GONDWANA EIGHT

ASSEMBLY, EVOLUTION AND DISPERSAL

PROCEEDINGS OF THE EIGHTH GONDWANA SYMPOSIUM
HOBART / TASMANIA / AUSTRALIA / 21-24 JUNE 1991

Gondwana Eight

Assembly, evolution and dispersal

Edited by

R.H.FINDLAY
Tasmania Department of Mines, Australia

R.UNRUG
Wright State University, USA

M.R.BANKS
University of Tasmania, Australia

J.J.VEEVERS
Macquarie University, Australia

CRC Press
Taylor & Francis Group
Boca Raton London New York

CRC Press is an imprint of the
Taylor & Francis Group, an **informa** business
A BALKEMA BOOK

CIP-DATA KONINKLIJKE BIBLIOTHEEK, DEN HAAG

Gondwana

Gondwana Eight : Assembly, Evolution and Dispersal : Proceedings of the eighth Gondwana symposium, Hobart, Tasmania, Australia, 21-24 June 1991 / ed. by R. H. Findlay ... [et al.]. - Rotterdam [etc.] : Balkema. - ill.
With index.
Subject headings: Gondwana / historical geology.

The texts of the various papers in this volume were set individually by typists under the supervision of each of the authors concerned.

Published by:
CRC Press/Balkema
P.O. Box 447, 2300 AK Leiden, The Netherlands
e-mail: Pub.NL@taylorandfrancis.com
www.crcpress.com – www.taylorandfrancis.com

ISBN 13: 978-90-5410-304-2 (hbk)

Gondwana Eight, Findlay, Unrug, Banks & Veevers (eds) © 1993 Balkema, Rotterdam. ISBN 90 5410 304 3

Table of contents

Evolution

Dispersal

Gondwana Eight, Findlay, Unrug, Banks & Veevers (eds) © 1993 Balkema, Rotterdam. ISBN 90 5410 304 3

Preface

The 8th International Gondwana Earth-science Symposium was held at the University of Tasmania, Hobart, Tasmania (Australia) between June 24th and June 28th 1991. 131 registrants from 18 nations produced a programme of 142 oral papers and 14 poster sessions. Additional specialist presentations were given at workshops held by the Mesozoic Stratigraphic Working Group of the Subcommission for Gondwana Stratigraphy (IUGS), at a workshop on the Gondwana-wide 500 Ma event, at the Prince Charles Mountains (Antarctica) workshop, and at meetings of International Geological Correlation Projects 237, 260, 272, 276 and 288. The Tasmania Department of Mines provided facilities for the IGCP260 workshop to study core from Tasmania's classical Gondwana sequence, the Parmeener Supergroup. Also held was an evening meeting for the lay public which was attended by 250 people and at which speakers presented aspects of Gondwana geology including the Antarctic-Australian Gondwana correlation.

Preceding the Symposium a week-long field excursion was held to study aspects of Tasmania's Parmeener Supergroup, which constitutes 'the Gondwanas' of Tasmania, and several delegates participated in the joint excursion run by the 7th Ordovician Symposium to Australia's Barrier Reef.

The 8th Gondwana Symposium formed a continuation of Tasmania's tradition of scientific support for Antarctic research; this tradition extends from the first decade of this century to the modern presence in Tasmania of the Commonwealth of Australia's Antarctic Division at Kingston, the Institute of Antarctic and Southern Ocean Studies at the University of Tasmania, and the Commission for Antarctic and Marine Living Resources in Hobart, together with major logistics-support and manufacturing facilities geared to Antarctic field-work. During the Symposium and in addition to other Antarctic papers, a day-long workshop was held on the geology of the Prince Charles Mountains, an area of the Antarctic craton which has been the focus of detailed work for several years by geologists from Australia and the USSR. Papers arising in particular from the Russian work in the Prince Charles Mountains accompany others in this volume which include new faunal discoveries in the Antarctic Peninsula, aspects of fission track geochronology related to the separation of Antarctica from Australia, and discussion of correlations between Antarctica and Sri Lanka and Australia. These aspects of the Symposium complement the work of the Antarctic Earth Science Symposium which was held in Japan in September 1991.

A new and developing aspect of Gondwana geology is the topic of Southeast Asia in Gondwana. Although this topic is not represented by a large number of papers, it is to be expected that future work, including that of IGCP321 (Gondwana Dispersion and Asian accretion in the Eastern Tethys and Western Circum-Pacific region), will lead to major developments in this field of Gondwana geology.

The Symposium also saw an increased interest in the mode of the assembly of Gondwana. This topic was the theme of 29 papers, in part sponsored by IGCP288 (Gondwanaland Sutures and Fold Belts), and was complemented by an additional 10 presentations during the Prince Charles Mountains Workshop. The papers on the assembly concluded with an open forum which was recorded and transcribed for publication in this volume by Prof. C.McA.Powell. Again, this topic represents a new focus for workers in Gondwana geology.

The conference's social events included a Symposium Dinner at which Prof. S.W.Carey reminded delegates that an open mind is a necessary attribute for successful geology, and also included a lunch of Tasmanian fare and wine during the mid-winter day-long field excursion to view the Gondwana aspects of the Hobart district. Also held were five 'working lunches' at a single venue to enable delegates to meet and discuss ideas in a relaxed atmosphere.

The present volume contains 56 papers, arranged according to the themes Gondwana assembly, evolution, and dispersal. In order to minimise production costs the authors were requested to present camera-ready copy for publication. This has led inevitably to various styles of type-face and minor differences in format which the senior editor attempted to minimise; the developing national differences in English usage were also recognised, although ambiguities and acronyms were discouraged strongly.

The editorial committee consists of R.H.Findlay (coordinator) and R.Unrug (Gondwana Assembly), M.R.Banks (Gondwana Evolution) and J.J.Veevers (Gondwana Dispersal); the final responsibility for production of the volume rests with R.H.Findlay.

Gondwana Eight, Findlay, Unrug, Banks & Veevers (eds) © *1993 Balkema, Rotterdam. ISBN 90 5410 304 3*

Acknowledgements

The organising and editorial committees thank the sponsors who contributed to the success of Gondwana Eight and to publication of the Abstracts and Symposium Volumes. Without this assistance, the conference could not have been held.

The Symposium was sponsored by: the Tasmanian State Government; the Federal Australian Department of Industry, Technology and Commerce through the Institution of Engineers; the Bureau of Mineral Resources, Geology and Geophysics; the Institute of Antarctic and Southern Ocean Studies, University of Tasmania; and the International Union of Geological Sciences. Additional financial support was provided to some delegates by IGCP288 (Gondwanaland Sutures and Fold Belts) and IGCP272 (late Palaeozoic – early Mesozoic events).

The Symposium committee consisted of: R.H. Findlay (Convener), C. Calver, M. Clarke, S. Forsyth and E. Williams (Tasmania Department of Mines); M.R. Banks, R. Lanyon and R. Varne (University of Tasmania); and N. Kemp (Tasmanian Museum).

The logistics load was shed on to a private-enterprise conference planning organisation, Conference Design Pty Ltd, who organised accommodation, the in-conference catering and some travel for delegates.

The field-excursions were organised by M. Clarke, M.R. Banks and N. Kemp (pre-conference excursion) and R. Varne (day excursion). The University of Tasmania's Students' Union provided excellent meals during the conference's 'working lunches'. Mrs June Pongratz, University of Tasmania, is thanked for her meticulous preparation of the Abstracts Volume and Programme.

Assembly

Gondwana Eight, Findlay, Unrug, Banks & Veevers (eds) © 1993 Balkema, Rotterdam. ISBN 90 5410 304 3

The Gondwana supercontinent: Middle Proterozoic crustal fragments, Late Proterozoic assembly, and unresolved problems

R.Unrug
Wright State University, Dayton, Ohio, USA

ABSTRACT: The Late Proterozoic supercontinent cycle includes the break-up of an earlier Middle Proterozoic supercontinent, and the accretion of several cratons and the large East Gondwana continent to form Gondwanaland. The accretion can be analysed in terms of plate tectonics. Early collisional events started at 870 Ma; the main tectonothermal events were in the 650-500 Ma range. The 500-450 Ma thermal event postdate the supercontinent assembly.

1 THE SUPERCONTINENT CYCLE

The formation of supercontinents containing most of, or all continental crust, followed by break-up and dispersal of fragments operates in a cycle that can be traced back to Middle Proterozoic. The supercontinent cycle provides the record of processes that control the formation and redistribution of continental crust. Through magmatism and orogeny it controls the major geochemical cycles, and in some phases creates environmental stress that stimulates bursts in biosphere evolution.

The dispersal phase of the most recent1supercontinent, the Late Paleozoic Pangea, is known in great detail. In contrast, the accretion phase of supercontinent formation remains poorly known.

Gondwanaland is an excellent research target for obtaining an improved understanding of supercontinent accretion. A well controlled reconstruction of Gondwanaland is available, and deep-level Proterozoic crustal structures are reasonably well exposed in mobile belts of ex-Gondwana continents.

The Gondwana Supercontinent formed by accretion of several cratons containing Archaean-Early Proterozoic nuclei, some fringed by Middle Proterozoic orogens, the large Eastern Gondwana continent, and a few smaller terranes (Fig. 1). The cratons are sutured by mobile belts of the Late Proterozoic Brasiliano-Pan African-Beardmore-Adelaidean orogenic cycle. The northern margin of Gondwana was dominated by calving of terranes now found in North America, Europe, Central and South-East Asia. A passive margin was present in the Transantarctic Mts. at the Pacific margin of Gondwana (Stump, 1987). Two large regions of juvenile crust formation in magmatic arcs: the Arabian-Nubian and the Avalonian were present at the northern margin of Gondwana.

2 MIDDLE PROTEROZOIC CRUSTAL FRAGMENTS IN GONDWANALAND

The Middle Proterozoic supercontinent that included a number of Archaean-Early Proterozoic cratons later incorporated in Gondwanaland, and possibly also the Laurentian craton (Moores, 1991; Dalziel, 1991; Hoffman, 1991), was sutured by The orogens formed at 1300-1000 Ma. This supercontinent was short lived, as its disintegration into lithospheric plates that interacted to form Gondwanaland began during the late stages of the Middle Proterozoic orogeny.

This is best documented at the margins of the Kalahari and Sao Francisco-Congo cratons, where a string of pull-apart basins is present in a rift zone induced by collision in the Namaqua orogen (Pfurr et al., 1990). The basins, filled with thick volcano-sedimentary sequences, some intruded by granites, formed at 1200-950 Ma. They extend from Namibia to Zambia and Mozambique (Borg, 1988, Böhme et al., 1990, Unrug, 1990.

The Amazonian and Sao Francisco-Congo cratons underwent three events of anorogenic magmatism, incipient rifting and some metamorphism of the rift basins fill, at 1700-1450 Ma, 1300-1100 Ma and 1000 Ma (De Almeida et al., 1981; Alvares and Maurin, 1991) but remained coherent cratonic units throughout the Gondwanaland cycle.

The West African, Amazonian, Rio de la

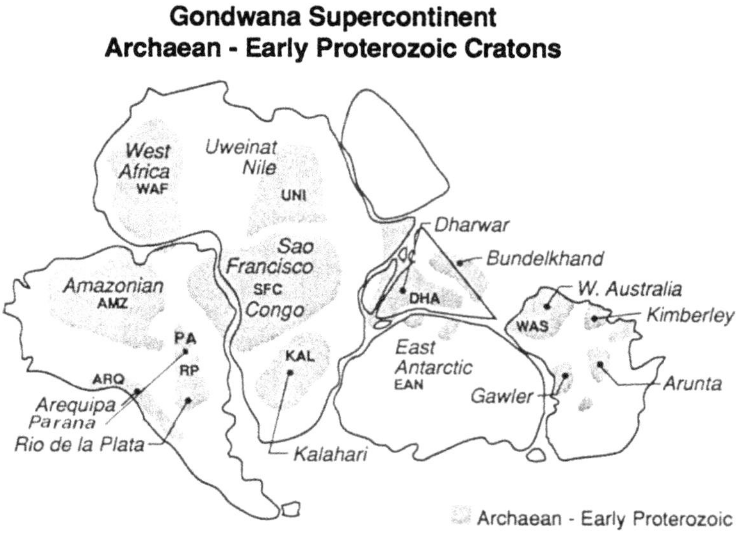

Fig. 1. Gondwana Supercontinent: Archaean-Early Proterozoic cratons. Tight fit Gondwana reconstruction after Lawver and Scotese (1987).

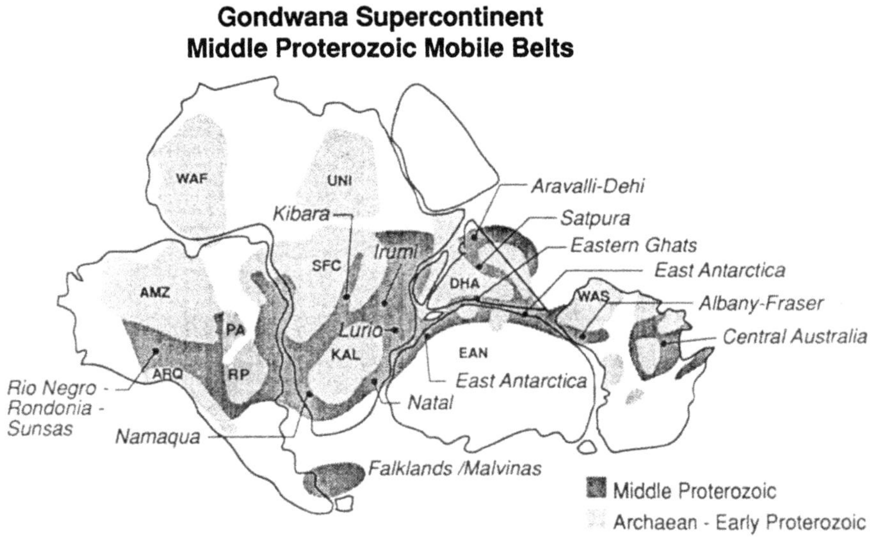

Fig. 2. Gondwana Supercontinent: Middle Proterozoic mobile belts plotted on tight fit Gondwana reconstruction after Lawver and Scotese (1987).

Plata, and Sao Francisco-Congo cratons, and the large Eastern Gondwana continent are the building blocks for the Late Proterozoic Gondwana assembly. Eastern Gondwana consisted of the Kalahari, Indian, West Australian and Eastern Antarctic cratons welded by Middle Proterozoic (1300 -1000 Ma) Natal-Lurio-Eastern Ghats-Albany/Fraser orogens and corresponding orogenic terranes of Eastern Antarctica (Fig. 2). These orogens are fringing the coastlines of ex-Gondwana continents which result from Mesozoic break-up. In the reconstruction of the Late Proterozoic upercontinent they extend across the large continental mass of Eastern Gondwana. Tectonic analysis confirms thus the

Gondwana Supercontinent
Late Proterozoic Mobile Belts

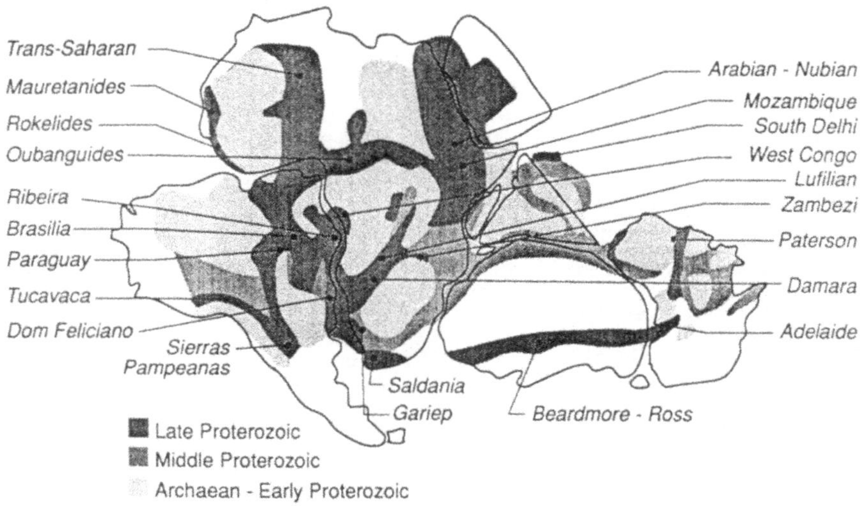

Fig. 3. Gondwana Supercontinent: Late Proterozoic mobile belts plotted on tight fit Gondwana reconstruction after Lawver and Scotese (1987).

existence of Eastern Gondwana defined on paleomagnetic data (McWilliams, 1981). In contrast, all Western Gondwana cratons are separated by Late Proterozoic orogens, and a coherent Western Gondwana never existed after the break-up of the Middle Proterozoic supercontinent.

Central Australia, the Jebel Uweinat terrane with poorly defined boundaries, and the Arequipa terrane of the Andes are other continental crust constituents of Gondwana. The Afif terrane of the Arabian shield, described as having continental affinities (Stoeser and Camp, 1985) is now considered to include onl8 one small area of 1628 Ma-old basement (Stoeser and Stacey, 1988).

3 LATE PROTEROZOIC ASSEMBLY OF THE GONDWANA SUPERCONTINENT

The Gondwana supercontinent assembled during the Brasiliano -Pan African-Beardmore-Adelaidean orogenic cycle lasting some 320 Ma from 870 to 550 Ma (Fig. 3). This orogenic cycle included several tectono-thermal events that can be analysed in terms of plate tectonics. The Late Proterozoic orogens of South America and Africa are collectively referred to as Brasiliano and Pan African, respectively.

The earliest events are known from Africa, in Central Hoggar at 870 Ma, and in Eastern Hoggar at 730 Ma (Caby, 1987). Slightly younger collisional orogenic phases at 820 Ma include arc-arc collision in the Arabian shield (Stoeser and Camp 1985), and continent-continent collision in the Sabachian event in the Mozambique belt of Kenya (Key et al. 1989) and in the Zambezi belt of Zambia (Hanson et al. 1988). Granulite facies metamorphism in the Mozambique belt of Tanzania, interpreted as indicative of peak crustal thickening is dated at 715 Ma (Maboko8et al., 1985).In Aïr the collision age is dated at 701 to 664 Ma (Black et al., 1990). Magmatic arc collisions are dated in western Adrar des Ilforas at 720 Ma, and in the western Pharusian belt of the Trans-Saharan Orogen at 700-680 Ma (Caby et al., 1981). The Bou Azzer ophiolite was emplaced at 685 Ma (Leblanc, 1981).

The main tectonothermal events of the Late Proterozoic orogenic cycle in Western Gondwana fall in the time range of 650-500 Ma. In Brasiliano orogens the early to syntectonic magmatism is dated at 765-550 Ma and the syn- to late-tectonic magmatism at 660-550 Ma (Leterrier et al. 1990). A syntectonic migmatite with a leucosome U-Pb zircon age of 512 Ma is reported by Machado et al. (1989).

In Africa, the Pan-African I at 650 Ma and Pan-African II at 550 Ma are recognized in the orogens fringing the West African craton in the south and west (Villeneuve and Dallmeyer, 1987). In the Mozambique belt of Kenya oblique continent-continent collision prior to 620 Ma was followed by transcurrent faulting at 620 and 580 Ma (Key et al., 1989). In the Lufilian fold belt deformation is dated at 670-602 Ma (Unrug, 1988). In the

Damara fold belt two tectonic events at 660-560 and 550-520 Ma are present (Hartnady et al., 1985).

In the Arabian Shield the arc accretion phase took place at 715-630 Ma (Stoeser and Camp, 1985).

In the south Delhi Foldbelt of India a tectonothermal event at 840-790 Ma is followed by a thermal event at 600-500 Ma (Volpe and McDougall, 1990).

In Australia, the Paterson Orogen marks a continent- continent collision at the NE margin of the West Australian craton in the 750-550 Ma range. Similar ages of metamorphism and deformation are recorded in the Pinjarra Orogen (Myers, 1990). The Adelaide Foldbelt developed from a rift formed around 800 Ma (Von der Borch, 1980). The Paterson and Adelaide regions may be linked by the Petermann Ranges Orogen of Central Australia where deformation is latest Proterozoic in age (Derrick, 1980).

Deformation and metamorphism in the Beardmore Orogen of the Transantarctic Mts. dated at 730-600 Ma, is followed by the thermal Ross Orogeny at 500 Ma (Stump, 1987).

Evolution of the Gondwanaland cycle sedimentary basins can be described in many cases in terms of rift basin evolution to passive margin-to deep-water basin. Evidence of destruction of oceanic crust during Gondwanaland assembly range from unequivocal ophiolites presence (Leblanc, 1981; Shackleton, 1988) to indirect arguments (Caby, 1987). However the potential for ophiolite preservation at the deep structural levels of Late Proterozoic orogens exposed by erosion is low (Shackleton 1988, Caby 1987), and in some Gondwanaland orogens the absence of classical ophiolites does not preclude ocean closure.

Most of Late Proterozoic Gondwana orogens include craton- vergent externide thrust and fold belts developed from passive margins, and internide zones with lower crustal rocks, crust generated granites, reworked older terranes and exotic terranes of uncertain provenance (Brito-Neves, 1986; 1990; Myers, 1990).

Large systems of shear zones (Fig. 4) hundreds of km wide and thousands of km long are present in all Late Proterozoic Gondwanaland orogens (e.g. Sadowski and Motidome, 1987; Sultan et al. 1987; Unrug, 1990). The strike-slip displacement on these shear zones ranges from tens to hundreds of km. Individual steep mylonite belts are up to several km thick. These shear zones are poorly understood, yet very important for kinematic analysis of Gondwanaland assembly.

Deformation in the Late Proterozoic mobile belts of Gondwanaland appeared to be completed by 550 Ma. The later widespread anorogenic magmatism and radiometric rejuvenation of Late Proterozoic rocks around 500 Ma are enigmatic. An interesting hypothesis (Sacci, 1989) relates the rejuvenation to large scale hydrothermal circulation and crustal granite melting in hot crust flooded by the large Cambrian transgression.

4 UNRESOLVED PROBLEMS

To improve the understanding of Gondwana Supercontinent assembly progress is necessary in the following areas:

- delineation of constituent cratons of the Gondwana Supercontinent and determination of the tectonic nature of craton boundaries;
- correlation of Proterozoic mobile belts, including fold and thrust belts, sutures, shear zones and major faults to determine their continuity on ex-Gondwanaland present continents; delineation of tectonostratigraphic terranes in the mobile belts;
- determination of kinematics of Proterozoic mobile belts: vergence of folds and thrust faults, sense of movements on transform and transcurrent faults, amount of crustal shortening;
- interaction of lithospheric plates during the Proterozoic and the nature of their boundaries;
- analysis of processes of repeated reactivation of large Proterozoic pericratonic dislocation zones.

REFERENCES

Alvarez, P. & J.C. Maurin 1991. Evolution sédimentaire et tectonique du bassin protérozoïque supérieur de Comba (Congo): stratigraphie séquentielle du Supergroupe Ouest-Congolien et modèle d'amortisement du décrochement dans le contexte de la tectogénèse panafricaine. Precambrian Research 50: 137-171.

Black, R., Liegeois, J.P., Navez, J. & Y. Vialette 1990. Early collision (700-665 Ma) after oceanic closure and Pan-African events (610-565 Ma) in Air, Niger. 'Abstracts, 15th Colloqium of African Geology'. CIFEG, Orléans: 248.

Böhme, F.W., Ahrendt, H. & K. Weber 1990. New aspects of the structural evolution and a revised stratigraphy in the southern margin zone of the Pan-African Damara Orogen, Namibia. 'Abstracts, 15th Colloqium of African Geology'.CIFEG, Orléans: 249.

Borg, G. 1988. The Koras-Sinclair-Ghanzi rift in southern Africa. Volcanism, sedimentation, age relationships and geophysical signature pf a late Middle Proterozoic rift system. Precambrian Research 38: 75-90.

Brito Neves, B.B. 1986. Tectonic regimes in the Proterozoic of Brazil. Proceedings, XII Simposio de Geologia do

Nordeste: 235-252.
Brito Neves B.B. 1990. Proterozoic passive margins of Brazil-Africa.'Abstracts, 15th Colloqium of African Geology'. CIFEG, Orléans: 168.
Caby, R. 1987. The Pan-African belt of West Africa from the Sahara desert to the Gulf of Benin. In J.P. Schaer & J. Rodgers (eds): 'The anatomy of mountain ranges'. Princeton University Press, Princeton, NJ.
Caby, R., Bertrand, J.M.L. & R. Black 1981. Pan-African closure and continental collision in the Hoggar-Iforas segment, central Sahara. In A. Kröner (ed.): 'Precambrian Plate Tectonics'. Elsevier, Amsterdam: 407-434.
Dalziel, I.W.D., 1991. Gondwana amalgamation and the origin of the southern Pacific ocean. American Geophysical Union Program and Abstracts Spring Meeting: 289.
De Almeida, F.F.M., Hasui, Y.; De Brito Neves, B.B. & R.A. Fuck 1981. Brazilian structural provinces: an introduction. Earth Science Reviews 17: 1-29.
Derrick, G.M. 1980. Solid Geology.'BMR Earth Science Atlas.' Bureau of Mineral Resources, Canberra.
Hanson, R.E., Wilson, T.J. & Wardlaw, M.S. 1988. Deformed batholiths in the Pan African Zambezi Belt, Zambia. Age and implications for regional Proterozoic tectonics. Geology 16, 1134-1137.
Hartnady, C., Joubert, P. & C. Stowe 1985. Proterozoic crustal evolution in southwestern Africa. Episodes 8: 236-244.
Hoffman, P.F. 1991. Did the breakout of Laurentia turn Gondwanaland inside out? Science, 252: 1409-1412.
Key, R.M., Charsley, T.J., Hackman, B.D., Wilkinson, A.F. & C.C. Rundle 1989. Superimposed Upper Proterozoic collision-controlled orogenies in the Mozambique orogenic belt of Kenya. Precambrian Research 44: 197-225.
Leblanc, M. 1981. The Late Proterozoic ophiolites of Bou Azzer (Morocco): evidence for Pan-African plate tectonics. In A. Kröner (ed.): 'Precambrian Plate Tectonics'. Elsevier, Amsterdam: 435-451.
Leterrier, J., De Sa J.E.F. & M.H. De Freitas Macedo 1990. Magmatic and geodynamic signature of the Brasiliano cycle plutonism in the Serido belt, NE Brazil. Sociedade Brasiliera de Geologia Anais 4: 1640-1655.
Maboko, M.A.H., Boelrijk, N.A.I.M., Priem, H.N.A. & E.A.T. Verdurmen 1985. Zircon U-Pb and biotite Rb-Sr dating of the Wami River granulites, Eastern Granulites, Tanzania: evidence of approximately 715 Ma ald granulite facies metamorphism and final Pan African cooling approximately 475 Ma ago. Precambrian Research 30: 361-378.
Machado, N., Schrank, A., De Abreu, F.R., Knauer, L.G. & P.A.A. Abreu 1989. Resultados preliminares da geochronologia U/Pb na Serra do Espinhaço meridional. 5-o Simposio Geologico Nucleo Minas Gerais/1-o Simposio Nucleo Brasilia, Anais. Sociedade Brasiliera de Geologia, Belo Horizonte, MG: 171-174.
McWilliams, M.O. 1981. Paleomagnetism and Precambrian tectonic evolution of Gondwana. In A. Kröner (ed.): 'Precambrian Plate Tectonics'. Elsevier, Amsterdam: 649-687.
Moores, E.M. 1991. Southwest U.S.-East Antarctic (SWEAT) connection: a hypothesis. Geology 19: 425-428.
Myers, J.J. 1990. Precambrian tectonic evolution of part of Gondwana, southwestern Australia. Geology 18: 537-540.
Pfurr, N., Ahrendt, H. & K. Weber 1990. U/Pb and Rb/Sr isotopic study of red granitic gneisses and associated metavolcanics in the area of the Rostock massifs, southern margin zone of the Damara orogen, Namibia and their implications for the lithostratigraphy of this crustal segment. 'Abstracts, 15th Colloqium of African Geology'. CIFEG, Orléans: 185.
Sacchi, R. 1989. On the Pan-African rejuvenation. Atti della Accademia delle Scienze di Torino 123, Geologia e Paleontologia: 1-13.
Sadowski, G.R. & M.M. Motidome 1987. Brazilian megafaults. Revista Geologica de Chile 37: 61-75.
Shackleton, R.M. 1988. Contrasting structural relationships of Proterozoic ophiolites in Northeast and Eastern Africa. In S. El-Gaby & R.O. Greiling (eds.): 'The Pan-African belt of Northeast Africa and adjacent areas'. F. Vieweg & Sohn, Braunschweig: 183-193.
Stoeser D.B. & V.E. Camp 1985. Pan-African microplate accretion of the Arabian Shield. Geological Society of America Bulletin 96: 817-826.
Stoeser D.B. & J.S. Stacey 1988. Evolution, U-Pb geochronology and isotope geology if the Pan-African Nabitah orogenic belt of the Saudi Arabian shield. In S. El-Gaby & R.O. Greiling (eds): 'The Pan-African belt of Northest Africa and adjacent areas'. F. Vieweg & Sohn, Braunschweig: 227-288.
Stump, E. 1987. Construction of the Pacific margin of Gondwana during the Pannotios cycle. In G. Mc. Kenzie (ed.): 'Gondwana Six: Structure, Tectonics and Geophysics'. American Geophysical Union, Geophysical Monograph 40, Washington, D.C.: 77-87.
Sultan, M., Arvidson, R.E., Duncan, I.D., Stern, R.J. & B. El Kaliouby, B. 1988. Extension of the Najd shear system from Saudi Arabia to the central Eastern Desert of Egypt based on integrated field and Landsat observations. Tectonics 7: 1291-1306.

Unrug, R. 1988. Mineralization controls and source of metals in the Lufilian fold belt, Shaba (Zaire) Zambia and Angola. Economic Geology 83: 1247-1258.
Unrug, R. 1990. Modes of Kibaran basement recycling in the Pan African Lufilian and Zambezi fold belts and the Mwembeshi-Chimaliro Dislocation zone, Zambia. Sociedade Brasiliera de Geologia Anais 6: 2672-2680.
Volpe, A.M. & J.D. MacDougall 1990. Geochemistry and isotopic characteristics of mafic (Phulad Ophiolite) and related rocks in the Delhi Supergroup, Rajasthan, India; implication for rifting in the Proterozoic. Precambrian Research 48: 167-191.
Villeneuve, M. & R.D. Dallmeyer 1987. Geodynamic evolution of the Mauritanide, Bassaride and Rokelide orogens (West Africa). Precambrian Research 37: 19-28.
Von der Borch, C.C. 1980. Evolution of Late Proterozoic to Early Paleozoic Adelaide foldbelt, Australia: comparison with post-Permian rifts and passive margins. Tectonophysics 70: 115-134.

ACKNOWLEDGEMENTS

I thank P.F. Hoffman for a review. This paper is an IGCP Project 288 contribution.

Gondwana Eight, Findlay, Unrug, Banks & Veevers (eds) © 1993 Balkema, Rotterdam. ISBN 90 5410 304 3

Late Proterozoic to early Paleozoic paleomagnetism and the formation of Gondwanaland

Z.X.Li & C.McA.Powell
Department of Geology, The University of Western Australia, Nedlands, W.A., Australia

ABSTRACT: Late Proterozoic to Early Paleozoic paleomagnetic data from the major Gondwanan continents have been reviewed to examine the tectonic relationship between the continents over this time interval, and, if possible, to determine when Gondwanaland was formed. The paleomagnetic data have been graded into five classes according to their quality/reliability, and only data of the top three classes have been used. Although reliable data for the Late Proterozoic interval are scarce, they indicate that (i), India and cratonic Australia had been together since at least 730 Ma; (ii), there could have been significant relative movements between the West Gondwanan cratonic blocks during the Late Proterozoic; and (iii), there was a wide ocean between eastern Africa and East Gondwanaland during the Cambrian. It is suggested that Gondwanaland was amalgamated during the Pan-African Orogeny.

1 INTRODUCTION

The time when Gondwanaland formed has long been a matter of debate. Many early workers, mostly paleomagnetists, suggested that Gondwanaland existed since at least 600-700Ma (McElhinny & Embleton 1976; Veevers & McElhinny 1976; Piper 1976; Morel & Irving 1978). McWilliams (1981) reviewed the then-available paleomagnetic data from Gondwanan continents, and concluded that before final suturing in Cambro-Ordovician time, West Gondwanaland (consisting of Africa, Arabia, South America, and possibly Madagascar) and East Gondwanaland (consisting of East Antarctica, Australia and India) had been two internally coherent tectonic units since at least mid-Proterozoic. Piper (e.g. Piper 1987) supported the Cambro-Ordovician birth for Gondwanaland, but proposed a much different, albeit conjoined, configuration for the Gondwanan continents as part of his Precambrian global supercontinent.

The range of opinion derived from paleomagnetic data alone can be attributed to both the insufficient data available, and ways the rapidly growing paleomagnetic data base has been used. In this study, we use a much more stringent quality filter than previously applied to assess the available paleomagnetic data from the Gondwanan continents. The assembly process of Gondwanaland is then re-examined in the light of the quality-filter selected data and geological information.

2 PALEOMAGNETIC DATA SELECTION

The quality of available paleomagnetic data varies greatly for diverse reasons, e.g. the lack of stepwise demagnetization for most of the early results, and the degrees to which the paleohorizontal and age of a remanent magnetization are constrained. It is therefore critical that each data set be evaluated carefully before being used to investigate tectonic implications.

Several data-quality classification schemes exist for paleomagnetic pole positions. They can be generalized into two types based upon the way in which they work. Incremental schemes (or point systems), such as those described by McElhinny & Embleton (1976) and Van der Voo (1990), set out a number

Table 1. Paleomagnetic data quality filter

Criteria		Pole quality and usefulness
(1) All specimens have been subjected to demagnetization using one or more demagnetizing techniques, i.e. thermal, alternating field and chemical demagnetizations. Magnetic remanent components have been distinguished. (2) Results (experimental) fully documented.	No →	'E' pole (Unacceptable)
↓ Yes		
(1) Reasonably large sample: ≥3 sites (or ≥12 samples) in sedimentary rocks, ≥3 layers of lava flows or dykes (with a minimum of 3 samples from each flow or dyke). (2) Reasonable precision ($\alpha_{95} \leq 30^o$). (3) A presumption of tectonic coherence with the craton or tectonic block concerned. (4) Paleohorizontal not suspect; Palaeohorizontal determined either by fold test or other geological test; or, structural correction makes no significant change to the remanentdirection.	No →	'D' pole (Very unreliable)
↓ Yes		
(1) Sufficiently large sample: ≥6 sites, or ≥24 samples for sedimentary rocks, ≥6 layers of intrusives or dykes. (2) Moderate to high precision ($\kappa \geq 10$ & $\alpha_{95} \leq 16^o$). (3) Age of magnetization established within a period or 60Ma, whichever is smaller.	No →	'C' pole (Acceptable, but use with caution)
↓ Yes		
Age of magnetization established within an epoch for the Phanerozoic or 30Ma, whichever is smaller.	No →	'B' pole (Reliable) — Key Pole
↓ Yes 'A' class pole (Best pole)	———	Key Pole

of parallel criteria, with the quality of a paleopole being measured by the number of criteria it meets.

Stepwise schemes, such as the one proposed by Briden & Duff (1981), assign reliability criteria with a range of priorities. The quality of a paleopole is assessed on a step-by-step basis, with a paleopole proceeding to a higher quality level only if it passes each step. The quality filter used in this study, as listed in Table 1, is modified from Briden & Duff's (1981) scheme, although some of the criteria in Van der Voo's (1990) scheme have been incorporated. By using this quality filter, we have tried to emphasize that a good control on the age of a given pole is meaningful only if the pole has no paleohorizontal ambiguity and meets certain statistical requirements.

All the published paleomagnetic data for the Late Proterozoic to Ordovician interval, to the authors' best knowledge, have been assessed according to this quality filter. Paleomagnetic poles of A- and B-class have a good age control and no paleohorizontal ambiguity. They are treated as key poles in defining the apparent polar wander path (APWP) of a continent. C-class poles generally have paleohorizontal control, but lack tight age control, or are statistically less reliable. They can be used as subsidiary poles in defining an APWP where key poles are available, or to indicate possible trends of an APWP when no key poles are available. D- and E-class poles are too unreliable to be used for APWP construction. From all the paleomagnetic data sets from the major Gondwanan continents in the studied time interval, about 30% of the paleopoles are of A- to C-class, and less than 7% of them can be regarded as key poles (Table 2).

To identify the tectonic unit which each paleopole represents, all the source regions for the selected paleopoles have been plotted

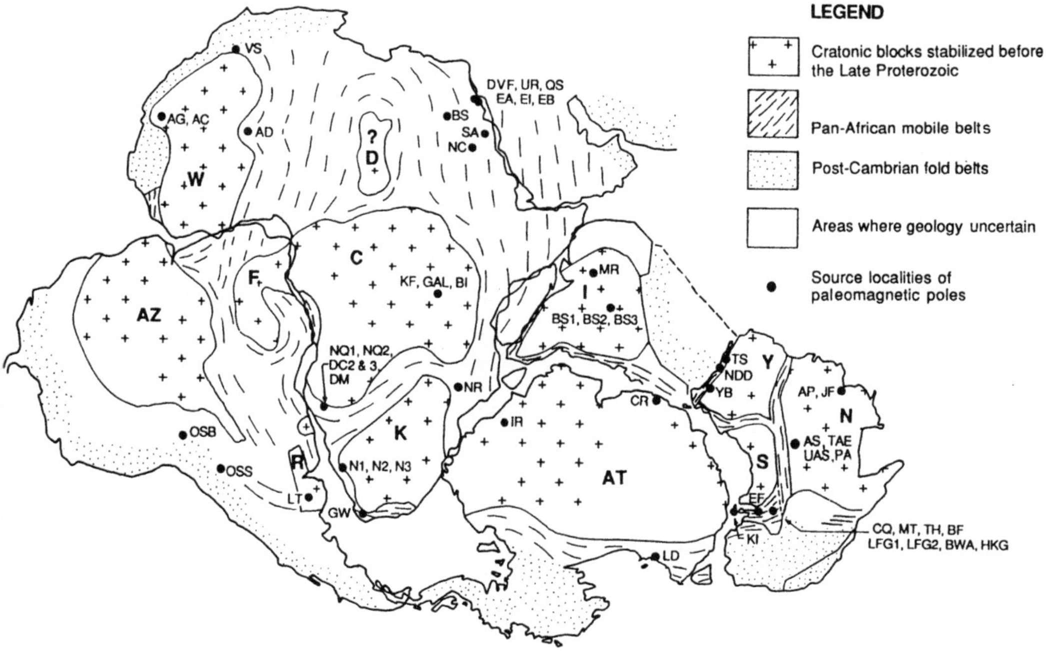

Fig. 1. Late Proterozoic tectonic framework of the Gondwanan continents (after Toens 1984) showing source localities of the selected paleomagnetic poles. Gondwanaland reconstruction is after Powell *et al.* (1988) and Scotese & McKerrow (1990). Cratonic blocks in Late Proterozoic: AT-East Antarctica; AZ-Amazonia; C-Congo; D-Chad; F-Sao Francisco; I-India; K-Kalahari; N-North Australia; R-Rio de la Plata; S-South Australia; W-West Africa; Y-West Australia.

on a schematic tectonic base map (Fig. 1). Data from the Salt Range region of the Indian subcontinent, the Himalaya region, the Tasman Fold Belt of Australia, and the Antarctica Peninsula, where post-Devonian rotations relative to the main continents have been suspected, have been excluded from this study.

All the A- to C-class paleopoles from the major Gondwanan continents are plotted in Fig. 2, with their source regions marked by different symbols, their quality marked by the sizes of the symbols and their ages marked by the line patterns within these symbols. The chosen polarities for the paleopoles are, to a certain degree, arbitrary, but angular distances between poles of similar ages have been kept to the minimum based on the minimum-energy principle (McElhinny *et al.* 1968).

3 TECTONIC SIGNIFICANCE

3.1 *Relationships between the Australian cratonic blocks*

Relationships between the Australian cratonic blocks during the Late Proterozoic have long been debated. Harrington *et al.* (1973) suggested, from geological evidence, that there was a rifting event (crustal extension of hundreds of kilometers) between different cratonic blocks during the earliest Paleozoic. Such a rifting model was supported by McWilliams (1981) based on a paleomagnetic analysis, but discounted by Rumph (1978) on geophysical evidence that there seems to be a continuous basement under the proposed rifting regions, and by Idnurm and Giddings (1988) based on a new paleomagnetic analysis. In contrast to the rifting model, Myers (1990) proposed that there was a major continental collision during Late Proterozoic between the North

Table 2 Selected Paleozoic paleomagnetic poles from Gondwanaland

Rock Formation	Pole Mnemonic	Age (Ma)	Plat (^{o}N)	Plong (^{o}E)	α_{95}(^{o}) (dp,	dm)	Qual	Reference.
Cratonic Australia:								
Jinduckin Formation	JF	O_1	-13	25	11		C	Luck (1970, 1972)
Tumblagooda Sandstone	TS	O?	-26.7	33.7	2	3	C	Schmidt & Embleton (1990)
Lake Frome Group	LFG2	ε_2	-31.4	26.9	5.1	10.1	C	Klootwijk (1980)
Lake Frome Group	LFG1	ε_2	-16	25	12.5		C	Embleton & Giddings (1974)
Areyonga Section	AS	ε_2	-32.5	11.5	5.1		C	Klootwijk (1980)
Billy Creek Fm, Wirrealpa Lst. & Aroona Creek Lst.	BWA	mid-ε	-37.4	20.1	7.2	14.4	B	Klootwijk (1980)
Camb. rocks in the Kangaroo Island	KI	ε_1	-33.8	15.1	6.2	12.3	C	Klootwijk (1980)
Hawker Group	HKG	ε_1	-26.7	2.3	8.1	14.3	C	Klootwijk (1980)
Todd River Dolomite, Allua Fm, & Eninta Sandstone	TAE	ε_1	-43.2	339.9	4.5	7.7	A	Kirschvink (1978a,b)
Antrim Plateau Volcanics	AP	earliest ε_1	-9	340	17		C	McElhinny & Luck (1970)
Northampton dolerite dykes	NDD	Pε/ε	-47.1	317.3	7.9		C	Embleton & Schmidt (1985)
Upper Arumbera Sandstone	UAS	Pε	-46.6	337.3	2.6	4.6	A	Kirschvink (1978a,b)
Upper Pertatataka Fm & lower Arumbera Sandstone	PA	latest Pε	-44.3	341.9	7.7	13.6	C	Kirschvink (1978a,b)
Brachina Formation	BF	latest Pε	-33	328	12	20	B	McWilliams & McElhinny (1980)
Tapley Hill Formation	TH	latest Pε	-39	28	8	15	C	McWilliams & McElhinney (1980)
Merinjina tillite	MT	latest Pε	-45	346	6	11	C	McWilliams & McElhinny (1980)
Copley Quartzite	CQ	latest Pε	-36	10	7	15	C	McWilliams & McElhinny (1980)
Elatina Formation	EF	680	-50.9	337.1	2.2		A	Embleton & Williams (1986), Schmidt *et al.* (1991)
YB dykes	YB	730-690	-19.9	282.0	28.1		C	Giddings (1976)
India: (Excluding data from the Salt Range and the Himalayas):								
Bhander Series	BS3	ε?	-51.3	042.7	11.2		C	McElhinney *et al.* (1978)
	BS2	ε?	-48.5	033.5	3	6	C	Klootwijk (1973)
	BS1	ε?	-31.5	019.0	3.1	5.9	C	Athavale *et al.* (1972)
Malani Rhyolites	MR	729±10	-80.5	223.5	8.0	11.5	B	Klootwijk (1975)
East Antarctica:								
Lamprophyre Dykes from Taylor Valley	LD	~470	-9.3	026.7	5.5	10.9	C	Manzoni & Nanni (1977)
Sør Rondane igneous intrusions	IR	~480	-28.5	009.5	6	7	C	Zijderveld (1968)
Wright Valley granitic rocks	WG	500-480	5.4	018.5			C	Funaki (1984)
Vanda porphyry of Wright Valley	VP	500-480	5.8	024.9			C	Funaki (1984)
Charnockitic rocks from Mirnyy Station	CR	520±25	-1.5	028.5	8	16	C	McQueen *et al.* (1972)
Africa:								
Salala ring complex	SA	460-466	40	330	8	12	B	Bachtadse & Briden (1989)
Graafwater Formation	GW	0_1	28	014	8.8		C	Bachtadse *et al.* (1987)
Qena-Safaga dyke swarm	QS	<late Pε >480	-87.3	124.2	5.0	4.8	C	Davies *et al.* (1980)
Um Rus dyke swarm	UR	<late Pε >464	-85.9	013.0	6.2	5.4	C	Davies *et al.* (1980)
Morocco Volcanics & Sed.	VS	532±18	46.9	042.2	6.8		C	Daly & Pozzi (1977)
Esh El Mellaha Range acidic dykes	EA	530?	-87.3	205.9	9.6	14.5	C	Abdullah *et al.* (1984)
Esh El Mellaha Range intermediate dykes	EI	530?	-81.6	300.8	3.3	5.2	C	Abdullah *et al.* (1984)

Esh El Mellaha Range basic dykes	EB	530?	-82.7	093.5	5.6	9.1	C	Abdullah *et al.* (1984)
Ntonya Ring Structure	NR	630-480	27.8	344.9	1.8		C	Briden (1968)
Atar Cliff Group	AC	<595±43	00.3	069.3	5.7	11.5	C	Perrin *et al.* (1988)
Nama Group overprint	N3	<N2	-04	335	12	16	C	Kröner *et al.* (1980)
Upper Nama Group	N2	ε_1	-05	091	9	16	C	Kröner *et al.* (1980)
Mulden Group	DM	560-550	-13	090	10	18	B	McWilliams & Kröner (1981)
Bir Safsaf dykes	BS	586±13	-80.0	069.7	9.2		C	Saradeth *et al.* (1989)
Nabati complex	NC	589±15	-67.9	134.1	18.8		C	Saradeth *et al.* (1989)
Dokhan Volcanic Formation	DVF	~600	-36.1	017.1	14.8	12.1	C	Davies *et al.* (1980)
Lower Nama Group	N1	553±13-686±32	-62	241	8	15	C	Kröner *et al.* (1980)
Adma Diorite	AD	616-590	-32.5	164.7	15.9		B	Morel (1981)
Otavi Group	DC2 & 3	<651±34	-55	044	20	24	C	McWilliams & Kröner (1981)
Nosib Group	NQ2	650-1000	-51	213	8	14	C	McWilliams & Kröner (1981)
Atar Group	AG	775±52	-34.0	066.9	4.6	8.6	C	Perrin *et al.* (1988)
Bukoban Intrusives	BI	860, 810	-11	281	12.1	21.6	C	Piper (1972)
Gagwe Amygdaloidal Lavas	GAL	960, 820	-29	283	7.7	14.2	C	Piper (1972)
Kigonero Flags	KF	1100-890	11	273	10.3	16.0	C	Piper (1972)
Nosib Group	NQ1	1000-900	28	323	11	18	C	McWilliams & Kröner (1981)
South America:								
Ordovician sediments at Salta & Jujuy	OSS	0	11	333	~10		B	Creer (1970)
Ord. sediments, Bolivia	OSB	0	4	302	16		C	Creer (1970)
La Tinta Formation	LT	709±24	-79.3	300.9	4.8		C	Valencio *et al.* (1980)

Pε - Proterozoic; ε – Cambrian; O - Ordovician

Australia block and the West Australia block (N & Y in Fig. 1). However, apart from overthrusting of supracrustal rocks by some tens of kilometers in the Petermann Ranges of Central Australia (Forman & Shaw 1973) and some Late Proterozoic-Early Cambrian tectonic reactivation elsewhere, no direct geological evidence of large-scale (hundreds of kilometers or more) Late Proterozoic crustal shortening has yet been reported in the mobile belts separating the three Australian cratonic blocks.

All but one C-class pole (pole AP) from Australia distribute along a simple trend (Fig. 2a), which starts from pole YB, the oldest, and traverses about 80° to the two Ordovician C-class poles (poles JF and TS in NW Africa). Furthermore, the four key poles with a latest Precambrian to Early Cambrian age (i.e. poles UAS and TAE from the North Australia block, and poles EF and BF from the Adelaide Fold Belt, the basement of which is believed to be the continuation of the South Australia block, form a very tight group. Taken at face value, this implies that no significant relative rotation has occurred between the Australian blocks since latest Proterozoic. However, reliable data of 900-700Ma age from both cratons are required to examine their relative positions during that interval.

3.2 *Formation of East Gondwanaland*

There are three C-class, supposedly Cambrian, poles (poles BS1, 2 and 3, all from the Bhander Series) and one B-class Late Proterozoic pole (pole MR) from India. Pole MR, dated as about 730Ma, falls close to C-class pole YB from the West Australia block, dated at about 730-690Ma. The other three poles fall close to the latest Proterozoic-Cambrian poles from Australia (Fig. 2a). There are only three C-class Cambro-Ordovician poles (poles CR, LD and IR) available from East Antarctica within the studied time interval. These poles are in good agreement with the two C-class poles (poles JF and TS) from cratonic Australia (Fig. 2a). It is therefore permissible to conclude that East Gondwanaland can be dated back to Cambro-Ordovician time, with the possibility of its being older than 730Ma.

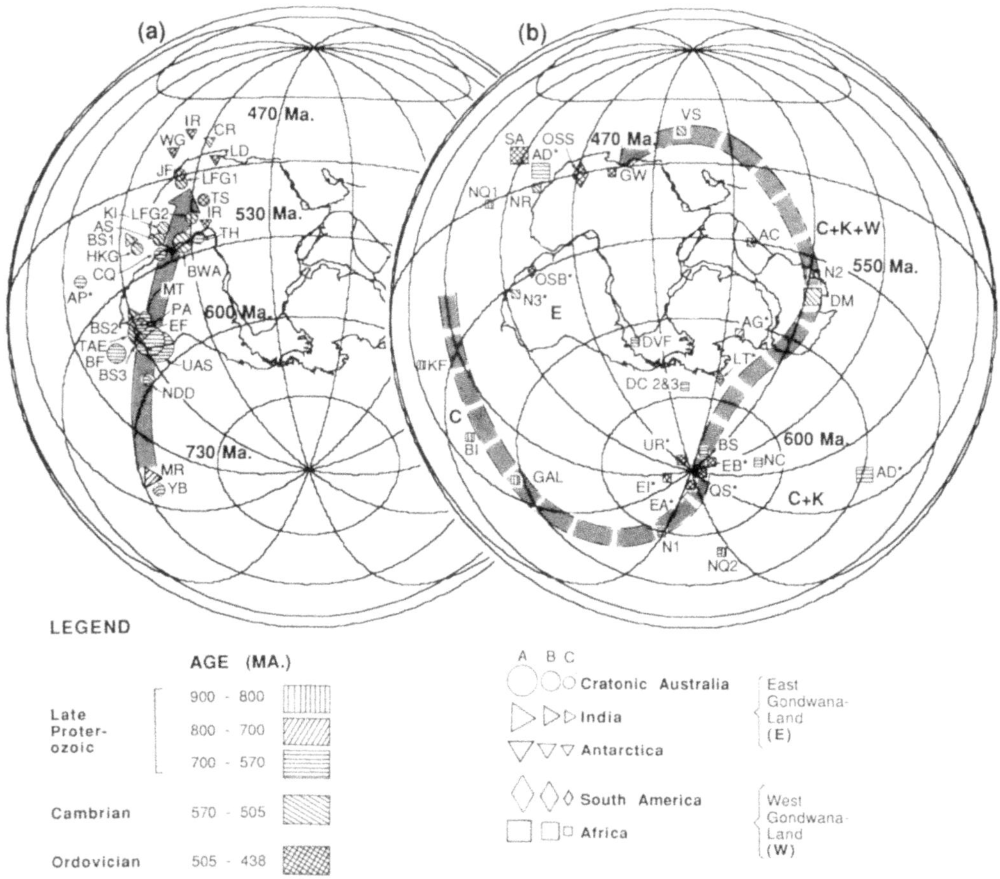

Fig. 2: Locations of selected paleomagnetic poles from (a) East Gondwanaland and (b) West Gondwanaland in Gondwanaland reference. The apparent polar wander path (APWP) of East Gondwanaland (E) is reasonably well constrained, but the APWP for the accreting "West Gondwanaland" is suggestive only. Poles with a star on their mnemonics are discarded during the drawing of the APWPs because of their aberrant positions, which could be due to either incorrect age assignment, or possible relative rotation of their source regions with respect to the other blocks. C, K and W stand for African cratonic blocks as shown in Fig. 1. Note that there is no A-class pole from Africa or South America.

3.3 *Tectonic relationships between the cratonic blocks of West Gondwanaland during the Pan-African Orogeny*

As shown in Fig. 1, there were at least five cratonic blocks in West Gondwanaland during the Late Proterozoic, separated by Pan-African orogenic belts. The dominant tectonic activity in these belts occurred between 800Ma and 600Ma.

All but one of the total of 13 Late Proterozoic paleopoles from West Gondwanaland are C-class poles. Therefore, strictly speaking, no legitimate test can be made of the tectonic relationships between these blocks in the Late Proterozoic. The data allow only a tentative discussion of the possible implications.

As the four oldest poles (poles NQ1, KF, BI and GAL) are all from the Congo block (Figs 1 & 2b), no comparison can be made for the 900-800Ma interval. Pole N1 from the Kalahari block falls close to the similarly-aged pole NQ2 from the southern margin of the Congo block. It is therefore possible that there has been no significant tectonic movement between these cratons since about 700Ma. On the other hand, the significant distances between these two poles and poles AG, from the West African craton, and LT, from southern South America, could imply that the West African block and southern South America were not tectonically coherent with the combined Congo-Kalahari block at that time. This is consistent with geological interpretations that there was oceanic subduction east of the West Africa block during the 700-600 Ma interval (Caby *et al.* 1981).

Poles with ages around 600Ma include DC2&3 from the southern margin of the Congo block, AD from the eastern margin of the West African block, and NC, BS and DVF from the western bank of the present Red Sea (Fig. 1). The last three poles, although from the same region, are scattered (Fig. 2b). However, their mean position is close to pole DC2&3. This could indicate that central-north Africa joined the combined Congo-Kalahari block by around 600Ma. Such an interpretation is supported by geological observations that plate subduction in the Arabian-Nubian region finished by about 600 Ma (e.g. Gass 1981; Vail 1988). Pole AD, dated at 616-590 Ma, from the Pan-African Fold Belt just east of the West African block, is significantly apart from the above cluster of poles. This could be due to local rotations around a vertical axes during the final stage of collision between the West Africa block and the combined central-east Africa. However, the close relationship between the antipole of AD and the Cambro-Ordovician poles of Australia and southern Africa (pole NR, referred to Late Cambrian age) leads us to suspect that this pole could also be of Cambrian age. The chemical origin of the corresponding remanence (Morel 1981) makes such a speculation feasible.

Poles with Early Cambrian ages include DM from the southern margin of the Congo block and N2 from the Kalahari block. Pole AC from the West African block falls close to these two poles (Fig. 2b). The loose control on the age of pole AC makes it feasible to be an Early Cambrian pole. If this inference is correct, it is the first time since the Late Proterozoic that there is an indication of tectonic coherence between the West African block and the other African blocks. Pole VS, dated as about 530Ma, can easily be accommodated between these Early Cambrian poles and the Cambro-Ordovician poles clustered around northwestern coast of Africa (see discussion below). No data of Early to Middle Cambrian age are available from South America.

There is no key-pole of Late Cambrian age from the West Gondwanan continents. Two key-poles of Ordovician age and two C-class Cambro-Ordovician poles nevertheless cluster around the northwestern margin of Africa (poles SA, NR, OSS and GW in Fig. 2b). Another five C-class poles (poles EB, EI, EA, UR and QS in Fig. 2b) from supposedly Cambro-Ordovician dykes on the western bank of the present Red Sea are suspected to represent a thermal overprint related to the opening of the northwestern end of the Red Sea. Alternatively, these dykes could be older than what people have believed. One important feature to note is that for the first time since the Late Proterozoic, there is a key-pole from South America (pole OSS) which falls close to contemporaneous poles from Africa. This implies that South America and Africa were joined no later than the Ordovician.

3.4 *Formation of Gondwanaland*

As discussed above, West Gondwanaland was formed by the convergence of allochthonous cratonic blocks, and accretion of island arcs (e.g. Shackleton 1986) during the Late Proterozoic. Therefore, Gondwanaland, in whatever configuration, could not have existed during that time interval. As Figure 2 indicates, paleopoles from East Gondwanaland and the accreting West Gondwanaland are significantly discordant before Cambro-Ordovician time. It is thus safe to conclude that Gondwanaland did not exist until the end of Cambrian or the Early Ordovician (~510Ma). This conclusion is consistent with opinions of some previous paleomagnetic workers (McWilliams 1981; Piper 1987), and is supported by recent geological work (e.g. Hoffman 1991; Unrug, 1992). It is notable that the time interval of the Pan-African Orogeny coincides with the assembly time of Gondwanaland. This leads us to postulate that the Pan-African Orogeny was part of the tectonic process through which Gondwanaland was formed.

Figure 3 shows a possible paleo-reconstruction of the Gondwanan continents during the Late Proterozoic-Ordovician interval. In the 720Ma reconstruction (Fig. 3a), the paleolatitudes of East Gondwanaland and the Congo block are plotted according to the paleomagnetic data shown in Fig. 2. East Gondwanaland is placed against the western margin of Laurentia according to a geologically-based reconstruction (Bell & Jefferson 1987; Moores 1991; Dalziel 1991; Hoffman 1991). However, the West Gondwanan blocks are placed adjacent to the opposite side of Laurentia. Although this configuration is quite different from that of Hoffman (1991), it also provides a roughly continuous Grenvillian (1.3-1.1 Ga) belt, one of the principal constraints Hoffman used in his reconstruction. However, our reconstruction implies a larger clockwise rotation for the nucleus of West Gondwanaland around a Euler pole adjacent to the southern end of Africa during the latest Precambrian-Cambrian interval (Fig. 3b-d). This sense of rotation is consistent with the reported Late Proterozoic, largely left-lateral, oblique collisions and post-collisional tectonic movements, in the Arabian-Nubian region and the Mozambique Belt (Stoeser & Camp 1985; Shackleton 1986; Berhe 1990).

One of the ambiguities in the above paleoreconstruction is the paleoposition of the Kalahari block. Most recent models based on geological data seem to favor an East Gondwanan affinity for the Kalahari block in the Late Proterozoic-Cambrian interval. In Hoffman's (1991) 700 Ma reconstruction, the Kalahari block is placed closer to the East Antarctica than to the Congo block during Late Proterozoic; Unrug (1992) also places the Kalahari block as part of his Late Proterozoic East Gondwanaland. Toens (1984) and Kaz'min (1988) believed that the Pan-African tectonic activity between the Kalahari block and East Antarctica was finished by about 600 Ma, and therefore that no significant relative movements could have occurred between them during the Cambrian. On the other hand, although the Khomas "Sea" between the Congo and Kalahari blocks was supposed to have been closed by the end of Proterozoic (Stanistreet *et al.* 1991), the related Damara Orogeny, which started from about 675Ma, lasted until mid-Cambrian (Hartnady *et al.* 1985). However, paleomagnetic data from the Kalahari block, although all of C-class, indicate a Congo block affinity (see discussion earlier). More reliable paleomagnetic data, particularly from the Kalahari block, are needed to reconcile this controversy.

Another question yet to be answered is whether there was still a large ocean between the western margin of East Gondwanaland and eastern Africa at the end of the Proterozoic. If the answer is yes, then where are the final suture or sutures?

As far as geological data are concerned, continent-arc or continent-continent collisions in both northeastern Africa (the Arabian-Nubian region) and the Mozambique Orogenic Belt to the south, had finished by 700-600 Ma (e.g. Gass 1981; Stoeser & Camp 1985; Shackleton 1986; Vail 1988; Key *et al.* 1989; Berhe 1990; Pallister *et al.* 1990). However, paleomagnetic data indicate a significant discordance between East Gondwanaland and the accreting West Gondwanaland during the Early Cambrian (Figs 2 & 3c). Although one can put the southern end of Africa close to East Antarctica by moving the continents along latitude lines (this would overlap West Gondwanaland with the equatorial Laurentia), there is always going to be a big gap between the west coast of India and northeast Africa. If such a wide ocean really existed during latest Proterozoic to Early Cambrian times, one has to find a Cambrian suture (or sutures) between East Africa and the western margin of East Gondwanaland.

There is some geological evidence for such a suture: Kaz'min (1988), based on a facies analysis of the northeastern African Horn in Somalia, suggested that there could have been a Cambrian suture east of the coast of Oman, eastern South Yemen and Somali (SU in Fig. 3c); Gass (1981) also reported geochemical evidence for the existence of subduction beneath the Arabian Shield during the 670-600 to 600-500 Ma interval. However, no direct exposure of such a suture has ever been reported along the southeastern coast of Africa, Madagascar, west coast of India, or East Antarctica.

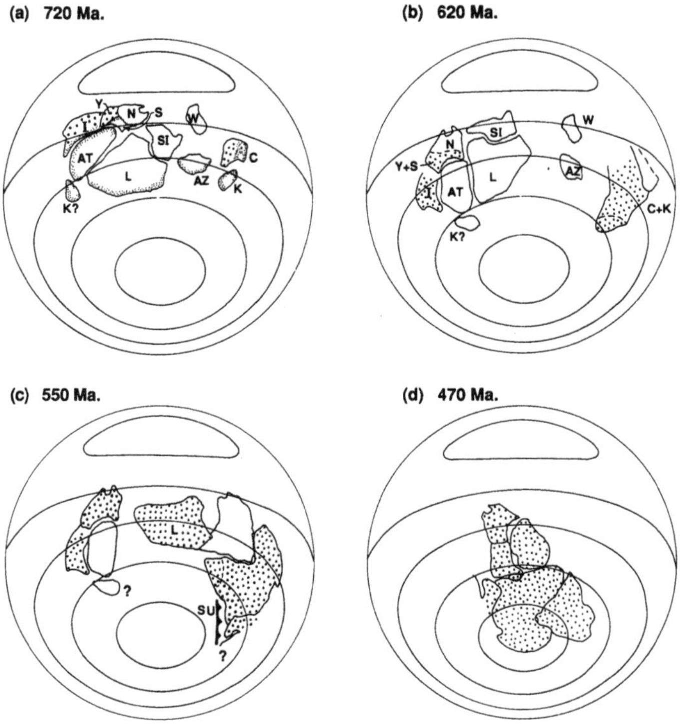

Fig. 3: Paleoreconstructions of the Gondwanan continents during the late Precambrian-Ordovician interval. Paleolatitudes of the coarsely stippled continents are plotted according to paleomagnetic data. Paleolatitude of Laurentia (L) and Siberia (SI) in (a) and (b) is roughly the same as in Hoffman (1991, fig. 1). Laurentia in (c) is plotted according its Early Cambrian pole compiled by Van der Voo (1990, table 7). Finely stippled regions in (a) are the Grenvillian belt following Hoffman (1991). SU is the Cambrian subduction zone suggested by Kaz'min (1988). Mnemonics for continents are the same as in Fig. 1.

Possible explanations for this ambiguity are: 1, There was a Cambrian suture between East Gondwanaland and East Africa, which has either been rifted away from the main continents during the splitting of Gondwanaland in the Mesozoic, or has been eroded away. 2. The relative movement between East Africa and the western margin of East Gondwanaland during the Cambrian was mainly transcurrent rather than convergent. If this were the case, a reconstruction similar to Fig. 8a or 8b of Powell *et al.'s* open forum discussion (1992) must be adopted; 3. The paleomagnetic data is erroneous. The last possibility is considered unlikely because the Early Cambrian paleolatitude of both East Gondwanaland and Africa is controlled by key-poles (Table 2 and Fig. 2).

4 CONCLUSIONS

The following conclusions can be drawn from this study:

1. East Gondwanaland was most likely formed before 730Ma.

2. Reliable paleomagnetic data from West Gondwanan cratonic blocks are too scarce to give a conclusive view about when they joined together. The available data (mostly of C-class) nevertheless indicate that there could have been significant relative movements between these blocks during the Late Proterozoic.

3. Paleomagnetic data available require a large amount (thousands of kilometers) of crustal shortening (or transcurrent movement) between Africa and East Gondwanaland during the Cambrian times,

but there is as yet little geological evidence for such a major tectonic event.

4. Gondwanaland was not formed until Cambro-Ordovician times (~510 Ma). The Pan-African Orogeny involved the assembly and consolidation of Gondwanaland.

ACKNOWLEDGMENTS

The authors wish to thank Drs B.J.J. Embleton, R. Unrug, A. Trench, J.P. Dobson and two anonymous referees for constructive comments, G. Gáal, L.R. Harris and J. Ridley for discussion on Pan-African mobile belts, D. Johnson, C.N.G. Dampney and J. Gillings for offering their INTERMAP graphic system, B. Agar for giving us his preprint on Arabian tectonics and metallogeny, and Ms B. Jess for drafting the diagrams. The work is supported by ARC grant No. A39030719.

REFERENCES

Abdullah, A., A.E.M. Nairn, D. Savino & K. Sprague 1984. Paleomagnetic study of some of the dikes from the Esh El Mellaha Range, Eastern Desert, Egypt. *Journal of African Earth Sciences* **2:** 267-275.

Athavale, R.N., A. Hansraj & R.K. Verma 1972. Palaeomagnetism and age of Bhander and Rewa Sandstones from India. *Geophysical Journal of the Royal Astronomical Society* **28:** 499-509.

Bachtadse, V. & J.C. Briden 1989. Palaeomagnetism of the Early to mid-Ordovician Salala igneous ring complex, Red Sea Hills, Sudan. *Geophysical Journal International* **99:** 677-685.

Bachtadse, V., R. Van der Voo & I.W. Halbich 1987. Paleomagnetism of the western Cape Fold Belt, South Africa, and its bearing on the Paleozoic apparent polar wander path for Gondwana. *Earth and Planetary Science Letters* **84:** 487-499.

Bell, R.T. & C.W. Jefferson 1987. An hypothesis for an Australia-Canadian connection in the Late Proterozoic and the birth of the Pacific Ocean. *Proceedings of Pacific Rim Congress* 87, Gold Coast, Queensland: 39-50.

Berhe, S.M. 1990. Ophiolites in Northeast and East Africa: implications for Proterozoic crustal growth. *Journal of the Geological Society, London* **147:** 41-57.

Briden, J.C. 1968. Paleomagnetism of the Ntonya Ring Structure, Malawi. *Journal of Geophysical Research* **73:** 725-733.

Briden, J.C. & B.A. Duff 1981. Pre-Carboniferous palaeomagnetism of Europe north of the Alpine Orogenic Belt. In: McElhinny, M.W. & D.A. Valencio (eds) Paleoreconstruction of the Continents. *American Geophysical Union Geodynamics Series* **2:** 137-149.

Caby, R., J.M.L. Bertrand & R. Black 1981. Pan-African ocean closure and continental collision in the Hoggar-Iforas segment, Central Sahara. In: Kröner, A. (ed) *Precambrian Plate Tectonics, Developments in Precambrian Geology* **4:** 407-434. Elsevier, Amsterdam.

Creer, K.M. 1970. A palaeomagnetic survey of South American rock formations. *Philosophical Transactions of the Royal Society of London* **A267:** 457-558.

Dalziel, I.W.D. 1991. Pacific margin of Laurentia and East Antarctica/Australia as a conjugate rift pair: Evidence and implications for an Eocambrian supercontinent. *Geology* **19:** 598-601.

Daly, L. & J.P. Pozzi 1977. Determination d'un nouveau pole paleomagnetique africain sur des formations cambriennes du Maroc. *Earth and Planetary Science Letters* **34:** 264-272.

Davies, J., A.E.M. Nairn & R. Ressetar 1980. The paleomagnetism of certain late Precambrian and Early Paleozoic rocks from the Red Sea Hills, Eastern Desert, Egypt. *Journal of Geophysical Research* **85:** 3699-3710.

Embleton, B.J.J. & J.W. Giddings 1974. Late Precambrian and Lower Palaeozoic palaeomagnetic results from South Australia and Western Australia. *Earth and Planetary Science Letters* **22:** 355-365.

Embleton, B.J.J. & P.W. Schmidt 1985. Age and significance of magnetizations in dolerite dykes from the Northampton Block, Western Australia. *Australian Journal of Earth Sciences* **32:** 279-286.

Embleton, B.J.J. & G.E. Williams 1986. Low palaeolatitude of deposition for the late Precambrian periglacial varvites in South Australia: implications for palaeoclimatology. *Earth and Planetary Science Letters* **79:** 419-430.

Forman, D.J & R.D. Shaw 1973.

Deformation of the crust and mantle in Central Australia. *Bureau of Mineral Resources, Bulletin* **144:** 20pp.

Funaki, M. 1984. Investigation of the paleomagnetism of the basement complex of Wright Valley, southern Victoria Land. *Journal of Geomagnetism and Geoelectricity* **36:** 529-563.

Gass, I.G. 1981. Pan-African (Upper Proterozoic) plate tectonics of the Arabian-Nubian Shield. *In:* Kröner, A. (ed) *Precambrian Plate Tectonics, Developments in Precambrian Geology* **4:** 387-405. Elsevier, Amsterdam.

Giddings, J.W. 1976. Precambrian palaeomagnetism in Australia I: basic dykes and volcanics from the Yilgarn Block. *Tectonophysics* **30:** 91-108.

Harrington, H.J., K.L. Burns & B.P. Thompson 1973. Gambier-Beaconsfield and Gambier-Sorell fracture zones and the movement of plates in the Australia-Antarctica-New Zealand region. *Nature* **245:** 109-112.

Hartnady, C., P. Joubert & C. Stowe 1985. Proterozoic crustal evolution in southwestern Africa. *Episodes* 8: 236-244.

Hoffman, P.F. 1991. Did the breakout of Laurentia turn Gondwanaland inside-out? *Science* **252:** 1409-1412.

Idnurm, M. & J.W. Giddings 1988. Australian Precambrian polar wander: a review. *Precambrian Research* **40/41:** 61-88.

Kaz'min, V.G. 1988. Tectonic development of the Mozambique Zone: from accretion to collision. *Geotectonics* **22:** 213-219.

Key, R.M., T.J. Charsley, B.D. Hackman, A.F. Wilkinson & C.C. Rundle 1989. Superimposed Upper Proterozoic collision-controlled orogenies in the Mozambique Orogenic Belt of Kenya. *Precambrian Research* **44:** 197-225.

Kirschvink, J.L. 1978a. The Precambrian-Cambrian boundary problem: paleomagnetic directions from the Amadeus Basin, Central Australia. *Earth and Planetary Science Letters* **40:** 91-100.

Kirschvink, J.L. 1978b. The Precambrian-Cambrian boundary problem: magnetostratigraphy of the Amadeus Basin, Central Australia. *Geological Magazine* **115:** 139-150.

Klootwijk, C.T. 1973. Palaeomagnetism of Upper Bhander Sandstones from central India and implications for a tentative Cambrian Gondwanaland reconstruction . *Tectonophysics* **18:** 123-145.

Klootwijk, C.T. 1975. A note on the palaeomagnetism of the late Precambrian Malani Rhyolites near Jodhpur - India. *Journal of Geophysics* **41:** 189-200.

Klootwijk, C.T. 1980. Early Palaeozoic palaeomagnetism in Australia. *Tectonophysics* **64:** 249-332.

Kröner, A., M.O. McWilliams, G.J.B. Germs, A.B. Reid & K.E.L. Schalk 1980. Paleomagnetism of Late Precambrian to Early Paleozoic mixtite-bearing formations in Namibia (South West Africa): the Nama Group and Blaubeker Formation. *American Journal of Science* **280:** 942-968.

Luck, G.R. 1970. The palaeomagnetism of some Cambrian and Ordovician sediments from the Northern Territory, Australia. *Geophysical Journal of the Royal Astronomical Society* **20:** 31-39.

Luck, G.R. 1972. Palaeomagnetic results from Palaeozoic sediments of Northern Australia. *Geophysical Journal of the Royal Astronomical Society* 20: 31-39.

Manzoni, M & T. Nanni 1977. Paleomagnetism of Ordovician lamprophyres from Taylor Valley, Victoria Land, Antarctica. *Pageoph.* **115:** 961-977.

McElhinny, M.W. & G.R. Luck 1970. The palaeomagnetism of the Antrim Plateau Volcanics of Northern Australia. *Geophysical Journal of the Royal Astronomical Society* **20:** 191-205.

McElhinny, M.W. & B.J.J. Embleton 1976. Precambrian and Early Palaeozoic palaeomagnetism in Australia. *Philosophical Transactions of the Royal Society of London* A. **280:** 417-431.

McElhinny, M.W., J.C. Briden, D.L. Jones & A. Brock 1968. Geological and geophysical implications of palaeomagnetic results from Africa. *Reviews of Geophysics* **6:** 201-238.

McElhinny, M.W., J.A. Cowley & D. Edwards 1978. Palaeomagnetism of some rocks from peninsular India and Kashmir. *Tectonophysics* **50:** 41-54.

McQueen, D.M., C.K. Scharnberger, L. Scharon & M. Halpern 1972. Cambro-Ordovician paleomagnetic pole position and Rubidium-Strontium total rock

isochron for charnockitic rocks from Mirnyy Station, East Antarctica. *Earth and Planetary Science Letters* **16:** 433-438.
McWilliams, M.O. 1981. Palaeomagnetism and Precambrian tectonic evolution of Gondwana. In: Kröner, A. (ed) *Precambrian Plate Tectonics, Developments in Precambrian Geology* **4:** 649-687. Elsevier, Amsterdam.
McWilliams, M.O. & A. Kröner 1981. Paleomagnetism and tectonic evolution of the Pan-African Damara Belt, South Africa. *Journal of Geophysical Research* **86:** 5147-5162.
McWilliams, M.O. & M.W. McElhinny 1980. Late Precambrian paleomagnetism of Australia: the Adelaide Geosyncline. *Journal of Geology* **88:** 1-26.
Moores, E.M. 1991. Southwest U.S.—East Antarctic (SWEAT) connection: a hypothesis. *Geology* **19:** 425-428.
Morel, P. 1981. Palaeomagnetism of a Pan-African diorite: a Late Precambrian pole for western Africa. *Geophysical Journal of the Royal Astronomical Society* **65:** 493-503.
Morel, P. & E. Irving 1978. Tentative paleocontinental maps for the Early Phanerozoic and Proterozoic. *Journal of Geology* **86:** 535-561.
Myers, J.S. 1990. Precambrian tectonic evolution of part of Gondwana, southwestern Australia. *Geology* **18:** 537-540.
Pallister, J.S., J.C. Cole, D.B. Stoeser & J.E. Quick 1990. Use and abuse of crustal accretion calculations. *Geology* **18:** 35-39.
Perrin, M., D.P. Elston & A. Moussine-Pouchkine 1988. Paleomagnetism of Proterozoic and Cambrian strata, Adrar de Mauritanie, cratonic Western Africa. *Journal of Geophysical Research* **93:** 2159-2178.
Piper, J.D.A. 1972. A palaeomagnetic study of the Bukoban System, Tanzania. *Geophysical Journal of the Royal Astronomical Society* **28:** 111-127.
Piper, J.D.A. 1976. Palaeomagnetic evidence for a Proterozoic super-continent. *Philosophical Transactions of the Royal Society of London.* A **280:** 469-490.
Piper, J.D.A. 1987. *Palaeomagnetism and the Continental Crust.* Open University Press, Milton Keynes.
Powell, C.McA. (Chair & transcriber) 1992. Assembly of Gondwanaland - open forum. (In: Findlay, R.H., M.R. Banks, J.J. Veevers & R. Unrug (eds.) *Gondwana 8: assembly, evolution and dispersal.* A.A. Balkema, Rotterdam.
Powell, C.McA., S.R. Roots & J.J. Veevers 1988. Pre-breakup continental extension in East Gondwanaland and the early opening of the eastern Indian Ocean. *Tectonophysics* **155:** 261-283.
Rumph, B. 1978. Regional gravity and magnetic data of the central Eromanga Basin area: Implications on the crustal structure, regional geology and tectonic history. M.Sc. thesis, University of Sydney (unpublished).
Saradeth, S., H.C. Soffel, P. Horn, D. Muller-Sohnius & A. Schult 1989. Upper Proterozoic and Phanerozoic pole positions and potassium-argon (K-Ar) ages from the East Sahara craton. *Geophysical Journal International* **97:** 209-221.
Schmidt, P.W. & B.J.J. Embleton 1990. The palaeomagnetism of the Tumblagooda Sandstone, Western Australia: Gondwana Palaeozoic apparent polar wandering. *Physics of the Earth and Planetary Interiors* **64:** 303-313.
Schmidt, P.W., G.E. Williams & B.J.J. Embleton 1991. Low palaeolatitude of Late Proterozoic glaciation: early timing of remanence in haematite of the Elatina Formation, South Australia. *Earth and Planetary Science Letters* **105:** 355-367.
Scotese, C.R. & W.S. McKerrow 1990. Revised world maps and introduction. In: McKerrow, W.S. & C.R. Scotese (eds) Palaeozoic Palaeogeography and biogeography, *The Geological Society Memoir* **12:** 1-21. The Geological Society, London.
Shackleton, R.M. 1986. Precambrian collision tectonics in Africa. In: Coward, M.P. & A.C. Ries (eds) Collision Tectonics, *Geological Society Special Publication* **19:** 329-349.
Stoeser, D.B. & V.E. Camp 1985. Pan-African microplate accretion of the Arabian Shield. *Geological Society of America Bulletin* **96:** 817-826.
Stanistreet, L.G., P.A. Kukla & G. Henry 1991. Sedimentary basinal response to a Late Precambrian Wilson Cycle: the Damara Orogen and Nama Foreland, Namibia. *Journal of African Earth*

Sciences **13:** 141-156.

Toens, P.D. 1984. The time-bound character of Uranium mineralising proccess, with special reference to the Proterozoic of Gondwana. *Precambrian Research* **25:** 13-36.

Unrug, R. 1992. The Gondwana supercontinent: Middle Proterozoic crustal fragments, Late Proterozoic assembly, and unresolved problems (In: Findlay, R.H., M.R. Banks, J.J. Veevers & R. Unrug (eds.) *Gondwana 8: assembly, evolution and dispersal.* A.A. Balkema, Rotterdam.

Vail, J.R. 1988. Tectonics and evolution of the Proterozoic basement of northeastern Africa. *In* El-Gaby, S. & R. O. Greiling (eds) *The Pan-African Belt of northeast Africa and adjacent areas.* Friedr. Vieweg & Sohn: Braunschweig/Weisbaden: 195-226.

Valencio, D.A., A.M. Sinito & J.F. Vilas 1980. Palaeomagnetism of Upper Precambrian rocks of the La Tinta Formation, Argentina. *Geophysical Journal of the Royal Astronomical Society* **62:** 563-575.

Van der Voo, R. 1990: Phanerozoic paleomagnetic poles from Europe and North America and comparisons with continental reconstructions. *Reviews of Geophysics* **28:** 167-206.

Veevers, J.J. & M.W. McElhinny 1976. The separation of Australia from other continents. *Earth-Science Reviews* **12:** 139-159.

Zijderveld, J.D.A. 1968. Natural remanent magnetizations of some intrusive rocks from the Sor Rondane Mountains, Queen Maud Land, Antarctica. *Journal of Geophysical Research* **73:** 3773-3785.

Gondwana Eight, Findlay, Unrug, Banks & Veevers (eds) © 1993 Balkema, Rotterdam. ISBN 90 5410 304 3

Upper Proterozoic – lower Paleozoic transgondwanic ruptures and events between North Africa and South America

B.A.J. Baldis, R.D. Martínez, M.E. Pereyra, A.M. Pérez, C.R. Villegas & P. Martínez de Giménez
Geotectonic and Latin American Regional Geology Division, Institute of Geology, National University of San Juan, Argentina

ABSTRACT: A synthesis of transgondwanic events between 700 and 450 Ma is presented on the basis of transcontinental correlation from northern Africa to South America. Alternate extension compression events, resulting in basin generation and granitoids intrusions interact with successive stages of transtensional processes, giving a complex pattern of basin configuration.

In recent years, several studies have considered the reconstructions of the Gondwanic basins between the ages of 700 to 450 Ma (Upper Precambrian - Lower Paleozoic) and have tried to integrate the controlling structures between Africa and South America (Kroner 1979, Fulfaro et al. 1982, Cordani et al. 1984, Caby 1989, and Baldis et al. 1990b-1991).

The main early Cambrian basins of Northern Africa, as proposed by Schandelmeier (1988), based mainly on the scheme of Klitzch (1970), originated by pull-apart processes and were then filled with Cambrian molasse sediments. They formed subparallel rifts which trend perpendicularly to the Trans African Lineament from the Nile Delta to the Chad depressions and have been named as Dakhla, Kufra, Erdis and Musbat basins occupying the western regions of Egypt, southwest Lybia, and northern Chad, in northeast Africa.

The Trans-African Lineament converges with the Central-African Lineament in the southern regions of Niger (Benin). Also it is well-known that these African lineaments and the Kandi-Hoggar Lineament correspond with the triangular-shaped, fractured region of Borborema in northeast Brazil (Nagy et al. 1976, Caby 1989). In western Africa, Destombes (1971) and Lecorche et al. (1989) have proved the existence of early Paleozoic basins associated with marginal rifts along the Mauritanides, and even in the areas where the Variscan deformation has destroyed most of the primary structural features, from Morocco to the Reguibat Uplift.

In South America, Cordani et al. (1984) have integrated the development of the Precambrian sequences with the tectonic events that resulted in the generation of many Brazilian basins; specifically those related with lower Paleozoic molasse fills and the later expansion during Siluro-Devonian times.

Three main Brazilian megastructures can be correlated with the African lineaments: the Araguaia fracture and mobile belt links to the Mauritanides; the Trans-Brazilian Lineament corresponds with the Kandi Lineament in Nigeria; the Patos and Pernambuco fractures relate to the Trans-African and Central-African Lineaments. Many small lower Cambrian molasse basins developed along these lineaments which converge towards a junction zone between the Monte do Carmo and Trans-Brazilian basins. Other important controlling elements are: the prolongation of the Araguaia Belt toward the Paraguay Belt; and the link from the Trans-Brazilian Lineament to the geographic origin of the Asuncion and Pay Ubre Uplifts. The integration map (Fig. 1) depicts all these lineaments, uplifts, fold and fracture belts, and all the other tectonic elements which are considered to have been active from 700 to 450 Ma.

In the South American central region, the most notable features consist of the subparallel controlling uplifts in the Amazon basin. The Carauari Uplift prevented the Pacific lower Paleozoic ingression (Baldis 1988) and it was later covered by the extensive Silurian-Devonian

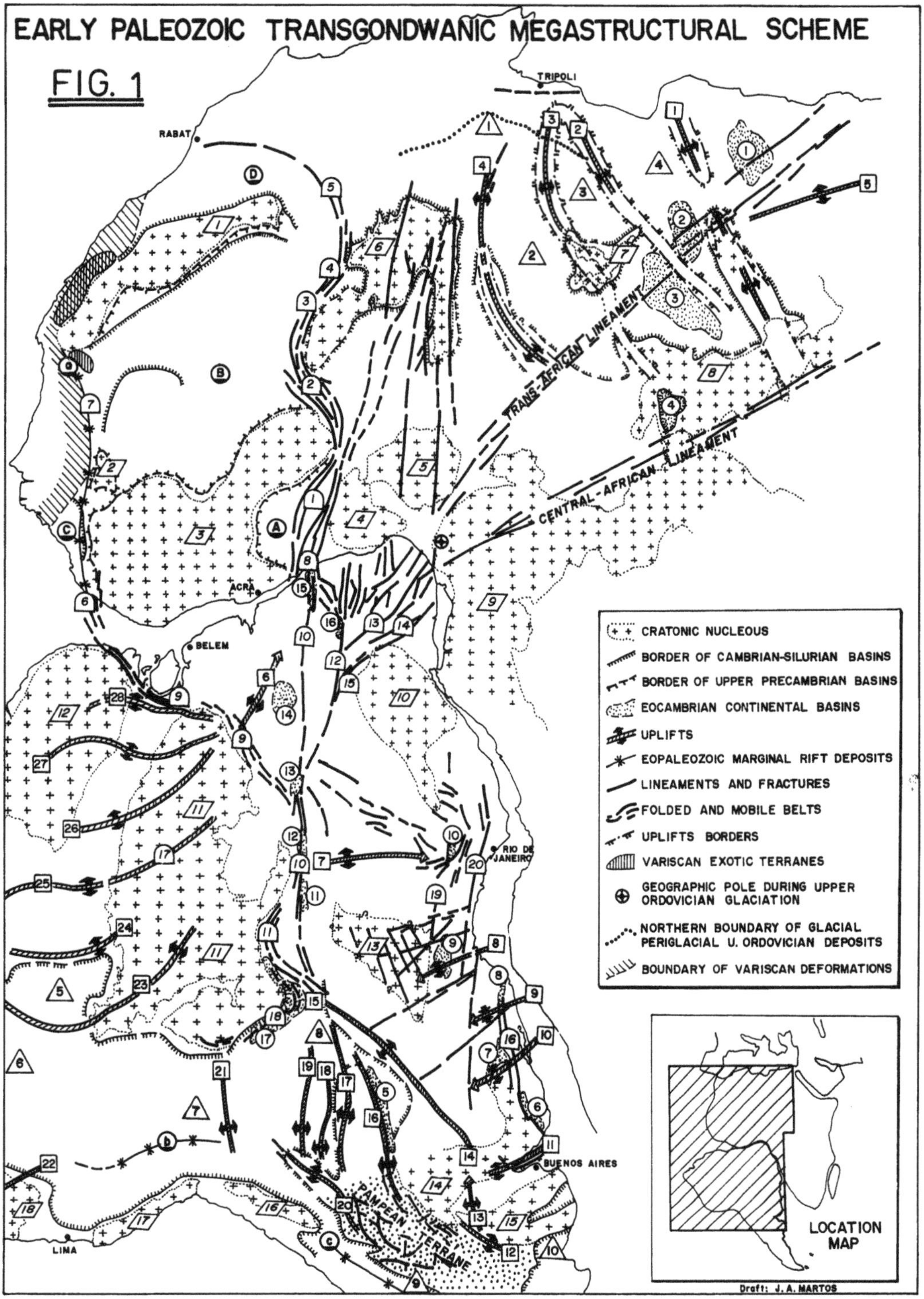
EARLY PALEOZOIC TRANSGONDWANIC MEGASTRUCTURAL SCHEME
FIG. 1
TRIPOLI
RABAT
ACRA
BELEM
RIO DE JANEIRO
BUENOS AIRES
LIMA
TRANS-AFRICAN LINEAMENT
CENTRAL-AFRICAN LINEAMENT
PAMPEAN TERRANE
CRATONIC NUCLEOUS
BORDER OF CAMBRIAN-SILURIAN BASINS
BORDER OF UPPER PRECAMBRIAN BASINS
EOCAMBRIAN CONTINENTAL BASINS
UPLIFTS
EOPALEOZOIC MARGINAL RIFT DEPOSITS
LINEAMENTS AND FRACTURES
FOLDED AND MOBILE BELTS
UPLIFTS BORDERS
VARISCAN EXOTIC TERRANES
GEOGRAPHIC POLE DURING UPPER ORDOVICIAN GLACIATION
NORTHERN BOUNDARY OF GLACIAL PERIGLACIAL U. ORDOVICIAN DEPOSITS
BOUNDARY OF VARISCAN DEFORMATIONS
LOCATION MAP
Draft: J. A. MARTOS

MOBILE BELTS AND MOBILE FRACTURE LINEAMENTS

1-Dahomey
2-Gourma
3-Adrar-DesIforas
4-Pharusian
5-Ougarta
6-Rockelides
7-Mauritania
8-Sobral
9-Araguaia-Tocantins
10-Transbrazilian
11-Paraguay
12-Pompeu
13-Paraiba
14-Pernambuco
15-Rio Preto M.B.
16-Canguzú
17-Aripuaná
18-Sunsas Belt
19-Jacutinga Fault
20-Corupá Fault

MOLASSIC EARLY PALEOZOIC BASINS

1-[illegible]
2-Kufra
3-Erdis
4-Musbat
5-Las Breñas
6-Piedra de Afilar
7-Camacuá
8-Itajaí
9-Castro
10-Eleuterio
11-Piranhas
12-Agua Bonita
13-Monte de Carmo
14-Transbrazilian
15-Jaibaras
16-Cococi
17-Sunsas

UPLIFTS

1-Calanscio-Uweniat-Hoggar
2-Haroudj
3-Trípoli-Tibesti
4-Tihemboka
5-Uweniat-Bir-Safsaf-Aswan
6-Xambioá
7-Goiania
8-Ponta Grossa
9-Uplift-Syncline of Torres
10-Río Grande
11-Martín García
12-Laboulaye
13-Camilo Aldao
14-Pay Ubre
15-Asunción
16-Charata-Las Breñas
17-Pampeana
18-Quirquincho
19-Michicola
20-Puneña
21-Northern Bolivian
22-Marañacu
23-Iquitos
24-Carauari
25-Purus
26-Itacoatiara
27-Monte Alegre
28-Gurupa

CRATONIC NUCLEOUS

1-Reguibat
2-Kenieba
3-Leo
4-Benin
5-Nigeria
6-Hoggar
7-Tibesti
8-Nubia
9-Zaire
10-San Francisco
11-Central Brazilian
12-Eastern Guyana
13-Cratonic Nucleous of Parana Basin
14-Río de la Plata
15-Tandil
16-Antofalla
17-Areqipa
18-Western Ecuador

CAMBRIAN-ORDOVICIAN-SILURIAN PLATFORM BASINS

1-Homra
2-Murzug-Djalo
3-Dor-el Gusa
4-Calanscio
5-Solimoes
6-Apure
7-Central Andes
8-Chaco
9-Cuyo
10-Ventana

UPPER PRECAMBRIAN BASINS

A-Volta
B-Taoudei
C-Bove
D-Tindout

MARGINAL RIFTS

a-Mauritanides-Bassarides Rift-Terrane (Early Paleozoic)
b-Precordillera Marginal Rift
c-Central Peruvian Early Paleozoic Rift

transgression which reached further into the basin. The Paraguai Mobile Belt links with the Sunsas Belt trough an area of molasse and associated platform, lower Paleozoic sediments. The Chaco region, which consist of northern Argentina, Bolivia and Paraguay is highly controlled by subparallel uplifts that began at the junction with the Trans-Brazilian Lineament. These uplifts describe a divergent, sigmoidal pattern which converges into a group of regional fractures that characterizes the borders and central regions of the Pampean terrane in central Argentina.

The early Paleozoic Andean basins were mainly subject to the control of marginal rifts located along the central Andes of Bolivia and Argentina (Damm et al. 1989) and at the Precordillera basin of western Argentina (Kay et al. 1984; Baldis et al. 1992).

The Paraguai-Pay Ubre Uplift bounds the Chaco and Parana basins. The Parana basin, in southeast Brazil shows a subsided central cratonic nucleus (Cordani et al. 1984; Baldis et al. 1990b) which controls several minor uplifts generated from the Rio de la Plata Craton and the Dom Feliciano Belt. Also, some Early Cambrian molasse basins have been generated under the tectonic control of the Dom Feliciano Belt as well as many transcurrent faults, such as the Jacutinga, Corupa and others. These molasse basins are smaller and more numerous than the ones generated through the same type of pull-apart transtensional processes in northern Africa (Schandelmeier 1988).

The overall structure forms a series of Transgondwanic structures with defined "bipolar" affects. This is referred to similar structural patterns at both extremes of the fractures systems. The common effects mentioned before are represented by:

- The existence of transtensional stages along megastructures with definite right-lateral displacements producing Lower Cambrian pull-apart basins aligned with a northeast trend in Egypt, Libya, and Chad (Schandelmeier 1988).
- The opening of the main ruptures in two "fan-shaped" systems: a) the African system, comprising the Trans-African Lineament, the Central-African Lineament, the Gourma-Ougarta Belt, the Hoggar-Tibesti, and the Rockelides Belt with the apex located on central-east Brazil (Fig. 1); and b) the South American system including the Paraguai, Trans-Brazilian, Pay Ubre, Sunsas, Las Breñas, Michicola and Quirquincho Belts and that could also comprise the Trans-Amazonian lineaments. These last ruptures seem to converge along the Amazonian cratonic area into the apex of the Tocantins-Araguaia mobile belt (Cordani et al. 1984).

From the synthesis of events presented by Soares (1988), and Pereyra & Villegas (1990), and our own compilations, several activation events can be postulated in association with the Transgondwanic megastructures of northern Africa and South America:

a) a first series of mobile events, possibly associated with Pan-African suturing processes, which resulted in the generation of several mobile belts between 850 and 600 Ma. The geometry of the ruptures is characterized by typical sigmoidal distribution patterns for the mobile belts and definite linear shapes for the fracture systems. It also is represented by cordilleran to island-arc type environments at the western part of the Pan-African belt which displays a series of synorogenic granitoids at about 600 Ma (Caby 1987).

b) a regional event, between 500 and 540 Ma, defined by transcurrent, right lateral displacements, along the main trans-continental fracture systems resulting in the opening of complemented by high level granitoid intrusions and alkaline dykes (Caby 1987, Schandelmeier et al. 1987; Schandelmeier 1988).

c) several successive processes of marginal rifting along the peri-Gondwanic platforms at about 500 Ma.

d) peri-Gondwanic rifting located transversely to the transcontinental structures and interior platforms at about 470 Ma. This event has been named as "Guandacol Phase" in the Andean region, and can be correlated with the "Late Brasiliano Cycle" in Brazil and tentatively compared to the very last stages of the "Late Pan-African Event" at some places in Africa.

Related to this sequence of events in South America, the tectonic behaviour of the suture of the Pampean terrane with the Rio de la Plata Craton in Argentina shows the cyclic effects of alternate compression and extension represented by sigmoidal ruptures within the Pampean terrane (Baldis et al. 1989). These ruptures are associated with basement uplifts related to Transgondwanic fracture and fold belts generated during the Brasiliano Cycle and reactivated up to the present.

The comparison of the tectonic sequences of events for several localities on the periphery of the Rio de la Plata Craton in southern South America shows a correlation of successive extension-compression periods. The region of analysis comprises areas of northwest and central-west Argentina, Uruguay, and southeast Brazil. The effects of the transcontinental control are represented in most places by correlated episodes of: rifting, generation of pull-apart basins, metamorphism to anphibolites facies, and are associated with mobile belts and granitoid intrusions.

The further understand all these tectonic events, a general stratigraphic correlation of several sequences, ranging from Middle Llanvirnian to Lower Caradocian ages along the South American Andean platforms, has resulted in the correlation of a distinctive tectonic event, at approximately 465 Ma (Baldis et al. 1992), named the Guandacol diastrophic phase. From the Los Llanos basin in Colombia, to southeast Bolivia, and continuing to the south with the northwest Argentina sequences of Zapla and Aguilar, the Western Puna sequences of Poquis, Aguada La Perdiz, Cordon de Lila in northeast Chile, and the classic Precordillera outcrops of Guandacol and Villicun, this is a distinctive feature characterizing this phase with marked unconformities in most of these place and transitional sequences in others. The lower Paleozoic reconstructions resulting from all these various data outline different tectonic models for each region of analysis (Baldis et al. 1992).

REFERENCES

Baldis, B.A. 1988. Posibilidades de Correlación del Ordovícico en Los Llanos Colombianos y la Alta Amazonia Brasileño-Ecuatoriana. Anais do VII Cong. Latinoam. de Geología, I, 500-505.

Baldis, B.A. 1989. Regmagenic fracturation in South America. Correlacion Geológica Nº 5, Tucumán, Argentina.

Baldis, B.A.; R.D. Martínez; M.E. Pereyra; & C.R. Villegas 1990a. Problems in Peri-Gondwanic Lower Paleozoic between Andes and Tethys. Memoire Geol. 8, Laussane, Switzerland.

Baldis, B.A.; R.D. Martínez; M.E. Pereyra; A.M. Pérez & C.R. Villegas 1990b. Fracturación Transversal del Continente y sus efectos desde Nord Africa hasta la Argentina Central. Pub. Fundación Univ. Nacional de San Juan.

Baldis, B.A.; R.D. Martínez; M.E. Pereyra & C.R. Villegas 1992. An Approach to the Megastructural Transgondwanic control during Early Paleozoic, from the Paleotethys to the Andes. Geologica Balcanica, 21, 1, 5-12.

Baldis, B.A.; R.D. Martínez; M.E. Pereyra, A.M. Pérez & C.R. Villegas 1992. Ordovician events in the South American Andean platform, in: Global perspective on Ordovician Geology. Webby & Laurie (eds.). Balkema Rotterdam.

Caby, R. 1987. The Pan-African belt of West Africa the Sahara Desert to the Gulf of Benin, in: Shaer, J.P., and Rogers, J. (eds.), Anatomy of mountain ranges: New Jersey, Princeton University Press, p. 129-170.

Caby, R. 1989. Precambrian terranes of Benin-Nigeria and northeast Brazil and the Late Proterozoic South Atlantic fit. in: Dallmeyer R.D. (ed.) Terranes in the Circum-Atlantic Paleozoic Orogens, G.S.A. Special paper 230. 145-150.

Cordani, U.G.; B.B.B. Neves; R.A. Fuck; R. Porto; A.T. Filho & F.M.B. Cunha 1984. Estudio preliminar de integracao de Pre-Cambriano com os eventos Tectonicos das bacias sedimentares Brasileiras. Rio de Janeiro. Petrobras-Cenpes-Sintep, Brazil.

Damm, K.W.; S. Pichowiak; R.S. Harmon; W. Todt; R. Omarini & H. Niemeyer 1989. The old Central Andes. A reflection of an active continental margin? G. S. A. Special Paper.

Destombes, J. 1971. L'Ordovician au Maroc Essai de synthese stratigraphique. Memoire du B.R.M. N 73.

Fulfaro, V.J.; A.P. Saad & R.B. Vianna, 1982. Compartimentacao tectonica de Bacia do Parana. Rev.Brasileira de Geoc., 12/4; 590-611.

Kay, S.M.; V.A. Ramos & R.W. Kay 1984. Elementos mayoritarios y trazas de las vulcanitas Ordovícicas de la Precordillera Occidental. Basaltos de Rift oceánico temprano (?), próximos al margen continental. IX Cong.Geol.Argentino, II:48-65.

Klitzsch, V.E. 1970. Die Stuckturgeschichte der Zentralsahara, Geologische Rundschau, V. 59; N 1-3, 459-525.

Kroner, M. 1979. Pan African plate tectonics and its repercussions on the crust of northeast Africa. 28th I.G.C. Abtr., 2; 273-274.

Lecorche, J. P.; R. D. Dallmeyer; M. Villeneuve 1989. Definition of tectonostratigraphic terranes in the Mauritanide, Bassaride, and Rokelides orogens, West Africa, in: Terranes in Circum-Atlantic Paleozoic Orogens, Dallmeyer, R. D. (ed.), G.S.A. Spec.paper, 230.131-144.

Pereyra, M.E.; C.R. Villegas 1990. Eventos periféricos al Cratón Río de la Plata entre los 500 a 600 Ma. XI Congr.Geolog. Argentino, II, 347-351.

Nagy. R.M.; M.A. Ghuma & J.J.W. Rogers, 1976. A crustal structure and lineament in North Africa. Tectonophysics, Amsterdam, 31, 167-172.

Schandelmeier, H.; A. Richter & U. Harms 1987. Proterozoic deformation of East Sahara Craton in SE Lybia, Southern Egypt and Northern Sudan. Tectonophysics, 140- 233-246.

Schandelmeier, H. 1988. Pre - Cretaceous intraplate basins of NE Africa. Episodes II, N 4, 270-274.

Soares, P.C. 1988. Tectonica colisional en torno de bloco Parana, Brasil. Anais do VII Cong. Latino-Americano de Belem, Brasil, I, 63-79.

Gondwana Eight, Findlay, Unrug, Banks & Veevers (eds) © 1993 Balkema, Rotterdam. ISBN 90 5410 304 3

Late Proterozoic tectonism in the interior of the São Francisco craton

Farid Chemale Jr
Department of Geology, UNISINOS, São Leopoldo, RS, Brazil

Fernando F. Alkmim & Issamu Endo
Department of Geology, UFOP, Ouro Preto, MG, Brazil

ABSTRACT: The São Francisco (SF) craton is underlain by Archean through Proterozoic continental crust, and is bordered all sides by Brasiliano mobile belts which verge toward the interior of the craton. In map view, the São Francisco craton is hook shaped. The N-S segment, here called Domain I, and the E-W trending segment, here called Domain II, join along a NNW-SSE doubly verging deformed belt, the ESPB belt, which includes the Espinhaço Setentrional, the Paramirim basement high and the western portion of the Chapada Diamantina. Our structural analysis data point out that Mid and Upper Proterozoic supracrustal rocks exposed in the interior of the craton were affected by Brasiliano Orogeny, in the style of thin-skinned foreland fold-thrust belt. They display the same style and orientation of structures than those that can be observed in the adjacent Brasiliano belts. The structural framework of the ESPB suggests that the SF craton was possible subdivided into two parts during Brasiliano. We could not recognize a record of Uruaçuano deformation (1.3 ba bp) in the Mid Proterozoic strata in the SF craton, which had been previously described.

1 INTRODUCTION

The São Francisco (SF) craton (Almeida 1967, 1977) of southeastern Brazil is one of four "cratons" in the country. These cratons are defined as regions which are underlain by Archean through Lower Proterozoic continental crust (granite-greenstone belts; high-grade terranes; and Early and Middle Proterozoic supracrustal sequences), and were largely unaffected by the Late Proterozoic Brasiliano Orogeny (Pan-African equivalent; around 600 Ma bp). The SF craton is a Brasiliano or Late Proterozoic structural (or geodynamic) unit, which, in map view, is hook shaped (Figure 1). The N-S segment, here called Domain I, and the E-W trending segment, here called Domain II, join along a NNW-SSE doubly verging deformed belt, which includes the Espinhaço Setentrional, the Paramirim basement high and the western portion of the Chapada Diamantina (Figure 1). For simplicity this structural zone will be refered to as the ESPB belt (Espinhaço Setentrional-Paramirim Block belt). Domain II was continous with the Congo-Zaire craton in Africa, prior to Mesozoic rifting.

Brasiliano mobile belts border all sides of the SF craton, except for the rifted Atlantic margin. The Brasília belt lies on the west, the Araçuaí on the east, the Rio Preto, the Riacho do Pontal and Sergipano on the north, the Mantiqueira on the southeast, the Rio Pardo on the south (Figure 1). All of these belts verge toward the interior of the craton.

The purpose of this paper is two fold. First, we will demonstrate that a large part of the Late Proterozoic strata in the interior of the craton was deformed during the Brasiliano Orogeny, and that the style of this

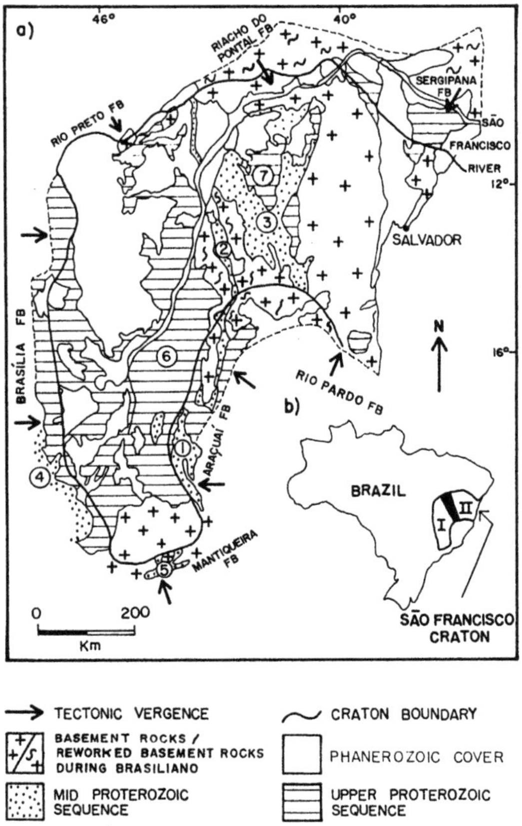

Fig.1 (a) Simplified geologic map of the SF craton showing Mid and Upper Proterozoic sequences, the basement rocks, and adjacent FTBs. b) Location of the SF craton with Domains I, II, and the ESPB belt (black) (modified from Mascarenhas et al. 1984). 1 = Espinhaço Meridional, 2 = Espinhaço Setentrional, 3 = Chapada Diamantina, 4 = Canastra Group, 5 = São João del Rei Group, 6 = SF basin, 7 = Irecê basin

deformation is characteristic of the foreland portion of fold-thrust belts (FTBs) . Second, we will show that deformation of Middle Proterozoic strata in the Domain II and in the ESPB belt ocurred during the Brasiliano orogeny, not during the 1.3 ba Uruçuano event as has been previously thought (Jardim de Sá et al. 1976; Almeida 1977; Brito Neves et al. 1979; Inda & Barbosa 1978; Moutinho da Costa & Inda 1982). Thus, the two domains of the SF craton were separated by an ensialic basin until the end of the Proterozoic. Our conclusions are based on detailed field-structural analysis, gravimetry, lithostragraphic data, and recently available geochronological data.

2 STRATIGRAPHIC FRAMEWORK OF THE SÃO FRANCISCO CRATON

Rocks older than 1.8 ba bp are exposed in the south part of Domain I, in the ESPB belt, and in the western portion of Domain II (Figure 1). These exposures are composed of Archean gneiss, granite-greenstone terrane and Early Proterozoic supracrustal sequences, which are the basement rocks for Middle and Upper Proterozoic sequences. According to Távora et al. (1967), Cordani (1973a, 1973b, 1978), and Inda & Barbosa (1978), basement rocks along the south border of Domain II and in the ESPB belt were sufficiently reheated during the Brasiliano Orogeny (600 Ma) to reset the K-Ar system. Based on radiometric data Cordani (1973b) has suggested thah the SF craton should be cut in two parts along the Paramirim Province which corresponds in this paper to the ESPB belt. However Almeida (1977) interpreted the Brasiliano K/Ar ages (650 to 450 Ma) as just a thermic event without tectonic significance.

Mid-Proterozoic strata in the interior of the SF craton include the Espinhaço Supergroup, a sequence of volcanogenic, alluvial, coastal eolian, deltaic and marine sediments (Figure 2). Espinhaço rocks are exposed in the core of doubly plunging anticlines in the eastern part of Domain I, along the Espinhaço Setentrional and in the Chapada Diamantina (Figure 1).

Much of Domain I encompasses the SF hydrographic basin, which was a locus of deposition during Late Proterozoic (Figure 1). Strata deposited during this time are known as the SF Supergroup.

The SF Supergroup is divided into two units, the Macaúbas Group (and correlatives the Jequitaí and Bebedouro Formations) (Figure 2) and the overlying Bambuí Group. Macaúbas Group sediments are representative of a widespread glaciation event (Dardenne 1978; Karfunkel & Hoppe 1988), which is dated at about 950 Ma. Shallow-marine strata, probably representative of cratonward expansion of Brasiliano marginal basins, during the post-glacial high sea level event (Alkmim et al. 1989), comprise the lower and middle parts of the Bambuí Group (Figure 2). These rocks are overlain by a shallow-marine to alluvial unit, the Três Marias Formation, which represents deposition in a foreland basin formed in advance of Brasiliano FTBs (Costa & Angeiras 1971; Marshak & Alkmim 1989; Ortu 1990) (Figure 2). SF Supergroup strata also crop out in the north-central and eastern border of the Chapada Diamantina (Figure 1).

Reactivation of subsidence in the SF craton occurred in Mesozoic time. The northern half and part of the western portion of the SF basin is now covered by Cretaceous sediments (Figure 1). Along the eastern border of Domain II, Neocomian rifting associated with opening of the Atlantic resulted in the formation of the Recôncavo, Tucano and Jatobá basins (Figure 1).

3 STRUCTURAL FEATURES OF DOMAIN I

The southern portion of Domain I, where the Upper Proterozoic supracrustal rocks are not covered by younger strata, can be divided into three zones (Alkmim et al. 1989): a) a Western zone which is the influence zone of the E-verging Brasília FTB; b) an Eastern zone which is the influence zone of W-verging Araçuaí FTB; c) a Central zone, subdivided into C1 and C2, corresponding to two regions, where the sediments of the SF Supergroup are not deformed (Figure 3). The Brasília and Araçuaí Influence Zones, W and E zones, respectively, are in fact the intermediate to external zones of those FTBs .

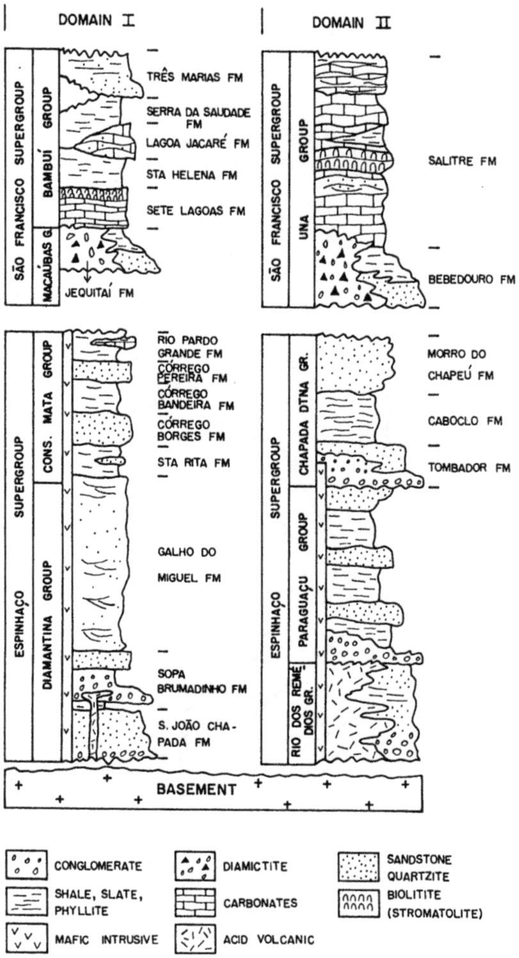

Figure 2. Stratigraphy columns of Mid and Upper Proterozoic strata in the interior of SF craton. a) Domain I (after Marshak & Alkmim 1989). b) Domain II (after Inda & Barbosa 1978; Pedreira 1988; Souza 1986; Monteiro et al. 1984).

3.1 Brasília Influence Zone

The western portion of Domain I displays the effects of two deformational events. The first event is extensional, and its occurrence within the SF basin is based primarily on gravimetric data reported by Lesquer et al. (1981). Gravity anomalies outline basement highs and lows whose shape is suggestive of rift-basin geometry. Structural and lithostra-

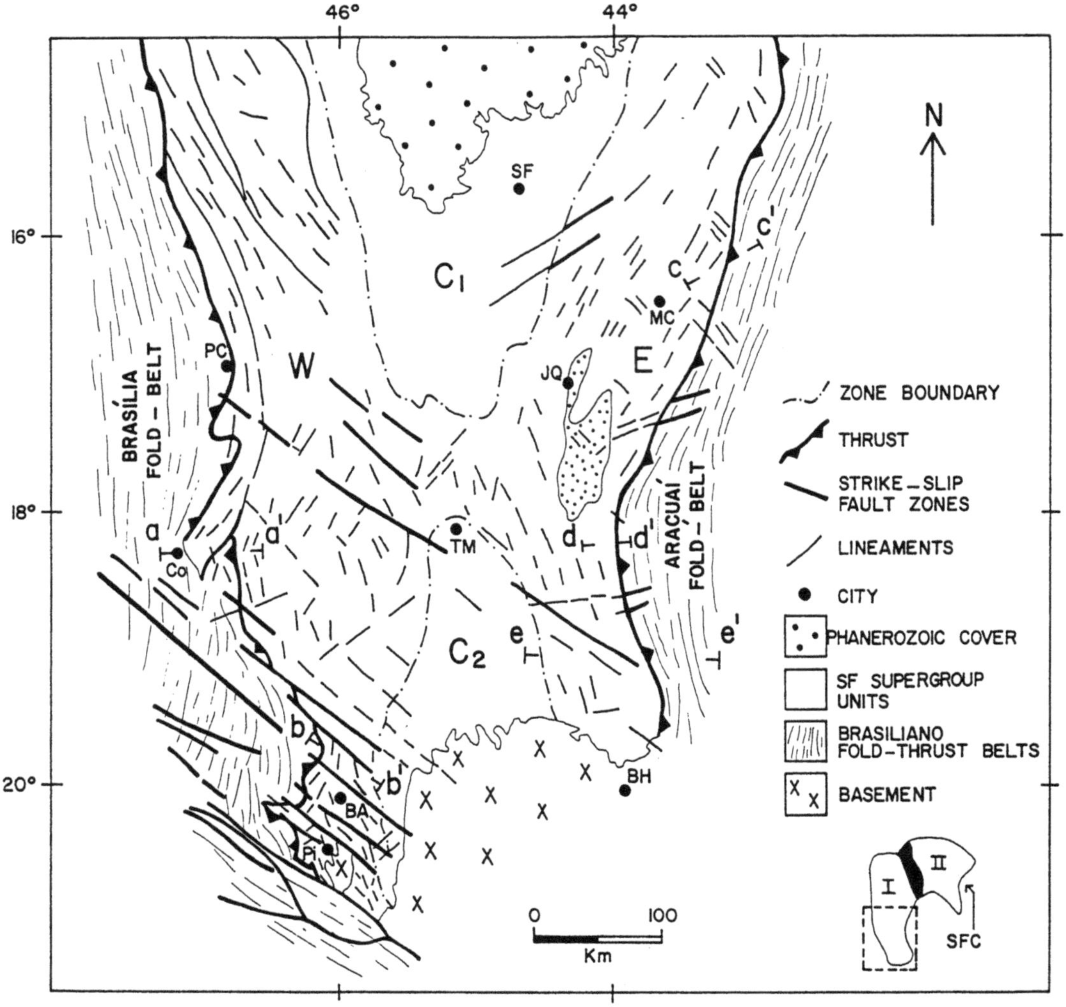

Fig 3 Structural map of the southern portion of Domain I (modified from Alkmim et al. 1989). W = Brasília Influence Zone; E = Araçuaí Influence Zone; C1 and C2 =indeformed zones. Cities: SF = São Francisco; MC = Montes Claros; JQ = Jequitaí; TM = Três Marias; BH = Belo Horizonte; PC = Paracatu; Co = Coromandel; Ba = Bambuí; Pi = Piumhí. Stippled area in the Araçuaí Influence Zone corresponds to weakly deformed Mid-Proterozoic units.

tigraphic field work, carried out by Muzzi Magalhães (1989), Bacellar (1989), and Pinho et al. (1989) suggest that both deposition and deformation of Bambuí Group lower and middle units were controlled by this basement topography.

The second event is compressional, and is manifested in Bambuí Group strata by the occurrence of detachment faults, thrust faults, east-verging flexural folds (Figure 4a and b), and in some places, close to the Brasília belt, westward-inclined cleavage with a down-dip mineral lineation. The intensity of this deformation diminishes eastward, and developed in ductile-brittle conditions. In the south corner of Domain I, the belt of compressional structures terminates against a series of NW-SE trending sinistral strike slip faults (Figure 3). Lithostratigraphic relations suggest these faults had

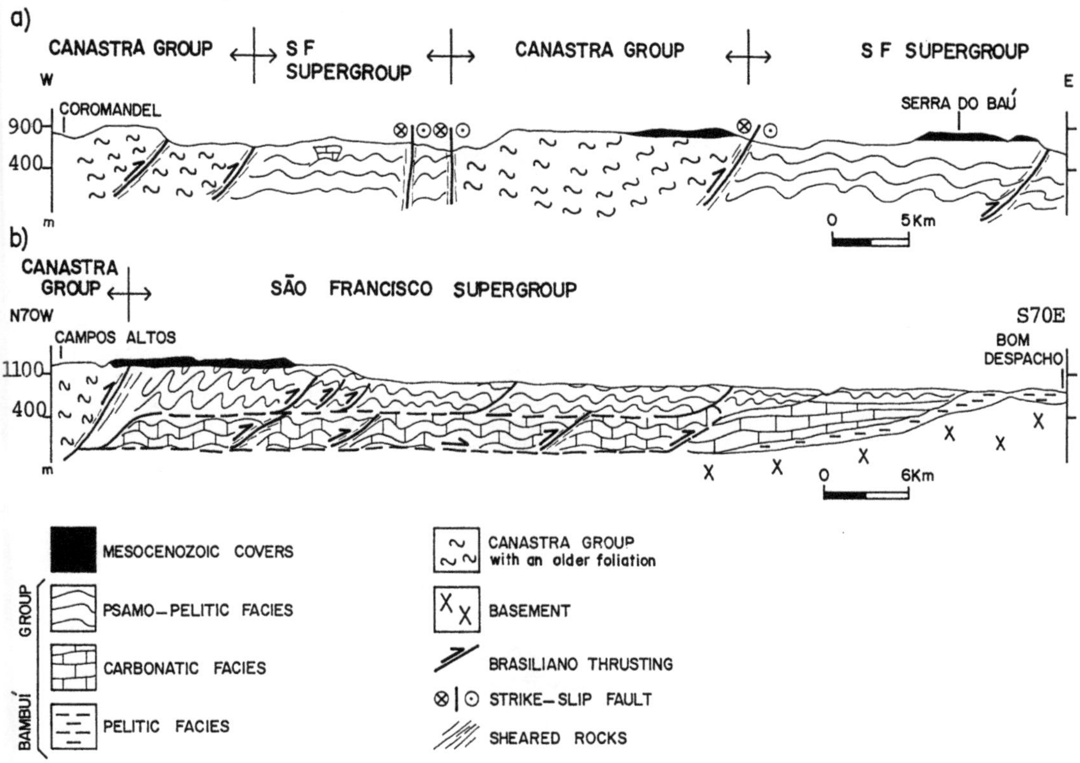

Fig. 4 Schematic cross sections along the western border of SF basin (for location see Figure 3). a) a-a′ section adapted from Bacellar (1989); b) b-b′ section adapted from Muzzi Magalhães (1989).

been active during the extensional event as well. Sinistral movements on these faults rotated N-S trending compressional fabrics to a trend of N60°W developed during the compressional event (Muzzi Magalhães 1989).

Our proposal that the structures affecting Bambuí strata in the western edge of Domain I are related to tectonism in the Brasília belt is based on correlation with structures of similar style and geometry within the Brasília belt. The Brasília belt adjacent to Domain I is underlain primarily by quartzite and phyllite/schist of the Canastra Group. The rocks developed a mylonitic foliation and a penetrative E-W mineral lineation prior to the deformation which affected the Bambuí Group. The age of the early foliation is related either to Uruçuano (1.3 ba) or to initial phase of Brasiliano Cycle. The extensional event is manifested by occurrence of normal faults which cut Canastra mylonite, and the compressional event is manifested by inversion of normal faults and by refolding and thrusting of the mylonite, carring eastward over strata of the Bambuí Group (Muzzi Magalhães 1989). Bonhomne (1976) and Iyer (1984) suggest that this thrusting took place around 600-450 Ma.

3.2 Araçuaí Influence Zone

On the eastern side of Domain I, the main compressional event is manifested in strata of the Bambuí and Macaúbas Groups by ocurrence of west-directed thrusts, west-verging inclined folds, and related structures (Figure 5 c, d and e), which are coeval to those structures of the second compressive event of the Brasília Influence Zone (Magalhães 1988). Deep-

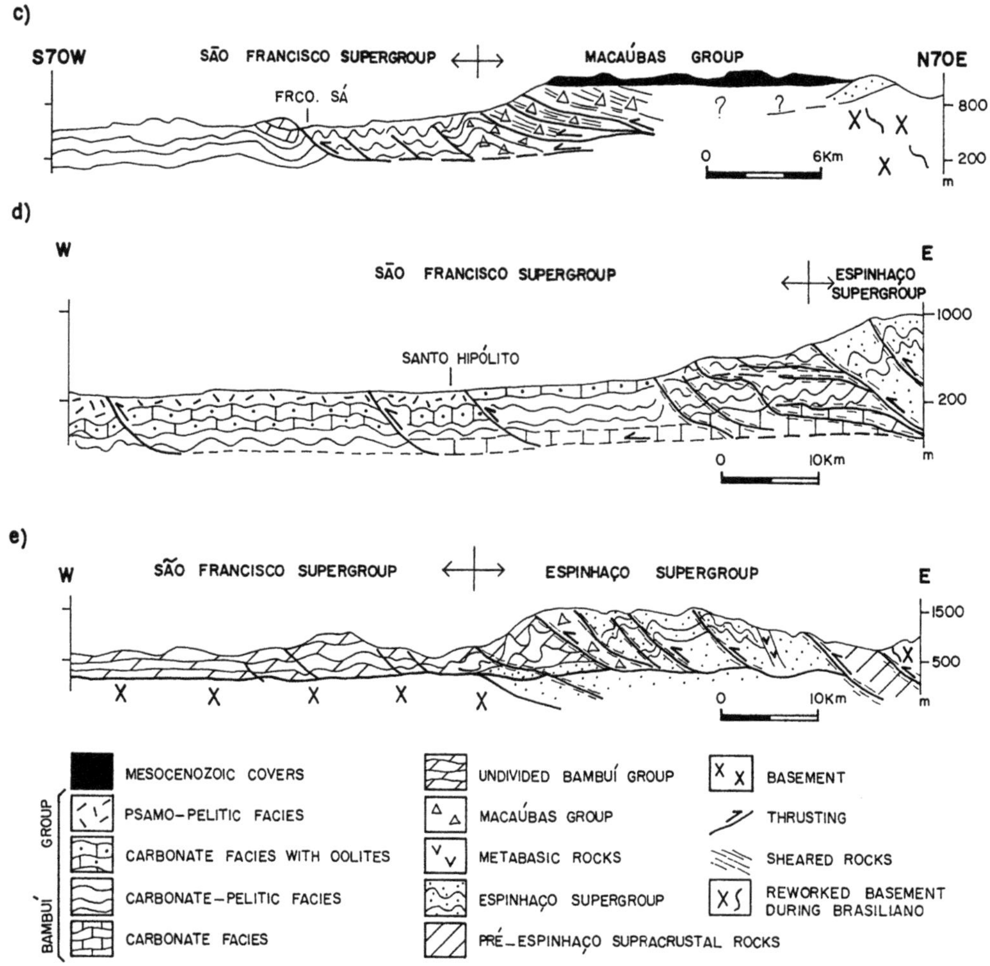

Fig. 5 Schematic cross sections along the eastern border of SF basin (for location see Figure 3). c) c-c′ section adapted from Oliveira 1989; d) d-d′ section adapted from Bacellar (1989); e) e-e′ section adapted from Dossin et al. (in prep.)

level exposures along Rio das Velhas and in the vicinity of Lagoa Santa, indicate that Macaúbas Group pinches out to the center of the domain, and that a subhorizontal detachment, defined by intensely sheared carbonate, occurrs at the base of the Bambuí Group, just above the basement/Bambuí contact (Figure 5e). N-S trending spaced cleavage in susceptile units of the Bambuí Group, as well as open folds, can be traced as far west as Sete Lagoas (Magalhães 1988 and Alkmim et al. 1989). The intensity of deformation in the SF Group increases progressively eastwards. The metamorphism on this side of Domain I, eastern border of SF basin, is of low greenschist facies and decreases westwards to very low grade (anchizone) in the direction of the basin center (Schöll 1973).

The Araçuaí belt adjacent to Domain I is represented by units of the Espinhaço Supergroup (Figure 5 e) which shows the same style and orientation of structures as those of the SF Supergroup (Figure 5 c, d and e). Thus, an older compressional event like the one manifested in the

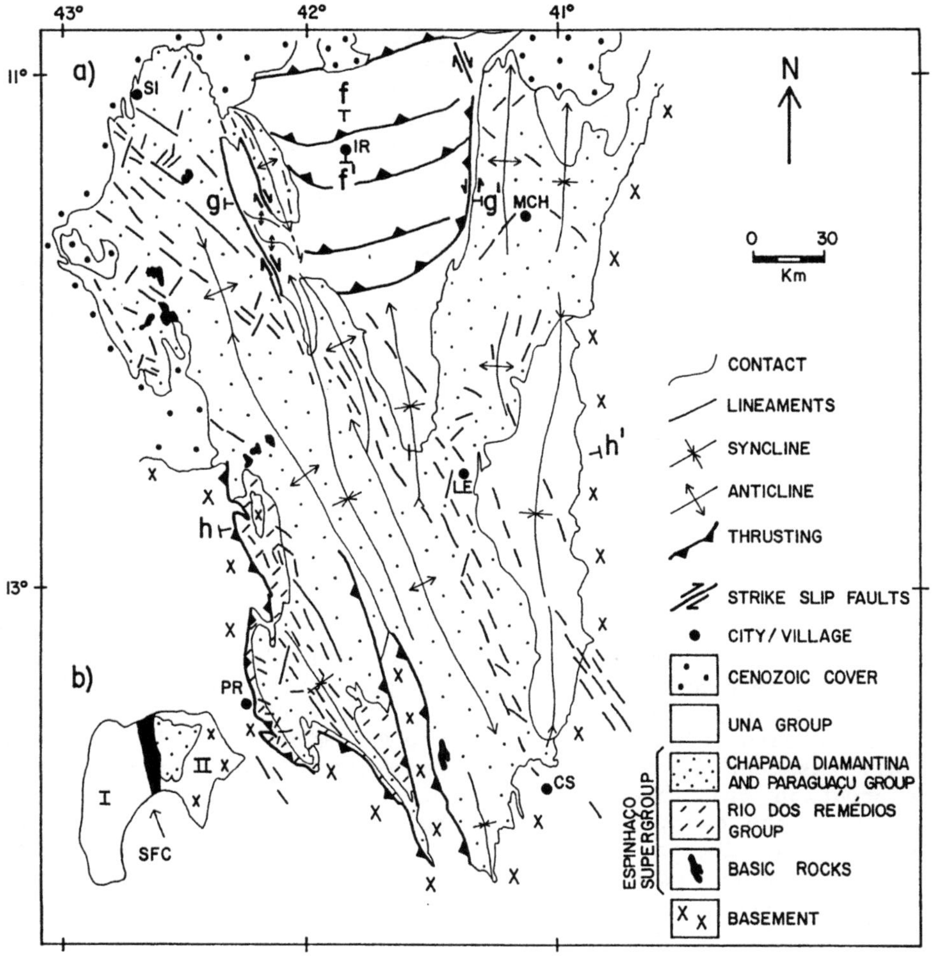

Fig. 6 a) Structural map of the Chapada Diamantina region (after Danderfer Filho (1990) and Lagoeiro (1990)). MCH = Morro do Chapéu; LE = Lençóis; CS = Contendas do Sincorá; IR = Irecê; SI = Santo Ignácio; PR = Paramirim b) Map of SF craton with Domains I, II, and ESPB belt (black). For Domain II: stippled = Mid and Upper Proterozoic sequences; cross = basement

Canastra Group has not been found in Espinhaço Supergroup units (Uhlein et al. 1986). In fact, the western boundary of the Araçuaí belt is a topographic/lithologic boundary, not a profound strain discontinuity. The structural boundary is a N-S thrust trace to the east of which are thrust slices containing resistant quartzite of the Espinhaço Supergroup and the underlying crystalline basement. Presumably, the detachment at the base of SF Supergroup roots beneath the thrust slices of Espinhaço strata (Marshak & Alkmim 1989).

3.3 The Undeformed Zones

Structures of the Araçuaí and Brasília influence zones overlap in a node at the center of Domain I (north of Três Marias, Figure 3); at the latitude of this node, Late Proterozoic strata (Bambuí Group) are deformed across the entire width of the Domain. But to the north and south of this node, there are intervals in which no deformation fabrics have been identified. These undeformed regions, called C1 and C2 subzones (Figure 3), appear to be related to pre-Brasiliano highs which controlled not only the deposition but also influenced markedly the inversion of the SF basin.

4 STRUCTURAL FEATURES OF DOMAIN II

In Domain II, Espinhaço and São Francisco Supergroups strata occur in a high plateau area called Chapada Diamantina in the State of Bahia. Both units are involved in a NNW-SSE striking fold-thrust system whose hinterland is part of the ESPB belt (Figure 6).

The structures in the Chapada Diamantina show the effect of two superimposed non-coaxial deformations. The first deformation results in a NNW-SSE trending open doubly plunging folds, high angle reverse faults and thrusts (Figures 6 and 7 g and h). Basement is involved in the deformation only near to the ESPB belt; to the east, structures are thin-skinned. Strain, magmatism and metamorphism increase progressively toward the west, and the structures verge toward the east (Danderfer Filho 1990) (Figure 7g and h).

The second event is related to E-W striking south-verging Riacho do Pontal FTB that occurs along the northern border of Domain II (Figure 1). Structures associated with the E-W belt are well displayed in the Irecê basin, a north plunging open syncline (Figures 6and 7f), in whose trough carbonates of the SF Supergroup were transported to the south over a detachment surface (Lagoeiro 1990). Inclined flexural to shear folds, thrusts, duplexes and high angle reverse faults show an outward vergence from the Irecê syncline (Figures 6 and 7f).

Danderfer Filho (1990) and Lagoeiro (1990) interpreted the NNW-SSE trending fold-thrust structures to represent the closure of an ensialic (rift) basin during the Brasiliano event. Their recognition of a Brasiliano age for this event is based on correlation of structures found in SF Supergroup strata with the structures involving Mid Proterozoic sediments. The E-W striking structures simply represent the influence zone of the Riacho do Pontal fold belt along the northern border of Domain II, which developed after the first event (NNW-SSE striking structures) during the Brasiliano Orogeny, too.

5 THE STRUCTURAL FEATURES OF ESPB BELT

The Espinhaço Setentrional-Paramirim Block belt (ESPB belt) is a intracratonic deformed zone that can be divided into three parts. The western part is the Espinhaço Setentrional, in which Espinhaço Supergroup strata (Figure 2) comprise a west verging FTB. The central area corresponds to the Paramirim block in which Archean basement and younger granitoids are exposed. The eastern part of the ESPB belt is the hinterland of the Chapada Diamantina belt (Figures 1 and 8). The Espinhaço Setentrional displays effects of two deformational events. The first event is compressional, coeval with the first event of Domain II, and represents initiation of a west-verging thrust belt (Section A-A′, Figure 8). This event is manifested by ocurrence of thrust faults, west-verging folds, intrastratal displacement along the bedding, eastward inclined cleavage. During this event, a N-S trending oblique shear zone with a sinistral displacement also developed. The Espinhaço Setentrional units have been metamorphosed from anchimetamorphism to greenschist facies, decreasing westwards. The second event resulted in formation of N-S striking normal mesofaults and shear zones, which are exposed only on the west side of Espinhaço Setentrional. The age of the extensional event is unclear.

Within the Paramirim block, the Central area of the ESPB zone, there are NNW-SSE high angle thrusts which verge toward east (Section A-A', Figure 8), and are connected with NNW-SSE trending structures of the eastern part of the ESPB belt wich continue into the units of Domain II (Section h, Figure 7).

The present geometry of the ESPB belt and adjacent areas is comparable to a large positive flower structure (Figure 8), which may represent the inversion of a pair of ensialic rift basins separated by a basement high, that corresponds to the Espinhaço aulacogen of Moutinho da Costa & Inda (1982). The basement high is the present Paramirim block. Though strata on either side of the Paramirim block have traditionally correlated, and both have been considered as Espinhaço Supergroup units, it is important to emphasize that the sediments in the two basins are quite dissimilar; in the Espinhaço Setentrional, Espinhaço strata are deep marine (turbidites) (Dominguez & Rocha 1989), whereas the Espinhaço of the Chapada Diamantina are shallow marine and fluvial. It suggests that the two basins were never connected.

During basin closure sinistral strike-slip and associated transpression appears to have led to development of flower structure, in which basement is thrust over the bordering strata. The gravimetric data also display a strong anomaly related to the ESPB (see Figure 8). In effect, the ESPB is a transition zone between the two Domains of the SF craton.

Previous authors (Jardim de Sá et al. 1976; Almeida 1977; Brito Neves 1979; Inda & Barbosa 1978; Mc-

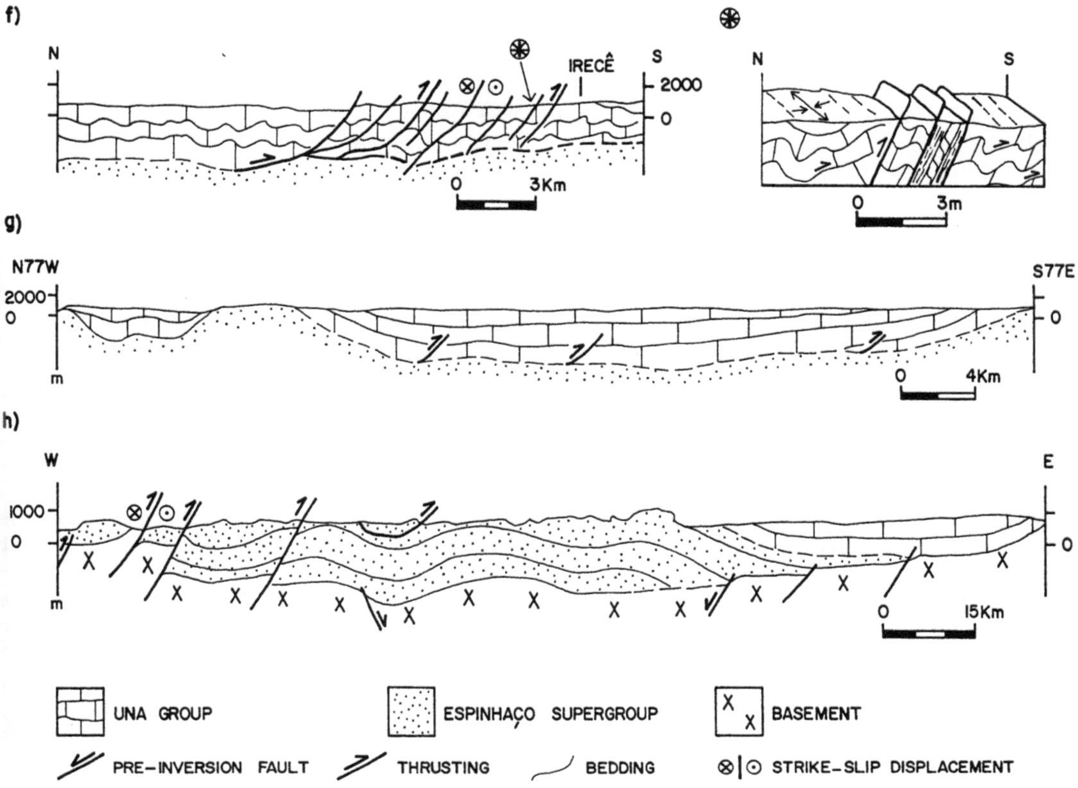

Figure 7. Cross sections along the Chapada Diamantina region (for location see Figure 6). f-f' and g-g' sections adapted from Lagoeiro (1990); h-h' section adapted from Danderfer Filho (1990)

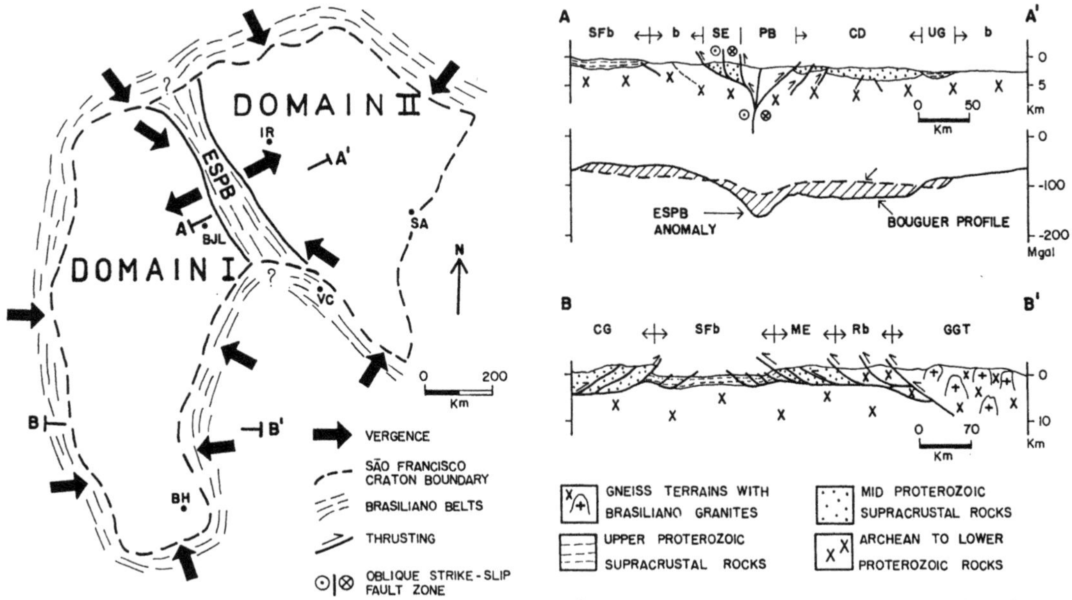

Fig. 8 Schematic map of the SF craton showing Brasiliano belts and its vergences. a) A-A′ simplified section along the ESPB belt showing the positive flower structure, and respective gravimetric profile (after Davino 1980). SFb = SF basin; b = basement SE = Espinhaço Setentrional; PB = Paramirim block; CD = Chapada Diamantina; UG = Una Group; SA = Salvador; IR = Irecê; BH = Belo Horizonte; BJL = Bom Jesus da Lapa; VC = Vitória da Conquista. b) B - B′ simplified section along the southern portion of Domain I. CG = Canastra Group; SFb = SF basin; ME = Espinhaço Meridional; Rb = basement affected by Brasiliano Orogeny; CGT = gneiss terrains with Brasiliano granites

Reath et al. 1981; Moutinho da Costa & Inda 1982) have considered the deformation of the ESPB belt to be Uruaçano (approximately 1.3 ba bp). Jardim de Sá et al. (1976) , Brito Neves (1979) and McReath et al. (1981) used Rb/Sr radiometric data of hydrotermalized metavolcanic rocks (Rio dos Remédios Fm., Figure 2) and weakly deformed shales (Cabloco Fm., Figure 2) to characterize the Uruaçano Orogeny in this belt. But, as pointed out by Cordani as long ago as 1973, the region was reactivated in Brasiliano time because the K-Ar system was reset at that time. In addition to this, based on structural and radiometric data Caby & Arthaud (1987) described one single structural-metamorphic event of Brasiliano age for 1.7 to 18 Ga orthogneissic Lagoa Real Complex, which is situated in the south part of the Paramirim Block. Furthermore we note that structures within the ESPB belt correlate with structures affecting SF Supergroup strata on either sides. Therefore, we propose that the closure of these basins is related to Brasiliano event. In this way, this region behaved as cratonic area during the Mid-Proterozoic era.

Since the ESPB belt divides the SF craton into two Domains, and it is a Brasiliano structure, we suggest a possible subdivision of the Synbrasiliano SF craton either in two blocks or in two more "cratons" (Cordani 1973a; Cordani 1973b), the Guanambi and Salvador "cratons", which correspond Domain I and II, respectively (Figure 8).

6 CONCLUSIONS

This paper emphazises the following

ideas:
(1) Latest Proterozoic strata within almost entire interior of the craton were deformed during the Brasiliano event, largely in the style of thin-skinned foreland fold thrust belts. These belts verge toward the interior of the craton.
(2) It is possible to define influence zones in the craton interior associated with marginal orogens.
(3) Using the age of Brasiliano Bambuí strata as a reference, it is possible to place an older limit on the marginal orogens.
(4) We provide new observations concerning the ocurrence of deformation in Bambuí strata bordering the Espinhaço Setentrional and in the Chapada Diamantina. Correlation of these structures with structures in the ESPB belt indicates that the region including the Espihaço Setentrional, Paramirim block and Chapada Diaman-tina is a closed ensialic basin (the Espinhaço aulacogen, as pointed out by Moutinho da Costa & Inda 1982) which was tectonized during the Brasiliano Orogeny.
(5) the ESPB belt is interpreted as a large positive flower structure of Brasiliano age which divides the SF craton in two distinct Domains, Domain I and II. These Domains could represent two different blocks of the Brasiliano SF craton.
(6) we could not recognize any evidence of an Uruçuano compressive deformation in the Mid- Proterozoic Espinhaço Supergroup rocks exposed in the interior of the SF craton.

Acknowledgments

We gratefully acknowledge graduate students André Danderfer Filho, Janio P. de Oliveira, Leonardo Lagoeiro, Lincoln Magalhães, Luis A. P. Bacellar and Pierre M. Magalhães for help in the field. Field work was supported by the Conselho Nacional de Desenvolvimento Científico e Tecnológico (CNPq) and the Petrobrás. Special thanks to Stephen Marshak for fruitful discussion and critical review of the earlier manuscript. The manuscript has benefitted from the review by Roland Trompette, too.

REFERENCES

Alkmim, F.F.; Chemale Jr., F.;Bacellar, L.A.P.; Oliveira, J.; Magalhães, P.M. 1989. Arcabouço estrutural da porção sul da Bacia do São Francisco. *An V Simp. Geol. Núcleo Minas Gerais-SBG*, Belo Horizonte:289-293.

Almeida, F.F.M. 1967. Origem e Evolução da Plataforma Brasileira. *Boletim DNPM*, Rio de Janeiro, 216:1-96.

Almeida, F.F.M. 1977. O Cráton São Francisco. *Revista Brasileira Geociências* 7: 285-295.

Bacellar, L.A.P. 1989. *Geologia estrutural do Supergrupo São Francisco ao longo da travessa* Coromandel-*Três Marias-Conselheiro Mata, MG*. M.Sc. Thesis Universidade Federal de Ouro Preto, Brazil.

Bonhomne, M.G. 1976. Mineralogie des fractions fines et datations rubidium-strotium dans le Groupe Bambuí, Brésil. *Revista Brasileira Geociências* 6: 39-43.

Brito Neves, B.B.; Kawashita, K.; Cordani, U.G.; Delhal, J. 1979. A evolução geocronológica da Cordilheira do Espinhaço - Dados novos e integração. *Revista Brasileira Geociências* 9: 71-85.

Caby, R. & Arthaud, M. 1987. Petrostructural evolution of the Lagoa Real Subalkaline Metaplutonic Complex (Bahia, Brasil). *Revista Brasileira de Geociências* 17: 636.

Cordani, U.G. 1973a. *Evolução geológica pré-cambriana da faixa costeira do Brasil, entre Salvador e Vitória*. Tese de Livre Docência Universidade de São Paulo, Brazil.

Cordani, U. G. 1973b. Definição e

caracterização do Cráton São Francisco. *An. 27. Congresso Brasileiro de Geologia*, 2, Aracaju: 142-145.

Cordani, U. G. 1978. Comentários filosóficos sobre a evolução geológica pré-cambriana. *Publ. Espec. SBG*, 3, Salvador:33-65.

Costa, L.A. M. & Angeiras, A. G. 1971. Tectonic zoning in the Epi-Baykalian of Central Brazil. *Anais da Academia Brasileira de Ciências* 41: 641.

Danderfer Filho, A. 1990. *Análise estrutural descritiva e cinemática do Supergrupo Espinhaço na região da Chapada da Diamantina (BA)*. M.Sc. Thesis, Universidade Federal de Ouro Preto, Brazil.

Dardenne, M. A. 1978. Síntese sobre a estratigrafia do Grupo Bambuí no Brasil Central. *An 30. Congresso Brasileiro Geologia*, 2, Recife: 607-610.

Davino, A. 1980. Delineamento dos limites e estruturas do Cráton Paramirim por gravimetria. *An 31. Congresso Brasileiro Geologia*, 5, Salvador: 2573-2582.

Dominguez, J.M.L. & Rocha, G.M.F. 1989. Sequências deposicionais do Espinhaço Setentrional na região sudeste do estado da Bahia: uma margem continental proterozóica. *An V Simpósio Geologia Núcleo Minas Gerais-SBG*, Belo Horizonte: 195-199.

Dossin, I.A.; Dossin, T.M. & Chemale Jr., F. Evolucão Tectônica Proterozóica do Espinhaço Meridional e Regiões Adjacentes. *Acta Geologica Leopoldensia* (in prep.).

Inda, H.A.V. & Barbosa, J.F. 1978. *Texto explicativo para o mapa geológico do Estado da Bahia, escala 1:100.000*. Salvador: Secretaria de Minas e Energia.

Iyer, S. S. 1984. A discussion on the lead isotope geochemistry of galenas from the Bambui Group, Minas Gerais - Brazil. *Mineral Deposita* 19:132-137.

Jardim de Sá E.F.; Brito Neves, B. B.; McReath, I.; Bartels, R.L. 1976. Geocronologia e o modelo tectono-magmático da Chapada Diamantina e Espinhaço Setentrional. *An 29. Congresso Brasileiro Geologia*, Belo Horizonte: 205-228.

Karfunkel, J. & Hoppe, A. 1988. Late Proterozoic Glaciation in Central-Eastern Brazil: shynThesis and model. *Paleogeography, Paleoclimatology & Paleoecology* 65: 1-21.

Lagoeiro, L. E. 1990. *Estudo da deformação nas sequências carbonáticas do Grupo Una na região de Irecê, BA*. M.Sc. Thesis, Universidade Federal de Ouro Preto,Brazil.

Lesquer, A.; Almeida, F.F.M.;Davino, A.; Lachaud, J.C. & Mailard, P. 1981. Signification structurale des anomalies gravimétriques de la partie sud du Craton du São Francisco (Brésil). *Tectophysics* 76: 273-293.

Magalhães, L. 1988. *Análise estrutural qualitativa dos sedimentos do Grupo Bambuí, região sudeste da Bacia do São Francisco*. M. Sc. Thesis, Universidade Federal de Ouro Preto, Brazil.

Marshak, S. & Alkmim, F.F. 1989. Proterozoic contraction/extension tectonics of the southern São Francisco region, Minas Gerais, Brazil. *Tectonics* 8: 555-571.

Mascarenhas, J.F.; Pedreira, A.J.; Motta, A.C. & Jardim de Sá, E.F. 1984. Província São Francisco. In: Almeida, F.F.M. & Hasui, Y. (eds.): *O Pré-Cambriano do Brasil*. São Paulo, Edgard Bluecher: 46-122.

McReath, I; Jardim de Sá, E. F. &

Fryer, B. J. 1981. As vulcânicas ácidas proterozóicas da região da Bacia do Paramirim, Bahia. In: Inda, H. A. V. & Duarte, F. B. (eds.): *Geologia e recursos minerais do Estado da Bahia*. Secret. Minas e Energia, Bahia: 121-132.

Monteiro, M.D.; Carvalho, M.P.& Conceição Filho, V.M. 1984. Caracterização faciológica e sistemas deposicionais do Grupo Chapada Diamantina. *An 33. Congresso Brasileiro Geologia*, 3, Rio de Janeiro,: 1090-1105.

Moutinho da Costa, L.A. & Inda,H.A. V. 1982. O aulacógeno do Espinhaço. *Ciências da Terra* 2: 13-18.

Muzzi Magalhães, P. M. 1989. *Análise estrutural qualitativa das rochas do Grupo Bambuí, na porção SE do São Francisco*. M. Sc. Thesis, Universidade Federal de Ouro Preto, Brazil.

Oliveira, J. R. P. 1989. *Comportamento estrutural dos sedimentos dos grupos Macaúbas e Bambuí numa porção do centro norte do Estado de Minas Gerais*. M. Sc. Thesis, Universidade Federal de Ouro Preto, Brazil.

Ortu, J. C. 1990. *Modelagem tectono-geofísica da porção sul da Bacia do São Francisco, Minas Gerais*. M. Sc. Thesis, Universidade Federal de Ouro Preto, Brazil.

Pedreira, A.J. 1988. Sequências deposicionais no Pré-cambriano: exemplo da Chapada Diamantina Oriental, Bahia. *An 35. Congresso Brasileiro Geologia*, 2, Belém: 648-659.

Pinho, J.M.M., Dardenne M.A. & Rigobello, A.E. 1989. Evolução tectônica da mineralização de zinco de Vazante. *An V Simpósio de Geologia Núcleo Minas Gerais-SBG*, Belo Horizonte: 275-276.

Schöll, W.U. 1973. *Sedimentologie und Geochemie der Bambuí Gruppe am SE-Rand des São Francisco-Beckens, Minas Gerais, Brasilien*. Ph.D. Thesis, Univ. Heidelberg, Germany.

Souza, W.S.T. 1986. Contribuição aos estudos de revisão estratigráfica do Pré-cambriano brasileiro a partir da utilização do conceito de unidade deposicional. *An 34. Congresso Brasileiro Geologia*, 3, Goiânia: 391-402.

Távora, F.I.; Cordani, U.G.;Kawashita, K. 1967. Determinações de idade potássio-argonio em rochas da região central da Bahia. *An 21. Congresso Brasileiro Geologia*, Aracaju: 214-224.

Uhlein, A., Trompette, R. & Silva, M. E. 1986. A Estruturação Tectônica do Supergrupo Espinhaço na Região de Diamantina (MG). *Revista Brasileira Geociências.*, 16:212-216.

Gondwana Eight, Findlay, Unrug, Banks & Veevers (eds) © *1993 Balkema, Rotterdam. ISBN 90 5410 304 3*

Orogenic belts, sutures and block faulting on the northwestern Gondwana margin

M.Villeneuve
Institut de Géodynamique, Sophia-Antipolis I, Valbonne, France
J.J.Cornée & J.Muller
URA 1208, Centre de Sédimentologie, Marseille, France

ABSTRACT: At least six major tectonic events and five sutures are distinguished on the northwestern margin of Gondwana. Three of these events, documented, are related to the Pan-African orogens (Pan-African I, Pan-African B2 and Pan-African II); one is linked to a possible Taconic tectonic event in northern Morocco, another, more poorly documented, fits with a Variscan orogeny in northeastern Morocco while the last one is linked to the main Hercynian tectonic event as the result of the collision between the North American continent and Gondwana. The pre-Alleghanian sutures delineate several crustal blocks which were strongly squeezed at the end of the Carboniferous time by the Hercynian collision between Gondwana and North America. The continuing activity within weakness zones of northwestern Gondwana, from late Precambrian until late Paleozoic, provides a good example of tectonic inheritance.

1 INTRODUCTION

According to recent palaeo-reconstructions, the northwestern margin of the African continent also corresponds to the northwestern margin of Gondwana (Dalziel, 1991; Li & Powell 1991; Hoffman 1991). This part of Africa consists of the Archean to Lower Proterozoic West African craton which is surrounded by several Pan-African belts. On the western side on the craton the Pan-African orogens are involved in a Hercynian fold belt. The studied area extends from Morocco to Liberia (Fig. 1). We distinguish at least 6 orogenic belts and 5 sutures ranging in age from Late Proterozoic to Carboniferous (Tabl. 1). The first one occured around 650 My and the last one (Hercynian) is linked to the Alleghanian orogen (330-280 My). In this study we only take into account the belts and sutures located on the eastern side of the Alleghanian suture (Fig. 7).

2 BELTS AND SUTURES

We will give now a brief review of the belts and sutures identified in this area and related to six main different tectonic events.

TABLE 1: Summary of the main orogenic events in the northwestern Gondwana margin during Late Proterozoic and Paleozoic times.

AGES My	EVENT	BELTS
330/280	ALLEGHANIAN	• MOROCCAN HERCYNIDES • DHLOU/ZEMMOUR BELT • MAURITANIDES
360/375	VARISCAN	• NORTHEASTERN MOROCCO (MIDELT BLOCK)
430/488	TACONIC	• NORTHERN MOROCCO (SEHOUL BLOCK)
550/530	PANAFRICAN II	• ROKELIDES
615/620	B2 PHASE	• MOROCCO (ANTI-ATLAS)
685/650	PANAFRICAN I	• BASSARIDES • MOROCCO (B1 PHASE, ANTI-ATLAS)
ARCHEAN AND EBURNEAN		

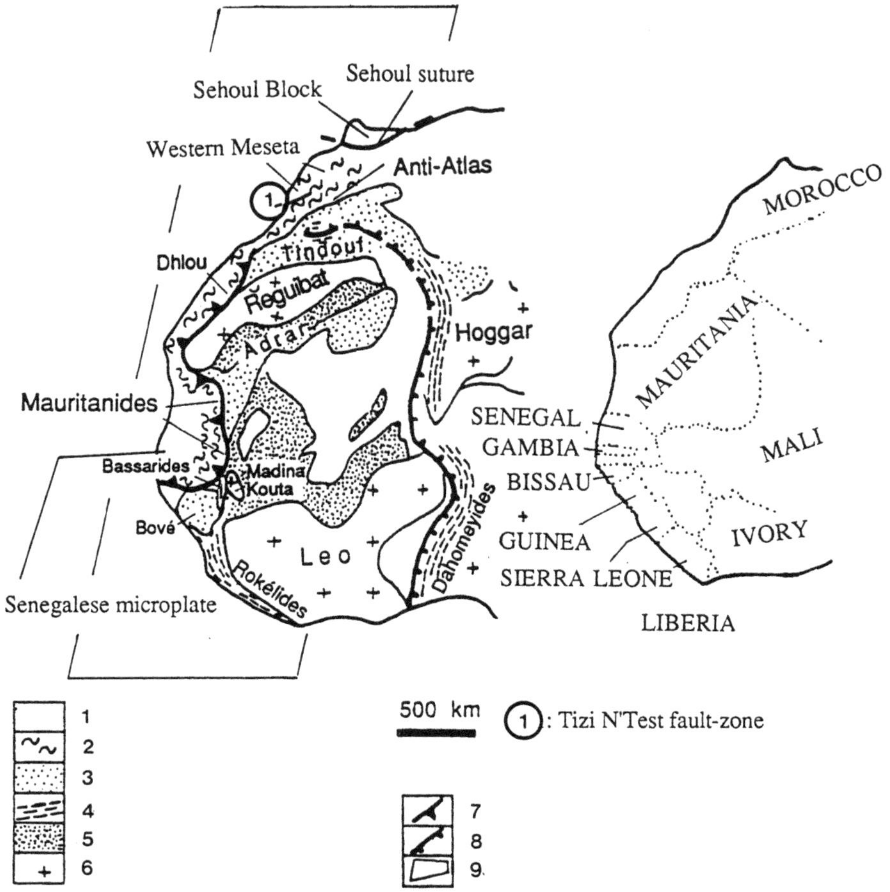

Fig. 1: West African geological sketch map and the studied area location. 1. Meso Cenozoic basins, 2. Hercynian belt, 3. Paleozoic sediments, 4. Pan-African belts, 5. Proterozoic sediments, 6. Archean and Eburnean basement, 7. Hercynian thrust, 8. Pan-African thrust, 9. Studied area .

2.1 Pan-African I tectonic event (Fig. 2)

In the Moroccan Anti-Atlas, an hypothesis of ophiolite obduction onto the West African Craton was supported by Leblanc (1975). This obduction was probably caused by the southward subduction of an oceanic crust beneath the craton. For Saqaque et al. (1989) the subduction would act northward. The subsequent metamorphism was dated from 685±15 My (Clauer 1976). Although the geodynamic model of this belt is still under discussion (Leblanc & Lancelot, 1980; Schermerhorn et al. 1986), this metamorphism is linked to the Pan-African B1 phase which falls in the same bracket as the Pan-African I Orogen of West Africa (Tab. 1). The ophiolite itself displays an age around 787±10 My (Clauer 1974). In Mauritania and Senegal, the Pan-African I orogenic event was first identified in the Bassaride belt (Villeneuve, 1982; 1984). Later, it was suspected to have been involved in the Mauritanide belt because of petrographical, geophysical and geochronological data (Dupont et al 1984; Villeneuve 1984; Dallmeyer & Lécorché 1989). Recently, Villeneuve (1991) proposed this belt extends westward, across the Gambia

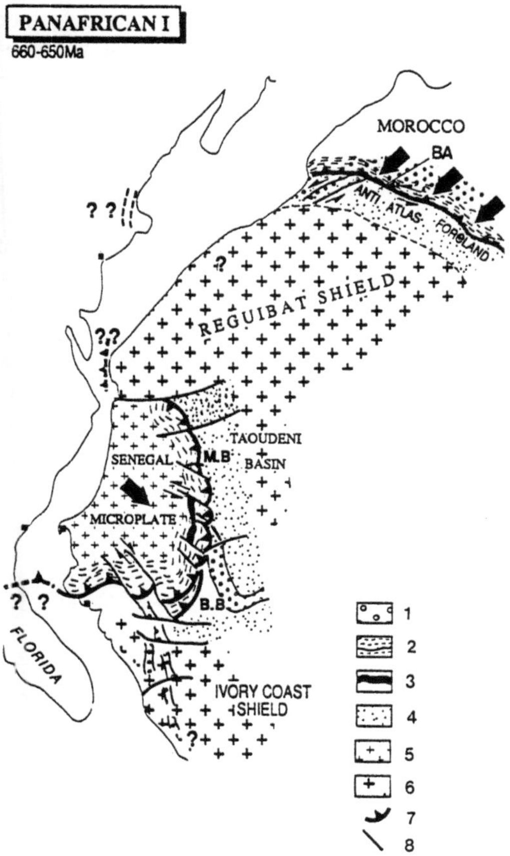

Fig. 2: Sketch map of the Pan-African I belts in West Africa. 1. Pan-African I molassic deposits, 2. Metamorphic zones, 3. Ophiolites, 4. Late Proterozoic deposits in the foreland, 5. Archean and Eburnean basement, 6. Pan-African I sutures, 7. Main panafrican faults, BA Bou Azzer, MB Mauritanide Belt, BB Bassaride belt. Arrows: supposed block motion.

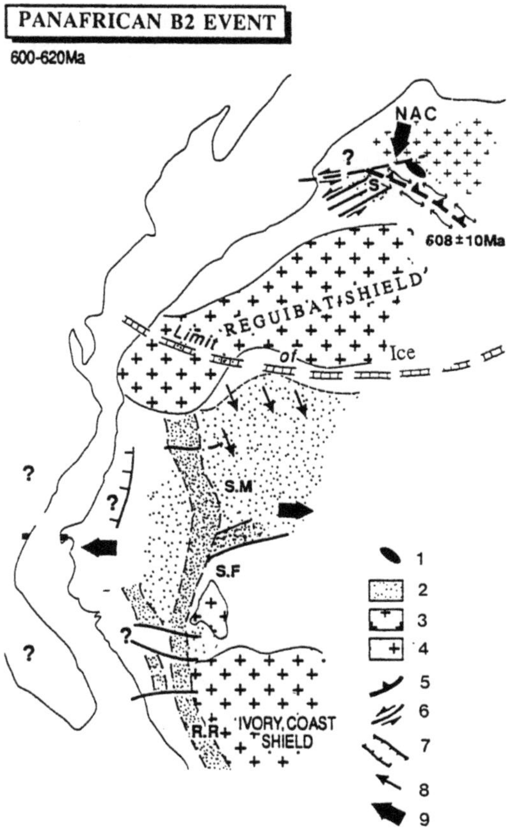

Fig. 3: Sketch map around 600 My. The Pan-African B2 tectonic event. 1. Bleida granodiorite, 2. Late Precambrian glacial deposits, 3. North Atlas Craton (CNA), 4. West African Craton, 5. Anti-Atlas suture, 6. Strike-slip fault, 7. Grabens, 8. Glacial flow direction, 9. Tectonic movement direction, S Siroua, SM Moudjeria trough, SF Faleme trough, R.R Rokel-River trough.

area (Gambia belt) rather than to the south (Guinea and Sierra Leone) as previously proposed (Villeneuve 1984; Bonvalot et al. 1991). Ophiolites occur in the Mauritanian area, associated with a magmatic calc-alkaline complex; a related Pan-African I suture is supported by gravity interpretation (Guetat 1981; Ponsard et al. 1988) which indicates a westward dip of this suture beneath the Senegalese microplate. Radiometric data (39Ar/40Ar on syntectonic muscovite) pointed to plate collision at about 650 My (Dallmeyer & Villeneuve 1987) with subsequent molassic deposits in Guinea and Mali (Villeneuve, 1991).

The geodynamic model first proposed by Villeneuve & Dallmeyer (1987) and then by Dallmeyer & Lécorché (1989), suggests a Red-Sea type rift system that opened at about 700 My and was closed at about 650 My.

2.2 Anti-Atlas Pan-African B2 tectonic event (Fig. 3)

In southern Morocco, sedimentary and volcano-sedimentary rocks were deposited after the Pan-African B1 metamorphism. These rocks were intruded by the Bleida Granodiorite which immediately predates the late Precambrian deformation (B2 phase, Leblanc 1975); the granodiorite is dated at 615±12 My (Ducrot 1979). On the other hand the overlying undeformed volcano-clastics are dated at between 618±22 My (Rb/Sr, Jeannette et al. 1981) and 578±5 My (U/Pb, Midfal & Peucat 1985). Consequently, the B2 Pan-African event is assumed to be 615-620 My old (Hassenforder, 1987). This tectonic event has been ascribed to the final collision between the West African Craton and an hypothetical northern craton. Left-lateral strike-slip movements on northeast trending faults located in the foreland were linked to this oblique collision (Hassenforder 1987). The deformed rocks are overlain by molassic and volcano-sedimentary rocks deposited into some northeast trending grabens and pull-aparts basins (Azizi et al. 1990; Hassenforder 1987).

In the southern countries of West Africa, no evidence of a compressional tectonic event has been found during the time of the B2 Pan-African tectonics in Morocco. This part of Africa was covered by sediments of glacial origin. Indeed, some of these sediments deposited in north-trending grabens along the future location of the Mauritanides belt show an extensional tectonic regime. The origin of these grabens is not definitely understood.

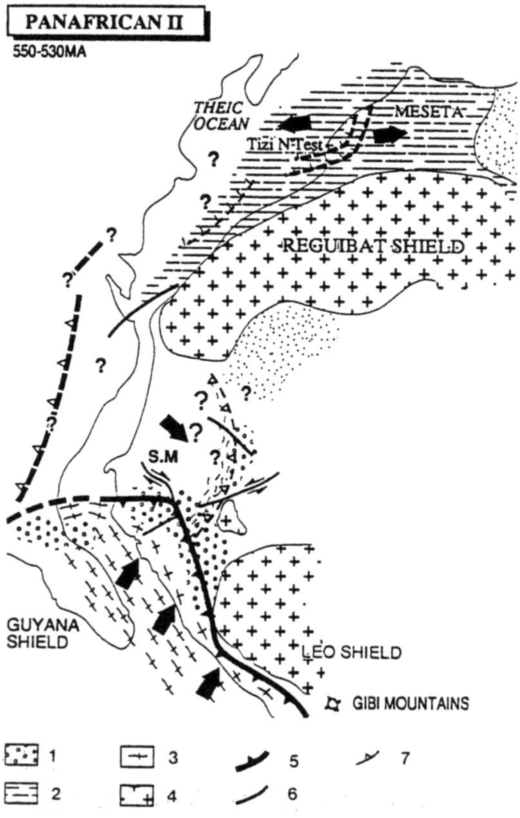

Fig. 4 : Sketch map of the Pan-African II belts in west Africa. 1. Pan-African II molassic deposits, 2. Mid-Cambrian marine sediments, 3. Pan-African II belt (Rokelide belt), 4. Archean and Eburnean basement, 5. Rokelide suture, 6. Pan-African II faults, 7. Thrusts, SM, Senegalese Microplate. Arrows: Supposed direction of plate motion.

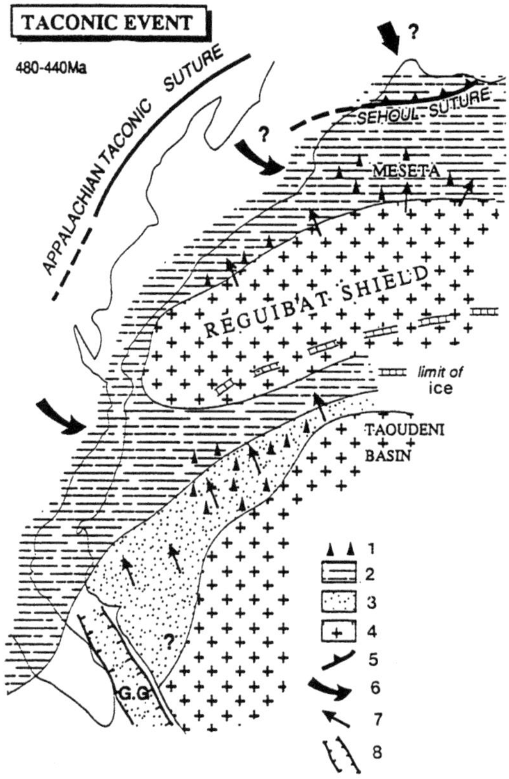

Fig. 5: Sketch map around 480-440 My. The Moroccan Taconic event. 1. Tillites, 2. Ordovician marine and epicontinental sediments, 3. Continental sediments, 4. Archean and Eburnean basement, 5. Suturing zone, 6, Direction of marine transgression, 7. Direction of the sediment transportation by fluvial streams, 8. Ordovician grabens, GG Guinean graben.

2.3 The Pan-African II tectonic event (Fig. 4)

This tectonic event, dated around 550-530 My by Dallmeyer & Villeneuve (1987), only occurred in the southern part of the studied area (Liberia to Gambia). It was induced by a northeastward movement of the Guyana shield. This movement is responsible for the Rokelide belt building (Allen 1969; Culver et al. 1979) with the thrust of the Guyana Shield over the West African Craton. The erosion of the Rokelide belt produced a lot of molassic sediments (Youkounkoun Group). The Rokelide suture was interpretated as an intracontinental suture by Thorman (1976) and Culver & Williams (1979). New gravimetric interpretations suppose a Pan-African I suture underneath the Rokelide Belt (Bonvalot et al., 1991). However, a Pan-African I rift branch could also be assumed to explain the origin of the Rokelide Belt .

In the northern part of the craton an extensional tectonic regime is indicated by the West Moroccan grabens that opened to the north of the Tizi-n'Test fault (Bernardin et al. 1988). Marine and epicontinental sediments, which strongly contrast with the molassic deposits of the southern Youkounkoun Group, were deposited in the West Moroccan grabens by the Early and Middle Cambrian.

2.4 The Taconic tectonic event (Fig. 5)

Throughout lower Paleozoic times, no compressional tectonic event occured in this part of Gondwana except in the northern part of Morocco. While in the southern part of Morocco marine clastic sediments were deposited during Ordovician times, in the northernmost part Cambrian and Tremadocian deltaic sediments were folded and metamorphosed before the emplacement of an undeformed granite dated from 430 My (Piqué et al. 1991). This metamorphic area has been called the Sehoul block (Fig. 1 and 5) which consists of the products of a tectono-metamorphic event that evidently occured between 488 and 430 My. The Sehoul suture was strongly reworked during the Alleghanian orogenic event and its initial geometry is unknown.

In the southern area of the craton, an extensional tectonic regime is suspected during the Ordovician time. This event is supported by a deep graben filled with Ordovician sandstones, between Guinea and Florida (Villeneuve et al. 1990). Every part of this area underwent the influence of the Late Ordovician glaciation as shown by outcrops of tillites across the West African Craton.

2.5 Variscan tectonic event (Fig. 6)

By Late Devonian to Early Carboniferous, most of central Morocco was uplifted above sea-level, except in some restricted areas such as the tectonically controlled Sidi Bettache basin (Piqué et al. 1991) in the central

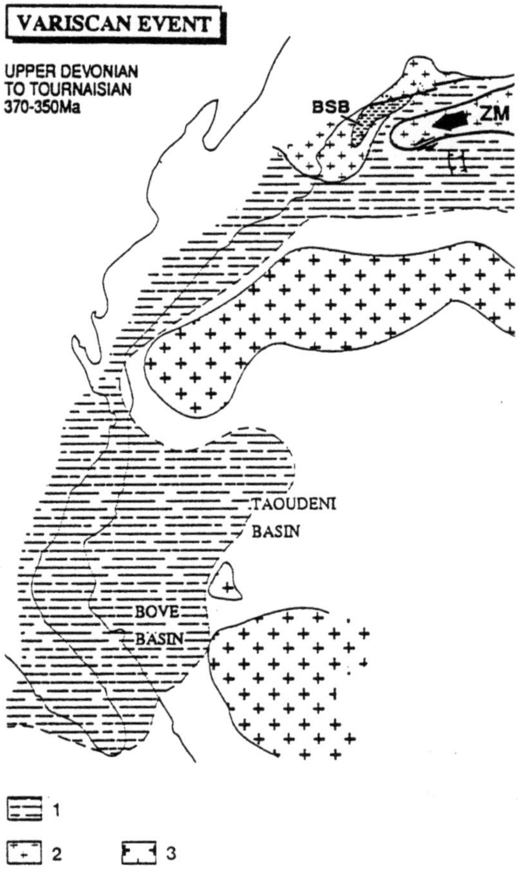

fig. 6: Sketch map around 370-350 My. The Variscan (or Acadian) tectonic event. 1. Marine and epicontinental Devonian sediments, 2. Midelt block of Morocco, 3. Archean and Eburnean basement. BSB Sidi Bettache basin, ZM Midelt Zone.

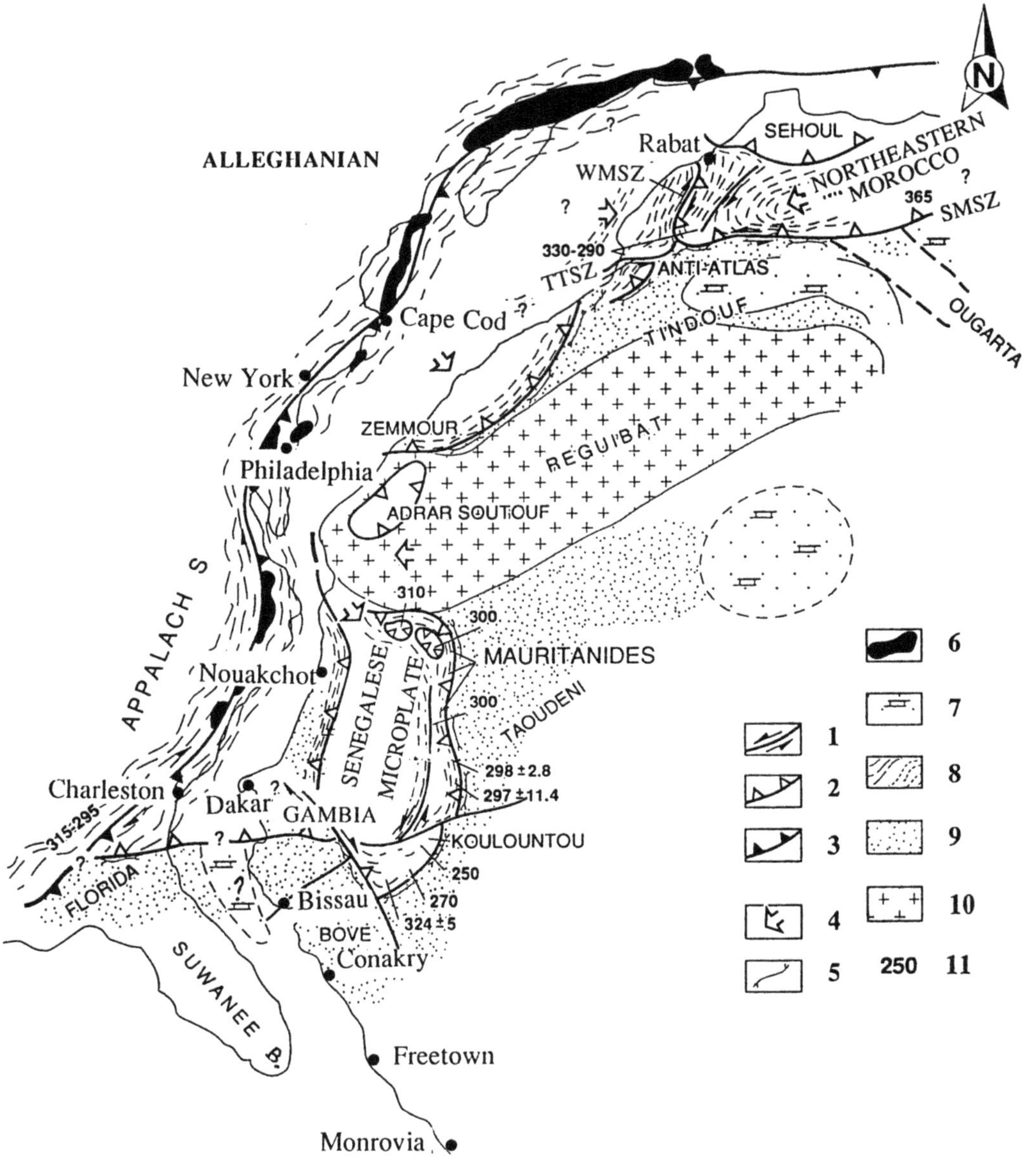

Fig. 7: Geological sketch map of the Hercynian belts (Alleghanian tectonic event). a. Magnetic anomalies fitting with the Alleghanian suture, b. Post-Visean deposits. 1. Strike-slip fault, 2. Hercynian thrusts, 3. Ante-Hercynian thrusts, 4. Supposed direction of plate motions, 5. Foreland folds, 6. Ophiolitic complexes, 7. Post-Namurian-Permian? deposits, 8. Hercynian fold belt, 9. Foreland sediments, 10. Basement craton, 11. Ages of the deformations, My, WMSZ West Meseta shear zone, TTSZ Tizi-n'Test shear zone, SMSZ South Meseta shear zone, RTF Rabat-Tiflet shear zone.

Moroccan area. As a result the region was separated by a Visean trough into two main blocks. In the eastern block, the synmetamorphic deformations occured around 366±7 or 368±8 My (K/Ar, Clauer et al. 1980; Huon 1985). From studies of these Variscan deformations in northeastern Morocco, and from the radiometric and geochemical data of the Moroccan Carboniferous and Permian granites, Piqué & Michard (1989) and Piqué et al. (1991) deduced an intracontinental west dipping suture zone along the easternmost boundary of Morocco (Fig. 7).

In the southern block, a Devonian regression lead to a Carboniferous emersion but no tectonic event was registered by this time. This area still remained a passive oceanic margin untill the Alleghanian tectonic event.

2.6 The Alleghanian tectonic event (Fig. 7)

The Alleghanian tectonic event, which occured at between 330 to 280 My, is the main regional tectonic event of the northwestern Gondwana margin. The whole area was affected by this event which was related to the collision between the Gondwana and the North American continent. According to Lefort (1984), the Alleghanian suture, underlined by gravity anomalies, runs along the Appalachian fold belt from Florida to Spain. Deformations at about 300 My occurred not only in the Appalachian belt, but also in Morocco, in the Zemmour area and around the Senegalese microplate.

In Morocco, the main Alleghanian deformations are concentrated at the boundaries of the different blocks, for example the SMSZ (South Meseta shear zone), the WMSZ (West Meseta shear zone), the TTSZ (Tizi-n'Test shear zone) and, of course, the RTF (Rabat-Tiflet fault). The SMSZ shows a left lateral strike-slip movement meanwhile the ZCMO and the ZDTT show a right lateral strike-slip movement (Cornée 1989; Piqué et al 1991). The FRT, which corresponds to the Taconic Sehoul suture, was thrust towards the south. Strike-slip faults are locally associated with strong metamorphism, magmatic intrusions and thrusts towards the west, the south and the southeast. These movements were interpreted by Cornée (1989) and Piqué et al (1991) as a consequence of the westward motion of the Midelt block (northeastern

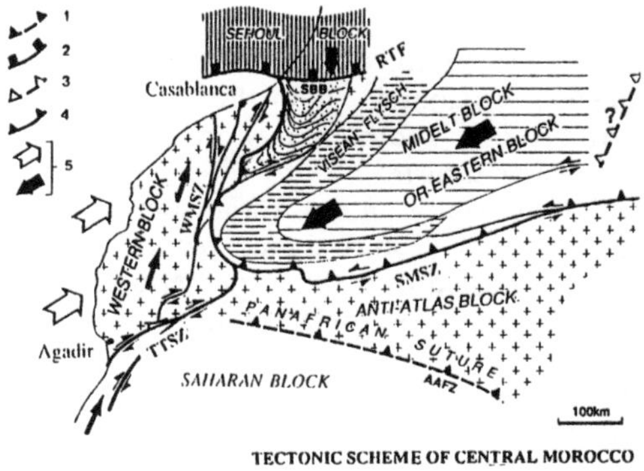

Fig. 8: Tectonic sketch map of central Morocco. 1. Anti-Atlas Pan-African I suture, 2. Sehoul Taconic suture, 3. Eastern Variscan suture, 4. Alleghanian suture, 5. Motion of the blocks. WMSZ, RTF, SMSZ and TTSZ see legend fig. 7. SBB: Sidi Bettache basin; AAFZ: Anti-Atlas Fault Zone

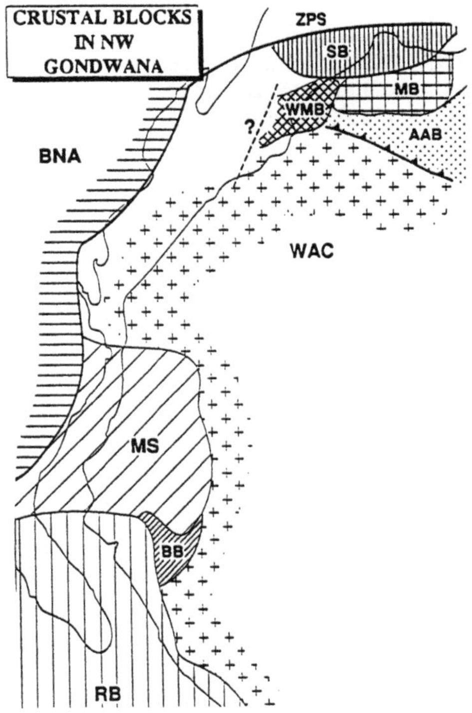

Fig. 9: Crustal blocks on the northwestern Gondwana margin. ZPS, South Portugal Block, BNA, North America Craton, SB: Sehoul Block, MB: Midelt Block, AAB: Anti-Atlas Block, WMB: West Meseta Block, WAC: West African Craton, MS: Senegalese Microplate, BB: Bassarides Block, RB: Rokelides Block.

Morocco).

To the south, in the Dhlou area, Sougy (1969) assumed that the folded Early Ordovician sediments were affected by Hercynian thrusts.

Farther south, in the Adrar Souttouf, Sougy and Bronner (1969) argued for a Proterozoic nappe thrusted over the Reguibat uplift during the Alleghanian tectonic event.

South of the Reguibat uplift, major Hercynian deformations were demonstrated either by field obervations or by geochronological data. The first tectonic evidence was cited by Tessier et al. (1961) in the Guelb el Hadej klippen. Since this time, several nappes or thrusts involving Paleozoic sediments have been documented around the Senegalese microplate. The most representative feature of the Hercynian tectonic event is the left-lateral Mont Oua Oua strike-slip fault (Le Page 1983). Beneath the Meso-Cenozoic sedimentary cover of the Senegalo-Mauritanian basin, some new seismic sections suggest some Hercynian thrusts along the Atlantic margin (Villeneuve et al. 1990) and radiometric ages from different methods indicate them to belong to the Alleghanian tectonic event (Dallmeyer & Villeneuve 1987; Dallmeyer & Lecorché 1989). East of the Appalachians, very few outcrops of Late Carboniferous or Permian molassic deposits have been found except in Morocco. Permian troughs are suspected by seismic interpretation in the middle of the Senegalese microplate and off-shore of the Bissau Atlantic margin.

3 BLOCK FAULTING MODEL AND TECTONIC INHERITANCE

The Pan-African and Paleozoic sutures setting on the northwestern Gondwana margin delineate several crustal blocks as well in the north as in the south.

3.1 The Moroccan block faulting (Fig. 8)

The different blocks and sutures of the central Moroccan area are represented in Figure 8. The most important faults, assumed to be crustal weakness zones, such as the TTSZ and the WMSZ, were probably active at the end of the Pan-African orogeny and could have been generated by this event. Thus, block tectonic regimes occurred in Morocco throughout Late Precambrian times until the Alleghanian period. The post-Pan-African deformations seem to be guided by this initial block faulting but motions were first influenced by the general geodynamic pattern which later often changed (Piqué et al 1991).

3.2 The West African block faulting (Fig.9)

Villeneuve (1991) supports a Pan-African I origin for the Senegalese microplate and the Bassaride block. The Rokelide block, including Florida, was formed at the end of the Pan-African II orogen, probably by reworking of some Pan-African I intracontinental faults. Thus, the Hercynian deformations of the Mauritanide belt seem to have been generated by a southeastern motion of the Senegalese microplate. Consequently, the Alleghanian suture in the Mauritanide belt most likely corresponds to the rejuvenation of the Pan-African I suture.

4 CONCLUSIONS

The northwestern Gondwana margin underwent multiple tectonic events, mainly during Late Proterozoic and Palaeozoic times. The different orogens are related to sutures. Some of the sutures are well understood, like the Pan-African I or Pan-African II sutures, but others are still poorly known or even uncertain such as the Moroccan Taconic or Variscan sutures. The concentration of numerous sutures reflects a long lived weakness zone fringed by oceanic domains such as the Iapetus or Theic oceans. The northwestern Gondwana margin underwent a main block faulting break, early in the Palaeozoic time and a much lesser break during the Hercynian orogeny. Consequently, the main thrusts of the Mauritanides or Moroccan belts should not be considered any longer as the eastern front of the Appalachian belt but only as rejuvenation of a major foreland fault system. We believe that these blocks, located on the eastern boundary of the Avalonian microcontinent, were squeezed at the time of the collision between the Gondwanian and the North American continents. Thereafter, the main part of the deformation was taken up in narrow block boundaries.

REFERENCES

Allen, P.M. 1969. The stratigraphy of part of an orogenic belt in western Sierra Leone, West Africa. *Geologische. Rundschau* 58: 588-620.

Azizi Samir, M.R., J. Ferrandini & J.L Tane 1990. Tectonique et volcanisme tardi-panafricains (580-560 My) dans l'Anti-Atlas central (Maroc): interprétation géodynamique à l'échelle du Nord-Ouest de l'Afrique. *Journal of African Earth Sciences* 10: 549-563.

Bernardin, C., J.J. Cornée, M. Corsini, S. Mayol, J. Muller & M. Tayebi 1988. Variations d'épaisseur du Cambrien moyen en Meseta marocaine occidentale: signification géodynamique des données de surface et de subsurface. *Canadian Journal of Earth Sciences* 25: 210-2117.

Bonvalot, S., M. Villeneuve & Y. Albouy 1991. Gravity interpretation of western Sierra Leone (West Africa): implications on the structure and evolution of the Rokelide orogenic belt. *Comptes Rendus de l'Académie des Sciences,* Paris 312: 841-848.

Clauer, N. 1974. Utilisation de la méthode Rb/Sr pour la datation d'une schistosité de sédiments peu métamorphisés: application au Précambrien II de la boutonnière de Bou-Azzer-El Graara (Anti-Atlas, Maroc). *Earth and Planetary Science Letters* 22: 404-41.

Clauer, N. 1976. *Géochimie isotopique du strontium des milieux sédimentaires. Application à la géochronologie de la couverture du craton Ouest Africain.* Thèse Doctorat en Sciences Strasbourg (Fr.): 1-227.

Clauer, N., D. Jeannette & D. Tisserant 1980. Datation isotopique des cristallisations successives d'un socle cristallin et cristallophyllien (Haute Moulouya, Maroc). *Geologische Rundschau* 169: 63-83.

Cornée, J.J. 1989 *Le haut Atlas occidental paléozoique: un reflet de l'histoire hercynienne du Maroc occidental. Stratigraphie, sédimentation et tectonique.* Thèse Doctorat Sciences Aix-Marseille III (Fr.): 1-901.

Culver, S.J. & H.R. Williams 1979. The Late Precambrian and Phanerozoic geology of Sierra-Leone. *Journal of the Geological Society of London* 136: 605-618.

Dallmeyer, R.D. & M. Villeneuve 1987. 40AR/39Ar. Mineral age record of a polyphase tectonothermal evolution in the southern Mauritanide orogen, southeastern Senegal. *Bulletin of the Geological Society of America* 98: 602-611.

Dallmeyer, R.D. & J.P. Lecorché 1989. 40Ar/39Ar polyorogenic mineral age record within the central Mauritanide orogen. *Bulletin of the Geological Society of America* 101: 55-70

Dalziel, I.W.D.1991. Pacific margins of Laurentia and East Antarctica-Australia as a conjugate rift pair: evidence and implications for an Eocambrian supercontinent. *Geology* 19: 598-601.

Ducrot, J. 1979. Datation à 615 Ma de la granodiorite de Bleida et conséquences sur la chronologie des phases tectoniques, métamorphiques et magmatiques panafricaines dans l'Anti-Atlas marocain. *Bulletin de la Société géologique de France.* XII, 21: 495-499.

Dupont, P.L., M. Villeneuve & H. Lapierre 1984. Mise en évidence de reliques océaniques au sein de la chaîne panafricaine des Mauritanides dans la région des Bassaris (Guinée-Sénégal). *Comptes Rendus de l'Académie des Sciences, Paris,* II, 299: 65-70.

Guetat, Z. 1981. *Etude gravimétrique de la bordure occidentale du craton ouest-africain. Essai de corrélations géologiques.* Thèse de 3° Cycle, Université de Montpellier (Fr.): 1-176.

Hassenforder, B. 1987. *La tectonique panafricaine et varisque de l'Anti-Atlas dans le massif du Kerdous (Maroc).* Thèse de Doctorat en Sciences Strasbourg (Fr.): 1-249.

Hoffman, P.F. 1991. Did the breakout of Laurentia turn Gondwanaland inside out ? *Science* 252: 1409-1412.

Huon, S. 1985. *Clivage ardoisier et réhomogénisation isotopique K/Ar dans des schistes paléozoiques du Maroc.* Thèse Université Strasbourg (Fr.): 1-124.

Jeannette, D., F. Benziane & A. Yazidi 1981. Lithostratigraphie et datation du Protérozoique de la boutonnière d'Ifni (Anti-Atlas, Maroc). *Precambrian Research* 14: 363-378.

Leblanc, M. 1975. *Ophiolites précambriennes et gîtes arséniés de cobalt (Bou-Azzer, Maroc).* Thèse de Doctorat Sciences Université de Paris VI: 1-220.

Leblanc, M. & J.R. Lancelot 1980. Interprétation géodynamique du domaine panafricain (Précambrien terminal) de l'Anti-Atlas (Maroc) à partir des données géologiques et géochronologiques. *Canadian Journal of Earth Sciences* 17: 142-155.

Lefort, J.P. 1984. Mise en évidence d'une virgation carbonifère induite par la dorsale Réguibat (Mauritanie) dans les Appalaches du Sud. Arguments géophysiques. *Bulletin de la Société géologique de France* 26: 1293-1303.

Le Page, A. 1983. *Les grandes unités des Mauritanides aux confins du Sénégal et de la Mauritanie. Evolution structurale de la chaîne du Précambrien supérieur au Dévonien.* Thèse de Doctorat en Sciences Université d'Aix-Marseille III: 1-518.

Li, Z.X & C.A. Powell 1991. A review of the Late Proterozoic to Late Paleozoic paleomagnetism of Gondwana continents, and its relevance to the formation of Gondwanaland. *8th International Symposium on Gondwana, Hobart, Australia,* abstract: 53.

Midfal, A. & J. Peucat 1985. Datation U/Pb et Rb/Sr du volcanisme acide de l'Anti-Atlas marocain et du socle sous-jacent dans la région de Ouarzazate. Apport au problème de la limite Précambrien/Cambrien. *Sciences Géologiques, Bulletin,* Strasbourg (Fr.), 38: 185-200.

Piqué, A. & A. Michard 1989. Moroccan Hercynides : a synopsis. The paleozoic sedimentary and tectonic evolution at the northern margin of West Africa. *American Journal of Science* 289: 286-330.

Piqué, A., J.J. Cornée, J. Muller & J. Roussel 1991. The Moroccan Hercynides. In: R. D. Dallmeyer & J.-P.Lécorché (eds).*The West African orogens and circum-Atlantic correlatives.* Berlin, Springer Verlag: 229-263.

Ponsard, J.F., J. Roussel & M. Villeneuve 1988. The Pan-African orogenic belt of southern Mauritanides and Northern Rokelides (Southern Senegal and Guinea, West Africa): Gravity evidence for a collisional suture. Journal of african Earth Sciences 7: 463-472.

Saqaque, A., H. Admou, J.A. Karson, K. Heffran & I. Reuber 1989. Precambrian accretionary tectonics in the Bou Azzer - El Graara region, Anti-Atlas, Morocco. *Geology*, 17: 1107-1110.

Schemerhorn, L.J.G., E. Wallbrecher, K.M. & Huch 1986. Der Subduktionskomplex, Granitplutonismus und Schertektonik im Grundgebirge des Sirwa-Doms (Anti-Atlas, Marokko). *Berliner Geowissenschaft Abhandlungen* 66: 301-332.

Sougy, J. 1969. Grandes lignes structurales de la chaîne des Mauritanides et de son avant-pays (Afrique de l'Ouest). *Bulletin de la Société géologique de France* 11: 133-149.

Sougy, J. & G. Bronner 1969. Nappes hercyniennes au Sahara espagnol méridional, (tronçon nord des Mauritanides). *5th. Colloque de géologie africaine*, Clermont-Ferrand, (Fr.), Géologie Minéralogie 19: 75-76.

Tessier, F., R. Dars & J. Sougy 1961. Mise en évidence de charriages dans la série d' Akjoujt ; *Comptes rendus de l'Académie des Sciences Paris,* 252: 1186-1188.

Thorman, C.H. 1976. Implications of klippen and a new sedimentary unit at Gibi Mountain, Liberia, West Africa, in the problem of Pan-african-Liberian age province boundary. *Bulletin of the Geological Society of America* 87: 851-856.

Villeneuve, M. 1982. Schéma lithostratigraphique des Mauritanides au sud du Sénégal et au Nord de la Guinée, d'après les données actuelles. *Bulletin de la Société géologique de France* 24: 249-254.

Villeneuve, M. 1984. *Etude géologique sur la bordure ouest du craton ouest africain. La suture panafricaine et l'évolution des bassins protérozoïques et paléozoïques de la marge NW du continent de Gondwana.* Thèse de doctorat d' Etat, Université d'Aix-Marseille III (Fr.): 1-552.

Villeneuve, M. 1991. Découverte de la molasse de démantellement de l'orogenèse du Panafricain I dans la chaîne des Bassarides (Afrique de l'Ouest). *Comptes rendus de l'Académie des Sciences Paris* II, 313: 557-564.

Villeneuve, M. & R.D. Dallmeyer 1987. Geodynamic evolution of the Mauritanide, Bassaride and Rokelide orogens (West Africa). *Precambrian Research* 37: 19-28.

Villeneuve, M., J. Rochet, M. Faye & A.W. Bally 1990. Héritages structuraux panafricains et hercyniens sur la marge Africaine de l'Atlantique central. Géodynamique des bassins et des marges continentales. *Réunion extraordinaire de la Société géologique de France*, Sophia-Antipolis (Fr.) , abstract: 40.

Gondwana Eight, Findlay, Unrug, Banks & Veevers (eds) © 1993 Balkema, Rotterdam. ISBN 90 5410 304 3

Geological and structural framework of the Zambezi Belt, northeastern Zimbabwe

C.M.Barton
British Geological Survey, St. Just, Exeter, UK

J.N.Carney & M.J.Crow
British Geological Survey, Keyworth, Nottingham, UK

J.A.Evans
NERC Isotope Geosciences Laboratory, Keyworth, Nottingham, UK

S.Simango
Geological Survey Department, Harare, Zimbabwe

ABSTRACT: The sector of the Zambezi belt described here lies at the northeastern corner of the Kalahari craton and is adjoined to the east by the Mozambique belt. It comprises a tectonically stacked gneissic sequence formed during the Lufilian-Zambezi (Late Proterozoic) and Pan-African (Late Proterozoic to early Palaeozoic) tectonic events. An initial Zambezi-Mozambique deformation (DZM1) imbricated the lower crust along a series of granulite facies mylonite zones. The granulitic sequences were then tectonically emplaced onto deformed Proterozoic craton margin sedimentary rocks during a DZM2 episode of thick-skinned crustal overthrusting which occurred at about 820 Ma. The DZM2 fabric orientations suggest that tectonic transport directions were both normal (from north to south) and oblique (from northwest to southeast) with respect to the craton margin. A third deformation (DZM3), dated 820 Ma or younger, is correlated with a period of westward convergence within the southern Mozambique belt. In Zimbabwe, DZM3 generated large-amplitude refolding and rethrusting of the DZM2 tectonic stack along structures which verge toward the northwest.

1 INTRODUCTION AND REGIONAL SETTING

The Zambezi belt forms one segment of a system of Late Proterozoic orogenic belts that wraps around and separates Archaean or Early Proterozoic cratonic nucleii (Figure 1). The orogenic belts are tectonic sutures and sites of displacement between the cratons during the various phases of assembly of the Gondwana supercontinent. In a typical cycle of orogenic activity, rifted continental or oceanic basins are closed and compressed into curvilinear belts in which may be juxtaposed lower crustal granulitic sequences, deformed and metamorphosed basement or cover rocks and syn- to post-tectonic granitoid intrusions.

For the region shown in Figure 1, radiometric and structural studies suggest that the amalgamation of terranes involved three overlapping and broadly-defined phases of deformation. The earliest phase equates with the Kibaran tectonothermal cycle between about 1100 and 950 Ma; structures of the Irumide belt (e.g. Daly, 1986) and the Chimwadzulu Zone against which the Zambezi belt terminates to the east (e.g. Andreoli, 1984) formed during this phase. Kibaran deformation reactivated the Early Proterozoic Ubendian belt as a major zone of strike-slip faulting (Daly, 1986). A second event, which corresponds to the main phase of deformation in the Zambezi belt of Zambia, was named the 'Lufilian-Zambezi' event by Daly (1986) and equates with the 'D1' phase of Hanson et al. (1988), who constrained its age to about 820 Ma in Zambia. The latest phase, the 'Pan-African tectonic cycle', occurred during the period between 800 and 500 Ma in Zambia (Daly, 1986; Johns et al., 1989). Both the Lufilian-Zambezi and Pan-African events have been identified by radiometric determinations in the northeastern Zimbabwe sector of the

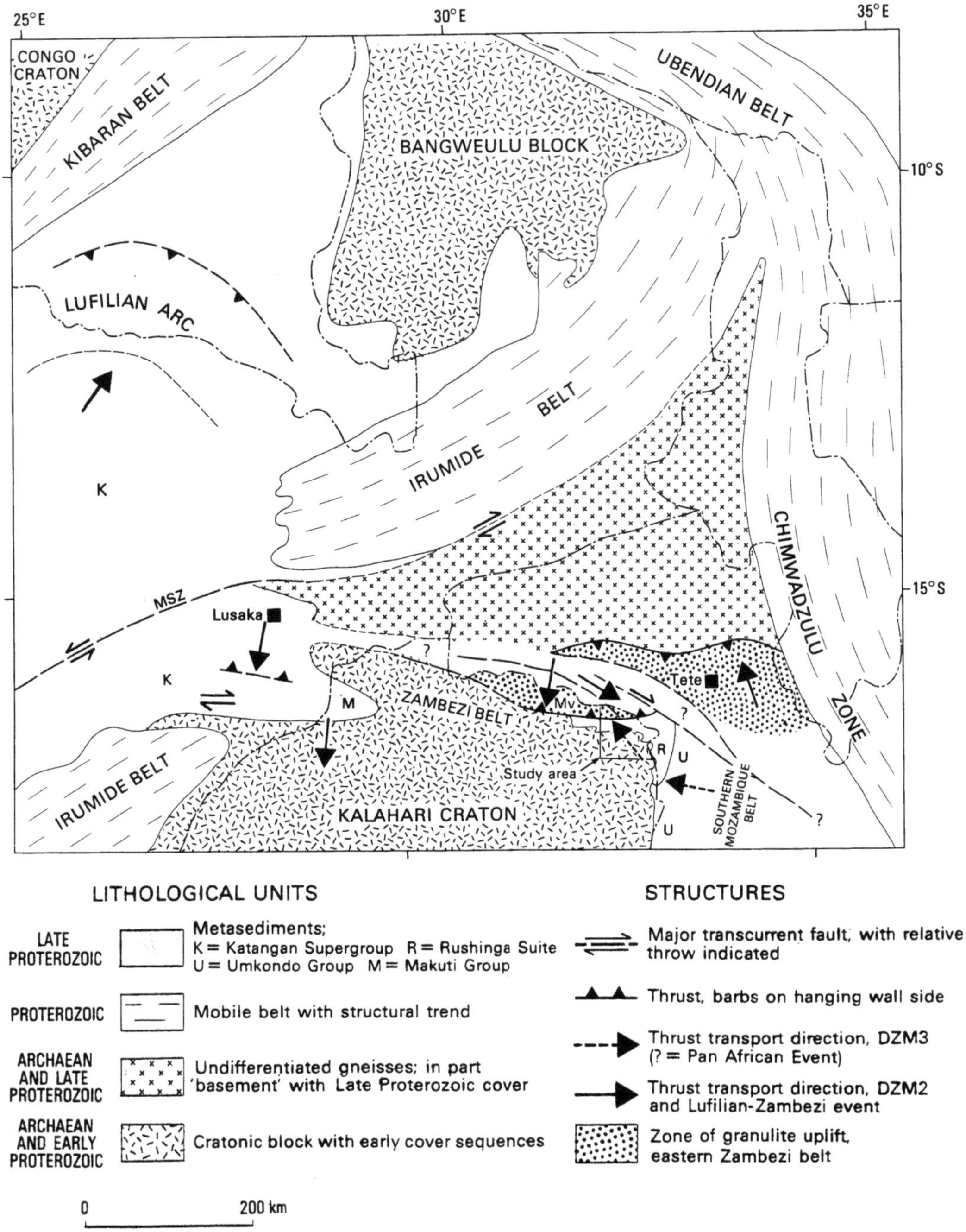

Fig.1 Regional tectonic setting of the Zambezi belt, and location of the study area. Mv, Mavuradonha Mountains; K, Katangan Supergroup; M, Makuti Group; R, Rushinga Metamorphic Suite; U, Umkondo System; MSZ, Mwembeshi Shear Zone. Structural and lithological data for Zambia and Mozambique is taken from Wilson et al. (this volume), Johns et al. (1989) and Barr and Brown (1987). NB: Considerable areas of the Zambezi belt between Tete and Lusaka are covered by Phanerozoic deposits, not shown on this map

Zambezi belt and are described in the present report.

The Zambezi belt occupies a pivotal position with respect to the Lufilian Arc and southern Mozambique belt (Figure 1); northerly structural trends of these adjacent belts contrast with the east-west trend of the Zambezi belt, suggesting that important changes in the timing and direction of tectonic movement must have occurred in this region during the assembly of Gondwana (Porada, 1989). For the western part of the Zambezi belt, in Zambia, previous workers have suggested that transcurrent movement along the northeasterly trending Mwembeshi Shear Zone accompanied early thrust deformation (e.g. Daly, 1988). Hanson et al. (1988) subsequently dated the fabrics developed along the Mwembeshi Zone at about 650 Ma; the data suggest that the Mwembeshi Zone is a Pan-African structure, younger in age than their 'D1' event and more probably associated with the oblique closure of basins forming the Damara branch of the orogenic system farther to the south-west (Porada, 1989). Transcurrent faulting within the Zambian sector of the Zambezi belt is nevertheless recognised by Hanson et al. (1988) and Wilson et al. (this volume), who proposed that the 'D1' deformation involved southwards-directed (craton-normal) thrusting as well as sinistral and dextral (craton-oblique) motions.

In the study area of northeastern Zimbabwe (Figure 2), we show that the Zambezi belt deformation involved three separate events. The most widespread, which we term DZM2, dates from about 820 Ma and is coeval with the 'D1' event of Hanson et al. (1988) in Zambia. The correlation forms the basis of a tectonic model that attempts to link deformation in the study area with that deduced for the Zambezi belt both in Zambia and Mozambique.

In the present study area deformation associated with the Zambezi belt occurs within three fault-bounded tectonostratigraphic units which were for convenience termed 'terranes' in earlier descriptions (e.g. Barton et al., 1991). The terranes are concentrically disposed around the north-eastern margin of the Archaean Kalahari craton and comprise: a *Migmatitic Gneiss Terrane* of Archaean cratonic gneisses reworked during Zambezi belt deformation, a *Marginal Gneiss Terrane* comprising a Proterozoic metasedimentary cover tectonically interleaved with Archaean basement during the Zambezi deformation, and an overlying *Zambezi Allochthonous Terrane* of granitic orthogneisses, granulites and metagabbros. A tabular-shaped granitoid batholith (termed the *Basal Rushinga Intrusive Complex*) occurs at the deformed interface of the Proterozoic cover and the reworked Archaean gneisses. In Zimbabwe, the northern margin of the Zambezi belt is defined by high-angle normal faults along the edge of an east-trending Mesozoic rift valley or graben.

This paper is concerned only with the Marginal Gneiss Terrane, Zambezi Allochthonous Terrane and the Basal Rushinga Intrusive Complex, the structural and metamorphic evolution of which are important for understanding the early tectonic history of Gondwana in central southern Africa.

2 SUMMARY OF THE MAIN ROCK UNITS

Full details of stratigraphical relationships in the Zambezi belt of north-eastern Zimbabwe (Figure 2) are given in Barton et al. (1991).

2.1 *Marginal Gneiss Terrane*

The Marginal Gneiss Terrane consists predominantly of a sequence of migmatitic and feldspathic gneisses, metaquartzites, metaconglomerates, marbles and pelitic gneisses known as the *Rushinga Metamorphic Suite*. The Rushinga suite is thought to be the remnants of a sedimentary cover to the northeastern margin of the Kalahari craton.

Within the Marginal Gneiss Terrane, a subordinate unit, termed the *Chimanda Metamorphic Suite* consists of decimetres-thick, tectonically-bounded gneissic slivers characterised by mylonitic and cataclastic fabrics which rework earlier tectonic foliations. The Chimanda suite gneisses form the hanging walls of thrusts which repeat the Rushinga suite succession; they are interpreted as slices of sheared Archaean basement incorporated tectonically into the Rushinga suite during the Zambezi belt deformation.

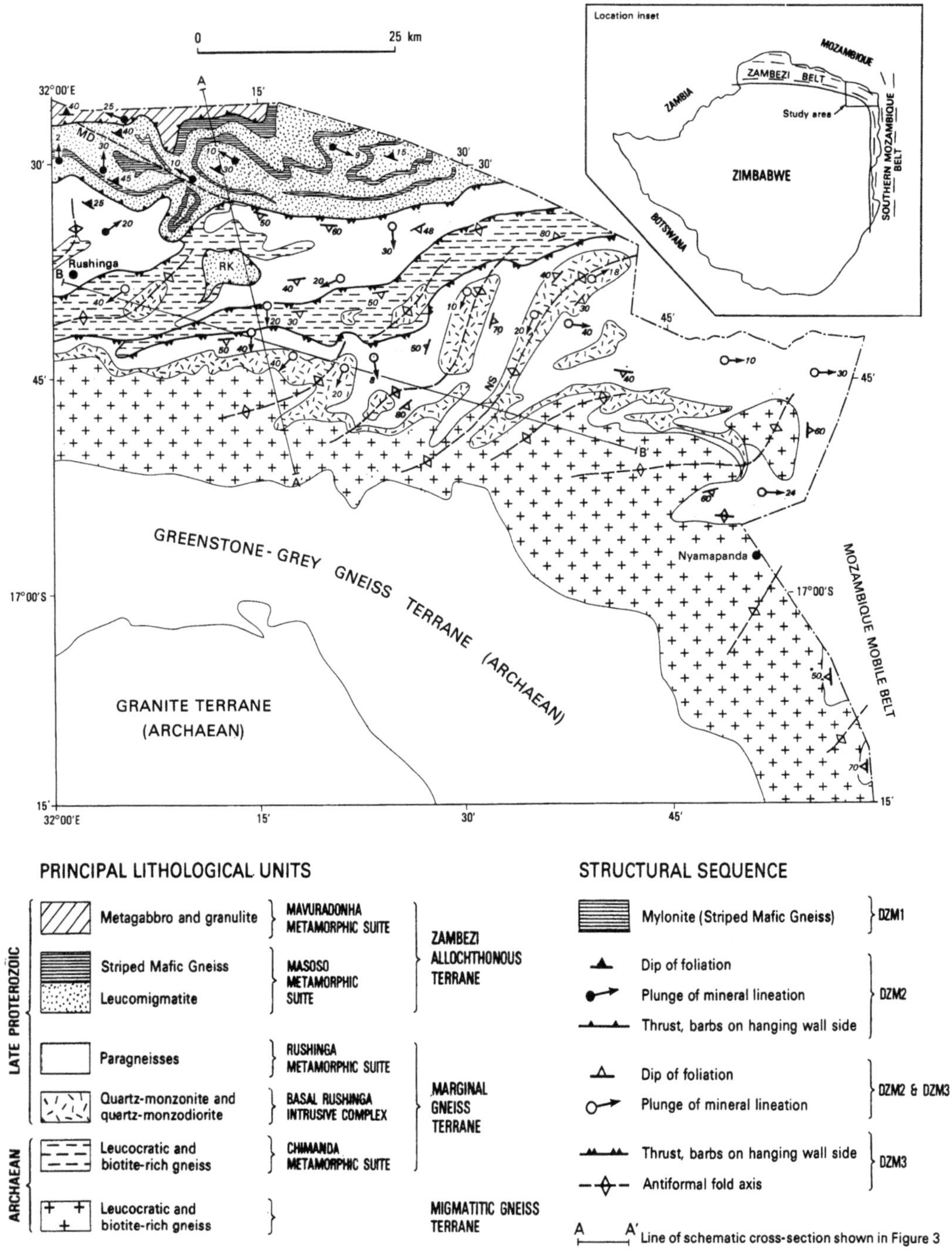

Fig.2 Distribution of the principal lithological and structural elements in the study area. MD, Mudzi Discontinuity; NS, Nyadire Synform; RK, Rusambo Klippe. The regional location of the area is shown in Figure 1

The base of the Marginal Gneiss Terrane is obscured by structural complexity and granitoid sheets of the Basal Rushinga Intrusive Complex. The terrane is overlain structurally by the rocks of the Zambezi Allochthonous Terrane.

The age of the Rushinga suite is uncertain. It is intruded by the Basal Rushinga Intrusive Complex, for which radiometric data suggest an age of about 829 Ma (see below) and provide a minimum age for deposition of the sedimentary precursors to the Rushinga suite. The Umkondo Group (or 'System') of metasedimentary rocks in eastern Zimbabwe (Figure 1) has been correlated with the Rushinga suite by Vail (1964), although Stocklmayer (1980) suggests that in Mozambique there may be an unconformity between the two sequences. The Umkondo Group is intruded by dolerite sills and contains basic volcanic intercalations towards the top. The sills and volcanic rocks have been dated at about 1080 Ma (Allsopp et al., 1989), which is therefore the depositional age of that part of the Umkondo Group. The Rushinga suite has also been correlated with the Makuti Group farther west, in the central Zambezi belt segment (Stocklmayer 1977, 1981) (Figure 1). The Makuti Group is thought to be an eastward extension of the Katangan Supergroup (Broderick 1980), which contains the Kafue metarhyolites which have yielded a U-Pb age of 881 $\pm$6 Ma (Wilson et al., this volume).

2.2 *Zambezi Allochthonous Terrane*

The Zambezi Allochthonous Terrane consists of a lower, predominantly granitic sequence, the Masoso Metamorphic Suite and an upper, gabbroic and granulitic sequence, the Mavuradonha Metamorphic Suite (Figure 2).

The *Masoso Metamorphic Suite*, which overlies tectonically the Rushinga Metamorphic Suite, is dominated by a monzo-granitic leucomigmatite. The leucomigmatite contains a pervasive segregation layering, and many leucosome ribbons folded into rootless recumbent isoclinal folds which are axial planar to the layering. In addition, 70 to 200m-thick, laterally extensive layers characterised by millimetre to decimetre-scale and regular interlayering of mafic and granitic material occur within the leucomigmatite. These rocks, known as Striped Mafic Gneisses, range in metamorphic rank from amphibolite facies in the east to garnet granulite facies in the west. Those in the west are strongly mylonitised at the garnet granulite facies, with a fabric defined by fine-grained mafic and leucocratic ultramylonitic layers which anastomose around tectonic pods formed by less-strained, coarse-grained garnet granulite protolith material (Carney et al., 1991).

The Masoso Metamorphic Suite is thought to have been emplaced as a granitic batholith which was tectonically imbricated along ultramylonite zones and metamorphosed at about 40 km depth in the lower crust during the earliest deformation in the Zambezi belt (DZM1). The Striped Mafic Gneiss layers represent early basic intrusive sheets that subsequently acted as zones of deformation and migmatisation along which the DZM1 mylonites were formed. Intrusive sheets of gabbro and monzodiorite are not migmatised and therefore were emplaced after the DZM1 event had largely ceased; they have coronitic textures indicative of crystallization at 25 km depth. Foliations and mineral elongation lineations are pervasively developed throughout the Masoso suite; in large part they are attributed to a DZM2 deformation which folded the DZM1 mylonite zones (Figure 2).

Analyses of Masoso suite leucomigmatites were carried out on bulked samples, in order to minimise the effects of element partitioning into migmatitic phases. The compositions, considered to be close to that of the protolith, show large contents of Ce, La, Nb, Y and Zr appropriate to the field of granites from within-plate tectonic settings (Harris et al., 1986). Mafic granulite layers and corona-textured metagabbros are subalkaline and similarly enriched in these elements, total Fe and titanium (Barton et al., 1991).

Rb-Sr whole rock data from the Masoso Metamorphic Suite give an errorchron age of about 820 Ma. The age is a fabric age, and most probably relates to the second, highly pervasive, deformational event (DZM2). It is interpreted as a minimum age for the formation of the Masoso suite (Evans in Barton et al., 1991).

The Masoso suite is structurally overlain by the *Mavuradonha Metamorphic Suite,* a sheet-like body of deformed mafic igneous rocks and granulites, with subordinate REE-enriched carbonate bodies, all characterised by extensive CO_2-Cl metasomatism. Metamorphic grade in the Mavuradonha suite ranges from greenschist to amphibolite facies for basal mylonitised gneisses to garnet granulite facies higher in the sequence.

Basal mylonites form the sole of a scapolitised metagabbro body several hundreds of metres thick and at least 50 km long. The upper part of this body is strongly foliated with screens of metagabbro alternating with amphibolite and garnet granulite. West of the study area, in the Mavuradonha Mountains, the upper part of the metagabbro is overlain by mylonitised garnet granulites which form the tectonised base of the highest structural unit of the Mavuradonha suite, a sequence of well-layered and foliated plagioclase-rich garnet granulites (observations by the present writers). The latter were termed 'anorthosite' by Bache et al. (1984).

The geochemistry of the Mavuradonha suite metagabbros (Barton et al., 1991) closely resembles that of present-day continental tholeiites, although it is difficult to be certain of the original tectonic setting as all of the rocks are plagioclase-rich and thus have formed from a highly fractionated protolith.

There are no age data available for the Mavuradonha suite. The Tete Complex of Mozambique (Figure 1) contains similar types of gabbroic and granulitic rocks with locally-developed carbonate-REE mineralisation (Andreoli, 1984), and may be a correlative. Radiometric ages from the Tete Complex suggest that the peak of granulite facies metamorphism occurred between 900 and 1000 Ma and was followed by scapolite metasomatism, uplift and thrust deformation dated at between at 850 and 650 Ma (Andreoli 1984). The latter age range includes the 820 Ma fabric age of the Masoso suite.

2.3 *Basal Rushinga Intrusive Complex*

The Basal Rushinga Intrusive Complex is a multiphase igneous complex that consists of plutons of quartz-monzonite and quartz-monzodiorite, with subordinate granite, quartz-diorite, quartz-syenite and anorthosite; the original volume is estimated to have been at least 2000 km^2. In large part, the igneous complex comprises the 'granodiorite gneiss' division of Talbot (1973), interpreted by that author as basement which has been remobilised to form mantled gneiss domes.

The Basal Rushinga Intrusive Complex occupies the lithological, structural and metamorphic break between the Archaean basement gneisses and the overlying paragneisses of the Rushinga suite (Figure 2). It has been extensively deformed and occupies the cores of large north-east trending folds within the Rushinga and Chimanda suites.

The margins of the complex are strongly gneissic with fabrics concordant with those in the host rocks. The gneissosity decreases in intensity away from the margins although local zones of ductile foliation occur throughout. Granoblastic textures, rather than original igneous ones, are predominant, however the presence of cognate dioritic xenoliths and of internal intrusive contacts attest to the magmatic affinities of these rocks. Also present throughout the complex is a preferred hornblende mineral lineation along a north-easterly trend, parallel with individual pluton long axes and to fold axes within the country rocks (Figure 2).

Geochemical data (Barton et al., 1991) show that individual plutons have similarities with the granitoid gneisses of the Masoso suite; they contain large though variable concentrations of Ce, La, Nb, Y and Zr, and resemble the mildly alkaline A-type granites from rift zones or areas of thinned continental crust. Initial strontium isotope ratios that range between 0.703 and 0.716 suggest that the igneous complex is isotopically heterogeneous, containing strontium derived from mantle as well as crustal sources (Evans in Barton et al., 1991).

3 STRUCTURAL EVOLUTION

In the study area, three phases of Zambezi deformation (DZM1-3) are recognised.

3.1 *DZM1: Deep-crustal deformation in the Zambezi Allochthonous Terrane*

DZM1 structures are recognised only in the Masoso Metamorphic Suite (Carney et al., 1991). They are preserved in the Striped Mafic Gneisses, and as melt-segregation layers in the host leucomigmatite. The Striped Mafic Gneiss layers show evidence for the participation of a granitic melt phase during the formation of typical mylonitic fabrics, and are interpreted as laterally-extensive zones of melt-enhanced deformation (cf. Hollister and Crawford, 1986) and extreme crystal plasticity, along which the granitic orthogneiss host rock was tectonically imbricated. There are no field data which identify the mylonites with a particular tectonic movement regime; foliations and lineations were extensively rotated during successive generations of the mylonitic foliation and, subsequent to mylonitisation, the fabrics were folded into regional-scale sheath-like structures.

The within-plate geochemical signature of the Masoso suite indicates an intracrustal setting for the deformation. The mylonites may represent flat-lying detachment zones formed during a phase of extension deep within the crust. However, we cannot exclude alternative interpretations that regard the mylonitisation as an early phase of crustal shortening and thickening within the Zambezi belt.

3.2 *DZM2: Regional overthrusting*

DZM2 is the principal deformation event of the Zambezi belt; the gabbroic slab and basal granulites of the Zambezi Allochthonous Terrane were uplifted and emplaced southward onto craton margin metasedimentary rocks of the Rushinga suite (Figure 3a). Although the DZM2 fabrics have been extensively reworked, obscuring original criteria for structural vergence, a broadly north-to-south tectonic transport direction is consistent with the original disposition of lithostratigraphic units along the northern craton margin. Structures of the DZM2 phase are best preserved within the Zambezi Allochthonous Terrane and adjacent northwestern part of the Marginal Gneiss Terrane. Farther south and east, the DZM2 fabrics were extensively rotated and re-foliated during DZM3 (Figure 2).

In the Zambezi Allochthonous Terrane, mylonitic and migmatitic layering of DZM1 age within the Striped Mafic Gneisses was folded into northwest trending sheath-like isoclinal structures by the DZM2 deformation. Penetrative foliations were imposed throughout the Masoso suite, and in part are defined by transposed and isoclinally folded, early melt segregation layers. The foliation surfaces contain a mineral elongation lineation which changes its orientation along the strike of the Allochthonous Terrane. In the sector extending from the Mavuradonha Mountains into the western part of the study area, the foliation dips north and the plunge of the lineation is down-dip. Farther east, however, the plunge of the lineation changes to northwest or southeast, subparallel to the strike of the foliation and to the axes of the sheath-like fold structures. The change in structural orientation takes place across a northwest to southeast trending gneissic shear zone termed the Mudzi Discontinuity (Figure 2). Gneisses within the Mudzi Discontinuity have highly ductile and locally blastomylonitic fabrics that include entrained DZM1 mylonitic layers, disconnected folds and narrow slivers. Foliations and mineral elongation lineations are parallel or subparallel with the strike of the gneissic shear zone.

The Mudzi Discontinuity is interpreted as a DZM2-age structural hinge zone along which the Zambezi belt thrust stack was segmented into thrust packages showing craton-normal (north to south) and craton-oblique (north-west to south-east) movement directions. To either side and within the Mudzi Discontinuity, the Striped Mafic Gneiss layers in the Masoso suite show similar garnet granulite facies metamorphic grades, suggesting that the transcurrent DZM2 structures were deeply rooted within the crust.

3.3 *DZM3: Folding and backthrusting*

During DZM3, the eastern craton-ward part of the Zambezi belt was folded, turned on edge and locally thrusted. In the west of the study area, the original DZM2 tectonic stacking sequence is

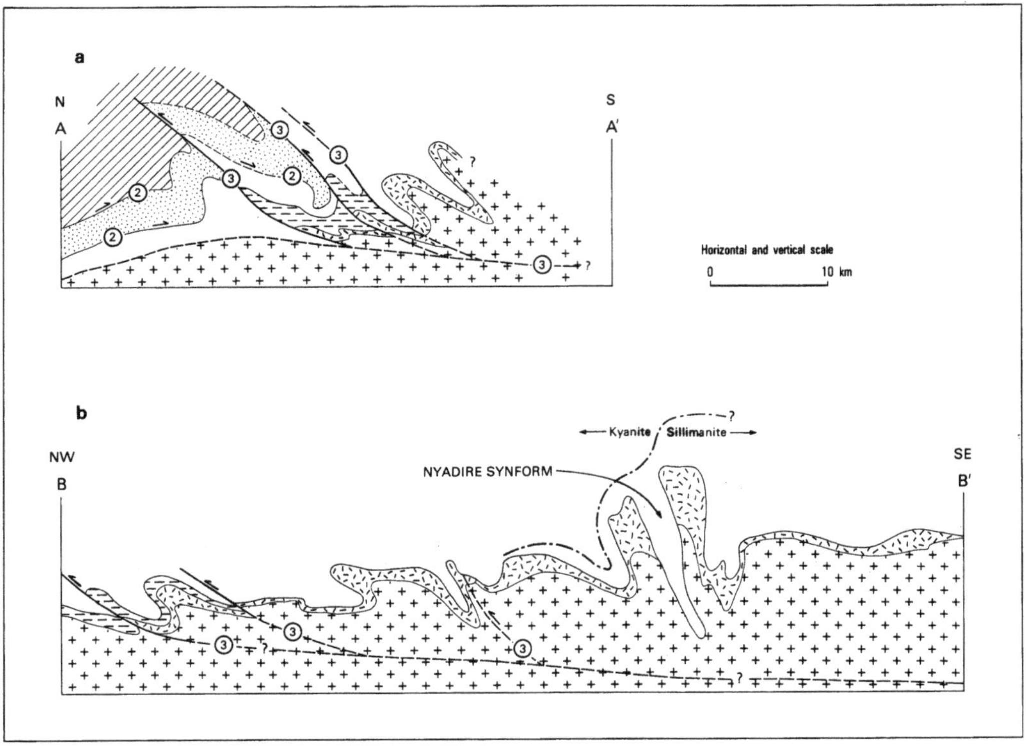

Fig.3 Schematic structural sections in the Zambezi belt (a) parallel to the DZM2 thrust direction and (b) parallel to the DZM3 thrust direction. The DZM deformational phases for major thrusts are shown by the circled numbers. An explanation of the rock units is given in Figure 2

preserved, in which the Rushinga suite is structurally overlain by the Zambezi Allochthonous Terrane. Farther east and south, however, the DZM2 fabrics and thrust zones are progressively rotated and overturned toward the north or north-west (Figure 3a). In that domain, the DZM3 and reworked DZM2 foliation surfaces are inclined to the southeast and contain a down-dip plunging mineral elongation lineation.

The major DZM3 structures are thrusts (Figure 2) that contain slivers of reworked Archaean basement gneiss (Chimanda Metamorphic Suite), and a set of large-scale northeast trending folds developed mainly in the southeast of the study area (Figures 3a&b). Fold axial surfaces dip consistently at moderate to large angles (more than 45°) toward the southeast. The folds range from extremely large-amplitude (10km) structures, developed in the Nyadire Synform (Figure 3b), to smaller amplitude and larger wavelength structures which occur to the northwest and southeast. Circular to ellipsoid-shaped fold interference structures formed at this time and include the Rusambo Klippe (Figure 2).

The orientation of DZM3 fold axes, thrust surfaces and foliations are all consistent with tectonic transport toward the northwest, a direction close to the reciprocal of that deduced from DZM2 movement history.

3.4 *Structures in the Basal Rushinga Intrusive Complex*

The absence of chilled margins and the prominence of gneissic and migmatitic fabrics at the margins of the intrusive complex indicate forcible emplacement. Granoblastic textures and ductile deformational fabrics are developed throughout the intrusive complex, and suggest recrystallization in a regime of high-temperature deformation.

The structural position of the intrusive complex at the sheared interface between the Rushinga suite and Archaean basement (Figures 2 and 3) suggests that it was emplaced during or soon after DZM2 thrusting. Foliations within the complex were folded together with those in the host rocks along DZM3 structural axes. The northeast trending linear fabric within the intrusive complex is parallel with both the elongation direction of individual plutons and DZM3 fold axes, suggesting that the lineation reflects a northeasterly tectonic extension direction.

4 METAMORPHISM

Metamorphic grade increases northwards, and structurally upwards, from the Marginal Gneiss Terrane into the Zambezi Allochthonous Terrane. The metamorphic inversion occurs at the structural base of the Zambezi Allochthonous Terrane, and is the result of DZM2 thrust-emplacement of the granulitic terrane onto the Rushinga suite metasedimentary cover sequence.

4.1 *Marginal Gneiss Terrane*

Metamorphic recrystallization, migmatisation and shear deformation of the Rushinga suite have destroyed most sedimentary structures and original grain boundaries. Pebble outlines are preserved in some quartzitic conglomerates in the eastern part of the sequence. Elsewhere in the Rushinga suite the presence of metamorphic clinopyroxene without garnet in mafic lithologies, and the assemblage Bi-Grt-Ky(Sill)-Qtz-Pl in pelites is indicative of crystallization at uppermost amphibolite facies (Ghent et al., 1982). Plagioclase in mafic and aluminous layers is Ab_{20} in composition and indicates at least the middle amphibolite facies of metamorphism. The kyanite-bearing assemblages reflect the high-pressure peak of metamorphism in this terrane; their crystallization was in response to shear deformation and tectonic burial of the gneisses beneath the Zambezi allochthon during DZM2 overthrusting.

In the pelitic rocks, kyanite occurs as rods which form part of a mineral elongation lineation. In the southern and western part of the Marginal Gneiss Terrane, however, kyanite is commonly fragmented and wrapped around by mica-quartz aggregates of a later generation; these textures indicate reworking of DZM2 mineral fabrics during the DZM3 deformational event.

Sillimanite both pseudomorphs kyanite and also grows across the DZM3 planar fabric, indicating that it is in part post-tectonic with respect to the latter. In the eastern gneisses of the Rushinga suite, sillimanite progressively replaces kyanite, and is the sole aluminosilicate phase east of 32° 30', suggesting that the high pressure (DZM2) phase of metamorphism in the Marginal Gneiss Terrane was succeeded in that area by a later high temperature or low pressure event. The zone of sillimanite metamorphism extends southwards from the study area for at least 250 km along the western margin of the southern Mozambique belt (Vail, 1966). The widespread growth of sillimanite could be a consequence of pressure relaxation that accompanied and immediately followed DZM3 folding and thrusting. However, in the study area the kyanite to sillimanite transition is most pronounced in the roof zone of the Basal Rushinga Intrusive Complex, which may suggest that DZM2 and DZM3 are not separated by a long period of time.

4.2 *Zambezi Allochthonous Terrane*

The allochthonous orthogneisses display two contrasting styles of granulite facies metamorphism, neither of which can be correlated with metamorphic events in the underlying Rushinga suite.

Within the *Masoso Metamorphic Suite*, coarse-grained shear pods enclosed in the Striped Mafic Gneiss layers preserve evidence for a mafic granulite protolith with the assemblage Grt-Cpx-Pl-Rt-Spn-Qtz, and with a blue Cl-rich tschermakitic amphibole preserved as a relict mineral. The textures are extensively modified by the development of symplectic intergrowths between clinopyroxene and plagioclase. The mylonitic garnet granulite host rocks represent primary DZM1 shear zones along which the protolith was dismembered; they contain essentially the same mineral assemblage but are extremely fine-grained and foliated, with garnets broken down mechanically into micro-porphyroclasts. Geothermometric and geobarometric studies (Carney et al., 1991) show that the coarse-grained pods and host mylonites equilibrated under the same P-T conditions, of about 800°C and 12 kbar. However, the blue amphibole in the coarse-grained pods is the relict phase of an event at higher pressures and temperatures that pre-dated the mylonitisation.

In the western part of the Masoso suite, the granulite assemblages contain late biotite and amphibole, indicating retrogression during the DZM2 overthrusting event. However, the hydrous phases are not extensively developed, suggesting that the mylonitic rocks in particular were resistant to retrogression. By contrast, east of the Mudzi Discontinuity, many Striped Mafic Gneiss layers are recrystallized to amphibolite facies assemblages, with the fine-grained mylonitic fabrics only sporadically preserved as relicts. This more extensively retrogressed domain corresponds to the part of the Zambezi belt most affected by the DZM3 fold and thrust systems (Figure 2).

In the Masoso suite metagabbros, reaction coronas are developed at olivine-plagioclase and clinopyroxene-plagioclase boundaries. The former consist of an inner zone of granular orthopyroxene succeeded outwards by granular clinopyroxene and then by garnet. The P-T conditions for the corona development are 625°-700°C and 7 $\pm$2 kbar, corresponding to about 25 km depth (Carney et al., 1991). These coronas are interpreted as cooling textures formed during uplift, rather than during isobaric cooling from the peak of granulite metamorphism. Their crystallization documents a certain stage reached during the DZM2 phase of uplift and tectonic emplacement of the Zambezi Allochthonous Terrane.

In the *Mavuradonha Metamorphic Suite*, the layered plagioclase-rich granulites, though highly foliated, show equilibrium textures throughout. Garnet and clinopyroxene form augen-like mafic aggregates and thin layers in a granoblastic-textured matrix of plagioclase and scapolite. Metamorphic P-T conditions of 710°-820°C and 11.4 kbar deduced for this assemblage (M.T. Styles pers. comm. 1985) are similar to those inferred for Masoso suite granulites suggesting a similar pre- to syn-DZM1 age of crystallization. Mylonitic low-angle shear zones separating the various gabbroic and granulitic components of the Mavuradonha suite have not been studied in detail, but are assumed to contain mineral assemblages related to late-DZM1 or DZM2 age recrystallization. Metagabbros in the Mavuradonha suite show original igneous textures overprinted by clusters of retrograde minerals that include garnet, hornblende, sodic plagioclase and scapolite. The survival of igneous textures suggests that the gabbros may have resisted anhydrous granulite-facies recrystallization, although we cannot rule out the possibility that they were emplaced at shallower crustal levels following uplift of the granulite sheet. The retrograde mineral assemblage suggests amphibolite-facies downgrading of the gabbros which possibly occurred during DZM2 thrust-emplacement.

5 GEOCHRONOLOGY OF STRUCTURAL AND METAMORPHIC EVENTS

There are limited geochronological data available for defining the structural and metamorphic events described above. In the Zambezi Allochthonous Terrane, samples of Masoso suite rocks with DZM1 mylonitic fabrics from the Mudzi Discontinuity (Figure 2) have yielded Rb-Sr whole rock errorchron ages of 823 $\pm$54 Ma and 823 $\pm$57 Ma, for Striped Mafic Gneiss and leucomigmatite respectively (Evans in Barton et al., 1991). These ages, in fabrics crystailised at granulite facies, may reflect the DZM1 deep crustal event, as suggested by Carney et al. (1991). However, the pervasiveness of the DZM2 event, and the presence of varying proportions of late-stage hornblende and biotite in the samples analysed, makes it more likely that the ages reflect DZM2 disturbances of the isotope systems. The age of the DZM1 granulite-facies event is therefore uncertain, although could correlate with the 900-

1000 Ma age for granulites in the Tete Complex of Mozambique (Andreoli, 1984).

The age of emplacement of the Basal Rushinga Intrusive Complex is given by an Rb-Sr errorchron age of 829 ±78 Ma (Evans in Barton et al., 1991).

The 823-829 Ma ages have a strong regional significance for Zambezi belt evolution. They are comparable with the 843 Ma age of the syntectonic Lusaka granite (Barr et al., 1978), which itself corresponds to an early phase of the Lusakan folding event (Daly, 1986). The results also correlate well with recent K-Ar ages from hornblendes from central Zambia which give ages of between 864 to 804 Ma (Ng'ambi et al., 1986). Also closely similar is the U-Pb zircon age of 820 ±7 Ma for the Ngoma Gneiss of Zambia. The Ngoma Gneiss is a syntectonic batholith emplaced within a crustal-scale ductile shear zone and its fabrics are considered to reflect the influence of the main-phase D1 deformational event in the Zambezi belt (Hanson et al., 1988).

Some Masoso suite gneisses, and feldspathic paragneisses from the Rushinga suite, show two broad clusters of younger Rb-Sr whole rock ages (Evans in Barton et al., 1991). The earlier of these, obtained from quartzo-feldspathic lithologies, is around 670 Ma; the large errors (between 44 and 300 Ma at 2-sigma) make the ages difficult to interpret; however, it could indicate extensive resetting of isotope systems by post-DZM2 events of the type that caused the overprinting of kyanite by sillimanite in the Rushinga suite. A second group of ages, of between 440 and 481 Ma, was obtained from micaceous lithologies from the Rushinga suite and is consistent with the lower blocking temperatures of those mineral assemblages consequent upon final uplift and cooling of the orogen (Evans in Barton et al., 1991).

6 DISCUSSION AND REGIONAL FRAMEWORK

6.1 *Tectonic setting of the eastern Zambezi belt*

Structural orientations, and the tectonic stacking order of rock units in the study area (Figure 2) suggest the closure of a basin orientated approximately east-west in which the mainly quartzose and quartzo-feldspathic sedimentary precursors of the Rushinga Metamorphic Suite were deposited. Sediment deposition occurred at some time during the interval between 1080 Ma (maximum age of the Umkondo Group) and 829 Ma (emplacement of the Basal Rushinga Intrusive Complex); a precise depositional age of 881 Ma would follow from a correlation of the Rushinga suite with the Katangan Supergroup of Zambia (Wilson et al., this volume), although such long-ranging correlations remain uncertain.

An intracontinental origin for the western sector of the Zambezi belt in Zambia has been proposed by Wilson et al. (this volume), based on the range of sedimentary rock types in the Katangan Supergroup and on the fact that the Zambezi belt, although cutting across the older Irumide belt, does not significantly displace the structural trends within it (Figure 1). In north-eastern Zimbabwe only a relatively small fragment of the Zambezi belt is seen and its geodynamic setting is difficult to determine. The deep-crustal gneisses (leucomigmatite of the Masoso suite) and syntectonic granitoids (Basal Rushinga Intrusive Complex) have geochemical signatures compatible with an origin in thinned continental crust, suggesting a comparison with other Proterozoic mobile belts formed in ensialic crustal settings (e.g. Kroner, 1977). On the other hand, the Mavuradonha suite metagabbros and garnet granulites are of a radically different geochemical type. While these rocks are evidently not true ophiolites, their presence in the highest unit of the thrust stack indicates that there exists a type of cryptic or geochemical suture within this part of the Zambezi belt.

6.2 *Gondwana assembly in the eastern Zambezi belt*

Our data show that Zambezi belt deformation in the study area involved three major fabric-forming events. The first of these, DZM1, occurred at about 40 km depth, imposing mylonitic and migmatitic fabrics on layered mafic and granitic orthogneisses that were metamorphosed at garnet granulite facies. This deformation produced a strongly layered lower crustal section, but its significance to the development of the Zambezi belt is obscure. It may date from a period of deep-seated extension of a crust thickened as a result of underplating by granitic and mafic magmas (leucomigmatite, Striped Mafic Gneiss and metagabbros of the Masoso suite, and possibly the metagabbros and basic granulites of the Mavuradonha suite). Such extension may have accompanied the formation of a proto-Zambezi belt basin. Alternatively, the DZM1 fabrics may record a deep-crustal response to a phase of crustal shortening and imbrication that was a prelude to later tectonic exhumation of the granulitic sections, as discussed by Carney et al. (1991).

The most widespread deformation (DZM2) occurred at about 820 Ma and is correlated with the D1 phase in Zambia (Hanson et al., 1988). In Zimbabwe, lower crustal granulites were emplaced as allochthonous tectonic slices (Zambezi Allochthonous Terrane) onto the margin of the Kalahari craton. The high-pressure peak of metamorphism in the Marginal Gneiss Terrane, involving the growth of kyanite in pelitic assemblages, and extensive reworking of Archaean basement gneisses, date from this episode. Two aspects of the DZM2 phase are noted here. First, the thrusts responsible for exhumation of the granulites penetrated deep within the crust, indicating that this deformation was thick-skinned in type. A similar style of deformation was recognised in the Zambian sector of the Zambezi belt by Wilson et al. (this volume), although there, the granulites exposed south of Lusaka are of a different type, and include kyanite eclogites (Vrana et al., 1975). Second, a change occurred in the direction of overthrusting along the strike of the Zambezi belt, from more or less perpendicular to the craton in the west (thrust transport from north to south) to oblique farther east (thrust transport from northwest to southeast). A transcurrent component to the deformation has also been deduced for Zambezi belt structures of D1 age in Zambia by Wilson et al. (this volume), and attributed to the lateral escape of crustal blocks in response to an impinging indenter of the type thought to have occurred in the India/Asia collision zone (e.g. Taponnier et al., 1982).

The DZM2 thrust structures strongly influenced the emplacement of the Basal Rushinga Intrusive Complex, which spread out as a sheet-like body along the sheared interface between the Rushinga Metamorphic Suite and Archaean craton. The intrusive sheet remained in a ductile state during folding and tectonic thickening within the cores of DZM3 anticlines. The syn-orogenic mode of emplacement of the intrusive complex is not inconsistent with its within-plate geochemical signature since Harris et al. (1986) suggest that magmas of this type could develop at any stage of a collision event if lithospheric structures allow mantle-derived magmas to be tapped. A deep-seated origin is indicated by the mantle affinities of some of the contained strontium in the intrusive complex; a geochemical contribution from crustal melting is also suggested by these data, however.

The DZM3 event involved fold and thrust structures indicative of compression directed toward the northwest (Figure 1). The style of this deformation (Figure 3b) suggests that the Zambezi belt acted as a foreland to the main phase of compression in the southern Mozambique belt. Highly ductile fabrics were formed during DZM3 and indicate that the Marginal Gneiss Terrane and Basal Rushinga Intrusive Complex were deformed at elevated temperature. The closure of basins along the southern Mozambique belt may have immediately followed the main-phase (DZM2) event in the Zambezi belt.

6.3 *Granulite correlations in the eastern Zambezi belt*

Andreoli (1984) and Barr and Brown (1987) relate the development of the Tete Complex to that of the granulites and mafic igneous rocks within the Kibaran-age Chimwadzulu Zone and its northerly and southerly extensions (Figure 1). In this paper, however, we have presented evidence for a lithological correlation between the Tete Complex and the Mavuradonha Metamorphic Suite of the Zambezi belt. The correlation is tentative and requires confirmation by further geochemical and isotopic study; it is supported by the similar structural histories, both that inferred here for the Mavuradonha suite and the scheme of evolution for the Tete Complex proposed by Andreoli (1984).

We envisage that the Tete and Mavuradonha sequences (and possibly also the Masoso suite) are lower crustal igneous and metamorphic complexes whose main crystallization occurred at about 1000 Ma. This lower crustal section was then tectonically exhumed at about 820 Ma during the Zambezi belt DZM2 deformation. Cross-sections drawn for the Tete Complex by Barr and Brown (1987) suggest it is bounded to the north by a south-dipping thrust, in contrast with the northward-dipping structure mapped below the Zambezi Allochthonous Terrane in Zimbabwe. A geometry in which DZM2-age thrusts have opposed transport directions (Figure 1), though speculative here, resembles that of the crustal-scale 'pop-up' structure containing garnet granulites and metagabbros of the Jijal and Chilas Complexes in Kohistan, Himalayas (Coward et al., 1986), suggesting that a model of collision-driven crustal thickening may be appropriate for the eastern part of the Zambezi belt.

Structures related to the transcurrent component of the DZM2 deformation may have segmented the granulite terranes between the Tete and northeastern Zimbabwe areas. As suggested in Figure 1, such structures may continue southeastwards, merging with and reactivating older Kibaran tectonic lines represented by the Chimwadzulu Zone and its extensions in Mozambique.

ACKNOWLEDGEMENTS: C.M.B., J.N.C. and M.J.C. publish with the permission of the Director of the British Geological Survey (NERC) and J.A.E. with the permission of the Head of the NERC Isotope Geosciences Laboratory. J.W. Baldock, S. Noble and three anonymous reviewers are thanked for their comments on an earlier version of this paper.

REFERENCES

Allsopp, H.L., Kramers, J.D., Jones, D.L. and Erlank, A.J. 1989. The age of the Umkondo group, eastern Zimbabwe, and implications for palaeomagnetic correlations. *South African Journal of Geology* 92: 11-19.

Andreoli, M.A.G. 1984. Petrochemistry, tectonic evolution and metasomatic mineralisation of Mozambique belt granulites from S Malawi and Tete (Mozambique). *Precambrian Research* 25: 161-186.

Bache, J.J., Dallas, S. and Milian, J.F. 1984. A preliminary report on the geology of the country around Centenary and Mount Darwin. *Annals of the Zimbabwe Geological Survey* IX (for 1983): 34-58.

Barr, M.W.C., Cahen, L. and Ledent, D. 1978. Geochronology of syntectonic granites from Central Zambia: Lusaka Granite and Granite NE of Rufunsa. *Annales de la Societe Geologique de Belgique* 100: 47-54.

Barr, M.W.C. and Brown, M.A. 1987. Precambrian gabbro-anorthosite complexes, Tete Province, Mozambique. *Geological Journal, Thematic Issue (Winter)* 22: 139-159.

Barton, C.M., Carney, J.N., Crow, M.J. and Simango, S. 1991. The geology of the country around Rushinga and Nyamapanda. *Zimbabwe Geological Survey Bulletin No.*92: 220pp.

Broderick, T.J. 1980. The geology of the Makuti Group as it occurs between Makuti and Vuti, Urungwe District. *Zimbabwe Geological Survey Annals* V (for 1979): 1-12.

Carney, J.N., Treloar, P.J., Barton, C.M., Crow, M.J., Evans, J.A. and Simango, S. 1991. Deep-crustal granulites with mylonitic and migmatitic fabrics from the Zambezi belt, northeastern Zimbabwe. *Journal of Metamorphic Geology* 9: 461-479.

Coward, M.P., Windley, B.F., Broughton, R.D., Luff, I.W., Petterson, M.G., Pudsey, C.J., Rex, D.C. and Asif Khan, M. 1986. Collision tectonics in the NW Himalayas. In Coward, M.P. and Ries, A.C. (eds). *Collision Tectonics*. Geological Society Special Publication 19: 203-219.

Daly, M.C. 1986. Crustal shear zones and thrust belts: their geometry and continuity in Central Africa. *Philosophical Transactions, Royal Society of London* A317: 111-128.

Daly, M.C. 1988. Crustal shear zones in Central Africa: a kinematic approach to Proterozoic tectonics. *Episodes* 11: 5-11.

Ghent, E.D., Stout, M. and Raeside, R.P. 1982. Plagioclase-clinopyroxene-garnet equilibria and the geobarometry and geothermometry of garnet amphibolites from Mica Creek, British Colombia. *Canadian Journal of Earth Sciences* 20: 699-706.

Harris, N.B.W., Pearce, J.A. and Tindle, A.G. 1986. Geochemical characteristics of collision zone magmatism. In Coward, M.P. and Ries, A.C. (eds). *Collision Tectonics*. Geological Society Special Publication 19: 67-81.

Hollister, L.S. and Crawford, M.L. 1986. Melt-enhanced deformation: a major tectonic process. *Geology* 14: 558-561.

Hanson, R.E, Wilson, T.J. and Wardlaw, M.S. 1988. Deformed batholiths in the Pan-African Zambezi belt, Zambia: Age and implications for regional Proterozoic tectonics. *Geology* 16: 1134-1137.

Johns, C.C., Liyungu, K., Mabuku, S., Mwale, G., Sakungo, F., Tembo, D., Vallance, G. and Barr, M.W.C. 1989. The stratigraphic and structural framework of eastern Zambia: results of a geotraverse. *Journal of African Earth Sciences* 9: 123-136.

Kroner, A. 1977. Precambrian mobile belts of S. and E. Africa-ancient sutures or sites of ensialic mobility? A case for crustal evolution towards plate tectonics. *Tectonophysics* 40: 101-135.

Ng'ambi, O., Boelrijk, N.A., Priem, H. and Daly, M.C. 1986. Geochronology of the Mkushi gneiss complex, central Zambia. *Precambrian Research* 32: 279-295.

Porada, H. 1989. Pan-African rifting and orogenesis in southern to equatorial Africa and eastern Brazil: *Precambrian Research* 44: 103-136

Stocklmayer, V.R.C. 1977. The geology of the area around Inyanga, with special reference to the Umkondo Group. Unpublished D.Phil. Thesis, University of Rhodesia: 323pp.

Stocklmayer, V.R.C. 1980. The geology of the Inyanga-North Makaha area. *Zimbabwe Geological Survey Bulletin* 89: 167pp.

Stocklmayer, V.R.C. 1981. The Umkondo Group. In Hunter, D.R. (ed.). *Precambrian of the Southern Hemisphere:* 556-561, Elsevier, Amsterdam.

Talbot, C. J. 1973. The geology at the eastern end of the Zambezi orogenic belt, Rhodesia. *Transactions of the Geological Society of South Africa* 76: 113-133.

Taponnier, P., Peltzer, G., Le Dain, A.Y., Armijo, R. and Cobbold, P.G. 1982. Propagating extrusion tectonics in Asia: new insights from simple experiments with plasticine. *Geology* 10: 611-616.

Vail, J.R. 1964. The Umkondo System: Late Precambrian sedimentation along the western edge of the Mozambique orogenic belt in Southern Africa. *XXII International Geological Congress, New Delhi,* Part X, Section 10: 439-455.

Vail, J.R. 1966. Zones of progressive regional metamorphism across the western margin of the Mozambique Belt in Rhodesia and Mozambique. *Geological Magazine* 103: 231-239.

Vrana, S., Prasad, R. and Fediukova, E. 1975. Metamorphic kyanite eclogites in the Lufilian arc of Zambia. *Contributions to Mineralogy and Petrology* 51: 1-22.

Wilson, T.J., Hanson, R.E. and Wardlaw, M.S. (this volume). Late Proterozoic evolution of the Zambezi belt, Zambia: implications for regional Pan-African tectonics and shear displacements in Gondwana. In Findlay, R.H.; Banks, M.R.; Veevers, J.J.; Unrug, R. (eds). *Gondwana 8: assembly, evolution and dispersal.* A.A. Balkema, Rotterdam.

Gondwana Eight, Findlay, Unrug, Banks & Veevers (eds) © 1993 Balkema, Rotterdam. ISBN 90 5410 304 3

Late Proterozoic evolution of the Zambezi belt, Zambia: Implications for regional Pan-African tectonics and shear displacements in Gondwana

Terry J.Wilson
Department of Geological Sciences, Ohio State University, Columbus, Ohio, USA
Richard E.Hanson
Department of Geology, Texas Christian University, Fort Worth, Tex., USA
Melissa S.Wardlaw
Department of Geological Sciences, San Diego State University, Calif., USA

ABSTRACT: The Pan-African Zambezi belt forms part of a Late Proterozoic orogenic system transecting southern Africa. Rifting to form the Zambezi supracrustal basin occurred at ca. 880 Ma, based on a U-Pb zircon age for metavolcanic rocks at the base of the supracrustal sequence. Orogenesis is dated at ca. 820 Ma, based on several isotopic ages from Zambia and Zimbabwe, including an igneous age for a syntectonic batholith injected into a transcurrent shear zone within the belt in Zambia. This shows that deformation in the Zambezi belt is nearly 200 Ma older than in the Damara belt along strike in Namibia, precluding synchronous closure of a single ocean basin along the Damara-Zambezi trend.

Orogenesis in the Zambezi belt involved penetrative deformation of the supracrustal basin infill together with extensive remobilization of underlying sialic basement. In Zambia, F_1 fold vergence and mineral lineation patterns in the supracrustal rocks indicate southward-directed thrust or reverse displacement. Transcurrent shearing is indicated by strike-parallel mineral stretching lineations in mylonitized granitic orthogneisses in the central part of the belt. In contrast, within a major 10-15 km wide shear zone that marks the southern margin of the Zambezi belt, mineral lineations and shear-sense indicators document dominantly dextral, north-over-south oblique shear. Thus, the Zambezi belt is characterized by a complex kinematic history, with compression accompanied by tectonic transport parallel to the orogen.

An older, Middle Proterozoic (Kibaran) orogenic province can be traced across the Zambezi belt, suggesting that the belt evolved as an intracontinental rift or narrow ocean basin with limited displacement of terranes on each side. This in turn implies that this portion of Gondwana was assembled before development of the Zambezi belt and was affected primarily by intraplate deformation during the Pan-African event. Thrusting and transcurrent shearing during main-phase Zambezi orogenesis at 820 Ma reflects compression and indentation along the Kalahari-Congo craton boundary. This may have been linked to coeval stretching in the Damara rift zone to the southwest. Recent models have proposed that the final welding of East and West Gondwana occurred by Cambrian collision along the Mozambique belt. If so, then any associated displacements along the Zambezi part of the Damara-Lufilian-Zambezi orogenic trend in southern Africa must have been accommodated by post-Zambezi transcurrent shear zones, such as the Mwembeshi dislocation.

1 INTRODUCTION

The Proterozoic tectonic mosaic of Africa is dominated by the network of Pan-African orogenic belts that surround and separate cratonic blocks. The development of these belts in the Late Proterozoic and early Paleozoic generally has been considered to be part of the culminating phase of the assembly of the Gondwana supercontinent. The orogenic system that transects south-central Africa, including the inland branch of the Damara belt, the Lufilian arc, and the Zambezi belt (Figure 1), constitutes a major sector of the Pan-African network. This orogenic system separates the Congo and Kalahari cratons and links the Mozambique belt on the eastern side of Africa with the coastal branch of the Damara belt and the West Congolian, Gariep, and Saldania belts along the western side (e.g. Hunter & Pretorius 1981; Hartnady et al. 1985; Porada 1989).

The transcontinental Damara-Lufilian-Zambezi orogenic system shows evidence of a complex structural and kinematic evolution that is still only partly understood. Thrust tectonics and nappe emplacement are characteristic of the entire system, although there are marked differences in vergence between the three individual belts. A striking feature of the system is the prominence of crustal-scale shear zones showing orogen-parallel tectonic transport. Coward (1983) related strike-slip transport on low-angle shear zones in the Damara belt to oblique convergence between the Congo and Kalahari cratons and suggested that the Damara belt may have developed during oblique motion along a transform-plate margin, similar to the 'strike-slip' orogen model

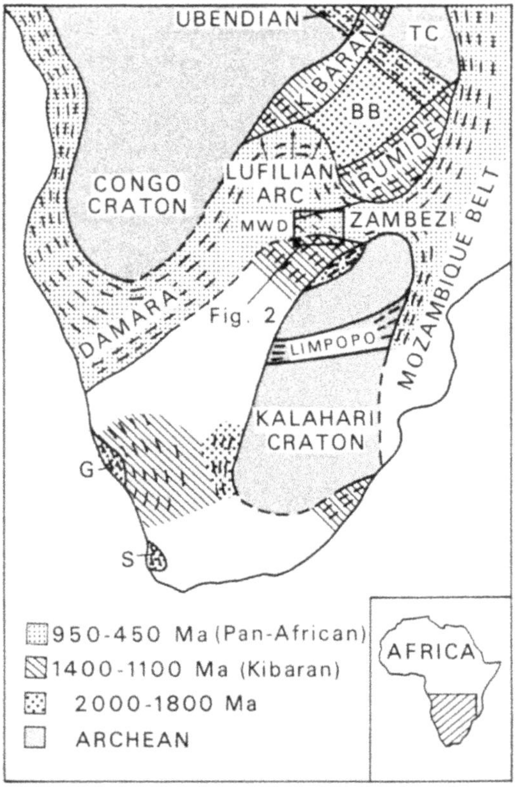

Figure 1. Relation of the Zambezi belt to the regional network of orogenic belts in southern Africa (after Hunter & Pretorius 1981). BB = Bangweulu Block; G = Gariep belt; MWD = Mwembeshi dislocation; S = Saldania belt; TC = Tanzania craton. Box shows approximate area of Figure 2. Structural trends shown schematically in orogenic belts.

for the Hercynian belt proposed by Badham (1982). Evidence for the existence of oceanic crust in the Damara belt led Hartnady et al. (1985) to suggest that a narrow, northeastward-tapering ocean basin (Adamastor Ocean) closed by convergent clockwise rotation of the Kalahari craton relative to the Congo craton. Models relating possible large-scale strike-slip motion to Congo-Kalahari convergence further northeast along the Damara-Lufilian-Zambezi trend have been formulated by Unrug (1983) and Coward & Daly (1984). Daly (1986a, 1988) proposed a kinematically linked system of Pan-African thrust belts and strike-slip shear zones across the width of the continent that operated over a period of ca. 400 Ma (see also Porada 1989). Sutton & Watson (1986) pointed out that such a continent-wide system of linked shear zones may indicate that this part of Africa underwent extensive internal deformation within a single plate during this time interval, rather than breaking up into discrete crustal blocks that were widely dispersed and then reassembled during amalgamation of Gondwana.

The Zambezi belt forms a key element within the Pan-African tectonic framework of southern Africa. Throughout its length, the belt is characterized by extensive tracts of older, remobilized sialic basement. Thick supracrustal sequences, dominated by metasedimentary rocks, unconformably overlie the basement and record a major period of basin development prior to Late Proterozoic orogenesis (de Swardt & Drysdall 1964; Stagman 1978; Cahen et al. 1984; Porada 1989; Hanson et al. 1992). In eastern Zimbabwe, the eastern end of the Zambezi belt merges with coeval, north-south structural trends of the Mozambique belt (Johnson & Vail 1965). The western end of the belt in Zambia is defined by the Mwembeshi dislocation, which is a sinistral strike-slip shear zone that separates the Zambezi belt from the Lufilian arc to the northwest (Figure 1) (de Swardt et al. 1965; Unrug 1983). Large parts of the Zambezi belt are hidden by Phanerozoic cover in Zimbabwe and adjacent parts of Mozambique, but in Zambia the entire cross-strike width of the belt is exposed, so that the nature, extent, and boundaries of the major lithotectonic units within the belt can be examined. In this paper we discuss aspects of the geology of this part of the Zambezi belt that bear on the Pan-African tectonic development of the region. We discuss constraints on the timing of events in the belt and summarize the results of kinematic studies of extensive ductile shear zones in the belt. Systematic measurements of mineral extension lineation attitudes in the shear zones indicate that regionally consistent lineation patterns are present. Regionally coherent stretching lineation patterns of this type are excellent markers of the geometry of tectonic transport and plate-scale kinematic patterns in orogenic belts (e.g. Shackleton & Ries 1984; Shackleton 1986; Brunel 1986; Ellis & Watkinson 1987).

2 THE ZAMBEZI BELT IN ZAMBIA

The main lithotectonic elements of the Zambezi belt in Zambia consist of wide zones of remobilized crystalline basement along the northern and southern margins, a supracrustal sequence exposed in the central part of the belt, and two major units of granitic orthogneiss in contact with the supracrustal sequence (Mpande and Ngoma Gneisses in Figure 2). Contacts between these units are parallel to a penetrative west-northwest foliation, which is present throughout the belt.

Along the northeastern margin of the belt in Zambia, the Zambezi structures are superimposed on a complex metamorphic basement that records a long history of Early and Middle Proterozoic tectonothermal activity (Johns et al. 1989). North of Lusaka, the northern margin of the belt truncates northeast-trending structures of the Middle Proterozoic (Kibaran) Irumide belt (Daly 1986b). Similar Middle Proterozoic northeast structural trends reappear to the south of the Zambezi belt in southern Zambia (Figures 1 and 2) within the Choma-Kalomo block (Hanson et al. 1988). Numerous workers have

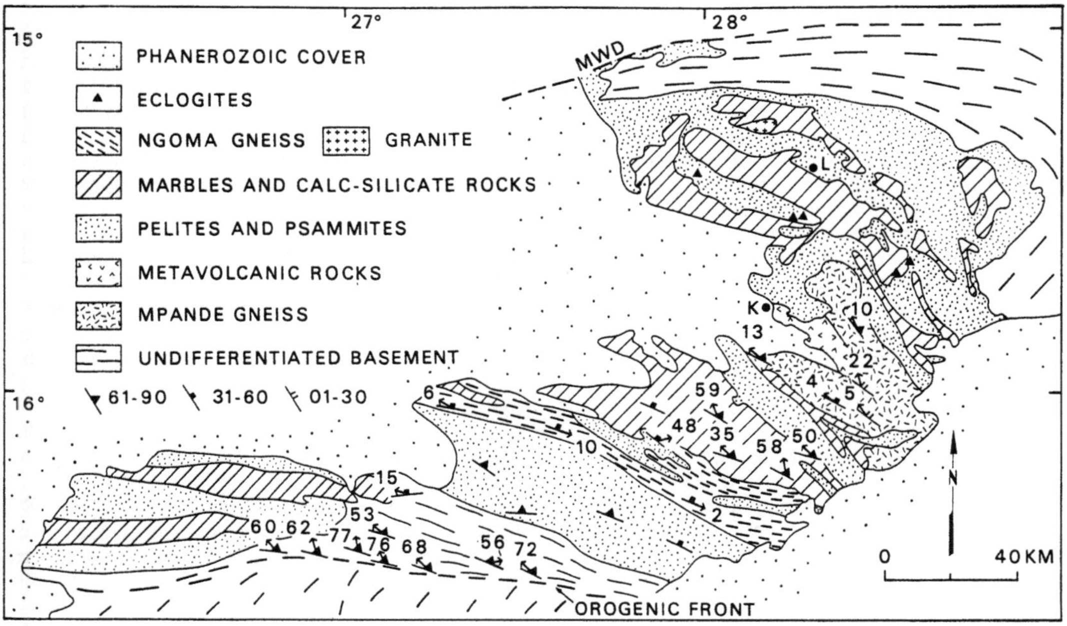

Figure 2. Geologic map of the Zambezi belt in Zambia (modified from Thieme & Johnson 1981). The Mpande Gneiss represents part of the basement of the belt, whereas the Ngoma Gneiss is a younger, syntectonic batholith. Representative structural data come from our own unpublished mapping and from Brown (1966, 1967) and Newton (1963). Attitude of Zambezi S_1 foliation planes shown by symbols defined in legend; arrows with numbers indicate trend and plunge of mineral elongation lineations. L = Lusaka; K = Kafue; MWD = Mwembeshi dislocation.

correlated this area in southern Zambia with the Irumide belt to the north (e.g. Drysdall et al. 1972; Shackleton 1973; Hunter & Pretorius 1981; Kröner 1983), and this has been supported by recent geochronologic studies (Hanson et al. 1988). Thus, tectonic models for the Zambezi belt must account for the fact that the belt appears to cut directly across the Irumide belt without displacing the older trends on each side.

2.1 Stratigraphic evolution

The Zambezi supracrustal sequence is believed to correlate, at least in a general sense, with the Katangan Supergroup in the Lufilian arc of Zambia and Zaire (de Swardt & Drysdall 1964; Drysdall et al. 1972; Unrug 1983). Detailed stratigraphic analysis of the Zambezi supracrustal sequence is precluded by the intense Pan-African tectonic overprint, but it is clear that the sequence must be several kilometers thick. General map patterns and structural relations permit recognition of a regional, three-part stratigraphy. The lowest unit of this tripartite sequence is a bimodal basalt-rhyolite pile which is assumed to have been deposited nonconformably on older basement represented by the Mpande Gneiss (discussed below), although the contact has been modified by subsequent thrusting and shearing (Smith 1963;

Table 1. Timing of events in the Zambezi belt

650 - 450 Ma	Late Pan - African thermal events, cooling
820 Ma	Main-phase deformation; syntectonic plutonism (Ngoma Gneiss)
	Deposition of Zambezi supracrustals
880 Ma	Rifting and volcanism
— — — unconformity — — —	
1100 Ma	Intrusion of Mpande batholith
1200 - 1230 Ma	Mid-Proterozoic tectonism in basement

Mallick 1966). A metarhyolite within the volcanic unit has yielded a U-Pb upper-intercept zircon age of 879 ± 19 Ma (M. Wardlaw, unpub. data), which dates the age of extrusion and gives the time at which the Zambezi supracrustal sequence began to accumulate (Table 1).

The bimodal volcanic unit is overlain by a thick unit

of psammites and pelites, which in turn is succeeded by an extensive unit of marbles and calc-silicate rocks (Figure 2; de Swardt & Drysdall 1964). Amphibolite-facies mineral assemblages are typical of most parts of the sequence (Ridgway & Ramsay 1986; Munyanyiwa 1988, 1990). Geochemical studies of the carbonate and calc-silicate rocks show that they were derived from a series of nearly pure dolostones, impure limestones, marls, and calcareous shales and sandstones (Munyanyiwa & Hanson 1988). Calc-silicate gneisses are one of the most common rock types and are characterized by a well developed S_0 banding that represents interbedding of more carbonate-rich versus more pelitic layers. Scapolite is common in the marbles and calc-silicate rocks, where it clearly forms part of the equilibrium mineral assemblages (Drysdall & Stillman 1966; Munyanyiwa 1985, 1990), and it is sporadically developed elsewhere in the belt as well (e.g. Simpson & Stillman 1963; Vrána et al. 1975). This widespread distribution of scapolite suggests that evaporites were originally present in the supracrustal sequence; their dissolution during metamorphism supplied the Cl and SO_3 necessary for scapolite stabilization relative to plagioclase (e.g. Mora & Valley 1989).

2.2 Mpande and Ngoma orthogneisses

The Mpande and Ngoma Gneisses form two extensive units of granite orthogneiss in the central part of the Zambezi belt (Figure 2) and were interpreted by previous workers as masses of older basement tectonically uplifted during Zambezi orogenesis (Smith 1963; de Swardt & Drysdall 1964; Brown 1966, 1967; Mallick 1966). Hanson et al. (1988) obtained a U-Pb zircon upper-intercept age of 1106 ± 19 Ma for the Mpande Gneiss. They considered the gneiss to represent a pre-tectonic batholith intrusive into the Zambezi supracrustal sequence, but subsequent work has shown that the gneiss is basement to the sequence (Hanson et al. 1992), as considered by earlier workers.

Further south, the Ngoma Gneiss forms an extensive unit that is dominated by quartzo-feldspathic mylonite. Zircons from a protomylonitic part of the unit have yielded a U-Pb zircon upper-intercept age of 820 ± 7 Ma, which is taken as the age of igneous crystallization of the granite protolith (Hanson et al. 1988). This relatively young age proves that the Ngoma Gneiss is not basement, as previously thought, but instead represents a granite batholith intruded into, and deformed with, the Zambezi supracrustal rocks. The age for the gneiss thus provides a younger age limit for deposition of the supracrustal sequence (Table 1). Structural evidence indicates that the Ngoma Gneiss is a syntectonic granite batholith intruded as a large concordant sheet along a major shear zone within the central part of the Zambezi belt (Hanson et al. 1988; Elliot et al. 1989).

2.3 Structural development

The structural style in the Zambezi belt in Zambia is dominated by intense shearing and widespread transposition of pre-existing banding or layering into the west-northwest Zambezi trend. In the supracrustal units, the transposition affects original sedimentary layering, whereas in the marginal zones of remobilized Kibaran basement, pre-existing tectonic foliation and banding are refolded and transposed. The penetrative Zambezi metamorphic foliation, termed here S_1, is thus marked by a composite banding and schistosity throughout much of the area. The schistosity in general developed under amphibolite-facies conditions, and intermediate-pressure, Barrovian-type regional metamorphism was associated with the deformation in most parts of the belt (Ridgway & Ramsay 1986; Munyanyiwa 1988, 1990). The Mpande and Ngoma orthogneisses contain a single foliation that is continuous with S_1 in the adjacent supracrustal rocks. In both these units, and in an extensive zone of sheared basement rocks that marks the southern boundary of the Zambezi belt, the Zambezi foliation is a mylonitic fabric. The belts of mylonite mark the sites of major ductile shear zones developed during main-phase Zambezi deformation.

Supracrustal rocks in that part of the belt north of the Mpande Gneiss contain both macroscopic and outcrop-scale southwest-vergent, isoclinal, overturned to recumbent folds with sheared limbs (de Swardt & Drysdall 1964; Mallick 1966; Matheson & Newman 1966). Crustal thickening during this compressional deformation explains the presence of a discontinuous belt of eclogites and talc-kyanite whiteschists located south of Lusaka (Vrána & Barr 1972; Vrána et al. 1975). These rocks shows that high pressures were attained during metamorphism in parts of the Zambezi belt in Zambia, which is comparable to high-pressure metamorphism in the eastern part of the belt in Zimbabwe (Treloar et al. 1990; Carney et al. 1991).

Within the supracrustal rocks south of the Mpande Gneiss, the general southwest vergence of structures continues, although dips are generally steeper. In the extensive calc-silicate gneisses in this area, domains of noncylindrical, open to tight folds with a sheath-like geometry (Figure 3a) grade within meters into zones where the S_0 sedimentary layering is completely transposed into the Zambezi S_1 foliation (Figure 3b). In the domains of open to tight folding, fold hinge lines define girdle patterns, and slip-line analysis based on fold asymmetry patterns (Hansen 1971) indicates north-over-south reverse movement (Figure 4a). As the bulk strain increases, fold hinge lines develop uniformly steep plunges parallel to the mineral elongation lineations (Figure 4b), indicating rotation of the hinge lines into parallelism with the extension direction during shearing. Mineral elongation lineations in the calc-silicate gneisses plunge steeply but generally are not down-dip, indicating a component of oblique movement (Figure 4a). The adjacent schist units show similar variations

A

B

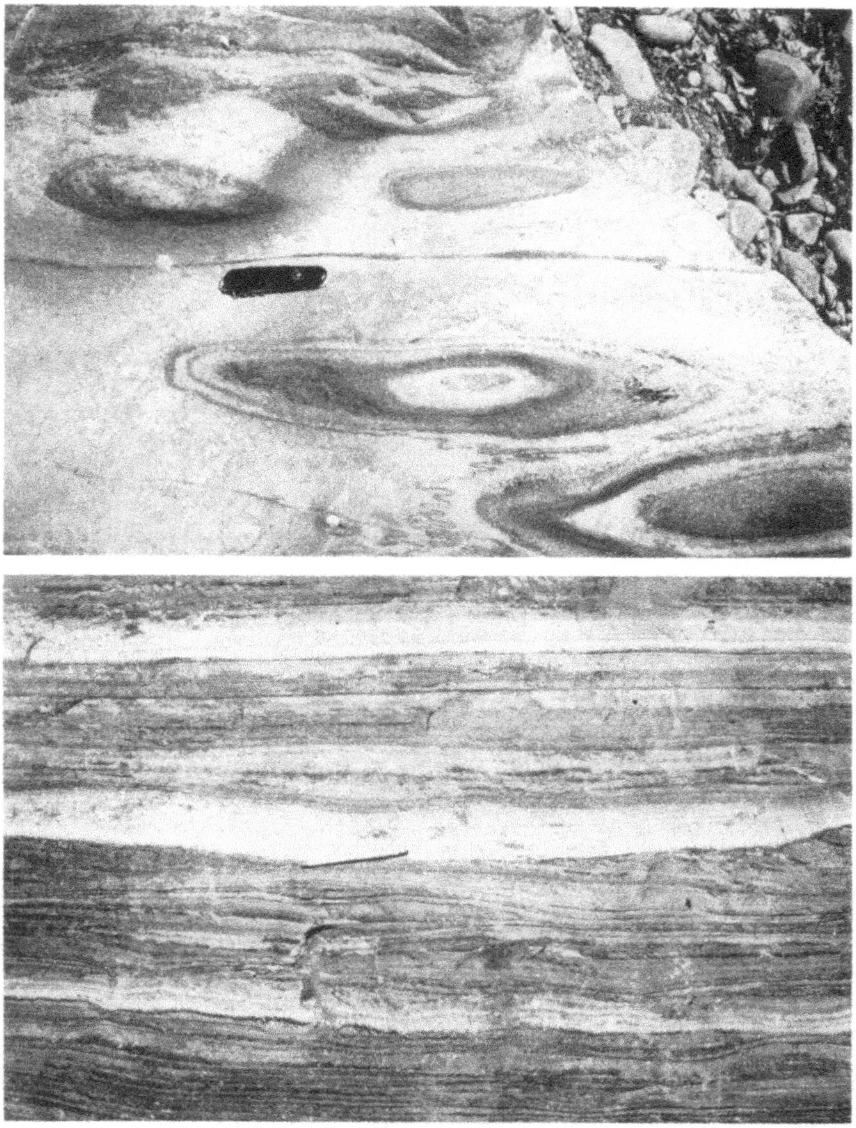

Figure 3. A: Outcrop exposure showing cross section through noncylindrical, southward-closing sheath-like folds in calc-silicate gneiss along the Munyameshi River within the central part of the Zambezi belt, documenting southward-directed reverse motion in the supracrustal rocks. Pen knife parallels the WNW - ESE S_1 foliation trend. Data shown in Figure 4a come from this locality and nearby areas.

B: Outcrop view of zone of nearly complete transposition of S_0 banding parallel to S_1 foliation in calc-silicate gneiss. Pencil points to remnant fold limb, where fold hinge has been eliminated by shearing. Data shown in Figure 4b come from slightly less transposed zones adjacent to this locality on the Munyameshi River.

in strain between domains of relatively open folding and domains of intense transposition, and mineral lineations on the S_1 foliation planes in the highly transposed zones again have steep plunges, consistent with southward reverse transport.

The Mpande and Ngoma Gneisses are characterized by a penetrative, protomylonitic to mylonitic L-S fabric. In contrast to the zones of remobilized basement along the northern and southern margins of the Zambezi belt, the Mpande Gneiss lacks a pre-Zambezi foliation, and it apparently represents a Middle Proterozoic batholith emplaced in the basement to the belt after Kibaran deformation had ceased in the region. The syntectonic Ngoma Gneiss

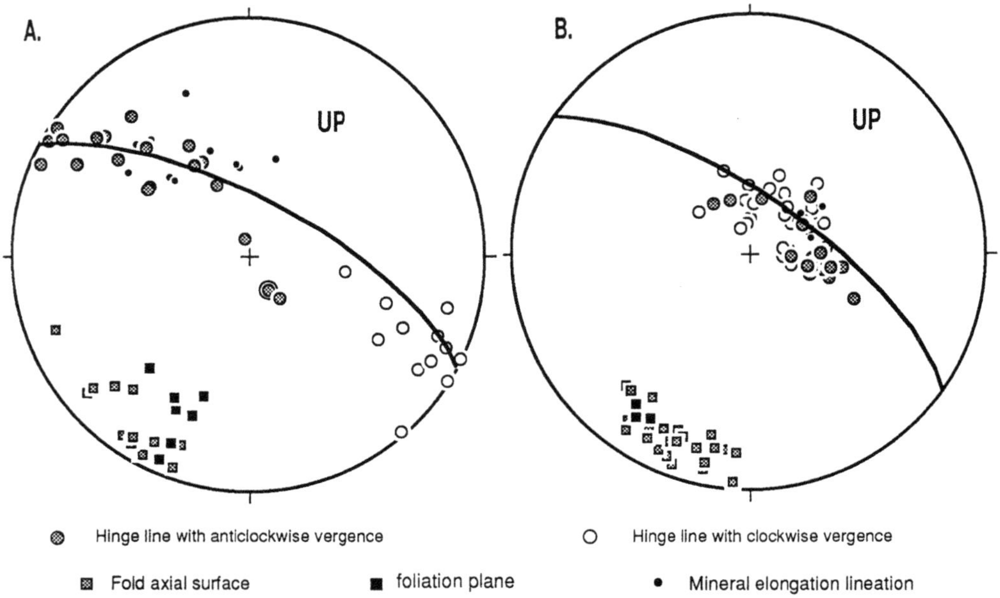

Figure 4. Structural data from calc-silicate gneisses in the Munyameshi River area within the central part of the Zambezi belt. Additional data are given in Munyanyiwa (1985). A: Fold hinge lines in domains of noncylindrical, open to tight folding define a great circle girdle; fold asymmetry pattern indicates north-side-up reverse shearing. B: Fold hinge lines in zones of more intense transposition show uniformly steep plunges parallel to the mineral elongation lineation, indicating rotation into the extension direction during shearing.

consists primarily of fine-grained mylonite and ultramylonite and marks one of the most important ductile shear zones within the Zambezi belt in Zambia. The S_1 foliation in both gneiss units generally has a moderate to shallow northeast dip, and shallow to subhorizontal mineral stretching lineations are pervasively developed (Figures 2, 5 and 6). This shows that the mylonitic gneiss units in the central part of the belt in Zambia represent low-angle, crustal-scale transcurrent shear zones and are not thrust slices. Examination of a small number of samples from the Mpande Gneiss has not revealed a consistent sense of displacement during shearing. In the Ngoma Gneiss, asymmetric K-feldspar augen (e.g. Simpson & Schmid 1983) indicate dominantly sinistral displacement (Elliot et al. 1989).

Along the northern margin of the Zambezi belt, mylonitization of remobilized basement gneisses is extensive (e.g. de Swardt & Drysdall 1964; Simpson 1967; Garrard 1968), but no detailed kinematic studies have been carried out in this region. The southern margin of the Zambezi belt consists of a zone ca. 30 km wide in which Kibaran crystalline basement rocks of the Choma-Kalomo block have been structurally transposed into the west-northwest Zambezi structural trend (de Swardt 1963; de Swardt & Drysdall 1964; Brown 1966; Wilson et al. 1985; Hanson et al. 1988; Hanson et al. 1992). The southern limit of intense Pan-African structural overprinting is defined by a zone 10-15 km wide dominated by mylonites that extends along the entire southern margin of the Zambezi belt in Zambia. This crustal-scale shear zone defines the orogenic front of the belt (Figure 2) (Daly 1986a). Between the extensive mylonites of the orogenic front and the contact with the Zambezi supracrustal rocks to the north, the basement shows variable degrees of shearing. Zones in which early banding in basement paragneisses is refolded and transposed by Zambezi F_1 folds grade across strike into mylonitic shear zones. South of the orogenic front, the Zambezi structural overprinting decreases abruptly and is represented primarily by narrow, discrete shear zones that wrap lozenge-shaped, less deformed masses of gneiss and granite.

The major shear zone defining the southern orogenic front contains steeply dipping (>65°) mylonitic orthogneisses together with quartz mylonites derived from silicification of granite orthogneiss during shearing. Slices of Zambezi supracrustal rock are tectonically interleaved with the transposed and mylonitized basement rocks, and intensely deformed and dismembered syntectonic pegmatites locally are abundant (Wilson & Hanson 1987, 1988; Wilson 1988). Mineral stretching lineations typically pitch 60-70° in the foliation planes (Figures 2 and 7). A variety of shear-sense indicators occur in the mylonitized rocks and include shear band foliations (White et al. 1980), trails of asymmetric pegmatite porphyroclasts and 'stringers' in mylonitized syntectonic pegmatites (Mawer 1987), and, on the microscopic scale, asymmetric augen,

Figure 5. Strike-parallel mineral elongation lineation (arrow; parallel to hammer handle) on steeply dipping foliation plane in quartzo-feldspathic mylonite within the Ngoma Gneiss south of Chivuna Mission.

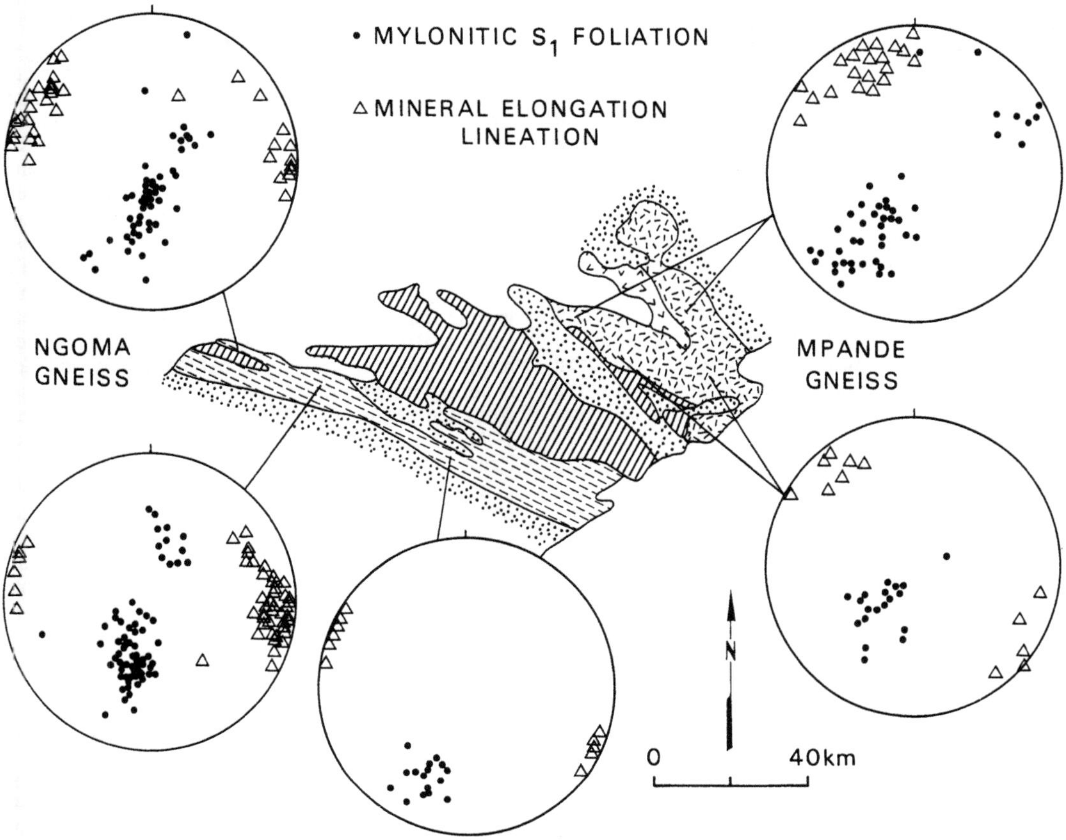

Figure 6. Attitudes of mylonitic foliation and mineral elongation lineations in shear zones within the Mpande and Ngoma Gneisses (equal area, lower hemisphere projections). Patterns as in Figure 2.

displaced broken grains, S-C fabrics, and oblique grain-shape fabrics (Simpson & Schmid 1983; Lister & Snoke 1984; Passchier & Simpson 1986). These indicate dominantly dextral, east-southeast-directed, north-over-south oblique shear.

2.4 Kinematic evolution

The consistent northeast dip of the foliation and fold axial surfaces in the Zambezi belt in Zambia, and the pronounced southwest fold vergence in the northern part of the belt around Lusaka, led previous workers to interpret the belt in Zambia as a southwest-verging thrust terrane (e.g. de Swardt & Drysdall 1964; Daly 1986a; Porada 1989). Daly (1986a) proposed a two-stage movement picture for the Zambezi belt, based on his view that two sets of mineral lineations of different ages occur in the belt. He suggested that early, west-southwest-directed thrusting resulted in substantial crustal thickening and produced high-pressure metamorphism. A later, south-directed thrusting event was interpreted as gravity spreading perpendicular to the margins of the orogen in response to the earlier crustal thickening. More recently, Hoffman (1991) has emphasized the role of transcurrent shearing in the Damara-Zambezi system, as part of an interpretation in which the development of Late Proterozoic orogenic belts in Africa is related to the breakup of a Proterozoic supercontinent and the subsequent assembly of Gondwana.

The structural data discussed herein indicate that kinematic models emphasizing either thrusting or predominantly transcurrent shearing are not applicable to the Zambezi belt as a whole. Major domains of nappe emplacement in the area around Lusaka, and reverse movement in the supracrustal rocks further south, alternate with extensive zones of sheared gneiss (Mpande and Ngoma Gneisses) which record strike-slip motion. The thrusting and transcurrent shearing, together with oblique reverse slip along the southern orogenic front, all appear to have been essentially coeval. This is indicated by the continuity of the S_1 fabric between domains with different directions of tectonic transport, and the lack of structural overprinting or observable changes in metamorphic grade across domain boundaries.

The present data thus require a complex kinematic model for the Zambezi belt in Zambia (Figure 8), involving southwest-vergent thrusting in the supracrustal rocks, transcurrent shearing in the Mpande Gneiss (of unknown direction), sinistral transcurrent shearing in the crustal-scale shear zone defined by the Ngoma Gneiss, and dextral, east-southeast-directed, oblique reverse shearing in the major mylonite zone along the southern orogenic front. We have found no evidence as yet in the areas we have studied for the two separate mineral lineations mentioned by Daly (1986a).

Major south-vergent ductile thrusting also occurred in the Zambezi belt further east along strike in Zimbabwe, where the southern margin of the belt against Archean and Lower Proterozoic basement is a zone of reverse-slip or thrust displacement (e.g. Barton et al. 1985; Daly 1986a; Treloar et al. 1990; Carney et al. 1991). Significant horizontal tectonic transport away from the core of the orogen is documented by the Urungwe klippe, which was emplaced some 40 km south of the orogenic front (Shackleton et al. 1966). Orpen et al. (1989) suggested that sinistral, strike-slip shearing with displacement on the order of 150 km occurred in the southern part of the belt in Zimbabwe. This was based on the separation between the northern end of the Great Dyke, which is intersected by the Zambezi belt, and an ultramafic complex further west, inferred by them to have been originally connected to the Great Dyke. Wilson & Prendergast (1989), on the other hand, attributed the marked swing of the northern edge of the Great Dyke to the east to reflect drag associated with dextral transcurrent shearing in the Zambezi belt. Barton et al. (this volume) provide evidence for transcurrent shearing in the belt in northeastern Zimbabwe and suggest that this may relate to a complex transpressive tectonic regime.

2.5 Timing of deformation

Snelling et al. (1972) and Barr et al. (1978) report several K-Ar and Rb-Sr mineral ages from the Zambezi belt in Zambia, which fall between 600-450 Ma. A smaller number of ^{40}Ar-^{39}Ar mineral ages fall in the same range (Hanson et al. 1988; Hanson et al. 1990). These mineral ages are believed to date late cooling following a post-tectonic thermal event in the region (Hanson et al. 1992). The best age for the timing of penetrative deformation and regional amphibolite-facies metamorphism in the Zambezi belt in Zambia is given by the U-Pb zircon igneous age of 820 ± 7 Ma for the syntectonic Ngoma Gneiss (Hanson et al. 1988). This is within error of a U-Pb zircon age of 846 ± 68 Ma for a small, syntectonic granite that intrudes the supracrustal sequence near Lusaka (Barr et al. 1978; the age and error have been recalculated from the original published data) and also is consistent with other limited geochronologic data from the region (see discussion in Hanson et al. 1992). Furthermore, thrusting and metamorphism of Zambezi supracrustal rocks in the eastern part of the belt in northeastern Zimbabwe occurred at ca. 820 Ma, based on Rb-Sr isochron ages discussed by Treloar et al. (1990), Carney et al. (1991) and Barton et al. (this volume). This indicates that orogenesis affected the entire length of the Zambezi belt in both Zambia and Zimbabwe at ca. 820 Ma.

3 DISCUSSION

3.1 Intracontinental origin of the Zambezi belt

In Zambia, it is possible to recognize a general depositional sequence within the Zambezi belt that begins with localized, bimodal volcanic rocks at the base, passes up into thick psammites and pelites, and is capped by extensive carbonate rocks which apparently originally contained evaporites, based on

Figure 7. Mineral elongation lineations with moderate westerly plunge (parallel to hammer handle) developed on steeply dipping mylonitic foliation planes within basement orthogneiss along the Mutama River. Such attitudes are dominant within the major shear zone that defines the southern orogenic front of the Zambezi belt (see Figure 2).

the abundance of scapolite in metamorphic assemblages. The nature of the rock types and the overall stratigraphic succession are most consistent with deposition in an intracontinental rift basin in which initial extension-related volcanism and terrigenous infilling was followed by incursion of shallow-marine waters (Porada 1989; Hanson et al. 1992). Available age constraints show that accumulation of the supracrustal sequence occurred in a relatively short interval, between 880 and 820 Ma (Table 1). Although extensive tracts of supracrustal rock are preserved in the belt, ophiolite suites or calc-alkaline arc rocks are absent. This makes the Zambezi belt similar to the Lufilian arc to the northwest in Zambia and Zaire, which is interpreted as the deformed infill of a major aulacogen or rift basin (Annels 1984; Unrug 1983, 1988; Porada 1989; Kampunzu et al. 1991).

An intracontinental setting for the Zambezi belt is supported by the fact that the belt cuts directly across the Irumide belt without significantly displacing the older, Middle Proterozoic trends on each side. This requires either that the depositional basin within which the Zambezi supracrustal sequence accumulated was floored by stretched and thinned continental crust, or that the basin developed as a narrow, Red-Sea-type ocean basin that closed directly back upon itself during Pan-African orogenesis (Hanson et al. 1988).

Closure of the basin occurred at ca. 820 Ma and involved penetrative deformation and regional metamorphism of the basin infill, extensive remobilization and mylonitization of the underlying sialic basement, and syntectonic granite intrusion. Coeval transcurrent shearing and oblique thrusting within the belt might be explained, at least in part, by closure of the basin oblique to its margins. Opposing displacement directions in the major shear zones within the belt, however, cannot be explained by a simple transpressive model. Instead, it appears that crustal blocks were being transported laterally in different directions parallel to the belt during overall northeast-southwest compression (Figure 8). This style of deformation is similar to the lateral escape of crustal blocks in response to an impinging indenter, of the type associated with the India/Asia collision zone (e.g. Tapponnier et al. 1982), although it is developed on a smaller scale in the Zambezi belt. The magnitude of tectonic transport that occurred parallel to the belt must have been limited, in view of the continuity of the Middle Proterozoic Irumide trends on each side of the Zambezi belt.

3.2 Relations to the regional Pan-African network

According to Porada (1989), the Zambezi supracrustal basin formed one element of a network of rift basins or narrow ocean basins that developed across West Gondwana (Africa and South America) beginning about 1,000 Ma. Orogenesis occurred at various points along this network over a time span of several hundred million years (Miller 1983; Cahen et al. 1984; Daly 1986a; Hanson et al. 1988). Coward & Daly (1984) suggested that the Zambezi belt represents an oblique collisional plate boundary between the Congo and Kalahari cratons. According to their model, the Mwembeshi dislocation, at the western end of the Zambezi belt, is a sinistral transform plate boundary between the two cratons

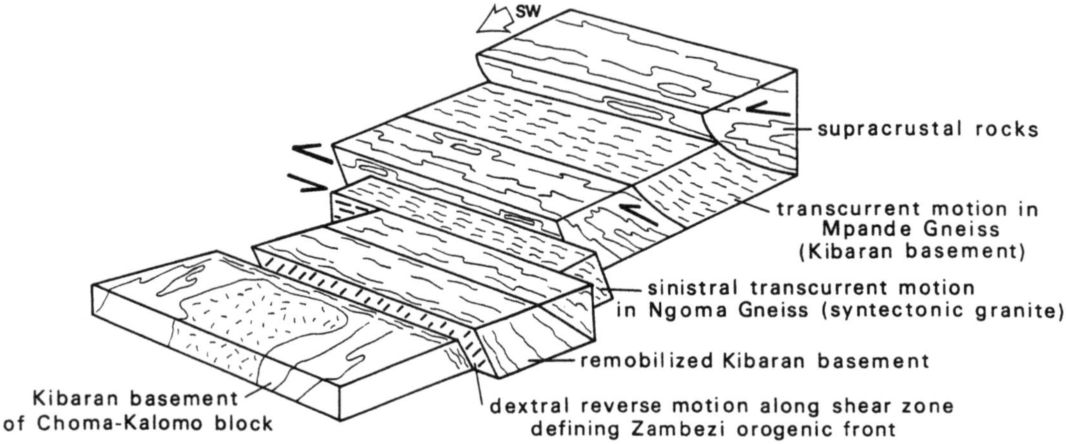

Figure 8. Kinematic model of block displacements during main-phase deformation in the Zambezi belt in Zambia. See text for discussion.

and connects the Zambezi belt with the Damara belt further to the southwest. This model requires that motion along the Mwembeshi dislocation is synchronous with orogenesis in both belts. New U-Pb zircon age data reported by Hanson et al. (1991), however, show that shearing along the Mwembeshi dislocation in the area west of the Zambezi belt affected rocks of latest Proterozoic - earliest Paleozoic age. This is much younger than main-phase Zambezi orogenesis at ca. 820 Ma and indicates that transcurrent shearing along the Mwembeshi cannot relate directly to penetrative deformation in the Zambezi belt.

Evidence for the formation of Late Proterozoic oceanic crust in the Damara belt is provided by the Matchless amphibolite, which shows geochemical affinities to mid-ocean-ridge tholeiites (Miller 1983; Breitkopf & Maiden 1988). Based on this, Hartnady et al. (1985) proposed that a northeastward-tapering ocean basin, termed the Adamastor Ocean, developed in the Late Proterozoic along the Damara-Zambezi trend. Spreading, and subsequent closure during Pan-African orogenesis, is thought to have occurred about a tectonic rotation pole located near the Zambezi belt (Figure 9). The Zambezi belt lies at the northeastern end of the proposed ocean and therefore would have experienced only a small degree of opening. A major difficulty with any model invoking single-stage ocean-basin opening and closure along the Damara-Zambezi trend is the marked difference in timing of orogenesis between the Damara and Zambezi belts (Hanson et al. 1988). The first significant deformation in the Damara belt occurred at ca. 650 Ma (Miller 1983), which is substantially younger than main-phase orogenesis in the Zambezi belt at 820 Ma.

Renne et al. (1990) used recent paleomagnetic data from terranes on each side of the Damara-Zambezi trend to support the existence of the Adamastor Ocean. They infer that the Congo-São Francisco craton was separate from the Kalahari craton prior to Pan-African orogenesis, and that clockwise rotational convergence of the cratons occurred to close the intervening ocean basin. They attribute the diachronous nature of deformation along the Damara-Zambezi trend to progressive, northeast-to-southwest, oblique convergence during ocean closure. However, given Late Proterozoic plate motions comparable to modern rates, the nearly 200 m.y. diachronicity presumably would require closure of an ocean basin of much greater extent than present paleomagnetic data suggest. In addition, the continuity of Middle Proterozoic trends on each side of the Zambezi belt argues against the formation of a major ocean basin in that region, as discussed above.

The intracontinental branch of the Damara belt was the site of rifting beginning at ca. 1,000 Ma, with the principal period of associated volcanism extending from ca. 840 to 730 Ma (Miller 1983). Deformation in the Zambezi belt thus occurred when continental stretching was still taking place in the Damara belt. At the same time, ocean spreading was apparently active along the site of the coastal branch of the Damara belt to form the "Pan-African" or "proto-" South Atlantic Ocean (Miller 1983; Porada 1989). Porada (1989) suggested that the Zambezi belt was produced by deformation along the northeastern, leading edge of the Kalahari craton due in part to motion resulting from this spreading. An additional component of motion would arise if there was an increase in stretching to the southwest within the developing intracontinental Damara rift zone, as suggested by the development of deep-water conditions in areas adjoining the Pan-African South Atlantic Ocean (Miller 1983). The resultant counterclockwise rotation of the Kalahari craton relative to the Congo craton may have produced oblique convergence and indentation where the Congo and Kalahari cratons were pinned at their eastern junction (Figure 9), accounting for compression and lateral tectonic transport in the Zambezi belt at ca. 820 Ma.

3.3 Relation to Gondwana assembly

Arguments for an essentially intracontinental mode of formation of the Zambezi belt are significant in terms of the process of Gondwana assembly, because this indicates that cratonic elements on each side of the belt had become amalgamated before Late Proterozoic rifting and underwent limited relative motion, without dispersal and reassembly, during the Pan-African period. The Middle Proterozoic (Kibaran) Irumide belt, which is crosscut by the younger Zambezi belt, is part of a major Middle Proterozoic orogenic province that can be traced into South Africa, where it is represented by the Namaqua-Natal belts (Hartnady et al. 1985; Hanson et al. 1988). This Middle Proterozoic province wraps around the western and southern margins of the Archean Zimbabwe and Kaapvaal cratons and originally linked with a coeval province in East Antarctica, which lay adjacent to southern Africa prior to Gondwana breakup (Figure 9) (Groenewald 1991; Hoffman 1991; Weber & Arndt 1991). It thus appears that a substantial continental mass was already in existence in this part of Africa and adjacent East Antarctica by the Late Proterozoic and was affected internally only by "intraplate" rotations and translations of relatively limited magnitude during the Pan-African event.

Increasing attention is being given to the manner in which the various Late Proterozoic/earliest Paleozoic orogenic belts of the southern continents relate to the assembly of Gondwana (e.g. Murphy & Nance 1991). Recent models position Laurentia as a keystone between East and West Gondwana in a Late Proterozoic supercontinent (Moores 1991; Dalziel 1991), which requires that the amalgamation of these two parts of Gondwana occurred along the Mozambique belt in the Cambrian, after Laurentia had broken away from East and West Gondwana. Hoffman (1991) suggested that convergence between the two parts of Gondwana occurred by rotation about a pole located near the Weddell Sea, in the time frame from ca. 650 to 550 Ma. In such a reconstruction, the Damara-Zambezi trend is located approximately along a small circle of rotation, and this is used to account for the postulated "dominantly transcurrent shear" within the belts (Hoffman 1991). However, the age of ca. 820 Ma for Zambezi orogenesis precludes a direct link with deformation in the Damara belt in the required time frame. Instead, any relative motion between the Congo and Kalahari cratons in the region of the Zambezi belt at 650 - 550 Ma must have been accommodated by younger shear zones, such as the Mwembeshi dislocation.

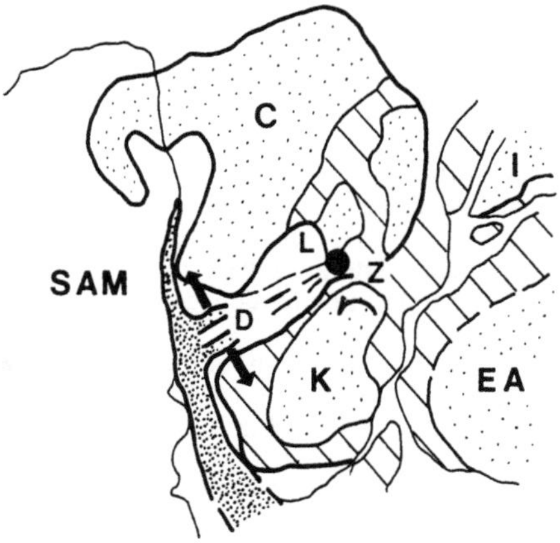

Figure 9. Model relating deformation in the Zambezi belt to the Pan-African tectonic evolution of southern Africa. Increased stretching in the southwestern part of the intracontinental branch of the Damara belt, together with spreading in the Pan-African South Atlantic Ocean, produced rotation and indentation along the eastern boundary between the Kalahari and Congo blocks, resulting in compression and shearing in the Zambezi belt. After Hartnady et al. (1985), Porada (1989), and Hoffman (1991). Dense stipple pattern = Pan African South Atlantic Ocean; diagonal line pattern = Middle Proterozoic (Kibaran) orogenic belts; C = Congo craton; K = Kalahari craton; SAM = South America; EA = East Antarctica; I = India; D = intracontinental Damara belt; L = Lufilian Arc; Z = Zambezi belt. Filled circle marks tectonic rotation pole for the Adamastor Ocean of Hartnady et al. (1985).

ACKNOWLEDGMENTS

This research was supported in part by National Science Foundation Grant INT 82-14234 to H.K. Brueckner and I.W.D. Dalziel, by University of Zambia Research Grants E71-18-939 and E71-18-940 to Hanson, Wilson, and T.C. Onstott, and by an Ohio State University Seed Grant award to Wilson. Parts of the field work were carried out with H. Brueckner, T. Onstott, H. Munyanyiwa, and numerous University of Zambia students. We are grateful to the Department of Geology and the School of Mines, University of Zambia, for substantial assistance during the field work.

REFERENCES CITED

Annels, A.E. 1984. The geotectonic environment of Zambian copper-cobalt mineralization. Journal Geological Society London 141: 279-289.

Badham, J.P.N. 1982. Strike-slip orogens - an explanation for the Hercynides. Journal Geological Society London 139: 495-506.

Barr, M.W.C., L. Cahen, & D. Ledent 1978. Geochronology of syntectonic granites from central Zambia: Lusaka Granite and granite NE of Rufunsa. Annales Societe Geologique Belgique 100: 47-54.

Barton, C.M., J.N. Carney, M.J. Crow & S. Simango 1985. Geology of the Archaean cratonic margin and the Zambezi-Mozambique orogenic belt in northeastern Zimbabwe (Rushinga, Mudzi and Mutoko districts). Annals of the Zimbabwe Geological Survey X (for 1984):1-39.

Barton, C.M., J.N. Carney, M.J. Crow, J.A. Evans & S. Simango 1992. Geology and structural framework of the Zambezi belt, northeastern Zimbabwe. In R.H. Findlay, M.R. Banks, R. Unrug & J. Veevers (eds), Gondwana 8 - Assembly, Evolution, and Dispersal. Rotterdam: Balkema (this volume).

Breitkopf, J.H. & K.J. Maiden 1988. Tectonic setting of the Matchless belt pyritic copper deposits, Namibia. Economic Geology 83: 710-723.

Brown, A.G. 1966. The geology of the Monze area: Explanation of Degree Sheet 1627, NW. Quarter. Geological Survey of Zambia Report No. 17: 38 p.

Brown, A.G. 1967. The geology of the Chikankata area: Explanation of Degree Sheet 1628, NW. Quarter. Geological Survey of Zambia Report No. 22: 42 p.

Brunel, M. 1986. Ductile thrusting in the Himalayas: Shear sense criteria and stretching lineations. Tectonics 5: 247-265.

Cahen, L., N.J. Snelling, J. Delhal & J.R. Vail 1984. The geochronology and evolution of Africa. Oxford: Clarendon Press.

Carney, J.N., P.J. Treloar, C.M. Barton, M.J. Crow, J.A. Evans & S. Simango 1991. Deep-crustal granulites with maylonitic and migmatitic fabrics from the Zambezi belt, northeastern Zimbabwe. Journal of Metamorphic Geology 9:461-479.

Coward, M.P. 1983. The tectonic history of the Damaran Belt: In R. McG. Miller (ed.), Evolution of the Damara orogen, South West Africa/Namibia. Special Publication Geological Society South Africa No. 11: 409-421.

Coward, M.P. & M.C. Daly 1984. Crustal lineaments and shear zones in Africa: Their relationship to plate movements. Precambrian Research 24: 27-45.

Daly, M.C. 1986a. Crustal shear zones and thrust belts: their geometry and continuity in Central Africa. Philosophical Transactions Royal Society London, A, 317: 111-128.

Daly, M.C. 1986b. The Irumide belt of Zambia and its bearing on collision orogeny during the Proterozoic of Africa. In Coward, M.P. & A.C. Reis (eds.), Collision tectonics. Geological Society London Special Publication 19: 321-328.

Daly, M.C. 1988. Crustal shear zones in central Africa: A kinematic approach to Proterozoic tectonics. Episodes 11: 5-11.

Dalziel, I.W.D. 1991. Pacific margins of Laurentia and East Antarctica-Australia as a conjugate rift pair: Evidence and implications for an Eocambrian supercontinent. Geology 19: 598-601.

de Swardt, A.M.J. 1963. Deformation of the Basement Complex associated with Lufilian folding south of Mapanza Mission, Northern Rhodesia. Transactions Geological Society South Africa 66: 75-92.

de Swardt, A.M.J. & A.R. Drysdall 1964. Precambrian geology and structure in central Northern Rhodesia. Geological Survey Northern Rhodesia Memoir 2: 82 p.

de Swardt, A.M.J., P. Garrard & J.G. Simpson 1965. Major zones of transcurrent dislocation and superposition of orogenic belts in part of central Africa. Geological Society America Bulletin 76: 89-102.

Drysdall, A.R., R.L. Johnson, T.A. Moore & J.G. Thieme 1972. Outline of the geology of Zambia. Geologie Mijnbouw 51: 265-276.

Drysdall, A.R. & C.J. Stillman 1966. Scapolite from the Katanga carbonate rocks of the Lusaka district. Geological Survey of Zambia Records 10: 20-24.

Elliot, R.J., T.J. Wilson & R.E. Hanson 1989. Late Proterozoic syntectonic plutonism in the Zambezi belt, Zambia. Geological Society of America Abstracts with Program 21: A69.

Ellis, M. & A.J. Watkinson 1987. Orogen-parallel extension and oblique tectonics: The relation between stretching lineations and relative plate motions. Geology 15: 1022-1026.

Garrard, P. 1968. The geology of the Chainama Hills area: Explanation of Degree Sheet 1528, NE Quarter. Geological Survey of Zambia Report No. 24: 80 p.

Groenewald, P.B., G.H. Grantham & M.K. Watkeys 1991. Geological evidence for a Proterozoic to Mesozoic link between southeastern Africa and Dronning Maud Land, Antarctica. Journal of the Geological Society, London, 148:1115-1123.

Hansen, E. 1971. Strain Facies. New York: Springer-Verlag.

Hanson, R.E., T.J. Wilson, H.K. Brueckner, T.C. Onstott, M.S. Wardlaw, C.C. Johns & K.C. Hardcastle 1988. Reconnaissance geochronology, tectonothermal evolution, and regional significance of the Middle Proterozoic Choma-Kalomo block, southern Zambia. Precambrian Research 42: 39-61.

Hanson, R.E., T.J. Wilson & H. Munyanyiwa 1992. Geologic evolution of the Late Proterozoic Zambezi belt in Zambia, In A.B. Kampunzu & J.T. Nanyaro (eds.), Proterozoic belts of Central-Eastern Africa. London: Pergamon Press (in press).

Hanson, R.E., T.J. Wilson & M.S. Wardlaw 1988. Deformed batholiths in the Pan-African Zambezi belt, Zambia: Age and implications for regional Proterozoic tectonics. Geology 16: 1134-1137.

Hanson, R.E., T.J. Wilson, M.S. Wardlaw & T.C. Onstott 1990. Nature and timing of depositional and orogenic events in the Late Proterozoic (Pan-African) Zambezi belt, Zambia. Geological Society of America Abstracts with Program 22: A264.

Hanson, R.E., M.S. Wardlaw, T.J. Wilson & G. Mwale 1991. U-Pb zircon ages from the Hook Granite Massif, Zambia: Implications for Pan-African Tectonics in south-central Africa. Geological Society of America Abstracts with Program 23(5): A97-A98.

Hartnady, C., P. Joubert & C. Stowe 1985.

Proterozoic crustal evolution in southwestern Africa. Episodes 8: 236-244.
Hoffman, P.F. 1991. Did the breakout of Laurentia turn Gondwanaland inside-out?. Science 252: 1409-1412.
Hunter, D.R. & D.A. Pretorius 1981. Structural framework (of southern Africa). In D.R. Hunter (ed.), Precambrian of the southern Hemisphere. Amsterdam: Elsevier: 397-422.
Johns, C.C., K. Liyungu, S. Mabuku, G. Mwale, F. Sakungo, D. Tembo & G. Vallance 1989. The stratigraphic and structural framework of Eastern Zambia: Results of a geotraverse. Journal of African Earth Sciences: 9: 123-136.
Johnson, R.L. & J.R. Vail 1965. The junction between the Mozambique and Zambezi orogenic belts, north-east Southern Rhodesia. Geological Magazine 102: 489-495.
Kampunzu, A.B., D. Kapenda & B. Manteka 1991. Basic magmatism and geotectonic evolution of the Pan-African belt in central Africa: evidence from the Katangan and West Congolian segments. Tectonophysics 190: 363-371.
Kroner, A. 1983. Proterozoic mobile belts compatible with the plate tectonic concept. In L.G. Medaris et al. (eds.), Proterozoic Geology: Selected papers from an international Proterozoic symposium. Geological Society of America Memoir 161: 59-74.
Lister, G.S. & A.W. Snoke 1984. S-C mylonites. Journal of Structural Geology 6: 617-638.
Mallick, D.I.J., 1966, The stratigraphy and the structural development of the Mpande Dome, southern Zambia: Transactions of the Geological Society of South Africa, v. 69, p. 211-230.
Matheson, G.D. & D. Newman, D. 1966. Geology and structure of the Lusaka area. Geological Survey of Zambia Records 10: 10-19.
Mawer, C.K. 1987. Shear criteria in the Grenville Province, Ontario, Canada. Journal of Structural Geology 9: 531-539.
Miller, R. Mc.G. 1983 The Pan-African Damara orogen of South West Africa/Namibia. In R. McG. Miller (ed.), Evolution of the Damara orogen, South West Africa/Namibia. Geological Society of South Africa Special Publication No. 11: 431-515.
Moores, E.M. 1991. Southwest U.S. - East Antarctic (SWEAT) connection: A hypothesis. Geology 19: 425-432.
Mora, C.I. & J.W. Valley 1989. Halogen-rich scapolite and biotite: Implications for metamorphic fluid-rock interaction. American Mineralogist 74: 721-737.
Munyanyiwa, H. 1985. The geochemistry and metamorphism of calc-silicate rocks, marbles and amphibolites in a portion of the Zambezi belt, southern Zambia [M.S. thesis]. Lusaka, Zambia, University of Zambia, 193 p.
Munyanyiwa, H. 1988. Metamorphism within the Zambezi belt, Zambia. Zambian Journal of Applied Earth Sciences 2: 28-40.
Munyanyiwa, H. 1990. Mineral assemblages in calc-silicates and marbles in the Zambezi mobile belt: Their implications on mineral-forming reactions during metamorphism. Journal of African Earth Sciences 10: 693-700.
Munyanyiwa, H. & R.E. Hanson 1988. Geochemistry of marbles and calc-silicate rocks in the Pan-African Zambezi belt, Zambia. Precambrian Research 38: 177-200.
Murphy, J. B. & R.D. Nance 1991. Supercontinent model for the contrasting character of Late Proterozoic orogenic belts. Geology 19: 469-472.
Newton, A.R. 1963. The geology of the country between Choma and Gwembe: Explanation of Degree Sheet 1627, SW and SE Quarters. Geological Survey Northern Rhodesia Report No. 8: 33 p.
Orpen, J.L., C.J. Swain, C. Nugent & P.P. Zhou 1989. Wrench-fault and half-graben tectonics in the development of the Palaeozoic Zambezi Karoo basins in Zimbabwe - the "Lower Zambezi" and "Mid-Zambezi" basins respectively - and regional considerations. Journal of African Earth Sciences 8: 215-229.
Passchier, C.W. & C. Simpson 1986. Porphyroclast systems as kinematic indicators. Journal of Structural Geology 8: 831-843.
Porada, H. 1989. Pan-African rifting and orogenesis in southern to equatorial Africa and eastern Brazil. Precambrian Research 44: 103-136.
Renne, P.R., T.C. Onstott, M.S. D'Agrella-Filho, I.G. Pacca & W. Teixeira 1990. $^{40}Ar/^{39}Ar$ dating of 1.0-1.1 Ga magnetizations from the São Francisco and Kalahari cratons: Tectonic implications for Pan-African and Brasiliano mobile belts. Earth and Planetary Science Letters 101: 349-366.
Ridgway, J. & C.R. Ramsay 1986. A provisional metamorphic map of Zambia - explanatory notes. Journal of African Earth Sciences 5: 441-446.
Shackleton, R.M. 1973. Correlation of structures across Precambrian orogenic belts in Africa. In D.H. Tarling & S.K. Runcorn (eds.), Implications of continental drift to the earth sciences, v. 2: London, Academic Press: 1091-1095.
Shackleton, R.M. 1986. Precambrian collision tectonics in Africa. In M.P Coward & A.C. Reis (eds.), Collision tectonics. Geological Society of London Special Publication 19: 329-349.
Shackleton, R.M. & A.C. Ries 1984. The relation between regionally consistent stretching lineations and plate motions. Journal of Structural Geology 6: 111-117.
Shackleton, R.M., J.R. Vail & D.S. Wood 1966. Preliminary report on the origin and significance of the Urungwe Klippe, Rhodesia. Annual Report Research Institute African Geology, University of Leeds, 10: 10-13
Simpson, C. & S.M. Schmid 1983. An evaluation of criteria to deduce the sense of movement in sheared rocks. Geological Society of America Bulletin 94: 1281-1288.
Simpson, J.G. 1967. The geology of the Chinyunyu area: Explanation of Degree Sheet 1529, NW Quarter. Geological Survey of Zambia Report

No. 19: 70 p.
Simpson, J.G. & C.J. Stillman 1963. Metamorphism and reaction phenomena in gabbros of the Mwembeshi and Lusaka areas. Geological Survey of Northern Rhodesia Records 9: 10-14.
Smith, A.G. 1963. The geology of the country around Mazabuka and Kafue: Explanation of Degree Sheets 1527, SE Quarter, and 1528, SW Quarter. Geological Survey Northern Rhodesia, Report 2: 32 p.
Snelling, N.J., R.L. Johnson & A.R. Drysdall 1972. The geochronology of Zambia. Records Geological Survey Zambia 12: 19-30.
Stagman, J.G. 1978. An outline of the geology of Rhodesia. Rhodesia Geological Survey Bulletin No. 80: 126 p.
Sutton, J. & J.V. Watson 1986. Architecture of the continental lithosphere. Philosophical Transactions Royal Society London, A, 317: 5-12.
Tapponnier, P., G. Peltzer, A.Y. Le Dain, R. Armijo & P. Cobbold 1982. Propagating extrusion tectonics in Asia: new insights from simple experiments with plasticine. Geology 10: 611-616.
Thieme, J.G. & R.L. Johnson 1981. Geological map of the Republic of Zambia. Geological Survey of Zambia, scale 1:1,000,000.
Treloar, P.J., J.N. Carney, M.J. Crow, J.A. Evans & C.N. Barton 1990. Pressure-temperature-time paths of granulite metamorphism and uplift, N.E. Zimbabwe. In D. Vielzeuf & Ph. Vidal (eds.), Granulites and crustal evolution. Netherlands: Kluwer Academic Publishers: 223-241.
Unrug, R. 1983. The Lufilian Arc: a microplate in the Pan-African collision zone of the Congo and the Kalahari cratons. Precambrian Research 21: 181-196.
Unrug, R. 1988. Mineralization controls and source of metals in the Lufilian fold belt, Shaba (Zaire), Zambia, and Angola. Economic Geology 83: 1247-1258.
Vrana, S. & M.W.C. Barr 1972. Talc-kyanite-quartz schists and other high-pressure assemblages from Zambia. Mineralogical Magazine 38: 837-846.
Vrana, S., R. Prasad & E. Fediukova 1975. Metamorphic kyanite eclogites in the Lufilian Arc of Zambia. Contributions to Mineralogy and Petrology 51: 1-22.
Weber, K. & N. T. Arndt 1991. Proterozoic link between Africa and Antarctica. Abstracts, Eighth International Symposium on Gondwana, University of Tasmania, Hobart, Tasmania, p. 90-91.
White, S.H., S.E. Burrows, J. Carreras, N.D. Shaw & F.J. Humphreys 1980 On mylonites in ductile shear zones. Journal of Structural Geology 2: 175-187.
Wilson, A.H. & M.D. Prendergast 1989. The Great Dyke of Zimbabwe - 1: Tectonic setting, stratigraphy, petrology, structure, emplacement, and crystallization. In M.D. Prendergast & M.J. Jones (eds.), Magmatic sulphides - the Zimbabwe volume. London: Institution of Mining and Metallurgy: 1-20.
Wilson, T.J. 1988. Textural transitions within quartzo-feldspathic mylonites in the Late Proterozoic Zambezi belt, Zambia. Geological Society of America Abstracts with Program 20: A214.
Wilson, T.J. & R.E. Hanson 1987. Structural evolution of the southern margin of the Pan-African Zambezi mobile belt. Geological Society of America Abstracts with Program 19: 252.
Wilson, T.J. & R.E. Hanson 1988. 'Tectonic conglomerates' developed from synkinematic pegmatites during ductile shearing, Zambezi belt, Zambia. Geological Society of America Abstracts with Programs 20: 395.
Wilson, T.J., R.E. Hanson & H. Brueckner 1985. Transposition, ductile shearing and noncylindrical folding of remobilized gneisses along the junction between two Proterozoic mobile belts in southern Zambia. Geological Society of America Abstracts with Program 17: 752.

Gondwana Eight, Findlay, Unrug, Banks & Veevers (eds) © 1993 Balkema, Rotterdam. ISBN 90 5410 304 3

Thermobarometry of mafic rocks within the Zambezi Mobile Belt, northern Zimbabwe

H. Munyanyiwa
University of Zimbabwe, Geology Department, Harare, Zimbabwe

ABSTRACT: The Zambezi belt links with the Late Proterozoic Pan-African network of mobile belts in central and southern Africa. In Zimbabwe, the belt forms an east-west trending zone dominated by medium- to high-grade quartzofeldspathic rocks along the northern margin of the Zimbabwe Archaean craton. The rocks are associated with concordant and discordant amphibolite bodies. Petrographic and phase chemistry data for these amphibolites reveal metamorphic conditions of 500-560°C, 5.6-6.8kbar for mafic rocks along the Zambezi Escarpment and 700-740°C, 6.8-7kbar for mafic rocks further north within the Chewore inlier. The P-T conditions are probably associated with main-phase deformation dated at c.820Ma. Widespread K-Ar and Rb-Sr mineral ages of c.500Ma, however are present within the Zambezi belt and elsewhere within central and eastern Gondwana. The mineral ages significantly postdate this main episode of penetrative deformation and regional metamorphism. The c.500Ma event possibly reflects a post-tectonic thermal episode in central and eastern Gondwana which is probably related to the late Precambrian/early Paleozoic tectonic assembly of Gondwana.

1 INTRODUCTION

The metamorphism of Late Proterozoic mobile belts of central and southern Africa is an essential component associated with the Proterozoic/early Paleozoic assembly of the Gondwana. Detailed knowledge of the metamorphic petrology of these belts is therefore important and contributes towards understanding the physical conditions associated with the tectonic assembly of these belts. This study is intended to present new petrographical and mineral chemistry data of the Zambezi belt metabasites in Zimbabwe located to the northwest of Guruve within the Zambezi Escarpment and within the Chewore inlier (Fig.1). The results of this study impose constraints on the metamorphic conditions in this remote and poorly understood segment of the Zambezi belt.

The regional geological setting of the Zambezi belt has been described by Johnson and Vail(1965), Unrug(1983), Coward and Daly(1984), Hanson et al.(1988) and Porada(1989). The Zambezi belt is part of a network of Pan-African mobile belts which is linked to the Mozambique belt to the east, the Lufilian arc to the north and the Damara belt to the southwest. In Zimbabwe, the belt was first described by Macgregor(1951), who pointed out that the belt forms an east-west trending medium- to high-grade metamorphic terrane which forms the northern margin of the Zimbabwe Archaean craton.

In Zimbabwe, the Zambezi belt is dominated by quartzofeldspathic gneisses and migmatites. These are intimately associated with concordant and discordant bodies of amphibolite, lenses of psammite, and marbles and pelites. Based on lithological and radiometric data, the migmatites and gneisses have been interpreted to

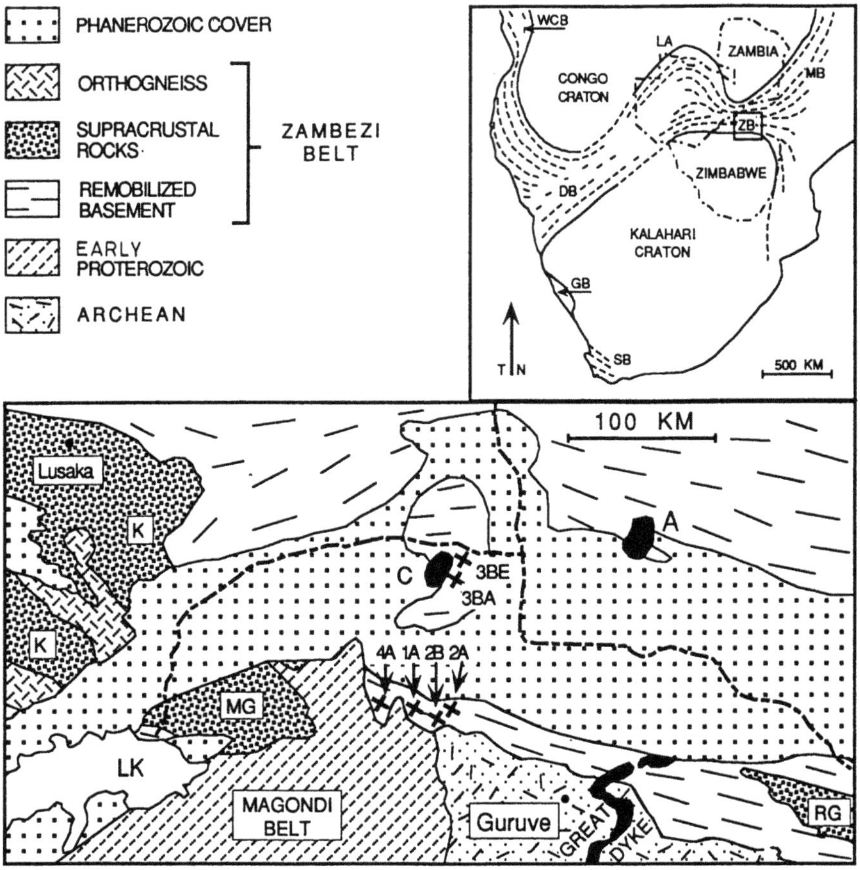

Fig. 1 Position of the Zambezi belt within the regional tectonic framework of southern Africa (After Hanson et al., in press). WCB, West Congolian belt; DB, Damara belt; LA, Lufilian arc; MB, Mozambique belt; GB, Gariep belt; SB Saldania belt. A portion of the regional geological map (location shown in box in tectonic map) shows: A, Atchiza Complex; C, Chewore complex; K, Katanga metasedimentary rocks; LK, Lake Kariba; MG, Makuti Group; RG, Rushinga Group (slightly modified after Hanson et al., in press). Localities of mafic rocks studied are also indicated. Samples 3BA and 3BE were collected within the Chewore Complex and samples 1A, 2A, 2B and 4A within the Zambezi Escarpment.

represent reworked older rocks of Early Proterozoic Magondi orogenic belt and the Archaean craton (Cahen et al., 1984; Hahn and Steiner, 1990).

The northwestern segment of the Zambezi belt in Zimbabwe is underlain by a thick supracrustal sequence of arkosic gneisses, metapelites, amphibolites and less abundant marbles and calc-silicates (Broderick, 1976; Fey and Broderick, 1990). These rocks have been termed the Makuti Group and have been correlated on lithological and radiometric grounds with metasedimentary rocks of central and southern Zambia (Fig.1; Broderick, 1976). The eastern part of the Zambezi belt in Zimbabwe consists of high-grade metapelites, granulites and orthogneisses (Bache et al., 1983). More specific details on the metamorphic geology of these rocks have been given by Treloar et al.(1990) and Carney et al. (1991).

2 GEOLOGIC SETTING

The geology of the northern margin of the Zimbabwe Archaean craton has been described by Hahn and Steiner (1990). The region is dominated by medium- to coarse-grained biotite gneisses, leucocratic gneisses and gneissic granite. The gneisses show banding on a cm-wide scale and locally grade into migmatites with well-developed compositional layering. The contacts between these rock types show prevalent lit-par-lit relationships. Bands of paragneisses, quartzites and mica schists crop out on the northern edge of the escarpment. These rocks have been termed the Guruve Metamorphic Complex and correlated on lithological grounds with the Makuti Group to the west and Rushinga Group to the east (Fig.1; Hahn and Steiner, 1990). Rocks of the Guruve Metamorphic Complex structurally overlie the migmatitic gneisses.

Medium- to coarse-grained amphibolites, the subject of the present study, occur as concordant lenses and dykes a few metres to several tens of metres wide within the migmatitic gneisses, paragneisses and pelitic schists. The contacts with the surrounding gneisses commonly are sharp and parallel to the foliation. In some outcrops, however, amphibolites cut obliquely across the foliation in the enclosing gneisses. Migmatites are locally present in the Zambezi belt but occur extensively in the Chewore inlier. The migmatites are coarse grained with migmatitic layering being composed of quartz, plagioclase and K-feldspar bands alternating with biotite-rich bands containing lesser hornblende and garnet.

3 PETROGRAPHY OF MAFIC ROCKS

The metabasites are represented by garnet amphibolites, biotite-hornblende schists, massive amphibolites and weakly foliated amphibolites. Garnet amphibolites occur extensively within the migmatites of the Chewore inlier. However, other garnet-bearing metabasites, such as the biotite-hornblende schists and weakly foliated amphibolites, occur locally along the Zambezi Escarpment.

3.1 Garnet amphibolites

The garnet amphibolites form concordant and discordant bodies up to 10m across within the migmatites. They have a weak foliation defined by hornblende and biotite visible only in thin sections. Plagioclase and hornblende are the major phases in these rocks. Garnet is found as porphyroblasts up to 2cm in diameter and contains abundant inclusions of quartz, rutile, calcite and opaque minerals. Zoisite and clinozoisite occur in trace quantities, whereas chlorite is present as a secondary mineral replacing biotite and garnet. Apatite occurs as an accessory mineral.

3.2 Biotite-hornblende schists

These rocks show a strong foliation defined by green hornblende and brown biotite flakes. Quartz and plagioclase are minor phases. Pale green chlorite is commonly intergrown with biotite and hornblende along the foliation. The textural features of chlorite strongly suggest that it is a primary metamorphic mineral. Epidote is found as clinozoisite showing yellowish-blue anomalous interference colours. Garnet occurs in trace quantities as anhedral crystals between hornblende. Euhedral pyrite is interstitially located between hornblende and biotite crystals.

3.3 Weakly foliated amphibolites

These rocks comprise mainly hornblende and plagioclase; both minerals are aligned along a weak foliation. Plagioclase is partially altered to fine-grained sericite and carbonate along the margins but twinning is preserved in most grains. The mineral further contains rare tiny inclusions of clinozoisite.

Table 1 Representative compositional parameters of minerals used in thermobarometry

	Garnet amphibolite (3BA)			biotite-hornblende schist (2A)			massive amphibolite (4A)			weakly foliated amphibolite (1A)	
	a	b	c	a	b	c	a	b	c	a	b
$(Mg/Mg + Fe)^{Grt}$	0.206	0.214	0.214	0.116	0.115	0.117					
$(Ca/(Ca + Mg + Fe + Mn))^{Grt}$	0.262	0.261	0.265	0.174	0.184	0.183					
$(Mg/Mg + Fe)^{Bio}$	0.552	0.542	0.548	0.468	0.471	0.487					
$(Mg/Mg + Fe)^{Hbl}$	0.482	0.483	0.480	0.384	0.380	0.373	0.518	0.502	0.494	0.556	0.545
$Na(M_4)^{Hbl}$	0.105	0.104	0.118	0.164	0.209	0.148	0.075	0.093	0.064	0.154	0.190
$Al(IV)^{Hbl}$	1.701	1.711	1.719	1.874	1.855	1.834	1.317	1.543	1.508	0.410	1.316
X_{Ca}^{Plg}	0.476	0.445	0.357	0.279	0.259	0.283	0.218	0.214	0.229	0.408	0.418
Ti^{Hbl}	0.132	0.136	0.138	0.045	0.044	0.041	0.065	0.061	0.060	0.082	0.068
Ti^{Bio}	0.389	0.389	0.387	0.169	0.164	0.175					

Bio- biotite, Grt - garnet, Hbl hornblende, $Na(M_4)$ - Na in M_4 Site in hornblende

Coarse-grained white mica (paragonite?) occurs in minor amounts and is commonly associated with altered plagioclase. Quartz, garnet and titanite are found in trace amounts.

3.4 Massive, garnet-poor amphibolites

The massive amphibolites are generally fine-to medium-grained rocks with granoblastic-polygonal textures defined by plagioclase, hornblende and quartz. Plagioclase grains are marginally replaced by calcite and sericite, as in the weakly foliated amphibolites. Biotite occurs in minor amounts as reddish-brown crystals whereas garnet is found as a trace mineral and is partially replaced by chlorite along cracks and rims. An opaque mineral occurs as anhedral skeletal crystals.

4 MINERAL CHEMISTRY

4.1 Analytical method

Electron microprobe analyses of amphiboles, plagioclase, garnet and biotite from the mafic rocks described above were conducted at the analytical laboratory, Geological Survey of Canada (Ottawa). A Cameca Camebax electron microprobe equipped with four wavelength-dispersive spectrometers was used. Accelerating voltage was 15KV with a beam current of 30nA and scanning time of 10s for peaks (5s for background). Natural silicates and oxides were used as standards. Fe^{3+}/Fe^{2+} ratios in amphiboles have been estimated by normalising total cations less(Ca + Na + K) to 13 (e.g. Leake, 1978). The I.M.A. nomenclature of amphiboles (Leake, 1978) is adopted. Only representative compositional parameters of the minerals used in thermobarometry below are presented (Table 1), but a full list of analyses is available from the author upon request.

4.2 Amphiboles

$Mg/(Mg+Fe^{2+})$ ratios in amphiboles in the analysed samples range from 0.38 to 0.56. Amphiboles in biotite-hornblende schists and garnet amphibolites have lower ratios than in the other rock types. This is attributed to higher Fe content in the host rocks, which in garnet amphibolites favoured the crystallisation of almandine-rich garnet. Amphiboles in biotite-hornblende schists have the highest Al, which also possibly reflects the protolith composition. Ti content of amphiboles in garnet amphibolites is higher than in the other samples. Based on the estimated Na(A)+K and Si cations per 23 oxygen atoms, the amphiboles vary in composition from tschermakite to tschermakitic, edenitic and pargasitic hornblendes and true hornblendes.

4.3 Plagioclase

In the garnet amphibolites from the Chewore inlier, plagioclase has high Ca/(Ca+Na) ratios of 0.36-0.48, as compared to 0.21-0.40 in other samples. Furthermore, plagioclase in the garnet amphibolites is reversely zoned with high calcic rims (An_{42-48}) relative to the low calcic cores (An_{36-37}). On the other hand plagioclase crystals in other mafic rocks are partially replaced by secondary carbonate and sericite as mentioned above. Consequently the chemical zonation in these plagioclase crystals does not reflect a primary metamorphic feature.

4.4 Garnets

Garnets are essentially unzoned with respect to the cations Mg, Fe, Ca and Mn. In general, the highest Mg-rich garnets are those from garnet amphibolites, whereas biotite-hornblende schists and garnets in weakly foliated amphibolites have relatively lower Mg(Table 1). Al in garnets from all samples is close to 2.0 per 12 oxygen atoms, which suggests that Fe^{3+} can only be present in trace amounts; all Fe in the mineral therefore is regarded as Fe^{2+}.

Table 2 Representative temperatures (°C) and pressures (kb) for the Zambezi belt mafic rocks estimated from thermobarometers

Sample	Grt-Bio[1]	Grt-Hbl[2]		T1	T2	T3	T4	T5	T6	PI
Garnet amphibolite (3BA)	1.281	3.611	4kb*	705	706	783	654	909	763	7.8
			8kb	723	722	783	654	930	763	7.7
Garnet amphibolite (3BE)	1.278	3.385	4kb	699	701	774	654	900	768	7.6
			8kb	716	716	774	654	921	768	6.8
Hornblende-biotite Schist (2A)	1.151	4.739	4kb	502	525	520	505	612	597	5.3
			8kb	517	539	520	505	628	597	5.7
Hornblende-biotite Schist (2B)	1.160	4.912	4kb	512	539	525	515	627	593	6.7
			8kb	526	552	525	515	644	593	6.2

T1 = Ferry and Spear (1978)

T2 = Indares and Martignole (1985)

T3 = Goldman and Albee (1977)

T4 = Thompson (1976)

T5 = Hoinkes (1986)

T6 = Graham and Powell (1984)

PI = Pressure estimates using the plg-amph-grt-qtz barometer of Kohn and Spear (1989)

K = 1, 2 K_D for Grt-Bio and Grt-Hbl exchange respectively

* = Temperatures were calculated at 4 and 8Kb but results show that thermometers with a correction factor are only slightly affected by pressure.

4.5 Biotites

Biotite is a major phase in the biotite-hornblende schists but is only found in minor amounts in the garnet amphibolites. Within the individual rock types, the mineral has a restricted compositional range. $Mg/(Mg+Fe^{2+})$ ratios are 0.54-0.55 in the garnet amphibolites and 0.47-0.49 in the biotite-hornblende schists (Table 1). Also, Ti contents of the latter samples are lower (Ti = 0.17-0.18) than the values of around 0.39 per 22 oxygen atoms in garnet amphibolites (Table 1).

5 METAMORPHISM

5.1 General statement

Equilibrium mineral assemblages of metabasites based on textural evidence may be summarised as follows:

1. hornblende + epidote + plagioclase + quartz+opaque + titanite
2. hornblende + epidote + plagioclase + quartz + biotite + chlorite + opaque + titanite
3. hornblende + zoisite/clinozoisite + plagioclase + quartz + garnet + biotite + rutile + opaque

The garnet amphibolites have rutile as a Ti-bearing phase in place of titanite whereas epidote is represented by traces of clinozoisite and zoisite. On the other hand, titanite, pistacite-rich epidote and primary chlorite occur in other mafic rocks described above. This suggests that the garnet amphibolites have experienced metamorphic conditions of upper amphibolite facies as compared to the other mafic rock suites. This is also reflected in the high Ca/(Ca+Na) ratio in plagioclase in the garnet amphibolites. It is generally acknowledged that the Ti content of biotite and amphibole increases with increasing metamorphic grade (e.g. Laird and Albee, 1981). The Ti content of biotite and amphibole from garnet amphibolites are higher than other mafic rocks from the Zambezi Escarpment, which further attests to high grade metamorphic conditions for the garnet amphibolites.

5.2 Metamorphic conditions

Microprobe analyses of amphibole, garnet, biotite and plagioclase have been used in combination with calibrated equilibria to arrive at estimates of metamorphic temperatures and pressures attained during metamorphism. The results of thermobarometry are discussed below and summarised in Table 2.

5.3 Temperatures

1. Grt-Hbl thermometry. The Fe-Mg exchange reaction between garnet and hornblende has been empirically calibrated as a geothermometer (Graham and Powell, 1984). The Kd values of 4.5-4.8 (where Kd is defined as (Fe, Grt/Mg, Grt)/(Fe, Hbl/Mg,Hbl) for coexisting garnet and hornblende in biotite-hornblende schists correspond to temperatures of 592 to 614°C. On the other hand, Kd values of 3.4-3.6 for coexisting garnet and hornblende in the garnet amphibolites correspond to higher temperatures of 730-760°C. As garnet and hornblende from these samples do not show any appreciable chemical zonation, the temperature ranges for each of these samples were calculated from cores and rims of the minerals.
2. Grt-Bio thermometry. The distribution of Fe and Mg between garnet and biotite has been studied by numerous workers resulting in the formulation of a number of garnet-biotite geothermometers (e.g. Goldman and Albee, 1977; Thompson, 1976; Ferry and Spear, 1978; Indares and Martignole, 1985; Hoinkes, 1986). Garnet amphibolites from the Chewore inlier reveal temperatures of 705-723°c at 4-8kb using the model of Ferry and Spear (1978). The temperatures are in good agreement with values calculated from Indares and Martignole (1985). The Goldman and Albee (1976) geothermometer gives higher

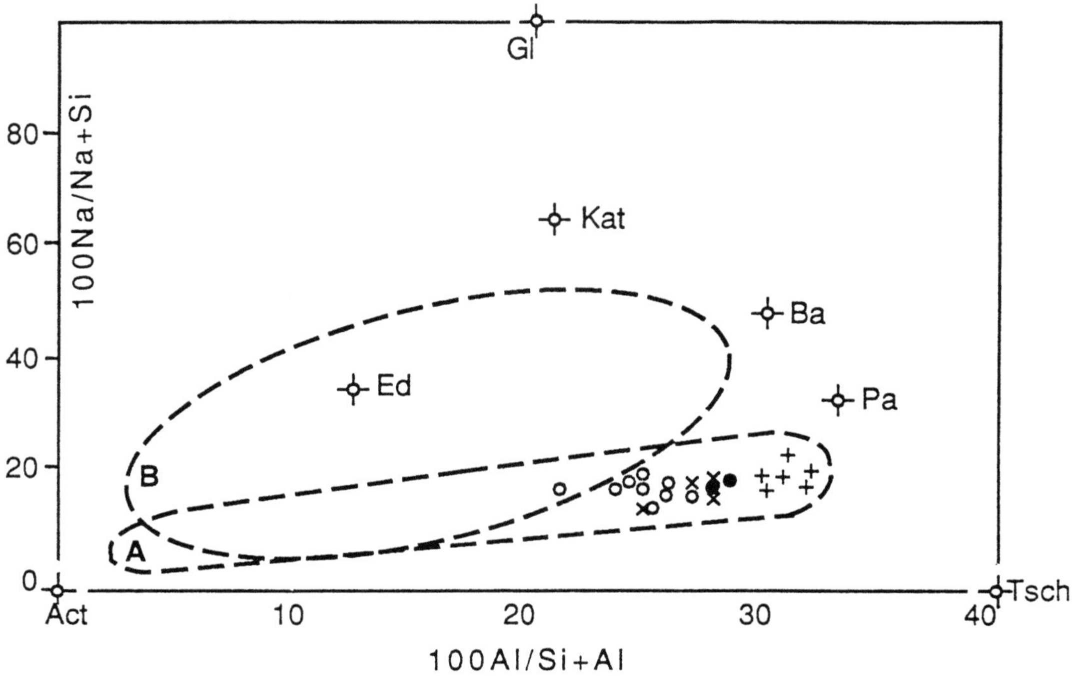

Fig. 2 Ratio diagram for calcic, sodic-calcic and sodic amphiboles in mafic rocks from the Zambezi belt. Fields A and B after Laird and Albee (1981). Field A is for medium-pressure amphiboles from Dalradian terranes and B is for high-pressure amphiboles from Sanbagawa and Franciscan terranes. All amphiboles from the Zambezi belt fall within the medium pressure field. Amphiboles from individual rock types are represented by different symbols as follows: X, garnet-amphibolites; +, biotite-hornblende schists; •, massive amphibolites; o, weakly foliated amphibolites. End-member compositions are Act, actinolite; Ba, barroisite; Ed, edenite; Gl, glaucophane; Kat, katopholite; Pa, pargasite; Tsch, tschermakite (-ф-).

temperatures of 774-784°C. Values calculated using the version of Hoinkes (1986) are unrealistically high (>900°C). For biotite-hornblende schists, the Ferry and Spear (1978) model yields lower values of 502-517°C, whereas that of Indares and Martignole (1985) yields slightly high temperatures of 525-552°C. Estimates using the Hoinkes thermometer are also unrealistically high, while the Goldman and Albee (1977) calibration for the same samples gives 520-557°C. The calibration of Thompson (1976) gave the lowest temperatures throughout; 650°C for garnet amphibolites and 505-515°C for biotite-hornblende schists.
3. Amp-Plg thermometry. An empirical amphibole-plagioclase is based on Na exchange between calcic amphibole and plagioclase (Spear, 1981). When applied to rocks of the present study this gives 625-700°C for garnet amphibolites from the Chewore inlier, 540-580°C for the biotite-hornblende schists and about 600°C for the massive amphibolites.

5.4 Pressures

1. Amphibole crossite content. The compositions of calcic and sodic amphiboles have long been used as indicators for metamorphic pressures (e.g. Raase, 1974; Brown, 1978; Laird and Albee 1981). In this study, the empirical barometer of Brown (1978, Fig.10), which relates Na(M4) and Al(IV) variations in amphiboles reveals

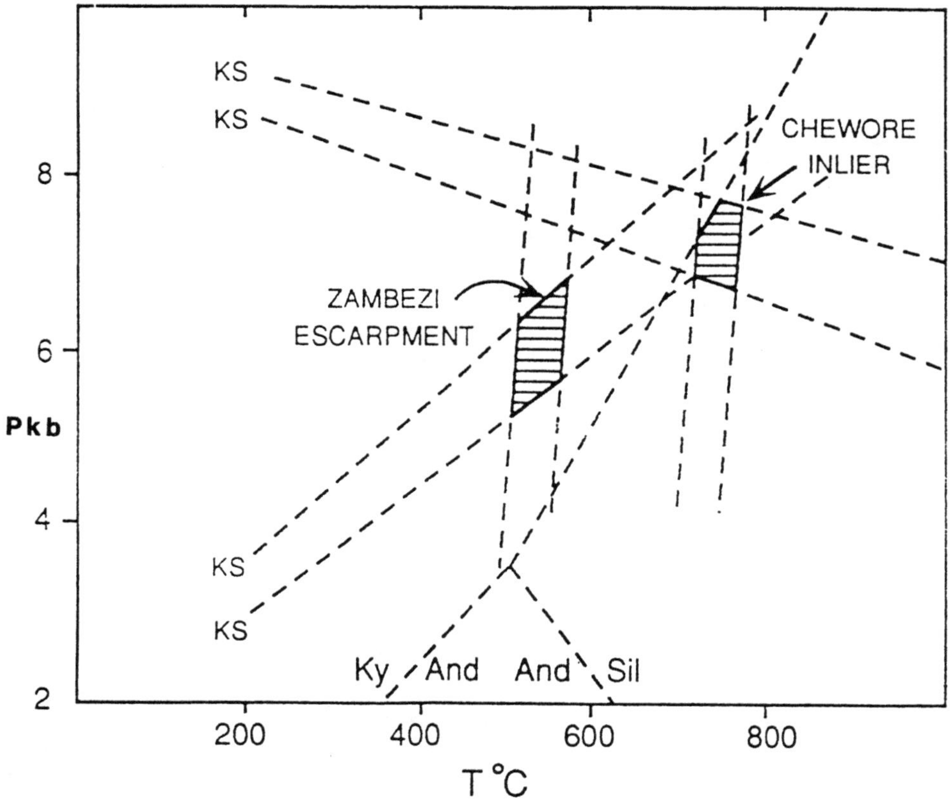

Fig. 3 P-T diagram showing metamorphic conditions for mafic rocks in the Zambezi belt determined from thermobarometry. KS represents equilibrium curves calculated from models A and B of Fend member calibrations of Kohn and Spear(1989). Upper and lower temperature limits have been estimated using Plg-Amp, Grt-Bio and Grt-Hbl thermometers text for discussion).

pressures of 5-6kbar. These moderate pressures are also reflected on a 100Na/(Ca + Na) versus 100Al/(Al+Si) plot (Fig.2), which shows the amphiboles to fall within the intermediate pressure field.

2. Plg-Amp-Grt-Qtz barometry. Kohn and Spear (1989) have formulated Mg and Fe end-member geobarometers based on equilibria among calcic amphibole, plagioclase garnet and quartz. Fe end-member calibration of the barometer gives 5.6-6.8kb for the biotite-hornblende schists and 7.0-8.0kb for garnet amphibolites. Applying the Mg end-member calibrations gave unrealistically high pressures (>10 kb) for the garnet-bearing mafic rocks. Because the pyrope contents of garnet are low, results of the Fe barometers are preferred, as it is possible that the solution model for Mg-garnet is inappropriate at low pyrope concentrations.

5.5 Summary

Temperatures estimated using different cation exchange thermometers and calibration models for garnet amphibolites from the Chewore inlier are higher than other amphibolite types from the escarpment. It is likely that the temperature differences between garnet amphibolites and other mafic rock types reflect real differences in metamorphic grade. Results from different cation-exchange equilibria and different calibrations thus yield conformable results. With exception of the Hoinkes (1986) calibration, which gives extreme temperatures (>900°C) in the case of garnet amphibolites, it seems justified to consider the mean temperature values form these thermometers as a best estimate (cf. Rameshwar Rao et al., 1991). This gives 720±20°C for the Zambezi Escarpment. Pressures are constrained by Plg-Amp-Grt-Qtz barometry and indicate 7.0-8.0 and 5.3-6.8kbr for the Chewore inlier and Zambezi Escarpment respectively (Fig.3). These pressures are preferred over those calculated from Brown's (1978) empirical barometer, because the latter calibration is greatly affected by Fe^{2+}/Fe^{3+} ratio and analytical accuracy.

6 REGIONAL TRENDS

6.1 Metamorphism

Regional metamorphic studies within the Zambezi belt in Zimbabwe are of limited extent (e.g. Treloar et al., 1990; Carney et al., 1991). Most information on the metamorphic geology is contained in Geological Survey of Zimbabwe bulletins. During mapping of the Chewore inlier, Both (1990) described garnet, staurolite, kyanite and sillimanite in pelitic and semipelitic schists. He further described garnet-sillimanite-biotite gneiss associated with augen gneiss and banded plagioclase-biotite-muscovite gneiss. Structural trends in both the schists and gneisses are northnorthwest and northwest. In the escarpment, Chenjerai (1987) and Hahn and Steiner (1990) have documented kyanite-, staurolite- and kyanite-bearing metapelites indicative of amphibolite facies metamorphism. These rocks have been correlated with the Rushinga and Makuti Groups and it seems probable that metamorphic mineral assemblages in them formed during the peak of Pan-African metamorphism. P-T conditions deduced for the area north west of Guruve within the Zambezi Escarpment fall within the kyanite stability field, which is consistent with the presence of kyanite in metapelites.

Amphibolite bands within migmatitic gneisses in the Chewore inlier reveal temperatures and pressures which fall mainly within the sillimanite field (Fig.3) This is also supported by he presence of sillimanite in some of the gneisses (Both, 1990). However, at this stage it is not conclusive that the high-grade migmatisation event in the Chewore area occurred during the Pan-African event or during an earlier orogeny. This question will be addressed when more structural and radiometric data become available. It should be pointed out, however, that high-grade metamorphic conditions documented for the Chewore inlier have been realised elsewhere within the Zambezi belt and have been ascribed to the Pan-African event (Broderick, 1976; Treloar et al., 1990; Carney et al., 1991).

On a regional scale and based on petrogrphic studies, Ridgway and Ramsay (1986) concluded that the Zambezi belt had experienced Barrovian-type metamorphism during the Pan-African event. Kyanite-staurolite-bearing metapelites are widespread throughout the Zambezi belt in southern Zambia and associated calc-silicate rocks have characteristic amphibolite-facies phases, which include diopside, dolomite, grossular, scapolite, plagioclase and epidote. In particular, scapolite and plagioclase in these rocks have uniform compositions of Me_{60-65} and An_{32-40} respectively over a large distance (Munyanyiwa and Hanson, 1988). Barrovian-type metamorphism has also been documented within the Makuti Group in Zimbabwe (Broderick, 1976).

The regional Barrovian-type metamorphism is, however, complicated by several small, sporadic bodies of eclogites derived from metagabbros south of Lusaka (Vrana, 1978) and by the presence of talc-kyanite-schists east of Lusaka (Vrana and Barr, 1972). Such rocks require pressures of at least 10kbar during metamorphism. This implies crustal thickening in certain segments of the Zambezi belt associated with Pan-African orogenesis. High-P conditions in excess of 10kbar also are recorded by granulites in northeastern Zimbabwe and are attributed to crustal thickening during Pan-African orogenesis (Treloar et al., 1990; Carney et al., 1991).

6.2 Geochronological constraints

Hanson et al.(1988) have dated the last major orogenic event affecting the Middle Proterozoic basement of the Zambezi belt at 1200-1230Ma. The main-phase Pan-African D1 event is dated at c.820Ma based on the U-Pb zircon igneous age of a syn-tectonic granite injected into a transcurrent shear zone within the central part of the belt in Zambia (Hanson et al., 1988). An age of c.820Ma is also consistent with other scattered geochronological results from the Makuti Group, which provide evidence of a tectonothermal event at c.850-800Ma (Loney, 1969). Furthermore, Treloar et al. (1990) and Carney et al.(1991) have documented an important event at c.820Ma in the easternmost part of the Zambezi belt.

Snelling et al.(1972), Cahen et al.(1984), Loney (1969) and Hahn and Steiner (1990) noted the 450-600Ma range of many K/Ar and Rb/Sr mineral ages within the Zambezi belt. As recognised by Hanson et al.(in press), these dates significantly post-date the main 820Ma tectonic event thus implying a widespread late Pan-African post-tectonic thermal episode within previous major zones of structural weakness. Hahn and Steiner (1990) have shown that the younger ages within the Zambezi belt become progressively older to the south. Late Precambrian to early Paleozoic ages have also been documented in Western Australia (Clarke, 1991), the East Antarctic Shield (Fitzsimons and Harley, 1991) and eastern India (Aswatha Narayana, 1964). The ages seem more widespread within central and eastern Gondwana than previously thought and are probably related to Late Precambrian - early Paleozoic tectonic assembly of Gondwana.

Acknowledgements

The project forms a contribution to IGCP 288 'Gondwana sutures and fold belts'. The University of Zimbabwe Research Board grant 3249 is gratefully acknowledged. The project was carried out within the framework of the NUFFIC-funded MINREST cooperative project between University of Zimbabwe and Free University of Amsterdam. Drs. S. Hanmer and R.Thieverge of the Geological Survey of Canada kindly assisted with the probe analyses. Thanks are due to Dr. R.E. Hanson and Prof. R. Unrug whose constructive comments improved the presentation of the manuscript.

7 REFERENCES

ASWATHA NARAYANA, U. 1964: Isotopic ages from the Eastern Ghats of India. Journal of Geophysical Research, 69: 3479-86.

BACHE, J.J.; DALLAS, S.; MILIAN, J.F.; 1983: A preliminary report on the geology of the country around Centenary and Mount Darwin. Annals of the Zimbabwe Geological Survey, 9: 34-58.

BOTH, F. 1990: The geology of the Chewore inliers (mid-Zambezi Valley): A preliminary report. Annals of the Zimbabwe Geological survey, 15: 6-10.

BRODERICK, T.J. 1976: Explanation of the geological map of the country east of Kariba. Rhodesia Geological Survey Short Report No. 43: 98p.

BROWN, E.H. 1978: The crossite content of Ca-amphibole as a guide to pressure of metamorphism. Journal of Petrology, 18: 53-72.

CAHEN, L.; SNELLING, N.J.; DELHAL, J.; VAIL, J.R. 1984: The geochronology and evolution of Africa. Oxford, Clarendon Press, 512p.

CARNEY, I.N.; TRELOAR, P.J.; BARTON, C.M.; CROW, M.J.; EVANS, J.A.; SIMANGO, S. 1991: Deep crustal granulites with migmatitic and mylonitic fabrics from the Zambezi Belt, northeast Zimbabwe. Journal of Metamorphic Geology, 9: 461-479.

CHENJERAI, K.G. 1987: A preliminary report on the geology north of Chenanga. Annals of he Zimbabwe Geological Survey, 13: 1-7.

CLARKE, G.L. 1991: Proterozoic

tectonic reworking in the Rudall Complex, Western Australia. Australian Journal of Earth Sciences, 38: 31-44.
COWARD, M.P.; DALY, M.C. 1984: Crustal lineaments and shear zones in Africa: Their relationship to plate movements. Precambrian Research, 24: 27-45.
FERRY, J.M.; SPEAR, F.S. 1978: Experimental calibration of the partitioning of Fe and Mg between biotite and garnet. Contributions to Mineralogy and Petrology, 66: 113-117.
FEY, P.; BRODERICK, T.J. 1990. Explanation of the geological map of the country east of Makuti, Hurungwe District. Zimbabwe Geological Survey Short Report No. 47: 84p.
FITZSIMONS, I.C.W.; HARLEY, S.L. 1991: Geological relationships in high grade gneiss of Brattsrand Bluffs coastline, Pyrdz Bay, east Antarctica. Australian Journal of Earth Sciences, 38: 497-519.
GOLDMAN, D.S.; ALBEE, A.L. 1977: Corrrelation of Mg/Fe partioning with $^{18}O/^{16}O$ partioning between quartz and magnetite. American Journal of Science, 277: 750-767.
GRAHAM, C.M.; POWELL, R. 1984: A garnet-hornblende geothermometer: Calibration, testing and application to the Pelona Schist, Southern California. Journal of Metamorphic Geology, 2: 13-31.
HAHN, L.; STEINER, L. 1990: Geology and Mineral Prospecting in the Makonde and Guruve Districts, Zimbabwe. Technical Cooperation Report, Hannover, 296p.
HANSON, R.E.; WILSON, T.J.; BRUECKNER, H.K.; ONSTOTT, T.C.; WARDLAW, M.S.; JOHNS, C.C. AND HARDCASTLE, K.C. 1988: Reconnaissance geochronology, tectonothermal evolution, and regional significance of the Middle Proterozoic Choma-Kalomo Block, Southern Zambia. Precambrian Research, 42: 39-61.
HANSON, R.E.; WILSON, T.J. and MUNYANYIWA, H. in press: Geologic evolution of the late Proterozoic Zambezi in Zambia. In: Kampunzu, A.B. (ed.), Proterozoic belts of Eastern and Central Africa. Pergamon, Oxford.
HOINKES, G. 1986: Effect of grossular-content in garnet on the partioning of Fe and Mg between garnet and biotite. Contributions to Petrology and Mineralogy, 92: 393-339.
INDARES, A.; MARTIGNOLE, J. 1985: Biotite-garnet geothermometry in the granulite facies: the influence of Ti and Al in biotite. American Mineralogist, 70: 272-278.
JOHNSON, R.L.; VAIL, J.R. 1965: The junction between the Mozambique and Zambezi orogenic belts, north-east Southern Rhodesia. Geological Magazine, 102: 489-495.
KOHN, M.J.; SPEAR, F.S. 1989: Empirical calibration of geobarometers for the assemblage garnet + hornblende + plagioclase + quartz. American Mineralogist, 74: 77-84.
LAIRD, J.; ALBEE, A.L. 1981: Pressure temperature, and time indicators in mafic schist: their application to reconstructing the polymetamorphic history of Vermont. American Journal of Science, 281: 127-175.
LEAKE, B.E. 1978: Nomenclature of amphiboles. Canadian Mineralogist, 16: 501-520
LONEY, P.E. 1969: The geology of the Kariba district, Rhodesia with special reference to geochronology and amphibolite petrochemistry. PhD. thesis, Leeds, University of Leeds, 110p.
MACGREGOR, A.M. 1951: Some milestones in the Precambrian of Southern Rhodesia. Proceedings of the Geological Society of South Africa, 54: 27-39.
MUNYANYIWA, H.; HANSON, R.E. 1988: Geochemistry of marbles and calc-silicate rocks in the Pan-African Zambezi belt, Zambia. Precambrian Research, 38: 177-200.
PORADA, H. 1989: Pan-African rifting and orogenesis in southern to equatorial Africa and eastern Brazil. Precambrian Research, 44: 103-136.
RAASE, P. 1974: Al and Ti contents of hornblendes: indicators of pressures and temperatures of regional metamorphism. Contributions to Mineralogy and Petrology, 45: 231-236.
RAMESHWAR RAO, D.; NARAYANA, B.L.; CHARAN, S.N.; NATARAJAN, R. 1991;

P-T Conditions and Geothermal Gradients of Gneiss-Enderbite Rocks: Dharmapuri Area, Tamil Nadu, India. Journal of Petrology, 32: 539-554.
RIDGWAY, J.; RAMSAY, C.R. 1986: A provisional metamorphic map of Zambia - explanatory notes: Journal of African Earth Sciences, 5: 4441-446.
SNELLING, N.J.; JOHNSON, R.L.; DRYSDALL, A.R. 1972: The geochronology of Zambia. Geological Survey of Zambia Records, 12: 19-30.
SPEAR, F.S. 1981: Amphibole-plagioclase equilibria: an empirical model for the relation albite + tremolite = edenite + quartz. Contributions to Mineralogy and Petrology, 72: 33-41.
TRELOAR, P.J.; CARNEY, J.N.; CROW, M.J.; EVANS, J.A.; BARTON, C.N. 199(P-T-t paths of granulite metamorphism and uplift, Zambezi belt, NE Zimbabwe. In VIELZEUF, D.; VIDAL, P.(eds.), Granulites and Crustal Evolution. Kluwer, Dordrecht, 223-241.
THOMPSON, A.B. 1976: Mineral reactions in pelitic rocks: II. Calculation of some P-T-X(Fe-Mg) phase relations. American Journal of Science, 276: 425-454.
UNRUG, R. 1983: Lufilian arc: A microplate in the Pan-African collision zone of the Congo and Kalahari cratons. Precambrian Research, 21: 181-196.
VRANA, S. 1978: Metamorphic pattern in Zambia and their bearing on problems of Zambian tectonic history-a comment. Precambrian Research, 5: 127-130.
VRANA, S.; BARR, M.W.C. 1972: Talc-kyanite-quartz schists and other high-pressure assemblages from Zambia. Mineralogical Magazine, 38: 837-846.

Gondwana Eight, Findlay, Unrug, Banks & Veevers (eds) © 1993 Balkema, Rotterdam. ISBN 90 5410 304 3

A review of the Precambrian geology of Sri Lanka and its comparison with Antarctica

M. Yoshida
Osaka City University, Japan

P.W. Vitanage
University of Peradeniya, Sri Lanka

ABSTRACT: The Precambrian rocks of Sri Lanka form principally three geologic units: the Wanni Complex, the Highland Group, and the Vijayan Complex. They were deformed and metamorphosed mainly during the Proterozoic although a relic Archaean history is known in part of the Highland Group. In East Antarctica between Enderby Land and central Queen Maud Land, are developed six Proterozoic unites. They have generally similar tectonothermal histories and form one geotectonic units (the Rayner Belt). All the Precambrian units of Sri Lanka are regarded as potential correlatives of the Rayner Belt. According to the distribution mode of different geologic units in Antarctica along with their structural relationships, Sri Lanka was probably joined to the Lützow-Holm Bay area of Antarctica. This involves the probable correlation of the Wanni Complex, the Highland Group, and the Vijayan Complex of Sri Lanka with the Prince Olav Complex, Ongul and Skallen groups, and the Yamato-Belgica Complex of Lützow-Holm Bay area respectively. In addition, lithology, structure, and tectonothermal history of the Ongul and Skallen groups are similar to those of the Highland Group, thus supporting this interpretation.

1. INTRODUCTION

The location of Sri Lanka in Gondwana has been one of the problems of various Gondwana reconstructions. Recently-presented Gondwana reconstructions locate Sri Lanka either at Lutzow-Holm Bay (e.g. Lawver and Scotese, 1987) or at the coast of central Queen Maud Land (e.g. De Wit et al., 1988). The palaeomagnetic pole positions of Sri Lanka during Mesozoic and early Palaeozoic times place Sri Lanka in the region between western Enderby Land and east-central Queen Maud Land, but do not identify a more specific location (Funaki et al., 1991, Yoshida et al., 1992). Petrological studies comparing the Precambrian geology of Sri Lanka, southern India, and Antarctica are given by Collerson and Sheraton (1986) and Grew and Manton (1986). In recent years, there has been much progress in understanding the tectonics, petrology and geochronology of both Sri Lanka and Antarctica (e.g., Yoshida et al., 1990, Shiraishi et al., 1987), and Yoshida et al.(1992) showed a Proterozoic to Mesozoic juxtaposition of Sri Lanka against the Lutzow-Holm Bay area in their East Gondwanian reconstruction based on geologic, palaeomagnetic, lineament, and other evidence. The present study reviews, discusses, and summarizes the Precambrian geology of Sri Lanka and Antarctica based on recent studies. The results presented here lend further support to the Gondwana reconstruction model proposed by Yoshida et al. (1992).

2 CHARACTERISTICS OF THE PRECAMBRIAN GEOLOGY OF SRI LANKA

The Precambrian of Sri Lanka is composed of the following five geotectonic units (Figure 1 and Table 1): (1) the Wanni Complex (Cordani and Cooray, 1989, equivalent to the former western Vijayan Complex) composed mostly of amphibolite facies gneisses with minor charnockites, associated with abundant granite and migmatite characterized by the predominance of pink K-feldspar; (2) the Arena Gneiss (Almond, 1991) composed mostly of hornblende-biotite gneisses of amphibolite facies grade occupying doubly plunging synforms in Highland Group granulites; (3) the Highland Group composed of various granulite facies supracrustal sequences and charnockites; (4) the Southwest Group composed mostly of pelitic gneisses and charnockites; and (5) the Vijayan Complex (equivalent to the former eastern Vijayan Complex) composed mostly of banded biotite gneisses and migmatites (summarized from

Table 1. Geologic characteristics of Sri Lanka and Luzow-Holm Bay area, Antarctica

	SRI LANKA	ANTARCTICA
	WANNI COMPLEX (including the Arena Gneiss)	OKUIWA GROUP (eastern part of the Prince Olav Complex)
Lithology	Pink banded biotite gneiss, biotite-hornblende gneiss, granitic gneiss, and augen gneiss, with minor amphibolite, quartzite, and calc gneisses.	Tonalitic to granitic gneises (mostly metapsammites) with much pink granitic injections. Minor amphibolite, biotite-hornblende gneiss, and rare calc gneisses.
Structures	Foliation is inclined moderately to steeply. Undulation and mineral lineations develop locally. Some intense folds develop, which are totally non-linear and noncontinuous.	Foliation is inclined moderately to steeply. Undulation and mineral lineations develop locally. Superimposed foldings develop which are non-linear and less continuous.
Metamorphism	Mostly amphibolite facies, with sporadic remnants of charnockites which occur extensively eastwards. Dominant migmatization associated with pink granite.	Medium-pressure amphibolite facies, including the transitional conditions to the granulite facies westwards. Dominant migmatization associated with pink granites.
	HIGHLAND GROUP (including the Southwest Group)	ONGUL AND SKALLEN GROUPS
Lithology	Metasediments including metapelite and metapsamite, marble and quartzite with abundant charnockites and some metabasites.	Metasediments including metapelites, metapsamites, marble, and quartzite with abundant charnockities and some metabasites.
Structure	Foliation is gently to moderately inclined. Lineation is dominant. Early isoclinal folds are superimposed by later open-tight and gentle folds.	Foliation is inclined gently to moderately. Lineation is dominant. Early isoclinal and recumbent folds are superimposed by later open-tight folds and gentle folds.
Metamorphism	Polymetamorphic composed of early medium-pressure granulite facies metamorphism and later amphibolite facies one.	Polymetamorphic composed of earlier high- to medium-pressure granulite facies metamorphism and later amphibolite facies one.
	VIJAYAN COMPLEX	YAMATO-BELGICA COMPLEX
Lithology	Banded biotite gneiss, hornblende-biotite gneiss, gneissose granite, and augen gneiss with minor amphibolite, calc gneiss and charnockites.	Abundant pink gneissose granite and migmatites, with minor charnockitic gneiss, amphibolite, and calc gneisses.
Structures	Foliation is inclined moderately to steeply. Lineation develops locally. Mega-structures are nonuniform, highly contorted, and nonlinear.	Foliation is inclined gently in the Belgica Mts. (BM) or moderately to steeply in the Yamato Mts (YM). Lineation develops locally. Macroscopic gentle fold superimposes over recumbent fold in BM.
Metamorphism	Amphibolite facies metamorphism with abundant migmatization associated with granitic rocks.	Early granulite facies and later amphibolite facies metamorphism of low-pressure type in YM and amphibolite facies in BM. Migmatization is dominant in both mountains.

Data are derived mostly from Kojima et al. (1981), Cooray (1984), Yoshida et al. (1983, 1990), and Shiraishi et al. (1983).

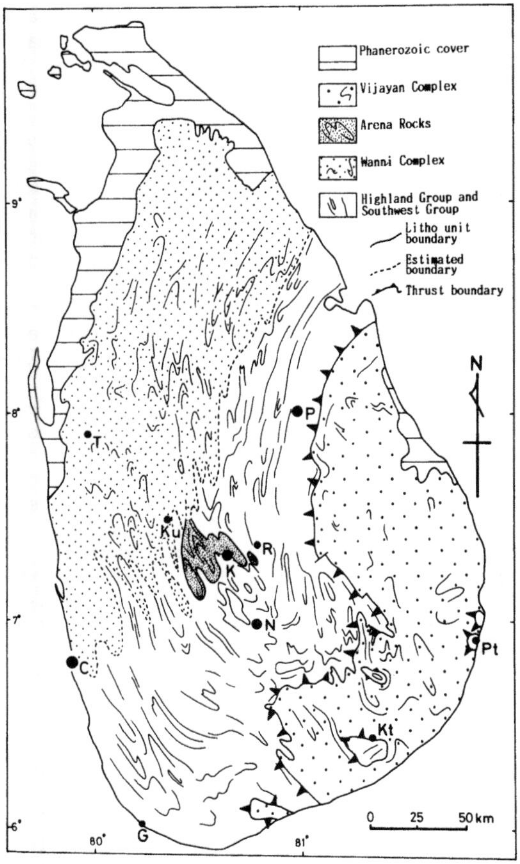

Figure 1. Geologic outline of Sri Lnka. Modified from the geological Map of Sri Lanka (Geological Survey Department of Sri Lanka, 1982). Solid line: Litho-unit boundary, Dotted line: Estimated boundary, Solid line with solid triangle: Thrust bundary. C: Colombo, G: Galle, K: Kandy, Kt: Kataragama, Ku: Kurunegala, N: Nuwara Eliya, P: Polonnaruwa, Pt: Pottuvilu, R: Rangala, T: Tonigala.

Cooray, 1978, 1984; Geological Survey Department of Sri Lanka, 1982; Cordani and Cooray, 1989, with some modifications). The Southwest Group is considered to differ only slightly from the Highland Group in their principal nature of metamorphism (Perera, 1984), structure (Yoshida et al., 1990), and geochronology (Milisenda et al., 1988) and therefore is included in the Highland Group in the following discussion.

2.1 The Highland Group and related rocks

The Precambrian rocks of Sri Lanka are considered to have originated during the Proterozoic, the Highland Group being older than the others according to zircon studies (Kroner et al., 1987) and Sm-Nd model ages (Milisenda et al., 1988). However, Archaean rocks are also known in some part of the Highland Group (Holland, 1985, Kagami et al., 1990). Structural and metamorphic studies have so far been centred mostly on the Highland Group terrane due to the lack of good exposures in the other terranes. Yoshida et al.(1990) studied the structure and metamorphism of the Highland Group and summarized the tectonothermal history mostly of this group. It is given below, supplemented by some recent works.

Charnockitic rocks of Archaean age occcur around Nuwara Eliya and yield Rb-Sr and Sm-Nd isochron ages of circa 2300 to 2500 Ma with a presumed crustal residence time of some hundreds of million years (Holland, 1985, Kagami et al., 1990). The circa 2400 Ma age is therefore considered to reflect a dominant tectonothermal event during the latest Archaean - earliest Proterozoic. Similar ages of detrital zircons from metaquartzite of the Highland Group determined by U-Pb ion microprobe studies (Kröner et al., 1987) may reflect some crust-forming event in the source area of the metaquartzite at this time. Some Sm-Nd model ages of Highland Group rocks ranging from about 2400 to 3200 Ma (Milisenda et al., 1988) are also considered to be the reflection of Archaean events which affected some portions of the Highland Group.

The youngest group of Sm-Nd model ages of the Highland Group is about 2200-2400 Ma; these ages are considered to reflect the crust-forming event at about this time. The next event is the high grade metamorphism of a considerable portion of the Highland Group; its age of about 2100 Ma is indicated by some Rb-Sr whole-rock isochron ages (Crawford and Oliver, 1969; Wickremansinghe, 1969; De Maesschalck et al., 1990). Rootless intrafolial folds with rotational microfabrics formed during this high-grade event, possibly associated with recumbent isoclinal folding. Some U-Pb zircon ages (Kröner et al., 1987; Hölzl et al., 1991) of around 1800 to 2000 Ma and a Rb-Sr whole-rock isochron age of circa 1800 Ma (Kagami et al., 1990) may reflect some subsidiary events including crustal formation associated with the metamorphism.

Some portions of the Highland Group, however, are considered to have formed sometime after 1800 Ma, deduced from the youngest ages of detrital zircons (Kröner et al., 1987). The next granulite facies metamorphism of the Highland Group is considered to have occurred at about 1100 Ma associated with isoclinal folds and flattening microfabrics normal to s-planes. Some of the recumbent and isoclinal folds identified as D_1-D_2 structures by Berger and Jayasinghe (1976) and Yoshida et al. (1990) formed during this metamorphic event or earlier. The circa 1100 Ma ages were given by a Rb-Sr whole-rock isochron age with a low Sr initial ratio (Kagami et al., 1990), Pb-Pb and U-Pb mineral isochron ages (Burton and O'Nions, 1990a), and U-Pb ion microprobe zircon ages (Kröner et al., 1987).

Overturned folding related to the eastward thrusting of the Highland Group over the Vijayan Complex is considered to have taken place after the major metamorphism at about 1100 Ma of both the Highland Group and the Vijayan Complex and before the major upright folding (Sandiford et al., 1989) at about 700 Ma. Some circa 800 Ma Rb-Sr whole rock isochron ages obtained from Vijayan Complex (De Maesschalck et al., 1990) are considered to date this thrusting event as will be discussed later. Following the above event, all Precambrian units underwent major upright folding under amphibolite facies and migmatitic conditions. This event is dated as circa 700 Ma according to Pb-Pb zircon ages of the pink granite surrounding Arena Gneiss (Kröner et al., 1987), U-Pb ion microprobe zircon age (upper intrcept of the incipient charnockite in Kurunegala (Baur et al., 1987), and rb-sr rock ages in the western part of Sri Lanka (Crawford and Oliver, 1969).

The last dominant tectonothermal event corresponds to ductile faults, gentle folds, intrusion of granite-pegmatite, and incipient charnockite formation under low-pressure amphibolite facies conditions. The age of this event is estimated to be at 450 -570 Ma according to K-Ar and Rb-Sr rock and mineral ages (Crawford and Oliver, 1969; Wickremansinghe, 1969), Rb-Sr, Sm-Nd, and Pb-Pb mineral isochron ages (Hölzl et al., 1991; Kagami et al., 1990; Burton and O'Nions, 1990a,b), and the age of dominant Pb loss on zircons (Kröner et al., 1987; Hölzl et al., 1989). This widespread and dominant tectonothermal event is considered to have obliterated most of the circa 700 Ma ages.

2.2 The Wanni Complex and Arena Gneiss

The Wanni Complex shows Sm-Nd model ages of 1300 - 2000 Ma (Milisenda et al., 1988). As to the age of metamorphism in the Wanni Complex, Crawford and Oliver (1969) gave a Rb-Sr whole-rock isochron dating of 986 Ma for a post-metamorphic granite, Cordani and Cooray (1989) reported some circa 1000 Ma Rb-Sr whole-rock ages for granitic rocks in the Wanni Complex, and Hölzl et al. (1991) obtained Rb-Sr and Sm-Nd mineral isochron ages of about 500 Ma and a zircon upper intercept age of 1074 Ma. All these data indicate that the Wanni Complex formed mostly at sometime around 1300 Ma, and was metamorphosed at least twice, i.e., sometime during between 1000 Ma and 1100 Ma, and at about 500 Ma.

The nature and history of the Arena Gneiss have been a matter of debate. Milisenda et al. (1988) reported Sm-Nd model ages of 1400 to 1600 Ma for metamorphites of Arena Gneiss; Pb-Pb zircon ages of circa 500 Ma, 700 Ma and 1100 Ma have been reported also (Kröner et al., 1987; Hölzl et al., 1991). These ages lie within an age range of the Wanni Complex which are distributed to the west of the Arena Gneiss, and differ from both the Highland Group and Vijayan Complex. These data, along with the lithologic similarity of the Arena Gneiss with the Wanni Complex as pointed out by Vitanage (1972) and the intimate distribution of the two units, are considered to indicate that the Arena Gneiss is the equivalent of the Wanni Complex.

Almond (1991) discussed the structural relationship between the Arena Gneiss and the structurally underlying Highland Group and concluded that they may be separated by either an unconformity or a thrust. Judging from the distinct difference between the earlier structures of the Arena Gneiss and the Highland Group as stressed by Yoshida (1989), the Arena Gneiss may be thrust over the Highland Group; the thrust is marked by gneissose pink granite in some arenas. However, an apparently transitional boundary between the Wanni Complex and the Highland Group is known in the northern part. Only part of the Wanni Complex, therefore, is thought to have been thrusted over the Highland Group and to have rested on it during the later tectonic event; this may be referred to the thrusting of the Highland Group over the Vijayan Complex. In Table 1, the Arena Gneiss is included in the Wanni Complex based on the above discussion.

2.3 The Vijayan Complex

The Vijayan Complex has systematically younger Sm-Nd model ages (mostly 1200 - 1400 Ma) than all other Precambrian units. De Maesschalck et al. (1990) also reported circa 800 Ma Rb-Sr whole-rock isochron ages from the Vijayan terrane and insisted that these ages represent the age of the Vijayan metamorphism. It should be noted, however,

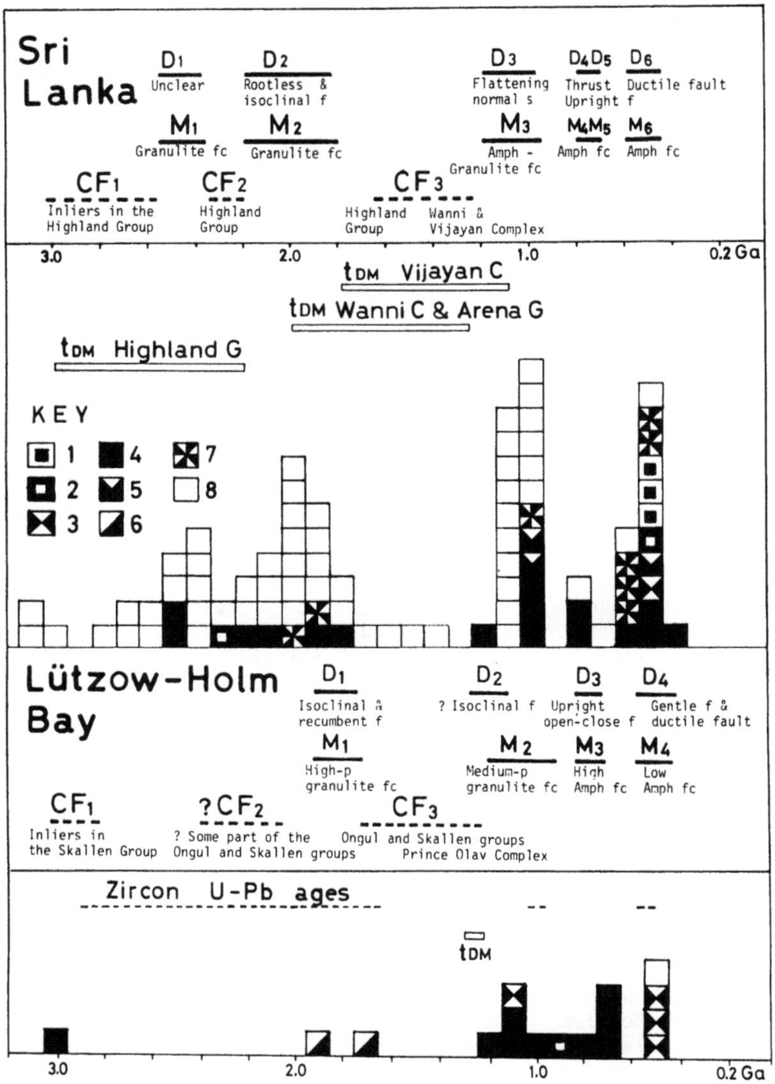

Figure 2. Comparison of Tectonothermal events of the Precambrian of Sri Lanka and Lutzow-Holm Bay area, Antarctica. s: s-plane, Amph: amphibolite, D: deformational event, f: folding, fc: facies, C: Complex, G: Group or groups, CF: Crust formation, M: metamorphic event, t_{DM}: Sm-Nd model age with the depleted mantle standard. Keys are 1: Sm-Nd mineral isochron, 2: Sm-Nd whole-rock isochron, 3: Rb-Sr mineral isochron, 4: Rb-Sr whole-rock isochron, 5: Pb-Pb mineral isochron, 6: Pb-Pb whole-rock isochron, 7: U-Pb discordia intercept, and 8: Pb-Pb mineral ages. Data source for radiometric ages includes: Saito et al., 1961; Maegoya et al., 1969; Crawford and Oliver, 1969; Wickremansinghe, 1969; Holland, 1985; Tanaka et al., 1985; Shibata et al., 1985, 1986; Kröner et al., 1987; Milisenda et al., 1988; Nakajima et al., 1988; Hölzl et al., 1991; Kagami et at., 1990; De Maesschalck et al., 1990; Fanning et al., 1991.

that all other geochronological studies have shown no evidence of this 800 Ma event; U-Pb zircon studies by Kröner et al. (1987) showed some metamorphic event of about 1100 Ma and 550 Ma; Crawford and Oliver (1969) and Cordani and Cooray (1989) reported several Rb-Sr rock and mineral ages of circa 1000 Ma and Rb-Sr whole-rock isochron ages of 1150 Ma (gneisses) and 1010 Ma (granites); and Hölzl et al. (1989) obtained a Rb-Sr mineral isochron age and zircon lower intercept ages of 450 - 500 Ma and zircon upper intercept ages of circa 1020 Ma. There is a possibility that the circa 800 Ma ages might reflect some local event, probably related to the thrusting of the Highland Group over the Vijayan Complex.

In the light of the above discussion, a model for the tectonothermal history of the Sri Lankan Precambrian may be summarized as in Fig. 2.

3 GEOLOGIC CHARACTERISTICS OF EAST ANTARCTICA FROM ENDERBY LAND TO CENTRAL QUEEN MAUD LAND AND COMPARISON WITH SRI LANKA

Along the coastal and near coastal areas of East Antarctica between Enderby Land and Queen Maud Land, sporadic outcrops enabled us to estimate the general geologic features of these areas. These areas are principally composed of Precambrian metamorphic and plutonic rocks, although Phanerozoic platform cover sequences have been supposed (Grikurov, 1978) to develop in some subglacial basins. In these areas, the following seven Precambrian units are distinguished from east to west (Fig. 3): Napier Complex, Rayner Complex, Prince Olav Complex, Ongul and Skallen groups, Yamato-Belgica Complex, Teltet-Vengen Group, and Nils Lalsenfjellet Group, geologic characteristics of these units being described below.

3.1 Enderby Land

The geology of Enderby Land was first widely studied by Soviet geologists (e.g., Ravich and Kamenev, 1975) and afterwards by geologists of other countries (e.g., Sheraton et al., 1987; Black et al., 1987). The Napier Complex is composed of Archaean high pressure - high temperature metamorphic rocks and can be omitted for the present discussion.

The Rayner Complex is composed of both reworked rocks of the Napier Complex and late Proterozoic igneous and metamorphic rocks which formed by middle to late Proterozoic magmatic and sedimentary processes (Black et al., 1987). A clockwise P-T-t path of the major metamorphism indicates an isothermal decompression, ranging from high- to medium-pressure granulite facies to low pressure amphibolite facies conditions. The age of the major metamorphism is dated at circa 1000 Ma. Later granite activity at circa 700 Ma, and pegmatite and associated alteration under greenschist facies conditions at circa 500 Ma have been identified. Some circa 700 Ma Rb-Sr isochron ages and U-Pb zircon ages obtained from the granitic rocks, however, have been thought to be meaningless, i.e. the interaction ages of circa 500 Ma low-grade metamorphism and circa 1000 Ma major metamorphism (Black et al., 1987).

3.2 Prince Olav Coast to Lützow-Holm Bay area

This area was first briefly studied by Soviet geologists (e.g., Ravich and Kamenev, 1975) and later studied in detail by Japanese workers (e.g., Yoshida et al., 1983). Geologic characteristics of this area are given in Table 1.

The Prince Olav Complex is equivalent to the amphibolite facies and amphibolite-granulite facies transition zone of the Lützow-Holm Complex of Shiraishi et al. (1987) and includes the Olav and Hinode serieses of Ravich and Kamenev (1975) and the Okuiwa Group of Yoshida (1978). They are mostly amphibolite facies metamorphites and magmatic rocks with dominant granite-migmatite bodies. The Okuiwa Group, which occurs in the southwestern part of Prince Olav Coast, includes dominantly granitic and migmatitic rocks and subordinate metasediments. The Prince Olav Complex area is considered to be polymetamorphic (Yoshida et al., 1983); the major regional metamorphism increases progressively southwestward from medium-pressure amphibolite facies to transitional zone with granulite facies, its progression southwestward into the Ongul and Skallen groups having been well documented (e.g., Shiraishi et al., 1987). The major metamorphism is believed to have occurred at circa 1100 Ma, as it is continuous with the major granulite facies metamorphism of the Lützow-Holm Bay area where the major metamorphism is considered to be about 1100 Ma (Yoshida et al., 1983). The 1100 Ma age is supported by Pb-Pb zircon ages recently reported by Fanning et al. (1991); Rb-Sr data yield whole-rock isochron ages of circa 700 Ma (Shibata et al., 1986 and Nakajima et al., 1988) and mineral isochron ages of circa 500 Ma(Shibata et al., 1985) and therefore late Proterozoic and Cambro-Ordovician resetting is considered to have occurred, associated with migmatitic and granitic activities (700 Ma) or pegmatite intrusions (500 Ma).

The Ongul and Skallen groups (Yoshida, 1978) are equivalent to the granulite facies rocks of the Lützow-Holm Complex of Shiraishi et al. (1987). These groups are composed of magmatic rocks and metasediments of possibly late and/or middle Proterozoic age, judging from their radiometric ages and strontium initial ratios. A late Archaean Rb-Sr whole-rock isochron age was reported from acid-intermediate granulites on the southern coast of Lützow-Holm Bay by Nakajima et al. (1988), and late Archaean to middle Proterozoic U-Pb zircon ages were recently obtained by Fanning et al. (1991). The Ongul and Skallen groups suffered the same progressive metamorphism as the Prince Olav Complex, attaining medium-pressure granulite facies with a pressure increase up to 9 kb observed towards the southwest (Yoshida, 1979; Katsushima, 1985). Relict high pressure granulite assemblages, which are considered to belong to an earlier metamorphism, also occur sporadically (Yoshida, 1979). Time of the major metamorphism is considered to be circa 1100 Ma, as estimated from two Rb-Sr whole-rock isochron ages with low Sr initial ratios (Shirahata, 1983; Shibata et al., 1986),a Rb-Sr mineral isochron age (Maegoya et al., 1968) and U-Pb zircon ages (Fanning et al., 1991). Subsequent metamorphic events are indicated at 700 Ma and 500 Ma by Rb-Sr and Pb-Pb data (Maegoya et al., 1968; Shibata et al., 1986, 1985 and Fanning et al., 1991 respectively). The 700 Ma modification produced an amphibolite facies overprint and concomitant migmatization, whereas the 500 Ma event produced granite-pegmatite intrusions.

3.3 Yamato and Belgica Mountains

The Yamato-Belgica Complex occupies all

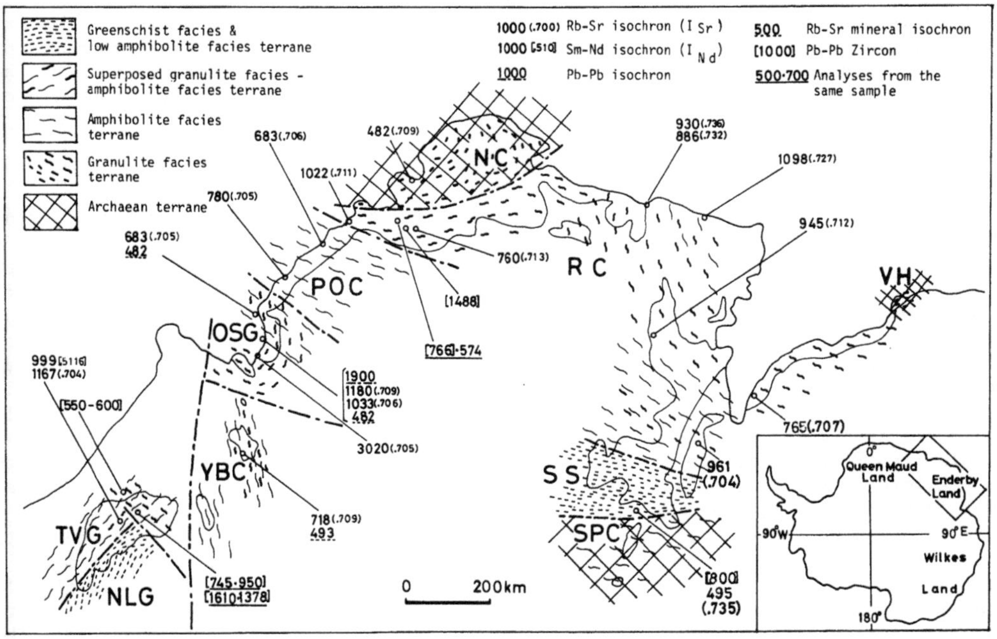

Figure 3. Geotectonic division and major radiometric ages of East Antarctica from Enderby Land to central Queen Maud Land. NC: Napier Complex, SS: Archaean low-grade supracrustals, SPC: Archaean Southern Prince Charles Mountains block, POC: Prince Olav Complex, OSG: Ongul and Skallen groups, YBC: Yamato-Belgica Complex, TVG: Teltet-Vengen Group, NLG: Nils Lalsenfjellet Group. Data source includes: Pasteel and Michot, 1970; Ravich and Kamenev, 1975; Kamenev and Ravich, 1979; Grikurov, 1979; Yoshida, 1978; Tingey, 1982; Yoshida et al., 1983; Shirahata, 1983; Shibata et al., 1985, 1986; Sheraton et al., 1987; Black et al., 1987; Shiraishi et al., 1987; Nakajima et al., 1988; Shiraishi and Kagami, 1989 and Shiraishi et al., 1991.

areas of these mountains and consists mainly of metasedimentary and magmatic rocks of probably late Proterozoic origin. Abundant migmatitic and metasedimentary gneisses of amphibolite facies grade, granites and syenites, and some granulite facies gneisses are distributed in the Yamato Mountains (Shiraishi et al., 1983), and metasediments and granitic gneisses of the amphibolite facies grade dominate in the Belgica Mountains (Kojima et al., 1981). Superposition of the later amphibolite facies metamorphism over the earlier granulite facies metamorphism was detected in the Yamato Mountains (Shiraishi et al., 1983). It is possible that the Yamato-Belgica Complex suffered a progressive amphibolite-granulite facies metamorphism which were superposed by amphibolite facies metamorphism associated with granite-migmatite activity. Time of the major granulite metamorphism may be either circa 700 Ma or earlier. Rb-Sr whole-rock isochron ages of circa 700 Ma with relatively high strontium initial ratios were obtained from Yamato Mountains by Shibata et al. (1986). These ages possibly correspond to the later amphibolite facies metamorphism because the Rb-Sr whole-rock plots on the isochron include rocks of the younger amphibolite-migmatite group. A Rb-Sr mineral isochron age and many K-Ar ages of circa 500 Ma are also known (Yanai and Ueda, 1974; Shibata et al., 1985). This age is considered to reflect a thermal event related to pegmatite activity.

3.4 Sør Rondane Mountains

This area was studied initially by Belgian workers (e.g., Van Autenboer and Loy, 1972) and recently by Japanese workers (e.g., Shiraishi et al., 1991). The Teltet-Vengen Group of the Sør Rondane Mountains is composed of metasediments and magmatic rocks of possible late Proterozoic age as deduced from preliminary Sm-Nd model ages (tDM) of 1150 - 1300 Ma (Shiraishi and Kagami, 1989). A late Archaean component is also present judging from presence of some old (up to circa 2700 Ma) detrital zircons (e.g., Pasteel and Michot, 1970). The Sør Rondane Mountains area suffered medium-pressure granulite facies metamorphism and late retrogressive amphibolite facies metamorphism and is intruded by various magmatic rocks. Preliminary Rb-Sr and Sm-Nd isochron ages (Shiraishi and Kagami, 1989) indicate that the main granulite facies metamorphism occurred at circa 1000 Ma and that the crustal residence time is very short. The time of dominant granitic and syenitic magmatic activity is considered to range from circa 600 - 500 Ma judging from Pb-Pb zircon and sphene ages (Pasteel and Michot, 1970) and a Rb-Sr whole-rock isochron age (Takahashi et al., 1990). The time of the amphibolite facies retrogression is unknown. It could be ascribed to the granitic-syenitic activities; but there is still another possibility that it took place at circa 650 Ma or a little earlier as reflected by a Pb-Pb age of some metamorphic zircons.

The Nils Lalsenfjellet Group is composed of sheared and retrogressed tonalitic rocks including some layers of more mafic rocks. Newly grown chlorite, epidote, actinolite and saussurite occur. Only limited petrological and geochronological works have been done on the group so far. Because of resistant zoned plagioclase and the original igneous fabrics of rocks of this group, they are considered not to be very old and possibly formed during the movement of the E-W fault running between the Teltet-Vengen Group and the Nils Lalsenfjellet Group (Shiraishi et al., 1991). However, two conformable ages, both a U-Pb zircon age (Pasteel and Michot, 1970) of 950 Ma and a Rb-Sr whole-rock isochron age (Takahashi et al., 1990) of about 950 Ma with a low Sr initial ratio from schistose tonalite of the Nils Lalsenfjellet Group indicate that the intrusion of the tonalite took place at about this time.

3.5 Late Proterozoic tectonothermal events

We can recognize a wide area mostly composed of middle to late Proterozoic sedimentary and magmatic rocks which suffered multiple tectonothermal events during early Proterozoic to earliest Phanerozoic times. They may be collectively termed as the Rayner Belt as proposed by Yoshida and Kizaki (1983). There is little evidence to distinguish the Proterozoic units according to their lithologic and tectonothermal characteristics along the coastal-near coastal areas of East Antarctica between Enderby Land and central Queen Maud Land, except for the Yamato-Belgica Complex and the Nils Lalsenfjellet Group. The Yamato-Belgica Complex suffered low-pressure metamorphism, while all others underwent the medium-pressure metamorphism characterized by a P-T-t path of isothermal decompression. The Nils Llalsenfjellet Group is distinct from others in its characteristic lithology.

The time of metamorphism and granite activity, which ranges from circa 1100 Ma to 500 Ma appears consistent throughout the Rayner Belt, although the intensity of the thermal events, which include granite production, during circa 700 Ma and circa 500

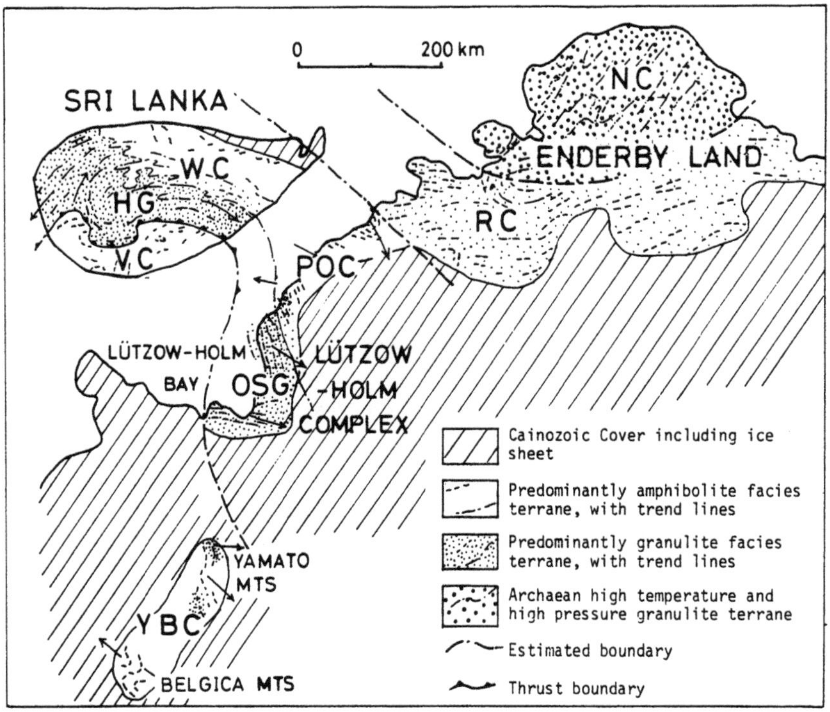

Figure 4. Juxtaposition of Sri Lanka with Antarctica (modified from Yoshida et al., 1992). Abbreviations are WC: Wanni Complex including Arena Gneiss, HG: Highland Group including Southwest Group, VC: Vijayan Complex. Other abbreviations are same as in Figure 3.

Ma appears to differ from place to place. The good whole-rock Rb-Sr isochron ages of circa 700 Ma from rocks collected between Prince Olav Coast and Yamato Mountains attests the strong isotopic homogenization during a major thermal event at this time. This contrasts with a less intense isotopic homogenization caused by a thermal event at 500 Ma which is indicated by Rb-Sr mineral and whole-rock ages, mineral isochron ages, and K-Ar mineral ages from the same area. Thus the 700 Ma activity is considered to have been dominant in areas from the Prince Olav Coast to Yamato Mountains through Lützow-Holm Bay, while it was less dominant in the Sør Rondane Mountains where only a Pb-Pb metamorphic zircon age of 745 Ma has been reported (Pasteel and Michot, 1970); we consider this to have resulted from the degree of resetting of the Rb-Sr clock that should have formerly dated the circa 1100 Ma regional metamrphic event.

4 JUXTAPOSITION OF SRI LANKA TO ANTARCTICA

The Highland Group of Sri Lanka is comparable to any one of the Rayner Complex, Ongul and Skallen groups, and the Teltet Vengen Group of Antarctica judging from their similarity in lithology, structure, age of original rocks, and type and age of metamorphism. There appears to be not so much evidence to identify the strongest candidate among the above three units. However, distinct differences in lithologies and characteristics of metamorphism and plutonism among geologic units in Sri Lanka are worth noting. In view of the above considerations, the Lützow-Holm Bay area is considered most suitable as a juxtaposing counterpart with Sri Lanka (Fig. 4, also see Table 1). The Prince Olav Complex - Ongul and Skallen groups - Yamato-Belgica Complex can be compared with the Wanni Complex - Highland Group - Vijayan Complex (Table 1). The Okuiwa Group of the Prince Olav Complex is composed of high-amphibolite facies metasediments with abundant pink granite-migmatite and compares closely with the Wanni Complex. The gradual increase in metamorphic grade of the Okuiwa Group towards the Ongul and Skallen groups is similar to the relationship of the Wanni Complex with the Highland Group.

The original lighology and complicated

tectonothermal history of the Ongul and Skallen groups compare well with those of the Highland Group. The occurrence of relict kyanite and staurolite in garnet porphyroblasts in garnet-sillimanite gneiss of both groups (e.g., Hiroi et al., 1983; Yoshida et al., 1990) shows a remarkable similarity in their respective metamorphic histories.

The Yamato-Belgica Complex may be correlated with the Vijayan Complex; they both are similar in lithology and metamorphic grade, although geochronologic results from both areas should be further examined. The boundary between the Ongul and Skallen groups and the Yamato-Belgica Complex is thought to be a fault, based on the distinct difference in the pressure type of metamorphism of the two units and from subglacial topography; this is comparable to the well documented tectonic boundary between the Vijayan Complex and the Highland Group.

5. SUMMARY

i) The Precambrian of Sri Lanka is principally composed of three geologic units. They are the Wanni Complex, Highland Group (including the Southwest Group) and Vijayan Complex from west to east. They have a prolonged tectonothermal history ranging from early Proterozoic to earliest Phanerozoic, except for some parts of the Highland Group which have further imprints of older events as old as late Archaean.

ii) The Precambrian sequence of East Antarctica from Enderby Land to central Queen Maud Land is composed of seven geologic units. They are the Napier Complex, Rayner Complex, Prince Olav Complex, Ongul and Skallen groups, Yamato-Belgica Complex, Teltet-Vengen Group, and Nils Lalsenfjellet Group from east to west. All these units except for the Napier Complex have a principally Proterozoic history and can be regarded as forming a broad tectonic belt to be termed the Rayner Belt. They exhibit almost similar tectonothermal history and isothermal decompressional P-T-t path, except for the Yamato-Belgica Complex which underwent low-pressure type metamorphism, and the Nils Lalsenfjellet Group which are distinct in lithology.

iii) Judging from the distribution mode and structural relationships among different geologic units of each area, the Lützow-Holm Bay area is considered to be the only appropriate location for juxtaposing Antarctica against Sri Lanka. Comparison of the Prince Olav Complex with the Wanni Complex, the Ongul and Skallen groups with the Highland Group, and the Yamato-Belgica Complex with the Vijayan Complex appears appropriate based on the similarity in lithology, tectonothermal history and structural relationships between the adjacent geologic units.

ACKNOWLEDGEMENT

The authors with to express their sincere gratitude to R.H. Findlay, J.M. Bertrand, and an anonymous reviewer for their valuable improvement on the early version of the manuscript, and to M. Santosh for the gramatical corrections. This work is a contribution to both IGCP 236 and 288.

REFERENCES

Almond, D.C. 1992. Arena Gneiss and Kandy Gneiss - a proposed subdivision of the Highland Series around Kandy, and its significance. Journal of the Geological Society of Sri Lanka 3: 41-50.

Baur, N., Kröner, A. & Todt, W. 1987. Structural control of progressive and retrogressive metamorphic reactions and U-Pb zircon isotopic systematics in high-grade Proterozoic gneisses of Sri Lanka. Terra Cognita 7: 133-134.

Berger, A.R. & Jayasinghe, N.R. 1976. Precambrian structure and chronology in the Highland Series of Sri Lanka. Precambrian Research 3: 559-576.

Black, L.P., Harley, S.L., Sun, S.S. & McChulloch, M.T. 1987. The Rayner Complex of East Antarctica: complex isotopic systematics within a Proterozoic mobile belt. Journal of Metamorphic Geology 5: 1-26.

Burton, K.W. & O'Nions, R.K. 1990a. The time scale and mechanism of granulite formation at Kurunegala, Sri Lanka. Contributions to Mineralogy and Petrology 106: 66-89.

Burton, K.W. & O'Nions, R.K. 1990b. Fe-Ti oxide chronometry: with application to granulite formation. Geochimica et Cosmochimica Acta 54: 2593-2602.

Collerson, K.D. & Sheraton, J.W. 1986. Bedrock geology and crustal evolution of the Vestfold Hills. In J. Pickard (ed.),Antarctic oasis, Terrestrial environments and history of the Vestfold Hills; 21-62. Sydney: Academic Press.

Cooray, P.G. 1978. Geology of Sri Lanka. Proceedngs of the third regional conference on the Geology and Mineral Resources of South-East Asia: 701-710. Bangkok: Arean Inst. Technol.

Cooray, P.G. 1984. An introduction to the geology of Sri Lanka (Ceylon). Colombo: National Museum of Sri Lanka Publication.

Cordani, U. & Cooray, P.G. 1989. Rb-Sr ages of granites and gneisses from the Precambrian of Sri Lanka. Journal of

the Geological Society of Sri Lanka 2: 35-43.
Crawford, A.R. & Oliver, R.L. 1969. The Precambrian geochronology of Ceylon. Geological Society of Australia Special Publication 2: 283-306.
De Maesschalck, A.A., Oen, I.S., Hebed, E. H., Verschure, R.H. & Arps, C.E.S. 1990. Rubidium-strontium whole rock ages of Kataragama and Pottuvil charnockites and East Vijayan gneisses: indication of a 2 Ga metamorphism in the Highlands of southeast Sri Lanka. Journal of Geology 98: 443-456.
De Wit, M., Jeffery, M. Bergh, H. & Nicolaysen, L. 1988. Geological map of Gondwana, scale 1:10,000,000, with explanations and references. Okulahoma: American Association of Petrolium Geologists.
Fanning, M., Hiroi, Y., Motoyoshi, Y., Shiraishi, K., Nakai, Y., Ellis, D.J. & Black, L. 1991. Evolution of the Lützow-Holm Complex, East Antarctica: geological, petrological and geochronological constraints. Abstracts, Sixth International Symposium on Antarctic Earth Sciences: 139. Tokyo:National Institute of Polar Research.
Funaki, M., Yoshida, M. & Vitanage, P.W. 1991. Natural remanent magnetization of some rocks from southern Sri Lanka. Proceedings of the NIPR Symposium on Antarctic Geosciences. 4: 231-240.
Geological Survey Department of Sri Lanka 1982. Geological map of Sri Lanka, 8 miles to one inch. Colombo: Geological Survey Department of Sri Lanka.
Grew, E.S. & Manton, W.I. 1986. A new correlation of sapphirine granulites in the Indo-Antarctic metamorphic terrain: late Proterozoic dates from the Eastern Ghats Province of India. Precambrian Research 33:123-137.
Grikurov, G.E. 1978. Tectonic map of Antarctica, 1:10,000,000. Leningrad: SEVMORGEA.
Hiroi, Y., Shiraishi, K., Yanai, K. & Kizaki, K. 1983. Aluminum silicates in the Prince Olav and Soya Coasts, East Antarctica. Memoirs of National Institute of Polar Research Special Issue 28:115-131.
Holland, J.G. 1985. The geochemistry and geological history of the Archaean and Proterozoic gneisses of Sri Lanka. In C.B. Dissanayake & P.G. Cooray (eds.), Recent advances in the geology of Sri Lanka: 75-86. Paris: International Center for Training and Exchanges in the Geosciences.
Hölzl, S., Köhler, H., Kröner, A., Jaeckel, P. and Liew, T.C. 1991. Geochronology of Sri Lankan basement. In A. Kroner (ed.), The crystalline Crust of Sri Lanka, Part I, Summary of Research of the German-Sri Lankan Consortium. Geological Survey Department of Sri Lanka, Professional Paper 5: 237-257.
Kagami, H., Owada, M., Osanai, Y. & Hiroi, Y. 1990. Preliminary geochronological study of Sri Lankan rocks. In Y. Hiroi & Y. Motoyoshi (eds.) Interim Report of Japan - Sri Lanka Joint Research: 55-70.
Kamenev, E.N. & Ravich, M.G. 1979. Map of metamorphic facies of Antarctica, 1:5,000,000. Leningrad: Research Institute of the Geology of the Arctic, Complex, "SEVMORGEO".
Katsushima, T. 1985. Granulite-facies rocks in several islands west of Langhovde, East Antarctica. Memoirs of National Institute of Polar Research 37: 95-110.
Kojima, H., Yanai, K. & Nishida, T. 1981. Geological map of Belgica Mountains, Antarctica. (scale 1:25,000), with an explanatory text. Antarctic geological map series, sheet 29. Tokyo: National Institute of Polar Research.
Kröner, A., Williams, J.S., Compton, W., Baur, N., Vitanage, P.W. & Perera, L.R.K. 1987. Zircon ion microprobe dating of high grade rocks in Sri Lanka. Journal of Geology 95: 775-791.
Lawver, L.A. & Scotese, C.R. 1987. A revised reconstruction of Gondwanaland. In D. McKenzie (ed.), Gondwana VI: Structure, tectonics, and geophysics. Geophysical Monograph 40: 17-23.
Maegoya, T., Nohda, S. & Hayase, I. 1968. Rb-Sr dating of the gneissic rocks from the east coast of Lutzow-Holm Bay, Antarctica. Memoirs of College of Science, University of Kyoto, Series B 35: 131-138.
Milisenda, C.C., Liew, T.C., Hofmann, A.W. & Kroner, A. 1988. Isotopic mapping of age provinces in Precambrian high-grade terrains: Sri Lanka. Journal of Geology 96: 608-615.
Nakajima, T., Shibata, K., Shiraishi, K., Motoyoshi, Y. & Hiroi, Y. 1988. Rb-Sr whole-rock age of metamorphic rocks from eastern Queen Maud Land, East Antarctica (2): Tenmondai Rock and Rundvagshetta (abstract). Proceedings of the NIPR Symposium on Antarctic Geosciences 2: 172.
Pasteel, P. & Michot, J. 1970. Uranium-lead radioactive dating and lead isotope study on sphene and K-feldspar in the Sør Rondane Mountains, Dronning Maud Land, Antarctica. Eclogea geologia Helvetia. 63:231-254.
Perera, L.R.K. 1984. Co-existing cordierite-almandine - a key to the metamorphic history of Sri Lanka. Precambrian Research 25: 345-364.

Ravich, M.G. & Kamenev, E.N. 1975. Crystalline basement of the Antarctic platform. New York: Halsfed Press.
Saito, N., Tatsumi, T. & Sato, K. 1961. Absolute age of euxenite from Antarctica. Antarctic Record (Nankyoku Shiryo) 12: 1057-1062.
Sandiford, M., Powell, R., Martin, S.F. & Perera, L.R.K. 1989. Thermal and baric evolution of garnet granulites from Sri Lanka. Journal of Metamorphic Geology 6: 351-364.
Sheraton, J.W., Tingey, R.J., Black, L.R., Offe, L.A. & Ellis, D.J. 1987. Geology of Enderby Land and western Kemp Land, Antarctica. BMR Bulletin 223, Canberra: Department of Resources and Energy, Bureau of Mineral Resources, Geology and Geophysics.
Shibata, K., Yanai, K. & Shiraishi, K. 1985. Rb-Sr mineral isochron ages of metamorphic rocks around Syowa Station and from the Yamato Mountains, East Antarctica. Memoirs of National Institute of Polar Research Special Issue 37: 164-171.
Shibata, k., Yanai, K. & Shiraishi, K. 1986. Rb-Sr whole-rock ages of metamorphic rocks from eastern Queen Maud Land, East Antarctica. Memoirs of National Institute of Polar Research Special Issue 43: 133-148.
Shirahata, H. 1983. Lead isotopic composition in metamorphic rocks from Skarvsnes, East Antarctica. In R.J. Oliver, P.R. James & J.B. Jago (eds.), Antarctic Earth Science: 55-58. Canberra: Australian Academy of Science.
Shiraishi, K. & Kagami, H. 1989. Preliminary geochronological study of granulites from the Sør Rondane Mountains, East Antarctica - a comparison of Rb-Sr and Sm-Nd ages - (abstract). Proceedings of the NIPR Symposium on Antarctic Geosciences 3: 152.
Shiraishi, K., Asami, M. & Ohta, Y. 1983. Geology and petrology of the Yamato Mountains. In R.L. Oliver, P.r. James, & J.B. Jago (eds.) Antarctic Earth Science: 50-53. Canberra: Australian Academy of Science.
Shiraishi, K., Hiroi, Y., Motoyoshi, Y. & Yanai, K. 1987. Plate tectonic development of late Proterozoic paired metamorphic complexes in eastern Queen Maud Land, East Antarctica. in G.W. McKenzie (ed.), Gondwana VI: Structure, tectonics and geophysics. Geophysical Monograph 40; 309-318.
Shiraishi, K., Asami, M., Ishizuka, H., Kojima, H., Kojima, S., Osanai, Y., Sakiyama, T., Takahashi, Y., Yamazaki, M. & Yoshikura, S. 1991. Geology and metamorphism of the Sør Rondane Mountains, East Antarctica. In M.R.A. Thomson (ed.), Geological evolution of Antarctica: 77-82. Cambridge: Cambridge Univ. Press.
Takahashi, Y., Arakawa, Y., Sakiyama, T., Osanai, Y. & Makimoto, H. 1990. Rb-Sr and K-Ar whole-rock ages of the plutonic bodies from the Sør Rondane Mountains, East Antarctica. Proceedings of the NIPR Symposium on Antarctic Geosciences 4: 1-22.
Tanaka, T., Nakajima, T. & Shiraishi,K. 1985. Higashi-Nankyoku Skarvsnes chiiki ni sansuru henseigan no samariumu-neojiumu nendai ni tuite (in Japanese) (Sm-Nd ages on the metamorphic rocks from Skarvsnes, East Antarctica. Abstracts, Sixth Symposium on Antarctic Geosciences: 20. Tokyo: National Institute of Polar Research.
Tingey, R.J. 1982. The geologic evolution of the Prince Charles Mountains. - An Antarctic Archaean cratonic block. In C. Craddock (ed.), Antarctic Geoscience: 455-464. Madison: University of Wisconsin Press.
Van Autenboer, T. & Loy, W. 1972. Recent geological investigations in the Sør Rondane Mountains, Belgicafjella and Sverdrupfjella, Dronning Maud Land. in R.J. Adie (ed.), Antarctic Geology and Geophysics: 561-571. Oslo: Universitets-forlaget.
Vitanage, P.W. 1972. Post-Precambrian uplifts and regional neotectonic movements in Ceylon. Proceedings of the Twenty fourth International Geological Congress 3: 642-653. Montreal.
Wichremansinghe, O.C. 1969. The geochemistry and geochronology of the charnockites and associated rocks of Ceylon. Thesis, University of Leeds, September, 1969.
Yanai, K. & Ueda, Y. 1974. Syowa kiti hukin san gansekino zettai nendai to sono igi (in Japanese) (Absolute ages and geological investigations on the rocks in the area of around Syowa Station, East Antarctica. Nankyokushiryo (Antarctic Record) 48: 70-81.
Yoshida, M. 1978. Tectonics and petrology of charnockites around Lützow-Holmbukta, east Antarctica. Journal of Geosciences, Osaka City University 21: 65-152.
Yoshida, M. 1979. Metamorphic conditions of the polymetamorphic Lützow-Holmbukta region, East Antarctica. Journal of Geosciences, Osaka City University 22: 97-140.
Yoshida, M. 1989. Comparative study of geologic structures of Precambrian rocks of Sri Lanka. Abstract volume, International Geological Correlation Programme, Project 236, Conference on Gondwana Fragments, 13-14. Nairobi: Geological Society of Kenya.
Yoshida, M. & Kizaki, K. 1983. Tectonic situation of Lützow-Holm Bay in East

Antarctica and its significance in Gondwanaland. In R.L. Oliver, P.R. James & J.B. Jago (eds.), Antarctic Earth Science: 36-39. Canberra: Australian Academy of Science.
Yoshida, M., Suzuki, M., Shirahata, H., Kojima, H. & Kizaki, K. 1983. A review of the tectonic and metamorphic history of the Lützow-Holm Bay region, East Antarctica. In R.L. Oliver, P.R. James & J.B,. Jago. (eds.), Antarctic Earth Science: 44-47. Canberra: Australian Academy of Science.
Yoshida, M., Kehelpannala, K.V. Hiroi, Y. & Vitanage, P.W. 1990. Sequence of deformation and metamorphism of granulites of Sri Lanka. Journal of Geosciences, Osaka City University 33: 69-107.
Yoshida, M., Funaki, M. & Vitanage, P.W. 1992. Proterozoic to Mesozoic East Gondwana: the juxtaposition of India - Sri Lanka - Antarctica. Tectonics 11: 392-404.

Gondwana Eight, Findlay, Unrug, Banks & Veevers (eds) © *1993 Balkema, Rotterdam. ISBN 90 5410 304 3*

Correlation of cratonic and orogenic provinces in southeastern Africa and Dronning Maud Land, Antarctica

P. Bruce Groenewald
Department of Geology, University of Natal, Pietermaritzburg, South Africa

ABSTRACT: Crustal evolution in western Dronning Maud Land, Antarctica, was essentially the same as that in southeastern Africa from the Archaean until Mesozoic, and suggests progressive assembly of this portion of Gondwana prior to 1000 Ma. Mid-Proterozoic supracrustal sequences of the cratonic Kaapvaal and Zimbabwe Provinces of Africa and the Grunehogna Province of Dronning Maud Land are similar in lithological succession, age and inferred depositional setting. Both continents are transected by 1 000 Ma orogenic belts. The Maudheim Province in Dronning Maud Land, an orogenic belt to the south and east of the Grunehogna Province, is comparable to the Mozambique and Natal provinces in southeastern Africa. All three have attributes of metamorphosed volcanic arc - marginal basin successions, and thermotectonic histories indicating that 1 000 Ma collision was a major stage in the assembly of Gondwana. The Maudheim and Mozambique Provinces were over-printed by a 600-450 Ma tectonothermal event, equivalent to the Pan-African and Ross orogenies, in which crustal extension with associated alkaline magmatism was followed by compressional tectonism. An early Phanerozoic trough parallel to the line followed by Mesozoic separation of Antarctica and Africa represents a failed rift in which vertical tectonism led to large-scale alluvial fan deposits. This suggests that segmentation of the supercontinent began 250 Ma before break-up. Although current geophysical data do not allow precise re-assembly of Africa and Antarctica, overlap of the modern continental margins is possible in southeastern Mozambique where the stratigraphy postdates break-up. Exclusion of these coastal plains from the continental margin allows closer re-assembly of Gondwana at 180 Ma, giving more precise alignment of the Natal, Maudheim and Mozambique belts and juxtapositioning of the Grunehogna and Kaapvaal Provinces.

1. INTRODUCTION

New information from Dronning Maud Land, Antarctica, permits more precise geological correlation of this region with southeastern Africa. Recent reconstructions of Gondwana juxtapose these regions at ca. 145 Ma on the basis of marine geophysical evidence (Martin & Hartnady 1986; Lawver & Scotese 1987). The evidence for correlating the stratigraphic and orogenic histories is reviewed and extended in an effort to enlarge on the suggestions of Yoshida & Kizaki (1983), Martin & Hartnady (1986), Grantham et al. (1988) and Groenewald et al. (1991). The difficulty of reasonably precise correlation stems partly from masking of the original contact areas, by a thick Cretaceous-Quaternary sequence in Mozambique and ice in Antarctica, and from the different levels of exposure resulting from diverse erosional histories.

Africa and Antarctica comprise cratonic nuclei separated by orogenic belts of various ages (Fig. 1). In southeastern Africa, the cratonic Kaapvaal and Zimbabwe Provinces consist of ancient granite-greenstone terrains overlain by volcano-sedimentary sequences ranging in age from late Archaean to mid-Proterozoic. In Dronning Maud Land, the Grunehogna Province is similar, having 3.0 Ga-old granites and 1.7 Ga-old volcano-sedimentary sequences. Around the cratons are mid-Proterozoic to early Phanerozoic tectonothermal provinces, some of which have characteristics of accreted terrains. Two distinct orogenies occurred, at ca. 1 000 Ma and 500 Ma.

Phanerozoic crustal evolution involved the development of major depositional basins and widespread volcanism which covered much of the older successions. In southeastern Africa, syn- depositional tectonism in the Karoo basin was related to early stages of Gondwana break-up when rifting and transcurrent faulting initiated partial separation of Antarctica from Africa.

2. CORRELATION OF CRATONIC SEGMENTS?

In western Dronning Maud Land, relatively undeformed and unmetamorphosed Archaean and Proterozoic rocks constitute the Grunehogna Province (Fig. 1). The Archaean rocks, occurring only in the Annandagstoppane area, are ca. 3 000 Ma granites (Barton et al. 1987). Some granitic intrusions in the eastern Kaapvaal and Zimbabwe Provinces are of similar age, but the limited data from Antarctica have been interpreted to indicate that the African granites have a distinct geochemical signature (Barton et al. 1987).

The Ritscherflya Supergroup, a supracrustal sequence of Proterozoic age, makes up the remainder of Grunehogna Province. The sedimentary rocks represent, from the base upwards, shallow marine, tidal flat, braided stream, and alluvial fan deposits (Ferreira 1986). Palaeocurrent directions are highly variable, but the depocentre was probably to the northeast. A sequence of continental tholeiitic basalts and basaltic andesites overlies the sedimentary rocks in Grunehogna Province (Watters et al. 1991). The geochronological data are equivocal, with reported ages ranging from 800 to 1 800 Ma (Moyes & Barton 1990). Watters et al. (1991) obtained Sm-Nd T_{chur} model ages in the range 1 300-1 620 Ma for the basalts, probably reflecting the crystallization age. These authors ascribed a Rb-Sr age of 876 Ma to later tectonothermal overprinting. The Ritscherflya sequence hosts extensive mafic and ultramafic intrusions which have been interpreted as being coeval with the volcanic part of the succession, and show a similar dispersion of isotopic ages as a consequence of hydrothermal alteration and crustal contamination (Krynauw et al. 1991).

The Umkondo, Waterberg, Soutspanberg and Ntingwe successions of southern Africa are similar in age and character to the Ritscherflya Supergroup. These are preserved near the margins of the Kaapvaal and Zimbabwe cratons, adjacent to the Mozambique, Limpopo and Natal orogenic provinces (Fig. 1). All contain immature fluvial to shallow marine sediments with distinct ferric pigmentation. Tholeiitic volcanic rocks are present only in the Soutpansberg and Umkondo Groups. Tankard et al. (1982) reported predominantly south to southeast palaeocurrent directions for the Soutpansberg and Waterberg Groups, which is opposite to those in the Ritscherflya. The sediments were deposited in sandy, braided alluvial plain settings, within either an aulocogen or a yoked basin. The Umkondo Group, situated on the eastern flank of the Zimbabwe Province, is similar to the Ritscherflya Supergroup in terms of sedimentary characteristics, which led Ferreira (1986) to suggest correlation between the two sequences. The Umkondo Group, as described by Button (1977), consists of sediments characteristic of tidal flat-lacustrine, progradational fan delta, and meandering river environments. Palaeocurrent directions are variable, but indicate a depocentre to the east. Correlation of the two groups is uncertain on both chronological and lithostratigraphic grounds. The age of 1 080 +140/-25 Ma for the Umkondo Group (Allsopp et al. 1989), obtained from mafic intrusions and volcanic rocks similar in character to those of the Ritscherflya Supergroup which yield older ages as outlined above, is significant although equivocal. Cheney et al. (1990) pointed out that little evidence for co-magmatic petrogenesis of the lavas and sills was presented, and the five samples of lava in the data set suggest an age of 1.9 Ga if considered separately. In addition, the Umkondo Group has stromatolitic carbonate units at the base which have not been reported from the Ritscherflya Supergroup. The latter contains tuffaceous rocks and conglomerates with jaspilitic clasts, none of which are known in the Umkondo Group.

Subtle differences in the lithologies could be accounted for by invoking either separate, overlapping sedimentary basins (Jansen 1982), or lateral variation within continuous unconformity bounded sequences of considerable extent, of which the present outcrop areas are preservational remnants rather than indicators of the original dimensions of the depositional basin (Cheney et al. 1990). The proximity of all the preserved mid-Proterozoic successions to orogenic provinces may suggest that pervasive crustal buckling occurred prior to the onset of the ca. 1 000 Ma Kibaran orogeny, a widespread and long-lived collisional tectonic event.

Although specific correlation between the Grunehogna and African cratons is not possible, the similarity in crustal evolution of the terranes is significant. The geometry of the orogenic provinces in the reconstruction (discussed below) supports the continuity of the

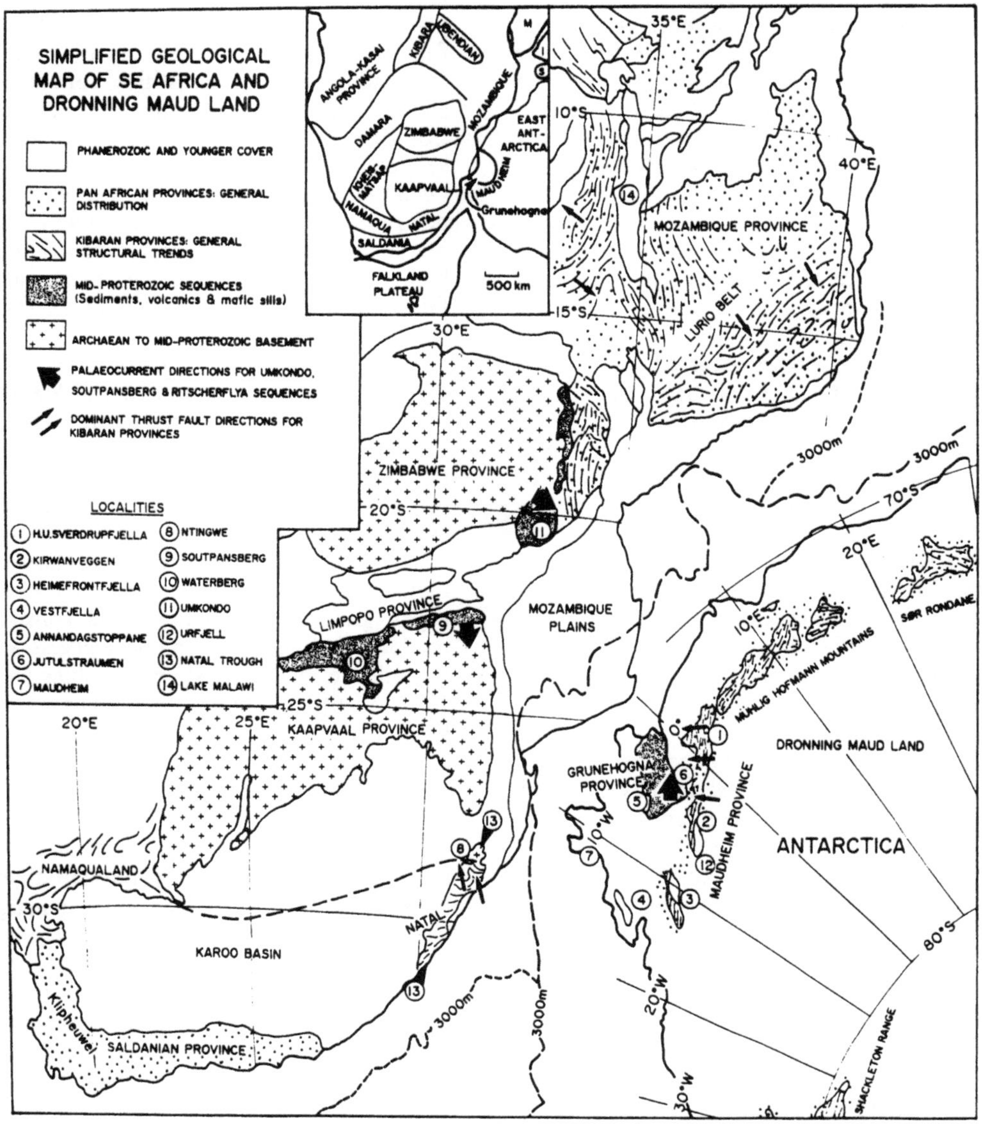

Figure 1. Simplified geological map of southeastern Africa and Dronning Maud Land, Antarctica, using the reconstruction of Martin & Hartnady (1986). The distribution of the tectonic provinces discussed in the text is shown in the inset on a segment of the reconstruction of Lawver & Scotese (1987). (Modified after Groenewald et al. 1991)

Grunehogna and Kaapvaal cratonic segments prior to the 1 000 Ma orogeny.

3. THE OROGENIC TERRANES

The continuity of the orogenic belts in Gondwana has been widely discussed and was fundamental to the earliest Gondwana reconstructions (see Craddock 1972; du Toit 1937). Linkage of those in southeastern Africa and Dronning Maud Land is supported by their general character and chronology, but no lithological successions or tectonic events have been correlated precisely as yet. The available data are synthesized here to assess this possibility.

3.1 *The 1000 Ma Mozambique, Maudheim and Natal Provinces*

A major orogenic cycle occurred about 1 000 million years ago and has been termed the Kibaran event (eg. Tankard et al. 1982; Clifford & Gass 1970). This is represented in Africa by the Kibaran (*sensu stricto*), Mozambique, Natal and Namaqua Provinces, by the Maudheim Province and widespread overprinting in East Antarctica, by the Musgrave and Frazer Provinces in Australia, and the Rondonides in South America. Correlation of the Mozambique, Maudheim and Natal Provinces has received some attention in recent publications (Grantham et al. 1988; Groenewald et al. 1991).

The Lurio Belt in the southeastern Mozambique Province (Fig. 1), described by Sacchi et al. (1984), comprises a migmatitic gneiss basement and a volcano- sedimentary cover sequence which underwent amphibolite facies metamorphism. Meta-quartzites and meta-carbonates are present, as are orthogneisses interpreted as being of calc-alkaline volcanic origin. The presence of interlayered mafic, ultramafic and sedimentary protoliths may indicate that parts of the Mozambique Province are ophiolitic in character. A nappe of granulite facies rocks, ranging from rhyolitic to ultramafic in composition, was thrust over the volcano-sedimentary succession. Several granites were intruded into this sequence.

Eastern Maudheim Province (Fig. 1) is underlain by the Sverdrupfjella Group (Roots 1953, 1969; Hjelle 1974; Wolmarans & Kent 1982; Grantham et al. 1988). In H.U. Sverdrupfjella, orthogneisses predominate adjacent to the Grunehogna cratonic province. They display layering on a variety of scales but are generally of rather monotonous, intermediate composition. These gneisses have calc-alkaline geochemical characteristics and are interpreted to be of volcanic origin. Farther east are paragneissic metacarbonates and pelites interlayered with grey orthogneisses. These are followed eastwards by predominantly quartzofeldspathic paragneisses which contain widespread, although subordinate, metapelites, iron-rich amphibolites, and quartzitic and semipelitic gneisses. Augen gneisses are the other major component in this area. This polarity of the lithostratigraphy, that is, lower grade metavolcanics close to the craton and higher grade mixed ortho- and para- gneisses farther from it, is considered important and will be discussed below.

Predominantly quartzofeldspathic gneisses are present in Kirwanveggen or central Maudheim Province. These have been interpreted as being of volcanic origin because low initial $^{87}Sr/^{86}Sr$ ratios (R_o) preclude any significant crustal prehistory (Wolmarans & Kent 1982). These authors also reported the sparse presence of calc-silicate and other metasedimentary lithologies as well as magnesium-rich rocks and augen gneisses. Heimefrontfjella consists predominantly of para- and orthogneisses with subordinate marbles and amphibolites (Juckes 1972; Fielitz & Spaeth 1991). The compositions described by these authors are very similar to those of Kirwanveggen, although shearing has been more intense in Heimefrontfjella.

The Natal Province consists of four zones (Matthews 1981). The orogenic belt abuts on the Kaapvaal Province in the northern marginal zone, which consists of a narrow southward dipping thrust belt of low grade metasediments. To the south of this is a wider, westward plunging nappe complex. The nappes consist of metamorphosed basic lavas, clastic and chemical meta-sediments which enclose several layered mafic-ultramafic intrusions, gneissic granites and diorites. The overthrusting was towards the north. Matthews (1972) considered this sequence to be ophiolitic in character, and proposed a subduction-obduction collision model. The northern zones are succeeded southwards by one in which large-scale synformal structures consisting of paragneissic sequences are separated by augen gneisses. Further south is an area in which granulite facies paragneisses, granites and charnockites predominate. Thomas (1989) divided Natal Province into a number of orogenic terranes on the basis of different metamorphic grades and subtle lithostratigraphic variations in the calc-alkaline granitoids and subordinate meta-sedimentary lithologies. He suggested that these terranes may represent repeated accretion on a relatively small scale, rather than a single collisional orogeny. Whichever model applies, the lithostratigraphy has a polarity in that metavolcanic rocks predominate adjacent to the craton, whereas metasediments predominate in the more distal region.

Syn-tectonic granites have been recognised in all of the 1 000 Ma provinces, and are of value in enabling radiometric dating to constrain tectonic events (see below). In the Maudheim Province, tabular pre- and syn- tectonic granites occur in H. U. Sverdrupfjella (Grantham et al. 1988), charnockitic granites in the Kirwanveggen (Wolmarans & Kent 1982), and relatively extensive, coarse-grained, gneissic granites in Heimefrontfjella (Juckes 1972). Similarly, in the Natal Province there are numerous granites and charnockites of pre-, syn- and late-tectonic age (Talbot & Grantham 1986; Thomas 1989). Domical and migmatitic granites about 1 000 Ma in age have been reported from the Mozambique Province (Sacchi et al. 1984; Costa et al. 1989).

The similar tectonic histories of the three orogenic provinces are compared in Table 1. Structural evolution in Lurio belt of the Mozambique Province, as documented by Sacchi et al. (1984), was dominated by thrust-nappe tectonics at ca. 1 000 Ma which produced strongly linear frontal ramps and imbricate stacks in areas of intense deformation. Elsewhere, recumbent folds are typical. The thrust system, which had a root zone in southern Malawi, involved tectonic transport towards the southeast. Outside of this system, northwest-trending folds predating the thrusting were superimposed on earlier structures. In the southeast, widespread gentle deformation with a west-northwest trend may represent a late stage of the thrusting event. This folding is more intense locally where it is accompanied by refoliation. Kibaran-age tectonics in Africa, as interpreted by Daly (1986a,b, 1988) and Shackleton (1986), involved considerable crustal shortening through northwest and southeast directed thrusting. This divergence was related to a postulated regional scale pop-up rooted in a mid-crustal shear or decoupling zone. The northwest verging structures are considered to represent the main thrust system whereas those to the southeast are thought to be backthrusts (Daly 1986b).

In Maudheim Province, the granulite facies paragneisses in H.U. Sverdrupfjella underwent two co-axial isoclinal folding events (Grantham et al. 1988), before being thrust northwest over the adjacent amphibolite facies succession (Groenewald & Hunter 1991). The thrust planes were subparallel to the general southeasterly to easterly dip in layering. Their local transgression across the tectonic foliation indicates that the thrusting postdated much of the folding. Subsequent deformation complicates the interpretation as open to tight folding has been superimposed on the earlier structures. Further west, in Heimefrontfjella, two generations of folding have been recognised. The earlier folds trend northwest with northeast vergence, and younger folds trend northeast and verge northwest (Spaeth & Fielitz 1987). An older granulite facies basement was recognized, but is separated from the amphibolite terrane by a major shear zone. Several flat-lying thrust faults are present in NE Heimefrontfjella.

The Natal orogenic province also underwent thrust and nappe dominated tectonism, as described by Matthews (1981). A zone of intense imbricate stacking along the boundary with Kaapvaal Province forms the thrust-front. Adjacent to this, southwards, is a sequence of large (10-35 km wide), northwards verging thrust nappes thought to comprise obducted ophiolites. Back-folding at the southern limit of these nappes has resulted in a steeply inclined thrust belt. The central part is characterised by upright synformal structures, whereas thrust faulting and isoclinal folding occurred in the south (Thomas 1989). Major shear zones transect the province.

High-grade metamorphism characterises all the 1 000 Ma-old provinces (Table 1). In southern Mozambique Province, amphibolite facies rocks predominate except in the overthrust segments which consist of granulite facies gneisses (Sacchi et al. 1984). Andreoli

Table 1. Summary and comparison of the geology of the Kibaran Provinces.

MOZAMBIQUE PROVINCE	MAUDHEIM PROVINCE	NATAL PROVINCE
LITHOSTRATIGRAPHY		
Gneiss-migmatite basement. Meta-sedimentary & meta-volcanic cover sequence. Minor carbonates, Fe-quartzites, greywackes. Dominantly calc-alkaline meta-volcanics. Overthrust granulite facies meta-sedimentary nappe in Lurio belt.	Migmatitic volcano-sedimentary succession. Basement - cover relations reported in Heimefront-fjella. Paragneiss from quartzite, greywacke, pelites and marbles. Orthogneiss: calc-alkaline volcanics. Abundant mafic dykes. Overthrust granulite facies paragneisses in the east.	Metavolcanic dominated northern zone. Ophiolitic affinity of amphibolites. Southwards - diverse metasediments (quartzites, calc-silicates, marbles, greywackes) Intermediate orthogneisses and various intrusive granitoids.
DEFORMATION		
Intense early folding and thrusting. Main thrusts to northwest in the west, to southeast in southern zone. Wide thrust belts and shear zones. Later more gentle folding, possibly accompanied by thrusting.	Two early, almost coaxial, isoclinal folding events accompanied by thrusting. Flattened foliations indicate M_1 predated F_1. Main thrusts & reverse faults towards northwest are later (Pan African ?) and associated with open to tight folding and major shear zones (?).	Intense early thrust and fold nappe development. Abundant evidence for thrusting towards north and northwest. Local backthrusts. Major shear zones. No significan post-Kibaran tectonism.
MAGMATISM		
Early deformed and relatively undeformed granites, latter with migmatitic aureoles. Late porphyritic granites are Pan-African in age.	Suites of megacrystic, tabular but highly deformed early granites (S-type and calcalkaline?). Charnockites in south. Later sheeted alkali and S-type granites are of Pan African age.	Abundant syn-kinematic granite & charnockite batholiths/sheets. S-type, calcalkaline and A-type bodies intruded at various stages of the tectonism.
METAMORPHISM		
Early regional mid-amphibolite facies in southern zone with klippen of granulites. In northwest widespread granulites, partly retrogressed. Peak: P=7-9kb, T=725-800°C.	Early amphibolite facies metamorphism in west, granulites above main thrusts to northeast and in the south. Near isothermal decompression. Late, widespread retrogression to epidote-amphibolite facies. Peak: P=9-11kb, T=800-850°C	Predominantly amphibolite facies in north, granulite facies in south. Local granulites throughout. No evidence for post-Kibaran retrogression/rehydration. Peak: P=5-7 kb, T=650-850°C

(1984) documented conditions of P=7-9 kb, T=725-800°C in southern Malawi. In Natal Province, amphibolite facies conditions predominated in the northern zones (Rhodes & Leith 1971), whereas granulite facies conditions applied in the southern zones (Talbot & Grantham 1987). In the Heimefrontfjella sector of Maudheim Province, Juckes (1972) found that the rocks had undergone almandine-amphibolite facies metamorphism, and Weber et al. (1987) identified a granulite facies basement underlying amphibolite facies cover. Wallace (in Wolmarans & Kent 1982) reported granulite and amphibolite facies assemblages in Kirwanveggen with an estimated P=6.4 kb and T=640°C for the peak of metamorphism. Groenewald & Hunter (1991) documented early granulite facies conditions of relatively high pressure (P=8-10 kb) and temperature (T=800°C), followed by amphibolite facies retrogression and rehydration (P=6 kb, T=600°C) in the main range of H.U. Sverdrupfjella. Lower grade conditions applied closer to the adjacent Grunehogna cratonic province, and amphibolite facies conditions were never exceeded in the western marginal area. Recent work in southern H. U. Sverdrupfjella has revealed that considerable tectonic inversion has resulted from thrust faulting, leading to the emplacement of the highest P/T rocks at the highest structural levels (Groenewald, Grantham, Krynauw and Strydom, unpublished data).

Available geochronological data from the various provinces reveal near contemporaneity of tectonothermal events (Table 2). In all the Kibaran-age provinces considered here, the marginal basin-volcanic arc successions accumulated before about 1 200 Ma, when metamorphism and deformation began. The orogeny was particularly prolonged, as revealed by the presence of syn- tectonic granites ranging from 1 163 to 850 Ma in age. Although these intrusions range from S-type to anorogenic alkaline in character, all show evidence of some metamorphic/deformational history. The ages may, therefore, represent metamorphic resetting rather than emplacement, especially as most were determined using the Rb-Sr isotopic system. Nonetheless, the initial $^{87}Sr/^{86}Sr$ ratios are particularly significant: of the isochrons reviewed in Table 2, many have R_o values between 0.703 and 0.705, indicative of a relatively short crustal residence time of the source rocks. In H.U. Sverdrupfjella, one of the S-type bodies (the Fugitive granite) probably originated during the peak of metamorphism and has yielded an age of 1163 Ma with a particularly low R_o (Groenewald & Moyes, unpublished data). Eglington et al. (1989) interpreted the isotopic data from Natal Province as indicating that the sequence was generated no earlier than 1 400 Ma and accreted onto the craton at about 1 000 Ma.

These common characteristics raise the question whether the three mobile belts constitute a single, extensive orogenic province. Thrust faulting in Natal Province and Dronning Maud Land involved tectonic transport towards the cratons, whereas that in the Mozambique Province was bi-directional. Thrusting in Mozambique and H.U. Sverdrupfjella emplaced granulite facies successions above amphibolites. Metamorphism shows the same general pattern of initial granulite and amphibolite grade, followed by amphibolite facies retrogression. Groenewald & Hunter (1991) presented a preliminary P-T-t path for H.U. Sverdrupfjella and argued that the metamorphism was related to plate collision. The lower P/T conditions in Natal may not be appropriate to collision orogeny, for which reason Tankard et al. (1982) suggested an ensialic rather than ensimatic marginal basin, although more than one accreted segment is recognised by Thomas (1989). Geochronological data suggest contemporaneity, although direct correlation of

Table 2. Geochronology of the Kibaran orogenic provinces. MSWD values in brackets, if known..

MOZAMBIQUE PROVINCE
Basement: Rb-Sr[1] 1200 Ma R_o=0.7060 (3 point isochron)
Rb-Sr[2] 1142±27 Ma R_o=0.7071 (0.04)
Cover: Rb-Sr[1] 950±40 Ma R_o=0.7091
Metamorphism: Rb-Sr[2] 963±23 Ma R_o=0.70896 (26.05)
Rb-Sr[2] 1003±35 Ma R_o=0.7097 (0.26)
Early granite:Rb-Sr[1] 1100 Ma R_o=0.7027
Rb-Sr[2] 1064±29 Ma R_o=0.7028 (0.19)
Rb-Sr[2] 1063±35 Ma R_o=0.7097 (0.26)
Rb-Sr[2] 1106±115 Ma R_o=0.7027 (1.94)
Amphibolites: Rb-Sr[3] 1391±88 Ma R_o=0.707 (0.3)
Granulites: Rb-Sr[3] 1073±45 Ma R_o=0.707 (0.59)

MAUDHEIM PROVINCE
Sverdrupfjella Group, Kirwanveggen
orthogneisses(?)[4]
zircon Pb-Pb zircon U-Pb:
1071±74 Ma (90) 1112±32 Ma (9.5)
1075±60 Ma (127) 1107±127 Ma (260)
1061±66 Ma (108) 1045±193 Ma (548)
Rb-Sr whole rock[4] 1015±24 Ma R_o=0.704 (1.04)
1164±78 Ma R_o=0.704 (11.2)
Rb-Sr whole rock-biotite ages[4] 460-485 Ma
Sverdrupfjella Group, H.U. Sverdrupfjella
Amphibolite facies orthogneisses
Rb-Sr whole rock[4] 1141±43 Ma R_o=0.708 (0.4)
1065±81 Ma R_o=0.707 (0.21)
Granulite facies paragneisses:
Rb-Sr whole rock[5] 1170±26 Ma R_o=0.7040 (9.76)
Early syn-orogenic granite:
Rb-Sr whole rock[5] 1163±99 Ma R_o=0.7036 (8.88)
Late syn-orogenic granitoids:
Rb-Sr whole rock[4] 519±17 Ma R_o=0.708 (1.7)
whole rock-biotite ages[5] 430-480 Ma
Heimefrontfjella[6]
Volcanism: 1100-1200 Ma
Metamorphism: 1000-1100 Ma

NATAL PROVINCE
All Rb-Sr whole rock[7]
Northern marginal area - metavolcanics
1240±23 Ma R_o=0.704 (2.2)
Northern marginal area - granites
1194±83 Ma R_o=0.7017 (0.8);
1067±20 Ma R_o=0.706 (1.72)
examples of granites farther south
calc-alkaline: 981±31 Ma R_o=0.7032 (11.3)
A-type: 1089±14 Ma R_o=0.7038 (1.73);
1003±29 Ma R_o=0.7054 (0.75)
transitional 1011±19 Ma R_o=0.7063 (2.36)
syn-orogenic metabasite 1024±32 Ma R_o=0.7026 (2.21)
tonalitic basement Nd T_{DM}=1405 Ma

Data source references:
1. Sacchi et al. 1984; 5. Moyes & Groenewald, (unpub.);
2. Costa et al. 1989; 6. Weber et al. 1987;
3. Cahen et al. 1984; 7. Eglington et al. 1989.
4. Moyes & Barton 1990;

lithostratigraphic units is not possible, probably as a consequence of differences in level of exposure. The lithologies in all three provinces are interpreted to be metamorphosed sequences of volcanic and sedimentary rocks typical of volcanic arcs and marginal basins. The ophiolites in the Natal and Mozambique Provinces are also evidence that accretion occurred through collision of crustal segments once separated by oceanic crust. The sense of polarity, in that metavolcanics predominate close to, and metasediments further away from, the cratons, is the same in each province and suggests the same style of accretion. These data are evidence that the Kibaran orogeny encompasses sub-provinces in east Africa, Antarctica and south Africa, was of considerable extent and represents major accretion at a relatively early stage in the construction of Gondwana.

3.2 *Pan-African and Ross orogenic events*

Tectonothermal activity of variable intensity affected much of Gondwana in the period 800-450 Ma, resulting in several fold belts. The Gariep, Damara and Saldanian orogenic provinces and pervasive thermal overprinting of larger tracts of Africa are ascribed to the Pan-African event, whereas the thermotectonic history of similar age in the Transantarctic mountains of Antarctica is termed the Ross Orogeny. In southeastern Africa, direct evidence for tectonothermal overprinting is restricted to the Mozambique orogenic province where Sacchi et al. (1984) recognised gentle refolding to be related to the emplacement of 500 Ma-old granites in the Lurio belt. Despite the limited refoliation and retrogression associated with this event, widespread resetting of isotopic systems occurred, particularly Rb-Sr and K-Ar, which resulted in this province being considered entirely of Pan-African-age in some early publications.

The Maudheim Province also underwent orogeny at this time. Many of the 500 Ma-old granitic intrusions in H.U. Sverdrupfjella were emplaced syntectonic to D3, which was characterized by open to tight folding and reverse or thrust faulting. The granites form subhorizontal to subvertical tabular bodies which are highly variable in thickness, and are associated with NW-directed thrust faulting. The Brattskarvet alkaline/peralkaline granitoids form the largest of these intrusions ($\approx$100 km^2 in area) and have yielded a well-constrained whole rock Rb-Sr age of 519$\pm$17 Ma (MSWD = 1.7, R_o = 0.708) (Moyes & Barton 1990). Other late granites, characterised by tourmaline or magnetite phenocrysts, are considered to be younger because they cut monzonitic dykes of the Brattskarvet suite. These have given a Rb-Sr whole rock - mineral age of 469$\pm$5 (MSWD=2.5, R_o=0.735) (Moyes & Barton 1990). Brattskarvet granitoids have a tectonic foliation in which biotite crystallized. This biotite, in conjunction with whole-rock data, has yielded Rb-Sr ages of 460-490 Ma in the Brattskarvet intrusion (Moyes & Groenewald, unpublished data), similar to biotite blocking temperature ages of 475 Ma from Kirwanveggen (Elworthy, in Wolmarans & Kent 1982). Most of the K-Ar age determinations by Russian workers in the 1960's ranged from 400-500 Ma throughout the region (Ravich & Solov'ev 1966), demonstrating that the thermal effect of this orogeny was pervasive isotopic resetting through retrogressive rehydration.

In Heimefrontfjella, Spaeth & Fielitz (1987) and Weber et al. (1987) recognized northeast verging folding with associated thrust faulting as being younger than the main 1000 Ma tectonism. This younger deformation was accompanied by extensive retrogression. Weber et al. (1987) interpreted these events to be of Pan-African-age, which is supported by the 450 Ma-age of a dyke displaced by thrust faulting (Juckes 1972). The great extent of this thermal event in Antarctica is revealed by the occurrence of gneisses of this age in the Shackleton Range, at the western limit of Dronning Maud Land, and in the Sør Rondane to the east (Fig. 1; Rex 1972; Picciotto et al. 1964), and the considerable extent of the Ross Orogeny defined by Craddock (1972).

There is no direct evidence of Pan-African tectonothermal effects in the Natal Province. Isotopic data reveal no ages younger than 850 Ma, nor is there reported evidence for deformation or retrogression after the main Kibaran orogeny. There is, however, an orogenic province of Pan-African-age in the southernmost part of Africa (Fig. 1), termed the Saldanian, in which thrust faulting, intense folding and granite emplacement occurred (see Tankard et al. 1982, for references). It would therefore seem likely that the Pan-African swathe of tectonism passed from the Mozambique Province through Maudheim Province, then trended southwest and did not extend westwards into Natal.

The recent reconstructions of Gondwana place the Falkland Plateau between the western limit of Dronning Maud Land and southeastern Africa. This is problematic in that the only reported Kibaran-age metamorphites there are located at Cape Meredith, the southern point of West Falkland (Rex & Tanner 1982). The Deep Sea Drilling Project intersected 535 Ma crust at the northeastern extremity of the submerged Falkland plateau (Beckinsdale et al. 1977). If the Falkland Islands and Plateau are placed in the position suggested by current reconstructions, then these two points have reversed positions in terms of age-relations: the 1 000 Ma Cape Meredith being juxtaposed to the Saldanian Province, and the 535 Ma plateau being close to the 1 000 Ma Natal Province. In view of the probable continuities outlined above, there is support for the proposal of Mitchell et al. (1986) that the Falkland Islands lie on a microplate which rotated through about 60° during the breakup of Gondwana.

4. PHANEROZOIC-PALAEOZOIC DEPOSITION AND TECTONISM

Several intracratonic depositional basins developed on Gondwana after the Ross/Pan African orogeny. Although the major basins such as the Cape and Karoo of southern Africa have preserved evidence for considerable syn-depositional tectonism, these are beyond the scope of the present review. Three lesser sequences are, however, of particular interest in the present context: the Klipheuwel Formation of southwest Cape, the Natal Group of Natal and the Urfjell Group of Kirwanveggen.

Sedimentation which occurred during the waning stages of the Saldanian orogeny in southwest Cape is preserved as the Klipheuwel Formation, a sequence of intertonguing mudstones, arkoses and conglomerates (Tankard et al. 1982). This succession has been interpreted to be the product of alluvial fans in a block-faulted terrain.

Early Phanerozoic sedimentary rocks in the Natal Group are similar to those of the Klipheuwel Formation. In northern Natal, the conglomerates, arkoses and siltstones are the products of an alluvial fan system (Hobday & von Brunn 1979). The fan was of considerable size and extent as indicated by the pronounced predominance of durable lithologies as very well rounded boulders, and the thickness of the succession which is up to 1 000m. Palaeocurrent directions are generally to the south, south-southeast and south-southwest which suggests that part of the provenance was to the east of the present coastline. The distribution of the Group allows recognition of an elongate basin parallel to the modern coastline of Natal (Fig. 1). This subsidence close to and aligned with the subsequent line of detachment in the breakup of Gondwana was interpreted by Hobday & von Brunn (1979) as evidence for an "early Paleozoic failed rift". Preliminary results of radiometric dating of the micas in the Natal Group suggest an age of 560 Ma for detrital grains and 490-260 Ma for the authigenic material (C.G.A. Marshall, R.J. Thomas & M.K. Watkeys, personal communication, April 1991). The total absence of source rocks for the detrital mica of Pan-African age (ca. 560 Ma) in the vicinity of the present exposures of the Group either raises doubts as to the validity of the age measured, or requires the presence of an appropriate terrane adjacent to the current NE margin of Natal.

The Urfjell Group in the Kirwanveggen area is a 1 650 m thick succession of arenaceous sedimentary rocks (Aucamp et al. 1972). This Group postdates the Maudheim orogeny since it rests uncomformably upon the deformed Sverdrupfjella Group. It displays evidence of gentle deformation, thus predating the undeformed Permian sediments in the same area. The Urfjell lithostratigraphy is similar to parts of the Natal Group in that it consists predominantly of quartz arenites, arkoses, conglomerates and siltstone. Very consistent southwest to south-southwest palaeocurrent directions (shown in Fig. 4 of Aucamp et al. 1972, but described as northeast to north-northeast in their text) and the poorly sorted nature of the conglomerates suggests a similar palaeogeographic setting to that of the Natal Group. Further work is needed to clarify this interpretation. The three early Phanerozoic sedimentary successions may be evidence of widespread and pronounced block-faulting, in view of their predominantly alluvial nature, contemporaneity and thickness. It is suggested that the final stages of the Pan-African orogeny were subtly extensional while the generally high associated heat-flow waned.

5. CONCLUSIONS

The similarity of the stratigraphy and orogenic evolution of the two continents supports the close reassembly proposed by Martin & Hartnady (1986), and has implications for the construction of Gondwana. The Mozambique Plains (Fig. 1) postdate break-up, and exclusion of these from the current coastal margin of southeastern Africa allows close juxtapositioning of western Dronning Maud Land and Africa, bringing the Grunehogna and Kaapvaal Provinces together. It has been argued that Archaean granitoid basement in Grunehogna Province is very different from the granites of similar age on the eastern margin of the Kaapvaal Province and had a different provenance. These features have been used to propose that Grunehogna Province was at one time a discrete microplate (Barton et al. 1987). The juxtapositioning of the Grunehogna and Kaapvaal segments in the reconstruction is bound by the probable continuity of the 1 000 Ma-old orogenic provinces discussed above, hence the proposed recognition of a microplate on the basis of geochemical differences derived from ten whole rock analyses from a limited outcrop area is difficult to accept.

Much of the Gondwana supercontinent, in particular the part adjacent to southeast Africa, was almost certainly complete 1 000 Ma ago, given the substantial similarities between the Kibaran-age provinces. Furthermore, the Mühlig Hofmann mountains, extending eastwards from the H. U. Sverdrupfjella to the Sør Rondane area (Fig. 1), have very similar lithostratigraphy, metamorphic history and age (Ravich & Solov'ev 1966; Asami et al. 1989; Shiraishi & Kagami 1989) to the Maudheim Province, which suggests an even greater extent for this orogeny. Tectonic events at ca. 1 000 Ma have also been recognised in Madagascar (Cahen et al. 1984). Should this correlation prove to be correct, the consequence would be the definition of the most extensive orogenic province on Earth. Considerably more work is obviously needed to assess this possibility.

The Pan-African event, in the areas discussed here, involved predominantly ensialic orogeny which ranged from extensional to compressive tectonism with considerable variation in heat-flow. This, and subsequent tectonism, involved jostling of crustal segments prior to and during the break-up of Gondwana as different crustal blocks were subjected to the vast stress field associated with drifting and the suppression of thermal activity in the mantle.

ACKNOWLEDGEMENTS

The Antarctic data reported here were accumulated during work on the South African National Antarctic Research Program, supported by the Department of Environment Affairs. Many discussions with D.R. Hunter, R. Unrug, G.H. Grantham, B. Corner, V. von Brunn, and M.K. Watkeys were enlightening. C.G.A. Marshall, A.B. Moyes and R.J. Thomas are thanked for their unpublished data. The travel funding provided by IGCP 288 and the Institute of Antarctic and Southern Ocean Studies, University of Tasmania, is acknowledged gratefully. P.F. Hoffman and an anonymous reviewer read the manuscript and their constructive comments are greatly appreciated.

REFERENCES

Allsopp, H.L., Kramers, J.D., Jones, D.L. & Erlank, A.J. 1989. The age of the Umkondo Group, eastern Zimbabwe, and implications for palaeomagnetic correlations. *South African Journal of Geology* 92: 11-19.

Andreoli, M.A.G. 1984. Petrochemistry, tectonic evolution and metasomatic mineralizations of Mozambique belt granulites from S. Malawi and Tete, (Mozambique). *Precambrian Research* 25: 161-186.

Asami, M., Makimoto, H. & Grew, E.S. 1989. Geology of the eastern Sør Rondane Mountains, East Antarctica. *Proceedings of Symposium on Antarctic Geoscience* 3:81--99. Tokyo: National Institute of Polar Research.

Aucamp, A.P.H., Wolmarans, L.G. & Neethling, D.C. 1972. The Urfjell Group, a deformed (?) early Palaeozoic sedimentary sequence, Kirwanveggen, Western Dronning Maud Land. *In* Adie, R.J. (ed.): *Antarctic Geology and Geophysics*: 557-562. Oslo:

Univertetsforlaget.
Barton, J.M., Klemd, R., Allsop, H.L., Auret, S.H. & Copperthwaite, Y.E. 1987. The geology and geochronology of the Annandagstoppane granite, western Dronning Maud Land, Antarctica. *Contributions to Mineralogy and Petrology* 97:488-496.
Beckinsdale, R.D., Tarney, J., Darbyshire, D.P.F. Hunn, M.J. 1977. Rb-Sr and K-Ar age determinations on samples of the Falkland Plateau basement at site 330. *Initial reports of the Deep Sea Drilling Project* 36:923-927. Washington: U.S. Government Printing Office.
Button, A. 1977. Stratigraphic history of the Middle Proterozoic Umkondo Basin in the Chipinga area, southeastern Rhodesia. University of Witwatersrand Economic Geology Research Unit, *Information Circular* 108.
Cahen, L., Snelling, N.J., Delhal, J. & Vail, J.R. 1984. *The geochronology and evolution of Africa.* Oxford: Clarendon.
Cheney, E.S., Barton, J.M., Jr. & Brandl, G. 1990. Extent and age of the Soutpansberg sequences of southern Africa. *South African Journal of Geology* 93:664-675.
Clifford, T.N. & Gass, I.G.(eds.) 1970. *African Magmatism and Tectonics.* London: Oliver & Boyd.
Costa, M., Ferrara, G., Tonarini, S. & Sacchi, R. 1989. Rb-Sr datings from the upper Proterozoic basement of Zambesia (Mozambique): a re-appraisal. *28 th International Geological Congress, Washington, D.C., Abstracts:* 1:331-332.
Craddock, C. 1972. Antarctic tectonics. In Adie, R. J. (ed.) *Antarctic Geology and Geophysics*: 449-455. Oslo: Univertetsforlaget.
Daly, M.C. 1986a. The intracratonic Irumide Belt of Zambia and its bearing on collision orogeny during the Proterozoic of Africa. *In* Coward, M.P. & Ries, A.C. (eds.) *Collision Tectonics.* Geological Society, London, Special Publication 19: 321-328.
Daly, M.C. 1986b. Crustal shear zones and thrust belts: their geometry and continuity in central Africa. *Philosophical Transactions, Royal Society of London, Series A* 317: 111-128.
Daly, M.C. 1988. Crustal shear zones in Central Africa: a Kinematic Approach to Proterozoic Tectonics. *Episodes* 11: 5-11.
du Toit, A.L. 1937. *Our Wandering Continents.* Edinburgh: Oliver & Boyd.
Eglington, B.M., Harmer, R.E. & Kerr, A. 1989: Isotope and geochemical constraints on Proterozoic crustal evolution in south-eastern Africa. *Precambrian Research* 45: 159-174.
Fielitz, W. & Spaeth, G. 1991. A structural study of Precambrian rocks, western Neuschwabenland, with special reference to the basic dykes. *In* Thomson, M.R.A., Crame, J.A. & Thomson, J.W. (eds.) *Geological Evolution of Antarctica:* 67-72. Cambridge: Cambridge University Press.
Ferreira, E.P. 1986. *'n Sedimentologies - stratigrafiese ondersoek van die sedimentêre gesteentes in die Ahlmannryggen, Antarctica.* Thesis, University of Stellenbosch, South Africa.
Grantham, G.H., Groenewald, P.B. & Hunter, D.R. 1988. Geology of the northern H. U. Sverdrupfjella, western Dronning Maud Land and implications for Gondwana reconstructions. *South African Journal of Antarctic Research* 18: 2-10.
Groenewald, P.B. & Hunter, D.R. 1991. Granulites of northern Sverdrupfjella, western Dronning Maud Land: metamorphic history from garnet-pyroxene assemblages, coronas and rehydration reactions. *In* Thomson, M.R.A., Crame, J.A. & Thomson, J.W. (eds.) *Geological Evolution of Antarctica*: 61-66. Cambridge: Cambridge University Press.
Groenewald, P.B., Grantham, G.H. & Watkeys, M.K. 1991. Geological evidence for a Proterozoic to Mesozoic link between Dronning Maud Land, Antarctica, and southeastern Africa. *Journal of the Geological Society of London* 148:1115-1123.
Hjelle, A. 1974. Some observations on the geology of H.U. Sverdrupfjella, Dronning Maud Land. *Norsk Polarinstitutt Arbot* 1972:7-22.
Hobday, D.K. & von Brunn, V. 1979. Fluvial sedimentation and paleo- geography of an early Paleozoic failed rift, southeastern margin of Africa. *Palaeogeography, Palaeoclimatology, Palaeoecology* 28:169--184.
Jansen, H. 1982. The geology of the Waterberg Basin in Transvaal, Republic of South Africa. *Memoirs, Geological Survey of South Africa* 71.
Juckes, L.M. 1972. The geology of northeast-

ern Heimefrontfjella, Dronning Maud Land. *British Antarctic Scientific Reports:* 65. London: British Antarctic Survey.

Krynauw, J.R., Watters, B.R., Hunter, D.R. & Wilson, A.H. 1991. A review of the field relations, petrology and geochemistry of the Borgmassivet intrusions in the Grunehogna province, western Dronning Maud Land, Antarctica. *In* Thomson, M.R.A., Crame, J.A. & Thomson, J.W. (eds.) *Geological Evolution of Antarctica*: 33-39 Cambridge: Cambridge University Press.

Lawver, L.A. & Scotese, C.R. 1987. A revised reconstruction of Gondwanaland. *In* Mckenzie, G.D. (ed.) *Gondwana Six: Structure, Tectonics and Geophysics.* American Geophysical Union Monograph 40: 17-33.

Martin, A.K. & Hartnady, C. 1986. Plate tectonic development of the southwest Indian Ocean: a revised reconstruction of East Antarctica and Africa. *Journal of Geophysical Research* 91: 4767-4785.

Matthews, P.E. 1972. Possible Precambrian obduction and plate tectonics in southeastern Africa. *Nature* 240:37-39.

Matthews, P.E. 1981. Eastern or Natal sector of the Namaqua-Natal mobile belt in Southern Africa. *In* Hunter, D.R. (ed.) *Precambrian of the Southern Hemisphere*: 705-714 Amsterdam: Elsevier.

Mitchell, C., Taylor, G.K., Cox, K.G. & Shaw, J. 1986. Are the Falkland Islands a rotated microplate? *Nature* 319:131-134.

Moyes, A.B. & Barton, J.M.Jr. 1990. A review of isotopic data from western Dronning Maud Land, Antarctica. *Zeitblatt Geologie und Palaontologie* 1:19-31.

Picciotto, E., Deutsch, S. & Pasteels, P. 1964. Isotopic ages from the Sør Rondane Mountains, Dronning Maud Land. *In* Adie, R.J. (ed.) *Antarctic Geology*: 570-578. Amsterdam: North Holland.

Ravich, M.G. & Solov'ev, D.S. 1966. *Geology and petrology of the mountains of central Queen Maud Land.* Jerusalem: Israel Programme for Scientific Translations.

Rex, D.C. 1972. K-Ar determinations on volcanic and associated rocks from the Antarctic Peninsula and Dronning Maud Land. *In* Adie, R.J. (ed.) *Antarctic Geology and Geophysics*: 133-136. Oslo: Univertetsforlaget.

Rex, D.C. & Tanner, P.W.G. 1982. Precambrian age for gneisses at Cape Meredith in the Falkland Islands. *In* Craddock, C. (ed.) *Antarctic Geoscience*: 107-108. Madison: University of Madison Press.

Rhodes, R.C. & Leith, M.J. 1971. Granitic rocks in the basement complex around Mapumulo, Natal, in the light of zircon studies. *Transactions of the Geological Society of South Africa* 74:57-68.

Roots, E.F. 1953. Preliminary note on the geology of western Dronning Maud Land. *Norsk Geologisk Tidsskrift* 32: 17-33.

Roots, E.F. 1969. *Geology of western Queen Maud Land.* Explanation of Plate VI, Folio 12, Antarctic map series, American Geographical Association.

Sacchi, R., Marques, J., Costa, M. & Casati, C. 1984. Kibaran events in the southernmost Mozambique belt. *Precambrian Research* 25:141-159.

Shackleton, R.M. 1986. Precambrian collision tectonics in Africa. *In* Coward, M.P. & Ries, A.C. (eds.) *Collision Tectonics.* Geological Society Special Publication 19: 329-349.

Shiraishi, K. & Kagami, H. 1989. Preliminary geochronological study of granulites from the Sør Rondane mountains, East Antarctica- A comparison of Rb-Sr and Sm-Nd ages. *Proceedings of Symposium on Antarctic Geoscience* 3:152. Tokyo: National Institute of Polar Research.

Spaeth, G. & Fielitz, W. 1987. Structural investigations in the Precambrian of western Neuschwabenland, Antarctica. *Polarforschung* 57:71-92.

Talbot, C. J. & Grantham, G. H. 1987. The Proterozoic intrusion and deformation of deep crustal 'sills' along the south coast of Natal. *South African Journal of Geology* 90:520-538.

Tankard, A.J., Jackson, M.P.A., Eriksson, K.A., Hobday, D.K., Hunter, D.R. & Minter, W.E.L. 1982. *Crustal Evolution of Southern Africa- 3.8 billion years of earth history.* New York: Springer Verlag.

Thomas, R.J. 1989. A tale of two tectonic terranes. *South African Journal of Geology* 92:306-321.

Watters, B.R., Krynauw, J.R. & Hunter, D.R. 1991. Volcanic rocks of the Proterozoic Jutulstraumen Group in western Dronning Maud Land, Antarctica. *In* Thomson, M.R. A., Crame, J.A. & Thomson, J.W. (eds.) *Geological Evolution of Antarctica*: 41-46.

Cambridge: Cambridge University Press.
Weber, K., Arndt, N.T., Behr, H.J., Emmerman, R., Schenk, V. & Tapfer,M., 1987. Age and geodynamic setting of the vulcanite series in Western Neuschwabenland. *5th International Symposium on Antarctic Earth Science, Cambridge, Abstracts:* 168.
Wolmarans, L.G. & Kent, L.E. 1982. Geological investigations in Western Dronning Maud Land, Antarctica- a synthesis. *South African Journal of Antarctic Research, Supplement 2.*
Yoshida, M. & Kizaki, K. 1983. Tectonic situation of Lutzow-Holm Bay in East Antarctica and its significance in Gondwanaland; *In* Oliver, R.L., James, P.R. & Jago, J.B. (eds.) *Antarctic Earth Science*: 36-39. Cambridge: Cambridge University Press.

Gondwana Eight, Findlay, Unrug, Banks & Veevers (eds) © 1993 Balkema, Rotterdam. ISBN 90 5410 304 3

A preliminary model for the 500 Ma event in the East Antarctic Shield

K. Stüwe & M. Sandiford
Department of Geology, Adelaide University, S.A., Australia

ABSTRACT: Most rocks from the East Antarctic Shield preserve structures, mineralogies and mineral compositions that record granulite facies metamorphic events during Proterozoic and Archaean times. However, there is independent evidence that the shield experienced a major tectono-thermal event during the Early Cambrian. This evidence includes the pervasive resetting of isotopic signatures of the Precambrian gneisses, of Cambrian ages for many undeformed granites, of greenschist metamorphic assemblages and shear zones of this age and Phanerozoic fission track ages. The event is also indirectly indicated by the nature of the Proterozoic events because most tectonic models derived for these events are insufficient to explain complete exhumation of these rocks. The combined evidence suggests that a tectono-thermal event during the Cambrian may have exhumed the East Antarctic Shield and adjacent parts of Gondwana from as deep as 10-15 kilometres. Nevertheless, there is no obvious pervasive deformation associated with this event. Substantial basaltic underplating during a period of minor lithospheric thinning may have been responsible for this exhumation. The model is consistent with the evidence and also with the occurrence of major convergent orogenies in adjacent parts of Gondwana. A simple one dimensional thermal model is presented that explores the thermal and isostatic consequences of the proposed model.

INTRODUCTION

Despite the availability of a large body of geochronological evidence for a major tectono-thermal event in the East Antarctic Shield during the Early Cambrian, little has been done to establish a model for this event and most previous studies have concentrated on interpretations of the Precambrian history. Nevertheless, many of the studies of the Precambrian geology record or infer a tectono-thermal event which occurred after Precambrian granulite facies metamorphism (see amongst others: Black et al. 1987; Harley 1987; Sheraton et al. 1984; Stüwe and Powell 1989a,b) This event is variably thought to have caused exhumation of the shield from as much as 15km depth, pegmatite and late granite intrusion, greenschist facies metamorphism (Stüwe and Oliver 1989) and shear zone formation (Sandiford 1985). The timing of this event is generally equated with the Early Cambrian isotopic ages repeatedly recorded on the East Antarctic Shield, i.e. the region between 40°E and 160°E (Table 1). Absolute ages for this event cover a range of dates between 500 Ma and 600 Ma (Table 1) but in the light of the uncertainties of especially the earlier published ages we will henceforth refer to this event as the 500 Ma event. The recorded evidence is extremely important to the Phanerozoic history of the shield as granitoid intrusion and major uplift imply substantial heating of the lower crust as well as crustal thickening; interpretations that are mostly inconsistent with the absence of pervasive deformation at this time. In this paper we present a model that can explain these features and is consistent with the geological evidence. A simple one-dimensional thermal model is presented in order to explore the thermal consequences of the model.

REVIEW OF THE EVIDENCE

Together with Southern Africa the East Antarctic Shield forms the central part of ancient Gondwana and is surrounded by a number of orogenic provinces that formed as the consequence of Cambrian collisional tectonics. These include the Ross Orogeny in the Transantarctic Mountains, the Delamerian Orogeny of South Australia and the Saldanian Orogeny underlying the Cape Fold Belt of southern Africa (Tankard et al. 1982; Kröner 1980). During this time, the Precambrian granulites of the East Antarctic Shield remained apparently unaffected by pervasive ductile deformation and substantial mountain-forming processes.

Within this enormous Precambrian granulite terrain, Grew (1982) distinguished three Proterozoic belts of metamorphism, plutonism and tectonism of different age. These are a 1500-1700 Ma belt in eastern Wilkes Land, a 1100-1400 Ma belt in the Bunger Hills and Windmill Islands and a 900-1100 Ma belt west of 80°W encompassing the Prydz Bay area and the Rayner Complex of Enderby Land (Grew 1982; Ellis 1983). All three belts are considered to at least partially rework older gneiss terranes because they contain small areas of Archaean high-grade metamorphic sequences that remain unaffected by the Proterozoic events. The largest of these Archaean areas are the Napier Complex of Enderby Land, the southern Prince Charles Mountains of MacRobertson Land and the Vestfold Hills in the Prydz Bay region. Rocks formed after the Proterozoic high-grade events are late Proterozoic sediments, as well as minor Permian sediments and Tertiary volcanics which are of no relevance to this paper. Despite the extreme lack of outcrop there is abundant evidence for a thermal event in Early

Table 1: Summary of some historically important radiometric dates showing the 500 Ma event in eastern Antarctica. Newer references are cited in the text. References as follows: (ARR75): Arriens 1975; (GRE78: Grew 1978; (GREW82): Grew et al. 1982; (PIE76): Pieters et al. 1976; (STA60,61): Starik et al. 1960, 1961; (PIC63): Piciotto and Coppez 1963; (RAV65) Ravich et al. 1965; (SHE83) Sheraton 1983; (TIN82): Tingey 1982; (KAM76): Kamenev 1972; (OLI83): Oliver et al. 1983. Additional summaries of absolute ages from the shield can be found by Stuiver and Braziunas (1985) and Grew (1982).

Enderby Land			
460±250Ma	67°43S 45°30E	Freeth Bay	(GRE78)
512±155Ma	67°40S 45°51E	Molodeznaya	(GRE78)
503Ma,445Ma	67°40S 45°53E	Thala Hills	(GRE78)
465Ma,530Ma	67°39S 45°58E	Thala Hills	(KAM72)
511±10Ma	67°30S 53°00E	Enderby Land	(PIE76)
MacRobertson Land			
535Ma	67°36S 62°53E	Mawson Station	(STA60,PIC63)
516Ma	67°36S 62°53E	Mawson Station	(STA60,PIC63)
504±20Ma	70°47S 67°53E	Fox Ridge	(TIN82,SHE83)
Ingrid Christensen and Wilhelm II Coast			
500Ma	79°00S 73°00E	Landing Bluff	(ARR75,)
504±17Ma	79°00S 73°00E	Sandefjord Bay	(SHE83A)
540Ma	69°24S 76°13E	Larsemann Hills	(RAV65)
500Ma	68°33S 78°15E	Vestfold Hills	(ARR75)
Queen Mary Coast and Wilkes Land			
520Ma	66°31S 93°00E	Haswell Isl.	(RAV64,RAV65)
Georg V Coast			
428±103Ma	67°00S 142°40E	Cape Denison	(OLI83)
487Ma	67°43S 146°40E	Cape Bage	(STA60,PIC63)

Cambrian times in large extents of all three Proterozoic mobile belts (Grew 1982). A summary of some of this evidence in geographical order from east to west along the coast line of eastern Antarctica (Fig.1) is given below.

At one of the easternmost exposures of the East Antarctic Shield, at *Cape Hunter* and *Cape Denison*, Stüwe and Oliver (1989) recognised a late greenschist facies metamorphic event which affected sediments deposited after Precambrian granulite facies metamorphism. They correlated the timing of this event with a 500 Ma disturbance of the isotopic signatures of zircons at *Cape Denison*, recognised by Oliver et al. (1983). Through structural evidence this greenschist facies metamorphism can be correlated with higher grade overprints in rocks of *King George V Land* further east. The inferred change of grade is consistent with the absence of such an event in *Adelie Land* further west. The eastwards increase in grade was interpreted as the consequence of differential uplift along north-striking shear zones exposing successively deeper levels to the east and closer to the Ross Orogeny.

In the *Bunger Hills*, no detailed radiometric work has been conducted which would give evidence for the Phanerozoic overprinting of the Precambrian granulites. However, Stüwe and Powell (1989b) recognized a pressure-temperature path for the Proterozoic granulite facies metamorphic event that involves compression of the terrane during the cooling period. Therefore, a later event must have exhumed these rocks. Moreover, Sheraton and Oliver (1991) have recognised possibly Cambrian overprinting of many granulite facies gneisses by greenschist metamorphic assemblages.

Most evidence for a 500Ma event in eastern Antarctica comes from the *Prydz Bay region* and the *Rayner Complex* further west (Fig.1). Here, isotopic evidence for an Early Cambrian event has been recorded by Sheraton et al. (1984) Collerson and Sheraton (1986), Black et al. (1987), Grew and Manton (1981) and a number of earlier Soviet and Australian authors (Table 1). The isotopic evidence is supported by the observation of greenschist facies metamorphic overprints in the Rauer Group (Harley 1987, 1988), by the shape of the Proterozoic pressure-temperature path in the Larsemann Hills area (Stüwe and Powell 1989a; Stüwe et al. 1989) and by the existence of undeformed granites at Landing Bluff (Arriens 1975). Alkaline dykes of approximately 500 Ma age intruded most of the Prydz Bay area, the Prince Charles Mountains and large parts of MacRobertson Land and Enderby Land (Sheraton and Black 1988). For the MacRobertson Land section of the Rayner Complex Black et al. (1987) suggest greenschist metamorphism during the Cambrian and Clarke et al. (1988) suggest exhumation after Proterozoic tectonism in order to explain the Proterozoic pressure-temperature path. In Enderby Land, at the western margin of the shields sector discussed here, Sandiford (1985) suggested exhumation of the rocks from as much as 20 km during an early Palaeozoic reactivation event and recognised amphibolite facies assemblages in reactivated shear zones that are considered to have formed around 500Ma.

In summary it can be said that practically all pressure-temperature paths derived for the Precambrian granulites of the East Antarctic Shield show evidence for termination of the Precambrian tectonic history at mid-crustal levels and that there is widespread radiometric evidence for magmatism and metamorphism in Early Cambrian times. Evidence from fission-track ages supports the interpretation of denudation of the shield in Cambrian times with cooling through the 400°C isotherm in Cambrian times and final cooling and exhumation no earlier than during the breakup of Gondwana in the Mesozoic (Arne et al. this volume). This, together with the evidence for pervasive resetting of K/Ar isotopic signatures (Table 1), the greenschist metamorphic overprinting and possibly heating to 600°C (Sandiford 1985) indicates that a major tectono-thermal event must have affected the shield during Early Cambrian times. Considering the present normal thickness of the crust underneath the East Antarctic Shield this tectono-thermal event must have been associated with crustal overthickening. Crustal overthickening and a temperature high enough to induce melting of the lower crust are two essential features that must be accounted for by a model that attempts to interpret the nature of this event.

In order to explain the nature of the 500 Ma event Sandiford and Wilson (1984) and Sandiford (1985) used the presence of beryl and/or tourmaline in many of the pegmatites that intruded the shield at this time to suggest that the magmas were derived from an undepleted continental source and that exhumation during Cambrian times may have occurred as the consequence of low angle thrusting. However, in the light of the steep geothermal gradients inferred for the event (all gradients between the suggested 300° - 600°C at 10-15 km depth are steep) it is unlikely that crustal and conductive processes alone could be responsible for the event. In contrast, Sheraton et al. (1984) have shown that most of the 500 Ma granites in the Prydz Bay region are probably derived from a U- depleted granulite facies source and may therefore be derived from previously metamorphosed rocks from within the shield, but they did not discuss reasons for crustal melting. A new model that is consistent with the data presented above and can account for high geothermal gradients in parts of the East Antarctic Shield during the 500 Ma event is presented below.

DISCUSSION OF THE MODEL

Any model that explains the 500 Ma event in the East Antarctic Shield must account for: (i) an increase of the buoyancy of the shield initiating at least 10 kilometres of erosion in the absence of pervasive ductile deformation; (ii) pervasive intrusion of granitoid magmas throughout the shield; (iii) heating of the rocks to upper greenschist and lower amphibolite facies conditions; and (iv) the large aerial extent over which these features are recognised. The large areas involved present a special problem which is difficult to reconcile with plate tectonic processes and this has been previously recognised by Perkins and Newton (1981). A different process that may account for these features is massive underplating of basaltic, asthenosphere-derived melts at the bottom of the cratonic crust. The importance of such a process in producing epirogenic uplift has been discussed by McKenzie (1984). White and McKenzie (1989) have shown that substantial amounts of such melts can be produced by partial melting of the asthenosphere if it is only slightly hotter than normal and if only minor thinning of the overlying continental lithosphere occurs. Such melts may underplate continental crust or extrude on the surface in the form of flood basalts as we see them in the Karoo of southern Africa or the Deccan Traps of India. In this section we will assume that massive basaltic underplating is the cause of the 500 Ma event in the East Antarctic Shield and investigate the necessary geometry and magnitude of such an event which could produce the observed isostatic and thermal effects. We will then discuss whether or not the estimated magnitude of the event is plausible in the light of the thermal effects in the East Antarctic Shield as summarised above.

We begin by estimating the melt thickness necessary to produce the total amount of exhumation inferred to have taken place during and subsequently to the 500 Ma event. The total amount of erosion, E, necessary to restore isostatic equilibrium after emplacing a basaltic underplate of the density ρ_u into lithosphere of undisturbed thickness is:

$$E = z_u \left(\frac{\rho_m - \rho_u}{\rho_m - \rho_c}\right)$$

if no thinning of the mantle lithosphere is considered and ρ_m and ρ_c are the densities of the asthenospheric mantle and the lithosphere respectively and zu is the thickness of the underplate. With $\rho_m = 3000$ kg m^{-3} $\rho_u = 2900$ kg m^{-3} and $\rho_c = 2800$ kg m^{-3} for the average densities this amounts to E = 0.75 z_u or an underplate thickness of 12.5 kilometres for an ultimate denudation of 10 kilometres. This estimate appears large and should be considered as an upper limit if not an overestimate, because it is known that final exhumation of the shield from unknown depths only occurred in association with the breakup of Gondwana in Mesozoic times (Arne et al. this volume).

On the other hand, it is likely that such massive intrusion into the continental lithosphere is associated with lithospheric deformation (McKenzie 1984) However, there is to date little published work from the East Antarctic Shield that discusses deformation associated with the 500 Ma event. Nevertheless, we will attempt to discuss briefly the type of deformation that may be expected to be associated with the emplacement of massive amounts of basaltic melts as suggested above. McKenzie (1984) pointed out that, unlike the suggestion of previous authors, there is no necessity for a direct correlation between the amount of extension of the lithosphere and the amount of melt produced, if the system is not invariant to temperature changes. McKenzie and Bickle (1988) however, considered such temperature changes and calculated the volumes of basaltic melt which would be produced in a 'slightly hotter than normal' asthenosphere during extension of the lithosphere. Using data of these authors it can be shown that, in order to produce a thickness of 10 - 15 km of basaltic magma, it is necessary to rise the 1500°C isotherm to a depth of 70 - 80 kilometres, or the 1480°C isotherm to 35-50 kilometres depth. Whereas the rising of the 1480°C isotherm to levels of the

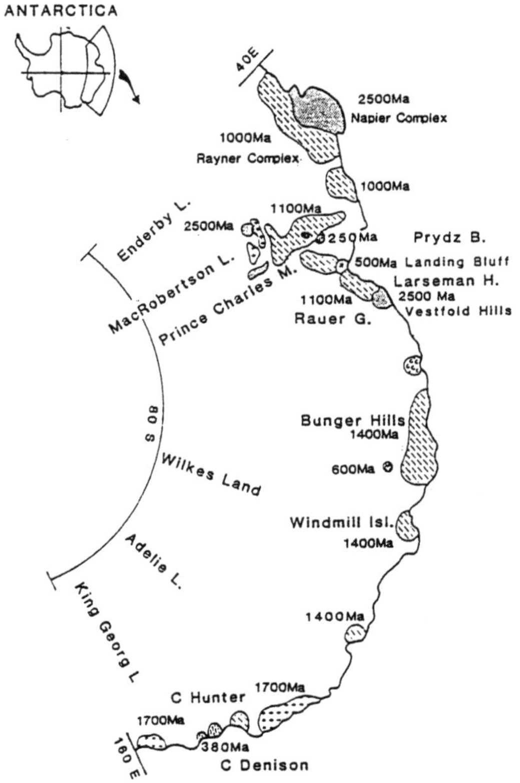

Figure 1: Location and outcrop map of the East Antarctic Shield with some radiometric ages for the ages of Proterozoic and Archaean metamorphism as well as for some 500 Ma intrusives. 500 Ma resetting ages of K/Ar signature are not plotted, but occur in large areas of the map area shown (see text for details). Dotted areas : Archaean metamorphics; diagonal dashes: middle Proterozoic metamorphics; crosses: early Proterozoic metamorphics; vertical dashes: Palaeozoic metasediments; x: Palaeozoic intrusives; c: Cenozoic volcanics of the Gausberg.

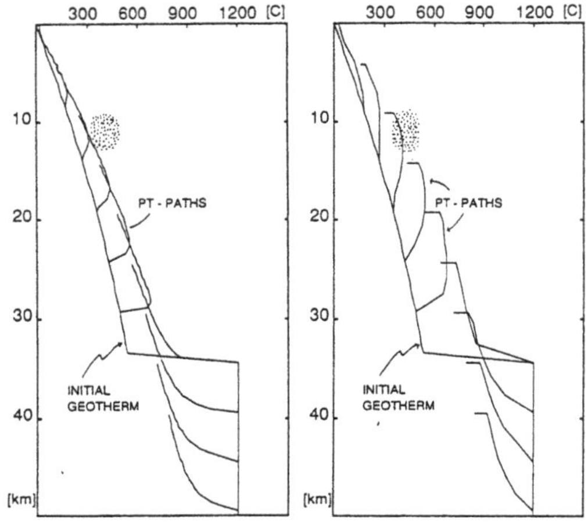

Figure 2: Pressure-temperature paths of rocks that are exhumed by erosion due to uplift caused by underplating of basaltic material at the bottom of the continental crust. *Left*: A slow erosion rate of 0.5 km my^{-1}, as may be realistic for the East Antarctic Shield (Stüwe, 1991) does not intersect the metamorphic pressure-temperature conditions suggested for the 500 Ma event (dotted area). *Right:* Erosion rates of 2 km my^{-1} can only barely advect pressure temperature paths into this field but are of an unlikely magnitude for the erosion rates following the 500 Ma event. Total run-time is 20 my.

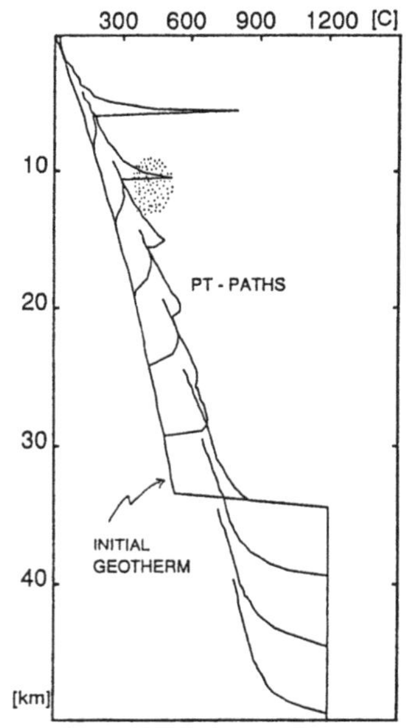

Figure 3: Pressure-temperature paths of rocks as in figure 2 but with consideration of partial melting in the section immediately above the underplate and emplacement of these melts at shallow crustal levels. Pressure-temperature paths of rocks at 10-15 kilometres depth in the Early Cambrian meet the metamorphic pressure temperature conditions suggested for the 500 Ma event (dotted area). Total run-time is 20 my.

Moho implies complete delamination of the subcrustal lithosphere, it is possible to envisage only minor thinning of the mantlepart of the lithosphere and a location of the 1580°C isotherm at 80 kilometres depth. White and McKenzie (1989) have shown that, in the central part of cells in the convecting asthenosphere, temperatures may be more than 300°C hotter than normal and it is therefore plausible that the generation of such large amounts of asthenosphere derived melts is not an unusual process. Moreover, secular changes of the convective vigour of the asthenosphere during the Cambrian (Sandiford 1989) may have been associated with steeper thermal gradients towards the centre of such cells than inferred for present day scenario of White and McKenzie (1989). If such convection cells operated under the East Antarctic Shield and were responsible for the generation of large amounts of basaltic underplate, then it is easy to argue that they are also responsible for the generation of the convergent forces in the orogenic belts that surrounded the shield.

In order to investigate the thermal consequences of the process discussed above, a simple one-dimensional thermal model was designed to explore the thermal evolution of the crust following massive basaltic underplating (Fig.2). We solved the one-dimensional form of the time dependent heat flow equation:

$$\frac{\delta T}{\delta t} = \kappa \frac{\delta^2 T}{\delta x^2}$$

with a Crank-Nicholson finite-difference approximation for an initial geotherm with normal geothermal gradient and being superimposed on the temperature profile of an instantaneously emplaced 20 kilometre thick magma body of 1200°C at the base of the crust (Fig. 2). The initial geotherm was calculated using 0°C and 30mW/m^2 as the boundary conditions and assuming low crustal heat production as appropriate for crust that has undergone multiple high-grade metamorphism. Subsequent erosion was implemented through a LaGrangian description. The results show that the thermal evolution of the upper crust is strongly dependent on the subsequent erosion rate (Fig.2). Only with erosion rates corresponding to unusual high erosion rates as observed in present mountain belts like New Zealand (Koons, 1987) can enough heat be advected to reach the pressure temperature field inferred for the 500 Ma event. However, Stüwe (1991) has shown that erosion rates of the East Antarctic Shield are likely to have been small because of the large, two dimensional extent of the area. Moreover, even for fast erosion rates, little thermal affect of the underplate is 'felt' by the rocks at 10-15 kilometres depth. This is not inconsistent with our knowledge that geothermal gradients of the inferred grade cannot be reached by conductive heat transfer alone (England and Thompson 1984; Sandiford and Powell 1990). Thus, it is necessary to discuss the advection processes that may be induced by the proposed underplating event. From figure 2 it can be seen that large parts of the lower crust will get hotter than 700°C during the cooling of the underplate and it is therefore likely that that they undergo partial melting. Such partial melts may segregate and be emplaced into the upper crust. This is consistent with our record of crustal melts in the form of pegmatites, granites and charnockites in large parts of the shield and it is suggested that advection of heat by melts derived from partial melting of the lower crust which has been heated by basaltic magma is an important contribution to the heating of the upper levels of the Cambrian crust in the East Antarctic Shield (Fig.3). Such a model is supported by our observation that there are few contact metamorphic aureoles around the 500 Ma old intrusives and more regional heating effects are therefore likely to be associated with this magmatism. In order to model these processes in a schematic manner we assumed 30 volume percent of melt as sufficient to segregate from the rock and assumed instantaneous emplacement of these melts into the upper crust (Fig.3). The subsequent temperature evolutions plotted in figure 3 indicate schematically that such processes are able to provide a substantial contribution to the heat budget of shallow crustal levels.

CONCLUSIONS

Underplating of basaltic, asthenosphere-derived magmas at the bottom of the cratonic crust of eastern Antarctica is a model that is consistent with the observed evidence for the 500 Ma event that has been pervasively imprinted on the Precambrian granulites of the shield. The model is also consistent with the simultaneous occurrence of convergent orogenies in other parts of Cambrian Gondwana. The steep geothermal gradients that have been inferred for the event in the East Antarctic Shield, however, can only be reached if heat advection occurred in addition to heat conduction. This presents no problem because advection of heat by crustal melts is expected to occur if substantial underplating causes partial melting of the lower crust. The record of the widespread occurrence of 500 my old crustal melts without distinct contact metamorphic aureoles supports the model.

ACKNOWLEGEMENTS

We would like to thank ANARE for logistic support during a total of 4 field seasons in which we had the opportunity to obtain data from a large section of eastern Antarctica. KS thanks ASAC for financial support during a Post Doctoral fellowship under which this work was conducted.

REFERENCES

Arne, D.C., Kelly, P.R., Brown, R.W. & Gleadow A.J.W. 1992. Reconnaissance apatite fission track data from the east Antarctic shield. In: Findlay R.H.; Banks, M.R., Unrug; R.; Veerers, J.J. (eds). Gondwana 8 - assembly, evolution and dispersal, A.A. Balkema, Rotterdam, 1992. this volume.

Arriens, P.A., 1975. The Pre-Cambrian geochronology of Antarctica. 1st Australian geological convention , Abs.vol., Adelaide. 1: 97-98.

Black, L.P., Harley, S.C., Sun, S.S. & McCulloch, M.T., 1987. The Rayner Complex of East Antarctica: complex isotopic systematics within a Proterozoic mobile belt. J. Met. Geol. 5: 1-26.

Clarke, G.L., Powell, R. & Guiraud, M., 1988. Low pressure facies metapelitic assemblages and corona textures from Mac Robertson Land, east Antarctica: The importance of Fe_2O_3 and TiO_2 in accounting for spinel-bearing assemblages. J. Met. Geol. 7:323-336.

Collerson, K.D. & Sheraton, J.W., 1986. Age and geochemical characteristics of a mafic dyke swarm in the Archaean Vestfold Block, Antarctica: inferences about Proterozoic dyke emplacement in Gondwana. J. Pet. 27: 853-886.

Ellis, D.J., 1983. The Napier and Rayner Complexes of Enderby Land, Antarctica - Contrasting styles of metamorphism and tectonism. In R.L. Oliver, P.J. James & J.B. Jago (eds.) Antarctic Geoscience. 20-24. Austral. Academy of Science, Canberra.

England P. & Thompson A.B., 1984. Pressure temperatur time paths of regional metamorphism I. Heat transfer during the evolution of regions of thickened continental crust. J. Pet. 25: 894-928.

Grew, E.S., 1978. Precambrian basement at Moledezhnaya Station, East Antarctica. Geol. Soc. Am. Bull. 89: 801-813.

Grew, E.S. & Manton, W.I., 1981. Geochronologic studies in East Antarctica: Ages of rocks at Reinbolt Hills and Moledezhnaya station. Antarctic J. United States , 16,5: 5-7.

Grew, E.S., 1982. The Antarctic margin. In A.E.M. Nairn & F.G. Stehli (eds.) The ocean basins and margins, V.6: The Indian Ocean: 697-755, Plenum Press, New York.

Grew, E.S., Manton, W. & Sandiford M.A., 1982. Geochronological studies in East Antarctica: ages of pegmatites in Casey Bay , Enderby Land. Antarctic J. of the United States. 17: 1-2.

Harley, S.L., 1987. Metamorphic evolution of granulites from the Rauer Islands, East Antarctica: decompression following Proterozoic collision. 5th Symposium of Antarctic Earth Science, Cambridge.

Harley, S.L., 1988. Proterozoic granulites from the Rauer Group, east Antarctica 1. PTt paths derived from mafic and felsic gneisses. J. Pet. 29:1059-1095.

Kamenev, E.N. 1972. Geology and structure of Enderby Land. In: R.J. Adie (ed.) Antarctic geology. 579-583. North Holland, Amsterdam.

Koons, P.O. 1987. Some thermal and mechanical consequences of rapid uplift: an example from the southern Alps, New Zealand. Earth Planet. Sci. Lett. 86: 307-319.

Kröner, A., 1980. Pan-African crustal evolution. Episodes, International Union Geological Sciences, 2: 3-8.

McKenzie , D. 1984. A possible mechanism for epirogenic uplift. Nature. 307: 616-618.

McKenzie, D. & Bickle, M.J. 1988. The volume and composition of melt generated by extension of the lithosphere. J. Pet. 29: 625-679.

Oliver, R.L., Cooper, J.A. & Truelove, A.J., 1983. Petrology and zircon geochronology of Herring Island and Commonwealth Bay and evidence for Gondwana reconstruction. In R.L. Oliver, P.R. James & J.B. Jago (eds.) Antarctic Earth Science. 64-68, Australian Academy of Science, Canberra.

Perkins III, D. & Newton R.C., 1981. Charnockite geobarometers based on coexisting garnet- pyroxene-plagioclase-quartz. Nature. 292: 144-146.

Piciotto, E. & Coppez, A., 1963. Bibliographie des mesures d'ages absolus en Antarctique. Annales Societe Geologique de Belgique, 85,8, B263-308.

Pieters P. E. and Wyborn D., 1976. Enderby Land and MacQuarie Island . Bur. Min. Res. report. 194.

Ravich, M.G., Klimov, M.V. & Soloviev, D.S., 1965 (Engl. transl. 1968). The Precambrian of East Antarctica. Transl. Science Research Institut of the Geology of the Arctic, Ministry Geol. USSR, Moscow, 475pp., Israel Program for Scientific Translations, Jerusalem.

Sandiford, M., 1985. The origin of retrograde shear zones in the Napier Complex: implications for the tectonic evolution

of Enderby Land, Antarctica. J. Struct. Geol. 7: 477-488.
Sandiford M., 1989. Secular trends in the thermal evolution of metamorphic belts. Earth Planet. Sci. Lett. 95: 85-96.
Sandiford, M. & Wilson, C.J.L., 1984. The structural evolution of the Fyfe Hills-Khmara Bay region, Enderby Land, East Antarctica. Austral. J. Earth Science, 31, 403-426.
Sandiford, M. & Powell, R., 1991. Some remarks on high temperature low pressure metamorphism in convergent orogens. J. Met. Geol. 9: 333-341.
Sheraton, J.W., 1983. Geochemistry of mafic igneous rocks of the northern Prince Charles Mountains. J. Geol. Soc. Australia. 30: 295-304.
Sheraton, J.W., Black, L.P. and McCulloch, M.T., 1984. Regional geochemical and isotopic characteristics of high-grade metamorphics of the Prydz Bay area: The extent of Proterozoic reworking of Archaean continental crust in East Antarctica. Precamb. Res. 26: 169-198.
Sheraton, J.W. & Black, L.P., 1984.Regional geochemical and isotopic characteristics of high grade metamorphics of the Prydz Bay area: The extent of Proterozoic reworking of Archaean continental crust in East Antarctica. Precamb. Res. 26: 169-198.
Sheraton, J.W. & Black, L.P., 1988. Chemical evolution of granitic rocks in the East Antarctic Shield, with particular reference to post-orogenic granites. Lithos. 21: 37-52.
Sheraton, J.W. & Oliver R.L., 1991. Geology and geochemistry of the Bunger Hills. Antarctic Science.
Starik, I.Y., Ravich, M.G., Krylov, A.Y., Silin, Y.I., Atrashenok, L.Y. & Lovtsyus, A.V., 1960 (engl.Trans. 1961). Novye dannye ob absolyutnom vokhraste porod Vostochnoy Antarktidy. Doklady Akademii Nauk SSSR 134, 1421-1423. (New data on the absolute ages of rock in eastern Antarctica. Doklady Acad. Sci. USSR 134, 956-958.).
Starik, I.Y., Krylov, A.Y., Ravich, M.G. & Silin, Y.I., 1961. The absolute ages of east Antarctic rocks. In Geochronology of Rock Systems, Annales of the New York Academy of Sciences, 91,2, 576-582.Sheraton, J.W., 1982. Origin of charnockitic rocks of MacRobertson Land. Antarctic Geoscience. 59: 489-497.
Stuiver, M. & Braziunas, T.F., 1985. Radiocarbon. Compilation of isotopic dated from Antarctica. Am. J. Sci. 27, 2A.
Stüwe, K., 1991. Flexural constraints on the denudation of asymetric mountain belts. J. Geophys. Res. 96:10401-10408.
Stüwe, K. & Oliver, R., 1989. Geological history of Adelie Land and King George V Land, Antarctica: Evidence for a polycyclic metamorphic evolution. Precamb. Res. 43: 317-334.
Stüwe, K. & Powell, R., 1989a. Low pressure granulite facies metamorphism in the Larsemann Hills area, East Antarctica; petrology and tectonic implications for the evolution of the Prydz Bay area. J. Met. Geol. 7: 465-483.
Stüwe, K. & Powell R., 1989b. Metamorphic evolution of the Bunger Hills, East Antarctica: evidence for substantial post-metamorphic peak compression with minimal cooling in a Proterozoic orogenic event. J. Met. Geol. 7: 449-464.
Stüwe, K., Braun, H.M. & Peer, H., 1989. Geology and Structure of the Larsemann Hills area, Prydz Bay, East Antarctica. Austral. J. Earth Sci. 36: 219-241.
Tankard, A.J., Jackson, M.P.A., Erikson, K.A., Hobday,D.K., Hunter, D.R. & Minter,W.E.L., 1982. Crustal evolution of southern Africa, 3.8 Billion years of Earth history. Springer, New York.
Tingey, R.J., 1982. The geologic evolution of the Prince Charles Mountains - an Antarctic Archaean cratonic block. In C. Craddock (ed.) Antarctic Geoscience. 455-464, Univ. Wisconsin Press, Madison , Wisconsin.
White, R.S., and McKenzie D.P., 1989. Magmatism at rift zones: the generation of volcanic margins and flood basalts. J. Geophys. Res. 94: 2647-2668.

Gondwana Eight, Findlay, Unrug, Banks & Veevers (eds) © 1993 Balkema, Rotterdam. ISBN 90 5410 304 3

Thrust patterns of the Ross/Delamerian orogens in northern Victoria Land (Antarctica) and southeastern Australia and their implications for Gondwana reconstructions

T. Flöttmann
Department of Geology and Geophysics, The University of Adelaide, S.A., Australia

G. Kleinschmidt & T. Funk
Johann Wolfgang Goethe Universität, Frankfurt am Main, Germany

ABSTRACT: Contractional faults characterize the boundaries of the terranes along the eastern craton edges of southeastern Australia and northern Victoria Land, Antarctica. East-directed displacement kinematics prevail at the eastern margins of the orogenic belts and similar foreland thrust belts mark the western orogen margins. The structural patterns represent similar imprints of Cambro-Ordovician tectonism, and therefore give evidence of a continuous convergent Palaeo-Pacific Gondwana margin in Australia and Antarctica. These patterns can be used to delimit Australian-Antarctic Gondwana reconstructions.

1 INTRODUCTION

Convergence at continental margins is typically reflected by structural imprints, such as folding or contractional faulting. Characteristically, these structures often exhibit divergent displacement patterns. In this paper we utilize structural patterns for a Gondwana correlation of southeastern Australia and northern Victoria Land. Possible Gondwana correlations involving also Tasmania have been discussed by Findlay, (1987a), Findlay et al. (1991), Laird et al. (1977) and Jago (1981).

The Wilson terrane, part of the Ross Orogen in northern Victoria Land (Antarctica) and the Delamerian Orogen which crops out in the Mt. Lofty Ranges and in the Glenelg river area (southeastern Australia) lie between the Proterozoic Antarctic and Australian cratons in the west and accreted terranes constituted by the Bowers and Robertson Bay terranes and the Lachlan Fold Belt in the east (Fig. 1). Both the Ross and Delamerian orogenies are characterized by Cambro-Ordovician magmatism and related low-pressure/high-temperature metamorphism. The broad lithological similarities of the orogens have been previously used for correlations of the position of Antarctica and Australia within Gondwana (Oliver et al. 1983; Stump et al. 1986).

The results presented here suggest that the edges of both orogens appear to be defined by major divergent contractional faults which provide constraints on the location and extent of the Palaeo-Pacific margin of Gondwana. The comparatively well defined, narrow crustal discontinuities give a more clear outline of the orogen widths and in addition kinematic indicators define the type of displacement during crustal convergence along the Palaeo-Pacific margin.

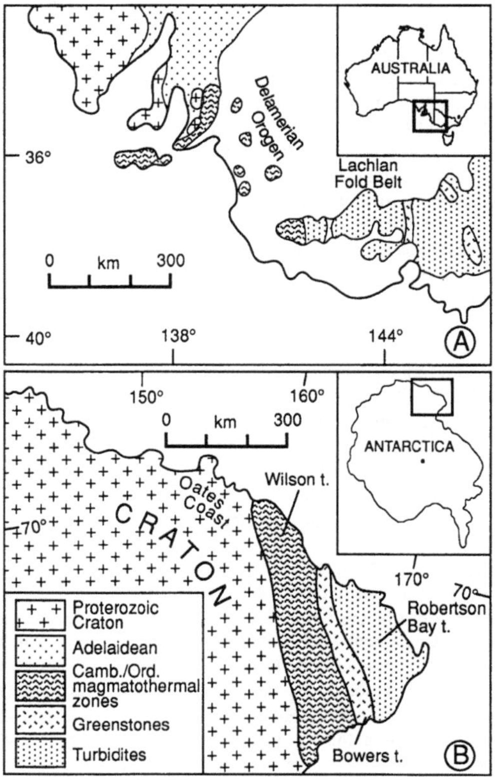

Figure 1: A) Geological setting of the Delamerian Orogen between the Australian craton and overlying Late Proterozoic sequences (Adelaidean) in the west and the greenstones and turbidite sequences of the Lachlan Fold Belt in the east ; B) setting of the Wilson terrane mobile belt between the Antarctic craton and the greenstone and turbidite sequences of the Bowers and Robertson Bay terranes, respectively.

2 NORTHERN VICTORIA LAND

2.1 Geological setting

Three contrasting tectonometamorphic units occur in northern Victoria Land (Fig. 2). The central Bowers terrane includes a

Cambrian oceanic volcanic arc (Weaver et al. 1984), the rocks of which are overlain by the mixed marine and fluvial sequences of the Cambro-Ordovician Mariner and Leap Year groups, respectively. The Bowers terrane is bound in the north by high angle thrust faults with mainly west-over-east directed displacement (Gibson & Wright 1985; Flöttmann & Kleinschmidt 1991a) against the Wilson terrane to the west and the Robertson Bay terrane to the east (Fig. 2). The central and southern Bowers - Robertson Bay terrane boundary is interpreted variably (Findlay 1987a, 1987b; Wright & Brodie 1987; Wright & Dallmeyer 1991). The latter terrane contains Cambrian - Lower Ordovician flysch-type sedimentary rocks deformed by open to very tight folds. The Devonian Admiralty intrusives occur predominantly in the Bowers and Robertson Bay terranes. The Wilson terrane shows relics of presumed Proterozoic granulites (Carmigiani et al. 1989). Subsequent low-pressure/ high-temperature overprinting led to abundant migmatization and local anatexis in the Wilson terrane (Babcock et al. 1987; Grew et al. 1984). The dominant gneisses are coarsely foliated and consist largely of quartz, plagioclase and biotite, and in addition microcline, cordierite and garnet may occur. In markedly migmatitic areas a regional foliation is made up by cm - dm thick segregations of mafic and felsic components. These are commonly cut by oblique bands of quartz-feldspar veinlets which may, at least in part be derived by partial melting. The less migmatic rock types usually contain biotite, quartz and garnet and the regional foliation is marked by biotite alignment. The general strike of the foliation is northnorthwest, the dips are on average between 30° and 60° to the northeast and to the southwest.

A belt of (?)Cambrian high-pressure, partly ultramafic rock relics (Grew et al. 1984; Grew & Sandiford 1984) crops out at the eastern margin of the Wilson terrane. Turbiditic sedimentary rocks occur predominantly at the flanks of the central Wilson terrane and form the Berg Group and the Priestley schists in the west and the Morozumi Range and McCain Bluff in the east (Flöttmann & Kleinschmidt 1991b). Rafts of pelitic sedimentary rock relics occur also within the high grade migmatites of the Wilson terrane.

The rocks of the Wilson terrane are extensively intruded by granites forming a magmatic arc (Babcock et al. 1987; Borg et al. 1987), with granites of I-type affinity in the east and S-type granitoids in the west (Borg et al. 1987; Vetter & Tessensohn 1987). Sedimentary rocks of the Berg Group (Klimov & Soloviev 1958; 1960), Priestley Formation (Lombardo et al. 1989) and the Morozumi Range (Engel 1984) experienced lower grade regional static metamorphism probably related to the granite intrusions.

The rocks of the Wilson terrane basement are multiply deformed (Kleinschmidt & Skinner 1981). Based on observations of the northern part of the terrane the earliest structural element is a metamorphic layering (S_1) which supposedly formed during a first deformation (D_1). During a second deformation D_2, S_1 was folded isoclinally. The F_2-axes plunge at 35° - 80° to the west to northwest. Semi-tight to tight folds of a third generation (F_3) refold the second generation folds. The migmatic layering is bent around the F_3-folds, which plunge shallowly at 5° - 30° to the northwest and southeast. F_3-fold asymmetries and vergences indicate east-directed displacement in the eastern parts of northernmost Victoria Land, whereas F_3-fold geometries in the western part of the Oates Coast suggest west directed displacement. Mobilisate veins associated with the migmatisation of the basement transect the F_3-fold hinges. It is thus concluded that the D_3-deformation was contemporaneous with the thermal peak of metamorphism. Locally large scale north-northwest - south-southeast trending F_4-folds are also developed. In contrast to the multiply deformed basement the flanking metasedimentary rocks in the Berg Mountains, McCain Bluff and Lonely One Nunatak exhibit only one generation of large scale northwest - southeast trending folds with wavelengths up to some 100m.

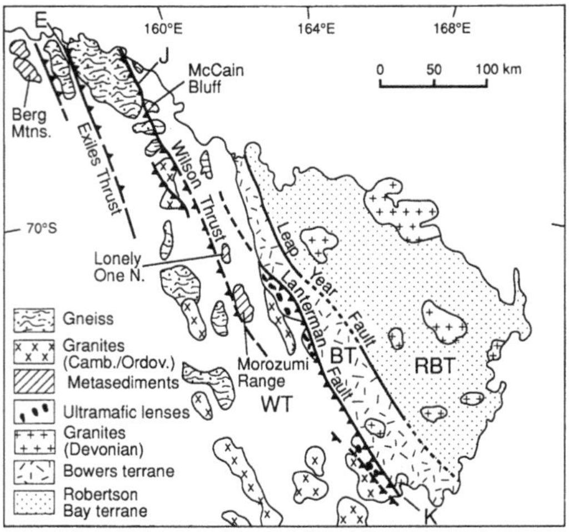

Figure 2: Divergent thrust displacement in northern Victoria Land, Antarctica along the western Exiles Thrust, and the eastern Wilson Thrust and the Lanterman Fault; note that the eastern Bowers terrane is also controlled by a fault, which is, however, not necessarily related to structures discussed here. Letters (E,J,K) refer to kinematics given in Fig. 5. (modified after Ganovex and Italantartide Teams, 1991 and own data)

2.2 Eastern Wilson terrane margin

The basement of the eastern Wilson terrane is cut by the north-northwest trending "Wilson Thrust", a ductile shear zone of up to several hundred meters thickness, which can be identified over more than 80 km (Fig. 2).

The Wilson Thrust consists of numerous highly strained mylonitic shear zones, which anastomose around phacoids of less deformed country rock. The width of the thrust zone varies from some metres to some hundreds of metres. The mylonites of the Wilson Thrust dip on average at 40° to the west - southwest. On the mylonite planes a mineral stretching lineation is formed by elongate feldspars, biotite and quartz. A consistent top-to-the-east-northeast sense of thrusting (Fig. 5) is recorded by several asymmetric rotation features such as S-C-surfaces, asymmetrically elongated feldspars and micas and non-conjugate ecc-cleavages (as described by Berthe et al. 1979; Simpson & Schmid 1983; Passchier & Simpson 1986; Platt & Vissers 1980). The mineral deformation and recrystallization behaviour indicates a deformational regime at a metamorphic grade consistent with the transition between greenschist and amphibolite facies (Flöttmann & Kleinschmidt 1991b). The east directed vergence shown by the F_3-fold-geometries in the eastern Wilson terrane suggests that the F_3-folds are kinematically related to thrusting. The thrust is intruded by syn-kinematic and crosscut by post-kinematic pegmatites, which are interpreted as late phase products of the granites (Flöttmann & Kleinschmidt 1991b), locally granites are deformed by thrusting. The Wilson Thrust is mainly an intra-basement structure, however, the Wilson Thrust juxtaposes the Wilson terrane basement complex and the eastern sedimentary rocks of McCain Bluff and the Morozumi Range.

The Lanterman Fault marks the boundary between the Wilson and the Bowers terranes (e.g. Gibson & Wright 1985), and shows principally similar kinematics along the northern segment, where a minor fault following the (concealed) main Lanterman structure shows west over east thrusting along steeply west dipping mylonites (Jordan 1981; Flöttmann & Kleinschmidt 1991a). Transpressional, top-towards-the-northeast-directed kinematics locally occur along the central (Gibson 1987) and southern parts of the Lanterman Fault (Fig. 5). Deformation occured in the very low to low grade transition of metamorphic grade (Flöttmann & Kleinschmidt 1991a).

2.3 Western Wilson terrane margin

A second major thrust system (Exiles Thrust) is also developed along the northern part of the western Wilson terrane (Fig. 2). Here, over a thickness of at least 400 metres hornblendic granites and aplites are extremely strained, and the principal mylonitic foliation dips at 30° - 65° to the northeast. The mylonites strike to the northnorthwest and S-C mylonites and asymmetric porphyroclast systems in combination with the mineral elongation lineation show that thrusting along the Exiles Thrust was directed towards the west. The overall sense of displacement of the Exiles Thrust is thus directly opposite to the displacement along the Wilson Thrust (Fig. 5).

The Exiles Thrust system generally marks the main boundary between the metasedimentary rocks of the Berg Group and Wilson terrane with the basement rocks thrust toward the craton over the metasedimentary rocks of the Berg Group. Aeromagnetic surveys (Damaske 1991) indicate that the Exiles Thrust coincides with the boundary of the Wilson terrane and the Antarctic craton.

The Priestley Formation which is of possible Cambrian age (Casnedi 1989) and the Berg Group rocks are interpreted to be lithologically similar. The Exiles Thrust lines up along strike with the boundary between the Priestley Formation and the Wilson terrane basement. However, in the southern segment of northern Victoria Land the contact relationships are obliterated by post-kinematic granites.

3 SOUTHEASTERN AUSTRALIA

3.1 Geological setting

Southeastern Australia is characterized by three different geological provinces, comprising from west to east: the Proterozoic craton (including the Middle to Late Proterozoic cover), the central Delamerian Orogen and the external Lachlan Fold Belt.

The Early to Middle Proterozoic metamorphic rocks of the Yorke and Fleurieu peninsulas represent the easternmost outcrops of the southeast Australian cratonic basement in mainland Australia (Fig. 1). The basement is overlain by the Late Proterozoic sequence of the Adelaidean basin which Preiss (1987) subdivides into: basal platform deposits which are followed by mixed carbonate and clastic sequences (Callana and Burra Groups), followed upwards by glaciomarine deposits (Umberatana Group) and a post-glacial transgressive sequence of the Wilpena Group. The Adelaidean basin is overlain by transgressive Middle Cambrian deposits (Normanville Group) and the (?)Middle Cambrian mostly turbididtic Kanmantoo Group (Daily & Milnes 1972), which crops out in the eastern Mt. Lofty Ranges. The sedimentary rocks of the Glenelg complex in western Victoria are considered as equivalent to the Kanmantoo Group (Vandenberg 1978). The Glenelg complex consists of a sequence of metapelites, amphibolites, local calc-silicates and cherts (Wells 1956). Serpentinites with an ophiolitic affinity (Turner et al. 1991) which occur within the Glenelg complex are absent in the Kanmantoo Group to the west.

The lateral extent of the Delamerian Orogen is indicated by Cambro-Ordovician granite intrusions occuring in the Glenelg complex rocks and in the Kanmantoo and Adelaidean Groups of the Mt. Lofty Ranges. Few outcrops show Delamerian magmatism in the central Murray Basin, in which Palaeozoic rocks are otherwise covered by younger sequences.

The syn-kinematic granites of the Delamerian Orogen are dominantly of I-type character, and elongated along a northerly trend (Figs. 3,4); late-kinematic granites have A-type characteristics (Foden et al. 1990). Metamorphism is generally of low grade, although higher grade amphibolite facies metamorphism and local anatectic melting are confined to the granite aureoles.

The contact relationships of the Delamerian Orogen and the Lachlan Fold Belt, which consists of continent-derived turbidites and fault-controlled Cambrian greenstones of volcanic arc provenance, are not clearly established. For most of its extent the boundary is obscured by Silurian rhyolites, which lie between the eastern Glenelg complex and the pre-Silurian sandstones of the Grampians Group which are intruded by Siluro-Devonian granites and are characteristic of the Lachlan Fold Belt.

3.2 Eastern Delamerian Orogen margin (Glenelg complex)

The structural record within the Glenelg complex (Fig. 3, see also Wells 1956) varies considerably with respect to different thermal and strain regimes. In the low grade metamorphic rocks, which occur predominantly in the western part, one fold generation (F_1) is developed, folding a roughly north - south trending bedding transposition foliation, which is often delineated by bedding-parallel quartz segregations. The F_1-axes trend generally northwest - southeast, and the wavelengths of folds vary from decimetre to metre scale. An axial planar foliation dips to the northeast, and contains biotite in progressively increasing amounts to the east.

The D_2-deformation is dominant in the upper greenschist - amphibolite facies metamorphic rocks, which are temporarily and spatially associated with the granitoids in the central and eastern part of the complex. F_2-fold-geometry is highly variable. In weakly strained areas the folds trend dominantly west - east to northwest - southeast plunging moderate to steeply to the east. F_1-folds may be recognized by disrupted, isoclinal hinges within the F_2-axial-plane- foliation. The F_2-

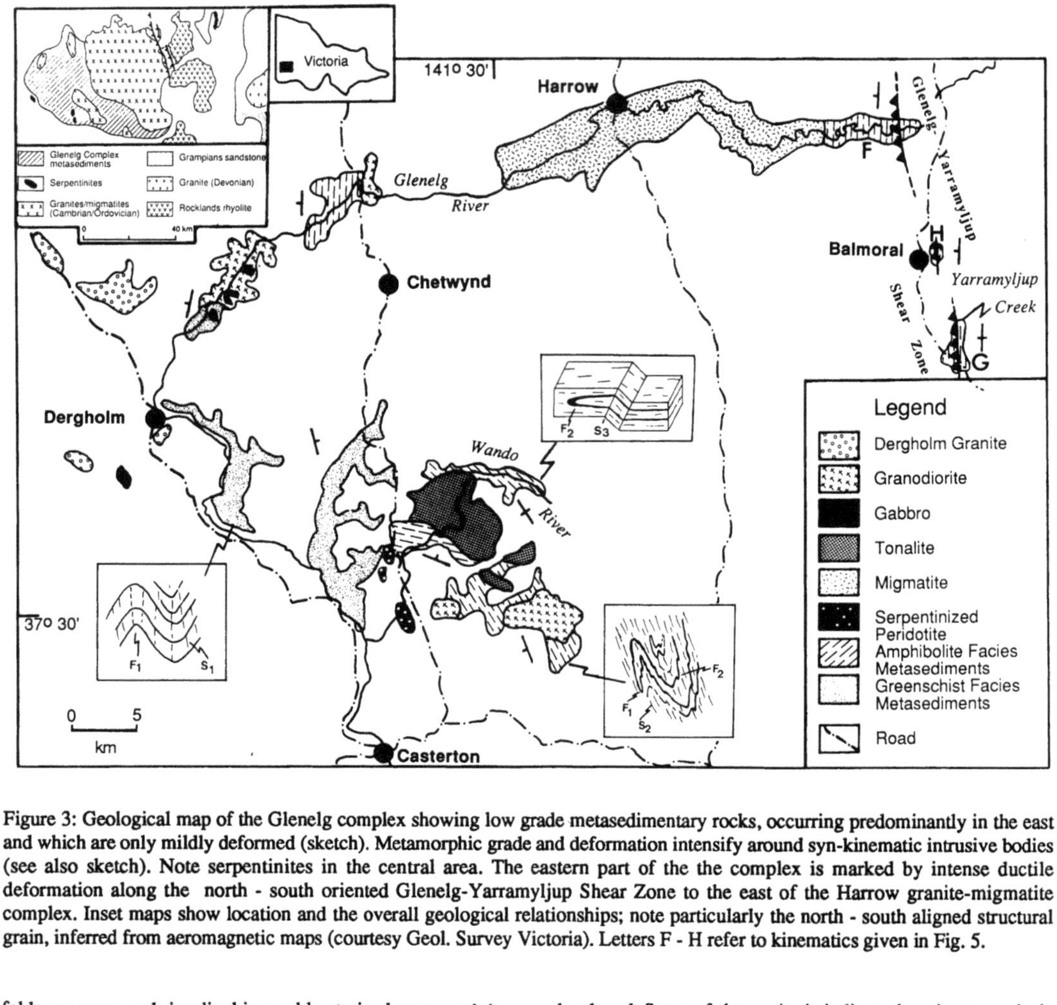

Figure 3: Geological map of the Glenelg complex showing low grade metasedimentary rocks, occurring predominantly in the east and which are only mildly deformed (sketch). Metamorphic grade and deformation intensify around syn-kinematic intrusive bodies (see also sketch). Note serpentinites in the central area. The eastern part of the the complex is marked by intense ductile deformation along the north - south oriented Glenelg-Yarramyljup Shear Zone to the east of the Harrow granite-migmatite complex. Inset maps show location and the overall geological relationships; note particularly the north - south aligned structural grain, inferred from aeromagnetic maps (courtesy Geol. Survey Victoria). Letters F - H refer to kinematics given in Fig. 5.

folds are open - sub-isoclinal in weakly strained areas, and the F_2-fold axes trend west - east to northwest - southeast with east-dipping foliation planes in the west and west-dipping foliations in the east. Syn-kinematic granites intruded the basement contemporaneously with the D_2-deformation. Along the highly deformed granite margins the D_2-foliation is the predominant fabric throughout the country rock and also penetratively developed in the granitoids (e.g. Wando Granodiorite). Generally the foliation of the granitoids strikes northnorthwest, dipping at 50 - 80° to the northwest. At the granitoid margins a penetrative mineral elongation lineation plunges northwest to west. Along the eastern part of the complex a zone of several kilometres width crops out, where pegmatite veining is very common. Many pegmatites are deformed by the D_2-deformation, but post-kinematic pegmatites also occur.

A north trending zone of intense ductile deformation occurs in the schists of the eastern Glenelg River. A closely spaced, locally mylonitic, foliation contains a northwest plunging mineral elongation lineation. Within the foliation the F_2-folds are usually isoclinal and the fold axes are oriented parallel to the stretching lineation, and incipient sheath folds are developed. Sense of shear criteria indicate thrusting towards the southeast (Fig. 5).

Some kilometres further east, at Yarramyljup Creek (Fig. 3) the thrust planes dip steeply towards the west-northwest. Asymmetrically stretched pegmatite boudins, boudinaged isoclinal F_2-fold relics and ecc-fabrics indicate sinistral strike-slip deformation. Both, the mineral elongation lineations and the F_2-fold axes are oriented subhorizontally, suggesting progressive rotation into parallelism with the extension lineation.

The kinematics along the margin of the eastern Glenelg complex, defined by the Glenelg-Yarramyljup Shear Zone, are overall characteristic of southeast directed transpression, which is also consistent with the geometry of the deformation features of the granites, suggesting syn-kinematic intrusion into a regime of oblique convergence. The abundance of pegmatites at the highly strained eastern margin of the Glenelg complex indicates a genetic relationship between deformation and the localization of pegmatite precipitating fluids. Cambro-Ordovician magmatism does not occur east of the Glenelg-Yarramyljup Shear Zone, which thus marks the principal eastern limit of the Delamerian Orogen.

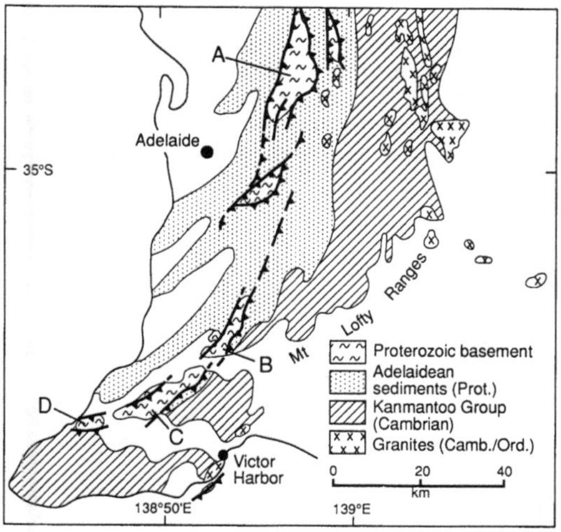

Figure 4: East dipping reverse faults mapped in the Mt.Lofty Ranges, South Australia displace the Kanmantoo and Adelaidean sequences affected by Delamerian thermal overprint related to granite intrusions. Letters A - D refer to kinematics given in Fig. 5.

3.3 Western Delamerian Orogen Boundary (Mt. Lofty Ranges)

The structure of the western margin of the Delamerian Orogen is dominated by west-verging large open folds and thrusts (Mancktelow 1990; Jenkins 1990; Steinhardt 1991) which deform the Adelaidean and the Kanmantoo sequences about generally northeast trending fold axes. The rocks around granite intrusives may attain sillimanite grade metamorphism and show similarly north trending open and upright folds (Offler & Fleming 1968). In more intensely deformed parts of the high grade metamorphic zone a prominent north - south trending, subhorizontal rodding fabric is developed in syn-kinematic granitoids such as the Rathjen Gneiss (Offler & Fleming 1968).

A series of north-northeast trending inliers of Proterozoic basement occur along the western margin of the Delamerian Orogen (Fig. 4). The basement inliers are juxtaposed against the Adelaidean and Kanmantoo sequences along major northeast trending thrust faults. Along the thrusts the basement is pervasively deformed. A downdip mineral elongation lineation (X-axis of the finite strain ellipsoid, see also Fig. 5) which dips at 30 - 50^{0} to the east - southeast is developed on the mylonite planes. The thickness of the mylonite ranges up to a hundred metres in strongly deformed zones (Steinhardt 1991). At the southernmost basement inlier the Kanmantoo Group is thrust against the Proterozoic basement, which shows a highly strained contact with the overturned stretched basal conglomerates of the Burra Group (Aldgate Sandstone). Kinematic indicators show generally northwest directed movement along the thrust (Steinhardt 1991). In mylonites within the basement rocks micas, as well as feldspars and quartz, show evidence for crystal-plastic deformation. Criteria indicative of the behaviour of deformation and crystallization (see Simpson 1985) such as dynamic quartz and muscovite recrystallization and brittle feldspar deformation suggest a middle to upper greenschist facies deformation regime according to Simpson (1985).

Further north, the basement inliers are in contact with the Adelaidean sequences, which form a southward thinning wedge between the Kanmantoo Group and the basement inliers (Fig. 4). Several intra-Adelaidean thrusts have been proposed by Jenkins (1990). Clarke & Powell (1989) have also suggested a thrust boundary between Kanmantoo and Adelaidean sequences.

Granitoids of Delamerian age at Victor Harbor have been deformed by ductile, northeast - southwest trending mylonite zones which dip at 40 - 60^{0} southeast and contain a downdip stretching lineation (Fig. 4). Rotation criteria indicate northwest directed reverse movement within the mylonites, which is consistent with the thrusting seen along the basement inliers. Along the northeasternmost basement inlier Mills (1973) reports a regional metamorphic overprint across thrust contacts. These relationships suggest that the west directed thrusting was roughly contemporaneous with the thermal imprint of the Delamerian Orogeny.

These features suggest that the boundary between the Delamerian Orogen and the Australian craton is marked by a foreland fold and thrust belt.

4 TIMING RELATIONSHIPS

Occurrence of Proterozoic crustal fragments in the Wilson terrane is supported by Nd-model ages from granulites of 2.0 - 1.8 Ga (Armienti et al. 1989), from granitoids (Borg et al. 1987) and also by a granite zircon ion microprobe study (Black & Sheraton 1990).

Adams (1986) suggests that the initial phases of the Ross Orogeny occured around 530 Ma as indicated by a Rb-Sr whole rock age from the Rennick schists (530 ± 20 Ma), which represent part of the precursor of the Wilson terrane gneisses. A similar age (544 ± 4) is obtained from ion microprobe studies of granite zircons (Black and Sheraton, 1990), which is interpreted to reflect the age of regional low-pressure/high-temperature metamorphism in the basement, whereas U-Pb monazite ages from gneisses suggest an age around 490 Ma for this metamorphic event (Adams 1986). The ages of the granite intrusions are loosely bracketed between 490 and 470 Ma (Adams 1986), which is in general accordance with an age (469 Ma) given by Black and Sheraton (1990).

Thrusting occured after the metamorphic peak since the migmatites produced during regional low-pressure/high-temperature metamorphism are cut through by the Exiles- and Wilson Thrusts. Locally the thrusts even transect granites which intrude in the migmatites. However, deformation along both thrusts was outlasted by thermal activity (Flöttmann & Kleinschmidt 1991b), suggesting that thrusting post-dates migmatisation and was roughly contemporaneous with the granite emplacement.

Deposition of the Kanmantoo Group took place during the Middle Cambrian, indicated by an ion microprobe U-Pb zircon age of 526 ± 4 Ma (Cooper et al. 1992) from a tuff layer in the Heatherdale Shale of the uppermost Normanville Group (Gatehouse et al., 1990) The highly deformed Rathjen Gneiss gives an age of 516 Ma (Sandiford et al. 1992; Foden et al. in prep.). In the Glenelg River area an Nd/Sm, 3-mineral whole-rock isochron from the syn-kinematic Wando Granodiorite gives an age of 516 Ma (Turner et al. in press). More importantly, the Hummocks ophiolitic serpentinite yields an

Nd-model age of 650 - 700 Ma thus giving evidence for the occurrence of Proterozoic oceanic crust in the Glenelg complex (Turner et al. in press). The age of the undeformed Dergholm granite is 485 ± 5 Ma (Turner et al. in press.), which is in accordance with ages from other late-kinematic Delamerian granites (Milnes et al. 1977; Sandiford et al. 1992; Foden et al. in prep.). K-Ar dating of late and post-kinematic pegmatites from the Glenelg-Yarramyljup Shear Zone gives a minimum age of 480 Ma for the thrusting (Turner et al. in press). K-Ar biotite and whole rock ages from schists and gneisses place the minimum age of metamorphism at about 475 Ma (Turner et al. in press).

Since thrust-related ductile shear zones in the southern Mt. Lofty Ranges deform Delamerian granites, but are conversely overprinted by the regional metamorphism of the Delamerian Orogeny the related thrusting is clearly also related to the Delamerian Orogeny. The syn-deformational age of the Wando Granodiorite may well reflect intrusion during the peak D_2-deformation. Retrograde thrusting along the Glenelg-Yarramyljup shear zone postdates or reflects late stages of the D_2-event, and therefore took place between 516 and 475 Ma.

5 DISCUSSION AND CONCLUSIONS

The identical structural setting of the contractional faults at the margins of the Wilson terrane and those delimiting the Delamerian Orogen is a strong pointer that these structures were once continuous between Australia and Antarctica. That is, these structures and the fold belts between them are regarded as key structures defining the early Palaeozoic fit between Antarctica and Australia.

The intra-basement setting and the general west-over-east displacement kinematics of the Wilson Thrust and Glenelg-Yarramyljup Shear Zone argues for a correlation of these structures (Fig. 6). The Glenelg-Yarramyljup Shear Zone may further coincide with the suture between the Delamerian Orogen and the Lachlan Fold Belt, however, it is equally possible that the (presumed) suture proper is, as in northern Victoria Land located slightly further east. The western boundary faults provide less compelling correlation evidence. In contrast to the clear structural contact between the craton and the Delamerian Orogen along the foreland thrust belt in the Mt. Lofty Ranges, the probably correlative Exiles Thrust does not necessarily mark the furthest western limit of the Wilson Terrane. The Exiles Thrust separates aeromagnetically contrasting terranes (Damaske 1991), and therefore may also coincide approximately with the boundary between the craton and the Wilson terrane.

Whereas both western margin foreland thrust belts show dominantly orogen-normal displacement, the eastern Delamerian margin is dominated by sinistral transpressional kinematics. At the northern Wilson terrane margin orogen-normal displacement dominates and dextral transpression occured at the southern segment of the Wilson terrane margin. The kinematic imprints at the Palaeo-Pacific margin can be unifyingly interpreted to reflect convergence with a central orogen-normal component of shortening. Orogen-parallel displacement components prevailed along the northern and a less prominently, at the southern margin segments. The greater width of the Delamerian Orogen could be a consequence of a greater proportion of orogen parallel displacement during convergence whereas the narrow width of the Wilson terrane is a direct consequence of its dominantly orogen-normal shortening.

Our structural findings are in general accordance with the lithological similarities of northern Victoria Land and southeastern Australia as outlined by Stump et al. (1986), Oliver et al. (1983), Grindlay and Davey (1982) and Findlay (1987a). We suggest a Gondwana correlation of Antarctica and

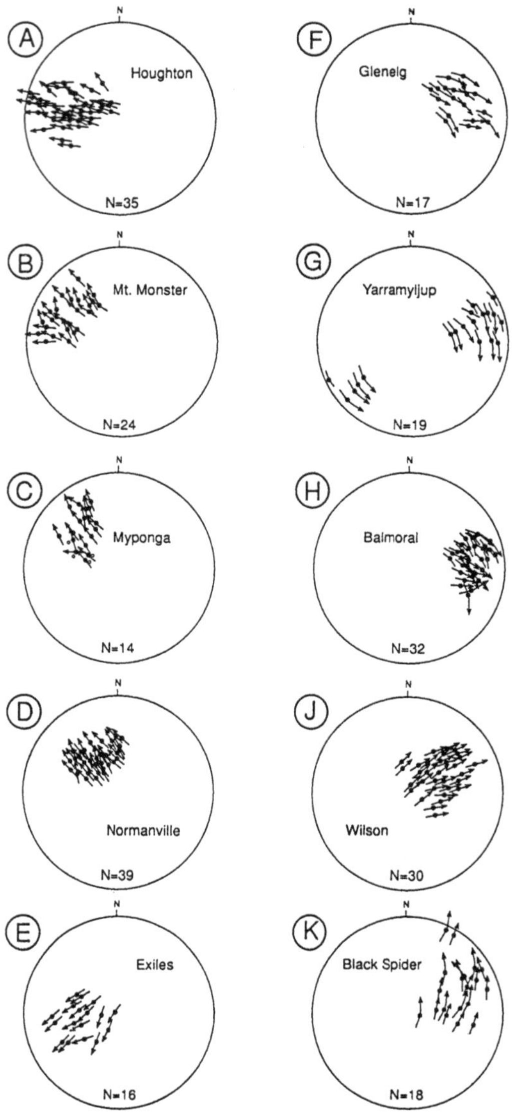

Figure 5: Synoptic diagram of major fault-system kinematics arranged from north to south along western boundary reverse fault systems (A - E) and eastern boundary fault systems (F - K), see maps Figs. 2 - 4 for location details. Schmidt net shows poles to mylonite planes (solid circles), lines show great circle trace of mutual poles to mineral elongation lineation and poles of corresponding mylonite planes, arrows give sense of displacement (lower hemisphere equal area projection).

Australia along the thrust systems of the eastern Ross/Delamerian Orogen margins. This is in agreement with the reconstruction based on the correlation of the easternmost Archean outcrops in both continents proposed by Oliver et al. (1983). The structures described here furthermore give evidence for laterally continuous convergence along the Palaeozoic margin of the Pacific.

REFERENCES

Adams, C.J. 1986. Age and ancestry of metamorphic rocks in the Daniels Range, USARP Mountains, Antarctica. In E. Stump (ed.), Geological Investigations in northern Victoria Land: 25-38. Washington: American Geophysical Union.

Armienti, P., C. Ghezzo, F. Innocenti, P. Mannetti, S. Rocchi, & S. Tonarini 1989. Genesis and evolution of Granite Harbour intrusives of the Wilson terrane: geochemical and isotopic evidence. Proceedings 2nd meeting Earth Science in Antarctica, Siena 27th - 28th September 1988, pre-print: 133-137.

Babcock, R.S., C.C. Plummer, J.W. Sheraton & C.J. Adams 1987. Geology of the Daniels Range northern Victoria Land, Antarctica. In E. Stump (ed.), Geological Investigations in northern Victoria Land: 1-24. Washington: American Geophysical Union.

Berthe, D., P. Choucroune & P. Jegouzo 1979. Orthogneiss, mylonite and non-coaxial deformation of granites: an example from the South Armorican shear zone. Journal of Structural Geology 1: 31-42.

Black, L.P. & J.W. Sheraton 1990. The influence of Precambrian source components on the U-Pb Zircon age of a Palaeozoic granite from northern Victoria Land, Antarctica. Precambrian Research 46: 275-293.

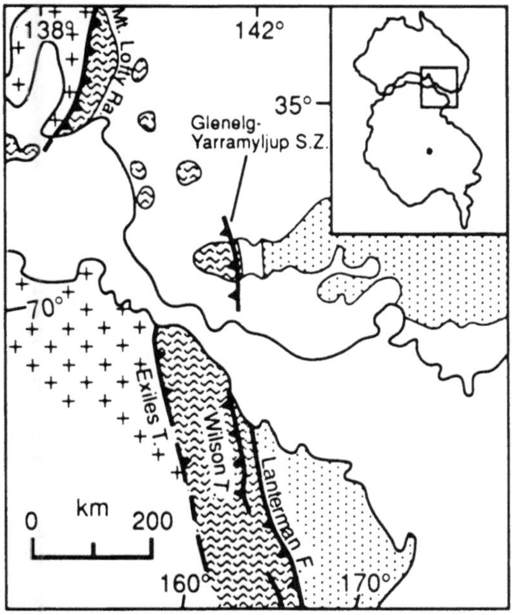

Figure 6: Synoptic map of divergent detachment patterns bracketing the Ross/Delamerian Orogens between the Australian and the Antarctic craton in the west and the accreted terranes in the east and the Gondwana correlation of Antarctica and Australia based thereon.

Borg, S.G., E. Stump, B.W. Chappel, M.T. McCulloch, D. Wyborn, R.L. Armstrong & J.R. Holloway 1987. Granitoids of northern Victoria Land, Antarctica: Implications of chemical and isotopic variations to regional crustal structure and tectonics. American Journal of Science 278: 127-169.

Carmigiani, L., L. Ghezzo, G. Gosso, B. Lombardo, M. Meccheri, A. Montrasio, P.C. Pertusati & F. Salvini 1989. Geology of the Wilson Terrane area between David and Mariner Glaciers, Victoria Land (Antarctica). Memorie della Societa Geologica Italiana, 43: 77-98.

Casnedi, R. 1989. Facies analysis of the Lower Palaeozoic in northern Victoria Land (Antarctica). 3rd Meeting Scienze della Terra in Antartide, Siena 4th - 6th October 1989, Abstracts: 37-39.

Clarke, G.L. & R. Powell 1989. Basement-cover interaction in the Adelaide Fold Belt South Australia: the development of an arcuate foldbelt. Tectonophysics 158: 209-226

Cooper, J.A., R.J.F. Jenkins, W. Compston & I.S. Williams 1992. Ion probe zircon dating of a tuff in the South Australian Lower Cambrian. Journal of the Geological Society of London 149: 185-192.

Daily, B. & A.R. Milnes 1972. Revision of the stratigraphic nomenclature of the Cambrian Kanmantoo Group, South Australia. Journal of the Geological Society of Australia 19: 197-202.

Damaske, D. 1991. Regional anomalies from aeromagnetic survey in Victoria Land. Memorie della Societa Geologica Italiana 46: 527-532.

Engel, S. 1984. Petrogenesis of contact schists in the Morozumi Range, North Victoria Land, Antarctica. Geologisches Jahrbuch B60: 167-185.

Findlay, R.H. 1987a. A review of the problems important for interpretation of the Cambro-Ordovician Palaeogeography of Northern Victoria Land (Antarctica), Tasmania and New Zealand. In G.D. McKenzie (ed.), Gondwana Six: Structure, Tectonics and Geophysics: 49-66. Washington D.C.: Geophysical Monograph Series 40.

Findlay, R.H. 1987b. Structural Geology of the Robertson Bay and Millen terranes, northern Victoria Land, Antarctica. In E. Stump (ed.), Geological investigations in northern Victoria Land: 91 - 114. Washington, D.C.: Antarctic Research Series 46.

Findlay, R.H., A.V. Brown & M.P. McClenaghan 1991. Confirmation of the correlation between lower Palaeozoic rocks in western Tasmania and northern Victoria Land, Antarctica and a revised tectonic interpretation. Memorie della Societa Geologica Italiana 46: 117-133.

Flöttmann, T. & G. Kleinschmidt 1991a. Kinematics of major structures of North Victoria and Oates Lands, Antarctica. Memorie della Societa Geologica Italiana 46: 273-282.

Flöttmann, T. & G. Kleinschmidt 1991b. Opposite thrust systems in northern Victoria Land, Antarctica: Imprints of Gondwana's Palaeozoic accretion. Geology 19: 45-47.

Foden, J.D., S.P. Turner & R.S. Morrison 1990. Tectonic implications of Delamerian magmatism in South Australia and western Victoria. In J.B. Jago & P.S. Moore (eds.), The evolution of a Late Precambrian - Early Palaeozoic rift complex: the Adelaide Geosyncline: 465-482. Brisbane: Geological Society of Australia Special Publication 16.

Ganovex and Italantartide Teams 1991. Preliminary geological-structural map of Wilson, Bowers and Robertson Bay terranes in the area between Aviator and Tucker Glaciers (northern Victoria Land, Antarctica). Memorie della Societa Geologica Italiana 46: 267-272.

Gatehouse, C.G., J.B Jago & J.A. Cooper 1990. Sedimentology and stratigraphy of the Carrickalinga Head Formation (low-stand fan to high-stand systems tract) Kanmantoo Group, South Australia. In J.B. Jago & P.S. Moore (eds.), The evolution of a Late Precambrian - Early Palaeozoic rift complex: the Adelaide Geosyncline: 351-368. Brisbane: Geological Society of Australia Special Publication 16.

Gibson, G.M. 1987. Metamorphism and deformation in the Bowers Supergroup: implications for terrane accretion in northern Victoria Land, Antarctica. In E.C. Leitch & E. Scheibner (eds.), Terrane accretion and orogenic belts: 207-219. Washington: Geophysical Monograph Series 42.

Gibson, G.M. & T.O. Wright 1985. Importance of thrust faulting in the tectonic development of northern Victoria Land, Antarctica. Nature 315: 480-483.

Grew, E.S., G. Kleinschmidt & W. Schubert 1984. Contrasting metamorphic belts in North Victoria Land, Antarctica. Geologisches Jahrbuch B60: 253-264.

Grew, E.S. & M. Sandiford 1984. A staurolite-talc assemblage in tourmaline- phlogopite-chlorite schists from northern Victoria Land, Antarctica, and its petrogenetic significance. Contributions to Mineralogy and Petrology 87: 337-350.

Grindley, G.W. & F.J. Davey 1982. The reconstruction of New Zealand, Australia and Antarctica. In C. Craddock (ed.), Antarctic Geoscience: 15-30. Madison: The University of Wisconsin Press.

Jago, J.B. 1981. Late Precambrian early Palaeozoic geological relationships between Tasmania and northern Victoria Land. In M.M. Cresswell & P. Vella (eds.), Gondwana Five: 199-205. Rotterdam: Balkema.

Jenkins, R.J.F. 1990. The Adelaide Fold Belt: tectonic reappraisal. In J.B. Jago & P.S. Moore (eds.), The evolution of a Late Precambrian-Early Palaeozoic rift complex: the Adelaide Geosyncline: 395-420. Brisbane: Geological Society of Australia Special Publication 16.

Jordan, H. 1981. Metasediments and metavolcanics in the Frolov Ridge, Bowers Mountains, Antarctica. Geologisches Jahrbuch B41: 139-154.

Kleinschmidt, G. & D.N.B. Skinner 1981. Deformation styles in the basement rocks of North Victoria Land, Antarctica. Geologisches Jahrbuch B41: 155-200.

Klimov, L.V. & D.S Soloviev 1958. A preliminary communication about geological observations in the eastern Antarctic. Information Bulletin of the Soviet Antarctic Expeditions 1: 27-30.

Klimov, L.V. & D.S Soloviev 1960. A correlation of the geological structures of the coastal areas of the Ross Sea and of Oates Land. Information Bulletin of the Soviet Antarctic Expeditions 16: 7-15.

Laird, M.G., R.A. Cooper & J.B. Jago 1977. New data on the lower Palaeozoic sequence of northern Victoria Land, and its significance for Australian-Antarctic relations in the Palaeozoic. Nature 265: 107-110.

Lombardo, B., B. Cappelli, L. Carmigiani, G. Gosso, I. Memmi, A. Montrasio, P.C. Pertusati, C.A. Ricci, F. Salvini & F. Talarici 1989. The metamorphic rocks of the Wilson Terrane between David and Mariner Glaciers, northern Victoria Land, Antarctica. Memorie della Societa Geologica Italiana 43: 99-130.

Mancktelow, N.S. 1990. The structure of the Adelaide Fold Belt, South Australia. In J.B. Jago & P.S. Moore (eds.), The evolution of a Late Precambrian - Early Palaeozoic rift complex: the Adelaide Geosyncline: 369-395. Brisbane: Geological Society of Australia Special Publication 16.

Mills, K.J. 1973. The structural geology of the Warren National Park and the western portion of the Mount Crawford State Forest, South Australia. Transactions of the Royal Society of South Australia 907: 281-315.

Milnes, A.R., W. Compston & B. Daily 1977. Pre- to syntectonic emplacement of early Palaeozoic granites in south eastern South Australia. Journal of the Geological Society of Australia 24: 87-106.

Offler, R. & P.D. Fleming 1968. A synthesis of folding and metamorphism in the Mt. Lofty Ranges, South Australia. Journal of the Geological Society of Australia 15: 245-266.

Oliver, R.L., J.A. Cooper & A.J. Truelove 1983: Petrology and zircon geochronology of Herring Island and Commonwealth bay and evidence for Gondwana reconstructions. In R.L. Oliver, P.R. James & J.B. Jago (eds.), Antarctic Earth Science: 64-68. Canberra: Australian Academy of Science.

Passchier, C.W. & C. Simpson 1986. Porphyroclast systems as kinematic indicators. Journal of Structural Geology 8: 831-843.

Platt, J.P. & R.L.M. Vissers 1980. Extensional structures in anisotropic rocks. Journal of Structural Geology 2: 379-410.

Preiss, W.V. 1987. The Adelaide Geosyncline - late Proterozoic stratigraphy, sedimentation, palaeontology and tectonics. Geological Survey of South Australia Bulletin 53: 1-438.

Sandiford, M., J. Foden, S. Zhou, & S.P. Turner. Granite genesis and the mechanics of convergent orogenic belts with application to the southern Adelaide Fold Belt. Transactions of the Royal Society of Edinburgh 83: 83-93.

Simpson, C. 1985. Deformation of granitic rocks across the brittle-ductile transition. Journal of Structural Geology 7: 503-511.

Simpson, C. & S.M. Schmid 1983. An evaluation of criteria to deduce the sense of movement in sheared rocks. Geological Society of America Bulletin 94: 1281-1288.

Steinhardt, C. 1991. The microstructural anatomy of a major thrust zone on Fleurieu Peninsula, South Australia. Australian Journal of Earth Sciences 38: 139-150.

Stump, E., A.J.R. White & S.G. Borg 1986. Reconstruction of Australia and Antarctica: evidence from granites and recent mapping. Earth and Planetary Science Letters 79: 348-360

Turner, S.P., C.J. Adams, T. Flöttmann & J.D. Foden, in press. The Glenelg River Complex. Australian Journal of Earth Sciences.

Turner, S.P., C.J. Adams & J.D. Foden 1991. Geochemistry and geochronology of the Glenelg River Complex, western Victoria. 8th International Symposium on Gondwana, Abstracts. Hobart: University of Tasmania: 86.

Vetter, U. & F. Tessensohn 1987. S- and I-type granitoids of North Victoria Land, Antarctica, and their inferred geotectonic setting. Geologische Rundschau 76: 233-243.

Vandenberg, A.H.M. 1978. The Tasman fold belt in Victoria. Tectonophysics 48: 267-297.

Weaver, S.D., J.D. Bradshaw & M.G. Laird, 1984. Geochemistry of Cambrian volcanics of the Bowers Supergroup and implications for the early Palaeozoic tectonic evolution of northern Victoria Land, Antarctica. Earth and Planetary Science Letters 68: 128-140.

Wells, B.E. 1956. Geology of the Casterton district. Proceedings of the Royal Society of Victoria 68: 85-110.
Wright, T.O. & C. Brodie 1987. The Handler Ridge Formation, a new unit of the Robertson Bay Group, northern Victoria Land, Antarctica. In G.D. McKenzie (ed.), Gondwana Six, Structure, Tectonics and Geophysics: 25-29 Washington D.C.: Geophysical Monograph 40.
Wright, T.O. & R.D. Dallmeyer 1991. The age of cleavage development in the Ross Orogen, northern Victoria Land, Antarctica: evidence from 40Ar/39Ar whole rock ages. Journal of Structural Geology 13: 677-690.

Gondwana Eight, Findlay, Unrug, Banks & Veevers (eds) © 1993 Balkema, Rotterdam. ISBN 90 5410 304 3

Structure and evolution of the Antarctic shield in Precambrian

E.N. Kamenev
North Branch for Marine Geologic Exploration 'Sevmorgeologia', Russia

ABSTRACT: The following major structural types are created by the Precambrian tectonic environments of the Antarctic shield: 1) Archean protocratonic blocks; 2) Proterozoic Wegener-Mawson mobile belt and 3) Late Proterozoic sedimentary-volcanogenic protoaulacogens. Each type reflects successive stages of evolution of the Antarctic shield in Precambrian. Protocratonic blocks were formed between 3900 and 2700 Ma, when the environments of the charnockite-enderbite nuclei and of the granite-greenstone terranes dominated. The Wegener-Mawson mobile belt strikes throughout East Antarctica from east to west. It was formed between 2500 and 500 Ma, when the environments of granulite-gneiss terranes dominated. A polycyclic evolution with four cycles of tectonic activity culminating at 2400, 1700, 1000 and 550 Ma is a very important feature of this mobile belt. Metamorphic prograde and retrograde zoning in the mobile belt is best explained by its nappe-block structure, caused by accretion, collision and amalgamation of protocratons. Proterozoic protoaulacogens were formed between 1200 and 500 Ma. Their formation marked the transition to the rift stage evolution of the Antarctic shield. This kind of structural interpretation of the Antarctic shield allows a more accurate definition of the position of Antarctica within Gondwana. In particular the Wegener-Mawson mobile belt, the Albany-Fraser Province of Australia, the Mozambique belt of Africa form part of a network of Proterozoic mobile belts.

1 INTRODUCTION

Between 30°W and 100°E East Antarctica is a giant Precambrian shield where cover sedimentary rocks have been eroded or have never accumulated. Mountain chains, isolated nunataks and ice-free hills facing the Indo-Atlantic coast of Antarctica are built up predominantly of intensively deformed Precambrian rocks with prevailing granulite and amphibolite facies assemblages. Recent aeromagnetic and bedrock topography data indicate those partly exposed coastal mountain areas may be continuous with the sub-ice Gamburtzev Mountains and other highlands in the Antarctic interior.

From the late 1950/s many of the individual well exposed parts of the Antarctic shield were subjected to detailed geologic and petrologic investigations.Following small scale mapping the main features of geologic structure and history of some vast mountaineous regions are known now also. During the last two decades, geochronological studies have made a major contribution to understanding the Precambrian history of Antarctica. These and other results from Antarctic Precambrian research allow the development of a general conception of structure and evolution of the Antarctic Precambrian shield as a whole.

2 PRECAMBRIAN TECTONIC ENWIRONMENTS

Because of the extremely long Precambrian geologic record, many geological features and events typical of Precambrian history, such as the order and peculiarities of deformational episodes, metamorphic and magmatic processes, lithology of individual units, metallogenic considerations, etc. have often been used to define and correlate paleotectonic environments of widely variable ages. A similar approach has been applied in Antarctic Precambrian studies carried out by the Australian, Belgian, British, German, Japanese, Norwegian, Soviet, South African and other expeditions since the early days of Antarctic geology research. As the result of these multinational cooperative efforts, it is now possible to recognize four major types of paleotectonic environments (Fig.1).

A - of belts of highly metamorphosed rocks (granulite-gneiss terranes) charac - characterized by intense and repeated tectonic deformations, granulite and high-grade amphibolite facies metamorphism, postkinematic subalkaline plutonic activity and the presence of some supracrustal rocks and mafic suites,including stratified mafic and anorthosite complexes;

B - of granite-greenstone belts (granite-greenstone terranes) built up of granite batholiths, domes, slabs and nappes in complex combination with nappes, slabs, piles of slabs and lenses of supracrustal sedimentary-volcanic rocks. The latter are usually characterized by a relatively weak, often zonal metamorphism ranging from the low-grade greenschist facies to the initial migmatitic grade of the amphibolite facies, and by intensive Alpine-type deformations;

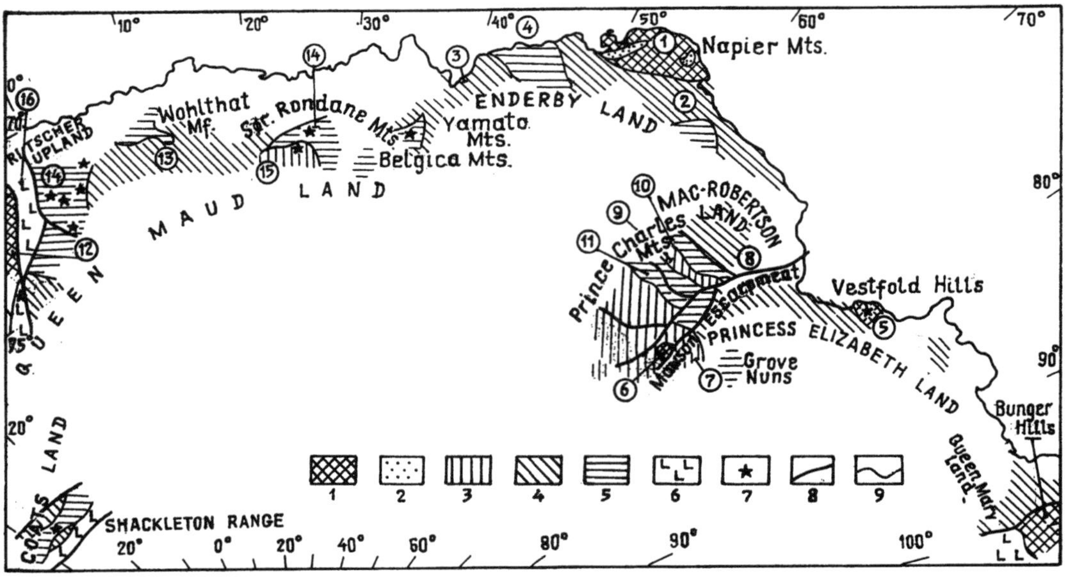

Fig.1 Distribution of the Precambrian tectonic environments in the exposed parts of the Antarctic shield. Compiled by Kamenev E.N. Explanations: 1- grey gneiss terranes (mostly charnockite-enderbite nuclei); 2- fragments of supracrustal sequences in the nuclei;3- granite-greenstone terranes; 4- charnockite-granulite belts; 5- granite-gneiss-schist belts; 6- sedimentary-volcanic troughs; 7- areas of tectono-thermal activations in the Phanerozoic; 8- axes of the rifts; 10- contours of the terranes with distinct tectonic environments. Numbers in circles: 1- Napier charnockite-enderbite nucleus; 2- Rayner charnockite-granulite belt; 3- Ongul charnockite-granulite belt; 4- Polkanov granite-gneiss-schist belt; 5- Vestfold charnockite-enderbite nucleus; 6- Mawson Escarpment charnockite-enderbite nucleus; 7- Ruker granite-greenstone terrane; 8- Beaver charnockite-granulite belt; 9- Fisher granite-greenstone terrane; 10-11- granite-gneiss-schist belts between Fisher and Ruker terranes (10) and between Beaver belt and Fisher terrane (11); 12- West Ritscher granite-grey gneiss nucleus; 13- Humboldt charnockite-granulite belt; 14- Sor-Rondane and Sverdrupfjella granite-gneiss-schist belts; 15- Ritscher sedimentary-volcanic trough

C - of linear zones or troughs with a thick sedimentary and volcanic fill undeformed or slightly deformed and metamorphosed under prehnite-pumpellyite to low-grade greenschist facies conditions and/or hydrothermally altered; mafic and subalkaline intrusions of subcrustal derivation are common;

D - of areas of tectono-thermal activation superimposed on all previously created Precambrian structures; they control distribution of the so-called post-cratonic basic, ultrabasic and alcaline magmatism mainly related to Phanerozoic rifting.

The paleotectonic environments of granulite-gneiss terranes are further sub-divided into three subtypes:

AI - of grey gneiss terranes (or nuclei of shield) built up predominantly of orthogneisses (charnockite-enderbite or granite-tonalite composition) with linear fragments and xenoliths of basic rocks and supracrustal folded sequences; the latter are typically composed of metavolcanic rocks and quartzites (including banded iron formations) and do not contain carbonate rocks.

A2 - of charnockite-granulite belts where charnockites are associated with abundant supracrustal rocks and anorthosites metamorphosed in the granulite facies; supracrustal sequences contain usually carbonate and calc-silicate rocks as well as a wide spectrum of metasedimentary and metavolcanic rocks;

A3 - of granite-gneiss-schist belts of predominantly supracrustal rocks metamorphosed in the high grade amphibolite facies; abundant migmatites and post deformational bodies and veins of granite and pegmatite are common; in places the granite-gneiss-schist belts represent transitional zones between the granite-greenstone and charnockite-granulite belts.

Archean plutonic orthogneiss bodies and supracrustal rock complexes are established on the Antarctic shield only in AI and B paleotectonic environments. A typical example of the Archean grey gneiss terrane is represented by the Napier Complex of Enderby Land (Kamenev, 1972; Kamenev, 1975; Ravich and Kamenev, 1975; Sheraton et al., 1980; James and Black, 1981; Sheraton, 1985; Sheraton et al., 1987; Ivanov and Kamenev, 1990). In this area gneissic enderbites, gneissic mesoperthite charnockites and garnet granite gneisses dominate and the portion of supracrustal sedimentary-volcanogenic rocks (Tula Supracrustals) is rather small.

The central part of the Napier Complex terrane consists of gneissic mesoperthitic charnockite intruded by garnetiferous granite gneiss. These rocks together compose deformed domal and oval structures inside and between which minor tectonic lenses and xenoliths of supracrustal, basic and ultrabasic rocks are present. Similar deformed domes and ovals of gneissic enderbites with layers and lens-like inclusions of pyroxene-plagioclase schists (metabasites) are widespread to the west and east of the main charnockite fields.The largest bodies (lenses, bands, layers, piles of beds, tectonic slabs, xenoliths) of the Tula Supracrustals are located between and within the gneissic enderbites.The Tula Supracrustals includes banded mesoperthitic garnet gneisses and migmatites, pyroxene-plagioclase schists, quartzites (with spinel, sapphirine, osumilite, cordierite, hypersthene, garnet), garnet -sillimanite and hypersthene-sillimanite plagiogneisses, meta-ultrabasites, meta-anorthosites and banded pyroxene-quartz-magnetite schists. The protoliths of this supracrustal rock suite are very similar to the ordinary,but slightly metamorphosed Archean greenstone sequences. The complicated multiple recumbent folds, tectonic slabs and nappes, tectonic layering which result from several deformations are very characteristic of the Tula Supracrustals. The layering was produced by intense tectonic activity dominated by horisontal movements involving major thrusting and overfolding and was accompanied by the emplacement of very large amounts of granitic material as sub-horisontal sheets. Schists, ortho- and paragneisses are the products of three stage of high grade granulite metamorphism ($T=1000^{o}$-$850^{o}C$, P=5-10 kbars) culminating about 3500, 3100 and 2800 Ma (James and Black, 1981: Sheraton, 1985; Sheraton et al., 1987; Belyatsky et al., 1990; Ivanov and Kamenev, 1990). There is some isotopic evidence of a prolonged (from 3900 Ma) pre-granulite facies crustal history for at least some Napier Complex rocks (Black et al.,1986). The P-T data suggest that this area became an essentially stabilized continental craton following deformational and metamorphic events at 2800 Ma (Harley,1983). Further, the Napier Complex was affected by younger events connected with crustal development of the neighboring Proterozoic charnockite-granulite belt (see Table 1, Napier protocratonic block) (James and Tingey, 1983; Sheraton et al.,1987; Ivanov and Kamenev, 1990). Other areas of AI environment of the Antarctic shield are represented by smaller fragments at the Vestfold Hills (Princess Elizabeth Land) (Collerson et al., 1983; Sheraton et al., 1984),possibly Bunger Hills (Queen Mary Land), Anandagstoppane Mts (Queen Maud Land) (Wolmarans and Kent, 1982) and the Read Mountains (Shackleton Range) (Pankhurst et al., 1983) (Fig.1). With the exception of the Vestfold Hills,only few geological events have been dated in these regions. At the Vestfold Hills, Archean and younger events were dated properly. Almost all dated Archean events in these places, particularly the 2800 Ma event (age of stabilization) coincide with some usually similar Archean events in the Napier Complex, although older 3100 Ma events are also known (see Tables 2 and 3, protocratonic blocks).

The only example of Archean granite-greenstone terrane is represented by the Ruker Complex of Southern Prince Charles Mountains (Tingey,1982; Tingey and Convine, 1982; Belyatsky and Kamenev, new unpublished data). This terrane comprises several individual greenstone belts intermingled with different kinds of granitoids.Large amounts of basic and acid volcanics, ultramafic rocks (komatiite type), different kinds of quartzites (including banded iron formation), tuffs and conglomerates, minor metapelites, marbles and gyprocks are characteristic of the supracrustal sequences of the Ruker Complex. Some of the granitoids represent reworked blocks of sialic basement, but more often they have been derived from pretectonic and syntectonic acid intrusives. The structure of the Ruker Terrane is dominated by tight folding, thrusting and shearing. The majority of the large structures are the tectonic slabs, nappes, folded nappes and domes with intensive internal folding, shearing and cleavage. Tectonic breccias, melange and conglomerates with tectonically rounded boulders and pebbles are very common. These features are the result of 2 or 3 deformations.

In the central part of the Ruker Complex the metamorphic grade corresponds to the greenschist facies, in an adjacent to the greenschists zone - to the epidote-amphibolite facies and in the marginal part - to the almandine-amphibolite and also cummingtonite-amphibolite facies. The metamorphic zonation is subparallel to the trend of the tectonic features.

Isotopic data indicate that evolution of the greenstone belts (volcanism, sedimentation, granite intrusion, deformation and metamorphism) occured between 3100 and 2700 Ma (Tingey,1982; Tingey and Convine, 1982; Belyatsky and Kamenev, unpublished data).

Initially, volcanic-sedimentary troughs developed upon and laterally to a much older granulite-gneiss terrane that had existed here since about 3300 Ma. Subsequent tectono-thermal activation of the Ruker Complex is indicated by the intrusion of metabasites (amphibolite dykes), progressive amphibolite facies metamorphism,and the intrusion of granite and pegmatite veins and plugs dated at 2400, 2100, 1500, 1300, 950 Ma. These events were contemporaneous with the crustal development of neighboring Proterozoic granulite-gneiss terrane (see Table 2, Southern Lambert Block) (Tingey and Convine, 1982 ; Belyatsky and Kamenev unpublished data).

Proterozoic supracrustal and plutonic rock complexes are established on Antarctic shield in A2, A3,B,C and D paleotectonic environments. The most widespread environments in the Proterozoic of the Antarctic shield are the granulite- gneiss terranes where tectonically intermixed charnockite-granulite and granite-gneiss-schist belts constitute polymetamorphic systems (Van Autenboer and Loy, 1972; Ravich and Kamenev, 1975; Grew, 1978; Tingey, 1982; Yoshida and Kizaki, 1983; James and Tingey, 1983; Sheraton et al.,1984; Bormann et al., 1986; Kojima and Shiraishi, 1986; Kamenev, 1987; Black et al., 1987; Ivanov and Kamenev, 1990). Although there are individual charnockite-granulite and granite-gneiss-schist belts, usually the charnockite-granulite belts are adjacent to the Archean enderbite-charnockite nuclei and contain their allochtonous residuals,where as the granite-gneiss-schist belts are adjacent to the granite-greenstone terranes. To distiguish them in a border zone is a very complicated problem that requires detailed studies of tectonic, metamorphic, igneous and geochronologic features.

Table 1. The dated geological events in the protocratonic block and the Wegener-Mawson Mobile Belt at the Enderby Land.

Ma	Napier Block (charnockite-enderbite nucleus)	Wegener-Mawson Belt (charnockite- granulite and granite-gneiss-schist belts)
0.5		0.5 ep.-amph. facies metamorphism locally leucogranites and pegmatites subalkaline granitoid plutons
0.6	0.6 pegmatites and leucogranites locally	
0.7		
		minor leucogranite intrusions
0.8	0.8 mylonites	0.8 D4, granulite and amphibolite facies metamorphism
0.9		
1.0		
1.1		1.1 D3, granulite facies metamorphism
1.2	1.2 dolerite dykes	1.2 dolerite dykes
1.3		
1.4		
		granitoids
1.5		1.5 anorthosites
1.6		
1.7		1.7 D2, granulite facies metamorphism
1.8		
1.9		1.9 granitoids
2.0		
2.1		
2.2	2.2 pegmatites locally	
2.3		
2.4	2.4 metabasite dykes subalkaline granitoid plutons	2.4 supracrustals, anorthosites
2.5	2.5 D4, granulite and amphibolite facies diaphtoresis locally	2.5 D1, granulite facies metamorphism
2.6		
2.7		
2.8	2.8 D3, granulite facies metamorphism	
2.9		
3.0		3.0 relics of basement
3.1	3.1 D2, granulite facies metamorphism	
3.2		
3.3		
3.4		
3.5	3.3 D1, granulite facies metamorphism granitoids	
3.6	3.6 supracrustals	
3.7		
3.8		
3.9	3.9 granitoids	

Data source includes: Grew,1978; Yoshida,1979; Sheraton et al.,1980; James and Black,1981; James and Tingey,1983; Harley,1983; Yoshida and Kizaki,1983; Sheraton,1985; Black et al.,1986; Black et al.,1987; Sheraton et al.,1987; Belyatsky et al.,1990; Ivanov and Kamenev,1990;

The rocks of Proterozoic granulite-gneiss terranes are typically represented by ortho- and paragneisses in near equal proportion. Metabasites, anorthosite plutons, metapelites, marbles and calc-silicate rocks are common within the sequences of dominantly biotite-hypersthene, biotite-garnet, amphibole and biotite gneisses. Large

Table 2. The dated geological events in the protocratonic blocks and the Wegener-Mawson Mobile Belt at the mountaneous frame of the Lambert Glacier and the Amery Ice Shelf

Ma	Westfold Block (charnockite-enderbite nucleus)	Wegener-Mawson Belt (charnockite-granulite, granite-gneiss-schist and granite- greenstone belts)	Sothern Lambert Block (relics of charnockite-enderbite nucleus & granite- greenstone belt)
		mylonites	
0.5		0.5 leucogranites, pegmatites subalkaline granitoids	0.5 pegmatites, leucogranites
0.6		D4, amph.fcs diaphthores. mylonites	
0.7		pegmatites	
0.8		0.8 granitoid plutons	
0.9			
1.0	1.0 mylonites		pegmatites
1.1		1.1 D3, granulite, amph. & ep.-amph. facies metamorphism	1.1 D5, amphibolite facies metamorphism locally
1.2		granitoids	pegmatites
1.3	1.3 basic dykes	1.3 supracrustals, basite plutons	1.3 D4, amphibolite facies metamorphism locally
1.4		1.4 granitoids	
1.5		1.5 supracrustals	
1.6			1.6 pegmatites
1.7		1.7 granitoids supracrustals	
1.8	1.8 basic dykes pegmatites	D2, amph. & granulite facies metamorphism	
1.9	D4, amphibolite metamorphism locally	1.9 granitoids ultrabasite protrusions	
2.0			
2.1		2.1 supracrustals	
2.2		2.2 granitoids	
2.3			pegmatites
	2.35 basic dykes	D1, metamorphism	D3, low grade metamorph.
2.4		2.4 basite plutons supracrustals	2.4 basic dykes
2.5	2.5 D3, granulite facies metamorphism granitoids		
2.6			
2.7			
2.8	2.8 D2, granulite facies metamorphism		2.8 greenschist & amphibolite facies metamorphism
2.9			
3.0	D1, granulite facies 3.0 metamorph., granitoids supracrustals	3.0 relics of basement	3.0 granitoids basite plutons
3.1			3.1 supracrustals
3.2			
3.3			3.3 D1, granulite facies metamorphism
3.4			
3.5			
3.6	3.6 granitoids		3.6 granitoids

Data source includes: Tingey,1982; Tingey and Convine,1982; Collerson et al.,1983; James and Tingey,1983; Ravich et al.,1984; Sheraton et al.,1984; Ivanov and Kamenev, 1990; Belyatsky and Kamenev, unpublished data.

Table 3. The dated geological events in the protocratonic block and the Wegener-Mawson Mobile Belt at the Queen Maud Land

Ma	Western Ritscher Block (grey gneiss nucleus)	Wegener-Mawson Belt (charnockite-granulite, granite-gneiss-schist and granite- greenstone belts)	Ritscher Protoaulacogene (sedimentary-volcanic trough)
		leucogranites, pegmatites	
0.5		0.5 subalkaline granitoids	
		D4, amphibolite facies metamorphism	subalkaline granitoids
0.6		locally	0.6 greenschist & epidote- amphibolite facies m-m
0.7			
			basic sills & dykes
0.8			0.8 supracrustals
0.9			
		granite plutons	
1.0		1.0 D3, metamorphism	
		basite plutons	
1.1		1.1 supracrustals	
			basic sills & dykes
1.2			1.2 supracrustals
1.3			
1.4			
1.5			
		D2,amph. & granul.fcs	
1.6		1.6 granitoids	
		supracrustals	
1.7			
1.8			1.8 basite plutons
1.9			
2.0		2.0 mylonites	
2.1			
2.2		2.2 D1, granulite facies metamorphism	2.2 basite plutons
2.3			
2.4		2.4 anorthosites	
2.5			
2.6			
2.7		2.7 relics of basement	
2.8	2.8 granitoids		
2.9			
3.0	3.0 granitoids		
3.1			
3.2	3.2 granitoids		

Data source includes: Pasteels et Michot, 1968; Van Autenboer and Loy,1972; Ravich and Kamenev, 1975; Wolmarans and Kent,1982; Asami and Shiraishi,1985; Bormann et al.,1986; Kojima and Shiraishi,1986; Kamenev,1987; Ivanov and Kamenev,1990.

amounts of migmatites, granite and pegmatite veins and plugs are very characteristic of these areas. The majority of postkinematic sub-alkaline plutonic bodies are concentrated there also.

There is evidence throughout the Antarctic Proterozoic granulite-gneiss terranes of polyphase metamorphism and multiple (from 2 to 4) deformation (Yoshida, 1979; Ravich et al., 1984; Asami and Shirashi, 1985; Black et al., 1987). As a result of this there was created a linear structure of these granulite-gneiss terranes with initial isoclinal folding overprinted by open folds, deformed dome-like structures, thrusting, shearing and overthrusting. Steeply dipping tectonic slabs and sheets are very typical for these terranes (Ivanov and Kamenev, 1990).

Isotopic data indicate that evolution of the Proterozoic granulite-gneiss terrains occured between 2500 and 500 Ma (Pasteels and Michot, 1968; Tingey, 1982; Yoshida and Kizaki, 1983; Tingey, 1983; Pankhurst et al., 1983;

Sheraton et al., 1984; Sheraton, 1985; Black et al., 1987; Belyatsky et al., 1990). The most common cycles of tectonic activity culminate at approximately 2400, 1700, 1000 and 550 Ma and the last two cycles are manifested clearly everywhere (see the Tables 1, 2, 3, Wegener-Mawson mobile belt).

Proterozoic granite-greenstone terranes occur in two localities: in the Sør-Rondane Mountains (Kamenev, 1987) and in the Prince Charles Mountains (Ivanov and Kamenev,1990). Evolution of both of these terranes occured during the Late Proterozoic between 1300 and 850 Ma (Pasteels et Michot, 1968 Belyatsky and Kamenev unpublished data). The better exposed rocks of the Fisher Terrane in the Prince Charles Mountains are entirely faulted against rocks of the Proterozoic granulite-gneiss terranes. The rocks of the Fisher Terrane are mostly volcanics and intrusives of mafic, intermediate and felsic composition metamorphosed to epidote-amphibolite facies grade. The metavolcanics are interleaved with minor intrusive bodies of porphyritic rocks, lenses of banded iron formation, calcareous tuffs, felsic tuffaceous pyroclastic rocks, and metacherts. Tectonic sheets of supracrustal units are intercalated with the slabs of coarse-grained plutonic rocks. The bodies of the latter extend colaterally with thrust faults and axial zones of folds in the enclosing metavolcanic sequences. The petrographic composition of the plutonic rocks varies from gabbro to granite.Mafic intrusions came first. Gabbro and diorite plutons are most abundant.The plutonic rocks show the same deformation and metamorphic history as the volcanic units. The structure of the Proterozoic granite-greenstone terranes is still unclear. There are piles of tectonic slabs, thrust fault zones, internal isoclinal folding, schistosity in metavolcanics, and shear zones inside the plutons. Locally there is evidence of two folding episodes. The strike of the tectonic features of the granite-greenstone terranes is similar to the strike of the adjacent granite-gneiss-schist belts.

The only good example of the Proterozoic troughs is the area of the Ritscher Upland in the Western Queen Maud Land (Wolmarans and Kent, 1982; Ivanov and Kamenev, 1990). The supracrustals of the Sandow and Amundsen Mounts in the Denman Glacier area represent the same environment probably. Thick sedimentary volcanic fill of the Proterozoic trough includes mainly the slightly metamorphosed lavas, ignimbrites and tuffs of acid, intermediate and basic compositions. Sedimentary terrigenous rocks are located in the lower parts of the section. The bedding is generally subhorizontal,although near the adjacent granite-gneiss-schist belt there are some gentle folds and monoclinal structures and dip of bedding may attain 50 degrees.

The supracrustals are intruded by sills and stocks of trapps and subalkaline granitoids. Some of the most thick trapp sills are layered.

Isotopic data indicate that the evolution of the trough area occured mostly between 1200 and 500 Ma. However there were precursory layered basic intrusions about 2200 and/or 1800 Ma (Wolmarans and Kent. 1982).

Summarizing the above the evolution of the Antarctic shield through the paleotectonic environments is integrated in Fig.2. This evolution,and the distribution of the proposed paleotectonic environments throughout the shield, support the existence of the following major structural types in the Precambrian Antarctic shield : 1) Archean protocratonic blocks (APBs) (fragments of protocratonic basement); 2) Proterozoic Wegener-Mawson Mobile Belt (PWMMB) (a branch of the ancient protogeosynclinal network) and 3) Late Proterozoic sedimentary -volcanogenic protoaulacogens or protorifts. Evidently each type reflects successive stages of evolution of the Antarctic shield in Precambrian.

3 ARCHEAN STRUCTURAL TYPES AND THEIR EVOLUTION

Archean protocratonic blocks are situated on the flanks of the Proterozoic Wegener-Mawson mobile belt which is described in the next section. These are (from east to west) Bunger (Queen Mary Land), Vestfold (Princess Elizabeth Land), Napier (Enderby Land) and West Ritscher (Queen Maud Land) APBs on the northern flank of the PWMMB, and South Lambert (Princess Elizabeth Land and McRobertson Land) and Read APBs (Shackleton Range) on the southern flank of the PWMMB (Fig.3). Coastal parts of northern APBs were activated intensively in the Proterozoic or bordered by narrow zones of the PWMMB.

We do not know much about the initial stages of evolution of the APBs: however,the most ancient isotopic ages of enderbites in Napier Block of about 3900 Ma (Black et al.) record a generation of Archean tectono-thermal cycles as old as the formation of the Earth crust. The products of the next stage of the APBs evolution (3600-3100 Ma) (James and Black, 1981; Collerson et al., 1983; Harley, 1983; James and Tingey, 1983; Sheraton, 1985; Sheraton et al., 1987; Belyatsky et al., 1990; Ivanov and Kamenev, 1990) are represented by ultrametamorphic plutonic domes formed of enderbitic and charnockitic gneisses and granite gneisses, garnetiferous granite gneisses, tonalitic and trondhjemitic gneisses and granite gneisses. Relics of the oldest greenstone belts, ultramafic bodies and mafic plutons occur between these domes and as xenoliths in the granitoids (AI environment). The oldest supracrustal belts are characterized by the absence of carbonate rocks.The highest geothermal gradient (Ravich and Kamenev,1975; Harley,1983; Sheraton et al.,1987; Ivanov and Kamenev,1990) is a very distinctive feature of this stage. Metamorphism corresponds to the high grade granulite facies. Relics of the greenstone belts show evidence of the metamorphic conditions in case of lower geothermal gradient.

Typical bimodal greenstone belts (B environment) were originated at the final stage of APB's creation (3100-2700 Ma) on the basement (protocratonic nuclei) formed of the products of the preceding stages (see Table 2,Fig.2).The protocratonic nuclei were enriched by new portions of granites and basites and retrogressively remetamorphosed (D environment). The greenstone belts of that stage are characterized by the wide distribution of basic and acid volcanics and volcanoclastics, occurrence of komatiites, chemical precipitates (quartzites, carbonate giprocks, carbonate rocks), terrigenous sediments (conglomerates, sandstones), and intrusion of granites. Metamorphism of the greenstone belts varied from greenschist facies to low grade amphibolite facies. After the formation of Late

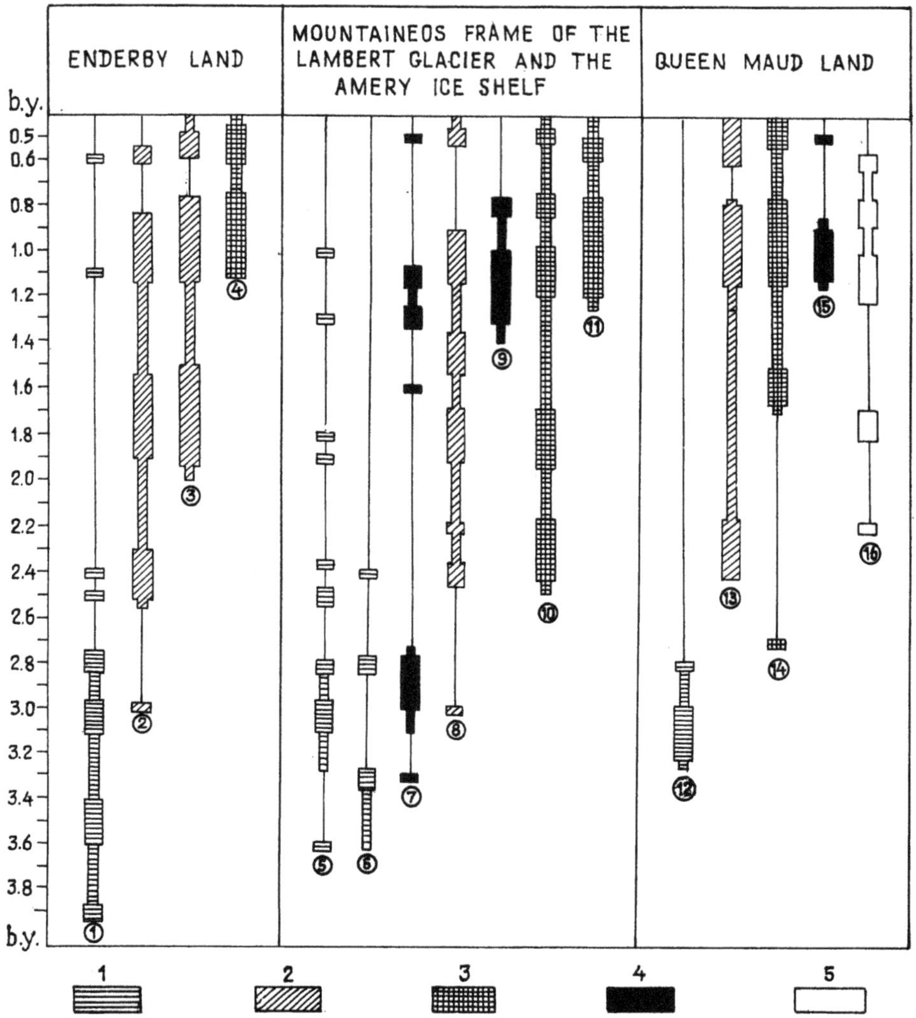

Fig.2 Presentment of the evolution of the Antarctic shield in Precambrian (through the main mountaineous regions). Compiled by Kamenev E.N. Explanations: 1- charnockite-enderbite or granite-grey gneiss nuclei; 2- charnockite-granulite belts; 3- granite-gneiss-schist belts; 4- granite-greenstone terranes; 5- sedimentary-volcanic troughs. Numbers in circles have the same meanings as on Fig.1. Wide boxes of the columns - well confirmed geological events, narrow boxes of the columns - less confirmed geological events; lines of the columns - no data

Archean greenstone belts, cratonic regime was persisted up to 2500 Ma (see Table 2,Fig 2). Younger geological events in the APBs were synchronous with the impulses of tectonic activity in the PWMMB and realized in local repeated metamorphism,and intrusion of granite,and basic and alkaline magmatism which constitutes the D environment.

4 PROTEROZOIC STRUCTURAL TYPES AND ELEMENTS: THE WEGENER-MAWSON MOBILE BELT

The PWMMB strikes throughout the whole shield from east to west and includes the APBs as marginal or internal basement blocks (Fig 3). The PWMMB evolved polycyclically from 2500 Ma and was not stabilized up to the Early Paleozoic. A full cycle of the evolution included supracrustal formations, basic and ultrabasic intrusions, granite formation, folding, metamorphism and postkinematic intrusion of subalkaline plutons, aplite, pegmatite and leucogranite veins. Four tectonic cycles at

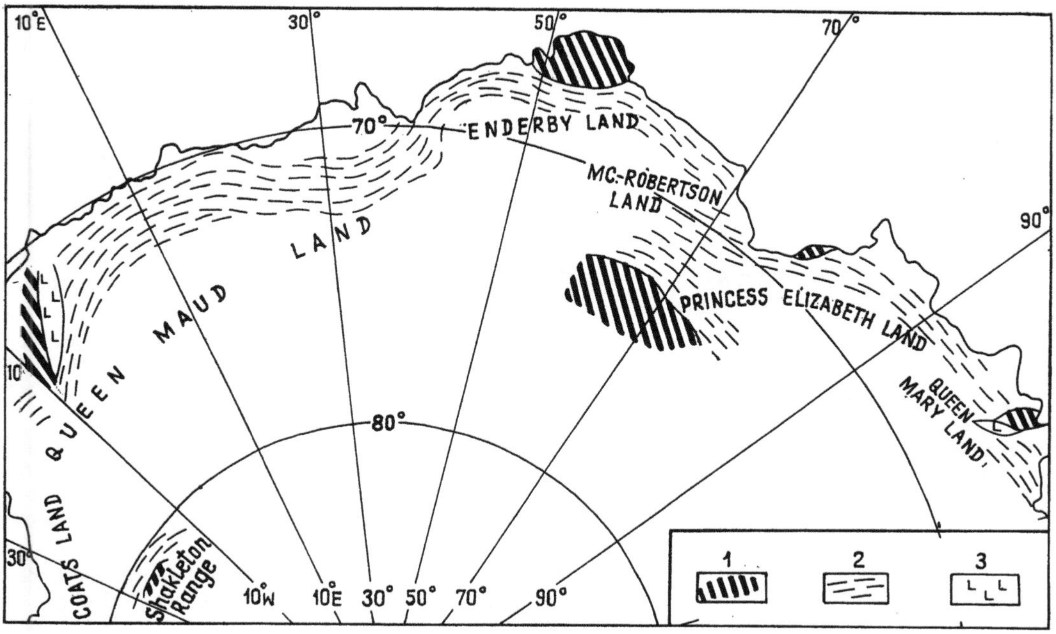

Fig.3 Main Precambrian structural types of the Antarctic shield. Compiled by Kamenev E.N. Explanations: 1 - Archean protocratonic blocks; 2 - Wegener-Mawson Mobile Belt; 3 - Late Proterozoic aulacogenes

approximately 2400, 1700, 1000 and 550 Ma are recognized in the A2, A3 and B environments of the PWMMB. Few of the cycles are manifested fully and uniformly through area of the PWMMB. Only that cycle which culminates at 1000 Ma is most evident and is completely developed almost everywhere from early to late stages.Only late stage of final cycle which culminated at 550 Ma is manifested through postkinematic intrusion of sub-alcaline plutons, aplite, pegmatite and leucogranite veins and local metamorphism. It appears that earlier stages of final cycle are absent. The whole PWMMB is characterized by ensialic evolution and the predominance of A2 and A3 environments. The B environment is rare, beeng seen only locally or as minor ultrabasic protrusions or chains of basic plutons.

The structure of the PWMMB is very complicated. Regional combinations of exposed metamorphic belts form distinct systems of nappes, slabs, overthrust sheets, recumbent folds, and blocks and domes of intrusive rocks, all separated by regional transcurrent or thrust faults. The sub-latitudinal distribution of the metamorphic zonality (progressive and retrogressive), which is subparallel to the strike of the PWMMB, the abundance of the thrusts and underthrusts, recumbent folds and nappes, the distribution of the isotopic ages, all support the concept that the PWMMB formed as a result of accretion, collision and amalgamation of the protocratons moving sublongitudinally. Evidently, the northern craton (or cratons) have been thrusted over the southern craton. As a result of that the northern craton have been subjected to retrograde metamorphism and subsequently eroded deeply, and southern craton suffered prograde metamorphism and preserved the greenstone belts from erosion.

Late Proterozoic aulacogens are located on the margins of the Antarctic shield and strike obliquely to the main strike of the PWMMB. They were formed between 1200 and 500 Ma (see Table 2 and 3, Fig.2), at the same time as were formed the Proterozoic greenstone belts of the PWMMB. The fill of both aulacogenes and Proterozoic greenstone belts is very similar but their structure is very different as stated above due to the differences of strain direction and intensities. The structure of aulacogenes indicates a decrease of tectonic activity in the marginal parts of the shield during Late Proterozoic times and they are associated with a tectonic regime which is transitional to the rift stage of development; this charcterizes the East Antarctic craton in the Phanerozoic (D environment) after 450 Ma (Ivanov and Kamenev,1990).

5 CONCLUSION

In the Gondwana reconstruction (Fig.4) the Proterozoic Wegener-Mawson Mobile Belt of the Antarctic shield is contiguous or continuous with the Proterozoic mobile belts of Australia (Albany-Fraser Province), Peninsular India (Eastern Ghats), Sri Lanka, Madagascar and Africa (Mosambique Belt). Evidently the PWMMB is a branch of the Gondwanian or even global system of the Proterozoic mobile belts. The Antarctic northern Archean protocratonic blocks are disposed inside such a mobile belt and may be the giant allochtonous relics of the single

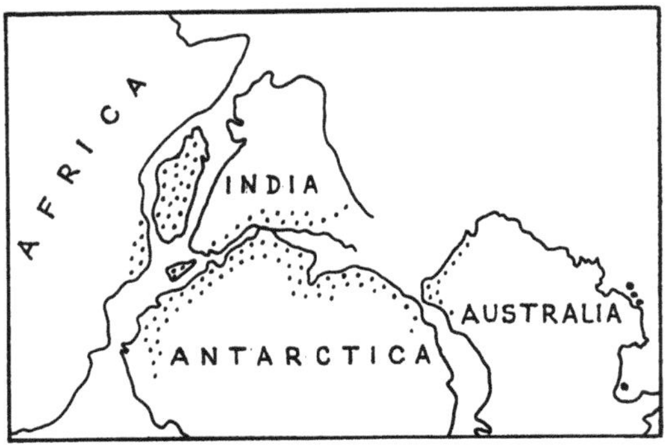

Fig.4 Proterozoic mobile belts (dotted areas) on the Gondwana reassembly. Gondwana reassembly by the participants of the Reunite Gondwanaland workshop at the University of Witwatersrand, Johannesburg 1979 (Cresswell and Vella, 1981)

Archean protocraton. Thus the breakup of the Gondwana supercontinent appears to have followed for the most part this Proterozoic mobile belt system. The first manifestations of such split seem to have been along the Late Proterozoic aulacogens.

REFERENCES

Asami M.,and Shiraishi K., 1985. Retrograde metamorphism in the Yamato Mountains, East Antarctica. Mem.Inst.Pol.Res.,N 37, p. 147-163.

Belyatsky B.V., Krylov D.P., Levsky L.K., Grikurov G.E., 1990. Zircon geochronology of the granulite complexes of Enderby Land, East Antarctica. Zbl.Geol. Paläont. Teil I (1/2). Stutgart,p.1-18.

Black L.P., Williams T.S., Compston W., 1986. Four zircon ages from one rock: the history of a 3930 Ma-old granulite from Mount Sones, Enderby Land, Antarctica. Contrib.Mineral.Petrol.,v.94, N 4, p.427-437.

Black L.P., Williams T.S., Sheraton J.W., 1987. The Rayner Complex of East Antarctica: complex isotopic systematics within a proterozoic mobile belt. J.Metamorphic Geology, v.5, N 1, p.1-26.

Bormann P., Bankwitz P., Bankwitz E., Damm V.,Hurtig E.,Kämpf H., PAECH H.-J., Schäfer U., Stackenbrandt H., 1986. Structure and development of the passive continental margin across the Princess Astrid Coast, East Antarctica. Journal of Geodynamics, 6, p.347-373.

Collerson K.D., Reid E., Millar D., McCulloch M.T., 1983. Lithological and Sm-Nd isotopic relationships in the Vestfold Block:implications for Archaean and Proterozoic crustal evolution in the East Antarctica. In:Oliver R.L., James P.R. and Jago J.B.(eds.), Antarctic Earth Science. Australian Academy of Science,Canberra, p.77-84.

Cresswell M.M. and Vella P., eds., 1981. Gondwana Five. A.A. Balkema, Rotterdam, 339 pp. Grew E.S., 1978. Precambrian basement at Molodezhnaya Station, East Antarctica. Geol.Soc.Am.Bull., v.89, p.801-813.

Harley S.L.,1983. Regional Geobarometry-geothermometry and metamorphic evolution of Enderby Land, Antarctica. In: Oliver R.L., James P.R. and Jago J.B. (eds.), Antarctic Earth Science. Australian Academy of Science, Canberra, p.25-30.

Ivanov V.L. and Kamenev E.N., (eds.), 1990. Geology and mineral resources of Antarctica (in Russian).Nedra, Moscow, 232 pp.

James P.R. and Black L.P., 1981. A review of the structural evolution and geochronology of the Archean Napier Complex of Enderby Land, Australian Antarctic territory. Geol. Soc. Austr., Spec.Pub., 7, p.71-83.

James P.R. and Tingey R.J., 1983. The Precambrian geological evolution of the East Antarctic metamorphic shield: a review. In:Oliver R.L., James P.R. and Jago J.B. (eds.), Antarctic Earth Science. Australian Academy of Science, Canberra, p.5-10.

Kamenev E.N., 1972. Geological structure of Enderby Land. In:Adie R.J.(ed.). Antarctic Geology and Geophysics, Universitetsforlaget, Oslo, p.579-583.

Kamenev E.N.,1975. The geology of Enderby Land (in Russian). Acad. Sci. USSR, Comm.Antarct.Res.Rep.14, p.34-58.

Kamenev E.N.,1987. The geology of Sør- Rondane Mountains (in Russian). Ministry of Geol.USSR. Geol. and geophys.res. in Antarctic. Leningrad, p.7-25.

Kojima S. and Shiraishi K., 1986. Note on the geology of the western part of the Sor-Rondane Mountains, East Antarctica. In: Yoshida Y.(ed.).Proceedings of the sixth symposium of Antarctic Geoscience. Nat.Inst.of Pol.Res., Tokyo, p.116-132.

Pankhurst R.J., Marsh P.D., Clarkson P.D., 1983. In: Oliver R.L., James P.R., and Jago J.B. (eds.). Antarctic Earth Science. Australian Academy of Science, Canberra, p.176-182.

Pasteels P. et Michot J., 1968. Nouveaux résultats géochronologiques obtenus par la méthode U-Pb sur des zircons des monts Sør-Rondane (Antarctique). Annales de la Societé Géologique de Belgique, t.91, p.283-303.

Ravich M.G. and Kamenev E.N., 1975. Crystalline basement of the Antarctic platform. Translated from Russian. John Willey and Sons, New York, 574 pp.

Ravich M.G., Solov'ev D.S., Fedorov L.V., 1984. Geological Structure of MacRobert- son Land (East Antarctica). Translated from Russian. Amerind Publishing Co. Pvt. Ltd., New Delhi, 274 pp.

Sheraton J,W., 1985. Geology of Enderby Land and western Kemp Land. 1:500 000 scale map. Bureau of Mineral Resources, Mercury Walch, Hobart.

Sheraton J.W., Black L.P., McCulloch M.T., 1984. Regional geochemical and isotopic charcteristics of high grade metamorphics of the Prydz Bay area: the extent of Proterozoic reworking of Archean continental crust in East Antarctica. Precambrian Research, N 26, p.169-198.

Sheraton J.W., Offe L.A., Tingey R.J. and Ellis D.J., 1980. Enderby Land, Antarctica - an unusual Precambrian high grade metamorphic terrain. Geol. Soc. Aust. J., 27, p.1-18.

Sheraton J.W., Tingey R.J., Black L.P., Offe L.A. and Ellis D. J., 1987. Geology of an unusual Precambrian high-grade metamorphic terrane - Enderby Land and Western Kemp Land, Antarctica. Austral. Bur.Miner.Res.Geol.Geophys.Bull., 223, Canberra, 51 pp.

Tingey R.J.,1982. The geologic evolution of the Prince Charles Mountains - an Antarctic Archean cratonic block. In: Craddock C.(ed.), Antarctic Geoscience. The University of Wisconsin Press, Madison, p.455-464.

Tingey R.J. and Convine J.W., 1982. Geology of the Southern Prince Charles Mountains. 1:500 000 scale map. Bur. Miner. Res., Mercury-Walch, Hobart.

Van Autenboer T. and Loy W., 1972. Recent geological investigations in the Sør-Rondane Mountains, Belgicafjella and Sverdrupfjella, Dronning Maud Land. In: Adie R.J. (ed.), Antarctic Geology and Geophysics. Universitetsforlaget, Oslo, p.563-571.

Wolmarans L.G. and Kent L.E., 1982. Geological investigations in Western Dronning Maud Land, Antarctica, a synthesis. South Afr.J.of Ant. res., suppl.2, 93 pp.

Yoshida M.,1979. Metamorphic conditions of the polymetamorphic Lutzow-Holmbukta region, East Antarctica. Journ. of Geosci., Osaka City Univ., vol.22, part 4, p.97-140.

Yoshida M. and Kizaki K., 1983. Tectonic situation of Lutzow-Holm Bay in East Antarctica and its significance in Gondwanaland. In: Oliver R.L., James P.R. and Jago J.B. (eds.). Antarctic Earth Science. Australian Acad. of Sci., Canberra, p.36-39.

Gondwana Eight, Findlay, Unrug, Banks & Veevers (eds) © 1993 Balkema, Rotterdam. ISBN 90 5410 304 3

High-grade metamorphic rocks of the northern Prince Charles Mountains, East Antarctica: Their subdivision and history

M. B. Sergeyev
VNII Okeangeologia, St. Petersburg, Russia

V. M. Kaulio
VIRG Rudgeofizika, St. Petersburg, Russia

ABSTRACT: Magnetic susceptibility weighed by rock volume can be used to separate the two rocks series: Larsemann Series has low values generally between 0.0001 to 0.001 SI, while Reinbolt Series generally has high values between 0.001 and 0.1 SI. The difference in magnetic susceptibility reflects a difference in the amount of magnetite, which is due to a different origin and pre-metamorphic history. The Larsemann Series is thought to be older supracrustal rocks, generally underlain by younger plutonic Reinbolt Series. The present horizontal distribution of rocks is attributed to vertical movements on E to ENE trending faults exposing Reinbolt Series in elongate fault blocks.

1 INTRODUCTION

The Prince Charles Mountains are one of the largest and most investigated massives of the East Antarctic shield. This region is mainly composed of Precambrian metamorphic rocks forming sublatitudiual belts. The grade of these rocks generally increases from greenschist facies in the south to granulite facies in the north.

The object of this study is the high-grade polymetamorphic complex that is exposed in the Northern Prince Charles Mountains (north of Fisher Massif). The rocks of this complex indicate at least two metamorphic events : the mineral assemblages of granulite facies co-exist with the later superimposed assemblages of upper amphibolite facies.

The Soviet geologists have divided this high-grade complex into two lithological units - the mainly aluminous Larsemann Series and the mainly charnockitic Reinbolt Series (Ravich & Fedorov 1982, Ravich et al. 1984). Each unit is more than 5 km thick. The Larsemann Series are mainly biotite-garnet gneiss and plagiogneiss, while the Reinbolt Series are mainly hypersthene gneiss and plagiogneiss. It is generally possible to determine the lithological series while at the outcrops by determining the predominant rock type.

It is very difficult to determine the lithological series at some outcrops. Firstly, some rocks have undergone intensive superimposed granitization and migmatization. Secondly, because most types of rocks (amphibolites, crystalline schists, sillimanite-bearing rocks and other rocks) are present in both series (although in different quantities), sometimes there are such combinations of rocks on outcrops that the determination of lithological series is highly difficult even in absence of superimposed processes. Moreover, even hypersthene gneisses and plagiogneisses are known in the Larsemann Series, and biotite-garnet gneisses and plagio-gneisses are present rarely in the Reinbolt Series. Therefore it is desirable to have some other criteria for the separation of these series.

Most investigators consider the rocks of the Reinbolt Series to be orthometamorphic. Gneisses and plagiogneisses of the Larsemann Series are usually considered parametamorphic rocks because of their essentially aluminous composition. These suppositions are reflected in the terms Larsemann Paragneisses (or Larsemann Supracrustals) and Reinbolt Orthogneisses that are sometimes used.

Although Soviet geologists have traditionally divided the high-grade complex of the Northern Prince Charles Mountains into two lithologic series, the modern spatial relationship between these series and, especially, their original relationship are not clear. The determination of the modern spatial relationship is difficult because contacts between the rocks of these two series are seldom exposed. The suppositions about the original relationship between these two formations are especially doubtful, and there are now at least four models based on different principles. They are discussed below in the section "Nature and original relationship between the Larsemann and Reinbolt Series".

So, in our opinion, it is very important :

-to find some additional criteria for the objective separation of the Larsemann and Reinbolt series ;

-to determine the modern spatial relationship between these two units ;

-to understand the original relationship between these two formations and to verify their nature.

This study is devoted to these problems.

2 THE REASONS OF THE USE OF PETROPHYSCAL METHODS, INITIAL DATA AND ROCKS' COLLECTION, METHODS OF MEASUREMENTS

Already during reconnaissanse geological and aeromagnetic investigations it was noted that the rocks of the Reinbolt Series usually had high magnetization, and the rocks of the Larsemann Series usually had low magnetization (Volnukhin & Kurinin 198O). This difference in magnetic susceptibility may be used to distinguish between these lithological series. Magnetic susceptibility is a very convenient criterion, because its measurement is simple, cheap and quick. But because of the difference in magnetization was determined using only a small collection of specimens, it was necessary to confirm it using a more representative collection.

This study is based on the observations and rocks' collections that were obtained by Soviet field parties from 1986 to 199O. During this period almost all areas of high-grade rocks in the Northern Prince Charles Mountains was investigated. Visits were mainly short and helicopter-borne, but some were longer and used field camps. Detailed measurements of magnetic susceptibility were made directly on the outcrops. This field work was followed in laboratory by petrophysical studies that included measurements of magnetic susceptibility and density of all collected samples (more than 25OO), and selective (4OO specimens) measurements of remanent and induced magnetization.

Magnetic susceptibility was measured on the specimens and on the outcrops using portable susceptibility meters of type KT-5 (made in Czeckoslovakia). The mean square error of measurements is 5.9 per cent. In the field detailed measurements were made along geological traverses with measurements at each lithology layer, or every 5O meters within uniform, thick layers. In every measurement point 3-5 observations were made, and the average recorded.

The petromagnetic map and histograms of Figures 1 and 2 are based only on representative specimens. We have considered as representative only the specimens of metamorphic rocks that were collected from layers over 2 meters thick. The specimens that were collected from thin layers, contact zones and zones of intensive alteration were considered separately, but were not used to prepare Figures 1 and 2.

The volume proportions of different types of rocks were estimated at each geological observation point along the traverses. After the measurement of magnetic susceptibility of representative specimens, the mean-weighted values of susceptibility was calculated for each geological observation point and shown on petromagnetic map (Fig. 1). We have not yet finished interpreting the observations for the whole area of the high-grade rocks. Results presented in this paper are mainly for the northwestern part of the Northern Prince Charles Mountains (the Porthos and Athos ranges, Stinear and ANARE nunataks), the area of helicopter geological survey of 1989-9O field season. In this area there has been relatively small amount of modification subsequent to granulite metamorphism, so mineral composition, concentrations of chemical elements and physical properties are similar to those at the end of the granulite metamorphism.

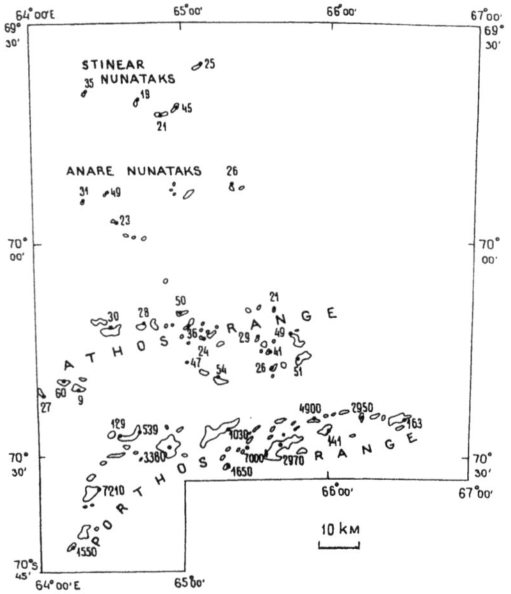

Fig.1 Petromagnetic map of the area of helicopter geolgical survey in 1989-9O season.

Points of geological observations are shown by dots. Numbers near the dots are the mean-weighted values of magnetic susceptibility of high-grade rocks (in $1O^{-5}$ SI)

3 MAGNETIC SUSCEPTIBILITY AS THE CRITERION FOR SEPARATION OF THE LARSEMANN AND REINBOLT SERIES

The area of helicopter geological survey of 1989-9O field season can be divided, using geological and petromagnetic data, into two zones. The southern zone (the Porthos range) is composed of the rocks of the Reinbolt Series, mainly hypersthene gneisses and plagiogneisses and also pyroxene crystalline schists. The northern zone (the Athos range, the Stinear and

ANARE nunataks) is composed of the rocks of the Larsemann Series, mainly aluminous plagiogneisses, gneisses and quartzite-gneisses.

The big difference in magnetic susceptibility between the rocks of the Larsemann and Reinbolt Series is illustrated on the petromagnetic map (Fig. 1) and histograms of mean-weighted values of magnetic susceptibility (Fig. 2a,b). The mean value of magnetic susceptibility of the rocks of the Reinbolt Series is 0.02580 SI, but of the Larsemann series is only 0.00034 SI. Such a big difference in magnetic susceptibility is attributed to a high proportion of magnetite in almost all rocks of the Reinbolt Series, and the usual absence of magnetic minerals in the rocks of the Larsemann Series. This is in good agreement with microscope observations.

Nevertheless, some rocks of the Reinbolt Series are low-magnetic (Fig. 2d). The low magnetization of most of these rocks is inferred to be due to late processes destroying magnetite.

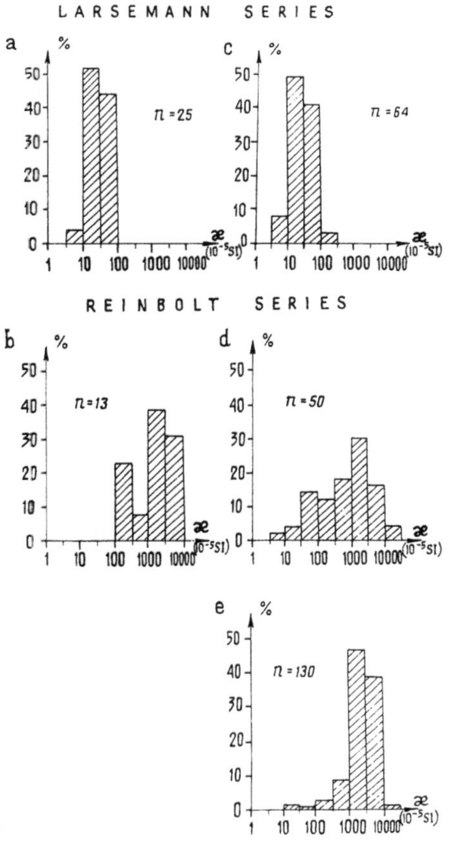

Fig.2 Histograms of magnetic susceptibility.
a,b - mean-weighted values for the separate points of geological observations; c,d - representative specimens; e - detailed measurements on Martin Massif.

This conclusion was checked by detailed field petromagnetic measurements on Martin Massif (central part of the Porthos range). In this massif post granulite facies processes are minor, so the concentration of magnetite and, consequently magnetic susceptibility must be near to their primary values. Almost all the measured magnetic susceptibility values are very high (Fig. 2e). This detailed study supports the view that the high concentration of magnetite and high magnetic susceptibility of the rocks of the Reinbolt series are their primary properties, and are not connected with post-granulite metamorphic processes.

In contrast, there are no representative specimens with high magnetic susceptibility collected from the rocks of the Larsemann Series (Fig. 2c). The low magnetic susceptibility is attributed to the present-day absence of magnetic minerals. The absence of these minerals is not thought to be due to post granulite facies processes because these processes are distributed here unevenly, and if the magnetic minerals had been present in the Larsemann Series rocks after the granulite metamorphism, they would have remained at least locally. But the magnetic minerals are absent now and they are thought to be absent in the Larsemann Series rocks by the time of granulite metamorphism.

It is very important to emphasize that the absence or presence of magnetite in these high-grade rocks is, generally, not connected with their chemical or mineral composition. For example, hypersthene plagiogneisses and gneisses in the Reinbolt Series, where they are predominant rocks, are usually highly magnetic. Rarely similar rocks occur in the Larsemann Series, but there they are non-magnetic. On our opinion, this indicates persuasively that the Larsemann and Reinbolt Series differ not only in lithology, but also in their mode of formation and/or their pre-metamorphic history.

Hence in the Reinbolt Series rocks pre-metamorphic conditions contributed to the origin and preservation of magnetite, while in the Larsemann Series rocks pre-metamorphic conditions were against the formation of magnetite, or favoured its destruction.

So, use of mean-weighted values of magnetic susceptibility for each point of geological observations as a criterion for distinguishing between the Larsemann and Reinbolt Series is based on two findings. First, the two formations differ greatly in magnetic susceptibility. Second, the difference in magnetic susceptibility of rocks is thought to be primary, and connected with different conditions of their primary formation and pre-metamorphic history, is not connected with later processes.

Although later processes lead to low-magnetic rocks among the rocks of the Reinbolt Series, they do not prevent to use the mean-weighted values of magnetic susceptibility to distinguish between the two units. The later processes are usually unevenly, and locally, distributed, and rocks effected by these processes form a minor proportion of all rocks.

It is illustrated by the difference between histograms of Fig. 2b and 2d.

The relationship given above is generally also correct for the high-grade rocks south of Porthos Range, in spite of later processes being more widespread in this area.

4 MODERN SPATIAL RELATIONSHIP BETWEEN THE LARSEMANN AND REINBOLT SERIES

As it is noted above, it is very difficult to determine the modern spatial relationship between the Larsemann and Reinbolt Series on the basis of the geological observations alone, because the direct contacts between the rocks of the two units are exposed rarely. However we conclude below that the Reinbolt Series underlies the Larsemann Series, using petrophysical, geological and aeromagnetic information.

Induced magnetization is greater than remanent magnetization (factor Q is usually O.2 - O.4) so it is the the main contributer to magnetic anomalies. It is possible to use for the interpretation of aeromagnetic data only the information on magnetic susceptibility (Fig. 3,4). The rocks of the Reinbolt Series usually have high susceptibility and, as expected over

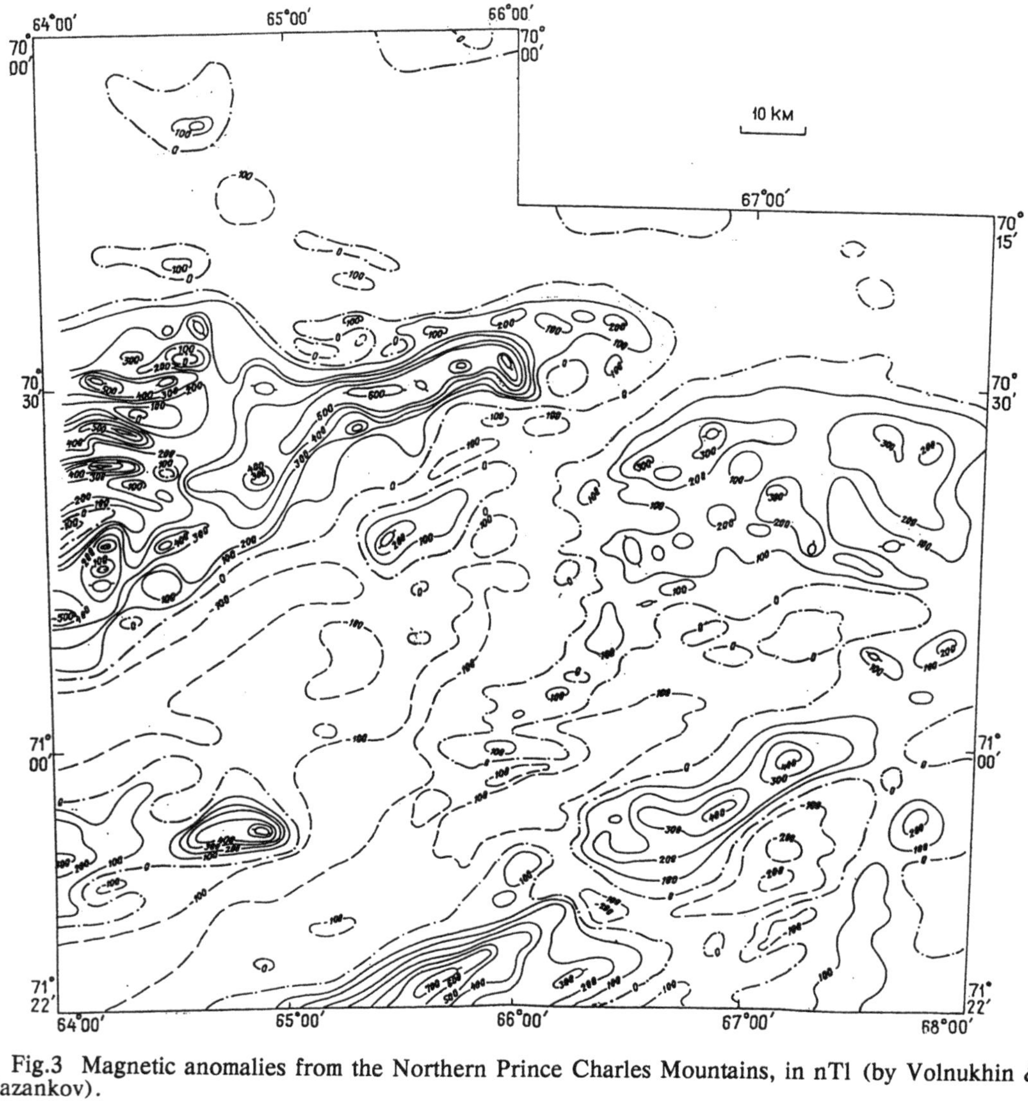

Fig.3 Magnetic anomalies from the Northern Prince Charles Mountains, in nTl (by Volnukhin & Kazankov).

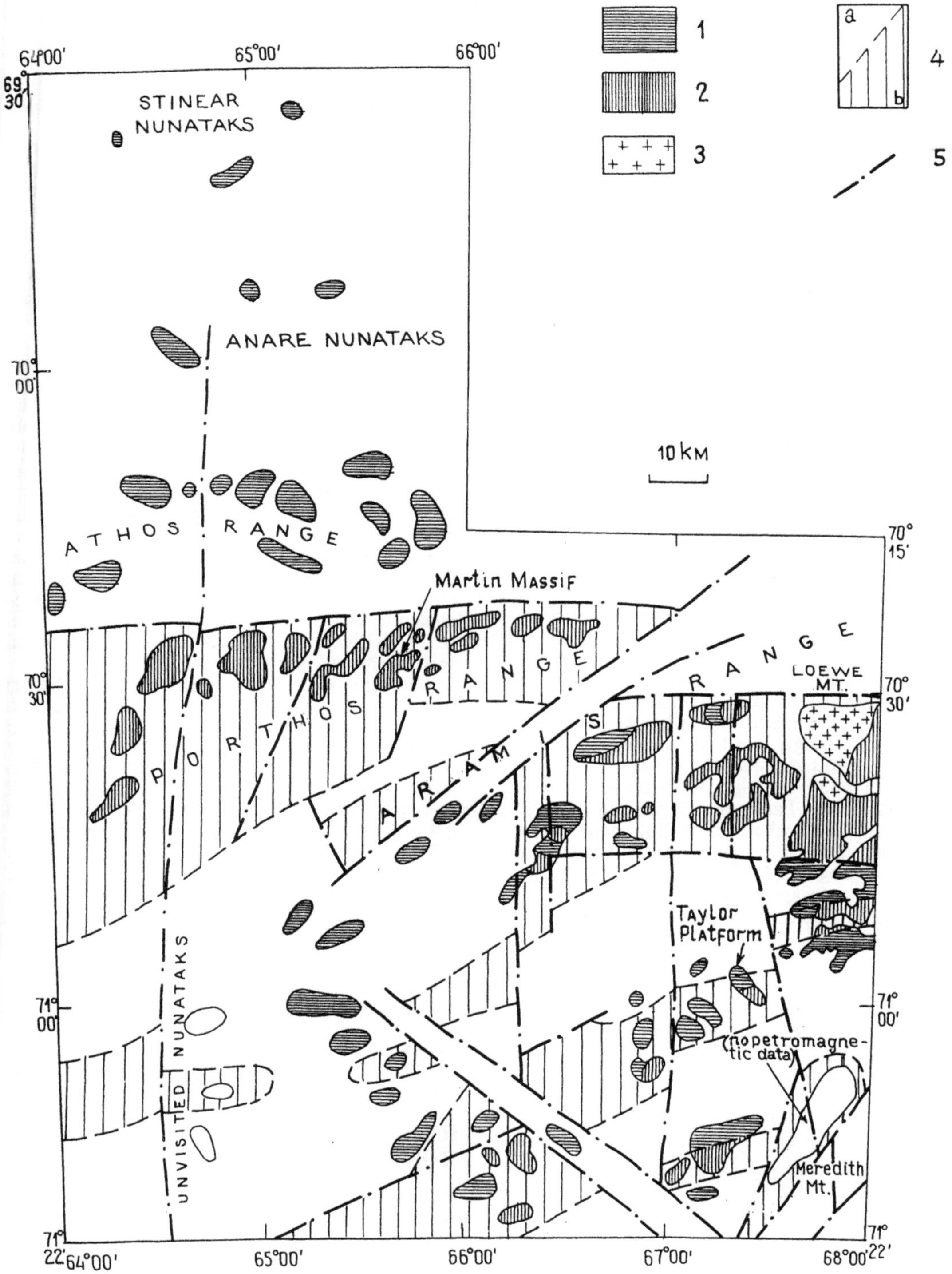

Fig.4 Zones of high and low magnetic field, and faults bounding block structures.
1 - Larsemann Supracrustals; 2 - Reinbolt Orthogneisses; 3 - charnockitoides of Loewe Mt.; 4 - zones of low (a) and high (b) magnetic field, and boundaries between these zones; 5 - faults defined on the basis of aeromagnetic data.

their outcrops are high-amplitude positive magnetic anomalies. High magnetic susceptibility is also typical for the intrusive (?) charnockitoides of Loewe massif and magnetic anomalies over this massif also are very positive. In contrast, the rocks of the Larsemann Series are of low magnetization and the associated magnetic anomalies must be near zero value. The few positive magnetic anomalies associated with outcrops of the Larsemann Series can be explaned by the presence of magnetic rocks of the Reinbolt Series, or intrusive (?) charnockitoides, below the non-magnetic rocks of the Larsemann Series.

The modern spatial relationship between the Larsemann and Reinbolt Series is illustrated by the results of interpretation of surface magnetic survey on Taylor platform (Fig. 5). The highly-magnetic unit (the Reinbolt series) is exposed in the south-east of Taylor platform and plunges gently to the north-west below the non-magnetic rocks of the Larsemann Series that are exposed in the north-west part of this platform. South-east of Taylor platform the highly-magnetic rocks of the Reinbolt Series are overlain by ice, morain and, in all probability, by rocks of the Larsemann Series.

We interpret the magnetic field in terms of a block faulting model (Fig. 4). The blocks without outcrops of the Larsemann Series (the blocks of the Porthos range, the block of Loewe massif) have risen the most. The blocks with outcrops of the both series have risen an intermediate amount. Magnetic anomalies over the blocks of these two types are strongly positive, because the highly-magnetic rocks are situated near or at the surface. Zones of relatively high magnetic anomaly with outcrops of only Larsemann Series are thought to have magnetic rocks at a shallow depth. Blocks with outcrops of the Larsemann Series, and overlying magnetic lows, are thought to have risen the least, and magnetic rocks may be only at great depth (e.g. blocks north of the Porthos range).

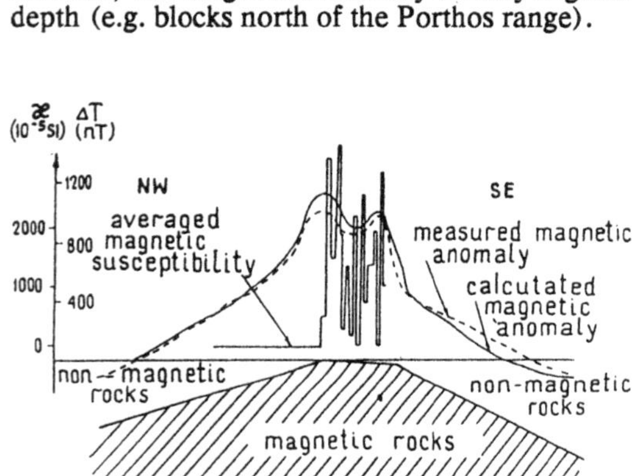

Fig.5 Profile across the Taylor platform and two-dimensional interpretation.

5 NATURE AND ORIGINAL RELATIONSHIP BETWEEN THE LARSEMANN AND REINBOLT SERIES

The petromagnetic data, on our opinion, could help to verify the nature and to understand the original relationship between two units divided inside the studied high-grade complex. As it is noted above, the rocks of the Larsemann Series are usually considered as parametamorphic, and the rocks of the Reinbolt Series are probably orthometamorphic. The primary relationship between these two units is a very controversial topic, so there are at least four different models.

The first model has never published, but it has been considered by investigators. It is supposed that the Larsemann and Reinbolt Series are of similar age, and of different lithology, both series forming thick alternated layers (Fig. 6a). After formation the rocks were multiply deformed, so the pattern of exposure is now complicated. However this model is not consistent with the petromagnetic and geophysical data. First, as it is noted above, the consistency in the difference in magnetic susceptibility of the Larsemann and Reinbolt Series, indicates that they are two separate units with different conditions of primary formation and/or pre-metamorphic history. Second, with this model it is difficult to explain the modern spatial relationship of the two series.

Second model was proposed by early Soviet investigatores (Ravich & Fedorov 1982, Ravich et al. 1984).The Reinbolt Series was considered as ancient basement, and to be overlain by supracrustal cover of the Larsemann Series (Fig. 6b). The authors' main argument for this model is that the Reinbolt Series rocks, as a rule, form granite-gneisses domes which are surrounded by folded rocks of the Larsemann Series. The modern spatial relationship inferred in this study are in agreement with this model, however it is unlikely that the basement rocks all had primary magnetite. Moreover, this model is in conflict with the later structural observations of N.N. Krasnikov (oral communication) that the Larsemann Series rocks underwent more stages of deformation than the rocks of the Reinbolt Series.

Some authors have also discussed the opposite interpretation (Fig. 6c). Though this model is in good agreement with above-mentioned structural observations, it conflicts with the modern spatial relationship determined in this work.

The authors of this paper propose a fourth model of original relationship between the Larsemann and Reinbolt Series. The supracrustal rocks of the Larsemann Series primarily overlay the more ancient basement that should be in part the precursor of the Reinbolt Series (Fig. 7a). Then, as a result of the rise of magmatic front (Fig. 7b) the rocks of the ancient basement and of the low part of the cover were reworked (Fig. 7c). The rise of magmatic front was connected with the same tectonic events that lead to main deformation and granulite metamorphism. These plutonic rocks intruded before and/or during the main stage of metamorphism, and the reworked basement, is the present Reinbolt Series.

This model is consistent with the behaviour of magnetite in metamorphic processes, and with the geological history of region inferred by Australian investigators (Tingey 1982).

The non-magnetic nature of rocks of the Larsemann series may be due to their supracrustal origin. Parametamorphic rocks are in most cases of low susceptibility, because magnetite is unstable in low-grade environment (Dortman 1984). Magnetite is destroyed in many ways in weathering and low-grade metamorphism of volcanic and sedimentary rocks. Increase in metamorphic grade, as a rule, is not accompanied by regeneration of magnetite, therefore high-grade supracrustal rocks usually do not contain magnetite. Hence, the petromagnetic data are consistent with a supracrustal origin for the Larsemann Series. Magnetite sometimes can form in high-grade supracrustal rocks during retrograde and ultrametamorphic processes, but in the Larsemann Series rocks these processes did not lead to wide forming of this mineral.

The highly-magnetic nature of the Reinbolt Series rocks is consistent with their supposed plutonic origin. If a plutonic rock was originaly void of magnetite, the orthometamorphic rock formed from it would be void of magnetite too. However, if a primary plutonic rock contains magnetite, then it will be stable, and continue to be present, thought high-grade metamorphism.

The presence of primary magnetite in the acidic-intermediate plutonic rocks indicates that these magmas were generated at large depth (Petcherskiy et al. 1975) . If magmas had been generated at relatively shallow depth, as a result of the melting of metasediments, then the plutonic rocks formed would be void of magnetite. Hence, the high concentration of magnetite in the rocks of the Reinbolt Series indicates, firstly, the magmatic origin of this formation, and, secondly, the deep level of origin of magma.

This interpretation of petromagnetic data is consistent with the conclusions of Australian investigators on the geological history of this region. On the basis of differences in Sr-87 / Sr-86 initial ratios in the rocks of this high-grade complex, it was concluded (Tingey 1982) that Late Proterozoic metamorphism involved the materials of two kinds : older crustal rocks and younger mantle derivatives. Probably, Larsemann Supracrustals are older crustal rocks, while all or part of Reinbolt Orthogneisses may be younger mantle derivatives. Nevertheless, the model presented in this paper needs further confirmation.

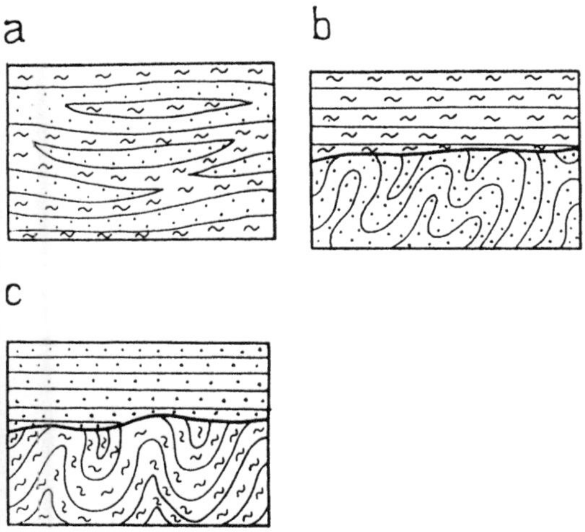

Fig.6 Models of original relationship between the Larsemann and Reinbolt Series. These models are not in agreement with petromagnetic data.

a - alternation of the very thick layers of different lithological composition; b - the Reinbolt Series is the more ancient basement, the Larsemann Series is the supracrustal cover; c - the Larsemann Series is the more ancient basement, the Reinbolt Series is the overlying formation.

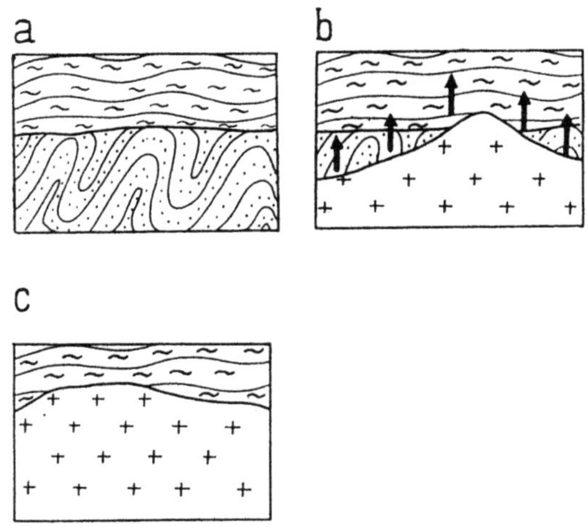

Fig.7 The model of original relationship between the Larsemann Supracrustals and Reinbolt Orthogneisses, that is proposed in this paper.

a - Larsemann Supracrustals on the primary basement; b - the rise of magmatic front accompanied by reworking of primary ancient basement and the lower part of Larsemann Supracrustals; c - Reinbolt Orthogneisses as the pseudobasement under more ancient Larsemann Supracrustals.

6 CONCLUSIONS

1. The mean-weighted magnetic susceptibility for each geological observations point may be used to distinguish lithological series when at the outcrops. The low values are typical for the Larsemann Series and the high values for the Reinbolt Series.

2. Rocks of the Reinbolt Series are generally situated below rocks of the Larsemann Series.

3. The studied region has a block tectonic structure. When petromagnetic, aeromagnetic and geological data are considered together, blocks can be mapped that have been relatively uplifted.

4. Differences in magnetic susceptibility reflects differences in the amount of magnetite. Magnetite is usually absent in the Larsemann Series, and abundant in the Reinbolt Series. This difference is attributed to the different conditions of primary formation and pre-metamorphic history of the rocks of the Larsemann and Reinbolt Series and not to post-granulite processes.

5. The absence of magnetite in the Larsemann Series is consistent with their supracrustal origin, while the large amount of magnetite in the Reinbolt Series is consistent with their deep-seated plutonic origin.

6. The geological model given above is consistent with the geological history of region given by Tingey (1982). In the context of Tingey's model the Reinbolt Series (all or part) represents the younger mantle derivatives, and the Larsemann Series represents the relatively older reworked crustal rocks.

REFERENCES

Dortman, N.B.(ed) 1984. Physical properties of rocks and minerals (Petrophysics). - publ. Nedra, Moscow. /in Russian : Fizicheskie svoistva gornih porod i poleznih iskopaemih (Petrofizika) /.

Petcherskiy, D.M., Bagin, V.I., Brodskaya, S.Y. & Sharonova, Z.V. 1975. Magnetism and conditions of formation of igneous rocks. - publ. Nauka, Moscow. /in Russian: Magnetizm i usloviya obrazovaniya izverzhennih gornih porod /.

Ravich, M.G. & Fedorov, L.V. 1982. Geologic evolution of MacRobertson Land and Princess Elizabeth Land,East Antarctica. -in: Antarctic Geoscience, ed. by C. Craddock, The University of Wisconsin Press, Madison, pp. 499-504.

Ravich, M.G., Solov'ov, D.S. & Fedorov, L.V. 1984. Geological Structure of Mac- Robertson Land (East Antarctica). - Amerind Publ. Co. Pvt. Ltd., New Delhi.

Tingey, R.J. 1982. The geologic evolution of the Prince Charles Mountains - an Antarctic Archean cratonic block.-in: Antarctic Geoscience, ed. by C.Craddock, The University of Wisconsin Press, Madison, pp. 455-464.

Volnukhin,V.S. & Kurinin, R.G. 1980. The physical properties of rocks of Lambert Glacier region. -in: Geophysical studies in Antarctica., NIIGA, Leningrad, pp. 52-58. /in Russian : Fizicheskie svoistva gornih porod raiona lednika Lamberta. -in: Geofizicheskie issledovaniya v Antarktide/.

Gondwana Eight, Findlay, Unrug, Banks & Veevers (eds) © 1993 Balkema, Rotterdam. ISBN 90 5410 304 3

The Kannack complex of the Vietnam Kontum massif of the Indochina block: An exotic fragment of Precambrian Gondwanaland?

M.B.Katz
University of New South Wales, Kensington, N.S.W., Australia

ABSTRACT: Precambrian granulite facies rocks occur over an area of 100 km in the eastern part of the Kontum Massif of Vietnam. This unusual occurrence of Precambrian high grade metamorphic rocks is separated from the closest similar rocks to the west in the Indian Eastern Ghats by over 2000 km. The intervening area is occupied by the Mesozoic-Cenozoic Burma-Malay orogen. In the reconstruction of Precambrian Gondwanaland a granulite facies mobile belt can be traced from Sri Lanka, India, Antarctica through to Australia. However a considerable gap remains between the Indian-Eastern Ghats section and the Albany-Fraser Belt of S.W. Australia. This gap has been suggested to be, in part, filled by similar rocks found in the Shillong Plateau, Assam. High grade metamorphic rocks of uncertain age also occur in the Mogok area in central Burma. The tectonic history of the Indochina Block in the Precambrian is debatable, however some latest Precambrian paleotectonic interpretations join this block and the Tibetan, Chinese and Iranian Blocks with Precambrian Gondwanaland. If the Kannack Complex is proved to be Archaean/Proterozoic, preliminary and incomplete information suggest that it could have been placed adjacent to the Eastern Ghats of India in the earliest Precambrian.

1 INTRODUCTION

The reconstruction of Gondwanaland by Lawver and Scotese (1987) has resulted in a reasonable fit of Africa-Madagascar-India-Sri Lanka with East Antarctica and Australia. In the reassembly, a granulite facies Precambrian mobile belt has been traced through Sri Lanka, the Eastern Ghats of India and East Antarctica to the Albany Fraser mobile belt of southwest Australia (Katz, 1989) and the extension of this belt into Madagascar and East Africa has also been suggested (Katz 1989; 1990). One of the major problems in this scheme is the apparent gap between the Eastern Ghats of India and the Albany Fraser Belt of Australia. This gap may have been filled by rocks found in the Shillong Plateau or the Mogok area of Burma during Precambrian times. Another possibility is that Precambrian fragmented continental blocks from Tibet and SE Asia (McKerrow and Scotese, 1990) could have been originally sited in this gap. One such SE Asian fragment, the Indochina Block, has similar granulite facies rocks located in the Kontum Massif of Vietnam (Fig. 1).

2 PRECAMBRIAN GONDWANALAND GRANULITE MOBILE BELT

The Sri Lanka - Albany Fraser Belt of southwest Australia is considered as a continuous metamorphic terrane in Precambrian Gondwanaland with a history dating back to the Archaean and it is distinguished by important charnockite-forming events in late Proterozoic times. Typical lithologies are two-pyroxene meta-mafic rocks, charnockites, garnet-sillimanite-cordierite gneisses, quartzites, marbles and diopside-scapolite-wollastonite calc-silicate rocks (e.g. Katz, 1969 - Sri Lanka). The prograde metamorphic assemblages imply low-medium pressure conditions under high temperatures. The rocks show multi-phase deformation, the development of L-S fabrics, shear zones and the emplacement of granites of Pan African age. The similarities in lithology,

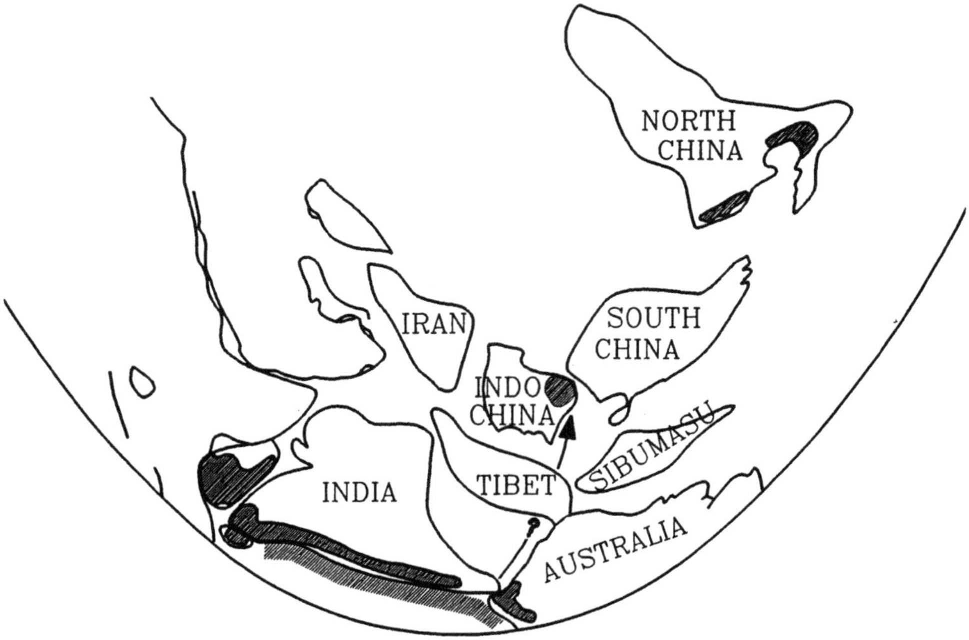

Fig. 1 Latest Precambrian Gondwanaland (after McKerrow and Scotese, 1989). Shaded areas represent granulite facies terrains.

metamorphism, structure, age relations, igneous activity and their apparent alignment in Precambrian Gondwanaland all point to a common origin.

Although the major fragments of this once-continuous belt are brought together in the reconstruction (Fig. 2), important minor fragments, especially between the Eastern Ghats and Albany Fraser Belts, may still be missing. The Eastern Ghats strikes east into the sediments of the Bengal Basin and thus may underly these rocks, possibly being exposed in the Shillong Plateau (e.g. Ghosh et al., 1991). Another small granulite facies fragment to the east may be found in the Mogok area of Burma (Iyer 1953). Other possibilities have to do with fragments that have been dispersed and displaced, such as the Indochina Block granulites that make up part of the Kontum Massif of Vietnam (Tran, 1990). Similar rocks are also found in the North China block fragment (e.g. Cheng et al. 1982) (Fig. 1).

3 INDOCHINA AND SE ASIA IN PRECAMBRIAN GONDWANALAND

There has been some debate on the position of SE Asia in Gondwanaland (e.g. Metcalfe, 1990, Burrett et al., 1990). Recent paleomap atlases (McKerrow and Scotese, 1990) speculate that several Asian blocks including Tibet, Indochina and South China (Fig. 1) make up part of Precambrian Gondwanaland. In relating these uncertain Asian blocks to the well defined fit of the rest of Gondwanaland certain preliminary assumptions are taken.

1 Whatever their exact past distribution they are considered to be part of Precambrian Gondwanaland.

2 The portions of these blocks that relate to the Gondwanaland granulite belt must be of similar metamorphic grade, structural history and at least of Proterozoic age.

3 In the breakup of Gondwanaland these marginal fragments had an independent history during their subsequent reassembly.

Using these preliminary criteria, based on very incomplete data, the only fragments that could be related to the granulite belt are the Vietnamese-Kontum Massif fragment of the Indochina block and similar fragments found in

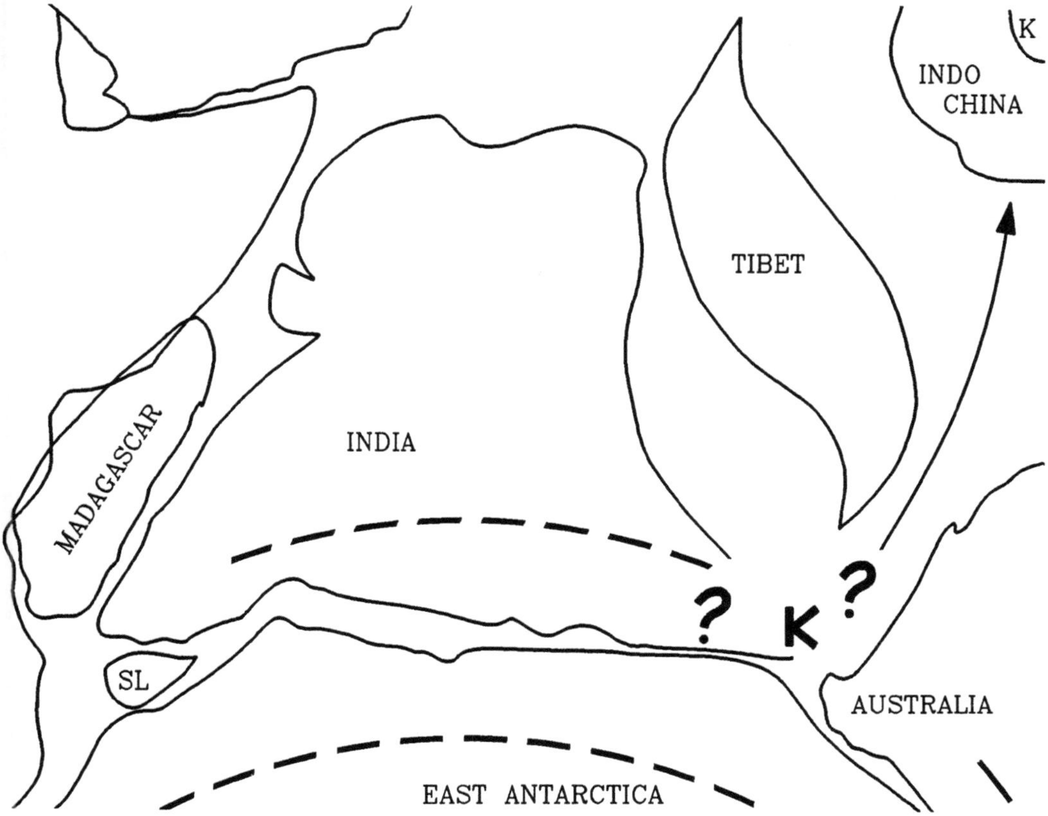

Fig. 2 Reconstruction of East Gondwanaland (after Lawver and Scotese, 1987). The Sri Lanka through to SW Australia granulite mobile belt is represented by the inter-arcs path. K - represents the speculative palaeo position of the Kontum massif in Archaean/Proterozoic times.

the North China block. Since the Kontum massif appears to be closest to where the gap occurs (Fig. 1), it is conceivable that in Proterozoic-Archean times its position could have been the missing piece between the Eastern Ghats and the Albany-Fraser Belt (Fig. 2).

REFERENCES

Burrett, C., Long, J. and Stait, B. 1990. Early-Middle Palaeozoic biogeography of Asian terranes derived from Gondwana. *Geological Society Memoir* 12: 163-174.

Cheng, Y., Bai, J. and Sun, D. 1982. The lower Precambrian of China. *Revista Brasileira Geociencias* 12: 65-73.

Ghosh, S., Chakraborty, S., Bhalla, J.K., Paul, D.K., Sarkar, A., Bishu, P.K. and Gupta, S.N. 1991. Geochronology and Geochemistry of Granite Plutons from East Khasi Hills, Meghalaya. *Journal Geological Society India.* 37: 331-342.

Iyer, L.A.N. 1953. The Geology and Gemstones of the Mogok Stone Tract, Burma. *Memoir Geological Survey, India.* 82:100 pp.

Katz, M.B. 1969. The Precambrian Metamorphic Rocks of Ceylon. *Geologische Rundschau,* 60: 1523-1549.

Katz, M.B. 1989. Sri Lanka - Indian Eastern Ghats - East Antarctica and the Australian Albany Fraser Mobile Belt: Gross Geometry, Age Relationships and Tectonics in Precambrian Gondwanaland. *Journal of Geology* 97: 646-648.

Katz, M.B. 1990. Continuity of Mobile Belts from Africa into East Precambrian Gondwanaland, 15th Colloquium of African Geology, *CIFEG Occasional Publication* 1990/20: 171.

Lawver, L.A. and Scotese, C.R. 1987. A Revised Reconstruction of Gondwanaland. American Geophysical Monograph, 40: 17-23.

McKerrow, W.S. and Scotese, C.R. (eds) 1990. Atlas of Paleozoic Basemaps. Paleozoic Paleogeography & Biogeography, *Geological Society Memoir*. No. 12.

Metcalfe, I. 1990. Allochthonous terrane processes in Southeast Asia. *Philosophical Transactions of the Royal Society London, A*, 331: 625-640.

Tran, Q.H. 1990. Archean Granulites of the Kontum Massif. (Abstracts) *3rd International Archaean Symposium*, Perth.

Gondwana Eight, Findlay, Unrug, Banks & Veevers (eds) © 1993 Balkema, Rotterdam. ISBN 90 5410 304 3

Correlations of tectonothermal events between the Central Indian Tectonic Zone and the Albany Mobile Belt of Western Australia

L. B. Harris
Department of Geology, The University of Western Australia, Nedlands, W.A., Australia

ABSTRACT: Correlations in terms of precursor lithologies and the nature and timing of tectonothermal events have been established between the Central Indian Tectonic Zone and the Albany Mobile Belt of Western Australia. Sedimentation and mafic intrusion took place within an intracratonic basin during the Late Archaean to Early Proterozoic. Collision between the Bundelkhand Block and the Deccan Protocontinent of India has been proposed for a major tectonothermal event in the Central Indian Tectonic Zone at ~1.6 Ga. As this event has not been recognised in the Albany Mobile Belt, it is suggested that either a possible transform margin existed along the western Yilgarn margin at this time, linking with compressional tectonics in the Gascoyne Orogen, or the suture may be situated within Antarctic terranes. Deformation involving dextral transpression, northwestwards thrusting and extension (orogenic collapse?) at ~1.1-1.2 Ga was accompanied by amphibolite to granulite facies metamorphism and the intrusion of syn- to post-tectonic dolerite dykes and granitoids. Late Proterozoic sedimentary sequences were deposited in both terranes, and deformed during reactivation of basement shear zones. Dolerite dykes intruding conjugate cataclastic shear zones are also common to both Central Indian Tectonic Zone and southwestern Western Australia (where an Early Cambrian age has been established). Correlations presented strengthen models for a Gondwana fit where northern India is juxtaposed against southwestern Australia.

1 INTRODUCTION

The Proterozoic Albany-Fraser Province of Western Australia comprises the Albany and Fraser mobile belts, along the southern and southeastern margins of the Archaean Yilgarn craton (Figures 1 and 2). Recent structural and geochronological data (Black *et al.* in press, Sheraton 1990) support correlations between the Albany-Fraser Province and the Bunger Hills and Windmill Islands regions of Antarctica, although there is still uncertainty in the exact fit, with either the Bunger Hills (Lovering *et al.* 1981) or Windmill Islands (Oliver *et al.* 1983) placed opposite Albany. On the basis of sea-floor spreading and palaeomagnetic data, Powell *et al.* (1988) juxtapose the eastern extremity of India against the margin of southwestern Australia. Similar relative positions are shown by DeWit *et al.* (1988). In a model by Ricou *et al.* (1990), the western margin of Western Australia abutted the eastern coast of India. In the models of Powell *et al.* (1988) and DeWit *et al.* (1988), correlations would be expected between northern Indian Proterozoic terranes and the Albany-Fraser Province; the model of Ricou *et al.* (1990) suggests possible correlations with the Gascoyne Province (northern Yilgarn margin).

Existing correlations (*eg.* Katz 1980) suggest a continuous, Proterozoic granulite belt linking the Fraser and Albany mobile belts of Western Australia, East Antarctic terranes, the Eastern Ghats (on the southeastern margin of the Central Indian Peninsular Block, Figure 1) and Sri Lanka. Proterozoic terranes within the Central Indian Tectonic Zone (Figure 1) have, however, been ignored in this and subsequent reconstructions (such as by Dalziel 1992). Due to this omission, and new data that the main tectonothermal events in the Fraser Mobile Belt at 1.6 and 1.3 Ga (D. Nelson, Geological Survey

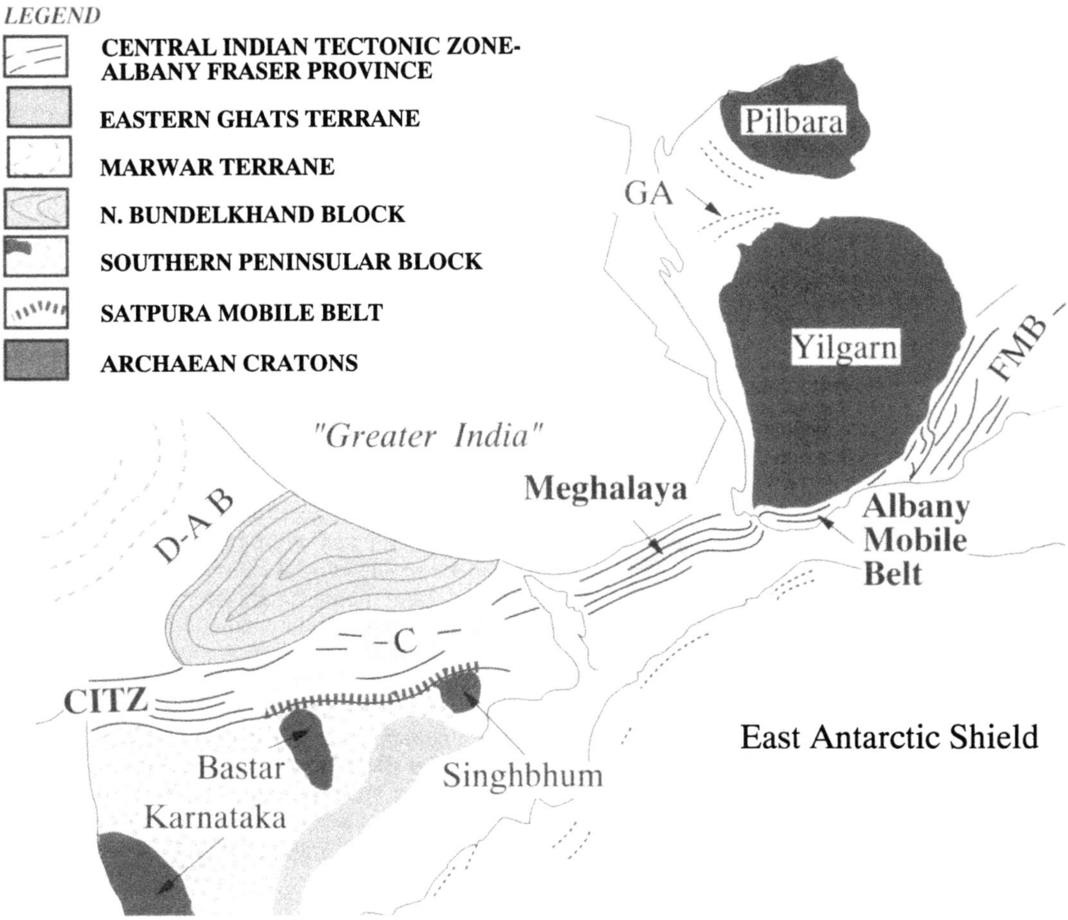

Figure 1. Gondwana reconstruction showing the main tectonic subdivision of the Indian and Australian Precambrian Shields. Note the continuity between the Albany Mobile Belt of Western Australia and the Central Indian Tectonic Zone. Plate reconstruction taken from Powell *et al.* (1988). Indian terrane boundaries from Radhakrishna (1989).

AMB = Albany Mobile Belt and FMB = Fraser Mobile Belt (comprising the Albany-Fraser Province); CITZ = Central Indian Tectonic Zone, with C = Chotanagpur Gneiss Terrane; D-A-B = Delhi-Aravalli belt and GA = Gascoyne Orogen.

of Western Australia *pers. comm.* 1991) have not been recognised in the Albany Mobile Belt (Black *et al.* in press), modifications to previous models must be sought regarding correlations between Indian and Western Australian orogenic belts.

When the tectonic subdivisions for Western Australia and India (taken from Radhakrishna 1989) are superposed on the Gondwana fit of Powell *et al.* (1988) in Figure 1, continuity is suggested between the Albany Mobile Belt and the Central Indian Tectonic Zone.

The aims of this paper are to:
(i) summarise the stages of sedimentary basin formation, igneous activity, metamorphism and deformation in the Albany Mobile Belt and Proterozoic terranes of the Central Indian Tectonic Zone;
(ii) demonstrate that these terranes may represent segments of a once continuous mobile belt (with common Early Proterozoic through to "Pan-African" events); and
(iii) thereby provide supporting geological evidence for the relative positions of the Indian

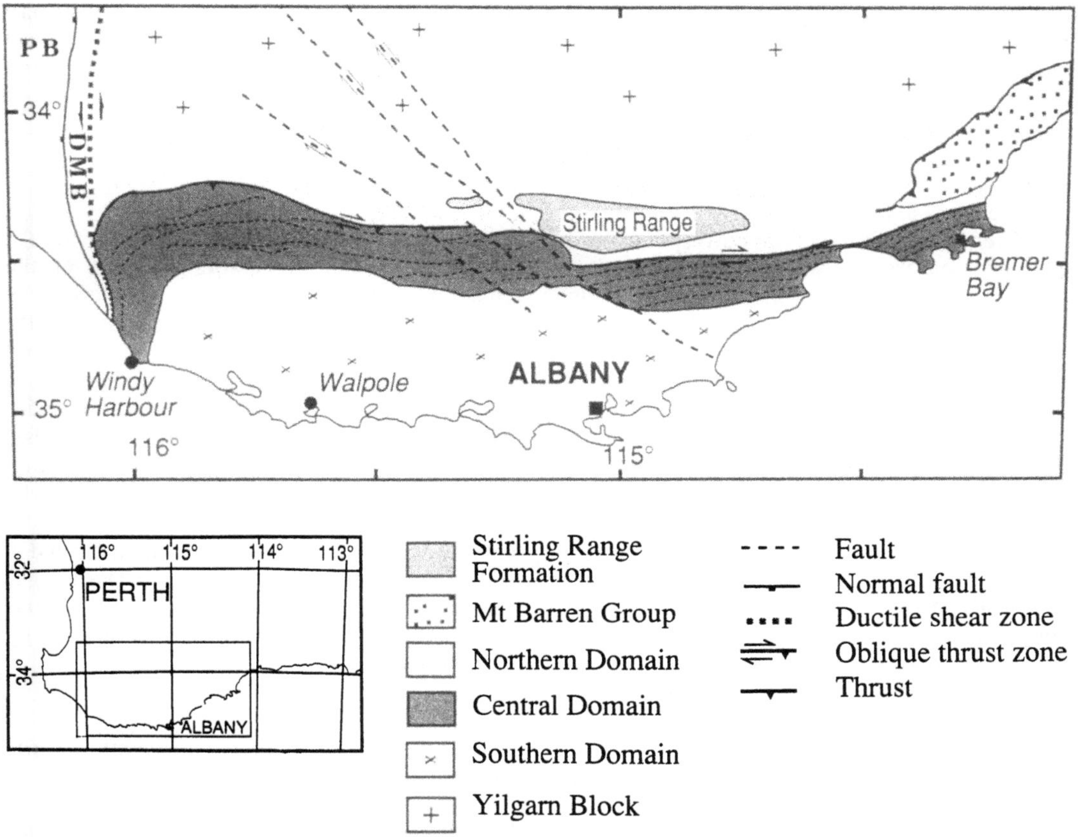

Figure 2. Simplified subdivision of the Albany Mobile Belt, Western Australia and its relationship to the Archaean Yilgarn Block.

Inset shows location within southwestern Australia. Note how domain boundaries are rotated into a North-South orientation by sinistral, "Pan-African" displacements along the Darling Mobile Belt (DMB). PB = Perth Basin.

and Australian shields in the Gondwana reconstruction of Powell *et al.* (1988).

2 ALBANY MOBILE BELT

2.1 *General geology*

The gneiss and granitoid complex of the Albany Mobile Belt (Figure 2) comprises Archaean orthogneiss and Early Proterozoic meta-sediments, dolerite dykes and meta-volcanics deformed and metamorphosed during the Middle Proterozoic. Three domains defined on the basis of lithology, metamorphic grade and aeromagnetic expression can be identified (modified after Beeson *et al.* 1988):

1. *The Northern Domain* is an ~25 km wide sequence of amphibolite facies gneisses. There is a general increase in deformation intensity and metamorphism southwards (except for localised high strain along internal oblique thrust zones) as illustrated by Beeson *et al.* (1988).

2. *The Central Domain* is an ~22 km wide zone of two-pyroxene granulites (characterised on aeromagnetic images by a very high magnetic expression) abutting the Northern Domain along a dextrally oblique thrust (associated with retrograde, amphibolite-facies metamorphism). Tectonic contacts and magnetic horizons within this domain show a remarkable lateral continuity (Figure 2) clearly defining the general east-west trend of the belt.

Towards the eastern margin of the Albany Mobile Belt, the Central Domain swings to an eastnortheast orientation, resulting in a change in tectonic regime to northwest directed thrusting. The structurally highest nappe of the Central Domain in the eastern Albany Mobile Belt preserves evidence for extension tectonics at a deep crustal level.

3. *The Southern Domain* corresponds to a zone characterised by widespread, syn- to late-tectonic granitoid intrusions. Granulite facies assemblages are locally preserved in gneisses and mafic rocks, although the general metamorphic grade of this domain is amphibolite facies.

In the west of the belt, the three domains have been rotated into a north-south orientation by sinistral transcurrent movements along the Darling Mobile Belt on the western margin of the Yilgarn craton (Figure 2) during a Pan-African tectonothermal event (Harris 1987).

The Mount Barren Group sequence of meta-conglomerates, quartzites, phyllites and schists, (metamorphosed up to the staurolite zone of the amphibolite facies), outcrops near the junction of the Albany and Fraser mobile belts. An Early to Middle Proterozoic age has previously been proposed for the entire Mt Barren Group (Thom *et al.* 1980). The presence of trace fossils (burrows at variable orientation to bedding) within quartzite beds, first described by Thom (1977), leads to the possibility that the northern, lower grade metasediments currently included in the Mt Barren Group represent a second, younger sequence. Further studies are in progress to ascertain the age of the northern-most sequence in the Mt Barren Group. An unconformity is observed between the Mount Barren Group and the Yilgarn Block in the northeast, whereas elsewhere contacts with the basement are northwest directed thrusts.

Lower greenschist facies quartzites, shales and phyllites of the Stirling Range Formation lie unconformably on the Northern Domain and southern Yilgarn Block margin (Figure 2). Although previously thought to be stratigraphically equivalent to the Mt Barren Group and deformed along with basement gneisses during the Middle Proterozoic, the discovery of body and trace fossil assemblages of Ediacaran affinities indicates a ~590-540 Ma depositional age (Cruse *et al.* in press). The Stirling Range Formation has been deformed due to subsequent reactivation of basement structures (Beeson 1992), for which preliminary results of Rb-Sr dating (currently in progress) suggest a Cambrian age (a detailed discussion of "Pan-African" and younger events is beyond the scope of this paper).

2.2 *Archaean precursors*

On the basis of Sm-Nd studies, Black *et al.* (in press) show that precursors to some of the gneisses in the Albany Mobile Belt formed at ~3.1 Ga. U-Pb zircon data by these authors also provide evidence for metamorphism and deformation at about 3110 ± 57 Ma for those rocks with igneous precursors. Granulite facies gneisses with mixed mafic, ultramafic and felsic igneous precursors in the western Albany Mobile Belt gave a T_{DM} (depleted mantle) model age of 3.2 Ga (Fletcher *et al.* 1983). Granite emplacement and metamorphism are indicated at ~2.7 Ga (Black *et al.* in press). These ages suggest reworking of the Western Gneiss Terrain of the Yilgarn Block as described by Myers (1990).

2.3 *Early Proterozoic components of the Albany Mobile Belt*

Sm-Nd model ages of ~2.0 Ga were obtained for components of the Albany Mobile Belt by Fletcher *et al.* (1983) and McCulloch (1987). Black *et al.* (in press), also on the basis of Sm-Nd model ages, provide evidence for a crustal formation event of Early Proterozoic (~2.1 Ga) age for parts of the Southern and Central domains. Fletcher *et al.* (1983) interpret their data in terms of a major crust-forming event ~2.0 Ga ago. McCulloch (1987) proposes that either there were either two Proterozoic crustal formation events or that mixing occurred between Middle Proterozoic and late Archaean crust. Black *et al.* (in press) support a combination of both options, where both Archaean (~3.1 Ga) and Early Proterozoic (~2.1 Ga) crust contributed to the formation of the Albany Mobile Belt.

Late Archaean to Early Proterozoic north-south extension in southwestern Australia is suggested by an approximately east-west striking dolerite dyke suite (as portrayed by Hallberg 1986 and Isles & Cooke 1990) for which dates of 2.4 Ga (Sm-Nd method; Fletcher *et al.* 1987) and 2.1 Ga (Rb-Sr method; Turek

1966) have been obtained. As the frequency of dyke emplacement within the Yilgarn Craton increases southwards (Hallberg 1986), crustal extension may have been greatest along the southern Yilgarn margin, leading to the initiation of the Albany Mobile Belt. Little has, however, been documented in detail regarding Early Proterozoic sedimentary sequences in this region. Meta-quartzites (and quartz-mylonites), mica schists, phyllites, talc schists and sillimanite ± cordierite-bearing paragneisses in the Albany Mobile Belt may represent deformed and metamorphosed sediments deposited during this episode. It is, however, also possible that some paragneisses may represent instead reworked Archaean meta-sediments of the Yilgarn Block Western Gneiss Terrane, and/or result from a younger cycle of Proterozoic sedimentation.

The source rocks of the sillimanite-bearing paragneisses have been dated at ~2.3 Ga and layered gneisses at ~2.2 Ga (Fletcher *et al.* 1983). Syn- to late-tectonic granites in the Albany Mobile Belt (described below) gave Early Proterozoic (~2.1 Ga) T_{DM} model ages which have been taken by Black *et al.* (in press) as indicating the age of precursor gneisses, melted during the main 1.1-1.2 Ga tectonothermal event.

2.4 *Middle Proterozoic tectonothermal events in the Albany Mobile Belt*

Differences in deformation regime result from the change in orientation of the Yilgarn Craton margin from a generally easterly orientation, in the central and western parts of the belt, to a more northeasterly orientation in the east (towards the junction with the Fraser Mobile Belt). Both sections of the Albany Mobile Belt will therefore be considered separately.

2.4.1 *Central and Western Albany Mobile Belt.*

The first Middle Proterozoic event recorded in the Albany Mobile Belt is the intrusion of the coarse-grained Albany enderbite at 1289±10 Ma (Pidgeon 1990). The timing of pegmatites and granite intrusions with respect to structural features has been used to constrain the tectonothermal evolution of the Albany Mobile Belt by Black *et al.* (in press). This study documents a complex, progressive tectonothermal event between 1190 to 1150 Ma.

A progressive southwards increase in strain is apparent, traversing from the Yilgarn Craton into the mobile belt (Beeson *et al.* 1988). Middle Proterozoic deformation is marked by west striking, dextral ductile shear zones which cross-cut the northnorthwest trending foliation typical of the Archaean gneisses of the Yilgarn Craton. The onset of these shear zones defines the northern margin of the belt. Dextral shear zones increase in intensity and width southwards until a penetrative Proterozoic metamorphic layering/shear foliation (dipping steeply south) is developed. With the southwards increase in strain, dolerite dykes have been rotated into parallelism with the foliation, foliated, boudinaged and metamorphosed up to granulite facies.

Three dominant end-member fold styles (representing successive stages of fold development during progressive deformation) are commonly observed in the Northern and Southern domains (outcrop of the Central Domain in this part of the belt is very poor and only limited structural studies may be carried out). The earliest folds are isoclinal, deforming compositional layering in gneisses and bedding in meta-sediments. These folds may show an axial planar foliation and have in turn been tightly to isoclinally, and coaxially refolded by a second set of folds. Northeast to east trending upright, generally northwest verging folds refold all previous structures. Ductile shear zones, within which shear criteria indicate dextral and reverse movement components, have also been observed in all domains. The orientation and spatial distribution of fold axes agrees with shear zone analysis and lineation trajectory mapping (*e.g.* Beeson *et al.* 1988) that the central and western parts of Albany Mobile Belt have been deformed in a dextral, transpressional regime, indicating northwest directed convergence across the belt (*i.e.* oblique to the Yilgarn craton margin).

The transpressional nature of deformation has resulted in a linear belt of deformation in which structures and lithologies show remarkable lateral continuity (Figure 2). Dextral transpression may be interpreted, within the Gondwana context, as indicating dextral displacement of the East Antarctic Shield with respect to the Yilgarn craton/Western Australian Precambrian Shield.

Conjugate ductile shear zones (northeast sinistral and westnorthwest dextral) cross-cut previous structures. Rotation of folds and foliations in the vicinity of these structures has resulted in further localised, broad, open folding. Brittle-ductile conjugate shear zones bisected by northwest striking K-feldspar + biotite + quartz pegmatites indicate deformation at a higher crustal level, with a continuation of northwest-southeast shortening, during final stages of this event.

2.4.2 *Eastern Albany Mobile Belt*

The eastern Albany Mobile Belt provides excellent examples of:
(i) basement-cover relationships and progressive deformation within a thrust belt, and
(ii) extensional deformation within a convergent tectonic setting.

A complex pattern of northwest directed thrusting and nappe emplacement, fold development, rotation and overprinting, along with intrusion of pegmatites and dolerite dykes, is observed in the eastern Albany Mobile Belt. Granulite facies gneisses of the Central Domain have been thrust northwestwards onto the Yilgarn Craton. Again, only 1190 to 1150 Ma ages have been recorded in basement gneisses (Black *et al.* in press). Northwestwards thrusting is compatible with dextral transpression in the rest of the belt as both result from northwest-southeast shortening at the southern Yilgarn Craton margin.

Discrete, northwest directed thrusts also affect Archaean granites, gneisses and greenstones in the south-eastern Yilgarn Block in proximity the Albany Mobile Belt. Metasediments of the Mount Barren Group have been emplaced as fold-thrust nappes over the Yilgarn craton and the basal sequence in the western Mt Barren Group represents an intense zone of ductile shearing and mylonitisation. A progressive pattern of fold evolution is again observed consistent with the northwest directed thrusting.

Within granulite facies gneisses of the eastern Central Domain, normal ductile shear zones (forming detachment structures with ramp-flat geometries) indicate at least one period of southeast directed extension, *i.e.* extension parallel to the regional bulk shortening direction deduced from other structures, as outlined above. Detachment faults generally have the same strike as thrusts. Earlier structures in gneisses above detachments are folded into open, generally upright, shallowly northeasterly plunging folds due to changes in dip of basal shear zones. Dolerite and granulite facies pegmatite dykes are generally localised along detachments and accompanying dilational sites, or intruded sub-parallel to axial surfaces of folds formed during extension. Minor folding of dolerite dykes (where folding cannot be explained in terms of changes in dip of detachments), and cross-cutting ductile and brittle-ductile shear zones, indicate continued northwest-southeast shortening, after formation of extensional structures.

Similar extension contemporaneous with, and parallel to, shortening in a collisional mountain belt has also been described in the Himalayan orogen (Burchfiel *et al.* 1992) where a gravitational collapse model was proposed; such a model is also suggested to explain extensional structures in the Albany Mobile Belt.

3 THE CENTRAL INDIAN TECTONIC ZONE

3.1 *Introduction*

A composite, eastnortheast striking orogenic province, referred to by Radhakrishna (1989) as the Central Indian Tectonic Zone, crosses the Indian continent (Figure 1). The Central Indian Tectonic Zone separates two continental blocks of Archaean to Early Proterozoic age (the Bundelkhand Block to the north and the Peninsular Block/Central Indian Shield to the south). The Central Indian Tectonic Zone records evidence of a complex tectonothermal history from the Early Proterozoic to Recent, and incorporates the Early Proterozoic Sausar-Sakoli-Singhbhum or Satpura orogenic belt on the northern margin of the Singhbhum Craton and gneisses of the Chotanagpur Terrane further north ("C" in Figure 1).

The evolution of different sections of the zone will be briefly summarised below for comparison with the Albany Mobile Belt.

3.2 *The Satpura Mobile Belt*

A linear belt of intense deformation, the Satpura

Mobile Belt (Figure 1) has been recognised along the southern boundary of the Central Indian Tectonic Zone (Yedekar *et al.* 1990; Sarkar *et al.* 1988; Bhattacharyya in press). Tectonothermal events described by these authors mark the start of the Proterozoic orogenic cycle in Central India. The two regions comprising this zone, the Central Indian Suture and the Singhbhum Shear Belt (along the northern margins of the Bastar and Singhbhum cratons respectively; Figure 1) will be described independently following current Indian terminology.

3.2.1 *Central Indian Suture*

The Central Indian Suture of Yedekar *et al.* (1990) is an Early Proterozoic terminal suture between the Bundelkhand Block and the Deccan Protocontinent (or Southern Peninsular Block of Radhakrishna & Ramakrishnan 1988, which incorporates the Bastar and Singhbhum cratons). This belt is also referred to as the Sausar-Gangpur-Singhbhum belt (Radhakrishna & Ramakrishnan 1988) and is situated south of Middle Proterozoic terranes of the Central Indian Tectonic Zone. The Central Indian Suture represents a major ductile shear zone extending for over 500 km, with widths up to 4 km of mylonites and phyllonites derived from granitic gneisses and their enclaves (Yedekar *et al.* 1990).

A model comprising southwards subduction beneath the Deccan Protocontinent is proposed by Yedekar *et al.* (1990). In this model, shelf sediments of the Sausar Group were deposited in an intracratonic basin during the Late Archaean to Early Proterozoic. Subduction commenced in the Early Proterozoic, with an initial plutonic phase up to 2.2 Ga, followed by arc volcanism and monzonitic intrusives culminating at ~2.1 Ga. Continent-arc and continent-continent collision, emplacement of granulites at a higher crustal level and migmatisation have been proposed at the cessation of subduction.

A rifting episode with volcanism has been described at ~1.7 Ga (Yedekar *et al.* 1990) for which both back-arc extension and post-orogenic extensional collapse models have been proposed. Terminal suturing, accompanied by upper amphibolite facies regional metamorphism, is presumed to have occurred at ~1.6 Ga (Rb-Sr ages cited by Sarkar *et al.* 1990 span a 1.7-1.5 Ga period).

Resetting ages of Rb-Sr and K-Ar isotopic systems at ~1006 Ma and ~840 Ma in eastnortheast trending Sausar Group gneisses from the northern Central Indian Suture (Sarkar *et al.* 1967) indicate a northwards change in the locus of tectonothermal activity for subsequent events.

3.2.2 *Singhbhum Shear Belt*

The geology of the Singhbhum Craton in Eastern India (Figure 1) has been summarised by Saha & Ray (1984), Bose (1986) and Bhattacharyya (in press). Tonalite-trondhjemite gneiss with Sm-Nd ages >3.8 Ga, along with the ~3.3 Ga Iron Ore Group of metasediments and metavolcanics (Moorbath & Taylor 1988), have been intruded by the Singhbhum granite at 3.3 Ga. Medium-grade metamorphism and deformation at 3.2 Ga has been described by Saha & Ray (1984). Potash-rich granites intruded at ~3.1 Ga with metamorphism closing at ~2.95 Ga ending this cycle. Further granitoid intrusion occurred between 2.2 and 2.5 Ga. Volcanics and sediments of the Dhanjori Group were deposited on the Singhbhum Craton margin, and gabbro, anorthosite, granite and ultramafic intrusion is recorded at ~2.1 Ga (Vohra *et al.* 1991).

The Singhbhum Shear Belt on the northern margin of the Singhbhum craton has been described by Sarkar *et al.* (1988) and Bhattacharyya (in press). The belt is again marked by a prominent east-west trending shear zone, which may therefore represent an eastern continuation of the Central Indian Suture. The following sequence of events have been interpreted for the Proterozoic evolution of the belt based on studies by Sarkar (1982):

1. Mica schists, phyllites and quartzites, meta-volcanics and sub-volcanics, cherts, limestones and calc-silicates of the Singhbhum Group indicate a period of sedimentation and volcanism within an early extensional tectonic regime in the interval 2.0 to 1.7 Ga.
2. Migmatites and orthogneisses formed during a crustal shortening event at ~1.6 Ga.
3. Two to three generations of approximately east-west trending, tight to isoclinal folds, accompanied by southwards directed reverse faults and shear zones, dated at ~1170 Ma.

4. Gravity sliding and folding (which may be interpreted as being associated with extension as the result of orogenic collapse, as recognised in the Albany Mobile Belt).

5. Intrusion of dolerite and epidiorite dykes along with cataclastic deformation. Interpretation of plate III of Sarkar (1982) suggests two generations of dykes controlled by conjugate shear fractures, bisected acutely by extensional fractures. The first dyke suite implies southsoutheast-northnorthwest extension, the second east-west extension. No ages may be directly attributed to intrusion of these dykes.

3.3 *Chotanagpur Gneiss Terrane*

The Chotanagpur Gneiss Terrane ("C" in Figure 1) of the Central Indian Tectonic Zone lies north of the Satpura Mobile Belt and comprises foliated gneiss and amphibolites, interleaved with supracrustal rocks. Radhakrishna (1989) considers this terrane to represent remobilised Archaean basement, Proterozoic granitic plutons and supracrustal rocks of diverse ages. Charnockitic rocks and granitic gneisses in the Chotanagpur Gneiss Terrane gave Rb-Sr ages of 1599±27 - 1590±30 Ma and 1522±21 Ma respectively (Barman *et al.* 1989; Pandey *et al.* 1986 a & b) which may represent cooling and/or deformation ages. Granite emplacement at 1590±30 Ma (Pandey *et al.* 1986a) took place at the close of this early, ~1600 Ma orogenic event.

Metasediments and migmatites of the east-west trending Bihar Schist Belt, within the Chotanagpur Gneiss Terrane, were isoclinally folded about easterly trending axes and cut by anatectic granites (Mazumder 1986; Radhakrishna 1989). Rb-Sr ages of 1242±34 and 1100± 20 Ma (Pandey *et al.* 1986 a & b) and 1086 Ma (Saha *et al.* 1987) have been recorded for late/post-kinematic granite intrusion and migmatites have been dated at 1119±24 Ma (Barman *et al.* 1989).

The Chotanagpur Gneiss Terrane has therefore undergone tectonothermal events at ~1.6 Ga and between ~1.0 and 1.2 Ga (*i.e.* the same timing of Proterozoic events recorded in the Singhbhum Shear Belt). Thermal resetting is also suggested by Saha *et al.* (1987) between 765 and 635 Ma, although the tectonic significance (if any) of these ages is unknown.

3.4 *Meghalaya region*

The Meghalaya Province of India is the closest area to the Albany Mobile Belt in the reconstruction of Powell *et al.* (1988). The Meghalaya region comprises multiply deformed metamorphic rocks of the Gneissic Complex (Mazumder 1986) overlain by less deformed, low-grade metasediments of the Shillong Group.

3.4.1 *Gneissic Complex of the Meghalaya Region.*

The Gneissic Complex of the Meghalaya region consists of amphibolite facies biotite ortho- and augen-gneiss, migmatites, amphibolites (representing early mafic intrusive sheets), schists and meta-quartzites intruded by syn-tectonic granitoids (*i.e.* similar lithologies to the Chotanagpur Gneiss Terrane). The Gneissic Complex shows a complex fold pattern as described by Barooah (1973), Murthy *et al.* (1976) and Mazumder (1986). Early broad, recumbent or reclined isoclinal folds have been refolded by tight to isoclinal, eastnortheast to northeast trending folds. These structures have been refolded by a third phase of open, north-south folds. A phase of mafic intrusion occurred at the end of the first two generations of folding.

Migmatites in the Nongpoh gneiss have been dated (using the Rb-Sr method) by Ghosh *et al.* (1991) at 1150±26 Ma. Further intrusive and/or resetting ages of 850 Ma and 505±5 Ma have been obtained from pegmatites by Van Breemen *et al.* (1989).

3.4.2 *Shillong Group metasediments*

Quartzites, siltstones, shales and phyllites of the Shillong Group unconformably overlie the Gneissic Complex (Mazumder 1986). Reliable isotopic data for the Shillong Group meta-sediments are lacking; only a K-Ar age on muscovite from a mica schist in the Shillong Formation of ~472 Ma has been determined by Sarkar (1968). This age may either date metamorphism, or records a subsequent thermal pulse.

Granitoids and syenites intruding the Shillong Group metasediments (described in detail by Mazumder 1986) have been analysed by Rb-Sr methods, however ages show a large range. The Nongpoh and Kyrdem plutons were dated at 550±15 Ma and 479±26 Ma respectively (Ghosh *et al.* 1991) and the Songsak Granite yielded an age of 500±40 Ma (Kumar 1990). These results suggest a period of granite intrusion at ~500 Ma. An "age" of 690±19 Ma was, however, obtained by Ghosh *et al.* (1991) for the South Khasi Batholith which intruded, metamorphosed and deformed Shillong Group sediments (Mazumder 1986). If this figure is correct, then sediments of the Shillong Group must have been deposited prior to 690±19 Ma, otherwise they may be as young as ~550 Ma (*i.e.* of comparable age to the Stirling Range Formation). It is possible that this oldest "age" may be in error due to contamination by gneisses and mafic amphibolites from the Gneissic Complex, resulting in a mixed isochron of no direct geological significance. Mazumder (1986) describes numerous gneissic inclusions and basic xenoliths at variable stages of assimilation within the South Khasi Batholith. He also states that biotite and sphene in the granites were developed due to assimilation reactions on basic xenoliths (Mazumder 1986, page 43). These, and other detailed petrological descriptions concerning the origin of feldspars provided by Mazumder (1986) suggest that the Rb-Sr method may not result in reliable ages for these plutons.

3.5 *Proterozoic gneisses of the Lesser Himalaya*

Proterozoic metamorphic rocks have been incorporated as tectonic slices in the lesser Himalaya during Tertiary orogenesis. Rb-Sr whole rock ages for gneisses of 1620±90 Ma (Ghose 1989) and 1050 Ma (Acharyya 1989) provide evidence for two tectonothermal events, comparable to those from within the Central Indian Tectonic Zone.

4 COMPARISONS BETWEEN THE CENTRAL INDIAN TECTONIC ZONE AND THE ALBANY MOBILE BELT

Common protolith lithologies and similarities in tectonothermal events (summarised in Table 1) suggest a comparable evolutionary history of the Central Indian Tectonic Zone and the Albany Mobile Belt.

4.1 *Archaean precursors*

Both the Satpura Orogen and Albany Mobile Belt contain gneisses derived from Archaean protoliths of similar (3.1-3.3 Ga) ages.

4.2 *Early Proterozoic extension*

Generally east-west trending dolerite dykes intruded both the Yilgarn Craton and Indian Archaean terranes (Ikramuddin & Stueber 1976; Murty *et al.* 1987) between 2.4 and 2.1 Ga. Sarkar *et al.* (1990) suggest that the Central Indian Tectonic Zone initiated as a rift system in the Early Proterozoic (leading to the Satpura Mobile Belt on its southern margin. A similar origin is proposed for the onset of the Albany Mobile Belt. It is therefore possible that Early Proterozoic meta-sediments in the Albany Mobile Belt are equivalent to meta-sediments within the Singhbhum Group where differences in mineralogy (*e.g.* sillimanite in gneisses of the Albany Mobile Belt and kyanite in the Singhbhum Belt) reflect differences in metamorphic grade. Meta-sediments, including iron-formations metamorphosed at granulite facies (Delor 1987), are present in both the Central Domain of the Albany Mobile Belt and the Singhbhum Group (Dhalbhum Formation of Sarkar 1982). A common early phase of crustal extension, resulting in sedimentation and mafic dyke intrusion, may be therefore be proposed.

4. 3 *The 1.6-1.5 Ga tectonothermal event*

An important tectonothermal event at ~1.6-1.5 Ga recorded in the Central Indian Tectonic Zone has not been recognised in the Albany Mobile Belt. Possible reasons for the apparent absence of this event in the Albany Mobile Belt are discussed in section 5 below.

4.4 *The 1.2-1.1 Ga tectonothermal event*

Ages and structures for the second deformation event in the Singhbhum belt (as outlined by Sarkar 1982) and from the Chotanagpur Belt equate with those for the main ~ 1.2 - 1.1 Ga

Table 1. Summary of the main events recorded within terranes of the Central Indian Tectonic Zone and the Albany Mobile Belt of Western Australia (see Figure 1 for location).

Note ages for precursor rocks reworked within the mobile belts are also given for comparison. Exact ages, references and interpretations are given in the text.

AGE (Ga)	Albany Mobile Belt	Meghalaya	Chotanagpur	Singhbum Belt	
	rifting (all terranes)				Mobile Belt
0.5	thermal event/dolerites	granite intrusion			Mobile Belt
	reactivation basement structures				Mobile Belt
	deposition Stirling Range Formation, Shillong Group, Upper Vindyhan				Mobile Belt
		metamorphism or resetting (Meghalaya to Singhbum Belt)			Mobile Belt
	sedimentation?		deposition Lower Vindyhan		Mobile Belt
1.0	deformation, metamorphism, granite intrusion (all terranes)				Mobile Belt
	enderbite intrusion		granite intrusion		Mobile Belt
1.5					Mobile Belt
		deformation, metamorphism, granite intrusion (Meghalaya to Singhbum Belt)			Mobile Belt
				sedimentation, volcanism	Mobile Belt
2.0					
	rifting, sedimentation	- - - - - -	- - - - - -	rifting, sedimentation	Precursor material
	dolerite dyke intrusion				Precursor material
2.5					Precursor material
	deformation, metamorphism, granite intrusion				Precursor material
3.0	mantle extraction ages, Western Gneiss Terrane			deformation, metamorphism, granite intrusion	Precursor material
				deformation, metamorphism, granite intrusion	Precursor material
3.5	xenocrystic zircons, Yilgarn				Precursor material

tectonothermal event in the Albany Mobile Belt. In all belts, folds trending both east-west and northeast-southwest were formed during progressive deformation. The "second" and "third", approximately coaxial, events of Sarkar (1982) may represent two stages in a single orogenic episode, similar to progressive fold and shear zone development in the Albany Mobile Belt. Refolding relationships, along with changes in trends of fold axes through ~ 45° to parallel the margins of the Singhbhum Shear Zone, suggest a strong dextral component of displacement along this structure (as described above for the Albany Mobile Belt). Extensional (orogenic collapse?) structures are common to both the Singhbhum Shear Zone and Albany Mobile Belt. Both areas are also cut by syn- to late-tectonic dolerite and granitoid intrusives and northnortheast striking, sinistral ductile shear zones.

The Meghalaya region also shows many similar features to the Albany Mobile Belt. Comparable lithologies are present within the "gneiss complexes" of both areas and Middle Proterozoic meta-sediments of the Mount Barren Group have counterparts within the Meghalaya region, in terms of both lithology and metamorphic grade. Both areas show at least two phases of isoclinal, inclined folds, followed by open folding. An excellent correlation can therefore be made between the two terranes, again suggesting a common ~1.1-1.2 Ga tectonothermal event.

4. 5 *Younger events*

No ages of ~850 Ma have been reported from the Albany Mobile Belt to correlate with pegmatite intrusion and/or resetting recorded in Meghalaya. However, 849±9 Ma alkaline ultrabasic dykes near the eastern boundary of the Yilgarn Craton and in the Fraser Mobile Belt (Robey *et al.* 1986) provide evidence for a Late Proterozoic igneous event in southwestern Australia, the extent of which is unknown. It is of interest to note that 832±40 Ma alkaline intrusives have also been reported from the Indian Precambrian Shield by Ikramuddin & Stueber (1976), suggesting that alkaline dyke intrusion may be more widespread than suggested by existing geochronology in this sector of Gondwana.

Both the Stirling Range Formation and Shillong Group are unconformable upon basement gneisses, show similar lithologies and thickness of sediments and were both metamorphosed to lower greenschist facies. Similar orientations and styles of folding have been described in both meta-sedimentary sequences. Structures with the same style and orientation are also present in sedimentary rocks of the Vindhyan Supergroup in northwestern India (Ghosh *et al.* 1981). In all of the above sequences, the *en échelon* stepping of folds and rotation of axes into parallelism with basement faults suggests that deformation of Middle Proterozoic to Early Cambrian strata may be related to dextral transpressional reactivation of basement structures.

No granitoids intrude the Stirling Range Formation, whereas the Shillong Group is intruded by a suite of ~500 Ma plutons. Isotopic resetting at ~500 Ma in the Albany Mobile Belt and western Yilgarn Craton (Libby & De Laeter 1987; De Laeter & Libby in press) may however reflect this event in Western Australia.

Late cataclastic shear zones controlling dolerite and acid dyke emplacement in the Singhbhum shear zone are identical in style and orientation with those in the Albany Mobile Belt which also control emplacement of dolerite dykes (Harris & Li 1991).

The Narmada-Son rift represents reactivation of the Proterozoic suture in the Central Indian Tectonic Zone (Sarkar *et al.* 1990); in an analogous manner, separation between Antarctica and Australia has taken place along Proterozoic trends in the Albany Mobile Belt, with the formation of the Bremer Basin of the Southern Rift System (Stagg & Willcox 1991).

5 PROBLEMS IN CORRELATION

There is no mention in existing geological literature as to the nature of foliations in the Gneiss Complex of Meghalaya, and whether there is evidence for non-coaxial deformation as seen in the Albany Mobile Belt. The timing of events within the Albany Mobile Belt is now well established (Black *et al.* in press), however little precise dating has been carried out from migmatites of the Gneiss Complex of Meghalaya. Further dating of pegmatites, intrusive into different structural sites, is therefore required to establish whether the events in the two areas are comparable.

As mentioned above, there is currently no evidence for a 800±100 Ma (Rb-Sr isotope) event (as described in Indian terranes by Sarkar *et al.* 1990) in the Albany Mobile Belt, and the tectonic significance of existing Indian ages is unknown.

Problems in correlation arise concerning the absence of a tectonothermal event at ~1.5-1.6 Ga in the Albany Mobile Belt, corresponding to the first event in the Singhbhum and Chotanagpur belts (Satpura Orogeny). Whilst this may be due to inadequate sampling in previous geochronological studies, the apparent absence of such a major event in the Albany Mobile Belt may provide indirect evidence for correlation with other Western Australian terranes at this time. Ages ranging from ~1.7 to 1.6 Ga have been recorded in sheared gneisses from the Gascoyne Province (Figure 1) on the northern Yilgarn craton margin (interpreted by Myers 1990 as a collisional orogen). Comparable ages have been recorded from drill-hole samples of basement gneisses from the Darling Mobile Belt beneath the Perth Basin (Figure 2) on the western Yilgarn margin (Fletcher *et al.* 1985). To explain the absence of ~1.5-1.6 Ga ages in the Albany Mobile Belt, it is proposed that the Darling Mobile Belt may have acted as a transfer/transcurrent shear zone linking collision in both the Gascoyne Province and Satpura Orogen. An alternative interpretation is that a terrane equivalent to the Central Indian Suture/Singhbum Shear zone exists in Antarctica, and that Middle Proterozoic events in the Albany Mobile Belt were situated north of this terrane.

Further dating (*eg.* by ion microprobe) is required to resolve the problems in determining the true age of plutons intruding the Shillong Group. If the 690±19 Ma age of Ghosh *et al.* (1991) obtained from the South Khasi Batholith indeed corresponds to the age of granitoid intrusion, then the Shillong Group and Stirling Range Formation would not represent stratigraphic equivalents as proposed herein (and the Shillong Group may then have been deposited during the Lower Vindyan cycle).

6 CONCLUSIONS

The correlations presented above allow a simplified history for Proterozoic to Cambrian tectonism in this sector of Gondwana to be proposed:

1. Emplacement of generally east-west trending dolerite dykes between 2.4 and 2.1 Ga in the Yilgarn craton and Indian Archaean terranes. Extension on the southern Yilgarn craton margin resulted in sediment deposition and emplacement of mafic intrusives. In India, deposition of the Sausar Group shelf sediments took place in an intracratonic basin during the late Archaean to Early Proterozoic.
2. Collision between the Bundelkhand Block and the Deccan Protocontinent resulted in a major tectonothermal event in the Central Indian Tectonic Zone at ~1.6 Ga. A transform margin is postulated along the Darling Mobile Belt, linking with collision on the northern Yilgarn margin in the Gascoyne orogen.
3. Dextral transpressional deformation and thrusting at ~1.1-1.2 Ga was accompanied by amphibolite to granulite facies metamorphism and granitoid intrusion, within an orogenic belt extending across India and continuing between Antarctica and the West Australian Precambrian Shield.
4. Deposition of the Lower Vindhyan sediments of the Vindhyan Supergroup (Ghosh *et al.* 1981) in India. No equivalent to the Lower Vindhyan has yet been confirmed in the Albany Mobile Belt.
5. Intrusion of pegmatites and isotopic resetting in Meghalaya at 800±100 Ma. Possibility of dyke intrusion in the Albany Mobile Belt (alkaline dykes in this time interval intrude the eastern Yilgarn Craton and the adjacent Fraser Mobile Belt).
6. Deposition of the Shillong Group sediments near the northern margin of the Central Indian Tectonic Zone in India and the Stirling Range Formation on the northern margin of the Albany Mobile Belt, unconformably upon basement gneisses during the Late Proterozoic (contemporaneous with deposition of the Upper Vindhyan in northwestern India).
7. Sinistral transcurrent displacement between India and western Australia along the Darling Mobile Belt during the Late Proterozoic.
8. Folding of the Shillong Group and Stirling Range Formation during dextral reactivation of basement structures (Early Cambrian?).
9. Early Cambrian cataclastic shear zones accompanied by intrusion of dolerite dykes and sills in southwestern Western Australia; equivalent dykes are interpreted in the Central

Indian Tectonic Zone based on similarities with Cambrian dykes in the Albany Mobile Belt, but radiometric or palaeomagnetic confirmation of their age is still required.

Due to the similarities described above, it may be concluded that the Albany Mobile Belt of southwestern Australia and the Central Indian Tectonic Zone constituted a continuous orogenic belt during the 1.1-1.2 Ga tectonic cycle common to many other areas of Gondwana. Both terranes continued to share a parallel evolution up to and including the Early Cambrian.

A 1.1-1.2 Ga tectonothermal event is also recorded in the Eastern Ghats (Grew & Manton 1986; Aftalion *et al.* 1988) on the southeastern margin of Peninsular India (Figure 1), the Bunger Hills and Enderby Land (Sheraton *et al.* 1980, 1990) regions of East Antarctica and in the Fraser Mobile Belt of Western Australia (Turek 1966; D. Nelson *pers comm.* 1991). One can therefore envisage a mobile belt between southwestern Australia and the Bunger Hills-Windmill Islands areas of Antarctica, bifurcating around Archaean and Early Proterozoic terrains of India. In this model, one arm includes the Eastern Ghats-Sri Lanka-East Antarctica, the other the Albany Mobile Belt-Central Indian Tectonic Zone. Dextral transpression within the central and western Albany Mobile Belt and Central Indian Tectonic Zone is coeval with thrust tectonics in the Eastern Ghats and eastern Albany Mobile Belt. North striking, sinistral ductile shear zones in the Fraser Mobile Belt (described by Harris & Delor 1991) formed in a conjugate relationship to the Albany Mobile Belt during this event.

The correlations established between northern India and Western Australia support the Gondwana fit of Powell *et al.* (1988). The fit proposed by Ricou et al. (1990) where the Indian continent lies against the western margin of Australia is not compatible with the conclusions presented here.

7 Acknowledgments

Fieldwork in the Albany Mobile Belt was funded through Australian Research Council and University of Western Australia Special Research grants. J. Beeson, C. Delor, T. Cruse and M. Pascoe have provided geological input during varying stages of this project. Geochronological studies in the Albany Mobile Belt were carried out with the assistance of the Bureau of Mineral Resources by L. Black. S.K. Mazumder is thanked for providing information on the Meghalaya region, and R.H. Findlay, S. Wetherley, D. Byrne and an anonymous reviewer are thanked for their helpful comments on earlier drafts of this paper.

REFERENCES

Acharyya, S.K. 1989. The Daling Group, its nomenclature, tectono-stratigraphy and structural grain: with notes on their possible equivalents. *In*: Daling Group and related rocks, *Special Publication of the Geological Survey of India* 22: 5-13.

Aftalion, M., D.R. Bowes, B. Dash & T.J. Dempster 1988. Late Proterozoic Charnockites in Orissa, India: a U-Pb and Rb-Sr study. *Journal of Geology* 96: 663-676.

Barooah, B.C. 1973. Tectonic patterns of the Precambrian rocks around Tyrssad, Meghalaya. *In*: Precambrian geology of the Peninsular Shield. *Miscellaneous Publication of the Geological Survey of India* 23: 485-495.

Barman, T.R., P.K. Bishui & A. Sarkar 1989. Dating of Early Precambrian granite-greenstone complex of the Eastern Indian Precambrian Shield with special reference to Chotanagpur granite gneiss complex. *Geological Survey of India Record* 123: 25-27.

Beeson, J. 1992. A field and experimental study of structures in the Albany Mobile Belt, Western Australia. *PhD thesis, The University of Western Australia (unpublished).*

Beeson, J., C.P. Delor & L.B. Harris 1988. A structural and metamorphic traverse across the Albany Mobile Belt, Western Australia. *Precambrian Research* 40/41: 117-136.

Bhattacharyya, D.S. in press. Precambrian metallogeny, tectonics and geochronology of the Singhbhum Cu-U belt, Eastern India. *Precambrian Research.*

Black, L., L.B. Harris & C.P. Delor in press. The timing of tectonothermal events in the Albany Mobile Belt, Western Australia. *Precambrian Research.*

Bose, M.K. 1986. Growth of Precambrian continental crust - a study of the Singhbhum

segment in the Eastern Indian Shield. *In*: S.M. Naqvi (ed.) Precambrian continental crust and its economic resources. *Developments in Precambrian Geology* 8: 267-286.
Burchfiel, B.C., C. Zhiliang, K.V. Hodges, L. Yuping, L.H. Royden, D. Changrong & X. Jiene 1992. The South Tibetan Detachment System, Himalayan Orogen: extension contemporaneous with and parallel to shortening in a collisional mountain belt. *Geological Society of America Special Paper* 269.
Cruse, T., L.B. Harris & B. Rasmussen in press. The discovery of Ediacaran trace and body fossils in the Stirling Range Formation, Western Australia: implications for sedimentation and deformation during the "Pan-African" orogenic cycle. *Australian Journal of Earth Sciences.*
Dalziel, I.W.D. 1992. Antarctica; a tale of two supercontinents. *Annual Revue of Earth and Planetary Science* 20: 501-526.
De Laeter, J.R. & W.G. Libby in press. 500 Ma biotite Rb/Sr dates on the Yilgarn Block near Harvey, Western Australia. *Australian Journal of Earth Sciences.*
De Witt, M., M. Jeffrey, H. Bergh & L. Nicoloysen 1988. Geological map of sectors of Gondwana. American Association of Petroleum Geologists.
Delor, C.P. 1987. Petrology of garnet-orthopyroxene assemblages from the Albany-Fraser Province (Western Australia). *Abstracts, Precambrian events in the Gondwana fragments.* International Geological Correlation Program project 236, Institute of Fundamental Studies, Kandy, Sri Lanka.
Fletcher, I.R., W.G. Libby, & K.J.R. Rosman 1987. Sm-Nd dating of the 2411 Ma Jimberlana dyke, Yilgarn Block, Western Australia. *Australian Journal of Earth Sciences* 34: 523-525.
Fletcher, I.R., S.A. Wilde W.G. Libby & K.J.R. Rosman 1983. Sm-Nd model ages across the margins of the Archaean Yilgarn Block. Western Australia - II: southwest transect into the Proterozoic Albany-Fraser Province. *Journal of the Geological Society of Australia* 30: 333-340.
Fletcher, I.R., S.A. Wilde & K.J.R. Rosman 1985. Sm-Nd model ages across the margins of the Archaean Yilgarn block, western Australia-III. Western Margin. *Australian Journal of Earth Sciences* 32: 73-82.
Ghose, A. 1989. Structural and metamorphic history of the crystallines of Lesser Himalaya, Uttar Pradesh. *In:* Daling Group and related rocks, *Geological Survey of India Special Publication* 22: 65-70.
Ghosh, D.B, M.K. Soni, K.K.K. Nair & M.M. Munshi 1981. Stratigraphy and sedimentation of the Vindhyans in the Lower Narmada Valley. *Geological Survey of India Miscellaneous Publication* 50: 157-161.
Ghosh, S., S. Chakraborty, J.K. Bhalla, D.K. Paul, A. Sarkar, P.K. Bishui & S.N. Gupta 1991. Geochronology and geochemistry of granite plutons from East Khasi Hills, Meghalaya. *Journal of the Geological Society India* 37: 331-342.
Grew, E.S. & W.I. Manton 1986. A new correlation of sapphirine granulites in the Indo-Antarctic metamorphic terrain: Late Proterozoic dates from the Eastern Ghats Province of India. *Precambrian Research* 33:123-137.
Hallberg, J.A. 1986. Post-cratonization mafic and ultramafic dykes of the Yilgarn Block. *Australian Journal of Earth Sciences* 34: 135-149.
Harris, L.B. 1987. A tectonic framework for the Western Australian Shield and its significance to gold mineralization. *In:* S.E. Ho & D. I. Groves (eds.), Recent Advances in Understanding Precambrian Gold Deposits. *Geology Department and University Extension, University of Western Australia, Perth, Publication* 11: 1-27.
Harris, L.B. & C.P. Delor 1991. Tectonic evolution of the Fraser Mobile Belt, Western Australia. *Geological Society of Australia Abstracts* 31: 24-25.
Harris, L.B. & Z.X. Li 1991. Tectonic controls on Cambrian dolerite dyke emplacement in the Albany Mobile Belt, Western Australia. *Abstracts, Eighth International Symposium on Gondwana, Hobart*: 41-42.
Ikramuddin, M. & A.M. Stueber 1976. Rb-Sr ages of Precambrian dolerite and alkaline dikes, southeast Mysore State, India. *Lithos* 9: 235-241.
Isles, D.J. & A.C. Cooke 1990. Spatial association between post-cratonization dykes and gold deposits in the Yilgarn Block, W.A. *In*: A.J. Parker, P.C. Rickwood & D.J. Tucker (eds.), *Mafic dykes and*

emplacement mechanisms: 57-162. Rotterdam: Balkema.
Katz, M.B. 1980. Schematic transform kinematics of the Australian Precambrian: An example of Precambrian intraplate tectonics. *Journal of Geology* 88: 467-473.
Kumar, S. 1990. Petrochemistry and geochronology of pink granite from Songsak, East Gara Hills, Meghalaya. *Journal of the Geological Society of India* 35: 39-45.
Libby, W.G. & J.R. De Laeter 1987. Biotite ages and cooling history at the western margin of the Yilgarn Block. *Geological Survey of Western Australia Annual Report for 1987*: 79-87.
Lovering, J.F., G.A. Travis, D.J. Comaford & P. R. Kelly 1981. Evolution of the Gondwana Archaean Shield: zircon dating by ion microprobe, and relationships between Australia and Wilkes Land (Antarctica). *In:* J.E. Glover & D.I. Groves (Eds.), Archaean Geology. *Geological Society of Australia Special Publication* 7: 193-203.
Mazumder, S.K. 1986. The Precambrian framework of part of the Khasi Hills, Meghalaya. *Geological Survey of India Record* 117: 1-59.
McCulloch, M.T. 1987. Sm-Nd isotopic constraints on the evolution of Precambrian crust in the Australian continent. Proterozoic lithospheric evolution. *Geodynamic Series* 17: 115-130.
Moorbath, S. & P.N. Taylor 1988. Early Precambrian crustal evolution in Eastern India: the ages of the Singhbhum Granite and included remnants of older gneiss. *Journal of the Geological Society India* 31: 82-84.
Murthy, M.V.N., S.K. Mazumder & N. Bhaumik 1976. Significance of tectonic trends in the geological evolution of the Meghalaya uplands since the Precambrian. *Geological Survey of India Miscellaneous Publication* 23: 471-484.
Murty, Y.G.K., V. Babu Rao, D. Guptasarma, J.M. Rao, M.N. Rao & S. Bhattacharji 1987. Tectonic, petrochemical and geophysical studies of mafic dyke swarms around the Proterozoic Cuddapah Basin, south India. *In:* H.C. Halls & W.F. Fahrig (eds), Mafic dyke swarms. *Geological Association of Canada Special Paper* 34: 303-316.
Myers, J.S. 1990. Western Gneiss Terrane. *In*: Geology and mineral resources of Western Australia. *Geological Survey of Western Australia Memoir* 3: 13-31.
Oliver, R.L., J.A. Cooper & J.A. Truelove 1983. Petrology and zircon geochronology of Herring Island and Commonwealth Bay and evidence for Gondwana reconstruction. *In:* R.L. Oliver, P. R. James & J.B. Jago (eds.), *Antarctic Earth Science:* 64-68. Australian Academy of Science.
Pandey, B.K., J.N. Gupta & Y. Lall 1986a. Whole rock and mineral Rb-Sr isochron ages for the granites from Bihar Mica Belt of Hazaribagh, Bihar, India. *Indian Journal of Earth Science* 12: 157-162.
Pandey, B.K., D.L. Upadhayay & K.K. Sinha 1986b. Geochronology of Jajawal-Binda-Nagnaha granitoids in relation to Uranium mineralisation. *Indian Journal of Earth Science* 13: 163-168.
Pidgeon, R.T. 1990. Timing of plutonism in the Albany Mobile Belt, southwestern Australia. *Precambrian Research* 47: 157-167.
Powell, C. Mc A., S.R. Roots & J.J. Veevers 1988. Pre-breakup continental extension in East Gondwanaland and the early opening of the eastern Indian Ocean. *Tectonophysics* 155: 261-283.
Radhakrishna, B.P. 1989. Suspect tectono-stratigraphic terrane elements in the Indian Subcontinent. *Journal of the Geological Society of India* 34: 1-24.
Radhakrishna, B.P. & M. Ramakrishnan 1988. Archaean-Proterozoic boundary in India. *Journal of the Geological Society of India* 32: 263-278.
Ricou, L.E., J., Besse, J. Marcoux & P. Patriat 1990. Une reconstruction du Gondwana révisée à partir de données pluridisciplinaires. *Comptes Rendues de l'Academie de Science, Paris* 311(4): 463-469
Robey, J.V.A., J.W. Bristow, M.R. Marx, J. Joyce, R.V. Danchin & F. Arnott 1986. Alkaline ultrabasic dikes near Norseman, Western Australia. *In:* J. Ross, A.L. Jaques, J. Ferguson, D.H. Green, S.Y. O'Reilly, R.V. Danchin & A.J.A. Janse (eds.), Kimberlites and related rocks Volume 1: their composition, occurrence, origin and emplacement. *Geological Society of Australia Special Publication* 14: 382-391.
Saha, A.K. & S.L. Ray 1984. The structural and geochemical evolution of the Singhbhum batholithic complex, India. *Tectonophysics* 105: 163-176.

Saha, A.K., S.S. Sarkar & S.S. Rej 1987. Petrochemical evolution of the Bihar Mica Belt granites, Eastern India. *Indian Journal of Earth Science* 14: 22-45.
Sarkar, A.N. 1982. Structural and petrological evolution of the Precambrian rocks in western Singhbhum, Bihar. *Memoirs of the Geological Survey of India* 113: 97 pp.
Sarkar, S.N. 1968. *Precambrian stratigraphy and geochronology of Peninsular India.* Dhanbad Publishers, India.
Sarkar, S.N., A. Gupta, & A. Basu 1988. North Singhbhum Proterozoic Mobile Belt, Eastern India. *In:* Precambrian ore deposits and tectonics, IGCP Project 247. Jadavpur University and Geological Survey of India: 31.
Sarkar, A., G. Sarkar, D.K. Paul & N.D. Mitra 1990. Precambrian geochronology of the Central Indian Shield - a review. *Geological Survey of India Special Publication* 28: 453-482.
Sarkar, S.N., E.K. Gerling, A.A. Pollkanov & F.V. Chukov 1967. Precambrian geochronology of Nagpur-Bhandara-Durg, India. *Geological Magazine* 104: 525-549.
Sheraton, J.W., L.P. Black, M.T. McCulloch & R.L. Oliver 1990. Age and origin of a compositionally varied mafic dyke swarm in the Bunger Hills, East Antarctica. *Chemical Geology* 85: 215-246.
Sheraton, J. W., L.A. Offe, R.J. Tingey & D. Ellis 1980. Enderby Land, Antarctica- an unusual Precambrian high grade metamorphic terrain. *Journal of the Geological Society of Australia* 27: 1-18.
Stagg, H.M.J. & J.B. Willcox 1991. Structure and hydrocarbon potential of the Bremer Basin, southwest Australia. *Bureau of Mineral Resources Journal of Australian Geology and Geophysics* 12: 327-337.
Thom, R. 1977. The evolution of Proterozoic rocks near the Fraser Front at Ravensthorpe, Western Australia. *Ph D thesis, University of London (unpublished).*
Thom, R., J.R. De Laeter & W.G. Libby. 1980. Rb-Sr dating of tectonic events in the Proterozoic Mount Barren Group near Hopetoun. *Annual Report of the Geological Survey of Western Australia* 1980: 109-112.
Turek, A. 1966. Rubidium - strontium isotope studies in the Kalgoorlie-Norseman area, Western Australia. *Ph D thesis, Australian National* University, *Canberra* (unpublished).
Van Breemen, O., D.R. Bowes, C.C. Bhattacharjee & P.K. Chowdhary 1989. Late Proterozoic-Early Palaeozoic Rb-Sr whole-rock and mineral ages for granite and pegmatite, Goalpara, Assam, India. *Journal of the Geological Society of India* 33: 89-92.
Vohra, C.P., S. Dasgupta, D.K., Paul, P.K. Bishui, S.N. Gupta & S. Guha 1991. Rb-Sr chronology and petrochemistry of granitoids from the south-eastern part of the Singhbhum Craton, Orissa. *Journal of the Geological Society India* 38: 5-22.
Yedekar, D.B., S.C. Jain, K.K.K. Nair & K.K. Dutta 1990. The Central Indian Collisional Zone. *Geological Survey of India Special Publication* 28: 1-42.

Gondwana Eight, Findlay, Unrug, Banks & Veevers (eds) © 1993 Balkema, Rotterdam. ISBN 90 5410 304 3

Southeast Asian terranes: Gondwanaland origins and evolution

I. Metcalfe
University of New England, Armidale, N.S.W., Australia

ABSTRACT: Southeast Asia is a complex assembly of allochthonous continental terranes now far removed from their original sites of origin together with island arcs, accretionary complexes and small ocean basins. The boundaries between these terranes are marked by major fault zones or by sutures recognised by the presence of ophiolites, mélanges and accretionary complexes. Stratigraphical, sedimentological, palaeobiogeographical and palaeomagnetic data suggest that most, and probably all, of the Southeast Asian continental terranes were derived directly or indirectly from Gondwanaland. These terranes were assembled between the late Palaeozoic and Cenozoic, but the precise times of rifting from Gondwanaland and timings of amalgamation and accretion are still contentious. The continental 'core' of the region comprises four principal continental blocks: South China, Indochina, Sibumasu and East Malaya. Palaeomagnetic and faunal data suggest that South China rifted from Gondwanaland in the Silurian or Devonian. Indochina and East Malaya may also have rifted from Gondwanaland in pre-Carboniferous times because their late Palaeozoic faunas and floras do not appear to show any Gondwanaland affinities and there are no Carbo-Permian glaciogenic sediments recorded on these blocks. Indochina and East Malaya also exhibit extremely similar late Palaeozoic and Mesozoic geology which suggests that they were already amalgamated by the late Palaeozoic. Amalgamation of South China and Indochina-East Malaya occurred during the Early Carboniferous along the Song Ma Suture to form "Cathaysialand". Cathaysialand, together with North China, formed a large continental region within the Palaeotethys during the Late Carboniferous and Permian. Palaeomagnetic data indicate that this continental region was in equatorial to low northern palaeolatitudes which is consistent with the tropical Cathaysian flora developed on these terranes.

Palaeozoic (Cambrian - Early Permian) faunas of Sibumasu show strong affinities with NW Australia and indicate that this block formed part of the north-eastern margin of Gondwanaland along with the small Hainanese terranes, the Lhasa Block and Changtang Block (part) of Tibet, which have similar stratigraphy and faunal affinities. This is supported by the presence of Late Carboniferous - Early Permian glacial-marine diamictites, Early Permian cool-water and low-diversity faunas and palaeomagnetic data suggesting southern hemisphere palaeolatitudes during Carboniferous and Early Permian times. A major rifting phase occurred on the north and east margin of Gondwanaland in the Early-to-Middle Permian, and palaeomagnetic and faunal data suggest that Sibumasu (as part of the elongate Cimmerian continent) rifted from NW Australian Gondwanaland at this time. Collision between Sibumasu and Indochina-East Malaya probably occurred within Palaeotethys in Early Triassic times to form the continental core of Southeast Asia. The Early Triassic suturing age is indicated by mélange ages, ages of collision-related plutons, palaeobiogeographic data and development of Late Triassic post-orogenic basins. The Early Triassic suturing of Sibumasu to Cathaysialand was followed by collision with North China in the Late Triassic, as evidenced by palaeomagnetic data, resulting in an overall consolidation of the Southeast Asian cratonic 'core'.

Other smaller terranes including West Burma, the Woyla terranes (Sikuleh, Natal), Semitau and SW Borneo were accreted to this core during the Jurassic and Cretaceous to form 'Sundaland'. The West Burma terrane and possibly some of the Woyla terranes, together with other unidentified crustal blocks may represent the material that provided sediment to the Triassic intracratonic basin of Timor and which rifted from NW Australia in the Middle Jurassic. The origin of the Southwest Borneo and Semitau terranes is still unclear. Palaeomagnetism indicates that they have been located in equatorial palaeolatitudes since the Cretaceous. Triassic and Jurassic faunal affinities are with Indochina, South China and Japan which suggests a possible Cathaysialand origin.

Eastern Southeast Asia is composed of continental fragments, rifted from the Australian and South China margins during the Late Mesozoic and Cenozoic, together with subduction complexes, island arcs and marginal seas. The spatial distribution of the small continental fragments in eastern Southeast Asia is a result of the interaction between the Eurasian, Indo-Australian and Philippine Sea plates during the Cenozoic. A series of palaeogeographic maps are presented showing interpreted palaeopositions of the various Southeast Asian terranes in the Palaeozoic, Mesozoic, and Cenozoic, in relation to Gondwanaland and Tethys, consistent with the current geological and geophysical constraints.

1 INTRODUCTION

Recent studies of Southeast Asian geology, in the light of plate tectonics and terrane analysis, have shown the region to be a giant 'jigsaw puzzle' of allochthonous continental blocks and fragments and other terranes of island arc and oceanic affinities set in a matrix of extended continental or transitional crust, accretionary complex material and young oceanic crust of 'marginal' basins (Burrett, 1974; Ridd, 1980; Mitchell, 1981; Stauffer, 1983; Sengör, 1984; Klimetz, 1987; Metcalfe, 1986, 1988, 1990; Figure 1). The principal continental terranes are separated and bounded by major geological discontinuities such as major strike-slip faults or narrow 'mobile belts' some of which are recognisable sutures with accretionary complexes and ophiolites representing former oceans. The concept of multiple Tethys oceans and tectonic collages developed by Sengör (1979 and subsequent papers) has led to questioning of the traditional reconstructions of the Permian Pangaea with a wide , empty, V-shaped Tethys ocean. Until very recently, most Palaeozoic and Mesozoic palaeogeographic reconstructions have shown Southeast Asia as being an integral part of Eurasia since the early Palaeozoic. Palaeogeographic reconstructions using palaeomagnetic data have also been made with Southeast Asia regarded as a single tectonic unit despite the mounting evidence that it is otherwise. Possible Southeast Asia - Gondwanaland connections were first suggested by Burton (1970) and Ridd (1971) and recent integrated studies (Audley-Charles, 1983, 1984, 1988; Metcalfe, 1986, 1988, 1990, 1991; Burrett et al., 1990) utilising stratigraphic, palaeontological and palaeomagnetic data indicate that most, and probably all, of Southeast Asia was derived directly or indirectly from Gondwanaland. The timings of rifting, drifting and amalgamation/accretion of the Southeast Asian terranes are still matters of current debate and this paper attempts to constrain those timings and presents palaeogeographic reconstructions for the Late Palaeozoic to Cenozoic which are considered consistent with the currently available geological and geophysical data.

2 THE SOUTHEAST ASIAN TERRANES

The continental terranes of Southeast Asia are illustrated in Figure 1. The continental 'core' of mainland Southeast Asia comprises four principal terranes, South China, Indochina, Sibumasu and East Malaya, which were assembled between the Early Carboniferous and Early Triassic (Metcalfe, 1986, 1988, 1990). Other authors have suggested that some of these terranes may be composite. Cameron et al. (1980) and Pulunggono and Cameron (1984) proposed two terranes west of the Bentong-Raub suture, the Malacca and Mergui microplates separated by a Triassic suture represented by the 'Mutus Assemblage' of Central Sumatra and the Semanggol Formation in N.W. Peninsular Malaysia (Figure 2A). The 'Mutus assemblage' was correlated with the Semanggol Formation on the basis of similar lithologies and a single K-Ar date of 222 Ma from undescribed tuff. However, the presence of ophiolites, accretionary complex material or other rocks suggestive of a suture within the Mutus assemblage or the Semanggol Formation has yet to be demonstrated. Furthermore, Hutchison (pers. comm.) demonstrates that the

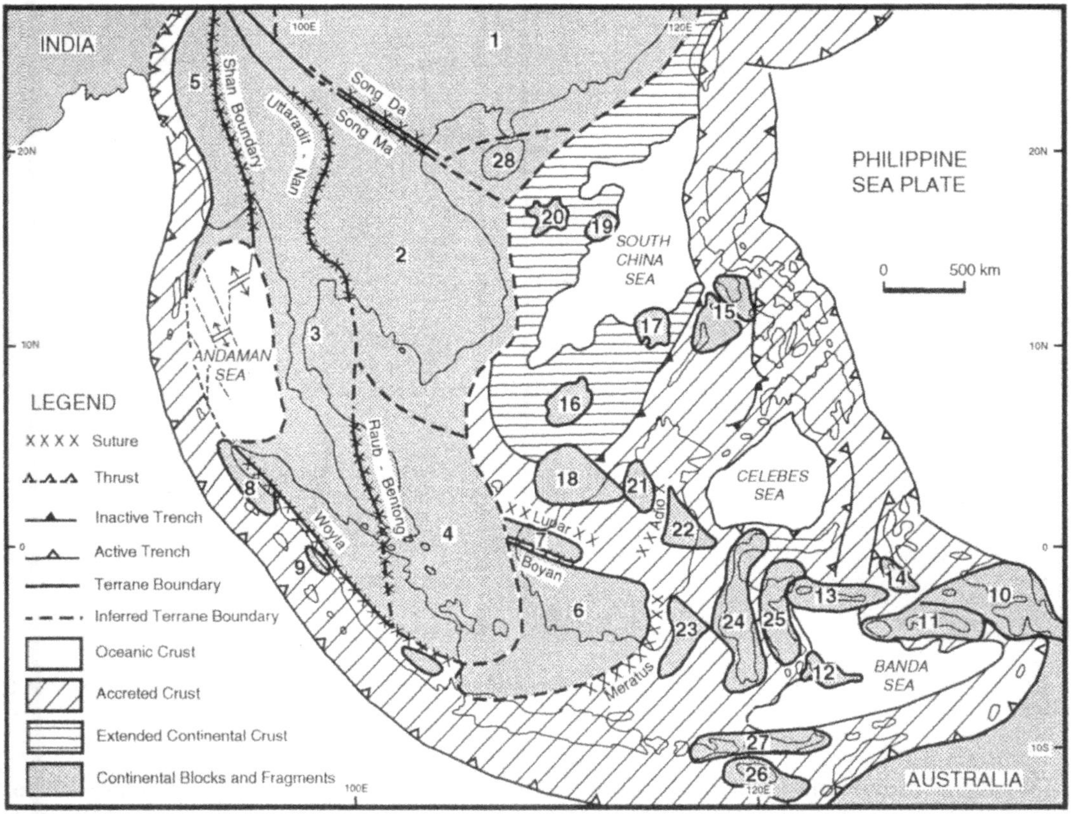

Fig. 1. Distribution of continental blocks and fragments (terranes) and principle sutures of Southeast Asia (modified after Metcalfe, 1990). 1. South China 2. Indochina 3. Sibumasu 4. East Malaya 5. West Burma 6. S.W. Borneo 7. Semitau 8. Sikuleh 9. Natal 10. West Irian Jaya 11. Buru-Seram 12. Buton 13. Bangai-Sula 14. Obi-Bacan 15. North Palawan 16. Spratley Islands-Dangerous Ground 17. Reed Bank 18. Luconia 19. Macclesfield Bank 20. Paracel Islands 21. Kelabit-Longbowan 22. Mangkalihat 23. Paternoster 24. West Sulawesi 25. East Sulawesi 26. Sumba 27. Banda Allochthon 28. Qiongzhong and Yaxian terranes of Hainan.

Mutus assemblage lithologies can be readily equated with the Lower Palaeozoic rocks of Peninsular Malaysia. The presence of pebbly mudstones in the basement of the Central Sumatran basin (Koning and Darmono, 1984) and south of Kuala Lumpur (Tjia and Anizan Isahak, 1987) which appear to be correlatives of the glacial-marine Singa Formation and Phuket Group also suggest that these regions form part of the Sibumasu terrane. The presence of a seperate 'Malacca microplate' as suggested by Cameron et al. (1980) and Pulunggono and Cameron (1984) therefore seems unlikely. Cooper et al. (1989) have also suggested the presence of a seperate 'Phuket terrane' (Figure 2B) characterised by Carbo-Permian glaciogenic pebbly mudstones. Again, the presence of a suture separating this terrane from Sibumasu has not been demonstrated and the presence of glacial-marine diamictites on other parts of the Sibumasu terrane negate the necessity for a 'Phuket terrane'. South China has been regarded as a composite terrane and the term as used in this paper includes the Yangtze and Huannan blocks of Sengör et al. (1988), see Metcalfe (1990) for discussion. The Indochina and East Malaya terranes probably formed a coherent tectonic block from the Late Palaeozoic onwards as evidenced by their similar Late Palaeozoic and Mesozoic geological histories.

A number of smaller terranes are found accreted to this Southeast Asian 'core'. On the west and southwest, they include the West Burma terrane (Mount Victoria Land of Mitchell, 1989, here renamed to avoid

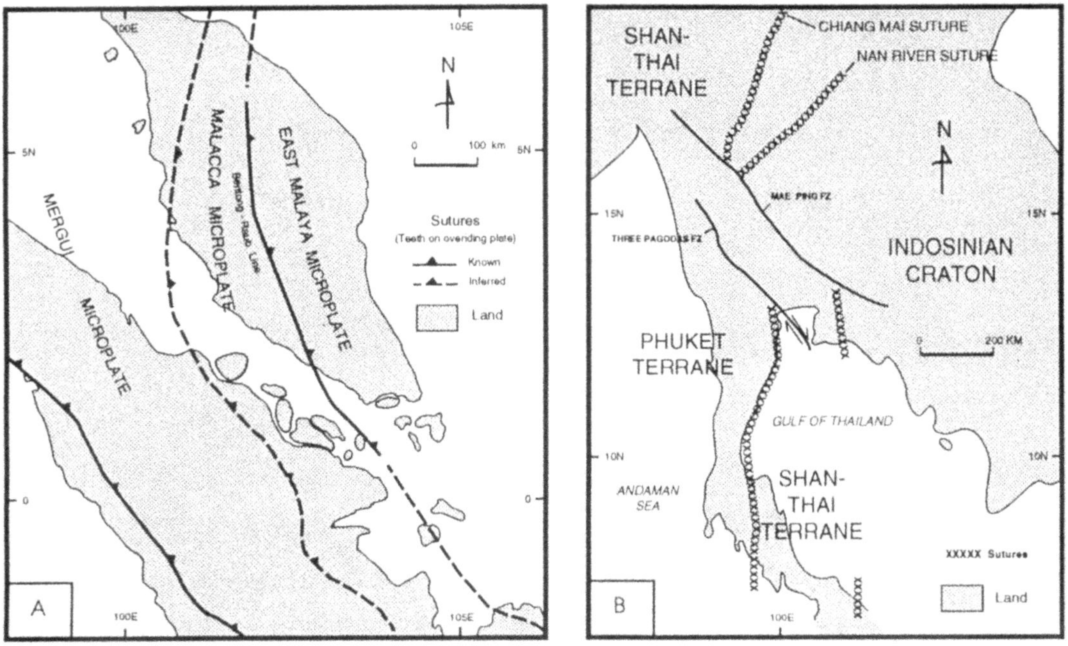

Fig. 2. Terranes proposed for parts of western mainland Southeast Asia by A: Cameron et al.(1980) and Pulunggono and Cameron (1984) and B: Cooper et al. (1989).

confusion with Victoria Land in Antarctica) and the Sikuleh and Natal terranes (Pulunggono and Cameron, 1984; Wajzer et al., in press). On the southeast they include the South West Borneo and Semitau terranes (Metcalfe, 1990) and on the northeast the small Quiongzhong and Yaxian terranes of Hainan Island (Yu, 1989; Yang et al., 1989). The other pre-Mesozoic terranes of the eastern part of the region (Figure 1) comprise a number of small continental terranes of South China or Australian affinities.

3 TERRANE ORIGINS

A growing body of evidence suggests that most, and probably all, of the Southeast Asian continental terranes were derived either directly or indirectly from Gondwanaland (Metcalfe, 1986, 1988, 1990). Studies of early Palaeozoic faunas of South China and the Quiongzhong terrane of Hainan and Early to Late Palaeozoic faunas of Sibumasu show that they have close faunal affinities with Gondwanaland faunas of the same age and in particular with those of north and northwest Australia (Figure 3) and indicate close continental proximity of these areas (Burrett and Stait, 1985; Burrett and Long, 1988, 1989; Burrett et al., 1990). The small Yaxian terrane of Hainan, however, exhibits a lower Palaeozoic stratigraphic sequence and faunas similar to those of the Georgina Basin and indicates an Eastern Australian origin (Yu, 1989). Early Palaeozoic trilobites recorded from the southern part of Hainan Island (Yaxian terrane) include forms endemic to the Georgina Basin of Australian Gondwanaland (Yu, 1989) and suggest that this small Hainanese terrane was derived from the Australian part of Eastern Gondwanaland. Most early to middle Palaeozoic palaeogeographic reconstructions (eg. Burrett et al., 1990; Nie et al., 1990) place Sibumasu adjacent to Australia and South China against the western Himalaya - Iran region, positions which are consistent with the available palaeomagnetic data. Nicoll and Totterdell (1990), using stratigraphical and biogeographic data, have suggested that North China was also adjacent to north western Australia in the Ordovician; and Burrett et al. (1990) placed North China adjacent to New Guinea in the

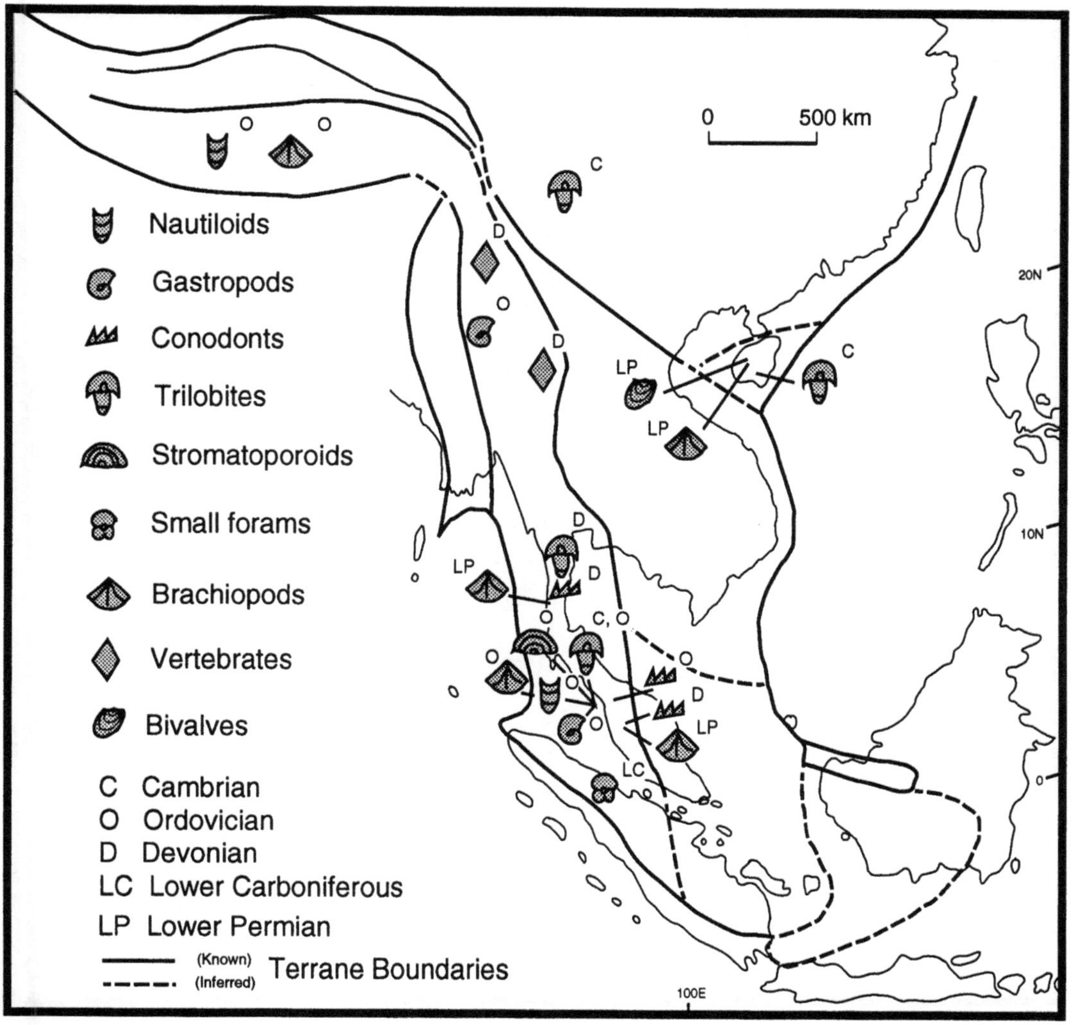

Fig. 3. Occurrence of Palaeozoic Gondwanaland faunal and floral elements in Southeast Asia. Compiled from Metcalfe (1988), Burrett et al. (1990) and references therein, Yu (1989) and Shi and Waterhouse (1991).

Cambrian and Ordovician, based on faunal evidence. The biogeography of early and middle Palaeozoic faunas of Indochina is poorly known but late Palaeozoic faunas have affinities with South and North China suggesting proximity at this time (Metcalfe, 1988). The presence of Late Carboniferous-Early Permian glacial-marine diamictites (Stauffer and Mantajit, 1981; Stauffer and Lee, 1989), Lower Permian cool-water faunas and $\delta^{18}O$ cool-water indicators (Waterhouse, 1982; Ingavat and Douglass, 1981; Rao, 1988; Fang and Yang, 1991; Figure 4) and distinctive Gondwanaland faunas and faunas with Australian affinities on Sibumasu (Burrett and Stait, 1985; Burrett et al., 1990; Archbold et al., 1982; Shi and Waterhouse, 1991; Figure 3), strongly suggest a NW Australian origin for the Sibumasu terrane. This is supported by late Palaeozoic and early Mesozoic palaeomagnetic data which indicate southern palaeolatitudes consistent with a position off NW Australian Gondwanaland in the Devonian, Carboniferous and Early Permian (Wu et al., 1989; Bunopas, 1982; Bunopas et al., 1989; Metcalfe, 1990; Huang and Opdyke, 1991). Lower Permian Gondwanaland faunas and diamictites interpreted as glacial-marine have

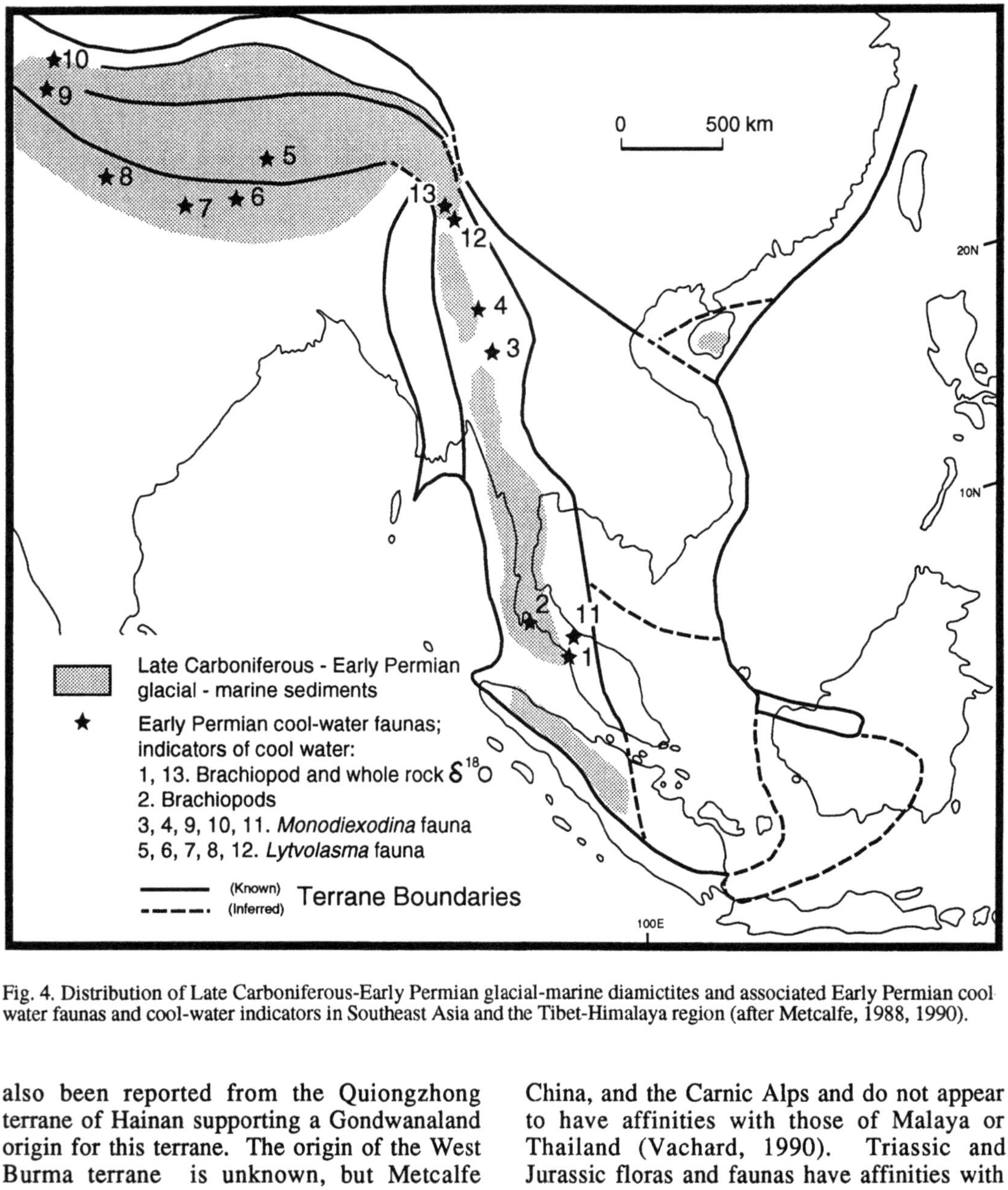

Fig. 4. Distribution of Late Carboniferous-Early Permian glacial-marine diamictites and associated Early Permian cool water faunas and cool-water indicators in Southeast Asia and the Tibet-Himalaya region (after Metcalfe, 1988, 1990).

also been reported from the Quiongzhong terrane of Hainan supporting a Gondwanaland origin for this terrane. The origin of the West Burma terrane is unknown, but Metcalfe (1990) has suggested that it may represent part of the continental sliver that must have rifted from NW Australia in the Jurassic, and which provided sediment derived from the north to the Triassic intracontinental basin in Timor (Audley-Charles, 1988). The origins of the South West Borneo and Semitau terranes are not known. Upper Carboniferous and Lower Permian fusulinid faunas represent the Eurasian-Arctic province and Lower Permian microfossils are similar to South and North China, and the Carnic Alps and do not appear to have affinities with those of Malaya or Thailand (Vachard, 1990). Triassic and Jurassic floras and faunas have affinities with South China, Indochina, Japan, and the Philippines (Kimura, 1984; Kobayashi and Tamura, 1984; Hayami, 1984.). As yet no faunas or floras indicating a Gondwanaland connection have been found on the South West Borneo or Semitau terranes. A number of small continental fragments now found around the margins of the South China Sea (Paracel Islands, Macclesfield Bank, Reed Bank, Spratley Islands-Dangerous Ground, North Palawan, Luconia, and Kelabit-Longbowan)

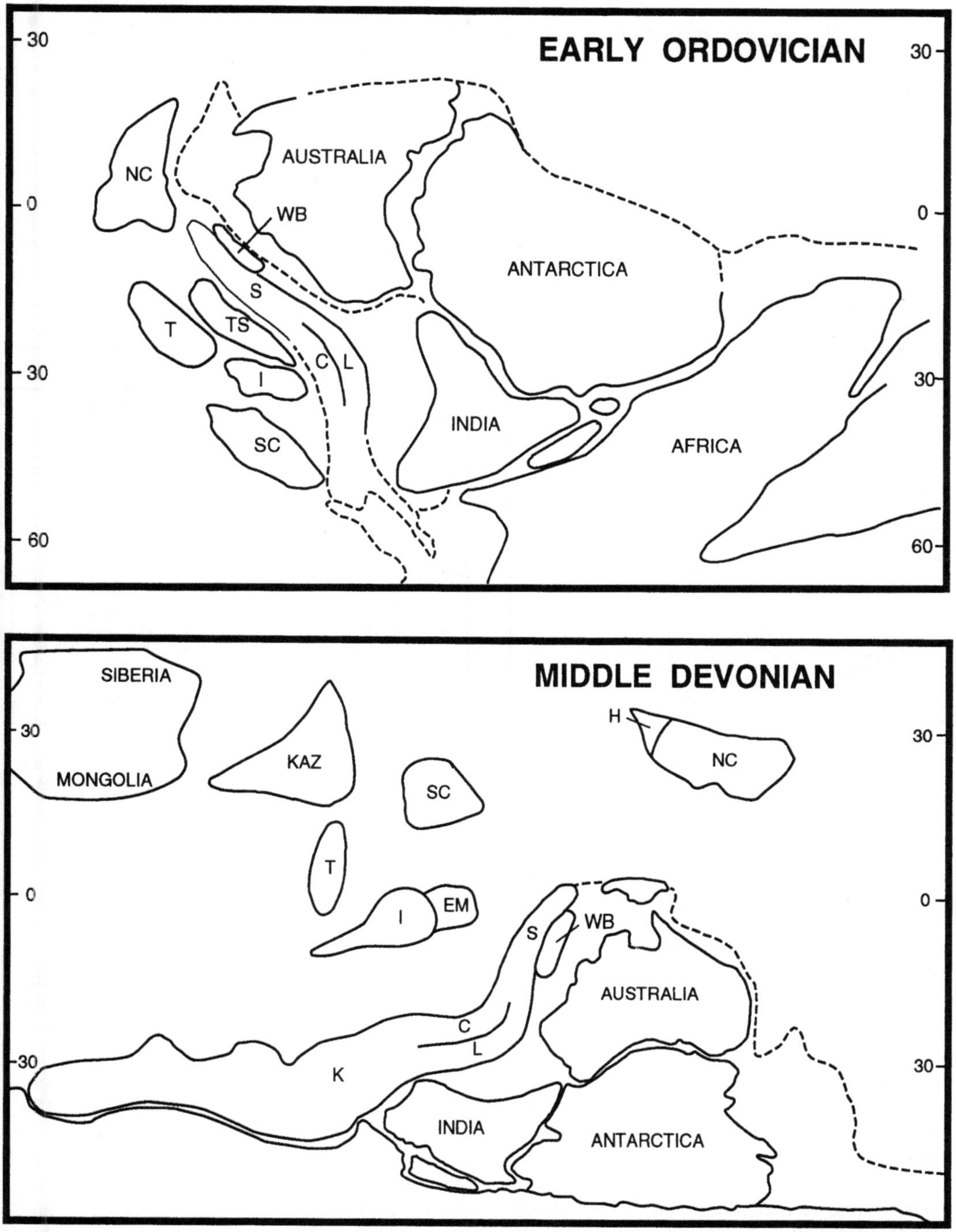

Fig. 5. Reconstructions of eastern Gondwanaland for the Early Ordovician and Middle Devonian, showing the postulated positions of the Southeast Asian terranes. Partly after Burrett et al. (1990). NC = North China SC = South China KAZ = Kazakhstan T = Tarim I = Indochina EM = East Malaya K = Kreios C = Changtang L = Lhasa S = Sibumasu WB = West Burma. H = Hexizoulang. Present day outlines are for reference only.

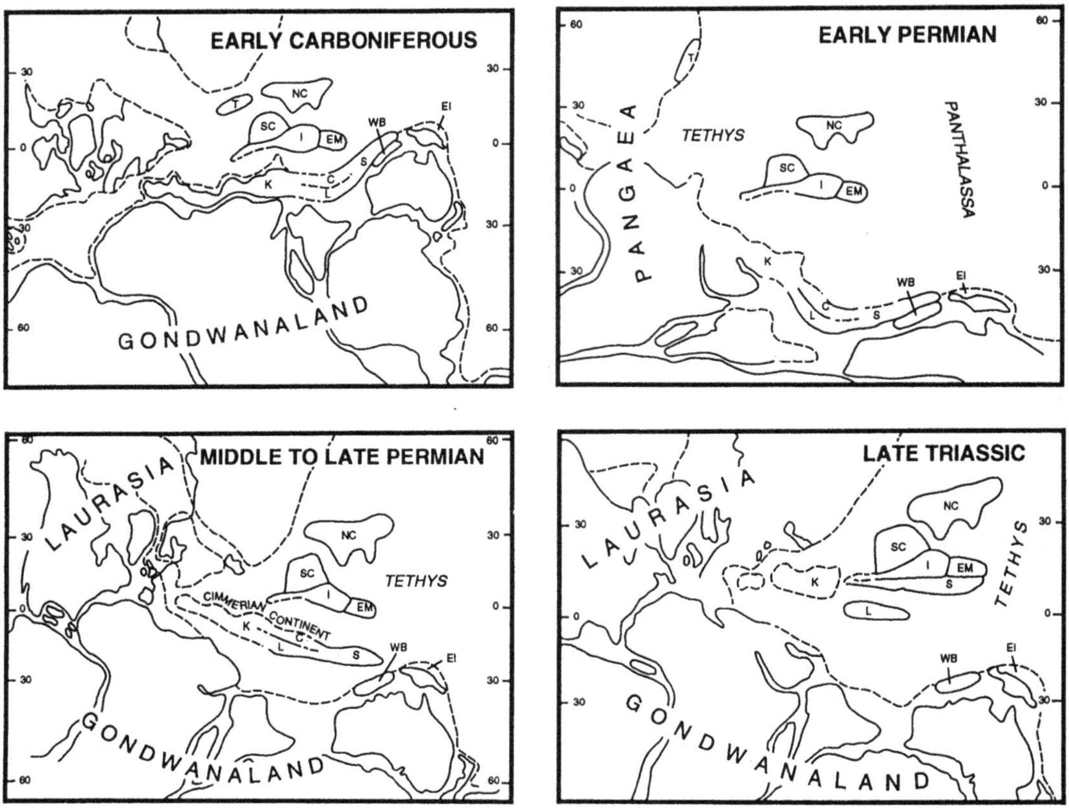

Fig. 6. Palaeogeographic reconstructions of the Tethyan region for Early Carboniferous, Early Permian, Middle-Late Permian and Late Triassic. Based on the reconstructions of Smith et al. (1981) and partly after Metcalfe (1988, 1990). Present day outlines are for reference only. EI = Eastern Indonesian terranes, other symbols as for figure 5.

are believed to have been derived from the South China and Indochina margins during the opening of the South China Sea (Holloway, 1981; Taylor and Hayes, 1980, 1983; Metcalfe, 1990). The rather meagre data on stratigraphy and faunas of these terranes is consistent with such a derivation (Metcalfe, 1990). The other eastern Southeast Asian terranes (West Irian Jaya, Buru-Seram, Obi-Bacan, Bangai Sula, Buton, Mangkalihat, West and East Sulawesi, Sumba and possibly the 'Banda Allochthon' of Audley-Charles) were derived from the Australian margin by rifting in the Jurassic and by north and westwards translation along major transcurrent faults due to the interaction of the Australian and Pacific plates. These small fragments show stratigraphies which are remarkably similar to that of the Australian margin (Pigram and Panggabean, 1984).

4 TERRANE EVOLUTION

4.1 *Palaeozoic*

The Sibumasu, Indochina, East Malaya, and South China terranes were located close to the Iran - Himalayan - NW Australian margin of Gondwanaland during the early Palaeozoic as indicated by faunal and stratigraphical comparisons and by palaeomagnetism. The proposed relative positions of these terranes to each other and other components of Gondwanaland in the Early Ordovician are shown in Figure 5A. Palaeomagnetic data suggest that South and North China, possibly along with the Tarim terrane, Indochina and East Malaya, rifted from Gondwanaland in the Silurian (Wu et al., 1989; Lin, 1987a,). The South China terrane was effectively isolated from other continental masses during the Devonian (Figure 5B) as shown by its high degree of Devonian faunal and floral endemism (Burrett et al., 1990). The South China and Indochina terranes are currently separated by northwest-southeast oriented mobile belts in North Vietnam within which Early Devonian, Early Carboniferous and Late Triassic orogenic events have been determined. Two sutures are currently recognised within this belt, the Song Ma and Song Da sutures of Palaeozoic and Mesozoic age, respectively. The age of the collision and amalgamation of Indochina and South China is still contentious. Hutchison (1989a, 1989b) and Gatinsky and Hutchison (1987) have proposed that the Early Carboniferous event represents the collision between Indochina and South China along the Song Ma suture to form the 'East Asia Continent' or 'Cathaysialand'. Alternatively, Sengör et al. (1988) regard this event as the collision between an arc fragment and an already accreted arc along a south-west dipping subduction zone and consider the final welding of Indochina to South China as having taken place in the Late Triassic along the Song Da suture. The large-scale folding and thrusting and nappe formation which took place in the Early to Middle Carboniferous suggests collision of something larger than an arc fragment and recognition of the Song Da Permo-Triassic sequences as probable rift basins (Tran, 1979; Hutchison, 1989a, 1989b), supports the amalgamation of Indochina and South China as occurring in the Early Carboniferous. Palaeomagnetic data indicate that the South China terrane was equatorial during the Early Carboniferous (Lin et al., 1985; Lin, 1987a; see Figure 6). No palaeomagnetic data are available for the Carboniferous of Indochina but Middle and Late Carboniferous and younger faunas and floras have affinities with South China. Palaeomagnetic data indicate that the North China was not attached to either South China or Siberia in the Carboniferous and that it lay in low northern palaeolatitudes (Lin, 1987a; Lin and Fuller, 1990; Figure 6).

A major rifting phase occurred on the margin of northeast Gondwanaland in the Early to Middle Permian (Powell, 1976; Bird, 1987; Audley-Charles, 1988) indicating that a substantial continental fragment or fragments separated from Gondwanaland at that time. Audley-Charles (1988) suggested that these terranes were Iran, North Tibet (Changtang) and Indochina, but I have argued (Metcalfe 1986, 1988, 1990, 1991) that the rifting terranes were in fact Sibumasu, Lhasa and Changtang (along with other terranes constituting the Cimmerian continent of Sengör). Palaeomagnetic data indicate that Sibumasu travelled rapidly from southern to northern palaeolatitudes in the Permo-Triassic (Figures 6 and 7). Similar data from the Changtang and Lhasa terranes of Tibet also indicate rapid northwards motion of these in the Permo-Triassic (Dewey et al., 1988; Lin and Watts, 1988). Post-Middle Permian faunas of Sibumasu show affinities with Indochina and South China and do not contain any Gondwanaland elements (Metcalfe, 1988). The Cimmerian continent probably broke up, both longitudinally and latitudinally, during its rifting from Gondwanaland, separating the Lhasa and Sibumasu-Changtang terranes and the Western Cimmerian Continent in Turkey and Iran (Kreios of Tollmann and Tollman, 1985). Continental palaeomagnetism indicates that Cathaysialand remained in low northern palaeolatitudes during this time (Figure 6). Sibumasu sutured to the Indochina and East Malaya portions of Cathaysialand along the Uttaradit-Nan and Raub-Bentong sutures in Permo-Triassic times. The precise age of suturing has been the subject of much recent debate and is still contentious, with ages of Lower Carboniferous (Helmke, 1985a; Altermann, in press), Permian (Helmke and Kraikhong, 1982; Helmke and Lindenberg, 1983; Helmke, 1983, 1985b), Early Triassic (Mitchell, 1989; Cooper et al., 1989; Metcalfe, 1990) and Late Triassic (Ridd, 1980; Mitchell, 1981; Sengör, 1984) being proposed. A latest Permian-earliest Triassic suturing age is here favoured. Harbury et al. (1990) present structural evidence which seems to preclude a Late Triassic collision, and a Carboniferous suturing is clearly untenable in view of the sedimentological, palaeontological and palaeomagnetic data (see above). The youngest mélange ages known along the suture are Late Permian or earliest Triassic (Metcalfe, 1989), the Main Range 'collisional' 'S' Type granites range from Late Triassic (230+/-9 Ma) to earliest Jurassic (207+/- 14 Ma) in age (Liew and Page, 1985; Darbyshire, 1987) and post-orogenic extensional basins developed in Thailand in the Late Triassic (Cooper et al., 1989). Thus, by Late Triassic times, Sibumasu had amalgamated with Cathaysialand and this amalgamated superterrane was located in low

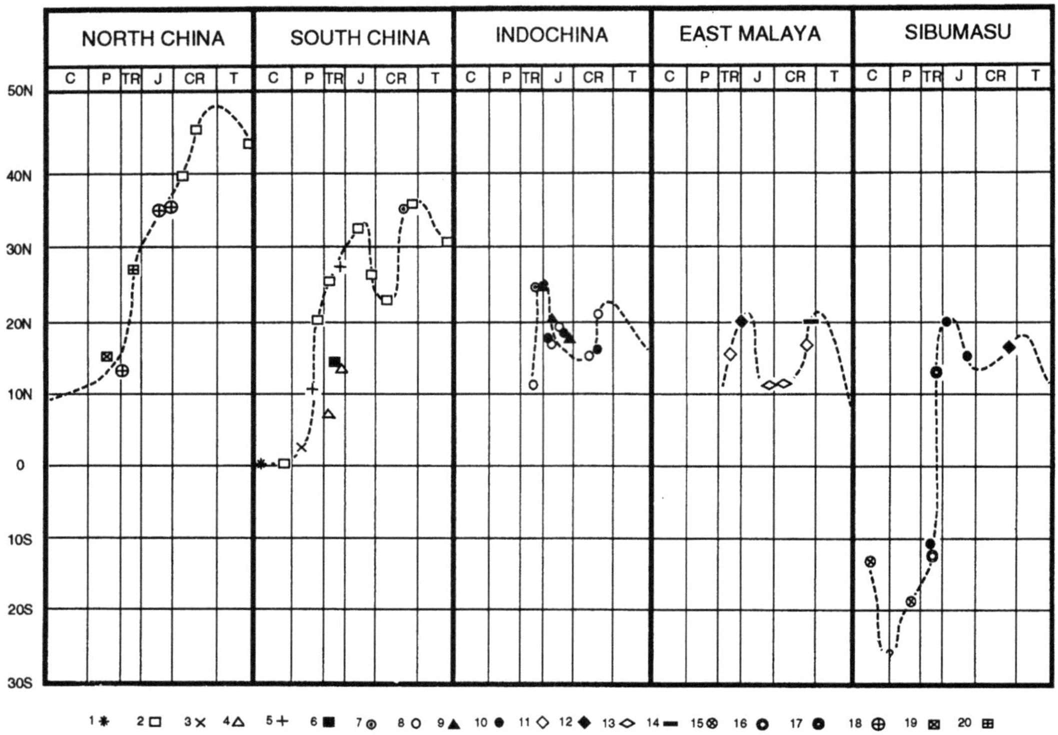

Fig. 7. Palaeolatitude plots for the North China, South China, Indochina, East Malaya and Sibumasu terranes from Carboniferous to Tertiary. Sources of data are 1. Lin (1987a); 2. Lin et al. (1985); 3. McElhinney et al. (1981); 4. Opdyke et al. (1986); 5. Sasajima and Maenaka (1987); 6. Chan et al. (1984); 7. Achache et al. (1983); 8. Maranate and Vella (1986); 9. Achache and Courtillot (1985); 10. Bunopas et al. (1989); 11. McElhinney et al. (1974); 12. E.A. Schmidtke and M. Fuller (personal communication); 13. Haile and Khoo (1980); 14. Haile et al. (1983); 15. Bunopas (1982); 16. Sasajima et al. (1978); 17. Haile (1979a); 18. Lin (1987b); 19. Zhao and Coe (1987); 20. Cheng et al. (1988).

northern palaeolatitudes (Figure 6). It is not yet known if the small Gondwanaland terranes of Hainan Island, which yield trilobites and other invertebrates endemic to Australia and Gondwanaland and which have Early Permian glacial-marine diamictites, were accreted to South China in the late Palaeozoic or Mesozoic.

4.2 *Mesozoic*

The Early Triassic suturing of Sibumasu to Indochina-East Malaya resulted in the generation of the Main Range 'S' Type granitoids of Peninsular Malaysia and Thailand, which were emplaced in the Late Triassic and earliest Jurassic some 30Ma after collision, consistent with observed and calculated time differences between generation and emplacement of collisional granites (England and Thompson, 1986). Subduction of the Neotethys (Tethys II) beneath the Sumatra - Burma portion of Sibumasu commenced following collision as evinced by the identification of Triassic ocean floor material in the Woyla Group accretionary complex by Wajzer et al. (in press) and Triassic sediments identified in the Andaman Islands by Sharma and Mehrotra (1984) and Sharma and Sarjeant (1987). The occurrence of Triassic sediments in the Andaman Islands has however been challenged recently by Kumar (1990).

Palaeomagnetic data (Zhao and Coe, 1987; Lin and Fuller, 1990) show that collision between, and amalgamation of South and North China along the Quinling-Dabie suture took place in the Late Triassic to Early Jurassic. This collision resulted in a spectacular anticlockwise rotation of North China and an overall consolidation of the Southeast Asian cratonic 'core'. The marine Triassic sediments of the region were deformed at this time by the

'Indosinian Orogeny' and the Triassic marine basins closed. Terrestrial deposits became widespread over large parts of the Southeast Asian craton in the Jurassic and Cretaceous.

Stratigraphic and structural evidence suggests that a second continental sliver separated from NW Australian Gondwanaland in the Late Jurassic (Falvey and Mutter, 1981) with the Cenotethys (Tethys III) opening behind it. Audley-Charles (1983, 1984, 1988) has suggested that this sliver comprised south Tibet (Lhasa terrane), Burma, Malaya, Sumatra, (Sibumasu) 'West Borneo', 'East Borneo', West Sulawesi and the 'Banda Allochthon'. From discussions above it is evident that this sliver did not include Sibumasu or the Lhasa Terrane which must have rifted in the Permian or Triassic. Following Metcalfe (1990), it is here suggested that the continental sliver included part of western Burma (West Burma terrane) and other continental material now represented by microcontinents incorporated in the Woyla terranes of Sumatra (Sikuleh and ?Natal terranes) and by the Mangkalihat and West Sulawesi terranes and the 'Banda Allochthon' (Figure 8A). Limited palaeomagnetic data from the Sikuleh terrane, which indicates palaeolatitudes of 26°S for the Late Triassic and 10°S in the Late Mesozoic (Haile, 1978) and from the West Sulawesi terrane which indicates southern palaeolatitudes in the Cretaceous (Sasajima et al., 1980), supports this view. This rifted sliver of continent travelled north by the opening of the Tethys III and by subduction of Tethys II (Mesotethys) beneath Eurasia during the Late Jurassic and Cretaceous (Figure 8B) and undoubtedly fragmented during this northward travel. The Lhasa Terrane collided with the Changtang and Sibumasu terranes in the Late Jurassic - Early Cretaceous along the Banggong Suture (Dewey et al., 1988). The West Burma terrane and the small continental terranes of the Woyla terranes continued their northward drift in the Cretaceous and had accreted to Sibumasu along the Sagaing (Early Cretaceous) and Woyla (Late Cretaceous) sutures (Metcalfe, 1990) by the Late Cretaceous (Figure 8C). The Semitau terrane also sutured to SW Borneo along the Boyan suture during the Late Cretaceous. India and Australia had also begun their northward drift at this time following the breakup of Gondwanaland.

4.3 *Cenozoic*

India travelled rapidly (10cm per annum) northeastwards relative to Eurasia during the Late Cretaceous and Early Eocene (between 84 and 45 Ma) and began its collision with Eurasia at about 45 Ma, after which it persued a northwards motion relative to Eurasia at a slower (5cm per annum) rate (Dewey et al., 1989). Australia also moved northwards more slowly at this time, being separated from India by major transform faults (Fig. 8).

Cenozoic modification of Southeast Asia involved substantial movements along, and rotations of strike-slip faults, rotations of continental blocks and the development and spreading of 'marginal' seas. These modifications were due to the combined effect of the interaction of the Eurasian, Pacific and Indo-Australian plates and the collision of India with Eurasia.

Tectonic models for the India-Eurasia collision propose three basic mechanisms for the northwards motion of India into Eurasia. The first model involves underthrusting of India beneath Eurasia (Argand, 1924; Powell and Conaghan, 1975; Zhao and Morgan, 1987); the second involves crustal shortening and thickening (Dewey and Burke, 1973; England and McKenzie, 1982; Dewey et al., 1988); and the third involves eastwards lateral extrusion (Molnar and Tapponier, 1975; Tapponier et al., 1986). Molnar and Tapponier (1975) and Tapponier et al. (1986) have proposed substantial extrusion of Southeast Asia to the southeast and clockwise rotations of crustal blocks consequent upon the collision of India with Eurasia. In this model, the opening of the South China Sea is regarded as a consequence of clockwise rotation and southeast extrusion of "Sundaland" relative to Indochina and South China. Whatever model is proposed, the collision of India with Eurasia undoubtedely had a significant effect on mainland Southeast Asia, and an analysis of strike-slip fault movements and block rotations in Southeast Asia provides important constraints on the tectonic models for the India-Eurasia collision.

Dewey et al. (1989), based on data from Tibet, the Himalayas and China, discussed arguments for and against the three models and concluded that models advocating underthrusting and extrusion were untenable. Furthermore, Dewey et al. (1988) and Dewey

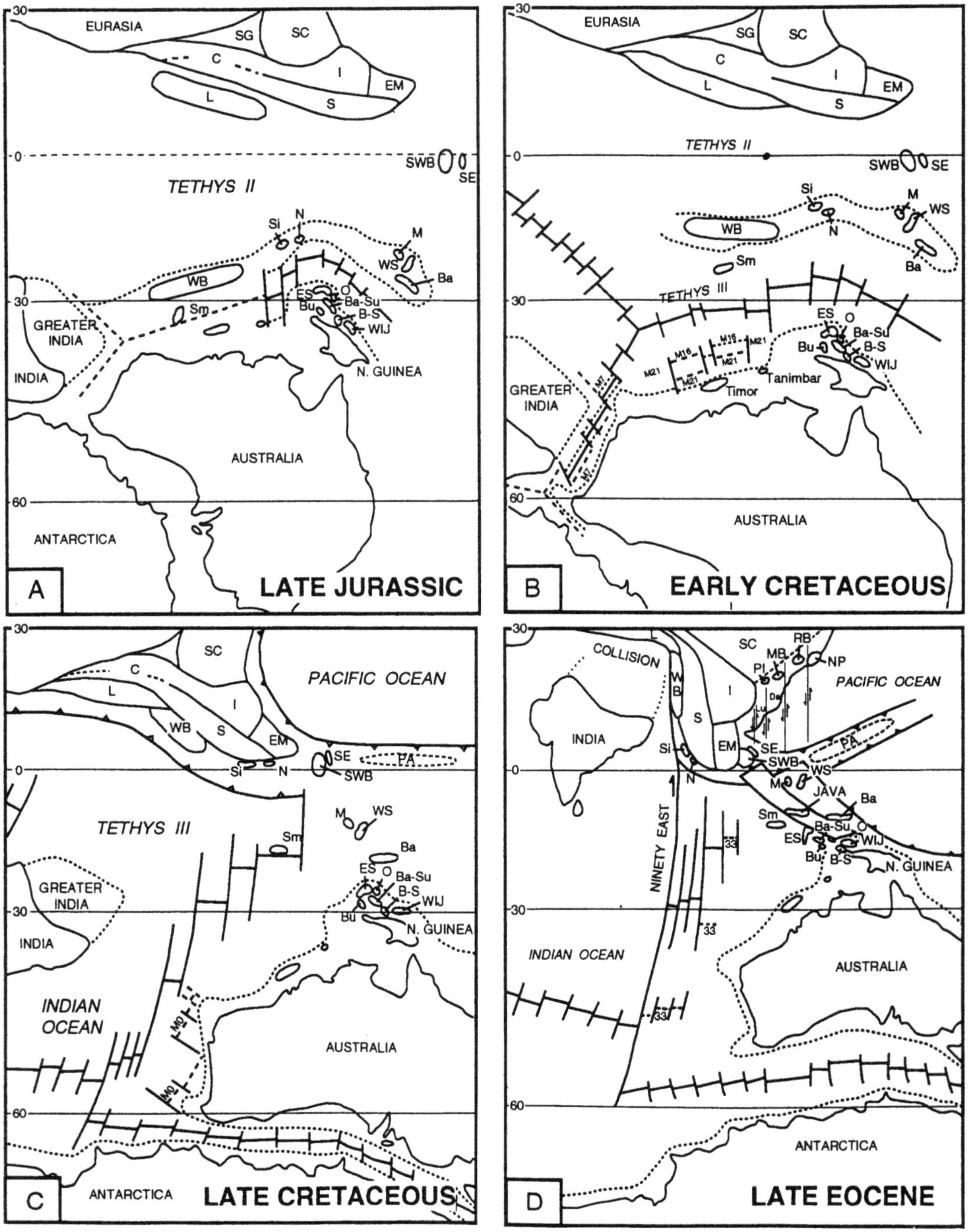

Fig. 8. Palaeogeographic reconstructions for Eastern Tethys in A: Late Jurassic, B: Early Cretaceous, C: Late Cretaceous and D: Late Eocene times. SG = Songpan Gangzi accretionary complex SWB = South West Borneo SE = Semitau Si = Sikuleh N = Natal M = Mangkalihat WS = West Sulawesi Ba = Banda Allochthon ES = East Sulawesi O = Obi-Bacan Ba-Su = Bangai-Sula Bu = Buton B-S = Buru-Seram WIJ = West Irian Jaya Sm = Sumba PI = Paracel Islands MB = Macclesfield Bank RB = Reed Bank NP = North Palawan Lu = Luconia Da = Spratley Islands-Dangerous Ground PA = Philippine Arc. M numbers represent Indian Ocean magnetic anomalies. Other terrane symbols as in figures 5 and 6. Partly after Smith et al. (1981), Audley-Charles (1988) and Audley-Charles et al. (1988). Present day outlines are for reference only.

et al. (1989) proposed that most of mainland Southeast Asia formed part of a zone of dextral megashear extending from Sumatra to the Tanlu Fault in China and that very little extrusion of China and Southeast Asia occurred during the intracontinental convergence between India and Eurasia during the last 45Ma. This is supported by recent work in eastern Tibet (England and Molnar, 1990) who demonstrated clockwise fault and block rotations in a zone of simple dextral megashear.

If mainland Southeast Asia did form part of a zone of dextral megashear in the Cenozoic,we would expect to see, either clockwise rotations of blocks and NW oriented sinistral faults, or anticlockwise rotations of blocks and NE oriented dextral faults within the dextral megashear boundaries.Evidence from the analysis of recent movements along strike-slip faults in eastern Tibet suggests that such a simple right-lateral megashear exists there and that "'extrusion' or 'expulsion' is an illusion" (England and Molnar, 1990). England and Molnar (1990) further suggest that the left-lateral faults and the blocks between them are rotating at 1-2 degrees per million years and that the east-west dimension of eastern Tibet is shortening at 10-20 mm per year. If we extend the current right-lateral shear zone in Tibet southwards into western Southeast Asia we should therefore see increasing amounts of clockwise block rotations and a narrowing of the megashear zone.

Data for the Cenozoic rotations of crustal blocks in mainland Southeast Asia are sparse (Figure 9). However, clockwise rotations are observed in Indochina and western Thailand and major progressive anticlockwise rotations are seen in Borneo and the Malay Peninsula. These anticlockwise rotations observed in Borneo and Malaya are at variance with the extrusion model. Current analysis of fault and block movements and kinematic modelling of the region (with J.F. Dewey) and palaeomagnetic studies (M. Fuller and co-workers) will hopefully resolve the conflict between predicted and observed Cenozoic deformation in mainland Southeast Asia in relation to the India-Eurasia collision.

Fragments of the eastern part of the continental sliver that rifted from Gondwanaland in the Jurassic (West Sulawesi, Mangkalihat and the Banda Allochthon) were translated north and west by major transcurrent faults as a result of the interaction of the westward moving Pacific and northwards moving Australian plates during the Cenozoic. This 'bacon-slicer' mechanism also translated small fragments of the New Guinea margin of Australia (East Sulawesi, Buton, Bangai-Sula, Obi-Bacan) to the north and west where they now form part of eastern Indonesia (Figure 8D).

5. ACKNOWLEDGEMENTS

I would like to thank Dr. Peter H. Stauffer, Prof. John J. Veevers and Dr. Clive F. Burrett for their constructive reviews of the paper. I would also like to thank the Department of Geology & Geophysics, University of New England for facilities provided.

REFERENCES

Achache, J. and Courtillot, V. 1985. A preliminary Upper Triassic palaeomagnetic pole for the Khorat plateau (Thailand): consequences for the accretion of Indochina against Eurasia. *Earth Planet. Sci. Lett.*, 73: 147-157.

Achache, J., Courtillot, V. and Besse, J. 1983. Palaeomagnetic constraints on the Late Cretaceous and Cenozoic tectonics of Southeastern Asia. *Earth and Planetary Science Letters* 63: 123-136.

Altermann, W. in press. Tracing Permo-Carboniferous volcanism in Central Thailand: Implications for a volcanic arc model. *Palaeogeography, Palaeoclimatology, Palaeoecology.*

Archbold, N.W., Pigram, C.J., Ratman, N. and Hakim, S. 1982. Indonesian Permian brachiopod fauna and Gondwana - South-East Asia relationships. *Nature* 296: 556-558.

Argand, E. 1924. La tectonique de l'Asie. *Proceedings 13th International Geological Congress, Brussels* (1924) 7: 171-372.

Audley-Charles, M.G. 1983. Reconstruction of eastern Gondwanaland. *Nature* 306: 48-50.

Audley-Charles, M.G. 1984. Cold Gondwana, warm Tethys and the Tibetan Lhasa block. *Nature* 310: 165.

Audley-Charles, M.G., 1988. Evolution of the southern margin of Tethys (North

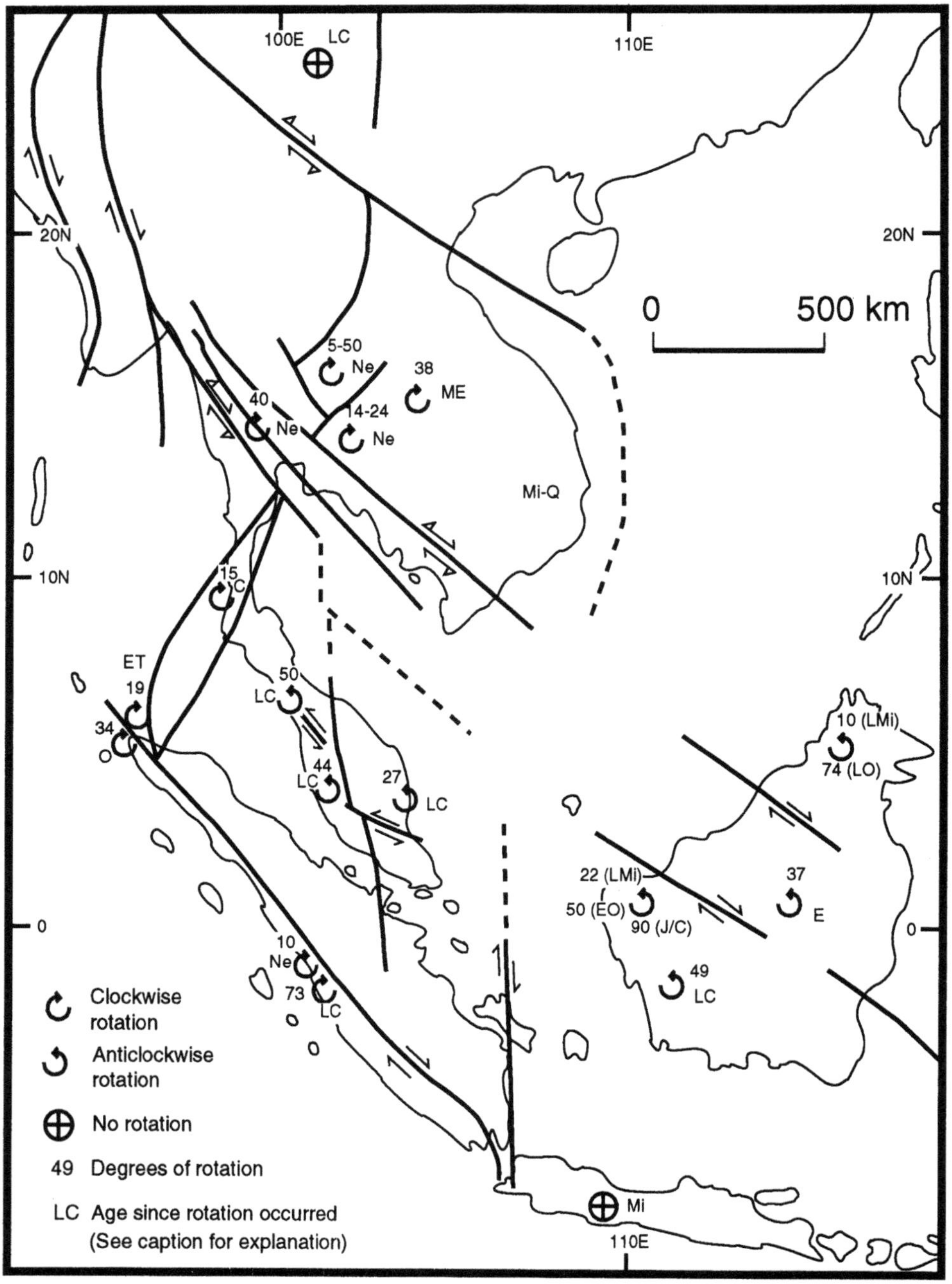

Fig. 9. Cenozoic palaeomagnetic rotation data for Southeast Asia. Heavy lines represent principal strike-slip faults with sense indicated. J = Jurassic, C = Cretaceous, L.C. = Late Cretaceous, ET = Early Tertiary, ME = Middle Eocene, EO = Early Oligocene, LO = Late Oligocene, O = Oligocene, Mi = Miocene, LMi = Late Miocene, Ne = Neogene, Q = Qua ternary. Sources of data as for figure 7, plus Chen and Courtillot (1989), Giang (1982), Haile (1974, 1978, 1979b), Haile and Briden (1982), Haile et al. (1977), Hirooka et al. (1980), Hirooka et al. (1976), Jarrard and Sasajima (1980), Kent et al. (1986), Kent et al. (1987), Lee et al. (1963), McCabe et al. (1988), Sasajima et al. (1980), Schmidtke et al. (1985) and Schmidtke et al. (1990).

Australian region) from Early Permian to Late Cretaceous. In: M.G. Audley-Charles, and A. Hallam (Editors), *Gondwana and Tethys*, Geological Society Special Publication No. 37, Oxford University Press, Oxford, pp. 79-100.
Audley-Charles, M.G., Ballantine, P.D. and Hall, R. 1988. Mesozoic-Cenozoic rift-drift sequence of Asian fragments from Gondwanaland. *Tectonophysics* 155: 317-330.
Bird, P. 1987. *The geology of the Permo-Triassic rocks of Kekneno, west Timor.* Ph.D. thesis, University of London, U.K.
Bunopas, S. 1982. Palaeogeographic history of Western Thailand and adjacent parts of Southeast Asia--a plate tectonics interpretation. *Geological Survey paper NO. 5, Department of Mineral Resources, Thailand*, 810pp.
Bunopas, S., Maranate, S. and Vella, P. 1989. Palaeozoic and Early Mesozoic rotation and drifting of Shan-Thai from Gondwana-Australia. *4th International Symposium on pre-Jurassic evolution of East Asia, IGCP Project 224, Reports and Abstracts*, 1: 63-64.
Burrett, C. 1974. Plate tectonics and the fusion of Asia. *Earth and Planetary Science Letters* 21: 181-189.
Burrett, C. and Long, J. 1988. Palaeobiogeographic and palaeomagnetic evidence for the derivation of the South East Asian and Chinese terranes from Gondwana. *3rd International Symposium on pre-Jurassic evolution of East Asia, IGCP Project 224, Abstracts:* 6.
Burrett, C. and Long, J. 1989. Palaeoposition of the Shan-Thai terrane in the Early-Middle Palaeozoic. *4th International Symposium on pre-Jurassic evolution of East Asia, IGCP Project 224, Reports and Abstracts* 1: 83.
Burrett, C., Long, J. and Stait, B. 1990. Early-Middle Palaeozoic biogeography of Asian terranes derived from Gondwana. *In*: McKerrow, W.S. and Scotese, C.R. (eds) *Palaeozoic Palaeogeography and Biogeography*, Geological Society Memoir No. 12: 163-174.
Burrett, C. and Stait, B. 1985. South-East Asia as part of an Ordovician Gondwanaland. *Earth and Planetary Science Letters* 75: 184-190.
Burton, C.K. 1970. The palaeotectonic status of the Malay Peninsula. *Palaeogeography, Palaeoclimatology, Palaeoecology*:: 51-60.
Cameron, N.R., Clarke, N.C.G., Aldiss, D.T., Aspden, T.A. and Djunuddin, A. 1980. The geological evolution of northern Sumatra, *Ninth Indonesian Petroleum Convention, Jakarta*, 53pp.
Chan, L.S., Wang, C.Y. and Wu, X.Y. 1984. Palaeomagnetic results from some Permian-Triass rocks from Southwestern China. *Geophysical Research Letters* 11: 1157-1160.
Chen,Y. and Courtillot, V. 1989. Widespread Cenozoic(?) remagnetisation in Thailand and its imlications for the India-Asia collision. *Earth and Planetary Science Letters* 93: 113-122.
Cheng, G., Bai, Y. and Sun, Y. 1988. Palaeomagnetic study on the tectonic evolution of the Ordos block, North China (in Chinese with English abstract). *Seismological Geology* 10: 81-87.
Cooper, M.A., Herbert, R. and Hill, G.S. 1989. The structural evolution of Triassic Intermontane Basins in Northeast Thailand. In: T. Thanasuthipitak, and P. Ounchanum (Editors) *Proceeding of International Symposium on Intermontane Basins: Geology and Resources:* .231-242, Chiang Mai University, Thailand.
Darbyshire, D.P.F. 1987. Geochronology of Malaysian granites. Natural Environment Research Council, Isotope Geology Centre Open File Report 88/3.
Dewey, J.F. and Burke, K.C.A. 1973. Tibetan, Variscan and Precambrian basement reactivation: products of a continental collision. *Journal of Geology* 81: 683-692.
Dewey, J.F., Cande, S. and Pitman, W.C. 1989. Tectonic evolution of the India/Eurasia Collision Zone. *Eclogae geologie Helvetica* 8:, 717-734.
Dewey, J.F., Shackleton, R.M., Chang Chengfa, and Sun Yiyin. 1988. The tectonic evolution of the Tibetan Plateau. *Philosophical Transactions of the Royal Society of London* A327: 379-413.
England, P.C. and McKenzie D.P. 1982. A thin viscous sheet model for continental deformation. *Geophysical Journal of the Royal Astyronomical Society* 70: 295-321.
England, P.C. and Molnar, P. 1990. Right-lateral shear and rotation as the explanation for strike-slip faulting in eastern Tibet. *Nature* 344: 140-142.

England, P.C. and Thompson, A. 1986. Some thermal and tectonic models for crustal melting in continental collision zones. *In*: Coward, M.P. and Ries, A.C. (eds), *Collision tectonics*, pp. 83-94, Geological Society Special Publication no. 19.

Falvey, D.A. and Mutter, J.C. 1981. Regional plate tectonics and the evolution of Australia's passive continental margins. *BMR Journal of Australian Geology & Geophysics* 6: 1-29.

Fang, N. and Yang, W. 1991. A study of the oxygen and carbon isotope records from Upper Carboniferous to Lower Permian in Western Yunnan, China. *In*, Ren, J. and Xie, G. (Eds), *Proceedings of First International Symposium on Gondwana Dispersion and Asian Accretion - Geological Evolution of Eastern Tethys*: 35-36, China University of Geosciences, Beijing.

Gatinsky, Y.G., and Hutchison, C.S. 1987. Cathaysia, Gondwanaland and the Palaeotethys in the evolution of continental Southeast Asia. *Bulletin of the Geological Society of Malaysia* 20: 179-199.

Giang, N.T. 1982. Paleomagnetic studies of Cenozoic basalts in Vietnam. *In* Briden, J.C. (ed) *Palaeomagnetic research in Southeast and East Asia*, U.N. Dev. Proj. Pub., pp.117-130. Committee for Coordination of Joint Prospecting for Mineral Resources in Asian Offshore Areas, Bangkok, Thailand.

Haile, N.S. 1974. Are the Kuantan dolerite dykes and extrusive basalt related to each other or to the Segamat basalt? Palaeomagnetic evidence. *Geological Society of Malaysia Newsletter* 52: 25-26.

Haile, N.S. 1978. Reconnaissance palaeomagnetic results from Sulawesi, Indonesia, and their bearing on palaeogeographic reconstruction *Tectonophsics* 46: 77-85.

Haile, N.S. 1979a. Palaeomagnetic evidence for rotation and northward drift of Sumatra. *Journal of the Geological Society of London* 136: 541-545.

Haile,, N.S. 1979b. Rotation of Borneo microplate completed by Miocene: Palaeomagnetic evidence. *Warta Geologi*, 5, 19-22.

Haile,, N.S., Beckinsale, R.D., Chakraborty, K.R., Abdul, H.H. and Tjahjo, H. 1983. Palaeomagnetism, geochronology and petrology of the dolerite dykes and basaltic lavas from Kuantan, West Malaysia. *Bulletin of the Geological Society of Malaysia* 16: 71-85.

Haile,, N.S. and Briden, J.C. 1982. Past and future palaeomagnetic research and the tectonic history of East and Southeast Asia. In Briden, J.C. (ed). *Paleomagnetic research in Southeast and East Asia*, U.N. Dev. Proj. Pub., pp.25-46, Committee for Coordination of Joint Prospecting for Mineral Resources in Asian Offshore Areas, Bangkok, Thailand.

Haile,, N.S. and Khoo, H.P. 1980. Palaeomagnetic measurements on Upper Jurassic and Lower Cretaceous sedimentary rocks from Peninsular Malaysia. *Bulletin of the Geological Society of Malaysia* 12: 75-78.

Haile,, N.S., McElhinny, M.W. and McDougall, I. 1977. Palaeomagnetic data and radiometric ages from the Cretaceous of West Kalimantan (Borneo), and their significance in interpreting regional structure. *Journal of the Geological Society of London* 133: 133-144.

Harbury, N.A., Jones, M.E., Audley-Charles, M.G., Metcalfe, I. and Mohamad, K.R. 1990. Structural evolution of Mesozoic Peninsular Malaysia. *Journal of the Geological Society of London* 147: 11-26.

Hayami, I. 1984. Jurassic marine bivalve faunas and biogeography in Southeast Asia. *Geology and Palaeontology of Southeast Asia* 25: 229-237.

Helmke, D. 1983. On the Variscan evolution of central mainland Southeast Asia. *Earth Evolution Sciences* 4: 309-319.

Helmke, D. 1985a. The orogenic evolution (Permo-Triassic) of central Thailand. Implications on palaeogeographic models for mainland SE-Asia. *Memoirs de Societe Geologie France, N.S.* 147: 83-91.

Helmke, D. 1985b. The Permo-Triassic 'Paleotethys' in mainland Southeast-Asia and adjacent parts of China. *Geologisches und Rundschau* 74: 215-228.

Helmke, D. and Kraikhong, C. 1982. On the geosynclinal and orogenic evolution of central and northeastern Thailand. *Journal of the Geological Society of Thailand* 5: 52-74.

Helmke, D. and Lindenberg, H.G. 1983. New data on the 'Indosinian' orogeny from Central Thailand. *Geologisches und*

Rundschau 72: 317-328.

Hirooka, K., Otofuji, Y., Sasajima, S., Nishimura, S., Musuda, Y., Thio, K.H., Dharma, A.and Hehuwat, F. 1980. An interim report of paleomagnetic study in Jawa Island. *Phys. Geol. Indonesian Island Arcs*, 67-71.

Hirooka, K., Sasajima, S., Nishimura, S., Thio, K.H. and Hehuwat, F. 1976. Preliminary report of the paleomagnetic study in Sunda Arc. In, *Proceedings of the 1st International Congress on Pacific Neogene Stratigraphy, Science Council of Japan*, Tokyo, 341-342.

Holloway, N.H. 1981. The North Palawan block, Philippines: its relation to the Asian Mainland and its role in the evolution of the South China Sea. *Bulletin of the Geological Society of Malaysia* 14: 19-58.

Huang, K and Opdyke, N.D. 1991. Paleomagnetic results from the Upper Carboniferous of the Shan-Thai-Malay block of western Yunnan, China. *Tectonophysics* 192: 333-344.

Hutchison, C.S. 1989a. The Palaeo-Tethyan realm and Indosinian orogenic system of Southeast Asia. *In*: Sengör, A.M.C. (ed) *Tectonic evolution of the Tethyan region*, pp.585-643, Kluwer Academic Publishers.

Hutchison, C.S. 1989b. *Geological Evolution of South-East Asia*. Oxford, Clarendon Press, 368pp.

Ingavat, R. and Douglass, R. 1981. Fusuline fossils from Thailand, Part XIV. The fusulinid genus *Monodiexodina* from Northwest Thailand. *Geology and Palaeontology of Southeast Asia* 22: 23-34.

Jarrard, R.D. and Sasajima, S. 1980. Palaeomagnetic synthesis for Southeast Asia: constraints on plate motions. In Hayes, D.E. (ed) T*he tectonic and geologic evolution of Southeast Asian seas and islands*, pp.293-316, Am. Geophys. Union Geophys. Monograph No. 23.

Kent, D.V., Xu, G., Huang, K., Zhang, W.Y. and Opdyke, N.D. 1986.Paleomagnetism of upper Cretaceous rocks from South China. *Earth and Planetary Science Letters* 79: 179-184.

Kent, D.V., Zeng, X., Zhang, W.Y. and Opdyke, N.D. 1987. Widespread Late Mesozoic to Recent remagnetisation of Paleozoic and lower Triassic sedimentary rocks from South China. *Tectonophysics* 139: 133-143.

Kimura, T. 1984. Mesozoic floras of East and Southeast Asia, with a short note on the Cenozoic floras of Southeast Asia and China. *Geology and Palaeontology of Southeast Asia* 25: 325-350.

Klimetz, M.P. 1987. The Mesozoic tectonostratigraphic terranes and accretionary heritage of South-Eastern Mainland Asia. *In*: Leitch, E.C. and Scvheibner, E. (eds) Terrane accretion and orogenic belts (Geodynamics Series 19), pp. 221-234. American Geophysical Union, Washington DC.

Kobayashi, T. and Tamura, M. 1984. The Triassic Bivalvia of Malaysia, Thailand and adjacent areas. *Geology and Palaeontology of Southeast Asia* 25: 201-228.

Koning, T. and Darmono, F.X. 1984. The geology of the Beruk Northeast Field, Central Sumatra: oil production from Pre-Tertiary basement rocks. *Proceedings Thirteenth Annual Convention*, pp.384-406, Indonesian Petroleum Association, Jakarta.

Kumar, A. 1990. Late Triassic dinoflagellate cysts and acritarchs from the Andaman Islands: Discussion. *Modern Geology* 14: 245-253.

Lee,C., Lee, H., Lio, H., Lio, G. and Yeh, S. 1963. Preliminary study of paleomagnetism of some Mesozoic and Cenozoic redbeds of South China. *Acta Geologica Sinica* 43: 261-264.

Liew, T.C. and Page, R.W. 1985. U-Pb Zircon dating of granitoid plutons from the West Coast Province of Peninsular Malaysia. *Journal of the Geological Society of London* 142: 515-526.

Lin, J. 1987a. Late Carboniferous paleogeographic reconstruction. *11th International Congress of Carboniferous Stratigraphy and Geology, Abstracts* 1: 283.

Lin, J. 1987b. The apparent polar wander path for the South China block and its geological significance (in Chinese with English abstract). *Scientia Geologica Sinica* 4: 306-315.

Lin, J. and Fuller, M. 1990. Palaeomagnetism, North China and South China collision, and the Tan-Lu fault. *Philosophical Transactions of the Royal Society of London* A331: 589-598.

Lin, J. and Fuller, M. and Zhang, W. 1985. Preliminary Phanerozoic polar wander paths for the North and South China Blocks. *Nature* 313: 444-449.

Lin, J. and Watts, D.R. 1988. Palaeomagnetic results from the Tibetan Plateau. *Philosophical Transactions of the Royal Society of London* A327: 239-262.

Maranate, S. and Vella, P. 1986. Paleomagnetism of the Khorat Group, Mesozoic, Northeast Thailand. *Journal of Southeast Asian Earth Sciences* 1: 23-31.

McCabe, R., Celaya, M., Cole, J., Han, H.C. and Ohnstad, T. 1988. Extension Tectonics: The opening of the north-south trending basins of Central Thailand. *Journal of Geophysical Res*earch 93: 11899-11910.

McElhinny, M.W., Haile, N.S. and Crawford, A.R. 1974. Palaeomagnetic evidence shows Malay Peninsula was not part of Gondwanaland. *Nature* 252: 641-645.

McElhinny, M.W., Embleton, B.J.J., Ma, X.H. and Zhang, Z.K. 1981. Fragmentation of Asia in the Permian. *Nature* 293: 212-216.

Metcalfe, I. 1986. Late Palaeozoic palaeogeography of Southeast Asia: some stratigraphical, palaeontological and palaeomagnetic constraints. *Geological Society of Malaysia Bulletin* 19: 153-164.

Metcalfe, I. 1988. Origin and assembly of Southeast Asian continental terranes. In: M.G. Audley-Charles, and A. Hallam (Editors), *Gondwana and Tethys*. Geol. Soc. London Spec. Pub. No. 37: 101-118.

Metcalfe, I. 1989. Triassic sedimentation in the Central Basin of Peninsular Malaysia. In: T. Thanasuthipitak, and P. Ounchanum (Editors), *Proceedings of International Symposium on Intermontane Basins: Geology and Resources*, pp. 173-186, Chiang Mai University, Thailand.

Metcalfe, I. 1990. Allochthonous terrane processes in Southeast Asia. *Philosophical Transactions of the Royal Society of London* A331: 625-640.

Metcalfe, I. 1991. Late Palaeozoic and Mesozoic palaeogeography of Southeast Asia. *Palaeogeography, Palaeoclimatology, Palaeoecology* 87: 211-221.

Mitchell, A.H.G. 1981. Phanerozoic plate boundaries in mainland Southeast Asia, the Himalayas and Tibet. *J. Geol. Soc. London*, 138: 109-122.

Mitchell, A.H.G. 1989. The Shan Plateau and Western Burma: Mesozoic-Cenozoic plate boundaries and correlation with Tibet. In Sengör, A.M.C. (ed) *Tectonic Evolution of the Tethyan Region*, pp.567-583, Kluwer Academic Publishers.

Molnar, P. and Tapponnier, P. 1975. Cenozoic tectonics of Asia: effects of a continental collision. *Science* 189: 419-426.

Nicoll, R.S. and Totterdel, J.M. 1990. Conodonts and the distribution in time and space of Ordovician sediments in Australia and adjacent areas. *Geological Society of Australia, Abstract Series Number 25, Gondwana: terranes and resources*: 46-47.

Nie Shangyou, Rowley, D.B. and Ziegler, A.M. 1990. Constraints on the locations of Asian microcontinents in Palaeo-Tethys during the Late Palaeozoic. *In*: McKerrow, W.S. and Scotese, C.R. (eds) *Palaeozoic Palaeogeography and Biogeography*, Geological Society Memoir No. 12: 397-409.

Opdyke, N.D., Huang, K., Xu, G., Zhang, W.Y. and Kent, D.V. 1986. Paleomagnetic results from the Triassic of the Yangtze Platform. *Journal of Geophysical Research* 91: 9553-9568.

Pigram, C.J. and Panggabean, H. 1984. Rifting of the northern margin of the Australian continent and the origin of some microcontinents in Eastern Indonesia. *Tectonophysics* 107: 331-353.

Powell, D.E. 1976. The geological evolution and hydrocarbon potential of the continental margin off north-west Australia. *Journal of the Australian Petroleum Exploration Association* 16: 13-23.

Powell, C.M. and Conaghan, P.J. 1975. Tectonic models of the Tibetan Plateau. *Geology* 3: 727-731.

Pulunggono, A. and Cameron, N.R. 1984. Sumatran microplates, their characteristics and their role in the evolution of the Central and South Sumatra basins. *Proc. 13th Ann. Convention Indonesian Petroleum Assoc.*, 1: 121-143.

Rao, C.P. 1988. Paleoclimate of some Permo-Triassic carbonates of Malaysia. *Sedimentary Geology* 60, 163-171.

Ridd, M.F. 1971. Southeast Asia as a part of Gondwanaland. *Nature* 234:531-533.

Ridd, M.F. 1980. Possible Palaeozoic drift of S.E. Asia and Triassiccollision with China. *Journal of the Geological Society of London* 137: 635-640.

Sasajima, S. and Maenaka, K. 1987. A paleomagnetic aspect on the assemblage of East Asian fragmented continents. *IGCP Project 224, Pre-Jurassic geological evolution of eastern continental margin of Asia, Rep*. no. 2: 139-150.

Sasajima, S., Nishimura, S., Hirooka, K., Otofuji Y., Vanleeuan, T. and Hehuat, F. 1980. Palaeomagnetic studies combined with fission-track datings on the western arc of Sulawesi, East Indonesia. *Tectonophysics* 64: 163-172.

Sasajima, S., Otofuji Y., Hirooka, K. Suparka, Suwijanto and Hehuwat, F. 1978. Palaeomagnetic studies on Sumatra Island and the possibility of Sumatra being part of Gondwanaland. *Rock Magnetism and Paleogeophysics*, 5, 104-110.

Schmidtke, E., Dunn, J.R., Fuller, M.D. and Haston, R. 1985. Preliminary paleomagnetic results from Sabah, E.Malaysia. *EOS* 66: 864.

Schmidtke, E.A., Fuller, M.D. and Haston, R.B. 1990. Paleomagnetic data from Sarawak, Malaysia (Borneo) and the Late Mesozoic and Cenozoic tectonics of Sundaland. *Tectonics* 9: 123-140.

Sengör, A.M.C. 1979. Mid-Mesozoic closure of Permo-Triassic Tethys and its implications. *Nature* 279: 590-593.

Sengör, A.M.C. 1984. The Cimmeride orogenic system and the tectonics of Eurasia. *Geological Society of America Special Paper* 195: 82pp.

Sengör, A.M.C., Altiner, D., Cin, A., Ustaomer, T. and Hsu, K.J. 1988.Origin and assembly of the Tethyside orogenic collage at the expense of Gondwana Land. In: M.G. Audley-Charles, and A. Hallam, *Gondwana and Tethys*. Geological Society Special Publication No. 37, Oxford University Press, Oxford, pp. 119-181.

Sharma, J. and Mehrotra, N.C. 1984. Discovery of Late Triassic sediments from the Andaman Islands; some new palynological evidence. *ONGC Bulletin* 21: 69-73.

Sharma, J. and Sargeant, W.A.S. 1987. Late Triassic dinoflagellate cysts and acritarchs from the Andaman Islands; India. *Modern Geology* 11: 255-264.

Shi, G.R. and Waterhouse, J.B. 1991. Early Permian brachiopods from Perak, west Malaysia. *Journal of Southeast Asian Earth Sciences* 6: 25-39.

Smith, A.G.,Hurley, A.M. and Briden, J.C. 1981. *Phanerozoic palaeocontinental world maps*. Cambridge: Cambridge University Press, 102pp.

Stauffer, P.H. 1983. Unravelling the mosaic of Palaeozoic crustal blocks in Southeast Asia. *Geologisches und Rundschau* 72:. 1061-1080.

Stauffer, P.H. and Lee, C.P. 1989. Late Palaeozoic glacial marine facies in Southeast Asia and its implications. *Geological Society of Malaysia Bulletin* 20: 363-397.

Stauffer, P.H. and Mantajit, J. 1981. Late Palaeozoic tilloids of Malaya, Thailand and Burma. *In*,. Hambrey, M.J. and Harland, W.H. (Eds), *Earth's pre-Pleistocene glacial record*, pp. 331-337, Cambridge.

Tapponnier, P., Peltzer, G. and Armijo, R. 1986. On the mechanics of the collision between India and Asia. *Geological Society of London Special Publication* 19: 115-157.

Taylor, B. and Hayes, D.E. 1980. The tectonic evolution of the South China Sea Basin. *In*: Hayes, D.E. (ed) *The tectonic and geologic evolution of Southeast Asian seas and islands*, Geophysical Monograph 23, pp.89-104, American Geophysical Union.

Taylor, B. and Hayes, D.E. 1983. Origin and history of the South China Sea Basin. *In*: Hayes, D.E. (ed) *The tectonic and geologic evolution of Southeast Asian seas and islands*, part 2, Geophysical Monograph 27, pp.23-56, American Geophysical Union.

Tjia, H.D. and Anizan Isahak. 1987. Permian glaciogenic deposits at Salak Tinggi, Selangor. *Geological Society of Malaysia Annual Conference 1987, Abstracts*:.24.

Tollmann, A. and Tollmann, E.K. 1985. Palaeogeography of the European Tethys from Palaeozoic to Mesozoic and the Triassic relations of the eastern part of Tethys and Panthalassa. *In*: Nakazawa, K. and Dickins, J.M. (eds) *The Tethys*, pp.3-22, Tokai University Press, Tokyo.

Tran, V.T. 1979. *Geology of Vietnam (North Part)*. General Geological Department, Hanoi, Vietnam.

Vachard, D. 1990. A new biozonation of the limestones from Terbat area, Sarawak, Malaysia. *CCOP Technical Bulletin* No. 20: 183-208.

Wajzer, M.R., Barber, A.J., Hidayat, S. and Suharsono. in press. Accretion, collision and strike-slipfaulting: The Woyla Group as a key to the tectonic evolution of NorSumatra. *Journal of Southeast Asian Earth Sciences*

Waterhouse, J.B. 1982. An early Permian cool-

water fauna from pebbly mudstones in south Thailand. *Geological Magazine* 119: 337-354.
Wu, F., Van derVoo, R. and Qizhong Liang. 1989. Devonian palaeomagnetism of Yunnan province accross the Shan Thai-South China suture. *Tectonics* 8: 939-952.
Yang Shufeng, Yu Zi-Ye, Guo Lingzhi and Shi Yangshen. 1989. The division and palaeomagnetism of the Hainan Island and plate tectonic significance (in Chinese with English abstract). Journal of Nanjing University (Earth Sciences Edition) 1: 38-46.
Yu Zi-Ye. 1989. Hainan Island - A terrane from Gondwana. *4th International Symposium on pre-Jurassic evolution of East Asia, IGCP Project 224 Reports and Abstracts* 1: 11-12.
Zhao, X. and Coe, R.S. 1987. Palaeomagnetic constraints on the collision and rotation of

Gondwana Eight, Findlay, Unrug, Banks & Veevers (eds) © 1993 Balkema, Rotterdam. ISBN 90 5410 304 3

Phytogeographic evidence for the palaeoposition of Madagascar

O. Hankel
Sonderforschungsbereich 69, Technische Universität Berlin, Germany

ABSTRACT: Middle and Late Triassic microfloras of Tanzania and Malagasy are used to test the validity of the different fits proposed for Madagascar's palaeoposition. Microfloras of this age show a pronounced regional differentiation in the Australian region where two different types have been distinguished, the Onslow Microflora of northwestern Australia and the Ipswich Microflora of southern and eastern Australia. This regional differentiation of Middle and Late Triassic microfloras exists possibly also in the East African region. Microfloral assemblages from the northern part of the Morondava Basin of Malagasy are of the Onslow-type while those from the southern part have been assigned to the Ipswich-type. In Tanzania, Middle and Late Triassic microfloras have been recorded from the upper Karoo sediments of the Luwegu Basin. These microfloras are of the Onslow-type.

It is known that the regional distribution of Onslow and Ipswich Microflora is climate controlled. During Triassic times the Onslow Microflora occurred in a narrow mid-latitudinal belt while the Ipswich Microflora occurred in higher latitudes of the southern hemisphere. Based on these relationships it is shown that neither a southern position of Madagascar nor a position in situ is consistent with the present data. However, discrepancies exist also with reconstructions in which Madagascar is placed far to the north opposite southern Somalia, Kenya and Tanzania. Such a placement shows the occurrence of the Onslow Microflora of Tanzania at the same palaeolatitude as the occurrence of the Ipswich Microflora of Malagasy. If it is true that microfloral assemblages of the Ipswich-type occur in SW Malagasy, then it seems more reasonable to place Madagascar some 400 - 500 km further to the south opposite southeast Kenya, Tanzania and northern Mozambique.

1. INTRODUCTION

In the past various data have been used to reconstruct the palaeoposition of Madagascar prior to the fragmentation of Gondwana. So far no attempt has been made to provide palaeopalynological evidence in support of these reconstructions. Such an attempt is made here. It is based on palynological data from the Middle and Upper Triassic of Tanzania and Malagasy (Fig. 1). Selected palynomorph species of biostratigraphic and phytogeographic significance are used to demonstrate the affinity of the Middle and Late Triassic microfloras of Tanzania and Malagasy with coeval microfloras of Australia.

The Middle and Late Triassic microfloras of Australia show a pronounced regional differentiation. Two different types have been distinguished, the Onslow Microflora of northwestern Australia and the Ipswich Microflora of southern and eastern Australia (Dolby & Balme 1976). The palynomorph species used here characterize the Onslow Microflora. Their association, however, is unknown from the Ipswich Microflora.

The presence or absence of the palynomorph species selected, therefore, may be used to assign the Tanzanian and Malagasian microfloras to the Onslow Microflora or Ipswich Microflora respectively. The assignment to the one or the other type is of relevance for the reconstruction of the original position of Madagascar relative to Africa since the distribution of the Onslow and Ipswich Microflora is known to be latitudinal controlled.

2. DATA BASE

In the present paper published and unpublished data are used. In Tanzania, Middle to Late Triassic microfloras are known from the upper Karoo of the Luwegu Basin (Hankel 1987). These assemblages have been recovered from surface sediments in the Mahogo area (samples TZ-34, TZ-156, TZ-65) (Fig. 2). Their sampling localities lie close to latitude 9°S. Middle to Late Triassic microfloras have been recorded later also from the subsurface of the Luwegu Basin. They have been recovered from upper Karoo sediments in well Liwale-1 from the interval 500 - 1728 m (Shell, unpubl. report; information kindly provided by W.M.L. Schuurman, 11.04.1991). The well Liwale-1 is located at about latitude 9°26'S.

In Malagasy, Middle to Late Triassic microfloras are known from the Morondava Basin (De Jekhowsky & Goubin 1964, Goubin 1965). These are assemblages of Goubin's Unit III which have been recorded from wells in the northern and southern part of the basin. In the north the assemblages of Unit III have been recovered from sediments of the Isalo I (Zone III NA) and the lower part of Isalo II (Zone III NB). The wells in the northern part of the basin are located between latitude 17°30'S and 19°S. In the south the assemblages of Unit III come from the Isalo I sediments (Zone III S). The wells examined in the southern part of the basin are located between latitude 22°S and 24°S.

In addition to the information provided by Goubin (1965) new data are available from the northern part of the Morondava Basin (Hankel, this work). Two microfloral assemblages have been recovered recently from Isalo sediments near Bemolanga. The samples come from well CDB-4 (depth 145 m: Isalo I)

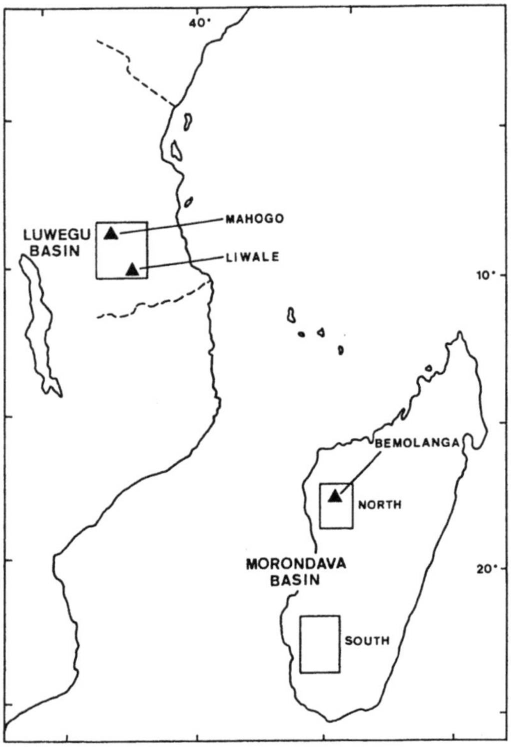

Fig. 1 Location of occurrences of Middle to Late Triassic microfloras in Tanzania and Malagasy.

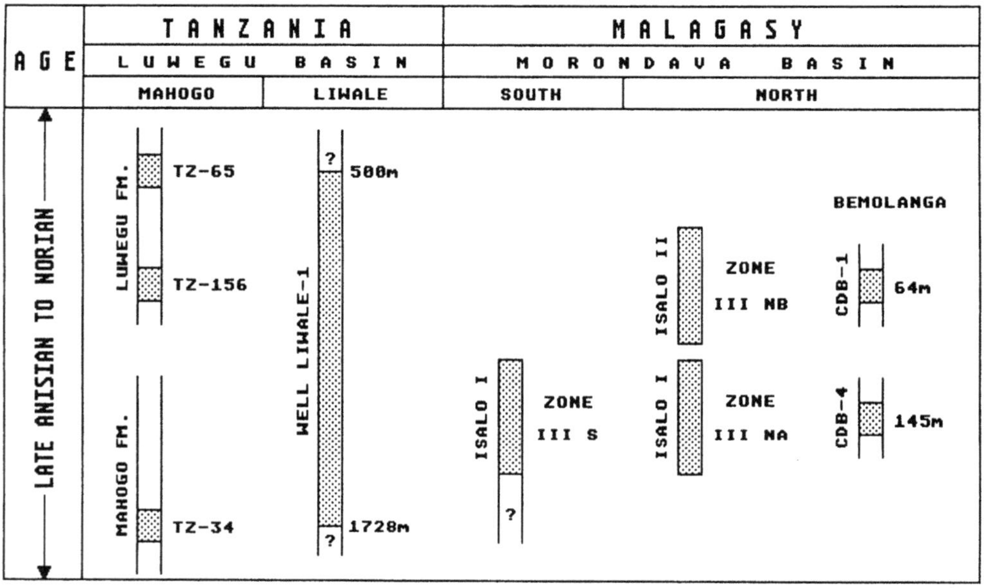

Fig. 2 Source and suggested age relationship of Middle and Late Triassic microfloras of Tanzania and Malagasy.

and CDB-1 (depth: 64 m: lower part of Isalo II). Material from these two wells has been used already by Goubin (1965) to define Zone III NA and III NB.

3. COMPOSITION

The distribution of selected palynomorph species in Middle to Late Triassic microfloral assemblages of Tanzania and Malagasy is shown in figures 3a-e. The tentative correlation and suggested age of these assemblages is also indicated in each figure.

In the Luwegu Basin of Tanzania all of the species selected are present in the assemblages that have been recovered from surface sediments in the Mahogo area (Hankel 1987) (Fig. 3a). Very similar in their gross composition are the microfloral assemblages recorded from well Liwale-1. The palynological data so far released by Shell do not include a complete list of the large number of taxa observed. The information available (W.M.L. Schuurman, personal communication), however, indicates that at least some of the most significant species are present in the assemblages of well Liwale-1 (Fig. 3b).

In Malagasy, microfloras of somewhat similar composition occur in the northern part

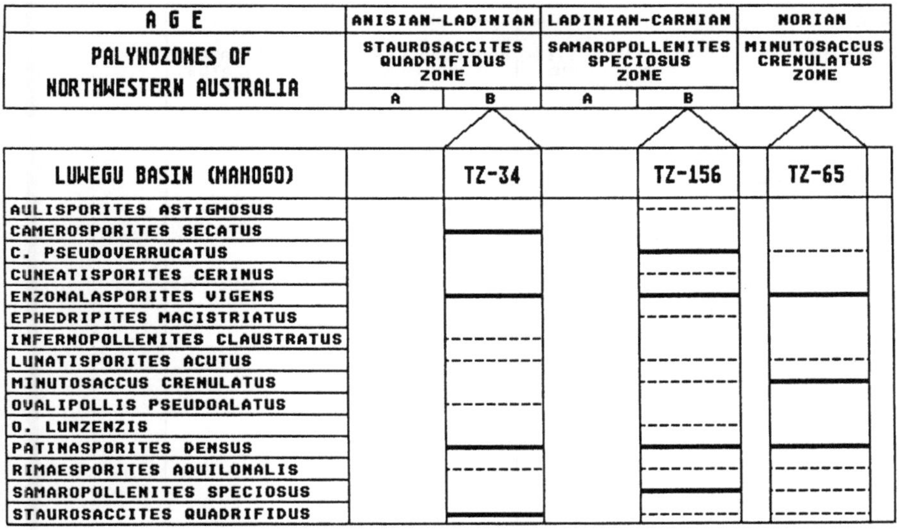

Fig. 3a

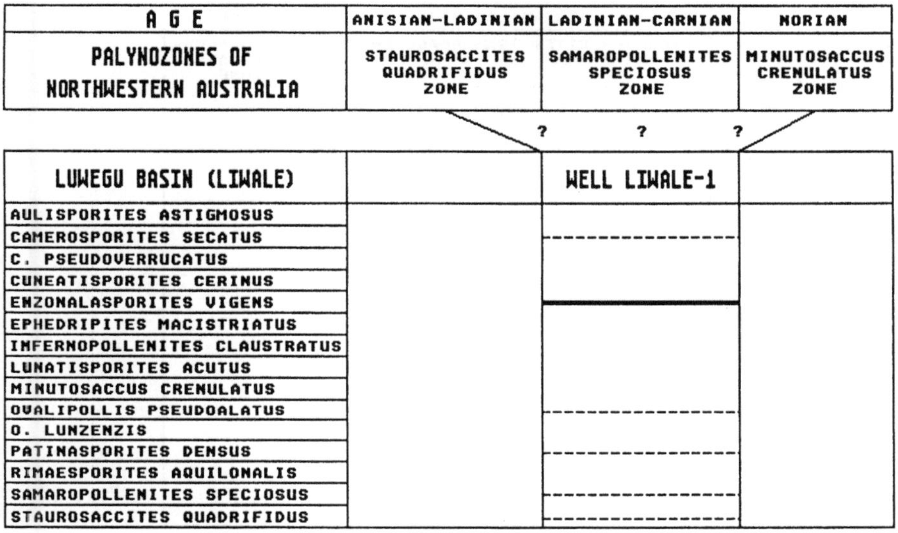

Fig. 3b

of the Morondava Basin. Of the selected species used here Goubin (1965) recorded only Samaropollenites speciosus and Rimaesporites aquilonalis from Zone III NA and III NB. Based on illustrations in De Jekhowsky & Goubin (1964) and Goubin (1965) it has been suggested that Aulisporites and Camerosporites are also present in these assemblages (Dolby & Balme 1976) (Fig. 3c). The new data available from wells near Bemolanga (Hankel, this work) reveal that most of the species which characterize the Tanzanian assemblages also occur in Goubin's northern Unit III (Fig. 3d). The assemblage from well CDB-4 comes from Zone III NA, the other assemblage from well CDB-1 is from Zone III NB.

No definite information exists from the southern part of the Morondava Basin. Most of the species used here are unknown from Zone III S (Fig. 3e). Only two elements are common to the northern and southern assemblages. These are Rimaesporites aquilonalis recorded by Goubin (1965) and a form illustrated in De Jekhowsky & Goubin (1964, Fig. 7, form 1053) which has been referred to Camerosporites by Dolby & Balme (1976). Contradictory information exists with regard to the presence of Samaropollenites. According to Goubin (1965) Samaropollenites

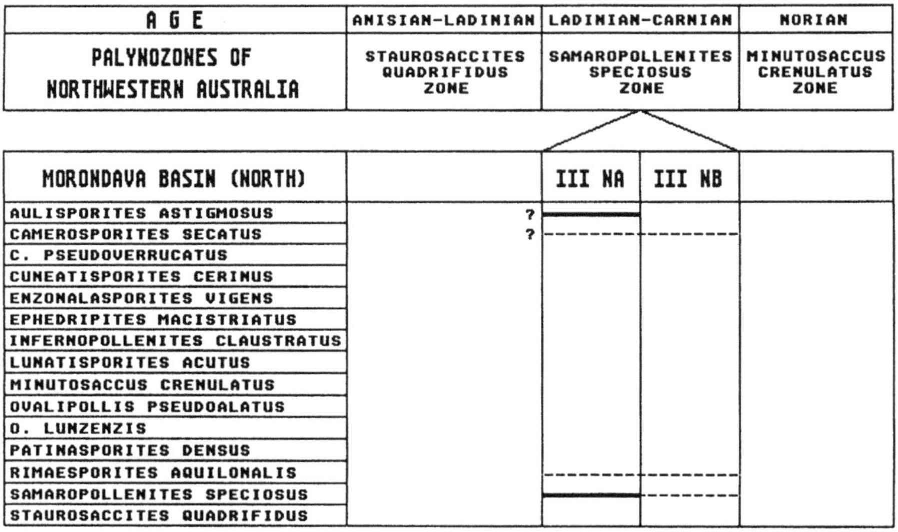

Fig. 3c

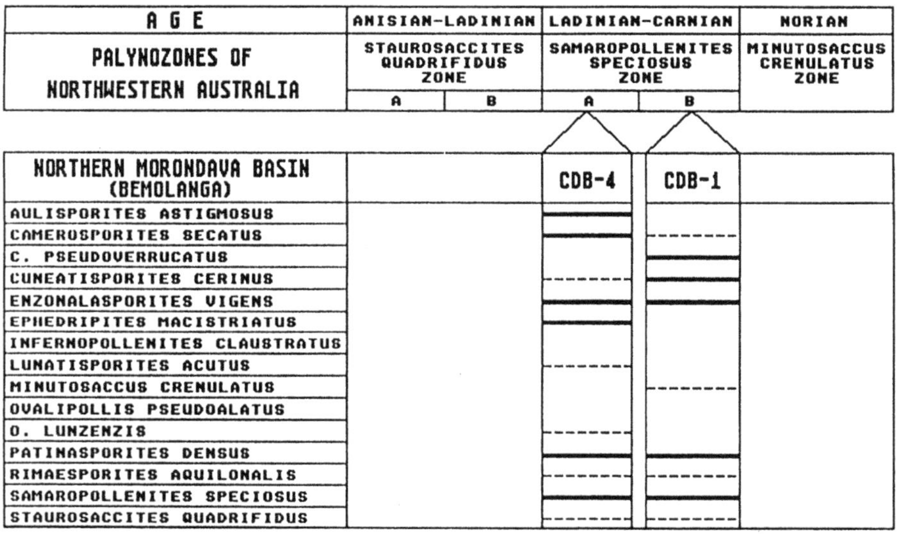

Fig. 3d

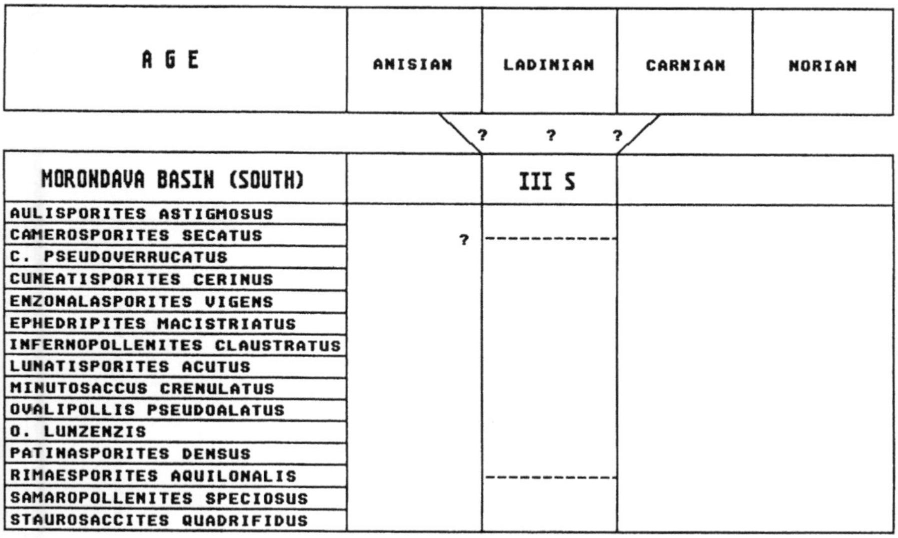

Fig. 3e

Fig. 3 Distribution of selected palynomorph species in Middle to Late Triassic palynofloras of Tanzania and Malagasy. Heavy line indicates common elements, broken line rare elements. The tentative correlation and/or suggested age of the microfloral assemblages is shown in each figure: a) Mahogo area, Luwegu Basin (Hankel 1987), b) Liwale area, Luwegu Basin (W.M.L. Schuurman, personal communication), c) Northern Morondava Basin (De Jekhowsky & Goubin 1964, Goubin 1965, Dolby & Balme 1976), d) Bemolanga area, Northern Morondava Basin (Hankel, this work), e) Southern Morondava Basin (De Jekhowsky & Goubin 1964, Goubin 1965).

speciosus is absent in Zone III S. Visscher & Van der Zwan (1981, Fig. 1), however, indicate the occurrence of an association of Camerosporites and Samaropollenites in SW Malagasy.

4. COMPARISON

All of the palynomorph species used here are characteristic elements of microfloral assemblages recorded from the Middle and Upper Triassic of the Carnarvon Basin of northwestern Australia (Dolby & Balme 1976) (Fig. 4). These are assemblages of the Staurosaccites quadrifidus Zone, the Samaropollenites speciosus Zone, and the Minutosaccus crenulatus Zone, which range in age from Anisian to Norian (Dolby & Balme 1976, Helby et al. 1987). Assemblages from these three zones have been referred to as the Onslow Microflora of northwestern Australia, while coeval microfloras of southern and eastern Australia which lack the characteristic elements have been distinguished as the Ipswich Microflora (Dolby & Balme 1976).

The assemblages from the Mahogo area in the Luwegu Basin, therefore, were assigned to the Onslow-type by Hankel (1987). Likewise, assemblages from well Liwale-1 clearly belong to the Onslow Microflora (W.M.L. Schuurman, personal communication). The microfloras recorded by Goubin (1965) from the northern part of the Morondava Basin have been referred to the Onslow-type by Dolby & Balme (1976). The new data reported here from wells near Bemolanga corroborate this assignment. Onslow-type assemblages have been recorded also from the South Rewa Basin of India (Maheshwari & Kumaran 1979, Kumaran & Maheshwari 1980).

It is difficult, however, at this time to make any unequivocal statement about the character of the palynofloras from the southern part of the Morondava Basin. De Jekhowsky & Goubin (1964) state that a relationship between Unit III in the northern part of the Morondava Basin and Unit III in the southern part is not evident. Based on the apparent absence of characteristic Onslow elements Dolby & Balme (1976) regarded the assemblages from southwestern Malagasy to belong to the Ipswich Microflora. This may be true. The possibility, however, also exists that assemblages of Zone III S belong to the Onslow Microflora. The arguments may be summarized as follows:

(1) The composition of the assemblages has not been documented completely by De Jekhowsky & Goubin (1964) and Goubin (1965). Only the main forms of each unit are recorded.

	TANZANIA	MALAGASY		INDIA	AUSTRALIA
	LUWEGU BASIN	MORONDAVA BASIN		SOUTH REWA BASIN	CARNARVON BASIN
		NORTH	SOUTH		
TAXA	1,2	4,5	3,4	6,7	8
AULISPORITES ASTIGMOSUS	x	x		x	x
CAMEROSPORITES SECATUS	x	x	?	x	x
C. PSEUDOVERRUCATUS	x	x			x
CUNEATISPORITES CERINUS	x	x			x
ENZONALASPORITES VIGENS	x	x		x	x
EPHEDRIPITES MACISTRIATUS	x	x			x
INFERNOPOLLENITES CLAUSTRATUS	x			?	x
LUNATISPORITES ACUTUS	x	x		x	x
MINUTOSACCUS CRENULATUS	x	x		x	x
OVALIPOLLIS PSEUDOALATUS	x				x
O. LUNZENZIS	x	x			x
PATINASPORITES DENSUS	x	x		x	x
RIMAESPORITES AQUILONALIS	x	x	x		x
SAMAROPOLLENITES SPECIOSUS	x	x		x	x
STAUROSACCITES QUADRIFIDUS	x	x		?	x

Fig. 4 Distribution of selected palynomorph species in Middle to Late Triassic palynofloras of Gondwana. Numbers refer to authors: (1) Hankel (1987), (2) W.M.L. Schuurman (personal communication), (3) De Jekhowsky & Goubin (1964), (4) Goubin (1965), (5) Hankel (this paper), (6) Maheshwari & Kumaran (1979), (7) Kumaran & Maheshwari (1980), (8) Dolby & Balme (1976).

(2) At least two elements characteristic of the Onslow Microflora are known from Zone III S (Rimaesporites aquilonalis, Camerosporites sp.).

(3) Assemblages from the Isalo I in the south (Zone III S) may be slightly older than those from the Isalo I in the north (Zone III NA). This could explain the absence of Aulisporites astigmosus and Samaropollenites speciosus in Zone III S since the two species are absent also in the oldest assemblage (TZ-34) recorded from the Mahogo area in the Luwegu Basin (Fig. 3a).

(4) Other Onslow elements may have been attributed to taxa which are not critical for distinguishing the Onslow and Ipswich character. The species Staurosaccites quadrifidus for example has been erected by Dolby in Dolby & Balme (1976). This form may have been incorporated by Goubin (1965) in Guttulapollenites hannonicus which is typical of the Late Permian of Malagasy (Zone I). The presence of Guttulapollenites hannonicus as a rare element of Zone III S has been indicated by Goubin (1965).

These are all speculations. They demonstrate, however, that both possibilities should be considered. Assemblages of Zone III S in the southern part of the Morondava Basin may belong to the Ipswich-type or the Onslow-type.

5. MIDDLE AND LATE TRIASSIC PHYTOGEOGRAPHY OF GONDWANA

Dolby & Balme (1976) have recognized that the regional differentiation of the Middle and Late Triassic flora of Australia is not an isolated phenomenon. They have shown that Middle and Late Triassic microfloras from other Gondwana fragments conform to the concept of the Onslow Microflora and Ipswich Microflora of Australia (Fig. 5a). As already noted Dolby & Balme (1976) assigned palynofloras from NW Malagasy to the Onslow-type and those from SW Malagasy to the Ipswich-type. This pattern in Malagasy was said to correspond with the pattern in Western Australia, where the Onslow Microflora occurs in the Carnarvon Basin and the Ipswich Microflora in the Perth Basin. Palynofloras from the Middle and Upper Triassic of Argentina, Antarctica and New Zealand have been referred to the Ipswich Microflora. Thus, Dolby & Balme (1976) have demonstrated that in an unfragmented Gondwana the Onslow Microflora occurs in a narrow belt along palaeolatitude 30°S, while in contrast localities of the Ipswich Microflora are all situated in higher palaeolatitudes. It has been suggested, therefore, that microfloral assemblages of the Onslow-type represent plant associations of warmer zones.

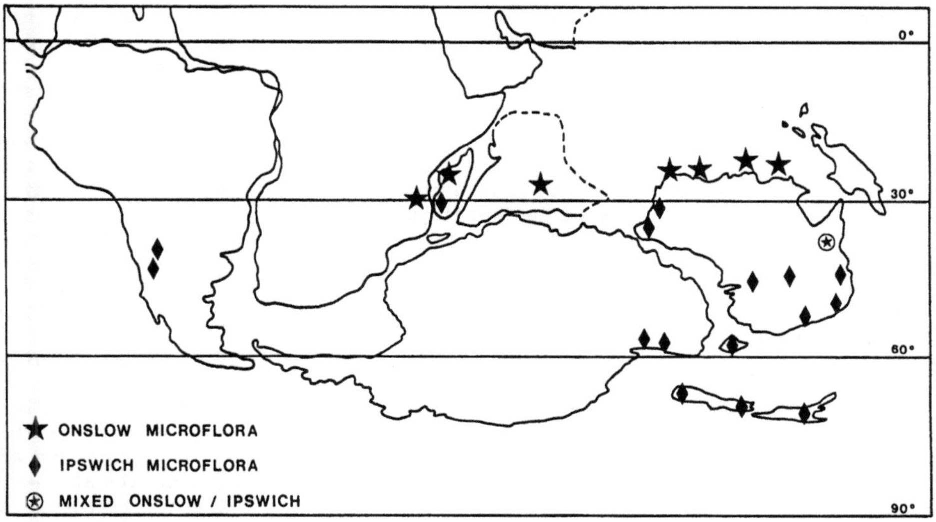

Fig. 5a

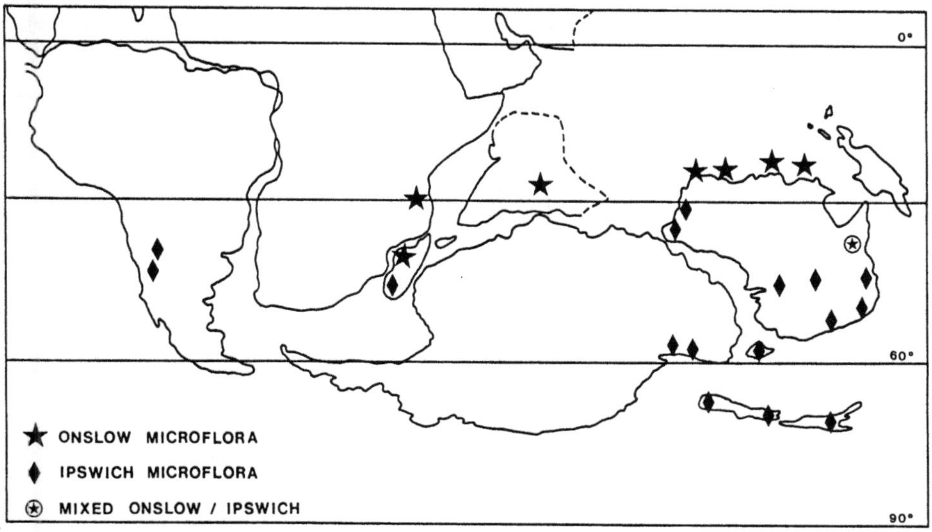

Fig. 5b

Fig. 5 Distribution of microfloral assemblages of the Onslow- and Ipswich-type in Gondwana after Dolby & Balme (1976), supplemented by records from India (Maheshwari & Kumaran 1979, Kumaran & Maheshwari 1980), Tanzania (Hankel 1987; W.M.L. Schuurman, pers. communication), Australia (De Jersey & McKellar 1981), Antarctica (Farabee et al. 1989), and New Zealand (De Jersey & Raine 1990). Occurrences are plotted on the southern hemisphere of the palaeocontinental world map of the Late Triassic of Smith et al. (1981). a) Madagascar in the northern position, b) Madagascar in the southern position.

On the basis of this phytogeographic model Dolby & Balme (1976) predicted the occurrence of Onslow-type assemblages in the South Rewa Basin of India and in the Karoo basins of Kenya and Tanzania. Apart from Kenya these predictions came true. The additional records of the Onslow Microflora from India and Tanzania convincingly demonstrate that during Middle and Late Triassic time the Onslow Microflora indeed occurred in a narrow mid-latitudinal belt which is now known to have extended from northwestern Australia through India and Malagasy to southern Tanzania. Visscher & Van der Zwan (1981) have shown that the regional differentiation of Ladinian and Carnian floras in the northern hemisphere corresponds principally to the pattern of the southern hemisphere.

6. DISCUSSION AND CONCLUSIONS

If it is accepted that the distribution of Onslow and Ipswich Microflora in Gondwana is latitudinal and thus controlled by climate, then the occurrences of Middle and Late Triassic microfloras of Tanzania and Malagasy may be used to test the validity of the different fits proposed for the palaeoposition of Madagascar.

A southern position of Madagascar opposite southern Mozambique conflicts with the present data. Such a placement (Fig. 5b) dislocates the occurrence of the Onslow Microflora of Malagasy from its mid-latitudinal position to the high palaeolatitudes characteristic of the Ipswich Microflora. The same applies to the in situ fit of Madagascar. From this point of view it would be unwise to postulate any southern position of Madagascar during Triassic time.

It is clear from the foregoing that the present data are consistent only with reconstructions in which Madagascar is placed in a position further to the north. The most favoured reconstruction shows Madagascar opposite southern Somalia, Kenya, and Tanzania. The consistency of this reconstruction with the present data depends on the phytogeographic character of the palynofloras of SW Malagasy which are either of the Ipswich-type or of the Onslow-type. This is demonstrated here using the reconstruction of the geological map of sectors of Gondwana by De Wit et al. (1988).

Discrepancies exist if the microfloral assemblages from the southern part of the Morondava Basin belong to the Ipswich Microflora as suggested by Dolby & Balme (1976). The reconstruction shows the occurrence of the Onslow Microflora of Tanzania at the same palaeolatitude as the occurrence of the Ipswich Microflora of Malagasy (Fig. 6a). Such a close coexistence of Onslow and Ipswich Microflora at the same palaeolatitude cannot be explained satisfactorily. Even if the microfloral assemblages of SW Malagasy

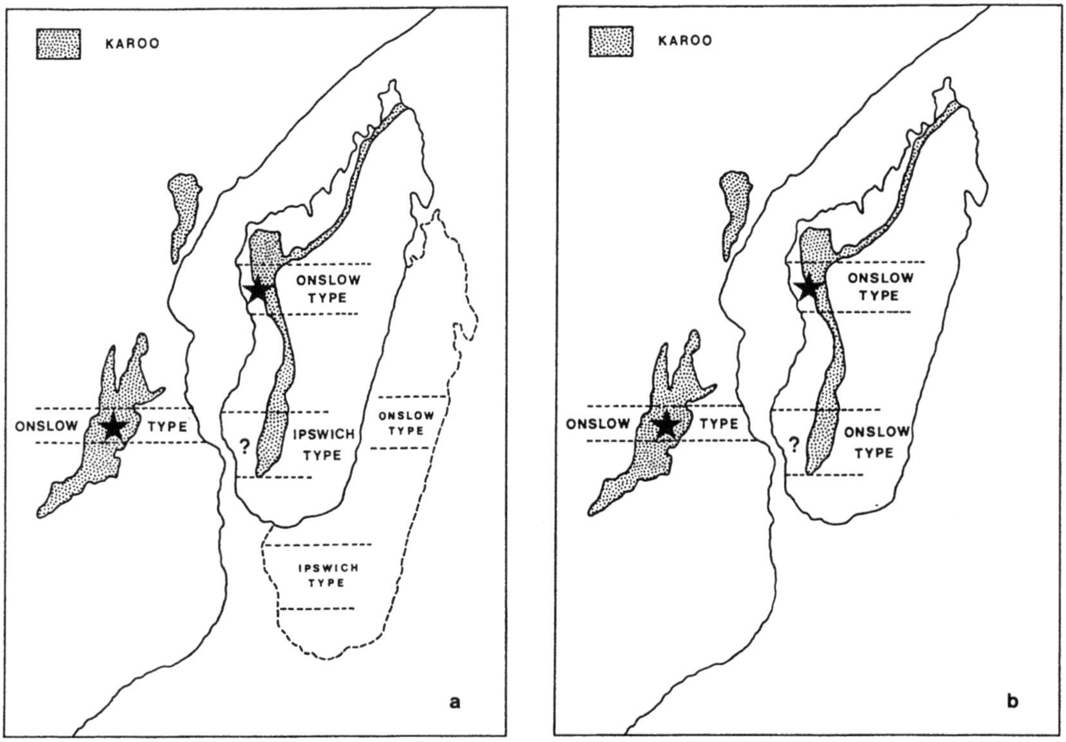

Fig. 6 Palaeoposition of Madagascar according to the geological map of sectors of Gondwana of De Wit et al. (1988) versus phytogeographic evidence from the Middle and Late Triassic of Tanzania and Malagasy. The two possibilities are shown: a) Ipswich-type assemblages in SW Malagasy (alternative fit indicated by broken line), b) Onslow-type assemblages in SW Malagasy.

should represent a mixed association of Onslow and Ipswich Microflora as it is known from the Galilee Basin of eastern Australia (De Jersey & McKellar 1981) it would be difficult to explain why such a mixed association does not occur in the Luwegu Basin of Tanzania. In this case it seems more reasonable, therefore, to place Madagascar some 400 - 500 km further to the south, so that the occurrence of the Onslow Microflora in northwestern Malagasy occupies the same palaeolatitudinal position as the occurrence of the Onslow Microflora of Tanzania. This alternative fit shows Madagascar opposite the present southeast Kenya, Tanzania, and northern Mozambique.

No discrepancies exist if palynofloras of southwestern Malagasy are of the Onslow-type (Fig. 6b). In this case the phytogeographic evidence from the Middle and Late Triassic is consistent with reconstructions which are based on evidence from the Davie Fracture Zone and magnetic anomalies in the western Somali Basin (Coffin & Rabinowitz 1987, Fig. 11). This, however, is no argument to neglect the other possibility discussed here.

7. REFERENCES

Coffin, M.F. & P.D. Rabinowitz 1987. Reconstruction of Madagascar and Africa: evidence from the Davie Fracture Zone and western Somali Basin. Journal of Geophysical Research 92, B9: 9385-9406.

De Jekhowsky, B. & N. Goubin 1964. Subsurface palynology in Madagascar: a stratigraphic sketch of the Permian, Triassic and Jurassic of the Morondava Basin. In A.T. Cross (ed.), Palynology in oil exploration. Society of Economic Palaeontologists and Mineralogists Special Publication 11: 116-130. Tulsa.

De Jersey, N.J. & J.L. McKellar 1981. Triassic palynology of the Warang Sandstone (Northern Galilee Basin) and its phytogeographic implications. In M.M. Creswell & P. Vella (eds.), Gondwana Five. Selected papers and abstracts of papers presented at the Fifth International Gondwana Symposium, Wellington, New Zealand: 31-37. Rotterdam: Balkema.

De Jersey, N.J. & J.I. Raine 1990. Triassic and earliest Jurassic miospores from the Murihiku Supergroup, New Zealand. New Zealand Geological Survey Paleontological Bulletin 62: 1-164. Lower Hutt.

De Wit, M., M. Jeffery, H. Bergh & L. Nicolaysen 1988. Geological map of sectors of Gondwana. The American Association of Petroleum Geologists, Tulsa.

Dolby, J.H. & B.E. Balme 1976. Triassic palynology of the Carnarvon Basin, Western Australia. Review of Palaeobotany and Palynology 22: 105-168.

Farabee, M.J., T.N. Taylor & E.L. Taylor 1989. Pollen and spore assemblages from the Falla Formation (Upper Triassic), central Transantarctic Mountains, Antarctica. Review of Palaeobotany and Palynology 61: 101-138.

Goubin, N. 1965. Description et répartition des principaux pollenites permiens, triasiques et jurassiques des sondages du bassin de Morondava (Madagascar). Revue de l'Institut Francais du Pétrole 20: 1415-1461.

Hankel, O. 1987. Lithostratigraphic subdivision of the Karoo rocks of the Luwegu Basin (Tanzania) and their biostratigraphic classification based on microfloras, macrofloras, fossil woods and vertebrates. Geologische Rundschau 76: 539-565.

Helby, R., R. Morgan & A.D. Partridge 1987. A palynological zonation of the Australian Mesozoic. In P.A. Jell (ed.), Studies in Australian Mesozoic palynology: 1-94. Sydney: Association of Australasian Palaeontologists.

Kumaran, K.P.N. & H.K. Maheshwari 1980. Upper Triassic sporae dispersae from the Tiki Formation. 2. Miospores from the Janar Nala section, South Rewa Gondwana Basin, India. Palaeontographica B173: 26-84.

Maheshwari, H.K. & K.P.N. Kumaran 1979. Upper Triassic sporae dispersae from the Tiki Formation. 1. Miospores from the Son river section between Tharipathar and Ghiar, South Rewa Gondwana Basin. Palaeontographica B171: 137-164.

Smith, A.G., A.M. Hurley & J.C. Briden 1981. Phanerozoic paleocontinental world maps. Cambridge: Cambridge University Press.

Visscher, H. & C.J. Van der Zwan 1981. Palynology of the circum-Mediterranean Triassic: phytogeographical and palaeoclimatological implications. Geologische Rundschau 70: 625-634.

Gondwana Eight, Findlay, Unrug, Banks & Veevers (eds) © 1993 Balkema, Rotterdam. ISBN 90 5410 304 3

Early Jurassic vertebrates from the Kota Formation, Pranhita-Godavari Valley, India

P. Yadagiri
Geological Survey of India, India

ABSTRACT: The Gondwana sequence of the Pranhita-Godavari Valley, Andhra Pradesh, India, comprises a thick succession of sediments ranging in age from Early Permian to Early Cretaceous. The Mesozoic Gondwana sediments contain a number of vertebrate-bearing zones. A rich vertebrate assemblage containing dinosaurs, fishes, microvertebrates and early mammals was recovered from the Kota Formation around Yamanpalli, based on which a Liassic age is assigned. Of two species of pholidophorid fishes, *Pholidophorus kingii* (Yadagiri and Prasad 1977) from India was more primitive than some Early Jurassic forms, viz. *Oreochima* from Antarctica and archaeomaenids from Australia. The discovery of dinosaurian fauna from Kota Formation has been a major breakthrough in the knowledge of Mesozoic terrestrial vertebrates from India. The excavation carried out near Yamanpalli has yielded a number of skeletal parts belonging to different dinosaurs including a new sauropod *Kotasaurus yamanpalliensis* (Yadagiri 1988). The discovery of early mammals from Kota is important in view of their paucity from Gondwanaland and sparse records from Early Jurassic sediments of the world. The report of two new genera of symmetrodonts *Trishulotherium kotaensis* and *Indotherium pranhita* (Yadagiri 1984) and an amphidontid symmetrodont *Nakunodon paikasiensis* (Yadagiri 1985) indicate complicated diversified nature of symmetrodont stock and has greatly expanded the biogeographic evaluation of the Mesozoic mammals. New data on the vertebrate fauna help in tracing the evolutionary history of early mammals and dinosaurs and also afford scope for palaeogeographic reconstruction. The significance of vertebrate fauna from Kota Formation is critically evaluated in the paper.

INTRODUCTION

The Gondwana sequence of the Pranhita-Godavari Valley represents a fairly continuous succession ranging in age from Lower Permian to Early Cretaceous. The Upper Gondwana rocks are classified into three divisions, viz. the Maleri, the Kota and the Chikiala formations based on lithological and faunal characters (King 1881). The Maleri Formation contains a Triassic fauna, whereas the Kota Formation contains a Liassic fauna. Jain et al. (1964) and Kutty (1969) have modified the stratigraphy in the light of lithology and additional vertebrate fossils.

The Maleri Formation consists of clays, lime pellet rocks and sandstones. The vertebrate fauna contains the dipnoan *Ceratodus*, the pleurocanth *Xenacanthus*, the labyrinthodont *Metoposaurus*, the rhynchosaur *Paradapedon*, a coelurosaur and a prosauropod dinosaur. The Dharmaram Formation comprises a succession of alternating sandstone and clay bands. The fauna from the Dharmaram Formation includes at least two dinosaurs, a plateosaurid and a prosauropod. The Kota Formation comprises mainly light brown sandstones and grits, clays and a prominent limestone zone. The basal limestone bed contains abundant ostracod shells cemented with lime mud. The marly bed immediately underlying this bed has yielded dinosaurs and early mammals. The succeeding limestone bed is massive, cream coloured and with a few chert veins. It is stromatolitic in places. The laminated limestone bed has yielded abundant fish fossils.

The Kota Formation has been assigned a Liassic age on the evidence of the fish fauna. Jain (1973), on the basis of hypsomid fishes suggested a Liassic age. The majority of pholidophorus species thrived

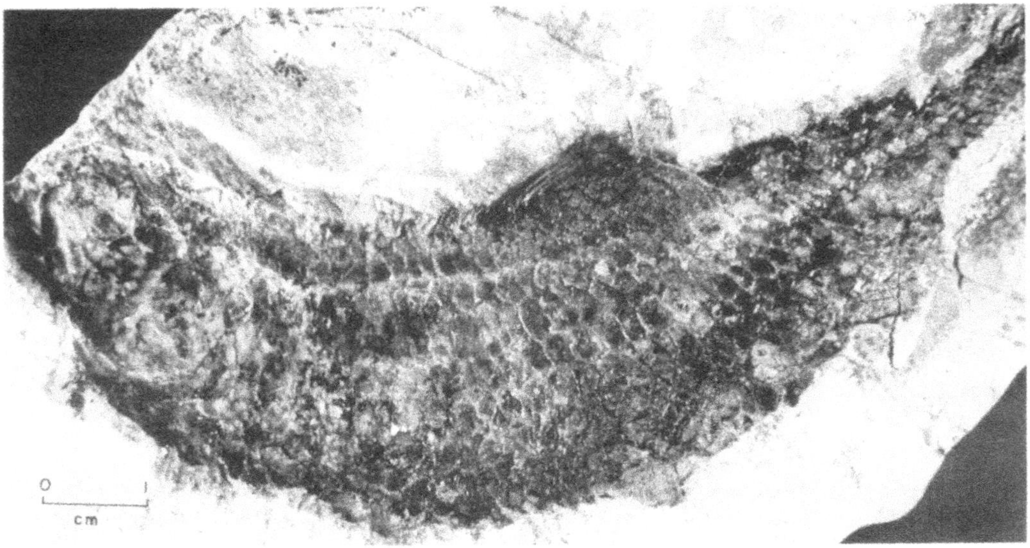

Figure 1 *Pholidophorus kingii*, lateral view.

under marine and estuarine conditions in the Liassic. The Pholidophorus first made its appearance in Triassic and was displaced by the teleosteans, at the end of the Jurassic period. The presence of Pholidophorus along with other ganoid fishes suggests an age of Lower Jurassic for the Kota Formation (Yadagiri and Prasad 1977).

KOTA VERTEBRATE FAUNA

The following vertebrate taxa have been described from the Kota Formation:

Fishes
- Semionotidae
  - *Lepidotes deccanensis* Egerton 1851
  - *Paradapedium egertoni* Jain 1973
  - *Tetragonolepis oldhami* Egerton 1851
- Pholidophoridae
  - *Pholidophorus kingii* Yadagiri & Prasad 1977
  - *Pholidophorus indicus* Yadagiri & Prasad 1977
- Coelacanthidae
  - *Indocoelacanthus robustus* Jain 1974

Reptiles
- Dimorphodontidae
  - *Campilognathoides indicus* Jain 1974
- Sauropod dinosaur
  - *Barapasaurus tagoreii* Jain et al. 1975
  - *Kotasaurus yamanpalliensis* Yadagiri 1988

Mammals
- Kuehneotheridae
  - *Kotatherium haldanei* Datta 1981
  - *Trishulotherium kotaensis* Yadagiri 1984
  - *Indotherium pranhitai* Yadagiri 1984
  - *Nakunodon paikasiensis* Yadagiri 1985

The Kota fauna represents aquatic, terrestrial and arboreal habitats.

FISHES

Lepidotes is the most commonly occurring member of the Piscine fauna, and five species were designated by Egerton (1851). A detailed study of the Kota material of *Lepidotes* (Yadagiri et al. 1980, Jain 1983) and an examination of the validity of characters used to distinguish species within the limits of genus, suggest that all Kota specimens should be placed in a single species, viz. *Lepidotes deccanensis*.

The semionotid *Dapedium* recognized in the 19th century were found morphologically different and a new genus *Paradapedium* was erected (Jain 1973).

Tetragonolepis, another semionotid, was redescribed by Jain (1973) under the species *Tetragonolepis oldhami*. Two other species of the genus were considered by him as not valid.

The coelacanth *Indocoelacanthus robustus* (Jain 1974) is estimated to be slightly larger than *Holophagus*. In the Kota specimen, the skull bones are heavily tuberculated except the frontals which are slightly ornamented.

Of the two species of pholidophorid fishes, namely *Pholidophorus kingii* and *P. indicus* (Yadagiri and Prasad 1977), the former (Fig. 1) is of smaller size (about 110 mm in length) with a fusiform body. The length of the head is almost equal to the depth of the body. The ratio of the length of head to the lengh of body is 1:4. The preoperculum has a very characteristic shape. The posterior margin of the preoperculum is deeply notched with straight sides and a convex base; on the body, there are four longitudinal rows of enlarged scales, the deepest being about twice as deep as broad. All scales have an even posterior margin. The dorsal fin is directly opposite to the pelvic fin.

Australian Lower Jurassic continental deposits have also yielded pholidophorid fishes. Based on the scale features, Griffith and Patterson (1963) have divided these fishes into two groups: (i) *Aetheolepis* and *Aphnelepis* which have rhombic scales that are somewhat deepened on the flank; and (ii) *Archaeomaene* and *Madariscus* which have thin cycloidal scales. Antarctica has also yielded pholidophorid fishes which were retrieved from Early Jurassic sedimentary interbeds in the Kirkpatrick Basalt of the Ferrar Group. Schaeffer (1972) described the form as *Oreochima ellioti*.

It is interesting to note that all the pholidophorid fishes from the three continents, viz. Australia, Antarctica and India were freshwater forms, and apparently were restricted to Early Jurassic times.

INTER-RELATIONSHIPS OF EARLY JURASSIC PHOLIDOPHORIDS

The pholidophoriform fishes are represented by three important families, viz. Pholidophoridae (Nybelin 1966), the Archaeomaenidae (Wade 1941), and the Ichthyokentemidae (Griffith and Patterson 1963) which are closely related. The Pholidophorids from Australia and Antarctica were assigned to the family Archaeomaenidae, whereas the Kota fishes were recognised under Pholidophoridae. A critical evaluation is attempted here of all the above fishes based on their morphological characters.

Archaeomaene and *Aphnelepsis* were first described by Woodward from the Upper Hawksbury–Wainamatta beds, Talbragar, New South Wales and were placed in the new family Archaeomaenidae by Griffith and Patterson (1963).

Aphnelepis is characterised by one and a half times in the maximum depth of the trunk and less than one quarter of the total length of the fish. Pelvic fins arise opposite the origin of dorsal. Scales ornamented with coarse crimpings. The scales are rhombic, somewhat deepened at the flank. Other two forms *Archaeomaene* and *Madariscus* have thin cycloidal scales.

The small (60 mm long) Antarctic pholidophoriform *Oreochima* was assigned to the family Archaeomaenidae by Schaeffer (1972). It differed from other genera in the family by having deep irregular serrations on the posterior borders of the extrascapular, dermosphenotic, preopercular and subopercular, in having a relatively smaller suborbital and in having the free edges of the rhomboidal flank scales strongly serrated.

DISCUSSION

Schaeffer (1972) on comparison of characters of Archaeomaenids with the same characters in the pholidophorids and ichthyokentemids suggested the primitive pholidophoriform condition as follows: premaxilla in contact below rostral, parietal rectangular, dermoteric with fairly broad anterior extension bordering frontal, circumorbital ring complete, with one enlarged infraorbital below single suborbital of nearly same size, maxilla arcuate, extending to middle or posterior border of orbit, preopercular narrow ventrally and extending dorsally to dermopteric, opercular-subopercular suture nearly horizontal, scales rhomboidal.

The following characters in the pholidophorids may be regarded as derived: nasals separated by frontals, dermipteric with narrow anterior extension, circumorbital probably incomplete dorsally, preopercular enlarged ventrally, does not extend to dermipteric, opercular triangular, suture with subopercular oblique.

In *Archaeomaene* and *Madariscus*, the primitiveness is evident in many characters. However, the short certical arm of the preoperculum and the cycloid scales in them are undoubtedly derived.

Oreochima shares most of the primitive pholidophoriform characters. It was assigned to the Archaeomaenidae on the basis of its snout pattern and shortened vertical ramus of its preopercular. The main differences between *Oreochima* and the Australian Archaeomaenids are the relatively smaller suborbital and denticulated dorsal bones and scales in *Oreochima*. *Oreochima* has scales similar to *Aetheolepis* and *Aphnelepis* which have rhombic scales that are somewhat deepened on the flank. However, it is more primitive than either in having a small suborbital.

The high proportion of shared primitive character states in the Pholidophorids, Ichthyokentemids and Archaeomaenids, makes it difficult to postulate the inter-relationships of these families. However, based on the same important characters, the primitiveness of individual genera may be pointed out.

The striking difference between the *Pholidophorus kingii* and *Oreochima ellioti* is the shape of the preopercular. In *Oreochima* it has short vertical arm which terminates near the ventral border of the suborbital and does not reach the dermopteric. The above character may be considered as derived. In *Pholidophorus kingii*, the preopercular is deeply notched with preopercular sensory canal running nearer to the anterior margin. The above character suggests the primitive pholidoform condition for *Pholidophorus kingii* than *Oreochima*. Other characters like body proportions, squamation and suborbital, etc., which are similar in the two forms are suggestive of primitive nature only. So, it can be stated that *Oreochima* although belonging to Archaeomaenidae, is primitive compared to *Aetheolepis* and *Aphnelepsis*. *Pholidophorus kingii* by its nature of characters like preopercular can be recognised as a more primitive form than both *Oreochima*, *Aetheolepis* and *Aphnelepis*.

DINOSAURS

A major gap in the understanding of the evolution of Sauropod dinosaurs has been due to lack of good fossil material from Lower Jurassic rocks, except the fragmentary material *Ohmdenosaurus* described by Wild (1978) from Europe. The discovery of dinosauran fauna from the Kota Formation has been a major breakthrough in the knowledge of Mesozoic terrestrial vertebrates in India. A Sauropod dinosaur, namely *Barapasaurus tagorei*, was described from Posampalli area in Maharashtra (Jain et al. 1980). The excavation carried out near Yamanpalli in Pranhita-Godavari Valley, has yielded a number of skeletal parts belonging to different dinosaurs (Yadagiri et al. 1979). In the collection a new Sauropod dinosaur *Kotasaurus yamanpalliensis* was recognised (Yadagiri 1988).

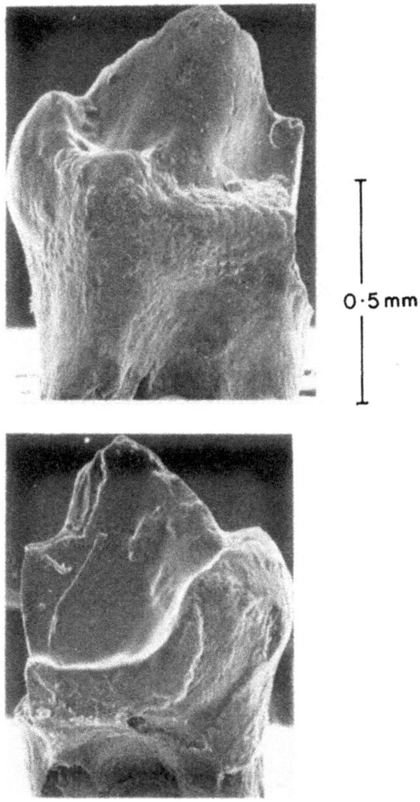

Figure 2 *Nakunodon paikasiensis* in (top) lingual and (bottom) buccal views.

The *Kotasaurus* has many sauropodian characters. Its level of development is intermediate between sauropods and prosauropods. It seems to possess a mosaic of prosauropod and sauropod features. An important character of *Kotasaurus* is its elongated and straight pelvis. The ilium of a sauropod is in general rounded in the margin with an extended pubic process. The ilium of *Kotasaurus* is entirely different. The dorsal margin of the ilium is straight and the anterior and posterior processes are enlarged and tapered. The ilium in general is comparable to prosauropodian pelvis.

The femur is very slender, the slenderness index being 12. The slenderness is least compared to other Jurassic Sauropods including the *Barapasaurus* which has 13.1. The fourth trochanter in the femur of *Kotasaurus* has an acuminate and slightly declined tip, a feature apparently not seen in later Sauropods, but almost universal among the prosauropods. The character of the astragalus in *Kotasaurus* which is ovoidal in shape with low mound in the centre and two

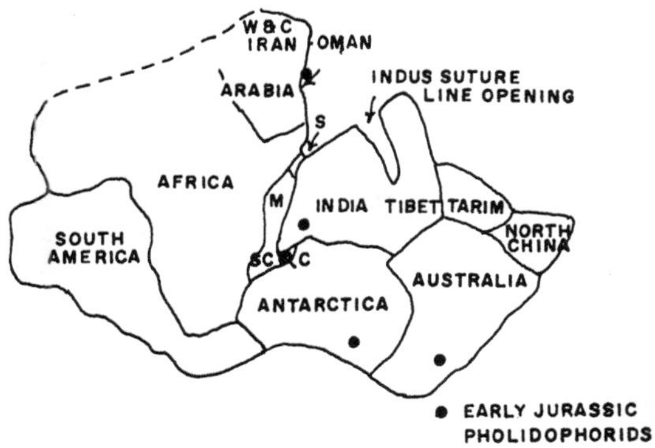

Figure 3 Gondwana at its maximum extent. This reconstruction (after Crawford 1974) satisfies the condition for the dispersal of the Pholidophorids.

depressions on either side, is suggestive of a prosauropod.

Some features of *Kotasaurus* strongly indicate a relationship with the family Melanorosauridae among the prosauropods; on the other hand, several characters indicate if not a close relationship at least a convergence between *Kotasaurus* and the Sauropod family. It differs from *Barapasaurus* in slenderness of femur, straight ilium, slender scapula, and low mound astragalus. The *Kotasaurus* is an intermediate form possessing the characters of prosauropods and sauropods.

EARLY MAMMALS

The mammalian history of the Jurassic period is poorly documented. Any new occurrence from the Early Jurassic, however scanty, may help in understanding the early evolutionary history of this important class.

A preliminary report of the Indian Jurassic mammals was published by Datta et al. (1978). Three new genera of symmetrodonts, Kuehneotheriid family, viz. *Kotatherium haldenei* (Datta 1981), *Trishulotherium kotaensis* and *Indotherium pranhitai* (Yadagiri 1984) have been described. Recently, an amphidontid symmetrodont *Nakunodon paikasiensis* was also reported (Yadagiri 1985) (Fig. 2). Detailed morphological studies are in progress to establish their affinities.

The discovery of an early mammalian fauna in the Kota Formation has greatly expanded the biogeographic evaluation of the Mesozoic mammals. Earlier reports were confined to Laurasia but the Kota mammals are the first record from Gondwana.

PALAEOBIOGEOGRAPHIC SIGNIFICANCE OF KOTA VERTEBRATES

The significance of Kota vertebrate fauna, as evidence of the Lower Jurassic palaeogeography, can be discussed on three different counts: (i) pholidophorid fishes; (ii) early mammals; and (iii) dinosaurs.

The find of pholidophorid fishes in India has great palaeogeographic significance in the sense that it adds a new evidence to evaluate the position of India, vis-a-vis the position of other continents of Gondwana before its fragmentation. Several approaches taken by different workers have resulted in varied configurations of Australia, Antarctica and India. Craddock (1977), while presenting a key paper on the evolution and fragmentation of Gondwanaland, in particular drew attention to some of the unsolved problems about the nature of the India–Antarctica–Australia relationship.

Tasch et al. (1975) brought out the significance of estherinids for continental fit. Jurassic beetles and palaeolimnadids from Antarctica and India, and Triassic palaeolimnadids and estherinids in India and Australia required nonmarine dispersal routes which suggest the proximity of the three continents. He, however, pointed out the configurations of Smith and Hallam (1967) and Veevers et al. (1971) are not adequate to explain the dispersal of the fauna in three continents.

Schaeffer (1972) mentioned that the presence of freshwater Archaeomaenidae in both Australia and Antarctica during the Jurassic was not surprising, as these two continents were in contact throughout the Mesozoic.

Crawford (1974) presented the concept of Greater Gondwanaland, based upon geophysical and structural studies, wherein he included part of the Asiatic continent as an integral portion of Gondwana. This was also corroborated by the presence of *Lystrosaurus* fauna in the Tienshan and the Turfan Basin of Sinkiang. The *Lystrosaurus* fauna were found in many parts of Gondwana, including India, Antarctica and Africa. The configuration suggested by Crawford (1974) (Fig. 3) with the position of Antarctica and Australia in contact with India including Tibet, is the plausible reconstruction which explains the distribution of pholidophorids in them during Early Jurassic times.

It is observed that a sea barrier is the main factor in limiting the spread of animal species on land. Seaways are also great barriers for freshwater fishes which can only gain access to other continents which are in close proximity. The significant factor to note is that all the pholidophorid fishes from three continents, viz. Australia, Antarctica and India were freshwater forms. They were restricted to early Jurassic horizons and there is no other record of pholidophorid fishes from these continents in any any other geological horizons. Thus, the occurrence of pholidophorids supports the view of Australia, Antarctica and India being in close proximity which was a necessity for the dispersal of these fishes. The fossil evidence is in conformity to the configuration suggested by Crawford (1974). As discussed earlier, the study of the pholidophoriform condition suggests that pholidophorids might have dispersed from India to Australia through Antarctica.

A major evidence in understanding the evolutionary transition from synapsids to mammals is being facilitated by the increased refinement of determination of the ages and correlations of local faunas that were considered of Late Triassic (Rhaetian) or Early Jurassic age (Lillegraven et al. 1979). Current research shows that the so-called "Rhaetic fissure fillings" of England and Wales, which have produced a wealth of specimens of early mammals and advanced synapsids, were deposited at various times during this interval. Most of the fillings yielding *Morganucodon* and *Kuehneotherium* are now throught to be of Liassic, probably Sinemurian, age (Kermack et al. 1981). In the light of the new findings, the Kayenta Formation of North America, the upper part of the Stormberg Group of South Africa, the Kota Formation of India and the Lufeng beds of China are now thought to be broadly correlative units and of Early Jurassic age. The aforesaid occurrences suggest that early mammals were widely dispersed through Laurasia and Gondwana before fragmentation of the supercontinent.

The presence of sauropod dinosaurs, namely *Barapasaurus* and *Kotasaurus*, in the Kota Formation suggests its close proximity to the South African region, as dinosaurs of the Late Triassic and Middle Jurassic period are found in the latter continent. The occurrence of *Rhoteosaurus* in the Middle Jurassic sediments of Australia also supports the view that during the Early to Middle Jurassic, pensinsular India was a land-locked part of Gondwanaland.

ACKNOWLEDGEMENTS

The author is thankful to Dr D.K. Ray, Director General, Geological Survey of India, for kind permission to publish. Thanks are due to Dr M.N. Balasubramanyan, Senior Deputy Director General, and Shri. D.C. Das Sarma, Director, Regional Palaeontological Laboratory, Geological Survey of India, Southern Region, for their encouragement.

REFERENCES

Bonaparte J. 1979. A Jurassic assemblage from Patagonia. *Science* 205: 1377–1379.

Cassiliano M.L. & Clemens W.A. 1979. Symmetrodenta. *In* Lillegravan et al. (eds), *Mesozoic Mammals*. University of California Press, Berkeley: 150–166.

Charig A.J., Attridge J. & Crompton A.W. 1965. On the origin of the sauropods and the classification of the Saurischia. *Proceedings of the Linnean Society of London* 176: 197–222.

Colbert E.H. 1975. *Wandering lands and animals*. Hutchinson, London: 223.

Colbert E.H. 1978. Mesozoic tetrapods and the northwards migration of India. *Journal of the Palaeontological Society of India* 20: 108–145.

Colbert E.H. 1984. Mesozoic reptiles, India and Gondwanaland. *Indian Journal of Earth Sciences* 11: 25-37.

Coombs W.P. 1975. Sauropod habits and habitats. *Palaeogeography, Palaeoclimatology and Palaeoecology* 17: 1–33.

Craddock C. 1977. The evolution and fragmentation of Gondwanaland. *In* Laskar, B. and Raja Rao, C.S. (eds), Fourth International Gondwana Symposium Papers. Hindustan Publishing Corp., Delhi: 1–17.

Crawford A.R. 1974. A greater Gondwanaland. *Science* 184: 1179–1184.

Datta P.M., Yadagiri P. & Rao B.R.J. 1978. Discovery of Early Jurassic micromammals from Upper Gondwana sequence of Pranhita-Godavari Valley, India. *Journal of the Geological Society of India* 19: 6468.

Datta P.M., Yadagiri P. & Rao B.R.J. 1981. The first Jurassic mammal from India. *Zoological Journal of the Linnean Society of London* 73: 307-312.

Egerton P.M. 1851. Descriptions of specimens of fossil fishes from the Deccan. *Quarterly Journal of the Geological Society of London* 7: 273.

Freeman E.F. 1979. A Middle Jurassic mammal from Oxfordshire. *Palaeontology* 22: 135–166.

Gardiner B.G. 1960. A revision of certain Actinopterygian and coelacanth fishes, chiefly from the Lower Lias. *Bulletin of the British Museum (Natural History), London* 4: 239–384.

Griffith A. & Patterson C. 1963. The structure and relationships of the Jurassic fish *Ichthyokentema purbeckensis*. *Bulletin of the British Museum (Natural History), Geology* 8: 1–43.

Jain S.L. 1973. New specimens of Lower Jurassic holostean fishes from India. *Palaeontology* 16: 149–177.

Jain S.L. 1974. *Indocoelacanthus robustus* n.gen., n.sp. (Coelacanthidae, Lower Jurassic), the first fossil coelacanth from India. *Journal of Palaeontology* 48: 49–62.

Jain S.L. 1983. A review of the genus *Lepidotes* (Actinopterygii: Semionotiformes) with special reference to the species from Kota Formation (Lower Jurassic), India. *Journal of the Palaeontological Society of India* 28: 7–42.

Jain S.L., Kutty T.S., Roychowdhury T. & Chatterjee S. 1975. The sauropod dinosaur from the Lower Jurassic Kota Formation of India. *Proceedings of the Royal Society of London* 188: 221–228.

Jain S.L., Kutty T.S., Roychowdhury T. & Chatterjee S. 1980. Some characteristics of *Barapasaurus tagorei*, a sauropod dinosaur from the Lower Jurassic of Deccan, India. *In* Laskar, B. and Raja Rao, C.S. (eds), Fourth International Gondwana Symposium Papers. Hindustan Publishing Corp., Delhi: 39–47.

Jain S.L., Robinson P.L. & Roy-Chowdhury 1964. A new vertebrate fauna from the Triassic of Deccan, India. *Quarterly Journal of the Geological Society of London* 18: 115–124.

Kermack K.A., Mussett F. & Rigney H.W. 1981. The skull of *Morganucodon*. *Zoological Journal of the Linnean Society of London* 71: 1–158.

King W. 1881. The geology of the Pranhita-Godavari Valley. *Memoir of the Geological Survey of India* 18: 1–151.

Kutty T.S. 1969. Some contributions to the stratigraphy of the Upper Gondwana Formations of the Pranhita-Godavari Valley, Central India. *Journal of the Geological Society of India* 10: 33–48.

Lillegraven J.A., Kraus M.J. & Brown T.M. 1979. Palaeogeography of the world, Mesozoic. *In* Lillegraven et al. (eds), *Mesozoic mammals: the first two thirds of mammalian history*. University of California Press, Berkeley: 277–308.

Lehman J.P. 1966. Actinopterygii. *In* Pivetea, J. (ed.), *Traite de Paleontologie, 4*. Masson et cie, Paris: 1–242.

Nybelin O. 1966. On certain Triassic and Liassic representatives of the family Pholidophoridae *s.s.* *Bulletin of the British Museum (Natural History) Geology* 11: 35–432.

Patterson C. 1968. The caudal skeleton in Lower Liassic pholidophorid fishes. *Bulletin of the British Museum (Natural History) Geology* 16: 201–239.

Rayner D.H. 1941. The structure and evolution of the Holostean fishes. *Biological Reviews, Cambridge* 16: 218–237.

Raath M.A. 1974. A new dinosaurus (Reptilia: Saurischia) from near the Triassic-Jurassic boundary. *Aynoldia* 50 (3): 1–37.

Rayner D.H. 1948. The structure of certain Jurassic Holostean fishes with special reference to the neurocyania. *Philosophical Transactions of the Royal Society, London (B)* 233: 287–345.

Schaeffer B. 1972. A Jurassic fish from Antarctica. *American Museum Novitates* 2495: 1–17.

Smith A.G. & Hallam A. 1970. The fit of southern continents. *Nature* 225: 139–144.

Tasch P., Sastry M.V.A., Shah S.C., Rao B.R.J., Rao C.N. & Ghosh S.C. 1975. Estheriids of the Indian Gondwanas: significance for the continental fit. *In* K.S.W. Campbell (ed.), *Gondwana Geology: papers from the Third Gondwana Symposium*. ANU Press, Canberra: 443–452.

Veevers J.J., Jones J.G. & Talent J.A. 1971. Indo-Australian stratigraphy and the configuration and dispersal of Gondwanaland. *Nature* 229: 383–388.

Wade R.T. 1941. The Jurassic fishes of New South Wales. *Journal and Proceedings of the Royal Society of New South Wales* 75: 71–84.

Wild R. 1978. Ein sauropoden rest (Reptilia, Saurischia) aus dem Posidenienschiefer (Lias, Toarcium) von Holzmaden. *Stuttgarter Beit. Naturk.* 41: 1–15.

Gondwana Eight, Findlay, Unrug, Banks & Veevers (eds) © 1993 Balkema, Rotterdam. ISBN 90 5410 304 3

Assembly of Gondwanaland – Open forum

C. McA. Powell
Geology Department, The University of Western Australia, Nedlands, W.A., Australia

1 INTRODUCTION

This 2-hour discussion followed three symposia on Palaeomagnetism (4 papers presented), Precambrian Mobile Belts/Sutures and Proterozoic Assembly of Gondwana (8 papers) and Late Proterozoic/Early Palaeozoic Tectonics (13 papers). Documentation of many of the answers to questions discussed are in the papers from these symposia. Speakers were limited to one overhead projection for each comment being made. A base map of Gondwanaland showing cratons which existed before 900 Ma, and the younger mobile belts (Fig. 1) was used as a prompt to guide discussion. The forum was chaired by C. McA. Powell.

Chris Powell, The University of Western Australia
The principal question we should attempt to address is when and how the supercontinent, Gondwanaland, was assembled, and there are four main themes we should explore. When I think of the kinds of answers that might be given, I wonder how I might answer the parallel question, 'When was Asia assembled?' Obviously, it has taken a lot of time, starting at least by the Triassic with various fragments of Gondwanaland being accreted to the eastern half of what is now Asia. India is just one of the late arrivals, and in another fifty million years or so Australia could be welded to far eastern Asia. Thus, the answer to the question of when Gondwanaland assembled could well be that it took quite a long time, some hundreds of millions of years, and that what we need to identify are stages in the growth of the Gondwanan nucleus.

The second theme which we need to address is: what is the oldest transcontinental Gondwanaland nucleus? Where is the oldest part which we are fairly happy was assembled in its Gondwanaland configuration? When did that nucleus come together? Something that will recur, the third theme, is what the palaeomagnetic constraints are on the existence of Gondwanaland or parts of it. Can we use palaeomagnetism to constrain some of our reconstructions? We saw in Ian Dalziel's keynote talk that palaeomagnetism is crucial in trying to put any particular reconstruction together. The last theme, which also relates to the keynote address, is whether we can reconcile the Eocambrian breakup of parts of the Neoproterozoic supercontinent with the coeval continuing amalgamation of blocks around its margins in northern Africa and Arabia.

Discussion opened up:

2. WHEN DID GONDWANALAND FORM?

2.1 *Helmut Wopfner, University of Cologne, Germany*

The first point I would like to make is that we normally cut off Gondwana at the northern margin of present-day Africa, and I think it's demonstrable that, during at least the Permo-Carboniferous time, Gondwana certainly did extend further to the north, so I think we have to specify a northern boundary between Gondwana and Laurasia.

Chris Powell: The focus of this symposium is on the formation of the entity we call Gondwanaland which, as Prof. John Veevers has noted in his keynote address, preceded

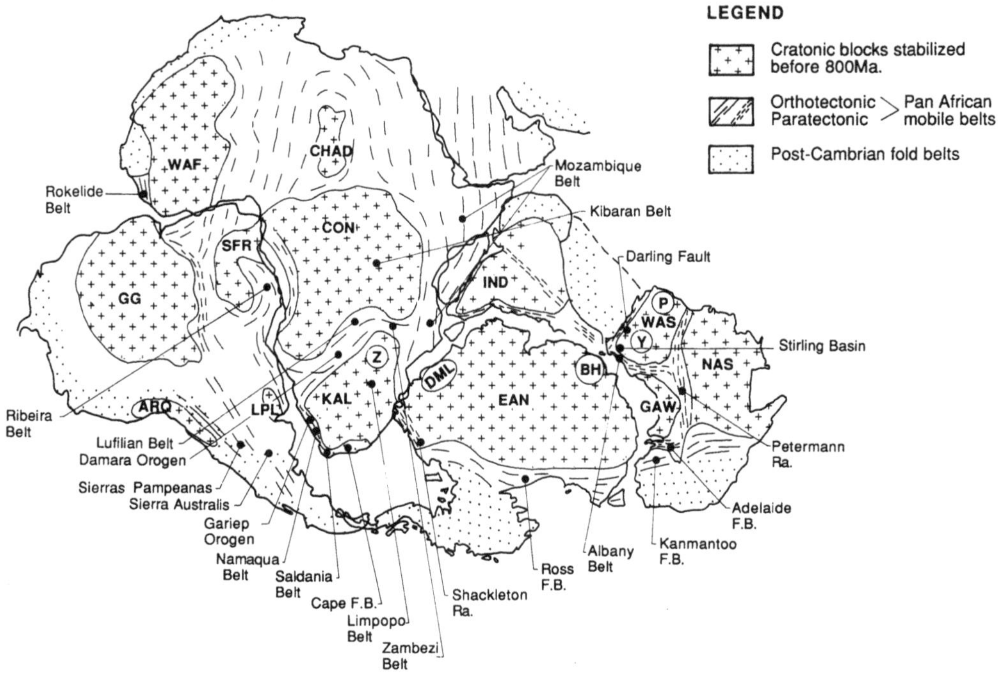

LPL	La plata Craton
GG	Guayana & Guopore Cratons
ARQ	Arequipa massif
WAF	West African Cratons
CHAD	Chad Craton
SFR	Saõ Francisco Craton
CON	Congo and other African Cratons
KAL	Kalahari Craton
EAN	East Antarctica Cratons
IND	Indian Cratons
WAS	West Australian Craton
NAS	North Australian Craton
GAW	Gawler Graton

(Y)	Yilgarn Craton
(P)	Pilbara Craton
(DML)	Dronning Maud Land
(Z)	Zimbabwe Craton
(BH)	Bunger Hills & Denman Glacier

Fig 1: Tectonic sketch of Gondwanaland showing cratonic blocks known to have stabilised before Late Proterozoic (~1000 Ma) and Pan African (820 to 500 Ma) mobile belts. Orthotectonic Pan African mobile belts contain fragments of ophiolite, which may indicate closure of former oceans; paratectonic Pan African mobile belts are reactivated older fold belts.

Pangea, to which I think you are referring. We are mainly concerned with the Late Precambrian and Cambrian aggregation of the various blocks. Perhaps someone who addressed the old transcontinental Gondwanaland nucleus, the 1100 to 1300 Ma entity, might like to make a few comments.

2.2 *Raphael Unrug, Wright State University, Ohio, USA*

The first and most important thing arising from this conference is welcome Laurentia and farewell the permanent Pacific Ocean! This model is extremely interesting and will stimulate of lot of work; it is testable. One comment I have is that we need to move Laurentia a little earlier, if you want to close, fan-like, Eastern and Western Gondwana, and the reason for that is that in Eastern Africa and the Zambezi Belt we have a very clear cut evidence of a tectonic event at 820 Ma, so perhaps the closure of the Mozambique Ocean is to some extent earlier. However, these are the fine points of the model that is generally testable. Talks presented during the symposium gave reasonably clear evidence that a large portion of Eastern Gondwana, including the Western Australian craton, East Antarctica, the Kalahari craton and most

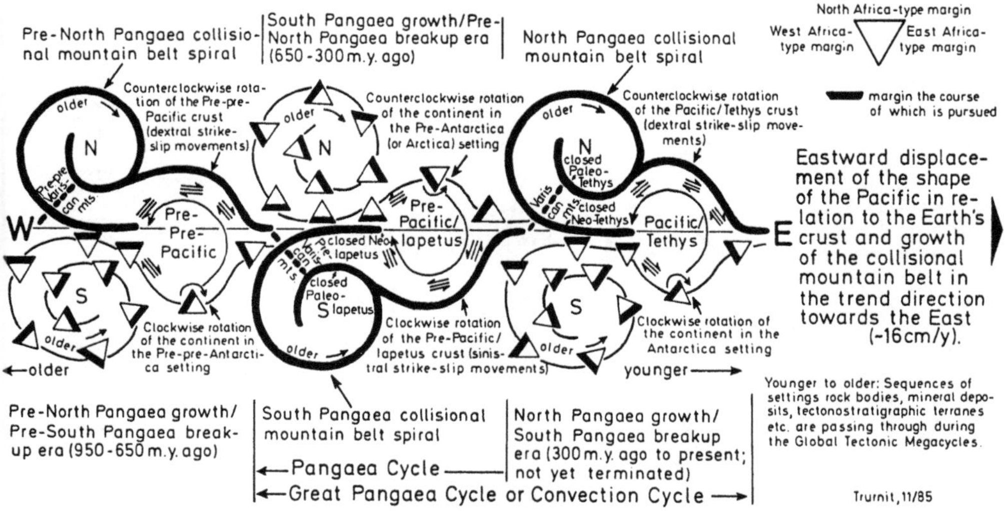

Fig 2: The Great Pangea Cycle or Convection Cycle: Two global tectonic megacycles of 360° per 200 to 250 m.y. duration make one Pangea Cycle; two Pangea Cycles amount to one Great Pangea Cycle (from Trurnit 1989).

probably India were fused or sutured in the 1300 to 1100 Ma events, and that they remained more or less stable until the Mesozoic break-up of the supercontinent. Nonetheless, there were a number of deformational events. In Western Australia, the Stirling basin of Late Proterozoic age was created at the junction of the Yilgarn craton and the Albany mobile belt, and it suffered some deformation in the latest Proterozoic. I think that right now we are entering a state of flux (of ideas) and there is some possibility for misunderstanding arising from terminology. We need to pay good attention to this. We probably need to differentiate, by different names, the supercontinents of various ages, and the same applies to the oceans. We had a very good suggestion from John Baldock that the lineage of the Greek Gods should be followed closely and the predecessor of Iapetus should be Uranus.

2.3 *Peter Trurnit, Federal Geological Survey of Germany, Hanover*

I agree with Professor Unrug that there should be names for the super-continents. Laurasia, I call North Pangea, and the mountain belt, from the North American Laramides (Cordilleras/ Rockies) eastwards via the Alpine mountains of the Rockies, Europe and Asia to the Himalayas, I call the North Pangea collisional mountain belt (Fig. 2). Gondwana existed when we had a South Pangea collisional mountain belt spiral. It assembled between 650 and 300 Ma. During that era the Appalachians/Caledonides mountain belt (a two-lap fragment from the South Pangea spiral - Fig. 2) formed. Today, we live in a North Pangea growth/South Pangea break-up era. During 650 to 300 Ma we had South Pangea growth/pre-North Pangea break-up era, and the 950 to 650 Ma interval was a pre-North Pangea growth/pre-South Pangea a break-up era.

Chris Powell: Are you suggesting that the Gondwanaland assembly began to form from 650 Ma onwards?

Peter Trurnit: Yes, the first suture was the Baikalian / Assyntic / Cadomian / Avalonian mountain belt during Vendian times.

Chris Powell: I believe there are palaeomagnetic constraints on that hypothesis.

2.4 *Dr. Zheng Xiang Li, The University of Western Australia*

There are basically two ways to use

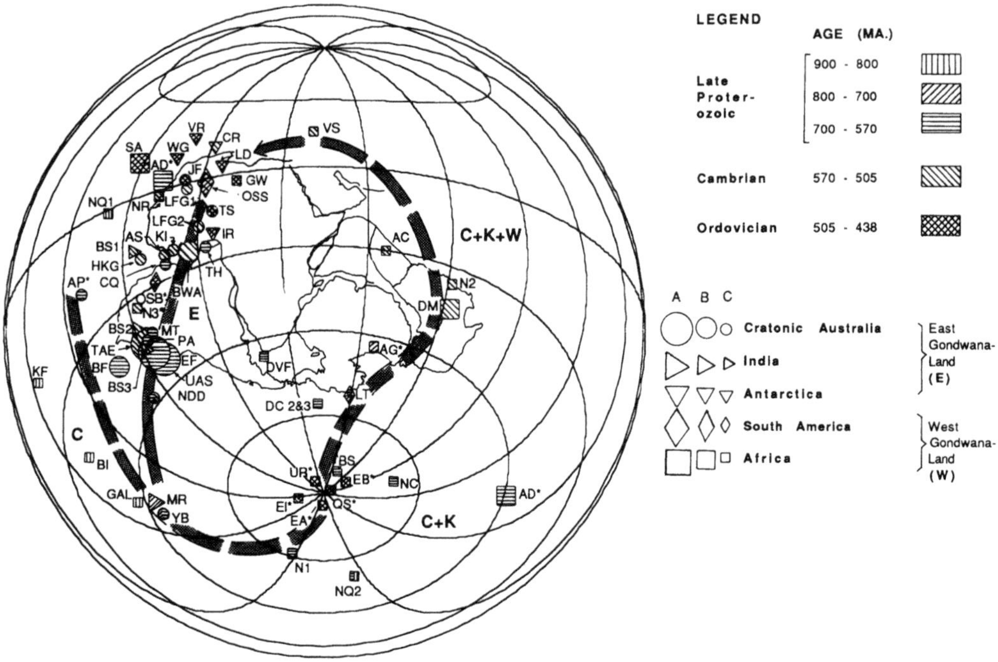

Fig 3: Palaeomagnetic poles of quality class C and better (Li and Powell, this volume) for Gondwanaland cratons in the Late Proterozoic to Ordovician interval. (a) East Gondwanaland palaeopoles, (b) West Gondwanaland palaeopoles. Apparent polar wander paths for East and West Gondwanaland do not converge until Cambro/Ordovician. Size of symbols relates to reliability of the palaeopoles, with the largest being most reliable.

palaeomagnetic data to test different models. One way is to use the apparent polar wander paths (APWPs). Suppose you have palaeomagnetic data from two continents. When the APWPs have similar shape, you can put the two continents together by rotating one continent towards the other until the APWPs match. The two continents form a single coherent entity for the time that the APWPs match. However, when they diverge as you trace the APWPs back in time, it means that the continents were separate. Unfortunately, it is not easy to use this method to tell when Gondwanaland came together, because the data are very scarce for some blocks.

A different method is commonly used where poles of the same age from two continents are used to rotate the continents so that the poles overlap. This puts both continents in their correct palaeolatitude, but you can never say conclusively from palaeomagnetic data whether the two continents were together. Palaeomagnetic data just do not have control; you can put the continents together or they could be thousands of kilometres apart. Additional geological evidence must be used to constrain the reconstructions.

My second point concerns when Gondwanaland formed. A plot of the palaeomagnetic data from around 800 to 500 Ma (Fig. 3) shows that most of the poles from East Gondwanaland are from Australia, with a few from India and Antarctica. Although the data from West Gondwanaland are very scarce, and come from different blocks, they are well away from East Gondwanaland. The palaeomagnetic poles come together only by Late Cambrian to Early Ordovician time, so from palaeomagnetic data alone we can say that the Gondwanaland we know of in the Palaeozoic did not exist in the Late Proterozoic.

Chris Powell: The palaeomagnetic evidence shows that East and West Gondwanaland were not in their accepted relationship to each other until the end of the Cambrian, although parts of them might have been together earlier. Which parts assembled earlier is a leading question, with the earlier Gondwana nucleus extending from Western Australia across East Antarctica and India, and into the Kalahari craton of southern Africa. Perhaps that southern part of Africa belonged with East Gondwanaland prior to other parts of Africa coming together. The Damara belt becomes important in this context, as it could represent the place where various blocks of Africa sutured with East Gondwanaland.

Does anyone who has worked in South Africa wish to comment on when they see the coherence of a block that might extend into India and hence across to Australia, because, from the talks, I understand that to be the earliest assembly that is trans-continental?

2.5 *Peter Gresse, Geological Survey of South Africa. S. Africa (written communication)*
Earliest Assembly: Underlying the Cape Fold Belt along the southern tip of Africa is the Pan-African Saldania belt which is in turn underlain by the Kibaran Namaqua-Natal belt. The geophysical signature of the Permo-Triassic Cape Fold Belt is characterized by the Beattie magnetic anomaly and the so-called Southern Cape Conductive Belt (SCCB). A recent interpretation by Prof. I.W. Hälbich *et al.* (University of Stellenbosch) suggests that these geophysical features define a number of southward-dipping Kibaran-age crustal slabs consisting of northward thrusted wedges or horses. At least one of these slabs consists of ultramafic material, possibly oceanic crust. They therefore suggest a Kibaran-age suture along the southern tip of Africa, verging towards the north, at about 1000 - 1100 Ma.

Subsequent rifting led to the formation of the Saldanian basin at ~900–800 Ma. The Saldanian belt closed at about 650 Ma by northward-directed compression and thrusting, possibly along a major décollement at the Saldanian/Kibaran interface. The Saldanian belt contains possible ophiolitic remnants (Bridgetown Volcanics), suggesting a second event of oceanization and closure with northward-directed compression towards the Kalahari craton. Syn- to late-tectonic granite intrusions range from 520 to 610 Ma.

During the Late Cambrian, the Cape trough developed as an ?aborted rift, again along exactly the same lines. It was subsequently deformed by north-verging folding and thrusting at ~250–230 Ma (Permo-Triassic) by collision or whatever mechanism.

These features favour an early Gondwana assembly at ~1100 Ma followed by two more identical events at ~650 Ma and ~250 Ma. The final breakup of Gondwana again exploited exactly the same pre-existing basement features.

2.6 *Masaru Yoshida, Osaka City University, Japan*
You have asked about the earliest event in the East Antarctica, India and Australia portion of East Gondwanaland. I have some data on the continuity of India, Antarctica and Sri Lanka at earlier times. My point is that the Archaean cratonic part in India and in Antarctica match in the Gondwanaland reconstruction. I also note that the boundary between the Archaean and Proterozoic can be traced from India to Antarctica. We have some palaeomagnetic data on Sri Lankian Early Phanerozoic rocks and Antarctic Phanerozoic rocks, as well as from Jurassic dolerites from Sri Lanka and western Queen Maud Land.

To tie all this data together we don't consider it necessary to have different assembly models. In each period, between latest Archaean to earliest Proterozoic and the Jurassic, all the reassemblies seems to be fundamentally similar. So it appears that this indicates that since the earliest Proterozoic there was no significant disruption, although there might have been a collisional event when there was reactivation of some zone, without substantial re-organisation of the assembly.

Chris Powell: So you would extend the early Gondwanaland nucleus back much earlier than 1300 Ma?

Masaru Yoshida: Yes.

2.7 *Robin Oliver, Adelaide University, South Australia*
With reference to recent dating by Lance Black and others quoted by John Sheraton from Canberra, in the Bunger Hills region of Antarctica, there is a range of model ages

around 2.6 Ga, a very prominent group of ages at 1.2 Ga and another around 1.0 Ga. These ages can be matched fairly closely in Western Australia, and the correlation between Australia and Antarctica can also be made or suggested on the basis of the continuation of the trend of the Darling Fault in Australia along the trend of the Denman Glacier. Furthermore, there is a 500 Ma event in the Bunger Hills, with granites and some alkaline mafic dykes of that age. This 500 Ma event is also present in southwestern Australia. Because the 500 Ma event is manifested in part by alkaline mafic dykes in the Bunger Hills, it could be argued that this is perhaps an expression of asthenospheric uplift associated with break-up of the earlier Proterozoic continent.

Chris Powell: These data support the idea that the modern fit of Australia and Antarctica is correct and also that these two continents were together for a long time. I would like to know how large this early nucleus was.

2.8 *Mike Katz, University of N.S.W., Australia*

I have been intrigued by all this for a number of years and my general comments come from the heart rather than from the brain. I think we are all more or less agreed that there are Archaean cratons as shown on the sketch map (Fig. 1). These cratons seem to be surrounded by mobile belts which have been active throughout Proterozoic time. There are African dates which virtually extend throughout the entire era, although there are places where the frequency of these dates in a histogram indicates there are major events. Now the question being asked is 'When did Gondwanaland come together?' Every mobile belt on the sketch map (Fig. 1) represents a tectonic event whether that event is a separation and later coming together of cratons, or some kind of *in-situ* process where these mobile belts can be considered, on a fractal scale, as being high strain zones around more rigid blocks. You can generate a lot of these mobile belts possibly *in-situ* by shearing around the margins of the cratons (Katz, 1985). There is good evidence that there was an East and West Gondwanaland and that the whole assembly come together around 500 Ma, and that's the Pan African event seen all over the place. When you ask 'When did Gondwanaland come together?', I think the answer is that there were bits and pieces of Gondwanaland that come together at various times.

Chris Powell: Where would you put the join between East and West Gondwanaland in view of the apparent continuity of the Middle Proterozoic continental nucleus from Western Australia to the Kalahari craton?

Mike Katz: In the literature, the Mozambique belt is identified as being the suture. In the past, I have been interested in trying to correlate older mobile belts from East to West Gondwanaland. One belt that I have considered is the Limpopo belt, going through Madagascar and into East Antartica along the Eastern Ghats. The Limpopo belt is Early Proterozoic, if not Archaean. Again there is a possibility that, like the Atlantic Ocean, things came apart and later came together more or less along the same line.

Chris Powell: Let me ask one more question: What is the significance of the Damara Belt, where there are a lot of 820 Ma ages? Is that possibly reactivated and taking up the on-strike continuation of the closing Mozambique zone?

Raphael Unrug: The answer to this question, as far as we know now, is that the two cratons (Congo and Kalahari) were continuous at around 1200 Ma. Then they started breaking apart; first there was a string of pull-apart basins, and then separation occurred. Chris Hartnady from Capetown suggests that the Kalahari craton rotated counter-clockwise with relation to the Congo to open a triangular basin. This counter-clockwise rotation would fit nicely with the northeastern corner of the Kalahari craton pushing into the Congo craton to give deformation in the Zambezi belt. Then the Damara and Lufilian belts, which are probably separated by a transform fault, started closing at 650 Ma. Collision in the Zambesi and Mozambique belts probably occurred earlier at 820 Ma (Fig. 4).

2.9 *Mike Rickard, Australian National University, Canberra, Australia*

It seems to me that this question is something

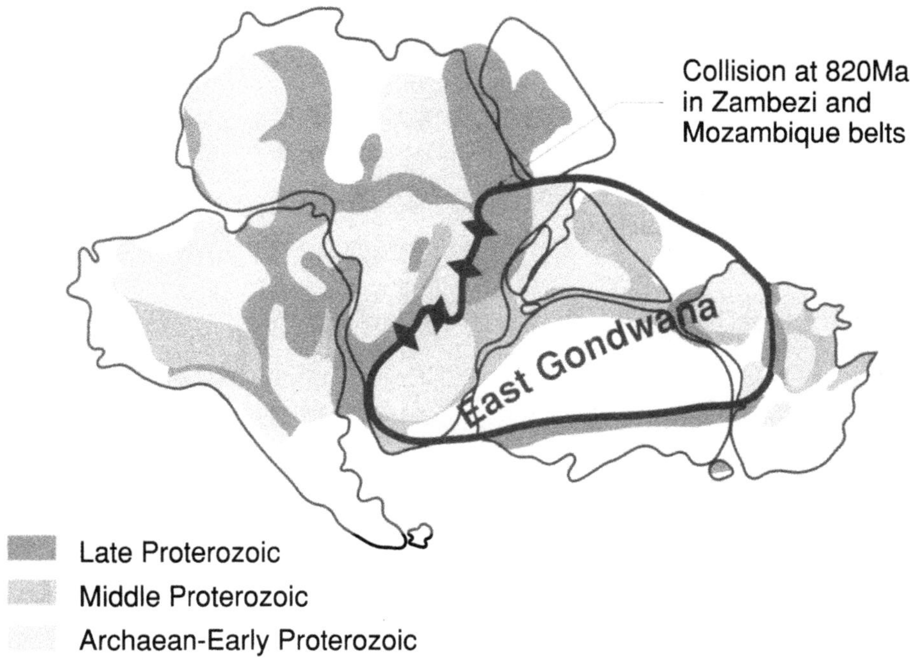

Fig 4: Gondwana Supercontinent showing Late Proterozoic mobile belts and the East Gondwanan continent which collided with African cratons at 820 Ma (after Unrug, 1991).

bigger than we can handle in the short time here. The answer to the problem depends on the nature of the mobile belts in the Proterozoic, which has long been a big argument: whether they are subduction-related or whether they are intracontinental, we can't really solve without any Archaean or Early Proterozoic palaeomagnetic results. It might be more profitable if we moved our attention to Cambrian and Palaeozoic for the purposes of this conference, i.e. the configuration of the Late Palaeozoic Gondwana Supercontinent.

Chris Powell: The reason we are concentrating on the Late Proterozoic assembly is that there are Cambrian and latest Precambrian palaeomagnetic data from East and West Gondwanaland showing a large relative movement between the two, and so having them continuous at 1100 to 1300 Ma poses some interesting questions. Perhaps, we have to do what Mike Katz suggested, and that is first open the mobile belt between the Congo and Kalahari cratons and then close it in the younger Precambrian time.

2.10 *John Baldock, British Geological Survey*

I have been privileged to work in various parts of Gondwana, much of East and Central Africa, the margin of South America and now on the basement of Britain. I want to make a few points. Recent research in the Zambezi belt is producing evidence of rocks with the 820 Ma ages including a lower crustal section containing highly metamorphosed allochthonous gabbros, which may represent transform ophiolites obducted on to the margin of the Zimbabwe craton. The 820 Ma dates from this belt, therefore, could represent an orogenic melt which is related to some form of subduction. The later ages that we get from the Mozambique belt, the 650 and 450 Ma dates, appear to be intracratonic. Whether that suggests there is again an opening and a closing of this belt is uncertain, but I don't believe that the main part of the Mozambique belt represents an opening of an oceanic basin between East and West Gondwana.

My second point concerns the northwestern margin of Gondwana (in this reconstruction), where we have a problem

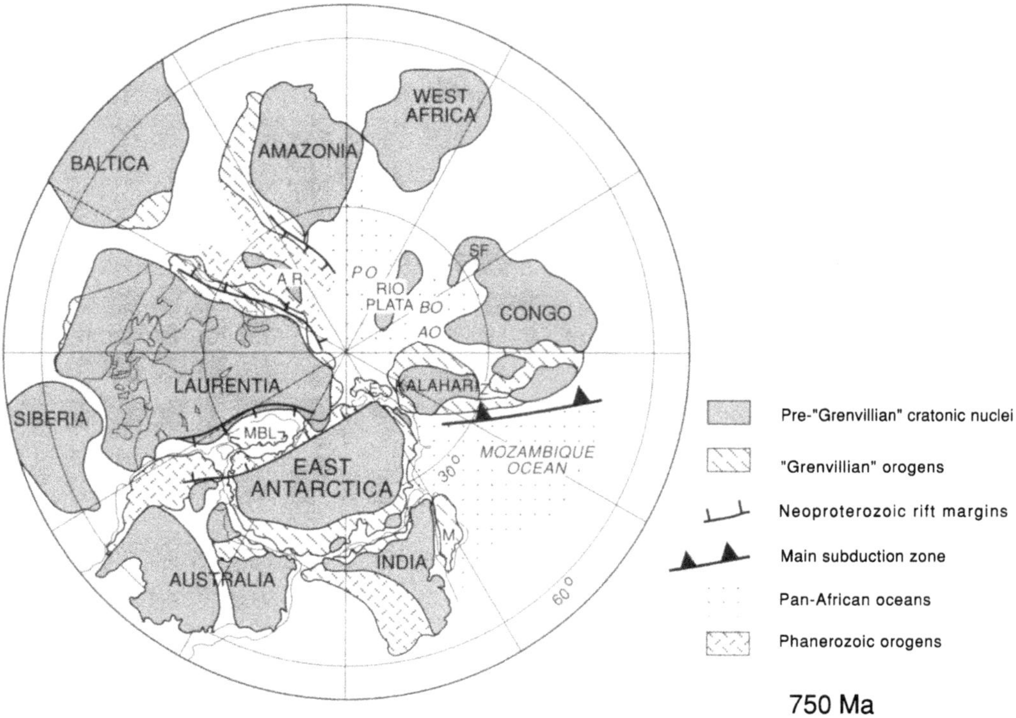

Fig 5: Hypothetical reconstruction of a Neoproterozoic supercontinent based on the North America-East Antarctica/Australia relationship suggested by Moores (1991) and amplified by Dalziel (1991). The relationship of North America and South America is that of Dalziel (1991) and Hoffman (1991). Positioning of Baltica and Siberia and identification of cratons and mobile belts follows Hoffman (1991). Palaeomagnetic poles and circles of confidence (alpha 95) are described in Dalziel (1991, fig. 3). From: Dalziel (1992, fig. 2).

with nomenclature which Raphael Unrug alluded to. The Iapetus Ocean, which represents a proto-Atlantic developed to the east of the Appalachians on the eastern side of Laurentia, is also shown as the oceanic crust to the northwest of Gondwana, and these two oceans may not be the same thing. We are in danger of using Iapetus in Ian Dalziel's sense for an opening ocean between the South American craton and the Laurentian craton, and in a different sense for opening an ocean to the northwest of Gondwana. I want to make a plea that we keep the nomenclature separate.

My final point is really a question to Ian Dalziel, 'Do the Precambrian massifs that are still left on the margin of South America (like the Arequipa massif) relate to part of Laurentia?'

3. ROLE OF LAURENTIA IN GONDWANALAND

3.1 *Ian Dalziel, University of Texas at Austin, U.S.A.*

There is nothing like destroying a map before it's even been published! With respect to John Baldock's points about Iapetus, one of the important things about these old oceans is that in the absence of any ocean floor, except as little bits of ophiolitic rock here and there, we are pushed into the position of talking about them purely as spaces between continental masses. So I am not particularly wedded to exactly what we call any ocean until we get into a reconstruction and see, let's say Baltica with respect to North America or North America with respect to South America. In this type of model the classical Iapetus ocean

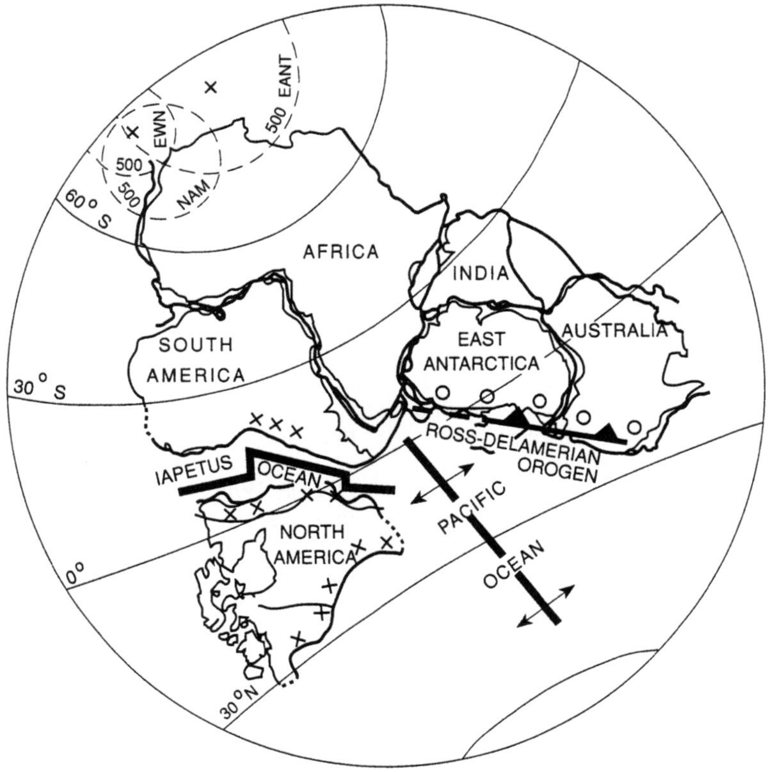

Fig 6: Palaeomagnetically controlled reconstruction of the Pacific and southern (i.e. South America-North America) Iapetus oceans on final amalgamation of the Gondwana supercontinent at the end of the Cambrian. From: Dalziel (1992, fig. 3).

might be between northeastern U.S.A. and Europe, Wilsonian Iapetus, but if we talk about the whole Appalachian belt, then the entire region between North America and South America would also be a part of Iapetus, if you think of Iapetus as being the ocean East of Laurentia. We probably need to do as you suggest, and bear in mind the ways in which these names have been used, but we can only talk about them in relation to the continental masses.

With regard to the question of the Arequipa massif, let me answer that in a general way by going back to the question our Chairman has asked, and looking at the Grenvillian belt. I am pretty sure that this is also how Paul Hoffman might suggest you look at it because this was the basis of the paper that he has published in Science after reviewing the Laurentia-East Antarctica story (Hoffman, 1991). If you say, regardless of the issue of what these Neo-proterozoic belts represent, that by the time there was cooling along these broadly Grenvillian belts (if you will permit me to use that term for 1.3 to 1.0 Ga mobile belts), the cratons bordering those belts had fused together. If you trace these Grenvillian belts (Fig. 5), then at least the Australian-East Antarctic-India area seems to have been fused by 1 Ga, notwithstanding important younger events at 500 Ma. On the model I have put forward (Fig. 5), North America (Laurentia) would also be part of that East Gondwanaland nucleus, and the rifting system on each side of the suggested juxtaposition would have cut across older belts like the Grenville front.

On the other hand, along the Atlantic margin of North America we have the suggested terminal Proterozoic rifting following a Grenvillian belt, which is the basis of Paul Hoffman's suggestion for embracing this in his reconstruction, because we have a Grenvillian-age belt to the west of the Trans-

Amazonian shield. So, if you say, 'Where is the early nucleus of Gondwanaland?', I might suggest including North America and Baltica which is tied by the Dahlslandian or Svecco-Norwegian province to the Grenville province (Fig. 5). Perhaps the early nucleus also included the Kalahari or Kaapvaal craton, which includes a little part of East Antarctica. It seems to me that, whereas in all those regions discussed above, terminal Proterozoic rifts are where oceans actually developed, in South America and Africa we have the Damara and several other Late Precambrian belts which have a lot of post-Grenvillian latest Precambrian orogenic activity within Gondwanaland. Along the Mozambique to Arabian margin of Africa, on the other hand, there seems to be very good evidence (ophiolitic fragments, blueschists and eclogites) of a lot of arc material being accreted in the latest Precambrian and perhaps into the Cambrian. There seems to me to be clear evidence of a suture there, and I would like to know how plausible are the broad notions that (1) East Gondwanaland including the Kalahari craton was one entity, (2) South America and Africa had some Late Precambrian mobile belts, which indicate suturing at a younger time, and (3) there is a lot of accreted material and a final suture along the Mozambique-East African margin (Fig. 5).

Chris Powell: What you are suggesting is that if the Kalahari craton were part of the East Antarctica craton, the southern part of Mozambique-East African subduction zone would deflect into part of the Damara belt, and that this would then satisfy constraints other people have outlined.

Ian Dalziel: That's close enough to what I was saying.

Chris Powell: Your reconstruction (Fig. 5) is a radically different looking Gondwanaland from previously postulated reconstructions, although it does not contravene the known constraints.

3.2 *John Long, West Australian Museum*

In 1980, Burrett and Richardson used computer reconstructions to examine, stage by stage, all the Cambrian trilobite data of the world (Burrett and Richardson, 1980). They examined the so-called endemic groups. In the earliest Cambrian, abundant *Redlichia* faunas occurred throughout Eastern Australia. They also occur in both North and South China, Tarim, Iran and Spain. The faunas seem to be marginal along the Gondwanaland edge, all in similar facies. In considering Ian Dalziel's reconstruction with North America adjacent to Eastern Australia (Fig. 5), perhaps we should look for *Redlichia* faunas on the western side of North America as a test for his model, and also test the model with data comparing shelly Early Cambrian faunas from Australia and China and parts of these northern Gondwana margins with North America.

Helmut Wopfner: I would like to point out that the two Richters (1941) have already drawn attention to the global distribution of *Redlichia*. They showed the *Redlichia* occurrences to extend from the Mediterranean, Palestine and Iran right across to South Australia. They noted a great similarity of the Middle Cambrian faunas over the entire extension of that belt, whereas the Middle Cambrian faunas of the Mediterranean differed considerably from those of Baltica. I think this is a very important aspect which should be considered in any kind of reconstruction. And that reverts back to my comment at the beginning of this session: What was Gondwana at that time and where was its "northern" limit?

3.3 *Dr. Jim Jago, University of South Australia*

Very briefly, what John Long said is correct. Redlichiid trilobites have also been found in southern France and North Africa. As far as I know redlichiids are not present in western North America and given that rocks of that age are pretty well known there this leads to a problem when you use Ian Dalziel's reconstruction (Fig. 5), although it depends to some degree on the time of the supposed breakup between North America and eastern Australia. The Cambrian faunas of Australia, Antarctica, New Zealand, China and Kazakhstan are all much the same, as oddly enough, are a few faunas from Alaska, but they are quite different from those of western North America. This poses a mjaor problem for this reconstruction (Fig. 5).

Ian Dalziel: In a paper published in South Africa (Hartnady, 1986), which I was not aware of until after I wrote my *Geology* paper (Dalziel, 1991), Chris Hartnady pointed out a considerable similarity between shallow-water near-coastal organisms of *Ediacaran* type found in the Death Valley area of southwestern USA, western Argentina, the Appalachians and southwest Africa, and suggested there could have been shallow-water continental-margin connections between those areas at that time. On the other hand, if we did have a reconstruction such as I have suggested, and if we generated the Pacific Ocean by spreading between North America and Australia/ Antarctica, then it probably happened pretty quickly. The reconstruction for 500 Ma (Fig. 6), based on palaeomagnetic data, shows a very big Pacific Ocean. The organisms are telling us something very clearly, and I think they are telling us that North America was isolated. If the 500 Ma reconstruction is correct then we did have a break between Australia-Antarctica and North America, but we may still have had a connection between South America and North America across what, if John Baldock will forgive me, is shown as a narrower Iapetus Ocean (Fig. 6).

Chris Powell: I would like to see that palaeontological information tested a little further. The timing of the breakup of Eastern Australia is controlled by relationships in the Adelaide and Kanmantoo Fold Belts. I understand from one of the talks given by Foden *et al.* (this symposium) that the age of the Truro Volcanics is quite young (~526 ± 4 Ma, Cooper *et al.* 1992) and this is at a level equivalent to the base of the Kanmantoo Group, which most people regard as the passive margin succession related to the continental breakup. This relationship is important, because the Ediacaran fauna pre-dates the breakup. Would anyone like to comment on the changes of palaeontological affinities from Ediacaran time into earliest Cambrian, and possibly talk about the significance and reliability of that Truro Volcanics date, because it is very important in the timing of events?

Jim Jago: As far as I know we cannot date the Truro Volcanics at Truro - I think the date mentioned comes from a tuff bed at Sellick Hill about 140 km away. This tuff bed is towards the top of the Heatherdale Shale (i.e. not far below the base of the Kanmantoo Group) and is in a similar, but not identical stratigraphic position to the Truro Volcanics.

3.4 *Thomas Flöttmann, Adelaide University, South Australia*

I don't know much about this date, except that it has been determined by Dr. John Cooper (Adelaide University) using the ion microprobe, and it is not from Truro itself but from a tuff bed in the Heatherdale Shale at Carrickalinga Head. Obviously, by around 500 Ma we have a compressional régime in the region.

Chris Powell: The importance palaeontologically is that if the age of the Heatherdale Shale, which is near the base of the divergent margin succession, is ~525 Ma, and that date is reliable, then there is not a lot of time between divergence and the 500 Ma reconstruction (Fig. 6). There should be some palaeontological connections and affinities; I imagine the Redlichids would not have been far from home at that stage if North America were the continent which broke away from Eastern Australia.

3.5 *Larry Harrington, Australia:*

Could I ask for a clarification of that age on the "Truro Volcanics", as it is critical in understanding when the Eo-Pacific Ocean opened.

Chris Powell (with assistance from the floor): Our current understanding is that the date of ~525 Ma is a yet-to-be published date, obtained using the ANU ion microprobe on samples from a tuffaceous band in the Heatherdale Shale, which underlies the Carrickalinga Head Formation. The latter is considered to be the proximal equivalent to the lower part of the Kanmantoo Group, which many people regard as the ocean-floored passive margin prism of sediment deposited in the wake of continental breakup in Eastern Australia. The tuff band is *not* in the Truro Volcanics, but in a unit potentially correlative with it. This relationship needs further investigation, irrespective of the reliability of the date, and we should not dwell too much on

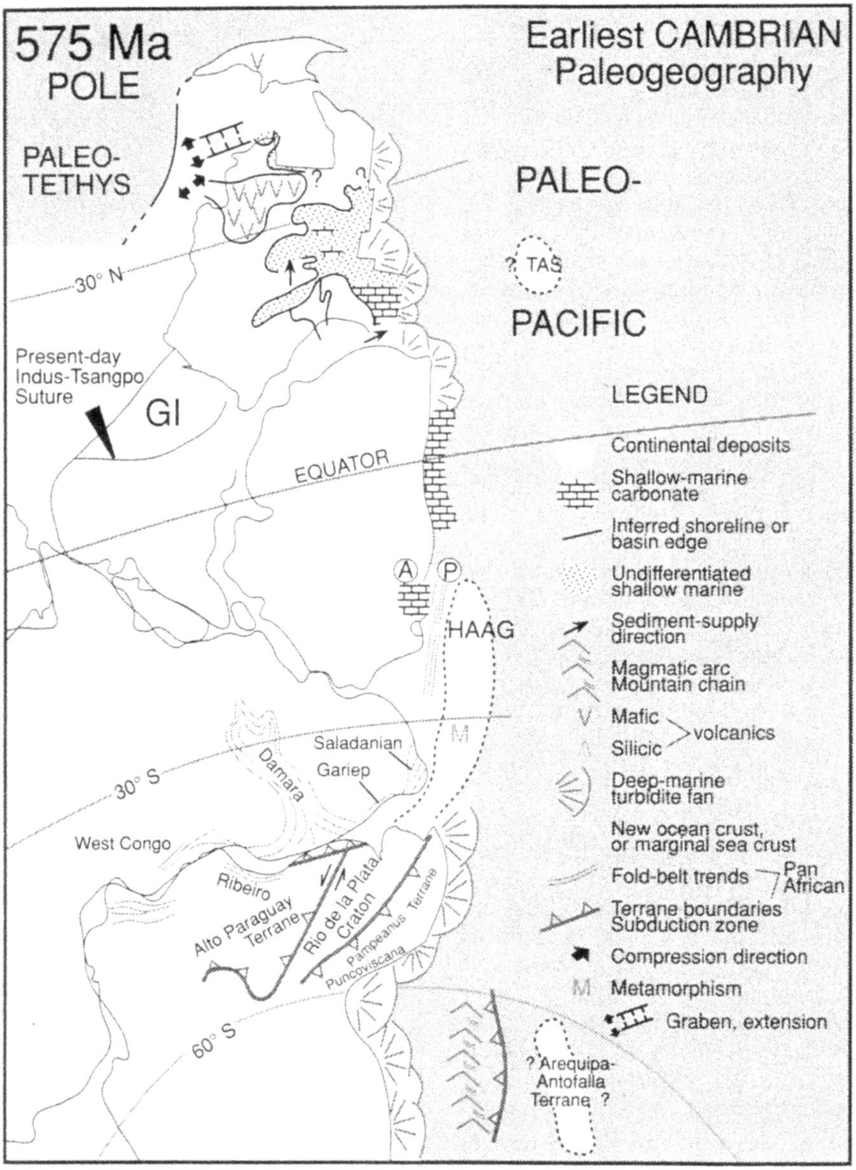

Fig 7: Earliest Cambrian palaeogeography of the Palaeo-Pacific margin located on a standard Gondwanaland reconstruction oriented so that the Palaeo-Pacific margin is along the length of the page. Oblique stereographic projection. Palaeolatitudes shown are derived from East Gondwanaland poles, principally from Australia (Li *et al.* 1990).

unpublished information. Suffice it to note that the continental breakup off Eastern Australia is commonly regarded as happening around the Precambrian/ Cambrian boundary and that shortly afterward (Late Cambrian) the region was deformed and uplifted. Thus, Ian Dalziel's reconstruction (Fig. 5) can be tested to see if rates of seafloor spreading are reasonable. For example, if there is only 25 Ma between breakup and the 500 Ma reconstruction, then even assuming a rapid rate of spreading, North America could only have rotated a few thousand km from Australia. Whether North America could go from off Eastern Australia prior to breakup to a position off South America within the time available needs to be checked. Ian, did you calculate the rates of opening in the Eo-Pacific Ocean?

Ian Dalziel: It's hard to calculate the rates of opening because of the uncertainty in the Cambrian time scale. I have done it as quantitatively as I can. If you put all the spreading into closure of a single Mozambique Ocean (cf. Figs. 5 and 6), you have to loose 8,000 km. In the reconstruction at the end of the Cambrian (Fig. 6), you have an 8,000 km-wide Pacific, and a 1,700 km wide (maximum) Southern Iapetus. The important point, then, is how long the Cambrian is. There are a lot of uncertainties, and I would like to ask the question of the audience here as to how long they think the Cambrian is. There are many rumours in the Northern Hemisphere about shortening the Cambrian. Dr. Mike McElhinny had a review of the Harland *et al.* (1990) timescale in which he commented that the Cambrian was probably significantly shorter. There was also an abstract by Dr. Bill Compston and his group on a 520 Ma late Tommotian tuff from North Africa, but there could, I understand, be problems with the ion-microprobe in Canberra when used in this mode. Without those numbers you cannot get a seafloor-spreading rate. You can only turn the information around the other way, and say that if you take present-day seafloor-spreading and convergence rates, and consider them reasonable, you need a certain length of time if you are not moving continents at non-uniformitarian rates.

3.6 *John Veevers, Macquarie University, Sydney:*

I would like to ask for more clarification, because, following on from what Ian Dalziel is saying, if this 525 Ma date pre-dates deposition of the main part of the Kanmantoo Group, then the inference is that the Eastern Australian margin opened only a little earlier. We then need to correlate that date with the stratigraphic time scale, because 525 Ma would be Late Cambrian on most current timescales (e.g. DNAG, Palmer 1983). That is not the point, however, because no fossils have been found in the Kanmantoo and it could be Late Cambrian. The problem is that from a lot of other information we can consider that breakup as being significantly earlier, so we come back to the timescale and look to an apparently very radical view that the Cambrian started about 530 Ma ago.

Chris Powell: We need clarification of the correlation between the biostratigraphic and radiometric timescales. Would anyone like to address the other part of the palaeontological connection, which is the distribution of the Ediacaran fauna in the reconstruction (Fig. 5) because that pre-dates the presumed break-up?

3.7 *Dr. Bob Findlay, Tasmania Department of Resources and Energy, Australia:*
Around the Late Proterozoic there is a continent-wide glaciation in Australia. Perhaps Ian Dalziel could comment on the age of his Neoproterozoic reconstruction, because in it the Equator runs through Australia?

Chris Powell: Before Ian Dalziel answers that question there are a couple of points I would like to add. First, the pole for the Late Proterozoic glaciation on the reconstruction (675 Ma, Fig. 5) has Australia in an equatorial position, with northern Africa and northern South America in the polar circle. This is a different latitudinal distribution from the younger interval around the Precambrian/Cambrian boundary for which palaeolatitudes on the map are drawn (Fig. 5). In younger interval, all the main parts of Neoproterozoic Gondwanaland are tropical, with no part lying poleward of 45° latitude. Possibly the global climatic gradients differed in the two configurations, with glaciation at 675 Ma occurring in Australia while it was near the Equator. Does anyone know what happened at the presumed polar latitudes when glaciation was occurring at the Equator?

Ian Dalziel: I have little to add, except to note that it all comes back to improving the palaeomagnetic resolution. All we are doing here is potentially providing a new framework in which to look at an existing problem.

Chris Powell: Are there not some glacials of Late Precambrian age in Africa?

Raphael Unrug: There are glacial deposits in

the Lufilian, and in the Damaran belt there are diamictites of disputed sedimentary origin, not necessarily glacial. But there are good dropstones in the Grand Conglomerate in Zambia and one of the explanations is that immediately preceding rifting there was uplift of the rift margins leading to local glaciation, with glaciers descending into the marine basin. In this way, vertical movements would be responsible for local centres of glaciation. This is different from the continent-wide glaciation Bob Findlay referred to.

3.8 *Peter Gresse (written communication):* Ediacaran Fossils, Precambrian/Cambrian Boundary, Late Proterozoic Glaciation and Post-550 Ma events. The Nama foreland basin was formed inboard of the Gariep and Damara orogens, i.e. on the W. margin of the Kalahari craton during collision and uplift in the orogens at ~650 Ma. The age of the Nama Group ranges from ~600 Ma at the base to ~540 Ma at the top. These sediments are syntectonic relative to the flanking orogens. The basal half of the Nama Group contain Ediacaran fossils - *Claudina* (in limestones) typical of the Late Precambrian. These sediments (shales and sandstones) also contain detrital micas dated at ~630 Ma and derived from the collision belts of the Gariep and the Damara. The upper half of the Nama Group contains the typical Cambrian trace fossil *Phycodes Pedum* and detrital micas dated at ~570 Ma. The whole Nama Group was deformed by cratonward-verging thrusting and folding from ~534 to 496 Ma. The Precambrian/ Cambrian boundary is therefore (in this case) >534 Ma and <570 Ma, i.e. ~540 - 550 Ma! The ±500 Ma event continued as uplift and cooling to ~460 - 480 Ma affecting the Damara, Gariep and Saldania belts as well as the Nama foreland basin.
Late Proterozoic glaciation occurs in the Damara and Gariep at ~700 and ~650 Ma and twice in the Nama Group, once at ~600 - 580 Ma and the other at the Precambrian/Cambrian boundary at ~540 - 550 Ma, both related to *glacial erosion features associated with limestones*.

3.9 *Hubert Miller, University of Munich, Germany:*
The relationship between Andean basement tectonics and the tectonics of Ribeira orogen (Brazil) and the Damara orogen (Africa) are very similar, so one should consider them in a unified sense. In NW Argentina, we have a southern part where metamorphism began at 580 - 570 Ma, whereas, in the north, sedimentation continued with a complete transition between the two areas. I think the transition reflects exposure of different tectonic levels. In the south, where the older part is exposed, there is a very thick succession of turbidity-current deposits, which are highly metamorphosed. In the north, younger parts of the turbidity-current succession are exposed. Judging from trace fossils, deposition probably lasted until 550-540 Ma, and metamorphism started at 535 Ma (K-Ar data). The first granite intrusions are in the north at 520 Ma; to south all plutonism is younger, between 480 and 450 Ma. In the Ribeira belt, newly published data (Söllner *et al.* 1991) indicate a very strong metamorphism up to granulite facies with charnockites at 590 to 560 Ma - essentially the same age as the NW Argentinan metamorphism. Intrusions at 520-480 Ma also correspond; an accompanying regional thermal overprinting at about 510 Ma marks the end of the orogenic history. Seen in their entirely, these data are very similar and as I remember the Damara orogen data are exactly the same, perhaps somewhat older in part. So, I return to the first theme, which is the question of when Gondwanaland formed. I think the Early Cambrian is when South America and Africa were joined. Irrespective of whatever happened in the Pacific Ocean, or in North America, there was a very pronounced subduction in NW Argentina and Bolivia where a small ocean must have existed.

4. EARLIEST CAMBRIAN PALAEOGEOGRAPHY OF THE PALAEOPACIFIC MARGIN

4.1 *Chris Powell:* I might take this opportunity to put up a map (Fig. 7) which shows the standard Late Palaeozoic-Early Mesozoic Gondwanaland reconstruction with palaeolatitudes and palaeogeography from the earliest Cambrian. The map is drawn to focus attention on the palaeo-Pacific margin of Gondwanaland. It shows that in Australia, and in Northern Victoria Land in Antarctica, there are platform limestones on the continents and deep-water turbiditic clastic prisms towards the ocean - indicating divergent or trailing-edge continental margins. In the Transantarctic Mountains there are widespread platform carbonates, but no evidence one way or the other of a continental margin. Further to the south, in the high-latitude regions on this reconstruction, there is a lot of tectonic activity to which Hubert Miller has just referred. The subduction and other processes relate to the amalgamation or consolidation of many terranes between the older Brazilian and African cratons. It appears as though the Pampeanus terrane, and everything to the west of it in modern-day coordinates, may still have been converging and colliding with the older Gondwanan nucleus extending from the Kalahari craton to Western Australia. In this reconstruction (Fig. 7), which is based on good palaeomagnetic data for Eastern Gondwanaland, the palaeo-Pacific margin of Gondwanaland is shown as extending through more than 110° of latitude. The reconstruction is quite different from the one shown by Ian Dalziel (Fig. 5), in which the correlative margin for Australia and East Antarctica would lie in western North America and its continuation along South America would have involved a sharp V-shaped re-entrant. Dalziel's reconstruction in the earliest Cambrian has the entire Palaeo-Pacific and Iapetus margins in the tropics, whereas this reconstruction shows that margin spanning equatorial to polar latitudes.

Hubert Miller: I wonder what was occurring between Africa and South America before, or around, the Precambrian/Cambrian boundary because, although the match of younger continental outlines of Africa and South America is the best of all the Gondwanan continents, the match of older geological features is not, apart from the very good fit of the Cape Fold Belt and the Sierra Australis. None of the Late Precambrian/Early Cambrian structures fit well. Very different groups are working in West Africa and Brazil, and few have seen each other's fold belts at first hand. I think it is possible there could be strike-slip faults which have disturbed the original connections between the two continents. Perhaps orogenies in Brazil and NW Africa were of transpressional type. There are extensive, very well-developed mylonite zones parallel to the continental margin, so we must think in terms of transpressional tectonics in this region.

4.2 *Michel Villeneuve, CNRS, France:*
I am very happy with this reconstruction (Fig. 7) because in the Rokelide belt (between Sierra Leone and Liberia) we have an intra-continental suture, probably dextral strike slip but with no oceanic crust. I would also like to see Florida attached to northwestern Africa as part of the Gondwanaland reconstruction.

Chris Powell: The shape and frame of this reconstruction (Fig. 7) was chosen for a different purpose, and we can add Florida as you suggest.

Zheng Xiang Li: I would like to make a comment on this map (Fig. 7), which has palaeolatitudes plotted from palaeomagnetic poles in East Gondwanaland only. The West Gondwanaland assembly has been put on mainly for illustration. According to the palaeomagnetic data (Li and Powell, this volume), West Gondwana had not joined with East Gondwana by that time (Early Cambrian). In a plot of palaeolatitude of Gondwanaland according to palaeomagnetic data (Fig. 8), East Gondwanaland is well controlled, but West Gondwanaland is controlled by only a few poles.

I would also like to make a point about the collision of East Africa with the western margin of East Gondwanaland. As palaeomagnetic data do not give a longitude control to a continent, and the polarity of the Early Cambrian poles of Africa is uncertain, there are many ways of plotting Africa in relation to East Gondwanaland. This figure (Fig. 8) just shows two of the possibilities. One would imply a sinistral movement along the collision zone, and the other implies a dextral movement.

From palaeomagnetism alone we can not distinguish between the two models.
Does the geology of the area enable these two closure models to be tested?

Chris Powell: If we are looking at an oblique closure of the ocean between the East and West Gondwanaland, is there any evidence of whether that might be a dextral or sinistral sense of closure, because in the palaeomagnetic data there is a requirement for a very big

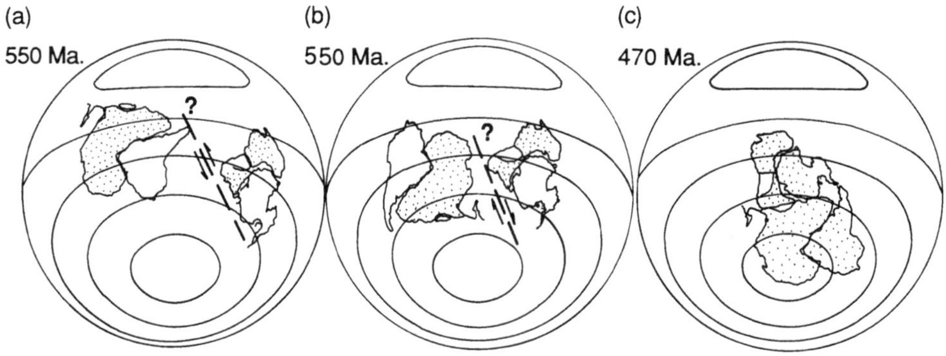

Fig 8: Palaeoreconstructions of Gondwanaland blocks according to palaeomagnetic data (Li & Powell, this volume). Stippled continents/cratons are those from which palaeomagnetic control is available; other continental blocks are located on geological grounds.

rotation?

Bob Findlay: From the palaeomagnetic data (Fig. 3), one can see that between 720 and 550 Ma Australia is at low latitude. The geological evidence, including striated pavements and tillites near Darwin, Northern Territory, and the Sturtian tillites near Adelaide, South Australia, indicates that Australia was glaciated at the same time during that interval. Can someone explain these observations?

Chris Powell: The problem of the Late Proterozoic glaciation is that it appears to have occurred in low latitudes. Glaciation in low latitudes is not unique to Australia. It could be related to a change in the inclination of the Earth's rotation axis to the ecliptic, so that the Equator does not receive as much solar radiation as at present.

4.3 *Johan Visser, University of The Orange Free State, South Africa*

We have done some work on the Late Proterozoic glaciations, and I have looked at them in Norway, Scotland, South America and in the Damara area. Almost everywhere they are low-latitude glaciations, and they are associated with carbonates. Up to now, this association has not been properly explained. The reason is that they are not typical lodgement tills; they appear to be some rain-out diamictites, and mostly debris flows, but with a clear glacial imprint on the stones.

4.4 *Ann Grunow, Ohio State University, U.S.A.*

I have a question about Ian Dalziel's reconstruction (Fig. 5). If we try to follow the Grenville Belt, bringing it down through Texas into the small area of Grenville rocks off East Antarctica (the Shackleton Range), is it possible that the small segment of East Antarctica in northern Dronning Maud Land was attached to Africa, instead of being part of East Antarctica in the Late Proterozoic? Later, during the closure of East Antarctica against Africa, could that little sliver have become attached to the East Antarctic craton, so that northern Dronning Maud Land is an exotic terrane accreted to East Antarctica? The other point is that this could provide an explanation for the Shackleton Range deformation, and also some of the plutonism found along the Dronning Maud Land and northern part of East Antarctica at that time.

Ian Dalziel: I think that idea needs to be explored. I don't have any firm answer because some of the evidence is hidden under the ice. I certainly don't believe there is any direct evidence of a suture, but, in taking a much closer look at the whole East Antarctica-Africa reconstruction and relationships from the Late Precambrian onwards, your suggestion is something to consider.

4.5 *David Macdonald, British Antarctic Survey*

I would like to make a couple of comments on the palaeogeography for the earliest Cambrian

(Fig. 7), with respect to what is now the Atlantic-ward end of the Transantarctic Mountains, including the Pensacola and the Ellsworth Mountains. The diagram implies that the area on the palaeo-Pacific side of the Transantarctic Mountains had silicic and mafic volcanism in the earliest Cambrian. This is not the case; the only volcanism in this region is Middle Cambrian or younger. The Union Glacier Volcanics in the Ellsworth Mountains are probably Middle Cambrian, the Gambacorta Formation in the Pensacolas, and the volcanic rocks of the Thiel Mountains are latest Cambrian or possibly earliest Ordovician. (*Editor's note:* Fig. 7 has been modified to remove the offending volcanics, which lay just above the P symbol, and south of the Equator). What you can say about the Cambrian palaeogeography, is that all through the Central Transantarctic Mountains along to the Pensacolas and Ellsworth Mountains there is a triad of volcanics, limestones and clastics, and the clastics tend to be reworked volcanics. The stratigraphy has a very rift-like signature to it, and my understanding of the geochemistry is that it does not have a subduction signature. If this is a rift margin, the simple-minded point is that something rifted off that margin during the Cambrian. Therefore, I don't think the passive margin sort of palaeogeography, shown for Eastern Australia and North Victoria Land is tenable for the Central Transantarctic Mountains area.

5. TASMANIA'S POSITION IN GONDWANALAND

5.1 *Chris Powell:* Since we are in Tasmania, it is appropriate that the last theme for discussion should be the position of Tasmania along the margin of Gondwanaland. Figure 7, which was drawn a couple of years ago, has Tasmania adrift in the Palaeo-Pacific in earliest Cambrian time. The evidence for that has, I think, been strengthened by presentations at this conference. Papers have shown that the Wilson terrane in North Victoria Land is directly on strike with, and has the same characteristics as, the Kanmantoo Fold Belt in Australia. Therefore, Tasmania, which now lies on the outboard or oceanward side of the Wilson-Kanmantoo belt is anomalous. Other than in Tasmania there is no known exposed Precambrian in the Tasman Fold Belt, though people debate whether Precambrian is present in the subsurface. The striking feature of the geological history of Tasmania, is the abundance of Late Precambrian dates indicating metamorphism, which is quite atypical of what we see in the Australian cratonic margin near Adelaide. The Adelaide region was a tectonically quiet margin from at least ~900 Ma onwards. The kind of metamorphic history in Tasmania is much more typical of what happened in southern South Africa and South America, where there was a lot of deformation and metamorphism in the 800 to 650 Ma interval, and probably younger. So I raise the questions, 'Where would you put Tasmania in a Late Precambrian reconstruction? Did it come from South America? Which features correlate and which don't?'

5.2 *Nick Turner, Tasmania Department of Resources and Energy, Australia*
I would just like to say that the stratigraphy of Tasmania in that interval looks quite like the palaeogeography shown in Fig. 7, with shelf carbonates overlying shallow-water clastics, and passing eastward into an extensive turbidite fan. The hinterland of that system, called the Rocky Cape Association in Tasmania, is quite similar to rocks that occur within the Adelaidean of South Australia. So, in at least those broad terms, there are substantial similarities.

Chris Powell: That point is accepted, but you could make the same correlation with any number of regions along the Transantarctic Mountains. It is the same platform succession, so where you place Tasmania along the margin is an open question. What is not the same is the 800 to 650 Ma metamorphic dates from the Precambrian slices that lie outboard, or east of, the Rocky Cape region. The core of Tasmania has a lot of metamorphic dates that we do not see in the Adelaide region.

Nick Turner: The Tyennan block in central Tasmania has a great range of metamorphic dates. I find it very difficult to be confident about when in that range a given tectonic event is occurring. Do you assume that every date you have in an extremely complex set of data actually measures a discrete geological event, or do you say that this big spread of data is approximating to some extended geological event?

Chris Powell: I do not suggest that there is a series of discrete geological events corresponding to each metamorphic date; what I do say is that there are dates which span the whole interval, and the same range of dates comes out of the South American and South African regions. If you look at the data at its broadest level, it indicates that deformation and metamorphism occurred throughout the 800 to 650 Ma interval. We have difficulty finding that kind of geological history in southeastern Australia, whereas we have no difficulty finding it in South America. That is the level of the correlation I am looking at.

Nick Turner: With the Tasmanian dates we have a small concentration around 800 Ma, and then a spread from 700 Ma through to 540 Ma.

Robin Oliver: We seem to be thinking of dates in the Adelaide region as starting with the Kanmantoo, but before the Kanmantoo we have the Adelaidean, a meta-sedimentary succession of Late Proterozoic age. There has been an age of 860 Ma determined on the Houghton Diorite, a constituent of the Baroma Complex, which is basement to the Adelaidean in the Adelaide region.

Chris Powell: You are not talking about regional metamorphic conditions, which is what the Tasmanian Precambrian rocks have - lots of deformation and metamorphism in mid-crustal levels. You do not see that in the Adelaide region, which, by most interpretations, is a relatively stable area with rifting in the latest Precambrian.

5.3 *Roye Rutland, Bureau of Mineral Resources, Canberra*

Just a minor comment on your proposition that the Tasmanian metamorphic ages do not occur elsewhere in Australia. These ~700 Ma metamorphic ages do occur in the Petermann Ranges in Central Australia, and I don't see any difficulty with seeing the Adelaidean as a marginal-basin passive margin with Tasmanian orogenic zone lying outboard of it (see Preiss *et al.*, 1981, esp. pp. 347-349). In the context of the spatial correlations between North America and eastern Australia now being suggested I would also like to draw attention to Reg Sprigg's original comparison (1952) between the Adelaidean and Beltian passive margins.

Chris Powell: The orogen Professor Rutland is referring to is a zone which runs right through Central Australia and along the northern margin of the Archaean Pilbara block. The geology of the broader region includes a platform succession, starting about 890 to 850 Ma, and including quartzites, carbonates and evaporites. The platform succession was disrupted by folding and thrusting, with a north-south shortening direction, between 700 and 600 Ma. An interpretation of the Western Australian part has been published recently by Myers (1990).

6. CONCLUDING REMARKS

6.1 *Prof. Dr. Ulrich Rosenfeld, University of Munster, Germany*

We have discussed today one of the most significant aspects of Gondwana geology but I want to draw your attention to the point that there are other problems. One is that we need very serious reinvestigations of Mesozoic volcanics. There are some areas in Gondwana where volcanism in the Early Mesozoic seems to be related to the onset of breakup. There is also volcanism related to Palaeozoic subduction, and we have no differentiation between the two.

The second problem is that we need more intense investigations of post-Gondwana sedimentary successions, especially in the Triassic and Lower Jurassic, for the same reasons. The Palaeozoic sedimentation finished with development of rift basins, and it seems that some of these rift basins were the very early start of breakup. In many regions, we have no differentiation between the last stages of the Palaeozoic subduction history and the younger breakup history.

Chris Powell: There are undoubtedly many problems we have not addressed and which will form the basis of future discussion sessions. This discussion session was set up to examine the problems associated with the assembly of Gondwanaland, and I believe we have identified some very important questions. The possibility that Gondwanaland did not come together until the end of the Cambrian seems to be very real, and it also seems, from the limited palaeomagnetic data, that there are very large rotations implied between Africa and East Gondwanaland in the latest Precambrian-Cambrian interval during the final merging. Additional possibilities are that parts of North America may fit into the palaeogeographic scheme. These are all questions I hope we can address more clearly when we next convene. In closing the open forum, I would like to thank all who have participated, and especially the convenors of Gondwana VIII who allocated this 2-hour session with no competing symposia. The success of this discussion has been underpinned by the fact that so many of the participants at Gondwana VIII were able to attend.

Finally, I would like to thank the convenors of the three symposia which preceeded this forum for their whole-hearted co-operation in arranging their sessions to lead into this free-ranging discussion. IGCP 288 was a co-initiator of the forum.

REFERENCES

Burrett, C. & R. Richardson 1980. Trilobite biogeography and Cambrian tectonic models. *Tectonophysics* **63:** 155-192.

Cooper, J.A., R.J.F. Jenkins, W. Compston & I.S. Williams, 1992. Ion-probe dating of a mid-Early Cambrian tuff in South Australia. *Journal of the Geological Society* (in press).

Dalziel, I.W.D. 1991. Pacific margin of Laurentia and East Antarctica/Australia as a conjugate rift pair: Evidence and implications for an Eocambrian supercontinent. *Geology* **19:** 598-601.

Dalziel, I.W.D. 1992. Antarctica: a tale of two supercontinents. *Annual Review of Earth & Planetary Science* (in press).

Harland, W.B., R.L. Armstrong, A.V. Cox, L.E. Craig, A.G. Smith & D.G. Smith 1990. *A Geologic Time Scale 1989* Cambridge, Cambridge University Press, 263p.

Hartnady, C. 1986. Was North America ('Laurentia') part of south-western Gondwanaland during the late Proterozoic era? *Suid-Afrikaanse Tydskrif vir Wetenskap* **82:** 251254.

Hoffman, P.F. 1991. Did the breakout of Laurentia turn Gondwanaland inside-out? *Science* **252:** 1409-1412.

Katz, M.B. 1985. The tectonics of Precambrian cratons - mobile belts: progressive deformation of polygonal miniplates. *Precambrian Research* **27:** 307-317.

Li, Z.X., C.McA. Powell, G.A. Thrupp & P.W. Schmidt 1990. Australian Palaeozoic palaeomagnetism and tectonics - II. *Journal of Structural Geology* **12:** 567-575.

Moores, E.M. 1991. Southwest U.S. - East Antarctic (SWEAT) connection: a hypothesis. *Geology* **19:** 425-428.

Myers, J.S. 1990. Precambrian tectonic evolution of part of Gondwana, southwestern Australia. *Geology* **18:** 537-540.

Palmer, A.R. 1983. The Decade of North American Geology 1983 Geological Time Scale. *Geology* **11:** 503-504.

Preiss, W.A., R.W.R. Rutland & B. Murrell. 1981. The Stuart Shelf and Adelaide Geosyncline. *In:* D.R. Hunter (ed) *Precambrian of the Southern Hemisphere,* 327-360. Elsevier, Amsterdam.

Söllner, F., B. Lammerer & K. Weber-Diefenbach 1991. Die krustenentwichlung in der Küstenregion nördlich von Rio de Janeiro/Brasilien. *Münchner Geologische Hefte* **4:** 101 S.

Sprigg, R.C., 1952. Sedimentation in the Adelaide Geosyncline and the formation of the continental terrace. *In:* Glaessner, M.F. & R.C. Sprigg (eds) *Sir Douglas Mawson Anniversary Volume,* 153-159. University of Adelaide, Adelaide.

Richter, R. & M. Richter 1941. Das Kambrium am Toten Meer und die älteste Tethys. *Abhandlungen der Senkenbergischen Naturforschenden Gesellschaft* No. **440:** 1-50, Frankfurt a.M.

Trurnit, T.P. 1989. Sequence of mineral deposits related to the theory of eastward migrating global tectonic megacycles. *Global Tectonics and Metallogeny* **3:** 125-158.

Unrug, R. 1991. The supercontinent cycle and Gondwanaland assembly: component cratons and suturing events timing. *J. Geodynamics* (submitted).

Evolution

Gondwana Eight, Findlay, Unrug, Banks & Veevers (eds) © 1993 Balkema, Rotterdam. ISBN 90 5410 304 3

Review of the Early Permian flora of the Nychum Volcanics north of Chillagoe, north Queensland

J.F. Rigby
Department of Resource Industries, Brisbane, Qld, Australia (Presently: School of Geology, QUT, Brisbane, Qld, Australia)

ABSTRACT: A Glossopteris flora containing predominantly Glossopteris nychumiensis sp. nov., from a tuff horizon in the Nychum Volcanics, is described. Other significant species found include Austroannularia subcircularis Rigby, Bergiopteris archangelskyi Rigby, Neomariopteris lobifolia (Morris) Maithy and Walkomiella australis (Florin) Florin. The enclosing volcanic rocks have been isotopically dated at approximately 270 Ma thus giving an age for the flora for purposes of correlation. Bergiopteris archangelskyi is restricted to the Nueva Lubecka Formation of probably Early Permian age, from Patagonia, other species are endemic, or widely distributed.

1 INTRODUCTION

Geological mapping of the area around Chillagoe, northern Queensland, was undertaken by a joint party of geologists from the Bureau of Mineral Resources, Canberra, and the Geological Survey of Queensland, Brisbane, during 1958 and subsequently. The area is approximately 140 km west of Cairns (Figure 1).

The geological parties recognised three outcropping areas which were grouped into a single stratigraphic unit and named the Nychum Volcanics (see Arnold & Fawckner, 1980). Recent investigations have shown these three areas need to be divided into separate stratigraphic units (Bultitude & Domagala, 1988). This is supported by fossil floral evidence in the case of two units. The strata exposed around Nychum homestead contains an Early Permian flora, whereas that exposed in the most northerly area, along the Mitchell River (see Figure 1, Black et al.; 1972) contains a Late Permian flora. The flora discussed below is derived from the strata near Nychum homestead (Figure 1).

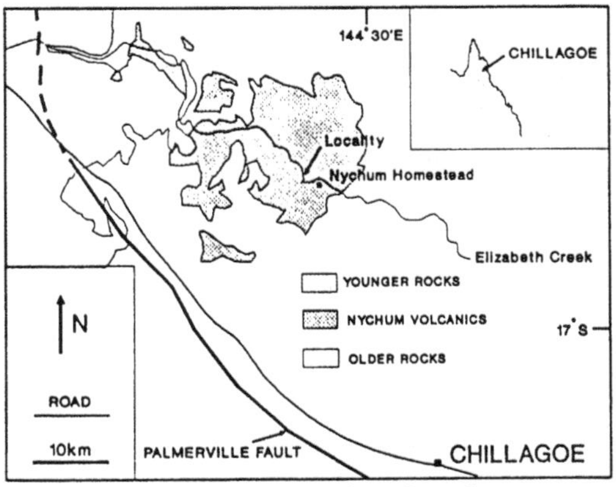

Figure 1. Locality map showing the collecting site.

2 PREVIOUS INVESTIGATIONS

White (1972) identified "Glossopteris indica Sch., Glossopteris communis Feist., Glossopteris fructification-Scutum sp., Cardiopteris polymorpha Goepp., Noeggerathiopsis hislopi (Bunb.), Palaeovittaria kurzi Feist., Gangamopteris cyclopteroides Feist., Walkomiella australis (Feist.) Florin, Sphenopteris polymorpha Feist., Annularia sp." from this area. None were described or figured; the list of names included the Laurasian Carboniferous genera Cardiopteris and Annularia mixed with typical Gondwana Permian genera. Black et al. (1972) concluded the age of the Nychum Volcanics was uppermost Carboniferous based on the fossils present and radiometric dating.

Balme (1973) drew attention to the use of the genus Cardiopteris for sterile fern material when Gondwanidium and some species of Neuropteridium, having similar foliage, were known from the Permian. Both of these generic names used in Gondwana countries are now known to be junior synonyms of Botrychiopsis Kurtz (Archangelsky & Arrondo, 1971). White (quoted by Bailey et al., 1982) accepted this and revised the identification of Cardiopteris polymorpha as Botrychiopsis plantiana (Carruthers) Archangelsky & Arrondo 1971. Bailey et al. (1982) gave a revised radiometric dating, proposing a Lower Permian age "close to 270 Ma", which is near the Artinskian-Kungurian boundary in Cowie & Bassett (1989) or Sakmarian in Harland et al. (1989).

Rigby (1988) gave a preliminary revised list of determinations: Paracalamites australis Rigby, Raniganjia sp., Neomariopteris lobifolia (Morris) Maithy, Botrychiopsis sp. nov., Glossopteris sp. A nov., Glossopteris sp. B nov.

This revision is based on the original collections made by officers of the Bureau of Mineral Resources, also the University of Queensland. The differences in names used by White (1972), Rigby (1988) and herein reflect taxonomic revision of various species, and undescribed specimens from the University of Queensland collection.

3 LOCALITY

All specimens discussed in this report come from the same locality, at or near locality 6 on the maps of Black et al. (1972, figure 1) and Bailey et al. (1982, figure 2). It is a plant-bearing, irregularly bedded, orange to buff tuff outcropping in Elizabeth Creek 2.5 km west-north-west of Nychum homestead, at grid reference 2226/81377, Bellevue 1:100 000 map sheet area (7764), at $16^{o}49\frac{1}{2}$'S, 144° 23'E.

One collection is housed by the Bureau of Mineral Resources (BMR), Canberra, it includes specimens CPC 27932 to CPC 27959 (formerly slabs numbered F 24227 to F 24243). The other is housed by the Geological Museum, University of Queensland (UQ) and includes specimens UQF 77357 to UQF 77364, UQF 79665, UQF 79666 and approximately 30 slabs under the locality number UQL 5492. The BMR collection was made by W.R. Morgan and I.H. Perrott in 1965 (Black et al., 1972, p. 191). Both collections were made at the same site, as slab UQF 77361 is counterpart to BMR specimens CPC 27932A/B and CPC 27933A. It is understood both collections were made at the same time.

4 CORRELATION AND DISCUSSION

Four species range throughout most of all of the Australian Permian, viz. Paracalamites australis, Neomariopteris lobifolia, Cordaites spatulata and Walkomiella australis.

Glossopteris nychumiensis is known only from this, the type locality. Austroannularia subcircularis was described from a specimen collected at an uncertain stratigraphic level in New South Wales (Rigby, 1990), possibly of Late Permian age. The only other occurrence of Austroannularia, viz. A. qubuensis, is in the Early Permian Qubu Formation of Tibet (Li et al., 1991). Bergiopteris archangelskyi was collected in the Nueva Lubecka Formation of Patagonia, Argentina. It occurs in the Ginkgoites exima Biozone of mid-Early Permian age (Archangelsky & Cúneo, 1984). The latter is the only precise correlation possible using this flora, both occurrences are probably of similar age, and palaeolatitude.

It is also significant that preliminary examination of other Permian floras from nearby areas of Queensland (work in progress) in the Mitchell River Volcanics and the Mt Mulligan Coal Measures lack the species Austroannularia subcircularis, Bergiopteris archangelskyi and Glossopteris nychumiensis. At present, it is thought this suggests these other floras may be significantly younger in age.

5 TAXONOMY

Austroannularia subcircularis Rigby 1990

Figure 2.2

1916. Annularia stellata (?), in Walkom: 233-234, pl. 25.

1990. Austroannularia subcircularis Rigby: 189-191.

This species is based on a specimen from the Permian of New South Wales collected many years ago, however, the name of the stratigraphic unit in which it was found had not been recorded.

White (1972) identified the BMR specimens as Annularia sp.. Rigby (1988) reidentified them as Raniganjia sp.. Rigby (1990) realized they differed significantly from Raniganjia and other hitherto named genera of Lower Gondwana sphenopsid leaf whorls so included them in the new genus Austroannularia.

Specimens: CPC 27944, CPC 27952A-E, CPC 27953, UQF 77359 (Figure 2.2, 2 specimens), UQL 5492 (4 specimens), total 12 specimens.

Paracalamites australis Rigby 1966

Three examples of this very widespread species occur on UQF 77359. By association at different localities, apparently identical stems probably belonged variously to a number of sphenopsid genera, here it was probably Austroannularia.

Neomariopteris lobifolia (Morris) Maithy 1974

Fragments of fern fronds all belong in this Australian-Antarctic species, which is widespread, and occasionally common. White (1972) identified the BMR specimens as Sphenopteris polymorpha.

SPECIMENS: CPC 27940, CPC 27942, CPC 27945, CPC 27950, UQF 77358, UQL 5492 (2 specimens), total 7 specimens.

Bergiopteris archangelskyi Rigby 1991

Figure 2.1

1982 Botrychiopsis valida, Archangelsky & Cúneo, p. 159-161, pls I-II.

1991 Bergiopteris archangelskyi Rigby, p. 146, fig. 3

Pinnule clusters, very much larger than those of Botrychiopsis valida (Feistmantel) Archangelsky & Arrondo 1971, occur. These specimens originally were identified as Cardiopteris polymorpha by White (1972).

Archangelsky & Cúneo (1982) dated the Argentinian occurrence of this species based on their identification of it as Botrychiopsis valida, as equivalent to the Indian Karharbari Formation. Based on Rigby's (1991) redetermination as Bergiopteris archangelskyi, this correlation no longer can be maintained. Archangelsky & Cúneo (1984) give the age of the Nueva Lubecka Formation in Patagonia as the older part of the Early Permian, possibly starting in the latest Carboniferous. The presence of a more warm climate genus, Bergiopteris, in southern Patagonia presents no problems due to the movement of Patagonia from adjacent to the coast of Peru in the Carboniferous to more or less its present position during the Late Permian (Valencio & Vilas, 1985) or Mesozoic (Martinez, 1980).

The palaeolocation of the occurrence of Bergiopteris in the Nychum Volcanics was in the lower latitudes during the Permian, therefore in a region of warmer climate which gives credence to biological and age correlation between the Nychum Volcanics and the Nueva Lubecka Formation.

SPECIMENS: CPC 27947, CPC 27948A-B (figure 2.1), CPC 27949, UQF 77359, UQF 77363, UQL 5492 (3 specimens).

Glossopteris nychumiensis sp. nov.

Figures 2.3 - 2.6

DIAGNOSIS: Estimated as medium to large leaf with maximum width of up to 50mm, apparently oblanceolate, acute rounded apex, tapering base. Midrib strong, clearly defined with longitudinally aligned anastomosing ribbing, density up to 6 veins per 1mm, ribs appear to be about half the width of secondary veins. Secondary venation of the communis type, gently arched, degree of arching 5°, marginal angle 50°-60°, density of secondary venation 20 per 10 mm; cross connections and anastomoses more common adjacent to midrib where the first series of areoles tend to be polygonal, sometimes no further cross connections between the central areoles and the margin through a region of strongly parallel venation; fibres absent; cuticles unknown as all type material are impressions.

HOLOTYPE: CPC 27934 (Figure 2.3).

PARATYPES: CPC 27932A,B,C (Figure 2.5 in part), CPC 27934, CPC 27936A-G, CPC 27937A-F, CPC 27938 (Figure 2.4), CPC 27943(?), CPC 27951A-B, UQF 77360 (Figure 2.6), UQF 77361, UQF 79665, UQF 79666, UQL 5492 (13 slabs). Total of 181 leaf fragments, no entire leaves known.

LOCALITY: See Introduction.

DESCRIPTION: The holotype is approximately the middle half of an oblanceolate leaf 48mm at its broadest, preserved length 95mm. This fragment shows the typical polygonal areola which occur along most of the midrib, but are infrequent towards the base (Figure 2.5). Only two complete apices were found, both were acute, rounded (Figure 2.4), a partial apex was slightly more acute.

One slab was covered with matted Glossopteris, all of which appear to belong in this species.

DISCUSSION AND COMPARISONS: All the specimens identified by White (1972, CPC specimens, each clearly labelled in the collection) as Glossopteris indica, G. communis, Gangamopteris cyclopteroides and Palaeovittaria kurzi appear to be different states of preservation of G. nychumiensis. The Gangamopteris form is an impression of the upper surface of a leaf where the midrib is masked as it projects only on the lower surface. She identified the leaves showing the lower surface as Glossopteris indica. These differ from the type specimen of G. indica (figures 40,41 of Rigby et al., 1980) in having more arched veins, and a more slender midrib, particularly basally where the leaf itself is more narrowly attenuate. Glossopteris nychumiensis has the more usual form of midrib for Glossopteris, namely a region of closely packed, coarser veins, so close they almost touch, with occasional cross connections between veins. Pigg (1990) found that anatomically the only difference between the secondary veins and vascular elements of the midrib was size, the latter were coarser.

White (1972) identified some leaves having imperfectly preserved midribs as Palaeovittaria kurzii, however all specimens herein identified as Glossopteris nychumiensis, including ones she identified as P. kurzii, have cross connections.

Leaves White (1972) identified as Glossopteris communis have a slightly more strongly arched secondary venation, but still lie within my circumscription of G. nychumiensis. The arching is not as pronounced, the venation is much more open and the leaf apex is more acute in our specimens than in the type of G. communis, figured by Chandra & Surange (1979, plate 1, figure 2).

Glossopteris sp.

Figure 2.7 (left side)

Specimen CPC 27957 is a fragment of a small, apparently obovate to elliptical leaf with an open, reticulate secondary venation, secondary veins usually more than 1mm apart. This is the specimen identified as G. browniana by White (1972). Glossopteris browniana has a marginal angle of 35^0-45^0, this leaf approaches 90^0.

Glossopterid fructification

White (1972) identified the fructification on specimen CPC 27939 as "Glossopteris fructification - Scutum sp", but insufficient is preserved to say whether it belongs in Scutum or in some other genus. It appears to have been an axially compressed, radially symmetrical organ.

Glossopterid scale leaves

Scale leaves similar to the scales of the female fructification Venustostrobus from India (Late Permian), Lidgettonia and Eretmonia both from India and Africa (Permian) occur on CPC 27955A ,B, CPC 27958, UQL 5492 (4 specimens). Generic identification depends on features not visible on these specimens.

Samaropsis sp.

Seeds on specimens CPC 27956 and UQL 5492 all belong to the same species; 4mm long by 2.5mm wide with a wing

0.4mm wide at base, 0.8mm wide in centre of seed, and 1.6mm wide at the apex. These dimensions differ from all other published descriptions of seeds from eastern Australia.

Samaropsis kurzii Leguizamon 1971 is twice as large and has a more uniformly wide wing throughout, not narrowing apically. It occurs in the Bajo de Veliz Formation, of (?) Late Carboniferous age, at Bajo de Veliz, northwestern Argentina.

Samaropsis criciumensis Millan 1967 is more obovate than elliptical, and the wing is more sharply curtailed at the base. It occurs in the Tatui Formation at Criciuma, southern Brazil, of Early Permian age.

Samaropsis goraiensis Surange & Lele 1957 is the same size, has a broader central body and narrower wing basally. It occurs in the Early Permian Talchir and Karharbari Formation of India.

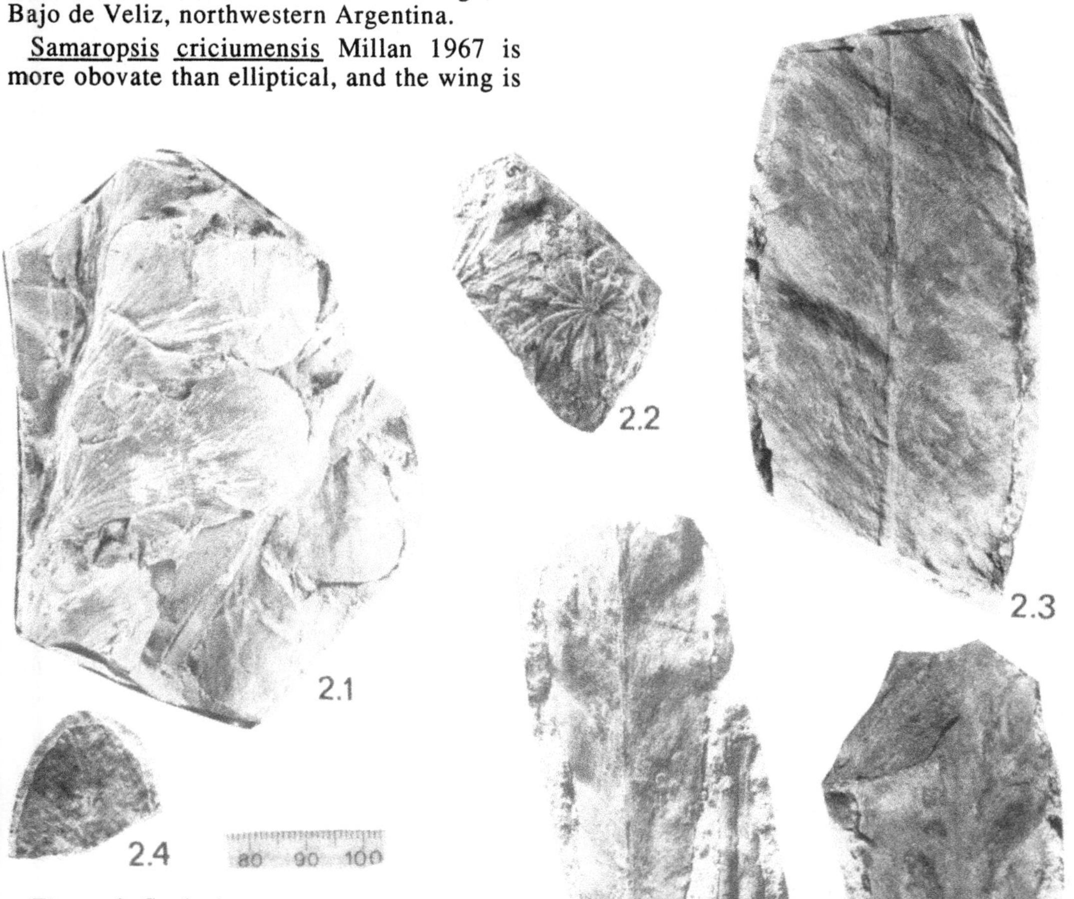

Figure 2. Scale in mm.
1. *Bergiopteris archangelskyi* Rigby. Fragments of two fronds superimposed. CPC 27948 A, B.
2. *Austroannularia subcircularis* Rigby. Complete whorl showing characteristic leaf drooping, and partial whorl. UQF 77359.
3-6. *Glossopteris nychumiensis* sp. nov.
3. Holotype, mid section of leaf. CPC 27954 A.
4-6. Paratypes.
4. Apical part of leaf. CPC 27938.
5. Paratypes. Leaf bases. CPC 27932 A, B.
6. Paratype. Lower part of leaf. UQF 73360.

Walkomiella australis (Feistmantel) Florin 1940

White (1972) identified a number of small fragments as this species, based on external morphology. This species is not common. It is linked with warm climates, mostly during the Late Permian.

SPECIMENS: CPC27941, CPC27946A,B.

Cordaites spatulata (Dana) Rigby, Maheshwari & Schopf 1980

White (1972) identified a few fragments of leaves as Noeggerathiopsis hislopii. Maheshwari & Meyen (1975) combined Noeggerathiopsis as a junior synonym of Cordaites keeping the Gondwana specimens in C. hislopii. Rigby et al. (1980) indicated that the species hislopii is a junior synonym of C. spatulata.

I have no supporting evidence, but I suspect when we eventually find the fructification of this species it will be one of the many so-called glossopterid fructifications. Pant & Verma (1964) found that the cuticles of Gondwanan species were very similar to Cordaites from Laurasia, however, no fructifications similar to Cordaianthus, the Laurasian Cordaites fructification, has ever been found in the Gondwanas.

SPECIMENS: CPC 27933A, B, CPC 27936, UQL 5492 (5 fragments).

ACKNOWLEDGEMENTS

Published with permission, Director-General, Department of Resource Industries, Queensland. J.M. Dickins and D.L. Strusz, Bureau of Mineral Resources, Canberra, for arranging the loan of specimens. J.S. Jell, Department of Geology and Mineralogy, University of Queensland, Brisbane, for arranging the loan of an undescribed collection of specimens. M.W. Sayers and M.D. McKillop, for photographic processing; Figure 1 was drafted by J.J. Draper; Figure 2 was prepared by J.W.Beeston; the manuscript was typed by Mrs D. Kinman; all from the Department of Resource Industries, Brisbane.

REFERENCES

Archangelsky, S. & Arrondo, O.G., 1971. Palaeophytologia Kurtziana III. 2. Estudio sobre el genero "Botrychiopsis" Kurtz (="Gondwanidium" Gothan) del carbonico y permico gondwanico. *Ameghiniana*, 8(3-4): 189-227.

Archangelsky, S. & R. Cúneo, 1982. Sobre la presencia del genero Botrychiopsis Kurtz en la formacion Neuva Lubecka, permico inferior de Chubut, Argentina. *Anais do II congreso Latinoamericano de Paleontologia*, 1: 157-167.

Archangelsky, S. & Cúneo, R., 1984. Zonacion del permico continental de Argentina sobre la base de sus plantas fosiles. *Memoria III Congreso Latino-americano de Paleontologia*: 143-153.

Arnold, G.O., & Fawckner, J.F., 1980. The Broken River and Hodgkinson Provinces. In Henderson, R.A. & Stephenson, P.J. (Editors): *The Geology and Geophysics of Northeastern Australia.* Geological Society of Australia, Queensland Division, Brisbane, 175-189.

Bailey, J.C., Morgan, W.R., & Black, L.P., 1982. Geochemical and istopic evidence for the age, orogenic setting and petrogenesis of the Nychum volcanic association, North Queensland. *Journal of the Geological Society of Australia*, 29: 375-393.

Balme, B.E., 1973. Correspondence. 'Age of a mixed Cardiopteris-Glossopteris flora'. *Journal of the Geological Society of Australia*. 20(1): 103-104.

Black, L.P., Morgan, W.R ., & White, M.E., 1972. Age of a mixed Cardiopteris-Glossopteris flora from Rb-Sr measurements on the Nychum Volcanics, north Queensland. *Journal of the Geological Society of Australia.* 19(2): 189-196.

Bultitude, R.J. & Domagala, J., 1988. Geology of the Bellevue 1:100 000 sheet area, northeastern Queensland - preliminary data. *Queensland Department of Mines Record* 1988/5: 1-121, 1 map .

Chandra, S. & Surange, K.R., 1979. Revision of the Indian species of Occurrence. *Monograph Birbal Sahni Institute of Palaeobotany*, 2: 1-291.

Cowie, J.W. & Bassett, M.G., 1989. International Union of Geological Sciences, 1989 global stratigraphic chart. Supplement to *Episodes*. 12(2): 1 page.

Florin, R., 1940. On Walkomia n. gen. A new genus of Upper Palaeozoic conifers from Gondwanaland. *Kungl Svenska Vetensskapsakakademiens Handlingar* (3) 18 (5): 3-23.

Harland, W.B., Armstrong, R.L., Cox, A.V., Craig,L.E., Smith A.G. & Smith, D.G., 1990. *A geologic time scale*. Cambridge Univ. Press, Cambridge. 263 pp.

Leguizamon, R.R., 1971. Palaeophytologia kurtziana III. 6. Révision de las semillas platispérmicas de Bajo de Los Velez (San Luis), ilustradas en el "Atlas de plantas fósiles de la República Argentina", (F. Kurtz, 1921). *Ameghiniana* 8: 259-264.

Li, X.-X., Rigby, J.F. & Wu,Y.-M. 1991. A biostratigraphical comparison between the Lower Gondwana floras of southern Xizang (Tibet) and Kashmir. *Gondwana Seven Proceedings*, 533-538.

Maheshwari, H.K. & Meyen, S.V., 1975. Cladostrobus and the systematics of cordaitalean leaves. *Lethaia* 8: 103-123.

Maithy, P.K., 1974. A revision of the Lower Gondwana Sphenopteris from India. *Palaeobotanist* 21(1): 70-80.

Martinez, C., 1980. Geologie des Andes Boliviennes. *Travaux et Documents de l'Orstom* 119: 1-352.

Millan, J.H., 1967. O gênero Samaropsis na flora do Gondwana do Brasil e dos outros países. *Boletim do Museo nacional do Brasil. Geologia* (n.s.) 32: 1-12.

Pant, D.D & B.K. Verma, 1964. The cuticular structure of Noeggerathiopsis Feistmantel and Cordaites Unger. *Palaeontographica* 115B: 21-44.

Pigg, K.B., 1990. Anatomically preserved Glossopteris foliage from the central Transantarctic Mountains. *Revue of Paleobotany and Palynology* 66: 105-127.

Rigby, J.F., 1966. The Lower Gondwana floras of the Perth and Collie Basins, Western Australia. *Palaeontographica* 118B: 113-152.

Rigby, J.F., 1988. The Permian flora of the Nychum Volcanics, North Queensland. *Geological Society of Australia Abstracts*. 21: 332.

Rigby, J.F., 1990. Austroannularia gen. nov., an asymmetrical Permian sphenopsid leaf whorl from Gondwanaland. *Palaeobotanist*. 37(2): 189-191.

Rigby, J.F. 1991. Bergiopteris and Botrychiopsis from the Late Palaeozoic of Gondwanaland. *XI Congrès International Stratigraphie et de Geologie du Carbonifère*, Compte Rendu 3: 143-148.

Rigby, J.F., Maheshwari, H.K. & Schopf, J.M., 1980. Revision of Permian plants collected by J.D. Dana during 1839 - 1840 in Australia. *Geological Survey of Queensland Publication* 376, *Palaeontological Papers* 47: 1-25.

Surange, K.R. & Lele, K.M., 1957. Studies in the Glossopteris flora of India - 6. Plant fossils from the Talchir Beds of South Rewa Gondwana Basin. *Palaeobotanist* 5 (2): 82-90.

Valencio, D.A. & Vilas, J.F., 1985. Evidence of a microplate in the southern Andes. *Journal of Geodynamics*. 2: 183-192.

Walkom, A.B., 1916. Note on a specimen of Annularia from near Dunedoo, New South Wales. *Memoirs of the Queensland Museum 5: 233-234*.

White, M.E., 1972. In Black, L.P.,Morgan, W.R., & White, M.E.: Age of a mixed Cardiopteris-Glossopteris flora from Rb-Sr measurements on the Nychum Volcanics, north Queensland. *Journal of the Geological Society of Australia*. 19(2): 190-191.

Gondwana Eight, Findlay, Unrug, Banks & Veevers (eds) © 1993 Balkema, Rotterdam. ISBN 90 5410 304 3

An unusual plant fossil from the Late Carboniferous of New South Wales

W. B. K. Holmes
'Noonee Nyrang', Wellington, N.S.W., Australia

ABSTRACT: An unusual plant fossil of unknown affinities is reported from the Late Carboniferous of New South Wales. *Burdekinia multiseptata* is a straplike organ, externally rather featureless, but characterised by a distinctive internal structure of regular transverse sphenopsid-like partitions which form rectangular sections. These sections are filled with a lattice-work of about six parallel rows of small chambers, the appearance of which resembles the internal structures formed by aerenchyma cells in the leaves of the extant marsh plants in the genus *Typha*.

1 INTRODUCTION

When compared with the rich pan-tropical floras preserved in the Carboniferous rocks of Europe and North America, the macrofloras of Gondwana during the Carboniferous Period are extremely lacking in diversity (Walkom, 1944; Gould 1975b; Morris 1975; Retallack, 1980; Rigby 1985). These floras reflect the deteriorating climate (Archangelsky 1981) and the isolation of Gondwana from Eurameria. Morris (1975) stated that for the whole of the Late Carboniferous period in New South Wales, only about ten species were present. Therefore the discovery of any new forms is considered noteworthy.

2 DESCRIPTIVE PALAEOBOTANY

Burdekinia multiseptata gen. et. sp. nov.
Figures A-G.

Combined diagnosis. Strap-like organs with parallel entire margins. Base, apex and lateral appendages if any, unknown. External surface smooth or with irregular longitudinal striations; well-spaced transverse grooves sometimes present but not obvious. Internal structure is a lattice-work formed from about six parallel longitudinal rows of small square or rectangular air chambers. At intervals, thicker transverse septa divide the organ into discrete rectangular sections.

Description. Some specimens are preserved as external surface impressions (Figs. A,B), some are casts which show details of the internal structure (Figs E-G) and occasional specimens show portions with both internal and external features (Figs C,D)
Specimens preserved as impressions of the external surface are from 6-25 mm in width and up to ca.0.5 metres in length. Specimens which show the internal structure are from 8-15 mm in width, with a maximum preserved length of 80 mm. All specimens have parallel entire margins and show no perceptible variation in width over the length preserved. No specimens show the base, the apex, branching or any lateral appendages. The external surface is smooth or with fine longitudinal striations, occasionally with more prominent irregular longitudinal flutings. Transverse grooves are sometimes present but are not conspicuous. The internal structure consists of about 6 parallel longitudinal rows of square or rectangular blocks each about 1 mm in length and enclosed by grooves ca.0.5 mm in width. Broader transverse grooves ca. 1 mm wide divide the organ into longitudinally oriented rectangular sections from 10-20 mm in length. On AMF 63824 (Fig.C) consecutive sections are 11, 12, 13, 12, 13 and 15 mm in length. On some of the broader specimens showing only the external surface, the transverse grooves are up to 50 mm apart.

Material. Holotype AMF 63824, Paratypes AMF 63825-33. Housed in the Plant Fossil Collection, Australian Museum, Sydney, N.S.W.

Type Locality. Vertical dipping beds of brown-grey weathering to white, siltstone exposed in a cutting on the western side of the Pacific Highway 8.7 km south of Bulahdelah on the southern side of Burdekin's Gap - stratigraphically near the top of the Isaac's Formation and below the Johnson's Peak Conglomerate - Late Carboniferous (Engel 1966). Grid Reference - 522984 Newcastle 1:250 000 Geological Series Sheet S1-56/2. Rigby (1985) recorded fragments of *Botrychiopsis ovata* and *Fedekurtzia* sp. from Burdekin's Gap but neither of these taxa are present at the type locality.

Derivation of Name. *Burdekinia* - from the type locality; *multiseptata* - referring to the partitions forming the internal lattice work structure.

3 DISCUSSION. I have used the noncommittal term "organ" to describe *B. multiseptata* as its relationship to a whole plant is not known. It may be a stem, a leaf or perhaps a root or rhizome. Specimens exhibiting internal structure, I believe, are the casts of the aerenchyma tissue formed by the entry of fine grained sediment into the airchambers after burial. Well known fossils having a similar origin include the internal casts of *Calamites* and other sphenopsid stems. Only the narrower examples of *B. multiseptata* show the internal structure. Perhaps the larger organs had better developed cortical tissues that were impervious to mud. Proof that the organs showing internal structure and those preserved only as external impressions belong to the same plant is demonstrated by two specimens (Figs C and D) which combine both types of preservation.

In life *B. multiseptata* must have been flattened in cross section. If originally cylindrical, on being compressed during fossilisation, the rows of airchambers would have been crushed and distorted. However all the available specimens show the rows of airchambers to be even in width and all the airchambers to be closely similar in dimensions.

The broader transverse septa suggest a sphenopsid relationship. However all sphenopsid stems are cylindrical in shape and none possess an internodal latticework of air chambers formed from aerenchyma tissue. Other fossils having transverse partitions are - *Artisia*, which is the contracted pith of *Cordaites* stems; and *Vertebraria*, the cylindrical root of the *Glossopteris* plant. In extant plants the grasses and many dicotyledonous angiosperms have nodal divisions of their stems.

Sifton (1945,1957) summarised the literature on airspace tissue in plants. The possession of aerenchyma is a striking characteristic of plants growing in marsh or aquatic habitats (Williams & Barber 1961). They noted that the mechanical requirement of anchoring a plant in a wet substrate is most economically met by having upper organs which are formed from a honeycomb structure to provide the greatest strength with the least amount of tissue. The presence of large airchambers in aerial leaves of marsh plants provides the greatest reservoir of oxygen enriched air and permits its free longitudinal diffusion to the submerged parts. Williams and Barber (1961) called these triple benefits accruing from the possession of aerenchyma, the "mechanical cum metabolic requirement", - a condition which at some stage in the establishment of any aquatic or marsh plant is paramount.

Leperonia articulata and members of the *Typha* genus are examples of present day marsh plants which exhibit aerenchyma structure similar to *Burdekinia multiseptata*. *L. articulata* has hollow terete stems with well developed septa at regular intervals. A thin layer of aerenchyma cells encloses air spaces inside the epidermis. The internal structure of the long strap-like leaves of *Typha* spp. consists of longitudinal parallel rows of 1 to 2 mm square air chambers, but there are no thicker transverse septa. No relationship is inferred between these plants and the fossil. Obviously they are all examples of parallel development of an internal structure that could cope with the physiological problems caused by a water-logged environment.

Nothorhacopteris argentinica leaves which occur on the same slabs as *B. multiseptata*, are often preserved in sediments associated with thin coal seams (Engel 1962). This indicates that the *Nothorhacopteris* plant grew in or around the coal swamps. Could *B. multiseptata* be the root organ of *Nothorhacopteris* analogous with *Vertebraria* and *Glossopteris* in the Permian coal swamps (Gould 1975a)?

Complete specimens or specimens preserved as coalified compressions or petrifactions may solve the position of this enigmatic fossil.

REFERENCES

Archangelsky, S. 1981. *Fedekurtzia*, a new Carboniferous frond from Gondwanaland and its fructification. American Journal of Botany 68 : 1130-1138.

Engel, B. 1962. Geology of the Bulahdelah-Port Stephens District. Journal Proceedings Royal Society, New South Wales 95 : 197-215.
Engel, B. 1966. Explanatory notes to accompany the Newcastle 1:250 000 Geological Sheet. Geological Survey of New South Wales.
Gould, R.E. 1975a. A preliminary report on petrified axes of *Vertebraria* from the Permian of Eastern Australia. In H.W.S. Campbell (ed.), Gondwana Geology. Canberra, ANU Press.
Gould, R.E. 1975b. The succession of Australian pre-Tertiary megafossil floras. Botanical Review 41 : 453-483.
Morris, N. 1975. The *Rhacopteris* flora in New South Wales. In H.S.W. Campbell (ed.) Gondwana Geology. Canberra, ANU Press.
Retallack, G.J. 1980. Late Carboniferous to Middle Triassic magafossil floras from the Sydney Basin. Bulletin Geological Survey of New South Wales 26 : 384-430.
Rigby, J.F. 1985. Aspects of Carboniferous palaeobotany in Eastern Australia. Dixième Congres International de Stratigraphie et de Geologie du Carbonifere Madrid 1983 Compte Rendu 4 : 307-312.
Sifton, H.B. 1945. Air space tissue in plants, 1. Botanical Review 11 : 108-143.
Sifton, H.B. 1957. Air space tissue in plants, 2. Botanical Review 23 : 303-312.
Walkom, A.B. 1944. Succession of the Carboniferous and Permian floras in New South Wales. Journal Royal Society of New South Wales 78 : 4-13.
Williams, W.T. & Barber, D.A. 1961. The functional significance of aerenchyma in plants. Society for Experimental Biology Symposium 15 : 132-144.

LEGEND FOR FIGURE.A-B, Impressions of external surface. A, AMF 78247, B, AMF 78248. C-D, Specimens showing both internal and external aspects. C, AMF 78249; D AMF 78250. E-G. Casts showing internal structure. E, AMF 78251, F, AMF 78252, G, Holotype AMF 78253. All illustrations natural size. The vertical or horizontal alignment of the illustrated fossils is not intended to represent the position of growth of the living plant.

Gondwana Eight, Findlay, Unrug, Banks & Veevers (eds) © 1993 Balkema, Rotterdam. ISBN 90 5410 304 3

Glossopterid megafossils in Permian Gondwanic non-marine biostratigraphy

Stephen McLoughlin
Department of Geology, University of Western Australia, Nedlands, W.A., Australia

ABSTRACT: Although glossopterids show strong species-level provincialism, widespread distribution of genera may enhance their role in Gondwana-wide correlations. Ovulate fructifications have diverse morphologies, well-defined stratigraphic ranges, and show a series of morphological transformations enabling construction of a tentative phylogenetic tree and generic key. Glossopteris leaves require extensive taxonomic revision but may be biostratigraphically valuable within individual basins. Strong provincialism may limit their use for inter-continental correlation. Other glossopterid organs (e.g., roots, scale leaves, pollenate fertiligers, and wood) are less stratigraphically useful owing to their relative uniformity throughout the Permian, across Gondwana, and between natural species.

INTRODUCTION

Glossopteris has long been noted as characterizing Permian Gondwanic strata (e.g., Feistmantel, 1890; Arber, 1905). The biostratigraphic potential of Ottokar Feistmantel's voluminous studies on glossopterids during 1876-1890 was, unfortunately, neglected for many years after Arber (1905) revised the Gondwanic Permian flora and reassigned Glossopteris leaves to just 13 broadly defined species. Most subsequent workers (until the 1960s) followed Arber's scheme incorporating superficially similar leaves from throughout the Permian in those few species. Intensive studies of fructifications and continuing re-evaluation of Glossopteris species since the 1950s have rekindled the biostratigraphic value of Gondwanic Permian plant megafossils. Nevertheless, certain limitations to their use remain. Foliage and reproductive organs are chiefly composed of delicate tissues and identifiable remains are predominantly restricted to non-marine, low energy, lowland depositional environments. Being commonly preserved as impressions, however, plant megafossils can often be identified from rocks too weathered or metamorphosed to yield palynomorphs; nor do they require lengthy processing or expensive equipment for analysis. Gondwanic (and other) basins were affected by lowering sea levels towards the end of the Permian (Ross & Ross, 1987a) and commonly hosted extensive non-marine deposits. Hence, plant megafossils may contribute significantly to biostratigraphic subdivision of these sequences. Their relevance is enhanced due to the occurrence of substantial economic coal reserves in Permian strata Gondwana-wide.

Correlations between terrestrial Gondwanic sequences and the Permian stages of Harland et al. (1990) remain equivocal. The relative ages of strata given here derive from numerous intra- and extra-Gondwanic palynological and invertebrate studies [e.g., Dickins (1976), Waterhouse (1976), Anderson (1977), Foster (1979, 1982), Archangelsky et al. (1980), Archangelsky & Marques-Toigo (1980), Balme (1980), Truswell (1980), Ross & Ross (1987b)].

LEAVES

Previous Australian Permian phytostratigraphic zonations (Gould, 1975; Retallack, 1980; Rigby, 1983) have offered only broad subdivisions of the system but further revision of the Australian glossopterids may provide the taxonomic basis for substantial refinement of these schemes. Workers elsewhere (Rösler, 1978; Anderson & Anderson, 1985) have also applied glossopterids to regional biostratigraphy. Leaves in the glossopterid alliance include Glossopteris Brongniart, Gangamopteris McCoy, Rubidgea Tate, Palaeovittaria Feistmantel, Belemnopteris Feistmantel, Euryphyllum Feistmantel, Rhabdotaenia Pant, Pteronilssonia Pant & Mehra, Surangephyllum Chandra & Singh, and Ilexoidephyllum Rohn & Rösler. Their affinities are based on similarities in form, venation, fructifications, and cuticular morphology (Pant, 1982). Gondwanic Cordaites (=Noeggerathiopsis) leaves may also belong to this alliance on the basis of their similarity to early non-anastomosing glossopterid leaves and their possible affinity to the fructification Arberia White emend. Rigby, here deemed to be ancestral to later undoubted glossopterid fertiligers.

Gangamopteris, at least in Australia, characterizes the Early Permian (Rigby, 1983). Elsewhere, Gangamopteris has occasionally been reported from the Late Permian (Srivastava, 1956; Maithy, 1974; Rösler, 1978) but critical examination of illustrated specimens of this age invariably suggests their reassignment to Glossopteris. Rubidgea and Euryphyllum, likewise restricted to the Early Permian (Lele, 1977), are uncommon and appear to be only slight morphological variants of Gondwanic Cordaites or Palaeovittaria leaves (Anderson & Anderson, 1985). Palaeovittaria lacks anastomosing veins and has a midrib only in the lower part of the leaf. Palaeovittaria-type leaves conceivably evolved more than once by coalescence and strengthening of the basal parallel veins in Cordaites- or Euryphyllum-type leaves. Hence, the Early Permian Palaeovittaria leaves from southern Africa may not be closely related to Indian forms which are restricted to the Late Permian.

Glossopterid leaf genera with more radically modified lamina and venation patterns (Belemnopteris, Surangephyllum, Ilexoidephyllum, Rhabdotaenia, and Pteronilssonia) characterize Late Permian strata, particularly the Capitanian-Lopingian Raniganj (India), Estcourt (South Africa), and Buckley Formations (Antarctica) and their equivalents. This Late Permian variation in glossopterid lamina morphology was concurrent with significant diversification amongst the fructifications.

Glossopteris is recognized from the base of the Permian although the presumably related taeniate disaccate pollen (Gould & Delevoryas, 1977) first enters the Gondwanic sequence in sediments assigned to the latest Carboniferous (Foster, 1983; Powis, 1984). The Late Carboniferous age assigned to Glossopteris by Black et al. (1972) has been adjusted to Early Permian on the basis of revised radiometric data (Bailey et al., 1982) and associated plant fossils (Rigby, this volume). Glossopteris reaches dominance over Gangamopteris during Kungurian time [Barakar Formation (India) and equivalents] and persists until its replacement by more xerophytic groups with the onset of redbed deposition [Rewan Group (Australia), Panchet Group (India) and equivalents] shortly before the end of the Permian. Reports of Glossopteris from younger rocks both within and beyond Gondwana (Johnston, 1887; du Toit, 1927; Harris, 1932; Delevoryas, 1969) have invariably been found to represent non-glossopterid or stratigraphically misplaced specimens (Delevoryas & Person, 1975; Meyen, 1987; Anderson & Anderson, 1989). Only in India have significant numbers of glossopterid leaves been reported from apparently Triassic strata (Pant & Pant, 1987) although they are not associated with other glossopterid organs and their stratigraphic positions are uncertain due to local faulting. Revisions of Glossopteris in India during the past two decades have established over 70 species. Most species are restricted to single formations (Chandra & Surange, 1979) thus providing the potential for zonation of the largely non-marine Permian sequences of peninsula India. Anderson & Anderson (1985), however, noted that a confused taxonomy has arisen through mis-identification of superficially similar leaves from different continents resulting in spurious stratigraphic correlations. Recent studies in the Bowen Basin, Queensland, revealed no Glossopteris species in common with the Indian or South African successions. While Glossopteris species may prove useful for local biostratigraphy, distant intra-Gondwanic correlations may have to rely on the regional and stratigraphic distribution of genera.

OVULATE FRUCTIFICATIONS

Recent collections from several Late Permian formations in the Bowen Basin (McLoughlin,

1990a, b, c) have revealed a wealth of diverse glossopterid fructifications. Glossopterid origins remain unclear although a cordaitalean ancestry has been suggested based on fructification structure (Schopf, 1976). A tentative phylogenetic tree is proposed for glossopterids (Text-fig. 1) based on ovulate fructification morphotrends throughout the Permian. This tree attempts to minimise repetitive morphogenesis of features and illustrates the development of four distinct groups (families of Anderson & Anderson, 1985). Chronostratigraphic ranges of fructification genera (Text-fig. 2) are based on an extensive review of the Gondwanic literature (McLoughlin, 1990c). Difficulties arise in correlating non-marine Gondwanic sequences with the standard international stages due to the absence of marine invertebrate index taxa (Anderson, 1973, 1977; Rösler, 1978; Foster, 1983) thus generic ranges may be subject to revision.

Arberia [syn. Dolianitia Millan (see Rigby, 1978)] is here regarded as the ancestral glossopterid fertiliger. Though not yet found attached to leaves, it has been recorded in association with Cordaites and Gangamopteris (Rigby, 1972; Anderson & Anderson, 1985). Its range is chiefly Asselian-Artinskian although a single modified species may have lingered into probable Ufimian times (upper Ecca Group equivalents) in southern Africa (Rösler, 1978; Rayner & Coventry, 1985). Planation and fusion of arberioid ovuliferous stalks may have produced loosely capitulate fructifications such as Ottokaria zeilleri Pant & Nautiyal. Ottokaria, the earliest capitulate glossopterid fructification, is chiefly an Early Permian genus although it ranges from the ?Asselian [Talchir Formation, India; Itararé Subgroup, South America (Virkki, 1938; Cazzulo-Klepzig et al., 1980)] to the ?Capitanian-Lopingian [Raniganj Formation, India; and Black Alley Shale, Australia (Banerjee, 1978; McLoughlin, 1990a)].

Plumsteadia Rigby, including the synonyms Derbyella White, Lanceolatus Plumstead, Cistella Plumstead, Gonophylloides Maheshwari, Fetura Benecke, Scopus Benecke, and Plumsteadiostrobus Chandra & Surange based on their equivalent compact elliptical polysperms with short stalks, bulbous seed scar imprints, and entire equi-width wings, is a common and possibly polyphyletic genus. It probably derived from Ottokaria-type fertiligers, arose in ?Sakmarian-Artinskian times (Archangelsky et al., 1981), and persisted to the ?Capitanian-Lopingian (Raniganj and Kamthi Formations, India) or perhaps the earliest Triassic (Panchet Formation of Bose et al., 1977).

Scutum Plumstead incorporates Pluma Plumstead, Venustostrobus Chandra & Surange, and Estcourtia Anderson & Anderson as all share ovate to elliptical polysperms with equi-width crenulate wings. Hirsutum Plumstead, the senior synonym of Elatra Appert, has elliptical to ovate receptacles with apically broadened wings. Scutum and Hirsutum appear during the Artinskian (Vryheid Formation, South Africa, and equivalents). Hirsutum disappears by the early Late Permian (Volkrust Formation, South Africa), whereas Scutum-type fertiligers persisted into the ?Capitanian-Lopingian [Raniganj Formation, India; Estcourt Formation, South Africa (Chandra & Surange, 1977; Anderson & Anderson, 1985)].

Capitanian-Lopingian times saw a diversification of capitulate fructification genera and species. Austroglossa Holmes and Satsangia Srivastava & Maheshwari are presently known only from Australia (Illawarra Coal Measures) and India (Gopad River area) respectively. The Triassic age for Satsangia campanulata Srivastava & Maheshwari 1973 should be regarded as questionable pending further stratigraphic studies on the type locality at Nidpur. Dictyopteridium Feistmantel ex Zeiller emend. McLoughlin [syn. Isodictyopteridium Rigby (McLoughlin, 1990a)] is potentially a useful index genus for Capitanian-Lopingian strata throughout Gondwana. Probably derived from elliptical Plumsteadia-type fructifications, Dictyopteridium is one of the most abundant and widely recognized glossopterid fruits with species in South Africa, India, Australia, and Antarctica. Senotheca Banerjee, recorded from India and Australia, is likewise a Capitanian-Lopingian genus although Zeiller's (1902) 'Cycadites (?) sp.', from the Early Permian Karharbari Formation, resembles a detached Senotheca polysperm. Zeiller's material from the Karharbari Coalfield was was collected by M.W. Ward and the precise formation of origin for his Senotheca-like specimen is questionable.

Individual Jambadostrobus Chandra & Surange polysperms resemble Plumsteadia suggesting a phylogenetic affinity. The Jambadostrobus fertiliger, however, differs in having several polysperms attached per leaf. Jambadostrobus fertiligers are here considered to

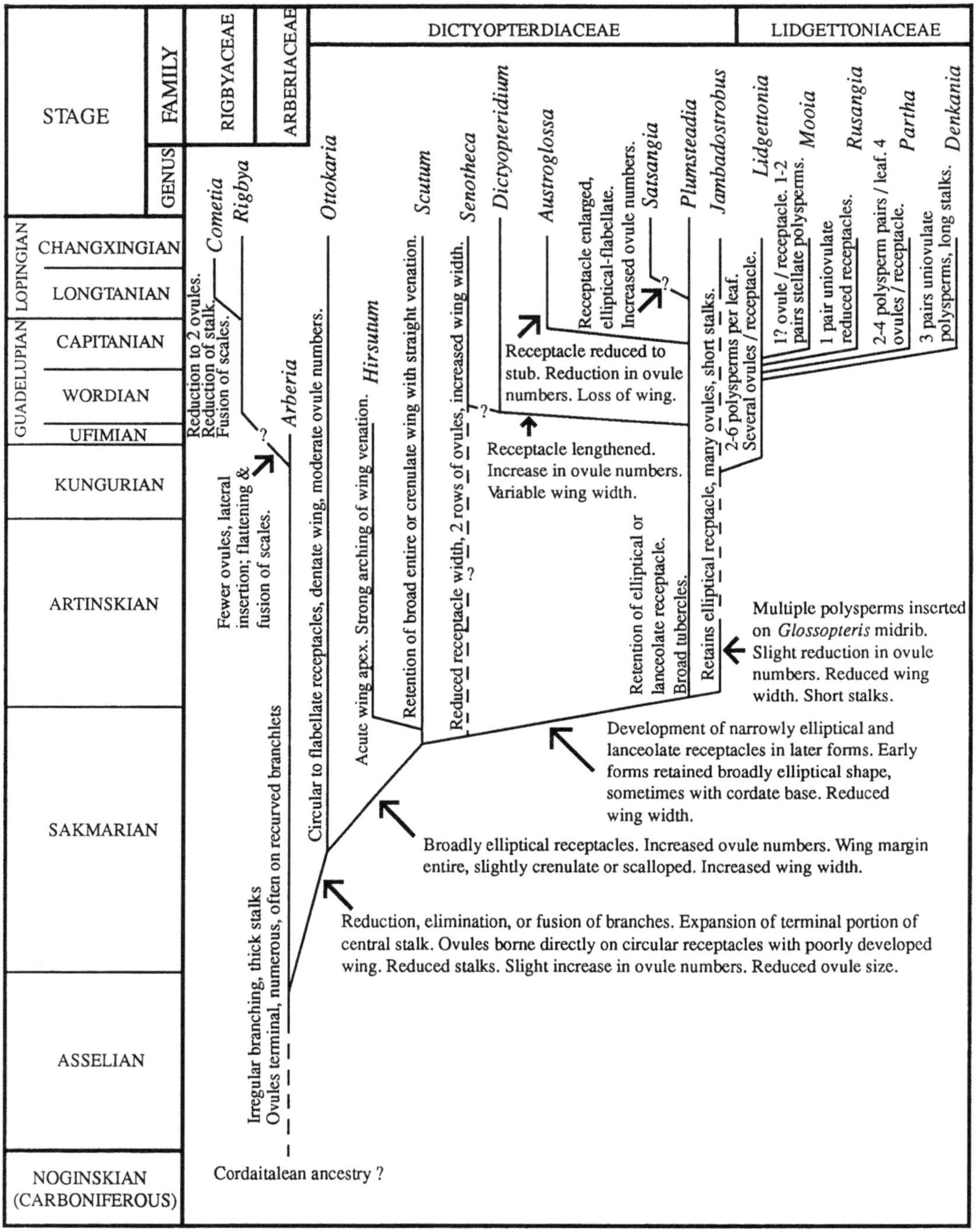

Text-figure 1. Tentative phylogenetic tree for glossopterids based on ovulate fructification morphotrends.

represent a link between solitary capitulate dictyopteridioid fruits and the multiple reduced peltate polysperms of lidgettonioid fructifications. Unfortunately the genus has only rarely been recorded from the Late Permian Raniganj Formation, India (Chandra & Surange, 1977), and the Early Permian Reids Dome Beds, eastern Australia (Rigby, 1983). Its stratigraphic

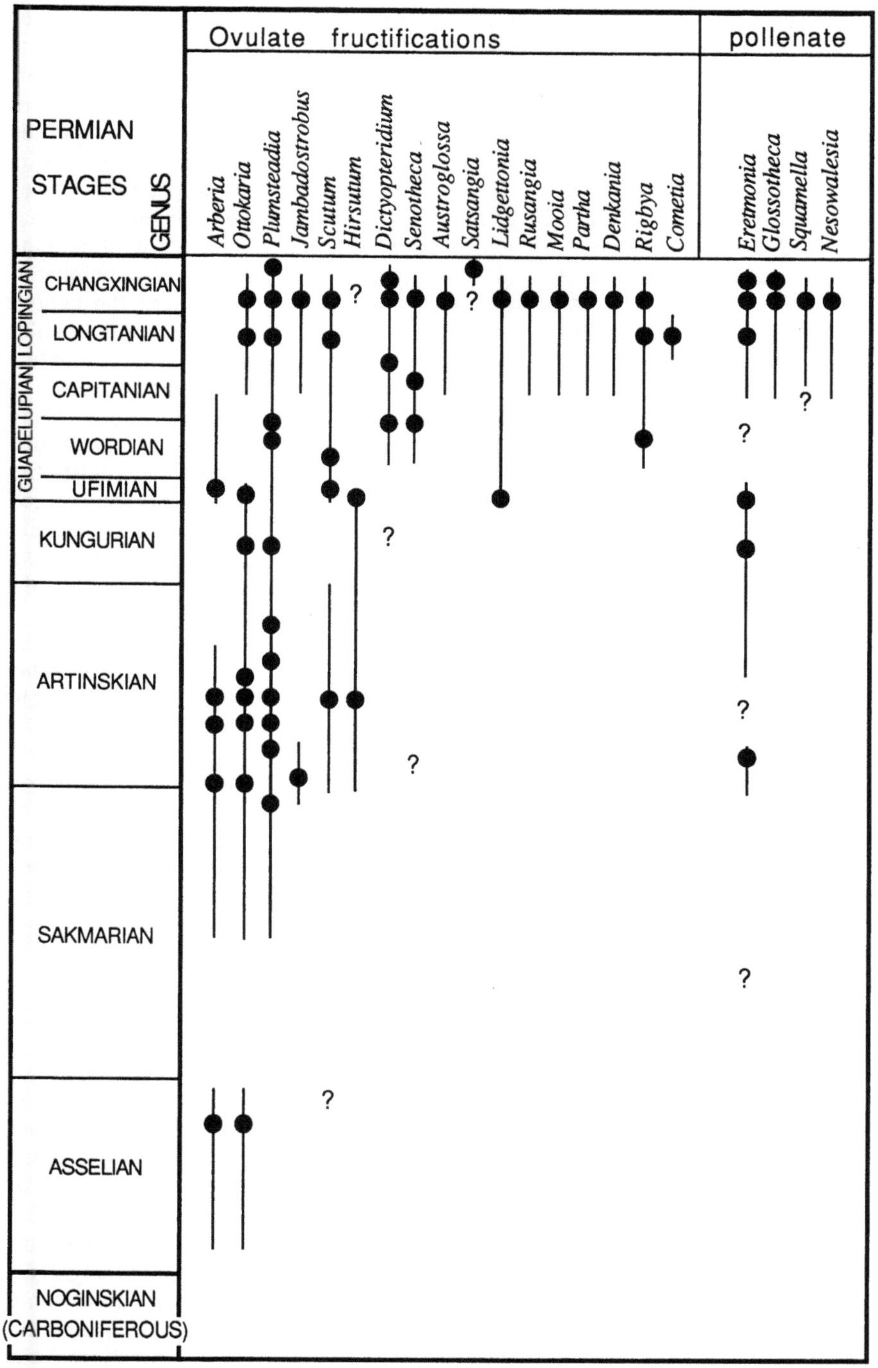

Text-figure 2. Chronostratigraphic ranges of Gondwanic Permian glossopterid fructifications [compiled from numerous sources (McLoughlin, 1990c)]. Many taxa are represented at one or few localities (suggested ages marked as ●) whereas host formations' chronostratigraphic ranges are indicated by solid lines.

utility is presently very limited.

Lidgettonia Thomas is first represented (by L. lawleyensis) in sediments referred to South African early Late Permian upper Ecca Group equivalents (Anderson & Anderson, 1985). The genus is more abundantly represented (by several species) in Capitanian-Lopingian strata of India, Australia, and South Africa (Surange & Chandra, 1974; White, 1978; Anderson & Anderson, 1985). Partha Surange & Chandra, Mooia Lacey, van Dijk, & Gordon-Gray, and Rusangea Lacey, van Dijk, & Gordon-Gray, represent morphological variants derived from the basic plan of the Lidgettonia fertiliger, differing in polysperm numbers per leaf, the size and number of seeds per fruit, polysperm shape, and the form of the attached leaf. These modified lidgettonioid fertiligers were a product of the Late Permian radiation amongst glossopterids and are confined to the Estcourt (South Africa), Raniganj, and Kamthi Formations (India), Newcastle Coal Measures (Australia), and their equivalents.

The lidgettonioid fruit Denkania Surange & Chandra has uniovulate long-stalked polysperms attached to the midline of reduced glossopterid leaves. Retallack & Dilcher (1981) used the morphological interpretation of Surange & Chandra (1973): cupulate polysperms with solitary enclosed ovules; to argue for a possible link between glossopterid fertiligers and early angiosperm multifollicular fruits. However, the three available specimens of Denkania indica Surange & Chandra (confined to the Raniganj Formation, India) are all preserved as indistinct impressions. Their interpretation as dorsiventral lobed scales, bearing solitary abaxial ovules, conforms with the structure of related genera and is compatible with the features seen in the impressions.

Rigbya Lacey, van Dijk, & Gordon-Gray (syn. Mudgea Melville), like Dictyopteridium, Senotheca, and Lidgettonia, is a relatively widespread and potentially useful late Guadelupian-Lopingian index genus occurring in Antarctica, South Africa, and eastern Australia (Plumstead, 1962; Schopf, 1967; Lacey et al., 1975; White, 1978; Rigby, 1978; McLoughlin, 1990c). Cometia McLoughlin, a fructification closely allied to the flabellate multi-lobed Rigbya but having only two ovulate scales, is currently known only from the Lopingian (Burngrove Formation) of eastern Australia. Although not yet found connected to glossopterid leaves, intimate association of these fruits with broad-meshed Glossopteris- or Surangephyllum-type leaves (Schopf, 1967, 1970; Anderson & Anderson, 1985; McLoughlin, 1990c) suggests their probable biological affinity. The relationships of the rigbyoid fertiligers to other glossopterids remains speculative. Their superficial similarity to some Early Permian Arberia specimens (Lacey et al., 1975) suggests development from that genus (Text-fig. 1). Anderson & Anderson (1985) proposed their derivation from Ottokaria-like forms by the suppression of the central receptacle and retention of the marginal ovules bordering the wing. Absence of the complete rigbyoid fertiliger remains a major handicap to the erection of glossopterid phylogenies. Were several rigbyoid polysperms to be found attached to a single leaf then an alliance with the lidgettonioid group could be envisaged (Maheshwari, 1990). A generic key (Appendix 1) is proposed for the ordination of the diverse ovulate fructifications and to rationalize the proliferation of generic names.

A number of other organs including Vannus Plumstead, Wankiea Lacey & Huard-Moine, Utkalia Chandra, and Khania Chandra & Singh, are possible glossopterid fructifications on the basis of local association with glossopterid leaves. Alternatively their multiple loose branching is quite unlike that of other glossopterid fruits or in the case of Vannus they may be immature dictyopteridioid fructifications bearing winged seeds and preserved in obscure orientations.

POLLENATE FRUCTIFICATIONS

Anderson & Anderson (1985) noted a dearth of pollenate glossopterid fructifications compared to ovulate fruits especially in the Early Permian. This may be due to their relatively delicate construction and the ready detachment of sporangia. Scale leaves like those of Eretmonia du Toit, Glossotheca Surange & Maheshwari, and Squamella White (but lacking sporangia) are very common throughout the Gondwanic Permian. Many if not all of these scales are probably the bracts of male fructifications from which the sporangia have been shed. Striate detached sporangia of these fertiligers are collectively referred to Arberiella Pant & Nautiyal although the enclosed pollen may be variable (Pant & Nautiyal, 1960; Gould & Delevoryas, 1977; Rigby & Chandra, 1990). Apart from their size and shape, scale leaves have few features to enable effective classification and thus they typically receive scant attention. Nesowalesia Pant, a cup-shaped organ and the aforementioned male genera have been chiefly reported from the Late Permian. Kendostrobus Surange & Chandra, bearing striate monolete Kendosporites Surange & Chandra spores, is probably a sphenophyllalean rather than glossopterid strobilus. The spores are morphologically similar to those recovered from Euramerian Carboniferous sphenophyllalean cones (Courvoisier & Phillips, 1975; Riggs & Rothwell, 1985).

SEEDS

Millan (1969) & Pant et al. (1985) have reviewed the geographic and stratigraphic distribution of Permian Gondwanic gymnospermic seeds. Few of these seeds are recorded attached to glossopterid fructifications (Rigby, 1972; Holmes, 1974; Lacey et al., 1975; Gould & Delevoryas, 1977), and many others are probably non-glossopterid. The dearth of seed species recorded from more than one Gondwanic region reinforces the notion of strong megafloral provincialism. The stratigraphic distribution of seeds in the Indian Permian (Pant et al., 1985) shows limited sharing of species between successive strata suggesting strong biostratigraphic utility. However, the few morphological characters available for their identification (especially when preserved as impressions), their small size, and the scarcity of past studies has resulted in seeds often being neglected in studies of fossil plant assemblages. Some seed species (e.g., Samaropsis pincombei Walkom, Maheshwariella spinocornuta Maheshwari & Tewari, and Alatocarpus johillense Lele) bear large or extravagantly ornamented wings and micropylar processes permitting easy identification. Such forms may be locally useful biostratigraphic indices but a workable zonation based on otherwise relatively similar seeds would require precise species delimitation and extreme care in identification.

ROOTS

Large-chambered roots (Vertebraria Royle emend. Schopf) and slender terminal rootlets (Lithorhiza Pant) are extremely abundant throughout the Permian of Gondwana and commonly outnumber Glossopteris leaves. Although palaeoenvironmentally useful for the recognition of palaeosols and ancient waterlogged environments, the roots of glossopterids show remarkably little morphological or anatomical variation between Gondwanic countries or throughout the Permian. Virtually all of the Gondwanic chambered roots can be assigned to the single species Vertebraria indica Royle which is an important Permian index fossil but is of little use for subdividing the Permian.

WOOD

Permian Gondwanic permineralized woods have received relatively little attention in comparison to other plant remains. Maheshwari (1972), Marguerier (1973), Mussa (1978), Prasad (1982, 1986) and Pant & Singh (1987) have provided important contributions to Gondwanic wood taxonomy and biostratigraphy. Classification of permineralized woods is chiefly dependent on the structure and arrangement of pits on xylem and vascular ray cells, the structure of the primary wood, and the presence or absence of resin canals and secretory cells. Hence, microscopic techniques are generally essential for species identification. Permineralized wood additionally has a greater potential for stratigraphic reworking than other plant remains. Numerous species have been described owing to the allocation of separate names to those woods variously possessing and lacking primary wood. However, many are described from few specimens and others are unrecognizable given certain states of preservation. The utility of fossil wood in Permian non-marine biostratigraphy has yet to be fully examined but at this juncture its potential use is deemed to be lower than that for leaves and fruits.

CONCLUSIONS

Although comprehensive taxonomic revision of glossopterid remains is required in several Gondwanic countries, leaf and fructification, and to a lesser extent seed and wood, species are potentially valuable for biostratigraphic zonation of individual Gondwanic basins or local provinces. Other glossopterid fossils (e.g., roots, male fruits, and sporangia) have, as yet, limited application to biostratigraphy owing to their relative uniformity throughout the Permian. Strong provincialism of glossopterids at species level may necessitate genus-level correlations between strata of remotely separated basins. The study of glossopterids and other stratigraphically significant elements of the Permian megafloras may provide a biozonation for terrestrial strata to complement marine invertebrate zonations. Such a biozonation would be valuable owing to the dearth of marine sediments towards the close of the Permian and the presence of extensive economically significant coal and hydrocarbon resources in Gondwanic strata of that age. A tentative phylogenetic tree and generic key are proposed for the ready identification of ovulate glossopterid fructifications and to clarify their morphological transformations throughout Permian times.

REFERENCES

Anderson, J.M. 1973. The biostratigraphy of the Permian and Triassic -part 2. *Palaeontologia Africana* 16: 59-83.

Anderson, J.M. 1977. The biostratigraphy of the Permian and Triassic. Part 3. A review of Gondwana Permian palynology with particular reference to the northern Karoo Basin, South Africa. *Memoirs of the Botanical Survey of South Africa* 41:1-67.

Anderson, J.M. & H.M. Anderson 1985. *Palaeoflora of southern Africa. Prodromus of South African megafloras Devonian to Lower Cretaceous.* Rotterdam: A.A. Balkema. 423 pp.

Anderson, J.M. & H.M. Anderson 1989. *Palaeoflora of southern Africa, Molteno Formation (Triassic) Volume 2. Gymnosperms (excluding Dicroidium).* Rotterdam: A.A. Balkema. 567 pp.

Arber, E.A.N. 1905. *Catalogue of the fossil plants of the Glossopteris flora in the Department of Geology, British Museum (Natural History)*: London. 255 pp.

Archangelsky, S. & M. Marques-Toigo 1980. La palinologia y el problema del limite Carbonico-Permico en el Gondwana Sudamericano. *Actas II Congeso Argentino de Paleontologia y Biostratigrafia y I Congreso Latinoamericano de Paleontologia. Buenos Aires, 1978*, 4: 221-229.

Archangelsky, S., C.L. Azcuy, I.D. Pinto, C.R. Gonzalez, M. Marques-Toigo, O. Rösler & R.H. Wagner 1980. The Carboniferous and early Permian of the South American Gondwana area: a summary of biostratigraphic information. *Actas II Congreso Argentino de Paleontologia y Bioestratigrafia y I Congreso Latinoamericano de Paleontologia. Buenos Aires, 1978*, 4: 257-269.

Archangelsky, S., A. Archangelsky & R. Cuneo 1981. Algunos elementos paleofloristicos de las Formaciones Piedra Shotel y Nueva Lubecka, Permico Inferior, Estancia la Casilda, Provincia de Chubut. *Ameghiniana* 18(3-4): 207-220.

Bailey, J.C., Morgan, W.R. & Black, L.P. 1982. Geochemical and isotopic evidence for the age, orogenic setting and petrogenesis of the Nychum volcanic association, north Queensland. *Journal of the Geological Society of Australia* 29: 375-393.

Balme, B.E. 1980. Palynology and the Carboniferous-Permian boundary in Australia and other Gondwana continents. *Palynology* 4: 43-55.

Banerjee, M. 1969. Senotheca murulidihensis a new glossopteridean fructification from India associated with Glossopteris taeniopteroides Feistm. In, Santapau, H.; Ghosh, A.K.; Roy, S.K.; Chanda, S; & Choudhuri, S.K. (eds) *J. Sen Memorial Volume*: 359-368. Calcutta: Botanical Society of Bengal.

Banerjee, M. 1978. Glossopteridean fructifications: 2. On the revision of Ottokaria Zeiller and the occurrence of O. raniganjensis Banerjee from the Raniganj Formation (Upper Permian) of India. *Indian Journal of Earth Science* 5(2): 129-140.

Black, L.P., W.R. Morgan & M.E. White 1972. Age of a mixed Cardiopteris - Glossopteris flora from Rb-Sr measurements on the Nychum Volcanics, north Queensland. *Journal of the Geological Society of Australia*19(2): 189-196.

Bose, M.N., J. Banerji & P.K. Maithy 1977. Some fossil plants from Ramkola-Tatapani Coalfield, Madhya Pradesh. *Palaeobotanist* 24(2): 108-117.

Cazzulo-Klepzig, M., M. Guerra-Sommer & G.E. Bossi 1980. Revisão fitoestratigráphica do Grupo Itararé no Rio Grande do Sul. 1. Acapamento Velho, Cambaí Grande, Budó e

Morro Papaléo. *Boletim I.G. Universidade de São Paulo* 11: 55-76.
Chandra, S. & K.R. Surange 1977. Cuticular studies of the reproductive organs of Glossopteris Part 3 - Two new female fructifications - Jambadostrobus and Venustostrobus - borne on Glossopteris leaves. *Palaeontographica* 164B: 127-152.
Chandra, S. & K.R. Surange 1979. *Revision of the Indian species of Glossopteris*. Lucknow: Monograph 2. Birbal Sahni Institute of Palaeobotany. 301 pp.
Courvoisier, J.M. & T.L. Phillips 1975. Correlation of spores from Pennsylvanian coal-ball fructifications with dispersed spores. *Micropalaeontology* 21: 45-59.
Delevoryas, T. 1969. Glossopterid leaves from the Middle Jurassic of Oaxaca, Mexico. *Science* 165: 895-896.
Delevoryas, T. & C.P. Person 1975. Mexiglossa varia gen. et sp. nov., a new genus of glossopteroid leaves from the Jurassic of Oaxaca, Mexico. *Palaeontographica* 154B: 114-120.
Dickins, J.M. 1976. Correlation chart for the Permian system of Australia. *Bureau of Mineral Resources Geology and Geophysics Australia, Bulletin* 156B: 1-26.
Du Toit, A.L. 1927. The fossil flora of the Upper Karoo beds. *Annals of the South African Museum* 22(2): 289-420.
Feistmantel, O. 1890. Geological and palaeontological relations of the coal and plant bearing beds of Palaeozoic and Mesozoic age in Western Australia and Tasmania; with special reference to the fossil flora. *Memoirs of the Geological Survey of New South Wales, Palaeontology* 3: 1-183.
Foster, C.B. 1979. Permian plant microfossils of the Blair Athol Coal Measures, Baralaba Coal Measures, and basal Rewan Formation of Queensland. *Publications of the Geological Survey of Queensland* 373, *Palaeontological Papers* 45: 1-244.
Foster, C.B. 1982. Spore-pollen assemblages of the Bowen Basin, Queensland (Australia): their relationship to the Permian/Triassic boundary. *Review of Palaeobotany and Palynology* 36: 165-183.
Foster, C.B. 1983. Review of the time frame for the Permian of Queensland. In; Foster, C.B. (ed.). *Permian Geology of Queensland*. 107-120. Brisbane: Geological Society of Australia, Queensland Division.
Gould, R.E. 1975. The succession of Australian pre-Tertiary megafossil floras. *Botanical Review* 41(4): 453-483.
Gould, R.E. & T. Delevoryas 1977. The biology of Glossopteris: evidence from petrified seed-bearing and pollen-bearing organs. *Alcheringa* 1: 387-399.
Harland, W.B., R.L. Armstrong, A.V. Cox, L.E. Craig, A.G. Smith, & D.G. Smith 1990. *A geological time scale 1989*. Cambridge: Cambridge University Press. 263 pp.
Harris, T.M. 1932. Fossil flora of Scoresby Sound, East Greenland. Part 2. *Meddelelser om Grønland* 85(3): 1-112.
Holmes, W.B.K. 1974. On some fructifications of the Glossopteridales from the Upper Permian of New South Wales. *Proceedings of the Linnean Society of New South Wales* 98(3): 131-141.
Johnston, R.M. 1887. Fresh contribution to our knowledge of the plants of Mesozoic age in Tasmania. *Papers and Proceedings of the Royal Society of Tasmania, 1886*: 160-185.
Lacey, W.S., D.E. van Dijk & K.D. Gordon-Gray 1975. Fossil plants from the Upper Permian in the Mooi River district of Natal, South Africa. *Annals of the Natal Museum* 22(2): 349-420.
Lele, K.M. 1977. Palaeoclimatic implications of Gondwana floras. *Geophytology* 6: 207-229.
McLoughlin, S. 1990a. Some Permian glossopterid fructifications and leaves from the Bowen Basin, Queensland, Australia. *Review of Palaeobotany and Palynology* 62(1/2): 11-40.
McLoughlin, S. 1990b. Late Permian glossopterid fructifications from the Bowen and Sydney Basins, eastern Australia. *Geobios* 23(3): 283-297.
McLoughlin, S. 1990c. *Palaeobotany and palaeoenvironments of Permian strata, Bowen Basin, Queensland*. PhD thesis. University of Queensland. 312 pp. (unpublished).
Maheshwari, H.K. 1972. Permian wood from Antarctica and revision of some Lower Gondwana wood taxa. *Palaeontographica* 138B: 1-43.
Maheshwari, H.K. 1990. The glossopterid fructifications: an overview. In; Douglas, J.G. & Christophel, D.C. (eds), *Proceedings of the 3rd International Organization of Palaeobotany Conference, Melbourne, 1988*: 11-15.

Maithy, P.K.S. 1974. The Lower Gondwana plants of India and their stratigraphic significance. *7th International Congress on Carboniferous Stratigraphy and Geology, Krefeld, 1971. Compte Rendu* 3: 385-390.

Marguerier, J. 1973. Paloxylogie du Gondwana Africain: Etude et affinities du genre Australoxylon. *Palaeontologia Africana* 16: 37-58.

Meyen, S.V. 1987. *Fundamentals of Palaeobotany*. London: Chapman & Hall. 432 pp.

Millan, J.H. 1969. The gymnospermic and platyspermic seeds of the Glossopteris flora from Brazil and correlated foreign regions. *Gondwana Stratigraphy. IUGS Symposium, Buenos Aires, 1967*. 107-122. Paris: UNESCO.

Mussa, D. 1978. Brasilestiloxylon e Solenobrasilioxylon, dois novos generos gondwanicos na Formacão Iraty, Estado de São Paulo, Brasil. *Boletim I.G. Universidade de São Paulo* 9: 118-126.

Pant, D.D. 1982. The Lower Gondwana gymnosperms and their relationships. *Review of Palaeobotany and Palynology* 37: 55-70.

Pant, D.D. & D.D. Nautiyal 1960. Some seeds and sporangia of the Glossopteris flora from Raniganj Coalfield, India. *Palaeontographica* 107B: 41-64.

Pant, D.D., D.D. Nautiyal & S.P. Tiwari 1985. On some Indian Lower Gondwana compressions of seeds. *Palaeontographica* 196B: 31-78.

Pant, D.D. & R. Pant 1987. Some Glossopteris leaves from Indian Triassic beds. *Palaeontographica* 205B: 165-178.

Pant, D.D. & V.K. Singh 1987. Xylotomy of some woods from Raniganj Formation (Permian), Raniganj Coalfield, India. *Palaeontographica* 203B: 1-82.

Plumstead, E.P. 1962. Fossil floras of Antarctica. *Trans-Antarctica Expedition Scientific Report* 9: 1-154.

Powis, G.D. 1984. Palynostratigraphy of the Late Carboniferous sequence, Canning Basin, W.A. In; Purcell, P.G. (ed.). *The Canning Basin, W.A.* Proceedings of the Geological Society of Australia/Petroleum Exploration Society of Australia Symposium, Perth: 429-438.

Prasad, M.N.V. 1982. An anatomical synopsis of Indian Palaeozoic gymnospermous woods. *Review of Palaeobotany and Palynology* 38: 119-156.

Prasad, M.N.V. 1986. Xylotaphoflora of the Kamthi Formation, Indian Lower Gondwana with remarks on the biostratigraphic importance of its taphoflora. *Palaeontographica* 201B: 111-134.

Rayner, R.J. & M.K. Coventry 1985. A Glossopteris flora from the Permian of South Africa. *South African Journal of Science* 81(1): 21-32.

Retallack, G. 1980. Late Carboniferous to Middle Triassic megafossil floras from the Sydney Basin. *Geological Survey of New South Wales Bulletin* 26: 384-430.

Retallack, G.T. & D.L. Dilcher 1981. Arguments for a glossopterid ancestry of angiosperms. *Palaeobiology* 7(1): 54-67.

Rigby, J.F. 1972. On Arberia White, and some related Lower Gondwana female fructifications. *Palaeontology* 15(1): 108-120.

Rigby, J.F. 1978. Permian glossopterid and other cycadopsid fructifications from Queensland. *Geological Survey of Queensland Publication* 367, *Palaeontological Papers* 41: 1-21.

Rigby, J.F. 1983. The role of Glossopteris in biostratigraphy - a preliminary assessment in the Reids Dome Beds. In; Foster, C.B. (ed.), *Permian Geology of Queensland*: 221-229. Brisbane: Geological Society of Australia, Queensland Division.

Rigby, J.F. (this volume). Review of the Early Permian flora of the Nychum Volcanics north of Chillagoe, north Queensland. in Findlay, R.H.; Banks, H.R.; Veevers, J.J.; Unrug, R. (eds).*Gondwana 8 - Assembly, Evolution, and Dispersal.* Rotterdam: A.A. Balkema.

Rigby, J.F. & S. Chandra 1990. The Late Permian sporangiate fructification Nesowalesia edwardsii Pant, 1977, from Katoomba, New South Wales, Australia. *Review of Palaeobotany and Palynology* 65: 175-178.

Riggs, S.D. & G.W. Rothwell 1985. Sentistrobus goodii n. gen. and sp., a permineralized sphenophyllalean cone from the upper Pennsylvanian of the Appalachian Basin. *Journal of Paleontology* 59(5): 1194-1202.

Rösler, O. 1978. The Brazilian eogondwanic floral succession. *Boletim I.G. Universidade de São Paulo* 9: 85-90.

Ross, C.A. & J.R.P. Ross 1987a. Late Palaeozoic sea level and depositional sequences. *Cushman Foundation for Foraminiferal Research, Special Publication* 24: 137-149.

Ross, C.A. & J.R.P. Ross 1987b. Biostratigraphic zonation of late Paleozoic depositional sequences. *Cushman Foundation for Foraminiferal Research, Special Publication* 24: 151-168.

Schopf, J.M. 1967. Antarctic fossil plants collected during the 1966-67 season. *Antarctic Journal of the United States* 2(4): 114-116.

Schopf, J.M. 1970. Gondwana Palaeobotany. *Antarctic Journal of the United States* 5(3): 62-66.

Schopf, J.M. 1976. Morphologic interpretation of fertile structures in glossopterid gymnosperms. *Review of Palaeobotany and Palynology* 21: 25-64.

Srivastava, P.N. 1956: Studies in the Glossopteris flora of India - 4. Glossopteris, Gangamopteris, and Palaeovittaria from the Raniganj Coalfield. *Palaeobotanist* 5(1): 1-45.

Srivastava, S.C. & H.K. Maheshwari 1973. Satsangia, a new plant organ from the Triassic of Nidhpuri, Madhya Pradesh. *Geophytology* 3(2): 222-227.

Surange, K.R. & S. Chandra 1973. Denkania indica gen. et sp. nov., a glossopteridean fructification from the Lower Gondwana of India. *Palaeobotanist* 20(2): 264-268.

Surange, K.R. & S. Chandra 1974. Lidgettonia mucronata sp. nov. : a female fructification from the Lower Gondwana of India. *Palaeobotanist* 21(1): 121-126.

Truswell, E.M. 1980. Permo-Carboniferous palynology of Gondwanaland: progress and problems in the decade to i980. *BMR Journal of Australian Geology and Geophysics* 5: 95-111.

Virkki, C. 1938. A Lower Gondwana flora of the Salt Range, Punjab. (Abstract). *Proceedings of the 25th Indian Science Congress*: 150-151.

Waterhouse, J.B. 1976. World correlations for Permian marine faunas. *Papers of the Department of Geology University of Queensland* 7(2): 1-232.

White, D. 1908. Report on the fossil flora of the coal measures of Brazil. In White, I.C. *Final Report Comissão de Estudos das Minas de Carvão de Pedra do Brazil, Rio de Janeiro* 3: 336-617.

White, M.E. 1978. Reproductive structures of the Glossopteridales in the Australian Museum. *Records of the Australian Museum.* 31(12): 473-504.

Zeiller, R. 1902. Observations sur quelques plantes fossiles des Lower Gondwanas. *Memoirs of the Geological Survey of India, Palaeontologia Indica,* N.S. 2: 1-40.

Appendix 1. Key to genera of ovulate glossopterid fructifications.

1a. Fructifications with 2 or more orders of dichotomous, pinnate, or irregular branching. Ovules borne terminally on straight or recurved ultimate branchlets. --- Arberia White emend. Rigby.

1b. Fructifications isobilateral, dorsiventral, consisting of variously fused or divided capitula or cupules. --- 2.

2a. Capitulum consisting of central receptacle bearing >10 ovules and an outer sterile transversely striate wing. --- 3.

2b. Capitulum consisting of single entire or multiple variously fused or divided peltate, flabellate, or linear scales or cupules. <10 ovules per capitulum. --- 10.

2c. Capitulum reduced to slightly thickened terminal stub on long stalk. 5-10 ovules per capitulum. --- Austroglossa Holmes.

3a. Receptacle circular, flabellate, or suborbiculate. Wing dentate or deeply lobate. --- Ottokaria Zeiller emend. Pant & Nautiyal.

3b. Receptacle ovate, elliptical, flabellate, lanceolate, or linear. Wing crenulate or entire. --- 4.

4a. Wing markedly broader near apex than base, apex often acuminate; wing venation strongly arched towards apex. --- Hirsutum Plumstead ex Anderson & Anderson.

4b. Wing about equi-width throughout; wing venation straight, perpendicular to margin. --- 5.

5a. Receptacle ovate to widely elliptical. Wing crenulate or rarely entire. --- Scutum Plumstead.

5b. Receptacle narrowly ovate, elliptical, wide flabellate, lanceolate, or linear. Wing entire or rarely undulose. --- 6.

6a. Receptacle linear strand-like, with two rows of oppositely arranged ovules, one on either side of receptacle. Wing at least twice as wide as receptacle --- Senotheca Banerjee.

6b. Receptacle narrowly ovate, elliptical, wide flabellate, lanceolate, rarely linear; ovules or ovule scars irregularly arranged on receptacle. Wing < twice as wide as receptacle. --- 7.

7a. Receptacle large (>4 cm wide), flabellate, with thick robust stalk (ca. 1 cm wide). --- Satsangia Srivastava & Maheshwari.*

7b. Receptacle narrowly ovate, elliptical, lanceolate, rarely linear (<4 cm wide); stalk short and relatively narrow (<5 mm wide). --- 8.

8a. Receptacle very narrowly elliptical, lanceolate, rarely linear, apex pointed acute, rarely rounded. Tubercles small, discrete, consisting of a small knob atop a low disc, 100-600 per receptacle. --- Dictyopteridium Feistmantel ex Zeiller emend. McLoughlin.

8b. Receptacle narrowly ovate, elliptical, rarely lanceolate, apex typically rounded acute or obtuse. Tubercles relatively large, closely spaced, consisting of a small knob atop a broad bulbous boss. Generally <300 tubercles per receptacle. --- 9.

9a. Solitary fructifications, dissociated, attached to, or axillary to Glossopteris or Palaeovittaria leaves. --- Plumsteadia Rigby.

9b. More than one receptacle adnate to Glossopteris leaf. --- Jambadostrobus Chandra & Surange.

10a. Capitulum flabellate, consisting of 2-9 variously fused or divided obovate to linear lobes. Single ovule or ovule scar on abaxial surface of each scale. Only solitary isolated fructifications known. --- 11.

10b. Capitulum consisting of single palmate, flabellate, or campanulate scale or cupule. 1-6 ovules per scale. Fructifications occur in 1-4 pairs, adnate to, or attached by long stalks to midrib of modified (reduced) glossopterid leaf. ---12.

11a. Two ovulate lobes per capitulum. --- Cometia McLoughlin.*

11b. 4-9 ovulate lobes per capitulum. --- Rigbya Lacey, van Dijk, & Gordon-Gray.*

12a. 1-4 pairs of flabellate scales with entire, crenulate, or dentate margins, attached by long pedicels to midline of modified glossopterid leaf. Each scale with 1-6 ovules. --- Lidgettonia Lacey, van Dijk, & Gordon-Gray.

12b. 1-4 pairs of reduced peltate to ovate scales or cupules with shallow to deeply notched margins, attached by short or long pedicels to midline of modified glossopterid leaf. Scales or cupules with 1-4 ovules. --- 13.

13a. 3-4 pairs of ovate scales with rounded apices or 3 apical lobes; scales borne on long (>2 cm) stalks. Single ovule borne on abaxial surface of each scale. --- Denkania Surange & Chandra.

13b. 1-4 pairs of peltate scales or cupules with shallow to deeply notched margins; scales borne on short (<2 cm) stalks. Each scale or cupule with 1-4 ovules on abaxial surface. --- 14.

14a. Peltate scales or cupules with 4 broadly rounded lobes each with a single ovule or ovule scar. 1-2 pairs of cupules per leaf. --- Partha Surange & Chandra.

14b. Peltate scales or cupules with 4-6 dentate lobes. 1 ovule per cupule. 1-2 pairs of cupules per leaf. --- 15.

15a. 1-2 pairs of large peltate scales with 6 dentate lobes. Attached leaf obovate to narrowly oblanceolate. --- Mooia Lacey, van Dijk, & Gordon-Gray.

15b. 1 pair of reduced peltate scales with 4-?6 dentate lobes. Attached leaf elliptical to obovate. --- Rusangea Lacey, van Dijk, Gordon-Gray.

* Denotes fructification not yet found directly attached to, but inferentially associated with, a glossopterid leaf.

Gondwana Eight, Findlay, Unrug, Banks & Veevers (eds) © 1993 Balkema, Rotterdam. ISBN 90 5410 304 3

Taxonomic diversity of Triassic seeds from India

Shyam C. Srivastava & S.R. Manik
Birbal Sahni Institute of Palaeobotany, Lucknow, India

ABSTRACT: Based upon the distinctive cutinized membranes, six new genera: Sahnispermum, Delevorvaspermum, Cupolaspermum, Urceolaspermum, Tayloriaspermum and Konaspermum have been identified from the plant bearing sediments of Nidpur. The Triassic bed occurs in a small fault-bounded outcrop containing exquisitely preserved plant organs in carbonaceous sandy shale of dark to light-grey colours. The study has been carried out on isolated seed compressions found in detached conditions. Generally, the fossil seeds yield three to five membranes which are outer-inner cuticle of the integument, nucellus and megaspore membrane. The seed taxa have been differentiated using the following parameters: cutinization of integument and nucellus, cell shape and size, surface striations, occurrence of papillae, thickness of anticlinal wall, presence or absence of inner cuticle of integument, presence or absence of megaspore membrane, structural organization of pollen chamber, extent of fusion of integument with nucellus and nature of micropylar hole. The relationship of genera have been determined on the circumstanstial evidence provided by morpho-taxonomical details of seed cuticle.

INTRODUCTION

The Triassic sequence in the Son Valley of South Rewa Gondwana Basin consists of three distinct formations. In ascending order, these are the Panchet (including Upper Pali), Tiki and Parsora formations (Roy Chowdhary et al, 1975). The Nidpur beds are considered to be the youngest unit and are exposed in the Gopad river (=Tiki Formation). The outcrop has revealed an extremely diverse plant assemblage including foliage , unattached fertile organs, scale leaves, and seeds. Palynomorphs have also been re-trieved from the fragile carbonaceous shale (Srivastava, 1988, Bharadwaj & Srivastava, 1969, and Tiwari & Ram Awatar 1990). All the major taxonomic groups are represented in the flora. Among these, Pteridospermophyta is the most diverse and dominant group, and is represented by the taxa Lepidopteris, Dicroidium and the microsporangiate organ Pteruchus. Of these taxa Dicroidium is overwhelmingly dominant. The preponderance of nonsriate bisaccate grains has also been recorded. Palaeofloristically a Middle Triassic age for the Nidpur beds is ascertained. Like other plant organs, seeds are also preserved in a variety of ways depending upon their structure and the conditions prevailing at the time of deposition. Seeds are represented as cast, impression, compression and petrifactions.

Morphologically, the seed consists of an envelope (integument) with a micropyle and mgasporangium (the nucellus) inside which there is megagametophyte consisting of nutritive tissue and archegonia.

From the Mesozoic of India, the fossil seed impression Samaropsis was recorded by Feistmantel (1880) from the Panchet Formation of Raniganj Coalfield. Later, Lele

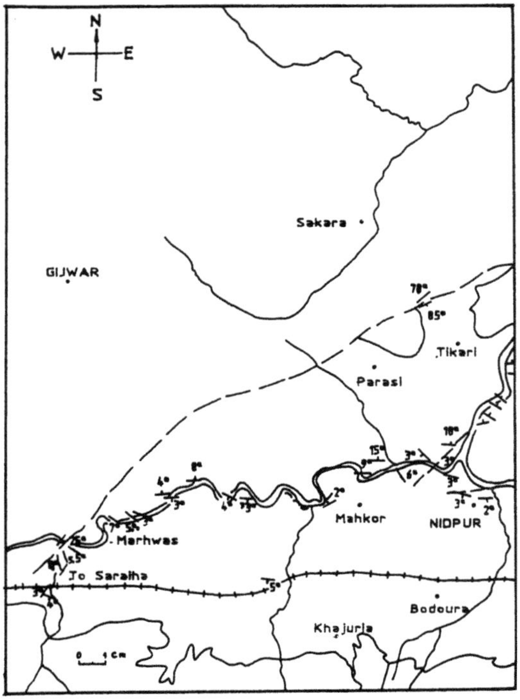

LOCATION-MAP (After Raja Rao,1983)

(1955,1962) furnished an illustrated account of seed impressions revealing representative of several species of *Samaropsis* and *Cardaicarpus* from the Parsora Formation of South Rewa Basin. Banerji et al (1976,1978)reported impressions of seed-like bodies from the Triassic of the Gopad River section near Nidpur, and also recorded *Spermatitis* spp. i.e. impressions from Upper Triassic exposures of the Janar Nala section near Harai. In addition some isolated compressed seeds identical to *Amphorispermum* were described by Bose and Banerji (1984) from Middle Upper Jurassic strata of Kachchh.

Recently, while further investigating the fertile structures from the Triassic beds of Nidpur, Pant and Basu (1977), Manik (1988) and Srivastava and Manik (1990) have reported seed compactions, namely, *Rugaspermum*, *Savitrispermum*, *Nidspermum*, *Rotundaspermum*, *Pyriformispermum*, *Pantiaspermum* and *Rostrumaspermum* and enriched our knowledge of the flora.

From further collections of seed compressions, six new genera have been instituted based upon their distinctive cuticular membranes, involving their exomorphic characters, viz., *Sahnispermum*, *Delevoryaspermum*, *Cupolaspermum*, *Urceolaspermum*, *Tayloriaspermum* and *Konaspermum*. Collectively these findings have demonstrated considerable taxonomic diversity among the Triassic seeds.

TAXONOMY

Sahnispermum gen.nov.
(pl.1 Figs.1,3 Text-figs.1A-D)
Diagnosis: Seed oval,platyspermic, overall outline undulating, slightly tapering towards micropylar end, outer integument papillate, stomatiferous, papillae overcarching stomatal pit,nuceller membrane enclosing megaspore sac, pollen chamber distinctly differentiated, composed of shorter polygonal cells, more or less dome-like,endwalls converging towards micropylar end, highly cutinized, micropylar opening protruding,appearing somewhat saucer-shaped.

Type Species: *Sahnispermum indicum*

Holotype: Slide No.BSIP 10633; Middle Triassic, Nidpur, India.

Derivatio nominis: To commemorate the Birth-Centenary Year 1991 of Professor Birbal Sahni.

Discussion

In the papillate nature of the outer integument, this taxon closely approaches the seed *Rugaspermum media* (Pant and Basu,1977) but it is readily distinguished from the latter by the absence of wrinkles or rugose outer integument. The seed cuticle of *S. indicum* also resembles that of *Peltaspermum thomasi*, (Townrow, 1957) where the papillae overarch the stomatal pit. The epidermal character of *S.indicum* is similar to that of *Bosea indica* (Srivastava, 1975) and the general epidermal feature compares favourably with *Lepidopteris indica*.

Pollen grains inside the pollen chamber of *S.indicum* belonging to *Weylandites*-complex have been

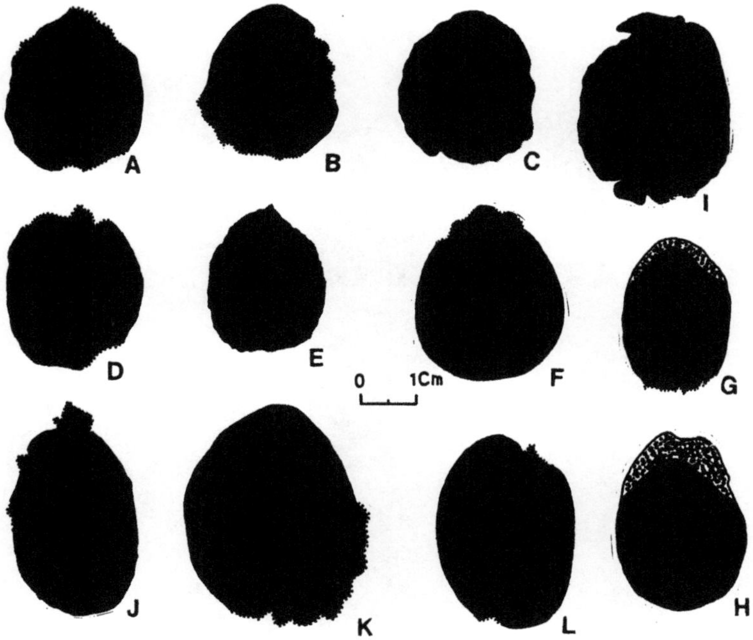

Text-fig.1 Sahnispermum indicum gen.et.sp.nov.; A,C & D,Carbonised seeds. Isotype-Nos.10636,10637,10639x5; B.Holotype No.10633x5.E.Delevoryaspermum nidpurensis gen.et.sp.nov.;10641x5. F.Cupolaspermum marhwaseanum gen.et. sp.nov.; 10642x5. G. Urceolaspermum gopadensis gen.et.sp.nov.; Isotype No.10647x5; H.Holotype No.10646x5. I,J & L. Konaspermum sidhiensis gen. et.sp.nov.; Isotype No.10650,10649,10653 x5. K. Holotype No.10648.

observed. These pollen have been reported in dispersed condition from the Nidpur shale by Bharadwaj and Srivastava (1969). This indicates costate pollen grains were typically involved in pollination.

As a consequence of their constant association in the Nidpur shale, and because of the similarity of their epidermal features, the vegetative leaf Lepidopteris indica, the pollen organ Bosea Indica and the seed taxon Sahnispermum indicum may have been derived from the same plant.

Affinities

The free integument from the nucellus upto the base of the seed, a characteristic feature of pteridosperms, reveals the relationship of Sahnispermum indicum with the Pteridospermales.

Delevoryaspermum gen.nov.
(Pl.1,Figs.4-5 Test figs.1-E,2,3)

Diagnosis: Seed oval having uneven outline, outer integument smooth, stomatiferous, pollen chamber broadly triangular, micropylar end bulging out somewhat like a protuberance, nucellar membrane interspersed with fine creases, megaspore centrally located.

Type species: Delevoryaspermum nidpurensis

Holotype: Slide No.BSIP 10640, Middle Triassic, Nidpur, India.

Discussion :

Delevoryaspermum, while resembling Savitrispermum and Nidispermum in its smooth surface, differs from the latter on bearing a stomatiferous integument. It differs from Sahnispermum indicum by the absence of papillae over the integument. In epidermal

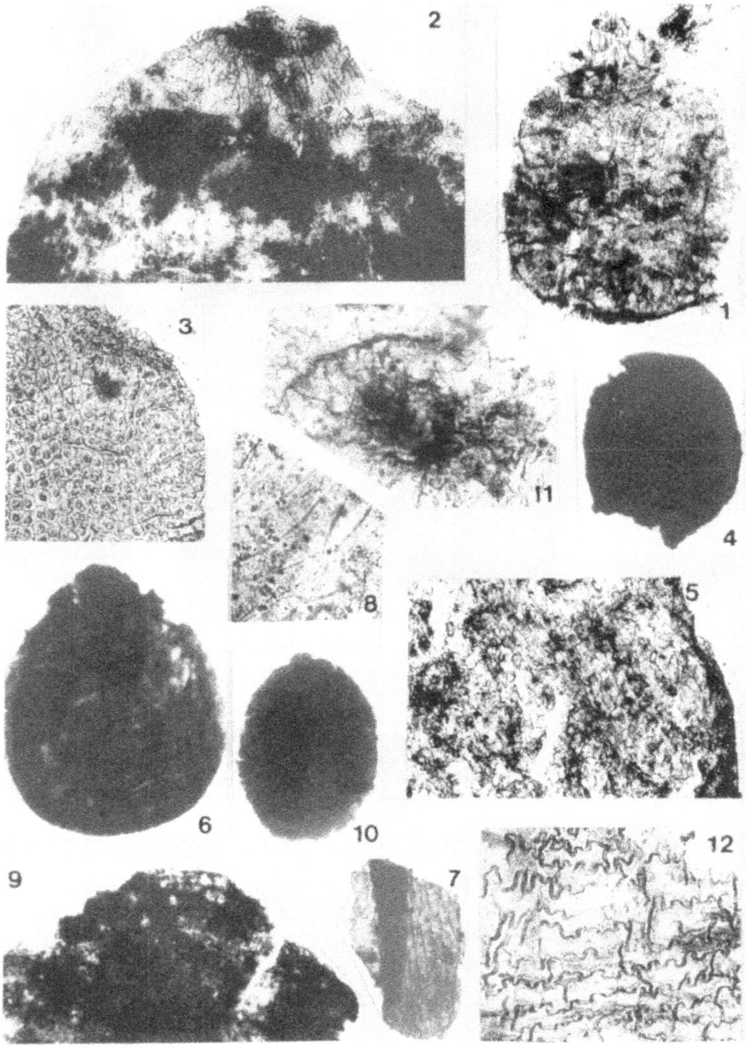

Plate-1 Sahnispermum indicum gen.et sp.nov. Fig.1. Seed after alkali treatment showing distinctly differentiated pollen chamber,micropylar opening and nucellar membrane associated with inner integument. BSIP Slide No.10635x25. Fig.2. Seed showing micropylar hole and slight depession at the distal end of nucellus forming pollen chamber. Holotype-BSIP Slide No.10635x50. Fig.3.Outer investment of seed with distinct papillate periclinal wall. BSIP slide No.10637x50. Delevoryaspermum nidpurensis gen.et sp.nov. Fig.4. Unmacerated seed dipped in glycerine. BSIP Specimen No.10640x10. Fig.5. Cellular structure of outer integument showing irregular distributed stomata. BSIP Slide No. 10640X50. Cupolaspermum marhwaseanum gen.et sp.nov. Fig.6. Seed after acid treatment showing broad chalazal end and obtusely pointed micropylar end. Holotype, BSIP Slide No.10642X20. Cellular structure of outer integument alongwith a portion of nucellar membrane, BSIP Slide No.10642X50. Fig.8. A portion of outer integument showing cellular details. BSIP Slide No.10642X250. Fig.9. Dome-shaped pollen chamber coalesced with outer integument. BSIP. Slide No.10642X50. Tayloriaspermum sinuosum gen.et.sp.nov. Fig. 10. Acid treated seed. Holotype BSIP Slie No.10644X20. Fig.11. Micropylar end of seed lying in flattened position showing micropylar hole associated with cellular remains of outer integument. BSIP Slide No.10645X150. Fig.12. Outer cuticle of integument showing sinuous cells. BSIP Slide No. 10644X150.

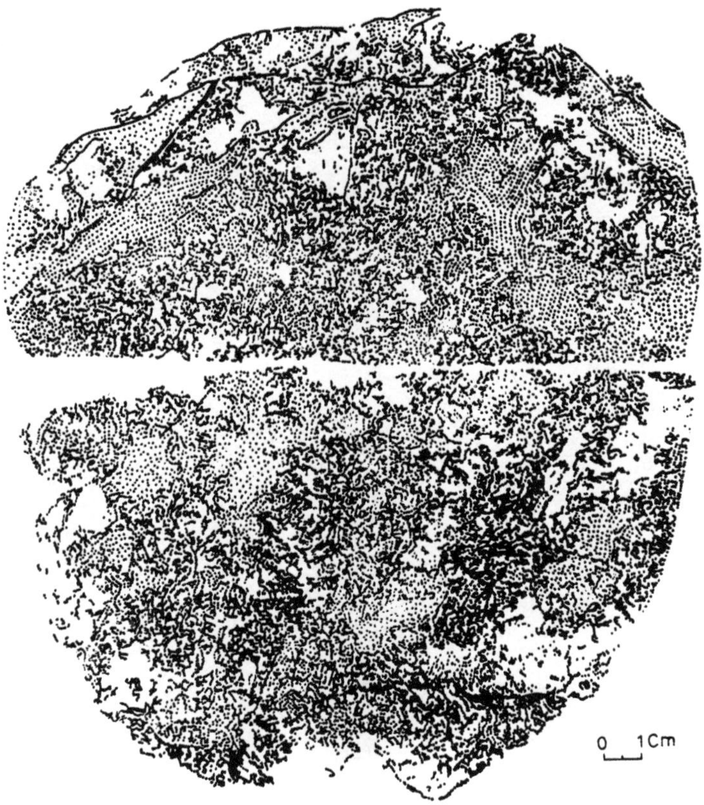

Text-fig.2. Delevoryaspermum nidpurensis gen.et. sp.nov.; A -Upper half portion of seed showing inegument associated with nucellus. BSIP Slide No.10641 X50.;A2. Portion of outer integument in Lower half of seed-BSIP Slide No.10641 X50.

characters, Delevoryaspermum shows similarity with Dicroidium nidpurensis (Bose and Srivastava,1971). Futher, when Delevoryaspermum nidpurensis is compared with the various species of Pteruchus described from the Nidpur beds, it is noted that the former comes close to P. nidpurensis in its smooth nature of the cell surface wall and in addition to this similarity both stomatal apparatus are alike.

Thus, it can be inferred that Delevoryaspermum nidpurensis could have been the seed of the Dicroidium nidpurensis plant because of the structural similarity of the outer integument with that of the leaf cuticle of D. nidpurensis.

This match of epidermal features between Delevoryaspermum nidpurensis and Pteruchus nidpurensis probably indicates that they can be ascribed to the Dicroidium nidpurensis plant.

Affinities :

Delevoryaspermum nidpurensis clearly demonstrates its relationship to the pteridospermales by its distinctive free outer integument upto the seed base, and by the elaborate pollen chamber.

Cupolaspermum gen.nov.
(Pl.1 Fig.6-9 Text-figs.1-F)

Diagnosis: Seed ovate; forming apically distinctive cupola bearing ornate or sculptured edge; outer integument smooth marginal flange-cells around micropylar hole, perforated generally elliptical in shape.

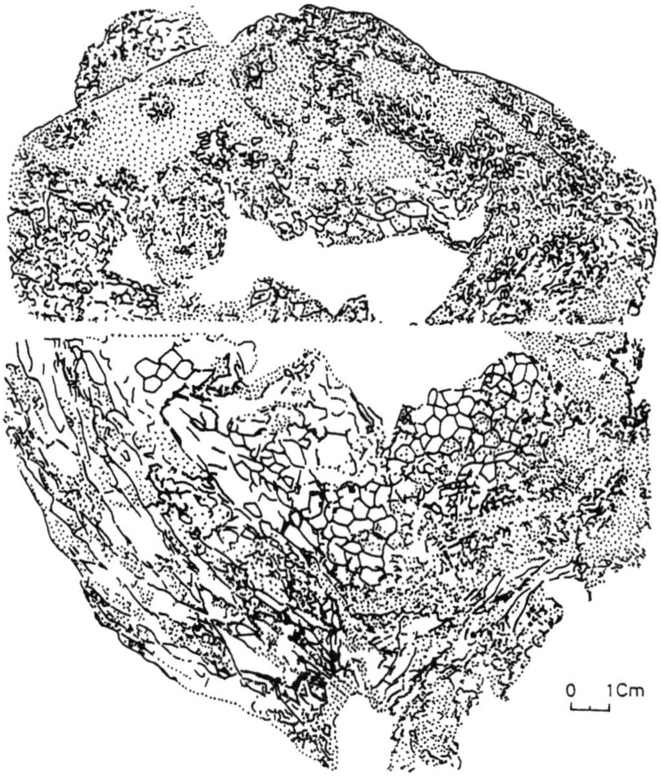

Text-fig.3. Delevoryaspermum nidpurensis gen. et. sp.nov.; A -Upper half of seed showing nucellus and micropylar region. BSIP Slide No.10641 X150; A2-Lower half of seed showing chalazal end. BSIP Slide No.10641 X150.

Type species : Cupolaspermum marhwaseaanum

Holotype: Slide No.BSIP 10642 Middle Triassic, Nidpur, India.

Discussion:

The typical cupola enables Cupolaspermum gen.nov. to be distinguished from the seed genera Rugaspermum, Savitrispermum, Nidispermum, Rotundaspermum, Pyriformispermum, Pantiaspermum and Rostrumaspermum but in this particular character this genus closely compares with Menaisperma greenlyii (Pettitt & Lacey 1972).

Affinities

Its well developed dome shaped massive pollen chamber indicates its botanical affiliation with the pteridosperms.

Urceolaspermum gen.nov.
(Pl.2 Figs.1-3 Text-figs.1,G-H)

Diagnosis: Seed pitcher shaped, micropylar hole sub-spherical bearing jagged edge; integument surface smooth; intimately adherent to nucellus, apically nucellus free from integument leaving crescent shaped pollen chamber, nucellus apex dome-shaped, nucellar membrane enclosing dark brown massive megaspore.

Type species : Urceolaspermum gopdensis.

Holotype: Slide NO. BSIP 10646 Middle Triassic, Nidpur, India.

Discussion :

Because of its pitcher-shaped character Urceolaspermum gopadensis can be readily differentiated

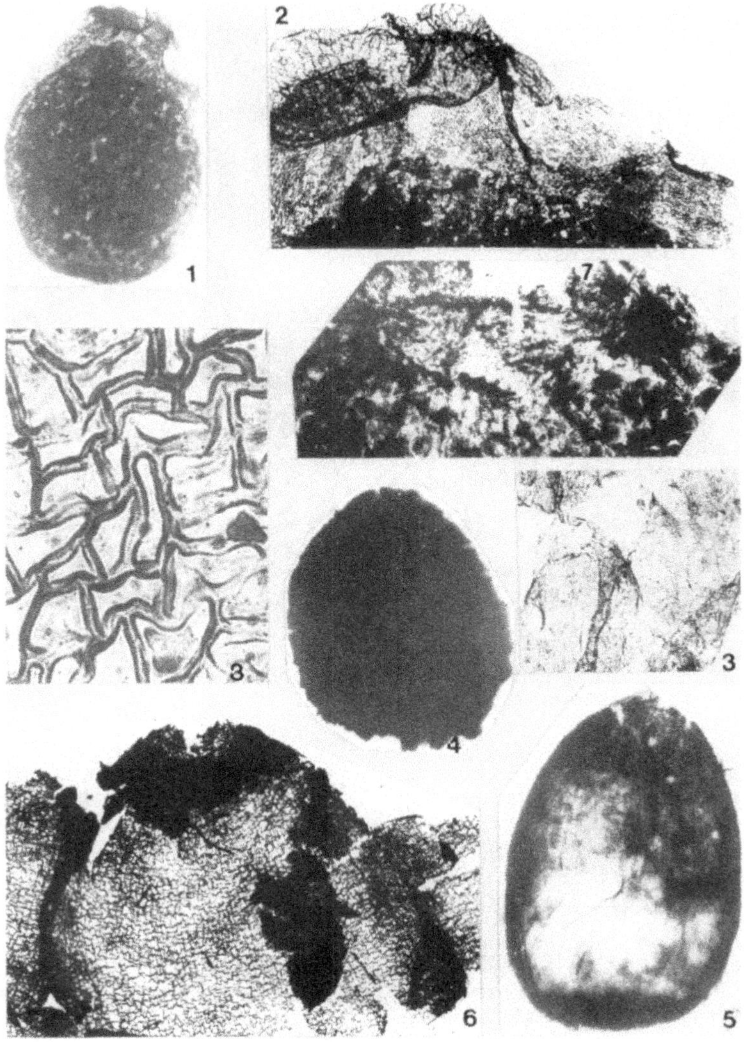

Plate-2: Urceolaspermum gopdensis gen.et.sp.nov. Fig.1. Flask-shaped seed having distinct micropylar opening after acid treatment. Holotype-BSIP Slide No. 10646 X 20. Fig.2. A portion of micropylar end showing details of micropylar region and distinct nucellar dome. BSIP Slide No. 10646 X 50. Fig.3. A portion of outer integument covering in the micropylar region showing varied cells. BSIP Slide No.10647, X 150. Konaspermum sidhiensis gen. et sp.nov. Fig.4. Unmacerated seed dipped in glycerine. Holotype BSIP Specimen No.10648 X 10. Fig.5. Seed after alkali treatment showing differentiating membranes. BSIP Slide No.10649 X20. Fig.6. Micropylar end of seed magnified to show heavy cutinization and cutinized concentric rings all around micropylar opening associated with distinctive angulate epidermis of outer integument. BSIP Slide No.10648 X50. Fig.7. Encircling cutinized rim all around distinct micropylar opening. BSIP Slide No.10649 X150. Fig.8. Epidermal structure of outer integument showing angular cells with thickly cutinized anticlinal walls. BSIP Slide No.10651 X300.

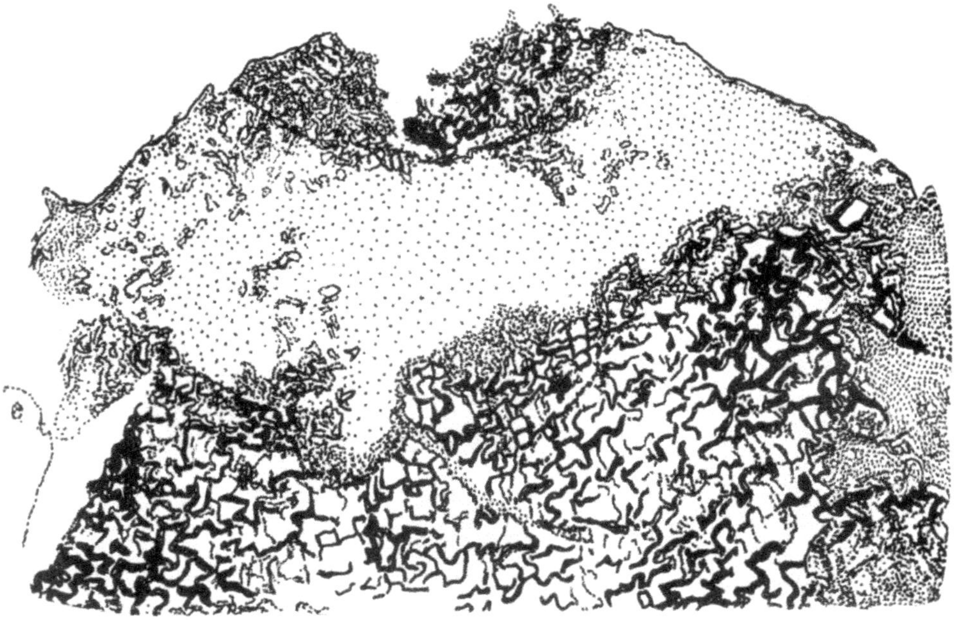

Text-fig.4. Konaspermum sidhiensis gen. et. sp.nov.: A. Micropylar region of seed showing strong cutinization and cells of outer cuticle of integument. BSIP Slide No.10648 x 50.

from all the known seed genera. However, in its smooth surface of epidermal cells, the taxon shows resemblance to Delevoryaspermum nidpurensis, Nidispermum glabrosum and Cupolaspermum marhwaseanum. In having nucellar apex free from the outer integument Allicospermum xystum (Harris,1935) from the Rhaetic of Greenland compares very closely with U. gopadensis. Both genera possess circular, fairly thick megaspores.However,A. xystum differs in shape and is much bigger in size.

Affinities :

The pollen chamber of Urceolaspermum gopadensis is much reduced compared to those of pteridospermus seeds and in this respect U. gopadensis shows greater similarity to the cycad s . Further the separation of the nucellar apex from the outer integument favours affinity with the Cycadales.

Tayloriaspermum gen.nov.
(Pl.1, Figs.10-12)

Diagnosis: Seed almond shaped, micropylar rim heavily cutinized, outer integument composed of sinuous cells a distance equivalent to three quarters the length of the nucellus towards the micropyle possessing an apical nucellar membrane separated from the outer integument, nucellar apex depressed, somewhat semilunar in appearance, pollen chamber extremely reduced appearing like a flattened apical bulge.

Type Species: Tayloriaspermum sinuosum

Derivatio nominis: After Professor Tom Taylor for his outstanding contribution to fossil seed morphology.

Holotype: Slide No.BSIP 10644 Middle Triassic, Nidpur, India.

Discussion:

Tayloriaspermum sinuosum differs from other recorded seed genera because it possesses sinuous cells. It is similar to U. gopadensis in possessing a nucellar apex separated from the outer integument . In the sinuous

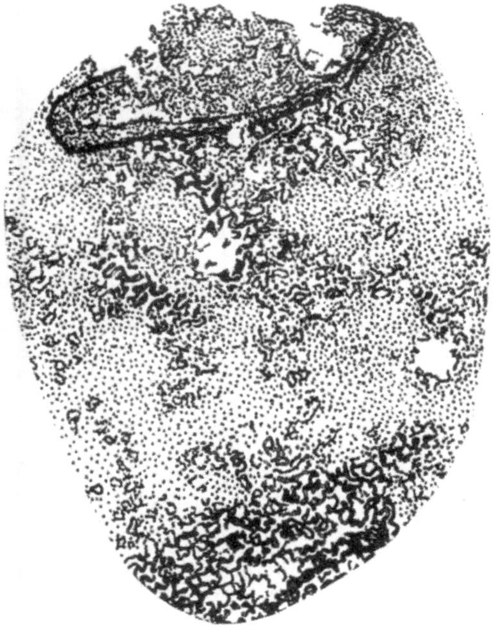

Text-fig.4B. Micropylar region of seedshowing circular opening and distinct nucellus. BSIP Slide No. 10649 X 50.

nature of the seed cuticle it resembles the leaf cuticle of Taeniopteris glandulata (Srivastava, 1971) from the Triassic of Nidpur.

Ctenis stewartiana (Harris, 1932) from the Rhaetic of East Greenland, resembles Tayloriaspermum sinuosum in the sinuous character of its integument. It also shows similarity with the leaf genera Nipaniophyllum raoi (Sahni,1948) and Yabeilla hirsuta (Sukh-Dev,1980) in bearing sinuous anticlinal walls thickened atthe curves.

Affinities:

Considering the extant Cycadales the leaves of Stangeria eriopus Pant and Nautiyal (1963) most closely resemble T. sinuosum because of their structural organisation of loop and crest. The leaves of S. eriopus differ however because of the presence of fine striations and plentiful hair bases. This seed genus shows affinity with the cycads due to differentiation of the outer integument from the nucellus at the apex. The nucellus remain free from the outer integument for a short distance, a feature, frequently encountered in cycadean seeds. Secondly,the extremely reduced pollen chamber and the nature of sinuous cells, justify the assignation of T. sinuosum to Cycadales.

Konaspermum gen. nov.
(Pl.2 Figs.4-8 Text-figs.4,A-B)

Diagnosis: Seed ovate, chalazal end broadly ellipsoidal,cutinized concentric rings encircling sunken micropylar opening, integument composed of angular cells, differentiated into cutinized and uncutinized areas giving a patchy or spotted appearance, inner side of anticlinal walls pronouncedly thickened, outer edge invariably shared by adjacent cells looking somewhat like a double wall, nucellus free from outer investment.

Type species: Konaspermum sidhiensis

Derivatio nominis: Hindi word "Kona" means angular.

Holotype: Slide No.BSIP 10648 Middle Triassic, Nidpur, India.

Discussion:

In possessing a depressed or sunken micropylar hole, spotted or patchy appearance and thickening of cell interiors, Konaspermum closely compares with Amphorispermum pullum (Harris, 1932,1964) but the later is conspicuously different in the absence of angular epidermal cells. In its angulate cells K. sidhiensis compares with seed-type-2 described from the Permian of Congo (Hoeg and Bose,1960).The other seeds described from the Triassic of Nidpur differs from Konaspermum by possessing a spotty and patchy integument consisting of angulate cells.

Affinities

Konaspermum sidhiensis belongs to the order Coniferales because the nucellar membrane is free from the integument over most of the seed and because there is no definite pollen chamber.

CONCLUSION

Study of these gymnospermous seeds has revealed a great diversity indicating all major taxonomic groups;Pteridospermales, Cyucadales and Coniferales flourished in the vicinity of Nidpur. Based on the resemblance of their epidermal features, the notable alliances are: the leaf Lepidopteris indica, the pollen organ Bosea indica containing Weylandites type pollen and the seed Sahnispermum indicum have been attributed to the Lepidopteris- type plant; the leaf Dicroidium nidpurensis, pollen organ Pteruchus nidpurensis and seed Delevoryaspermum nidpurensis have been ascribed to the plant Dicroidium nidpurensis. Amongst the cycadean seeds, Tayloriaspermum sinuosum shows its affinity with Taeniopteris glandulata. In additon Savitrispermum crateriformis, the seed taxon represented in the Gondwanic continent is quite prolific in the Nidpur sediments and based upon epidermal features has shown similarity with seed fructification Umkomasia. The present study of fossil seeds provide a significant key to our understanding of the structure and differentiation of the various groups of plants represented in the flora. The systematic taxonomy further strengthens the view that the Nidpur area was a sheltered fresh water lake or estuary bordered by marshy habitat supporting the growth of cryptogamic plants(Algae,Bryophyta etc.). Also, the features such as trichome, papillae, heavily cutinized cell walls, sinuous cellular outlines must have developed in response to the fluctuating climatic conditions prevalent during the Triassic period at Nidpur. Thus, this investigation of fossil seeds from the Triassic of India has contributed to an acceptable reconstruction of a palaeoenvironment in which in situ burial of vegetation took place.

REFERENCES

Banerji,J., Maheshwari,H.K. & Bose M.N.(1976). Some plant fossils from the Gopad river section near Nidpur Sidhi District, Madhya Pradesh, Palaeobotanist 23: 59-71.

Banerji, J., Kumaran, K.P.N. & Maheshwari, H.K. (1978). Upper Triassic sporae dispersae from the Tiki Formation-1, Megaspores from the Janar Nala Section, South Rewa Gondwana Basin, Palaeobotanist 25:1-26.

Bhardwaj,D.C. & Srivastava, Shyam C. (1969). A Triassic mioflora from Inia. Palaeontographica B125: 119-149.

Bose,M.N. & Banerji,J. (1984). The fossil floras of Kachchh-I Mesozoic megafossils. Palaeobotanist 33:1-89.

Bose,M.N. & Srivastava,Shyam C. (1971). The genus Dicroidium from the Triassic & Nidpur, Madhya Pradesh, India. Palaeobotanist, 19: 41-51.

Bose,M.N. & Srivastava,Shyam C. (1972). Lepidopteris indica sp. nov. from the Lower Triassic of Nidpur, Madhya Pradesh. Journal of the Palaeontological Society of India, 15: 64-68.

Feistmantel,O. (1880). The fossil flora of the Gondwana System; the flora of the Damuda and Panchet Divisions. Memoirs of the Geological Survey of India, Palaeontologia Indica, 3:1-77.

Harris,T.M. (1932). The fossil flora of Scoresby Sound East Greenland Pt.2: Description of seed plants incertae sedis together with a discussion of certain cycadophyte cuticles. Meddelelser om Gronland Kjobenhavn, 85: 1-112.

Harris,T.M. (1935). The fossil flora of Scoresby Sound, East Greenland, Pt. 4, Ginkgoales, Coniferales, Lycopodiales, and isolated fructifications. Meddeleser om ronland Kjobenhavn, 112:1-176.

Harris,T.M. (1964) The Yorkshire Jurassic flora II Caytoniales, Cycadales and Pteridopserms. British Museum (Natural History) London: 1-191.

Hoeg,O. & Bose,M.N. (1960). The Glossopteris flora of the Belgian Congo, Annales du Musee r. du Congo, Belge Serie, 8vo, 32: 1-109.

Lele,K.M. (1955). Plant fossil from Parsora in the South Rewa Gondwana Basin, India. Palaeobotanist, 4:23-24.

Lele,K.M. (1962). Studies in the

Indian Middle Gondwana flora 3: Platyspermic seeds and megaspore impressions from the South Rewa Gondwana Basin, Palaeobotanist, 11:13-18.

Manik,S.R. (1988). Some new genera of Triassic seeds.Palaeobotanist 36: 197-200.

Pant,D.D. & Basu, N. (1977). On some seeds, synangia and scales from the Triassic of Nidpur, India. Palaeontographica, B163: 162-178.

Pant,D.D. & Nautiyal,D.D. (1963). Cuticleand epidermis of recent cycadales, leaves, sporangia, seeds.Senckenbergiana biologica, 44: 257-348.

Pettitt,J.M. & Lacey,W.S. (1972). A Lower Carboniferous seed compression from North Wales. Review of Palaeobotany and Palynology, 14: 159-169.

Roychowdhury,M.K., Sastry,M.V.A., Sah,S.C., Singh,G. & Ghosh,S.C. (1975).Triassic floral succession in the Gondwana of Peninsular India: pp.149 - 158. In: Campbell, K.S.W. (Eds.) Gondwana Geology, 3rd International Gondwana Symposium, Canberra:1-705.

Sahni,B. (1948). The Pentoxyleae a new group of Jurassic gymnosperms from the Rajmahal Hills of India. Botanical Gazette 110: 47-48.

Srivastava,Shyam C. (1971). Some Gymnospermic remains from the Triassic of Nidpur, Sidhi District, Madhya Pradesh. Palaeobotanist 18: 280-296.

Srivastava,Shyam C. (1975). A new microsporangiate fructification from the Triassic of Nidpur, India. Palaeobotanist 22:19-22.

Srivastava,Shyam C. (1988). Stratigraphic positon and age of plant bearing Nidpur beds. Palaeobotanist 36: 154-160.

Srivastava,Shyam C. & Manik,S.R. (1990) Rostrumaspermum venkatachalae gen.et.sp.nov., an archegoniate seed from Triassic of Nidpur, India. In: Jain,K.P. & Tiwari,R.S. (eds.)-Proceedings of the Symposium Vistas in Indian Palaeobotany, Palaeobotanist,38: 98-104. BSIP, Lucknow.

Sukh-Dev, (1980). Occurrence of genus Yabeiella Oishi in the Jabalpur Formation. Palaeobotanist 26: 257-260.

Tiwari,R.S. & Ram-Awatar (1990). Palyno dating of Nidpur beds, Son Graben, Madhya Pradesh, In: Jain,K.P. & Tiwari, R.S. (eds) Proceedings of the Symposium "Vistas in Indian Palaeobotany" Palaeobotanist 38: 105-121. BSIP Lucknow.

Townrow, J.A.(1957): On Dicroidium probably pteriospermous leaf, and other leaves now removed from this genus. Transactions of the geological society of South Africa 60: 21-56.

Gondwana Eight, Findlay, Unrug, Banks & Veevers (eds) © 1993 Balkema, Rotterdam. ISBN 90 5410 304 3

Advent of angiospermous pollen in India and its spatial relationship in Gondwanaland

Archana Tripathi
Birbal Sahni Institute of Palaeobotany, Lucknow, India

ABSTRACT: Recent palynological researches in the subsurface Mesozoic sediments comprising Intertrappean sediments (Rajmahal Formation) in the northern part of Rajmahal Basin have revealed the presence of early angiospermous pollen. Reticulate monosulcate angiospermous pollen appears first in the Indian Early Cretaceous. The distribution pattern of early angiospermous pollen in Gondwanaland suggests a concurrent appearance of reticulate-monosulcate forms in various parts of the supercontinent during Late Neocomian to Aptian time.

1 INTRODUCTION

The Gondwana sediments of the Rajmahal Basin are a part of Rajmahal-Malda-Purnea master basin which extends over a large area of Bengal and north Bihar. They are mostly covered by the Rajmahal traps and alluvium. The stratigraphic succession in Rajmahal Basin (Table 1) includes the Dubrajpur and Rajmahal Formations representing the Upper Gondwana (Triassic to Lower Cretaceous). The palaeobotanical studies of Upper Gondwana sediments in this area have enabled identification of various megafloral and palynological assembalge zones (Sengupta, 1988; Banerji, 1990; Tiwari et al., 1984; Tiwari & Tripathi, (in press). Palynologically a Late Jurassic to Early Cretaceous age has been assigned to the upper part (Infratrappean) of the Dubrajpur Formation while Early Cretaceous to Rajmahal Formation (Tiwari et al., 1984; Tiwari & Tripathi, (in Press). The radiometric dating of Rajmahal volcanics gives a 105 to 115 Ma age (McDougall & McElhinny, 1970; Agarwal & Rama, 1976; Baksi et al., 1987). However, at present the earliest volcanic event in the Rajmahal Basin is not precisely dated because the lower traps are geographically neither synchronous nor well correlated.

2 PALYNOASSEMBLAGE IN THE RAJMAHAL BASIN

During the palynological study of subsurface bore-core material (bore-hole RJNE-32; Tripathi & Tiwari, 1991) rich assemblage from five intertrappean horizons have been recorded showing early angiospermous pollen together with other typical Early Cretaceous palynomorphs. The palynoassemblage from the Infratrappean (uppermost part of Dubrajpur Formation) broadly correlates with the Callialasporites segmentatus Assemblage-Zone having a Tithonian tagging (Venkatachala et al., 1972; Tiwari & Tripathi, (in press) The palynoflora of this level is dominated by Araucariacites and the other taxa present are Podocarpidites, Callialasporites, Gleicheniidites, Baculatisporites, Osmundacidites, Santhalisporites, Cibotiumsporites, Podosporites, Classopollis, Coptospora, Cooksonites and Cicatricosisporites. No angiospermous pollen have been recorded at this level despite thorough searches.

The palynoflora of the Intertrappean beds show close similarity with the Microcachryidites antar-

Table 1 : Geological succession in Rajmahal Basin (adapted after Raja Rao,1987).*levels identified on the basis of palynoflora.

Series	Formation	Lithology
Recent to subrecent	Alluvium	Loose soil, silt and clay
Upper Tertiary		Coarse to medium grained sst., gravel, pebble beds, silt, clay
----------Unconformity----------		
Lower Cretaceous	Rajmahal traps and Intertrappean	Flows of basalt, pitchstone and Intertrappean beds (sst.shale,ash).
----------Unconformity----------		
*Upper Jurassic/ Lower Cretaceous		Pebbly and coarse to fine grained sst., siltstone, clays, grey to pink shales
*Upper Triassic	Dubrajpur	
*Upper Permian		
----------Unconformity----------		
Lower Permian	Barakar	Coarse to medium grained and pebbly sst, grey shale, clay and coal
	Talchir	Tillite, fine to medium grained sst, olive green shale
----------Unconformity----------		
Precambrian		Basic rocks, amphibolites, quartzite limestone, gneisses and granite

cticus Assemblage-Zone of Venkatachala et al. (1972); also Tiwari & Tripathi (in press). The taxa Araucariacites/Podocarpidites dominate the assemblage. The pteridophytic component is represented by a large variety of palynomorphs. Additionally, some angiospermous pollen grains have been recorded from the 1st, 4th and 5th Intertrappean beds, the frequency of which is less than one percent of the total population. They have been assigned to Retimonocolpites peroreticulatus, R. dividus, R. sp. and R. sp.1, Liliacidites sp., aff.Clavatipollenites sp.1., Asteropollis vulgaris, cf. Stephanocolpites sp., cf. Stellatopollis sp., Type A, Type B and Tricolpites variabilis. The first appearance of these taxa observed by Tripathi & Tiwari (1991) reveals that the various types discretely appear at different levels in the sequence (Table 2).

The presence of the index species Coptospora microgranulosa, Foraminisporis asymmetricus, Triporoletes reticulatus, Januasporites spinulosus, Aequitriradites spinulosus, A. verrucosus, Stereisporites antiquasporites, Podosporites microsaccatus etc. together with angio-

Table 2 - Distribution of angiospermous pollen through Rajmahal Formation in bore-hole RJNE 32,Rajmahal Basin

FORMATION	B.H.RJNE 32	ANGIOSPERMOUS PALYNOTAXA
RAJMAHAL	TRAP VI	
	INTERTRAP.	Tricolpites variabilis,Asteropollis vulgaris,Retimonocolpites dividus,Retimonocolpites sp.1,Clavatipollenites hughesii
	TRAP V	
	INTERTRAP.	cf. Stellatopollis sp.,Clavatipollenites hughesii,Retimonocolpites sp.,Liliacidites sp.,cf. Stephanocolpites sp.
	TRAP IV	
	INTERTRAP.	
	TRAP III	
	INTERTRAP.	
	TRAP II	
	INTERTRAP.	aff.Clavatipollenites sp,Type B,Type A, Retimonocolpites peroreticulatus
	TRAP I	
DUBRAJPUR	INFRATRAP.	NO ANGIOSPERMOUS POLLEN

spermous pollen reflect an Early Cretaceous (Late Neocomian to Aptian) aspect. The conclusion is corroborated by the radiometric dates (105-115 Ma) of basaltic flows (McDougall & McElhinny, 1970, Agarwal & Rama, 1976; Baksi et al., 1987).

3 ANGIOSPERMOUS POLLEN MORPHOLOGY

Angiosperm pollen are characterized by: (i) a columellate pollen wall structure differentiable into ectexine consisting of a foot layer, columella, and tectum and nonlaminated endexine; and (ii) aperture types - monosulcate, colpate, porate and colporate (Muller, 1970; Walker & Doyle, 1975). The palaeobiologists's efforts to decipher evolutionary aspects of early angiosperm have led them to study fossil angiospermous pollen wall structure through electron microscopy. Such investigations have revealed the ultrastructure of various components of the pollen wall, i.e. muri, columella, foot-layer, and endexine. The tricolpate pollen have been considered as true angiosperm pollen. Doubts have been raised from time to time for monosulcate pollen type with columellate exine structure (i.e. pollen having both gymnospermous and angiospermous features). They have been described as angiospermid and may represent a transitional stage between a gymnosperm and angiosperm pollen (Hughes, 1976,1984). The monosulcate pollen with reticulate columellate exine structure, viz.

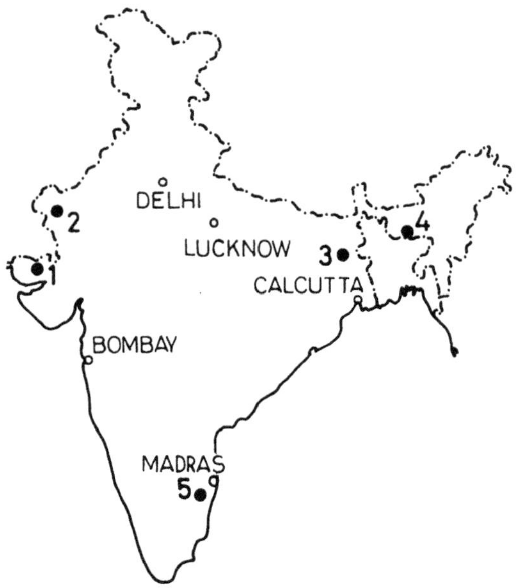

Fig.1 Areas of early angiospermous pollen record during Late Neocomian to Turonian in India. The index numbers 1-5 are given for the data base. 1.Mathur & Mathur, 1972; 2.Mathur et al., 1970; 3.Vishnu-Mittre 1956; Tripathi & Tiwari, 1991; 4.Nandi, 1984, 1990; Kar & Singh, 1986; 5.Jain & Taugourdeau-Lantz, 1973; Venkatachala & Sharma,1974a, 1974b.

Clavatipollenites, Retimonocolpites, Liliacidites, Stellatopollis have been recorded from the Barremian to Aptian (Van Campo, 1971; Doyle et al., 1975; Walker & Walker, 1984; Burger, 1990). The reports of such palynomorphs from Israel (Brenner, 1984), England (Hughes & McDougall, 1987), Morocco (Gubeli et al.,1988) and Italy (Trevisan, 1988) have suggested their inception during Hauterivian to Valanginian times. The find of Lethomasites from the Lower Potomac Group by Ward et al. (1989) adds a new group comprising granular monosulcate pollen, to the early angiosperm record. Recently pollen with similar pollen wall morphology have been recorded from the Late Triassic (Cornet, 1989; Doyle, 1978; Litwin, 1985; Zavada, 1990; Vasanthy et al., 1990) although an angiospermous affinity has not been attached to such forms.

Ultrastructural studies could not be attempted at present due to the paucity of specimens in the Rajmahal Basin. However, light microscopy revealed twelve different morphological groups (after Tripathi & Tiwari, 1991).

1. Coarsely reticulate-noncolumellate-monosulcate, eg. Retimonocolpites peroreticulatus (Fig.-2.f,g).
2. Finely reticulate-noncolumellate-monosulcate, eg. Retimonocolpites sp. (Fig.2.e).
3. Finely reticulate-monosulcate; reticulation absent on bordering region of sulcus and terminal areas, eg. Liliacidites sp. (Fig.-2.a).
4. Finely reticulate-columellate, monosulcate eg. Retimonocolpites sp.1, Retimonocolpites dividus (Fig.2.d).
5. Finely reticulate-semitectate--columellate-monosulcate, eg. Clavatipollenites hughesii (Fig.-2.b,c).
6. Tectate-columellate-reticulate--tricolpoidate, eg. cf.Stephanocolpites sp. (Fig.2.i).
7. Tectate-semiperforate-columellate-trichotomosulcate, eg. aff.-Clavatipollenites sp.1.
8. Tectate-columellate-reticulate--trichotomosulcate, eg. Asteropollis vulgaris (Fig.2.k).
9. Tectate-clumellate-reticulate--tricolpate,eg.Tricolpites variabilis (Fig.2.j).
10. Semitectate-columellate-trichotomosulcate, eg. Type A.
11. Monosulcate pollen bearing bacula, coni and clavae probably with reticulation (reticulation could not be confirmed under LM), eg. cf. Stellatopollis sp. (Fig.2.h).
12. From the Nipania chert near Amrapara village in Rajmahal Basin a juglandoid polyporate grain with 10-12 simple pores has been recorded by Vishnu-Mittere (1956). The exine stratification of this type is not distinct.

It is worth mentioning that most of the pollen are medium sized, except cf. Stephanocolpites whcih is small and aff. Clavatipollenites sp.1, Liliacidites sp. and Retimonocolpites sp.1 which fall in the large size group of Walker & Doyle (1975).

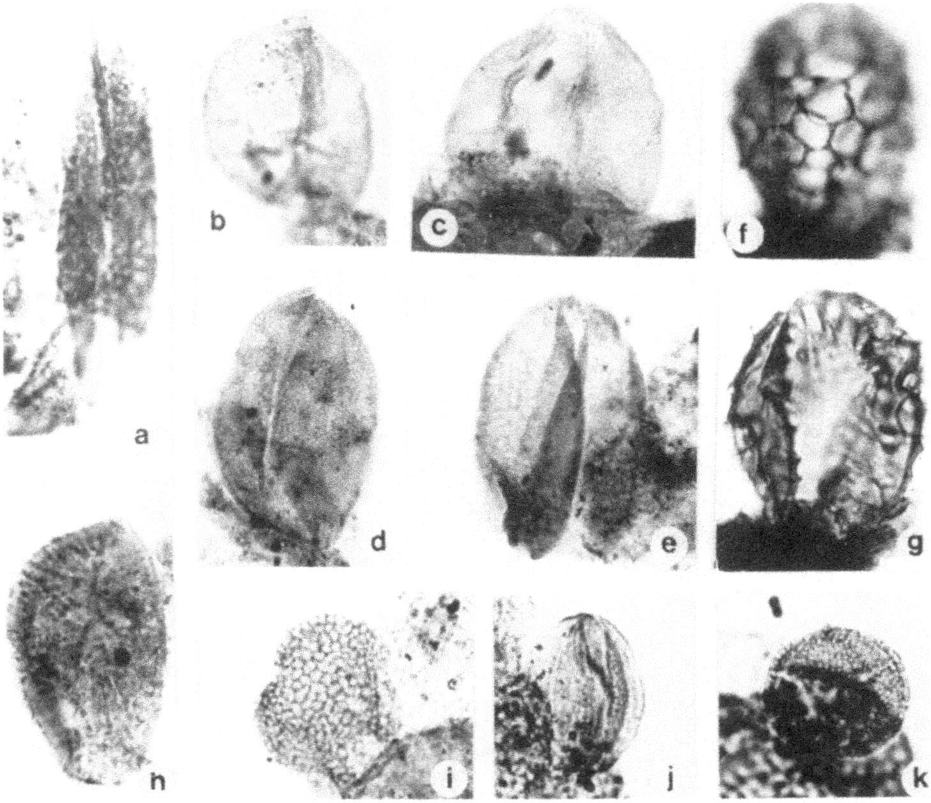

Fig.2. a-k Early angiospermous pollen from Rajmahal Formation Early Cretaceous sediments from Rajmahal Basin, India. (All photomicrog.aphs x 1000) (a) Liliacidites sp.; (b,c) Clavatipollenites hughesii Coup. 1958; (d) Retimonocolpites dividus Pierce 1961; (e) Retimonocolpites sp.; (f,g) Retimonocolpites peroreticulatus (Bren) Doyle et al. 1975; cf. Stellatopollis sp.; (h) cf. Stephanocolpites sp.; (i) Tricolpites variabilis (Pierce) Sriv, 1975; (j) Asteropollis vulgaris (Groot) Singh, 1963.

4 EARLY ANGIOSPERM POLLEN RECORD FROM INDIA

Extensive palynological analyses of subsurface and surface sediments in different parts of India during the last three decades, has revealed rich palynofloras in the Cretaceous sequence. A synthesis of Cretaceous palynological data reveals that the early angiosperm pollen are very rare during the Early Cretaceous but they diversify in the Late Cretaceous (Table 3, Fig.1). It is also evident that the earliest Indian record of angiosperm pollen yet found is from the Rajmahal Intertrappean sediments of Late Neocomian to Aptian age (Tripathi & Tiwari, 1991). They are represented by reticulate monosulcate pollen (Retimonocolpites spp. and cf. Stellatopollis sp.), semitectate-columellate finely reticulate monosulcate pollen (Clavatipollenites hughesii), semitectate-reticulate-columellate trichotomosulcate-pollen (aff. Clavatipollenites sp.1, Asteropollis vulgaris), tectate-columellate-reticulate tricolpoidate pollen (cf.Stephanocolpites sp.), and tectate-columellate-reticulate

Table 3 - Early angiospermous pollen record from india.Data base as given in Fig.1.The arrow indicates the age ranges of the taxa listed in front of the arrow.

AGE		GUJARAT,RAJASTHAN	RAJMAHAL	MEGHALAYA	CAUVERY
TURONIAN				Preudoplicapollis, Plicapollis, Atlantopollis, Complexiopollis, Tricolporopollenites, Tricolpopollenites, Liliacidites, Retimonocolpites, Clavatipollenites	
CENOMANIAN		Triporopollenites, Triporites, Retitetracolpites, Punctatricolpites, Retitricolpites, Tricolpites, Clavatipollenites			Scabrastephanocolpites, Tricolporopollenites, Gothanipollis, Ilexpollenites, Tricolpites, Liliacidites
ALBIAN		Striatricolporites, Striopollenites, Tricolporopollenites, Tricolpopollenites, Retitricolpites, Tricolpites, Clavatipollenites	Sporojuglandoidites	Stephanoporopollenites, Psilatricolporites, Psilastephanocolpites, Palmaepollenites, Palmidites, Liliacidites	Tricolpopollenites, Gothanipollis, Ilexpollenites, Tricolpites, Liliacidites
APTIAN			cf.Stellatopollis Tricolpites, Asteropollis, Liliacidites, cf Stephanocolpites, Clavatipollenites, aff Clavatipollenites, Retimonocolpites		tricolpate form
BARREMIAN					
NEOCOM	HAUTERIVIAN				
	VALANGINIAN				
	BARRIASIAN				

Table 4 - Relationship of Early Cretaceous angiospermous pollen from India, and various suites proposed by Burger (1981) on the basis of first appearence of aperture types. Suite I-monocolpate; Suite II-tricolpate; Suite III-tricolporate; Suite IV-triporate; the details of numbers 1-5 are as given in Fig.1.

Age		1,2 Gujarat Rajasthan	3 Rajmahal	4 Meghalaya	5 Cauvery
Cenomanian		III,IV		IV	
Albian	L			II,III	III,IV
	M				
	E				
Aptian		I,II			II
Barremian			I,II		
NEOCOM	Hauterivian				
	Valanginian				
	Barriasian				

tricolpate pollen (Tricolpites variabilis). Besides these, some monosulcate pollen are also recorded, which do not show a reticulate exine but have an infrastructured exine or a fine granulate or rugulate surface pattern.

The tectate-columellate-reticulate tricolpate pollen (eg. Tricolpites) is recorded from Cauvery Basin at a level dated as Aptian to Lower Albian on the basis of microplanktons, pollen and spores (Jain & Taugourdeau-Lantz, 1973. A variety of tricolpate pollen including Tricolpites spp. and Tricolpopollenites spp. are represented in Albian sediments of Kachchh, Rajasthan and Cauvery Basins (Mathur et al., 1970; Mathur & Mathur, 1972; Lukose, 1974; Venkatachala & Sharma, 1974b). The latter authors and Kar & Singh (1986) have described monosulcate, tricolpate and tricolporate pollen from Cauvery Basin and Meghalaya respectively and dated the assemblage as Late Albian to Cenomanian. From post Albian sediments pollen with varied aperture types (monosulcate, tricolporate and porate) and of exine structure are on record from Kachchh, Cauvery and Meghalaya (Mathur et al., 1970; Venkatachala & Sharma, 1974a,1974b; Nandi, 1984,1990).

In summary occurrence of angiosperm pollen evidently appeared in India during the Late Neocomian to Aptian and were mostly represented by reticulate monosulcate pollen. The other morphotypes appeared successively tricolpate in Early to Middle Albian, tricolporate and triporate in Late Albian to Cenomanian. Referring to the zonation scheme proposed by Burger (1981) based on the first appearance of various aperture types following relationship with Indian occurrences is suggested in Table 4.

5 RELATIONSHIP

Published data (Table 5) suggests that the earliest unquestionable angiosperm pollen Clavatipollenites hughesii appears in ?Valanginian to Haute-

Table 5 - Early Cretaceous angiospermous pollen record from Gondwanaland. Data base is from South America (Archengelsky & Gamero,1967, Valkeima & Salas,1975 in Burger,1981): Africa (Doyle et al.,1977, Schrank,1983,1987, Gubeli et al.,1988, Trevisan,1988, Burger,1981); Antarctica (Dettmann & Thomson,1987, Drinnan & Crane,1989, Truswell,1989: Australia (Burger,1988,1990, Dettmann,1986, Drinnan & Crane,1989): India (Data as in Fig.1). The arrow indicate age ranges of taxa.

AGE	SOUTH AMERICA	AFRICA	INDIA	ANTARCTICA	AUSTRALIA
ALBIAN	Stephanocolpites, Tricolpites; reticulate tricolpate	Tricolpites vulgaris; Liliacidites,Clavatipollenites	Gothanipollis, Ilexpollenites; Tricolpopollenites minutus, T.cf.micromunus, Tricolpites reticulatus, Clavatipollenites couperii	Asteropollis asteroides, Tricolpites sp.; Clavatipollenites	Tricolpites variabilis, Retimonocolpites peroreticulatus
APTIAN	Clavatipollenites hughesii	Tricolpites sp., Retimonocolpites dividus	tricolpate form; Tricolpites variabilis, Asteropollis vulgaris, Retimonocolpites dividus, Retimonocolpites sp.1, cf.Stellatopollis sp., cf. Stephanocolpites sp., Liliacidites sp.,Clavatipollenites hughesii, Retimonocolpites sp.,Type A ,Type B, aff. Clavatipollenites sp.1, Retimonocolpites peroreticulatus		Asteropollis asteroides, Clavatipollenites spp; Clavatipollenites hughesii
BARREMIAN	Clavatipollenites , Liliacidites	Retimonocolpites peroreticulatus, Asteropollis vulgaris,Liliacidites sp.,Clavatipollenites hughesii, cf.Stellatopollis barghornii			Clavatipollenites sp.
NEOCOM HAUTERIVIAN					
NEOCOM VALANGINIAN	Liliacidites type	Clavatipollenites hughesii, reticulate columellate monosulcate			
BERRIASSIAN					

rivian strata of Africa; although some reticulate-columellate monosulcate pollen have been recorded from the Barremian of Israel yet the identity of this form could not be confirmed as no illustrations or descriptions are given (Trevisan, 1988). In Indian Early Cretaceous assemblages Retimonocolpites peroreticulatus is first to appear in the Late Neocomian to Aptian (Rajmahal palynosequence). Its earliest record in Africa is from the Barremian to Early Aptian (Doyle et al., 1977, Gubeli et al., 1988) and in Australia from the Middle Albian (Drinnan & Crane, 1989). The next forms to appear in the Rajmahal sequence are aff. Clavatipollenites sp.1 and Type A (triangular trichotomosulcate). So far, similar pollen have not been reported from outside India. The third successive appearance is of pollen Type B (tectate-columellate-reticulate) the generic assignment of which is not possible due to incomplete nature of the specimen. The types Clavatipollenites hughesii, Retimonocolpites sp., Liliacidites sp. and cf. Stephanocolpites sp. appear next in the succession. Exact matching of these forms with records from other parts of Gondwanaland is difficult. However, at generic level, Liliacidites appears in the Barremian of South America (Archangelsky, 1980) and Africa (Doyle et al., 1977). Liliacidites type pollen are on record from Barriasian to Barremian in South America (Archangelsky, 1980). Similarly, the genus Retimonocolpites is observed for the first time in the Barremian of Africa (Doyle et al., 1977). The taxon Stephanocolpites is first registered in the Middle Albian of South America (Burger, 1981). The next organization to appear in the succession of Indian palynoflora is cf. Stellatopollis sp. Representative of Stellatopollis is first recorded from the Barremian of Africa (Doyle et al., 1977). The group Retimonocolpites dividus, Retimonocolpites sp.1, Asteropollis vulgaris and Tricolpites variabilis proceeds the occurrence of cf. Stellatopollis sp. Retimonocolpites dividus appears in the Aptian of Africa. The type Asteropollis vulgaris is first recorded in the Barremian (Schrank, 1983,1987) and Tricolpites variabilis is reported from the Early to Middle Albian of Africa (Gubeli et al., 1988). It is derived from the above discussion that the inception of early angiosperm pollen reticulate monosulcate and aff. Clavatipollenites groups of pollen took place at Barremian to Aptian time simultaneously in different parts of the Gondwanaland.

6 DISCUSSION

Various theories have been putforth for the origin, radiation and migration of angiosperms (Heer, 1863 in Thomas & Spicer, 1987; Saporta, 1877 in Thomas & Spicer, 1987; Takhtajan, 1969; Axelrod, 1952,1959,1970; Brenner, 1976; Hughes, 1961,1976; Stebbins, 1974 in Friis et al., 1987; Raven & Axelrod, 1974; Burger, 1990).

In relation to the present study the role of India as passage enroute from Africa to Australia is vital. This aspect has been dealt by Raven and Axelrod (1974) who proposed a probable route for migration of early angiosperms from West Gondwana to Australasia via India and Antarctica. Lately Burger (1990) has doubted the feasibility of such pathways because the data available had provided evidence for the earliest record of angiosperms in India from the Late Albian (Venkatachala & Sharma, 1974b, Srivastava, 1983). Evidently the story has changed now as the confirmed occurrence of angiospermous pollen from the Late Neocomian to Aptian has been documented in the present work. The conclusions that the earliest angiospermous pollen appeared almost at the same time in Africa, India and Australia cannot be avoided. In sequence of logical derivations it is further evident that the propositions for migrations of angiosperms from pole to equator (Dettmann, 1986) or from equator to pole from one centre (Axelrod, 1959; Hughes, 1961) seem to be improbable at the present juncture. The answers to the radiation and migratory routes have to be sought in the pollen stalks of Pre-Cretaceous time which bear angiospermoid exine structure (Cornet, 1989; Zavada, 1990; Vasanthy et al., 1990). The simultaneous occurrence of definite angiospermous pollen at the end Neocomian further gives an insight to the fact that such morphology must have a long preceding history which relate to the sudden proliferation of the angiosperms in the Upper Cretaceous.

7 CONCLUSIONS

(i) Advent of early angiospermous pollen in India is recorded during Late Neocomian to Aptian time.
(ii) Reticulate monosulcate pollen is the likely candidate for earliest angiospermous morphology.
(iii) The reticulate monosulcate group of angiospermous pollen show their appearance simultaneously in different parts of Gondwana during the Late Neocomian to Aptian.

8. ACKNOWLEDGEMENTS

The author is thankful to Dr. B.S. Venkatachala, Director, Birbal Sahni Institute of Palaeobotany, Lucknow for his kind permission to communicate this paper. Thanks are extended to Dr. R.S. Tiwari, Head, Dept. of Pre-Gondwana & Gondwana Palynostratigraphy, Birbal Sahni Institute of Palaeobotany, Lucknow for the constant encouragement and fruitful discussions during the progress of this work. I wish to gratefully acknowledge the Council of Scientific & Industrial Research, New Delhi, the Hon. Chief Minister, Government of Uttar Pradesh and Birbal Savitri Sahni Foundation, Lucknow for their financial support which made it possible for me to participate in this Symposium at Hobart, Tasmania, Australia.

REFERENCES

Agarwal, J.K. & Rama, F.A. 1976. Chronology of Mesozoic volcanics of India. Proc. Indian Acad. Sci. 84A(4):157-179.

Archangelsky, S. 1980. Palynology of the Lower Cretaceous of Argentina. Proc. 4th Int. Palynol. Conf. Lucknow 2, pp.425-428.

Archangelsky,S. & Gamerro,J.C.1967. Spore and pollen types from the Lower Cretaceous in Patagonia Argentina. Rev. Palaeobot. Palynol. 1:211-217.

Axelrod, D.I. 1952. A theory of angiosperm evolution. Evolution 6:29-60.

Axelrod, D.I. 1959. Poleward migration of early angiosperm flora Science 130:203-207.

Axelrod, D.I.1970. Mesozoic palaeogeography and early angiosperm history. Bot. Rev. 36:277-319.

Baksi,A.K., Ray Barman,T., Paul, D.K. & Farrar, E. 1987. Widespread Early Cretaceous flood basalt volcanism in eastern India. Geochemical data from the Rajmahal-BengalSylhet traps. Chemical Geology 63:133-141.

Banerji, J. 1990. Plant fossils from Dubrajpur Formation, Bihar and their significance in stratigraphy. Proc. Symp. 'Vistas in Indian Palaeobotany' Palaeobotanist 38:122-130.

Brenner, G.J. 1976. Middle Cretaceous floral provinces and early migration of angiosperm. Origin and early evolution of angiosperms, Columbia Univ. Press, New York, pp.23-47.

Brenner, G.J. 1984. Late Hauterivian angiosperm pollen from the Helez Formation Israel. 6th Internat. Palynol. Conf., Calgary, 1984, Abstracts, p.15.

Burger, D. 1981. Observation on the earliest angiosperm development with special reference to Australia. Proc. 4th Int. Palynol. Conf. Lucknow, 3:418-428.

Burger, D. 1988. Early Cretaceous environments in the Eromanga Basin. Palynological evidence from GSQ Wyandral Core hole. Assoc. Australas. Palaeontol. Mem. 5:173-186.

Burger, D. 1990. Early Cretaceous angiosperm from Queensland Australia. Rev. Palaeobot. Palynol. 65:153-164.

Cornet, B. 1989. Late Triassic angiosperm like pollen from the Richmond Rift Basin of Virginia, USA. Palaeontographica B 213(1-3):37-87.

Dettmann, M.E. 1986. Early Cretaceous palynoflora of subsurface strata correlative with the Koonwara Fossil Bed, Victoria.

Assoc. Australas. Palaeontol. Mem. 3:79-110.

Dettmann, M.E. & Thomson, M.R.A. 1987. Cretaceous palynomorphs from James Ross Island area, Antarctica-a pilot study. Br. Antarct. Surv. Bull. 77: 13-59.

Doyle, J.A. 1978. Origin of angiosperms. Ann. Rev. Ecol. Syst. 9:365-392.

Doyle, J.A., Biens, P., Döerenkamp, A. & Jardiné, S. 1977. Angiosperm pollen from the Pre-Albian Lower Cretaceous of Equatorial Africa. Bull. Cent. Rech. Explor. Prod. Elf-Aguitaine 1(2):451-473.

Doyle, J.A., Van Campo, M. & Lugardon, B. 1975. Observations on exine structure of Eucommiidites and Lower Cretaceous angiosperm pollen. Pollen Spores 17: 429-486.

Drinnan, A.N. & Crane, P.R. 1989. Cretaceous palaeobotany and its bearing on the biogeography of a Austral-angiosperms. In T.N. Taylor & E.L Taylor (eds), Antarctic Palaeobiology: Its role in the reconstruction of Gondwana. Springer Verlag, New York Inc., pp. 192-219.

Friis, E.M., Chaloner, W.G. & Crane, P.R. 1987. Introduction to angiosperms. In E.M. Friis et al. (eds.), The origin of angiosperms and their biological consequences. Cambridge University Press, Cambridge, pp.1-15.

Gubeli, A.A., Hochuli, P.A. & Wildi, W. 1988. Lower Cretaceous turbiditic sediments from the Rift Chain (Northern Morocco). Palynology, stratigraphy and palaeogeographic setting. Geologische Rundschau 73(3): 1081-1114.

Hughes,N.F. 1961.Fossil evidence and angiosperm ancestry. Sci. Prog. 49:84-102.

Hughes, N.F. 1976. Palaeobiology of angiosperm origins. Cambridge. Univ. Press, Cambridge.

Hughes, N.F.1984. Mesosperm palynological evidence and ancestors of angiosperms. Ann.Miss. Bot. Gard. 71:539-598.

Hughes, N.F. & McDougall, A.B. 1987. Records of angiospermid pollen entry into the English Early Cretaceous succession. Rev. Palaeobot. Palynol. 5-: 255-272.

Jain, K.P. & Taugourdeau-Lantz, J. 1973. Palynology of Dalmiapuram grey shale, Dalmiapuram Formation, District Trichnopoly, South India-1. Taxonomy. Geophytology 3(1):52-68.

Kar, R.K. & Singh, R.S. 1986. Palynology of the Cretaceous sediments of Meghalaya, India. Palaeontographica B 202(1-6): 83-153.

Litwin, R.J. 1985. Fertile organs and in situ spores of ferns from the Late Triassic Chinle Formation Arizona and New Mexico with discussion of the associated dispersed spores. Rev. Palaeobot. Palynol. 44: 101-146.

Lukose, N.G. 1974. Palynology of the subsurface sediments of Manhera Tibba structure, Jaisalmer, Western Rajasthan, India. Palaeobotanist 21(3):285--297.

Mathur, Y.K. & Mathur, K. 1972. Angiospermous pollen and associated fossil from the Mid-Cretaceous subsurface sediments of Rajasthan, India.J. Palynol. 8:89-96.

Mathur, Y.K., Soodan, K.S., Mathur, K., Bhatia,M.L., Juyal, N.P. & Pant, S. 1970. Microfossil evidences on the presence of Upper Cretaceous and Palaeocene sediments in Kutch. Bull. ONGC 7(2):109-114.

McDougall, I. & McElhinny, M.W. 1970. The Rajmahal Traps of India K-Ar ages and palaeomagnetism. Earth and Planat.Sci. Letters 9:371-378.

Muller. J. 1970. Palynological evidence on early differentiation of Angiosperms. Biol. Rev. 45:417-450.

Nandi, B. 1984. Palynostratigraphy of the Gumaghat Formation, Meghalaya India with special reference to the significance of the Normapolles Groups. A.K. Ghosh Commem. Vol.Evolutionary Botany & Biostratigraphy.Curr. Trends Life Sci. 10:521-540.

Nandi, B. 1990. Early angiosperm pollen grains from Meghalaya and Late Cretaceous event stratigraphy. Contributions. Sem. Cum Workshop IGCP-216 and 245, Chandigarh, 1990:62-63.

Raja Rao, C.S. 1987. Coal resources of Bihar(excluding Dhanbad District). Bull.Geol.Surv.India

Ser. A No.45 IV(1):1-336.
Raven, P.H. & Axelrod, D.I. 1974. Angiosperm biogeography and past continental movements. Ann.Miss.Bot.Gard.61:539-673.
Schrank, E. 1983. Scanning electron and light microscopic investigation of angiosperm pollen from the Lower Cretaceous of Egypt. Pollen Spores 25(2):213-242.
Schrank, E. 1987. Palaeozoic and Mesozoic palynomorphs from northeast Africa (Egypt and Sudan) with special reference to Late Cretaceous pollen and dinoflagellates. Berliner geowiss, Abh.(A) 75.1:249-310.
Sengupta, S. 1988. Upper Gondwana stratigraphy and palaeobotany of Rajmahal Hills, Bihar, India Geol. Surv. India Palaeont. Indica 48:1-182.
Srivastava, S.K. 1983. Cretaceous phytogeoprovinces and palaeogeography of the Indian plate based on palynological data. Proc. Symp. Cretaceous of India, Lucknow, Indian Assoc. Palynostratigr., pp.141-157.
Takhtajan, A. 1969. Flowering plants:Origin and Dispersal. Oliver and Boyd, Edinburgh.
Thomas, B. & Spicer, R.A. 1987. The evolution and palaeobiology of land plants. Croom Helm London & Sydney.
Tiwari, R.S., Kumar, P. and Tripathi, Archana 1984. Palynodating of Dubrajpur and Intertrappean beds in subsurface strata of north-eastern Rajmahal Basin. Proc. 5th Indian Geophytol. Conf. Lucknow, pp.207-225.
Tiwari, R.S. & Tripathi, A. in press . Marker Assemblage Zones of spores and pollen species through Indian Gondwana sequence. In Birbal Sahni Centenary Volume.
Trevisan, L. 1988. Angiospermous pollen (monosulcate-trichotomosulcate phase) from the verv early Lower Cretaceous of southern Tuscany (Italy): some aspects. 7th Internat. Palynol. Conf., Brisbane, Abstracts, p.165.
Tripathi, A. & Tiwari, R.S. 1991. Early Cretaceous angiospermous pollen from the Intertrappean beds of Rajmahal Basin, Bihar, India. Palaeobotanist 39(1): 50-56.
Truswell, E.M. 1989. Cretaceous and Tertiary vegetation of Antarctica:a palynological perspective. In T.N. Taylor & E.L. Taylor (eds), Antarctic Palaeobiology:its role in the reconstruction of Gondwana. Springer-Verlag, New York Inc.,pp.71-88.
Van Campo, M. 1971. Precision nouvelles sur les structures comparees des pollen de Gymnosperms et Angiospermes. Compt. Rendu Hebd. Seances Acad. Sci. S.D. 272:2071-2074.
Vasanthy, G., Venkatachala, B.S. & Pocock, S.A.J. 1990. The evolution of angiospermid pollen characteristics: conjectures and queries. Proc. Symp.'Vistas in Indian Palaeobotany'. Palaeobotanist 38:131-146.
Venkatachala, B.S. & Sharma, K.D. 1974a. Palynology of the Cretaceous sediments from the subsurface of Vridhachalam area, Cauvery Basin. Geophytology 4(2):153-183.
Venkatachala, B.S. & Sharma, K.D. 1974b. Palynology of the Cretaceous sediments from the subsurface of Pondicherry area, Cauvery Basin. New Botanist 1 (3 & 4):170-200.
Venkatachala, B.S., Sharma, K.D. & Jain, A.K. 1972. Palynological zonation of Jurassic-Lower Cretaceous sediments in the sub-surface of Cauvery Basin. Proc. Seminar Palaeopalynology and Indian Stratigraphy, Calcutta, pp. 172-187.
Vishnu-Mittre, 1956. Sporojuglandoidites jurassicus gen. et. sp.nov. a sporomorph from the Jurassic of Rajmahal Hills. Bihar. Palaeobotanist 4:151-152.
Walker, J.W. & Doyle, J.A. 1975. The bases of angiosperm phylogeny: Palynology. Ann. Miss. Bot. Gard. 62:664-723.
Walker, J.W. & Walker, A.G. 1984. Ultrastructure of Lower Cretaceous angiosperm pollen and the origin and early evolution of flowering plants. Ann. Miss. Bot. Gard. 71:464-521.
Ward, J.V., Doyle, J.A. & Hotton, C.L. 1989. Probable granular, magnolid angiosperm pollen from the Early Cretaceous. Pollen Spores 31(1-2): 113-132.
Zavada, Michael, S. 1990. The ultrastructure of three monosulcate pollen grains from the

Triassic Chinle Formation, Western United States. Palynology 14:41-51.

Gondwana Eight, Findlay, Unrug, Banks & Veevers (eds) © 1993 Balkema, Rotterdam. ISBN 90 5410 304 3

Paleoclimatic significance and age constraints of the Carboniferous coals of Paganzo basin, western Argentina

Oscar R. López-Gamundí
Texaco Inc., Frontier Exploration Department, Bellaire, Tex., USA

Silvia N. Cesari & Carlos O. Limarino
Departamento de Ciencias Geológicas, Universidad de Buenos Aires, Ciudad Universitaria, Argentina

ABSTRACT: Glacial deposits are extensively distributed in the Paganzo basin of western Argentina. Both palynological and megafloristic remains suggest a Middle Carboniferous age, partially coincident with the period of glaciomarine sedimentation that peaked between Late Namurian and Westphalian in the adjacent Calingasta-Uspallata basin. These glacial deposits are succeeded by cross-bedded sandstones, mudstones, carbonaceous shales and coals deposited in either fluvial or deltaic settings. The coal-bearing strata are characterized by the presence of plant remains of the NBG (*Nothorhacopteris*, *Botrychiopsis* and *Ginkophyllum*) megafloristic assemblage and spores of the *Raistrickia - Plicatipollenites* subzone (*Potoniesporites - Lundblandispora* zone), both indicative of a Late Westphalian-Stephanian age. Due to their narrow chronostratigraphic range and wide areal distribution, the coals constitute an excellent marker for intrabasinal correlation. The litho- and chronostratigraphic proximity of the glacial sediments and superjacent coals suggest that widespread blanket peats formed during the waning period of, or immediately after, glaciation under cold to temperate, humid climatic conditions. Autocyclic factors exerted a significant control on lateral thickness changes and quality of Paganzo coals. Allocyclic climatic factors account for the observed synchronous occurrence of coals throughout the basin.

INTRODUCTION

A recently proposed model of peat forming conditions for Gondwanan coals (Martini and Glooschenko, 1985) has revived some questions upon the paleoclimatic constraints of coal beds. The classic model of a humid and tropical environment has been contrasted to a model in which the environment is humid and cold. The "cold" model stems from well-documented examples from the Quaternary glacial and postglacial sediments of the northern hemisphere. Analogies to this have been proposed between the peat-forming conditions currently active in some sub-Arctic areas of Canada and the origin of the Permian coal deposits of Australia. This actualistic approach has been also extended to interpret the paleoclimatic conditions under which the Gondwanan coal deposits from Australia (Hobday, 1987; Martini and Johnson, 1987; Martini and Banks, 1989), India (Raja Rao and Mitra, 1985), South Africa (Hobday, 1987; Kreuser, 1987) and South America (Martini and Rocha-Campos, 1989) formed.

In the Paraná basin of central and southern Brazil an interglacial coal-bearing association of thin coals and diamictites has been identified in the lower portion of the Itararé Formation (Martini and Rocha-Campos, 1989). Another sedimentary association, of deltaic affinities, occurs in the postglacial Early Permian Rio Bonito Formation.

Coal-bearing strata of the Paganzo basin in western Argentina (Fig. 1) have a peculiar areal and temporal distribution: they overlie glacial or coeval deposits and their chronostratigraphic range lies within

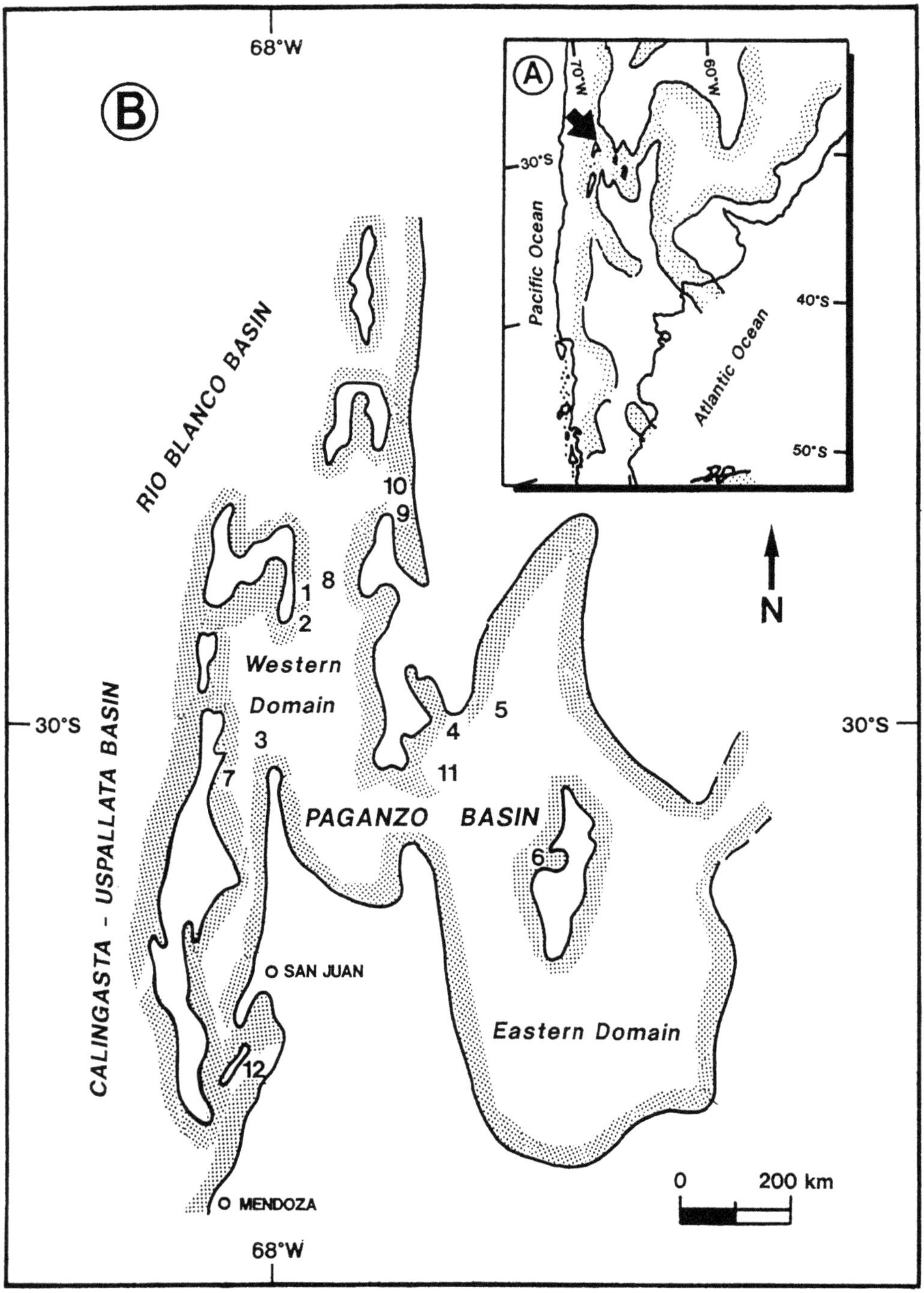

Figure 1. A: Generalized Carboniferous paleogeography of southern South America. Paganzo basin indicated by arrow. B: Carboniferous paleogeography of central western Argentina (adapted from Salfity and Gorustovich, 1981) with location of principal occurrences of coal-bearing strata indicated by numbers from 1 through 12. These numbers refer to localities cited in Tables 1 and 2.

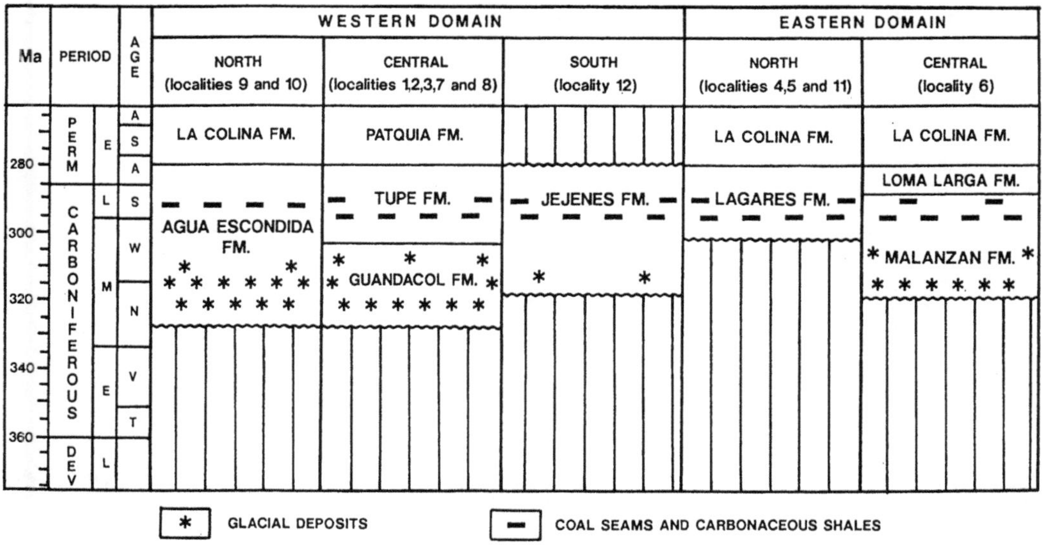

Figure 2. Stratigraphic chart for the coal-bearing units of the Paganzo basin, western Argentina. Localities as in Figure 1.

Middle Carboniferous times (Fig. 2). Most commonly, these coal-bearing beds rest on glaciterrestrial (glacilacustrine fine-grained deposits with abundant ice-rafted debris and associated diamictites with faceted and striated clasts) or glaciomarine (rain out tillites and dropstone shales) deposits of Middle Carboniferous age (paleoclimatic stage II, López-Gamundí et al., 1992). Their litho- and chronostratigraphic proximity to glacial deposits indicate a close paleoclimatic linkage with the Gondwanan glaciation. The purpose of this paper is to document this relationship in the context of the paleoclimatic evolution of this part of Gondwanaland.

STRATIGRAPHIC SETTING AND DISTRIBUTION OF THE COAL-BEARING STRATA

The Paganzo basin of western Argentina (Fig. 1), with an area of approximately 150,000 km ^{2}, is characterized by the widespread presence of deposits of Carboniferous and Permian age. Two sedimentary domains have been recognized in the basin (Fig. 1, B). The eastern domain is characterized by non-marine sediments represented by dominant fluvial deposits and subordinate lacustr ine sediments. Fluvial and deltaic facies are common in the western domain; paleogeographic reconstructions suggest a connection with the marine deposits of the Calingasta-Uspallata basin (Fig. 1, B).

The coal seams and carbonaceous mudstones and shales of the Paganzo basin occur exclusively at a specific stratigraphic interval within the Late Carboniferous Tupe Formation or correlatives, being especially abundant in the lower part of these units (Table 1 and Fig. 2).

Autocyclic controls are important in terms of the thickness and paleoenvironmental context of the coal-bearing strata. Coal thickness ranges between 0.5 and 1.5 m, exceptionally reaching 2 m (Alvarez Rojo, 1987). In the eastern domain coals are associated with low-sinuosity river deposits. In this part of the basin sedimentation commenced probably earlier in postglacial times during the Late Carboniferous; coal-bearing strata rest directly on granitic and metamorphic basement rocks in some localities. Lacustrine deposits are associated with carbonaceous mudstones in some localized paleovalleys of the southeast part of the basin (Malanzán area, locality 6 in Fig. 2). Recent geochemical work (Villar et al., 1991) indicates

Table 1. Middle Carboniferous coals and carbonaceous mudstones of the Paganzo basin (numbers in parentheses refer to location in Figure 1)

Locality	Formation	Facies type
La Victoria Mine (1)	Tupe	swamp (fluvial)
La Negra Mine (2)	Tupe	swamp (fluvial)
La Ciénaga Mine (3)	Tupe	swamp (fluvial)
Las Mellizas Mine (4)	Lagares	swamp (fluvial)
A.P. Green Mine (5)	Lagares	swamp (fluvial)
Malanzán (6)	Malanzán	lacustrine
Rio Francia (7)	Tupe	swamp (fluvial)
Las Pircas Mine (8)	Tupe	swamp (fluvial)
Tambillos Mine (9)	Agua Colorada	swamp (fluvial)
Casa Blanca (10)	Agua Colorada	lacustrine
Lagares (11)	Lagares	swamp (fluvial)
Jejenes (12)	Jejenes	swamp (fluvial)

that the coals and carbonaceous shales of the eastern domain show thermal maturities spanning from marginally mature to mature. Vitrinite reflectance values (RO%) from 22 selected samples range between 0.4 and 1.1, with a mean of 0.8. Thermal alteration index (TAI) values vary between 2^+ and 3^+. Rank stages range from sub-bituminous (dull and brown coals) to bituminous (hard coal); ash content is generally high (Table 2). Some dull-brown coals are associated with millimeter-thick (exceptionally centimeter-thick) laminae of bright-brown coal composed of vitrinite.

In the western domain, Mid-Carboniferous glacial deposits, corresponding to the Guandacol Formation and equivalents (Limarino and Cesari, 1988), grade upward to fluvial (meandering rivers), often deltaic, facies consisting of cross-bedded channel sandstones and rippled fine sandstones and mudstones indicative of a floodplain environment. Coals and carbonaceous shales occur mostly within the floodplain (swamp) facies with associated plant bioturbation and paleosols (Table 1). In the Huaco area (La Ciénaga Mine, Fig. 1, B and Table 1), the coal seams are interbedded with deltaic floodplain deposits and underlie a thin marine interval of fossiliferous calcareous mudstones (Limarino et al., 1987). Vitrinite and fusinite have been reported as the main kerogen types in the coals from the southern part of the western domain (Cuerda and Furque, 1983). Their association with paleosols confirm their autochthonous nature (Bossi and Andreis, 1985).

AGE CONSTRAINTS AND PALEOBIOLOGICAL ASPECTS

The glacial deposits underlying the fluvial and deltaic coal-bearing strata in the western Paganzo basin are characterized by palynological and megafloristic remains that suggest a Middle Carboniferous (Namurian-Westphalian) age. *Cristatisporites*-like spores and plant remains such as *Malanzania* and *Bumbudendron* indicate an age which is partially coincident with the period of glaciomarine sedimentation in the adjacent Calingasta - Uspallata basin.

An important assemblage described as the Tupense Flora (Archangelsky, 1971) and biostratigraphically known as the NBG (*Nothorhacopteris*, *Botrychiopsis* and *Ginkgophyllum*) zone is found in the coal beds. This megaflora consists of lycophytes, sphenophytes, progymnosperms, pteridosperms and Cordaitales and is associated with a microfloristic assemblage with higher diversity in species. A stable paleocommunity, made up of lycophytes, sphenophytes, progimnosperms and gymnosperms has been derived from pollen diagrams (Cesari, 1986).

The Late Carboniferous age assigned to the coals and carbonaceous shales is indirectly confirmed by the presence of marine fossils of the Fauna Intermedia (González, 1985) in the calcareous mudstones of the Tupe Formation. These mudstones are exposed in the Huaco section in the western part of the Paganzo basin and lie only 20 m above the coal beds (Ottone and Azcuy, 1986).

Table 2. Characteristics of selected coals of the Paganzo basin (information compiled from Yacimientos Carboníferos Fiscales and Dirección General de Minería y Geología)

	Sierra de Maz	Huaco	Famatina
Localities in Fig. 1	(1,2 & 8)	(3)	(9 & 10)
RANK STAGE S:sub-bituminous (brown coal) B:bituminous (hard coal)	S	S - B	S
ASH CONTENT (% in weight)	20 - 30	43	25
SULPHUR CONTENT (% in weight)	0.34 - 2.78	0.29	1.83
CALORIFIC VALUE (kcal/kg)	6,000-7,000	4,400	3,800

The age of the Zona Intermedia has been considered to be Late Carboniferous and equivalent to the NBG zone (Archangelsky et al., 1987).

PALEOCLIMATIC CONSTRAINTS

Evidence gathered from the Paganzo and adjacent Calingasta-Uspallata basins suggest that glaciation in western Argentina peaked during the Mid-Carboniferous. Glaciterrestrial facies associations prevail in the Paganzo basin (Limarino and Gutiérrez, 1988), whereas glaciomarine conditions are predominant in the Calingasta-Uspallata basin (López-Gamundí, 1987). Immediately after the glaciation, an expansion of marine environments has been documented in several basins of western Argentina (López-Gamundí, 1989). This postglacial expansion of the marine realm could have led to high humidity in relatively high latitudes (Frakes, 1979). During this rapid postglacial climatic amelioration, peat development took place under cool-temperate conditions. In addition, although with wide variations at any latitude, there is a tendency to find higher peat accumulation rates at lower latitudes (Barber, 1981).

Locally, paleoclimatic conditions are also indicated by the presence of thin levels of kaolinitic claystones (tonsteins of Di Paola, 1972 and Limarino, 1988) intimately related to the coal beds which suggest chemical alteration of K-feldspar and muscovite from the granitic basement under humid conditions.

The disappearance of the coals in the Paganzo basin can be at least partially related to climatic changes experienced in western Argentina by the latest Carboniferous. By Stephanian times, sedimentological and paleontological information indicates that a significant humidity decrease, probably coupled with a temperature increase, led to a marked seasonality evinced by the presence of red beds. By early Permian times, the association of red beds with playa evaporites and eolian sediments characterizes the sedimentary fill in the Paganzo basin (López-Gamundí et al., 1992). This overall paleoclimatic trend suggests that peat-forming conditions were unlikely by the latest Carboniferous.

An additional factor that could have exerted an influence on the disappearance of the coals is that coal accumulation rates may outstrip subsidence rates in cold climates (Fig. 3), in particular in basins with low subsidence rates (passive margins and pericratonic basins).

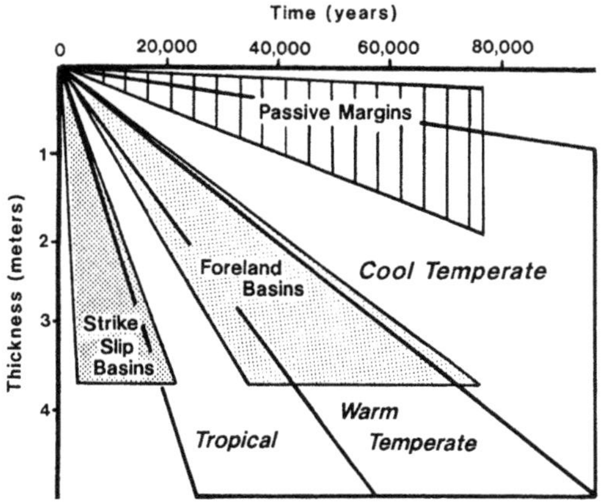

Figure 3. Tectonic subsidence rates for three different type basin types superimposed on coal accumulation rates of different climatic regimes (from McCabe, 1991). Note that spectrum of coal accumulation rates under cool to temperate conditions tend to outstrip subsidence rates in basins of mild tectonism. See text for discussion.

Under these circumstances subsidence is too slow for mires to accumulate thick peat beds. Furthermore, such a slow subsidence makes clastic inundation of the mires more likely (McCabe, 1991). Taking into consideration the thinness of the coal and carbonaceous mudstone beds of the Paganzo basin, which invariably are covered by fluvial deposits, it can be argued that the end of the peat-forming conditions was the result of the combined effect of low subsidence rates under cool temperate conditions superimposed on a general trend of climatic change characterized by increasing seasonal variations and accompanying water table fluctuations.

CONCLUSIONS

The association of coal seams with paleosols and tonstein levels and the evidence of significant vegetation growth suggest conditions of high humidity in the Paganzo basin for Late Carboniferous times. The modest amount of megaflora species in the coals and carbonaceous shales of the Paganzo basin indicates the lack of tropical or subtropical paleoclimatic conditions. Extremely cold paleoclimatic conditions can be discarded due to the lack of glacial deposits and presence of abundant *Eusphenopteris* and *Diplothnema* fronds related to pteridosperms and subarborescent lycophytes of up to 30 cm in diameter.

The chronostratrigraphic proximity of the glacial deposits and the overlying coal-bearing sediments in the Paganzo basin suggest that glacially-induced climatic changes such as increased rainfall and moderate temperature amelioration, typical of early postglacial periods, exerted a significant control on the peat-forming conditions.

The northern Canadian mires are intimately controlled by the migration of the mixing zone between arctic (polar) and temperate air masses. This interaction produces a significant amount of rainfall (Teraesmae, 1977). In addition, a rapid temperature amelioration coupled with increased rainfall characterizes the postglacial Holocene period in Canada and Britain (Hazeldine, 1988). This climate change led to the formation of blanket peat deposits during this postglacial period. Following this modern analogue, and even considering the low chronostratigraphic resolution of the biostratigraphic data available to date, it can be speculated that the coal-bearing strata of the Paganzo basin provide sufficient evidence to suggest moderately cold to temperate paleoclimatic conditions during the waning period, and/or immediately after, the Carboniferous glaciation of western South America.

Due to their narrow chronostratigraphic range and wide areal distribution the coal and carbonaceous shales of the Paganzo basin constitute an excellent marker for intrabasinal correlation. In addition, the unique paleoclimatic conditions under which they formed shed light on the strong influence exerted by the Gondwanan glaciation on coal occurrence.

REFERENCES

Alvarez Rojo, G., 1987. Depósitos carbonosos. In S.Archangelsky (ed.) El Sistema Carbonífero en la Argentina, Academia Nacional de Ciencias, Córdoba: 323-326.

Archangelsky, S., 1971. Las Tafofloras del Sistema Paganzo en la República Argentina. Anales Academia Brasileira de Ciencias, Rio de Janeiro: 43 (supl.): 67-88.

Archangelsky, S., C.L. Azcuy, C.R. González and N. Sabattini, 1987. Correlación general de las biozonas. In S.Archangelsky (ed.) El Sistema Carbonífero en la Argentina, Academia Nacional de Ciencias, Córdoba: 281-291.

Barber, K.E., 1981. Peat stratigraphy and climatic change. Rotter dam: Balkema.

Bossi, G. and R.R. Andreis, 1985. Secuencias deltaicas y lacustres del Carbonífero del Centro-Oeste argentino. Proceedings 10th International Congress on Stratigraphy and Geology of the Carboniferous, 3: 285- 309.

Cesari, S.N., 1986. Palinología de la Formación Tupe (Paleozoico superior), Sierra de Maz, provincia de la Rioja. Parte III. Análisis cuantitativo y conclusiones paleoecológicas. Ameghiniana 23: 197-202.

Cuerda, A. and G. Furque, 1983. Depósitos carbónicos de la Precordillera de San Juan. Parte II. Quebrada La Deheza. Asociación Geológica Argentina, Revista 38: 381-391.

Di Paola, E., 1972. Características litoestratigráficas de la Formación Lagares (Carbónico) en Paganzo-Amaná, provincia de La Rioja, República Argentina. Asociación Argentina Mineralogía, Petrografía y Sedimentología, Revista 3: 99-116.

Frakes, L.A., 1979. Climates througthout geologic time. Amsterdam: Elsevier.

González, C.R., 1985. Esquema bioestratigráfico del Paleozoico superior marino de la cuenca Calingasta-Uspallata-Iglesia, República Argentina. Acta Geológica Lilloana, S.M. de Tucumán, 16: 231-244.

Hazeldine, R.S., 1988. Coal reviewed: depositional controls, modern analogues and ancient climates. In: M.K. Whateley and K.T. Pickering (eds) Deltas: Sites and Traps for Fossil Fuels, Geological Society London, Special Publication No 41: 289-308.

Hobday, D.K., 1987. Gondwana coal basins of Australia and Africa: tectonic setting, depositional systems and resources. In: A.C. Scott (ed.) Coal and Coal-bearing Strata: recent Advances Geological Society of London, Special Publication No 32: 219-234.

Kreuser, T., 1987. Late Paleozoic Glacial Sediments and Transition to Coal-Bearing Lower Permian in Tanzania. Facies 17: 149-158.

Limarino, C.O., 1988. Paleoambientes sedimentarios y paleogeografía de la seccion inferior del Grupo Paganzo en el Sistema del Famatina. Anales Academia Nacional Ciencias Exactas, Físicas y Naturales, Buenos Aires 39: 110-119.

Limarino, C.O. and S.N. Cesari, 1988. Paleoclimatic significance of the lacustrine Carboniferous deposits in northwest Argentina. Palaeogeography, Paleoclimatology, Palaeoecology 65: 115-131.

Limarino, C.O. and P. Gutiérrez, 1988. Diamictites in the Agua Colorada Formation. A new evidence of Carboniferous glaciation in northwest Argentina. Journal of South American Earth Sciences 3: 9-20.

Limarino, C.O., H. Sessarego, S.N. Cesari and O.R. López-Gamundí, 1987. El perfil de la Cuesta de Huaco, estratotipo de referencia (hipoestratotipo) del Grupo Paganzo en la Precordillera Central. Anales Academia Nacional de Ciencias Exactas, Físicas y Naturales, Buenos Aires 38: 81-109.

López-Gamundí, O.R., 1987. Depostional models for the glaciomarine sequences of Andean Late Paleozoic basins of Argentina. Sedimentary Geology 52: 109-126.

López-Gamundí, O.R., 1989. Postglacial transgressions in Late

Paleozoic basins of western Argentina: A record of glacieustatic sealevel rise. Palaeogeography, Palaeoclimatology, Palaeoecology 71: 257-270.

López-Gamundí, O.R., C.O. Limarino, and Cesari, S.N., 1992. Late Paleozoic paleoclimatology of central west Argentina. Palaeogeography, Palaeoclimatology, Palaeoecology 91: 305-329.

Martini, I.P. and M.R. Banks, 1989. Sedimentology of the cold-climate, coal-bearing, Lower Permian "Lower Freshwater Sequence" of Tasmania. Sedimentary Geology 64: 25-41.

Martini, I.P and W. Glooschenko, 1985. Cold climate peat formation in Canada and its relevance to lower Permian coal measures of Australia. Earth Science Reviews 22: 107-140.

Martini, I.P. and D.P. Johnson, 1987. Cold-climate, fluvial to paralic coal-forming environments in the Permian Collinsdville Coal measures, Bowen Basin, Australia. International Journal of Coal Geology 7: 365388.

Martini, I.P. and A.C. Rocha-Campos, 1989. Permo-Carboniferous interglacial and early postglacial coal-bearing sequences of the Paraná basin, Brazil. Abstracts 28th International Geological Congress, Washington D.C. 12: 378-379.

McCabe, P.J., 1991. Tectonic controls on coal accumulation. Bulletin Societé geologique de France 162: 277-282.

Ottone, E.G. and C.L. Azcuy, 1986. El perfil de la quebrada La Delfina, provincia de San Juan, Argentina. Asociación Geológica Argentina, Revista 41: 124-136.

Raja Rao, C.S. and N.D. Mitra, 1985. Late Paleozoic Climate in the Indian Segment of Gondwanaland. 9th International Congress on Stratigraphy and Geology of the Carboniferous, Vol.5: 415-422.

Salfity, J.A. and S.A. Gorustovich, 1981. Paleogeografía del de la cuenca del Grupo Paganzo (Paleozoico Superior). Asociación Geológica Argentina, Revista 38: 437-453.

Teraesmae, J., 1977. Post-glacial history of Canadian muskeg. In N.W.Radforth and C.O. Brawner (eds) Muskeg and the Northern Environment in Canada. University of Toronto Press 9-30.

Villar, H.J., O.R. López-Gamundí and P.K. Siegele, 1991. Organic matter characterization of Carboniferous continental strata of the Paganzo basin, western Argentina. 10th International Congress on Stratigraphy and Geology of Carboniferous and Permian Systems (in press).

Gondwana Eight, Findlay, Unrug, Banks & Veevers (eds) © 1993 Balkema, Rotterdam. ISBN 90 5410 304 3

Upper Carboniferous-Permian volcano-sedimentary formations in Vietnam and adjacent territories

Phan Cu Tien
Geological Survey of Vietnam, Hanoi, Vietnam

ABSTRACT: The Upper Carboniferous–Permian formations of Vietnam are differentiated from region to region and related to several notable events in the geological development. Upper Carboniferous–Permian limestones are widely distributed in large areas in northeastern north-central Vietnam, whereas volcano-sedimentary formations are restricted to the Songda depression and Muongte in northwestern Vietnam, the Alin and Daklin areas in north central and the Namdu archipelago in the Gulf of Thailand in the south.

In the Songda depression, the Upper Carboniferous–Permian sequence is characterized by basalts, tuffs, sandstones, shales, limestones. This sequence is overlain discordantly by Upper Permian (Djulfian) mafic volcanics. In the northeast of Vietnam there is a similar discordant relationship between the Upper Carboniferous–Permian rocks and the Djulfian bauxite-bearing formations. In the Muongte, Daklin, Alin areas and Namdu archipelago, the volcano-sedimentary formations of Upper Carboniferous–Permian age are composed of andesites, trachytes, rhyolites, tuffs, sandstones, cherts, limestones and siltstones. The equivalent formations are observed in Xiengkhoang and Salavan in Laos and Stungtreng and Tbeng Meanchay in Cambodia. Limestones containing a Misellina–Neoschwagerina assemblage, correlated with the topmost parts of the above-mentioned formations, are widely known in Vietnam and adjacent territories. The Late Carboniferous–Permian intrusive rocks consist of a peridotite–wehrlite association and a diorite–grandiorite–granite association. At the end of the Permian, there was formed a gabbro–gabbrodiabase–diabase association which is related closely to mafic volcanics of the same age. The Late Carboniferous–Permian volcano-sedimentary formations and intrusive associations represent a tectonic event equivalent in age to the European Hercynides/Variscides. Late Permian (Djulfian)–Triassic formations developed during active tectonism in Vietnam and adjacent territories and seem to be related to plate tectonic rifting.

Mineralization within the Late Carboniferous–Permian sequence of the Songda depression is relatively abundant with nickel–copper, copper, gold, stratiform pyrites, rare metals and gold, iron ore, ilmenite, lead, zinc, bauxite, coal are related to those formations of Daklin, Alin, Muongte and some other area.

INTRODUCTION

The Late Carboniferous–Permian is a notable stage in the geological development of Vietnam and adjacent territories. In the former literature on geology of Vietnam, the Upper Carboniferous–Permian continental and subcontinental sequences were classified into Lower Indosinias (Fromaget 1941, 1952, 1954, Saurin 1956, Workman 1977, Fontaine and Workman 1978). On the basis of the geological research in the north of Vietnam, the Upper Carboniferous–Permian formations were regarded as a part of the monotonous limestone formation of Carboniferous–Permian age. However in this sequence there were recognised some stratigraphical discontinuities inferred to be of Visean, Namurian age (Dovjikov et al. 1965, Phan Cu Tien 1977, Nguyen Van Liem 1978). A subsequent stratigraphical correlation of the Upper Carboniferous–Permian sequences in Vietnam was made during the compilation of the geological map of Cambodia, Laos and Vietnam (Phan Cu Tien et al. 1988, 1991). This map shows the various volcanic and sedimentary facies of

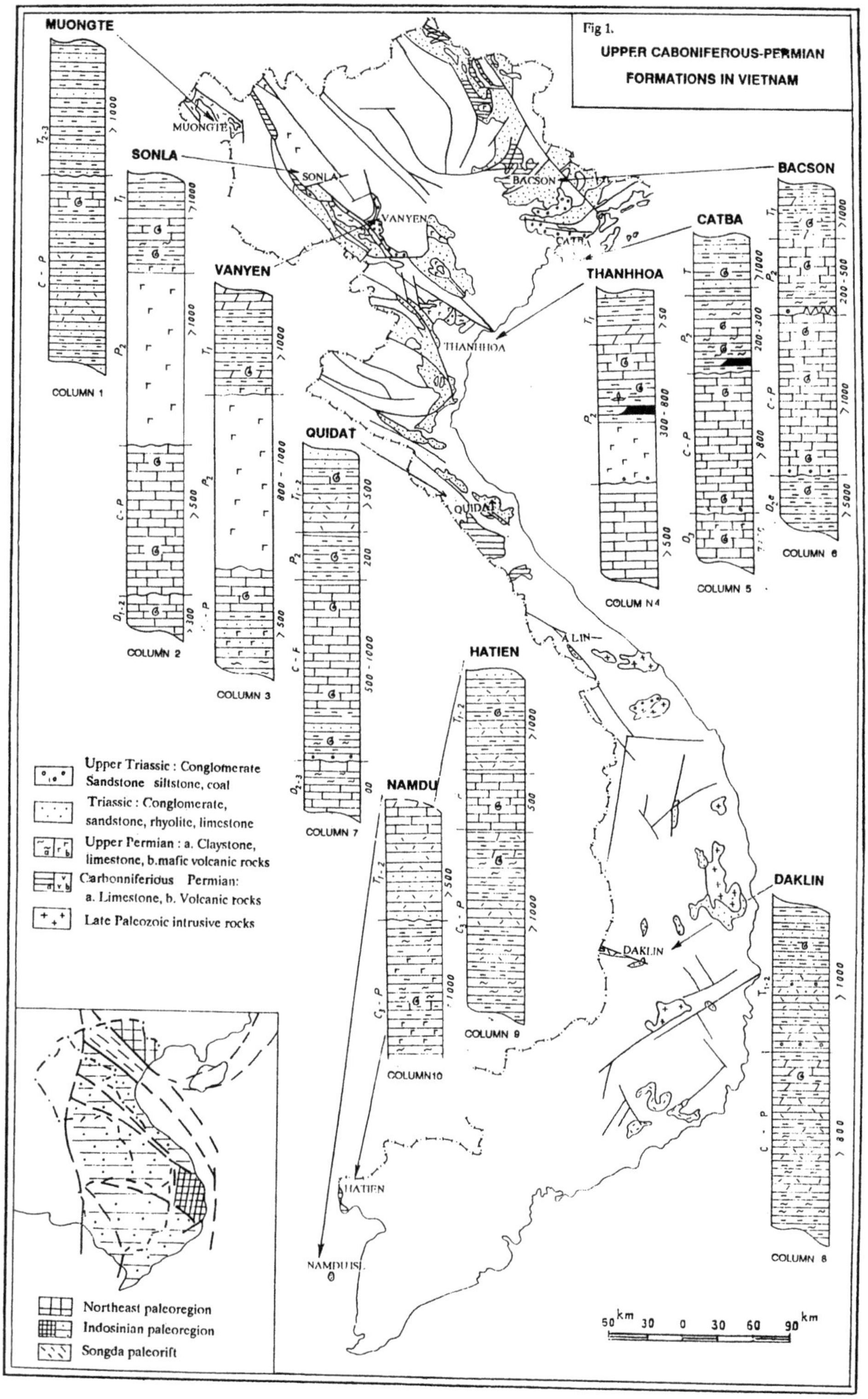

Fig 1. UPPER CABONIFEROUS-PERMIAN FORMATIONS IN VIETNAM

these formations and allows Upper Carboniferous–Permian stratigraphic correlations to be made within Southeast Asia and between Vietnam and South China and Australia (Phan Cu Tien 1990). These correlations permit the definition of the Indosinian (Early Kimmeridgian) tectonics in Vietnam and adjacent territories.

UPPER CARBONIFEROUS–PERMIAN FORMATIONS IN VIETNAM

The Upper Carboniferous–Permian formations are distributed widely in the nearly all part of northeast, northwest, northern central Vietnam. Those formations are also known in some exposures in southern central, south Vietnam and the Gulf of Thailand (Fig. 1).

In Bacson, Catba in the northeast, Sonla in the northwest and Quidat in northern central Vietnam, the Upper Carboniferous–Permian (Guadalupian) sequence is composed only of limestones. These rocks rest concordantly on similar sediments of Lower–Middle Carboniferous age. In some places such as Catba and Halang in the northeast and Quidat in northern central Vietnam, the Lower–Middle Carboniferous limestones overlie manganese-bearing sediments containing *Quasiendothyra kobeitusana*, *Quasiendothyra communis* or *Palmatolepis giges*, *Polygnathus normalis* (Fig. 1, column 5).

The collected faunas of the Upper Carboniferous–Permian limestone sequence are abundant and comprise Coral, Crinoid, Rugosa, Brachiopod, Gastropod and especially Foraminifera belonging to the horizons of *Triticites*, *Schwagerina* of Upper Carboniferous age and *Robustoschwagerina*, *Misellina–Parafusulina*, *Cancellina*, *Neoschwagerina–Verbeekina* of Lower or "middle" Permian age (Fig. 1, column 6 and 7). In northeast Vietnam, Upper Carboniferous–Permian limestones are overlain by bauxite-bearing formations and limestones and cherts containing *Leptodus*, *Codonofusiella dunbari*, *Nankinella inflata*, *Reichelina pulchra*, *Paleofusulina prisca* of the uppermost part of the Permian (Djulfian) (Fig. 1, columns 5 and 6). In the Songda depression, the equivalent formation is formed of two parts. The lower part is characterised by olivine basalt, tholeiite, diabase, tuffs and coal-bearing terrigenous sediment. In the upper part, limestones and interbeds of chert contain *Leptodus*, *Codonofusiella*, *Paleofusulina* as with those rocks in the northeast of Vietnam. The discordance between Djulfian volcano sedimentary formations and Guadalupian limestones is clearly and widely distinguished in northeast Vietnam and the peripheries of Songda depression. In north central Vietnam, Guidalupian limestones containing *Neoschwagerina craticulifera*, *Verbeekina verbeeki* are overlain by volcano-sedimentary formations of Triassic age (Fig. 1, column 7). There exist limestones containing *Nankinella* sp. and shale beds containing *Leptodus nobilis*, *Meekeila keuichowensis* which have been confirmed as the equivalent of Djulfian sequences (Ruzhensey et al. 1965). However note that the stratigraphical relationships between these sediments and older formations are not distinguished. Upper Carboniferous–Permian volcano-sedimentary formations are distributed in some separated areas in Vietnam such as the central part of the Songda depression and Muongte area in northwest Vietnam, Alin and Daklin areas in central Vietnam. In the central part of the Songda depression, there occur typically mafic volcanic and sedimentary formations; the other regions are characterised by intermediate, acidic and sedimentary sequences. In the Songda depression, the Upper Carboniferous–Permian volcano-sedimentary formations are distributed in narrow belts and composed of porphyrite, basalt, albitized basalt porphyrite, tuffs, intercalated beds of sandstone in the lower part and limestone beds in the upper one. The thickness of this formation is about 150 m but reaches 500 m in some places.

These mafic volcanics of Upper Carboniferous–Permian and Upper Permian (Djulfian) age are mainly distinguished on the basis of their stratigraphic position in geological sections and the collected fossils in interbeds of limestone and shale. These fossils are *Triticites schwageriniformis*, *Misellina ovalis*, *Propinacoceras* aff. *aktubense*, *Dytomopyge* in Upper Carboniferous–Permian sequence and *Nankinella inflata*, *Paleofusulina prisca* and *Leptodus nobilis* in Djulfian sequence (Fig. 1, columns 2 and 4). Generally, the mafic volcanic rocks are characterised by their low content of SiO_2 and Al_2O_3 and high content of TiO_2 (2.9–3.2), Fe, MgO and alkali.

In central Vietnam, the Upper Carboniferous–Permian volcano–sedimentary formations are restricted to the northern edge and widely distributed on the southern one of the Kontum uplift (Fig. 2). They are composed of shale, chert intercalated with limestone, porphyrite, andesite, rhyolite and tuffs; the total thickness is about

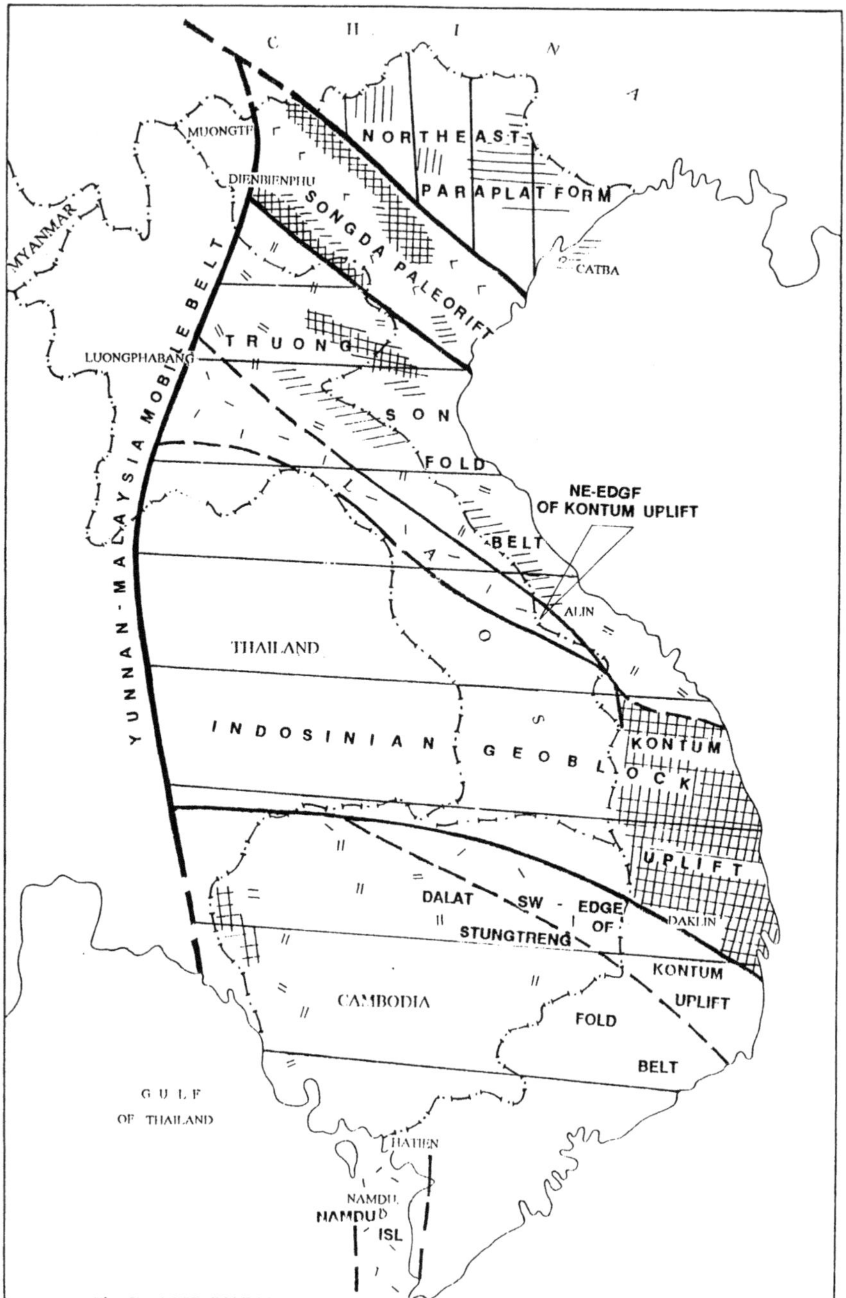

Fig. 2. LATE PALEOZOIC STRUCTURES AND PALEOGEOGRAPHY

Symbol	Legend	Symbol	Legend
r r	Mafic volcanic belt	(vertical and horizontal lines)	Precambrian uplift
(dashes)	Intermediate, acid volcanic belt	// \\	Paleozoic fold belt
(horizontal lines)	Epicontinental carbonate cover	(line)	Fault

500–1500 m. The collected fossils in this formation are *Schwagerine, Pseudofusulina, Verbeekina, Parafusulina, Minojapanella*. The stratigraphical relationship between this formation and the limestones containing *Parafusulina, Neoschwagerina, Lepidolina* in Hatien area seems to be concordant.

Volcanic rocks of Upper Carboniferous–Permian age in the edges of the Kontum uplift are similar to andesites classified by Deli (1933) which contain a high content of Fe (especially FeO, 5.5%) and Na_2O, low amounts of SiO_2, Al_2O_3, Cr_2O_3, H_2O and correspond to calc–alkaline volcanic rocks of basalto-andesite association.

Volcano-sedimentary formations of Upper Carboniferous–Permian age are also distinguished in western northwest Vietnam, in the Muongte area. These formations are composed of conglomerate, sandstone, shale, andesite, rhyolite, tuffs, limestone and marl containing *Misellina ovalis, Parafusulina japonica* and *Neoschwagerina craticulifera*. The thickness ranges from 500 m to 2000 m.

In the edges of the Kontum uplift and western northwest Vietnam, the relationship between the Upper Carboniferous-Permian and related stratigraphical subdivisions is not distinguished. Volcano-terrigenous formations of Triassic age rest on the Upper Carboniferous–Permian volcanic rocks at Daklin area and Namdu Island.

The Late Carboniferous–Permian intrusive rocks are divided into two associations: (1) peridotite–wehrlite association of Permian age, and (2) diorite–grandiorite–granite association of Late Permian to Early Triassic age. The rocks of the wehrlite–peridotite association include amphibole gabbro, gabbrodiabase and plagioclase peridotite. These usually occur as small bodies ranging from 2–3 meters to hundreds of meters thick, closely related to mafic volcanic rocks. These have been mapped along the central part of the Songda depression and also along the Dienbienphu–Luong Phabang fault. They are rich in FeO (0.8–1.4%), MgO (23.7–37.8%) and poor in alkali. Au, Ag, Pb, Zn, Hg and Ba mineralization is related to these rocks which occur as structurally-controlled deposits.

The diorite–grandiorite–granite association is found in the Muongte area, the Kontum uplift and along the Songma fault. This association is characterised by gabbro, gabbrodiorite, quartz diorite forming the earliest phase of intrusion, subsequent biotite–hornblende granodiorites, and granite aplite, granodiorite porphyries, diorite porphyries, which constitute the third intrusive/extrusive phase. These intrusive rocks are characterised by total content of alkali less than 5–7% in gabbro diorites, diorites, granodiorites and a bit higher in the granites. The K/Ar and U/Pb radiochronologic age of these rocks varies from 240 to 280 m.y. (Tran Duc Luong, Nguyen Xuan Bao et al. 1982, Phan Cu Tien et al. 1989, 1991)

GEOLOGICAL CORRELATION OF THE UPPER CARBONIFEROUS–PERMIAN FORMATIONS

We are able to interpret the geology of the Upper Carboniferous–Permian formations as below (Fig. 2).

The northeast Vietnam paraplatform (Dongbac paraplatform) of the South China plate is separated from the northern central fold belt (Truongson Fold Belt) of the Indosinian plate by the Songda paleorift. The latter seems to be part of an intercontinental paleorift which was formed initially during Late Carboniferous time. This was reactivated to form a depression with sub-oceanic crust in the Permo–Triassic. The Songda paleorift is separated from the Muongte area of the Yunnan–Malaysia mobile belt along the intercontinental Dienbienphu–Luong Phabang fault (Fig. 2). In the Muongte area, geosynclinal development had been continuing from the early Paleozoic until the Late Carboniferous time. Here, the calc-alkaline sub-volcanic association and carbonate, terrigenous and siliceous sediments together with the gabbrodiorite–grandiorite association of Late Carboniferous–Permian age are similar to sequences which are interpreted as marking a subduction zone of the Yunnan–Malaysia mobile belt. The oceanic crust of this mobile belt is formed by an ophiolite belt expected along the Luong Phabang–Uttradit lineament (Thanasuthipitak et al. 1978, Hamilton 1979).

During Late Carboniferous–Permian time, along the northeastern and southwestern edges of the Kontum uplift there existed an island-arc following a NW–SE trend. This is characterised by the calc-alkaline basalt–andesite association and intercalated beds of the terrigenous carbonate-siliceous sediments. The Late Carboniferous–Permian diorite–granodiorite–granite association is distributed widely in the Kontum uplift and as well at its southwestern and northeastern edges. This association represents the effects of the late

	Northeast Vietnam	North west Vietnam (Songda depression)	Northern central Vietnam	Southern central and south Vietnam
Upper Triassic	Hongai and Vanlang Suites: Coal-bearing sediments containing Floras and some Bivalves Basal conglomerates	Suoibang Suite: Coal-bearing sediment containing Juravites, Halobia, Burmesia, Parathisbites, Floras. Basal conglomerates	Nongson, Dongdo Suites: Coal-bearing red couloured formation containing Bivalves, Floras Basal conglomerates	?
Upper Triassic	Mauson Suites: Sandstones, shales, marls containing Freshwater Bivalves	Nammu, Songboi Formations: Siltstones, limestones, volcanics containing Halobia, Daonella, Discotropites	? Quylang Suite: Silstones, limestones containing Costatoria, Trigonodus	Chauthoi and Hongnghe Formation: Limestones, shales, volcanics containing Balatonites, Daonella
Middle Triassic	Nakhuat Suite: Volcanics, tuffs, siltstones, shales, marls containing Kellnerites, Costatoria, Trigonodus ?	Namtham Suite: Shale, marl, limestones containing Daonella, Posidonia Donggiao Suite: Limestones containing Cuccoceras, Paraceratites	Dongtrau Suite: Sandstones, shales, volcanics, limestones, Achrochordiceras, Costatoria Basal conglomerates	? Manggiang Formation: volcanics, tuffs, sandstones, siltstones containing Entolium
Lower Triassic	Langson and Songhien Suites: Siltstones, sandstones, marls, volcanics, tuffs, shales containing Claraia, Glyptophiceras, Flemingites, Columbites, Kashmirites	Conoi Suite: Sandstones, siltstones, limestones, volcanics, tuffs containing Claraia, Lytophiceras, Eumorphotis, Tirolites, etc.	?	Songsaigon Suite: Sandstones, shales marls containing Otoceras, Ophiceras, Claraia, Eumorphotis, Gyronrites
Upper Permian	Dongdang Suite: Limestones, shales containing Paleofusulina, Leptodus Basal beds of bauxite, breccias Eroded surface	Yenduyet Suite: Cherts, shales containing Leptodus, Gigantopteris, limestones containing Paleofusulina, Reichelina Camthuy Formation. Basalts, porphyrites	Limestones containing Codonofusiella, Pachyphloia, shales containing Leptodus, Meekella ?	Tathiet Formation: Limestones containing Paleofusulina, Neoendothyra, Reichelina ?
Lower-Middle Permian	Upper part of Bacson Series: Limestones containing Neoschwagerina, Verbeekina	Upper part of Bacson Series and Bandiet Formation: Limestones containing Triticites, Verbeekina. Basalts, porphyrites, tuff in Songda depression	Limestones containing Robustoschwagerina, Verbeekina, Neoschwagerina	Hatien Formation, Daklin Formation: limestones containing Neoschwagerina, Parafusulina, and andesites, rhyolites tuffs
Upper Carboniferous	Lower and Middle part of Bacson Series: Limestones containing Dainella, Endothyranopsis, Milerella, Profusulinella, Fusulinella, Obsoletes, Triticites, Schwagerina.	?	Muonglong Formation: Limestones containing Endothyranops, Profusulinella, Fusulinella, Obsoletes	?
Middle Carboniferous		Lower and Middle of Bacson Series: Limestones containing Endothyranopsis, Milerella, Profusulinella, Fusulina, Obsoletes, Triticites, Schwagerina.		?
Lower Carboniferous	Basal conglomerates	Basal conglomerates	Lakhe Suite: Limestones, cherts, coal shales containing Dainella etc. Basal conglemerates	Cherts, sandstones, shales
Upper Devonian	Toctat Suite: Manganese-bearing limestones, cherts containing Quasiendothyra, Palmatolepis	Bancai Suite: Manganese, limestones, cherts containing Quasiendothyra, Palmatolepis	Limestones containing Palmatolepis, Polygnatus, Cyrtospirifer	

Fig.3 GEOLOGICAL EVENTS IN THE LATE PALEOZOIC - EARLY MESOZOIC IN VIETNAM

Paleozoic collision in the Indosinian plate, which is equivalent in age with the Hercynian orogeny defined in the European regions.

The Late Permian–Triassic stage was a tectonomagmatic event in Vietnam and the adjacent territories. In the Songda paleorift and the southwestern edge of the Kontum uplift, major subsidence followed erosion during Early or "middle" Permian times. In nearly all parts of the territory of Vietnam, the intracontinental rifts were initiated. They are the Anchau and Songhien rifts in the northeast Vietnam paraplatform and the Samneua and Nongson Rifts in the Truongson Fold Belt. The major geological event in Djulfian time is observed in Vietnam and almost all adjacent territories in Southeast Asia as well (Dickins 1985, Supachai & Sungat 1978, Tan & Khoa 1978). This event is characterised by the discontinuity between the limestones of the Bacson Series of Carboniferous–Permian (Guadalupian) age or older formations, and Djulfian mafic volcanic rocks or bauxite, coal-bearing terrigenous and carbonate rocks of the overlying sequences (Fig. 3). The next major geological event in the Carboniferous–Triassic history of Vietnam occurred during the Norian stage with the appearance of basal conglomerates and coal-bearing sediments of Norian–Rhaetian age. The geological events in Early and Middle Triassic times were described in some previous geological literature (Dovjikov et el. 1965, Fontaine &Workman 1978); however these events are locally restricted (Phan Cu Tien 1990). The related Upper Permian (Djulfian)–Triassic mineralisation is diverse with nickel–copper, copper, gold, stratiform pyrites, asbestos, coals in the Songda Paleorift and gold, iron ore, ilmenite, zinc, bauxite in the northeast Vietnam paraplatform and the Truongson Fold Belt.

REFERENCES

Dickins J.M. 1985. Late Paleozoic and early Mesozoic "orogeny" in eastern Australia. Advances in the study of the Sydney Basin. Proceeding of the 19th Symposium, Dept. Geology, University of Newcastle: 7–8.

Fontaine H. & D.R. Workman 1978. Review of the geology and mineral resources of Kampuchea, Laos and Vietnam. Proceedings of the 3rd Conference of the Geology of Southeast Asia (GEOSEA), Bangkok: 359–608.

Fromaget J. 1941. L'Indochine francaise, sa structure geologique, ses roches, ses mines et leur possible avec la tectonique. Bulletin du Service Geologique de l'Indonchine. Volume XX: Faschio (number) 2. Hanoi.

Fromaget J. 1952. Etudes geologiques sur le Nord–Ouest du Tonkin et le Nord du Haut Laos. Bulletin du Service Geologique de l'Indochine, Volume XXIX: Faschio (number) 6.

Fromaget J. 1954. La 2[e] edition de la carte geologique au 2000.000e de l'Indochine. International Geological Congress, Algeria 1952. Section 13: Faschio (number) 14: 395–399.

Hamilton W. 1979. Tectonics of the Indonesian Region. United States Geological Survey Professional Paper 1078.

Hung D. 1986. Permian stratigraphy in Vietnam and its correlation with equivalent formations in Indochina. Proceedings of the 1st Congress of the Geology of Indochina (CIG), Hanoi 1: 98–100.

Junnavat S. & S. Prijasui 1978. Triassic rocks in Thailand. Proceedings of the 3rd Conference of the Geology of Southeast Asia (GEOSEA), Bangkok: 735–737.

Khuc V. 1986. Triassic in Vietnam. Proceedings of the 1st Congress of the Geology of Indochina, (CIG), Hanoi: 111–126.

Liem N.V. 1978. On the Carboniferous system in north Vietnam. Biology -Geography Volume 16: 78–85. (In Vietnamese)

Liem N.V. 1985. The upper Paleozoic in Vietnam. Science and Technology Publishing House, Hanoi. (In Vietnamese).

Luong T.D. & N.X. Bao (eds) et al. 1982. Geological map of Vietnam, 1:1.500.000 scale. General Department of Mines and Geology, Hanoi.

Ruzhensev G. (ed.) et al. 1965. The development and changes of marine organisms at the boundary between Paleozoic and Mesozoic. Science Publishing House (Moscow). (In Russian).

Saurin E. 1956. Indochine. In "Lexique Stratigraphique Internationale" 3 (6a). Centre Nationale de la Recherche Scientifique. Paris.

Tan B.K. & T.T. Khoa 1978. Review of the development in the geology and mineral resources of Malaysia and Singapore. Proceedings of the 3rd Conference Geology of Southeast Asia (GEOSEA), Bangkok: 655–671.

Thanasuthipitak T. et al. 1978. A review of the upper Paleozoic rocks of Thailand. Proceedings of the 3rd Conference of the Geology of

Southeast Asia (GEOSEA), Bangkok : 775–782.
Tien P.C. 1977. Upper Permian – Lower Triassic formations in Northwest Vietnam. In "The Geological Problems in Northwest Vietnam". Science and Technology Publishing House, Hanoi: 109–151. (In Vietnamese).
Tien P.C. (ed.) et al. 1988. Geological map of Kampuchea, Laos and Vietnam, 1:1,000,000 scale. General Department of Mines and Geology, Hanoi.
Tien P.C. (ed.) et al. 1989. Geology of Kampuchea, Laos and Vietnam. Explanatory note to the geological map of Kampuchea, Laos and Vietnam. 1: 1,000,000 scale. General Department of Mines and Geology, Hanoi.
Tien P.C. 1990. Geological events in late Paleozoic–early Mesozoic in Cambodia, Laos and Vietnam and their correlation. Geology of Cambodia, Laos and Vietnam. Volume 2. Geological Survey of Vietnam. Hanoi: 29–40.
Tien P.C. (ed) et al. 1991. Geology of Cambodia, Laos and Vietnam. Second edition with some improvements of the explanatory notes to the geological map of Cambodia, Laos and Vietnam at 1,000,000 scale. Geological Survey of Vietnam, Hanoi.

Gondwana Eight, Findlay, Unrug, Banks & Veevers (eds) © 1993 Balkema, Rotterdam. ISBN 90 5410 304 3

Early Permian brachiopods from the eastern Himalaya: Their provincial relationships with Australia

N.W.Archbold
Deakin University, Rusden Campus, Clayton, Vic., Australia

Trilochan Singh
Wadia Institute of Himalayan Geology, Dehra Dun, India

ABSTRACT: The Early Permian (Latest Tastubian - Sterlitamakian) brachiopod fauna of the Garu Formation, eastern Himalaya is reassessed. Faunal relationships are closest with other Himalayan regions and closer to western Australian correlative faunas than to eastern Australian faunas.

1 INTRODUCTION

Rigorous description and taxonomic assessment of Permian brachiopod faunas is essential before the implication of such faunas for palaeobiogeographical models can be tested. As part of a project on the comparison of Himalayan and Australian Permian brachiopod faunas, the two authors directly compared specimens from Early Permian eastern Himalayan brachiopod faunas with specimens from the Early Permian of eastern and western Australia. Recent statistical studies (Shi & Waterhouse, 1991) have indicated some differences in terms of the faunal affinities of the Early Permian Indo-Himalayan assemblages with respect to Australian faunas and hence the Garu Formation fauna provides a useful test case in terms of the current understanding of faunal relationships.

The distinctive Early Permian brachiopod fauna from the Siang District of the eastern Himalaya (Figure 1) was described originally by Singh (1978a, 1978b) and has been revised by the authors on the basis of additional collections and subsequent knowledge of the faunas of Australia. Specimens were found in the concretions present in the black carbonaceous and brown ferruginous shale of the Garu Formation and from within the shales themselves.

The Garu Formation has yielded a diverse fauna of brachiopods, bivalves, gastropods and cephalopods (Singh, 1973, 1978a, 1978b, 1978c, 1978d) and a microflora (Singh, 1979). Singh (1981) noted the distinctive Early Permian aspect of the fauna and inferred a Sakmarian to Artinskian age for the fauna. Archbold (1982) indicated that the age of the fauna may either straddle the Tastubian - Sterlitamakian boundary or be entirely Sterlitamakian.

The Garu Formation has yielded the ammonoid described by Singh (1978d) as *Uraloceras* cf. *irwinense* Teichert & Glenister (1952). This distinctive species is known from the Woolaga Limestone Member (Perth Basin, western Australia) and low in the Callytharra Formation (Carnarvon basin, western Australia) and has recently been referred to *Svetlanoceras* (Glenister et al., 1990). *Svetlanoceras irwinense* is an advanced member of the genus (Glenister et al. 1990) and the occurrence of an advanced *Metalegoceras* low in the Callytharra Formation (Glenister & Furnish, 1961) indicates a range of Latest Tastubian to Early Sterlitamakian for *Svetlanoceras irwinense*.

The brachiopod fauna of the Garu Formation shows significant links with that of the Sterlitamakian Callytharra Formation of western Australia and correlative faunas and hence the preferred age of the Garu fauna is in support of the ammonoid data.

2 GARU FORMATION BRACHIOPOD FAUNA

The brachiopod fauna of the Garu Formation

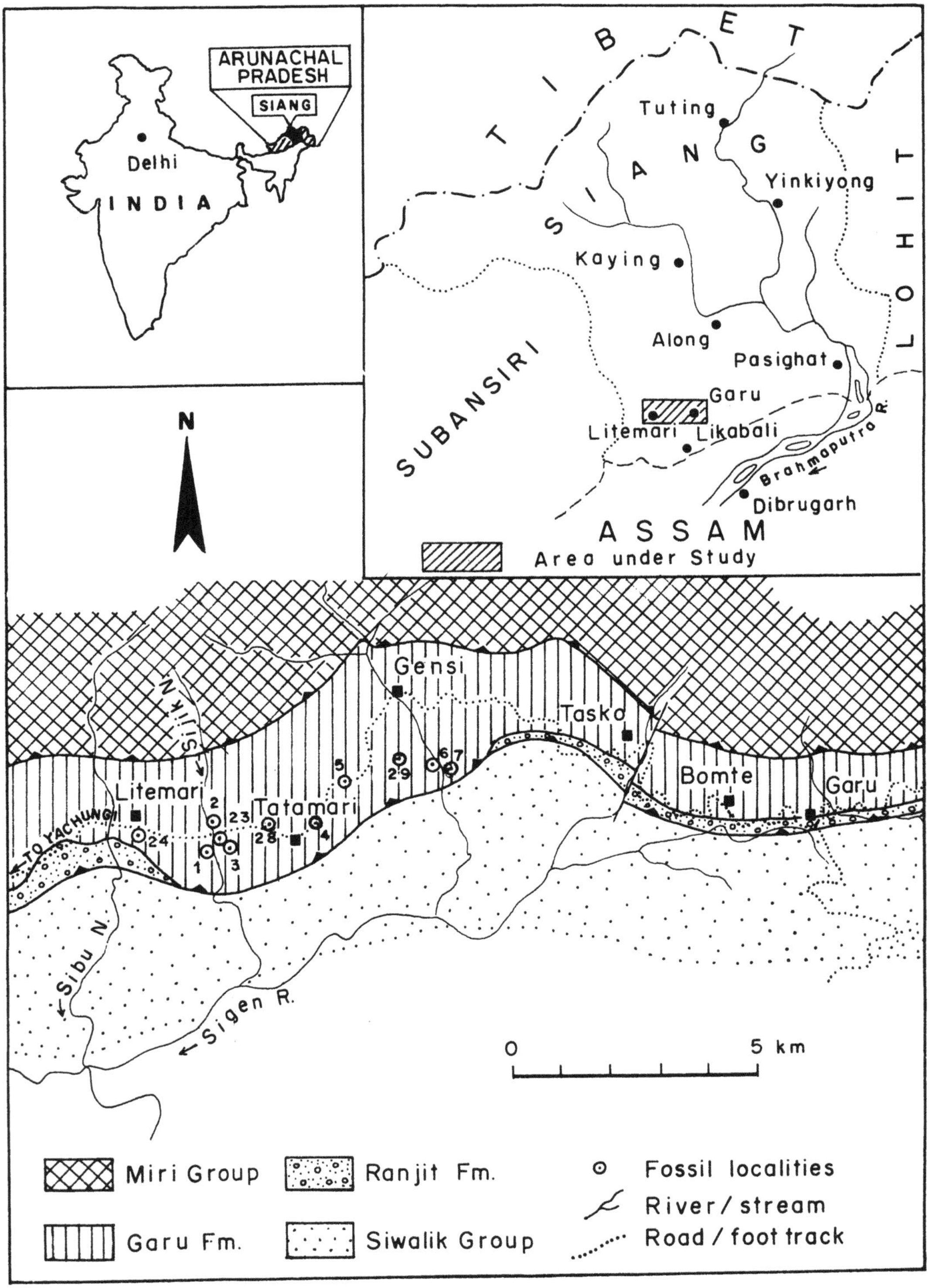

Figure 1. Map of Siang District, eastern Himalaya, showing Garu Formation outcrop and localities.

presently understood, consists of the following forms:
Tivertonia sp. nov.
Strophalosiidae gen. et sp. indet.
Costatumulus sp. nov.
Cyrtella cf. *nagmargensis* Bion
Subansiria spp.
Tomiopsis sp. nov.
Fredericksia? sp .and
Trigonotreta sp. nov.

The current understanding of the palaeobiogeographical relationships of each form are briefly summarized.

2.1 *Tivertonia sp. nov.*

The genus *Tivertonia* is well known from the Early Artinskian faunas of the Bowen, Yarrol and Sydney Basins of eastern Australia (Archbold 1983a, 1986a) and older Sakmarian occurrences from the Sydney Basin are currently being studied by the author with Dr. D. J. C. Briggs. The genus probably occurs in the Early Sakmarian of Argentina (Archbold 1983a). Several of the *Neochonetes* specimens from the Early Permian faunas of Chumik, Northwest Himalaya (Gaetani et al., 1990) indicate the presence of *Tivertonia*. The genus is not known from western Australia.

2.2 *Strophalosiidae gen. et sp. indet.*

Generic assignment of the material is not clear but *Strophalosia* species are well known from the early Permian of both western and eastern Australia (Archbold 1986b: Clarke, 1990) and the Indo-Himalayan region (Waterhouse & Gupta, 1978; Waterhouse & Rao, 1989).

2.3 *Costatumulus sp. nov.*

Traditionally species of *Costatumulus* would be placed in *Cancrinella*, but *Costatumulus* is useful for those species that lack dorsal spines and possess fine, low ventral rugae. The new species is close to the western Australian Sterlitamakian species *C. irwinensis* (Archbold, 1983b). *Costatumulus* is known from younger Artinskian faunas of eastern Australia (Briggs, 1991).

2.4 *Cyrtella cf. nagmargensis Bion*

This characteristic Early Permian western Himalayan species recalls species from western Australia (Thomas, 1971) and Timor (Archbold & Barkham, 1989) as well as Afghanistan (Termier et al., 1974). Comparable large Early Permian representatives of the genus are poorly known from eastern Australia but are probably present in Tasmania (Clarke, 1990), described under the generic name *Pseudosyrinx*.

2.5 *Subansiria spp.*

This distinctive genus is restricted to the eastern Himalayas (Singh, 1978a, 1978b) and Tibet (Hu, 1983).

2.6 *Tomiopsis sp. nov.*

This species is the first fully documented (including micro-ornament) representative of the genus from the Indo-Himalayan region. Its affinities are with western Australian and eastern Australian correlative species and species from the early Permian of Afghanistan (Termier et al., 1974) and the Transbaikal region (Kotlyar, 1968).

2.7 *Fredericksia? sp.*

A single specimen of the Garu fauna exhibits features comparable with very rare material, tentatively referred to the genus, from the Early Permian of western Australia (Archbold & Thomas, 1986a).

2.8 *Trigonotreta sp. nov.*

Although the genus is well developed in eastern Australia (Clarke, 1979) this new species is closer to species from western Australia (Archbold & Thomas 1986b) in terms of ontogenetic development. *Trigonotreta* probably has a wide peri-Gondwanan distribution but species require modern description.

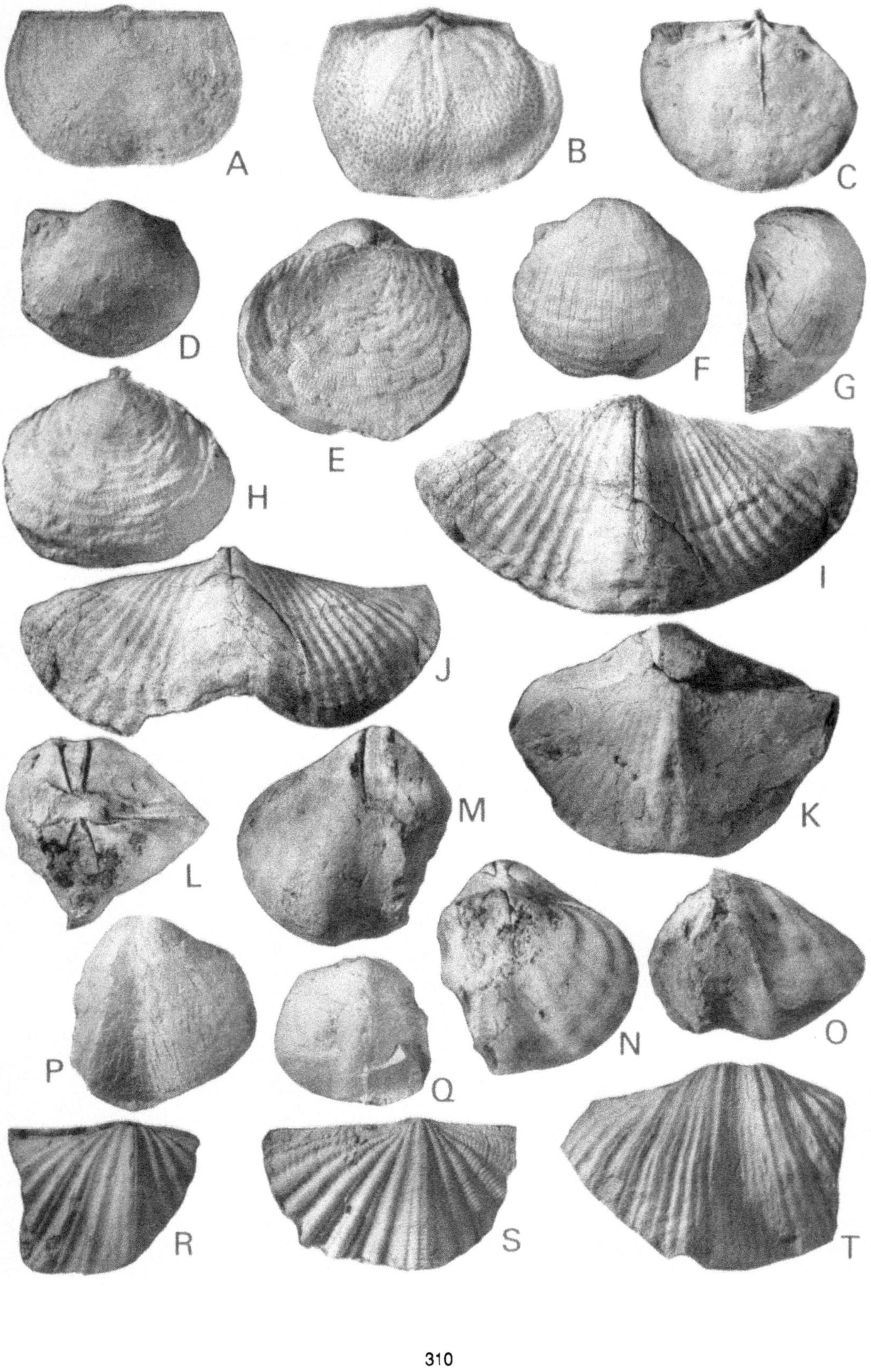
A
B
C
D
E
F
G
H
I
J
K
L
M
N
O
P
Q
R
S
T

Figure 2. Illustrations of specimens of brachiopods from the Garu Formation. A, B, *Tivertonia* sp. nov, dorsal external mould and ventral internal mould, x2, x1.8. C, Strophalosiidae gen. et sp. indet., latex cast of dorsal interior, x2.2. D-H, *Costatumulus* sp. nov., D, ventral valve in ventral view, x1.2; E-G, decorticated shell in dorsal, ventral and lateral views, x1.4, x1.2, x1.2; H, latex cast of dorsal interior, x 1.4. I, J, *Cyrtella* cf. *nagmargensis* Bion, dorsal internal mould in dorsal and anterior views, x0.85. K, *Subansiria* sp., decorticated shell in dorsal view, x1. L-O, *Tomiopsis* sp. nov., incomplete internal mould of shell in posterior, ventral, dorsal and anterior views, x1. P,Q, *Fredericksia*? sp., decorticated shell in ventral and dorsal views x1.5, x1. R-T, *Trigonotreta* sp. nov., R, S, internal mould of dorsal valve and latex cast of dorsal exterior, x 1.2. T, ventral valve in ventral view, x0.9.

3 CONCLUSIONS

Reassessment of the brachiopod fauna from the Garu Formation indicates the care required when using palaeontological data for palaeobiogeographical comparisons. Initially (Singh, 1978b, 1981), the fauna indicated some links with eastern Australian Early Permian faunas. However, subsequent work on western Australian faunas followed by collaboration by the two authors has demonstrated significant (and stronger) links with western Australian faunas. This conclusion tends to contradict reviews that imply relationships between the Early Permian faunas of the Indo-Himalayan and western Australian regions are not as strong as relationships between the Indo-Himalayan and eastern Australian regions during the Late Sakmarian.

ACKNOWLEDGEMENTS

We are grateful to the Wadia Institute of Himalayan Geology and the Australian Department of Science (Bilateral Science and Technology Programs) for funding the visit of T. S. to Australia. N. W. A.'s work on Late Palaeozoic brachiopod faunas is supported by the Australian Research Council. Our thanks to Mrs. C. Serpell for typing the manuscript.

REFERENCES

Archbold, N. W. 1982. Correlation of the Early Permian faunas of Gondwana: Implications for the Gondwanan Carboniferous-Permian boundary. *Journal of the Geological Society of Australia* 29: 267-276.

Archbold, N. W. 1983a. Permian marine invertebrate provinces of the Gondwanan Realm. *Alcheringa* 7: 59-73.

Archbold, N. W, 1983b. Studies on Western Australian Permian brachiopods 3. The family Linoproductidae Stehli 1954. *Proceedings of the Royal Society of Victoria* 95: 237-254.

Archbold, N. W. 1986a. *Tivertonia yarrolensis* (Maxwell). (Chonetidina, Brachiopoda) from the Permian of the Sydney Basin, Australia. *Alcheringa* 10: 413-415.

Archbold, N. W. 1986b. Studies on western Australian Permian brachiopods 6. The genera *Strophalosia* King, 1844, *Heteralosia* King, 1938 and *Echinalosia* Waterhouse, 1967. *Proceedings of the Royal Society of Victoria* 98: 97-119.

Archbold, N. W. & Barkham, S. T. 1989. Permian Brachiopoda from near Bisnain Village, West Timor. *Alcheringa* 13: 125-140.

Archbold, N. W. & Thomas, G. A. 1986a. Permian Ingelarellidae (Brachiopoda) from Western Australia and Argentina. *Journal of Paleontology* 60: 581-605.

Archbold, N. W. & Thomas, G. A. 1986b. *Neospirifer* and *Trigonotreta* (Spiriferida, Brachiopoda) from the Permian of western Australia. *Alcheringa* 10: 125-161.

Briggs, D. J. C. 1991. Stratigraphical distribution and biostratigraphical significance of the Permian Productidina and Strophalosiidina (Order Productida) of Eastern Australia. In, MacKinnon, D. I., Lee. D. E. & Campbell, J. D. (eds) *Brachiopods Through Time, Proceedings of the 2nd International Brachiopod Congress:* 367-374. Rotterdam: Balkema.

Clarke, M. J. 1979. The Tasmanian Permian spiriferid brachiopods *Trigonotreta stokesi* Koenig, 1825, *Grantonia hobartensis* Brown, 1953 and *Spirifer tasmaniensis* Morris, 1845. *Journal of Paleontology* 53: 197-207.

Clarke, M. J. 1990. Late Palaeozoic (Tamarian; Late Carboniferous - Early Permian) cold-water brachiopods from

Tasmania. *Alcheringa* 14: 53-76.
Gaetani, M., Garzanti E & Tintori, A. 1990. Permo-Carboniferous stratigraphy in SE Zanskar and NW Lahul (NW Himalaya, India). *Eclogae Geologicae Helvetiae* 83(1): 143-161.
Glenister, B. F., Baker, C., Furnish, W. M. & Thomas, G.A. 1990. Additional Early Permian ammonoid cephalopods from Western Australia. *Journal of Paleontology* 64: 392-399.
Glenister, B. F. & Furnish, W. M. 1961. The Permian ammonoids of Australia. *Journal of Paleontology* 35: 673-736.
Hu, C. M. 1983. New genera and species of spiriferacean brachiopods in the Late Carboniferous to Early Permian from Duoma District, Rutog, Xizang (Tibet), China. *Earth Science, Journal of Wuhan College of Geology* 1983 (1): 105-117.
Kotlyar, G. V. 1968. Pozdnepaleozoiskie predstaviteli roda *Tomiopsis* Zabaikal'ya i severnoi Mongolii. *Ezhegodnik Vsesoyuznoe Paleontologicheskoe Obshchestva* 18: 223-243.
Shi, G. R. & Waterhouse, J. B. 1991. Sakmarian (Early Permian) brachiopod biogeography and major associations as related to terrane drift. In, MacKinnon, D. I., Lee, D. E. & Campbell, J. D. (eds) *Brachiopods Through Time, Proceedings of the 2nd International Brachiopod Congress:* 355-365. Rotterdam: Balkema.
Singh, T. 1973. Note on the Upper Palaeozoic fauna from Subansiri District, Arunachal Pradesh. *Himalayan Geology* 3: 401-410.
Singh, T. 1978a. A new species of spiriferoid genus Subansiria from Subansiri District, Arunachal Pradesh. *Contributions to Himalayan Geology* 1: 162-164. Delhi: Hindustan Publ. Corp. (India).
Singh, T. 1978b. Brachiopods from Permian formation of Siang District, Arunachal Pradesh. *Contributions to Himalayan Geology* 1: 171-188. Delhi: Hindustan Publ. Corp. (India).
Singh, T. 1978c. Lower Permian gastropods and bivalves from Eastern Himalaya, India. *Recent Researches in Geology* 7: 276-310. Delhi: Hindustan Publ. Corp. (India).
Singh, T. 1978d. Lower Permian cephalopods from Eastern Himalaya, India. *Himalayan Geology* 11: 271-286.
Singh, T. 1979. Palynostratigraphy of the Permian rocks of Siang District, Arunachal Pradesh. In *Metamorphic Rock Sequences of the Eastern Himalaya*, P. K. Verma, (ed.): 100-113. Calcutta: K. P Bagchi & Coy.
Singh, T. 1981. Age and faunal affinity of the Garu Formation, Arunachal Pradesh. *Himalayan Geology* 11: 271-286.
Teichert, C. & Glenister, B. F. 1952. Lower Permian ammonoids from the Irwin Basin, Western Australia. *Journal of Paleontology* 26: 12-23.
Termier, G., Termier, H., De Lapparent, A. F & Marin, P. 1974. Monographie du Permo-Carbonifere de Wardak (Afghanistan Central). *Documents des Laboratoires de Geologie de la Faculte des Sciences de Lyon, Hors Serie*, 2: 1-167.
Thomas, G. A. 1971. Carboniferous and Early Permian brachiopods from Western and Northern Australia. *Bulletin of the Bureau of Mineral Resources, Geology and Geophysics* 56: 1-227.
Waterhouse, J. B. & Gupta, V. J. 1978. Early permian fossils from the Bijni Tectonic Unit, Garhwal Himalaya. *Recent Researches in Geology* 4: 410-437. Delhi: Hindustan Publ. Corp. (India).
Waterhouse, J. B. & Rao, A. R. 1989. Early Permian brachiopod and molluscan species from the Bap Formation of Peninsula India. *Palaontologische Zeitschrift* 63 : 25-39.

Gondwana Eight, Findlay, Unrug, Banks & Veevers (eds) © *1993 Balkema, Rotterdam. ISBN 90 5410 304 3*

A zonation of the Permian brachiopod faunas of western Australia

N.W.Archbold
Deakin University, Rusden Campus, Clayton, Vic., Australia

ABSTRACT: Permian brachiopod faunas of western Australia are abundant and diverse. A zonation scheme of some 17 zones is proposed with the nominate species of each zone being a particularly characteristic species of each assemblage. Early Permian zones are primarily based on the sequence of faunas developed in the Carnarvon Basin. Late Permian zones are based on the more limited development of Late Permian faunas of the Canning Basin.

1 INTRODUCTION

Permian brachiopod faunas of western Australia are noteworthy for their diversity and abundance. Previous reviews of the brachiopod faunas have high-lighted their provincial relationships (Teichert, 1951; Thomas, 1969; Archbold & Thomas, 1985). The co-existence of diverse faunas in association with significant ammonoid occurrences, as summarized in Table 1, results in the western Australian Permian sequences being of world significance for global Permian correlations.

Brachiopod assemblages are distributed throughout the marine Permian sequences of the Perth, Carnarvon and Canning Basins in Western Australia and the Bonaparte Gulf Basin (partially in the Northern Territory), Figure 1. The most complete Early Permian sequence is to be found in the Carnarvon Basin with important Late Permian sequences in the Canning Basin. The most recent reviews of the Permian stratigraphy of the basins are to be found as follows: Perth Basin, Playford et al. (1976); Carnarvon Basin, Hocking et al. (1987); Canning Basin, Towner & Gibson (1983); onshore Bonaparte Gulf Basin, Dickins et al. (1972).

Correlations of the major sequences have traditionally relied on critical ammonoid data (Glenister & Furnish,1961; Glenister et al.,1973; Glenister et al.,1990a; Glenister et al.,1990b) as summarized in Table 1.

The foundation study for the biostratigraphical zonation of the western Australian Permian was that of Dickins (1963) who recognised six faunal stages (informally named Oppel-zones) based primarily on the stratigraphical distribution of bivalve and gastropod taxa. For the present summary, brachiopod zones are formally named. The zones are essentially assemblage zones with the nominate species being a particularly characteristic species of each assemblage.

2 THE BRACHIOPOD ZONATION

The extensive brachiopod faunas of western Australia have been the subject of many studies and continue to be investigated by the present author and Dr. G. A. Thomas. The western Australian Permian successions continue to yield new assemblages but the majority of the brachiopod faunas have now been described, or are in preparation and, as a result, it is now timely to consider a refined biostratigraphy based on the stratigraphical distribution of the Brachiopoda.

All the brachiopod zones, except for the *Neochonetes (Sommeriella)* sp. nov. A Zone, are in objective stratigraphical sequence, as discussed below, and details of occurrence and correlation are provided with each zonal description.

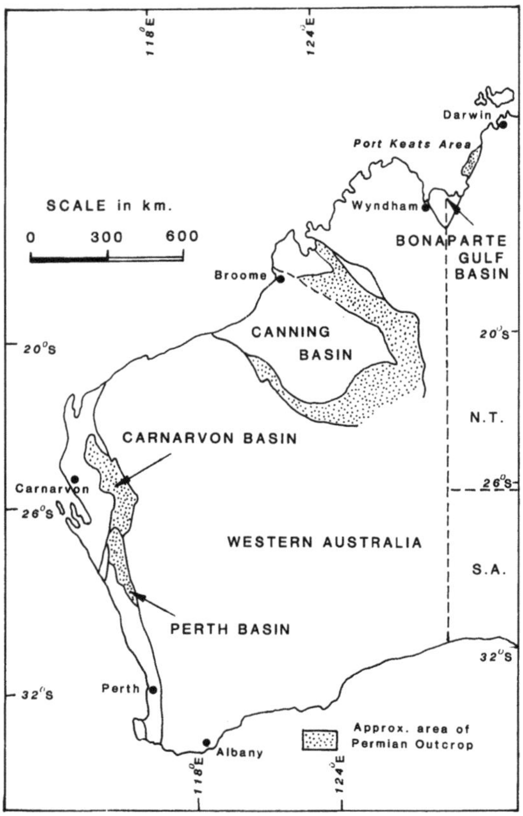

Figure 1. Permian sedimentary basins of western Australia.

2.1 *Lyonia lyoni Zone*

Lyonia lyoni (Prendergast) is a distinctive linoproductid found throughout most outcrops of the Lyons Group, the basal stratigraphical unit of the Carnarvon Basin Permian sequence. Fossil localities are scattered throughout the sequence (Dickins & Thomas,1959) but localities low in the Lyons Group yield brachiopods such as *Lyonia lyoni* (Prendergast), *Rhynchopora* sp. and *Kiangsiella* sp. (see Thomas, 1969) in addition to the bivalve *Deltopecten lyonsensis* Dickins. Localities in the middle of the Lyons Group demonstrate the incoming of additional elements including *Cyrtella australis* Thomas, *Grumantia* cf. *costellata* Clarke, *Tomiopsis notoplicatus* Archbold & Thomas and *Trigonotreta lyonsensis* Archbold & Thomas and the persistence of *Lyonia lyoni*. These localities low and in the middle of the Lyons Group are grouped together in the *Lyonia lyoni* Zone the top of which is defined as the incoming of *Trigonotreta occidentalis* Thomas and *Eurydesma playfordi* Dickins.

The age of the zone is most likely to be Early Tastubian, possibly ranging down into the Late Asselian. Correlation with the *Lyonia* sp. Zone of eastern Australia (Briggs, 1987) is indicated.

The zone is associated with sediments indicative of widespread glaciation.

2.2 *Trigonotreta occidentalis Zone*

The base of this zone is marked by the incoming of *Trigonotreta occidentalis* Thomas and the distinctive bivalve *Eurydesma playfordi* Dickins into localities high in the Lyons Group (Dickins & Thomas, 1959). Both species are also present in a conglomerate at the base of the Callytharra Formation (Thomas 1976) in the Merlinleigh Sub-basin of the Carnarvon Basin. This basal conglomerate, considered by Hocking et al. (1980) to be the equivalent of the Carrandibby Formation of the Byro Sub-basin, marks the highest occurrences of *Grumantia* cf. *costellata* Clarke, *Tomiopsis notoplicatus* Archbold & Thomas, *Eurydesma playfordi* Dickins and *Deltopecten lyonsensis* Dickins and provides the youngest evidence of glaciation in the western Australian Permian successions. Brachiopods that persist from the *Lyonia lyoni* zone are *Lyonia lyoni* (Prendergast) and *Cyrtella australis* Thomas. *Permorthotetes crespinae* Thomas and *Linoproductus* sp. also come from the Upper Lyons Group. The Carrandibby Formation (restricted) yields only one brachiopod species - *Tomiopsis notoplicatus* Archbold & Thomas.

Correlative units in western Australia include the middle part of the Holmwood Shale, including the Woolaga Limestone Member, of the Perth Basin on the basis of stratigraphical position below the Fossil Cliff Member. No brachiopods are reliably known from the middle of the Holmwood Shale. The Wye Worry Member of the Carolyn Formation, Grant Group, Canning Basin has yielded two localities with marine faunas. One fauna, described by Dickins et al. (1978) includes *Deltopecten lyonsensis* and a possible *Eurydesma* suggesting a link with the *Trigonotreta occidentalis* Zone. A new assemblage from varved shales includes a new species of *Neochonetes (Sommeriella)*,

Heteralosia sp. nov., *Costatumulus* sp. and fragments of *Trigonotreta*. The new species of *Neochonetes (Sommeriella)* is also possibly present in the Woolaga Limestone (Perth Basin) and the conglomerate at the base of the Callytharra Formation (Carnarvon Basin) although both occurrences require confirmation.

A small fauna described by Waterhouse (in Foster & Waterhouse, 1988) from an isolated section within the Grant Group on the Barbwire Terrace, Canning Basin is reassessed herein to consist of *Neochonetes (Sommeriella)* sp. nov., *Costatumulus capillata* (Waterhouse) and *Heteralosia* sp. nov. as well as various unfigured brachiopod fragments. This assemblage may correlate with either the *Lyonia lyoni* Zone or the *Trigonotreta occidentalis* Zone.

Correlation of the *Trigonotreta occidentalis* Zone with the "*Trigonotreta campbelli*" Zone of eastern Australia (Briggs 1987) is indicated. The age of the *Trigonotreta occidentalis* zone is Late Tastubian occurring as it does stratigraphically immediately below an Early Sterlitamakian ammonoid fauna. Apparent correlatives in the Perth Basin include a Late Tastubian ammonoid fauna.

2.3 *Strophalosia irwinensis Zone*

This zone is characterized by one of the most diverse brachiopod assemblages in western Australia. The brachiopod fauna of the Callytharra Formation, Carnarvon Basin, typifies the zone and consists of the following species : *Permorthotetes callytharrensis* Thomas, *Permorthotetes camerata* Thomas, *Arctitreta plicatilis* (Hosking), *Tornquistia occidentalis* Archbold, *Neochonetes (Sommeriella) prattii* (Davidson), *Strophalosia irwinensis* Coleman, *Heteralosia etheridgei* (Prendergast), *Aulosteges baracoodensis* Etheridge, *Aulosteges spinosus* Hosking, *Stictozoster senticosa* (Hosking), *Comuquia australis* Archbold, *Lethamia obscurus* Archbold, *Dyschrestia micracantha* (Hosking), *Callytharrella callytharrensis* (Prendergast), *Costatumulus irwinensis* (Archbold), *Globiella foordi* (Etheridge), *Stenoscisma* sp. nov., *Camerisma callytharrensis* (Hosking), *Myodelthyrium dickinsi* (Thomas), *Cyrtella australis* Thomas, *Latispirifer callytharrensis* Archbold & Thomas, *Neospirifer foordi* Archbold & Thomas, *Neospirifer hardmani* (Foord), *Neospirifer* sp., *Fusispirifer carnarvonensis* Archbold & Thomas, *Trigonotreta neoaustralis* Archbold & Thomas, *Elivina hoskingae* Archbold & Thomas, *Spirelytha fredericksi* Archbold & Thomas, *Phricodothyris occidentalis* Archbold & Thomas, *Tomiopsis woodwardi* Archbold & Thomas, *Tomiopsis* sp., *Fredericksia*? sp. nov. A, *Gjelispinifera decipiens* (Hosking), *Lamnaespina papilionata* (Hosking), *Callispirina* sp. nov., *Hustedia* sp. nov., *Cleiothyridina baracoodensis* (Etheridge) and *Fletcherithyris* cf *hardmani* Campbell.

Seventeen of the Callytharra Formation species, including *Strophalosia irwinensis,* are shared with the fauna of the Fossil Cliff Member of the Holmwood Shale, Perth Basin (Archbold, 1988a) which also includes *Cyrtella* sp. and *Taeniothaerus irwinensis* Coleman. The Cuncudgerie Sandstone, southern Canning Basin also yields a correlative fauna (Archbold, 1990) which includes : *Permorthotetes lindneri* Thomas, *Tornquistia subquadratus* Archbold, *Strophalosia irwinensis* Coleman, *Aulosteges* cf *baracoodensis* Etheridge, *Taeniothaerus* cf *irwinensis* Coleman, *Costatumulus irwinensis* (Archbold), *Cyrtella koopi* Archbold, *Myodelthyrium dickinsi* (Thomas), *Trigonotreta neoaustralis* Archbold & Thomas and *Spirelytha* cf *fredericksi* Archbold & Thomas.

Despite strong links in the correlation of the fauna of the Callytharra Formation with the fauna of the Nura Nura Member, basal Poole Sandstone, on the basis of Mollusca (Dickins, 1963; Glenister et al. 1973), the brachiopods of this northern Canning Basin unit are less reliably known. At present, adequately documented species include: *Streptorhynchus* sp., *Permorthotetes lindneri* Thomas, *Neochonetes (Sommeriella) prattii* (Davidson), *Neospirifer foordi* Archbold & Thomas, *Cratispirifer nuraensis* Archbold & Thomas, *Myodelthyrium dickinsi* (Thomas), *Tomiopsis woodwardi* Archbold & Thomas and *Fletcherithyris* cf *hardmani* Campbell. The correlation of the two units is accepted herein, following earlier studies.

The age of the *Strophalosia irwinensis* Zone is regarded as being Sterlitamakian in view of the Callytharra Formation and the Nura Nura Member both yielding advanced species of *Metalegoceras* as discussed by Glenister et al.

(1973). Ammonoids from the Nura Nura Member are more advanced than those from the Fossil Cliff Member and some of those from the Callytharra Formation (Glenister et al., 1973; Glenister et al., 1990a) and hence it is possible that the Nura Nura fauna may only represent the younger part of the Sterlitamakian. Stratigraphical data would also support this proposition in view of the distinct unconformity at the base of the Nura Nura Member. Nevertheless the current macrofaunal data still supports a broad correlation of the three units as advocated by Thomas & Dickins (1954).

2.4 *Neochonetes (Sommeriella) sp. nov. A Zone.*

This zone occurs in the basal beds of the High Cliff Sandstone, stratigraphically above the *Strophalosia irwinensis* Zone, with an apparent erosional disconformity separating the two zones. The assemblage includes *Neochonetes (Sommeriella)* sp. nov., *Aulosteges* sp., *Taeniothaerus* sp. nov., *Costatumulus* sp. nov. A, *Neospirifer* sp. nov., neospiriferid gen. et sp. nov. (allied to *Crassispirifer*), *Cyrtella*? sp. nov., *Tomiopsis rarus* Archbold & Thomas, *Cleiothyridina* sp., *Gilledia woolagensis* Campbell and *Hoskingia* sp. nov., and is currently being described by the author.

No direct correlatives of this zone are known elsewhere in western Australia and for the present the zone is regarded as corresponding to the break in sedimentation at the base of the Wooramel Group in the Carnarvon basin. *Neospirifer* sp. nov. appears to be related to a poorly known species from the *Strophalosia jimbaensis* Zone of the Carnarvon Basin.

2.5 *Strophalosia jimbaensis Zone.*

Distinctive, but poorly preserved assemblages from the Wooramel group of the Carnarvon Basin include *Streptorhynchus*? sp., *Permorthotetes* sp., *Neochonetes (Sommeriella) hockingi* Archbold, *Strophalosia jimbaensis* Archbold, *Aulosteges* sp., *Fusispirifer* sp., *Neospirifer* sp. and *Cleiothyridina* sp. (Archbold 1991). These assemblages, together with the larger but poorly known fauna of the Jimba Jimba Calcarenite of the Wooramel Group, are grouped together within this zone. The Jimba Jimba Calcarenite has yielded the two distinctive species *Strophalosia jimbaensis* Archbold and *Globiella flexuosa* (Waterhouse). The preferred age of the *S. jimbaensis* Zone is Aktastinian (Archbold, 1991). No direct correlatives of the zone are known elsewhere in western Australia but the precise time relationships with the *Neochonetes (Sommeriella)* sp.nov. A Zone of the Perth Basin are open to question.

2.6 *Echinalosia prideri Zone*

This zone is based on the assemblage from the lower part of the Madeline Formation, Carnarvon Basin which includes: *Permorthotetes* sp., *Streptorhynchus* sp., *Kiangsiella condoni* Thomas, rugosochonetid gen. et sp. nov., *Neochonetes (Sommeriella) robustus* Archbold, *Echinalosia prideri* Coleman, *Aulosteges lyndonensis* Coleman, *Costatumulus* sp., *Pseudosyrinx*? *sinuosa* Thomas and *Neospirifer plicatus* Archbold & Thomas. Molluscan assemblages from the top of the Wooramel Group (Dickins, 1956) are included within this zone. The fauna of the Mingenew Formation, Perth Basin (Archbold, 1988a) is probably a correlative of the *Echinalosia prideri* Zone.

Localities low in the Noonkanbah Formation, Canning Basin with *Kiangsiella* cf. *condoni* and *Costiferina wadei* (Prendergast) may indicate a correlative horizon.

2.7 *Wyndhamia colemani Zone.*

This zone is based on assemblages from the upper part of the Madeline Formation, Carnarvon Basin, which include: *Streptorhynchus* sp., *Wyndhamia colemani* Archbold, *Heteralosia* sp., *Aulosteges ingens* Hosking, *Costatumulus* sp., *Crassispirifer rostalinus* (Hosking), *Fusispirifer byroensis* (Glauert), *Tomiopsis pauciplicatus* Archbold & Thomas, *Spiriferella* sp., *Spirelytha* sp., *Cleiothyridina* sp. and *Hoskingia trigonopsis* (Hosking). Correlatives of this zone elsewhere in western Australia are not firmly established but may include some localities of the Mingenew Formation in the Perth Basin and the oldest described outcropping Permian fauna of the Bonaparte Gulf Basin.

2.8 *Fusispirifer byroensis Zone.*

This zone overlaps with the *Wyndhamia colemani* Zone and extends into younger strata (Mallens and Bogadi Sandstones). The brachiopod fauna is restricted and includes *Fusispirifer byroensis* (Glauert), *Crassispirifer rostalinus* (Hosking), *Hoskingia trigonopsis* (Hosking) and *Aulosteges ingens* (Hosking). No correlatives are known elsewhere in western Australia.

2.9 *Tornquistia magna Zone.*

A poorly known fauna from the Bulgadoo Shale includes *Tornquistia magna* Archbold, *Neochonetes (Sommeriella)* sp., *Fredericksia*? sp. nov. B and *Hoskingia kennediensis* Campbell. Despite being poorly known, the fauna indicates a distinct zone which is named. No obvious correlatives are known from the Perth or Canning Basins.

2.10 *Fusispirifer cundlegoensis Zone*

This zone is typified by the diverse fauna of the Cundlego Formation consisting of *Permorthotetes* sp., *Streptorhynchus hoskingae* Thomas, *Tornquistia* cf. *tropicalis* Grant, *Demonedys granti* Archbold, *Neochonetes (Sommeriella) tenuicapillatus* Archbold, *Quinquenella australis* Archbold, *Heteralosia prendergastae* Coleman, *Wyndhamia multispinifera* (Prendergast), *Retimarginifera perforata* Waterhouse, *Spiriferella cundlegoensis* Archbold & Thomas, *Fusispirifer cundlegoensis* Archbold & Thomas, *Hoskingia kennediensis* Campbell and *Gilledia* cf. *homevalensis* Campbell. The laterally equivalent Quinnanie Shale assemblage consists of *Semilingula occidentaustralis* (Archbold), *Streptorhynchus hoskingae* Thomas, *Neochonetes (Sommeriella) tenuicapillatus* Archbold, *Fusispirifer quinnaniensis* Archbold & Thomas, *Spiriferella cundlegoensis* Archbold & Thomas and *Spirelytha* sp. nov.

A few species of this zone range into the *Fusispirifer wandageensis* Zone above. Correlatives appear to exist at certain levels and localities within the Noonkanbah Formation of the Canning Basin. These appear to be localities higher in that formation than localities with *Costiferina wadei* Prendergast but probably lower in the formation than classical localities such as Mount Marmion which yield a fauna more closely correlated with that of the Wandagee Formation of the Carnarvon Basin.

2.11 *Fusispirifer wandageensis Zone*

This zone is represented by the diverse assemblage of the Wandagee Formation, Carnarvon Basin which includes *Streptorhynchus hoskingae* Thomas, *Tornquistia gregoryi* Archbold, *Neochonetes (Sommeriella) tenuicapillatus* Archbold, *Heteralosia complectens* (Etheridge), *Wyndhamia* sp., *Lialosia kimberleyensis* (Prendergast) *Taeniothaerus coolkiliensis* Coleman, *Taeniothaerus miniliensis* Coleman, *Taeniothaerus teicherti* Coleman, *Dyschrestia colemani* Archbold, *Retimarginifera perforata* Waterhouse, *Costatumulus bellus* (Etheridge), *Neospirifer amplus* Archbold & Thomas, *Neospirifer postplicatus* Archbold & Thomas, *Crassispirifer pinguis* Archbold & Thomas, *Spiriferella australasica* (Etheridge), *Spirelytha miloradovichi* Archbold & Thomas, *Tomiopsis teicherti* Archbold & Thomas, *Hustedia*? *basedowi* (Etheridge), *Cleiothyridina macleayana* (Etheridge), *Yochelsonia thomasi* Stehli, *Hoskingia wandageensis* Campbell and *Gilledia* cf. *homevalensis* Campbell.

Localities high in the Noonkanbah Formation such as Mount Marmion are clearly correlative with the Wandagee assemblage yielding many of the same productid and spiriferid species.

The *Fusispirifer wandageensis* Zone may extend into the earliest Kungurian Stage - the ammonoid data is equivocal.

2.12 *Neochonetes (Sommeriella) sp. nov. B Zone.*

The Nalbia Sandstone of the Carnarvon basin yields a distinctive small assemblege with a few species persisting from the Wandagee Formation. The assemblage includes *Neochonetes (Sommeriella)* sp. nov. B, *Svalbardia narelliensis* Archbold, *Wyndhamia* sp., *Lialosia kimberleyensis* (Prendergast), *Magniplicatina* sp. nov., *Spirelytha miloradovichi* Archbold & Thomas and *Hoskingia* cf. *wandageensis* Campbell. It appears likely that localities high in the

Noonkanbah Formation, Canning Basin, with *Svalbardia narelliensis* are correlative with the Nalbia Formation. The zone is regarded as being Early Kungurian in age.

2.13 *Svalbardia thomasi Zone.*

This short duration zone from the Baker Formation, Carnarvon Basin includes the following brachiopods: *Permorthotetes teicherti* Thomas, *Svalbardia thomasi* Archbold, the highest occurrence of *Lialosia kimberleyensis* (Prendergast), *Aulosteges*? sp., *Fusispirifer* sp. and *Cleiothyridina* cf. *macleayana* (Etheridge). Ammonoid data (Archbold & Dickins, 1991) indicate a Late Kungurian (Irenian) age. No correlatives of this zone have been specifically identified from elsewhere in western Australia.

2.14 *Neochonetes (Sommeriella) afanasyevae Zone.*

The Coolkilya Sandstone (excluding the top beds) has yielded a diverse assemblage: *Permorthotetes teicherti* Thomas, *Streptorhynchus johnstonei* Thomas, *Derbyia*? sp., *Neochonetes (Sommeriella) afanasyevae* Archbold, *Chonetinella* sp., *Aulosteges* sp., *Dyschrestia* sp., *Retimarginifera waterhousei* Archbold, *Costatumulus*? *coolkilyaensis* (Archbold), *Fusispirifer kennediensis* Archbold & Thomas, *Spiriferella etheridgei* Archbold & Thomas, *Spirelytha stepanoviana* Archbold & Thomas, *Tomiopsis globosus* Archbold & Thomas, *Cleiothyridina* sp. and ?*Yochelsonia thomasi* Stehli.

Correlatives include the fauna of the Lightjack Formation, Canning Basin which has yielded *Permorthotetes guppyi* Thomas, *Streptorhynchus perfidiabadensis* (Etheridge), *Neochonetes (Sommeriella) afanasyevae* Archbold, *Heteralosia* sp., *Costatumulus*? sp., *Costatispirifer gracilis* Archbold & Thomas, *Trigonotreta lightjacki* Archbold & Thomas, *Tomiopsis globosus* Archbold & Thomas, *Tomiopsis balgoensis* Archbold & Thomas and *Yochelsonia stehlii* Campbell. Also correlative are the beds at Fossil Head, Bonaparte Gulf Basin, with *Neochonetes (Sommeriella) afanasyevae* and *Streptorhynchus perfidiabadensis*.

2.15 *Fusispirifer coolkilyaensis Zone*

A restricted zone from the top beds of the Coolkilya Sandstone has yielded *Fusispirifer coolkilyaensis* Archbold and *Taeniothaerus* sp. as well as distinctive bivalves (Archbold & Skwarko, 1988). No direct correlatives are known elsewhere in western Australia.

Ammonoid data from the Coolkilya Formation could be compatible with an Early Ufimian age for both Coolkilya zones (ie. an equivalent of the Early Roadian of North America) although shown as a little older in Table 1.

2.16 *Late Permian Zones*

The Late Permian *Liveringia magnifica* Zone and *Waagenoconcha imperfecta* Zone were documented by Archbold (1988b) and will not be discussed here. Both zones are recognised in the Canning and Bonaparte Gulf Basins and ammonoid data (Table 1) indicates a Djhulfian age for the *Waagenoconcha imperfecta* Zone. Shared brachiopod species with the slightly older *Liveringia magnifica* Zone indicates a close time relationship for these zones.

ACKNOWLEDGEMENTS

Many years of discussions about western

Table 1. Correlation chart of the major Permian sequences of western Australia showing the proposed brachiopod zonation and critical ammonoid data. Ammonoids are indicated by black circles and the following numbers:

1. *Juresanites jacksoni*, 2. *Svetlanoceras irwinense*, 3. *Metalegoceras kayi*, 4. *Mescalites* sp., 5. *Metalegoceras* n. sp., 6. *Pseudoschistoceras simile*, 7. *Bamyaniceras* sp., 8. *Neocrimites* sp., 9. *Bamyaniceras* n. sp., 10. *Bamyaniceras australe*, 11. *Paragastrioceras wandageense*, 12. *Daubichites goochi*, 13. *Popanoceras* sp., 14. *Agathiceras applanatum*, 15. *Metalegoceras striatum*, 16. *Metalegoceras australe*, 17. *Thalassoceras wadei*, 18. *Propopanoceras ruzhencevi*, 19. *Cyclolobus persulcatus*

Table 1. Explanation opposite.

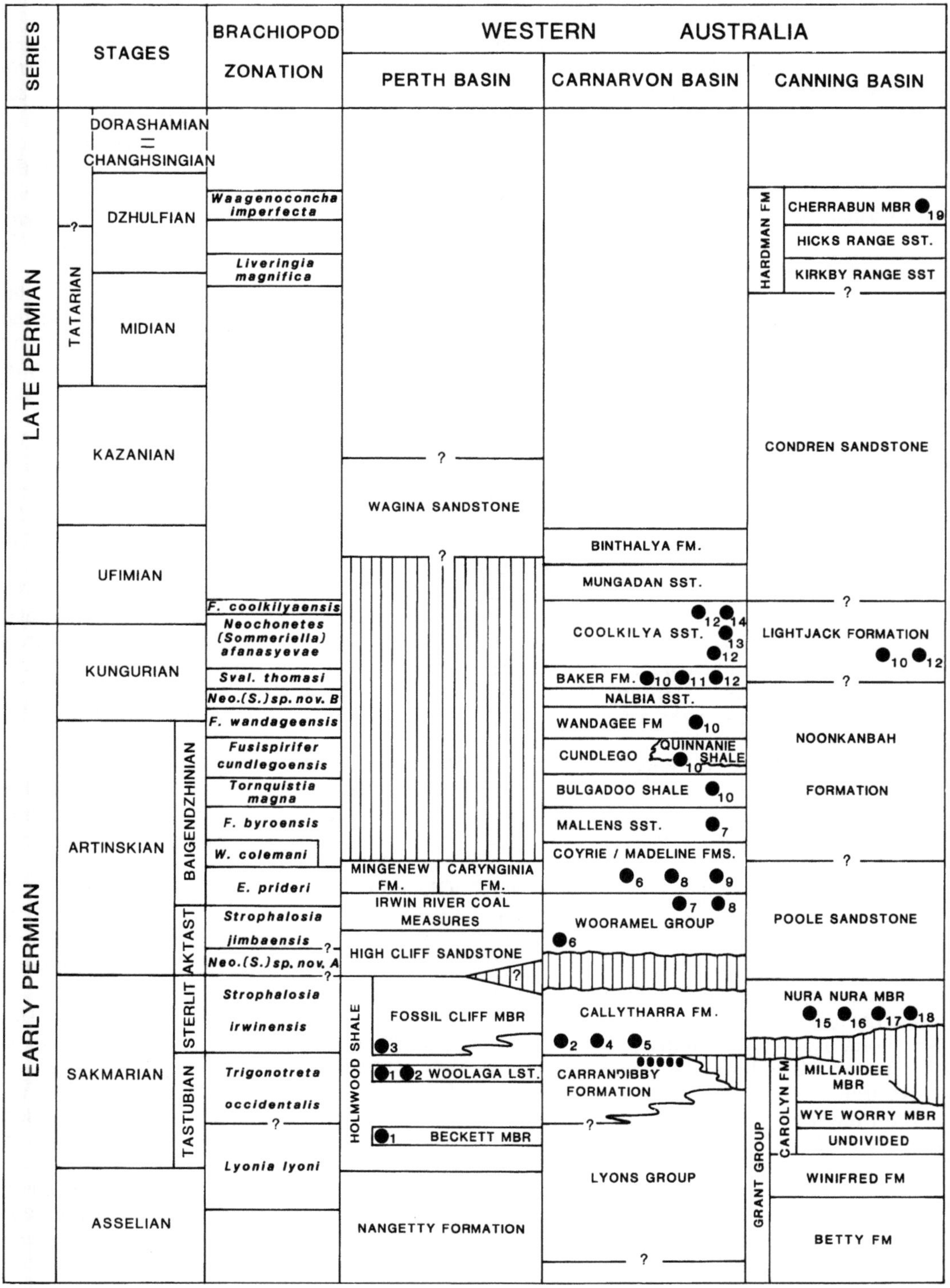

Australian Permian faunas with Dr. J. M. Dickins and Dr. G. A. Thomas are deeply appreciated although responsibilities for the opinions expressed herein rest solely with the author. Mrs. C. Serpell is thanked for typing the manuscript. The authors work on Late Palaeozoic brachiopod faunas is supported by the Australian Research Council.

REFERENCES

Archbold, N. W. 1988a. Permian brachiopod faunas of the Perth Basin - a study of progressive isolation. *Geological Society of Australia, Abstracts* 21: 46-47.

Archbold, N. W. 1988b. Studies on Western Australian Permian brachiopods 8. The Late Permian brachiopod fauna of the Kirkby Range Member, Canning Basin. *Proceedings of the Royal Society of Victoria* 100: 21-32.

Archbold, N. W. 1990. Studies on Western Australian Permian brachiopods 9. The Sterlitamakian brachiopod fauna of the Cuncudgerie Sandstone, Canning Basin. *Proceedings of the Royal Society of Victoria* 102 :1-13

Archbold, N. W. 1991. Studies on Western Australian Permian brachiopods 10. Faunas from the Wooramel Group, Carnarvon basin. *Proceedings of the Royal Society of Victoria* 103: 55-66.

Archbold, N. W. & Dickins, J. M. 1991. Australian Phanerozoic Timescales 6. Permian. *Bureau of Mineral Resources, Geology and Geophysics, Record* 1989/36 :1-17.

Archbold, N. W. & Skwarko, S. K. 1988. Brachiopods and bivalves of Kungurian (Late Early Permian) age from the top of the Coolkilya Sandstone, Carnarvon Basin, Western Australia. *Report of the Geological Survey of Western Australia* 23 :1-15.

Archbold, N. W. & Thomas, G. A. 1985. Permian Brachiopoda from Western Australia: A review in time and space. *Biostratigraphie du Paleozoique* 4 :431-438.

Briggs, D. J. C. 1987. Permian productid zones of New South Wales. *Advances in the Study of the Sydney Basin. Proceedings of the Twenty First Symposium, Department of Geology, University of Newcastle* : 135-142.

Dickins, J. M. 1956. The Permian marine macro-fossils of the Wooramel and Byro Groups of the Wooramel River Area, Western Australia. *Bureau of Mineral Resources, Geology and Geophysics, Record* 1956/113 :1-21.

Dickins, J. M. 1963. Permian pelecypods and gastropods from Western Australia. *Bulletin of the Bureau of Mineral Resources, Geology and Geophysics* 63: 1-203.

Dickins, J. M., Roberts, J. & Veevers, J. J. 1972. Permian and Mesozoic geology of the northeastern part of the Bonaparte Gulf Basin. *Bulletin of the Bureau of Mineral Resources, Geology and Geophysics* 125 :75-102.

Dickins, J. M. & Thomas, G. A. 1959. The marine fauna of the Lyons Group and the Carrandibby Formation of the Carnarvon Basin, Western Australia. *Report of the Bureau of Mineral Resources, Geology and Geophysics* 38 :65-96.

Dickins, J. M., Towner, R. R. & Growe, R. W. A. 1978. A Permian cold water marine fauna in the Grant Formation of the Canning Basin, Western Australia. *Journal of the Palaeontological Society of India* 20 :275-278.

Foster, C. B. & Waterhouse, J. B. 1988. The *Granulatisporites confluens* Oppel-zone and Early Permian marine faunas from the Grant Formation on the Barbwire Terrace, Canning Basin, Western Australia. *Australian Journal of Earth Sciences* 35 :135-157.

Glenister, B. F., Baker, C., Furnish, W. M & Dickins, J. M. 1990a. Late Permian ammonoid cephalopod *Cyclolobus* from Western Australia. *Journal of Paleontology* 64 :399-402.

Glenister, B. F., Baker, C., Furnish, W. M. & Thomas, G. A. 1990b. Additional Early Permian ammonoid cephalopods from Western Australia. *Journal of Paleontology* 64 :392-399.

Glenister, B. F. & Furnish, W. M. 1961. The Permian ammonoids of Australia. *Journal of Paleontology* 35 :673-736.

Glenister, B. F., Windle, D. L. & Furnish, W. M. 1973. Australian Metalegoceratidae (Lower Permian ammonoids). *Journal of Paleontology* 47 :1031 -1043.

Hocking, R. M., Moors, H. T & Van de Graaff, W. J. E. 1987. Geology of the Carnarvon Basin, Western Australia. *Bulletin of the Geological Survey of Western Australia* 133 :1-289.

Playford, P. E., Cockbain A. E. & Low, G. H.

1976. Geology of the Perth Basin. *Bulletin of the Geological Survey of Western Australia* 124 :1-311.
Teichert, C. 1951. The marine Permian faunas of Western Australia : An interim review. *Palaontologische Zeitschrift* 24 :76-90.
Thomas, G. A. 1969. The Permian brachiopod faunas of Western Australia. In, *Gondwana Stratigraphy* (International Union of Geological Sciences Symposium). U. N. E. S. C. O., Paris 217-234.
Thomas, G. A. 1976. Faunal evidence of climatic change in the Early Permian of the Carnarvon Basin, Western Australia. *Abstracts 25th International Geological Congress* 1 :318-319.
Thomas, G. A. & Dickins, J. M. 1954. Correlation and age of the marine Permian formations of Western Australia. *Australian Journal of Science* 16 : 219-223.
Towner, R. R. & Gibson, D. L. 1983. Geology of the onshore Canning Basin, Western Australia. *Bulletin of the Bureau of Mineral Resources, Geology and Geophysics* 215 :1-51.

Gondwana Eight, Findlay, Unrug, Banks & Veevers (eds) © 1993 Balkema, Rotterdam. ISBN 90 5410 304 3

Megousia solita from the Permian of New Zealand: Implications for terrane histories

D.J.C. Briggs
School of Earth Sciences, University of Melbourne, Parkville, Vic., Australia

H.J. Campbell
Department of Scientific and Industrial Research, Geology and Geophysics, Lower Hutt, New Zealand

ABSTRACT: In this paper we describe the distinctive late Early Permian anidanthid brachiopod *Megousia solita* Waterhouse and an associated linoproductid brachiopod *Terrakea* sp. from a fossiliferous block in the Stephens Formation on Stephens Island, and record the presence of comparable material in a limited collection from the upper Takitimu Group in Southland. We also briefly review other anidanthid occurrences from the Permian of New Zealand. The occurrences of *M. solita* have significant biostratigraphical, biogeographical and tectonostratigraphic implications for the Dun Mountain-Maitai and Brook Street terranes.

MEGOUSIA SOLITA IN EASTERN AUSTRALIA

Megousia solita was first described from the type and other localities in the lower Wandrawandian Formation of the southern Sydney Basin (Waterhouse, 1968a). The species has subsequently been recorded in mainland eastern Australia from the lower Branxton Subgroup in the northern Sydney Basin (McClung, 1980, fig. 19.2), from the lower and middle Wandrawandian Formation of the southern Sydney Basin (Runnegar, 1980, fig. 20.1), and from the lower South Curra Limestone of the Gympie Block (Waterhouse and Balfe, 1987). The South Curra Limestone record however is based on material which we assign to *Anidanthus springsurensis* (Booker), and the anidanthid in the middle Wandrawandian Formation is distinguishable as a separate species of *Megousia* (Briggs, in prep). In addition *M. solita* has been identified by one of us (DJCB) from various localities in the Elderslie Formation, Fenestella Shale, and undifferentiated Branxton Subgroup in the northern Sydney Basin, and from Bureau of Mineral Resources locality TH 407 in the Eight Mile Creek beds near Warwick in the southern New England Orogen. *M. solita* is not known from the northern New England Orogen or the Bowen Basin, probably due to widespread unfavourable facies and stratigraphic breaks in late Early Permian successions in these areas.

The biostratigraphic range of *M. solita* is at present best established from the occurrences in the Sydney Basin, where the species is confined to an interval corresponding to the *Echinalosia maxwelli* and *Echinalosia* n. sp. A Zones of Briggs (1991). In the northern Sydney Basin, this interval includes the greater part of the Elderslie Formation and the overlying "Fenestella Shale" in the Lochinvar Anticline area, and correlative strata (lower Branxton Subgroup) in the Belford Dome. In the southern Sydney Basin, the same interval is represented by the Snapper Point Formation and basal Wandrawandian Formation, although *M. solita* has been found to date only in the latter unit, probably due to the unfavourable (shallow-water) facies of the Snapper Point Formation. Immediately older strata in the Sydney - Bowen Basin and the New England Orogen, assigned to the upper *Echinalosia preovalis* Zone of Briggs (1991), contain the anidanthid *Anidanthus springsurensis* (Booker), while immediately younger rocks, of the *Echinalosia discinia* Zone, mostly lack anidanthids, but in the middle Wandrawandian Formation contain the unnamed new species of *Megousia* mentioned above.

The possible presence of *M. solita* in the Permian of Tasmania is suggested by Clarke and Banks' (1975) report of *"Anidanthus springsurensis-solitus"* in their Faunizones 5, 7 and 8, but the material on which this record is based has not been described to date. Clarke and Farmer (1976) subsequently recorded the range of *Anidanthus springsurensis* as Faunizone 5, and *"Anidanthus sp."* as Faunizones 7-8. Faunizones 7-8 are considered to be essentially correlative with the range of *M. solita* in the Sydney Basin (Briggs, 1991, fig. 4).

In terms of the international correlation adopted by Briggs (1991), the range of *M. solita* is early to mid-Kungurian (late Early Permian using the traditional two-fold division of the System). This is consistent with the apparent similarity of *M. solita* to a number of early Kungurian and "Leonardian" species noted in the systematic description below.

MEGOUSIA SOLITA FROM THE STEPHENS FORMATION

M. solita is here figured and described from collection GS 12643, from the fossiliferous sandstone block within the Stephens Formation (Dun Mountain - Maitai terrane) that was reported and figured by Campbell et al. (1984). (GS numbers refer to the collections of the New Zealand DSIR Geology and Geophysics, Lower Hutt). Figure 1 shows the position of the fossil occurrence in relation to the stratigraphic succession exposed on Stephens Island. The age of the Stephens Formation has been the subject of considerable controversy, and is either very late Permian, or Early Triassic (see Campbell et al., 1984, p. 288).

Associated species in the fossiliferous block identified by one of us (H.J.C.) are as follows: *Paraconularia* sp.; *Euryphyllum* sp.; Bryozoa; Chonetida gen. indet.; Aulostegidae gen. indet.; *Terrakea* sp.; *Cleiothyridina* sp.; *Aperispirifer* cf. *wairakiensis* Waterhouse; *Tomiopsis* aff. *confusa* Waterhouse; *Glendonia microstriatus* (Waterhouse); *Fletcherityris* cf. *biundata* McCoy; *Gilledia* sp.; Terebratulacea indet.; Rostroconchia indet.; Atomodesmatinae indet.; *Etheripecten* sp.; *Deltopecten?* sp.; *Streblopteria* sp.; *Myonia* sp.; Euomphalidae gen. et sp. indet.; *Peruvispira* cf. *acuta*; *Platyschisma* sp.; *Keeneia?* sp.; Crinoidea.

The productid recorded as *Terrakea* sp. is also described below. Although imperfectly preserved, the material displays more spines over the ventral valve ears than are seen in *Terrakea concava* Waterhouse, which occurs in the upper part of the range of *M. solita* in the Sydney Basin, within the uppermost *E. maxwelli* Zone and the *E.* n. sp. A Zone. The material appears comparable to an undescribed species of *Terrakea* with similar ear spines, known from limited material from the lower part of the range of *M. solita*, in the Elderslie Formation (lower *E. maxwelli* Zone).

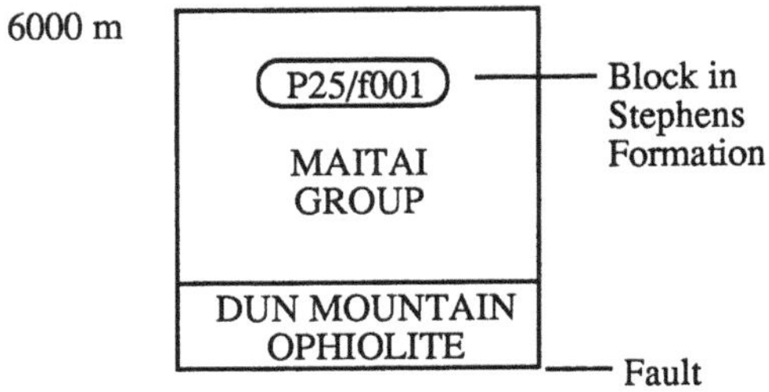

Figure 1. Occurrence of *Megousia solita* in Dun Mountain - Maitai terrane.

Based on provisional faunal identifications, Campbell et al. (1984) concluded that the fauna of the sandstone block was comparable at the generic level with faunas recorded from elsewhere in the Stephens Formation, and that there were a number of possible specific links with those faunas. They assigned the fauna to the *Aperispirifer nelsonensis* Zone of Waterhouse, and considered it to be the youngest fauna known from the Stephens Formation. Our identification of *Megousia solita* however demonstrates that the fauna in the block is late Early Permian in age. It follows that the fossiliferous sandstone can not be contemporaneous with the enclosing strata, and that the presence of a Permian fauna within it has no bearing on the debated Permian vs Triassic age of the Stephens Formation.

ANIDANTHID CF. *MEGOUSIA SOLITA* FROM THE UPPER TAKITIMU GROUP

M. solita may also be represented by a single dorsal valve internal mould from collection GS 7807, collected by J.B. Waterhouse from coarse volcaniclastic sandstone at locality D44/f9878, in rocks of the Brook Street terrane in the Wairaki Hills, Southland (Figure 2). Although previously mapped by Mutch (1972) as undifferentiated Productus Creek Group, the fossiliferous unit was thought by Waterhouse

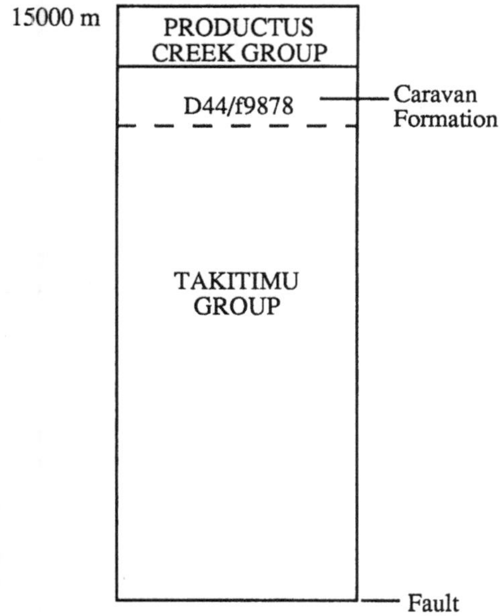

Figure 2. Occurrence of anidanthid cf. *Megousia solita* in Brook Street terrane.

(1982a) to be referable to the underlying Takitimu Group. Landis (1987) named the unit yielding the fossils the Caravan Formation, which he regarded as separate from the Takitimu Group, but which we treat as the uppermost unit of that Group. Associated species in the collection include *Terrakea* sp. of Waterhouse (1982a, p. 48, pl. 10h,j), which seems closely comparable to the *Terrakea* from the block on Stephens Island, along with *Glendonia microstriatus*, *Peruvispira* sp., Euomphalidae gen. et sp. indet, and indeterminable brachiopod, bryozoan, bivalve and crinoid material. Attempts by one of us (H.J.C.) to collect additional material from the locality have so far been unsuccessful.

The presence of *M. solita* would imply a late Early Permian age for the Caravan Formation, and the associated *Terrakea* sp. appears to support this, suggesting correlation specifically with the lower part of the range of *M. solita* in the Sydney Basin. This correlation in turn implies an age immediately older than the *E. maxwelli* -*T. concava* fauna of the apparently overlying lower Productus Creek Group, and immediately younger than the remainder of the Takitimu Group (now correlated with the upper *E. preovalis* Zone (Briggs, in prep.)), as was also suggested by Waterhouse (1982a). The fauna thus seems to argue against the existence of large-scale structural complications in this part of the succession.

OTHER NEW ZEALAND ANIDANTHID OCCURRENCES

A specimen figured by Waterhouse (1982b, plate 22c) as ?*Pseudomonotis* sp. C appears from the illustration to be an external mould of an anidanthid dorsal valve. The specimen is recorded as being from GS 9276 (Takitimu Group) in the plate caption, but in the description the specimen is said to be from GS 9697 in the Mangarewa Formation. Further work is required to establish the identity of the specimen, and to resolve the confusion regarding its provenance.

The *Anidanthus* reported by Campbell et al. (1984) from an isolated fauna (D42/f54A) in the Eglinton Volcanics in the Dunton Range, Southland, is now considered to be a productid of uncertain affinity, but not an anidanthid.

TECTONOSTRATIGRAPHIC IMPLICATIONS

The late Early Permian age of the sandstone block from Stephens Island is intriguing, as it seems to indicate that the sandstone is significantly older than the entire known sedimentary sequence (Maitai Group) in the Dun Mountain-Maitai terrane. This implies that the block was derived from another terrane. We know of no unit that is precisely comparable lithologically, but note a general similarity to volcaniclastic sediments in the upper Takitimu Group of the Brook Street terrane. This link is strengthened by our identification of comparable *Terrakea* sp. and possible *Megousia solita* from the limited fauna from the Caravan Formation, as well as the evidence from the richer overlying and underlying units that the uppermost Takitimu Group should be about the same age as the fossiliferous sandstone of the block. The matrix of the collection from the Caravan Formation is a poorly sorted coarse tuffaceous sandstone of comparable lithofacies to the matrix of the fossiliferous sandstone block from Stephens Island. Petrographically the two lithologies differ, but not to the extent that derivation from a common terrane is precluded. If, as seems likely, the block on Stephens Island was derived from the Brook Street terrane, its occurrence as a detrital block in the Dun Mountain - Maitai terrane implies close proximity of the two terranes during the period of Stephens Formation deposition, in either Late Permian or Early Triassic time.

SYSTEMATIC PALAEONTOLOGY

Family Anidanthidae Waterhouse, 1968a
Genus *Megousia* Muir-Wood and Cooper, 1960

Diagnosis: Anidanthidae with shallowly concave, alate dorsal valve bearing concentric ornament of free-standing lamellae over disc.

Type species: Megousia auriculata Muir-Wood and Cooper, 1960 (see also Cooper and Grant, 1975).

Remarks: In proposing the genus *Megousia*, Muir-Wood and Cooper (1960, p. 309) listed a number of apparent differences in external and internal morphology from *Anidanthus* Whitehouse, including a "more alate shell outline with greater development of ears, often forming crescentic extensions" and other external and internal features. Subsequently the presence of large costellate "ears" on the dorsal valve was taken to be the main feature distinguishing *Megousia* from *Anidanthus* by workers such as Waterhouse (1968a,b), Grigorieva and Kotlyar (in Sarytcheva, ed., 1977) and Cooper and Grant (1975). The separate validity of *Megousia* was therefore called into question when Gupta and Waterhouse (1979, p. 9) pointed out that large costellate ears are present in *Anidanthus* as well. Waterhouse (1986, p. 61) considered that other apparent differences between the respective type species of the two genera were due to differences in preservation, or were of only specific importance, and concluded that *Megousia* should be synonymized with *Anidanthus*. However the genus *Megousia* is retained here for alate anidanthids in which, as in the type species, the dorsal valve concentric ornament consists of sharp free-standing lamellae, rather than the very low, step-like crests seen in *Anidanthus* and *Protoanidanthus* Waterhouse, or the numerous very short, close lamellae seen in *Kuvelousia* Waterhouse.

Megousia solita Waterhouse, 1968a
Figure 3: 1-10

1968a *Megousia solita* Waterhouse, p. 1172, pl. 154, figs 1-6, 8-10.
1984 *Anidanthus* sp.; Campbell et. al., fig. 6:11-13.

Holotype: ANU F17752, from "Ulladulla Formation" (lower Wandrawandian Formation) at Berrara, New South Wales; figured by Waterhouse, 1968a, pl. 154, figs 1,9.

Diagnosis: Large, strongly transverse *Megousia*. Dorsal valve with shallowly concave disc and large, flat, capillate alae close to plane of commissure; disc bears ornament of low recumbent lamellae posteriorly and high, thick, free lamellae on anterior half.

Material: 42 moulds, almost exclusively of separate valves, in very coarse sandstone, grading into granule conglomerate, from block in conglomerate of Stephens Formation on Stephens Island (GS12643; P25/f001). The material comprises moulds of 5 incomplete ventral valve exteriors, 23 ventral valve interiors, 9 dorsal valve exteriors, three incomplete dorsal valve interiors, and the dorsal surface of one conjoined shell.

Dimensions (in mm)

	L_{vv}	L_{dv}	W_{vv}	W_{dv}	Ht_{vv}	D_{dv}	Hge
BR1641	18.0	14.0	23.0	-	8.5	-	22.0
BR2308	16.5	12.5	28.0	-	8.5	-	28?
BR2309	18.5	14.0	28.0	-	13.0	-	28?
BR2310	19.0	14.5	32.0	-	9.5	-	32.0
BR2311	11.5	9.5	17.0	-	6.0	-	16.0
BR2312	16.5	13.0	26.0	-	8.5	-	26.0
BR2313	10.5	9.0	16.0	-	6.0	-	16?
BR1314	12.5	9.5	21.0	-	7.0	-	21?
BR1642	-	11.0	-	28?	-	3.0	24?
BR2315	-	11.0	-	32?	-	3.5	30?
BR2316	-	11.5	-	33?	-	1.0	26.5
BR2317	-	13.5	-	35?	-	3.0	-
BR2318	-	14.0	-	30.0	-	4.0	-
BR2319	-	11.0	-	33?	-	2.5	26?

L_{vv} length of ventral valve
L_{dv} length of dorsal valve (estimated in ventral valves from length of commissure]
W_{vv} width of ventral valve
W_{dv} width of dorsal valve (exceeds W_{vv} at maturity due to alae)
Ht_{vv} height of ventral valve
D_{dv} (external) depth of dorsal valve
Hge length of hinge

Description: External. Adult shell small to medium sized: maximum length 19 mm, maximum width 28 mm. Ventral valve plan outline strongly transverse throughout growth: L/W_{vv} 0.54-0.68, average 0.64; umbo projects up to 0.24 of shell length behind hinge. Hinge equal to or very close to maximum width of ventral valve. Longitudinal profile uncoils evenly from umbo, apart from abrupt straightening into anterior segment ("trail"). Transverse profile moderately arched, with broad, flat median part curving evenly into gently sloping flanks. Ears gently convex. Sulcus consistently present, but very weak in some specimens; deepest in relatively weakly transverse shells; extends from near umbo to anterior margin. Umbo strongly incurved. Ventral valve ornament poorly preserved. Costellae appear mostly fine, about 10-12/5 mm on anterior part of valve; roughly equal in width to intercostal sulci. Spines and growth ornament not preserved.

Dorsal valve strongly transverse (L/W_{dv} 0.33-0.47, average 0.38). Disc shallowly and variably concave (depth/W 0.03-0.13, average 0.09); separated from broad flat ears by change of slope and (usually) by sudden termination of concentric lamellae. Very low, broad median fold consistently present, generally not accompanied by significant backward arching of lamellae. Alae almost flat, lying very close to plane of commissure, broadly U-shaped in plan view, projecting laterally and slightly anteriorly. Concentric ornament of disc consists of sharp terraces up to 3-4 mm from umbo, fine sharp oblique blades on mid-disc, and coarse erect blades, each about 1.0 mm thick, anteriorly. Costellae strong on posterior disc and on dorsal surfaces of lamellae on rest of disc. Alae weakly costellate over most of surface, smooth close to hinge; normally lack concentric ornament, but in one specimen bear 5 lamellae continuous with those of disc.

Internal. Ventral valve umbonal region moderately strongly thickened, with umbonal walls reaching up to 1.5 mm in thickness. Adductor scars smooth, weakly raised up to 1.0 mm above floor of valve; posteriorly placed, with mid-point located opposite posterior margin of diductor scars. Diductor scars depressed, weakly subdivided into radial bands. Small pits dotted over interior in front of adductor scars. Traces of very fine sharp pustules preserved close to anterior margin.

Dorsal valve. Cardinal ridges form flattened subhorizontal shelf well behind hinge, joining at base of cardinal process. Cardinal process with large, club-like outer walls flaring widely from massive median lobe formed by fusion of inner walls. Massive thickened visceral septum runs from process, passing into short high brachial septum at front of adductor scars. Anterior margin of disc displays abrupt, roughly right-angled geniculation, bordered by distinct low marginal ridge at maturity. Brachial ridges conspicuous, laterally placed, occupying middle half of disc. Pustules fine over anterior half of disc. Alae smooth internally.

Comparisons: The material closely resembles eastern Australian populations of the species, except that some of the latter include larger individuals up to 27 mm long. The coarse free lamellae on the dorsal valve distinguish the species from all older Australian anidanthid species (included in the genera *Anidanthus* and *Protoanidanthus*. An undescribed younger species of *Megousia* from the middle Wandrawandian Formation differs by having finer and more closely spaced dorsal lamellae.

M. yakutica Likharev of Ustritsky and Chernyak (1963, p. 89, p. 10, figs 3-14, pl. 11, fig. 1) from the lower Baikur Suite of Taimyr seems similar in shape and ornament. The figured specimens are relatively small compared with *M. solita*, and high free lamellae develop only on the distal part of the dorsal valve. Specimens from the Vilski Horizon (Filippovian) of North Timan figured by Barkhatova (1970, p. 144, pl. 6, figs 6-15) as *Pseudomarginifera kulikii* (Fredericks) are very similar in size and form to *M. solita*, and appear to have regularly spaced lamellae anteriorly. However these seem to be lower than those of *M. solita*, although the preservation of the specimens and the reproduction of the figures are inadequate to allow certainty. *Megousia kulikii* of Ifanova (1972, p. 114, pl. 5, figs 1-14) from the Ayachyaginsk and Rudnitz Subsuites of the Lekvorkutsk Suite (Kungurian) of the Pechora Basin agrees with *M. solita* in overall form, and in the internal and external character of the dorsal valve.

Megousia harlandi Gobbett and *M. kulikii* of Gobbett (1963) from the Spirifer Limestone of Spitzbergen are similar in shape to *M. solita*, but the details of their dorsal ornaments are obscure. *M. weyprechti* (Toula) of Gobbett (1963) from the "Brachiopod Chert", higher in the sequence at Spitzbergen, has a much larger, deeply sulcate ventral valve and an externally flat dorsal valve, and following Waterhouse (1968a) can be included in *Kuvelousia.*

Of the various species described by Cooper and Grant (1975) from the Glass Mountains of Texas, *M. solita* is closest to species from the Cathedral Mountains and Road Canyon Formations, in which the dorsal valve lamellae are numerous and densely crowded. Most similar is *M. umbonata* Cooper and Grant from the Cathedral Mountain Formation, which has very similar dorsal ornament to *M. solita.* The type species, *M. auriculata* Cooper and Grant from the Road Canyon Formation, differs by its small size, extremely strongly transverse dorsal valve, and highly elongate alae. The younger species, *M. flexuosa* Cooper and Grant, *M. mucronata* Cooper and Grant, and *M. definita* Cooper and Grant, all based on type material from the Word Formation, have relatively few lamellae, confined to the distal part of the disc.

Anidanthid cf. *Megousia solita*
Figure 3: 11-12

Material: Internal mould of a dorsal valve, BR 2320, from collection GS 7807, from locality D44/f9878, Caravan Formation, Takitimu Group, Wairaki Hills, Southland.

Description: The solitary specimen consists of an internal mould of a dorsal valve identified as an anidanthid on the basis of its short (possibly worn or broken) alae, and provisionally compared to *Megousia* on the basis of its quadrate shape and strongly transverse outline. Visible internal features are compatible with those seen in *M. solita*, and include a low, stout median septum, laterally placed brachial ridges (strongly raised anteriorly in this specimen), traces of fine endospines over the anterior part, and in particular a pair of stout cardinal ridges projecting strongly posteriorly, and separated from the hinge by a distinct groove. The cardinal process appears to be largely embedded within these cardinal ridges, but may be partly worn or broken away. The diagnostic external ornament of *M. solita* is not preserved, but no other productid species is known from the Austrazean Permian that shares comparable internal features.

Family Linoproductidae Stehli, 1954
Genus *Terrakea* Booker, 1930

Terrakea sp.
Figure 3: 13-19

1984 *Terrakea* sp.; Campbell et al., fig. 6:3-4

Material: About 14 specimens, preserved as moulds of separate valves in in very coarse sandstone, grading into granule conglomerate, from block in conglomerate of Stephens Formation on Stephens Island (GS12643; P25/f001).

Dimensions (in mm)

	L_{vv}	L_{dv}	W	Ht_{vv}	D_{dv}	Hge
BR1633	22.5	19.5	28.0	11.0	-	28?
BR1634	-	23.0	30.0	-	11.0	29.5
BR2321	21.0	17.0	25.0	13.0	-	-

Description: External. Ventral valve relatively roundly quadrate, with broad, evenly convex middle part (lacking sulcus) and short, steep flanks. Maximum length probably about 36 mm (estimated from length of longest dorsal valve of 30.0 mm). Disc covered by numerous distinct, markedly elongate spine ridges: at around 2 cm from umbo, ridges typically about 0.6 mm wide and 3.0 mm long, giving rise to low-angle recumbent spines about 0.5 mm in diameter. These spines arise in fairly regular quincunx, about 3 mm apart in rows about 3 mm apart. Intervening costellae much less prominent than spine ridges, and poorly preserved; fine (3-4 per 2 mm). Ornament of anterior part of mature valve not seen. Ear spines up to 0.6 mm in diameter; form small, moderately dense group, with 2-3 rows of 2-3 spines in each, extending over most of surface of ear.

Dorsal valve with very weakly concave disc and moderately long trail (11mm long in one specimen), separated on anterior and anterolateral margins by abrupt, rounded geniculation. Plan outline of disc, and of valve, marked by short, straight anterior segment, very broadly rounded anterolateral corners, wide hinge, and large ears; ears and broad convexities in front of ears occupy relatively large sectors of disc. Trail bears coarse, low-angle recumbent spines up to 1.0 mm in diameter, mostly grouped in two rows. Loose group of coarse spines (0.4 mm diameter) over ear; spines on rest of disc relatively fine and poorly preserved.

Internal. Largest ventral valve (BR 1633, which however is considerably smaller than estimated maximum size based on dorsal valves) very weakly thickened internally, and weakly impressed adductor scars. Dorsal valve interior very weakly thickened: in one large valve (30 mm long), concentric wrinkles of exterior are clearly reflected internally; adductor scars weakly impressed. Elongate pustules form broad anterior zone, coarsest on geniculation.

Comparisons. The relatively wide hinge, forming the widest part of the shell, and the relatively broad sectors occupied by the ears and their bordering slopes on the dorsal valve, together suggest comparison with only two other described *Terrakea* species, both of late Early Permian age. *Terrakea concava* Waterhouse, first described from the Productus Creek Group of Southland, New Zealand, is similar in size, overall shape, and most aspects of its ornament, but differs in the presence on the ventral valve of relatively few posterolateral spines, which do not spread onto the surface of the ears. In this respect *Terrakea* sp. of Waterhouse (1982a, p. 48, pl. 10h,j; see also Fig. 3:20 herein), from GS 7807 in the Caravan Formation at the top of the underlying Takitimu Group, appears to agree closely with the Stephens Island material. Unfortunately the material from GS 7807 is mostly fragmentary and immature, so that it can not be confidently assigned to the same species. *Terrakea* with similar features of shape and ornament is known from the Elderslie Formation (e.g. at University of New England locality L990), and will be described elsewhere (Briggs, in prep.).

Older species of *Terrakea* such as *T. pollex* Hill and *T. dickinsi* Dear also have ears occupying relatively broad sectors of the dorsal valve, but differ in having a slightly shorter hinge, distinctly narrower than shell width. In most Late Permian species of *Terrakea* the ears and their bordering slopes occupy somewhat narrower sectors of the dorsal valve.

ACKNOWLEDGEMENTS

Work by DJCB was carried out at the University of Melbourne during a three-year ARC funded project with Dr. Neil Archbold. The study has benefited from discussions with Professor C.A. (Chuck) Landis.

REFERENCES

Barkhatova, V.P., 1970. Biostratigraphy of the Carboniferous and Lower Permian of Northern Timan. *Trudy Vsesouznogo Neftyanogo Nauchno-issledovatelskogo Geologorazvedochnogo Instituta (VNIGRI)*, 283: 228 pp. [In Russian].

Briggs, D.J.C., 1991. Stratigraphical distribution and biostratigraphical significance of the Permian Productidina and Strophalosiidina (Order Productida) of eastern Australia. *In* MacKinnon, D.I., Lee, D.E. and Campbell, J.D., (eds), *Brachiopods through time, Proceedings of the 2nd International Brachiopod Congress, Dunedin, New Zealand*: 367-374. Balkema, Rotterdam.

Briggs, D.J.C., in prep. Permian Productidina and Strophalosiidina of the Sydney - Bowen Basin and the New England Orogen: Systematics and significance for Australasian Permian biostratigraphy. *Association of Australasian Palaeontologists, Memoir.*

Campbell, H.J., A.C. Coleman, M.R. Johnston. & C.A. Landis, 1984. Geology of Stephens Island and the age of the Stephens Formation. *New Zealand Journal of Geology and Geophysics*, 27: 277-289.

Clarke, M.J. & M.R. Banks, 1975. The stratigraphy of the lower (Permo-Carboniferous) parts of the Parmeener Supergroup, Tasmania. *In* Campbell, K.S.W., (ed.), *Gondwana Geology: Papers from the Third Gondwana Symposium, Canberra*: 453-467. ANU Press, Canberra.

Clarke, M.J. & N. Farmer, 1976. Biostratigraphic nomenclature for Late Palaeozoic rocks in Tasmania. *Papers and Proceedings of the Royal Society of Tasmania*, 110: 91-109.

Cooper, G.A. & R.E. Grant, 1975. Permian brachiopods of West Texas, III. *Smithsonian Contributions to Paleobiology*, 19. 1298 pp., pls 192-502.

Gobbett, D.J., 1963. Carboniferous and Permian brachiopods of Svalbard. *Norsk Polarinstitutt Skrifter* 127. 201 pp., 25 pls.

Gupta, V.J. & J.B. Waterhouse, 1979. Permian invertebrate faunas from the Lamnimargus himalayensis Zone of Spiti and Ladakh regions, North West India. *Contributions to Himalayan Geology* 1: 5-19.

Ifanova, V.V., 1972. Permian brachiopods of the Pechora Basin. In Ifanova, V.V. & Semenova, E.G., *Middle Carboniferous and*

Permian brachiopods from east and north of the European part of the USSR: 72-176 "Nauka", Moscow [In Russian].

Landis, C.A., 1987. Permian-Jurassic rocks at Productus Creek-Letham Ridge, Southland. *Geological Society of New Zealand, Miscellaneous Publication* 37C, 1987: 89-110.

McClung, G., 1980. Permian biostratigraphy of the northern Sydney Basin. *Geological Survey of New South Wales, Bulletin* 26: 360-375.

Muir-Wood, H.M. & G.A. Cooper, 1960. Morphology, life habits and classification of the Productoidea (Brachiopoda). *Geological Society of America, Memoir* 8: 447 pp., 135 pls.

Mutch, A.R., 1972. Geology of the Morley Subdivision. *New Zealand Geological Survey Bulletin* 76: 1-104.

Runnegar, B.N., 1980. Biostratigraphy of the Shoalhaven Group. *Geological Survey of New South Wales, Bulletin* 26: 376-382.

Sarytcheva, T.G. (ed.), 1977. Upper Palaeozoic productids of Siberia and the Arctic). *Trudy Paleontologicheskogo Instituta, Akademia Nauk SSSR* 161: 192 pp., 27 pls [In Russian].

Ustritsky, V.I. & G.E. Chernyak, 1963. Biostratigraphy and Brachiopoda of the Upper Palaeozoic of Taimyr. *Trudy Nauchno-Issledovatelskogo Instituta Geologii Arktiki* 134: 1-235. [In Russian].

Waterhouse, J.B., 1968a. New species of *Megousia* Muir-Wood and Cooper and allied new genus from the Permian of Australia and North America. *Journal of Paleontology*, 42: 1171-1185, pls 154-156.

Waterhouse, J.B., 1968b. Redescription of the Permian brachiopod *Anidanthus springsurensis* Booker. *Transactions of the Royal Society of New Zealand, Geology* 5: 235-243, pl. 1.

Waterhouse, J.B., 1982a. New Zealand Permian brachiopod systematics, zonation, and paleoecology. *New Zealand Geological Survey, Paleontological Bulletin* 48.

Waterhouse, J.B., 1982b. Permian Pectinacea and Limacea (Bivalvia) from New Zealand. *New Zealand Geological Survey, Paleontological Bulletin* 49.

Waterhouse, J.B., 1986. Late Palaeozoic Scyphozoa and Brachiopoda (Inarticulata, Strophomenida, Productida and Rhynchonellida) from the southeast Bowen Basin, Australia; with a contribution from D.J.C. Briggs. *Palaeontographica, Abt A*, 193: 1-76, pls 1-15.

Waterhouse, J.B. & P.E. Balfe, 1987. Stratigraphic and faunal subdivisions of the Permian rocks at Gympie. *In* Murray, C.G., and Waterhouse, J.B. (eds), *1987 Field Conference, Gympie District*: 20-33.Geological Society of Australia, Queensland Division, Brisbane.

Figure 3. All illustrations x 1.5

1-10: *Megousia solita* Waterhouse. 1-3, external moulds of three dorsal valves, BR2315, 2322, and 2316 respectively. 4-5, posterior and ventral views of ventral valves internal mould, BR2310. 6-7, ventral and anterior views of incomplete ventral valve internal mould BR2324. 8, latex cast of part of dorsal valve, BR2323. 9-10, anterior and ventral views of latex cast of incomplete dorsal valve BR2325.

11-12: Anidanthid cf. *Megousia solita*. Ventral and anterior views of latex cast of dorsal valve, BR2320.

13-19: *Terrakea* sp. 13, latex cast of umbonal part of ventral valve, BR2321. 14-16, posterolateral, ventral and anterior views of ventral valve internal mould, BR1633. 17-19, ventral, anterior and lateral views of dorsal valve external mould, BR 1634.

20: *Terrakea* sp. Mould of posterolateral part of ventral valve, BR2327.

1-10,13-19 from GS 12643, P25/f001, Stephens Formation, Stephens Island. 11,12,20 from GS 7807, D44/f9878, Caravan Formation, Takitimu Group, Wairaki Hills.

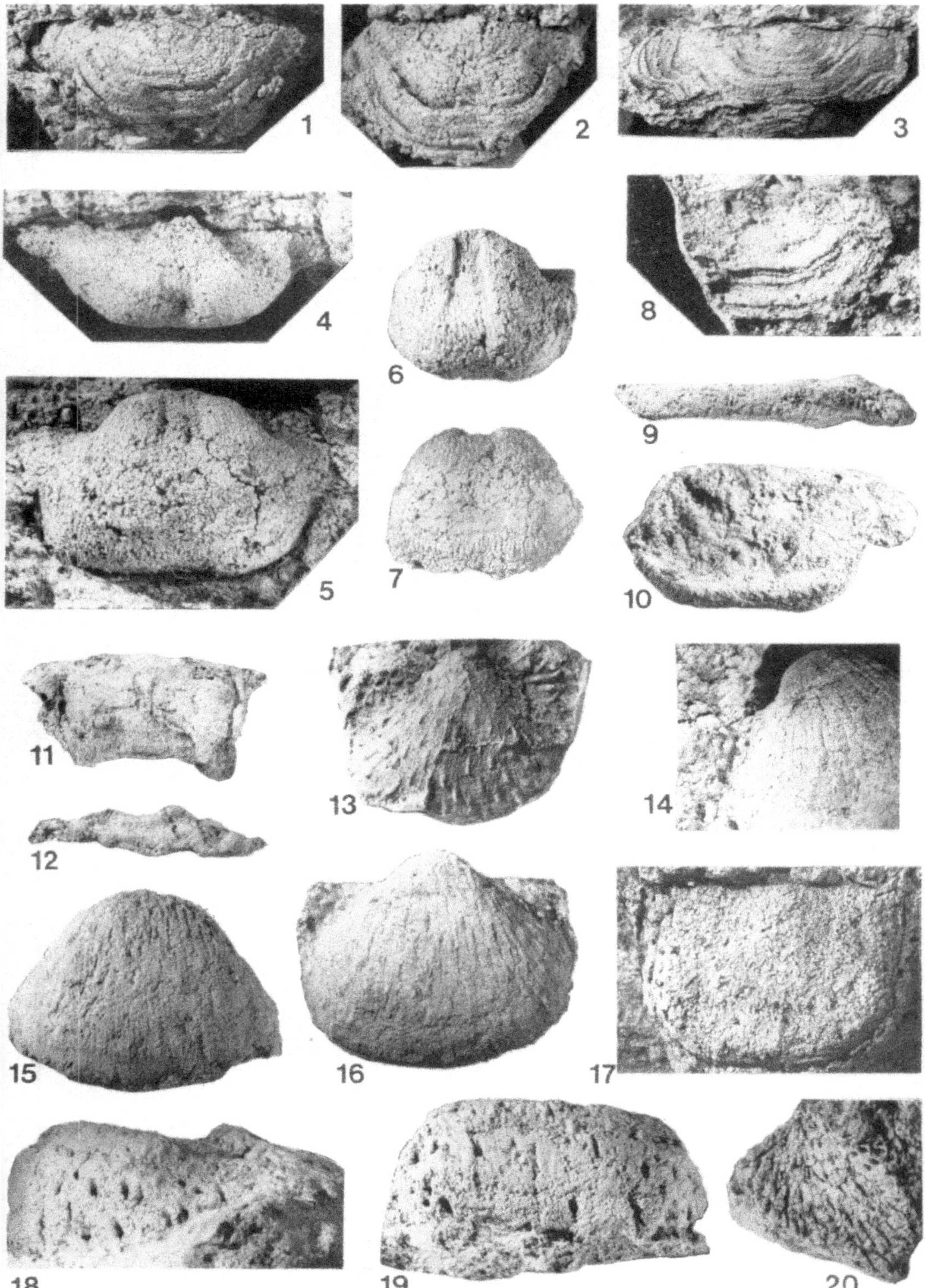
1
2
3
4
5
6
7
8
9
10
11
12
13
14
15
16
17
18
19
20

Gondwana Eight, Findlay, Unrug, Banks & Veevers (eds) © 1993 Balkema, Rotterdam. ISBN 90 5410 304 3

Some climatic and tectonic implications of the Permian marine faunas of Peninsular India, Himalayas and Tibet

J.M.Dickins
Bureau of Mineral Resources, Canberra, A.C.T., Australia

S.C.Shah
Geological Survey of India, Calcutta, India

N.W.Archbold
Victoria College, Rusden Campus, Clayton, Vic., Australia

Jin Yugan
Nanjing Institute of Geology and Palaeontology, People's Republic of China

Liang Dingyi & Liu Benpei
China University of Geosciences, Beijing, People's Republic of China

ABSTRACT: The earliest Permian (Asselian - Sakmarian) colder water *Eurydesma* fauna has long been known to be overlain by later warmer water faunas in the Himalayas (including Tibet). More recently early Permian colder marine water faunas have been recognized in the Lhasa Block and in north-west Tibet including *Eurydesma* itself in north-west Tibet where they are also overlain by faunas regarded as warm water. Further north and east only warmer water faunas are recognized. The explanation of the change in water temperature known to be characteristic of the Lower Permian only in this region and the adjacent southern part of the Arabian Peninsula and possibly Afghanistan and the Pamir poses a fascinating problem.

In this paper Asselian - Sakmarian (Tastubian), Sakmarian (Sterlitamakian) - Lower Artinskian, Upper Artinskian - Kungurian, Kazanian - Dzhulfian and Dorashamian (Changhsingian) faunas are reviewed in terms of their climatic and tectonic implications.

1 INTRODUCTION

The distribution of the cold water *Eurydesma* fauna in the southern continents (South Africa, Australia, South America) and India (more recently found in Tibet and New Zealand) and the occurrence of the *Glossopteris* flora in these regions played an important part in developing the concept of Gondwana. Consideration of the flora is, however, outside the scope of the present paper, although it is of interest, that the occurrence of the *Glossopteris* flora is equivocal, see inter alia Dickins & Shah (1981), Dickins (1985a), Chatterjee & Hotton (1986), in that *Glossopteris* and the *Glossopteris* flora is now known to occur apparently outside of what is generally regarded as Gondwana (e.g. eastern Siberia).

In the last two decades, firstly based on palaeomagnetic work and later following the development of plate tectonics, the position of India and the relationships of its faunas (and floras) with the Himalayas and adjacent regions have become of critical importance for judging the validity of Gondwana reconstructions. A large body of palaeontological work now indicates that India was never far away from Asia during the Palaeozoic and Mesozoic. This includes work by Teichert (1971), Dickins (1972), Stehli (1968, 1973), Ahmad (1978) and a recent review by Chatterjee & Hotton (1986) as well as many other papers. The present review of the Permian marine invertebrate faunas of this region has been undertaken to provide information on this problem as well as to elucidate other problems such as the tectonic, magmatic and climatic development.

The faunas are considered in

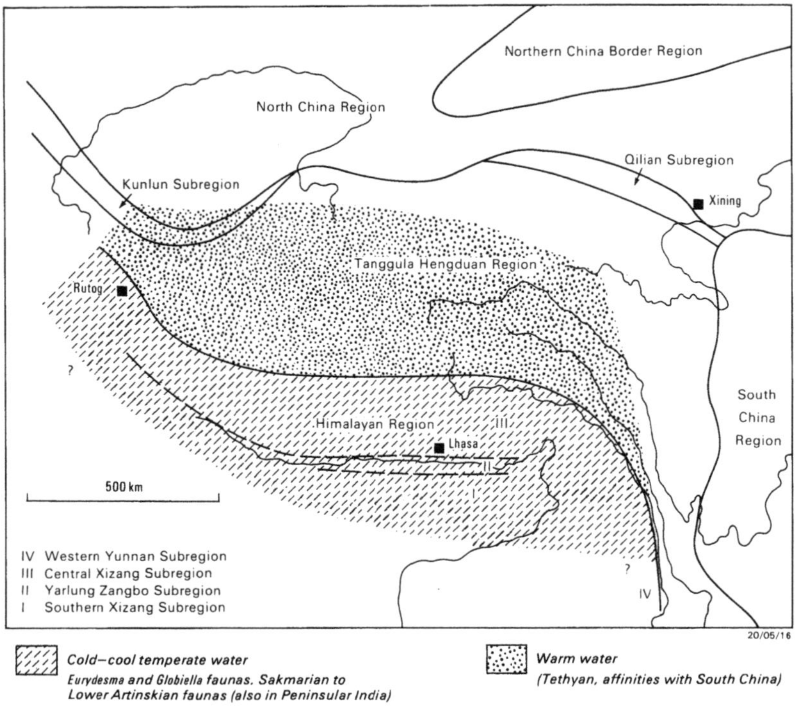

Fig. 1. Faunal distribution for Asselian-Lower Artinskian (Regional subdivisions - also in Figs. 2-4 - after Yin Yugan, 1985). The boundary between cold-cool temperate and the warm water faunas passes north of Rutog whereas the tectonic line (the Bangongghu-Nujiang Line) passes to the south - Fig. 3.

intervals which correspond essentially to those of Archbold & Gupta (1986) and Dickins and Shah (1987).

2 FAUNAS

2.1 Asselian to Sakmarian (Tastubian)

a) Cold - *Eurydesma* fauna

The earliest Permian faunas in Peninsular and Himalayan India and southern Tibet are cold water and, indeed contain *Eurydesma* itself, see Dickins & Shah (1981), Dickins (1985 a and b), Jin Yugan (1985). This is reflected, for example, in the brachiopods, gastropods and pelecypods and is associated with the absence of characteristic warm water forms such as the fusulinids. According to the regional subdivisions for Tibet used by Jin Yugan (1985) the cold water fauna occurs throughout the Himalayan Region of southern Tibet - see Fig.1.

In the Tanggula-Hengduan Region to the north, marine faunas of this level are dominated by fusulinids, algae and colonial corals and have a provincial affinity with South China (Jin Yugan, 1985). In the Kunlun Subregion in the northwest and the Qilian Subregion in the northeast, warm water Lower Permian including Asselian deposits with *Pseudoschwagerina* and faunally related to southern China, are recognized. (A twofold subdivision of the Permian System into Lower and Upper is used here corresponding to the standard twofold subdivision in the Urals area and the Russian Platform.

Glacially derived deposits are present in the Himalayan Region but apparently not to the north.

b) Cold or cool temperate - *Globiella* fauna

In earlier publications *Globiella* is referred to as *Stepanoviella* (Dickins & Shah, 1981; Jin Yugan, 1985). The work, however, of Archbold (1983) has shown that the genus present is more satisfactorily regarded as *Globiella* than as *Stepanoviella*. The *Globiella* fauna is found in Peninsular and Himalayan India and in southern Tibet - Himalayan Region and the Lhasa Block (Dickins & Shah, 1981; Jin Yugan, 1985).

This fauna can be regarded as cold or cool temperate. It is found in deposits overlying glacially derived sediments, contains *Globiella* and has a somewhat more diverse fauna. Therefore the fauna and enclosing sediments seems to suggest some amelioration in climate compared to the underlying interval. In Tibet this fauna does not appear to occur further to the north or east.

2.2 Sakmarian (Sterlitamakian) to Lower Artinskian

Cool temperate

These faunas are found in the upper beds at Badhaura, Peninsular India, in the upper part of the Agglomeratic Slate in Kashmir of Himalayan India, apparently in many parts of the Himalayas and in northwest Tibet near Rutog in the Himalayan Region of Tibet (Dickins & Shah, 1979; Dickins, 1985a). The fauna of Trilochan Singh (1987) from Arunachal Pradesh, Eastern Himalaya, belongs here.

They correspond to the faunas of Stages B and C of Dickins (1963) in Western Australia, where faunally the two stages have much in common (Archbold, et al., in press). Archbold (1991) has recently described brachiopods from Stage C which help to differentiate the two stages. In Western Australia *Globiella* occurs in Stages B and C and it would not be surprising if it ranged into equivalent levels in India and Tibet. The fauna can probably be best regarded as cold temperate based on the presence of some cold water forms, the entry of genera not represented lower in the sequence and the relative diversity of its fauna (Dickins, in press).

In the Rutog area in the Qudi Formation, overlying glacial deposits and the *Eurydesma* fauna, is a fauna with *Oriocrassatella* and other Gondwana related forms regarded above as cool temperate, and a fusulinid fauna (Liang Dingyi et al., 1983; Nie Zetong & Song Zhimin, 1983a). The fusulinid fauna occurs in the topmost part of the formation overlying the brachiopod and pelecypod fauna (the *Oriocrassatella* fauna) in the lower part. A number of fusulinids are present which are regarded as Sakmarian or Artinskian and appear to indicate a warm water environment. As indicated in the next section, the Tunlonggongba Formation overlying the Qudi Formation contains a fusulinid fauna related to that of South China of Chihsian (Artinskian to Kungurian) age (Nie Zetong & Song Zhimin, 1983b). The fauna can be regarded as warm temperate or even warm water.

A closely related fauna is found in the Haushi Formation of Oman (Dickins & Shah, 1979; Dickins, 1985a) where it is overlain by an Upper Permian warm water fauna in the Lusaba Formation.

This interval contains the youngest marine faunas found in Peninsular India, so far confined to the Badhaura area of north-western Peninsular India.

2.3 Upper Artinskian to Kungurian

Warm

Faunas of this interval seem less widespread than the underlying or overlying faunas and are not represented in Peninsular India. That of the Amb Formation of the Salt Range, Pakistan is well known. This interval is present in Ladakh (Dickins & Shah, 1979) and probably elsewhere but at present is difficult to identify because of lack of adequate systematics

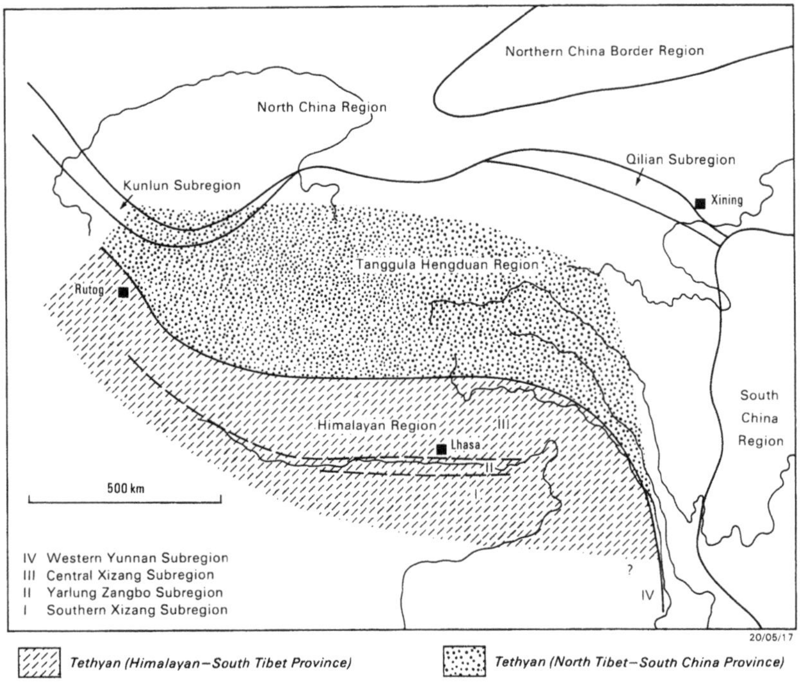

Fig. 2. Faunal distributions for Upper Artinskian-Kungurian. The boundary between the Himalayan-South Tibet Province and the North Tibet-South China Province is shown north of Rutog whereas the tectonic line (the Bangongghu-Nujiang Line) passes to the south - Fig.3.

and biostratigraphy, essential for understanding the development of the Himalayas. The interval is represented in northwest Tibet (Rutog area) in the Himalayan Region of Tibet. In this area the Tunlonggonba Formation overlying the Qudi Formation contains a fusulinid fauna from which seven genera are recorded and is roughly compared to the Chihsia Formation of South China (Nie Zetong & Song Zhimin, 1983b). Occurring with the fusulinid fauna is a distinctive coral fauna with *Lytvolasma* and a brachiopod fauna with *Juresania* (Liang Dingyi et al., 1983). This fauna is apparently of Artinskian to Kungurian age and is not a cold or cool temperate Gondwana fauna. In a general way it appears to be warm water and its possible meaning is discussed later in this paper.

In the Tanggula-Hengduan Region the faunas appear to be closely related to those of South China and are warm water (Jin Yugan, 1985).

Problems in the scope of the Chihsia are discussed elsewhere (Dickins, 1991). Briefly, the Chihsia in its original definition and traditional use in China corresponds the lower part of the Upper Permian of the twofold subdivision but together with the overlying Maokou makes up the traditional Lower Permian of China. Faunas older than the type Chihsia and younger than the Maping, however, have sometimes been included with the Chihsia. The faunas from the Tunlonggonba Formation compared with the Chihsia apparently correspond to this latter usage.

The Amb Formation is separated by unconformity from the underlying and overlying beds and in the Himalayan Region including Tibet this interval seems to be a time of regression. From its fauna and by comparison with the underlying deposits the Amb Formation appears to be warm water (Dickins & Shah, 1987).

2.4 Kazanian to Dzhulfian

Warm

These faunas are not present in Peninsular India but are widespread in the Himalayan Region: in Pakistan, India and Tibet. They can be recognized from Arabia (see above) to Salt Range, Pakistan, and along the Himalayan chain. They are found for example in displaced blocks in the Yarlung Zangbo Subregion of the Himalayan Region of Tibet (see Jin Yugan, 1985). They are limited above by faunas of Dorashamian age or where the faunas are not present, a hiatus separates this interval from overlying rocks.

The interval contains fusulinids, conodonts, colonial corals and a diverse fauna, all indicative of warm water. Even where a less diverse fauna is present such as in the Zewan of Kashmir, apparently because of the particular local environment, the genera present give positive indication of a similar water temperature as found where the fauna has more numerous genera and species. The Tethyan character of the faunas of this interval and the water temperature have recently been discussed by Kapoor & Tokuoka (1985). Despite suggestions that in places this interval may have had a cooler environment, for example in Nepal (Waterhouse, 1978), the evidence for warm water seems convincing (see also Dickins, 1985a). The suggestions of cooler climate seem to result from insufficiently taking into account the possibility that the lower diversities reflect less favorable local environments such as for example near shore conditions with a clastic sediment input. Such a clastic input characterizes, apparently, at least partly, the Nepal sequence, which may explain the low diversity rather than the water temperature.

The recent detailed description of this interval in the Salt Range by the Pakistani-Japanese Research Group (1985) serves as an important reference for this interval.

The Longge Formation in the Rutog area contains a diverse fusulinid fauna of this age including *Neoschwagerina* and *Verbeekina* which can be compared and correlated with the Maokou of South China (Nie Zetong & Song Zhimin, 1983c).

The Tanggulla-Hengduan Region contains much flysch and pyroclastic volcanic material. The marine faunas are similar to those of South China.

2.5 Dorashamian (Changhsingian)

Warm water

The fauna described by Grant (1970) from the base of the Kathwai Member of the Mianwali Formation apparently belongs here. The beds (or lenses) containing this fauna would probably best be separated as a unit distinct from the Kathwai Member according to the work of one of us (J.M.D) and the work of the Pakistani-Japanese Research Group (1985, p.264). The fauna is apparently distinctly older than the *Ophiceras* Zone of the Kathwai Member and the unit is separated from the lithologically different beds below by a disconformity. The fauna from Member E1 of the Khunamuh Formation of Kashmir (Nakazawa and Kapoor, 1981) may also belong here. This fauna is in need redescription in the light of recent taxonomic and biostratigraphical work. A small fauna from Selong in southern Tibet also seems to belong here according to the information supplied by Wang et al. (1989).

This interval is probably represented in the Tanggula-Hengduan Region according to the fusulinids *Reichelina changhsingensis* and *Paleofusulina* (see Jin Yugan, 1985, pp. 35 & 36). It has a warm water environment. The localities at which it occurs are not readily distinguishable faunally.

Although Waterhouse & Gupta (1986) have described a *Comelicania* fauna from Spiti which would be of Dorashamian age, its authenticity is questionable (Brock et al., 1991).

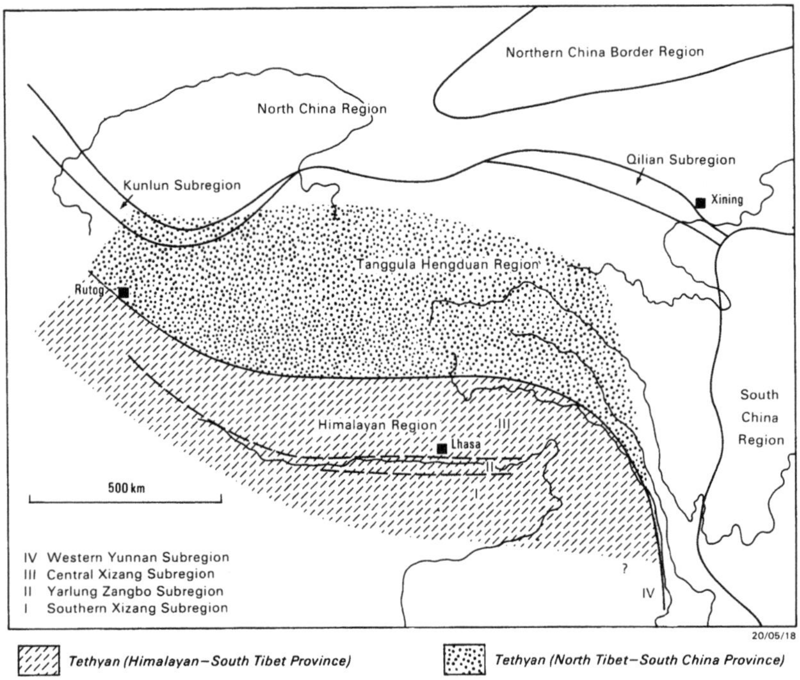

Fig. 3. Faunal distributions for Kazanian-Dzhulfian. The boundary between the Himalayan-South Tibet and the North Tibet-South China Provinces as shown is close along its length to the tectonic boundary (the Bangongghu-Nujiang Line) and is south of Rutog, an important change from the earlier part of the Permian illustrated in Figs. 1 and 2. Also note that at no time in the Permian does the Yarlung Zangbo structural or tectonic feature ("Line", "Suture" or Subregion) form a faunal boundary.

3 SYNTHESIS OF FAUNAL DATA AND COMPARISON WITH ADJACENT REGIONS.

3.1 Asselian-Lower Artinskian (2.1 and 2.2)

The faunas of the southern Arabian Peninsula (Oman), Peninsular India and the Himalayan Region (the Salt Range, Pakistan, Himalayan India and the Himalayan Region of southern Tibet including the Rutog area) all have closely related faunas of this age. They have a distinct Gondwana aspect and notably include elements with affinities to Australia. Similar relationships may be present in Afghanistan (Termier et al., 1974) and the Pamirs (Grunt & Dmetriev, 1973). Similar faunas are not known to the west, north or east. To the contrary, faunas are known to the north in the Tanggula-Hengduan Region and northern China which have affinities with the warm water faunas of South China. No similar faunas are known in Thailand or the Malayan Peninsula. The suggestion that the pebbly mudstone facies of Peninsular Thailand and western Malaysia is glacially derived seems to have been adequately dealt with by Altermann (1986). Either no glacial deposits are present, or, if present, they are not adequately documented. All or much of the pebbly mudstone represents debris flow deposit associated with active tectonism.

As discussed earlier in the paper, these faunas seem to range from cold at the base (associated with glacial deposits) to cold temperate at the top. As elsewhere there is no definite indication of glacial activity after the Asselian (Dickins, 1985b), although cold conditions with ice may have persisted longer as in eastern Australia.

3.2 Upper Artinskian-Kungurian (2.3)

Faunas of this interval have not been found in the southern Arabian Peninsula or in Peninsular India. They lack elements with distinct Gondwana affinities. In this context the faunas of Timor are not regarded as Gondwana as all the known faunas have Tethyan elements suggesting warmer water than coeval Australian faunas with the exception of the Upper Permian western Australian faunas which are themselves Tethyan in character.

In the Himalayan Region (Salt Range, Himalayan India and southern Tibet) and the Rutog area these faunas overlie cold to cold temperate faunas with Gondwana affinities. At Rutog the fusulinids have affinities with South China and other Tethyan regions but the brachiopods and corals have distinctive characters. The fauna from the Amb Formation of the Salt Range seems to be less diverse than South China and therefore in Fig. 2, these faunas are separated as the Himalayan-South Tibet Province.

The faunas to the north in the Tanggula-Hengduan Region have strong affinities with South China and are included in the North Tibet-South China Province.

Both provinces seem to be warm water and the differences between the two could reflect water temperature differences within the warm water tropical zone, differences in sedimentary conditions or physical barriers to migration.

3.3 Kazanian to Dzhulfian (2.4)

Although these faunas have been separated into two provinces, the distinction may result from different sedimentary conditions or (perhaps less likely) be related to barriers in migration. The diversity and the content of the fauna present in the Salt Range are as indicative of warm water conditions as anywhere in Tethys at the time. The fusulinid fauna seems to contain genera characteristic of other parts of Tethys such as central and western Asia and southern Europe as well as China (Pakistani-Japanese Research Group, 1985).

In the Rutog area the fusulinid fauna of this interval is also characteristic of many parts of Tethys including South China and is associated with the older fauna which in Fig. 2 is assigned to the Himalayan-South Tibet Province.

For this interval especially, the dividing line between the Himalyan-South Tibet Province and the North-Tibet South China Province may not be nearly as distinct as suggested in the figures.

3.4 Dorashamian (Changhsingian) (2.5)

As indicated earlier in the paper, the faunas of this interval are all warm water and at present, no provincial subdivision seems apparent.

4. TECTONIC AND MAGMATIC DEVELOPMENT

For the tectonic framework of the Himalayas and Tibet, reference is particularly made here to Wang Hongzhen (1983) and Stocklin (1980).

On the basis of the faunas in the Permian, the Yarlung Zangbo Subregion, "Line" or "Suture" does not form a dividing line between Gondwana (cooler water) and Tethyan (warmer) water faunas. Both are found on either side of this subregion or line (Figs 1-3) no matter which part of the Permian is considered. A greater change is associated with the boundary between the Himalayan Region of Tibet and the Tanggula-Hengduan Region ("Bangongghu-Nujiang Line"). The Asselian - Lower Artinskian cold to cold temperate water Gondwana faunas found to the south are largely absent to the north and east. The relationships in the Rutog area, however suggest that any faunal dividing line may not necessarily correspond to the tectonic line. In the Asselian-Lower Artinskian and possibly the Upper Artinskian, the faunal boundary is north of the Bangongghu-Nujiang Line, which runs south of Rutog, whereas in the Kazanian-Dzhulfian the faunal boundary is shown south of Rutog (see Figs 1-3).

The boundary between the Himalayan and the Tanggula-Hengduan Regions (the Bangongghu-Nujiang Line) appears to correspond to an important tectonic line during the Permian - and also the Triassic (Stocklin, 1980: Wang Hongzhen, 1983: Huang Zhixun, 1987). It corresponds broadly to the boundary of the island arc of Jin Yugan (1985, Figs. 9A-D) and to the mobile belt of Fig.4 of this paper. From the volcanic activity of the Lower Permian (largely basaltic ,Jin Yugan, 1985) and of the Upper Permian (also intermediate and acidic) this belt was active throughout the Permian and into the Triassic. This belt appears to be similar to the Eungella-Cracow Mobile Belt separating the Bowen Basin from the Permian-Triassic Basins to the east in Queensland and other similar belts of this time. Like the Eungella-Cracow Mobile Belt, the Tibetan Belt may have been submergent or largely submergent at times and at others largely emergent (Dickins & Malone, 1973).

The tectonic-magmatic picture, however, cannot have been as simple as shown in Fig. 4, because of, for example, Lower Permian basaltic magmatism found in the Kashmir area and along the length of the Himalayan mountain chain (see Dickins, 1987).

5. CONCLUSIONS AND SOME SUGGESTIONS.

Many workers on the faunas of the region, together with workers on sedimentation and tectonics, have supported the proximity of Peninsular India to Asia during the Palaeozoic and Mesozoic with comprehensive evidence. The review of Teichert (1971) alone offered ample evidence on this. Important information has continued to accumulate - for recent work see Chatterjee & Hotton (1986), Dickins (1987),Dickins & Shah (1987), Saxena et al. (1985), Jin Yugan (1985), and Waterhouse (1983).

The data considered here indicate quite clearly that during the Asselian-Lower Artinskian a single faunal province existed incorporating Peninsular India and the Himalayan Region as far north as the Central Xizang Subregion (Fig. 1).

This province apparently included the southern part of the Arabian Peninsula and may have extended to Afghanistan and Central Asia but was distinct from other Gondwana provinces of the time (Dickins, 1985a; Dickins and Shah, 1981, 1987).

In these circumstances, the splitting off of part of this province from Gondwana and its movement northward during the Permian, as sometimes postulated, seems rather unlikely - the whole province would have had to have moved as a whole. In view of the faunal relationships outlined here and the Permian history of the region, this too is hardly likely (see also Dickins & Shah,1981, 1987).

Jin Yugan (1985), on the basis of the proximity of India to Asia, has set out a realistic palaeogeographic scenario (his Figs. 9A-D), with a supply of sediment coming from a continental source both to the south (Peninsular India) and the north (Central Asia) into a trough (Tethys).

In the earliest Permian, a cold current may have been directed into this trough from the south and a warm current from the east separated by a developing mobile belt emergent or partly emergent in the middle part of Tethys (Fig.4). This belt may have swung south through Thailand and the Malay Peninsula separating the cold and warm currents. Overall warming took place during the Permian (Dickins, 1985; Kapoor &

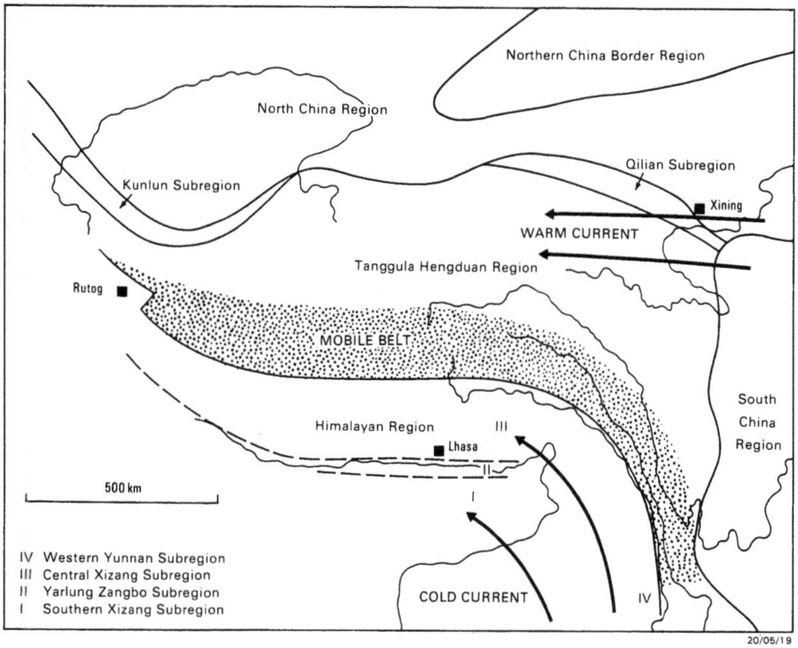

Fig. 4. Hypothetical current directions for Asselian-Lower Artinskian.

Tokuoka, 1985) but some water temperature differences may have remained until about the mid-Permian. By this time the earlier water temperature differences were apparently eliminated because of the overall world warming and the development of the mobile belt which resulted in the break down of previous water current and faunal migration barriers.

REFERENCES

Ahmad, F. 1978. "Gondwanaland", the concept that failed. Seventh Birbal Sahni Memorial Lecture, Birbal Sahni Institute of Palaeobotany, Lucknow.

Altermann, W. 1986. The Upper Palaeozoic pebbly mudstone facies of Peninsular Thailand and western Malaysia - continental margin deposits of Palaeoeurasia. Geologische Rundschau 75 (2): 371-381.

Archbold, N.W. 1983. Studies on Western Australian Permian brachiopods 3: The family Linoproductidae Stehli 1954. Proceedings of the Royal Society of Victoria 95: 237-254.

Archbold, N.W. 1991. Studies on Western Australian Permian brachiopods 10. Faunas from the Wooramel Group, Carnarvon Basin. Proceedings of the Royal Society of Victoria 103: 55-66.

Archbold, N.W, J.M.Dickins & G.A. Thomas (in press). Correlation and age of the Western Australian marine faunas. In S.K.Skwarko (ed.), Permian Palaeontology of Western Australia, Perth, Geological Survey of Western Australia.

Archbold, N.W. & V.J.Gupta 1986. Permian brachiopod faunas of the Himalayas and Western Australia: a comparison. Bulletin of the Indian Geologists Association 19 (2): 81-96.

Brock, G.A., R.Mawson, R.C.Talent, M.J.Engelbretsen & J.A. Talent 1991. Spurious and scambled data: V.J.Gupta's impact on the Prototethyan/Tethyan database. In T. Kotaka, J.M. Dickins, K.G. McKenzie, K. Mori, K. Ogasawara and G.D. Stanley Jr. (eds.), Shallow Tethys 3, Saito Ho-on Kai, Sendai, Japan, Special Publication 3: 477-480.

Chatterjee, S. & N. Hotton 1986. The paleoposition of India. Journal of Southeast Asian Earth Sciences, 1 (3): 145-89.

Dickins, J.M. 1963. Permian pelecypods and gastropods from Western Australia. Bulletin Bureau of Mineral Resources, Australia, 63: 1-203.
Dickins, J.M. 1972. Permian molluscs and continental drift. Australian Natural History 17: 270-273.
Dickins ,J.M. 1985a. Palaeobiofacies and palaeobiography of Gondwanaland from Permian to Triassic. In K. Nakazawa & J.M. Dickins (eds.) The Tethys, Tokyo, Tokai University Press: 83-92.
Dickins, J.M. 1985b. Late Palaeozoic glaciation. BMR Journal of Australian Geology & Geophysics 9: 163-169.
Dickins, J.M. 1987. Tethys - A geosyncline formed on continental crust? In K.G. McKenzie (ed.) Shallow Tethys 2: 149-158. Rotterdam, Balkema.
Dickins, J.M. 1991. Permian of Japan and its significance for world understanding. In T. Kotaka, J.M. Dickins, K.G. McKenzie, K. Mori, K. Ogasawara and G.D. Stanley Jr. (eds.), Shallow Tethys 3, Saito Ho-on Kai, Sendai, Japan, Special Publication 3: 343-351.
Dickins, J.M. in press. Permian climate of Western Australia. In S.K.Skwarko (ed.) Permian Palaeontology of Western Australia, Perth, Geological Survey of Western Australia.
Dickins, J.M. & E.J. Malone 1973. Geology of the Bowen Basin, Queensland. Bulletin Bureau of Mineral Resources Australia 130.
Dickins, J.M. & S.C. Shah 1979. Correlation of the marine Permian sequence of India and Western Australia. Fourth International Gondwana Symposium 2: 387-408. Calcutta, 1977, Delhi, Hindustan Publishing Corporation (India).
Dickins ,J.M. & S.C. Shah 1981. Permian paleogeography of Peninsular and Himalayan India and the relationship with the Tethyan region. In M.M. Cresswell & P. Vella (eds.) Gondwana Five: 79-83. Rotterdam, Balkema.
Dickins, J.M. & S.C. Shah 1987. The relationship of the Indian and Western Australian Permian marine faunas. In G.D. McKenzie (ed.) Gondwana Six, American Geophysical Union, Geophysical Monograph 41: 15-20.
Grant, R.E., 1970. Brachiopods from Permian-Triassic boundary beds and age of Chhidru Formation, West Pakistan. In B. Kummel & C. Teichert (eds.) Stratigraphic Boundary Problems: Permian and Triassic of West Pakistan. University of Kansas, Department of Geology Special Publication 4: 117-151.
Grunt, T.A. & Dmitriev, V.V. 1973. Permian brachiopods of Pamir. Academy of Sciences of the USSR, Trudy of the Palaeontological Institute, 136: 1-209.
Huang Zhixun 1987. On the paleobiogeography and paleostructure of Carboniferous-Permian in North Qinghai-Xizang (Tibet) Plateau. 11th International Congress Carboniferous Stratigraphy and Geology, Abstracts, 2: 368-369.
Jin Yugan 1985. Permian Brachiopoda and paleogeography of the Qinghai-Xizang (Tibet) Plateau. Paleontologia Cathayana 2: 19-71.
Kapoor, H.M. & T. Tokuoka 1985. Sedimentary facies of the Permian and Triassic of the Himalayas. In K. Nakazawa & J.M. Dickins (eds.) The Tethys: 23-58. Tokyo, Tokai University Press.
Liang Dingyi, Nie Zetong, Guo Tieying, Xu Baowen, Zhang Yizhi, & Wang Weipin 1983. Permo-Carboniferous Gondwana-Tethys facies in Southern Karakorum and Xizang. Earth Science-Journal of Wuhan College of Geology, 1983 1: 9-27.
Nakazawa, K. & H.M. Kapoor (eds.) 1981. The Upper Permian and Lower Triassic faunas of Kashmir. Palaeontologia Indica, New Series 46: 1-204.
Nie Zetong & Song Zhimin 1983a. Fusulinids of Lower Permian Qudi from Rutog of Xizang (Tibet), China. Earth Science-Journal of Wuhan College of Geology, 1983 1: 29-42.
Nie Zetong & Song Zhimin 1983b. Fusulinids of Lower Permian Tunlonggongba Formation from Rutog of Xizang (Tibet), China. Earth Science - Journal of Wuhan

College of Geology, 1983 1: 43-55.

Nie Zetong & Song Zhimin 1983c. Fusulinids of Lower Permian Maokouian Longge Formation from Rutog of Xizang (Tibet), China. Earth Science - Journal of Wuhan College of Geology, 1983 1: 57-68.

Pakistani-Japanese Research Group 1985. Permian and Triassic Systems in the Salt Range and Surghar Range. In K. Nakazawa & J.M. Dickins (eds.) The Tethys: 221-312. Tokyo, Tokai University Press.

Saxena, M.N., V.J. Gupta, A.A. Meyerhoff & N.W. Archbold 1985. Tectonic and spatial relations between India and Asia since Proterozoic time. In V.J. Gupta et al. (eds.) Geology of Western Himalayas, Contributions to Himalayan Geology 3: 187-207. Delhi, Hindustan Publishing Corporation (India).

Stehli, F.G. 1968. Taxonomic diversity gradients in pole location - the recent model. In E.T.Drake (ed.) Evolution and Environment: 163-227. Yale University Press.

Stehli, F.G. 1973. Permian brachiopods. In A.Hallam (ed.), Atlas of Palaeogeography: 143-149. Amsterdam, Elsevier.

Stocklin, J. 1980. Geology of Nepal and its regional framework. Journal Geological Society of London 137 (1): 1-34.

Teichert, C. 1971. Marine fossil invertebrate faunas of the Gondwana Region. In S.H. Haughton (ed.) Proceedings and Papers of Second Gondwana Symposium, South Africa 1970: 125-138.

Termier, C., H. Termier, A.F. De Lapparent & P. Marin 1974 Monographie du Permo-Carbonifere de Wardak (Afghanistan Central). Documents des Laboratoires Geologie de la Faculte Sciences de Lyon, H.S. 2: 1-167.

Trilochan Singh 1987. Paleoclimatic significance of fauna and mioflora of the Garu Formation (Permian), Arunachal Pradesh, Eastern Himalaya. In G.D. McKenzie (ed.) Gondwana Six: Stratigraphy, Sedimentology and Paleontology. American Geophysical Union, Geophysical Monograph 41: 191-194.

Wang Hongzhen 1983. On the geotectonic units of Xizang(Tibet) Region. Earth Science - Journal of Wuhan College of Geology, 1983, 1: 1-8.

Wang Yigang, Chen Chuzhen, Rui Lin, Wang Zhihao and He Jinwen 1989. On the definition of Permian - Triassic boundary. Journal of Stratigraphy 13 (3): 205-212.

Waterhouse J.B. 1978. Permian Brachiopoda and Mollusca from North-west Nepal. Palaontographica Abt.A.160: 1-175.

Waterhouse, J.B. 1983. How wide was the Tethyan "Ocean". In V.J.Gupta (ed.), Contributions to Himalayan Geology 2: 246-249. Delhi, Hindustan Publishing Corporation.

Waterhouse, J.B. & V.J. Gupta 1986. A significant new Permian athyrid brachiopod from the uppermost Gungri Formation (Kuling or Productus Shale) at Spiti, and its implications for classification and correlation. Bulletin of the Indian Geologists Association 19 (1): 45-56.

Gondwana Eight, Findlay, Unrug, Banks & Veevers (eds) © *1993 Balkema, Rotterdam. ISBN 90 5410 304 3*

Gondwana sediments (Permian) of Arunachal Himalaya: Stratigraphic status and depositional environment

Trilochan Singh
Wadia Institute of Himalayan Geology, Dehra Dun, India

ABSTRACT : In the Arunachal Himalaya three formations, viz., Rangit Formation, Garu Formation and Bhareli Formation represent the depositional realm of the Lower Gondwana Subgroup. Lithologically, these formations are characterised, respectively, by : a) polymictic tillite; b) black carbonaceous and brown ferruginous shale with concretions containing marine fossils; and c) arkosic, micaceous and quartzitic sandstone, associated with slaty shale containing a *Glossopteris* flora and thin coal bands. The enclosed marine fauna, flora and microflora confirm a Permian age for these formations.

Gondwana sedimentation in the Arunachal Himalaya started under glacio-marine conditions with deposition of polymictic tillite near the margin of the Gondwana continent. A marine transgression led to formation of a marginal swamp region characterised by lagoonal and basinal deposits, culminating in deposition of black shales. Frequent marine incursions gave rise to the development of the concretions and faunal coal balls. It is likely that the basins were also fed by some channels supplying fresh water terrestrial sediments under fluctuating colder and warmer climatic conditions. During the same time mafic volcanism took place under sub-aerial conditions in the eastern part of the area. Following regression, a thick sequence of fresh water sandstone and shale deposited in a deltaic regime to produce ultimately an alluvial plain.

1 INTRODUCTION

Godwin-Austen (1875) and La Touche (1885) were first to record the existence of rocks of the Damuda Series belonging to the Lower Gondwana Subgroup in the Dafla Hills (Subansiri District) and the Aka Hills (Kameng District) of the Arunachal Himalaya, respectively. Jacob & Banerjee (1954) reported the Glossopteris fronds from the Sela sub-Agency (Kameng District) and established the existence of Permian Gondwana sediments. The first marine fossils, were collected by MaClaren (1904) from rolled boulders at the mouth of Subansiri Gorge. Diener(1905) described these and assigned them an Anthracolithic (Permo-Carboniferous) age. Since then the Permian sediments of the Arunachal Himalaya have been referred to the Lower Gondwana Subgroup.

The Lower Gondwana sediments, in Arunachal, occur as a linear belt all along the frontal margin of the Lesser Himalaya (Fig. 1). These are thrusted over the sediments of the Siwalik Group in the south and in turn are over thrusted by the metasediments of Miri and Bomdila groups in the north.

The sediments under discussion comprise both continental and marginal marine (coastal) rocks and are traceable from the Kameng District in the west to the Siang District in the east (Fig. 2). The continental facies attains its maximum thickness in the Kameng District, whereas the marine influence increases towards the east. The nature of contact between these two units is uncertain, though it appears to be tectonic. Significantly, the basic rocks (Abor Volcanics : Coggin-Brown 1912) also occur associated with the Lower Gondwana Subgroup. The main volcanic body is located in the Siang District, and it can be observed in the Pasighat-Rotung-Pangin, Pangin-Dosing-Yinkiyong, and Yinkiyong - Maryang -

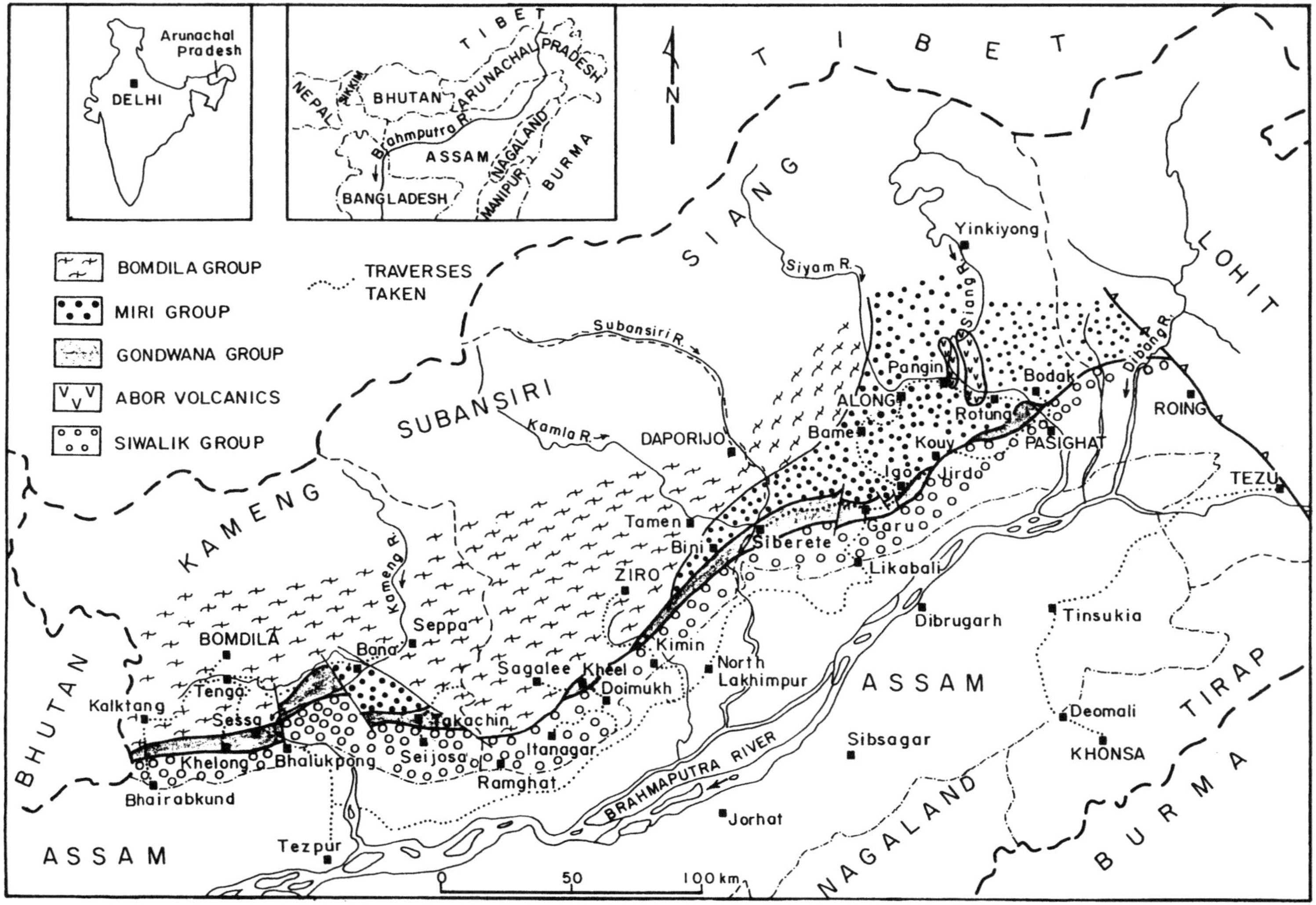

Fig. 1. Geological map of the southern parts of the Arunachal Himalaya, showing Lower Gondwana Group.

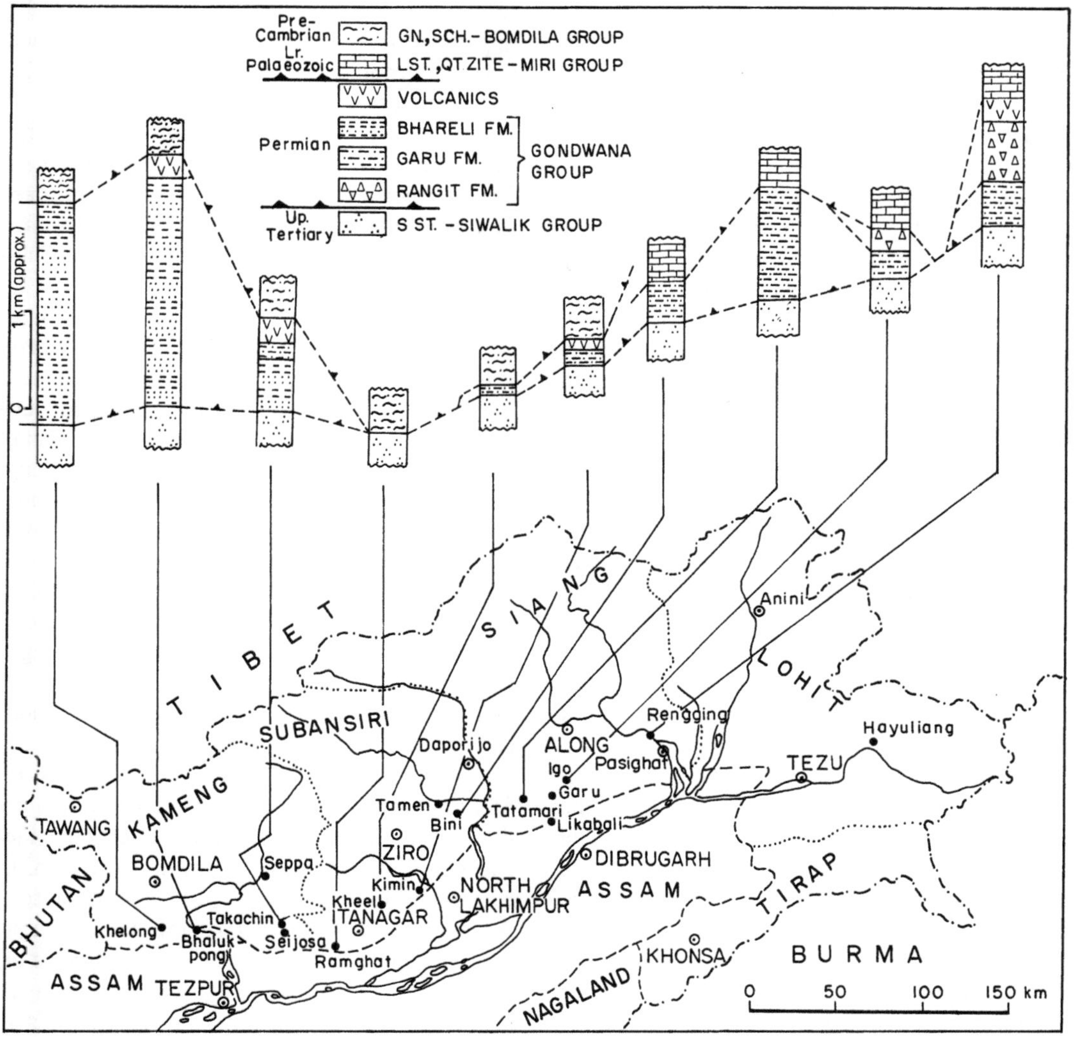

Fig. 2. Lithologs showing distribution of Rangit, Garu and Bhareli Formations in different sectors of the Arunachal Himalaya.

Bodak sections, where it is associated with the rocks of the Miri Group as well. Prominent exposures of volcanic rocks are also present in Kimin - Lichi - Ziro Section in the Subansiri District (Fig. 2).

2 STRATIGRAPHIC STATUS

In the Arunachal Himalaya the Lower Gondwana Subgroup consists of three tectonically juxtaposed formations : a) Rangit Formation; b) Garu Formation; and c) Bhareli Formation (Fig. 2). The stratigraphic contacts between these formations are not clear due to the tectonic nature of superposition. Therefore, the stratigraphy has been reconstructed on the basis of the faunal and floral contents.

2.1 Rangit Formation

Although the Rangit Formation is exposed throughout in isolated patches, its best exposures can be seen on the Pasighat - Pangin and Garu-Basar road sections in the Siang District, and on the Khuppi - Bana road section in the Kameng District. It bears similarity with the Pebble Slate unit of the Sikkim - Darjeeling Himalaya. The true thickness of the Rangit Formation could not be ascertained due to lack of bedding characteristics.

The basal unit of this formation is represented mainly by bluish-grey muddy, fine grained, matrix supported para-conglomerate containing poorly sorted clasts. The composition and texture of para-conglomerate varies from place to place. The size of the clasts varies from granule to pebble to boulder, and the shape varies from angular to rounded. The clasts are poorly sorted and show no preferred orientation. Quartzite clasts dominate over the other clasts, which include those of limestone, sandstone, shale, slate, phyllite, schist, gneiss, chert, and a few of granite. Most of the limestone clasts are dissolved leaving empty spaces. A few clasts have been replaced by limonitic matter. The pebbles have striations and are angular. These rocks are regarded as polymictic tillite. The unit is structureless and does not show any bedding characteristics. In places it is highly jointed.

Upward in the sequence impersistent greenish-grey shales with intercalated thin bands of fine grained greenish siltstone are exposed. The greenish-grey shales often contain burrows. These are overlain by the buff-grey earthy mudstone containing scattered clasts of varied lithology. However, the size of the clasts is relatively smaller than those found in the basal polymictic tillite of the formation. Another notable feature of the mudstone is the presence of calcareous concretions of varying size, which at places measure up to a metre in diameter.

Tandon & Kumar (1980) carried out the sedimentological studies of the basal polymictic tillite (= upper diamictite) and the buff grey earthy mudstone (=lower diamictite) associated with greenish grey shales and siltstone of the Siang District, and have also shown difference in clast composition, size and shape of the two units.

Although Acharyya et al. (1975) have reported the occurrence of squashed brachiopods, bivalves (*Eurydesma* sp., *Schizodus* sp.), and other fragmentary fossils from the Rangit Formation of Khuppi area in the Kameng District, a thorough search by this author failed to produce any macrofossil in the Rangit Formation. However, the upper mudstone unit of this formation has yielded a rich palynological assemblage dominated by radial monosaccates (Singh 1979; Srivastava & Dutta 1979). The assemblage comprising *Parasaccites, Plicatipollenites, Virkkipollenites, Potonieisporites, Caheniasaccites, Pilasporites, Limitisporites*, is comparable to the Talchir palyno-assemblage of the Lower Gondwana Subgroup of Peninsular India.

The upper mudstone unit of the Rangit Formation is best developed in the Siang District, where it is exposed along the Tatamari-Garu-Rilu, and Pasighat-Rengging road sections. Isolated outcrops of this unit also occur near Litemari in Siang, Kheel in Subansiri, and Takachin in Kameng districts. This unit had earlier been placed in the Garu Formation by Kumar & Singh (1974). However, on the basis of palynoflora recorded by Singh (1979) and Srivastava & Dutta (1979) from this unit being equivalent to the Talchir Formation, Srivastava et al. (1987) included it in the Rangit Formation.

The polymictic tillite of the Rangit Formation had been earlier named as the Rangit Pebble Slate by Acharyya (1971) after the Rangit Valley in Sikkim, though he considers the type section of this formation to be the Rakti River Section in the Darjeeling Himalaya (Acharyya 1971, p.63). Similarly in the Arunachal Himalaya too Acharyya et al. (1975) gave the name Rangit Pebble Slate to this unit, after its resemblance to that exposed in the Sikkim-Darjeeling Himalaya. However, since this unit does not show any slaty character uniformly and also the clasts are not always of pebble size, it is suggested that in order to avoid confusion with the lithological character of the unit, the name Pebble Slate should not be used and instead it may be referred to as 'Rangit Formation'.

2.2 Garu Formation

The name 'Garu Formation' is used here following Kumar & Singh (1974) with some modification. Originally Kumar & Singh (1974) divided the Garu Formation into Rilu, Bomte and Sikki Abu Members. Here, the classification is revised, and the scope of the Garu Formation is restricted to the Bomte Member of Kumar & Singh (1974) whereas the Rilu and Sikki Abu Members are placed under the Rangit Formation. Although Acharyya et al. (1975) had named the marginal marine sediments of Arunachal Pradesh as Rilu Formation, the name Garu Formation has priority for these rocks.

This unit is formed of black carbonaceous and brown ferruginous shales with small lenses and thin bands of fine grained sandstone. Associated with these are small thin bands and lenses of coal. A few tuffaceous beds also occur in this formation. The shale unit is characterised by the presence of concretionary nodules of various shapes and sizes which occur scattered throughout and are commonly fossiliferous. These concretions, formally referred to as 'faunal coal balls' by Anand - Prakash et al. (1988), have yielded well preserved fossils. A buff-grey to white micaceous sandstone forms a large lensoid body within the black carbonaceous and brown ferruginous shale, and has yielded well preserved bryozoans.

Fossils are very common in the Garu Formation; they occur mainly in the concretions found in the black carbonaceous and brown ferruginous shale unit, and also within the shale and sandstone. The fossils have been described earlier from a number of localities in the Subansiri and Siang districts (Singh 1973, 1976, 1978 a, b, c) and consist of : bryozoans (*Fenestella*, *Polypora*), brachiopods (*Lissochonetes*, *Linoproductus*, *Subansiria*, *Ambikella*), bivalves (*Quadratonucula*, *Phestia*, *Deltopecten*), gastropods (*Warthia*, *Platyteichum*), cephalopods (*Sueroceras* and *Uraloceras*) and crinoidal remains. A rich palyno-assemblage has also been recovered from black shales (Singh 1979). This assemblage has *Callumispora* in abundance alongwith *Parasaccites*, *Indotriradites*, *Vesicaspora*, and is considered equivalent to the Lower Karharbari palyno-assemblage of Peninsular India. Another palyno-assemblage, in which *Scheuringipollenites* attains overall dominance whereas radial monosaccate pollen and trilete spores decrease considerably was reported from higher up in the sequence by Srivastava et al. (1987) who considered it equivalent to the Lower Barakar palyno-assemblage of Peninsular India. The faunal and palyno-assemblages together indicate a Sakmarian to Artinskian age (Singh 1981) for the Garu Formation. Recently, this author recovered foraminifera and ostracodes from these sediments. The foraminifera are represented by endothyrids and arenaceous forms of Lower Permian age. These microfossils will be described elsewhere.

The Garu Formation is best exposed in the Siang District in Siberrete-Litemari-Tatamari-Gensi-Garu-Daring Section. However, representatives are also exposed in other sections of Siang, Subansiri and Kameng districts. Thin coal seams associated with the sediments of the Garu Formation are found near Takaso and Tatamari villages in Garu- Gensi-Tatamari Section in the Siang District, and near Petepara in Kimin-Lichi-Ziro Section in the Subansiri District.

2.3 Bhareli Formation

The Bhareli Formation is well exposed in almost all the sections in the Kameng District, viz. Bhairabkund - Kalaktang Section, Khelong - Sissni - Tenga Section, Bhalukpong - Sessa - Bomdila Section, N/Phu - Khuppi - Bana Section, and Seijosa -Takachin Section. It is, however, completely absent towards the east in the Siang District. The name used here has been followed after Acharyya et al. (1975).

This is characterised by fresh-water sediments formed of alternating beds of sandstones and slaty shales, with thin beds of coal in between. The sandstones are light-grey to bluish-grey, medium to coarse grained, sometimes gritty and hard in nature. They are usually highly indurated and almost quartzitic, but in places are also micaceous with large detrital grains of muscovite; they also contain carbonaceous matter. These beds are thinly laminated often showing cross-lamination. The thickness of sandstone and shale beds varies laterally. The shales are mostly slaty and sometimes contain pyrite. The carbonaceous shale and coal beds occur in between the sandstone and slaty shales, mostly in form of lenses.

The slaty shales often yield well preserved plant remains. Although a number of fossils have been reported from the Bhareli Formation, more recently Singh & Bajpai (1990) described the flora which includes equisetalean axes,*Phyllotheca* sp.,

Glossopteris spp., and *Vertebraria* sp., and considered this floral assemblage equivalent to that of the Raniganj Formation of Peninsular India. Evidence supporting this correlation was also provided by Misra et al. (1987), who on the basis of coal petrography correlated the Permian coal beds of the Kameng District with the Raniganj Formation of Peninsular India.

3 DEPOSITIONAL ENVIRONMENT

The deposition of the Lower Gondwana Subgroup in the Arunachal Himalaya started under frigid climatic conditions producing the basal polymictic tillite unit of the Rangit Formation, deposited over the uneven basement of the Miri Group of rocks (Fig. 9 I). The clasts present in the polymictic tillite unit indicate their derivation mainly from the Miri and Bomdila Groups exposed in the surrounding areas. Further, the occurrence of angular fragments (Fig. 3) particularly of limestone, shale, sandstone and slate / phyllite indicate a short distance of transportation. Many of the quartzitic clasts show joint planes / fractures having sharp edges, which do not continue in the matrix, thus suggesting retention of an inherited character. Some of the clasts may be displaced along the joint / fracture plane developed within the clasts (Figs. 4-7). The surfaces of most of the clasts show variable amounts of polishing. Several clasts show development of facets and striations (Fig. 8) on the surface. The polymictic tillites show a general lack of stratification and slump structures. According to Acharyya (1973) the polymictic tillite represents submarine mud-flow deposit in the tectonically active molasse furrows with associated subaerial intermediate to basic volcanic activity; though minor additional influence of glaciers depending upon the palaeo- topography was also envisaged by him. However, Tandon & Kumar (1980) have given a model with a glacial influence. Subsequently, in the present study the widespread distribution of these para-conglomeratic rocks with lateral facies changes, marine intercalations and various above mentioned features observed in the clasts indicate that these were brought to the depositional site, i.e. the marginal shore by the glaciers present in the vicinity.

Towards the final phase of the glacio-marine conditions, greenish-grey shales and fine siltstones were deposited in a locally reducing environment. The following mudstone unit has yielded a rich palyno-flora equivalent to the Talchir Formation of Peninsular India, and suggests a cool environment (Singh 1987a).

After the deposition of the Rangit Formation the sediments of the Garu Formation were laid down (Fig. 9 II). The fauna and palyno-flora of the Garu Formation are indicative of cool climatic conditions (Singh 1987a). The predominance of molluscs over the brachiopods suggests sub-littoral conditions. The faunal coal balls (fossil-bearing concretionary nodules) present in the shale unit indicate marginal swampy conditions with marine incursions at frequent intervals (Anand-Prakash et al. 1988). The black shales indicate a reducing environment and the succeeding brown ferruginous shales are suggestive of an oxygenic environment. The cyclic nature of climatic variation from reducing to oxidising was reasonably responsible to generate such a suit of lithological components. Further, the isotopic data (^{18}O and ^{13}C) of the fossil-bearing concretionary nodules (Singh & Srivastava : in prep.) suggests their burial under warmer diagenetic conditions as the ^{18}O and ^{13}C values range from -6 to -28 and -10 to -27, respectively, although the fauna present in them is indicative of a colder climate. It is, thus, interpreted that the Garu Formation was deposited in the lagoonal or partially enclosed basins fed by some channels supplying fresh water sediments to the swampy area, under fluctuating cold and warmer climatic conditions.

During the deposition of the Garu Formation, mafic volcanic activity took place in eastern part of the area. The volcanic suit of rocks is best exposed in the Siang District. The volcanics have typical continental thoeiilite chemistry in terms of rather high and variable Fe/Mg and incompatible trace element abundances and ratios (Bhat 1984; Bhat and Ahmad 1990). Bhat (1984) and Bhat and Ahmad (1990) suggested that the Abor Volcanics erupted sub-aerially in a rift setting. The intertrappean sediments found in Yinkiyong - Maryang Section have yielded palynomorphs indicating an early Permian age (Prasad et al. 1989).

Deposition of the Bhareli Formation took place under alluvial plain conditions as indicated by the

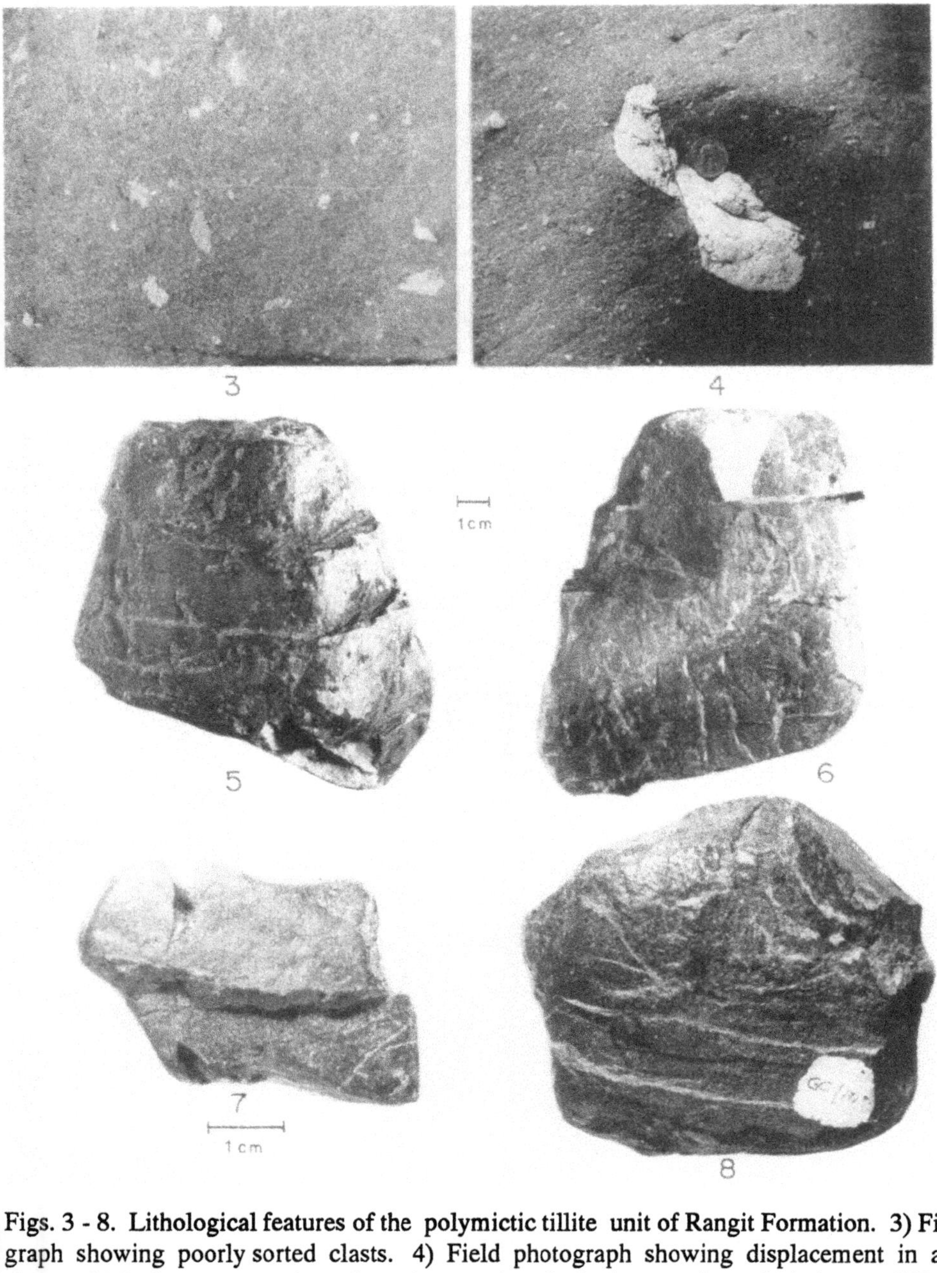

Figs. 3 - 8. **Lithological features of the polymictic tillite unit of Rangit Formation.** 3) Field photograph showing poorly sorted clasts. 4) Field photograph showing displacement in a quartzite clast, which does not continue into the matrix. 5, 6) A quartzite pebble in two different views showing displacement along the fracture plane. 7) A quartzite clast showing displacement. 8) A quartzite pebble showing striations.

thick sequence of alternating shale and sandstone showing laminated and cross laminated character (Fig. 9 III). The presence of *Glossopteris* flora in the slaty shales further suggests a terrestrial regime under cold climatic conditions. Misra et al. (1987), on the basis of petrographic and rank data of Permian coal from the Kameng District, have further suggested that the coal seams were

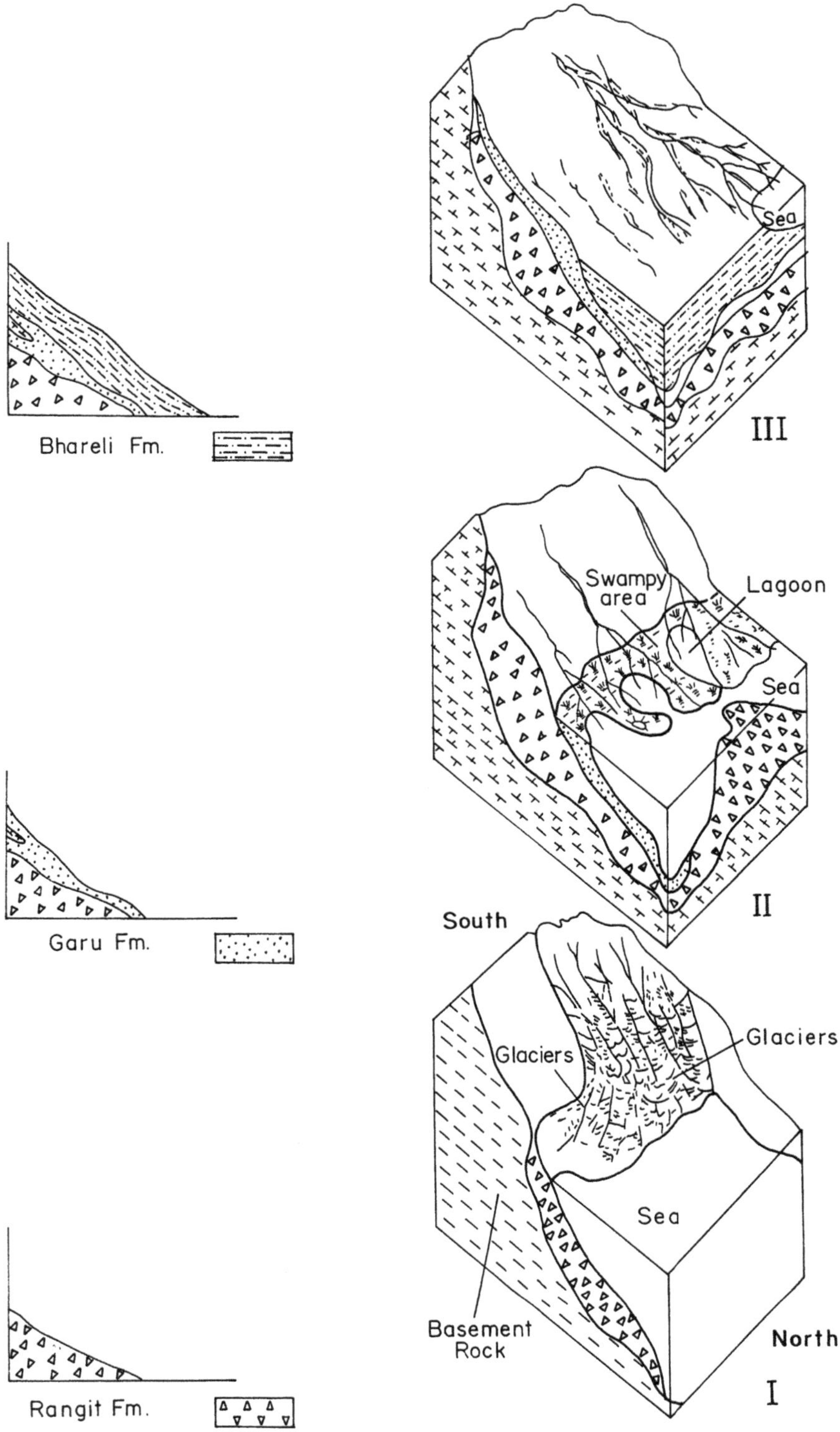

Fig. 9. Models depicting successive environmental conditions during the deposition of I) Rangit, II) Garu, and III) Bhareli Formation.

Table 1. Relationship of Permian sequence of Arunachal Himalaya with contemporaneous sequences in the other parts of the Indian subcontinent and Australia.

India			Pakistan	Nepal	Tibet		Australia		
Arunachal Himalaya	Kashmir	Peninsular India	Salt Range	Tansen	South	North	Perth Basin	Carnarvon Basin	Canning Basin
Bhareli Fm	Nishat Bagh bed	Raniganj Fm		Plant-bearing bed	Qubu Fm				
	Pyroclastic Div	Lower Barakar Fm	Amb Fm				Fossil Cliff Fm	Callytharra Fm	Nura Nura Member
Garu Fm				Barahakshetra Fm	Caya Fm	Qudi Fm / Ria Fm			
	Upper Diamictite Div	Lower Karharbari Fm	Dandot Fm				Holmwood Shale	Lyons Group	
Rangit Fm	Lower Diamictite Div	Talchir Fm	Tobra Fm		Jilong Fm		Nangetty Glacial Fm		

deposited in a deltaic regime rather than in a typical fluvial sedimentation cycle.

4 RELATIONSHIP WITH CONTEMPORANEOUS SEQUENCES OF GONDWANALAND

The various sequences equivalent to the Permian sediments of the Arunachal Himalaya are summarised in Table 1. Singh (1987b), while discussing the Permian biogeography, analysed various Permian outcrops of the Indian subcontinent on the basis of marine fauna and established that the Permian sediments of the Himalaya, Salt Range, South Tibet and Peninsular India belong to the same faunal province, the Gondwana Province. Srivastava et al. (1987) opined that although the Permian sediments of the Eastern Himalaya contrasted lithologically with the intracratonic continental sediments of Peninsular India, their floras are closely similar. A close similarity between the Arunachal Himalayan fauna and other early Permian faunas of the Indian subcontinent and Australia was also noted (Singh 1981). More recent studies have indicated that the Aruanchal fauna is also significantly close to that of Western Australia (Archbold & Singh : this volume). The Permian sequence of the Arunachal Himalaya with its characteristic basal polymictic tillite overlain by rocks containing an Early Permian cold water marine fauna, followed by beds containing *Glossopteris* flora also shows similarity with that of the South Tibetan sequence (Singh : in prep.).

5 CONCLUSIONS

The Lower Gondwana Subgroup of the Arunachal Himalaya shows a wide spectrum of litho-facies, comprising : a) polymictic tillite, b) black carbonaceous and brown ferruginous shale with concretions containing marine fossils, and c) arkosic, micaceous and quartzitic sandstone and slaty shale containing *Glossopteris* flora, referable to Rangit, Garu and Bhareli formations, respectively. The reconstruction of the palaeoenvironment suggests that deposition of the Gondwana Group in the Arunachal Himalaya started with the polymictic tillite under the glacio-marine conditions near the margin of the Gondwana continent. The area then became partly connected with the sea, and formed lagoons. The swampy area was also fed by fresh-water channels under fluctuating colder and warmer climatic conditions. Mafic volcanism occured under sub-aerial conditions during the same time in the eastern part of the area. As the sea regressed, the area was transformed into an alluvial plain giving rise to the deposition of fresh water sediments in the deltaic regime along the shore region.

ACKNOWLEDGEMENTS

The author is indebted to Prof. I. B. Singh of Lucknow University, Dr. J. M. Dickins of B.M.R., Canberra, Dr. V. Raiverman, Dehra Dun, and his colleagues Dr. R.A.K. Srivastava and Dr. Kishor Kumar for their valuable comments and suggestions. Thanks are due to the Director, Wadia Institute of Himalayan Geology for providing the necessary facilities.

REFERENCES

ACHARYYA, S. K. 1971. Rangit Pebble Slate - a new formation from Darjeeling foothills. *Indian Minerals* 25 : 61-64.

ACHARYYA, S.K. 1973. Late Palaeozoic glaciation vs volcanic activity along the Himalayan chain with special reference to the Eastern Himalaya. *Himalayan Geology* 3 : 209-230.

ACHARYYA, S.K., S.C. GHOSH, R.N. GHOSH & S.C. SHAH 1975. The continental Gondwana Group and associated marine sequences of Arunachal Pradesh (NEFA), Eastern Himalaya. *Himalayan Geology* 5 : 60-82.

ANAND-PRAKASH, T. SINGH & S.C. SRIVASTAVA 1988. Occurrence of faunal coal balls in Gondwana sediments (Permian) of Arunachal Himalaya, India. *International Journal of Coal Geology* 9 : 305-314.

ARCHBOLD, N.W. & T. SINGH This Volume. Early Permian brachiopods from the Eastern Himalaya and their provincial relationships with Australia. In R. H. Findlay, M. R. Banks, J. J. Veevers, R. Unrug (eds.), *Gondwana 8 : Assembly, Evolution and Dispersal.* A.A. Balkema : Rotterdam.

BHAT, M. I. 1984. Abor Volcanics : Further evidence for the birth of the Tethys ocean in the Himalayan segment. *Journal Geological Society, London* 141 : 763-775.

BHAT, M. I. & T. AHMAD 1990. Petrogenesis and the mantle source characteristics of the Abor Volcanic rocks, Eastern Himalayas. *Journal of the Geological Society of India* 36 : 227-246.

COGGIN-BROWN, J. 1912. A geological reconnaissance through the Dihong Valley, being the geological results of Abor Expedition 1911-1912. *Records of the Geological Survey of India* 42 (4) : 231-253.

DIENER, C. 1905. Note on the Anthracolithic fauna from the mouth of the Subansiri Gorge, Assam. *Records of the Geological Survey of India* 32 (2) : 189-198.

GODWIN-AUSTEN, H.H. 1875. Notes on the geology of part of the Dafla Hills, Assam. *Journal of the Asiatic Society of Bengal* 44-II : 35-41.

JACOB, K. & T. BANERJEE 1954. The occurrence of *Glossopteris* fronds in the North East Frontier Tracts, with a brief review of the Gondwanas of North Eastern India. *Proceedings National Institute of Science, India* 20 : 53-61.

KUMAR, S. & T. SINGH 1974. Lithostratigraphy of the southern parts of Siang District, Arunachal Pradesh. *Himalayan Geology* 4: 648-656.

LA TOUCHE, T.D. 1885. Notes on the geology of the Aka Hills. *Records of the Geological Survey of India* 17 (2) : 121-124.

MACLAREN, J. M. 1904. The geology of Upper Assam. *Records of the Geological Survey of India* 31 (4) : 179-204.

MISRA, B.K., M. AHMED & G.K.B. NAVALE 1987. Petrological, chemical and depositional aspects of Eastern Himalayan coals from Elephant Flat area, Kameng District, Arunachal Pradesh, India. *International Journal of Coal Geology* 8 : 279-297.

PRASAD, B., A. K. DEY, P. K. GOGOI & A. K. MAITHANI 1989. Early Permian plant microfossils from the intertrappean beds of Abor Volcanics, Arunachal Pradesh, India. *Journal of the Geological Society of India* 34 : 83-88.

SINGH, T. 1973. Note on the Upper Palaeozoic fauna from Subansiri District, Arunachal Pradesh. *Himalayan Geology* 3 : 401-410.

SINGH, T. 1976. Stratigraphy and Palaeontology of the marine Permian Formation, Siang District, Arunachal Pradesh. Unpublished Ph. D. Thesis, University of Delhi : Delhi.

SINGH, T. 1978 a. Lower Permian gastropods and bivalves from Eastern Himalaya, India. In *Recent Researches in Geology* 7 : 276-310. Hindustan Publishing Corporation (India) : Delhi.

SINGH, T. 1978 b. Brachiopods from Permian Formation of Siang District, Arunachal Pradesh. In V.J. Gupta (ed.), *Contribution to Himalayan Geology* 1 : 171-188. Hindustan Publishing Corporation (India) : Delhi.

SINGH, T. 1978 c. Lower Permian cephalopods from Eastern Himalaya, India. *Himalayan Geology* 8 : 178 -193.

SINGH, T. 1979. Palynostratigraphy of Permian rocks of Siang District, Arunachal Pradesh. In P.K. Verma (ed.), *Metamorphic Rock Sequences of the Eastern Himalaya* : 100 -113. K.P. Bagchi & Company : Calcutta.

SINGH, T. 1981. Age and faunal affinity of the Garu Formation, Arunachal Pradesh. *Himalayan Geology* 11 : 271-286.

SINGH, T. 1987 a. Palaeoclimatic sigificance of fauna and mio-flora of the Garu Formation, Eastern Himalaya. In G. D. Mckenzie (ed.), *Gondwana Six : Stratigraphy, Sedimentology and Palaeontology, Geophysical Monograph* 41 : 191-194. American Geophysical Union: Washington.

SINGH, T. 1987 b. Permian biogeography of the Indian subcontinent with special reference to the marine fauna. In G. D. Mckenzie (ed.), *Gondwana Six : Stratigraphy, Sedimentology and Palaeontology, Geophysical Monograph* 41 : 239-249. American Geophysical Union: Washington.

SINGH, T. In Preparation. Stratigraphic record of northern margin of the Gondwanaland in Indian subcontinent.

SINGH, T. & U. BAJPAI 1990. On some plant fossils from Gondwana equivalent sediments of Eastern Himalaya. *The Palaeobotanist* 37 : 284-291.

SINGH, T. & R.A.K. SRIVASTAVA In Preparation. Genesis of biogenic concretions from Permian sediments of Arunachal Himalaya - geochemical and isotopic evidences.

SRIVASTAVA, S.C., ANAND-PRAKASH & T. SINGH 1987. Permian palyno-fossils from the Eastern Himalaya and their genetic relationship. *The Palaeobotanist* 36 : 326-338.

SRIVASTAVA, S.C. & S.K. DUTTA 1979. A note on the palynology of the Gondwanas of Siang District, Arunachal Pradesh. *Geophytology* 7 : 281-283.

TANDON, S. K. & S. KUMAR 1980. Stratigraphic sedimentation of the Upper Palaeozoic diamictites of Eastern Himalaya with special reference to the diamictites of Siang Division. In K. S. Valdiya & S. B. Bhatia (eds.), *Stratigraphy and correlations of Lesser Himalayan Formations* : 255-266. Hindustan Publishing Corporation (India) : Delhi.

Gondwana Eight, Findlay, Unrug, Banks & Veevers (eds) © 1993 Balkema, Rotterdam. ISBN 90 5410 304 3

Permo-Triassic sedimentation within the Lambert Graben, northern Prince Charles Mountains, East Antarctica

John A.Webb
Department of Geology, La Trobe University, Bundoora, Vic., Australia
Chris R.Fielding
Department of Geology and Mineralogy, University of Queensland, Qld, Australia

ABSTRACT: The only substantial occurrence of Palaeozoic strata in East Antarctica is the Amery Group, which crops out around Beaver Lake in the northern Prince Charles Mountains, on the western margin of the Lambert Graben. The uppermost unit of the Amery Group is the Flagstone Bench Formation; this is over 550 m thick, and apparently conformably succeeds the underlying mid to Late Permian Bainmedart Coal Measures. The Flagstone Bench Formation is dominated by trough cross-bedded, coarse-grained quartzo-feldspathic sandstones, apart from the Jetty Member (here defined), which lies more or less in the middle of the formation, and contains abundant red-purple and olive coloured siltstones, frequently displaying large scale desiccation mud-cracks. The sandstone unit beneath the Jetty Member contains a mid to Late Permian microflora (Stage 5), and overlying the Jetty Member are siltstone beds with a Late Triassic megaflora containing *Dicroidium zuberi, Dicroidium crassinervis* forma *stelznerianum* and a ?voltzialean conifer (all illustrated herein). Except for the Jetty Member, the Flagstone Bench Formation was deposited by a northward-flowing braided river system; this ran axially down the Lambert Graben, which was continuous with the Mahanadi Graben in India at the time. Palaeocurrent directions within the Jetty Member are predominantly towards the east, and probably reflect tectonic uplift of the western margin of the graben. The transition within the Amery Group from a coal measure sequence to a formation containing red-beds probably reflects the change from a cool, humid climate to a warmer, perhaps drier one.

1 INTRODUCTION

Within the East Antarctic Craton there are few outcrops of Palaeozoic or younger strata, and from 0° longitude to 150° east the only substantial occurrence of Palaeozoic rocks is in MacRobertson Land, within the Prince Charles Mountains. These strata are Late Permian sandstones, conglomerates, shales and coals of the Amery Group; younger early Mesozoic sedimentary rocks were unknown prior to this study. By contrast, on the other side of the East Antarctic Craton, in the Transantarctic Mountains, thick sequences of Devonian to Jurassic siliciclastics are exposed (Barrett et al. 1986; Barrett 1991). The presence of Triassic strata in the Prince Charles Mountains had been suspected previously (Ravich et al. 1977), but was not confirmed until Triassic plant fossils were discovered there by the authors during the 1989/1990 summer field season. The Triassic megaflora was recovered from the uppermost formation of the Amery Group; this unit was relatively poorly known, so a detailed stratigraphic and sedimentological study was undertaken.

2 REGIONAL SETTING

Permian and Triassic strata in the northern Prince Charles Mountains crop out around the shores of Beaver Lake, which lies on the western side of the Lambert Glacier (Fig.1). This glacier, generally recognised as the largest valley glacier in the world, is as much as 80 km wide and more than 500 km long. Geophysical evidence, both within the glacier area (Fedorov

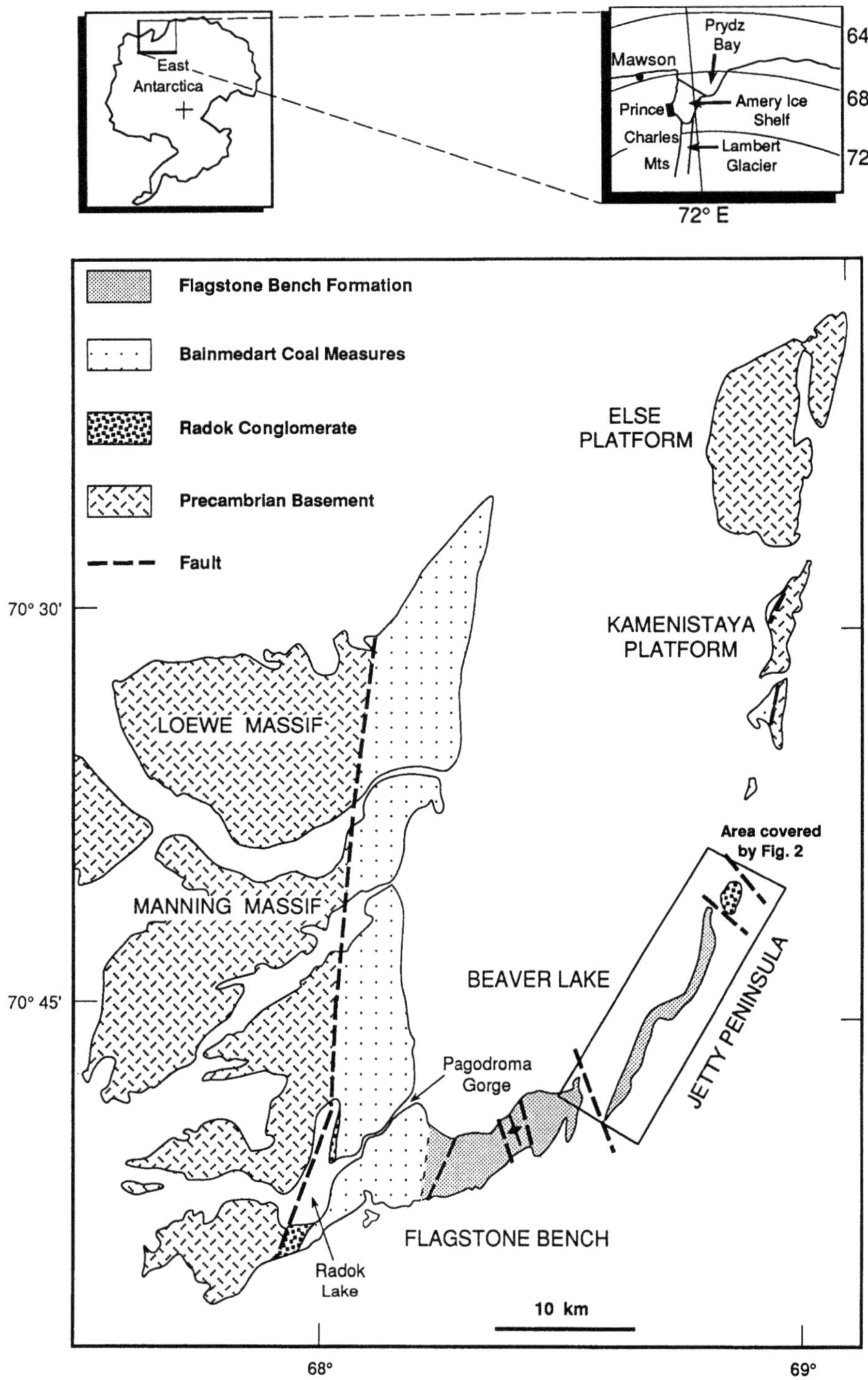

Fig. 1. Geological map of the Beaver Lake area, northern Prince Charles Mountains, showing outcrop areas of the three formations of the Permo-Triassic Amery Group. Note that most of the area underlain by Bainmedart Coal Measures on the western side of Beaver Lake is covered by moraine.

et al. 1982) and in Prydz Bay to the north (Stagg 1985), indicates that the Lambert Glacier occupies a structural trough, the Lambert Graben.

The basement rocks of the graben are part of the extensive Precambrian metamorphic terrane of the East Antarctic Craton. In the Prince Charles Mountains, along the western flanks of the Lambert Graben, the metamorphic grade increases northwards from greenschist to granulite facies (Tingey 1982). Within the northern Prince Charles Mountains the basement rocks are mostly felsic granulite and gneiss (McKelvey & Stephenson 1990).

The Lambert Graben contains a considerable thickness of sedimentary rocks, as shown by the geophysical data (Fedorov et al. 1982; Stagg 1985). The Permian outcrops at Beaver Lake, and Permian erratics in moraine near Mt Rymill, 250 kilometres to the south (Ruker 1963), clearly indicate the presence of late Palaeozoic strata throughout the graben. Drilling in Prydz Bay, at the northern extent of the graben, revealed the presence of Early Cretaceous fluviatile sediments overlying probable Amery Group equivalents (Turner & Padley 1991; Turner 1991).

3 AMERY GROUP

The late Palaeozoic strata of the Beaver Lake area have been collectively assigned to the Amery Group (Crohn 1959; Mond 1972), which is subdivided into three units: Radok Conglomerate, Bainmedart Coal Measures, and Flagstone Bench Formation, in ascending stratigraphic order (Ravich 1974; Ravich et al. 1977; McKelvey & Stephenson 1990). Brief descriptions of the lower units precede a detailed discussion of the Flagstone Bench Formation.

3.1 Lower Amery Group

The Radok Conglomerate crops out in two areas: around Radok Lake, where it is faulted against Precambrian basement, and on the northern tip of Jetty Peninsula, as a small block upfaulted into younger strata of the Amery Group (Fig. 1). The Radok Conglomerate consists of over 400 metres of interbedded greenish grey matrix-rich sandstones and conglomerates, as well as siltstones and occasional thin coal seams. The latter, identified during mapping by the authors, are restricted to the top portion of the unit. Clasts in the conglomerates are up to 1.5 metres in diameter and are composed mainly of felsic granulite, identical to the dominant lithology of the Precambrian basement to the west. The Panorama Point Beds (McKelvey & Stephenson 1990) probably represent a finer grained lateral facies equivalent of the Radok Conglomerate. Spore and pollen assemblages from the Radok Conglomerate indicate a mid to Late Permian age (Stage 5 - Baigendzhinian to Tatarian; Kemp 1973; Dibner 1978; Playford 1990).

The uppermost part of the Radok Conglomerate is transitional into the overlying Bainmedart Coal Measures, but a clear-cut boundary between the two formations is defined by a thin pebble to boulder conglomerate bed (the Dart Fields Conglomerate Member of McKelvey & Stephenson 1990), which lies at the base of the Bainmedart Coal Measures. Locally there is a very low angle unconformity between the Radok and Dart Fields Conglomerates (McKelvey & Stephenson 1990).

The Bainmedart Coal Measures, which are probably about 1100 metres thick, crop out extensively along the eastern side of Radok Lake, throughout Pagodroma Gorge and along the western side of Beaver Lake (Fig. 1). The small outcrops of Amery Group strata on the western side of Kamenistaya Platform, to the north of Jetty Peninsula (Fig. 1), probably also belong to this unit (Grew 1985; D. Cantrill, pers. comm.) The Bainmedart Coal Measures are dominated by coarse-grained light yellow-brown quartzose and feldspathic sandstone, with minor proportions of siltstone, coal and claystone. The sandstone bodies are trough cross-bedded and relatively even grained, but at their tops they generally fine upwards through rippled medium-grained sandstones into thin siltstones with abundant rootlets. These siltstones directly underlie coal seams, which are numerous (over 60), but mostly thin (less than 0.5 m), although there are several seams over two metres thick. The coals are of high volatile bituminous rank (Bennett & Taylor 1972). Glossopterid plant fossils and a diverse palynomorph assemblage from the siltstone layers indicate a mid to Late Permian age (Stage 5; Balme & Playford 1967; Kemp 1973; White

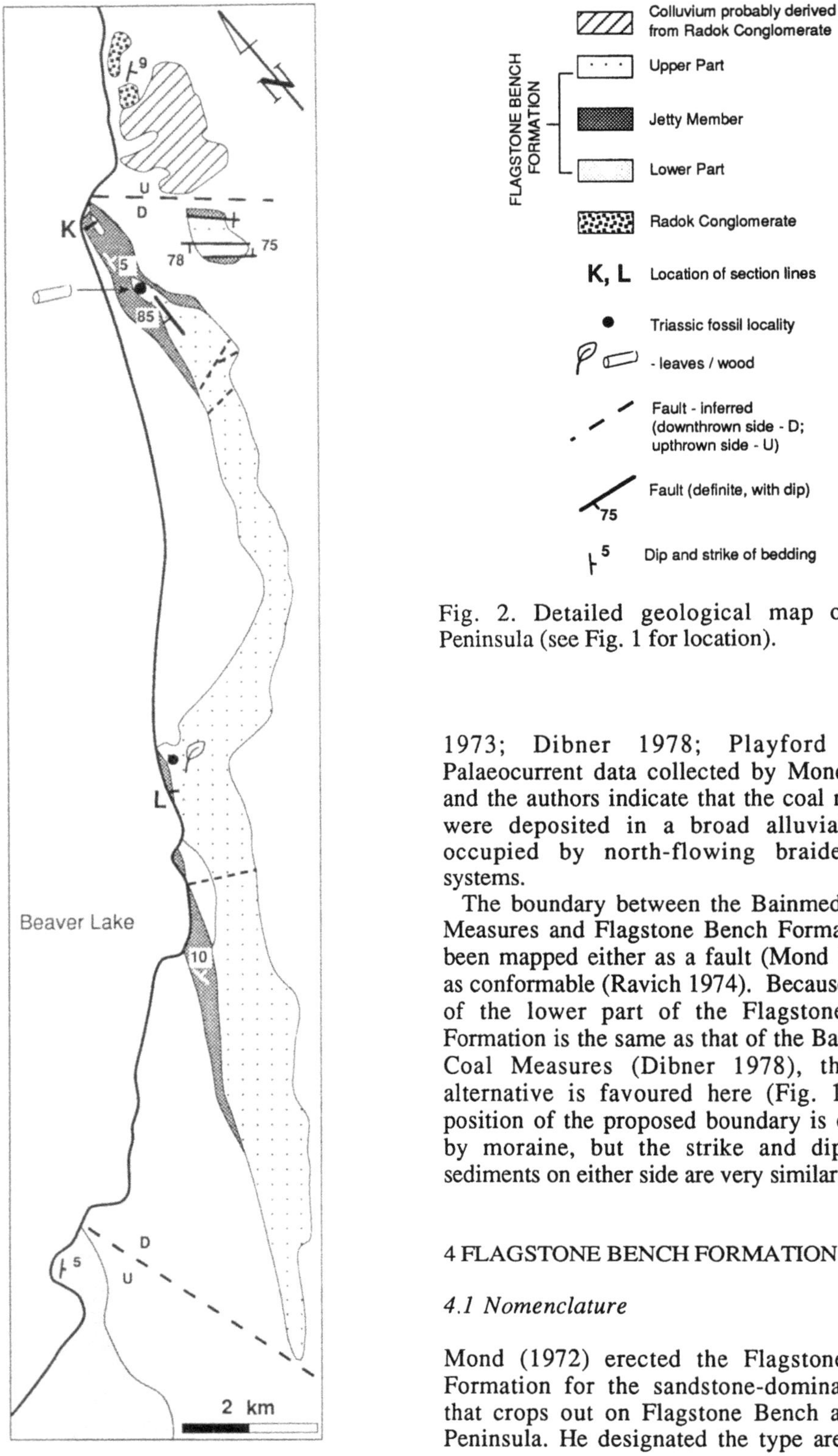

Fig. 2. Detailed geological map of Jetty Peninsula (see Fig. 1 for location).

1973; Dibner 1978; Playford 1990). Palaeocurrent data collected by Mond (1972) and the authors indicate that the coal measures were deposited in a broad alluvial valley occupied by north-flowing braided river systems.

The boundary between the Bainmedart Coal Measures and Flagstone Bench Formation has been mapped either as a fault (Mond 1972) or as conformable (Ravich 1974). Because the age of the lower part of the Flagstone Bench Formation is the same as that of the Bainmedart Coal Measures (Dibner 1978), the latter alternative is favoured here (Fig. 1). The position of the proposed boundary is obscured by moraine, but the strike and dip of the sediments on either side are very similar.

4 FLAGSTONE BENCH FORMATION

4.1 Nomenclature

Mond (1972) erected the Flagstone Bench Formation for the sandstone-dominated unit that crops out on Flagstone Bench and Jetty Peninsula. He designated the type area as the

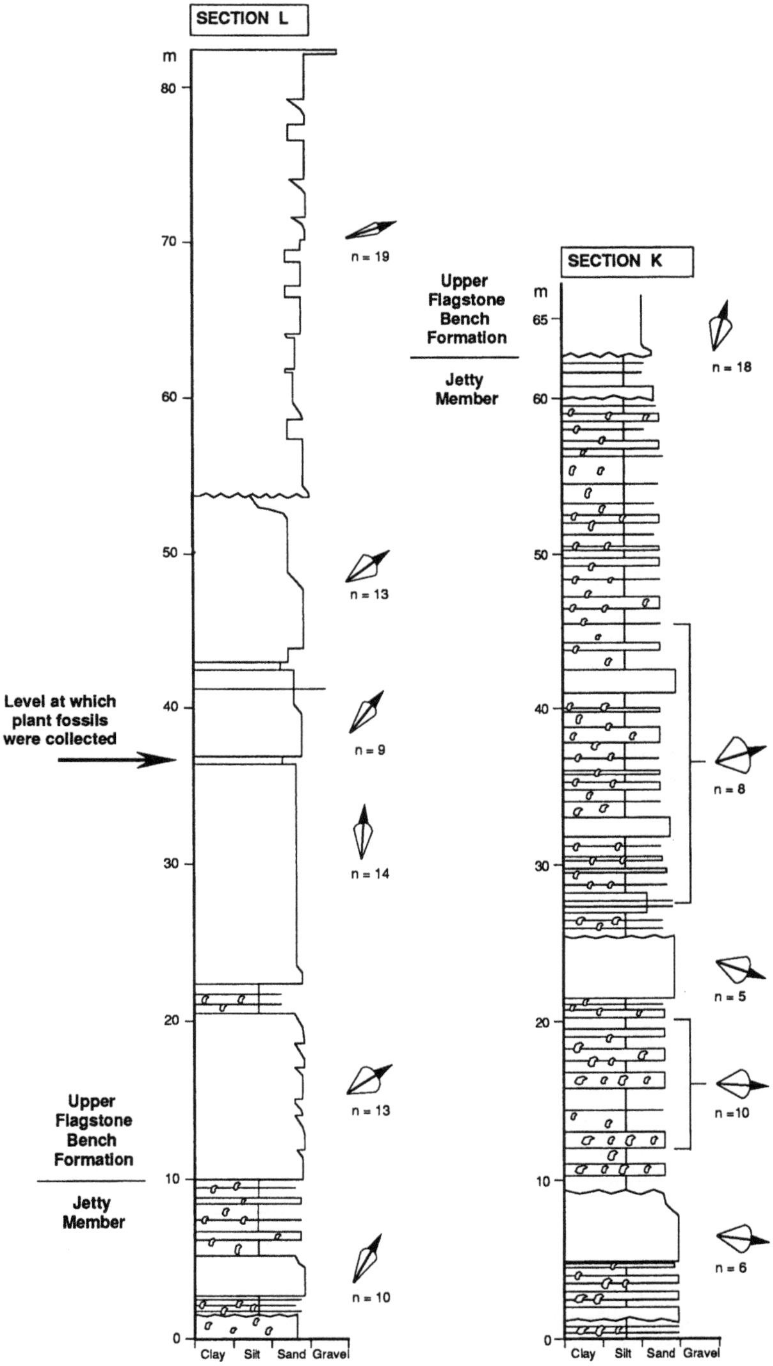

Fig. 3. Detailed stratigraphic sections through the Jetty Member and the upper part of the Flagstone Bench Formation; locations shown on Fig. 2. Arrow indicates stratigraphic level of siltstone containing plant fossils; this bed is absent in this section due to scouring at base of overlying sandstone. Colour mottling of siltstones indicated by irregular patches; horizontal scale indicates grain size. Palaeocurrents shown as mean and 95% uncertainty (obtained using Batschelet chart) for sandstone bodies or groups of beds; n = number of observations.

northeast part of Flagstone Bench. Traube (1991) separated out part of the sedimentary sequence on Jetty Peninsula as the Jetty Formation, but did not describe the unit in detail or erect a type section. Nevertheless his stratigraphic subdivision appears to be valid, although we prefer to regard the newly defined unit as a member within the Flagstone Bench Formation (see below). A detailed description of the Jetty Member and type section are given in Appendix 1.

4.2 Distribution

The Flagstone Bench Formation crops out extensively on Flagstone Bench and Jetty Peninsula, and is very well exposed in the cliffs along the edge of Beaver Lake (Figs 1 and 2). The beds dip gently eastward except on the central part of Flagstone Bench, where there is broad open folding. The sequence is cut by a number of steeply dipping or subvertical faults with variable amounts of displacement (Fig. 2).

Drilling in Prydz Bay, at the seaward of the Lambert Graben, revealed a sequence of continental red-beds that Turner (1991) believed were Amery Group equivalents or younger. These strata are similar to the Jetty Member (see below) in colour, sedimentary characteristics and petrography, and may be correlative.

4.3 Stratigraphic sequence and age

On Flagstone Bench the formation consists of at least 400 metres of coarse-grained, light yellow-brown quartzo-feldspathic sandstones, which have abundant ferruginous patches and layers. These sandstones form prominent vertical cliffs. Siltstones are rare and coal is absent. Ravich et al. (1977) discovered poorly preserved conchostracans at one locality, and Dibner (1978) recovered a moderately well-preserved, diverse palynoflora from one of the siltstone horizons. The species identified date the bed as Stage 5 (mid to Late Permian), the same palynological stage present in the underlying Bainmedart Coal Measures (Dibner 1978; Playford 1990). The similarity in age between these two units implies that their mutual boundary is conformable, as previously discussed.

On Jetty Peninsula two separate units can be distinguished (Fig. 2). The lowermost, the Jetty Member, consists of at least 60 m of interbedded sandstones and siltstones with abundant red-purple and olive-coloured alluvial palaeosols showing large desiccation mud-cracks (Fig. 3, section K). This unit forms steeply sloping, gullied topography. Overlying the Jetty Member is a sequence of more than 70 m of light yellow-brown feldspathic sandstones and very rare siltstones (Fig. 3, section L). These strata form prominent vertical cliffs, and are almost indistinguishable from the Flagstone Bench Formation sediments on Flagstone Bench, except that they are less ferruginous and contain occasional pieces of poorly preserved silicified gymnospermous wood. A siltstone bed just above the top of the Jetty Member yielded a well preserved Late Triassic megaflora (described below).

Thus the Jetty Peninsula sequence is considerably younger than the Flagstone Bench Formation on Flagstone Bench, and a major fault must separate the two, with at least 150 m of dip-slip displacement. Early Triassic sediments presumably underlie the Jetty Member, which may be partly Middle Triassic in age, and the entire sequence appears to be conformable. Given the lithological similarity of the sandstones above and below the Jetty Member, it seems appropriate to retain all these strata within the Flagstone Bench Formation, the total thickness of which exceeds 550 metres.

4.4 Triassic megaflora

A dark grey siltstone bed, approximately 25 metres above the top of the Jetty Member (Fig. 2), yielded a small collection of well preserved plant fossils (Fig. 4). The flora is dominated by *Dicroidium zuberi*; the specimens display a considerable range of pinnule shape (Figs 4b,c,f,g), similar to that ascribed to this species by Anderson & Anderson (1983). Numerous specimens of an undescribed ?voltzialean conifer are also present (Figs 4d,e); this superficially resembles *Voltziopsis angusta*, but has a different cuticle morphology. In addition, there are fragmentary cycad leaves, specimens of *Pteruchus* cf. *barrealensis* (probably the male fruiting body of *Dicroidium zuberi*), and a single specimen (Fig. 4a) that is referable to

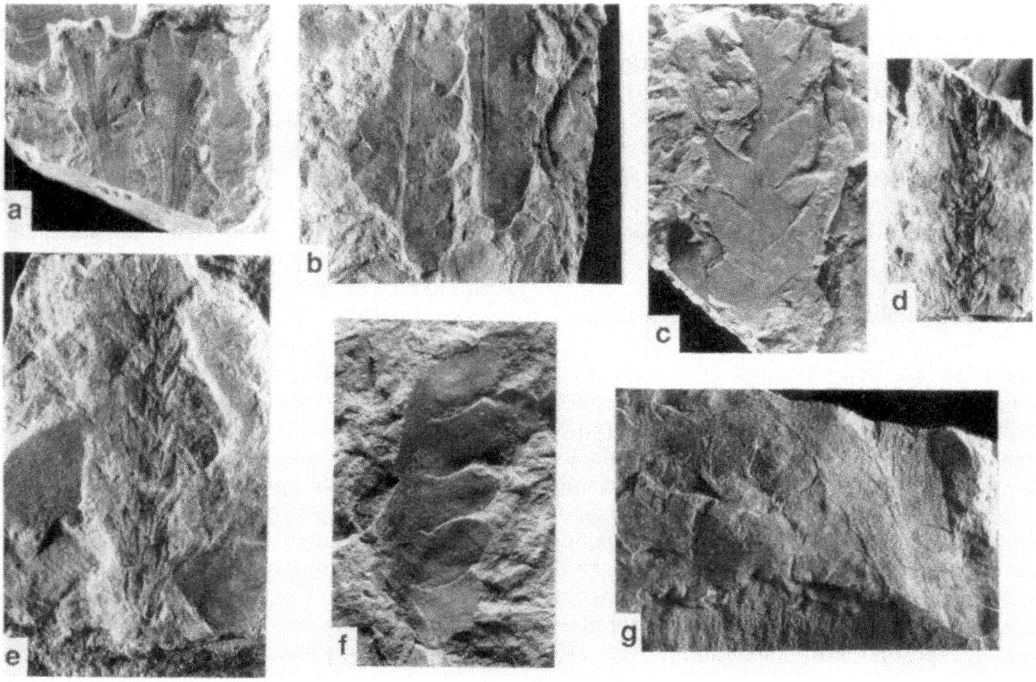

Fig. 4. Megaflora from upper part of Flagstone Bench Formation; locality shown on fig. 2. *Dicroidium zuberi*: (b) x 0.8, (c) x 0.5, (f) x 1.1 - note details of venation, (g) x 1.0 - note forking rachis, and bipinnate arrangement of leaves on left hand side of photograph. ?voltzialean conifer: (d) x 1.9, (e) x 1.7. *Dicroidium crassinervis* forma *stelznerianum*: (a) x 1.0.

Dicroidium crassinervis forma *stelznerianum* (of Anderson & Anderson 1983; = *Johnstonia stelzneriana* variety *stelzneriana* of Retallack 1977).

Dicroidium zuberi ranges in age from Early to Late Triassic (Smithian to Norian: Retallack 1977; Anderson & Anderson 1983). *Dicroidium crassinervis* forma *stelznerianum* has a more restricted time range of Middle to Late Triassic (Anisian-Ladinian to Norian: Retallack 1977; Anderson & Anderson 1983). A preliminary study of the microflora from the siltstone bed indicates that it is probably of Late Triassic (Carnian) age (C. Foster, pers. comm.).

Triassic floras are well known from elsewhere in Antarctica, and have been collected from the Lashly Formation of South Victoria Land (Plumstead 1962; Rigby 1985) and the Fremouw and Falla Formations of the Shackleton and Beardmore Glacier regions in the central Transantarctic Mountains (Townrow 1967; Smoot et al. 1985; Taylor et al. 1986; Smoot et al. 1987; Meyer-Berthaud & Taylor 1989; Pigg 1990). Of these, the flora most similar to that from the Flagstone Bench Formation is from an erratic of probable Falla Formation siltstone collected near Mount Bumstead (Townrow 1967); it contains *Dicroidium zuberi* and probably *Dicroidium crassinervis* forma *stelznerianum*, along with *Dicroidium elongatum* forma *elongatum* (*Dicroidium* identifications revised using Anderson & Anderson 1983). Palynomorphs from the Falla Formation indicate a Late Triassic (Carnian/Norian) age (Farabee et al. 1989), so the upper part of the Flagstone Bench Formation appears to be a time equivalent of the Falla Formation.

The Triassic localities in the Transantarctic Mountains and Prince Charles Mountains seem to be characterized by moderate diversity floras (around 5 species), dominated by one or more elements. The best studied Triassic locality, a Middle Triassic silicified peat horizon in the Fremouw Formation, contains one species of *Dicroidium*, along with a cycad, gymnospermous wood, various reproductive structures, and fungi (Taylor et al. 1986). The Mt Bumstead locality has four, perhaps five species of *Dicroidium*, as well as a ?gingkoalean and a ?cycad (Townrow 1967; Anderson & Anderson 1983). The diversity of the Jetty Peninsula flora, with its two species of *Dicroidium*, as well as a conifer and a cycad, is therefore typical of Triassic Antarctic floras.

4.5 Lithologies and sedimentary structures

Except for the Jetty Member, the rocks of the Flagstone Bench Formation are light yellow-brown sandstones with abundant ferruginous layers. Grain size ranges from coarse sandstone to granule conglomerate (Fig. 3). In the upper part of the formation the bases of sandstone beds occasionally contain well-rounded pebbles and cobbles of quartzite, as much as 35 centimetres in diameter. The sandstones are composed of moderately well sorted, angular to rounded grains with subordinate very fine grained matrix. Compositionally these sediments are subarkoses, although a few have sufficient matrix to be called feldspathic greywackes (using the classification of Folk 1974). The dominant framework grains are metamorphic quartz, and occasionally display syntaxial overgrowths. There is abundant sericitized K-feldspar, predominantly orthoclase, with lesser amounts of microcline; many of the orthoclase and microcline grains are perthitic. The occasional lithic fragments are felsic granulite. Muscovite flakes are a widespread but minor component, and there are trace amounts of heavy minerals (zircon, tourmaline, garnet).

The matrix is composed mainly of very fine grained kaolinite, in some places intergrown with another clay mineral (?illite) or very fine grained iron oxide. This material is probably depositional in origin, as it contains a few angular silt-sized quartz grains. However, there are often coarser grained kaolinite patches within the matrix; these probably formed as a result of diagenetic recrystallization. The boundaries of quartz grains in contact with the matrix are commonly diffuse and embayed, probably reflecting limited resorption of the grains by the matrix. The sandstones display considerable evidence of compaction in the form of interpenetrant grains and muscovite flakes which have been kinked around clasts.

The sandstones below the Jetty Member contain much biotite. The grains of biotite have mostly altered to colourless chlorite; this alteration has caused many of the grains to split and buckle, and has released a large amount of iron oxide, which now occurs as patches of ferruginous cement surrounding the altered biotites. Biotite is absent from the sandstones above the Jetty Member, and there is much less iron oxide in these rocks.

The sandstones of the Jetty Member itself are pale olive. They are rich in feldspar and have a high percentage of matrix; they vary from arkoses to feldspathic greywackes. Biotite is absent, and iron oxide is present either as very fine grains intergrown with the kaolinite matrix, or as scattered patches of cement often associated with altered K-feldpar crystals.

Ferroan calcite is sporadically present in sandstones throughout the Flagstone Bench Formation, either as irregular, often poikilitic, patches apparently replacing the kaolinite matrix, or associated with iron oxide cements. Small amounts of siderite also occur.

Above and below the Jetty Member the sandstones are thick-bedded, and dominated internally by trough cross-bedding. The sandstone beds are typically amalgamated into multi-storey bodies 10-30 m thick (Fig. 3), which show no consistent vertical grainsize trends. Individual cosets are as much as 3 metres in thickness, and are in many cases laterally continuous in a down-palaeocurrent direction over 10s to 100s of metres. Complex hierarchical relationships are well preserved; for example small-scale ripple cross-strata migrate down the foresets of some large dunes. Convolute lamination is occasionally present within cross-strata cosets, probably reflecting soft sediment deformation by current drag. Palaeocurrent measurements indicate directions predominantly towards the northnortheast and northeast (Fig. 3; also noted by Traube 1991). There are only minor fine-grained partings, and these are usually pale red-purple, pale olive or blue-grey, apart from the less oxidized dark grey fossiliferous siltstone in the upper part of the Flagstone Bench Formation.

In the Jetty Member the sandstones are thinner bedded and for the most part single-storey bodies (Fig. 3). They are dominated internally by trough cross bedding, in sets up to 0.4 m thick. Ripple cross-laminations are also abundant. Palaeocurrent data indicate predominantly eastward directed flow (Fig. 3; see also Traube 1991). Pale red-purple siltstones constitute a significant proportion of the Jetty Member, and show abundant, well-preserved polygonal desiccation cracks. These range up to twenty centimetres across. The infiltration of pale olive sands into the cracks during the deposition of overlying sediments give the siltstone horizons a mottled red-green appearance. Poorly preserved root channels are occasionally present.

Traube (1991) identified a breccia containing "exceptionally coarse debris" within the Jetty Member; a close examination of the Jetty Peninsula cliffs has failed to find this unit. The limestone noted by Traube (1991) at the top of the Jetty Member is best regarded as a calcareous palaeosol.

The Prydz Bay red-beds described by Turner (1991) closely resemble the Jetty Member in that both contain abundant red siltstones with numerous mottled palaeosols, sometimes showing desiccation mud-cracks. However, the sandstones within the Prydz Bay sequence contain abundant biotite, and are therefore more similar to the sandstones of the Flagstone Bench Formation below the Jetty Member. Nevertheless, it is probable that the Prydz Bay red-beds can be correlated with the Jetty Member; any petrographic dissimilarities may reflect minor differences in source area.

5 DISCUSSION

5.1 Depositional environment of the Flagstone Bench Formation

In the lower and upper Flagstone Bench Formation, the amalgamated multi-storey character of the sandstones is suggestive of accumulation in a major, low-sinuosity braided river system. This is supported by the strongly unimodal palaeocurrent distribution (Fig. 3) and the lack of preserved siltstones. The Jetty Member (and Prydz Bay red-beds), with their higher proportions of fine-grained material, probably represent sediment accumulation in less swiftly flowing, shallower braided channels draining an extensive alluvial plain.

5.2 Permo-Triassic palaeogeography and tectonism of the Lambert Graben

The Flagstone Bench Formation was deposited within the Lambert Graben, which probably occupied the full extent of the present-day Lambert Glacier valley. This is shown by the geophysical evidence (Federov et al. 1982; Stagg 1985) and the easterly palaeocurrents of the Jetty Member, which indicate that the alluvial valley extended well to the east of Beaver Lake. During the Permian and Triassic the Lambert Graben was continuous with the Mahanadi Graben in central eastern India (Stagg 1985). Consistent palaeocurrent directions and stratigraphic sequences in both areas indicate that a major braided river system flowed axially northwards down the graben from Antarctica into India, a distance of almost 1000 km. In Antarctica the river deposited the Bainmedart Coal Measures and Flagstone Bench Formation, while in India very similar sediments of approximately the same age accumulated (Raniganj Coal Measures and Mahadeva sequence respectively; Casshyap & Tewari 1987).

The easterly palaeocurrents within the Jetty Member are almost perpendicular to the northnortheasterly palaeocurrents of the underlying and overlying braided river sandstones, which represent axial flow down the graben. As the sedimentary sequence of the Jetty Member is typical of distal alluvial fan deposits (see Nilsen 1982), this unit may represent the distal portion of an alluvial fan originating on the western margin of the Lambert Graben in mid-Triassic times. The temporary change in regional gradient in this area during the mid-Triassic probably reflects tectonic uplift along the western boundary fault. Veevers (1991) noted that a mid-Triassic compressive event was widespread throughout Gondwana, and may mark the maximum coalescence of Pangaea.

On the western side of the East Antarctic Craton, the late Palaeozoic - early Mesozoic basins of the central Transantarctic Mountains accumulated similar stratigraphic sequences to the Lambert Graben. Late Permian coal measures (Buckley Formation) are overlain by sandstone-dominated units, sometimes with minor red-bed siltstones, deposited by braided rivers (Fremouw Formation: Collinson et al. 1981; Barrett et al. 1986; Collinson & Isbell

1986; Barrett 1991). However, the western edge of the Antarctic craton was associated with abundant volcanic activity, particularly in the Triassic, and the units of this age all have a prominent volcaniclastic component (Barrett 1991). This side of Antarctica may have been a convergent plate boundary at the time (Barrett 1991). By contrast, the Lambert Graben lacks volcaniclastics and represents an extensional intracratonic rift.

5.3 Permo-Triassic palaeoclimate in East Antarctica

Triassic palaeoclimates throughout the world were warmer and drier than the cool, humid climates of the Permian; the early part of the Triassic features a dominance of red-bed deposition, and in the Middle Triassic evaporite deposition was widespread at equatorial latitudes, indicating very arid conditions (Frakes 1979). These climatic changes are clearly shown by the Amery Group sequence, even though the Antarctic continent lay almost entirely in its present polar position throughout the Permian and Triassic (Barrett 1991). Compared to the underlying Bainmedart Coal Measures, the Flagstone Bench Formation lacks coal seams, and has fine-grained sediments that are generally oxidized red in colour and contain mud-cracked palaeosols. This suggests that the basin became better drained with time, probably as the result of a change to a warmer, perhaps seasonally drier climate. Furthermore, the palynomorphs from the lower part of the Flagstone Bench Formation have different proportions of species compared to the underlying coal seam sediments, even though they belong to the same palynological zone (Dibner 1978), indicating that a substantial change in the flora had taken place. This change must have occurred towards the end of the Late Permian. The Triassic climate in East Antarctica probably never became very arid; there is little evidence of pedogenic calcrete formation (Turner 1991), and the presence of abundant kaolinite suggests a warm, humid, well-drained environment (Robert & Kennett 1992).

The climatic changes from the Permian to Triassic may have been less marked in the Transantarctic Mountains, where Triassic red-beds are poorly developed or absent (Collinson et al. 1981; Barrett 1991). However, elsewhere in Gondwana, the climatic break is well defined. For example, in eastern Australia, within the Bowen, Sydney and Tasmanian Basins, coal-bearing units of Stage 5 age (Blackwater Group, Illawarra Coal Measures and Cygnet Coal Measures respectively) are overlain by Late Permian - Early Triassic sequences containing red-beds (Rewan Group, Narrabeen Group and Early Triassic quartz sandstone sequence respectively; Foster 1979; Herbert & Helby 1980; Day et al. 1983; Forsyth 1989).

6 CONCLUSIONS

A detailed study of the Flagstone Bench Formation, the uppermost unit of the Amery Group in the northern Prince Charles Mountains, East Antarctica, has shown that it ranges in age from Late Permian to Late Triassic. The formation is sandstone-dominated apart from the Jetty Member, which lies in the middle of the formation and consists of abundant red siltstones with mud-cracked palaeosols. The Flagstone Bench Formation was deposited by a very extensive braided river system almost 1000 km long which flowed northwards down the Lambert Graben into the Mahanadi Graben in central eastern India. A tectonic episode in the mid-Triassic uplifted the western margin of the Lambert Graben and initiated development of an easterly directed alluvial fan that deposited the Jetty Member. The transition within the Amery Group from the Bainmedart Coal Measures to the Flagstone Bench Formation probably reflects the change from a cool, humid climate to a warmer, perhaps drier one.

7 APPENDIX

Jetty Member (originally proposed by Traube 1991 as Jetty Formation)

Derivation of name: From Jetty Peninsula.

Type section: Cliffs on the northern tip of Jetty Peninsula (section K; Fig. 3).

Distribution: Well exposed in the cliffs along the western side of Jetty Peninsula (Fig. 2).

Thickness: >60 m (base not exposed).

Boundaries: Upper - conformable; base of first thick light yellow-brown sandstone unit. Lower - not exposed.

Stratigraphic relationships: Overlain by >70 m of Late Triassic light yellow-brown sandstones of the Flagstone Bench Formation. Separated by a fault from the older Late Permian light yellow-brown sandstones of the Flagstone Bench Formation on Flagstone Bench; these strata are >400 m thick.

Lithology: Interbedded sandstones and siltstones with abundant mottled red-purple and olive-coloured alluvial palaeosols often characterized by large-scale desiccation mudcracks. Sandstones range from arkoses to feldspathic greywackes, with abundant kaolinitic matrix, and occasional patches of ferruginous cement.

Age: Probably Middle Triassic, at least in part; a siltstone bed about 25 m above the top of the Jetty Member contains a Late Triassic megaflora.

Environment of deposition: Probably channels and floodplains of the distal portion of an alluvial fan, with predominantly easterly palaeocurrent directions.

ACKNOWLEDGEMENTS

We would like to express our appreciation for the logistic and financial support given by the Australian Antarctic Division. In particular we wish to thank the field leader, Martin Betts, and chief helicopter pilot, Pip Turner, on the 1989-90 Prince Charles Mountains Program. David Cantrill and Andrew Drinnan assisted with the identification of the conifer from the Jetty Member, and provided information from their field work carried out on the Jetty Peninsula during the 1991-1992 field season. Alguird Rahdon helped considerably with the petrological descriptions. Clinton Foster (Bureau of Mineral Resources) provided preliminary results from his study of the microflora from the fossiliferous siltstone in the upper Flagstone Bench Formation.

REFERENCES

Anderson, J.M. & H.M. Anderson 1983. *Palaeoflora of southern Africa; Molteno Formation (Triassic)*. Rotterdam: Balkema.

Balme, B.E. & G. Playford 1967. Late Permian plant microfossils from the Prince Charles Mountains, Antarctica. *Revue Micropaleontologie* 10: 179-192.

Barrett, P.J. 1991. The Devonian to Jurassic Beacon Supergroup of the Transantarctic Mountains and correlatives in other parts of Antarctica. *In* Tingey, R.J. (ed.), *The geology of Antarctica*: 120-152. Oxford: Clarendon Press.

Barrett, P.J., D.H. Elliot & J.F. Lindsay 1986. The Beacon Supergroup (Devonian - Triassic) and Ferrar Group (Jurassic) in the Beardmore Glacier area, Antarctica. *In* Turner, M.D. & J. Splattstoesser (eds.), Geology of the central Transantarctic Mountains, *Antarctic Research Series* 36: 339-428. Washington: American Geophysical Union.

Bennett, A.J.R. & O.H. Taylor 1972. Coals from the vicinity of the Prince Charles Mountains. *In* Adie, R.J. (ed.), *Antarctic geology and geophysics*, 591-598. Oslo: Universitetsforlaget.

Casshyap, S.M. & R.C. Tewari 1987. Depositional model and tectonic evolution of Gondwana basins. *The Palaeobotanist* 36: 59-66.

Collinson, J.W. & J.L. Isbell 1986. Permian - Triassic sedimentology of the Beardmore Glacier region. *Antarctic Journal of the United States* 21: 29-30.

Collinson, J.W., K.O. Stanley & C.L. Vavra 1981. Triassic fluvial depositional systems in the Fremouw Formation, Cumulus Hills, Antarctica. *In* Cresswell, M.M. & P.P. Vella (eds.): *Gondwana Five* , 141-148. Rotterdam: Balkema.

Crohn, P.W. 1959. A contribution to the geology of the western part of the Australian Antarctic Territory. *Bureau of Mineral Resources, Geology and Geophysics, Australia, Bulletin* 52: 1-103.

Day, R.W., W.G. Whitaker, C.G. Murray, I.H. Wilson & K.G. Grimes 1983. Queensland geology. *Geological Survey of Queensland, Publication* 383: 1-194.

Dibner, A.F. 1978. Palynocomplexes and age of the Amery Formation deposits, East Antarctica. *Pollen et Spores* 20: 405-422.

Farabe, M.J., T.N. Taylor & E.L. Taylor 1989. Pollen and spore assemblages from the Falla Formation (Upper Triassic), central Transantarctic Mountains, Antarctica. *Review of Palaeobotany and Palynology* 61: 101-138.

Fedorov, L.V., G.E. Grikurov, R.G. Kurinin & V.N. Masolov 1982. Crustal structure of the Lambert Glacier area from geophysical data. *In* Craddock, C. (ed.), *Antarctic Geoscience*, 931-936. Madison: University of Wisconsin Press.

Folk, R.L. 1974. *Petrology of sedimentary rocks*. Austin: Hemphills.

Forsyth S.M. 1989. Upper Parmeener Supergroup. *In* Burrett, C.F. & E.L. Martin (eds.), *Geology and mineral resources of Tasmania. Geological Society of Australia Special Publication* 15: 309-334.

Foster, C.B. 1979. Permian plant microfossils of the Blair Athol Coal Measures, Baralaba Coal Measures and basal Rewan Formation of Queensland. *Geological Survey of Queensland, Publication* 372: 1-244.

Frakes, L.A. 1979. *Climates through geologic time*. Amsterdam: Elsevier.

Grew, E.S. 1985. Field studies on the Jetty Peninsula (Amery Ice Shelf area) with the 30th Soviet Antarctic Expedition. *Antarctic Journal of the United States* 20: 52-53.

Herbert, C. & Helby, R. (eds.) 1980. A guide to the Sydney Basin. *Geological Survey of New South Wales, Bulletin* 26: 1-584.

Kemp, E.M. 1973. Permian flora from the Beaver Lake area, Prince Charles Mountains, Antarctica. 1. Palynological examination of samples. *Bureau of Mineral Resources, Geology and Geophysics, Australia, Bulletin* 126: 7-12.

McKelvey, B.C. & N.C.N. Stephenson 1990. A geological reconnaissance of the Radok Lake area, Amery Oasis, Prince Charles Mountains. *Antarctic Science* 2: 53-66.

Meyer-Berthaud, B. & T.N. Taylor 1989. The structure and affinities of woody stems from the Triassic of Antarctica. *American Journal of Botany* 76 (supplement): 170-171.

Mond, A. 1972 Permian sediments of the Beaver Lake area, Prince Charles Mountains. *In* Adie, R.J. (ed.), *Antarctic geology and geophysics*, 585-589. Oslo: Universitetsforlaget.

Nilsen, T.H. 1982. Alluvial fan deposits. *In* Scholle, P.A. & D. Spearing (eds.), *Sandstone depositional environments*, 49-86. Tulsa: American Association of Petroleum Geologists.

Pigg, K.B. 1990. Anatomically preserved *Dicroidium* foliage from the central Transantarctic Mountains. *Review of Palaeobotany and Palynology* 66: 129-145.

Playford, G. 1990. Proterozoic and Palaeozoic palynology of Antarctica: a review. *In* Taylor, T.N. & E.L. Taylor (eds.) *Antarctic paleobiology - its role in the reconstruction of Gondwana*, 50-70. New York: Springer-Verlag.

Plumstead, E.P. 1962. Fossil floras of Antarctica. *Transactions of Antarctic Expeditions, Scientific Reports* 9: 1-132.

Ravich, G.M. 1974. Razrez permskikh uglenosnykh otlozheniy rayona oz. Biver (gory Prins-Charlez, vostochnaya Antarktida). *Antarktika* 13: 19-25.

Ravich, G.M., Yo.G. Gor, A.F. Dibner & O.V. Dobanova 1977. Stratigrafiya verkhnepaleozoiskikh uglenocnykh otlozheniy vostochnoy Antarktidy (rayon ozera Biver). *Antarktika* 16: 62-75.

Retallack, G.J. 1977. Reconstructing Triassic vegetation of eastern Australasia: a new approach for the biostratigraphy of Gondwanaland. *Alcheringa* 1: 247-277.

Rigby J.F. 1985. Some Triassic (middle Gondwana) floras from South Victoria Land, Antarctica. *In* Hornibrook Symposium, Abstracts, 78-79. Christchurch: Department of Scientific and Industrial Research, New Zealand.

Robert C. & J.P. Kennett 1992. Paleocene and Eocene kaolinite distribution in the South Atlantic and Southern Ocean: Antarctic climatic and paleoceanographic implications. *Marine Geology* 103: 99-110.

Ruker, R.A. 1963. Geological reconnaissance in Enderby Land and the southern Prince Charles Mountains, Antarctica. *Bureau of Mineral Resources, Geology and Geophysics, Australia, Record* 1963/154.

Smoot, E.L., T.N. Taylor & J.W. Collinson 1987. Lower Triassic plants from Antarctica: diversity and paleoecology. *VII Simposium Argentina Paleobotanica i Palynologia, Actas*: 193-196.

Smoot, E.L., T.N. Taylor & T. Delevoryas 1985. Structurally preserved plants from Antarctica. I. *Antarcticycas* gen. nov., a Triassic cycad stem from the Beardmore Glacier area. *American Journal of Botany* 72: 1410-1423.

Stagg, H.M.J. 1985. The structure and origin of Prydz Bay and MacRobertson Shelf, East Antarctica. *Tectonophysics* 114: 315-340.

Taylor T.N., E.L. Smoot & J.W. Collinson 1986. Paleoenvironment of Lower Triassic plants from the Fremouw Formation, Antarctica. *Antarctic Journal of the United States* 21: 26-27.

Tingey, R.J. 1982. The geological evolution of the Prince Charles Mountains - an Antarctic Archean cratonic block. *In* Craddock, C. (ed.), *Antarctic geoscience*, 455-464. Madison: University of Wisconsin Press.

Townrow, J.A. 1967. Fossil plants from Allan and Carapace Nunataks, and from the upper Mill and Shackleton Glaciers, Antarctica. *New Zealand Journal of Geology and Geophysics* 10: 456-473.

Traube, V.V. 1991. Sedimentology and paleoenvironments of the Beacon strata at Beaver Lake, Prince Charles Mountains. *Abstracts, 6th International Symposium on Antarctic Earth Sciences*: 611-614.

Turner, B.R. 1991. Depositional environment and petrography of preglacial continental sediments from Hole 740A, Prydz Bay, East Antarctica. *In* Barron, J., B. Larsen, et al. (eds.), *Proceedings of the Ocean Drilling Program, Scientific Results* 119: 45-56.

Turner, B.R. & D. Padley 1991. Lower Cretaceous coal-bearing sediments from Prydz Bay, East Antarctica. *In* Barron, J., B. Larsen, et al. (eds.), *Proceedings of the Ocean Drilling Program, Scientific Results* 119: 57-60.

Veevers, J.J. 1991. Gondwana facies of the Pangaean Supersequence. *Abstracts, 8th International Symposium on Gondwana*: 87-88.

White, M.E. 1973. Permian flora from the Beaver Lake area, Prince Charles Mountains, Antarctica. 2. Plant fossils. *Bureau of Mineral Resources, Geology and Geophysics, Australia, Bulletin* 126: 13-18.

Gondwana Eight, Findlay, Unrug, Banks & Veevers (eds) © 1993 Balkema, Rotterdam. ISBN 90 5410 304 3

Time control in the Permian of the Sydney-Bowen Basin and the New England Orogen

David J.C. Briggs
School of Earth Sciences, University of Melbourne, Parkville, Vic., Australia

ABSTRACT: Permian tectonic events in the Sydney - Bowen Basin and New England Orogen are discussed based on a compilation of time-control data. Extensional features formed throughout most of the region in the Sakmarian, contemporaneously with island-arc volcanism preserved in the Gympie terrane, and possible continental-arc volcanism in the northern Bowen Basin. Volcanism in the Gympie arc was apparently terminated by a late Sakmarian to early Artinskian landward shift of the subduction zone, and may have initiated following an earliest Permian seaward shift of the zone. Probable continental-arc volcanism and plutonism resumed in the New England Orogen in the mid-Artinskian, and persisted almost to the end of the Permian, interrupted by a late Artinskian to early Kungurian deformation. In the late Late Permian, the extinct Gympie arc, mantled by volcaniclastic sediment and overlying slowly-deposited limestone, collided with the remainder of the Orogen, contemporaneously with widespread compressive deformation in the latter.

The Permian stratigraphy of the Sydney - Bowen Basin and the New England Orogen is remarkably complex, reflecting an intricate and continually changing palaeogeography that in turn reflects a history of varied tectonic, eustatic and climatic events. Even considering only units containing marine macrofaunas, some 153 named and unnamed lithostratigraphic units of formational rank are recognized in the region, compared with about 67 such units in the remainder of the Permian of Australia. This exceptional stratigraphic complexity stems from the location of the region on a rapidly evolving active continental margin throughout the Permian.

This paper is based on a compilation of data relevant to time control on this stratigraphy, including new data derived from taxonomic work on marine macrofossils by the author (Briggs, in prep.). As a framework for this compilation, the region was divided into 16 latitudinal bands, each corresponding to one series of laterally adjoining 1:250,000 sheet areas, and each of these bands was subdivided into between three and 18 structural domains. For each of these domains, Permian stratigraphic units, igneous intrusions, and tectonic events were listed, and all available data relevant to their age were compiled. These data comprise biostratigraphic and lithostratigraphic correlations, and radiometric dating.

Biostratigraphic correlations in the region presently rely mainly on marine invertebrate macrofossils (mostly brachiopods and molluscs) and palynomorphs. Marine macrofossils from the region have been studied in recent decades by workers including Dickins, Runnegar, Dear, Armstrong, McClung, Waterhouse and Parfrey, as well as by the present writer, whose work has concentrated on brachiopods of the suborders Strophalosiidina and Productidina. (Order Productida). Brachiopods of these suborders have particular biostratigraphic value because individual species are known to have wide geographical distributions, reasonably wide substrate tolerances, and short time ranges (Briggs, 1991). Palynomorphs have been studied by a number of workers, and a widely used scheme of numbered interval zones (APP1-6), based on successive incomings of selected index species, has been developed (Price et al., 1985; see also Draper et al., 1990). Marine microfossils have received less

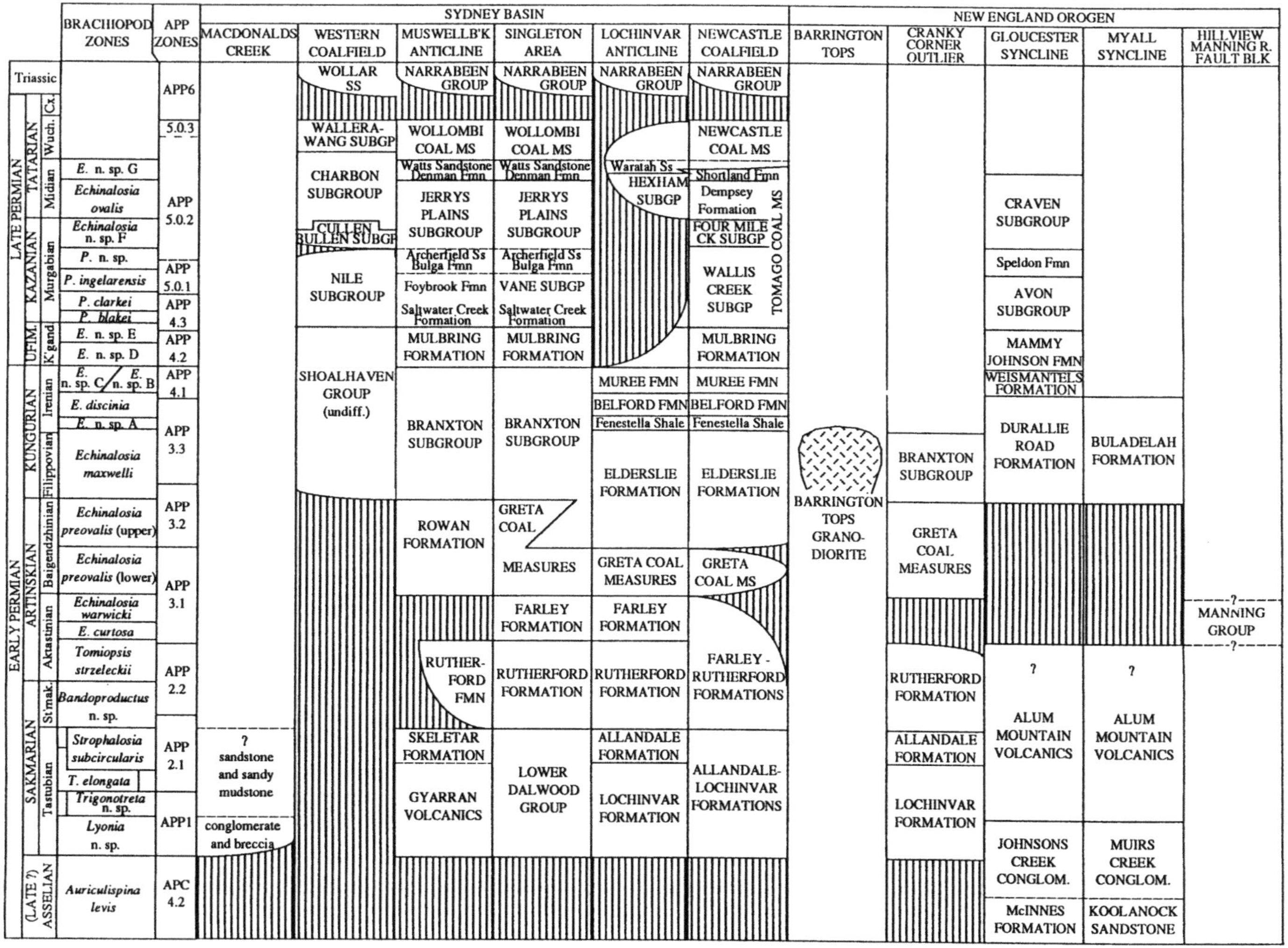

Figure 1. 32°00'S - 33°00'S: Permian units of the Dubbo, Singleton and Newcastle 1:250,000 Sheet areas. Intra-Permian unconformities vertically shaded.

attention so far, although the foraminifera have considerable biostratigraphic potential, and a zonal scheme has been outlined by Palmieri (1990). Plant macrofossils are relatively poorly known, and at present have only broad-scale biostratigraphic value.

Lithostratigraphic evidence has been widely used for intrabasinal correlations, especially in Late Permian non-marine successions where biostratigraphic resolution is poor. Examples include correlation of individual tuff beds, intensely tuffaceous intervals, coal seams, and transgressive intervals. Correlations based on geophysical data are of increasing importance, particularly in the southwestern Bowen Basin.

Radiometric dates in the region are mostly too imprecise to be useful for detailed stratigraphic correlations. Nevertheless they provide important (though at present often equivocal) evidence for relating stratigraphic successions to dated igneous intrusions and metamorphic events in adjoining orogenic areas.

Based on these biostratigraphic, lithostratigraphic and radiometric data, correlation charts have been prepared for each of the 16 latitudinal bands covering the region. Figures 1 - 3 show three of these charts, for the bands between latitudes of 32°00'S and 33°00'S, 29°00'S and 30°00'S, and 25°00'S and 26°00'S respectively. The chronological framework used is modified from the scheme presented by Briggs (1989), which related the macrofaunal and palynological zones of eastern Australia to a succession of depositional sequences found to be correlateable both locally within the Sydney - Bowen Basin, and internationally (based largely on ammonoid evidence from correlative depositional sequences recognized by the author in Western Australian successions) with the global sequence stratigraphic curve of Ross and Ross (1987). The full set of 16 correlation charts and a detailed discussion of the basis for these correlations have been submitted for publication elsewhere (Briggs, in prep.).

It must be stressed that these 16 correlation charts represent only an interpretation of the available data, and that evidence for the precise correlation of many units remains equivocal. Nevertheless, the development of a detailed, unified time framework, applicable to marine strata, non-marine strata and radiometrically-dated intrusions and tectonic events, has been of value in suggesting new hypotheses concerning the tectonic evolution of the region as a whole, as will be set out below.

REGIONAL STRUCTURE

Historically the boundary between the Sydney - Bowen Basin and the New England Orogen corresponds to a marked change in the mode of preservation of the Permian System, from an essentially continuous, westward-thinning wedge in the Sydney - Bowen Basin, to an array of isolated synclines, fault blocks and outliers in the New England Orogen. This fragmentation characterizing the Orogen is the result of mid-Permian, Late Permian and post-Permian deformation and plutonic intrusion. Along most of its length the western boundary of the New England Orogen lies along or just east of a series of major east over west thrust structures, the Burunga, Moonie - Goondiwindi, and Mooki - Hunter faults, whose course coincides with a fundamental crustal boundary, the Burunga - Mooki geosuture (Finlayson et al., 1990). Permian strata have been removed from large parts of the western New England Orogen by erosion, but where they are preserved they generally resemble, and almost certainly were originally continuous with, strata preserved in adjacent parts of the Sydney - Bowen Basin. This original continuity constrains the age of onset of major thrust movement along the Burunga - Mooki geosuture in at least some areas to no earlier than late Late Permian.

Excluding the Gympie terrane (discussed below), the New England Orogen may be divided into a western and an eastern part, according to the kind of sub-Permian basement present. In the western part, here referred to as the Yarrol - Tamworth belt, Permian successions rest unconformably to apparently conformably on Devonian to (mostly) Carboniferous arc and forearc sequences. In the eastern part, here referred to as the Wandilla - Tablelands belt, basement consists mostly of deformed Devonian to Carboniferous accretionary prism rocks, apart from three blocks (the Hastings Block, Emu Creek Block, and Alice Creek inlier) believed by many workers (e.g. Murray et al., 1987; Korsch and Harrington, 1987) to be tectonically emplaced forearc material derived from the Yarrol - Tamworth belt. Where exposed in the northern and southern New England Orogen, the boundary between the two belts corresponds to the Yarrol and Peel-Manning fault systems respectively.

The Gympie terrane, on the eastern margin of the New England Orogen, is widely regarded

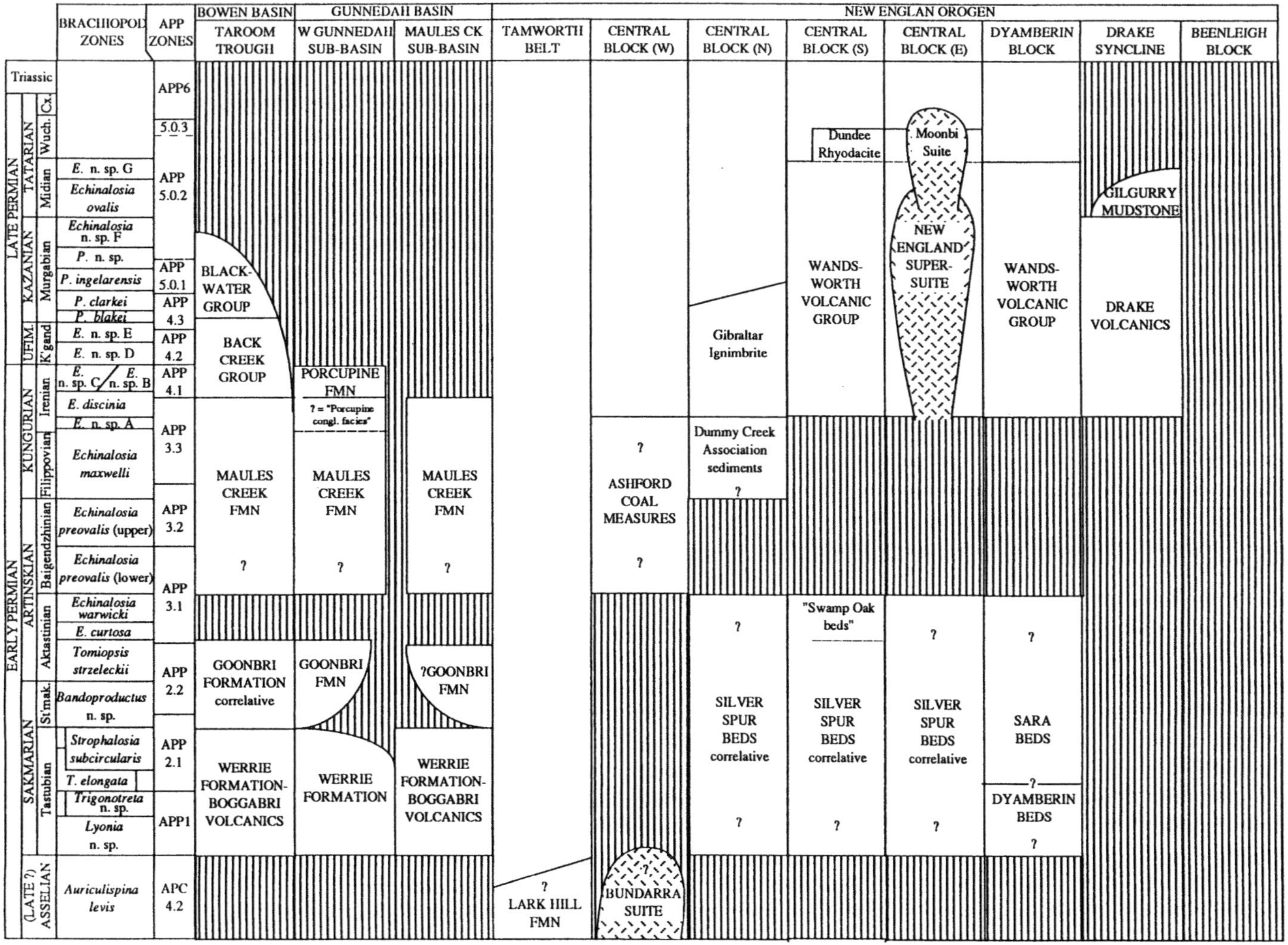

Figure 2. 29°00'S - 30°00'S: Permian units of the Moree, Inverell, Grafton, and Maclean 1:250,000 Sheet areas. Intra-Permian unconformities vertically shaded.

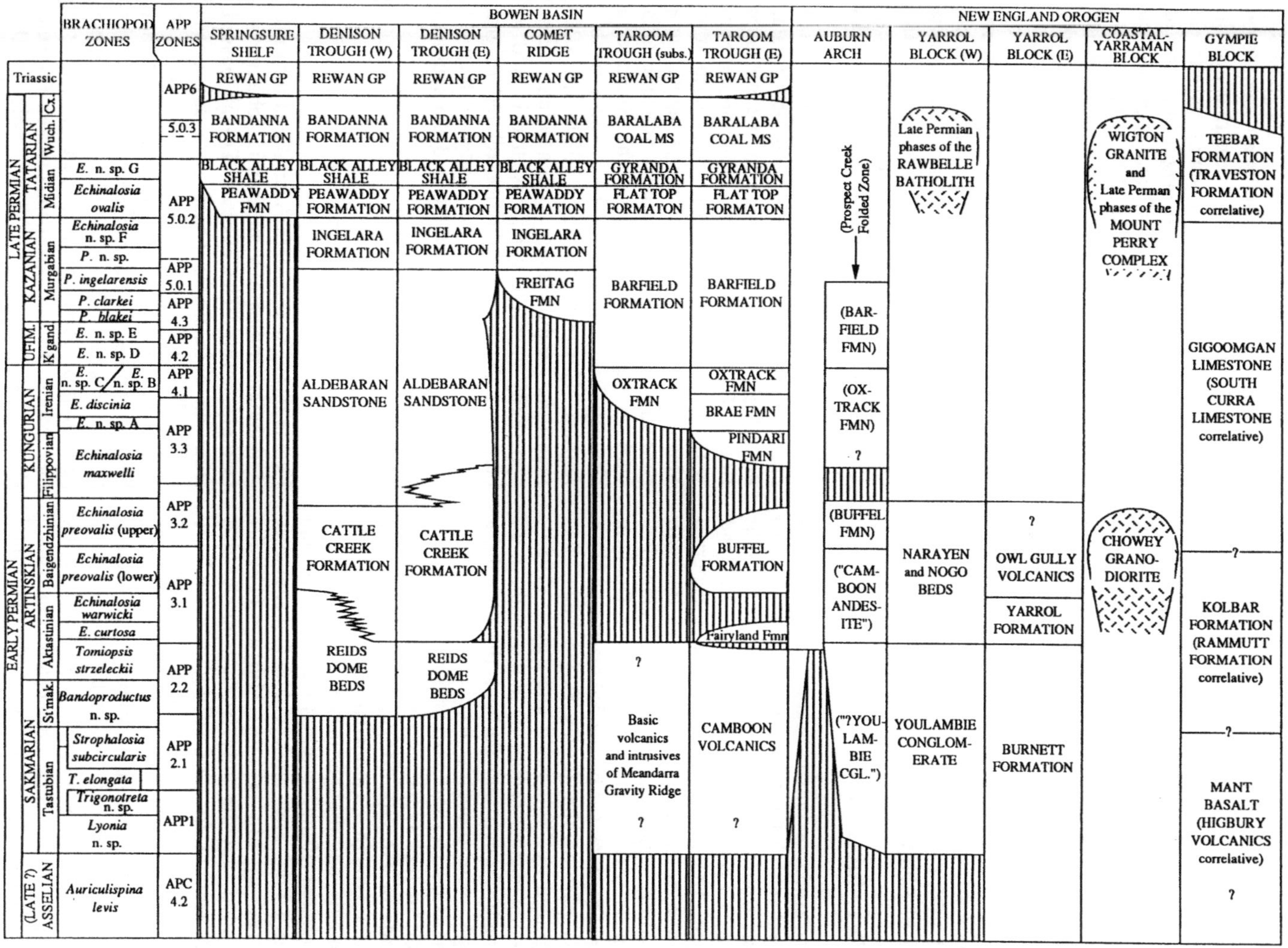

Figure 3. 25°00'S - 26°00'S: Permian units of the Eddystone, Taroom, Mundubbera and Maryborough Sheet areas. Bracketed units occur in adjacent sheet areas. Intra-Permian unconformities vertically shaded.

as allochthonous to the remainder of the Orogen, following Harrington (1983) (see also Murray, 1990a). Rocks of the Gympie terrane record a Permian history utterly unlike that of the rest of the Orogen; nevertheless there are relationships of timing that appear to have tectonic significance, and are perhaps the most intriguing results of this study. The Gympie terrane will be considered separately here following discussion of the rest of the region.

SYDNEY - BOWEN BASIN, YARROL - TAMWORTH BELT AND WANDILLA - TABLELANDS BELT.

Asselian - Sakmarian (early Early Permian)

Early Permian events indicating an extensional tectonic setting have been reported by various authors from widely separated parts of both the Sydney - Bowen Basin and the New England Orogen. These events seem to have occurred within a relatively confined time interval, coinciding approximately with the Sakmarian. An initial extensional phase in the Denison Trough, represented by the Sakmarian (APP2.1 - 2.2) part of the Reid's Dome beds, was recognized by Ziolkowski and Taylor (1985). Contemporaneous sediments to the south, in the Arbroath Trough and perhaps the Gilgandra Trough, were probably deposited in a similar setting. However the most striking extensional feature in the Sydney - Bowen Basin is the Meandarra Gravity Ridge, lying along the Taroom Trough - Gunnedah Basin - Sydney Basin axis, and thought to have been produced by incipient rifting, leading to the emplacement of intrusions and thick volcanic sequences (Murray et al., 1989). Murray (1990b) concluded that the only tectonic model that could account for this rifting was back-arc extension in response to a subduction-related volcanic arc to the east. A series of Early Permian volcanic units is known from outcrop and drilling in the vicinity of the course of the Meandarra Gravity Ridge, and as suggested by Murray (1990b) it seems reasonable to attribute these units to the same phase of extensional volcanism. These volcanic units include flows in the Lochinvar and Allandale Formations and the Gyarran Volcanics of the Sydney Basin, the Werrie Formation and Boggabri Volcanics of the Gunnedah Basin, and (although not included by Murray, 1990b) the Camboon Volcanics of the southeastern Bowen Basin. Although mostly rather broadly dated, these units all seem to be of roughly similar, mostly Sakmarian age, in places possibly ranging as young as earliest Artinskian. An extensional origin for some of these units has previously been suggested on geochemical grounds (Flood et al., 1988).

An extensional origin has also been suggested for various Early Permian sequences in the southern New England Orogen, including rapidly-accumulated sediments in the Manning Block, the Nambucca Block, the northern Central Block, and along the Peel Fault (Leitch, 1988). Sakmarian marine faunas are known from these units in the Manning Block, the western Nambucca Block, and the northern Central Block, and from several of the blocks along the Peel Fault.

Based on their position beneath Artinskian marine units, and on limited radiometric data, the Lizzie Creek Volcanics in the northern Bowen Basin seem roughly contemporaneous with the volcanics associated with the Meandarra Gravity Ridge, perhaps suggesting a continuation of the extensional volcanic axis well to the north of where the Meandarra Gravity Ridge ceases to be a recognizable feature. However Murray (1990b, p. 201) mentions unpublished geochemical data, apparently on these volcanics, that is said to clearly indicate a subduction-related origin. This would imply a major change in tectonic setting between the northern and the southern Bowen Basin at this time.

Artinskian (mid-Early Permian)

Events throughout eastern Australia in the early Artinskian indicate a substantial change in tectonic setting. The Denison Trough passed from an extensional phase to a sag phase (APP3.1 - 3.2) at this level, leading to a spreading of the area of sediment accumulation beyond the margins of the extensional troughs (Ziolkowski and Taylor, 1985). Comparable spreading of depocentres is seen elsewhere in the Sydney - Bowen Basin and the New England Orogen at this level, for example in the northern Hastings Block, where the Artinskian Yessabah Limestone in places is disconformable on the Carboniferous (Lennox and Roberts, 1988).

Widespread Early Permian volcanics in the New England Orogen to the east of the Bowen

Basin have been confused in the past with the Sakmarian volcanics of the Sydney - Bowen Basin, but marine faunas of the *Echinalosia warwicki* and *E. preovalis* Zones found within and below these units indicate that they are of entirely younger, mid- to late Artinskian age. These Artinskian volcanics include the "Camboon Andesite" in the Prospect Creek Folded Zone and the Gogango Overfolded Zone, the "Lizzie Creek Volcanics" and volcanics in the Carmila beds in the Connors Arch region, the Owl Gully Volcanics in the Yarrol Block, and the unnamed Early Permian volcanics in the Drake Syncline and Emu Creek Block. Day et al. (1983) grouped all of these units, along with the Camboon Volcanics of the southeastern Bowen Basin, as the "Camboon Volcanic Arc". The replacement name "Owl Gully Volcanic Arc" is suggested here because the Camboon Volcanics in their type area belong instead to the older (Sakmarian), apparently extensional volcanic phase of the Sydney - Bowen Basin. The Cedarton Volcanics and the Permian volcanics within the Amamoor beds (Gympie Block) may also belong with this group, as suggested by Murray et al. (1990a), but marine faunas from these units are not precisely dated at present.

The tectonic setting of these Artinskian volcanics units needs to be established more rigourously. The subduction-related arc origin implied by Day et al. (1983) is favoured, but at present supporting geochemical evidence is available only for the doubtfully included Cedarton and Amamoor volcanics (Murray 1990a). If the Artinskian volcanic units as a whole are correctly regarded as representing a continental-margin arc, they indicate resumption of plate convergence along most of the eastern margin of the New England Orogen by the mid-Artinskian.

Late Artinskian - early Kungurian deformation

Indications of a "mid-Permian" deformation in the southern New England Orogen have been recognized by various authors (e.g. Leitch, 1974) based on the contrast between deformed Early Permian strata and relatively undeformed Kungurian and younger volcanic successions now included in the Wandsworth Volcanic Group of Barnes et al. (1991). The "mid Permian" deformation can now be dated as late Artinskian to early Kungurian based on the presence of faunas as young as Artinskian in the Silver Spur beds (*Echinalosia warwicki* Zone) and the Early Permian sediments in the Drake Syncline (lower *Echinalosia preovalis* Zone), and faunas as old as late Kungurian in sedimentary interbeds in the lower part of the Wandsworth Volcanic Group (see below). Roughly correlative breaks are present between the Moah Creek beds and the underlying Rookwood Volcanics, and between the Boomer Formation and the underlying Rannes beds, in the northern New England Orogen, and between the Pindari Formation and the underlying Buffel Formation in the Bowen Basin. In the northern Denison Trough Ziolkowski and Taylor (1985) recognized a change from a sag phase to a compressional phase at about this level.

Korsch and Harrington (1987) noted that bedding and faulted margins of many outliers of the Silver Spur beds appear to have been deformed parallel to the curve of the Texas Orocline, and inferred that the deformation of the Silver Spur beds occurred at the same time as the formation of the Orocline. If accepted, their argument implies that formation of the Orocline, and hence emplacement of the Emu Creek Block into the Wandilla-Tablelands belt, must be attributed to the "mid-Permian" deformation. The "mid-Permian" deformation would therefore represent a period of large scale horizontal tectonic movements within the New England Orogen, which seem difficult to satisfactorily explain at present.

Kungurian (late Early Permian) to Late Permian

In the southern New England Orogen, resumption of extensive calc-alkaline volcanism of the Wandsworth Volcanic Group in the early late Kungurian (*Echinalosia discinia* Zone) is dated by marine faunas low in the Emmaville Volcanics near Glenmore, in the Rhyolite Range beds near Warwick, and in the Cataract River Member low in the Drake Volcanics near Drake. Younger marine faunas from the Drake Volcanics and Gilgurry Mudstone indicate that volcanism in that section spans to just above the *Echinalosia ovalis* Zone. The Dundee Rhyodacite and associated units are entirely terrestrial, and evidently post-date these marine-influenced units. Published radiometric dates on the Dundee Rhyodacite and the probably co-magmatic Moonbi Suite suggest either a latest Permian or earliest Triassic age. However the existence of a highly tuffaceous interval in the late Late Permian of the Sydney - Bowen Basin (see below), and the absence of a comparable interval in the Early Triassic Narrabeen Group, strongly favour a Late Permian age. The wholly terrestrial nature of this phase of the volcanism may reflect the combined effects of late Late Permian tectonic uplift and sea-level fall (see below).

Late Permian plutons of the New England Batholith are spatially associated with the Wandsworth Volcanic Group, suggesting a genetic relationship. However radiometric dates (Shaw et al., 1991) and intrusive relationships (Barnes et al., 1991) indicate that most of the exposed plutons were intruded late in the development of the volcanic pile, and only the youngest volcanic units (the Dundee Rhyodacite, Tent Hill Volcanics and parts of the Coombadjha Volcanic Complex) show clear geochemical links with the exposed plutonic rocks (Barnes et al., 1991). Presumably the plutons associated with earlier phases of the volcanism either did not intrude as high as present-day surface levels, or perhaps were largely "cannibalized" by younger intrusons.

In the northern New England Orogen, thick volcanic successions exposed on either side of the Esk Rift (the Cressbrook Creek Group and equivalents) are poorly dated, but presumably include equivalents of the Wandsworth Volcanic Group, and perhaps older Permian volcanic units as well. Further north, Late Permian plutons, including the Permian phases of the Rawbelle Batholith and Mount Perry Complex, the Wigton Granite, etc. indicate a northern continuation of the same Late Permian igneous province, although here any co-magmatic volcanics have been largely or entirely removed by erosion.

During this time interval thick sedimentary successions accumulated in the Sydney - Bowen Basin. Interbedded tuffs first appear in the Sydney Basin within the *E. discinia* Zone, in the upper Wandrawandian Formation, closely matching the age of the oldest volcanics of the Wandsworth Volcanic Group. Throughout the Sydney - Bowen Basin tuffs progressively increase in abundance upwards to the level of the *Echinalosia ovalis* Zone, above which there occurs throughout the basin an intensely tuffaceous interval, followed in places by a thin interval of non-tuffaceous sediments to the top of the Permian. The intensely tuffaceous interval includes units such as the Newcastle Coal Measures and equivalents in the Sydney Basin, the "tuffaceous stony coal facies" in the Black Jack Formation of the Gunnedah Basin, and the Kaloola Member and correlatives in the Bowen Basin. These tuffs are closely contemporaneous with volcanics of the Wandsworth Volcanic Group, with the earlier interval of scattered tuffs being directly correlated with the older part of the Group (Drake and Emmaville Volcanics), and the younger, intensely tuffaceous interval apparently correlating with the Dundee Rhyodacite and associated units.

The tectonic setting of the magmatism responsible for the New England Batholith/ Wandsworth Volcanic Group has received little attention to date, but a continental-margin volcanic arc (as suggested by Weber and Scheibner, 1977) seems most likely. Evidence discussed below in connection with the Gympie terrane independently supports the existence of a convergent plate margin to the east of the New England Orogen during the Late Permian.

Murray (1990b) assigned the Bowen Basin during this stage of its development to a (post-extensional) sag phase, passing in the Late Permian and Early Triassic into a fold-thrust belt foreland basin phase with the onset of compressive deformation in the New England Orogen. A reasonable alternative might be to regard the basin in the Kungurian and especially the Kazanian as a retroarc foreland basin, related to loading of the adjacent orogen by the Late Permian volcanic pile.

Late Permian deformation

A major Late Permian deformation in the New England Orogen is indicated by a concentration of radiometric dates of around 250-260 Ma on metamorphic units throughout the Orogen, as well as by effects on contemporaneous successions in the Sydney - Bowen Basin. While it may be argued that some of the 250-260 Ma radiometric dates might record uplift and cooling of substantially older metamorphics, rather than true metamorphic ages, in either case they indicate a major tectonic event. A date of 255 Ma on amphibolite-facies metamorphics associated with the Yarrol Fault system dates large-scale westward thrusting of ultramafics over the Yarrol Block (Murray, 1983). Dates obtained by Leitch and McDougall (1979) on metasediments in the Nambucca Block were taken to indicate a deformation around 250-255 Ma, at the time regarded as mid-Permian. Dates averaging 260 Ma have been obtained on muscovite and glaucophane from the Rocksberg Greenstone (Murray, 1987).

Roberts and Engel (1987) attributed the Late Permian regression in the Sydney Basin, represented by the Singleton Supergroup and equivalents, to uplift associated with contemporaneous deformation in the New England Orogen, and Fielding et al. (1990) emphasized the resurgence of volcanism on the eastern basin margin and the onset of compressive tectonics as the major influences on the Late Permian regression in the Bowen Basin. However the Late Permian regression does not necessarily reliably date deformation in the New England Orogen, as its timing appears to be strongly influenced by eustatic sea-level fall. Correlations with the global sea-level curve published by Ross and Ross (1987) indicate that the onset of Late Permian coal measure sedimentation in the Sydney Basin (Singleton Supergroup and equivalents) and the Gunnedah Basin (Black Jack Formation) can be correlated with a marked eustatic drop in the global curve at the top of the Roadian. Similarly, the termination of marine conditions in the southwestern Bowen Basin, and the termination of distinct marine influence in coal measures of the Sydney Basin (near the base of the Newcastle Coal Measures and correlatives), seem to correlate with a further sharp fall in the global curve at the top of the Capitanian.

A better indication of the timing of deformation is provided by the influx into the basin of coarse detritus derived from the orogen. Such an influx of detritus (including Woolomin-type chert clasts derived from the Wandilla - Tablelands belt) occurs in the northern Sydney Basin in the Late Permian Newcastle Coal Measures and correlative Wollombi Coal Measures, and in the Gunnedah Basin in the correlative interval within the Black Jack Formation. Alluvial fan conglomerates occur on the eastern margin of the Bowen Basin in the Early Triassic Rewan Formation (Fielding, 1990), and alluvial fan deposits are also preserved locally within the underlying Late Permian Baralaba Coal Measures (Fielding et al. 1990). Further east in the Rockhampton area, the Dinner Creek Conglomerate records an influx of coarse detritus of poorly constrained Late Permian age.

At a somewhat lower level, an angular break below the Hexham Subgroup of the Tomago Coal Measures may provide the first indication of the onset of deformation in the adjacent orogen, at a level roughly equivalent to the top of the *Echinalosia* n. sp. F. Zone.

GYMPIE TERRANE

The expression "Gympie terrane" as used here includes only the coherent, Permian succession of the Gympie Group, comprising the Highbury Volcanics, Rammutt Formation, South Curra Limestone and Tamaree Formation in the Gympie area, and the Mant Basalt, Kolbar Formation, Gigoomgan Limestone and Teebar Formation in the Biggenden area. The Gympie terrane in this sense comprises only part of the Gympie Block or Province of Day et al. (1978), which also includes Carboniferous subduction-complex rocks, deformed Permian rocks lacking a coherent stratigraphy, and younger, Triassic sediments and metasediments.

The Highbury Volcanics and Mant Basalt form the oldest exposed parts of the Gympie terrane. Harrington (1983), in suggesting the possibly allochthonous origin of the Gympie succession, compared these units with volcanics of the Brook Street terrane of New Zealand. Subsequent studies have added support to the postulated allochthonous origin of the Gympie succession, perhaps most significantly by demonstrating that some of the basal volcanics have geochemical and lithological attributes suggestive of an immature intra-oceanic island-arc environment (Murray, 1990a). What has not been confirmed however is the suggested correlation with the Brook Street terrane. While volcanics of the Brook Street terrane display lithological and geochemical similarities to the Highbury Volcanics, these only indicate formation in a similar tectonic environment; biostratigraphic evidence demonstrates that the Brook Street volcanics are largely, and probably entirely, younger than the volcanics of the Gympie terrane. The age of the Brook Street volcanics is best established from the succession in the Takitimu Hills of Southland; here marine fossils from levels scattered throughout the upper two-thirds of the succession show closest links with the late Artinskian upper *Echinalosia preovalis* Zone of eastern Australia, apart from a small fauna from the Caravan Formation at the top of the succession that is apparently early Kungurian (Briggs and Campbell, this volume). In contrast a pre-Artinskian (Sakmarian and possibly older) age for the volcanics low in the Gympie succession is indicated by early Artinskian marine faunas (*Tomiopsis strzeleckii* Zone and *Echinalosia warwicki* Zone) in the upper part of the Rammutt Formation.

With very few exceptions, the Permian faunas described from the Gympie terrane (mostly from the Rammutt Formation and South Curra Limestone levels) are shared with the rest of eastern Australia, and this faunal link implies reasonably close geographical proximity. It may therefore be significant that cessation of volcanic activity on the Gympie arc closely coincided in time with the termination of extensional volcanism in eastern Australia, and shortly preceded the resumption of postulated continental-arc volcanism along the Owl Gully volcanic arc in the New England Orogen. Unless this time relationship is coincidental, it suggests that the two arc systems are genetically related. Specifically, it suggests:

1. that the shift of arc volcanism from the Gympie island arc to the Owl Gully volcanic arc may have been caused by a continent-ward shift in the subduction zone in the late Sakmarian or early Artinskian, from an oceanic setting seaward of the Gympie arc to a continental-margin setting seaward of the New England Orogen.
2. that the back-arc spreading proposed by Murray (1990b) as the cause of incipient rifting along the Meandarra Gravity Ridge may have occurred in response to the Gympie island arc.

If this second conclusion is correct, the initiation of the Gympie arc would probably have shortly preceded the Sakmarian extensional phase in eastern Australia, and so would probably have occurred in the earliest Permian. Initiation of the island arc may therefore have closely followed the termination of Carboniferous continental-arc volcanism in the New England Orogen, dated at about 300 Ma by Murray et al. (1987). This again suggests a simple shift in the location of subduction, this time away from the continental margin to an oceanic setting. However this suggestion is more speculative, because the time of origin of the Gympie arc is not constrained by palaeontological or other direct means.

Apart from dacitic volcanics, representing the waning phases of arc volcanism, the Rammutt (=Kolbar) Formation consists largely of volcaniclastic conglomerates, sandstones and mudstones (Murphy et al., 1976), presumably locally derived by erosion of the arc. The transition from the terrigenous Rammutt/ Kolbar Formations to the biogenically-derived South Curra/Gigoomgan Limestones therefore indicates cessation of the arc as a clastic source, and effectively dates the submergence of the arc in the area. This transition is dated as late Artinskian by the presence of a fauna of the upper *Echinalosia preovalis* Zone in the basal South Curra Limestone.

Brachiopod faunas from the top of the South Curra and Gigoomgan Limestones are substantially younger than the basal faunas, and belong to the Kazanian *Echinalosia* n. sp. F Zone (Briggs, in prep.). This confirms that the limestones span a large portion of the Permian, as originally concluded by Runnegar and Ferguson (1969), though this was later questioned by Waterhouse and Balfe (1987). Accumulation of such a small thickness of limestone (about 135 m in the Gympie area) in such a prolonged interval of time may be consistent with accumulation on a submerged, extinct island arc remote from terrigenous sediment input. The absence of known marine macrofossils between the upper and lower faunas may be an effect of the extremely slow sediment accumulation rates on macrofossil preservation.

Pursuing further the idea of accumulation of the South Curra/Gigoomgan Limestones on a submerged, extinct arc, the abrupt transition to probable turbidites of the overlying Tamaree/ Teebar Formations presumably records arrival of the extinct arc at the Australian continental margin. This event is dated by the Kazanian faunas from the top of the South Curra/ Gigoomgan Limestones. The Tamaree/ Teebar Formations contain meagre Permian faunas, and are unconformably overlain by Early Triassic units comprising conglomeratic sediments at the base followed by deep-water marine sediments. This Triassic succession (Keefton +Traveston/ Brooweena Formations) is here interpreted as having accumulated over the seaward side of the accreted arc. The introduction of continentally derived clasts in the conglomeratic units at the base of the Triassic succession (Murphy et al., 1976, p. 32) greatly strengthens the case for regarding these

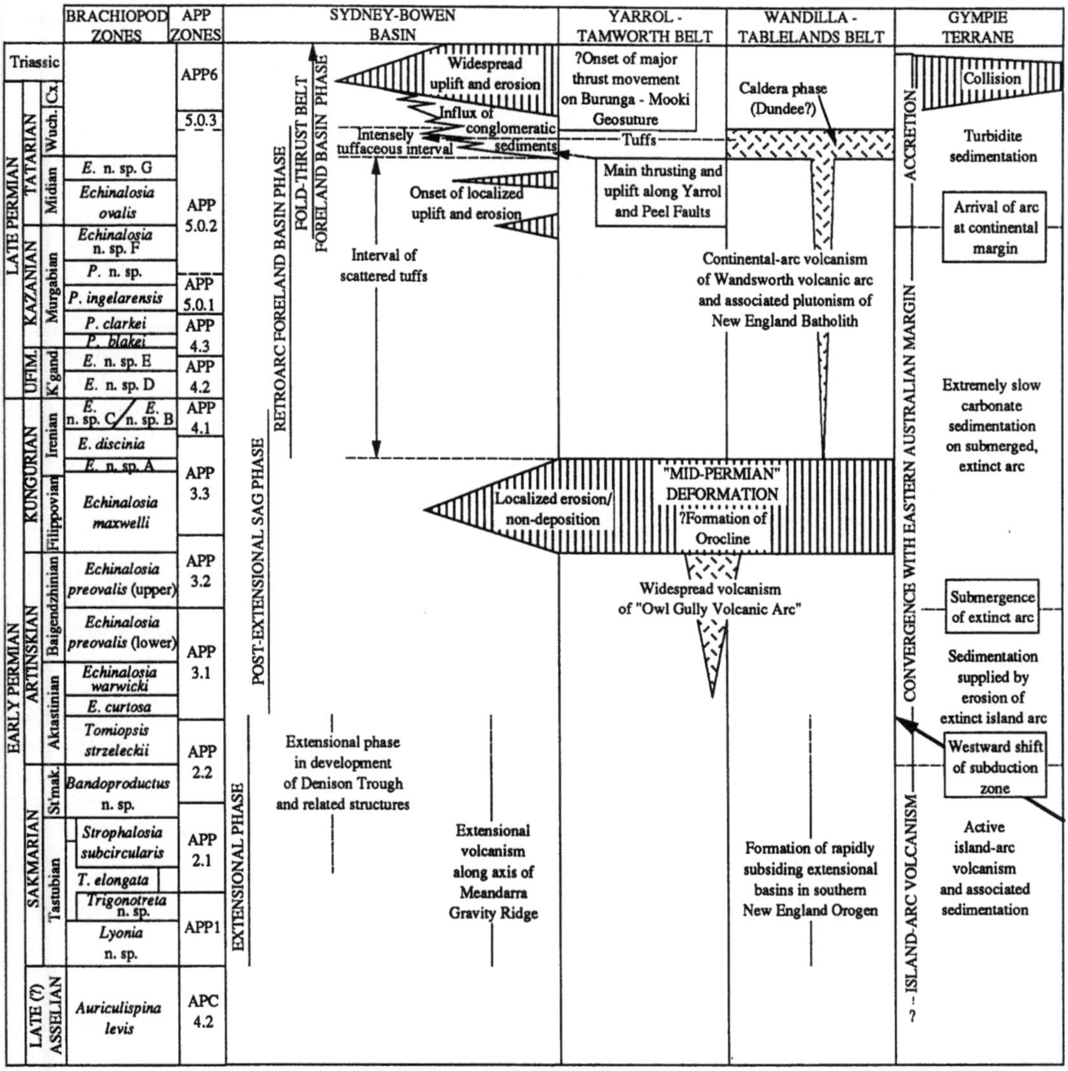

Figure 4. Summary of time relationships in the Permian of the Sydney - Bowen Basin and New England Orogen.

sediments as syn- or post-accretionary. Continued plate convergence during the Early Triassic could account for the contemporaneous deformation of parts of the Triassic sediment wedge before the intrusion of the Goomboorian Diorite at about 240 Ma, which otherwise presents a problem of timing (see Murray, 1987, p. 5).

The model presented here differs considerably from that proposed by workers such as Murray (1990a) and Harrington and Korsch (1985), who envisaged the Gympie Block, including the Early Triassic succession, as accreting to the rest of the New England Orogen in the mid-Triassic. Dating the collision as late Late Permian suggests that it may have contributed to the Late Permian deformation of the New England Orogen and thus to the effects of this deformation on the adjacent Sydney - Bowen Basin, including the conversion of the latter from either a sag or retroarc foreland basin to a fold-thrust belt foreland basin.

CONCLUSIONS

Figure 4 summarizes the main conclusions of the present study.

ACKNOWLEDGEMENTS

This paper was prepared with financial support from an Australian Research Council project with Dr. Neil Archbold (Project A38830210). Field work in New South Wales was supported by a CSIRO - University of Melbourne Collaborative Research Grant. Dr Pat Conaghan (Macquarie University) and two anonymous reviewers provided many helpful suggestions.

REFERENCES

Barnes, R.G., R.E. Brown, J.W. Brownlow, & W.J. Stroud, 1991. Late Permian volcanics in New England - the Wandsworth Volcanic Group. *Geological Survey of New South Wales, Quarterly Notes* 84.

Briggs, D.J.C., 1989. Can macro-palaeontologists, palynologists and sequence stratigraphers ever agree on eastern Australian Permian correlations? *Advances in the Study of the Sydney Basin, 23rd Newcastle Symposium, Proceedings:* 11-17.

Briggs, D.J.C., 1991. Stratigraphical distribution and biostratigraphical significance of the Permian Productidina and Strophalosiidina (Order Productida) of eastern Australia. *In* MacKinnon, D.I., D.E. Lee, & J.D. Campbell (eds), Brachiopods through time. *Proceedings of the 2nd International Brachiopod Congress, Dunedin, New Zealand:* 367-374. Balkema, Rotterdam.

Briggs, D.J.C., in prep. Permian Productidina and Strophalosiidina from the Sydney-Bowen Basin and the New England Orogen: Systematics and significance for Australasian Permian biostratigraphy. *Association of Australasian Palaeontologists, Memoir.*

Briggs, D.J.C., & H.J. Campbell, 1992. *Megousia solita* from the Permian of New Zealand: implications for terrane histories. In Findlay, R.H., Banks, M.R., Veevers, J.J., Unrug, R. (eds), *Gondwana 8: assembly, evolution and dispersal.* Balkema, Rotterdam.

Day, R.W., C.G. Murray, & W.G. Whitaker, 1978. The eastern part of the Tasman Orogenic Zone *In* Scheibner, E. (ed.), The Phanerozoic structure of Australia and variations in tectonic style. *Tectonophysics* 48: 327-364.

Day, R.W., W.G. Whitaker, C.G. Murray, I.H. Wilson, & K.G. Grimes, 1983. Queensland Geology. *Geological Survey of Queensland Publication* 383.

Draper, J.J., V. Palmieri, P.L. Price, D.J.C. Briggs, & S.M. Parfrey, 1990. A biostratigraphic framework for the Bowen Basin. *Bowen Basin Symposium 1990, Proceedings*: 26-35.Geological Society of Australia, Queensland Division, Brisbane.

Fielding, C.R., 1990. Tectonic evolution of the Bowen Basin, eastern Queensland. *Pacific Rim 90 Congress*, 2: 183-189. Australian Institute of Mining and Metallurgy, Brisbane.

Fielding, C.R., A.J. Falkner, J. Kassan, & J.J Draper, 1990. Permian and Triassic depositional systems in the Bowen Basin. *Bowen Basin Symposium 1990, Proceedings*: 21-25. Geological Society of Australia, Queensland Division, Brisbane.

Finlayson, D.M., K.D. Wake-Dyster, J.H. Leven, D.W. Johnstone, C.G. Murray, H.J. Harrington, R.J. Korsch, & P. Wellman, 1990. Seismic imaging of major tectonic features in the crust of eastern Australia. *Tectonophysics*, 173: 211-230.

Flood, R.H., S.J. Craven, D.C. Elmes, R.J. Preston, & S.E. Shaw, 1988. The Warrigundi Igneous Complex: volcanic centres for the Werrie Basalt, N.S.W. *In* Kleeman, J.D. (ed.), *New England Orogen Tectonics and Metallogenesis*: 166-171. Department of Geology and Geophysics, University of New England, Armidale.

Harrington, H.J., 1983. Correlation of the Permian and Triassic Gympie terrane of Queensland with the Brook Street and Maitai terranes of New Zealand. In *Permian Geology of Queensland*: 431-436. Geological Society of Australia, Queensland Division, Brisbane.

Harrington, H.J., & R.J. Korsch, 1985. Deformation associated with the accretion of the Gympie terrane in eastern Australia. In Leitch, E.C., (ed.), *Third Circum-Pacific Terrane Conference Extended Abstracts, Geological Society of Australia, Abstracts* 14: 104-108.

Korsch, R.J. & H.J. Harrington, 1987. Oroclinal bending, fragmentation and deformation of terranes in the New England Orogen, eastern Australia. *In* Leitch, E.C. & E. Scheibner (eds), *Terrane accretion and orogenic belts. American Geophysical Union Geodynamics Series* 19: 129 - 139.

Leitch, E.C., 1974. The geological development of the southern part of the New England Fold

Belt. *Journal of the Geological Society of Australia* 21: 133-156.

Leitch, E.C., 1988. The Barnard Basin and the Early Permian development of the southern part of the New England Fold Belt. *In* Kleeman, J.D. (ed.), *New England Orogen Tectonics and Metallogenesis*: 61-67. Department of Geology and Geophysics, University of New England, Armidale.

Leitch, E.C., & I. McDougall, 1979. The age of orogenesis in the Nambucca Slate Belt: A K-Ar study of low-grade regional metamorphic rocks. *Journal of theGeological Society of Australia,* 26: 111-119.

Lennox, P.G., & J. Roberts, 1988. The Hastings Block - a key to the tectonic development of the New England Orogen. *In* Kleeman, J.D. (ed.), *New England Orogen Tectonics and Metallogenesis*: 68-77. Department of Geology and Geophysics, University of New England, Armidale.

Murphy, P.R., H. Schwarzbock, L.C. Cranfield, I.W. Withnall, & C.G. Murray, 1976. Geology of the Gympie 1:250,000 Sheet area. *Geological Survey of Queensland Report* 96.

Murray, C.G., 1983. Permian Geology of Queensland. In *Permian Geology of Queensland*: 1-32. Geological Society of Australia, Queensland Division, Brisbane.

Murray, C.G., 1987. Review of the geology of the Gympie district, southeast Queensland, with particular reference to stratigraphic, structural and tectonic problems. *In* Murray, C.G., & J.B. Waterhouse (eds), *1987 Field Conference, Gympie District*: 1-19. Geological Society of Australia, Queensland Division, Brisbane.

Murray, C.G., 1990a. Comparison of suspect terranes of the Gympie Province with other units of the New England Fold Belt, eastern Australia. *Pacific Rim 90 Congress, Proceedings* 2: 247-255. Australasian Institute of Mining and Metallurgy, Brisbane.

Murray, C.G., 1990b. Tectonic evolution and metallogenesis of the Bowen Basin. *Bowen Basin Symposium, 1990, Proceedings*: 201-212. Geological Society of Australia, Queensland Division, Brisbane.

Murray, C.G., C.L. Ferguson, P.G. Flood, W.G. Whitaker, & R.J. Korsch, 1987. Plate tectonic model for the Carboniferous evolution of the New England Fold Belt. *Australian Journal of Earth Sciences,* 34: 213-236.

Murray, C.G., E. Scheibner, & R.N. Walker, 1989. Regional geological interpretation of a digital coloured residual Bouger gravity image of eastern Australia with a wavelength cutoff of 250 km. *Australian Journal of Earth Sciences,* 36: 423-449.

Palmieri, V., 1990. Permian foraminifera of Australia. *Geological Society of Australia, Abstracts* 25: 53-54.

Price, P.L., J. Filatoff, A.J. Williams, S.A. Pickering, & G.R. Wood, 1985. Late Paleozoic and Mesozoic palynostratigraphic units. *CSR Oil and Gas Division Palynology Facility Report 274/25. Queensland Department of Resource Industries Open File Report* CR 14012.

Roberts, J. & B.A. Engel, 1987. Depositional and tectonic history of the southern New England Orogen. *Australian Journal of Earth Sciences,* 34: 1-20.

Ross, C.A., & J.R.P. Ross, 1987. Late Palaeozoic sea levels and depositional sequences. *Cushman Foundation for Foraminiferal Research, Special Publication* 24: 151-168.

Runnegar, B.N., & J.A. Ferguson, 1969. Stratigraphy of the Permian and Lower Triassic marine sediments of the Gympie district, Queensland. *University of Queensland, Department of Geology, Papers* 7(9).

Shaw, S.E., P.J. Conaghan, & R.H. Flood, 1991. Late Permian and Triassic igneous activity in the New England batholith and contemporaneous tephra in the Sydney and Gunnedah Basins. *Advances in the Study of the Sydney Basin, 23rd Newcastle Symposium, Proceedings*: 44-51.

Waterhouse, J.B., & P.E. Balfe, 1987. Stratigraphic and faunal subdivisions of the Permian rocks at Gympie. In Murray, C.G., & J.B. Waterhouse (eds), *1987 Field Conference, Gympie District*: 20-33. Geological Society of Australia, Queensland Division, Brisbane.

Weber, C.R. & E. Scheibner, 1977. The origin of some Permo-Triassic metal zones in the New England region, New South Wales. *Geological Survey of New South Wales, Quarterly Notes* 26: 1-14.

Ziolkowski, V., & R. Taylor, 1985. The structure of the north Denison Trough. *Geological Society of Australia, Abstracts* 17: 129-135.

Gondwana Eight, Findlay, Unrug, Banks & Veevers (eds) © 1993 Balkema, Rotterdam. ISBN 90 5410 304 3

The Sakmarian-Kungurian palæogeography of Australia

Albert T. Brakel & Jennifer M. Totterdell
Bureau of Mineral Resources, Geology & Geophysics, Canberra, A.C.T., Australia

ABSTRACT: Three palæogeographic maps of Australia are presented for the Sakmarian - Kungurian epochs.

The early - middle Sakmarian interval saw the deglaciation of the continent, except for small local glaciers, accompanied by a rise in sea-level. In some basins, the sheer volume of sediment carried in more than counterbalanced the sea-level rise, and led to coastal outbuilding. The most striking feature along the eastern seaboard was a huge belt of mostly back-arc volcanism extending between Newcastle and Cairns.

By the late Sakmarian - early Artinskian, Australia was no longer affected by glaciation or associated isostatic effects, although the climate was still cold enough to provide seasonal ice-rafting. Features of the interval were a fall in sea-level, continuing volcanicity in the east, and the widespread development of coal measures.

During the late Artinskian - Kungurian, some valley glaciers re-appeared, most volcanism ceased, large lakes formed in the Cooper Basin, and there was renewed transgression. Terrane accretion and tectonism in the east ended the time slice.

INTRODUCTION

This paper presents some of the results of the BMR - APIRA Palaeogeographic Maps Project, for Australian Permian palynological Stages 3a to 4. Maps for three time slices have been compiled using the methodology of Cook (1988), and correspond in age to early - middle Sakmarian, late Sakmarian - early Artinskian, and late Artinskian - Kungurian respectively (Table 1). The time boundaries have been chosen to correspond to significant sedimentary or biological events. The lower boundary of the oldest time slice discussed is the beginning of Stage 3a, when an increase in abundance and diversity of gymnosperm pollen recorded an amelioration of the climate towards the end of the main Late Palæozoic glaciation; the upper boundary is marked by sedimentation changes in several regions. Local hiatuses and the replacement of dominantly terrestrial by dominantly marine regimes in many eastern Australian depocentres define the top of the middle time slice. The incoming of Fauna IV in much of eastern Australia, after a major tectonic episode, forms the upper limit of the youngest interval.

Other Permian time slices, which have been briefly described elsewhere, are one for the earliest Permian (Brakel & Totterdell, 1990), and three for the Late Permian (Brakel & Totterdell, 1988).

EARLY - MIDDLE SAKMARIAN

After a previous glacial maximum (Brakel & Totterdell, 1990), the Australian continent became deglaciated (Fig. 2), except for some valley glaciers in northern Victoria, that shed ice-rafted debris into the nearby sea (O'Brien, 1986), and in the Oakover Valley area on the southern Canning Basin margin (Hickman, 1978).

As sea-level rose because of the melting of the ice caps, marine transgressions occurred in the Bonaparte, Carnarvon, and northern Perth Basins in Western Australia, and in southern Victoria, Tasmania, the eastern Murray infrabasinal area (extending to the Oaklands Basin), the Temi

TABLE 1: Stratigraphic units in time slices, for selected basins

Time Slice	Epoch	Fauna	Palynological Stage	Perth Basin	Carnarvon Basin	Canning Basin	Cooper Basin	Bowen Basin	Sydney Basin	Tasmania Basin
Ł Art. - Kung.	KUNGURIAN ARTINSKIAN	III II	Stage 4	? Caryngina Fm	Coolkilya Sst Byro Group	L.Lightjack Fm ? Noonkanbah Fm	Roseneath Shale Epsilon Fm Murteree Shale	Gebbie Fm Tiverton Fm	Branxton Fm Greta CM Farley Fm	L.Malbina Fm Cascades Gp
Ł Sak.- E Art.	ARTINSKIAN	I	Stage 3b	Irwin R. CM High Cliff Sst Fossil Cliff Member	Wooramel Group Calytharra Fm	Poole Sst	Patchawarra Fm	Lizzie Creek Volcanics	Rutherford Fm	Faulkner Group
E - Ł Sak.	SAKMARIAN		Stage 3a	Holmwood Shale	Carrandibby Fm (of Lyons Gp)	upper Grant Group	upper Tirrawarra Sst		Allandale Fm upper Lochinvar Fm	Bundella Mudstone U.Woody Id Siltstone

Figure 1. Locations of places mentioned in the text.

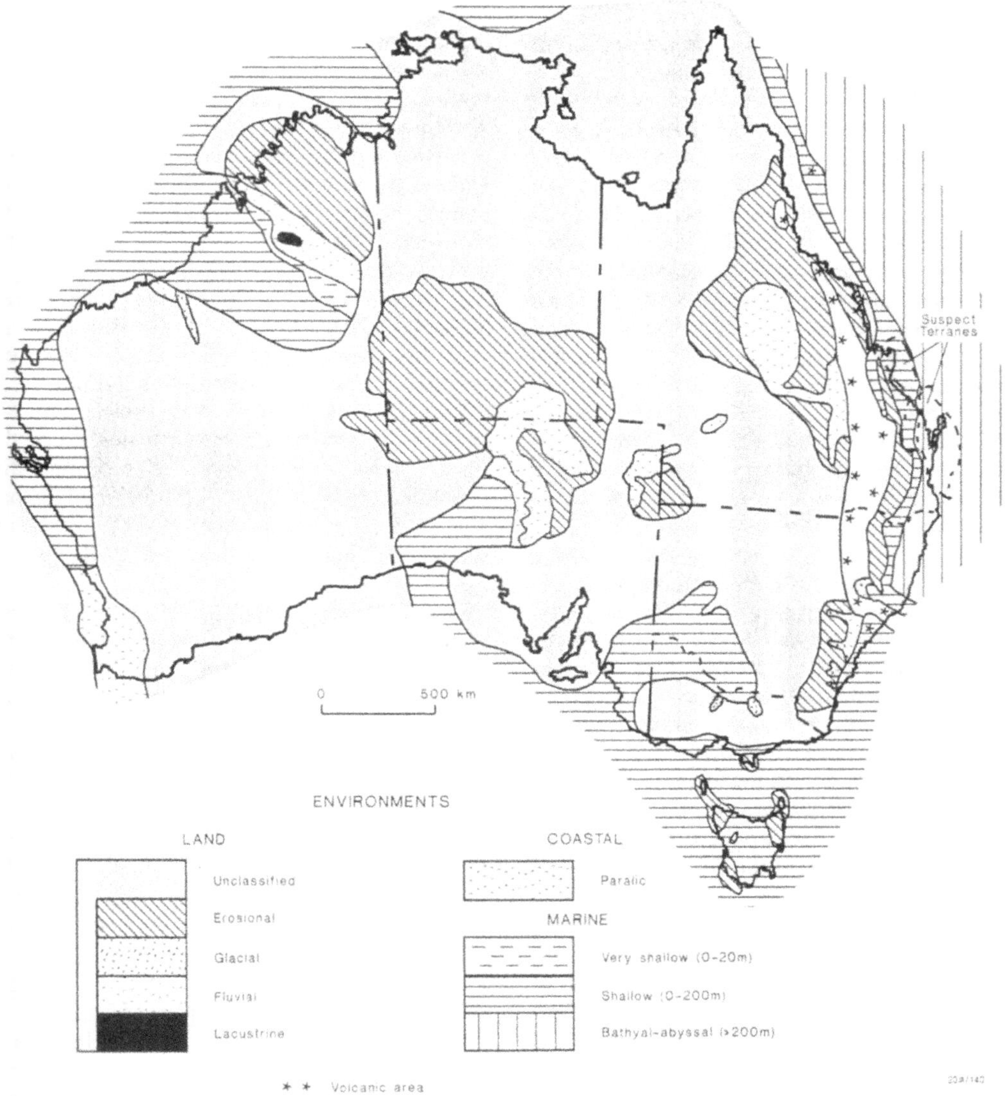

Figure 2. Palæogeography of Australia during the early - middle Sakmarian.

Basin, and to a lesser extent the Sydney Basin. But in the Canning and Arckaringa Basins the influx of sediment on their northeastern sides more than kept pace with the rising sea-level, and actually caused the shorelines to prograde. An apron of fluvial and some lacustrine and deltaic deposits along the northeastern Canning Basin (Crowe & Towner, 1976) was fed by an erosional region to the north, and similarly, a fluvial and deltaic apron along the northeastern Arckaringa Basin was supplied from highlands to the east (Townsend, 1976). In the Pedirka Basin, initially low energy, fine-grained fluvials gave way to higher energy sand-dominated fluvial aggradation by the end of the time interval (Youngs, 1976). This sedimentary influx into the central Australian basins was probably caused by recycling of glacial detritus left behind on the land surface, and by increased erosion in highlands rising by isostatic rebound after the removal of the ice load.

The southern Perth Basin was a graben between the Yilgarn Block and "Greater India", and the site of fluvial peatlands. Most sediment was derived from the eastern side, with little input from the western side (van de Graaff, 1981), but a source in adjoining Antarctica to the south may also have been significant. A sedimentary basin contiguous with the Perth Basin occupied a substantial portion of the southwestern Yilgarn Block; only remnants such as the Collie Basin now remain (Wilson, 1989). Apart from the Naturaliste Plateau and the Leeuwin Block, the land shown west of the Perth Basin was rifted away when India and Australia separated in the Early Cretaceous. The northern Perth Basin and the Carnarvon Basin comprised a cold-water marine gulf, supporting a fauna which was of low diversity compared to that of the following time slice (Thomas, 1976). Most of the Canning Basin, as well as the Bonaparte Basin, experienced shallow marine conditions. The environment prevailing at this time over the Scott Plateau off the northwestern coast is conjectural.

Seismic sections in the Arafura Basin show a substantial thickness of sediments between the sub-Jurassic unconformity and the youngest Palæozoic (palynological Stage 2) found by drilling (Bradshaw et al., 1990). This undrilled succession, which itself contains a sequence boundary, probably spans the entire Permian period, and possibly much of the Triassic as well. The seismic character of the beds is similar to shallow marine correlatives in the Bonaparte Basin (J. Bradshaw, BMR, pers. comm.).

The marine sediments at the top of the sequence in the Bacchus Marsh area of Victoria could be attributed to the higher sea-level resulting from the melting of the ice caps, before the crust had time to adjust fully by isostacy to the removal of the ice load. If so, they would record a very short span of time, of some 10 000 -20 000 years, before continuing isostatic rebound caused the flooded land to re-emerge. Early Permian marine sediments in the Duck Bay 1 well in eastern Victoria (Bowen & Thomas, 1988) may also have originated in such an event. Tasmania, where an archipelago of sizable islands had appeared, must have been similarly affected, but the larger marine areas remained as such even after rebound.

Marine transgression in the southern Sydney Basin took place across an uneven land surface, and turbulent littoral conditions denote a coast facing the open ocean. Large erratics were dropped from seasonal ice. In the north of the basin, the Allandale Formation records a sublittoral facies, locally developed around basaltic islands. No rocks of this age are preserved on the western side, and the land here may have been a sediment by-pass area.

The basaltic volcanism in the northern Sydney Basin was but the southern end of a huge bimodal volcanic belt extending as far as Cairns in northern Queensland. The belt had been in existence during the previous glacial maximum, when it reached farther south within the western Sydney Basin. It was referred to as the Ayr Volcanic Rift by Scheibner (1976), who attributed it to rifting in front of a subduction zone lying off the eastern coast, although the Lizzie Creek Volcanics and equivalents appear to represent a short-lived volcanic arc (Day et al., 1983), east and north of back-arc extension volcanics.

Near Taree, a formerly deep marine area was filled in to become shallow marine, but further north deep water turbidite sedimentation continued. In the Gympie Terrane, the Highbury Volcanics are amygdaloidal in places, suggesting shallowing, but the actual water depths are uncertain.

Carbonates were deposited within the generally clastic sequences on the shallow marine Yarrol Shelf. Parts of the nearby Gogango zone also continued as shallow marine, flanked to the east by fan deltas (Youlambie Conglomerate) derived from the adjacent eroding uplifted highland. Conditions in the Strathmuir Synclinorium and Grantleigh Trough were fluvial.

A coal-forming fluvial setting, including alluvial fan and lacustrine facies, prevailed in the SW Bowen Basin (Reids Dome beds)(Draper & Beeston, 1985). The eastern Galilee Basin was partly occupied by rivers and some lakes, but without peat development (Hawkins, 1978), and a shallow repository also began forming in the Lovelle Depression to the west (Hawkins & Harrison, 1978). In the Cooper Basin, a fluvial sequence ranged vertically and laterally from braided (Tirrawarra Sandstone) to meandering streams (Patchawarra Formation), with the latter facies containing the most coal. The lithofacies and palæogeographic maps by Thornton (1979) imply an outlet to the north, but this is uncertain; outflow may have gone either into the southern Galilee Basin, or into the Murray marine embayment. Outflow into the Arckaringa Basin is less likely in view of the highland area between the two basins, and closed internal drainage analogous to the modern Okavango Delta can be discounted because of the lack of carbonates or other evaporite minerals in the sediments.

The major erosional areas of the continent were in the Kimberley, central Australia, northern and

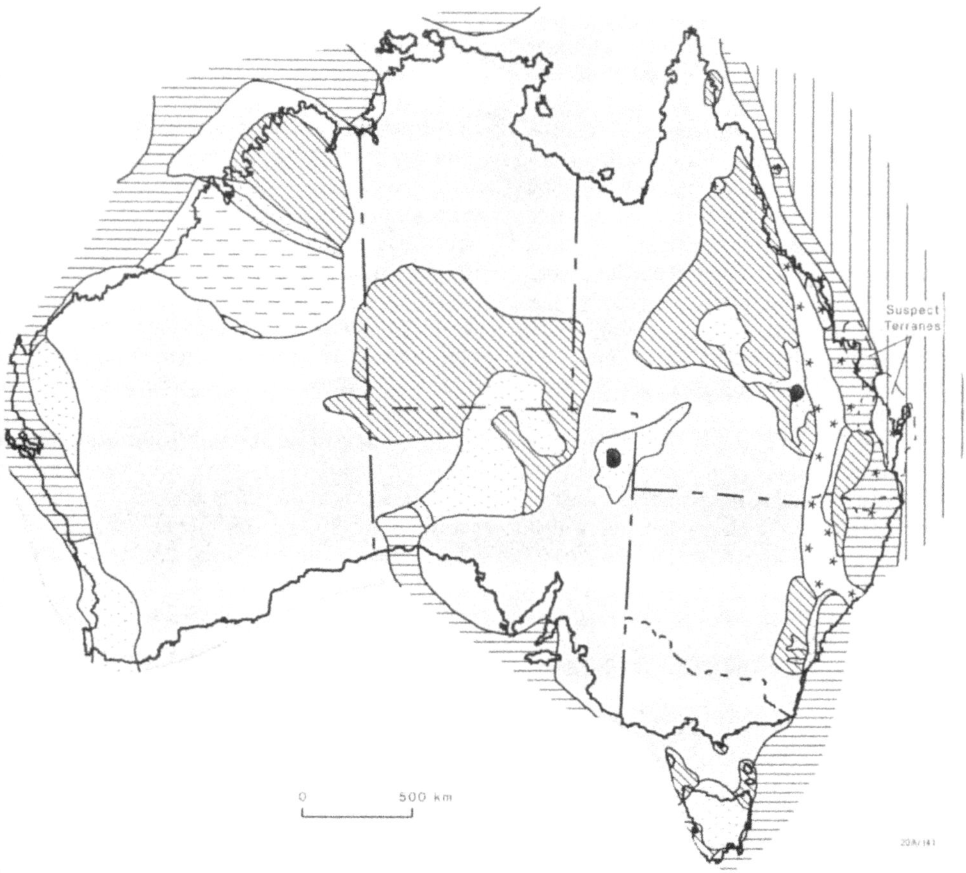

Figure 3. Palæogeography of Australia during the late Sakmarian - early Artinskian. Legend as for Figure 2.

central Queensland, western New England, and southeastern New South Wales. The deglaciated central Australian highlands, rising by isostatic rebound, became subject to water erosion, and probably remained so for the rest of the Permian. West of the Sydney Basin, probable early Permian palæokarst, pre-dating the Snapper Point Formation, has been recognized (Osborne & Branagan, 1985).

LATE SAKMARIAN - EARLY ARTINSKIAN

By the late Sakmarian, the continent was no longer affected by glaciation or associated isostatic effects (Fig. 3). There was a general retreat of the sea, which was particularly marked in southeastern Australia. Deposition ended in the Murray Basin and Victoria, while in Tasmania fluvial sediments, including coal measures, prograded across the state towards the southeast, except for some erosional areas, some east coast marine embayments, and a probable lagoonal complex extending northwards from Hobart. The sea was expelled from the Arckaringa Basin by the continued infilling by clastics derived from the east and from the large central Australian erosional area via the Pedirka Basin. Both basins developed coal measures in the later half of the time interval, as clastic input declined and peatlands were able to form.

Coal measures, mostly fluvial, are a feature of the time slice, as they were also deposited in the Denison Trough (Bowen Basin) and the Galilee, southern Cooper, Collie, and southern Perth Basins. The Denison Trough and Cooper Basins also contained lakes. The Nappamerri Trough of the Cooper Basin has a concentration of channel belt sandstones, because it subsided slightly faster than adjacent areas and attracted the drainage, whereas slower subsidence is indicated for the Patchawarra Trough (Wells & O'Brien, 1989). The Aramac Coal Measures collected in both depocentres of the Galilee Basin, with clastic components being fed in from neighbouring highs (Hawkins, 1978). The Sydney Basin was not coal-forming at this time, except for small, sporadic, uneconomic lenses in its southern portion, which represent marshes behind the fringing barrier bars of a fan delta complex (Evans et al., 1983). Seasonal ice-rafting shows that the climate was still moderately cold. In the rest of the basin, shallow marine conditions continued; a paralic zone presumably existed along the shore line. Minor uneconomic coal also occurs in the meandering stream facies of the upper Carmila beds in the Strathmuir Synclinorium of central eastern Queensland.

Eruptions continued along of the vast eastern Australian volcanic zone, and the Werrie Basin (NSW) was filled by basalts as the belt widened eastwards. Volcanism in the New South Wales portion of the belt ended sometime during the interval, however, and a period of intense subaerial weathering ensued (Brownlow, 1980). Activity persisted longer in the Queensland section of the belt, but did not last into the next time slice. Some dominantly felsic activity (Nychum Volcanics) occurred in north Queensland (Bailey et al., 1982).

On the shallow marine Yarrol Shelf, limestone now dominated over clastic deposition, and a wide range of marine fauna flourished. In the Candlelight Embayment northwest of Rockhampton, however, submarine eruptives of spilitic pillow lavas and pyroclastics came to interfinger with mudstones. The nearby Berserker Graben was possibly initiated as an actively subsiding fault trough at the eastern edge of the shelf (Day et al., 1983); its early fill was dominated by felsic and intermediate volcanics, with accompanying marine clastics.

The marine region of eastern New England shallowed to become a carbonate-clastic province, as indicated by units such as the Yessabah Limestone and Warbro Formations (Hastings Block) (Lindsay, 1969), and the Kimbriki Formation and Cedar Party Limestone (Manning district) (Engel & Laurie, 1978). The Nambucca beds and western equivalents still suggest deep water, but age control for them is poor, and they may be older than this time slice. Water depths over the Gympie Terrane are uncertain; the Highbury Volcanics were still erupting.

The northern Perth Basin early in the interval received a richly fossiliferous, shallow marine carbonate-lutite sequence (the Fossil Cliff Member), followed by the shallow marine High Cliff Sandstone which may have been partly littoral in origin (Playford et al., 1976). The siliciclastics coarsen towards the Darling and Urella Faults, perhaps being a tectonic signature of activity on these structures during or before the emplacement of these sediments. At the close of the interval, the fluvial facies of the Irwin River Coal Measures extended across the region following a marine regression.

The eastern Carnarvon Basin experienced a series of diverse events, recording an interplay of marine and paralic lithotopes due to fluctuating shorelines. At first, low clastic influx on a wide shelf gave rise to the highly fossiliferous carbonate-lutite associations of the Carrandibby and Callytharra Formations (Lavering, 1985); the latter was deposited in warm to temperate conditions (Crespin, 1958; Thomas, 1976). After a withdrawal of the sea exposed the southeastern basin to a period of karst-forming subaerial weathering, renewed sea-level rise and transgression set the stage for the outbuilding of a large delta complex (the Wooramel Group). This delta was initially sand-rich, but after inundation by further sea-level rise, a mud-prone deltaic progradation was established, only to be drowned in turn as marine shelf sedimentation was finally restored. The Wooramel delta complex had more than twice the area of the Mississippi delta according to Lavering's reconstruction, and must have been fed by a comparably large drainage system. The catchment area probably extended to the central Australian highlands, and may have drained part of Antarctica as well. Offshore from the delta, shelf conditions prevailed.

The Canning Basin began the interval with an erosional hiatus, caused either by uplift (Crowe et al., 1983), or by a eustatic low stand; if the latter, eustacy may be implicated in some sedimentation changes elsewhere at this time, but evidence for this is lacking in many regions. Most of the basin was then transgressed, resulting in a widspread nearshore to very shallow marine, dominantly sandy sequence (the Poole Sandstone). Its basal Nura Nura Member in the northwest tends to be calcareous and coquinitic, however, and preserves a faunal assemblage which lived in

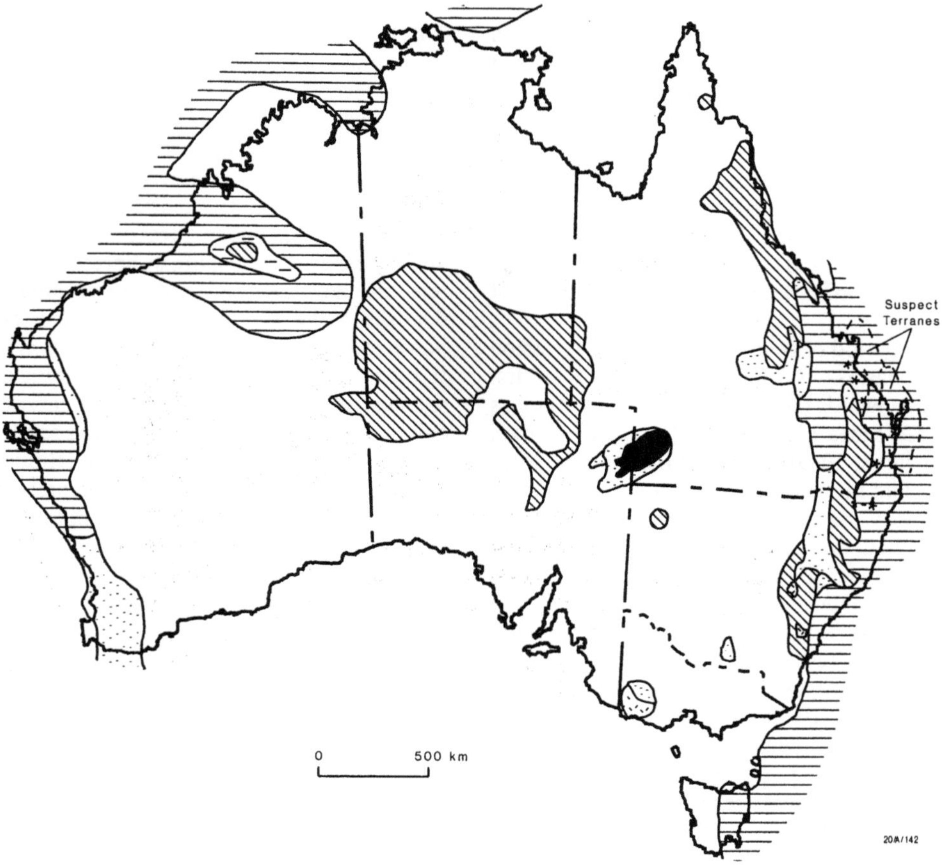

Figure 4. Palæogeography of Australia during the late Artinskian - Kungurian. Legend as for Figure 2.

warm to temperate open seas (Forman & Wales, 1981). On the northern margin a migrating barrier bar and lagoonal zone developed between the sea and an apron of coalescing fan deltas constructed at the distal end of braided fluvial systems draining the Kimberley highlands to the north.

A prograding shoreline of sandy deltas with minor coal was constructed along the eastern Bonaparte Basin (Dickins et al., 1972). Most of the basin continued to be shallow marine, receiving fine, carbonaceous, prodeltaic clastics, and some limestone (Laws & Brown, 1976). Hughes (1978) shows a basement high near Newby 1 on which no Permian was detected by seimic work; this may have formed a small erosional area throughout most of the Permian. Undrilled equivalents of the Bonaparte succession in the Arafura Basin, for reasons given earlier, were probably also shallow marine.

As this was a time of lowered sea-levels in Australia, the Scott Plateau was probably emergent.

LATE ARTINSKIAN - KUNGURIAN

A chilling of the climate brought the local return of glaciation to western Victoria and the uplands along the western margin of the Sydney Basin (Fig. 4). Evidence for this in Victoria comprises tillite and varved claystone (Bowen & Thomas, 1988), in a sequence which contains a palynological Stage 4 assemblage (Kemp et al., 1977). Alpine glaciers waxed amid rugged topography along the western Sydney Basin (Dulhunty, 1964), and these glaciers were the probable source of icebergs that rafted 4-5-tonne erratics into the marine Branxton Formation to the east, where dropstones were toppled onto a quiet sea floor in places covered with the delicate fronds of the bryozoan *Fenestella* (David, 1907).

Eruptions along the great Ayr Volcanic Rift had by now ended. The only extant activity was the andesitic submarine outpourings in the Yarrol Basin, minor submarine andesite in the Gympie Terrane, felsic to mafic volcanism in northeastern New England, and at least partly subaerial rhyolitic eruptions in the Yarraman block west of Brisbane.

It was a period of renewed marine transgression in many basins. In Tasmania, the sea at first advanced towards the north and west as two lobes during Cascades Group time (Banks & Clarke, 1987; Clarke, 1989). Later in the interval, after a hiatus had interrupted deposition in all but the easternmost parts of the basin, inundation overwhelmed the eastern half of the state, and eventually reached as far as the present west coast. The greatest subsidence and sedimentation was in a trough along the Tamar Fracture System, with sediment feeding in from source areas lying to the southwest, northwest, and northeast, at different times (Banks & Clarke, 1987). The sequences contain cold-water faunas, dropstones presumably carried in by seasonal ice, and limestones. The average sea water temperature of -1.8°C was comparable to that for the modern Antarctic shelf (Rao & Green, 1982), so similar sea-ice cover would have been present.

An apron of coalescing fan deltas, constructed south of a rising hinterland, formed a prograding coastline in the northern Sydney Basin early in the period (Branagan et al., 1976), while a similar line of alluvial fans and distal alluvial plains extended to the northwest, in the Gunnedah Basin (Hamilton & Beckett, 1984). The latter was divided into two sub-basins by the north-south Boggabri Ridge. Peat, aggrading in raised bogs on the fan surfaces, gave rise to the important seams of the Greta Coal Measures and their equivalents. Subaerial deposition was ended by a relative sea-level rise in Branxton Formation times, which also submerged the western margin of the Sydney Basin. Palynology shows that flooding was diachronous, and penetrated into the Gunnedah Basin from the south. The small Gloucester Basin north of Newcastle was also affected by this marine encroachment, and preserves a complex of nearshore marine, barrier and lagoonal facies of a wave-dominated (and coal-bearing) fan delta system that was sourced from the east and north (Lennox & Wilcock, 1985).

The only rocks of this age in the marine region of eastern New England contain Fauna II assemblages (Runnegar, 1970), diagnostic of the earliest part of the interval. These represent a continuation of the shallow marine environment of the previous time slice. The later transgression is not recorded, either because the orogen was already experiencing uplift, or because subsequent tectonic movements caused any ensuing marine deposits to be eroded. The Gympie Terrane, however, witnessed continuous deposition. Fauna II times were marked by the shallowing-upward shelf succession of the Rammutt Formation, which culminated in high-energy sublittoral sands at the base of the South Curra Limestone (Runnegar & Ferguson, 1969). A rapid deepening, probably coinciding with the transgression noted in other basins, led to the deposition of calcareous siltstone, before shallowing once again, with calcarenite aggradation forming the bulk of the formation.

Shallow marine conditions continued in the Berserker Graben and in the Yarrol Basin. In the latter however, most of the rocks constitute a thick submarine volcanic pile. Initially, the sea spread westwards from the Yarrol Basin across the Bowen Basin, and remained there. The Buffel Formation limestones in the southeast Bowen contain robust shelly fossils indicative of high-energy, and therefore very shallow, nearshore waters. There is a major hiatus in this eastern region, presumably because the marine Fauna III rocks deposited here were removed during later tectonism. On the western flank of the basin, the large Aldebaran Sandstone delta, supplied by detritus eroded off the Anakie High and probably the Galilee Basin area, at first filled the Denison Trough, and then spread across the confining Comet Ridge to the east (Brakel, 1989a). The northern end of the basin was occupied by the alluvial fan of the Collinsville Coal Measures, advancing mainly from the west before the marine transgression had reached the site, and eventually meeting the sea as a fan delta. The Calen Basin in the Strathmuir Synclinorium, meanwhile, collected coal measure sediments in a predominantly lower delta plain setting (Brakel, 1989b).

Marine advances also feature on the western side of the continent, both in the Bonaparte Basin and along the northern side of the Canning Basin. The Broome Arch in the central Canning was uplifted, however, and part of it became emergent. Most of the Canning Basin remained an extensive shallow sea, with clastic and some carbonate deposition (Noonkanbah Formation); the palæotemperature of the water was 24°-26°C (Forman & Wales, 1981). An erosional break between the Noonkanbah and Lightjack Formations in the Balgo Hills (Yeates et al., 1975) may be the local expression of a basin-wide sequence boundary. Above this boundary, a shallowing of the sea to less than 60 m is apparent (Forman & Wales, 1981).

The Carnarvon Basin experienced fluctuating shallow marine conditions, ranging from quiet lower offshore to wave-dominated upper shoreface zones (Moore et al., 1980), with storm activity prominent at various times (Moore & Hocking, 1983; Hocking et al., 1985). Scattered erratics near the base of the interval suggest the presence of floating ice; the climate became cool to cold for the first third of the interval (lower Byro Group) before an amelioration (Crespin, 1958; Moore et al., 1980). The cooling is consistent with the re-appearance of glaciers in SE Australia, but intriguingly, it is in marked contrast to the warm conditions in the nearby Canning Basin, implying that the latter was influenced by a subtropical ocean current that did not reach farther south.

Only the restricted shallow marine Carynginia Formation, at the base of the time slice, is preserved in the northern Perth Basin, and no significant areal change in the marine facies is indicated. However, the long post-Carynginia hiatus implies a subsequent erosional event, possibly co-incident with the tectonic hiatus at the end of the time slice in eastern Australia (see below). Fluvial coal measures continued to form in the Collie and southern Perth Basins. Border faults, affecting sedimentation, were active in the Collie Basin, and may have restricted the area of deposition to within the presently preserved basin (Park, 1982), but Wilson (1989) believes that the wider area of deposition of earlier times may have persisted.

No record of deposition is preserved in the Arckaringa and Pedirka Basins, but regardless of whether sedimentation continued in those basins during this time slice, or the basins had filled to become sedimentary by-pass areas, the earlier drainage patterns almost certainly lingered on.

The southern Cooper Basin was twice the site for the development of large lakes, of about 250 by 150 km in size, in both cases extending diachronously westwards. This suggests blocking of the drainage at the northeastern end of the basin so that the lakes were backfilled towards the southwest (Stuart, 1976), perhaps because of ponding as a result of global eustatic highstands, or tectonic movement along the Karmona trend. Although some workers have postulated restricted sea conditions, there is no evidence for marine environments in the Cooper Basin, or for any marine connection to the open ocean. The first lake was filled by deltas and the reclaimed territory converted to peatlands, which were in turn inundated when the second lake formed and encroached westwards (Thornton, 1979).

Highlands still existed in central Australia, and in the western New England -- Yarrol region between the Sydney - Bowen trough and the eastern sea. A small intermontane basin near Ashford accumulated coal measures, which were preserved by later faulting (O'Brien, 1989). Highlands also persisted from west of the Bowen Basin to beyond Cairns in north Queensland, as a lingering effect of earlier igneous and tectonic activity. In a disruptive end to the time slice, an episode of major uplift and folding affected a large portion of eastern Australia, generating hiatuses from the Sydney to the Bowen Basin. The formerly marine area in most of the Bowen Basin underwent a short-lived subaerial exposure and erosion, the thickness of sequence removed being greatest along part of the eastern margin and on the Roma and Springsure Shelves, where the Fauna III biozone, which accounts for the last three-quarters of the time slice, is missing. The Sydney Basin embayment shallowed due to regression. The long hiatus in the Galilee Basin, where there are no rocks at all of this age, could be a reflection of the same event. All this coincided with folding and serpentinite emplacement east of the Bowen Basin (Malone et al., 1969), both of which may have been caused by the docking of the Curtis terrane.

ACKNOWLEDGMENTS

The maps presented in this paper form part of the Australian Palaeogeographic Maps Project, which was undertaken by the Australian Bureau of Mineral Resources (BMR) and the Petroleum Division of the Australian Mineral Industries Research Association (AMIRA), with the support of a number of petroleum and mineral exploration companies, and input by all the State Geological

Surveys. We would also like to acknowledge the contributions to the project of many scientists, both within BMR and elsewhere, who are much too numerous to list individually here. The figures were drawn by Bruce Cotton of the BMR Cartographic Services Unit. This paper is published with the permission of the Executive Director, BMR Geology & Geophysics, Canberra.

REFERENCES

Bailey, J.C., W.R. Morgan, & L.P. Black 1982. Geochemical and isotopic evidence for the age, orogenic setting and petrogenesis of the Nychum volcanic association, North Queensland. *Journal of the Geological Society of Australia* 29: 375-393.

Banks, M.R., & M.J. Clarke 1987. Changes in the geography of the Tasmania Basin in the late Paleozoic. *In* McKenzie, G.D. (Editor) - Gondwana Six: stratigraphy, sedimentology, and paleontology. *American Geophysical Union, Geophysical Monograph* 41: 1-14.

Bowen, R.L., & G.A. Thomas 1988. Permian. *In* Douglas, J.G., & J.A. Ferguson (Editors) - Geology of Victoria. *Geological Society of Australia, Victorian Division*: 195-212.

Bradshaw, J., R.S. Nicoll & M. Bradshaw 1990. The Cambrian to Permo-Triassic Arafura Basin, northern Australia. *The APEA Journal* 30(1): 107-127.

Brakel, A.T. 1989a. Permian coal measures in the exposed part of the Bowen Basin. *In* Harrington, H.J., & others - Permian coals of eastern Australia. *Bureau of Mineral Resources, Australia, Bulletin* 231: 69-91.

Brakel, A.T. 1989b. Calen Basin. *In* Harrington, H.J., & others - Permian coals of eastern Australia. *Bureau of Mineral Resources, Australia, Bulletin* 231: 107-110.

Brakel, A.T., & J.M. Totterdell 1988. The Late Permian palaeogeography of the Australian continent. *22nd Newcastle Symposium on Advances in the study of the Sydney Basin, Proceedings*: 59-67. *Department of Geology, University of Newcastle, N.S.W.*

Brakel, A.T., & J.M. Totterdell 1990. Glaciation and paleogeography of Australia at the start of the Permian. *In* Gondwana: terranes and resources. 10th Australian Geological Convention, Hobart, 1990, *Geological Society of Australia, Abstracts* 25: 63-64.

Branagan, D.F., C. Herbert & T. Langford-Smith 1976. An outline of the geology and geomorphology of the Sydney Basin. *Science Press, Department of Geology and Geophysics, University of Sydney, N.S.W.*

Brownlow, J. 1980. Early Permian sediments in the Maules Creek district, New South Wales. *Coal Geology (Journal of the Coal Geology Group, Geological Society of Australia)* 2(4): 125-160.

Clarke, M.J. 1989. Lower Parmeer Supergroup. *In* Burrett, C.F., & E.L. Martin (Editors) - Geology and mineral resources of Tasmania. *Geological Society of Australia, Special Publication* 15: 295-309.

Crespin, I. 1958. Permian foraminifera of Australia. *Bureau of Mineral Resources, Australia, Bulletin* 48.

Cook, P.J. 1988. Palaeogeographic atlas of Australia. Volume 1 - Cambrian. *Bureau of Mineral Resources, Canberra.*

Crowe, R.W.A., & Towner, 1976. Environmental interpretation of the Permian Nura Nura Member of the Poole Sandstone, Noonkanbah Sheet area, Canning Basin: a gradation between fluviatile and shallow-water marine facies. *Geological Survey of Western Australia, Annual Report for 1975*, 59-62.

Crowe, R.W.A., R.R. Towner & D.L. Gibson 1983. Permian and Mesozoic geology of the Derby and Mount Anderson 1:250 000 Sheet area, Western Australia. *In* Towner, R.R. & D.L. Gibson - Geology of the onshore Canning Basin, Western Australia. *Bureau of Mineral Resources, Australia, Bulletin* 215: Appendix 4 (Micro-fiche).

David, T.W.E. 1907. The geology of the Hunter River coal measures, New South Wales. *Geological Survey of New South Wales, Memoir* 4.

Day, R.W., W.G. Whitaker, C.G. Murray, I.H.

Wilson & K.G. Grimes 1983. Queensland Geology: a companion volume to the 1:2 500 000 scale geological map (1975). *Geological Survey of Queensland, Publication* 383.

Dickins, J.M., J. Roberts & J.J. Veevers 1972. Permian and Mesozoic geology of the northeastern part of the Bonaparte Gulf Basin. *In* Geological Papers 1969. *Bureau of Mineral Resources, Australia, Bulletin* 125: 75-102.

Draper, J.J., & J.W. Beeston 1985. Depositional aspects of the Reids Dome beds, Denison Trough. *Queensland Government Mining Journal* 86: 200-210.

Dulhunty, J.A. 1964. Our Permian heritage in central eastern New South Wales. *Royal Society of New South Wales, Journal and Proceedings* 97: 161-166.

Engel, B.A., & J.R. Laurie 1978. A new species of the Permian trilobite *Doublatia* from the Manning District, New South Wales. *Alcheringa* 2: 49-54.

Evans, P.R., R.J. Seggie, & M.J. Walker 1983. Early Permian stratigraphy and palaeogeography, southern Sydney Basin. *17th Newcastle Symposium on Advances in the Study of the Sydney Basin, Abstracts* (supplement). *Dept. of Geology, University of Newcastle, N.S.W.*

Forman, D.J., & D.W. Wales (Compilers) 1981. Geological evolution of the Canning Basin, Western Australia. *Bureau of Mineral Resources, Australia, Bulletin* 210.

Hamilton, D.S., & J. Beckett 1984. Permian depositional systems in the Gunnedah region. *Geological Survey of New South Wales, Quarterly Notes* 55: 1-19.

Hawkins, P.J. 1978. Galilee Basin - a review of petroleum prospects. *Queensland Government Mining Journal* 79: 96-112.

Hawkins, P.J., & P.L. Harrison 1978. Stratigraphic and seismic investigations in the Lovelle Depression, western Galilee Basin. *Queensland Government Mining Journal* 79: 623-650.

Hickman, A.H. 1978. Nullagine, Western Australia (SF51-5): 1:250 000 Geological Series. *Geological Survey of Western Australia, Explanatory Notes.*

Hocking, R.M., S.J. Williams, P.S. Moore, P.D. Denman, I.H. Lavaring, & W.J.E. van de Graaff 1985. Kennedy Range, W.A.: 1:250 000 Geological Series. *Geological Survey of Western Australia, Explanatory Notes.*

Hughes, R.J. 1978. The geology and mineral occurrences of Bathurst Island, Melville Island, & Coburg Peninsula, Northern Territory. *Bureau of Mineral Resources, Australia, Bulletin* 177.

Kemp, E.M., B.E. Balme, R.J. Helby, R.A. Kyle, G. Playford, & P.L. Price 1977. Carboniferous and Permian stratigraphy in Australia and Antarctica: a review. *BMR Journal of Australian Geology & Geophysics* 2: 177-208.

Lavering, I.H. 1985. Permian environments and palaeogeography of the onshore Carnarvon Basin and implications for petroleum prospectivity. *PESA Journal* 7: 40-46.

Laws, R.A., & R.S. Brown 1976. Bonaparte Gulf Basin - southeastern part. *In* Leslie, R.B., H.J. Evans, & C.L. Knight (Editors) - Economic Geology of Australia and Papua New Guinea. 3. Petroleum. *Australasian Institute of Mining & Metallurgy, Monograph* 7: 200-208.

Lennox, M., & S. Wilcock 1985. The Stroud-Gloucester Trough and its relation to the Sydney Basin. *19th Newcastle Symposium on Advances in the Study of the Sydney Basin, Proceedings*: 37-41. *Department of Geology, University of Newcastle, N.S.W.*

Lindsay, J.F. 1969. Stratigraphy and structure of the Palaeozoic sediments of the lower Macleay region, north-eastern New South Wales. *Royal Society of New South Wales, Journal and Proceedings* 102: 41-55.

Malone, E.J., F. Olgers, & A.G. Kirkegaard 1969. The geology of the Duaringa and Saint Lawrence 1:250,000 Sheet areas, Queensland. *Bureau of Mineral Resources, Australia, Report* 121.

Moore, P.S., P.D. Denman, & R.M. Hocking 1980. Sedimentology of the Byro Group (Lower Permian), Carnarvon Basin, Western Australia.

Geological Survey of Western Australia, Annual Report for 1979: 55-64.

Moore, P.S., & R.M. Hocking 1983. Significance of hummocky cross-stratification in the Permian of the Carnarvon Basin, Western Australia. *Journal of the Geological Society of Australia* 30: 323-331.

O'Brien, P.E. 1986. Stratigraphy and sedimentology of Late Palaeozoic glaciomarine sediments beneath the Murray Basin, and their palaeogeographic and palaeoclimatic significance. *BMR Journal of Australian Geology & Geophysics* 10: 53-63.

O'Brien, P.E. 1989. Ashford coal field. *In* Harrington, H.J., & others - Permian coals of eastern Australia. *Bureau of Mineral Resources, Australia, Bulletin* 231: 67-68.

Osborne, R.A.L., & D.F. Branagan 1985. ?Permian palaeokarst at Billys Creek, New South Wales. *Royal Society of New South Wales, Journal and Proceedings* 118: 105-111.

Park, W.J. 1982. The geology of the Muja Sub-Basin a model for the Collie Basin, Western Australia. *In* Mallett, C.W. (Editor) - *Symposium on Coal Resources: origin, exploration and utilization in Australia, Proceedings, Melbourne, November 1982. Coal Geology (Journal of the Coal Geology Group, Geological Society of Australia)* 4(2): 319-340.

Playford, P.E., A.E. Cockbain, & G.H. Low 1976. Geology of the Perth Basin Western Australia. *Geological Survey of Western Australia, Bulletin* 124.

Rao, C.P., & D.C. Green 1982. Oxygen and carbon isotopes of Early Permian cold-water carbonates, Tasmania, Australia. *Journal of Sedimentary Petrology* 52: 1111-1125.

Runnegar, B.N. 1970. The Permian faunas of northern New South Wales and the connection between the Sydney and Bowen Basins. *Journal of the Geological Society of Australia* 16: 697-710.

Runnegar, B., & J.A. Ferguson 1969. Stratigraphy of the Permian and Lower Triassic marine sediments of the Gympie district, Queensland. *University of Queensland Papers, Department of Geology*: 6 (9).

Scheibner, E. 1976. Explanatory notes on the tectonic map of New South Wales, scale 1:1 000 000. *Geological Survey of New South Wales.*

Stuart, W.J. 1976. The genesis of Permian and Lower Triassic reservoir sandstones during phases of southern Cooper Basin development. *The APEA Journal* 16: 37-48.

Thomas, G.A. 1976. Faunal evidence of climatic change in the Early Permian of the Carnarvon Basin. *25th International Geological Congress, Sydney, Abstracts* 1: 318-319.

Thornton, R.C.N. 1979. Regional stratigraphic analysis of the Gidgealpa Group, southern Cooper Basin, Australia. *Geological Survey of South Australia, Bulletin* 49.

Townsend, I.J. 1976. Stratigraphic drilling in the Arckaringa Basin, 1969-1971. *Geological Survey of South Australia, Report of Investigations* 45.

van de Graaff, W.J.E. 1981. Paleogeographic evolution of a rifted cratonic margin: S.W. Australia - Discussion. *Palaeogeography, Palaeoclimatology, Palaeoecology* 34: 163-172.

Wells, A.T., & P.E. O'Brien 1989. Late Permian Gidgealpa Group, Cooper Basin of South Australia and Queensland. *In* Harrington, H.J., & others - Permian coals of eastern Australia. *Bureau of Mineral Resources, Australia, Bulletin* 231: 149-169.

Wilson, A.C. 1989. Palaeocurrent patterns in the Collie Coal Measures - the implications for sedimentation and basin models. *In* Professional Papers. *Geological Survey of Western Australia, Report* 25: 85-91.

Yeates, A.N., R.W.A. Crowe, R.R. Towner, L.A.I. Wyborn, & V.L. Passmore 1975. Notes on the geology of the Gregory Sub-basin and adjacent areas of the Canning Basin, Western Australia. *Bureau of Mineral Resources, Australia, Record* 1975/77.

Youngs, B.C. 1976. The geology and hydrocarbon potential of the Pedirka Basin. *Geological Survey of South Australia, Report of Investigations* 44.

Gondwana Eight, Findlay, Unrug, Banks & Veevers (eds) © 1993 Balkema, Rotterdam. ISBN 90 5410 304 3

Mesozoic volcanism in the central Transantarctic Mountains, Antarctica: Depositional environment and tectonic setting

D.H.Elliot
Byrd Polar Research Center & Department of Geological Sciences, Ohio State University, Columbus, Ohio, USA

D.Larsen
Byrd Polar Research Center & Department of Geological Sciences, Ohio State University, Columbus, Ohio, USA
(Presently: Department of Geology, University of New Mexico, Albuquerque, N.Mex., USA)

ABSTRACT: Pyroclastic and related rocks of silicic and basaltic composition form an important part of the uppermost Beacon Supergroup in the central Transantarctic Mountains. The Upper Triassic to Lower Jurassic Falla Formation consists of a sandstone-carbonaceous shale sequence succeeded by a sequence increasingly dominated by silicic tuffs. This volcaniclastic sequence is overlain by basaltic pyroclastic and laharic deposits of the Prebble Formation. The pyroclastic eruptions were abruptly followed by early Middle Jurassic flood-basalt eruption (Kirkpatrick Basalt). The pyroclastic deposits were the result in part of Plinian and phreatomagmatic processes. Monoclinal structures, indicative of basement uplifts, affected the sequence in the Early to Middle Jurassic. Development of topography is also indicated by thick sequences of mass-flow (laharic) deposits, the occurrence of arkosic grits in the lower part of the volcaniclastic sequence and ponded lavas in the overlying basalt sequence. The depositional and tectonic setting for these rocks is inferred to be a tectono-magmatic rift. In Victoria Land silicic volcanism and vertical movements occurred before onset of basaltic activity. Silicic magmatism is interpreted to be the result of anatexis caused by emplacement of basaltic magmas into the lower crust, followed by rise of those magmas to the surface. Magmatic activity in the Jurassic was associated with vertical tectonism that was expressed by basement uplifts and development of significant topography.

1 INTRODUCTION

Flood basalts commonly form as precursors to continental fragmentation and seafloor spreading (White & McKenzie 1989). Those of the Jurassic Ferrar Group in the Ross Sea sector of Antarctica, however, were not associated with sea-floor generation and therefore only the early stages of continental break-up are represented in the rock record. Silicic volcanism accompanying break-up is well known and documented in the Lebombo belt in southern Africa where it occurs with thick basalt sequences of Jurassic age (Cleverly et al. 1984; Eales et al. 1984), and in Patagonia (Gust et al. 1985) where it is Middle to Late Jurassic and spatially separated from the younger (Late Jurassic to Early Cretaceous) continental flood basalt province of the Parana (Piccirillo et al. 1988). Less well known is the Jurassic silicic volcanism in the Transantarctic Mountains (Larsen 1988; Elliot 1991) and the granitic plutonism in West Antarctica (Storey et al. 1988) (Figure 1).

This paper discusses the Jurassic silicic and basaltic volcanism in the central Transantarctic Mountains, and the tectonic setting in which this magmatic activity occurred.

2 STRATIGRAPHY OF THE UPPER PART OF THE BEACON SUPERGROUP, CENTRAL TRANSANTARCTIC MOUNTAINS

The most complete of the Antarctic Gondwana sequences crops out in the Beardmore Glacier region which is one of three localities where the capping tholeiitic flood-basalts are preserved (Figure 1). In that region (Figure 2) the upper part of the Permian to Triassic-Jurassic Victoria Group of the Beacon Supergroup

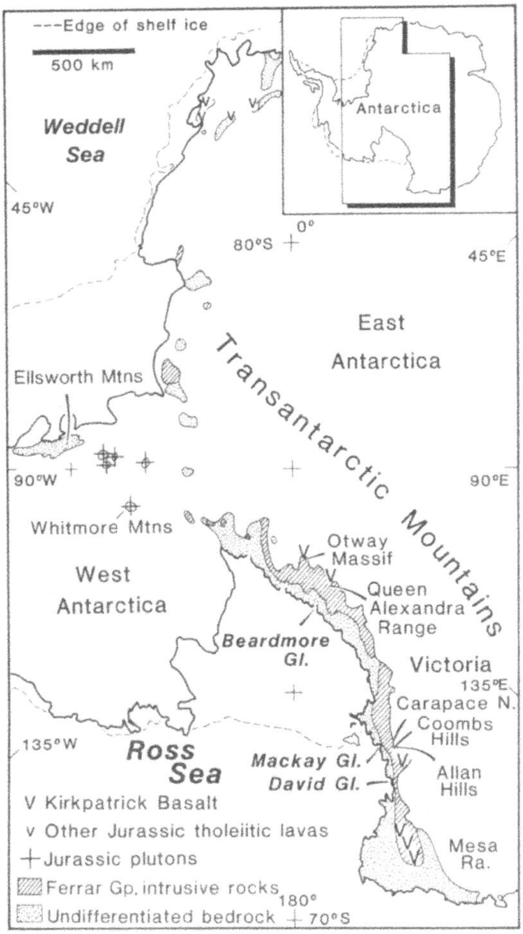

Fig. 1. Distribution of Mesozoic tholeiites and silicic rocks in the Transantarctic Mountains and West Antarctica

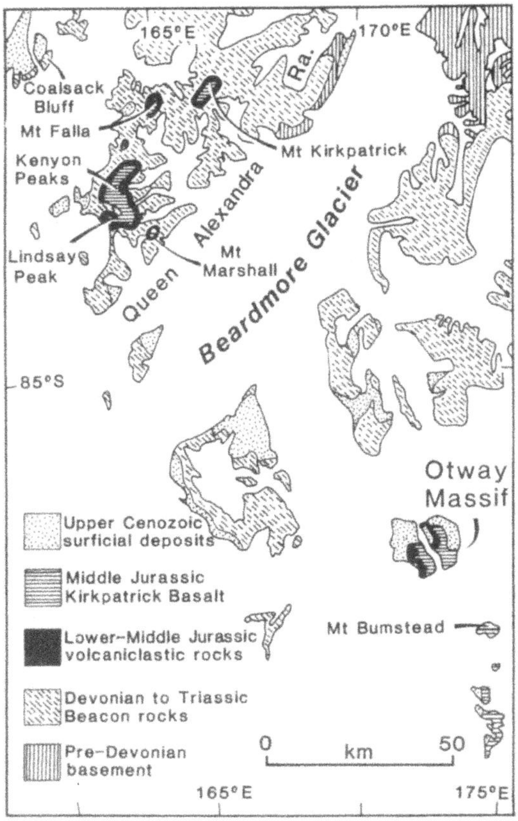

Fig. 2. Geologic sketch map of the Beardmore Glacier region, central Transantarctic Mountains. The Lower to Middle Jurassic rocks include the upper volcaniclastic part of the Falla Formation and the Prebble Formation, both assigned to the Beacon Supergroup. Ferrar Dolerite sills are coextensive with the Beacon Supergroup.

(Barrett et al. 1986) includes Upper Triassic *Dicroidium*-bearing beds (lower part of the Falla Formation) overlain by a volcaniclastic sequence of probable Early to Middle Jurassic age (upper part of the Falla Formation and the Prebble Formation). These volcaniclastic beds are capped by the lower Middle Jurassic tholeiitic lavas of the Kirkpatrick Basalt (Elliot et al. 1985; Barrett et al. 1986).

The interval from the plant-bearing beds to the basalts is best exposed at Mt Falla (Figure 3). There the lower part of the Falla Formation is a 296-m-thick cyclic sequence of channel sandstones and associated finer grained beds which commonly include carbonaceous mudstone or shale (recently measured stratigraphic thicknesses differ slightly from those of Barrett et al. 1986). The sandstones are feldspar and lithic-rich quartz sandstones, and the lithic grains are dominated by aphanitic silicic volcanic rock fragments. The volcaniclastic upper part of the Falla Formation, 235 m thick, has been divided into three informal members (Larsen 1988). The lowest (Unit 1, Figure 3) is a sequence of four sandstone units each grading up into beds with variable amounts of volcanic and tuffaceous material. The sandstones are similar in composition to those in the underlying carbonaceous sequence, but contain a higher proportion of volcanic rock fragments and include fragments with felted and trachytic textures. The finer-grained beds, particularly those in the lowest part of the unit, contain sparsely distributed glass shards. The middle volcaniclastic member (Unit 2, Figure 3) is a distinctive resistant tuff sequence with two tuffaceous sandstone intervals. These sandstones are matrix-rich and shard/pumice-rich, and contain volcanic rock fragments similar to those in Unit 1 but in lesser amounts. The upper volcaniclastic member (Unit 3, Figure 3) consists of a slope-forming sequence of tuffs which include beds with accretionary lapilli.

The slope-forming tuff sequence is abruptly overlain by thick laharic beds of the Prebble Formation (Figure 3). The laharic deposits, in general, are massive, matrix-supported, poorly sorted, and contain mostly lapilli-sized clasts although boulders or blocks up to 75 cm across occur in one 2 m thick lens-shaped interval. The lens has gradational but abrupt contacts and laterally the clast size decreases and the contacts become more diffuse. Other lahar deposits in the Prebble Formation show inverse grading or upward increase in the abundance of clasts. The clasts include basalt, dolerite, silicic tuff, sandstone, siltstone, and carbonaceous shale. The uppermost part of the sequence, mostly talus-covered, contains bedded tuffs and lapilli tuffs.

The subdivision of the upper volcaniclastic part of the Falla Formation at the type section is less obvious at all other outcrops examined, in part due to poor exposure. The Prebble Formation everywhere separates the volcaniclastic beds of the Falla Formation from the tholeiitic lavas. The Prebble Formation elsewhere is lithologically more varied, and includes basaltic pyroclastic breccia and lapilli tuff, silicic tuff, and volcanic sandstone. At the Otway Massif (Figure 2) it consists almost entirely of laharic deposits. The volcaniclastic upper part of the Falla Formation is 235 m thick at the type locality, but thins southward to the Marshall Mountains where it is about 120 m thick. The Prebble Formation ranges in thickness from 14 m in the Marshall Mountains to 75 m at Mt Falla, 145 m at Mt Kirkpatrick and more than 360 m on the Otway Massif.

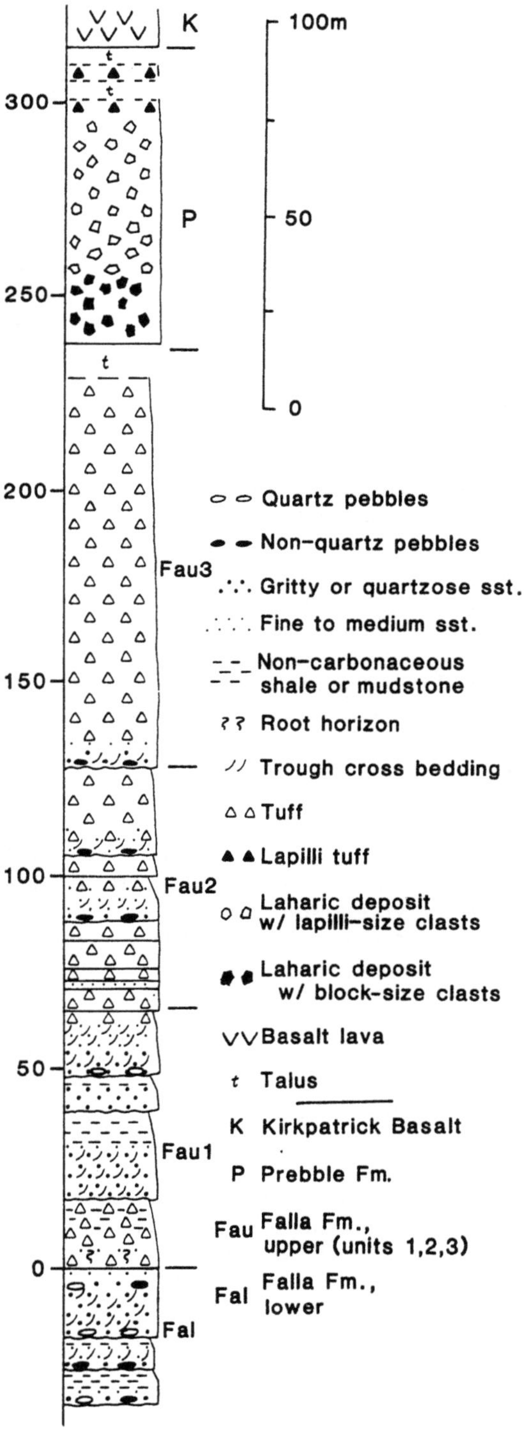

Fig. 3. Composite stratigraphic column for the upper volcaniclastic part of the Falla Formation and the Prebble Formation at Mt Falla.

3 PETROGRAPHY

The silicic tuffs and tuff clasts in the Falla and Prebble formations, respectively, show a range of petrographic compositions dependent mainly on the amount and type of lithic material. The tuffs containing the least amount of material of possible detrital, xenocrystic or xenolithic origin were analyzed geochemically (see later) and their petrography is summarized here. These tuffs are crystal-vitric ash deposits. Quartz predominates over feldspars (andesine-oligoclase and minor sanidine) and is accompanied by trace amounts of biotite, hornblende, pyroxene, garnet, zircon, sphene and opaques; with one exception, crystals form less than 7% of the rock. Lithic grains are dominated by microcrystalline and cryptocrystalline volcanic rock fragments and pumice; trace amounts of schistose metamorphic grains, plutonic rock fragments, and fine-grained sedimentary rock fragments accompany those of volcanic origin. Again with the one exception, lithic grains, excluding pumice, form less than 12% of the rock. Lithic and crystal grain sizes are generally less than 0.25 mm, and are comparable to those of the shards. Tricuspate and platy glass shards are common to abundant (20 to 50%) and are replaced by zeolite which is principally clinoptilolite. The interstitial matrix, forming 40 to 75% of the rock, consists of very fine grained shard, crystal and lithic debris together with varying amounts of phyllosilicate as dispersed shreds; silicification may be present in many of the samples.

The basaltic pyroclastic rocks in the Prebble Formation are all extensively to extremely altered, and commonly strongly oxidized so that hematite is pervasive; few original textural features remain. The breccia samples contain fragments of basaltic composition having microlites of plagioclase and pyroxene set in an irresolvable matrix and, in some instances, numerous vesicles filled by secondary minerals consisting of phyllosilicates, zeolites or calcite. The matrix to the clasts contains scattered grains of intermediate plagioclase, with lesser amounts of quartz, and trace amounts of alkali feldspar, pyroxene and garnet; the interstitial material to the mineral grains is irresolvable. Trace amounts of sedimentary and metamorphic rock fragments are also present. The vent breccia is very similar to the extrusive breccias.

4 VOLCANOLOGY AND DEPOSITIONAL SETTING

The fine-grained silicic tuffs of Units 1 and 2 of the upper Falla Formation are interbedded with trough cross-bedded fluvial sandstones. Pedogenic modification of the tuffaceous beds is suggested by root traces, argillaceous films or linings (possible ped structure) and mottled coloring (Retallack 1983, 1988). On the basis of their fine grain size and good sorting, these tuffs are interpreted as fallout from Plinian eruptions or its derivatives reworked by water (Larsen 1988). The coarser-grained, more poorly-sorted tuffs of Unit 3 of the upper Falla Formation and the Prebble Formation exhibit evidence for phreatomagmatic processes during eruption: beds as much as 20 cm thick and containing abundant accretionary lapilli up to 10 mm across occur in several sections and armored lapilli up to 4 mm across are sparsely distributed. Armored lapilli are indicative of hydrovolcanic processes but accretionary lapilli may have a variety of origins. The grain size and sorting of these tuffs, as well as their occurrence interbedded with laharic and breccia deposits, suggests that they probably originated as fallout from volcanoes within the Beardmore Glacier region.

The coarse-grained basaltic pyroclastic breccias and lapilli tuffs, together with the laharic deposits, indicate sources within several tens of km, though exceptional lahars are inferred to have traveled more than 100 km (Crandell 1971). Lahars can form either as mass-flows resulting from the eruptive process or by collapse of loose volcanic material on the flanks of a volcano (Rodolfo 1989). The primary or secondary origin of many of the laharic deposits is difficult to assess because of the apparent absence of key indicators, such as armored clasts or accretionary lapilli, and the abundance of non-volcanic debris. At the Otway Massif, the laharic deposits have a stratigraphic thickness of at least 360 m, the lower contact with the Falla Formation not being exposed in this region. The section appears to consist of only five separate mass-flow deposits. Clast sizes range up to several meters in the lowest exposed beds and the maximum size decreases both up-section and also within individual deposits. The upper contact is a flat surface with observed relief of no more than 7 m, and

Table 1. Representative analyses (recalculated to 100%) of pyroclastic rocks from the Beardmore Glacier region.

	1	2	3	4	5	6	7	8	9	10	11
SiO_2	67.23	71.97	73.94	74.82	75.21	76.29	78.05	71.73	52.67	59.63	56.21
TiO_2	0.47	0.43	0.37	0.32	0.35	0.43	0.35	0.20	0.67	0.60	0.63
Al_2O_3	14.75	14.69	13.58	13.81	13.92	12.31	12.32	16.39	16.18	14.08	14.81
Fe_2O_3T	5.38	3.24	3.65	2.51	3.03	2.39	2.06	1.84	12.26	8.49	9.19
MnO	0.19	0.08	0.11	0.06	0.04	0.07	0.05	0.03	0.10	0.12	0.16
MgO	2.72	0.96	0.65	0.68	0.79	0.69	0.58	1.08	7.15	6.54	6.59
CaO	1.97	2.20	2.53	2.40	1.76	2.46	2.43	4.46	6.61	3.89	4.79
Na_2O	1.46	2.98	2.16	2.37	3.32	3.49	2.56	3.37	3.51	5.22	5.92
K_2O	5.78	3.41	2.92	2.96	1.52	1.82	1.56	0.89	0.76	1.38	1.60
P_2O_5	0.05	0.04	0.08	0.07	0.05	0.04	0.03	0.01	0.08	0.06	0.10
LOI	5.37	6.43	6.74	7.02	5.70	4.72	6.19	10.48	4.90	4.63	4.51
Rb	141	174	93	154	86	108	78	91	134	65	31
Sr	134	282	288	401	176	58	187	94	85	46	29
Zr	186	203	51	162	181	197	156	130	192	130	129
Y	20	27	37	19	27	28	14	18	20	18	20

Fe_2O_3T = total Fe as Fe_2O_3. LOI = Loss on ignition (in weight percent).
Oxides in weight percent; trace elements in ppm.
Analyses 1-7: tuffs in the upper Falla Formation; 8: tuff clast in the Prebble Formation; 9,10: pyroclastic breccia in the Prebble Formation; 11: vent breccia in the Prebble Formation.

is overlain by basalt lavas. The beds directly under the basalts include tuffs and volcanic sandstones, but the maximum observed thickness is only about 7 m. The great thickness and apparently near planar upper surface of the Prebble Formation beds suggests confining topography, which is supported by the presence of ponded lavas in the overlying basalts. The accumulation of that thickness (360+m) of debris does not appear possible without such conditions because otherwise topography would be developed in the loose debris. This suggests the laharic deposits may be the result of redeposition of volcanic debris, xenoclastic fragments brought up by eruptive events, and possibly bedrock derived from the confining topographic highlands. The laharic deposits may include beds formed directly by phreatomagmatic activity.

Direct evidence for constructional volcanic landforms is confined to the eastern side of Kenyon Peaks where a volcanic neck cuts the volcaniclastic beds of the Falla and Prebble Formations. The Prebble Formation immediately beneath the basalts dips as much as 15° to the southeast and the basalts clearly lap onto those beds, demonstrating relief of about 30 m. The attitude of those beds precludes a direct relationship to the volcanic neck, although both consist mainly of altered basaltic material.

5 GEOCHEMISTRY

Interpretation of geochemical analyses of the pyroclastic rocks (Table 1) is hindered by the problems of xenolithic and xenocrystic material and secondary alteration. Most tuffs contain xenoliths of, for example, polycrystalline quartz or quartz-mica-schist. Furthermore almost all tuffs examined so far show zeolitization and clay mineralization. The

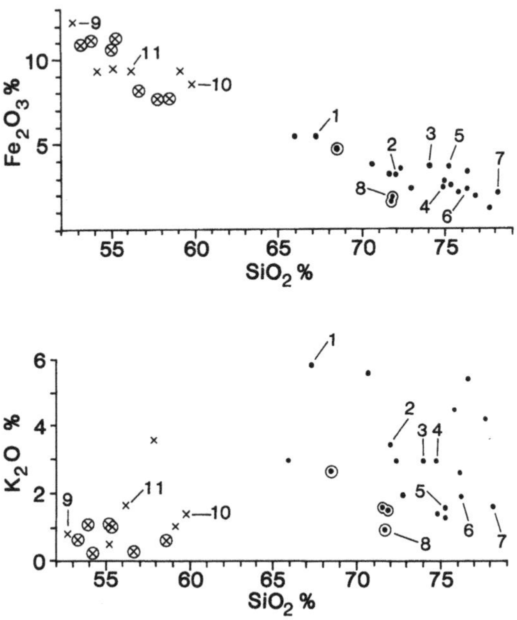

Fig. 4. Variation diagrams to illustrate the compositions of the pyroclastic rocks in the Beardmore Glacier region. Analyzed samples are keyed to Table 1. Symbols. × - basaltic to andesitic rocks; ⊗ - basaltic to andesitic clasts; • - dacitic to rhyolitic tuffs; ⊙ - dacitic to rhyolitic tuff clasts.

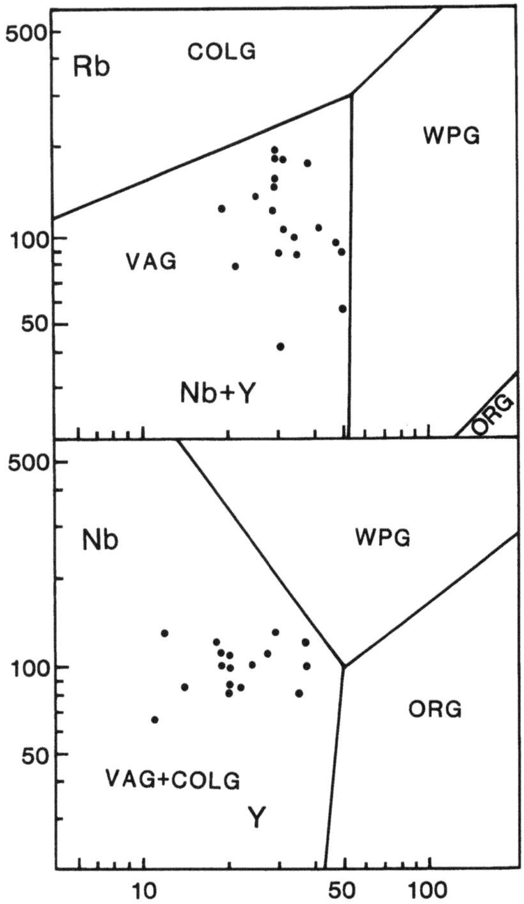

Fig. 5. Discriminant diagrams of Pearce et al. (1984) for silicic tuffs from the Falla and Prebble Formations; A: Rb-(Nb+Y); B: Nb-Y. Fields: VAG - volcanic arc granite; COLG - syn-collision granite; ORG - ocean ridge granite; WPG - within plate granite.

scatter in much of the chemical data, most obviously in the alkalis (up to 9% Na_2O), suggests alteration did not occur isochemically. These problems undoubtedly overwhelm effects such as possible aeolian fractionation.

The analytical results suggest the tuffs range from dacite to high silica rhyolite (classification of Ewart 1979) having SiO_2 = 65 - 77%. MgO and Fe_2O_3T show fair correlations with SiO_2 (Figure 4); other major and trace elements show less systematic relationship to SiO_2, probably due to element mobility during secondary alteration. On chondrite-normalized incompatible element diagrams, the tuffs have patterns similar to crustal rocks in that they show relative depletions in Nb, Sr, P and Ti. The Ga-Al discriminant diagrams (Whalen et al. 1987) for I, S and A-type granites (White & Chappell 1983) demonstrate the tuffs are not A-type. On the Nb-Y, Rb-(Y+Nb), and Ta-Yb discrimination diagrams of Pearce et al. (1984), the tuffs fall in the volcanic arc or volcanic arc plus syn-collision granite fields (Figure 5). The high initial strontium isotope ratio (0.7118) obtained for five tuffs of the Falla Formation (Faure & Hill 1973) suggests the magmas were not I-type, provided the initial ratios are not a function of Sr mobility and isotope exchange associated with zeolitization and secondary alteration.

The Prebble Formation at Lindsay Peak and Kenyon Peaks includes rocks that are distinctly different. Like the tuffs, these rocks are strongly altered but the geochemical data clearly set them apart as basaltic to andesitic in

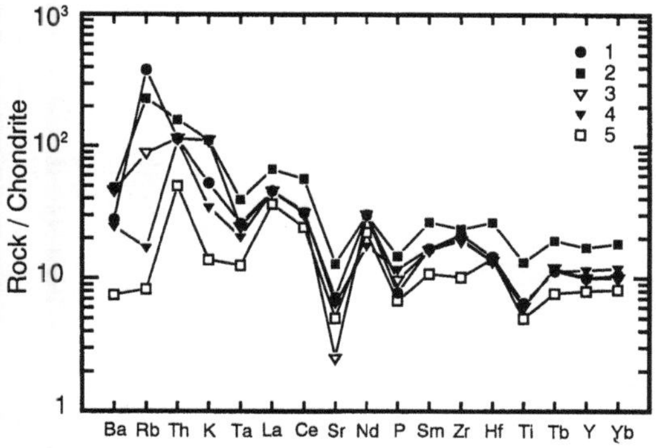

Fig. 6. Incompatible element diagram (normalized to chondrite except for K, P, Rb which are normalized to primitive mantle) for representative samples of the basaltic rocks in the Prebble Formation and the overlying Kirkpatrick Basalt to illustrate their geochemical similarities. 1 - altered basaltic rock (Prebble Fm.); 2 - basalt lava (Kirkpatrick Basalt); 3 - basaltic clast in vent breccia (Prebble Fm.); 4 -basaltic lapilli tuff (Prebble Fm.); 5 - dolerite clast in laharic deposit (Prebble Fm.). Normalizing factors from Thompson et al. (1982) and Sun (1980).

composition (SiO_2 = 51 - 59%). In addition, the laharic deposits carry basaltic clasts which are petrographically and geochemically similar to Ferrar Group rocks. Mafic volcanic rocks, basaltic clasts in the laharic deposits, pyroclastic breccias, clasts in the volcanic neck, and Ferrar Group rocks show similar patterns on incompatible element variation diagrams, most strikingly in the depletions in P, Ti, and Sr (Figure 6). As would be expected with altered rocks, the LIL elements are more scattered. These basaltic to andesitic rocks in the Prebble Formation are tholeiitic in composition.

The geochemical data indicate the Falla Formation tuffs are dominantly silicic, whereas the Prebble Formation pyroclastic rocks, although including a few fine-grained tuffs of silicic composition, are dominantly basaltic to andesitic. The geochemical data, and field relations, strongly suggest that the basaltic pyroclastic rocks are part of the Ferrar Group. The sparse record of silicic magmatism during deposition of the Prebble Formation is followed by an equally sparse record during accumulation of the overlying basalt sequence (Elliot et al. 1991).

6 STRUCTURE

At two localities, the Falla Formation is mildly deformed. At Lindsay Peak, volcanic sandstones of the upper Falla Formation are gently folded, with dips increasing from near horizontal at the west end of the outcrop to about 20° to the southeast at the east end (Figure 7). The dipping sandstones are overlain by silicic tuffs which in turn are in contact with volcaniclastic strata of the Prebble Formation; Kirkpatrick Basalt lavas overlie both Prebble and Falla Formation beds. These field relations are interpreted as showing monoclinal warping of the Falla Formation to produce local topography, infilling of topography by pyroclastic rocks of the Prebble Formation, erosion to produce a regionally planar surface, and finally accumulation of flood basalts. Field relations constrain this monocline to the Early (or early Middle) Jurassic. Structural relief on the monocline is more than 100 m. In the saddle to the east of Lindsay Peak, monoclinal flexure of the lavas faces east and has structural relief of about 200 m. At the Mt Falla monocline, rocks younger than the middle volcaniclastic member of the upper part of the Falla Formation are absent, however the presence of breccia identical to that of the Prebble Formation in the hinge line (D.H. Elliot & R.E. Hanson, unpublished field data 1991) indicates that flexuring was coeval with deposition of the Prebble Formation. Structural relief on this monocline is about 100 m. Structural relations at Coalsack Bluff have also been interpreted as a faulted monocline (Collinson & Elliot 1984) which pre-dates dolerite emplacement; probably it was broadly coeval with the other monoclines. The strike of all these monoclines is northnortheast.

7 TECTONIC SETTING

The monoclinal flexures are attributed to basement uplifts. The nature of the Prebble Formation at the Otway Massif, in particular, suggests topography confined a thick (360+ m) sequence of laharic deposits thus forming a planar surface onto which the lavas were erupted. The thickness of individual flows, more than 170 m at Mt Bumstead (Figure 2), and more than 100 m at the Otway Massif and in the southern Queen Alexandra Range, also

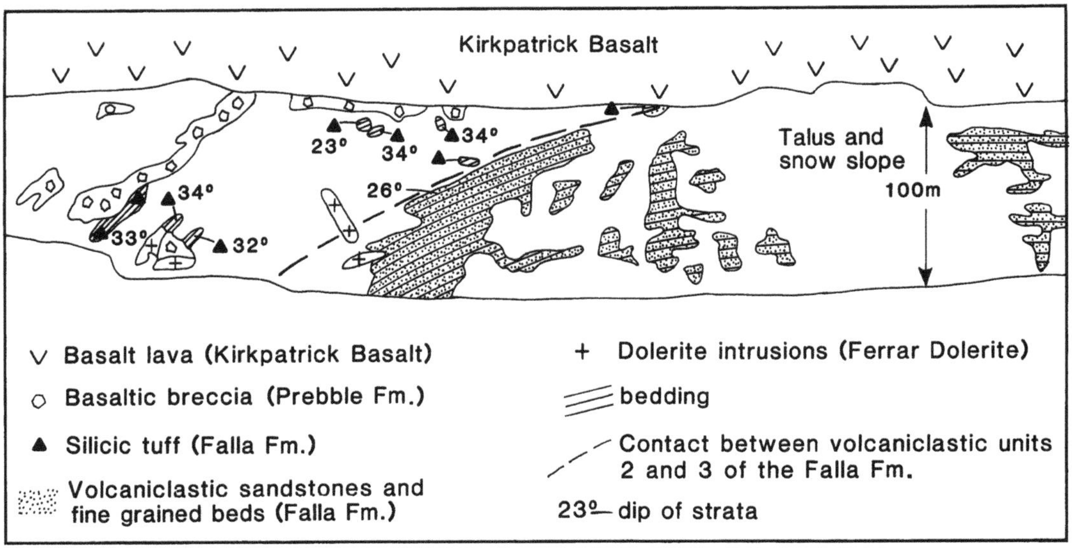

Fig. 7. Sketch of the field relations at Lindsay Peak (view toward the southsouthwest). The apparently irregular base of the Kirkpatrick Basalt is mainly a result of an irregular cover of talus on the contact. The sedimentary beds exhibit an increasing dip toward the southeast (from right to left in the sketch); the overlying tuffs exhibit variable attitudes, dipping toward the east quadrant, due to rotation of beds caused by dolerite emplacement.

requires that they were ponded by topography (Barrett et al. 1986). The occurrence of coarse-grained quartz grits with large (up to 2 cm) angular K-feldspar grains in the volcaniclastic part of the Falla Formation at Mt Falla and near Tempest Peak suggests a proximal non-volcanic source. The thickness (> 1400 m) of the Victoria Group on the polar plateau margin suggests the strata extended more than 100 km onto the craton. The influx of coarse arkosic grits with large angular K-feldspar grains implies a major change in sediment transport distance and rates of source region erosion. Basement uplifts exposing scarps of Precambrian metamorphic and igneous rocks would provide a suitable source terrain. The monoclinal structures and the evidence for development of topography leads to the conclusion that the regional setting was a volcano-tectonic rift. This interpretation, based on paleovolcanological, stratigraphical and petrological data, is afforded strong support by structural data (Wilson in press a) suggesting extension and normal faulting associated with Ferrar Dolerite dike emplacement.

The limits of this inferred local rift system are uncertain. Variations in the thicknesses of the lower non-volcaniclastic and upper volcaniclastic parts of the Falla Formation are difficult to demonstrate because of the lack of continuous sections. The total stratigraphic interval between the base of the Falla Formation and the base of the Kirkpatrick Basalt is 200 m or less on the western flank of the southern Marshall Mountains, whereas it is 600 m at Mt Falla and about 430 m at Mt Kirkpatrick. Although no precise figure can be given for the Mt Marshall area, the sequence appears to be thicker than 200 m. In the Marshall Mountains the volcaniclastic upper Falla Formation is about 150 m thick whereas it is 235 m at Mt Falla and 180 m at Mt Kirkpatrick. The thickness of the volcaniclastic part of the Falla Formation is less variable than that of the underlying non-volcanic part of the formation and, therefore, although simple thinning of the clastic sequence cannot be excluded, an important disconformity may be present in the middle of the formation. Such a disconformity could be related to regional warping before or at about the time of initiation of volcanism, and pre-dates the known monoclinal structures and inferred basement uplifts. The warping, however, did not change the regional northwest trend of fluvial paleocurrent flow (Barrett et al. 1986; Elliot unpublished field data 1991). The orientation of the monoclines is northnortheast and transverse to the regional paleocurrents and the present trend of the Transantarctic Mountains, but parallel to one of the fault and dike arrays identified by Wilson (in press a) who has interpreted the structural data as vertical shortening and two directions of horizontal extension (parallel and perpendicular to the range). Thus, vertical movements began before deposition of the arkosic beds in the lower volcaniclastic member of the upper Falla Formation; they are expressed by monoclinal structures that are younger than the middle volcaniclastic member at Mt Falla, and are younger than at least some of the upper volcaniclastic member of the Falla Formation at Lindsay Peak (Figure 8). Such movements continued during, or after, eruption of the basalts as shown by the monocline east of Lindsay Peak. The inferred age of the rift system spans the Early to Middle Jurassic.

8 VICTORIA LAND SEQUENCES

Rocks that occupy the interval between Upper Triassic *Dicroidium*-bearing beds and basalts of Middle Jurassic age are exposed in the Mesa Range region of north Victoria Land, and the area between the Mackay and David Glaciers in south Victoria Land (Figure 1). In north Victoria Land the uppermost Beacon Supergroup strata, the quartz-rich sandstones of the Section Peak Formation (Collinson et al. 1986), crop out in the vicinity of the Mesa Range. The Section Peak Formation does not include volcaniclastic beds but volcanic rock fragments are present and a few sandstones contain altered shards. The only observed contact with the overlying Exposure Hill Formation (Elliot et al. 1986a) is conformable. The Exposure Hill Formation is a volcaniclastic unit comprising breccias, tuffs, laharic deposits and volcaniclastic sandstones. Compositionally it is dominated by basaltic material although a silicic component is present in many beds. The occurrence of blocky sideromelane clasts suggests phreatomagmatic activity was associated with deposition of some of the units. The overlying lavas of the Kirkpatrick Basalt (Elliot et al. 1986b) include thick (100+ m) ponded flows.

The Lashly Formation, the uppermost unit of the Beacon Supergroup in south Victoria Land (Barrett & Kohn 1975), contains volcaniclastic

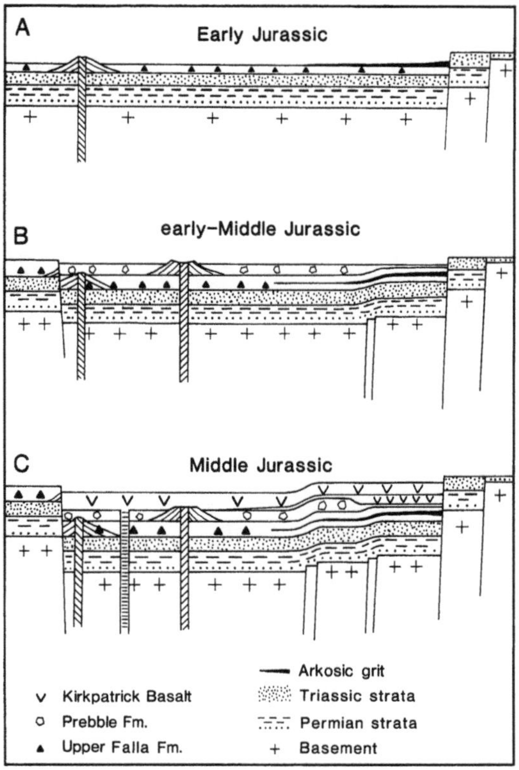

Fig. 8. Diagrammatic interpretation of the evolving rift basin in the Beardmore Glacier region. A: Early Jurassic - during deposition of the arkosic grits in the lower volcaniclastic member of the Falla Formation; B: early Middle Jurassic - during deposition of the Prebble Formation from basaltic volcanic centers such as at Kenyon Peaks; C: Middle Jurassic - after monoclinal flexuring of the Kirkpatrick Basalt.

sandstones but tuff beds comparable to those in the volcaniclastic part of the Falla Formation have not been found. However, silicic tuff megaclasts occur in the Mawson Formation (Bradshaw 1987), which overlies the Lashly Formation at Coombs Hills. Nowhere else is the basal contact exposed. The Mawson Formation, a basaltic breccia unit, overlies an unconformity which at Allan Hills cuts down at least 500 m through beds of Triassic age into the Permian Weller Coal Measures (Collinson et al. 1983). Beds under the Kirkpatrick Basalt at Carapace Nunatak consist of a 100+ m sequence of bedded volcaniclastic sandstones

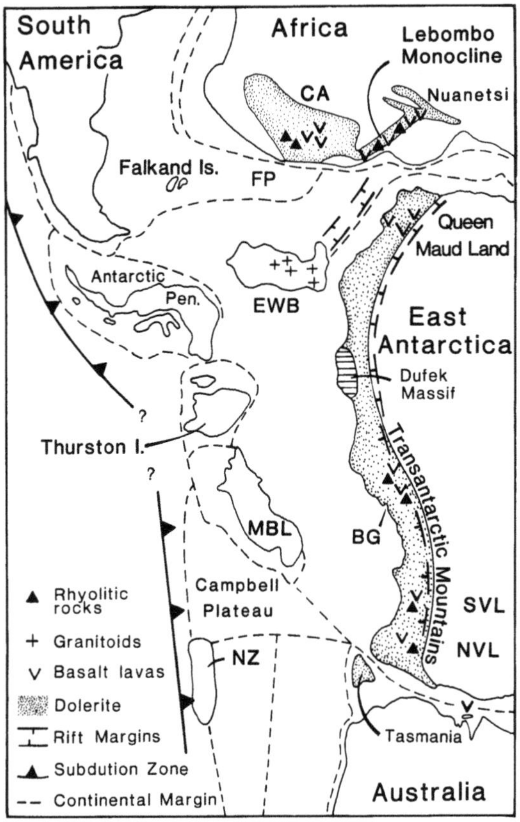

Fig.9. Gondwanaland reconstruction at 175 Ma, modified from Grunow et al. (1991) and Lawver et al. (1991). The age of the Antarctic and Australian rocks illustrated is about 175-180 Ma. Principal episodes of tholeiitic and related magmatism in southern Africa were 190-195 Ma in the Karoo Central Area (including intermediate to silicic rocks; see Marsh & Eales, 1984) and the southern Lebombo belt, and 175-180 Ma in the Nuanetsi region, north Lebombo belt and the Karoo Central Area (hypabyssal intrusions only in the Central Area). Only one margin is indicated for the Jurassic tectono-magmatic rift system in Antarctica because of uncertainties in the location of the other margin. The complete rift off Queen Maud Land is the site of the Explora Wedge, a dipping reflector sequence (Kristoffersen & Hinz, 1991) which is possibly correlative with the Lebombo Monocline. Crustal thinning between East Antarctica and Marie Byrd Land, Thurston Island and the Antarctic Peninsula was probably initiated during the Early to Middle Jurassic. New Zealand (NZ) is oriented so that the present east coast was facing the paleo-Pacific Ocean. BG: Beardmore Glacier; CA: Central Area of the Karoo; EWB: Ellsworth-Whitmore Mountains block; FP: Falkland Plateau; MBL: Marie Byrd Land; NVL: north Victoria Land; NZ: New Zealand; SVL: south Victoria Land.

(Bradshaw 1987), but elsewhere the basalts directly overlie breccias of the Mawson Formation. The unconformity at Allan Hills demonstrates vertical tectonism and a long period of erosion before emplacement of the breccias and overlying basalt lavas.

These sequences indicate that silicic volcanism, similar to that in the central Transantarctic Mountains occurred in Victoria Land. However, it is recorded in south Victoria Land only as clasts in younger basaltic breccias and in north Victoria Land only as a component in sandstones and tuffaceous rocks. Further, the south Victoria Land geology points to vertical tectonism before onset of basaltic magmatism, and the thick lavas in north Victoria Land indicate the presence of topography.

9 REGIONAL RELATIONS

The trend of Mesozoic tholeiitic magmatism across the center of Antarctica and parallel to the margin of the East Antarctic craton suggests a simple linear belt (Figure 9). Ford & Kistler (1980) noted the possibility that the Dufek Gabbro and Ferrar Group rocks mark, respectively, the locus and the failed arm of a triple junction. The outcrop distribution, however, is partly a function of ice cover. Little evidence exists for extensive sub-ice outcrop of the Beacon Supergroup and Ferrar Group except for a segment between about 135°W to 165°E on the polar plateau side of the Transantarctic Mountains (Drewry 1972, 1983). The distribution of the Ferrar Group in Victoria Land gives only a continent-scale indication of the trend of possible rift-related structures. Structural studies on Beacon and basement rocks indicate that in both south Victoria Land and the Beardmore Glacier region extension directions associated with Jurassic tectonism were oriented northeast-southwest and northwest-southeast (Wilson in press a). The trend of basement faulting in the Beardmore Glacier region, inferred from monoclinal structures, is north-northeast and sub-parallel to one of the fault arrays identified by Wilson (in press a), and presumably reflects more fundamentally significant structural trends in the upper crust than those displayed in the supracrustal rocks.

Cenozoic structures associated with uplift of the Transantarctic Mountains do not coincide with the Jurassic directions (Wilson in press b); no major displacements have been found which unequivocally cut only Beacon Supergroup strata. It is not possible to say whether the present outcrop pattern and inferred rift system marks the axis of rifting, one of a set of en echelon rifts which together constitute the rift system, or is only a region with relatively minor faulting on the flank of a major rift system. The apparent absence of major faults would suggest that the magmatism occurred on the flanks of the principal rift system which would have been located in what is now West Antarctica.

The granitic plutons in the Ellsworth-Whitmore Mountains block (Storey et al. 1988; Pankhurst et al., 1991) are part of the Early to Middle Jurassic episode of silicic magmatism in Antarctica which, in the Ross Sea sector of the Transantarctic Mountains, is represented by pyroclastic rocks. The Ellsworth-Whitmore Mountains crustal block, however, was located farther toward the Weddell Sea (Figure 9) during the Early to Middle Jurassic (Grunow et al. 1991; Lawver et al. 1991), and therefore the granitic plutonism of that region was unlikely to have been directly related to the volcanism in the Beardmore Glacier region and Victoria Land. Granite emplacement was coeval with the principal episode of Gondwanaland break-up magmatism (Elliot in press); the plutons are regarded as extension-related (Storey et al. 1988) and belong to the same tectono-magmatic province. In South Africa, Lower Jurassic rhyolitic rocks are prominent in the Lebombo Monocline and occur in a few scattered outcrops southwest of Lesotho as intrusions in beds stratigraphically below the basalts (Marsh & Eales 1984).

10 GEOLOGIC HISTORY

Sedimentation in the Permian to Triassic foreland basin of the Victoria Group (Collinson 1991) was brought to a close late in the Triassic or early in the Jurassic with the onset of silicic volcanism. Although the Lower to Upper Triassic Fremouw Formation and the lower part of the Falla Formation contain volcanic debris and tuffs, the volcanism was distal, and ash falls relatively uncommon. Sandstone-carbonaceous shale deposition of the Falla Formation was gradually overwhelmed by input of silicic debris consisting of: 1) ashfall deposits from Plinian eruptions; 2) local phreatomagmatic fallout deposits; and 3) reworked volcanic material in epiclastic and tuffaceous sandstones and siltstones. A marked southwestward thinning of the lower non-volcaniclastic part of the Falla Formation may express a regional warping associated with the onset of silicic volcanism, although this warping did not change the dominant northwesterly paleocurrent flow. The volcaniclastic sequence was interrupted in its lower part by input of locally-derived coarse arkosic grits which reflect basement uplifts beyond the outcrop area. Basement movements were later expressed within the outcrop area as monoclinal structures. This vertical tectonism created topography before eruption of basaltic breccias and accumulation of laharic deposits. During the same time interval vertical movements in south Victoria Land created topography on which similar coarse basaltic beds were deposited. Explosive basaltic activity was brought to an abrupt halt with the outpouring of flood basalts which accumulated to several hundred meters thickness. The ponded nature of many of the flows demonstrates that continuing tectonism maintained topography which confined the thick lavas, and this is supported by the monoclinal warping of lavas east of Lindsay Peak. Contemporaneously with eruption of lavas, thick dolerite sills and the layered basic intrusion of the Dufek Massif were emplaced. Silicic volcanism continued at a low level during eruption of lavas; the contemporaneous emplacement of granitoids occurred in the spatially-separated Ellsworth-Whitmore Mountains block.

The spatial coincidence and partial overlap of the basaltic and silicic magmatism, particularly in the Beardmore Glacier region, suggests they are related (Elliot 1991). The silicic magmas may represent products of anatexis caused by initial emplacement into the lower crust of mantle-derived basaltic magmas, which also introduced a thermal perturbation that was reflected as vertical tectonism at the surface. Subsequent migration of basaltic magmas to the surface led to explosive activity as the hot magmas interacted with groundwaters in the developing rift basins. The construction of volcanic landforms by this process was soon overwhelmed by the relatively quiet eruption of massive flood basalts which were also confined to the rift basins.

Acknowledgements: This work was supported by NSF grant DPP 84-19529 to DHE. We wish to thank our colleagues Tom Fleming and David Buchanan for their assistance with the fieldwork, and the U.S. Navy VXE-6 pilots and crew who enabled us to easily reach remote localities. We thank two anonymous reviewers and the editors for constructive comments.

REFERENCES

Barrett, P.J. & B.P. Kohn 1975. Changing sediment transport directions from Devonian to Triassic in the Beacon Super-Group of South Victoria Land. In: Campbell, K.S.W. (ed.) *Gondwana Geology*. Canberra: Australian National University Press: 15-35.

Barrett, P.J., D.H. Elliot & J.F. Lindsay 1986. The Beacon Supergroup (Devonian-Triassic) and Ferrar Group (Jurassic) in the Beardmore Glacier area, Antarctica. In: Turner, M.D & J.F. Splettstoesser (eds.), *Geology of the central Transantarctic Mountains*. American Geophysical Union, Antarctic Research Series 36: 339-428.

Bradshaw, M.A. 1987: Additional field interpretation of the Jurassic sequence at Carapace Nunatak and Coombs Hills, south Victoria Land, Antarctica. *New Zealand Journal of Geology and Geophysics* 30: 37-49.

Cleverly, R.W., P.J. Betton & J.W. Bristow 1984. Geochemistry and Petrogenesis of the Lebombo Rhyolites. *Special Publication of the Geological Society of South Africa* 13: 171-194.

Collinson, J.W. 1991. The palaeo-Pacific margin as seen from East Antarctica. In: Thomson, M.R.A., J.W. Crame & J.W. Thomson (eds.), *Geological evolution of Antarctica*. Cambridge: Cambridge University Press: 199-204.

Collinson, J.W. & D.H. Elliot 1984. Geology of Coalsack Bluff, Antarctica. In: Turner, M.D. & J.F. Splettstoesser (eds.), *Geology of the Central Transantarctic Mountains*. American Geophysical Union, Antarctic Research Series 36: 97-102.

Collinson, J.W., D.C. Pennington & N.R. Kemp 1983. Sedimentary petrology of Permian-Triassic Fluvial rocks in the Allan Hills, central Victoria Land. *Antarctic Journal of the U.S.* 18(5): 20-22.

Collinson, J.W., D.C. Pennington & N.R. Kemp 1986. Stratigraphy and petrology of Permian and Triassic fluvial deposits in northern Victoria Land, Antarctica. *In: Stump, E. (ed.), Geological investigations in northern Victoria Land*. American Geophysical Union, Antarctic Research Series 46: 211-242.

Crandell, D.R. 1971. Postglacial lahars from Mount Rainier volcano, Washington. *U.S. Geological Survey Professional Paper* 677: 1-75.

Drewry, D.J. 1972. Subglacial morphology between the Transantarctic Mountains and the South Pole. In: Adie, R.J. (ed.), *Antarctic Geology and Geophysics*, Oslo, Universitetsforlaget: 693-703.

Drewry, D.J. 1983. *Antarctica: glaciological and geophysical folio*. Cambridge: Scott Polar Research Institute.

Eales, H.V., J.S. Marsh & K.G. Cox 1984. The Karoo Igneous Province: an Introduction. *Special Publication of the Geological Society of South Africa*, 13: 1-26.

Elliot, D.H. 1991. Triassic-Early Cretaceous evolution of Antarctica. In: Thomson, M.R.A., J.A. Crame & J.W. Thomson (eds.), *Geological Evolution of Antarctica*. Cambridge, Cambridge University Press: 541-548.

Elliot, D.H. in press. Jurassic magmatism and tectonism associated with Gondwanaland break-up: an Antarctic perspective. In: Storey, B., T. Alabaster & R.J. Pankhurst (eds.), *Magmatism and the causes of continental break-up*. Geological Society of London, Special Publication 68.

Elliot, D.H., R.J. Fleck & J.F. Sutter 1985. Potassium-Argon age determinations of Ferrar Group rocks, central Transantarctic

Mountains. In: Turner, M.D. & J.F. Splettstoesser (eds.), *Geology of the Central Transantarctic Mountains*. American Geophysical Union, Antarctic Research Series 36: 197-224.

Elliot, D.H., M.A. Haban & M.A. Siders 1986a. The Exposure Hill Formation, Mesa Range. In: Stump, E. (ed.), *Geological investigations in northern Victoria Land*. American Geophysical Union, Antarctic Research Series 46: 267-278.

Elliot, D.H., M.A. Siders & M.A. Haban 1986b. Jurassic tholeiites in the region of the upper Rennick Glacier, North Victoria Land. In: Stump, E. (ed.), *Geological investigations in northern Victoria Land*. American Geophysical Union, Antarctic Research Series 46: 248-265.

Elliot, D.H., J. Bigham & F.S. Jones 1991. Interbeds and weathering profiles in the Jurassic basalt sequence, Beardmore Glacier region, Antarctica. In: Ulbrich, H. & A.C. Rocha Campos (eds.), *Gondwana Seven*, Instituto de Geociências, Universidade de São Paulo: 289-302.

Ewart, A. 1979. A review of the mineralogy and chemistry of Tertiary-Recent dacitic, latitic, rhyolitic, and related salic volcanic rocks. In: Barker, F. (ed.), *Trondhjemites, dacites, and related rocks*. Amsterdam: Elsevier: 13-122.

Faure, G. & R.L. Hill 1973. Age of the Falla Formation (Triassic), Queen Alexandra Range. *Antarctic Journal of the U.S.* 8(5): 264-266.

Ford, A.B. & R.W. Kistler 1980. K-Ar age, composition, and origin of Mesozoic mafic rocks related to Ferrar Group, Pensacola Mountains, Antarctica. *New Zealand Journal of Geology and Geophysics* 23: 371-390.

Grunow, A.M., D.V. Kent & I.W.D. Dalziel 1991. New paleomagnetic data from Thurston Island: implications for the tectonics of West Antarctica and Weddell Sea opening. *Journal of Geophysical Research* 96: 17935-17954.

Gust, D.A., K.T. Biddle, D.W. Phelps & M.A. Uliana 1985. Associated Middle to Late Jurassic volcanism and extension in southern South America. *Tectonophysics* 114: 293-313.

Kristoffersen, Y. & K. Hinz 1991. Evolution of the Gondwana plate boundary in the Weddell Sea area. In: Thomson, M.R.A., J.A. Crame & J.W. Thomson (eds.), *Geological Evolution of Antarctica*. Cambridge, Cambridge University Press: 225-230.

Larsen, D. 1988. The petrology and geochemistry of the volcaniclastic upper part of the Falla Formation and Prebble Formation, Beardmore Glacier area, Antarctica. Unpublished M.Sc.thesis, The Ohio State University, Columbus, U.S.A.: 273 p.

Lawver, L.A., J.-Y. Royer, D.T. Sandwell & C. R. Scotese 1991. Evolution of the Antarctic continental margin. In: Thomson, M.R.A., J.A. Crame & J.W. Thomson (eds.), *Geological evolution of Antarctica*. Cambridge: Cambridge University Press: 533-539.

Marsh, J.S. & H.V. Eales 1984. The chemistry and petrogenesis of igneous rocks of the Karoo Central Area, southern Africa. *Special Publication of the Geological Society of South Africa*, 13: 27-67.

Pankhurst, R.J., B.C. Storey & I.L. Millar 1991. Magmatism related to the break-up of Gondwana. In: Thomson, M.R.A., J.A. Crame & J.W. Thomson (eds.), *Geological evolution of Antarctica*. Cambridge: Cambridge University Press: 573-579.

Pearce, J.A., N.B.W. Harris & A.G. Tindle 1984. Trace element discrimination diagrams for the tectonic interpretation of granitic rocks. *Journal of Petrology* 25: 956-983.

Piccirillo, E.M., A.J. Melfi, P. Comin-Chiaramonti, G. Bellieni, M. Ernesto, L.S. Marques, A.J.R. Nardy, I.G. Pacca, A. Roisenberg & D. Stolfa 1988. Continental flood volcanism from the Paraná Basin (Brazil). In: Macdougall, J.D. (ed.), *Continental flood basalts*. Kluwer Academic Publishers: 195-238.

Retallack, G.J. 1983. Late Eocene and Oligocene paleosols from Badlands National Park, South Dakota. *Geological Society of America Special Paper* 193: 83 p.

Retallack, G.J. 1988. Field recognition of paleosols. *Geological Society of America Special Paper* 216: 20 p.

Rodolfo, K.S. 1989. Origin and early evolution of lahar channel at Mabinit, Mayon Volcano, Philippines. *Geological Society of America Bulletin* 101: 414-426.

Storey, B.C., M.J. Hole, R.J. Pankhurst, I.L. Millar & W. Vennum 1988. Middle Jurassic within-plate granites in West Antarctica and their bearing on the break-up of Gondwanaland. *Journal of the Geological Society of London* 145: 999-1007.
Sun, S-s. 1980. Lead isotope study of young volcanic rocks from mid-ocean ridges, ocean islands, and island arcs. *Philosophical Transactions of the Royal Society of London* A297: 409-445.
Thompson, R.N., A.P. Dickin, I.L. Gibson & M. A. Morrison 1982. Elemental fingerprints of isotopic contamination of Hebridean Palaeocene mantle-derived magmas by Archean sial. *Contributions to Mineralogy and Petrology* 79: 159-168.
Whalen, J.B., K.L. Currie & B.W. Chappell 1987. A-type granites: geochemical characteristics, discrimination and petrogenesis. *Contributions to Mineralogy and Petrology* 95: 407-419.
White, A.J.R. & B.W. Chappell 1983. Granitoid types and their distribution in the Lachlan Fold Belt, southeastern Australia. In: Roddick, T.A. (ed.), *Circum-Pacific Plutonic Terranes*. Geological Society of America Memoir 159: 21-34.
White, R. & D. McKenzie 1989. Magmatism at rift zones: the generation of volcanic continental margins and flood basalts. *Journal of Geophysical Research* 94: 7685-7729.
Wilson, T.J. in press a. Jurassic faulting and magmatism in the Transantarctic Mountains: implications for Gondwana breakup mechanisms. In: Findlay, R.H., M.R. Banks, J.J. Veevers & R. Unrug (eds.), *Gondwana 8: assembly, evolution, dispersal*. Rotterdam:A.A. Balkema.
Wilson, T.J. in press b. Mesozoic and Cenozoic kinematic evolution of the Transantarctic Mountains. In: Yoshida, Y. (ed.), *Recent Progress in Antarctic Earth Science*. Terra Scientific Publishing Company Tokyo.

Gondwana Eight, Findlay, Unrug, Banks & Veevers (eds) © 1993 Balkema, Rotterdam. ISBN 90 5410 304 3

Disturbance of Rb-Sr and K-Ar isotopic systems in the Kirkpatrick Basalt, north Victoria Land, Antarctica: Implications for middle Cretaceous tectonism

Thomas H. Fleming, David H. Elliot & Kenneth A. Foland
Department of Geological Sciences & Byrd Polar Research Center, Ohio State University, Columbus, Ohio, USA
Lois M. Jones
Department of Geology, Kansas State University, Manhattan, Kans., USA
John R. Bowman
Department of Geology and Geophysics, University of Utah, Salt Lake City, Utah, USA

ABSTRACT: Chemical and isotopic variations in an Fe-rich lava flow which caps the Kirkpatrick Basalt lava sequence in north Victoria Land provide evidence for middle Cretaceous alteration. $^{40}Ar/^{39}Ar$ incremental heating analyses of plagioclase separates yield good plateaus, firmly establishing an age of 177 $\pm$2 Ma for the flow and providing a minimum age for the entire lava sequence. On a Rb-Sr isochron diagram, whole-rock samples from this flow define a 103 Ma array whereas magnetically separated fractions from a single sample fall along a different 165 Ma trend. The whole-rock Rb-Sr isotopic data are interpreted to reflect an episode of alteration during the middle Cretaceous. Disturbance of the K-Ar system at this time is reflected in low dates resulting from Ar loss. Several paleomagnetic studies of the Ferrar Group in north Victoria Land report anomalous pole positions which may also reflect alteration during the middle Cretaceous. This alteration event appears to be associated with the development of the Rennick Graben which probably formed as a failed rift arm at about the same time that sea-floor spreading was initiated between Australia and Antarctica. Aqueous convective systems generated within the graben during the middle Cretaceous are the most logical explanation for the paleomagnetic pole positions, Sr isotopic variations, and range of K-Ar dates previously reported for the Ferrar Group in north Victoria Land.

1 INTRODUCTION

In north Victoria Land, Jurassic tholeiites of the Ferrar Group are represented by the Exposure Hill Formation, Kirkpatrick Basalt and Ferrar Dolerite (Elliot et al., 1986a, 1986b). These rocks are confined to the region of the Rennick Graben, a structure that may have developed during the separation of Australia and Antarctica in the late Mesozoic (Roland & Tessensohn, 1987). Paleomagnetic studies of rocks from the Ferrar Group in north Victoria Land document virtual geomagnetic pole (VGP) positions significantly different from those of the Ferrar Group in other regions of the Transantarctic Mountains. The difference in VGP has been interpreted to reflect either a Cretaceous age for the basalts in north Victoria Land, a magnetic overprint resulting from a middle Cretaceous thermal event (Delisle, 1983; Delisle & Fromm, 1984, 1989), or incomplete averaging of paleosecular variation by the lavas in north Victoria Land (McIntosh et al., 1986). This paper presents new data confirming a Jurassic age for the eruption of the Kirkpatrick Basalt in the Mesa Range, presents evidence for disturbance of Rb-Sr and K-Ar isotopic systems of the lavas by a middle Cretaceous event, and discusses the implications of such an event.

2 REGIONAL GEOLOGY

The lower Paleozoic and older basement rocks in north Victoria Land comprise three, or perhaps as many as five, terranes which are separated by major northwest trending faults (Tessensohn et al., 1981; Bradshaw et al., 1985) (Figure 1). The igneous and metamorphic Wilson terrane, formerly the margin of the East Antarctic craton, is

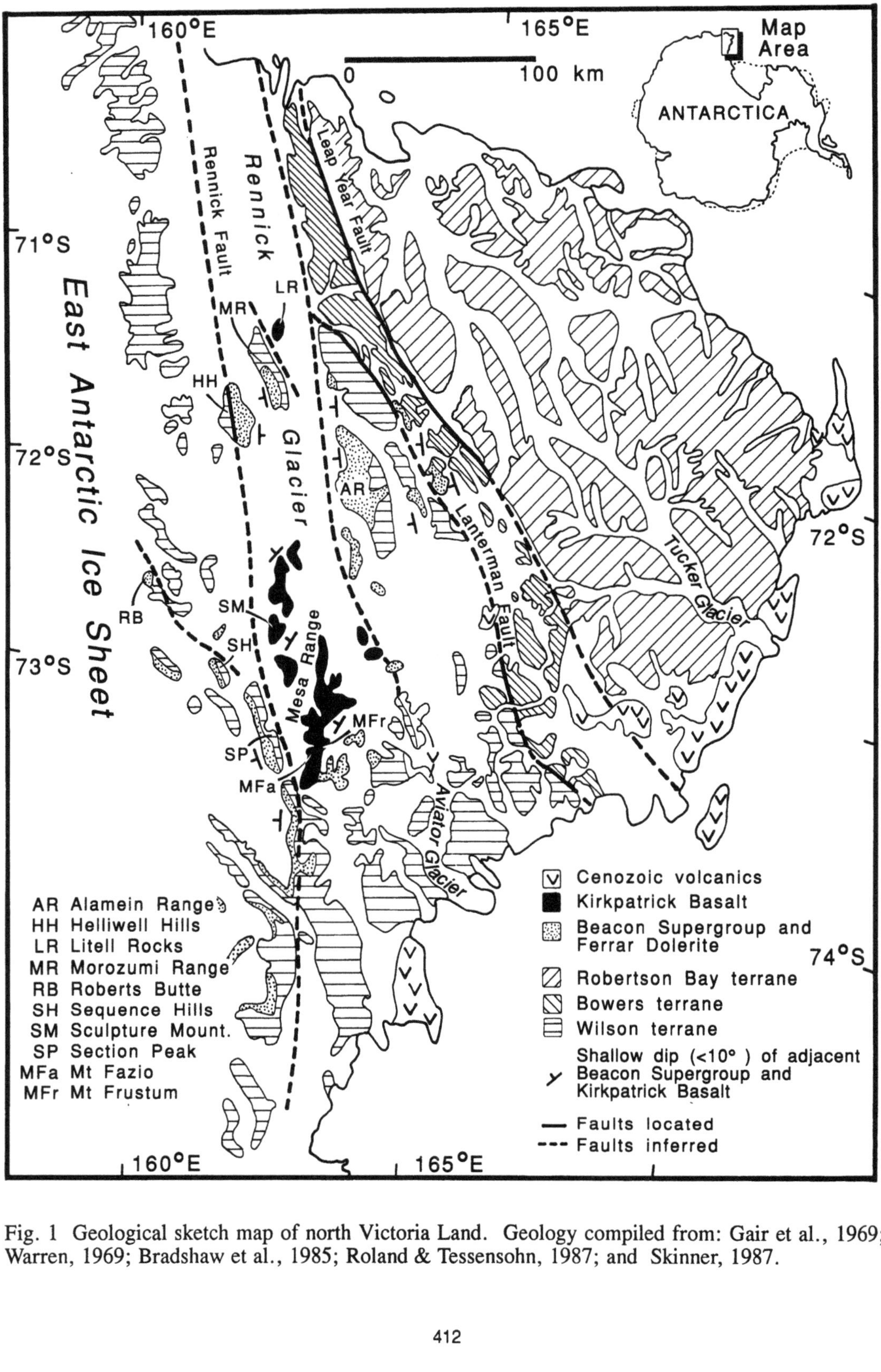

Fig. 1 Geological sketch map of north Victoria Land. Geology compiled from: Gair et al., 1969; Warren, 1969; Bradshaw et al., 1985; Roland & Tessensohn, 1987; and Skinner, 1987.

separated from the Bowers terrane to the northeast by the Lanterman Fault. The Bowers terrane, consisting mainly of clastic sedimentary sequences, is separated by the Leap Year Fault from the Robertson Bay terrane which lies yet farther to the northeast and consists primarily of turbidites. The Beacon Supergroup, of Late Carboniferous to Triassic age, overlies the basement of the two westernmost terranes and consists of about 750 meters of glacial and fluvial deposits (Collinson et al., 1986). Jurassic dolerite sills of the Ferrar Group intrude the basement of the Wilson and Bowers terranes and the overlying Beacon Supergroup, but have not been found in the Robertson Bay terrane. As much as 780 meters of basalts, which are chemically related to the sills, overlie the Beacon Supergroup sedimentary sequence (Elliot et al., 1986a). Basalt outcrops are confined to the Rennick Graben, which has developed since the Jurassic (Roland & Tessensohn, 1987); dolerite sills, however, extend beyond the inferred limits of the graben. The only younger rocks, other than glacial deposits, are Upper Cenozoic alkaline volcanic and plutonic rocks (Kyle, 1990). Along the Ross Sea coast, between the Aviator and Tucker Glaciers, about 10 km of uplift in the last 50 Ma has been inferred from fission track data, whereas farther west, along the Polar Plateau, uplift has been interpreted to be on the order of 3-4 km (Fitzgerald & Gleadow, 1988).

The chemistry of the lava sequence in north Victoria Land has been described by Mensing et al. (1984) and Siders and Elliot (1985). The capping flow unit in the Mesa Range has a unique iron-rich chemistry referred to as the Scarab Peak chemical type (SPCT) (Figure 2). Detailed chemical analysis of this flow shows extreme homogeneity except for small variations in the alkalis that are attributed to alteration. The unique chemistry and field characteristics of this flow have allowed first-order correlation of the lava sequence throughout the Mesa Range (Elliot et al., 1986a). The capping flow at Litell Rocks, approximately 150 km to the north, has a virtually identical chemical composition (Skinner et al., 1981). Flows which underlie the SPCT, together with the dolerite sills, are referred to as the Mt. Fazio chemical type (MFCT). These rocks, chemically distinct from the SPCT flow unit, tend to form coherent trends on elemental variation diagrams (Figure 2). Volumetrically, the MFCT rocks represent the bulk of the Ferrar Group in north Victoria Land.

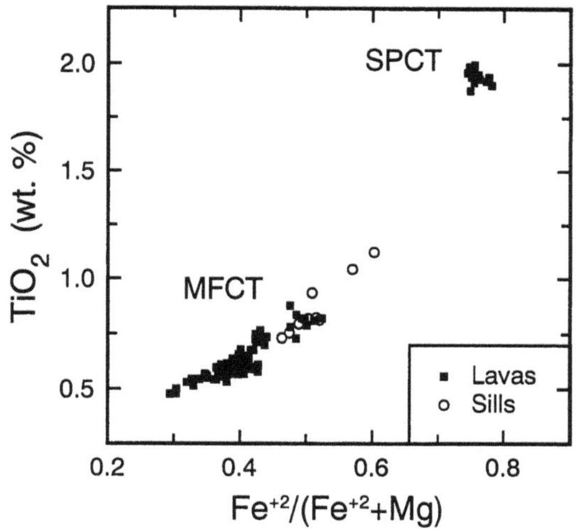

Fig. 2 Chemical variation diagram showing the distinct chemistry of the Scarab Peak chemical type (SPCT) and the Mt. Fazio chemical type (MFCT). Data from: Siders, 1983; Haban, 1984; Mensing et al., 1984; and Fleming, 1987.

Nearly all the rocks of the Ferrar Group in north Victoria Land show some signs of alteration, causing difficulty in petrogenetic interpretations of the Sr and O isotope data (Fleming et al., 1989, 1992). Quartz, chalcedony, carbonate, stilbite, apophyllite, heulandite and natrolite occur as fracture and amygdule fillings in the Mesa Range (Elliot et al., 1986a) and prehnite, babingtonite and rarely epidote are found at Litell Rocks (Skinner et al., 1981). Amygdaloidal zones of flow tops, commonly weathered to a very light brown color, may be either massive or rubbly and show extensive oxidation and secondary mineralization as listed above. Away from the amygdaloidal zones, basalts are less altered but invariably show some alteration of the fine-grained or glassy matrix. Typically, titanomagnetite is partially converted to titanomaghemite. In general, however, the microphenocrysts of plagioclase and pyroxene remain unaltered. Chemically, the alteration is revealed by a high degree of scatter of the more mobile elements (e.g., K, Na, Rb, and Sr) on variation diagrams, whereas the less mobile elements (e.g., Zr, Ti, Y, REE, Cr) show good correlations (Fleming et al., 1989, 1992). The high loss-on-ignition values and $H_2O(+)$ contents also reflect substantial alteration and hydration. The MFCT rocks are generally more altered than their SPCT counterparts because of their more mafic character (MgO = 4.5-7.5% compared to MgO = 2.2%, respectively). The sills are the exception, owing to their lack of amygdaloidal zones that would readily transmit fluids, the absence of glass commonly found in the lavas, and their less mafic chemistry (MgO = 3.7-5.2%). Overall, the alteration indicates that rocks of the Ferrar Group have been affected by significant amounts of hydrothermal fluids sometime in their history; however, field observations cannot determine whether the alteration occurred shortly after eruption or at a significantly later date.

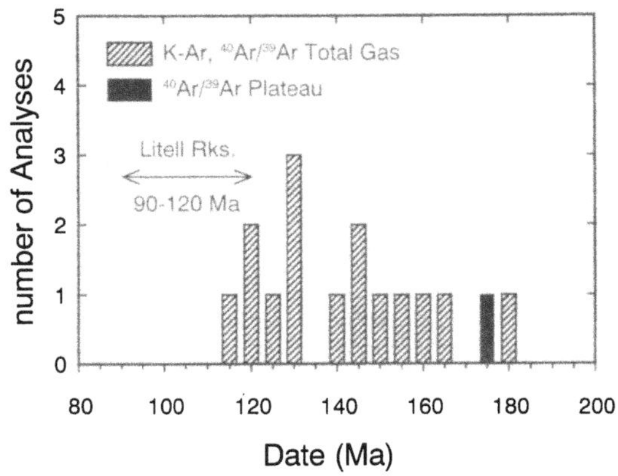

Fig. 3 Histogram of previously published K-Ar and $^{40}Ar/^{39}Ar$ dates from the Ferrar Group in north Victoria Land. Data from: Dow & Neall, 1972; Kreuzer et al., 1981; Elliot & Foland, 1986; McIntosh et al., 1986; and Brotzu et al., 1988.

3 AGE OF THE BASALTS

Conventional K-Ar determinations of lavas and sills from north Victoria Land yield dates ranging from about 90 to 180 Ma (Figure 3). The crystallization age of the rocks has been generally considered to be 180 ±5 Ma with the range of dates attributed to partial ^{40}Ar loss from glassy or altered matrix material (Elliot & Foland, 1986). A $^{40}Ar/^{39}Ar$ incremental-heating analysis of a whole-rock sample collected near the base of the lava sequence produced a discordant age spectrum, but with a 174 ±1 Ma plateau consisting of two increments (McIntosh et al., 1986), which is in general agreement with the interpretation of the conventional K-Ar data by Elliot and Foland (1986).

The capping SPCT flow has yielded discordant K-Ar dates of 178 ±3 Ma and 143 ±3 Ma (Elliot & Foland, 1986) and a discordant $^{40}Ar/^{39}Ar$ incremental-heating spectrum with a total gas age of 153 Ma (McIntosh et al., 1986). Elliot and Foland (1986) interpreted the 178 Ma K-Ar date as a minimum crystallization age for the flow. Mensing et al. (1988) have inferred, however, that the uppermost unit in the Mesa Range may be approximately 92 Ma based on whole-rock Rb-Sr data. This interpretation requires that the older K-Ar ages for the SPCT flow result from excess argon. In addition, a sill with chemical and Sr isotope compoition identical to the SPCT flow in north Victoria Land caps the lava sequence at Gorgon Peak in south Victoria Land and has given a $^{40}Ar/^{39}Ar$ plateau age of 163 ±2 Ma (Kyle et al., 1981; P.R. Kyle personal communication, 1991). From this apparent age, the sill has been considered to be about 15 Ma younger than the associated lava flows which underlie it. Based upon the previous work, the age of the SPCT flow in north Victoria Land is poorly constrained and previous workers have favored significantly different age assignments. Nevertheless, the age of the SPCT flow is important not only for interpreting the geologic history of north Victoria Land but also for understanding the petrogenesis of the Ferrar Group.

To clarify the age of the SPCT basalts, $^{40}Ar/^{39}Ar$ incremental-heating analyses have been performed on plagioclase and matrix concentrates magnetically extracted from a sample of the uppermost lava flow at Mt. Frustum in the Mesa Range. This sample previously yielded the conventional K-Ar whole-rock date of 143 ±3 Ma reported by Elliot and Foland (1986). The sample contains unaltered pyroxene and plagioclase phenocrysts in a partially devitrified and altered glassy matrix. Incremental heating of matrix-dominated fractions gives a steady increase in apparent age from about 110 Ma to a maximum of 175-177 Ma (Figure 4). The integrated

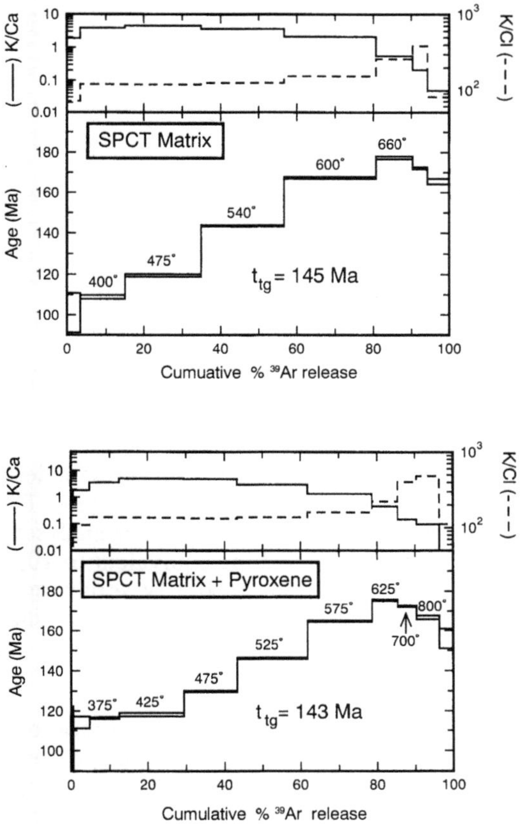

Fig. 4 $^{40}Ar/^{39}Ar$ age spectra for matrix-rich fractions of sample 81-7-2 from the Scarab Peak chemical type (SPCT) flow at Mt. Frustum. The total-gas ages are identical to the conventional K-Ar date reported for the whole-rock by Elliot & Foland, 1986. The discordant spectra have apparent age minima of ≈110 Ma and maxima at ≈ 175 Ma.

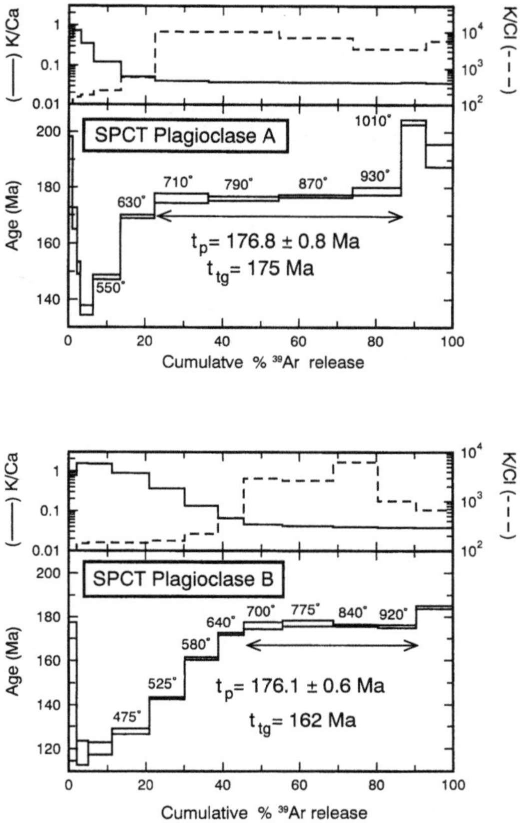

Fig. 5 $^{40}Ar/^{39}Ar$ age spectra for plagioclase-rich fractions of sample 81-7-2 from the Scarab Peak chemical type (SPCT) flow at Mt. Frustum. Fraction A contains ≈97% plagioclase and Fraction B ≈91% plagioclase. The remaining component in both fractions is matrix material attached to plagioclase microphenocrysts. This matrix material produces the low temperature discordance in the age specta. The plateau dates of the two spectra are analytically indistinguishable.

total-gas ages are within uncertainty of the conventional K-Ar date of the whole-rock. These matrix-rich fractions lose more than 90% of their Ar at laboratory heating temperatures below 700°C.

The $^{40}Ar/^{39}Ar$ results for plagioclase fractions markedly contrast with those for the matrix concentrates. The two plagioclase concentrates show a gradual increase in apparent age during low temperature increments but have well defined plateaus of 177 ±2 Ma at temperatures between 700-930°C (Figure 5). K/Ca and K/Cl ratios measured from Ar isotopes produced during irradiation indicate that the plateaus represent release of Ar from plagioclase whereas the low temperature discordance reflects minor matrix impurities with higher K and Cl contents. Because the matrix has high concentrations of K relative to plagioclase, even minor amounts result in low temperature discordance. The discordance is more marked in the separate which is less pure (Figure 5; plagioclase B). An entirely consistent framework to understand the $^{40}Ar/^{39}Ar$ ages is given by the relations between the apparent age, K/Ca and K/Cl ratios and the purity of the separates. Young ages are due to ^{40}Ar loss from the matrix components.

The 177 ± 2 Ma date from the plagioclase separates is considered a highly reliable age for crystallization of the flow. It also provides a minimum age for the entire lava sequence in north Victoria Land. It does not, however, constrain the age of onset of magmatism and further work is required to evaluate the duration of magmatic activity represented by the Ferrar Group. The low temperature $^{40}Ar/^{39}Ar$ steps and the previously reported young K-Ar dates (90-145 Ma) are both consistent with an episode of ^{40}Ar loss from glassy matrix material, most logically during the middle Cretaceous.

4 Rb-Sr ISOTOPIC DATA

Fleming et al. (1989, 1992) have shown that, on an Rb-Sr isochron diagram, the compositions of 25 whole-rock samples from the SPCT unit fall along a line with a slope equivalent to 103 Ma (Figure 6). This is similar to the 93 Ma Rb-Sr array described by Mensing et al. (1988) for the same unit. In contrast, magnetically separated fractions from a single sample which was also used for the $^{40}Ar/^{39}Ar$ analyses described above do not fall along the 103 Ma line but rather fall along a 165 Ma trend (Figure 6) (Fleming et al., 1989, 1992). Neither of these linear Rb-Sr trends is a valid isochron, sensu stricto, because the scatter of data about a regression line exceeds the deviations which can be attributed to analytical error. Such trends have been termed "errorchrons" to indicate uncertainty for the apparent age due to "geological error" from violation of assumptions of the Rb-Sr method (Brooks et al., 1972). The 103 Ma array contrasts with the 165 Ma array and indicates that these samples have not behaved as closed systems since crystallization.

Rather than reflecting a crystallization age as previously suggested by Mensing et al. (1988), the apparent middle Cretaceous Rb-Sr trend for the SPCT whole-rocks is interpreted to reflect Rb mobility in an originally homogeneous flow due to alteration of the abundant glassy and fine-grained matrix. The simplest interpretation is that Rb was redistributed in the matrix at about 100 Ma. The 165 Ma array is also considered to be an artifact of this alteration and believed to be a mixing line resulting from different proportions of altered matrix and unaltered plagioclase microphenocrysts in the magnetically separated fractions; therefore, it indicates neither the age of the rocks nor the age of a possible alteration event.

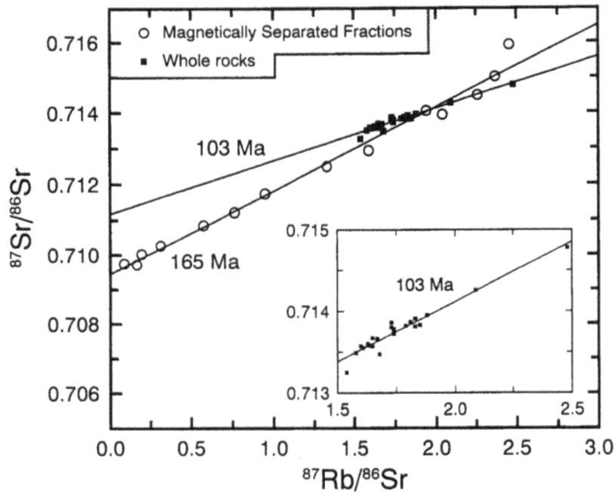

Fig. 6 Rb-Sr isochron diagram for samples from the Scarab Peak chemical type (SPCT) flow in the Mesa Range. Whole-rock analyses form an array with a slope equivalent to 103 Ma. A series of magnetically separated fractions from sample 81-7-2 give an array with a slope equivalent to 165 Ma.

5 PALEOMAGNETIC DATA

The majority of lavas and sills from the Ferrar Group in north Victoria Land record a virtual geomagnetic pole (VGP) offset from VGPs of rocks from the Ferrar Group at other localities farther south in the Transantarctic Mountains (McIntosh et al., 1982; 1986) (Figure 7). Several studies have suggested that the rocks may carry a significant component of secondary chemical remanent magnetization (CRM). Delisle (1983) showed the lavas at Litell Rocks have a VGP which lies near the apparent polar wander path (APW) for Australia (corrected for drift) at approximately 100 Ma and suggested the lavas either have a middle Cretaceous age or a middle Cretaceous magnetic overprint. Delisle and Fromm (1984, 1989) provided further paleomagnetic data in support of a middle Cretaceous alteration event; they also

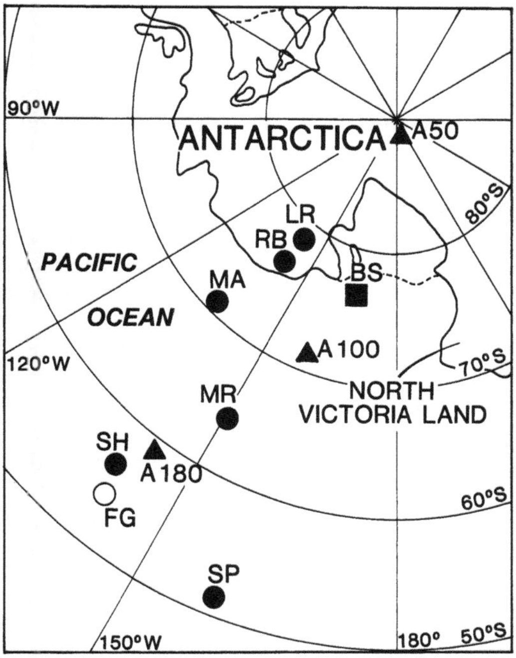

Fig. 7 Virtual geomagnetic pole (VGP) positions for rocks of the Ferrar Group from north Victoria Land (LR - Litell Rocks; MA - Morozumi Range and Alamein Range; MR - Mesa Range; RB - Roberts Butte; SH - Sequence Hills; SP - Section Peak) (Delisle & Fromm, 1983, 1989; McIntosh et al., 1986) compared with those from other areas of the Transantarctic Mountains (FG). The inferred overprint pole position from the Bowers Supergroup is also indicated (BS) (Schmierer & Burmester, 1986). The 180 Ma, 100 Ma and 50 Ma magnetic pole positions for Australia (A180, A100, A50, respectively) (Embleton & McElhinny, 1982) are shown rotated to the Antarctic frame of reference based on the reconstructions of Veevers and Eittreim (1988).

argued that the paleomagnetic data may record a second event near the Cretaceous/Tertiary boundary. Cherry and Noltimier (1984, 1985) identified a significant component of CRM resulting from low temperature oxidation of titanomagnetite to titanomaghemite in lavas from the Mesa Range. In addition, from examination of the paleomagnetic variations in an 86 meter thick flow at Mt. Fazio in the Mesa Range, they concluded that the CRM component may significantly post-date the primary thermo-remanent magnetization (TRM). Cherry (1984) also found evidence for a CRM component in the lavas at Sculpture Mountain in the Mesa Range. Conversely, McIntosh et al. (1986) found only slight evidence for maghemitization of the basalts and suggested the southerly VGP from the Mesa Range lavas is unlikely to be due to an unremoved CRM. They preferred to attribute it to incomplete averaging of secular variations of the earth's magnetic field. Our own observations by reflected light microscopy and electron microprobe analysis suggest that maghemitization of titanomagnetite is common, but variable in extent, in the Mesa Range basalts. It seems likely, therefore, that the southerly VGP of rocks from the Ferrar Group in north Victoria Land is caused by a middle Cretaceous overprint. A more precise age than "middle Cretaceous" is not warranted, however, because of large uncertainties for the overprint pole position and for the APW for Antarctica which, by necessity, is reconstructed from the Australian APW for that period.

A few rocks from the Ferrar Group in north Victoria Land appear to retain their original TRM magnetization unaffected by a CRM component. Sills west of the Mesa Range at Section Peak (McIntosh et al., 1986) and Sequence Hills (Delisle & Fromm, 1989) record VGP positions which are substantially further north than other rocks of the Ferrar Group in north Victoria Land and which are similar to those of Jurassic igneous rocks in other parts of the Transantarctic Mountains (Figure 7). Delisle and Fromm (1989) noted that the lower three flows examined by McIntosh et al. (1986) also have VGPs more typical of the Ferrar Group in other areas of the Transantarctic Mountains.

The paleomagnetic signature of basement rocks of the Bowers terrane may also record an alteration event during the middle Cretaceous. The Bowers Supergroup, in part representing a Cambro-Ordovician arc complex, is presently exposed in a narrow belt between the Lanterman and Leap Year faults (Laird et al., 1982; Weaver et al., 1984; Tessensohn, 1984; Wodzicki & Robert, 1986). Schmierer and Burmester (1986) have identified a paleomagnetic component in these rocks which lies close to the 100 Ma poles for Australia (corrected for drift). They have interpreted this component as a middle Cretaceous magnetic overprint (Figure 7). Because the rocks have been metamorphosed only to the prehnite-pumpellyite facies, probably during the early Paleozoic Ross Orogeny, the middle Cretaceous magnetization must have been acquired at relatively low temperatures (≤ 200°C).

6 DISCUSSION

Rb-Sr, K-Ar and paleomagnetic results from the basalts and dolerites suggest a disturbance in the middle Cretaceous. The currently available data, however, do not provide good constraints on the temperature attained during this disturbance. Although elevated temperatures are more likely to promote chemical exchange, the extremely low grade (or lack) of metamorphism in most rocks of the Beacon Supergroup (Collinson et al., 1986) and Ferrar Group (Elliot et al., 1986a) suggests that temperatures were less than 300°C. Moreover, early Paleozoic K-Ar dates from crystalline basement rocks of the Wilson and Bowers terranes are relatively undisturbed, which further supports this conclusion (Kreuzer et al., 1981; Adams et al., 1982; Vetter et al., 1983). Redistribution of Rb and loss of Ar from the glassy matrix material in the basalts could easily have occurred at very low temperatures (<100°C), particularly if facilitated by ground water circulation.

Titanomaghemite, which carries the middle Cretaceous magnetic signature in the Ferrar Group in north Victoria Land (Delisle & Fromm, 1989), may also be formed at near-surface temperatures (O'Reilly, 1984). Extensive exchange with meteoric water at low temperatures (0-150°C) is shown by oxygen isotope relations in the SPCT flow. Specifically, there is a 2 per mil difference between $\delta^{18}O$ of plagioclase (+5.5 per mil) and matrix separates (up to +7.5 per mil); also $\delta^{18}O$ of the whole-rock samples ranges from +5.8 to +8.2 per mil which is argued to reflect varying degrees of ^{18}O enrichment (Fleming et al., 1992). The timing of this O isotope exchange is difficult to assess. Some exchange probably occurred in the Jurassic as a result of hydrothermal processes associated with emplacement of the Ferrar Group. Additional exchange was likely during the middle Cretaceous because maghemite, which carries the paleomagnetic component of that age, forms by oxidation during interaction with water. The possibility that the range in $\delta^{18}O$ values has resulted from exchange with water during the Late Cenozoic can be eliminated because of the highly ^{18}O-depleted composition of precipitation during that period.

The K-Ar and paleomagnetic results indicate that the processes which have altered the rocks did not operate uniformly throughout the lava sequence, particularly in the Mesa Range. At Mt. Frustum, texturally identical samples, collected only a few meters apart in the same flow, show dramatically different K-Ar ages of 143 ±3 Ma and 178 ±3 Ma (Elliot & Foland, 1986). For rocks in such close proximity, this difference suggests localized effects such as might be developed by fluid/rock interaction (Elliot & Foland, 1986). It is likely, therefore, that the migration of fluids has played a more important role in controlling the alteration process than simple re-heating. The range of VGPs from lavas in the Mesa Range also suggests local variations in alteration. The lowest flows in the sequence, which would be expected to reach the highest temperatures (Flows 1, 2, and 3 of McIntosh et al., 1986), appear to be least affected by overprinting. This may reflect either local fluid control of the alteration process or the original magnetic mineralogy and texture of the flows.

A limited number of published fission track dates from basement rocks (Fitzgerald & Gleadow, 1988) may provide some constraint on the areal extent of the event. The presence of older apatite fission track dates (196-310 Ma) in the basement rocks immediately west of the Mesa Range indicates the region was not heated above ≈120°C at any time since about 200 Ma. Paleomagnetic study of sills at Sequence Hills and Section Peak, also just west of the Mesa Range, have given Jurassic pole positions (McIntosh et al., 1986; Delisle & Fromm, 1989), similarly suggesting the hydrothermal event may not have affected this area. To the north, at Helliwell Hills, 100 Ma fission track dates fall off the regional uplift trend formed by apatites farther east. The Helliwell Hills dates have been interpreted as uplift ages of a crustal block involved in less uplift than rocks farther to the east (Fitzgerald & Gleadow, 1988). As an alternative, these dates might reflect cooling ages resulting from a 100

Ma thermal event in crust which has not undergone significant uplift. Rocks found to the south and directly west of the Mesa Range, which have given older dates and which may be on the same structural block on the western flank of the Rennick Graben support this possibility.

In summary, the simplest interpretation of the K-Ar, Rb-Sr and paleomagnetic data is an episode of chemical mobility and remagnetization in the middle Cretaceous. The timing of the alteration, whether as a single episode, as several episodes or continuously, is not firmly established. However, the convergence of dates derived from the Rb-Sr whole-rock "errorchron", the youngest K-Ar dates, the low temperature $^{40}Ar/^{39}Ar$ incremental heating steps and the VGPs suggest an event at about 100 Ma. The cause of alteration was most likely fluid interaction involving groundwater at relatively low temperatures (< 150°C). It is reasonable to assume the fluids were "hydrothermal" because even slightly elevated temperatures would promote alteration reactions and would drive the circulation systems.

This middle Cretaceous (≈ 100 Ma) event, whatever the temperature attained, was probably linked to the breakup between Australia and Antarctica. Seafloor spreading between these two continents is thought to have been initiated at about 95 Ma (Cande & Mutter, 1982; Veevers, 1986) and continental extension probably started as early as 165 Ma (Powell et al., 1988; Williamson et al., 1990). Tectonic and thermal activity within the eastern part of the Australian-Antarctic rift at about 90-100 Ma is recorded by fission track and paleomagnetic data along the southern continental margin of Australia (Gleadow & Lovering, 1978; Gleadow & Duddy, 1981; Ferguson et al., 1983; Moore et al., 1983, 1986; Klootwijk, 1985; Foster & Gleadow, 1991) and includes middle Cretaceous (98 ±3 Ma) alkaline intrusions in Tasmania (McDougall & Leggo, 1965; Ford, 1989). Similar thermal and tectonic effects are recorded on the flanks of the rift between Australia and New Zealand (Morley et al., 1980; Schmidt & Embleton, 1981; Middleton & Schmidt, 1982; Seward, 1989) associated with sea-floor anomalies as old as C33 (≈ 82 Ma) (Weissel & Hayes, 1977).

The Mesa Range, however, is over 350 km from the continental margin and, consequently,

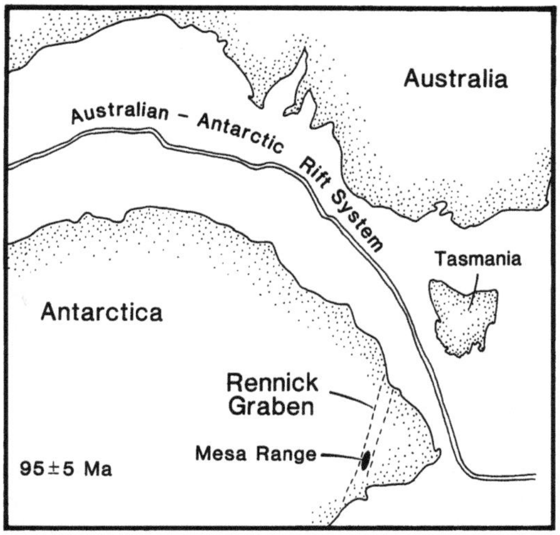

Fig. 8 Schematic diagram showing the relationship of the Mesa Range and Rennick Graben to the opening of the Southern Ocean in the middle Cretaceous. The continental reconstruction is from Veevers and Eittreim (1988).

unlikely to have been directly affected by thermal activity initiated within the main rift zone associated with the dispersal of Antarctica and Australia. A more local source is required to explain a thermal event in the Mesa Range. The Rennick Graben within which the Jurassic lavas appear to be confined is thought to be a failed rift arm developed during continental breakup (Figure 8) (Roland & Tessensohn, 1987). A thermal event is commonly associated with such rifts and could have raised temperatures significantly within the rift, or driven hydrothermal systems responsible for the low-temperature redistribution of Rb, loss of Ar, and formation of titanomaghemite. The axis of the Rennick Graben is presently occupied by the Rennick Glacier. The boundaries of the graben are poorly known, but appear to be defined by systems of step-like faults marking its shoulders. Tectonic activity related to graben formation is recorded by folded and faulted rocks of the Beacon Supergroup and Ferrar Group (Roland & Tessensohn, 1987), clearly dating it as post-Jurassic. The structures are mainly extensional but also locally compressional, possibly related to strike-slip motion along the eastern margin of the graben. Graben formation appears to have also reactivated older faults in the basement which separate the major terranes. It is not clear how the formation of the Rennick Graben relates, if at all, to the breakup structures postulated by Veevers et al. (1991, Figs 5 and 6) for the Tasmania region. The structural trend of the Rennick Graben lies parallel to several fracture zones in the Southern Ocean (Gahagan et al., 1988), the most prominent of these being the Tasman Fracture Zone. The trend of these fracture zones, however, reflects northward plate motion during a relativly fast period of spreading (27mm/yr) which followed the reoganization of spreading centers at 49 Ma (anomaly A21) (Veevers, 1990). The the initial stages of sea floor spreading, between 95 Ma and 49 Ma, were significantly slower (4.5mm/yr) and plate motions probably had a more southeasterly orientation (Veevers, 1990) which would have been oblique to the trend of the Rennick Graben. It is possible that the Rennick Graben could be somewhat older than middle Cretaceous and that the hydrothermal systems simply reflect processes occurring immediately before the start of seafloor spreading at 95 Ma.

Several lines of evidence suggest that the thermal activity within the Rennick Graben may have been more intense toward the north, in the direction of the continental margin. The basalts at Litell Rocks, which are much closer to the continental margin, have a higher-temperature secondary mineral assemblage, including prehnite and occasionally epidote, and lack the zeolites (Skinner et al., 1981) which are typically found in basalts farther south (Elliot et al., 1986a). Basalts at Litell Rocks also yield a younger range of K-Ar dates (90-120 Ma; Kreuzer et al., 1981), compared to Mesa Range basalts to the south (113-178 Ma; Elliot & Foland, 1986), which may reflect more complete loss of Ar during the middle Cretaceous. Northward in the graben, VGPs measured in dolerite and basalt also become increasingly disparate from a Jurassic position (Delisle & Fromm, 1984). VGP positions typical of the Ferrar Group farther south have been obtained from rocks at Sequence Hills and the lower three flows in the southern part of Mesa Range, whereas at Litell Rocks the VGP has a middle Cretaceous position. This change may reflect a smaller component of CRM magnetization in rocks further south within the graben.

7 CONCLUSIONS

$^{40}Ar/^{39}Ar$ data for plagioclase separates from the capping flow unit in north Victoria Land establish an age of 177 ± 2 Ma which is also a minimum age for the entire Kirkpatrick Basalt lava sequence in north Victoria Land. The younger dates recorded by Rb-Sr and K-Ar systems may result from an alteration event in the middle Cretaceous ($\approx$100 Ma). These isotopic data support the interpretation of the paleomagnetic data by Delisle and Fromm (1984, 1989) suggesting a mild thermal event during the middle Cretaceous. This episode of alteration is most likely associated with development of the Rennick Graben which formed as a failed rift arm during the breakup of Australia and Antarctica. Hydrothermal systems developed within the graben during the Cretaceous probably caused the observed paleomagnetic pole positions, Sr and O isotopic variations, and young K-Ar ages.

Acknowledgements: The authors would like to thank F. A. Hubacher for technical assistance with the $^{40}Ar/^{39}Ar$ analyses. Thanks are also extended to M. A. Siders and M. A. Haban, who initially characterized the geochemistry of the Ferrar Group in the Mesa Range making much of this work possible. This work has been supported, in part, by National Science Foundation grants DPP-8021401, DPP-8318957, DPP-9022603 and EAR-8904948.

REFERENCES

Adams, C.J.D., J.E. Gabites, A. Wodzicki, M.G. Laird, & J. D. Bradshaw 1982. Potassium-argon geochronology of the Precambrian-Cambrian Wilson and Robertson Bay Groups and Bowers Supergroup, northern Victoria Land, Antarctica. In: Craddock, C. (ed.) *Antarctic Geoscience*. University of Wisconsin Press, Madison, 543-548.

Bradshaw, J.D., S.D. Weaver & M.G. Laird 1985. Suspect terranes and Cambrian tectonics northern Victoria Land, Antarctica. In: Howell, D.G. (ed.) *Tectonostratigraphic Terranes of the Circum-Pacific Region*. Circum-Pacific Council for Energy and Mineral Resources, Houston, Texas, 467-479.

Brooks, C., S.R. Hart & I. Wendt 1972. Realistic use of two-error regression treatments as applied to rubidium-strontium data. *Reviews of Geophysics and Space Physics* 10:551-577.

Brotzu, P., G. Capaldi, L. Civetta, L. Melluso & G. Orsi 1988. Jurassic Ferrar Dolerites and Kirkpatrick Basalts in northern Victoria Land (Antarctica): stratigraphy, geochronology and petrology. *Memorie Della Societa Geologica Italiana* 43: 97-116.

Cande, S.C. & J.C. Mutter 1982. A revised identification of the oldest sea-floor spreading anomalies between Australia and Antarctica. *Earth and Planetary Science Letters* 58: 151-160.

Cherry, E. M. 1984. Paleomagnetic Response to Hydrothermal Alteration: Kirkpatrick Basalt Group, Victoria Land, Antarctica. Unpublished M.S. Thesis, The Ohio State University 155 p.

Cherry, E.M. & H. C. Noltimier 1984. Paleomagnetic response to hydrothermal alteration in an 86 m basalt flow. *Eos* 65:197.

Cherry, E.M. & H.C. Noltimier 1985. Paleomagnetism and hydrothermal alteration: Kirkpatrick Basalt Group Victoria Land. *Antarctic Journal of the U.S.* 20: 32.

Collinson, J.W., D.C. Pennington & N.R. Kemp 1986. Stratigraphy and petrology of Permian and Triassic Fluvial deposits in northern Victoria Land, Antarctica. In: Stump, E. (ed.): *Geological Investigations in Northern Victoria Land.* American Geophysical Union, Antarctic Research Series 46: 211-242.

Delisle, G. 1983. Results of paleomagnetic investigations in Northern Victoria Land. In: Oliver R.L., P.R. James P.R. & J. B. Jago (eds.): *Antarctic Earth Science.* Cambrige University Press, Cambrige, 148-149.

Delisle, G. & K. Fromm 1984. Paleomagnetic investigation of Ferrar Supergroup rocks, north Victoria Land, Antarctica. *Geologisches Jahrbuch* Reihe B 60:41-55.

Delisle, G. & K. Fromm 1989. Further evidence for a Cretaceous thermal event in north Victoria Land. *Geologishes Jahrbuch* Reihe E 38: 143-151.

Dow, J.A.S. & V.E. Neall 1972. Summary of the geology of lower Rennick Galcier, northern Victoria Land. In: Adie, R. J. (ed.): *Antarctic Geology and Geophysics.* Oslo, 339-344.

Elliot, D.H., M.A. Siders & M.A. Haban 1986a. Jurassic tholeiites in the region of the upper Rennick Glacier, north Victoria Land. In: Stump, E. (ed.): *Geological Investigations in Northern Victoria Land.* American Geophysical Union, Antarctic Research Series 46: 249-265.

Elliot, D.H.; M. A. Haban & M.A. Siders 1986b. The Exposure Hill Formation, Mesa Range. In: Stump, E. (ed.): *Geological Investigations in Northern Victoria Land.* American Geophysical Union, Antarctic Research Series 46: 267-278.

Elliot, D.H. & K.A. Foland 1986. K-Ar age determinations of the Kirkpatrick Basalt, Mesa Range. In: Stump, E. (ed.): *Geological Investigations in Northern Victoria Land.* American Geophysical Union, Antarctic Research Series 46: 279-288.

Embleton, B.J.J. & M.W. McElhinny 1982. Marine magnetic anomalies, paleomagnetism and the drift history of Gondwanaland. *Earth and Planetary Science Letters* 58: 141-150.

Ferguson, K.U., P.R. Kelly, A.J.W. Gleadow & J.F. Lovering 1983. Australia-Antarctic breakup: apatite fission track evidence from south and western Australia and East Antarctica. In: Oliver, R.L., P.R. James & J.B. Jago (eds.) *Antarctic Earth Science.* Australian Academy of Science, Canberra, 582.

Fleming, T.H. 1987. The role of fractional crystallization in the petrogenesis of the Kirkpatrick Basalt, north Victoria Land, Antarctica, based on major element, trace element and mineral chemistry. Unpublished M.S. Thesis, Ohio State University, Columbus, Ohio, 399 p.

Fleming, T.H., D.H. Elliot, L.M. Jones & J.R. Bowman 1989. Secondary alteration of iron-rich tholeiitic rocks of the Kirkpatrick Basalt, northern Victoria Land. *Antarctic Journal of the U.S.* 24:37-40.

Fleming, T.H., D.H. Elliot, L.M. Jones, J.R. Bowman & M.A. Siders (1992) Chemical and isotopic variations in an iron-rich lava flow from the Kirkpatrick Basalt, north Victoria Land: implications for low-temperature alteration. *Contributions to Mineralogy and Petrology* 111, (in press).

Fitzgerald, P.G. & A.J.W. Gleadow 1988. Fission-track geochronology, tectonics and structure of the Transantarctic Mountains in northern Victoria Land, Antarctica. *Isotope Geoscience* 73: 169-198.

Ford, R.J. 1989 Cretaceous alkaline rocks. In: Burrett, C.F. & E.L. Martin (eds.) *Geology and Mineral Resources of Tasmania* Geological Society of Australia Special Publication 15: 381-383.

Foster, D.A. & A.J.W. Gleadow 1991. Apatite fission track thermochronology of western Victoria, Australia, and implications for the architecture of continental rifting. Abstract Volume for the 8th International Symposium on Gondwana, Hobart, Tasmania, 31.

Gahagan, L.M., C.R. Scotese, J.Y. Royer, D.T. Sandwell, J.K. Winn, R.L. Tomlins, M.I. Ross, J.S. Newman, R.D. Muller, C.L. Mayes, L.A. Lawver, C.E. Heubeck 1988. Tectonic fabric map of the ocean basins from satellite altimetry data. *Tectonophysics* 155: 1-26.

Gair, H.S., A. Sturm, S.J. Carryer & G. W. Grindley 1969. The geology of northern Victoria Land. Antarctic Map Folio Series, folio 12, plate XII, American Geographical Society, New York.

Gleadow, A.J.W. & J. F. Lovering 1978. Geochronology of King Island, Bass Straight, Australia: relationship to continental rifting. *Earth and Planetary Science Letters* 37; 429-437.

Gleadow, A.J.W. & I.R. Duddy 1981. Early Cretaceous volcanism and the early breakup of southeastern Australia: evidence from fission track dating of volcanoclastic sediments. In: Cresswell, M.M. & P. Vella (eds.) *Gondwana V.* A.A. Balkema, Rotterdam, 295-300.

Haban, M.A. 1984. Mineral chemistry and petrogenesis of the Ferrar Supergroup, north Victoria Land, Antarctica. Unpublished M.S. Thesis, Ohio State University, Columbus, Ohio, 160 p.

Klootwijk, C. 1985. Paleomagnetism of the Tasman Fold Belt: indication for mid-Carboniferous large-scale southward dispalacement of the New England Region. In: Extended Abstracts: 3rd Circum-Pacific Terrane Conference, Geological Society of Australia, Sydney, 124-127.

Kyle, P.R., D. H. Elliot & J.F. Sutter 1981. Jurassic Ferrar Supergroup tholeiites from the Transantarctic Mountains, Antarctica, and their relationship to the initial fragmentation of Gondwana. In: Cresswell, M.M. & P. Vella (eds.) *Gondwana V.* A.A. Balkema, Rotterdam, 283-287.

Kyle, P.R. 1990. McMurdo Volcanic Group - western Ross Embayment: introduction. In: LeMasurier, W.E. & J.W. Thompson (eds.) *Volcanoes of the Antarctic Plate and Southern Oceans.* American Geophysical Union Antarctic Research Series, 48: 19-25.

Kreuzer, H., A. Hohndorf, H. Lenz, U. Vetter, F. Tessensohn, P. Muller, H. Jordan, W. Harre & C. Besang 1981. K-Ar and Rb-Sr dating of igneous rocks from north Victoria Land. *Geologisches Jahrbuch* Reihe B 41:267-273.

Laird, M.G., J.D. Bradshaw, A. Wodzicki 1982. Stratigraphy of the upper Precambrian and lower Paleozoic Bowers Supergroup, northern Victoria Land, Antarctica. In: Craddock, C. (ed.) *Antarctic Geoscience.* University of Wisconsin Press, Madison, 535-542.

McDougall, I., & P.J. Leggo 1965. Isotopic age determination on granitic rocks from Tasmania. *Journal of the Geological Society of Australia* 12: 295-332.

McIntosh, W.C., P.R. Kyle, E.M. Cherry & H.C. Noltimier 1982. Paleomagnetic results from the Kirkpatrick Basalt Group, Victoria Land. *Antarctic Journal of the U.S.* 17: 20-22.

McIntosh, W.C., P.R. Kyle & J.F. Sutter 1986. Paleomagnetic results from the Kirkpatrick Basalt Group, Mesa Range, north Victoria Land. In: Stump, E. (ed.): *Geological Investigations in Northern Victoria Land.* American Geophysical Union, Antarctic Research Series 46:289-303.

Mensing, T.M., G. Faure, L.M. Jones, J.R. Bowman & J. Hoefs 1984. Petrogenesis of the Kirkpatrick Basalt, Solo Nunatak, north Victoria Land, Antarctica. *Contributions to Mineralogy and Petrology* 87: 101-108.

Mensing, T.M., G. Faure, L.M. Jones & J Hoefs 1988. Evidence for Cretaceous volcanism in northern Victoria Land, Antarctica, and implications for a mantle enrichment event during the Ross Orogeny based on the chemical and isotopic data from the Kirkpatrick Basalt in the Mesa Range. Abstract Volume for the 7th Gondwana Symposium, Sao Paulo, Brazil, 98.

Middleton, M.F & P.W. Schmidt 1982. Paleothermometry of the Sydney Basin. *Journal of Geophysical Research* 87: 5351-5359.

Moore, M.E., A.J.W. Gleadow & J.F. Lovering 1983. Breakup of Gondwana:

apatite fission track dating evidence from southeast Australia. In: Oliver, R.L., P.R. James & J.B Jago (eds.) *Antarctic Earth Science*. Australian Academy of Science, Canberra, 583.

Moore, M.E., A.J.W. Gleadow & J.F. Lovering 1986. Thermal evolution of rifted continental margins: new evidence from fission tracks in basement apatites from southeastern Australia. *Earth and Planetary Science Letters* 78: 255-270.

Morley, M.E., A.J.W. Gleadow & J.F. Lovering 1980. Evolution of the Tasman Rift: apatite fission track dating evidence from the southeastern Australian continental margin. In: Cresswell, M.M. & P. Vella (eds.) *Gondwana V*, 289-293.

O'Reilly, W. 1984. *Rock and Mineral Magnetism*. Blackie and Son Ltd., London 220 p.

Powell, C.McA., S.R. Roots & J.J. Veevers 1988. Pre-breakup continental extention in East Gondwanaland and the early opening of the eastern Indian Ocean. *Tectonophysics* 155: 261-283.

Roland, N.W. & F. Tessensohn 1987. Rennick Faulting - an early phase of Ross Sea rifting. *Geologishes Jarhbuch*, Reihe B 66: 203-229.

Schmidt, P.W. & B.J.J. Embleton 1981. Magnetic overprinting in southeastern Australia and the thermal history of its rifted margin. *Journal of Geophysical Research* 86: 3998-4008.

Schmierer, K. & R. Burmester 1986. Paleomagnetic results from the Cambro-Ordovician Bowers Supergroup, northern Victoria Land, Antarctica. In: Stump, E. (ed.): *Geological Investigations in Northern Victoria Land*. American Geophysical Union, Antarctic Research Series 46:69-90.

Seward, D. 1989. Cenozoic basin histories determined by fission-track dating of basement granites, South Island, New Zealand. *Isotope Geoscience* 79: 31-48.

Siders, M.A. 1983. Intraflow variability, chemical stratigraphy and petrogenesis of the Kirkpatrick Basalt from the Mesa Range area, north Victoria Land, East Antarctica. Unpublished M.S. thesis, Ohio State University, Columbus, Ohio, 202 p.

Siders, M.A. & D.H. Elliot 1985. Major and trace element geochemistry of the Kirkpatrick Basalt, Mesa Range, Antarctica. *Earth and Planetary Science Letters* 72: 54-64.

Skinner, D.N.B. 1987. Terra Nova Bay and the Deep Freeze Range - the southern allochthonous border of north Victoria Land, Antarctica. *Memorie Della Societa Geologica Italiana* 33: 41-58.

Skinner, D.N.B., F. Tessensohn & U. Vetter 1981. Lavas in the Ferrar Group of Litell Rocks north Victoria Land, Antarctica. *Geologisches Jahrbuch*, Reihe B 41: 251-259.

Tessensohn, F., K. Duphorn, H. Jordan, G. Kleinschmidt, D.N.B. Skinner, U. Vetter, T.O. Wright & D. Wyborn 1981. Geological comparison of basement units in northern Victoria Land, Antarctica. *Geologisches Jahrbuch*, Reihe B 41: 31-88.

Tessenshon, F. 1984. Geological and tectonic history of the Bowers structural zone, north Victoria Land, Antarctica. *Geologisches Jahrbuch*, Reihe B 60: 371-396.

Veevers, J.J. 1986. Breakup of Australia and Antarctica estimated as mid-Cretaceous (95 $\pm$5 Ma) from magnetic and seismic data at the continental margin. *Earth and Planetary Science Letters* 77: 91-99.

Veevers, J.J. 1990. Antarctica-Australia fit resolved by satellite mapping of oceanic fracture zones. *Australian Journal of Earth Sciences* 37:123-126.

Veevers, J.J. & S.L. Eittreim 1988. Reconstruction of Antarctica and Australia at breakup (95 $\pm$5Ma) and before rifting (160 Ma). *Australian Journal of Earth Sciences* 35: 355-362.

Veevers, J.J., C. McA. Powell & S.R. Roots 1991. Review of seafloor spreading around Australia. I. Synthesis of the patterns of spreading. *Australian Journal of Earth Sciences* 38:373-389.

Vetter, U., N.W. Roland, H. Kreuzer, A. Hohndorf, A. Lenz & C. Besang 1983. Geochemistry, petrography, and geochronology of the Cambro-Ordovician and Devonian-Carboniferous granitoids of northern Victoria Land, Antarctica. In: Oliver, R.L., P.R. James & J.B. Jago (eds.) *Antarctic Earth Science*. Australian Academy of Science, Canberra, 140-143.

Warren, G. 1969. Geology of the Terra Nova Bay - McMurdo Sound area Victoria Land. Antarctic Map Folio Series, folio 12, plate XIII, American Geographical Society, New York.

Weaver, S.D, J.D. Bradshaw & M.G. Laird 1984. Geochemistry of Cambrian volcanics of the Bowers Supergroup and implications for the early Paleozoic tectonic evolution of

northern Victoria Land, Antarctica *Earth and Planetary Science Letters* 68:128-140.
Weissel, J.K. & D.E. Hays 1977. Evolution of the Tasman Sea reappraised. *Earth and Planetary Science Letters* 36: 77-84.
Williamson, P.E., M.G. Swift, G.W. O'Brien, & D.A. Falvey 1990. Two-stage early Cretaceous rifting of the Otway Basin margin of southeastern Australia: implications for the rifting of the Australian southern margin. *Geology* 18: 75-78.
Wodzicki, A. & R. Robert 1986. Geology of the Bowers Supergroup, central Bowers Mountains, northern Victoria Land. In: Stump, E. (ed.): *Geological Investigations in Northern Victoria Land.* American Geophysical Union, Antarctic Research Series 46:39-68.

Gondwana Eight, Findlay, Unrug, Banks & Veevers (eds) © 1993 Balkema, Rotterdam. ISBN 90 5410 304 3

Biofacies and biostratigraphic constraints on regression in the uppermost Fossil Bluff Group (Aptian-Albian), Alexander Island, Antarctica

Simon R.A. Kelly
British Antarctic Survey, High Cross, Cambridge, UK

ABSTRACT: An estimated 7 km of fore-arc basin sediments are now placed in the Fossil Bluff Group (Late Jurassic-Early Cretaceous) of Alexander Island. In the present study only the uppermost strata of Aptian to Albian age are examined, comprising the Pluto Glacier and Neptune Glacier formations. Environments range from offshore marine shelf to onshore fluvial. In shallowing order six biofacies are identified: heteromorph ammonite, *Aucellina* bivalve, inoceramid bivalve, trigoniid bivalve biofacies, *Cruziana* ichnofacies, and forest. Regression commenced in the Aptian and climaxed in the Late Albian when gymnosperm forests were established in a fluvio-lacustrine environment. The uppermost sediments record Late Albian transgression. Biostratigraphic control is by molluscan macrofaunas.

1 INTRODUCTION

The Fossil Bluff Group of Alexander Island comprises Mesozoic fore-arc basin deposits cropping out between the LeMay Group accretionary prism to the west, and the island arc volcanic and intrusive complex of the Antarctic Peninsula to the east (e.g. Nell and Storey 1991). Although the lithostratigraphy (Butterworth et al. 1988) and biostratigraphy (Crame and Howlett 1988) of the Fossil Bluff Group in the outcrop north of Fossil Bluff is now quite well understood, detailed information on higher strata exposed mainly to the south of Fossil Bluff (Fig. 1), is only starting to appear (Kelly and Moncrieff, in press). It is on these higher strata, which are of Aptian and Albian age, from the Pluto Glacier and Neptune Glacier formations, that this article concentrates. The aim is to describe the biofacies which reflect regression, and discuss the faunal elements which provide the dating framework (Fig. 2). The upper part of the Fossil Bluff Group provides a striking example of a regressive cycle, culminating in the emergence of land and afforestation. At least thirteen standing forests have been seen in one section, and individual, rooted, in situ trunks measure up to 5 m high.

The clastic lithologies present in the upper part of the Fossil Bluff Group range mainly from siltstones through to conglomerates. There is a high volcaniclastic component, and coarser clasts include both plutonic and volcanic types (Horne 1969a) and large mudstone intraclasts indicate penecontemporaneous reworking. There is common evidence of down-slope slumping (Horne 1969b) and there are several major disturbed zones (Macdonald and Butterworth 1986). Movement is generally towards the west with sources from the Antarctic Peninsula volcanic arc (Horne 1969b; A. Moncrieff pers. comm.). The sedimentary setting of the terrestrial interval is comparable to the Late Jurassic Huriwai braidplain delta of New Zealand (Ballance 1989) and the Recent South American Pacific margin (Kuenzi et al. 1979).

1.1 LITHOSTRATIGRAPHY

The formal lithostratigraphy for the upper part of the Fossil Bluff Group, namely the Neptune

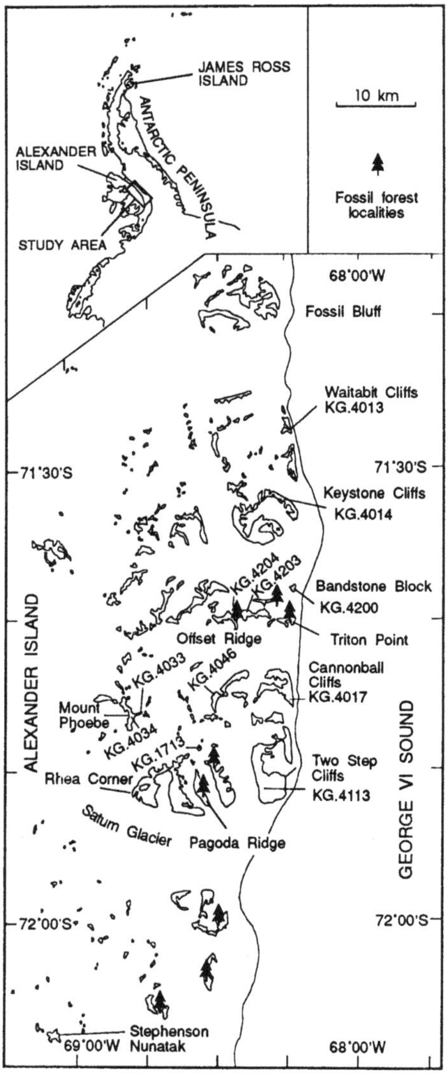

Fig. 1. Map showing location of Alexander Island and the Antarctic Peninsula.

Glacier Formation, follows that of Moncrieff and Kelly (in press). Three units are recognized above the Pluto Glacier Formation (Fig. 2). The marine deposits between the Pluto Glacier Formation and the fossil forests are termed the Deimos Ridge Member; the emergent fossil forest levels, the Triton Point Member; and the marine strata above the forests, the Mars Glacier Member. These units are described in more detail below.

The geology of the coastal outcrops south of Fossil Bluff, including the upper part of Waitabit and Keystone cliffs (Fig. 1), was summarized by Taylor et al. (1979), and comprises mainly the Pluto Glacier Formation. The fossil forest levels of the Triton Point Member, were described by Jefferson (1982 1983), although their precise stratigraphic relationship to underlying and overlying marine strata, the Deimos Ridge and Mars Glacier Members, is only now being clarified.

Exposures from the nunataks to the south of Saturn Glacier are referred to below, informally, as the 'southern nunataks'. Their biota is still known only from reconnaissance level work (eg. Thomson 1974 1981; Jefferson 1982; Doyle 1987), and their detailed lithostratigraphy remains unstudied.

1.2 BIOSTRATIGRAPHY

The biozonation of the lower part of the Fossil Bluff Group is based on molluscan macrofossils, principally ammonites, and secondarily belemnites and bivalves (Howlett 1989). These groups will also be used for the higher strata, but no biozonation scheme is available yet. There has been no satisfactory recovery of microflora, despite repeated attempts, from the Fossil Bluff Group because of thermal metamorphism or burial diagenesis relating to the nearby arc (J.B. Riding, pers. comm. 1991).

Crame and Howlett (1988) dated the Pluto Glacier Formation to the north of Fossil Bluff as Aptian. They placed the Barremian\Aptian boundary within the underlying Spartan Glacier Formation. Thomson (1974) dated the bulk of the Keystone Cliffs section, above the major disturbed zone, as of possible Barremian to Early Albian age. This interval is regarded here as being entirely within the Pluto Glacier Formation and is of Aptian age. The ammonite identifications provided conflicting ages which Thomson (1983) discussed in more detail. Crame (1981, 1982, 1983, 1985) described and discussed inoceramids including the *I. anglicus* Woods group, the *Birostrina concentrica* (Parkinson) group as well as the *Anopaea mandibula* Savel'yev group from the middle and upper parts of the Pluto Glacier Formation. Evidence from the first two groups suggested,

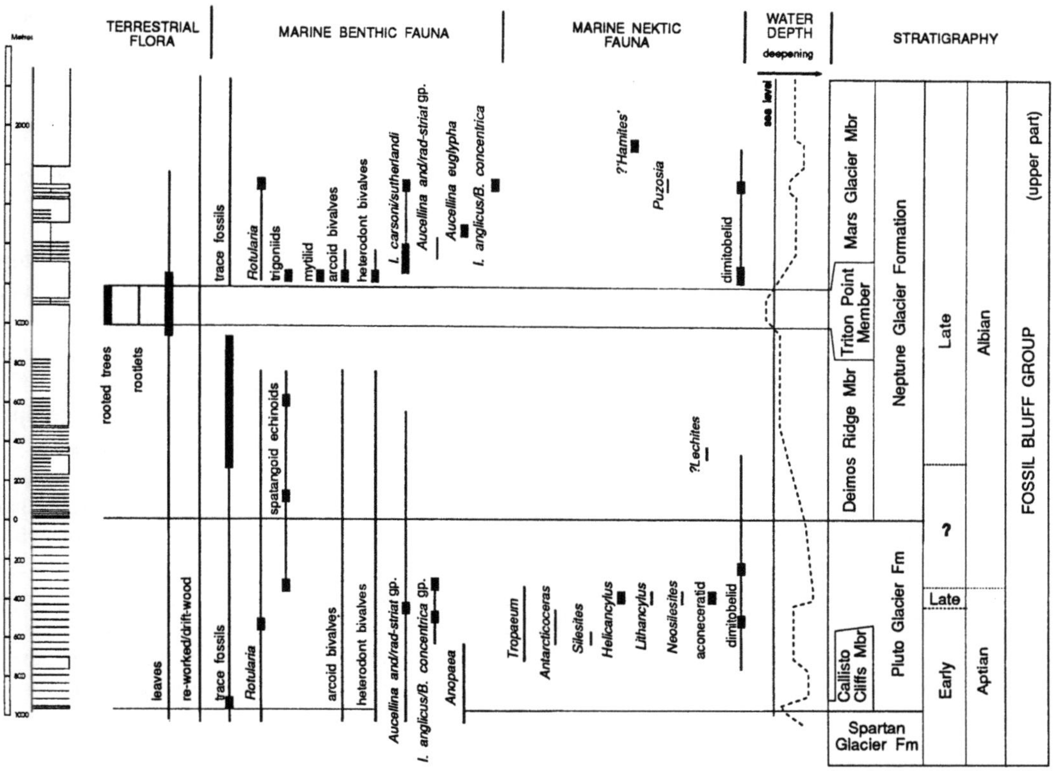

Fig. 2. Generalized section through upper part of Fossil Bluff Group, showing lithostratigraphy, selected biotic distributions, sea-level change and stages. —— = present; ▬ = adundant. For lithostratigraphy see also Butterworth et al. (1988) and Moncrieff and Kelly (in press).

initially, the presence of Middle to Late Albian or even Cenomanian strata. However, new ammonite evidence, including the occurrence of the heteromorphs *Helicancylus* cf. *patagonicum* (Stolley) and *Tropaeum* (see below), confirms an Aptian age for most of this unit. A sandy and conglomeratic upper part of the Pluto Glacier Formation, occuring at Rhea Corner, contains Albian *Dimitobelus diptychus* (M'Coy) (Doyle 1987), as well as reworked *Birostrina concentrica* (Parkinson) and *Eotetragonites*. New Late Albian faunas, figured here (Fig. 4), indicate marine strata overlying the fossil forests and which continue to the top of the Fossil Bluff Group.

Brief discussions of the stratigraphic implications of principal biota are contained in the Appendix 1 (See Fig. 2).

2 BIOFACIES

Six principal biofacies are encountered in the upper part of the Fossil Bluff Group, representing a regressive cycle from offshore marine, through to marginal marine and terrestrial environments (Fig. 5). The triangular diagram is analogous to the figure created by Kauffman (e.g. Kauffman 1984) for transgression in the Cretaceous Western Interior sea-way in the U.S.A.

The key section for understanding the order of facies in the Fossil Bluff Group is on Offset Ridge (Fig. 1) where the progression of facies showing transgression above the fossil forest is particularly clearly exposed. Butterworth and Macdonald (1991) identified facies ranging from deep marine to very shallow shelf in the lower part of the Fossil Bluff Group. Deep marine facies are not recognized here in the upper part of the Fossil Bluff Group. The slope deposits that they recognized are probably equivalent to the deeper shelf biofacies (heteromorph and *Aucellina* biofacies) identified here.

2.1 HETEROMORPH AMMONITE BIOFACIES

Because of post-mortem drift, this biofacies is only loosely defined. Ammonites occur occasionally in most of the marine facies apart from the trace-fossil-dominated, *Cruziana* ichnofacies. However, heteromorph ammonites occur fairly commonly at some levels in fine siltstones in the upper part of the Pluto Glacier Formation (*Helicancylus*) and high in the Mars Glacier Member ('*Hamites*') with little associated benthos. These could represent the deepest/quietest biofacies indicating the deeper parts of the shelf, but it cannot be precluded that this facies occurred when anoxia inhibited benthos. Batt (1989) regarded 'U' shaped heteromorphs (e.g. *Helicancylus*) as pelagic, with limited mobility. The more planispiral oxycones (e.g. aconeceratids) he regarded as more active swimmers.

2.2 *AUCELLINA* BIVALVE BIOFACIES

The epifaunal, byssate *Aucellina andina/radiatostriata* gp. dominates certain mudstones and siltstones in the upper part of the Pluto Glacier Formation (Fig. 3A,F). *Aucellina* cf. *euglypha* Woods (Fig. 4E) has a more restricted range within the lower part of the Mars Glacier Member. Other molluscan benthos is limited but may include occasional taxodont, heterodont, and pectinid bivalves. The trace-fossil *Chondrites* is locally common with occasional *Planolites*. Jeletzky (1983) noted that *Aucellina* coquinas occurred in silty sandstones on the Falkland Plateau, which he interpreted as having been deposited on a shelf as opposed to bathyal environment. Wood and Morter (1983) identified conspicuous alternations of *Aucellina*- and inoceramid-dominated facies in the north European mid-Cretaceous, showing that the taxa were generally not in abundance together, but one always dominated. The *Aucellina* biofacies occupied the middle to deeper parts of the shelf.

2.3 INOCERAMID BIVALVE BIOFACIES

As the sand content in the sediment increases, aucellinids give way to the large epifaunal, byssate/reclining inoceramids which are abundant in the silty sandstones towards the top of the Pluto Glacier Formation. *Anopaea* occurs low in the Pluto Glacier Formation, while *Birostrina concentrica* (Fig. 4J) and *I. anglicus* groups appear in the upper part. Above the forest levels the *I. carsoni*/*I. sutherlandi* groups follow the trigoniid biofacies in the Mars Glacier Member, firstly in graded sandstones, and then in sandy siltstones. *B. concentrica* also appears slightly higher in close association with *I. ipuanus* Wellman (Fig. 4H). In the DSDP bore-holes on the Falkland Plateau (Jeletzky 1983), inoceramids never dominate over *Aucellina*, suggesting that here conditions were too deep for them to have prospered. The inoceramid biofacies occupied the subtidal to middle shelf position.

2.4 TRIGONIID BIVALVE BIOFACIES

In the upper part of the Fossil Bluff Group, trigoniid bivalves occur only in one bed near the base of the Mars Glacier Member. They are associated with other, mainly thick-shelled, high-energy dwelling bivalves, as a death assemblage of isolated valves in current-stable position. The associated lithology is a compact sandstone with the shell accumulations on low-angle bedding foresets. At least four genera of trigoniids are represented (Fig. 4), as well as cucullaeids and heterodonts which were all active, shallow-burrowing suspension feeders. There are also abundant small mytilids (?*Eyrena*) which were byssally attached. Elsewhere in the Fossil Bluff Group, trigoniid bivalves (mainly *Myophorella*) are only significant close to the Jurassic-Cretaceous boundary. Butterworth and Macdonald (1991) indicated that these occurred in a deep marine assemblage which is in contrast to more generally accepted ideas that they are most typical of moderately high energy, sandy,

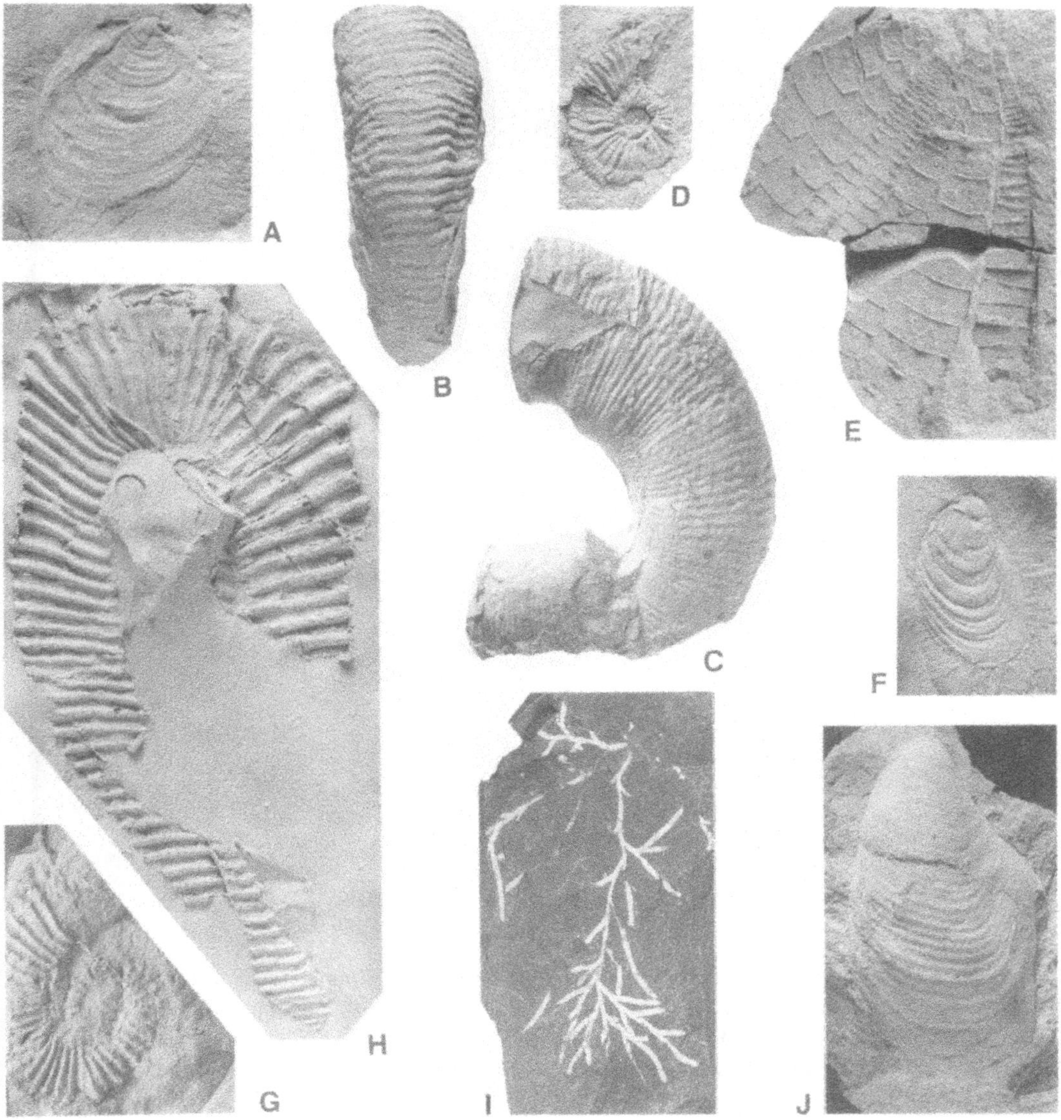

Fig. 3. Biota from the Pluto Glacier Formation, Aptian, SE Alexander Island: A, F, *Aucellina andina/radiatostriata* gp., A, KG.4034.74, right valve, internal mould, x0.7; F, KG.4034.49, left valve, internal mould x0.7. B,C, *Tropaeum* cf. *obesum* Klinger and Kennedy, KG.1413.222, lateral and ventral views, x0.3. D. *Neosilesites* sp. KG. 4046.28, lateral view, x1.2. E, echinoid, KG.4014.352, external mould, x1.2. G,H, *Helicancylus* cf. *patagonicum* (Leanza), G, KG.404.31, early coiled whorls, x0.7; H, KG.4034.25 hooked crozier, x0.7. I, *Chondrites* ichnsp., KG.4033.175, x0.7. J, *Birostrina concentrica* (Parkinson) gp., KG.4019.120, left valve, internal mould, x0.4. All specimens, except I, coated with ammonium chloride; A,D,F latex casts; B,C silicone rubber cast.

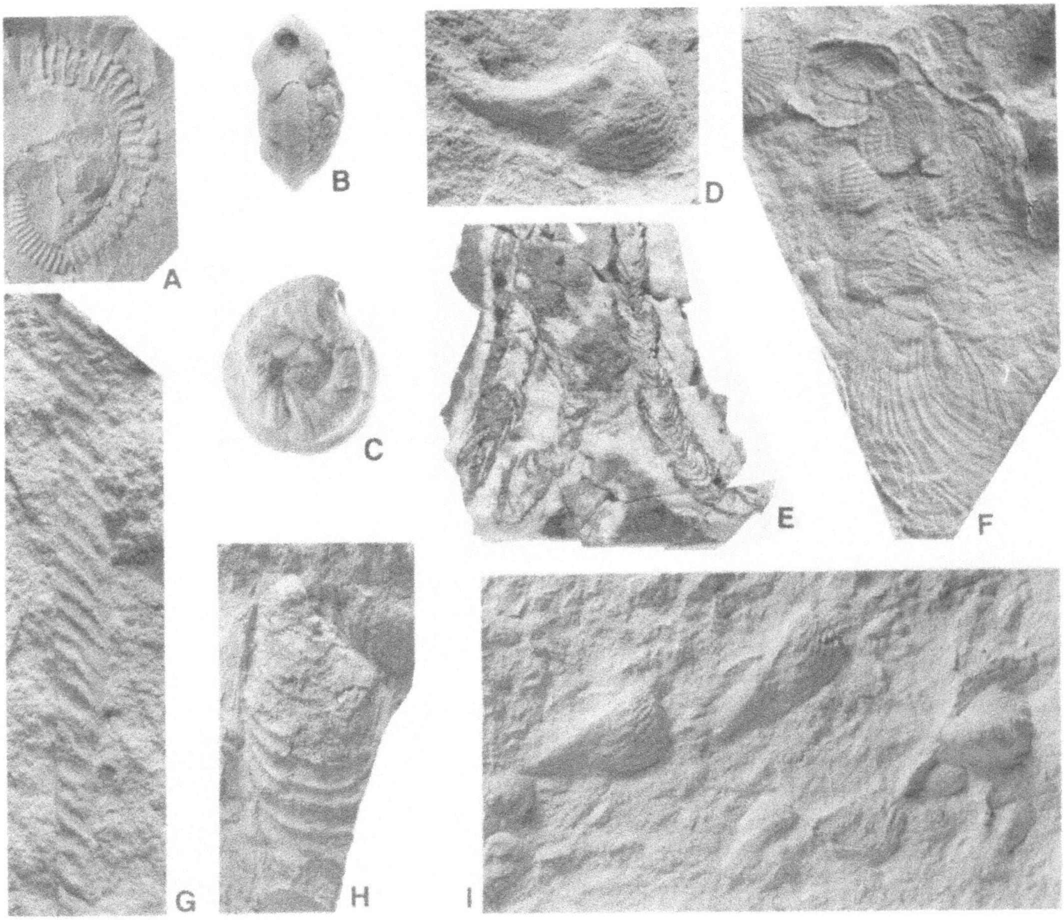

Fig. 4. Biota from the upper part of the Fossil Bluff Group: A, *Hamites virgulatus* (Brongniart), KG.4113.92, Mars Glacier Member, internal mould, x1.2. B,C, *Rotularia* sp. KG.4017.115, Deimos Ridge Member, x1.2. D, *Nototrigonia* (*Nototrigonia*) *spedeni* Fleming, KG.4203.157, Mars Glacier Member, x1.2. E, spatangoid echinoid-type burrow, KG.4200.96, Deimos Ridge Member, x0.25. F, *Aucellina* cf. *euglypha* Woods, KG.54203.26, crushed left and right valves, x1.2. G, *Lechites gaudini* (Pictet and Campiche), KG. 1713.1, Deimos Ridge Member, x1.2. H, *Inoceramus ipuanus* Wellman, KG.4017.134, Mars Glacier Member, internal mould, x0.4. I, bedding surface with *Nototrigonia ponticula* Skwarko (centre and left), *Eselaevitrigonia* aff. *meridiana* (Woods) (lower left) and *Pisotrigonia capricornia* (Skwarko) (right), KG.4203.175-176, Mars Glacier Member, x0.4. All specimens coated with ammonium chloride; F, latex cast; G,I, silicone rubber cast.

subtidal environments (e.g Stanley 1977). The associated assemblages with *Myophorella* mentioned by Crame and Howlett (1988) would suggest relatively shallow water biofacies, perhaps just subtidal. It is possible that the trigoniid and trace-fossil biofacies represent unstabilized and stabilized sediment respectively rather than one being deeper than the other. Shallow burrowing and some byssate bivalves are capable of being reworked during life.

2.5 *CRUZIANA* ICHNOFACIES

Trace-fossils are widespread in the Fossil Bluff Group. The *Cruziana* ichnofacies (Frey and Seilacher 1980), identified here, occurs in sandstones with a variable mud component, although *Cruziana* itself does not occur. The trace-fossils include *Ophiomorpha*, *Planolites*, *Teichichnus*, *Thalassinoides*, and large meniscate back-filled burrows typical of spatangoid echinoids (e.g. Bromley and Asgaard 1975). *Chondrites* is more typical of the muddier parts of the sequence, and also occurs in the inoceramid and aucellinid biofacies. The presence of *Ophiomorpha* suggests the proximity of the sandy shore *Skolithos* ichnofacies, but the *Cruziana* ichnofacies may range from infralittoral and shallow circalittoral to offshore enviroments.

The degree of burrowing in the Cruziana ichnofacies ranges from isolated examples to homogenisation complete enough to render bedding and biogenic structures almost invisible. Pure, thick-bedded sandstones appear to have been inhospitable to organisms, or have no sedimentary contrasts to show up the burrows. The *Cruziana* ichnofacies is most characteristic of the Deimos Ridge Member. It also occurs at the base of the Pluto Glacier Formation (the Callisto Cliffs Member, Fig. 2), and locally at the base of the Mars Glacier Member. This facies is generally devoid of body fossils.

2.6 FOREST BIOFACIES

This facies is restricted to Triton Point Member, best exposed on Pagoda Ridge and Triton Point (Fig. 1). It is characterized by standing, rooted trees and rootlet horizons. The trees are principally gymnosperms, including forms intermediate between '*Araucaria*' and '*Cupressus*' types (Jefferson 1982). Some occur as silicified casts. The lithologies include thin, depleted soil horizons, and thick, often dark sandstones which have swamped the forests and buried them rapidly. Jefferson (1983) recognized that finer grained facies contained a varied range of foliage suggesting probable precursor ferns growing on lake margins. This facies thickens from about 100m at its northernmost outcrop on Offset Ridge, to over 500m on the north side of Saturn Glacier (Moncrieff and Kelly in press).

3 REGRESSION/TRANSGRESSION

Butterworth and Macdonald (1991) indicated a broad shallowing event from Late Jurassic through to Albian times, for the northern part of the Fossil Bluff Group outcrop. The highest strata to which they refer include the base of the Pluto Glacier Formation, here thought to be Aptian in and equivalent to the sandy unit at the base of the generalized section (Fig. 2). The biotic facies sequence in the upper part of the Fossil Bluff Group clearly indicates a further broad cycle of regression followed by transgression. Haq et al. (1987) showed that from Aptian to Albian times there was a progressive increase in coastal onlap worldwide. In the Fossil Bluff Group there are distinct highstands in latest Aptian and in Late Albian. This is also the pattern in Australia (Frakes et al. 1987. In the South Atlantic, Reyment and Mörner (1977) indicated Aptian and late Middle Albian transgressions. The former related to early opening of the South Atlantic, and the latter to global eustasy. In comparison to the Fossil Bluff Group, the Aptian transgression may be a common feature, but there is no direct evidence for Middle Albian transgression, only of Late Albian transgression. In common with the Fossil Bluff Group, Hallam et al. (1986) showed that the local sea-level curve of the Chilean Andean back-arc basin did not match the global sea-level rise of this time period because of local tectonic interference. The regression in Alexander Island must be related partly to delta growth/switching from rivers draining the Antarctic Peninsula during the Late Aptian-Albian (Moncrieff pers. comm. 1991). The precise eustatic effect can only be estimated when there is improved precision on the dating. The closest sea-level curve to that shown for Alexander Island (Fig. 2) is that of Jeletzky (1983) for the Falkland Plateau in the South Atlantic, with marked shallowing events in the latest Barremian/Early Aptian.

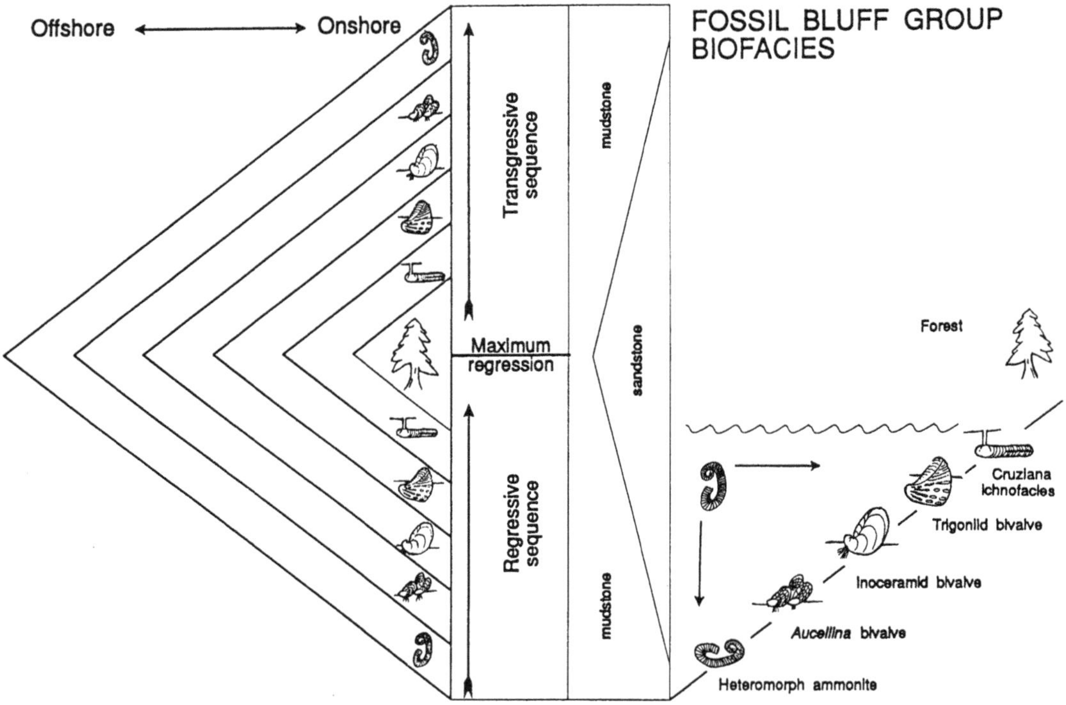

Fig. 5. Summary of the sequence of Fossil Bluff Group biofacies during regression and transgression.

4 CONCLUSIONS

Six biofacies (Fig. 5), ranging from offshore through marginal marine to terrestrial delta-top are identified in the upper part of the Fossil Bluff Group. They illustrate a major regressive event culminating in the formation of extensive forests, of Late Albian age. The regression is believed to have been caused mainly by delta switching rather than eustatic processes. New biostratigraphic dating places the Pluto Glacier Formation largely within the Aptian stage, with only the uppermost part being Albian. The top of the Fossil Bluff Group is Late Albian age (Fig. 2). Although biostratigraphic dating in the Fossil Bluff Group has improved, it is still imprecise and only reliable to sub-stage level. The absence of palynofloras, and the poor records of other microfossil groups, means that molluscan macrofossil dating remains the best tool.

Evidence presented here allows closer correlation of the upper part of the Fossil Bluff Group with Aptian and Albian strata in the northern Antarctic Peninsula. There is considerable biostratigraphic overlap between the top of the Fossil Bluff Group, of the fore-arc basin of Alexander Island, with the base of the Gustav Group, in the back-arc basin of James Ross Island (Thomson 1984). Aptian and Albian faunas of Alexander Island also can now be correlated more confidently with elements of contemporary faunas from Patagonia, southern Africa, Australia and New Zealand.

5 ACKNOWLEDGEMENTS

The author thanks: N. Armstrong and M. Hignall (formerly B.A.S.) for assistance in the field; J.A. Crame, P.A. Doubleday, D.I.M. Macdonald, A.C.M. Moncrieff, and M.R.A. Thomson (B.A.S.), M.B. Aguirre-Urreta (University of Buenos Aires), H.G. Owen (British Museum, Natural History, London), P.G Quilty (Australian Antarctic Division) for valuable discussions or comments on early draft typescripts; and C.J. Gilbert (B.A.S.) for photography.

REFERENCES

Aguirre-Urreta, M.B. 1986. Aptian ammonites from the Argentinian austral basin. The Subfamily Helicancylinae Hyatt 1894. Annals of the South African Museum 96: 271-314.

Aguirre-Urreta, M.B., Buatois, L.A., Chernoglasov, G.Ch.B. and Medina, F.A. 1990. First Polychelidae (Crustacea, Palinura) from the Jurassic of Antarctica. Antarctic Science 2: 157-162.

Ballance, P.F. 1989. The Huriwai braidplain delta of New Zealand, a late Jurassic, coarse grained, volcanic-fed depositional system in a Gondwana forearc basin. In: Nemec, W. and Steel, R. (eds): Fan deltas: sedimentology and tectonic setting. 430-444. Glasgow and London: Blackie and Son.

Batt, R.J. 1989. Ammonite shell morphotype distributions in the Western Interior Greenhorn Sea and some paleoecological implications. Palaios 4: 32-42.

Bromley, R.G. and Asgaard, U. 1975. Sediment structures produced by a spatangoid echinoid: a problem of preservation. Bulletin of the Geological Society of Denmark 24: 261-281.

Butterworth, P.J, Crame, J.A., Howlett, P.J. and Macdonald, D.I.M. 1988. Lithostratigraphy of Upper Jurassic-Lower Cretaceous strata of eastern Alexander Island. Cretaceous Research 9: 249-264.

Butterworth, P.J. and Macdonald, D.I.M. 1991. Basin shallowing from the Mesozoic Fossil Bluff Group of Alexander Island and its regional tectonic significance. In: Thomson, M.R.A, Crame, J.A., and Thomson, J.W. (eds): Geological evolution of Antarctica. 449-453. Cambridge, Cambridge University Press.

Cooper, M.R. 1988. A new species of trigoniid bivalve from the Lower Cretaceous (Albian) of Zululand, South African Journal of Geology 91: 326-8.

Cooper, M.R. 1991. Lower Cretaceous Trigonioida (Mollusca, Bivalvia) from the Algoa Basin, with a revised classification of the order. Annals of the South African Museum 100: 1-52

Crame, J.A. 1981. The occurrence of *Anopaea* Bivalvia: Inoceramidae) in the Antarctic Peninsula. Journal of Molluscan Studies 47: 206-219.

Crame, J.A. 1982. Late Mesozoic bivalve biostratigraphy of the Antarctic Peninsula Region. Journal of the Geological Society of London 139: 771-778.

Crame, J.A. 1983. Lower Cretaceous bivalve biostratigraphy of Antarctica. Zitteliana 10: 399-406.

Crame, J.A. 1985. Lower Cretaceous inoceramid bivalves from the Antarctic Peninsula region. Palaeontology 23: 475-525.

Crame, J.A., Howlett, P.J. 1988. Late Jurassic and Early Cretaceous biostratigraphy of the Fossil Bluff Formation, Alexander Island. British Antarctic Survey Bulletin 78: 1-35.

Day, R.W. 1967. *Lithancylus australis* sp. nov., a new ammonite from the Aptian of Queensland. Journal of the Geological Society of Australia 14: 19-22.

Day, R.W. 1969. The Lower Cretaceous of the Great Artesian Basin. In: Campbell, K.S.W. (ed.) Stratigraphy and palaeontology. Essays in honour of Dorothy Hill. 140-173. Canberra: Australian National University Press.

Dettmann, M.E. and Thomson, M.R.A. 1987. Cretaceous palynomorphs from the James Ross Island area, Antarctica. British Antarctic Survey Bulletin 77: 13-59.

Doyle, P. 1987. The Cretaceous Dimitobelidae (Belemnitida) of the Antarctic Peninsula region. Palaeontology 30: 147-177.

Edwards, A.R., Hornibrook, N. de B., Raine, J.I., Scott, G. H., Stevens, G.R., Strong, C.P. and Wilson, G.J. 1988. A New Zealand Cretaceous-Cenozoic Geological time scale. New Zealand Geological Survey Record 35: 135-149.

Fleming, C. A. 1987. New Zealand Mesozoic bivalves of the Superfamily Trigoniacea. New Zealand Geological Survey Palaeontological Bulletin 53: 1-104.

Frakes, L.A., Burger, D., Apthorpe, M., Wiseman, J., Dettmann, M., Alley, N., Flint, R., Gravestock, D., Ludbrook, N., Backhouse, J., Skwarko, S., Schneibnerova, V., Mcminn,

A., Moore, P.S., Bolton, B.R., Douglas, J.G., Christ, R., Wase, M., Molnar, R.E., Mcgowran, B., Balme, B.E. and Day, R.A. 1987. Australian Cretaceous shorelines. Palaeogeography, Palaeoclimatology, Palaeoecology 59: 31-48.
Frey, R.W. and Seilacher, A. 1980. Uniformity in marine invertebrate ichnology. Lethaia 13: 183-207.
Hallam, A., Biro-Bagoczky, L. and Perez, E. 1986. Facies analysis of the Lo Valdés Formation (Tithonian-Hauterivian) of the high Cordillera of central Chile, and the palaeogeographic evolution of the Andean Basin. Geological Magazine 123: 425-435.
Haq, B.U., Hardenbol, J. and Vail, P.R. 1987. Chronology of fluctuating sea levels since the Triassic. Science 235: 1156-1167.
Henderson, R. A. 1973. Clarence and Raukumara Series (Albian-?Santonian) Ammonoidea of New Zealand. Journal of the Royal Society of New Zealand 3: 71-123.
Henderson, R. A. 1990. Late Albian ammonites from the Northern Territory, Australia. Alcheringa 14: 109-148.
Horne, R.R. 1969a. Morphology, petrology and provenance of pebbles from Lower Cretaceous conglomerates of south-eastern Alexander Island. British Antarctic Survey Bulletin 21: 51-60.
Horne, R.R. 1969b. Sedimentology and palaeogeography of the Lower Cretaceous depositional trough of south-eastern Alexander Island. British Antarctic Survey Bulletin 22: 61-76.
Howlett, P.J. 1989. Late Jurassic-early Cretaceous cephalopods of eastern Alexander Island, Antarctica. Special Papers in Palaeontology 41: 1-72.
Jefferson, T.H. 1982. Fossil forests from the Lower Cretaceous of Alexander Island, Antarctica. Palaeontology 25: 681-708.
Jefferson, T.H. 1983. Palaeoclimatic significance of some Mesozoic Antarctic fossil forests. In: Oliver, R.L., James, P.R. and Jago, J.B. (eds): Antarctic Earth Science. 593-598. Canberra: Australian Academy of Science.
Jeletzky, J.A. 1983. Macroinvertebrate paleontology, biochronology, and paleoenvironments of Lower Cretaceous and Upper Jurassic rocks, deep sea drilling hole 511, eastern Falkland Plateau. Initial Reports of the Deep Sea Drilling Project 71: 951-975.
Kauffman, E.G. 1984. Paleobiogeography and evolutionary response dynamic in the Cretaceous Western Interior seaway of North America. Geological Association of Canada, Special Paper 27: 273-306.
Kelly, S.R.A. and Moncrieff, A.C.M. In press. Molluscan constraints on the age of the Cretaceous fossil forests of Alexander Island, Antarctica. Geological Magazine.
Klinger, H.C., Kennedy, W.J. 1977. Cretaceous faunas from Zululand, South Africa, and southern Mozambique. The Aptian Ancyloceratidae (Ammonoidea). Annals of the South African Museum 73: 215-359.
Kuenzi, W.D., Horst, O.H. and Mcgehee, R.V. 1979. Effect of volcanic activity on fluvial-deltaic sedimentation in a modern arc-trench gap, southwestern Guatemala. Geological Society of America Bulletin 90: 827-38.
Ludbrook, N.H. 1966. Cretaceous biostratigraphy of the Great Artesian Basin in South Australia. Bulletin Geological Survey South Australia 40: 1-223, 28 pls.
Macdonald, D.I.M. and Butterworth, P.J. 1986. Slope collapse deposits in a Mesozoic fore-arc basin, Antarctica. Abstracts 12th International Sedimentological Congress, Canberra, 193-194.
Medina, A., Rinaldi, C.A., Del Vallé, R.A. and Baldoni, M. 1982: Edad de la Formacion Lower Kotick Point en la Isla James Ross, Antartida. Ameghiniana 19: 263-272.
Moncrieff, A.C.M. and Kelly, S.R.A. In press. Lithostratigraphy of the uppermost Fossil Bluff Broup (Early Cretaceous) of Alexander Island, Antarctica. Cretaceous Research.
Nell, P.R. and Storey, B.C. 1991. Strike-slip tectonics within the Antarctic Peninsula fore-arc. In: Thomson, M.R.A, Crame, J.A. and Thomson, J.W. (eds): Geological evolution of Antarctica. 443-448. Cambridge: Cambridge University Press.
Reyment, R.A. and Mörner, N.-A. 1977. Cretaceous transgressions and regressions exemplified by the South Atlantic. Palaeontological Society of Japan, Special Papers 21: 247-261.

Riccardi, A.C., Aguirre-Urreta, M.B. and Medina, F.A. 1987. Aconeceratidae (Ammonitina) from the Hauterivian-Albian of southern Patagonia. Palaeontographica Abteilung A 196: 1-5-185, 13 pls.
Skwarko, S. 1966. Cretaceous stratigraphy and palaeontology of the Northern Territory. Bureau of Mineral Resources, Geology and Geophysics, Bulletin 73: 1-135, 14 pls.
Speden, I.G. 1977. The Tatai Series (Early Cretaceous) and the elimination of the Motoiwian Stage. New Zealand Journal of Geology and Geophysics 20: 537-562.
Stanley, S. 1977. Co-adaptation in the Trigoniidae, a remarkable family of burrowing bivalves. Palaeontology 20: 869-899.
Taylor, B.J., Thomson, M.R.A., Willey, L.E. 1979. The geology of the Ablation Point to Keystone Cliffs area, Alexander Island. British Antarctic Survey Scientific Report 82: 1-65.
Thomson, M.R.A. 1974. Ammonite faunas of the Lower Cretaceous of south-eastern Alexander Island. British Antarctic Survey Scientific Report 80: 1-44.
Thomson, M.R.A. 1981. Antarctica. In: Reyment, R.A. and Bengtson, P. (eds.): Aspects of mid-Cretaceous regional geology. 269-296. London: Academic Press.
Thomson, M.R.A. 1983. "European" ammonites in the Lower Cretaceous of Antarctica. Zitteliana 10: 407-412.
Thomson, M.R.A. 1984. Preliminary ammonite zonation of the mid-Cretaceous rocks of James Ross Island. British Antarctic Survey Bulletin 64: 85-91.
Wood, C.J. and Morter, A.A. 1983. Alternation of *Aucellina* and *Inoceramus* dominated faunas in Europe. Zitteliana 10: 515-529.

APPENDIX: Ammonite and bivalve faunas

1 AMMONITES

Crioceratitids?

Antarcticoceras antarcticum Thomson (1974) was described from the lower part of the section at Keystone Cliffs, and from Stevenson Nunatak at similar levels to *Silesites*, and is of probable Early Aptian age. Despite new collections of specimens, its higher taxonomic placement remains uncertain. However, its style of tuberculation does show similarities with *Helicancylus* cf. *patagonicum* (Stolley) which occurs c330 m below the top of the Pluto Glacier Formation. *Antarcticoceras* retains its open coiling throughout life, but *Helicancylus* develops the distinctive crozier form (Fig. 3H).

Ancyloceratids

Helicancylus patagonicum (Stolley) group (Fig. 3G,H) occurs c330 m below the top of the Pluto Glacier Formation (Kelly and Moncrieff in press). This group has been shown to be of Late Aptian age in Patagonia (Aguirre-Urreta 1986), and provides correlation with part of the Gustav Group on James Ross Island (Thomson 1984).

Large *Australiceras* and *Tropaeum* were recorded by Thomson (1974, 1983). *Tropaeum* of over 20 cm diameter were collected from the upper part of the Waitabit section (uncrushed, Fig. 3 B,C) and mid-way on the Keystone Cliffs section (crushed) in the present study. They retain relatively fine ribbing throughout. However near the base of the undisturbed sequence on Keystone Cliffs, a specimen (KG.4014.20), formerly about 40 cm diameter, shows initial fine ribs followed by a change to very coarse simple ribbing. While such finely ribbed *Tropaeum* are most likely referable to the *T.* sp. gr. *bowerbanki* var *densistriatum-hillsi* (Klinger and Kennedy 1977, fig. 45), of the Aptian I or II of Zululand, they could also be referred to *T. obesum* (Klinger and Kennedy 1977, figs 35-42) suggesting correlation with the Late Aptian (IV). One very crushed specimen of ?*Tropaeum* was figured by Kelly and Moncrieff (in prep.), which occurred at a similar horizon to the *Helicancylus* near the top of the Pluto Glacier Formation.

A single incomplete specimen (KG.1658.14) is referred to *Ancyloceras* (*Adouliceras*) sp. with its distinctive paired lateral/ventrolateral tubercles and traces of a crozier form. Unfortunately its precise horizon in the southern nunataks is not yet established, but its presence suggests correlation with the Early Aptian I or II of South Africa. *Lithancylus* (?) was recorded loose from Waitabit Cliffs by Thomson (1974). It also appears at the *Helicancylus* level in the Pluto Glacier Formation at Keystone Cliffs (Kelly and Moncrieff in prep.). *Lithancylus* is a cosmopolitan Aptian genus (Day 1967).

Lechitids

Thomson (1981) first reported *Lechites* from east of Mt. Phoebe from strata now referred to the Deimos Ridge Member. It is the highest ammonite occurrence below the forests. The specimens show coarse, prorsiradiate ribs (Fig. 4F). It appears to be *Lechites* cf. *gaudini* (Pictet and Campiche), comparable to forms from Northern Territory, Australia (Henderson 1990) of Late Albian age.

Hamitids

Fine ribbed hamitids were first recognized from an unknown level in the southern nunataks by Thomson (1981). The highest ammonites from the Fossil Bluff Group occur 360 m below the top at Two Step Cliffs where Kelly and Moncrieff (in press) identified a more coarsely ribbed *Hamites* cf. *virgulatus* (Brongniart) (Fig. 4A), known from the Late Albian of Northern Territory, Australia (Henderson 1990).

Aconeceratids

Small aconeceratids, referable to *Sinzovia* and *Aconeceras*, occur associated with *Helicancylus* in the upper part of the Pluto Glacier Formation. Riccardi *et al.* (1987) identified *Sanmartinoceras* and *Sinzovia piatnitzkyi* in the Late Aptian of Patagonia. The latter are quite close to Pluto Glacier Formation forms.

Silesitids

Thomson (1974, 1981) described and figured species including *Silesites antarcticus* Thomson from the Pluto Glacier Formation at Keystone Cliffs and Waitabit Cliffs, and also from the southern nunataks, which are of probable Early Aptian age. High on Keystone Cliffs and east of Mt. Phoebe, Kelly and Moncrieff (in press) recorded *Neosilesites* sp. (Fig. 3D), of Late Aptian aspect. However, on James Ross Island, Thomson (1984) recorded *Silesites* from the Kotick Point Formation, which he believed was Albian because of association with Albian palynomorphs (Dettmann and Thomson 1987). This level has been given an Aptian age (Medina et al. 1982; Aguirre-Urreta et al. 1990).

Puzosiinids

Puzosia sp. was collected from the inoceramid rich horizon (including *B. concentrica*) low in the Cannonball Cliffs section, in the Mars Glacier Member. Both were large and still septate at diameters of c160 mm and c270 mm respectively. Whilst puzosiinids range from the Hauterivian to Maastrichtian, *Puzosia* itself is an Albian to Turonian genus.

2 BIVALVES

Inoceramids

The inoceramids relevent to the present study belong in four groups (Fig. 2):

1. *Anopaea*. First recognized in Alexander Island by Crame (1982) who identified *A.* cf. *mandibula* (Savel'yev), originally from the Albian of the Caucasus, from the upper part of the Pluto Glacier Formation. This species appears together with *Antarcticoceras* and *Silesites* low in the section at Keystone Cliffs, and appears to be Early Aptian. One of the nearest austral representatives is *Anopaea constrictus* (Etheridge) (Day 1969) which occurs in the Early Albian, of Queensland.
2. *Birostrina concentrica* Parkinson group. Representatives placed in this group first appear in the upper part of the Pluto Glacier Formation, and bear superficial similarities to the European forms. It was described as being typical of the Middle to Late Albian and possibly extending into the Early Albian and Cenomanian (Crame

1985). However, the first appearance on Alexander Island is within Aptian strata and below or within levels where Late Aptian helicancylids occur. Higher levels above the fossil forests may contain more typical examples of the *B. concentrica* group, e.g. the horizon at which *Puzosia* also occurs on Cannonball Cliffs.

3. *Inoceramus anglicus* Woods group occurs also in the upper part of the Pluto Glacier Formation (Crame 1985) and is generally accepted as being a Middle to Late Albian form. However, like the earlier occurrences of the *B. concentrica* group, the Alexander Island specimens must be of Aptian age.

4. *I. carsoni* M'Coy/*I. sutherlandi* M'Coy group. Although known previously from James Ross Island (Crame 1985), it is now recorded above the fossil forests on Alexander Island. These species are both typical of the Late Albian in Queensland, where they occur in association with *Labeceras* and *Myloceras* (Ludbrook 1966). Also occuring in this group are forms similar to *Inoceramus ipuanus* Wellman (Fig. 4), from the Motuan of New Zealand which Edwards et al. (1988) regarded as Late Albian.

Aucellina

There is limited published information on the aucellinid bivalves of the upper part of the Fossil Bluff Group. The *A. andina-radiatostriata* group (Fig. 3A,F) occurs extensively within Aptian and Albian strata on Alexander Island (Crame 1982; Crame and Howlett 1988). This is probably the group which is widespread in the upper part of the Pluto Glacier Formation, and possibly up into the lower part of the Mars Glacier Member. *A.* cf. *euglypha* Woods (Fig. 4F) occurs higher in the Mars Glacier Member. Hitherto, *A. euglypha* sensu stricto has only been known from the Motuan of New Zealand (Speden 1977), subsequently regarded as Late Albian (Henderson 1973; Edwards et al. 1988).

Trigoniids

In the upper part of the Fossil Bluff Group, trigoniid bivalves only occur at one horizon. At this level they are common, in a diverse, shallow-water assemblage, just above the fossil forests (Kelly and Moncrieff in press) in the Mars Glacier Member. The assemblage includes *Eselaevitrigonia* aff. *meridiana* (Woods), *Nototrigonia* (N.) *spedeni* Fleming (Fig. 4D), *N.* (*N.*) *ponticula* Skwarko, *N.* (*Callitrigonia*) sp. and *Pisotrigonia capricornia* (Skwarko), species typical of the mid-Cretaceous in Australasia. In New Zealand *E. meridiana* is characteristic of the Ngaterian stage and *N. (N.) spedeni* in the Motuan (Fleming 1987). *N. (N.) ponticula* occurs in the late Neocomian of Northern Territory, Australia (Skwarko 1966). *Pisotrigonia* also compares with forms from southern Africa (Cooper 1988), but the fauna generally contrasts with earlier Cretaceous trigoniids such as the Valanginian assemblage from southern Africa (Cooper 1991).

Gondwana Eight, Findlay, Unrug, Banks & Veevers (eds) © *1993 Balkema, Rotterdam. ISBN 90 5410 304 3*

The Late Ordovician Ice Age: New evidence from the Argentine Precordillera

W. Buggisch
Geological Institute, University Erlangen, Germany

R. Astini
Universidad Nacional de Córdoba, Argentina

ABSTRACT: Diamictites in the Don Braulio section of the Sierra de Villicúm (Eastern Precordillera of San Juan, Argentina) are of glacial origin. The Late Ordovician age is well documented by fossils in the underlying and overlying strata. A comparison with other described glacigenic sediments confirms that southern Europe and parts of central Europe were attached to Gondwana during this early Palaeozoic glaciation. The new discovery of the southwesternmost occurrence of glacigenic sediments in South America implies a pole position close to the boundary between Africa and Southern America Gondwana during Late Ordovician times.

1 INTRODUCTION

The Late Ordovician continental glaciation of northern Africa in the vicinity of the Hoggar Mountains has been well documented since the work of Debyser et al. (1965) and Beuf et al. (1971). Due to rarity of fossils the age is not easily determinable in this area. The glacial and periglacial deposits are overlain by "terminal Ashgill to Lower Silurian" fine-laminated sandstones (Legrand 1985).

Striated glacial pavements and glaciomarine sediments are well dated as Upper Ashgill by the occurrence of brachiopods, rare trilobites and bivalves in the central Anti-Atlas (Destombes 1968, Destombes et al. 1985). More distal glaciomarine diamictites are reported from the High Atlas and the Moroccan Meseta ("argille microconglomeratique").

Further evidence of the Late Ordovician glaciation is published from western (Deynoux & Trompette 1981, Tucker & Reid 1981) and southern Africa (Rust 1981), Arabia (McClure 1978), Europe (Carls 1975) and South America (Schlaginweit 1943). Due to the lack of fossils, the age of most of the South American glacial deposits is poorly known. Therefore, the newly discovered Late Ordovician diamictites of the Don Braulio section in the eastern Argentine Precordillera (Peralta & Carter 1990; this paper) are of particular interest because they are intercalated between fossiliferous marine sediments.

2 THE ARGENTINE PRECORDILLERA

The Argentine Precordillera of Mendoza, San Juan and La Rioja (Fig. 1) exhibits an almost complete sequence of Palaeozoic rocks. During Cambrian and Early Ordovician times a shallow water carbonate platform was built up in the eastern Precordillera whereas deep water sediments were deposited in the west. On top of the Arenig San Juan Limestone a widespread regression led to nondeposition and erosion as known by lack of Middle and Upper Ordovician strata in many sections of the central eastern Precordillera. Siliciclastic sediments dominate from Late Ordovician to Permian times. Olistostromes of Silurian (and Devonian?) age interrupt the continuous sequences in some places (Rinconada Formation, Mogotes Negros Formation). Upper Devonian strata are completely missing. This is caused by tectonic movements and probably by the Famennian glacioeustatic sea level decline due to the accumulation

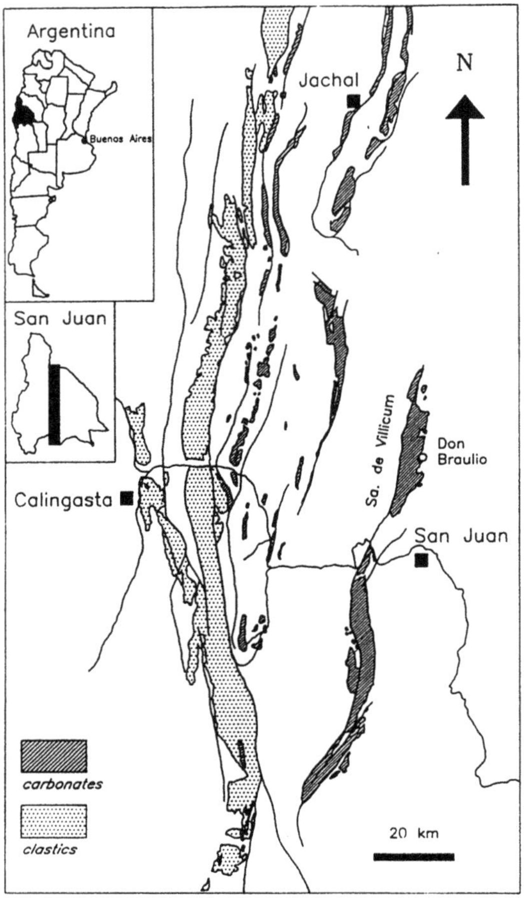

Fig. 1: Distribution of Ordovician sediments in the Precordillera of San Juan (Argentina) and location of the Don Braulio Section.

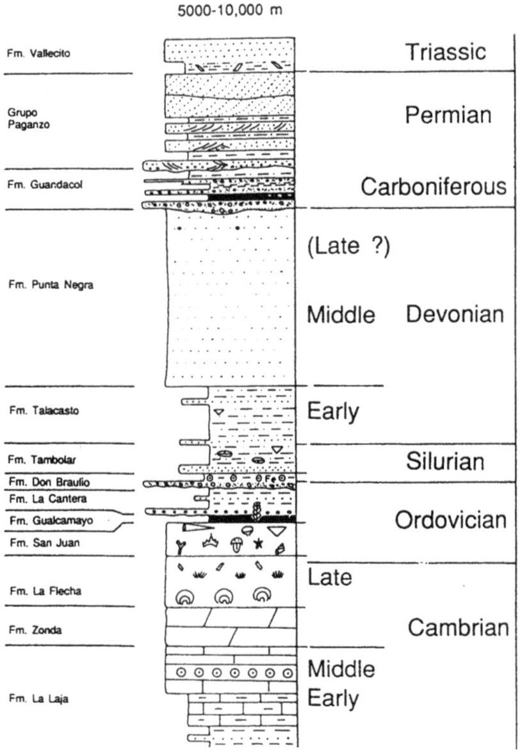

Fig. 2: Composite stratigraphic section of the Palaeozoic rocks in the eastern Precordillera.

of the Late Devonian Amazonas ice shield.

Because of synsedimentary block-faulting, Carboniferous sediments of various ages were deposited in individual basins. Fossiliferous marine black shales and siltstones interfinger with glacigenic and fluvial sediments which are the result of several Late Carboniferous ice shield advances and subsequent deglaciations. The Carboniferous sediments grade up into deltaic and fluvial siltsones and sandstones and finally into eolian dunes which form typical red bed deposits.

Though the entire succession of Palaeozoic strata is interrupted in many areas, it is possible to reconstruct an almost complete composite columnar section (Fig. 2) in order to deduce the history of climate of Southern America (Buggisch et al., in prep.). An important outcrop in this connection is the Don Braulio area at the eastern flank of the Sierra de Villicúm (Fig. 1 and 3) where an almost complete section of Ordovician strata is exposed.

2.1 *The Don Braulio Section*

Baldis et al. (1982) divided the Ordovician rocks of this section into four formations: San Juan, Gualcamayo, La Cantera and Don Braulio. The San Juan Formation consists of a thick sequence of Arenig shallow water platform carbonates. The transition to the Gualcamayo Formation is characterized by the interbedding of dark pelites and marly limestones from which conodonts of the Amorphognathus variabilis zone (Late Arenig to Early Llanvirn) are described (Sarmiento 1985). The thick black graptolite shales and intercalated limestone nodules (about 25 -30 m thick) of the Gualcamayo Formation include rich graptolite and trilobite faunas of Llanvirn age.

Conglomerates at the base of the La Cantera Formation cut with erosional contact into the underlying black shales. These channel-fill deposits correspond to a rapid sea-level decline which might be related to tectonic activities during the "Guandacol phase". The subsequent sea-level rise led to the deposition of more than 80 m of shales with intercalations of platy fine-grained sandstones in which parallel lamination, cross-beds and ripples are common. Slumpfolds indicate a westwardly inclined relief. Based on graptolites (Peralta 1985; 1986; 1990) the age of the La Cantera Formation is Llandeillo to Early Caradoc.

Middle Caradoc to Middle Ashgill fossils are unknown from the Don Braulio section. Therefore, after a considerable break (?) in sedimentation and erosion, the Don Braulio Formation rests almost conformably on the La Cantera Formation.

The Don Braulio Formation (Fig. 4) with a thickness of 50 to 60 m consists of two members. The lower member begins with about 10 to 15 m of mud-supported diamictites. The proportion of clasts ranges from 5 to 20%. They consist of sandstones, limestones, basic volcanic rocks, quartzites and quartz. Most of the clasts were probably derived from a local source area with mainly Ordovician bedrock; exotic crystalline erratics have not been observed. Although most of the clasts are fairly well-rounded, facetted and flatiron shaped rocks also occur. Striated clasts are common and exhibit up to three different directions of striae (Fig. 5).

Astini (1992) observed boulders with similar orientation of striations in the basal part of the lower member which he interpreted as relicts of a boulder pavement. Reworked bryozoan colonies and shell debris of brachiopods (Sanchez et al., 1991) in the silty and muddy matrix prove glacimarine conditions for most of the pebbly mudstones.

A channel about 100-200 m wide and up to 5 m deep filled with conglomerates and sandstones cuts into the pebbly mudstone. The conglomerates consist of grain-supported clasts with diameters up to 20 cm. The modal composition is similar to that of the clasts of the underlying pebbly mudstones, from which they were probably derived. The conglomerates are overlain by coarse, partly cross-bedded sandstones which grade up into medium-grained arenites.

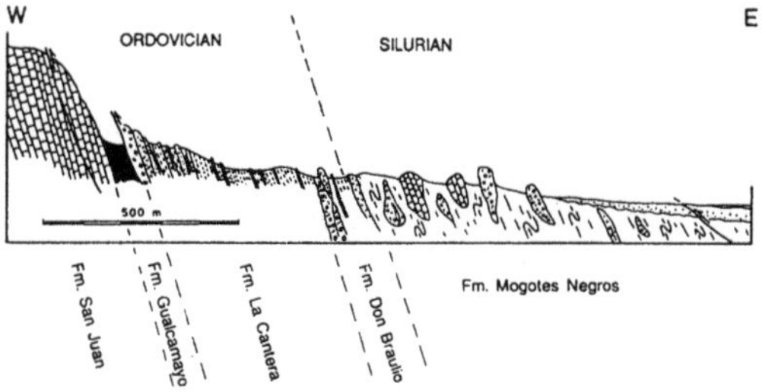

Fig. 3: Cross-section through the Don Braulio area at the eastern border of the Sierra de Villicúm.

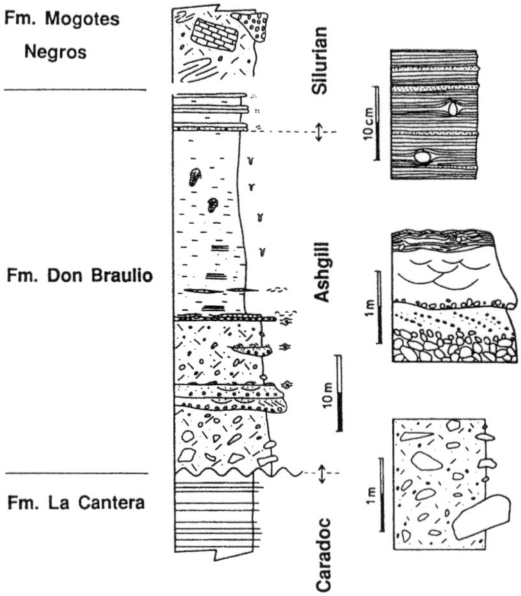

Fig. 4: Columnar section of the glacigenic Don Braulio Formation with the most important facies units (on the right).

Laminated dark grey and green shales and siltstones with dropstones and flaser bedding are exposed above the channel deposits.

The glacigenic sequence of the lower member is overlain by highly bioturbated, partly calcareous, sand- and siltstones of the upper member with abundant brachiopods, bivalves, trilobites, and graptolites.

The upper member of the Don Braulio formation terminates with yellow and green silt- and fine-grained sandstones which are characterized by the occurrence of Fe-ooids and "phosphorite" pebbles.

The Don Braulio Formation is overlain by a more than 1 km thick olistostrome sequence, the Mogotes Negros Formation.

The Palaeozoic rocks dip 40 to 60 toward the east. They are discordantly overlain by Late Tertiary coarse-grained clastic sediments.

2.2 *The age of the Don Braulio Formation*

The age of the Don Braulio Formation is confirmed by brachiopods (Levy & Nullo 1974, Benedetto 1985, 1986), trilobites (Baldis & Blasco 1975), bivalves (Sanchez 1990), graptolites (Peralta 1986) and chitinozoans (Volkheimer et al. 1980). A rich typical Hirnantian fauna is described from the Don Braulio Formation by these authors which indicates a (latest Early to) Late Ashgill age for the glacigenic lower member. Most of the upper member (Dalmanitina beds after Peralta 1990) is also of Late Ashgill age. Only the uppermost levels yielded graptolites and chitinozoans exclusively of Early Silurian (Llandovery) age.

2.3 *Interpretation of the depositional environment of the Don Braulio Formation*

As shown by facetted and striated clasts the pebbly mudstones at the base of the Don Braulio Formation are distinctly of glacigenic origin. Whilst pavement striations give evidence of a grounded glacier or ice sheet, fragments of bryozoan colonies and shell debris of brachiopods prove floating and glacimarine sedimentation. Therefore, the massive diamictites with obscure bedding are interpreted as meltout till of a partly grounded but mostly floating glacier.

The massive and trough-crossbedded conglomerates consist of outwash gravel from the underlying tills reworked by meltwater currents.

The decreasing proportion of clasts in the pebbly mudstones of the upper part of the lower member of the Don Braulio Formation, and the presence of dropstones, indicate a more distal glacimarine environment.

Fig. 5: Striated pebble in the basal diamictite of the Don Braulio Formation.

The glacio-eustatic sea-level rise caused the transgressive sequence of the upper member of the Don Braulio Formation, a rise partly compensated by isostatic uplift after deglaciation. Fe-oolites and the formation of phosphorites verify wide-spread upwelling conditions during a high sea level stand.

Coarse-grained, poorly-sorted, more or less massive conglomerates occur in several outcrops of Ashgill age (e. g. at the Cerro del Fuerte section located in the south of Jachal, between Los Azules and La Chilca Formations). After Baldis et al. (1982) the conglomerates are indicative of a Late Ordovician tectonic "Fase Villicúmica". Based on the new results it is much more probable that the Late Ordovician sedimentation in the eastern Precordillera is controlled by glacio-eustatic sea-level changes.

3 OTHER LATE ORDOVICIAN GLACIAL DEPOSITS (Figs. 6 and 7)

3.1 *Southern America*

Within the "Silurian" Cancaniri Formation of Bolivia, Argentina and Peru Schlaginweit (1943) named the glacial horizon "Zapla Formation". The Cancaniri Formation lies unconformably

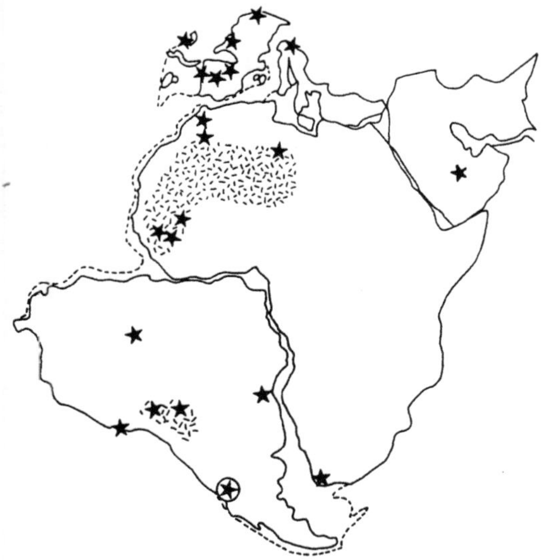

Fig. 6: Partial Ashgill Gondwana assemblage with the location of described glacigenic units (asterisks).

upon several different levels of Ordovician rock units up to Caradoc age. Ashgill strata below it are not known. Crowell et al. (1981) considered the formation as largely belonging to the Llandovery and Wenlock series. More recent results (Bahlburg 1991) ascribe a Late Ashgill to Wenlock (?) age to the glacial units (Monaldi & Bosco 1987). In the light of this new discovery a Late Ordovician age is most probable for these units.

The stratigraphic position of the Iapo tillite (Maack 1951; 1957) of central-eastern Parana (Brazil) is also poorly known (Rocha-Campos 1981a, b). It is underlain by Ordovician rhyolites and overlain by Devonian sandstones (Furnas Formation).

3.2 *Southern Africa*

Deposits of the Winterhoek Glaciation are marginal sediments of an extended Ordovician ice sheet centered in central Africa. Striated pavements, roches moutonees, tillites and glacioaquatic sediments characterize the Pakhuis Formation which is overlain by "lagoonal" mudstones of the Cedarberg Formation with a Hirnantia fauna (Tankard et al. 1982).

3.3 *Central Africa*

Late Ordovician glacimarine sediments are known from Sierra Leone (Tucker & Reid 1981). The rocks rest with a marked unconformity upon Precambrian sediments, granites and gneisses. The glacigenic sediments are intruded by Karoo dolerites. The age is unknown, nevertheless the glacigenic units are correlated with the Late Ordovician "Saharan Glaciation" (Tucker & Reid 1973).

3.4 *Northern Africa*

Upper Ordovician glacigenic sediments are well known from the Hoggar Mountains (Algeria) and were first reported by Debyser et al. (1965). The age is not very well established as in all glacigenic units. The youngest fossils beneath the basal unconformity are of Caradoc age, the transgressive beds above the Tamadjert Formation are of Middle to Late Llandovery age. Therefore, the Hoggar Mountains glaciation is of Late Ordovician (to Early Silurian) age (Biju-Duval et al. 1981).

This age determination is supported by data from the Moroccan Meseta and from the Anti-Atlas (Destombes in Destombes et al. 1985). The Upper Second Bani Formation (Upper Ashgill) containing glacigenic units is overlain by beds with Hirnantian faunas. Also in the central Hercynian Massifs of the Meseta glacioaquatic sequences are widespread in the "argille microconglomeratique".

3.5 *Arabia*

The diamictites of northern central Arabia are of glacial origin (Young 1981, Vaslet 1990). The glacial deposits cut with erosive contact into Cambro-Ordovician sediments: the Cambro-Arenig Saq Sandstone and the Llanvirn to Caradoc (Ashgill ?) Quasim Formation. Caradoc age is indicated by graptolites for the Ra'an shale which in turn is overlain by Cruziana and Scolithus containing sandstones (Quawarah Member: Ashghill ?). Two advances of the ice shield are documented in the basal tillites of the Ashgill Zarqua and Sarah Formations. Trilobites,

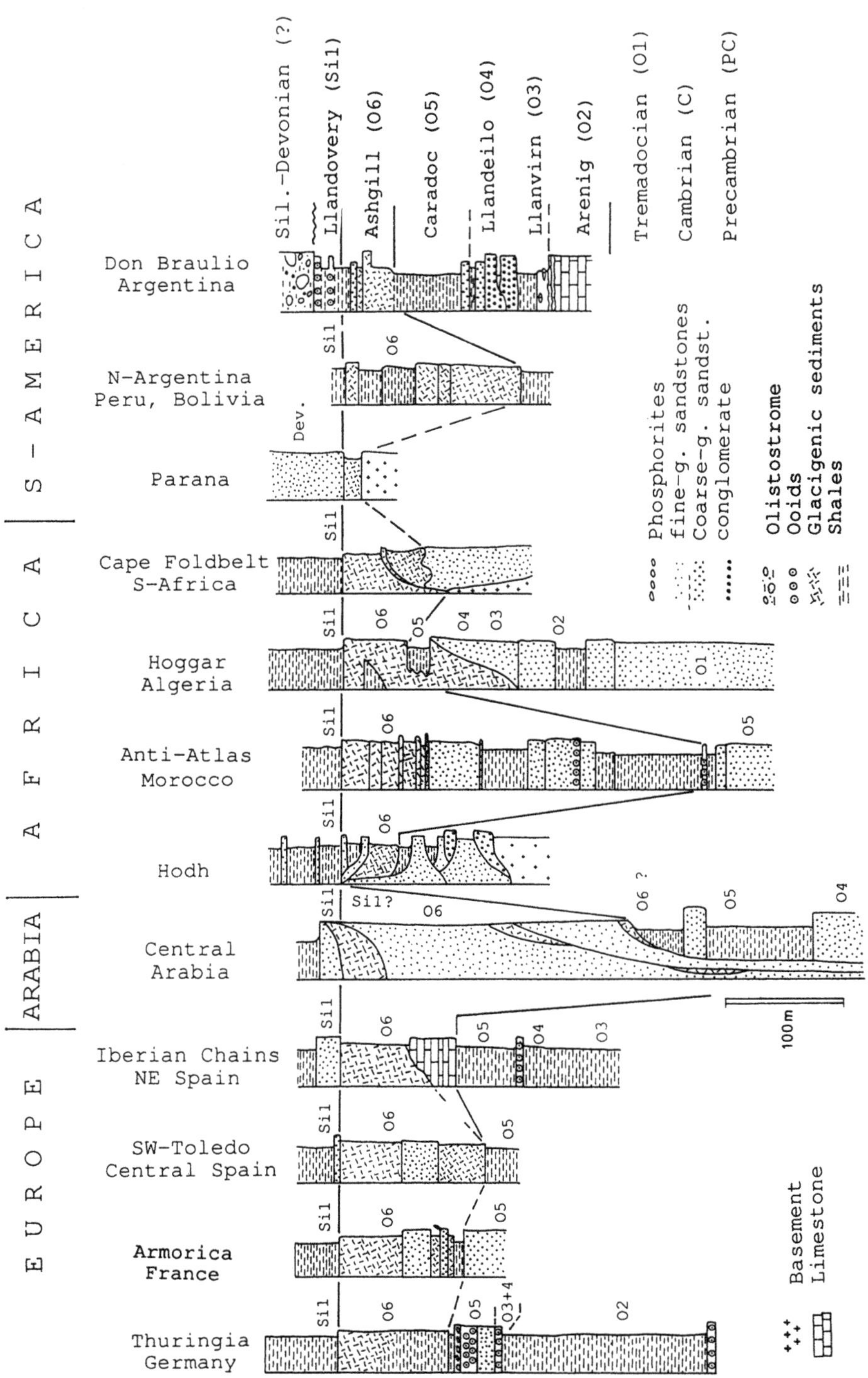

Fig. 7: Simplified columnar section with Late Ordovician glacigenic units and their stratigraphic correlation (for data sources see text).

acritarchs and graptolites indicate Landovery age of the following Qusaiba shales. The upper boundary of the glacigenic unit is considered to be a disconformity.

3.6 *Europe*

The glacial origin of the clasts in the Thuringian "Lederschiefer" in Germany was proposed by Deubel (1929). The stratigraphic age of the clast-bearing unit is probably Ashgill and might extend into the lowermost Silurian [younger than Caradoc and older than Early Silurian (Blumenstengel 1965, Steiner & Falk 1981, Fuchs 1990)].

Several outcrops of Late Ordovician glacigenic rocks are known from Normandy (Armorican Massif, France) where glaciomarine para-tills of Ashgill age occur (Doré 1981).

Similar shallow marine glacigenic sediments are widespread in the Iberian Peninsula (Valongo Anticline/northwest Portugal; Toledo Mountains/central Spain [Robardet 1981] and eastern Iberian chains in northeast Spain [Carls 1975]). Fossils were not found in the diamictites, but according to various faunas in the underlying and following strata (Late) Ashgill (to very Early Silurian ?) age is well confirmed.

A Hirnantia-fauna is described by Jäger et al. (1975) from the Carnic Alps (Southern Alps/Austria). The Ashgill (to Llandovery) sediments at the Hohen Trieb and the Rauchkofel contain rock-fragments up to 5 cm in diameter composed of shales, cherts, and limestones as well as quartz and feldspar grains in a fine-grained matrix. This lithofacies resembles the "argille microconglomeratique" of the Meseta of Morocco. It represents probably distal glaciomarine sediments of the Late Ordovician glaciation (Vai 1971).

4 CONSEQUENCES FOR LATE ORDOVICIAN GONDWANA RECONSTRUCTIONS

The new discovery of Ashgill tillites in the Don Braulio section (Precordillera/Argentina) is of particular interest, because it contradicts the older hypothesis, that the South American glaciation is younger than the North African ice sheet (Hambrey & Harland 1981). On the contrary, all known palaeontological data support the synchronism of the Late Ordovician tillites of Gondwana and Europe (Burrett 1973). Therefore, Europe south of the Variscan "Mid-German Crystalline Rise" was attached to Gondwana during early Palaeozoic times, as it is shown in more recent reconstructions (Scotese & Barrett 1990).

A Late Ordovician pole position close to the boundary between Southern America and Africa would be consistent with the distribution of glacial sediments in Gondwana. This is in accordance with recent reconstructions of the Palaeozoic polar wandering path (Bachtadse & Briden 1990; Kent & Van der Voo 1990).

ACKNOWLEDGEMENT

These investigations were supported by the Volkswagen Stiftung (AZ: I/63074) and by the Deutsche Forschungsgemeinschaft (AZ: Bu 312/17). We are also indebted to the anonymous reviewers for several suggestions that significantly improved the manuscript.

REFERENCES

ASTINI, R.A. 1992: Paleoambientes sedimentarios y secuentias depositacionales del Ordovícico clástico de la Precordillera Argentina. Unpub. PhD Thesis, 851 pp., Córdoba.

BACHTADSE, V.; BRIDEN, J.C. 1990: Palaeomagnetic constraints on the position of Gondwana during Ordovician to Devonian times. *In* McKerrow, W.S.; Scotese, C.R. (eds.): *Palaeozoic palaeogeography and biogeography*. Geological Society of London Memoir 12: 43-48.

BALDIS, B.; BERESI, M.; BORDONARO, O.; VACA, A. 1982: Síntesis evolutiva de la Precordillera Argentina. *V Congreso Latinoamericano de Geología, Actas 4*: 399-445.

BALDIS, B.; BLASCO, G. 1975: Premieros trilobites ashgillianos del Ordovícico sudamericano. *I Congreso Argentino de Palaeontología y Bioestratigrafía, Actas 1*: 33-48.

BAHLBURG, H. 1991: Die Entwicklung mariner Sedimentbecken am paläozoischen Kontinentalrand Südamerikas im Bereich der südlichen Zentralanden (NW-Argentinien, N-Chile). *Zeitschrift deutsche geologische Gesellschaft 142:* 131-148.

BENEDETTO, J.L. 1985: El hallazgo de la típica fauna Hirnantia en el Ashgilliano tardío de la sierra de Villicúm, San Juan. *Reunión de Comunicaciones Palaeontológicas:* 56-57.

BENEDETTO, J.L. 1986. The first typical Hirnantia fauna from South America (San Juan Province, Argentine Precordillera). *In* Rachboef, R.R.; Eming, C.C. (eds.). Les brachiopodes fossiles et actuelles. *Biostratigraphie du Paléozoique 4*: 439-447.

BEUF, S.; BIJU-DUVAL, B.; DE CHARPAL, O.; ROGNON, P.; GARIEL, O.; BENNACEF, A. 1971: *Les grès du Paléozoique inférieur au Sahara.* Sédimentation et discontinuités, évolution structurale d'un craton. Paris (Edit. Technip.).

BIJU-DUVAL, B.; DEYNOUX, M.; ROGNON, P. 1981: Late Ordovician tillites of the central Sahara. *In* Hambrey, M.J.; Harland, W.B. (eds.): *Earth pre-Pleistocene glacial record.* Cambridge University Press, Cambridge: 99-107.

BLUMENSTENGEL, H. 1965: Zur Ostracodenfauna eines Kalkgerölls aus dem Lederschiefer (Ordovizium). *Freiberger Forschungshefte C 182:* 63-78.

BURRETT, C.F. 1973: Ordovician biogeography and continental drift. *Palaeogeography, Palaeoclimatology, Palaeoecology 13*: 161-201.

CAPUTO, M.V.; CROWELL, J.C. 1985: Migration of glacial centers across Gondwana during Paleozoic Era. *Geological Society of America Bulletin 96*: 1020-1036.

CARLS, P. 1975: The Ordovician of the Eastern Iberian Chains near Fombuena and Luesma (Prov. Zaragoza, Spain). *Neues Jahrbuch für Geologie und Paläontologie Abhandlungen 150*: 127-146.

CROWELL, J.C.; SUÁREZ SORUCO, R.; ROCHA-CAMPOS, A.C. 1981: The Silurian Cancaniri (Zapla) Formation of Bolivia, Argentina and Peru. *In* Hambrey, M.J.; Harland, W.B.: *Earth prePleistocene glacial record.* Cambridge University Press, Cambridge: 902-907.

DEBYSER, J.; DE CHARPAL, O.; MERABET, O. 1965: Sur le charactère glaciaire de la sédimentation de l'Unité IV au Sahara central. *C.r. Séanc. Acad. Sci. Paris 261*: 5575-5576.

DESTOMBES, J. 1968: Sur la présence d'une discordance généràl de ravinement d'âge Ashgill supérieur dans l'Ordovicien terminal de l'Anti-Atlas (Maroc). *C.r.Séanc. Acad. Sci., Paris, 267*: 565-567.

DESTOMBES, J.; HOLLARD, H.; WILLEFERT, S. 1985: Lower Palaeozoic rocks of Morocco. *In* Holland, C.H. (ed.): *Lower Palaeozoic of north-western and west-central Africa.* John Wiley & sons, Chichester: 91-336.

DEUBEL, F. 1929: *Erläuterungen zur Geologischen Karte von Preussen und benachbarten deutschen Ländern*, Lieferung 301, Blatt Gräfenthal, 139 pp., Preuss. Geologische Landesanstalt, Berlin.

DEYNOUX, M.; TROMPETTE, R. 1981: Late Ordovician tillites of the Taoudeni basin, west Africa. *In* Hambrey, M.J.; Harland, W.B.: *Earth pre-Pleistocene glacial record.* Cambridge University Press, Cambridge: 89-96.

DORÉ, F. 1981: The late Ordovician tillite in Normandy (Amorican massif). *In* Hambrey, M.J.; Harland, W.B.: *Earth pre-Pleistocene glacial record.* Cambridge University Press, Cambridge: 582-584.

FUCHS, A. 1990: Bemerkungen zur paläogeographischen Position des Ordoviziums von Thüringen - abgeleitet aus Conodonten-Untersuchungen in der Schmiedefelder Folge (Stratigraphie, Palökologie, Farbindizes. *37. Jahrestagung der Gesellschaft Geologischer Wissenschaften der DDR in Jena*, Kurzreferate: 18-19.

HAMBREY, M.J.; HARLAND, W.B. 1981: Summary of Earth's pre-Pleistocene glacial record. *In* Hambrey, M.J.; Harland, W.B.: *Earth pre-Pleistocene glacial record.* Cambridge University Press, Cambridge: 943-954.

JÄGER, H.; HAVLICEK, V.; SCHÖNLAUB, H.P. 1975: Biostratigraphie der Ordovizium/Silur-Grenze in den Südalpen - ein Beitrag zur Diskussion um die Hirnantia-Fauna. *Verhandlungen Geologische Bundes-Anstalt Wien 1975 (4)*: 271-289.

KENT, V.D.; VAN DER VOO, R. 1990: Palaeozoic palaeogeography from palaeomagnetism of the Atlantic-bordering continents. *In* McKerrow, W.S.; Scotese, C.R. (eds.): Palaeozoic palaeogeography and biogeography . *Geological Society of London Memoir 12*: 49-56.

LEGRAND, PH. 1985: Lower Palaeozioc rocks of Algeria. *In* Holland, C.H. (ed.): *Lower Palaeozoic of north-western and west-central Africa.* John Wiley & sons, Chichester: 5-89.

LEVY, R.; NULLO, F. 1974: La fauna del Ordovicico (Ashgilliano) de Villicum, San Juan, Argentina. *Ameghiana 9*: 173-194.

MAACK, R. 1951: Vestígios pre-devonianos de glacicaos e la sequência das camadas devonianas do Estado Paraná. *Arg. Biol. Tecn 5/6*: 197-230.

MAACK, R. 1957: Über Vereisungsperioden und Vereisungsspuren in Brasilien. *Geologische Rundschau 45*: 547-595.

MCCLURE, H.A. 1978: Early Paleozoic glaciation in Arabia. *Palaeogeography Palaeoclimatology, Palaeoecology 25*: 315-326.

MONALDI, C.R.; BOSCO, M.A. 1987: Dalmanitina (Dalmanitina) subandina nov. sp. (Trilobita) en la formación zapla del norte Argentino. *IV Congreso Latinoamericano de Palaeontología, Bolivia* I: 149-157.

PERALTA, S.H. 1985: Graptolitos del Llandoveriano inferior en el Paleozoico de la sierra de Villicum, Precordillera Oriental de San Juan. *Acta de las Primeras Jornadas sobre Geología de Precordillera, Monografías y Reuniones Asociación Geológica Argentina A* 2: 134-138.

PERALTA, S.H. 1986: La graptofauna ordovicica de la Formación Agua de Jaguel (Uspallata, Prov. de Mendoza). *2da Reunión Argentina de Sedimentología,* 150-154; Buenos Aires.

PERALTA, S.H. 1990: Nuevas graptofaunas de la Formación La Cantera, Sierra de Villicum, Precordillera Oriental de San Juan, Argentina. *Acta V Congreso Argentino de Paleontología y Bioestratigafía, Correlación Geológica 7*, San Miguel de Tucuman.

PERALTA, S.H.; CARTER, C. 1990: La glaciación gondwánica del Ordovícica tardío: evidencias en fangolitas guijarrosas de la Precordillera de San Juan, Argentina. *Actas XI Congreso Geológico Argentino II*: 181-185.

ROBARDET, M. 1981: Late Ordovician tillites in the Iberian Peninsula. *In* Hambrey, M.J.; Harland, W.B.: *Earth pre-Pleistocene glacial record.* Cambridge University Press, Cambridge: 585-589).

ROCHA-CAMPOS, A.C. 1981a: Late Ordovician(?)-early Silurian Trombetas Formation, Amazonas Basin, Brazil. *In* Hambrey, M.J.; Harland, W.B.: *Earth pre-Pleistocene glacial record.* Cambridge University Press Cambridge: 896-901.

ROCHA-CAMPOS, A.C. 1981b: Early Palaeozioc Iapó Formation of Parana, Brazil. *In* Hambrey, M.J.; Harland, W.B.: *Earth pre-Pleistocene glacial record.* Cambridge University Press, Cambridge: 908-909.

RUST, C.I. 1981: Early Palaeozoic Pakhuis tillite, south Africa. *In* Hambrey, M.J.; Harland, W.B.: *Earth pre-Pleistocene glacial record.* Cambridge University Press, Cambridge: 113-117.

SANCHEZ, M.T. 1990: Bivalvos del Ordovicico Medo-Tardío de la Precordillera de San Juan (Argentina). *Ameghiana 27* (3-4): 251-261.

SANCHEZ, M.T.; BENEDETTO, J.L.; BRUSSA 1991: Late Ordovician stratigraphy, palaeoecology and sea level changes in the Argentine Precordillera. *In* Barnes, C.R.; Williams, C.H. (eds.). Advances in Ordovician Geology. *Geological Survey Canada Paper 90-9*: 245-258.

SARMIENTO, G.N. 1985: La biozona del Amorphognathus variabilis - Eoplacognathodus pseudoplanus (Conodonta), Llanvirniano inferior, en el flanco oriental de la sierra de Villicum. *Primeras Jornadas sobre la Geología de Praecordillera, Acta* I: 119-123.

SCHLAGINWEIT, O. 1943: La posición estratigrafica del yacimiento de hierro de Zapla y la difusión de horizonte glacial de Zapla en la Argentina y en Bolivia. *Revista Minera, Geológica y Mineralógica. Buenos Aires 13* (4): 115-127.

SCOTESE, C.R.; BARRETT, S.F. 1990: Gondwanas movement over the South Pole during the Palaeozoic: evidence from lithological indicators. *In* McKerrow, W.S.; Scotese, C.R. (eds.): *Palaeozoic palaeogeography and biogeography. Geological Society of London Memoir 12*: 75-85.

STEINER, J.; FALK, F. 1981: The Ordivician Lederschiefer of Thuringia. *In* Hambrey, M.J.; Harland, W.B.: *Earth pre-Pleistocene glacial record.* Cambridge University Press, Cambridge: 579-581.

TANKARD, A.J.; JACKSON, M.P.A.; ERIKSSON, K.A.; HOBDAY, D.K.; HUNTER, D.R.; MINTER, W.E.L. 1982: *Crustal evolution of southern Africa.* Springer, New York: 523 pp.

TUCKER, M.E.; REID, P.C. 1973: The sedimentology and context of late Ordovician glacial marine sediments from Sierra Leone, West Africa. *Palaeogeography, Palaeoclimatology, Palaeoecology 13*: 289-307.

TUCKER, M.E.; REID, P.C. 1981. Late Ordovician glaciomarine sediments, Sierra Leone. *In* Hambrey, M.J.; Harland, W.B.: *Earth pre-Pleistocene glacial record.* Cambridge University Press, Cambridge: 97-98.

VAI, G.B. 1971: Ordovicien des Alpes Carniques. *Mémoire BRGM* 73: 437-448.

VASLET, D. 1990: Upper Ordovician glacial deposits in Saudi Arabia. *Episodes 13 (3)*: 147-161.

VOLKHEIMER, W.; BALDIS, D.E.; BALDIS B.A. 1980: Quitinozoos de la base del Silúrico de la Sierra de Villicúm (Provincia de San Juan, Republica Argentina). *Revista Museo Argentino de Ciencias Naturales Bernardino Rivadavia (Paleontología)*, 2: 125-135.

YOUNG, G.M. 1981: Early Palaeozoic tillites of the northern Arabian Peninsula. *In* Hambrey, M.J.; Harland, W.B.: *Earth pre-Pleistocene glacial record.* Cambridge University Press, Cambridge: 338-340.

Gondwana Eight, Findlay, Unrug, Banks & Veevers (eds) © 1993 Balkema, Rotterdam. ISBN 90 5410 304 3

A reconstruction of the late Palaeozoic ice sheet on southwestern Gondwana

Johan N.J.Visser
Department of Geology, University of the Orange Free State, Bloemfontein, South Africa

ABSTRACT: Analysis of the palaeogeography, spatial and temporal distribution of glacial erosion features and lithofacies, and plate tectonic history of the region enables the reconstruction of the Permo-Carboniferous ice sheet at specific time intervals on southwestern Gondwana. During the Early to Mid-Carboniferous small mountain ice centres formed along the northern sector of the tectonic arc and in South America. Growth of ice centres in South America, Africa and Antarctica since the Mid-Carboniferous resulted in the formation of a supercontinental ice sheet having dimensions of about 10 000 km by 7 000 km. The major ice-spreading centre straddled Africa and Antarctica. Possible global sea-level rise led to the transformation of those sections of the ice cover located on the shelves to unstable marine ice sheets towards the end of the Carboniferous. Large sections of South America and part of the Sowegon basin thus became ice free during the Early Permian. Sea-level and climatic changes at the end of the Early Permian caused the collapse of the ice sheet and only small highland ice caps were maintained in Africa and possibly Antarctica in the Late Permian.

1 INTRODUCTION

The Permo-Carboniferous glaciation was one of the most dramatic events in earth history and is still a strong geological argument for the dispersion of Gondwana during the Mesozoic. Evidence for the glaciation on the Gondwana continents consists of remnants of erosion features (glaciated surfaces and valleys) and sedimentary records. Recording the distribution of the erosion features and sedimentary sequences, supplemented by studies on geochemistry, palaeontology and tectonic setting of the region, enables a reconstruction of the ancient glaciation.

This in itself is a very simplistic approach because the dynamic behaviour of an ice sheet is time constrained, the preservation potential of glacial features and sediments varies, and the distinction among glacial, glacially related and non-glacial sediments is difficult. The configuration of an ice sheet, which is determined by the ice budget, changes consistently and therefore the relative ages of the glacial erosion features and sedimentary units are crucial in the reconstruction of the ice sheet (Fig. 1). The preservation potential of glacial erosion features and sediments depends on the bedrock topography, syndepositional subsidence, the destruction of parts of Gondwana during breakup and post-breakup erosion of the land surfaces. On highlands all evidence of glaciation is usually removed by post-glacial erosion, whereas the glacial record is more complete on the shelves (Visser 1989a) or in rapidly subsiding continental basins. Careful interpretation of glacial sediments can assist in the positioning of the ice front (cf. Frakes & Crowell 1968). However, special care must be taken with ice-rafted stones which may be deposited hundreds of kilometres from the ice front. Sedimentation also records the last stage of the glaciation and there may be no way to determine when the glaciation actually began (Crowell & Frakes 1975).

The objective of this paper is to present maps illustrating the distribution of the Gondwana ice sheet from the Late Carboniferous to the Early Permian on southwestern Gondwana. Southwestern Gondwana is here defined as consisting of part of South America, southern and central Africa, Falkland Islands and part of Antarctica. Although peninsular India was not part of southwestern Gondwana it is included on some maps to give perspective to the reconstruction of the ice sheet. The maps are diagrammatic and interpretative and the prime data upon which they are based are from publications by the author unless otherwise stated.

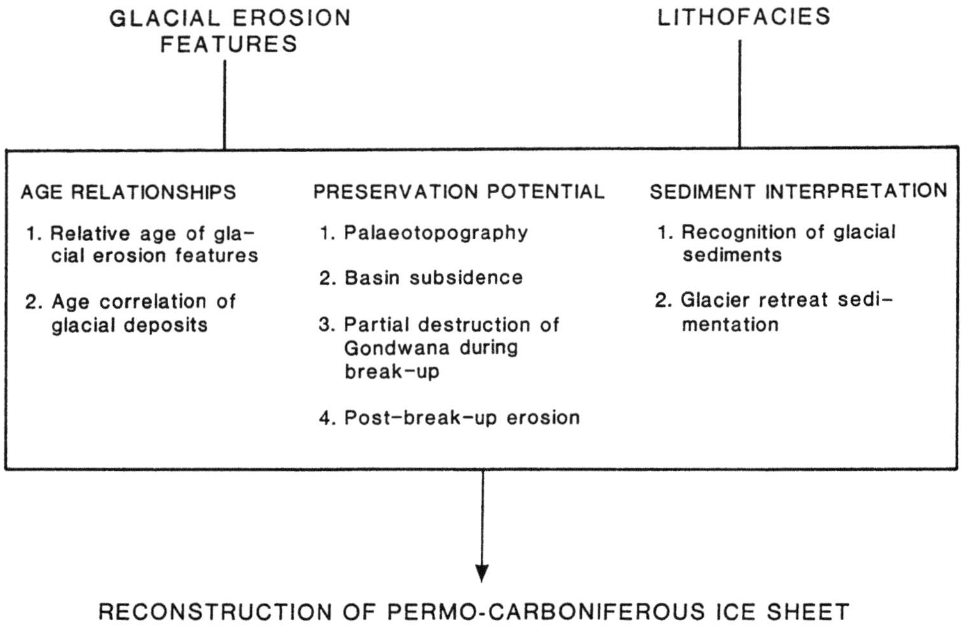

Fig. 1 The parameters and their limitations for the reconstruction of ancient ice sheets.

2 PAST RECONSTRUCTIONS OF THE ICE SHEET

There is an intimate relationship between the reconstruction of Gondwana and that of the Late Palaeozoic ice sheet, because the one was commonly used as a parameter for the reconstruction of the other. However, after the concepts of palaeomagnetism and plate tectonics were generally accepted, modern reconstructions of Gondwana enhanced ice sheet reconstruction (cf. Norton & Sclater 1979, Lawver & Scotese 1987, De Wit et al. 1988).

Du Toit (1937) was probably the first to publish a map of the "Great Ice Cap" overlying Gondwana during the Late Carboniferous. Crowell and Frakes (1975) excellently depicted the ice lobes radiating from subcontinental ice centres which migrated as the clustered Gondwana continents drifted across the south rotational pole. Although Crowell and Frakes (1975) discussed the time factor in the glacial facies model, their reconstruction of the ice lobes was, unfortunately, not always time constrained.

Martin (1981) did not publish a map illustrating the distribution of the Late Palaeozoic ice sheet, but his excellent description thereof based on the age of the deposits, the palaeogeography, palaeo-ice flow data and the lithofacies contributed to our knowledge of the glaciation. He concluded that the Gondwana glaciation cannot be explained by the assumption of moderately sized upland ice caps alone. Visser (1983, 1989b) published maps of southern Africa illustrating the extent of the ice sheet during the Late Carboniferous (maximum glaciation) and Early Permian (major deglaciation). Making use of a palaeotopographic reconstruction of the region, it was possible to show the distribution of large embayments with ice lobes, ice streams and ice domes. He suggested a maximum ice thickness of about 4000 m for the ice sheet while flowing over solid bedrock, but this thickness probably decreased to about 1000 m later when it rested on deformable substrate.

Veevers and Powell (1987) showed the distribution of at least four ice-spreading centres for southwestern Gondwana during the period 290 to 300 Ma, as well as the inferred duration of the glacial episode for each Gondwana subcontinent.

This brief review of the Permo-Carboniferous ice sheet in southwestern Gondwana clearly shows the need for refinement in age and topographic control, and in stratigraphic and lithofacies interpretation in order to better understand ice sheet dynamics during this time.

3 PALAEOGEOGRAPHY OF SOUTH-WESTERN GONDWANA

Southwestern Gondwana was dominated by a

combined foreland-back arc basin (Sowegon basin) reaching from Antarctica to South America. This north-trending basin was bounded in the east by a continental highland which extended into central Africa (Martin 1981). Valleys draining into the Sowegon basin were cut into the highland (Visser 1987). In the west the Sowegon basin was bounded by a tectonic arc formed by the subduction of the palaeo-Pacific plate underneath the southwestern Gondwana plate (Smellie 1981). The tectonic arc was probably discontinuous, allowing access of marine conditions into the basin. The Sowegon basin consisted of a shelf bordering the eastern highlands and a back-arc trench. Remnants of this proto-basin can be found in the Beacon basins and the Ellsworth Mountains of Antarctica, the Karoo and Kalahari basins of southern Africa, the Falkland Islands and the Chaco-Paraná, Tarija, Paganzo, Calingasta-Uspallata, Sierra de la Ventana and central Patagonia basins of South America. The Tarija, Paganzo, Calingasta-Uspallata and central Patagonia basins probably had a back-arc trench setting.

Several questions on the palaeogeography which had implications for the reconstruction of the ice sheet, remain unanswered. (i) Was the African highland continuous with the Antarctic highland or was there a depression along which the continents later split (cf. Hobday & Von Brunn 1979)? (ii) Did the Antarctic highland extend into India? (iii) What influence did tectonic instability of the basin have on the palaeogeography and consequently on the configuration and history of the ice sheet?

4 AGE OF THE GLACIATION

Alledged glacial deposits of Late Devonian to Mid-Carboniferous age are reported from South America (Caputo & Crowell 1985, Charrier 1986, Lopez-Gamundi 1989, 1992). Most of those deposits occur within the trench (Fig. 2) and formed primarily by resedimentation of debris left by mountain glaciers (cf. Frakes & Crowell 1969, Charrier 1986). The time of onset of the major glaciation on the Gondwana continents is unknown, although it probably started with the Namurian regression (Veevers & Powell 1987, Fig. 2). The oldest preserved glacial sediments deposited on the shelf during the continental glaciation are probably Westphalian in age

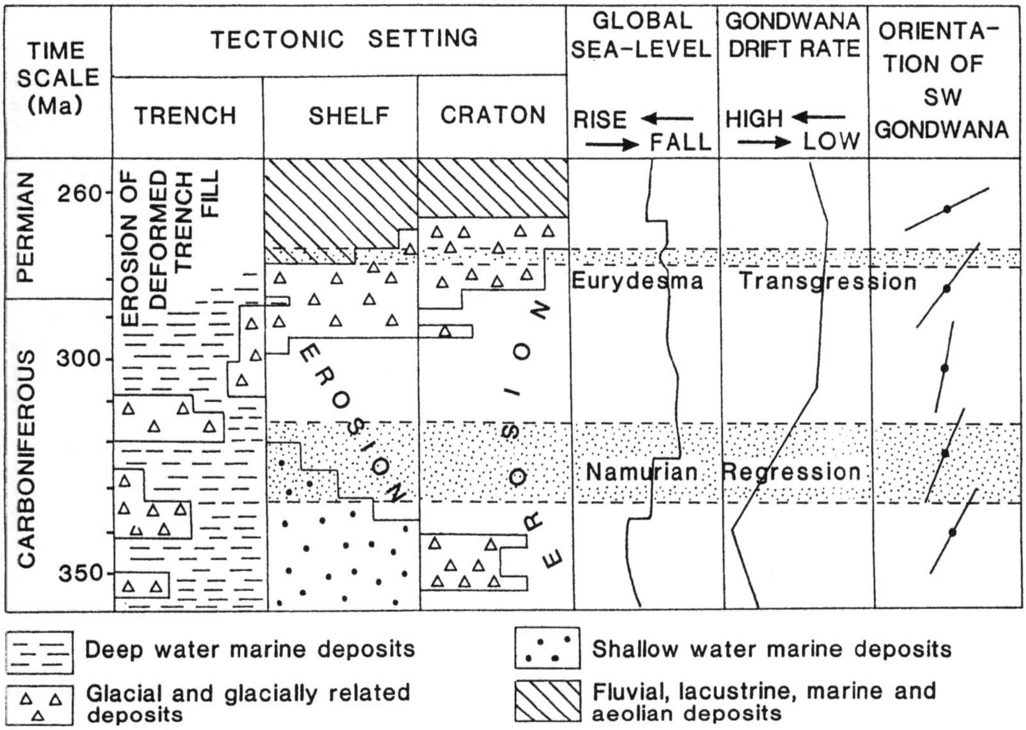

Fig 2 Tectonic setting of the glacial deposits in southwestern Gondwana during the Carboniferous and Permian. Global sea-level modified after Vail et al. (1977), Leggett et al. (1981) and Johnson et al. (1985).

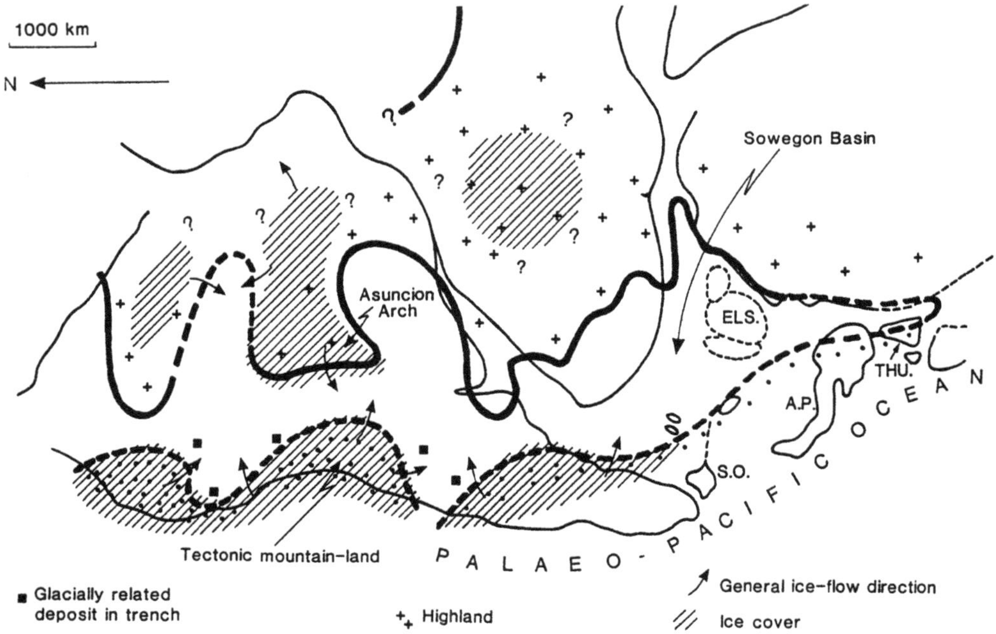

Fig. 3 The Early to Mid-Carboniferous (±340 Ma) ice cover in southwestern Gondwana. Palaeo-ice flow directions after Charrier (1986) and Franca & Potter (1991). Gondwana reconstruction after De Wit et al. (1988). ELS. = Elsworth Block, THU. = Thurston Block, A.P. = Antarctic Peninsula, S.O. = South Orkneys

(Martin 1981, Visser 1990a). As the ice retreated onto the highlands during the Sakmarian-Artinskian, open marine conditions prevailed on the shelves, although palynology in the Paraná Basin indicates the possible continuation of glacial conditions well into the Kungurian (Franca & Potter 1991). Ice caps with outlet glaciers infilling drowned glacial valleys remained on the highlands until the end of the Early Permian (Kungurian) (Visser 1990a).

5 EARLY TO MID-CARBONIFEROUS MOUNTAIN ICE CENTRES

Early to Mid-Carboniferous (Tournaisian-Namurian) diamictites from South America (Caputo & Crowell 1985, Charrier 1986, Lopez-Gamundi et al. 1992), having only slight glacial imprint, were deposited primarily in the back-arc trench and small intermontane basins. The sediment sources were small mountain ice centres and glaciers along the tectonic arc (cf. Lopez-Gamundi 1989) and uplifted cratonic regions like the Asunción Arch (Charrier 1986, Fig. 3). There is no evidence for an ice cover over the continental shelves during the Early Carboniferous. The facies distribution of the diamictites (rain-out and debris-flow deposits) and the palaeogeography suggest the presence only of mountain ice centres in South America, but small intermittent ice covers may also have been present over elevated regions in Africa (Fig. 3). These conclusions are substantiated by the presence of relatively high global sea-levels during the Early Carboniferous (Fig. 2).

Veevers and Powell (1987) drew a distinction between this glacial episode and the main Permo-Carboniferous glaciation. Palaeontological evidence suggested a relatively cool to cold climate for southwestern Gondwana since the Late Devonian to the Early Permian (Visser 1991a) because the south pole was located over Gondwana during the Carboniferous and Early Permian (Caputo & Crowell 1985). Therefore these mountain ice centres probably initiated the growth of the large ice sheet over the supercontinent during the Namurian. The maritime setting of the mountains also enhanced the formation of those initial ice centres (Charrier 1986).

6 ICE SHEET DISTRIBUTION DURING THE LATE CARBONIFEROUS

During the Namurian some ice caps in South America expanded into ice-spreading centres from which ice flowed into the Paraná Basin. The possible highland ice cover over Africa

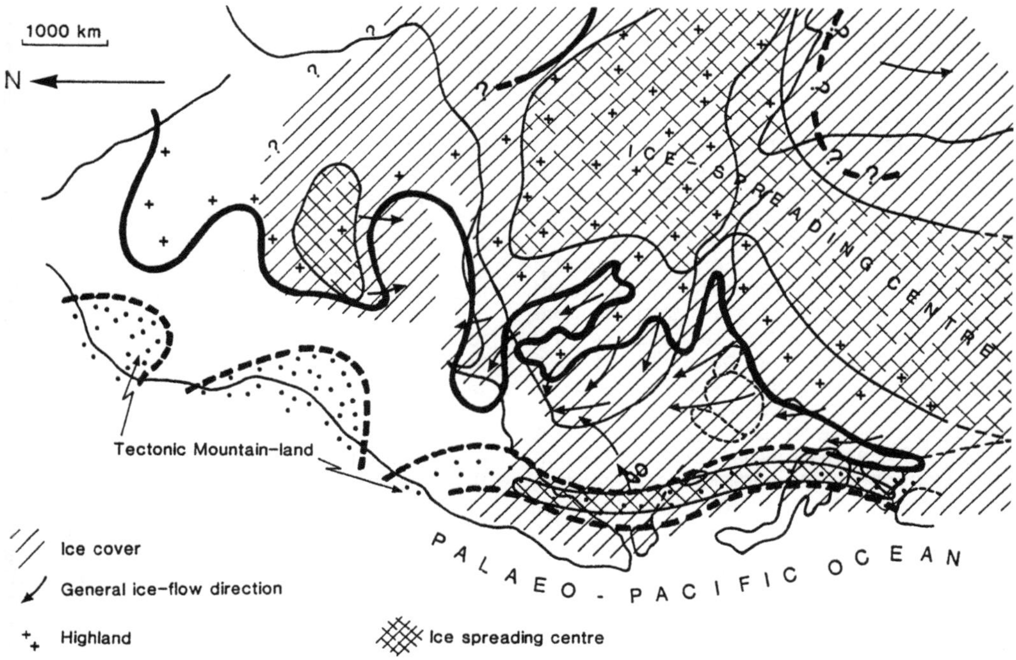

Fig. 4 The Late Carboniferous (±300 Ma) ice cover in southwestern Gondwana. Palaeo-ice flow directions after Martin (1981), Visser (1987), Hall et al. (1988) and Franca & Potter (1991).

also experienced rapid extensive growth, whereas a new ice-spreading centre was initiated over Antarctica. These ice centres soon coalesced to form a large supercontinental ice sheet with lobes flowing down the escarpments onto the shelf regions where they caused erosion of the exposed bedrock (Fig. 4). Palaeo-ice flow directions and clast compositions suggest the presence of a subsidiary ice-spreading centre over the tectonic arc. During maximum glaciation partly grounded ice lobes covered extensive shelf areas in the Sowegon basin as well as in India. Where the ice front terminated at the shelf break, spill-over of sediment into the back-arc trench took place. In most of the shelf sequences thick, widespread, waterlain diamictites first appear at about the time of the Carboniferous-Permian boundary suggesting the formation of marine ice sheets. Thus a global sea-level rise (cf. Ross & Ross 1985) or loading of the shelf by the grounded ice which caused crustal depression, resulted in the transformation of the ice lobes into marine ice sheets during the latest Carboniferous.

Crowell and Frakes (1975) suggested the presence of separate ice sheets for Gondwana. In this reconstruction a continuous ice cover for at least southwestern Gondwana is postulated, although no record of glacial deposition over the entire area exists. However, transport directions of up to 2000 km for some clasts, the presence of regional ice-spreading centres over continental highlands and the maintaining of large marine ice sheets all point to an ice mass of supercontinental dimensions of the order of about 10 000 km by 7000 km. These figures compare favourably with the 6000 km for the present Antarctic ice cover during maximum glaciation. Changes in the ice budgets of the spreading centres resulted in the interference of ice lobes entering the Sowegon basin. Thus although a continuous ice cover is suggested, the dynamics of the ice-spreading centres within the ice sheet changed consistently depending on overall climatic conditions.

Uncertainty prevails over the presence of an ice cover over part of the tectonic arc in South America and the continuation of the African ice-spreading centre into Antarctica. Because the south pole drifted away from South America during the Late Carboniferous (cf. Caputo & Crowell 1985), it is suggested that melting of the ice caps over the northern part of the tectonic arc set in during the Mid-Carboniferous leaving those mountains largely ice free (Lopez-Gamundi 1992). Gondwana reconstructions and the bedrock geology suggest a continuation of the East African highland into Antarctica and therefore an ice-spreading centre straddling the two continents is postulated.

7 ICE SHEET DISTRIBUTION DURING THE EARLY PERMIAN

The subsiding shelf regions with their marine ice sheets became the main depocentres for glacial sedimentation during the Early Permian (cf. Martin 1981). Marine ice sheets are inherently unstable and deposition occurred primarily during mass wasting of the ice fronts (cf. Visser 1989b, Anderson & Thomas 1991). Towards the end of the Early Permian the marine ice sheets completely disintegrated leaving small ice caps on the highlands. The collapse of the marine sections of the Gondwana ice sheet can be attributed to a continuous rise in global sea-level brought about by the progressive deterioration of the terrestrial part of the ice sheet (cf. Anderson & Thomas 1991) as well as isostatic depression of the lithosphere by sediment and shifting water loads (cf. Carter et al. 1986). In those areas where the ice front had retreated up to the palaeo-escarpments, deposition by tidewater glaciers became predominant. Valleys cut into palaeo-escarpments and fault-controlled basins on the cratonic interior became secondary depocentres for glacial sediment accumulation.

During this phase the ice sheet had dimensions of about 8000 km by 4000 km with a major ice-spreading centre over Antarctica and subsidiary centres over Africa and the southern part of the tectonic arc (Fig. 5). Large parts of the Sowegon basin were free of a permanent ice cover. Topographic control on the configuration of the thinning ice sheet became more important with mass wasting of the ice front occurring rapidly along topographic depressions.

The Early Permian reconstruction of the topography in South America and India poses a problem. The highlands were probably largely denuded by glacial erosion and inundated during regional subsidence which, in the case of South America, may be attributed to a thermal episode resulting in lithospheric flexure in the foreland basin (cf. Kay & Mpodozis 1991).

8 LATE EARLY PERMIAN ICE CAPS IN SOUTHERN AFRICA

Highland ice caps had predominated the scene in southwestern Gondwana since the Late Sakmarian (cf. Veevers & Powell 1987). Because of the lack of preservation of glacial sediment in a highland setting a glaciological map of southwestern Gondwana cannot be constructed, but a detailed map of southern Africa is available (Fig. 6). Local highland ice caps waxed and waned during the Artinskian and Kungurian (Visser 1990a) with their influence mainly manifestated in the

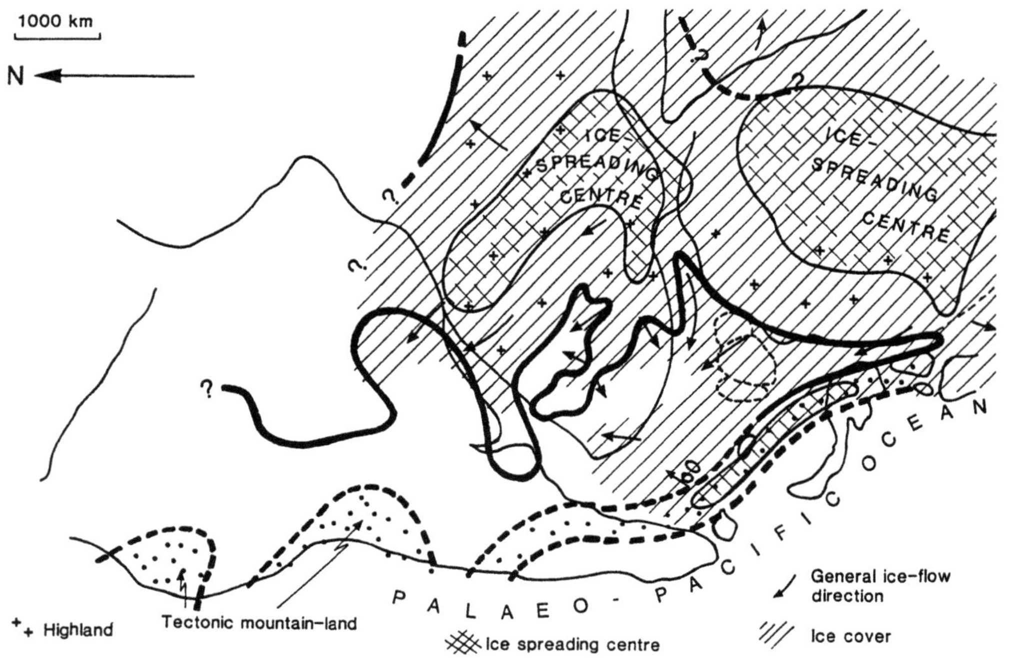

Fig. 5 The Early Permian (±275 Ma) ice cover in southwestern Gondwana. Palaeo-ice flow directions after Martin (1981), Visser (1987), Hall et al. (1988) and Franca & Potter (1991).

geochemistry of the basin sediments (cf. Visser & Young 1990), deposition of organic sediments (LeBlanc Smith & Eriksson 1979) and rare dropstones in argillites (Visser 1990b). A global sea-level rise during the Artinskian as well as regional subsidence caused the inundation of large sections of the highlands by a transgressing sea and by the Late Permian probably all ice on southwestern Gondwana had melted.

9 CONCLUSIONS

Small local ice centres formed in South America and the adjacent tectonic arc during the Early Carboniferous when the south pole was situated in that region. A large part of southwestern Gondwana had a permanent ice cover from about the Mid-Carboniferous to late in the Early Permian. The ice sheet was centred on the highlands and reached sea-level in the large Sowegon basin and along the Tethyan coast. The growth of ice-spreading centres during the Mid-Carboniferous can be attributed to the migration of the pole across Gondwana, but this is an oversimplification as the causes were probabably much more complex.

The drop in global sea-level, commonly attributed to the growth of the Gondwana ice sheet during the Mid-Carboniferous (cf. Veevers & Powell 1987, Fig. 2), was much larger than could be expected with ice sheet growth. It is suggested that this large eustatic change was caused by high rates of plate motion resulting in the uplift of large sections of the supercontinent. Exposure of the continental shelves increased the albedo of the earth's surface, which together with the

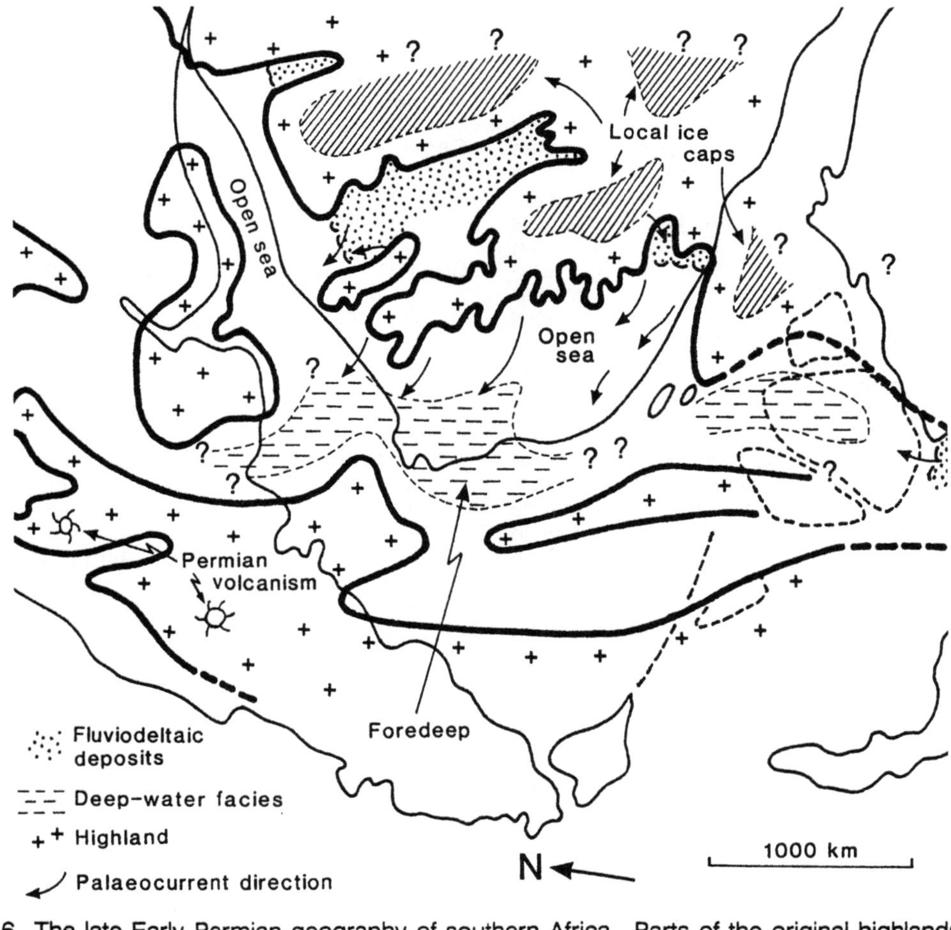

Fig. 6 The late Early Permian geography of southern Africa. Parts of the original highlands had been inundated by the Permian sea, while only small local ice caps remained. Palaeocurrent directions and basin tectonics after Visser (1991a, 1991b). Distribution of deep-water facies partly after Frakes & Crowell (1968).

reorientation of the palaeo-Pacific margin of Gondwana towards a more longitudinal position and the foreland-back arc tectonic setting of southwestern Gondwana (Fig. 2), may have been the triggering mechanisms for ice sheet growth. Topography, rather than climatic conditions, was probably more important for glaciation in the intermediate latitudes and continentality may have led to increased ice accumulation. Therefore, the large, uplifted, continental interior of southwestern Gondwana during the Mid-Carboniferous formed the ideal setting for the initiation of the large Permo-Carboniferous ice sheet.

Growth of the ice sheet was probably checked by the configuration of southwestern Gondwana, the availability of moisture and subsidence of the shelf margins due to isostactic loading by the ice. With the margins of the ice sheet becoming buoyant, its expansion was halted, but it remained stable, except for some interglacials, for almost 30 Ma. Major glacial deposition on the shelves only commenced when the ice sheet became unstable at the time of the Carboniferous-Permian boundary.

Complete collapse of the ice sheet in the Sakmarian-Artinskian (cf. Crowell & Frakes 1975) can be attributed to a global temperature increase resulting in rising global sea-level, reorientation of southwestern Gondwana to a more latitudinal position (Fig. 2) and basin subsidence resulting in relatively high sea-levels. Over large parts of southwestern Gondwana marine shales overlie glacial sediments without any break in sedimentation. Inundation of the land areas, contributing to the complete collapse of the marine ice sheets because their grounding lines are highly sensitive to buoyancy (MacAyel 1979, Visser 1990b, Anderson & Thomas 1991), may have already started at the time of the Permian-Carboniferous boundary and the marine transgression reached its maximum extent in the Late Permian. Subsidence could have been caused by isostatic depression due to lithospheric loading below the ice sheet (cf. MacAyel 1979, Carter et al. 1986) or to lithospheric flexure and deformation by tensional stress within the supercontinent (cf. Franca & Potter 1991). Glaciological models in which the control on ice sheet retreat by sea-level, not climate, are now being tested in areas of Late Pleistocene glaciation (Eyles & McCabe 1989).

The spatial and temporal behaviour of the Permo-Carboniferous ice sheet in southwestern Gondwana was controlled by several interrelated factors (topography, sea-level, atmospheric and oceanic circulation) which can be directly or indirectly attributed to plate tectonics. An understanding of the plate tectonic history of the region, in addition to the analysis of glacial erosion features and lithofacies, are thus essential for the reconstruction of the Gondwana ice sheet.

ACKNOWLEDGEMENT

The Foundation for Research Development (FRD) and the University of the Orange Free State are thanked for financial support of the Dwyka Project. I am grateful to John Crowell and Lawrence Frakes for helpful comments and suggestions on the original manuscript.

REFERENCES

Anderson, J. B. and Thomas, M. A. 1991. Marine ice-sheet decoupling as a mechanism for rapid, episodic sea-level change: the record of such events and their influence on sedimentation. Sedimentary Geology 70: 87-104.

Caputo, M. V. and Crowell, J. C. 1985. Migration of glacial centres across Gondwana during the Paleozoic Era. Geological Society of America Bulletin 96: 1020-1036.

Carter, R. M., Carter, L. and Johnson, D. P. 1986. Submerged shorelines in the SW Pacific: evidence for an episodic post-glacial transgression. Sedimentology 33, 629-649.

Charrier, R. 1986. The Gondwana glaciation in Chile: Description of alleged glacial deposits and paleogeographic conditions bearing on the extensive ice cover in southern South America. Palaeogeography, Palaeoclimatology, Palaeoecology 56: 151-175.

Crowell, J. C. and Frakes, L. A. 1975. The late Palaeozoic glaciation. In Campbell, K. S. W. (ed): Gondwana Geology. Australian National University Press, Canberra: 313-331.

De Wit, M., Jefferey, M., Bergh, H. and Nicolaysen, L. 1988. Geological map of sectors of Gondwana reconstructed to their disposition - 150 Ma. American Association of Petroleum Geologists & University of the Witwatersrand.

Du Toit, A. L. 1937. Our Wandering Continents. Oliver & Boyd, Edinburgh: 366 pp.

Eyles, N. and McCabe, A. M. 1989. The late Devensian (<22, 000 YBP) Irish Sea basin: The sedimentary record of a collapsed ice sheet margin. Quaternary Science Review 8: 307-351.

Frakes, L. A. and Crowell, J. C. 1968. Late Palaeozoic glacial facies and the origin of the south Atlantic basin. Nature 217: 837-838.

Frakes, L. A. and Crowell, J. C. 1969. Late

Paleozoic glaciation: I, South America. Geological Society of America Bulletin 80: 1007-1042.
Franca, A. B. and Potter, P. E. 1991. Stratigraphy and reservoir potential of glacial deposits of the Itareré Group (Carboniferous-Permian), Paraná Basin, Brazil. Bulletin of the American Association of Petroleum Geologists 75: 62-85.
Hall, B. A., Borns, H. W., Chaudhuri, A. K. and Rudra, D. K. 1988. The age(s) and character of the Talchir glaciation, Peninsular India: A new working hypothesis. Abstracts of the 7th Gondwana Symposium, Sâo Paulo, Brazil: 73.
Hobday, D. K. and Von Brunn, V. 1979. Fluvial sedimentation and paleogeography of an early Paleozoic failed rift, southeastern margin of Africa. Palaeogeography, Palaeoclimatology, Palaeoecology 28: 169-184.
Johnson, J. G., Klapper, G. and Sandberg, C. A. 1985. Devonian eustatic fluctuations in Euramerica. Geological Society of America Bulletin 96: 567-587.
Kay, S. M. and Mpodozis, C. 1991. Late Paleozoic to Triassic Gondwana magmatism and tectonism along the South American margin (28° to 33°). Abstracts of the XII International Congress on the Carboniferous-Permian, Buenos Aires: 59.
Lawver, L. A. and Scotese, C. R. 1987. A revised reconstruction of Gondwanaland. In McKenzie, G. D. (ed): Gondwana Six: Structure, Tectonics and Geophysics. Geophysical Monograph of the American Geophysical Union 40: 17-23.
LeBlanc Smith, G. and Eriksson, K. A. 1979. A fluvio-glacial and glaciolacustrine deltaic depositional model for Permo-Carboniferous coals of the northeastern Karoo Basin, South Africa. Palaeogeography, Palaeoclimatology, Palaeoecology 27: 67-84.
Leggett, J. K., McKerrow, W. S., Cocks, L. R. M. and Rickards, R. B. 1981. Periodicity in the early Palaeozoic marine realm. Journal of the Geological Society of London 138: 167-176.
Lopez-Gamundi, O. R. 1989. Postglacial transgressions in late Paleozoic basins of western Argentina: a record of glacioeustatic sea level rise. Palaeogeography, Palaeoclimatology, Palaeoecology 71: 257-270.
Lopez-Gamundi, O. R., Limarino C. O. and Cesari, S. N. 1992. Late Paleozoic paleoclimatology of central west Argentina. Palaeogeography, Palaeoclimatology, Palaeoecology 91: 305-329.
MacAyel, D. R. 1979. A catastrophe model of the paleoclimate. Journal of Glaciology 24: 245-256.
Martin, H. 1981. The late Palaeozoic Gondwana glaciation. Geologische Rundschau 70: 480-496.
Norton, J. O. and Sclater, J. G. 1979. A model for the evolution of the Indian Ocean and the breakup of Gondwanaland. Journal of Geophysical Research 84: 6803-6830.
Ross, C. A. and Ross, J. R. P. 1985. Late Paleozoic depositional sequences are synchronous and worldwide. Geology 13: 194-197.
Smellie, J. L. 1981. A complete arc-trench system recognized in Gondwana sequences of the Antarctic Peninsula region. Geological Magazine 118: 139-159.
Vail, P. R., Mitchum R. M. and Thompson, S. 1977. Seismic stratigraphy and global changes of sea level, Part 4: Global cycles of relative changes of sea level. In Payton, C. E. (ed): Seismic Stratigraphy - Applications to Hydrocarbon Exploration. Memoir of the American Association of Petroleum Geologists 26: 83-97.
Veevers, J. J. and Powell, C. M. 1987. Late Paleozoic glacial episodes in Gondwanaland reflected in transgressive-regressive depositional sequences in Euramerica. Geological Society of America Bulletin 98: 475-487.
Visser, J. N. J. 1983. Glacial-marine sedimentation in the late Paleozoic Karoo Basin, southern Africa. In Molnia, B. F. (ed): Glacial-Marine Sedimentation. Plenum, New York: 667-701.
Visser, J. N. J. 1987. The palaeogeography of part of southwestern Gondwana during the Permo-Carboniferous glaciation. Palaeogeography, Palaeoclimatology, Palaeoecology 61: 205-219.
Visser, J. N. J. 1989a. The distinction between ancient mountain and shelf glaciations. Abstracts of the 28th International Geological Congress, Washington D. C.: 3-303.
Visser, J. N. J. 1989b. The Permo-Carboniferous Dwyka Formation of southern Africa: deposition by a predominantly subpolar marine ice sheet. Palaeogeography, Palaeoclimatology, Palaeoecology 70: 377-391.
Visser, J. N. J. 1990a. The age of the late Palaeozoic glacigene deposits in southern Africa. South African Journal of Geology 93: 366-375.
Visser, J. N. J. 1990b. Influence of basin subsidence on the disintegration of the marine ice sheet in the Dwyka Basin, South Africa. Abstracts of the 13th International Sedimentological Congress, Nottingham: 575.
Visser, J. N. J. 1991a. Geography and climatology of the late Carboniferous to Jurassic Karoo Basin in southwestern

Gondwana. Annals of the South African Museum 99: 415-431.
Visser, J. N. J. 1991b. Basic tectonics in southwestern Gondwana during the Carboniferous and Permian. Abstracts of the Conference on Inversion Tectonics of the Cape Fold Belt, University of Cape Town: 15-16.
Visser, J. N. J. and Young, G. M. 1990. Major element geochemistry and palaeoclimatology of the Permo-Carboniferous glacigene Dwyka Formation and post-glacial mudrocks in southern Africa. Palaeogeography, Palaeoclimatology, Palaeoecology 81: 49-57.

Gondwana Eight, Findlay, Unrug, Banks & Veevers (eds) © 1993 Balkema, Rotterdam. ISBN 90 5410 304 3

Palaeogeography of Tasmania's Permo-Carboniferous glacigenic sediments

Stephen J. Hand
School of Earth Sciences, Macquarie University, N.S.W., Australia

ABSTRACT: The Permo-Carboniferous glacigenic deposits of the Tasmania Basin were deposited entirely under glacimarine conditions. Individual deposits have been interpreted, as either having been deposited from both an ice sheet or from fjord glaciers. An ice sheet appears to have been present over much of Tasmania until the final stages of glacial retreat by which time fjord glaciation had become dominant.

The glacigenic sediments cannot be subdivided biostratigraphically, nor can they be biostratigraphically separated from the lower part of the Woody Island Siltstone and correlates which overlie the glacigenic sediments upon a boundary which is probably diachronous. Marine conditions probably commenced in the Tasmania Basin due to isostatic loading during ice advance, the later stages of ice advance coinciding with a marine regression before the advancing ice front. The onset of marine conditions must have either coincided with, or preceded the deposition of the oldest preserved glacimarine sediments. The lower part of the Woody Island Siltstone and correlates in those areas beyond the maximum extent of glaciation may therefore be slightly older than the oldest preserved glacigenic sediments. This may be observed to be the case in at least one outcrop. Subsequent to the glacial maximum glacial retreat resulted in a marine transgression over those areas previously covered by ice.

1 INTRODUCTION

Tasmania's Permo-Carboniferous glacigenic sediments crop out over the western two-thirds of the State (Fig.1). The sediments occupy the basal position in the Carboniferous to Triassic Parmeener Supergroup (Fig.2) in the Tasmania Basin. A large number of geographically separate deposits occur, some of which were recognised quite early in the state's history (e.g. Stephens 1869) and have been studied by David (1907), Ahmad (1957), Carey & Ahmad (1961) and most recently by Banks & Clarke (Banks 1981, Banks & Clarke 1987, Clarke 1989). A number of different formation names have been used for these glacigenic sediments: the Wynyard Formation for deposits in northwestern Tasmania, the Stockers Formation for deposits in northern and central Tasmania and the Truro Formation for deposits in southeastern Tasmania. The present study was undertaken in response to the rapidly expanding literature on glacigenic sediments, and in particular glacimarine sediments, over recent years. All the lettered areas in Fig.1 have been examined by the author. For those not lettered, previous studies (Banks 1981, Banks & Clarke 1987) have been used.

Several techniques have been used in the present study which have not been applied to these rocks previously. Clast orientation fabric has proved a useful tool in distinguishing texturally similar diamictites of differing genesis (Lindsay 1968, Mark 1973, 1974, Lawson 1979, Domack & Lawson 1985, Dowdeswell & Sharp 1986). The technique also gives ice movement directions from the imbrication direction of clasts in tillite (Flint 1947, p.107) and less reliably in debris flows (Lawson 1979).

A study of basement palaeotopography has demonstrated the probable existence of fossil U-shaped valleys within which the glacigenic sediments were deposited.

Although clast composition has been used

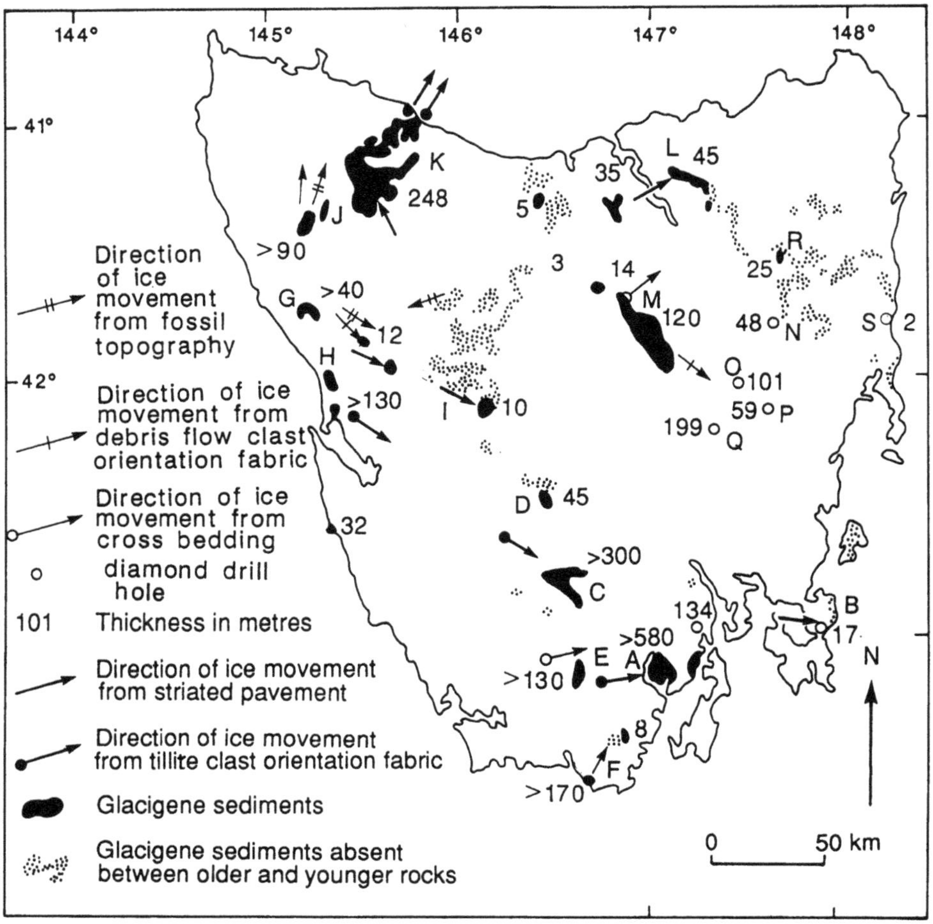

Fig.1 Geographic distribution of Permo-Carboniferous glacigene sediments in Tasmania: **A**-Cygnet, **B**-Eaglehawk Neck DDH, **C**-Maydena, **D**-Florentine Valley, **E**-Picton River, **F**-South Coast, **G**-Zeehan, **H**-Strahan, **I**-King William Saddle, **J**-Savage River, **K**-Wynyard, **L**-Lilydale, **M**-Western Tiers, **N**-Bonneys Plains DDH, **O**-Ross No.2 DDH, **P**-Ross No.1 DDH, **Q**-Tunbridge Tier DDH, **R**-Strickland Gorge, **S**-Douglas River. cf. Banks (1981), Banks and Clarke (1987).

Ice movement directions (IMDs) from striated pavements and imbrication direction of tillite clast orientation fabrics are generally in close agreement and are thought to be highly accurate (Holmes 1941, Mark 1973, 1974, Lawson 1979). Imbrication directions from debris flow clast orientation fabrics generally indicate the direction of flow (Lindsay 1968, Lawson 1979) but must be used with caution as flow direction of debris flows may differ greatly from IMD. IMDs interpreted from debris flow fabrics have only been used in the absence of better indicators of IMD and are the mean of several fabric measurements. Fossil topography refers to U shaped bedrock depressions, infilled with glacigenic sediments. These are interpreted as fossil glacial valleys and the orientation of their long axis therefore represents the IMD.

previously in Tasmania to establish provenance (Banks 1981), it is used in this study on a much greater scale.

The sedimentary rocks of Tasmania's Permo-Carboniferous glaciation have been described with the aid of the glacimarine sedimentation model of Powell (1984). The reasons for using a model of marine glacial sedimentation rather than one of terrestrial glacial sedimentation are given below. Powell

(1984) categorised the known variations of the glacimarine environment into eight glacimarine regimes, each with its own sedimentary characteristics (see Powell 1984 table V). These regimes are as follows. Regime 1 is where a valley glacier terminates in a floating ice shelf and has an alternately melting/freezing base. Regime 2 is identical except that the base of the glacier is permanently frozen. Regimes 3 and four are the same as regimes 1 and 2 respectively with the exception of the glaciers being ice sheets rather than valley glaciers. Regimes 5 and 6 are valley glaciers terminating in tidewater fronts with a melting/freezing base and a frozen base respectively and regimes 7 and 8 are ice sheets with tidewater fronts with a melting/freezing base and a frozen base respectively. Facies associations observed in the Tasmanian Permo-Carboniferous sediments have been compared with those documented by Powell (1984) from Quaternary (including modern) glacimarine environments. In a number of areas more than one Permo-Carboniferous glaciation has been recognised .

2 GENERAL ENVIRONMENT

The Permo-Carboniferous glacigenic sediments include a number of lithotypes and fossils that characterise the general environment. The high proportion of diamictites which can be interpreted as debris flow deposits is indicative of abundant meltwater (Powell 1984), as is the presence of thick intervals of glacial outwash, (conglomerates and sandstones). Dropstone diamictites and/or dropstone muds are present in the majority of deposits and require a relatively large open body of water, either lacustrine or marine, before deposition can take place. Tidal rhythmites (Mackiewicz et al. 1984) are present in over half of the deposits observed. The deposition of these sediments is dependent on considerable tidal fluctuation. They therefore indicate marine conditions as even lakes of the size of the Great Lakes in North America "are virtually tideless" (Press and Siever 1978 p.248). Lithological indicators therefore suggest a marine environment of deposition for most of Tasmania's Permo-Carboniferous glacigenic sediments and deposition in one or more large open bodies of water, possibly the sea, for the remainder of the deposits.

Fossils are present in a number of the glacigenic deposits. Marine macrofossils (see below) are present in a dropstone diamictite at the South Coast (Fig.1, locality F) and fragmentary marine macrofossils are present at Maydena (Fig.1, locality C) (Runnegar 1969, Jago 1972). A fossil insect, Psychroptilus burrettae, the plants Botrychiopsis plantiana and Aphlebia sp. and the probable arthropod track Tasmanadia twelvetreesi have been found in rhythmites in the Hellyer Gorge area south west of Wynyard (Fig.1, locality K) (Gulline 1967, Gould 1975, Riek 1976). Trace fossils are present in rhythmites in drill core from Eaglehawk Neck and from Ross No.1 (The Quoin) (Fig.1, localities B and P). These fossils were identified (T. Naing pers. comm. 1990) as being characteristic of shallow marine environments. Spinose acritarchs, representing brackish to marine conditions, occur in palynomorph samples from Zeehan and Wynyard (Truswell 1978). Thus fossil evidence, with the exception of that from the Hellyer Gorge, indicates marine or near marine conditions. Combining the results from fossils and lithology, over half of the glacigenic deposits have direct evidence of marine deposition and those which don't, contain evidence of deposition in a large body of open water. The glacigenic sediments are overlain by marine siltstones and mudstones (the Woody Island Siltstone and correlates) which can be demonstrated in places to be laterally equivalent to the glacigenic sediments (see below). Therefore it is considered likely that all the glacigenic sediments are marine rather than most being marine and some being lacustrine. However, in the case of a number of deposits, notably those in the Hellyer Gorge area, a lacustrine environment cannot be ruled out. In any case the specific environments of deposition have been interpreted using the glacimarine sedimentation model of Powell (1984).

3 THE GLACIGENIC SEDIMENTS

Cygnet (Fig.1, A): Three glaciations are recognisable in outcrop and subcrop of the Truro Fm. around Cygnet. The first glaciation may have been of regime 1 or 5, the second and third glaciations were of regime 5.

Eaglehawk Neck DDH (Fig.1, B): No tillite is present in the Eaglehawk Neck DDH core. However, debris flow diamictites, rhythmites and other sediments characteristic of glacimarine regime 5 are present. No evidence was present in these sediments suggesting more than a single

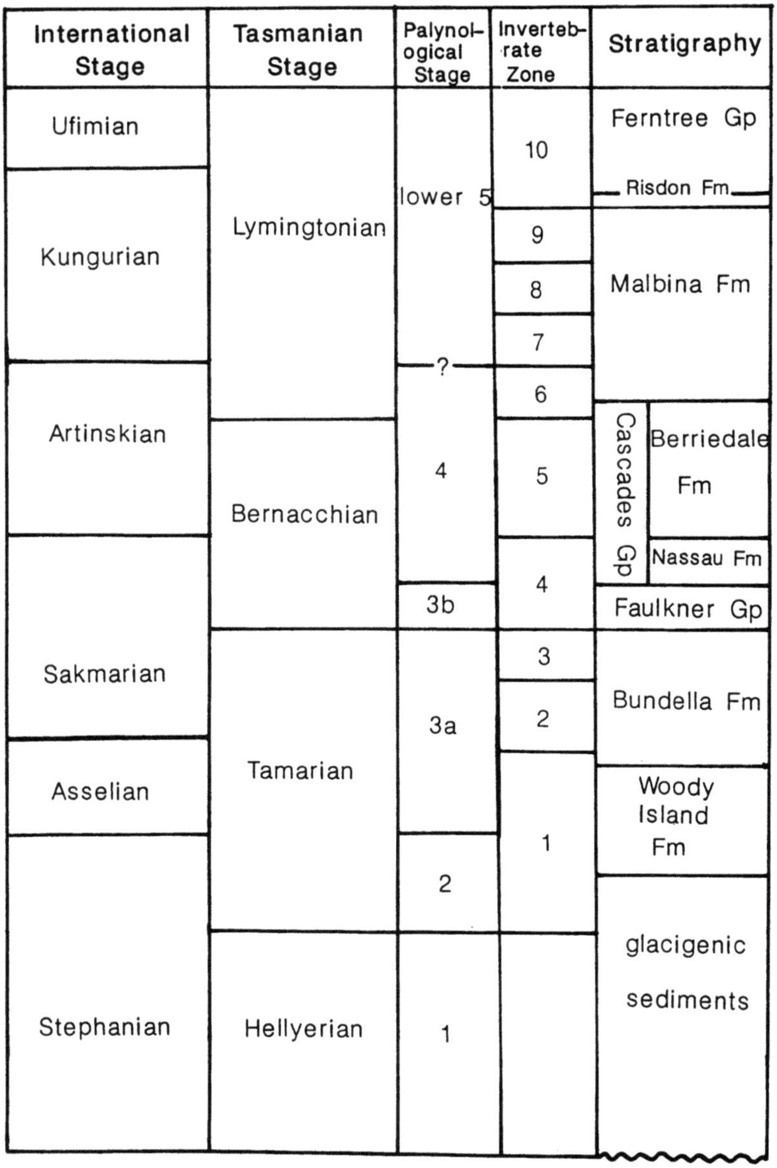

Fig.2 Generalised stratigraphic column for the lower (Carboniferous and Permian) part of the Parmeener Supergroup (after Banks & Clarke 1987, Clarke 1989). See Fig.5 for a revised interpretation of the stratigraphic relationships between the glacigenic sediments and the overlying Woody Island Fm.

glaciation.

Maydena (Fig.1, C): The glacigenic sediments at Maydena appear to represent only one glaciation. However the presence of reworked glacigenic material indicates a previous glacial episode. The sediments comprise a single retreat sequence of a glacier of regime 7. Rocks of the Woody Island Siltstone at Maydena were observed to conformably overlie the glacigenic sediments on a possible diachronous boundary. The siltstones both overlie the glacigenic sediments and occupy a laterally equivalent position immediately to their southeast. No faulting is present which might otherwise explain this outcrop pattern.

Florentine Valley (Fig.1, D): Although outcrop in the Florentine Valley is limited to one good exposure, this contains a range of lithofacies which can be interpreted as having been deposited by a glacier of regime 3. Only one glaciation is recognisable but this may possibly be due to the paucity of outcrop.

Picton River (Fig.1, E): Only the upper part of the glacigenic succession can be observed in outcrop at the Picton River. One glaciation is recognised, either of regime 3 or 7. Sediments of the Woody Island Siltstone crop out immediately to the east of the glacigenic sediments. The siltstones are found to occupy the same and slightly lower stratigraphic positions as the glacigenic sediments. As no faulting is observed which might otherwise explain this phenomenon it is possible that the Woody Island Siltstone overlies the glacigenic sediments on a diachronous boundary similar to that observed at Maydena.

South Coast (Fig.1, F): Extensive coastal outcrop at Shoemaker Bay on the south coast of Tasmania consists of sediments of regime 5. Only one glaciation is manifest in these sediments.

Zeehan (Fig.1, G): Two major areas of outcrop occur north of Zeehan: the Heemskirk Road section (Fig. 2A) and the Pieman Dam section. The former consists of lithofacies of regimes 1 or 5, and the latter could belong to any of regimes 1, 3, 5 or 7. The sheet like geometry of the Pieman Dam section suggests deposition from an ice sheet, i.e. either regime 3 or 7. Both areas are interpreted as having undergone only one glaciation.

Strahan (Fig.1, H): Outcrops at Strahan occur at a number of widely scattered localities, and comprise lithofacies which suggest deposition from a glacier of regime 1, 3 or 5. Only one glaciation is recognised, but because of the problems of correlation between outcrops, the possibility of more than one glaciation cannot be discounted.

King William Saddle (Fig.1, I): Insufficient outcrop exists at King William Saddle to establish the glacial regime.

Savage River (Fig.1, J): One glaciation of regimes 1 or 5 is recognisable in the glacigenic sediments north of Savage River (Fig. 2B).

Wynyard (Fig.1, K): These constitute the most geographically extensive of Tasmania's Permo-Carboniferous glacigenic deposits, three glaciations being recognisable. All three glaciations were of regime 5 (Hand 1988). An additional glaciation prior to that responsible for the first glaciation is inferred from the presence of clasts of reworked glacigenic sediment in rocks deposited as a result of that glaciation. A clearly diachronous interfingering boundary is present between the glacigene sediments and the overlying Woody Island Siltstone equivalent, the Inglis Siltstone on the old Murchison Highway on the north side of the Hellyer Gorge.

Lilydale (Fig.1, L): Exposures around Lilydale consist of a small number of poor quality outcrops. Insufficient information could be gained from these outcrops to reconstruct the glacial regime although the linear geometry of the outcrops, transverse to the local ice movement direction suggests deposition from an ice sheet.

Western Tiers (Fig.1, M): One glaciation of regime 7 is recognisable in the glacigenic sediments of the Western Tiers.

Bonneys Plains, Ross No.1 (The Quoin), Ross No.2 and Tunbridge Tier DDH's (Fig.1, N,O,P,Q): These four drill holes prove the subsurface extension of the Stockers Formation, which crops out in central and northern Tasmania. Only one glaciation can be inferred from the lithofacies succession in each of these cores, although in the case of both the Tunbridge Tier and Ross No.2 DDH's both a glacial advance and retreat phase is preserved. Deposition in these two areas is interpreted as having occurred just beyond the furthest extent of glacial ice. The thicknesses of sediment preserved show a thinning to the east. The glacigenic sediments at both Tunbridge Tier and Bonneys Plains were deposited from glaciers of regime 5 whereas the sediments at both the Ross DDH's were deposited from glaciers of either regimes 1 or 5.

Strickland Gorge (Fig.1, R) The Strickland Gorge section just north of Ben Lomond is unique in that Woody Island Siltstone equivalent appears to both underlie and overlie the glacigene sediments. The basal unconformity is not exposed in this area, the lowermost sediments exposed being dropstone free black mudstone. The mudstone interfingers with glacigene sediments and has been deformed by an overriding tillite. Immediately overlying this tillite is undeformed mudstone, otherwise identical to that below. This unit contains fragmentary brachiopods which could not be identified, as well as a single glossopteris stem. Not enough outcrop is available in the Strickland Gorge to work out the glacial regime.

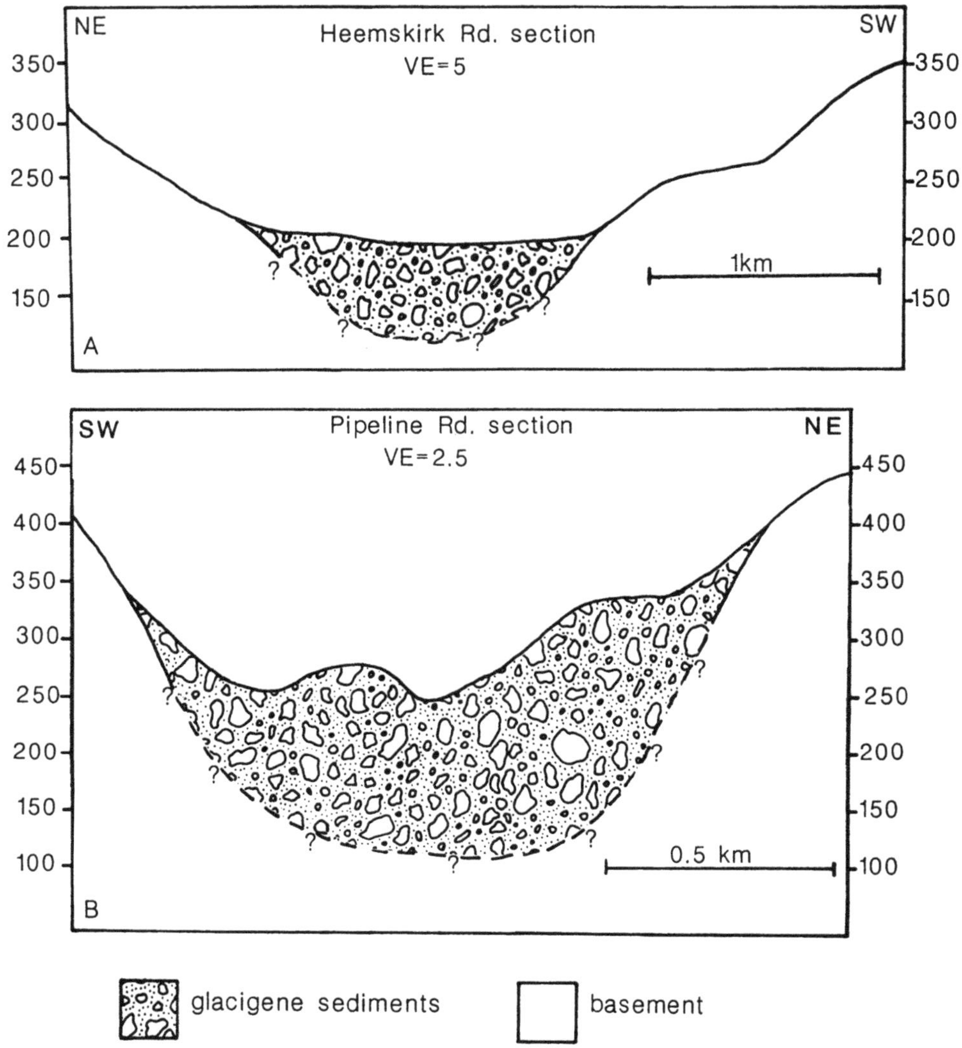

Fig.3 Basement topography of the Heemskirk Rd. section north of Zeehan (55GCP596673-55GCP565653) (A) and the Pipeline Rd. section north of Savage River (55GCQ521166-55GCQ505163) (B). The basement lows in which the glacigenic sediments sit are interpreted as having been fjords, the axes of which trend 132 and 020 respectively.

Douglas River (Fig.1, S) This drill hole contains the eastern-most glacials in Tasmania the 2m of core being insufficient to identify glacial regime.

4 BASEMENT TOPOGRAPHY

Study of the topography of the basal contact of the Permo-Carboniferous glacigenic sediments is possible for the majority of the deposits although in some cases, such as the Truro Formation around Cygnet, the deposits are entirely bounded by younger rocks. In most cases basement topography is very slight but at two localities, namely along the Heemskirk Road north of Zeehan and along the Pipeline Road north of Savage River pronounced basement topography is evident despite the effects of erosion(Fig.3). The geometry of the basal contact in both these areas indicates the presence of U shaped valleys during deposition of the glacigenic sediments although the possibility of these areas being fault controlled cannot be entirely ruled out. The lithofacies observed in these areas indicate glacimarine deposition which would indicate that if these were U shaped valleys they were actually fjords.

5 PROVENANCE

A study of clast types was undertaken in order to determine the provenance of debris included in the glacigenic sediments. Although these data are not detailed herein, the results have been incorporated into the palaeogeographic reconstructions (Fig.4), being particularly useful in determining ice movement directions over areas where outcrops of glacigene sediments may have existed prior to erosion.

6 BIOSTRATIGRAPHIC CONTROL ON TIMING OF GLACIATIONS

The only fossils which apparently allow time differentiation within the glacigene succession are microflora. The succession contains microflora of Stages 1 and 2 (Truswell 1978). Kemp et al. (1977) state that both of these stages are Late Carboniferous, the end of Stage 2 marking the Permo-Carboniferous boundary. Archbold (1982) placed the Carboniferous-Permian boundary at the end of Stage 1 (as Banks & Clarke (1987) have done, see Fig.2). Foster (in Foster & Waterhouse, 1988 p.143) states that "Stage 1 is probably a facies controlled palynoflora...". Stages one and two might therefore be partially time equivalent. Lithostratigraphic correlation between microflora sites has not been currently achieved. However, due to constant reworking during glacial advance, sediments deposited from a glacier are unlikely to be preserved except in a glacial retreat phase (Powell 1981). Therefore it might be expected that microfloral sites nearer ice entry points to Tasmania would yield stage 2 microflora while those sites far from ice entry points yielded stage 1 microflora. In fact no such correlation is observed which may be explained in one or more of three ways, 1) Stage 1 microfloral sites near ice entry points (e.g. Strahan(Fig.1, H)) may be in sediments which have not been reworked during glacial advance, 2) Stage 1 microflora have been reworked and redeposited during stage 2 time, or 3) Stage 1 is facies controlled and therefore the real relationship between stages 1 and 2 is unknown. Whatever the reason or reasons, it appears that Tasmania's Permo-Carboniferous glacigenic sediments cannot be subdivided palynostratigraphically.

Marine macro-fossils are present in two of the glacigenic deposits. At Maydena (Fig.1, C), fragmentary specimens of Trigonotreta, Deltopecten, Pyramus and stenoporids are present (Runnegar 1969, Jago 1972). At Shoemaker Bay on the South Coast (Fig.1, F) abundant well preserved specimens of Trigonotreta, Keenia, Eurydesma and stenoporids are present. The invertebrate fossils at both these localities fall into faunizone 1 of Clarke & Banks (1975) and are thought to be latest Carboniferous or earliest Permian in age (Clarke 1989).

Other macrofossils include the marine trace fossils Tasmanadia and Gyrochorte, an insect, Psychroptilus burrettae, and the plants Botrychiopsis plantiana and Aphlebia sp. Of these, only the plants yield biostratigraphic data suggesting a Late Carboniferous to earliest Permian age (Gould 1975).

The age ranges of the fossils from the glacigene sediments suggests a Late Carboniferous (Stephanian) to earliest Permian (Asselian) age. However, the bulk of the preserved glacigenic sediments are likely to represent the final glacial retreat and therefore it is possible that the onset of glaciation in Tasmania occurred earlier than the Late Carboniferous.

7 PALAEOGEOGRAPHIC RECONSTRUCTIONS

The palaeogeographic history of Tasmania's deglaciation during the Stephanian and/or Asselian is presented in Fig.4. The major assumption which has been made in the construction of these figures is that deglaciation of Tasmania occurred as a single smooth retreat. This is the simplest case and is assumed to be correct in the absence of conflicting stratigraphic evidence.

Elevated regions have been postulated in some of the areas where glacigenic sediments are absent between older and younger rocks. Among these areas are those where rocks older than the glacigene sediments crop out at higher elevations

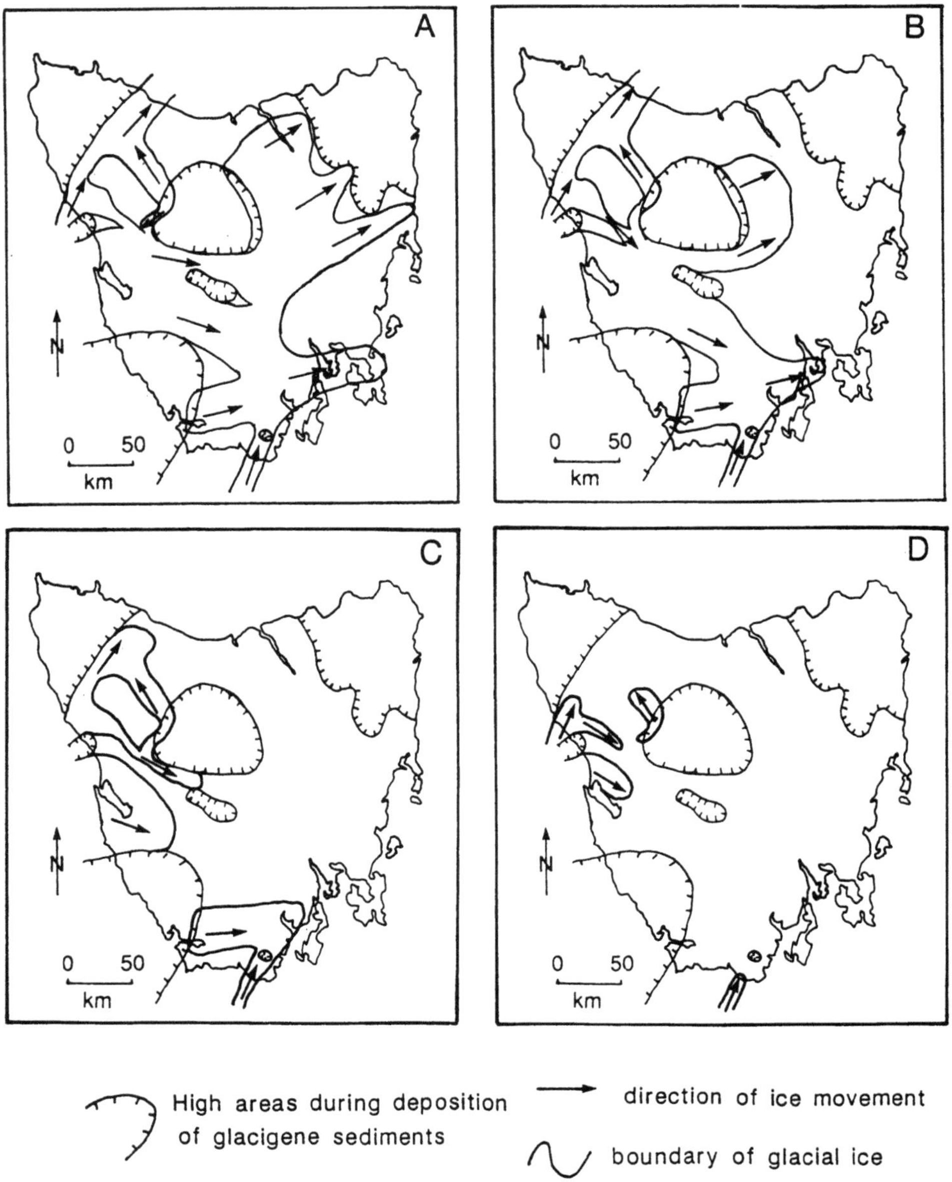

Fig.4 Palaeogeographic reconstructions of Tasmania at four separate times during the deposition of the glacigenic sediments. (A) shows the palaeogeography at the glacial maximum while (B), (C) and (D) show the palaeogeographies at successive stages during the glacial retreat. No attempt has been made to put an absolute time scale on these reconstructions due to the paucity of biostratigraphic controls.

than adjacent glacigene sediments. Others are where the ice movement directions suggest that ice might otherwise have moved over these areas. Other regions where glacigenic sediments are absent in the stratigraphic record probably represent areas beyond the furthest extent of glacial deposition.

Elevated regions are thought to have existed in southwest and northwest Tasmania due to the lack of deposition of glacigenic sediments and younger rocks of the Parmeener Supergroup in these areas. The Devonian Heemskirk Granite on the West Coast is thought to have been a nunatak. Ice movement directions at Strahan to the immediate southeast of the Heemskirk granite are from the northwest suggesting ice movement over the Heemskirk granite. The absence of granite clasts in sediments at Strahan suggests that ice did not move over the Heemskirk Granite and may have been deflected by the granite body towards the south as shown in Fig.4.

At Savage River and at Zeehan (Fig.1) the presence of fossil fjords is known from the palaeotopography; in other areas where fjord style sedimentation has occurred the size and, in some cases, orientation of fjords is based on supposition. A "U" shaped topographic low in Pre Permo-Carboniferous bedrock, possibly a fossil fjord, exists north of the Douglas River borehole (Fig.1, S) (C. Calver pers. comm. 1992) but has not been included on Fig.4A.

It is probable that ice entered Tasmania in three areas (Fig.4). The ice deviated around the Heemskirk Granite which was probably a nunatak, thus splitting the entry of ice across what is now Tasmania's West Coast into two corridors, one north and one south of the Heemskirk Granite. Ice entering through the northern of these corridors diverged into two separate streams immediately beyond the nunatak, one glacier flowing to the northeast towards Wynyard, and the second flowing southeast and probably coalescing with ice entering Tasmania through the other major corridor, south of the Heemskirk Granite Nunatak (Fig.4). The third corridor of ice entry into Tasmania was across the South Coast in the vicinity of Shoemaker Bay. Ice flowed northeast to coalesce with another glacier flowing east-northeast from uplands to the west (Fig.4).

At the glacial maximum ice covered around half of Tasmania. In several areas, most notably at Maydena (Fig.1, C) , correlates of the Woody Island Siltstone overlie the glacigenic sediments on a probable diachronous boundary. Open marine and glacimarine environments co-existed and as the ice retreated a marine transgression occurred. Given the ambiguity in interpretation of palaeontological data (see above) there is no direct evidence to suggest when non-glacial marine sedimentation commenced in Tasmania with relation to the deposition of the glacial sediments. No sediments, glacial or non-glacial, can be

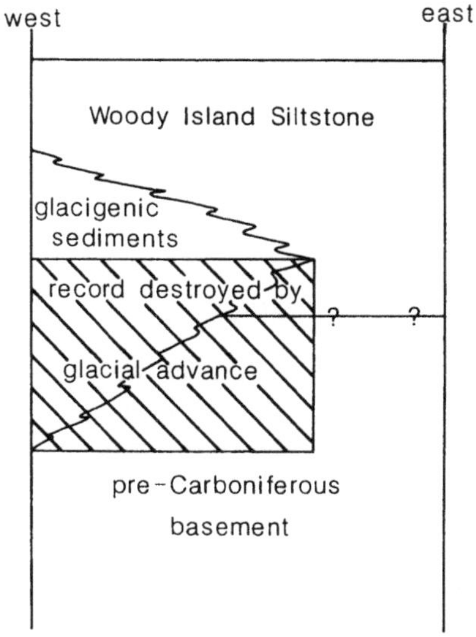

Fig.5 Stratigraphic column showing the proposed revision of the depositional sequence in the Tasmania Basin (cf. Fig.2). Sediments recording the onset of glaciation are unlikely to have been preserved due to ice advance and reworking. If such sediments do exist they have not been recognised. The onset of glaciation may therefore have occurred before the onset of marine conditions in the Tasmania Basin. The bulk of sedimentation, however, occurred after the onset of marine conditions in the Tasmania Basin and was thus glacimarine.

demonstrated to have been deposited in western Tasmania prior to the glacial maximum. It is assumed that the majority, perhaps all of such sediments have been reworked by subsequent ice advance. Therefore whenever glacigenic and non-glacigenic marine sediments are observed in the same section the latter overlies the former. However, glacimarine sediments were deposited into a marine basin and hence marine conditions must have existed in Tasmania at latest by the glacial maximum. It is highly likely that marine conditions existed prior to the glacial maximum and that the glacial advance represents, at least in its latter stages, a marine regression. Assuming this to have been the case the oldest rocks in the Parmeener Supergroup would be the basal beds of the Woody Island Siltstone and its correlates in northeastern Tasmania(in areas beyond the maximum extent of glaciation) (Fig.5). It is suggested that the onset of marine conditions was the result of isostatic loading of the Tasmania

Basin by and during ice advance. Hence, marine conditions probably first occurred part way through the glacial advance.

As the glaciers receded, a marine transgression occurred, finally resulting in marine conditions over most of Tasmania (Banks & Clarke 1987).

ACKNOWLEDGEMENTS

I am extremely grateful to Dr P.J. Conaghan, Dr M.R. Banks and an anonymous reviewer who read this manuscript and offered useful suggestions on how it could be improved.

REFERENCES

Ahmad, N.R. 1957. The environment of the Permian glacial sedimentation of Tasmania. Unpublished Ph.D. thesis: University of Tasmania

Archbold, N.W. 1982. Correlation of Early Permian Faunas of Gondwana: Implications for the Gondwana Carboniferous-Permian Boundary. Journal of the Geological Society of Australia 29: 267-276.

Banks, M.R. 1981. Late Palaeozoic tillites of Tasmania. In Hambrey, M.J. and Harland, W.B. (eds.), Earth's pre-Pleistocene glacial record: 495-501. Cambridge: Cambridge University Press

Banks, M.R. and Clarke, M.J. 1987. Changes in the Geography of the Tasmania Basin in the Late Palaeozoic. In McKenzie, G.D. (ed.) Gondwana six: Stratigraphy, Sedimentology and Paleontology: 1-14 Washington: American Geophysical Union

Carey, S.W. and Ahmad, N.R 1961. Glacial marine sedimentation. In Raasch, G.O. (ed.), Geology of the Arctic Vol.2: 865-894 Toronto: University of Toronto

Clarke, M.J. 1989. Lower Parmeener Supergroup. in Burrett, C.F. and Martin, E.L. (eds.), Geology and mineral resources of Tasmania: 295-309 Geological Society of Australia Special Publication 15

Clarke, M.J. and Banks, M.R. 1975. The stratigraphy of the lower (Permo-Carboniferous) parts of the Parmeener Super-group, Tasmania. In Campbell, K.S.W. (ed.) Gondwana Geology: 453-467 Canberra: Australian National University Press

David, T.W.E. 1907. The Permo-Carboniferous glacial beds at Wynyard near Table Cape. Tasmania. Australian Association for the Advancement of Science 11: 274-279

Domack, E.W. and Lawson, D.E. 1985. Pebble fabric in an ice rafted diamicton. Journal of Geology 93: 577-591

Dowdeswell, J.A. and Sharp, M.J. 1986. Characteristics of pebble fabrics in modern terrestrial glacigenic sediments. Sedimentology 33: 699-710

Flint, R.F. 1947. Glacial geology and the Pleistocene epoch. New York:John Wiley and sons.

Foster, C.B. and Waterhouse, J.B. 1988. The Granulatisporites confluens Oppel-zone and Early Permian marine faunas from the Grant Formation on the Barbwire Terrace, Canning Basin, Western Australia. Australian Journal of Earth Sciences. 35: 135-157

Gould, R.E. 1975. The succession of Australian pre-Tertiary megafossil floras. Bot. Rev. 41(4): 453-483

Gulline, A.B., 1967. The first proved Carboniferous deposits in Tasmania. Australian Journal of Science 29 (9): 332-333

Hand, S.J. 1988. Palaeogeography of the Wynyard Formation. Unpublished Honours Thesis: University of Tasmania.

Holmes, C.D. 1941. Till fabric. Geological Society of America Bulletin 52: 1299-1354

Jago, J.B. 1972. Geology of the Maydena Range. Papers and Proceedings of the Royal Society of Tasmania 106: 45-56

Kemp, E.M., Balme, B.E., Helby, R.J., Kyle, R.A, Playford, C. & Price, P.L. 1977. Carboniferous and Permian palynostratigraphy in Australia and Antarctica: a Review. BMR Journal of Australian Geology and Geophysics 2: 177-208

Lawson, D.E. 1979. A comparison of the pebble orientations in ice and deposits of the Matanuska Glacier, Alaska. Journal of Geology 87: 629-645

Lindsay, J.F. 1968. The development of clast fabric in mudflows. Journal of Sedimentary Petrology 38 (4): 1242-1253

Mackiewicz, N.E., Powell, R.D., Carlson, P.R., and Molnia, B.F. 1984. Interlaminated ice-proximal glacimarine sediments in Muir Inlet. Alaska. Marine Geology 57: 113-147

Mark, D.M. 1973. Analysis of axial orientation data including till fabrics. Geological Society of America Bullettin 84: 1369-1374

Mark, D.M. 1974. On the interpretation of till fabrics. Geology. 2: 101-104

Powell, R.D. 1981. A model for sedimentation by Tidewater Glaciers. Annals of Glaciology 2: 129-134

Powell, R.D. 1984. Glacimarine processes and inductive lithofacies modelling of ice shelf and tidewater glacier sediments based on Quaternary examples. Marine Geology 57: 1-52

Press, F. and Siever, R. 1978. Earth (2nd edition) San Francisco: W.H. Freeman and Co.

Riek, E.F. 1976. Neosecoptera, a new insect suborder based on specimen discovered in the Late Carboniferous of Tasmania. Alcheringa 1(2): 227-234

Runnegar, B.N. 1969. The Permian Faunal succession in Eastern Australia. Geological Society of Australia Special Publication 2: 73-98

Stephens, T. 1869. Remarks on the geological structures of part of the north-west coast of Tasmania, with special reference to the Tertiary marine beds near Table Cape. Proceedings of the Royal Society of Tasmania 5: 17-21

Truswell, E.M. 1978. Palynology of the Permo-Carboniferous in Tasmania: an interim report. Tasmanian Department of Mines, Geol. Surv. Bull. 56.

Gondwana Eight, Findlay, Unrug, Banks & Veevers (eds) © 1993 Balkema, Rotterdam. ISBN 90 5410 304 3

Facies analysis of glacial marine pebbly mudstones in the Tasmania Basin: Implications for regional paleoclimates during the late Paleozoic

Eugene W. Domack, Lewis A. Burkley & Cynthia R. Domack
Geology Department, Hamilton College, Clinton, N.Y., USA

Maxwell R. Banks
Geology Department, University of Tasmania, Hobart, Tas., Australia

ABSTRACT: Uppermost portions of the late Paleozoic glacigenic section in Tasmania consist of uniformly massive to poorly stratified diamictites that pass abruptly into overlying marine rocks. The lower facies of the succeeding marine section is an organic-rich mudstone that contains a variable ice-rafted component characterized by ubiquitous sand-sized material and scattered pebbles and cobbles that decrease in abundance upwards. These dropstones were modified by fluvial or beach processes and show little evidence for transport by glacial ice. The basal mudstone also contains Tasmanite oil shale. The Tasmanite beds record a basin-wide period of enhanced primary productivity coupled with ice rafting. Overall sedimentation rates were initially low and organic carbon was being preserved apparently without bottom water anoxia. Input of meltwater and terrigenous sediment was limited to the heads of restricted embayments where transport and deposition may have been influenced by Coriolis forces. The lower interval of the marine sedimentation records polar to sub-polar glacial marine conditions characterized by very cold waters, seasonal sea ice and associated ice-marginal phytoplankton blooms and shore-ice rafting. Temperate conditions were established later as marked by the presence of plant macrofossils and an apparent increase in sedimentation rates. A great deal of terrigenous sediment was introduced into the basin from a distal and/or subaqueous delta system that prograded from the southeast. Periods of decreased sedimentation (terrigenous input) are marked by pebble and fossil horizons. These may indicate intervals of renewed sub-polar conditions and intensified bottom current activity.

1 INTRODUCTION

As with other late Paleozoic glacial sequences across Gondwanaland the succession of glacial deposits in Tasmania is punctuated by a sharp contact between a diamictite and an overlying black mudstone. Such a marked lithologic transition is important in that it is widely believed to represent the end of glacial climates within the regional extent of the basins in which it is found. In order to understand the causes of regional deglaciation it is first necessary to establish the paleo-environmental context of the black mudstones and their facies relationship to underlying glacigenic deposits. From a stratigraphic viewpoint the timing of ice recession as recorded in these rocks is also important. Either the glacial recession was synchronous across Gondwanaland during the Late Carboniferous to Early Permian or it was asynchronous, and occurred as the landmass drifted across polar-centric climatic zones. Thus, resolution of the stratigraphic age of comparative sequences must also focus upon the "post-glacial" lithologies; namely the black mudstone/shale.

Contemporary glacial marine settings are found in a wide range of climatic zones, from temperate oceanic, sub-polar, to polar (Anderson and Domack, 1991). Different mechanisms drive the regional glacial recession within each of these climatic zones. Polar conditions (as exemplified at the periphery of the Antarctic continent) contain glacial systems, such as ice shelves, that respond primarily to sea-level fluctuations whereas temperate glaciers also respond to climatic forcing; that is warming and cooling and/or consequent changes in precipitation. Therefore, in order to understand the driving mechanisms for ancient deglaciations, such as those within the late Paleozoic basins of Gondwanaland,

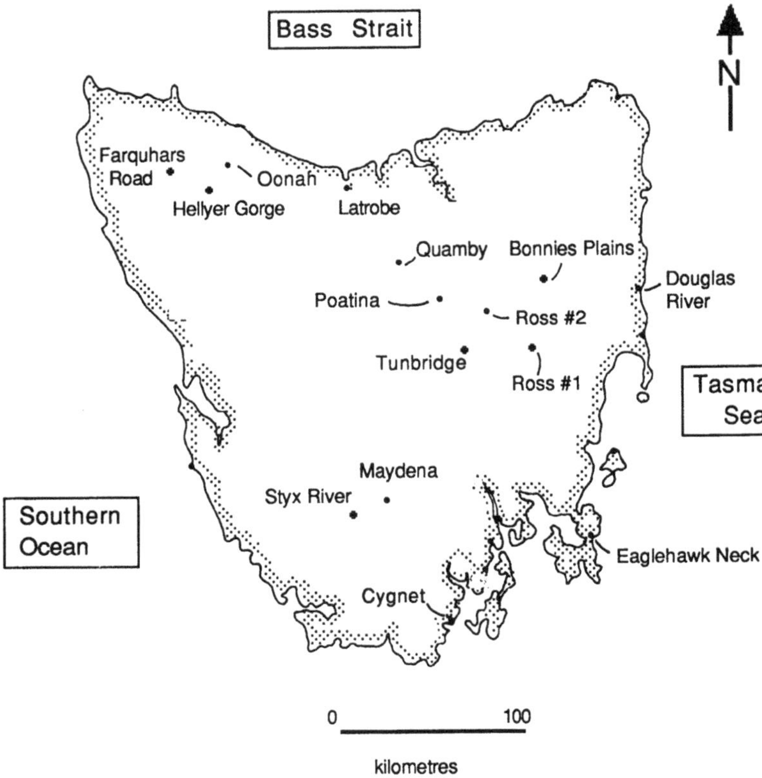

Fig. 1 Tasmania with locations mentioned in text.

representative sequences first need to be placed into a correct paleoclimatic setting.

The rocks involved within Tasmania are demonstrably marine and form the lower part of a relatively complete Upper Carboniferous to Permian sequence referred to as the Parmeener Supergroup (Banks and Clarke, 1987). Glacigenic beds form a thick succession (>200 m) that includes both structureless, massive diamictite and interbedded conglomerates, sandstones, and laminated mudstones. These lithologies locally form such units as the Stockers, Truro, and Wynyard Tillite. However, the paleo-environments for this sequence range from glacial (ice contact) to lacustrine, fluvial and glacial marine. Dark mudstones are found directly above the glacigenic sequence and contain both dropstones and marine fossils; which are suggestive of, at least, temporary glacial marine conditions during the glacial to marine transition. Such rocks are assigned to one of several units depending upon the location within Tasmania; these are the Inglis Formations, Quamby Formations, and Woody Island Formations.

This study began in 1988 and focused upon the sedimentologic, stratigraphic, and geochemical characteristics of the black mudstone/diamictite association. Eight outcrop and seven borehole sections (Figure 1) were examined, logged, and sampled for thin-section, clast shapes, and organic carbon analyses.

2 OBSERVATIONS AND RESULTS

2.1 *General Lithologic Succession and Regional Variations*

In most cases the contact of the black mudstone with the glacigenic diamictites is extremely sharp, as especially evident within sections from the central portion of the basin (Tunbridge, Ross #2, Figure 1). Outcrop relationships demonstrate transitional or interbedded rock types. For instance, at Oonah (Figure 2) structureless to horizontally foliated diamictites, interpreted as tillite, are overlain by a thin, stratified diamictite

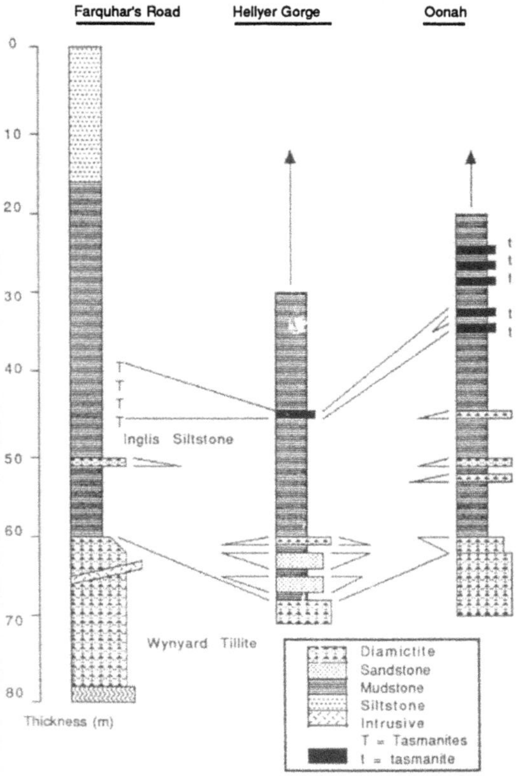

Fig. 2 Measured outcrop sections from the northwestern sub-basin in Tasmania. See Fig.1 for locations.

with dropstone clusters or pebble nests. Thin (1-0.5 m) diamictite beds alternate with dark grey pebbly mudstone over the succeeding 10 meters. Eventually the mudstones predominate. The thin diamictites are lenticular with flat and even lower surfaces and irregular but sharp upper contacts with the mudstone. Stratification within the overlying mudstones, where evident, tends to on-lap onto the irregular upper surfaces of the diamictite interbeds. Similar interbedded transitions were observed within the Ross #1 borehole section (Figure 1). At other localities the transitional contact is observed as a gradual decrease in pebble abundance as the poorly sorted matrix of diamictite gives way to the fine matrix of a pebbly mudstone (eg. near Cygnet).

Characteristic of the lower parts of the mudstone is a decrease in clast abundance and size from the diamictite contact upwards. The mudstone is characterized by a massive to hackly fracture surface. Bedding is most often massive but thin laminae were observed within a thin (0.5-1.0 m) zone at Oonah (Figure 2). Two thin, sandstone interbeds were observed at Hellyer Gorge (Figure 2) and these were characterized by well-preserved climbing ripple stratification.

Within the drill-cores the stratification is also massive with rare ill-defined, very thin laminae or fine-scale lenticular bedding. The only exception to this is the drill-core at Bonney's Plains (Figure 1). Within this section there is no structureless mudstone; rather, marine units directly above the glacigenic interval are characterized by thin bedded to inter-laminated sandstone and mudstone. Primary bedding is well preserved and includes both flaser, lenticular and horizontal to cross-laminations. Composite fining of the entire section takes place over a 40-50 m interval, where the thickness, grain size and abundance of sandstone interbeds decreases at the expense of mudstone. Several poorly sorted sandstones and sandy conglomerates also occur sporadically throughout the section. Large, sand-filled, horizontal burrows are common but dropstones and macrofossils are rare. The section at Bonney's Plain is distinctly different from all other drill-cores and outcrops within the Tasmania basin and therefore must represent unique environmental conditions.

2.2 *Tasmanites and Tasmanite Interval*

Evidence of the green alga Tasmanites has been found in every section examined, with the exception of the exposures of the Woody Island Fm (siltstone) in the far southeast of Tasmania and within the Bonney's Plains borehole. The first occurrence of Tasmanites in most sections is of isolated spores dispersed within the mudstone matrix. Near Latrobe, at the bend of the Mersey River, isolated Tasmanites can be found within the matrix of diamictite, below the contact with the mudstone. A similar occurrence was reported for sections near Beaconsfield (Truswell, 1978). Where the concentrations of spores exceeds about 10% a Tasmanite interval is produced. Such interbeds usually occur as two separate layers about 1 m apart, but near Oonah up to 5 such layers occur (Figure 2). In outcrop the Tasmanite is a hard durable layer that possesses a strong foliation due to the compression of the Tasmanites spores. Isolated clasts (dropstones) are always present. Fresh exposures of the Tasmanite are characterized by a dark, greenish-bronze color whereas weathered exposures are

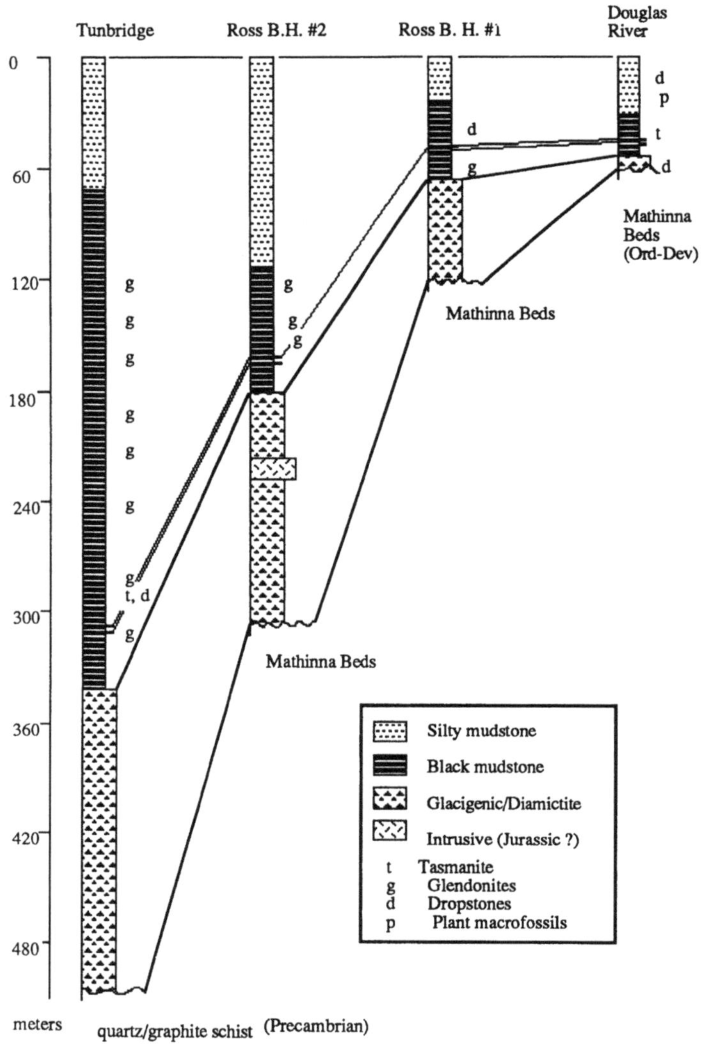

Fig. 3 Borehole sections from the east central portion of the main Tasmania basin. See Fig.1 for locations.

dull brown.

The Tasmanite beds occupy a consistent stratigraphic level across the Tasmania Basin where they are found some 10 to 30 m above the basal contact with diamictites. The lateral continuity of the Tasmanite beds is both remarkable and somewhat variable. Within the east central portions of the main basin two separate intervals can be correlated across some 125 km (Figure 3). Banks and Clarke (1987) originally suggested that the Tasmanite beds were restricted to the margins of the basin, especially near what would have been islands of crystalline basement.

In the far northwest we observed that the Tasmanite beds were far more abundant along the eastern side of a structural trough, or sub-basin. Toward the west, stratigraphic sections contain either a single Tasmanite bed or a zone of dispersed Tasmanites (Figure 2). The Tasmanite is also absent from the well-stratified and arenaceous sequence within the borehole at Bonney's Plains, and thus lateral variability of the Tasmanite facies seems to be controlled by local conditions rather than simply distance to the paleo-shoreline.

Dropstones of gravel to sand size are especially abundant within the Tasmanite as are pyrite nodules and burrows. Sedimentary structures are difficult to discern because of the strong foliation and bioturbation but some faint evidence for cross-laminations is observable (Denwer, 1987). Contacts with adjacent mudstone may be gradational or sharp with some deformation due to loading occurring along the lower contacts.

2.3 *Upper Mudstone Facies and Transitions with Overlying Unit*

In most sections the upper parts of the mudstone are transitional, with the overlying units consisting of abundantly fossiliferous and erratic-rich mudstones, limestones, and thin diamictite (Bundella Fm., Darlington Limestone, Golden Valley Grp.; Banks and Clarke, 1987). In the central parts of the basin, such as at Poatina (Figure 1), the transition is marked by an increase in the abundance and continuity of pebble layers within the mudstone. Isolated stones are common at all levels within the mudstone, but they form discontinuous layers and eventually occur as continuous layers higher in the sequence. Large and robust bivalves are commonly associated with the more continuous pebble layers as is a significant quantity of moderately to poorly sorted sand. Bioturbation is well preserved within such layers and shows evidence of down-mixing of the arenaceous material into underlying mudstone (Figure 4).

In the southern part of the basin the mudstone progressively coarsens upward into siltstone and, eventually, very fine-grained silty sandstone. Bioturbation (eg. Zoophycos) almost completely masks original bedding. Some medium scale cross-bedding (amplitudes of 15-20 cm) was observed near Cygnet (Figure 1) where a westerly paleocurrent direction was measured.

Fig. 4 Portion of borehole section from upper facies of mudstone illustrating coarse layers and bioturbation into underlying fine-grained mudstone.

2.4 *Clasts and Dropstones*

As mentioned previously, pebble-sized clasts are almost always present within the mudstone unit though the vertical abundance varies. Dropstones are difficult to recognize due to the structureless nature of the mudstone but can be seen in borehole sections where very thin laminae are preserved. Thin sections, especially within the Tasmanite horizons, also demonstrate the dropstone character of outsized clasts. The abundance of dropstones is discussed above, but it should be noted that sections along the margins of the basin such as at Douglas River (Figure 5) and Latrobe contain dropstones throughout the section. The size of clasts is restricted (Figure 6) to pebbles generally smaller than 5 cm. This compares to a more even distribution of clast sizes within the underlying diamictites (Figure 6). Clast form and texture determined by visual inspection, measures of Krumbein Roundness (Krumbein, 1941) and axial ratios are illustrated in Figures 7a and 7b. These data demonstrate that there are differences between the character of the clasts within the glacigenic (diamictite) and overlying mudstones. Of the 2000 clasts examined by us from the mudstone none showed any striations. This contrasts with the clasts from the diamictites in which on average about 20% were striated, including those extracted from the drill cores. There is a tendency for a greater proportion of spherical clasts in the mudstone, compared to those from the diamictite (Figure 7a,b). There is

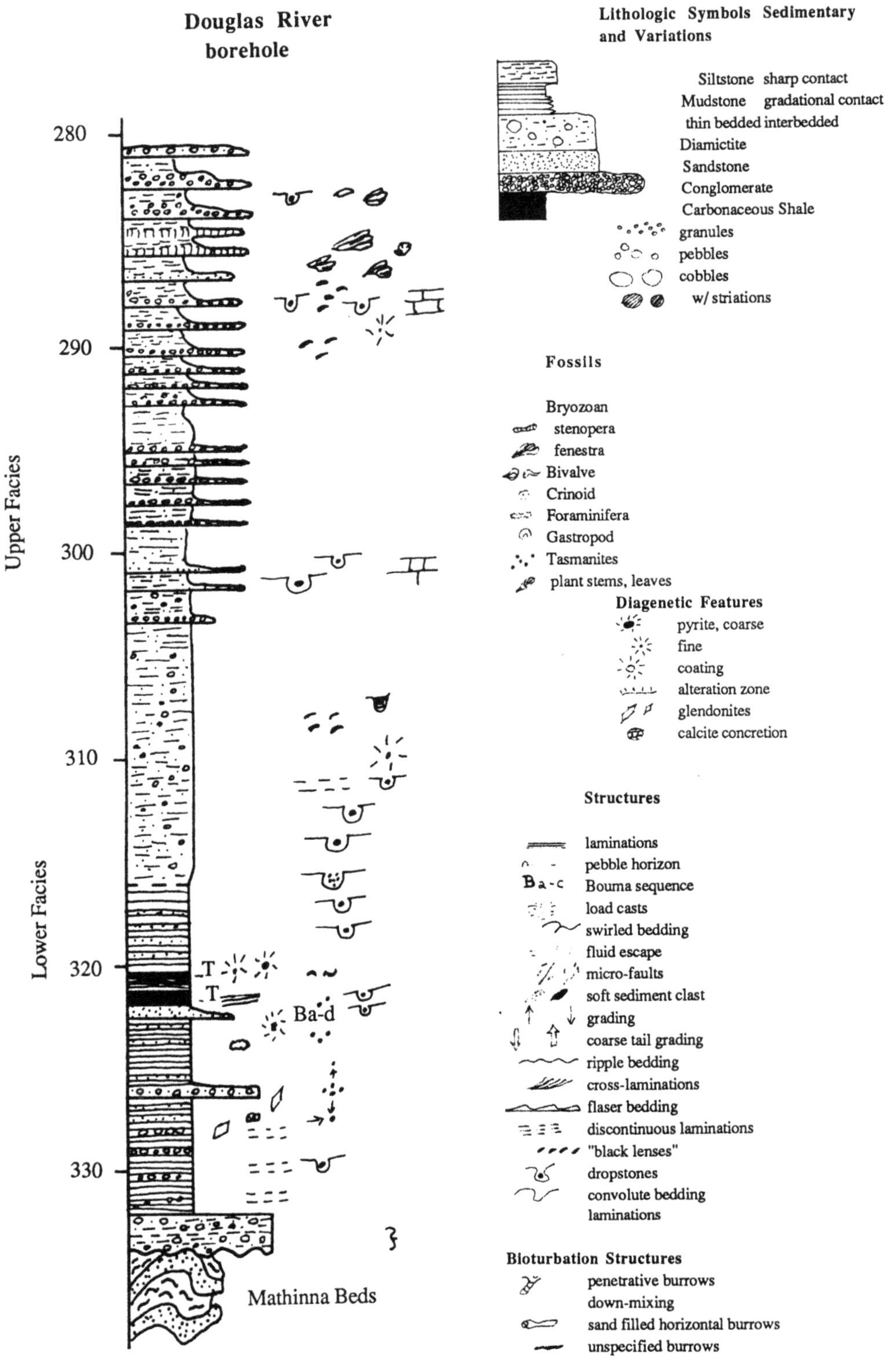

Fig. 5 Detailed measured section of the Douglas River borehole. See Fig. 1 for location.

also a noticeable increase in size and angularity of the dropstones within the formational units immediately above the mudstone; many of these are also striated.

2.5 *Early Diagenetic Features*

Of the many diagenetic/authigenic features found within the mudstone, two are of note and include the presence of glendonites and pyrite. Glendonites are calcite pseudomorphs after the mineral ikaite, a hydrated form of calcium carbonate (Suess et al., 1982; Shearman and Smith, 1985). Within the mudstone section the glendonites occur in a predictable and cyclical fashion being found generally above the Tasmanite horizon (Figure 3). Their first occurrence is marked by small (granule size) individual blades or very tiny rosettes. These become progressively larger (several cms to 10 cm) and more numerous upwards. The glendonites almost always occur as clusters or along horizons within the outcrops. Eventually the size and abundance of the glendonites decreases, in phase with the coarsening of the mudstone matrix from silty mudstone to muddy siltstone. The glendonites in the uppermost parts of the section are often corroded and show signs of having been exhumed and partially eroded prior to sediment consolidation. Glendonites are not found within the Bonney's Plains borehole but have been found within all other sections including exposures in the southeast.

Pyrite is ubiquitous within the mudstone and occurs in a variety of forms and associations. Coarse pyrite (cm sized) is almost always in the form of irregular nodules that have replaced organic material (coprolites ?). Fine pyrite is found disseminated throughout the mudstone matrix as individual grains or framboidal clusters. Pyrite coatings can also be found along the margins of bivalve shells, glendonites, and dropstones. An unusual occurrence is found within "quartzite" dropstones, where pyrite occurs both as a coating around the stone and within the matrix of the stone itself (borehole descriptions from Mersey district). This suggests that the "quartzite pebbles" were probably unlithified sand clots that were later lithified, along with the mudstone, but not before diagenetic pyrite was formed. Pyrite can also be found as laminae or very thin beds within the organic-rich parts of the lower mudstone.

Large concretions (up to 0.5 m diameter) also occur in outcrops of the mudstone and these are either calcitic or sideritic in composition.

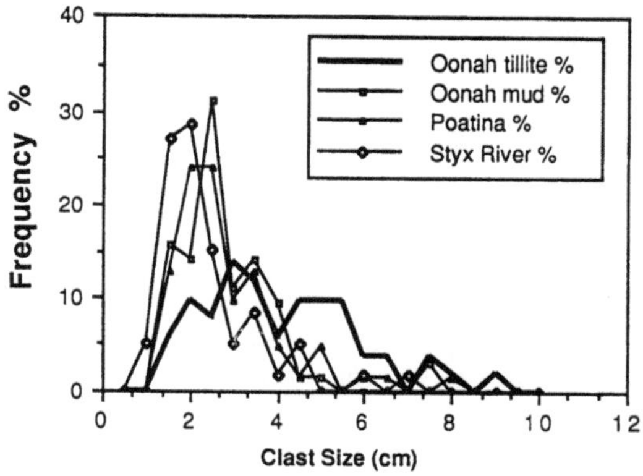

Fig. 6 Size frequency distribution of dropstone size for selected outcrop localaties and lithologies. See Fig. 1 for locations.

2.6 *Plant Fossils*

Spores can be extracted from the mudstone although the abundance, preservation and diversity vary depending upon the stratigraphic position. A poorly preserved, low diversity microflora is reported from the sections below, within, and just above the Tasmanite within the Douglas River borehole, Hellyer Gorge, Latrobe (Calver et al., 1983), and Quamby Brook (Truswell, 1978). Rare monosaccates and Microbaculispora are present but are corroded, thin, and coated with pyrite. The co-occurrence of acritarchs, reworked from older Paleozoic successions, suggests that the spores themselves may be reworked. Nevertheless, the microfloras are assigned to a Stage 2 assemblage and therefore indicate that the age of the mudstone sequence is latest Carboniferous to earliest Permian (Calver et al., 1984).

Plant macrofossils can be found within the mudstone sequence at a number of localities including: Quamby Brook, Styx River, and within the Douglas River Borehole. Leaf impressions and carbon films of Gangamopteris sp. were found within massive dark grey siltstone at Quamby Brook some 10s of meters above the Tasmanite horizon. Stems and very small leaf impressions were observed some 10s of meters above the basal contact near Styx River. In the Douglas River borehole carbonized plant debris occurs within the upper portions of the mudstone sequence (Figure 5).

2.7 *Total Organic Carbon (TOC)*

Samples for total organic carbon were taken every 30 cm in the Douglas River Borehole and every 50 cm in the Ross #1 borehole. Acid soluble components were removed and powdered samples were burnt in oxygen in a LECO induction furnace. Results are reported as weight % total organic carbon and are illustrated in Figures 8 and 9. Of note is the overall trend of decreasing TOC upwards through the mudstone. Maximum TOC is associated with intervals of Tasmanites in cores both with levels of greater than 20% TOC found within the Tasmanite in Douglas River, 17% in Ross #2, and greater than 2% in the Tasmanites zone in Ross #1 and Tunbridge. The lowest values of TOC are just under 0.3 % and are found within the upper facies of the mudstone.

The mudstone can be divided into two basic units based upon the TOC content. A lower section is characterized by a high amplitude of, and frequent fluctuations in, its TOC content and is found within the first 18 m of the mudstone in the Douglas River borehole and the first 20 m in the Ross #1 borehole. Although the TOC fluctuates a great deal, the overall TOC within this lower interval remains relatively constant, except for the major increases associated with the Tasmanite layers. Above the lower facies, the TOC contents tend to decrease with fluctuations being smaller and more widely spaced (Figures 8 and 9). These results are consistent with preliminary measurements on the Tunbridge and Ross #2 boreholes which, so far, show identical TOC patterns.

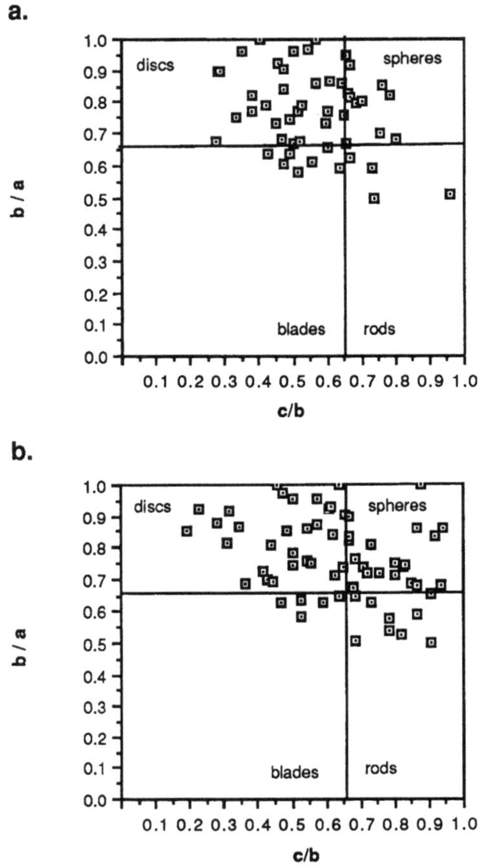

Fig. 7 Principle axial ratios for clasts from diamictite at Oonah (a) and mudstone at Oonah (b).

3 DISCUSSION AND INTERPRETATION

3.1 *Deglaciation*

We interpret the contact between the thick glacigenic sequence and the overlying mudstone as representing a very rapid change in depositional conditions; one that involved an ice sheet that was, in part, marine based. Decoupling of a marine-based ice-sheet from its bed would produce a transition from sub-glacial to deep marine environments within a matter of a few hundred years, as demonstrated from dated Quaternary sequences, along the Antarctic continental shelf (Domack et al., 1991). The glacigenic units just below the transition consist of massive diamictite that varies in thickness from a few meters (Douglas River; Figure 6) to over several tens of meters. The diamictites contain reworked bivalves that indicate derivation, in part, from marine sediments. Fluid-escape pipes are also present and are essentially dewatering structures that may suggest a subaqueous and debris flow origin for some of the diamictite. The overall structureless character of the diamictite suggests a lack of current sorting and/or amalgamation of individual sediment flows or tillites. An abundance of striated clasts attests to glacial deposition or the direct derivation from a glacial source. The abrupt and sharp contacts with the overlying mudstone are preceded by a noticeable

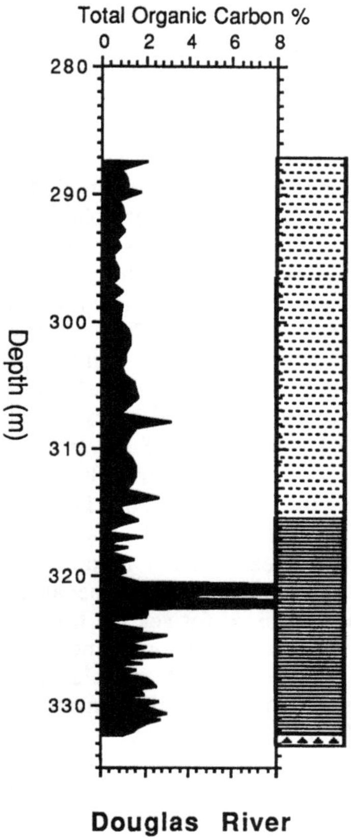

Fig. 8 Total organic carbon contents for the Douglas River borehole. Sample interval is 30 cm. See Fig 5 for details of measured section. Rock type symbols are the same as in Figs. 2 and 3.

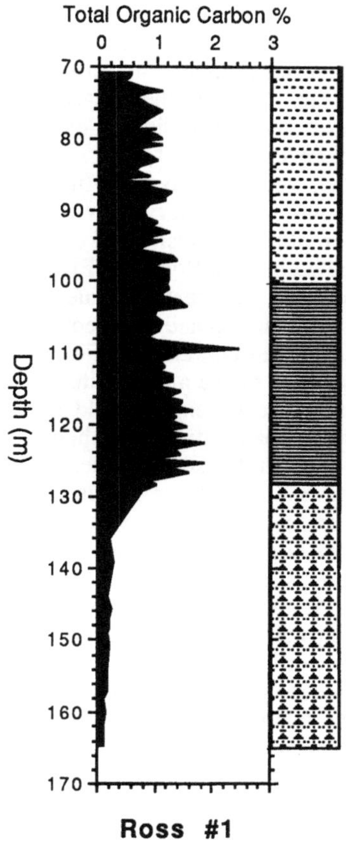

Fig. 9 Total organic carbon contents for the Ross #1 borehole. Sample interval is 50 cm. Rock type symbols are the same as in Figs. 2 and 3.

reduction in clast size within the matrix of the underlying diamicton. This could be a reflection of either a greater distance for debris flow run-out, as if from more water-saturated mud-rich flows and hence producing a smaller clast size. Alternatively, if undermelt is invoked as a depositional process (as would occur near the grounding line of an ice shelf) then the reduction in clast size could be a reflection of upper levels of debris within the base of the ice such as would be expected with increasing distance from the grounding line (Drewry and Cooper, 1981). In any event, the transition from glacigenic to marine environments was especially rapid in the center of the basin where contacts are extremely sharp.

Along the basin margins the transitional interbedding of diamicton with mudstone as seen at Oonah and within the Ross #1 borehole, most likely reflects the slowed recession of the ice margin, due in part to shallow water. This would allow for the development of sediment gravity flows that were generated near the grounding lines of the tidewater glaciers, as evinced by the lenticular diamictites and interbedded pebbly mudstones near Oonah.

3.2 *Paleoenvironment of Mudstones*

The uniform fine texture, dropstones and lack of any bottom current structures suggests that lower parts of the mudstone facies represent open, glacial marine conditions of water depths of at least several hundred meters. It should be kept in mind that the preserved section represents the most basinward conditions. Extreme shallow-water environments are not likely to have survived erosional conditions during the past 280 million years. Close to the fringes of the main basin, restricted sub-basins or even fjords are likely to have existed. Those sub-basins that were silled, would have been in-filled quickly by ice-proximal, sedimentary gravity flows and meltwater plume deposits, like the sections preserved within the Bonney's Plains borehole. Where unsilled, sedimentation rates would have been less pronounced and sediment gravity flows would have been conducted efficiently out into deeper portions of the main basin. Deflection of turbid (meltwater) plumes would have been to the westerly paleo-direction, because of the Coriolis Force in the southern hemisphere. This would have allowed for the interbedding of biogenic and terrigenous facies along the eastern margins of large sub-basins and the dominance of terrigenous facies along the western margins of the sub-basins; such as is shown by the facies relationships of pebbly mudstone and Tasmanite within the northwestern sub-basin (Figure 2).

The widespread distribution of the Tasmanite within the main basin (Figure 3) indicates that conditions responsible for its accumulation were also widespread. Enhanced preservation of organic carbon, as due to bottom water anoxia, cannot explain the Tasmanite or the generally high TOC values within the mudstone because of the presence of in-situ faunas and bioturbation.These characteristics are not commonly associated with anoxic bottom water conditions (Savdra et al., 1984). It is possible that the Tasmanite intervals represent condensed sections as would be developed during the maximum post-glacial flooding within a transgressive system (Haq et al., 1987). However it is clear that productivity models must also be considered in order to explain the Tasmanites and resulting Tasmanite intervals. Further, the productivity model must be temporally constrained as the Tasmanite is not found throughout the entire section.

Phytoplankton blooms in high latitude seas are limited by a pronounced seasonal variation in solar luminosity (Smith and Sakshaug, 1990). Therefore one must consider a pulsed system of phytoplankton supply for the Tasmania Basin as the paleo-latitude of the region has been estimated at between 70-80° S. As the Tasmanite intervals are several tens of cm thick the conditions of Tasmanites productivity must have persisted for at least many decades or a few centuries. This is because there is an upper limit to seasonal production particularly in polar seas. Hence the flux of preserved organic carbon to the bottom would also be limited.

A number of factors can be considered important in controlling excessive productivity within marine waters and these include: physical stability of the surface layer (photic zone) and nutrient supply, such as terrigenous and allochthonous nutrient loading (Paerl, 1988). Nutrient supply can be associated with upwelling and terrestrial runoff. Physical stability is enhanced under favorable wind regimes and within the low-salinity meltwater lense associated with seasonal recession of sea ice (Smith and Nelson, 1985;Sullivan et al., 1988). Such a sea-ice model is particularly attractive in explaining the Tasmanites abundances because it would involve physical stability over a widespread area. Such a model can also be limited in a temporal sense because climatic conditions required for sea ice formation (sub-polar to polar conditions) must eventually change. Some nutrient models can also be temporally constrained. For instance, nutrient supply from terrestrial runoff can be enhanced during the early stages of deglaciation due to retention of nitrogen within polar ice sheets (Zeller et al., 1988) but the resulting blooms would be limited to coastal or nearshore environments. A model involving upwelling is more difficult to present because it depends upon an understanding of the nature of deep water formation and local factors, such as wind direction and submarine topography, that would contribute to upwelling.

Of the above models, productivity associated with sea-ice marginal zones is the best explanation for the Tasmanite intervals. There is additional evidence that favours sea-ice cover during the early stages of marine conditions within the basin. As noted previously, poorly sorted sand and a restricted size of dropstones (granules and small cobbles) would indicate sea-ice as a rafting mechanism (Gilbert, 1990). The well rounded, spherical shapes and lack of striations on the dropstones would also point to a source related to beach and/or fluvial systems, consistent with anchor and shore fast ice processes in sub-polar systems. The presence of unlithified sand clasts that, in a frozen state, acted like dropstones also suggests shore ice incorporation and eventual ice rafting of sorted sediment.

The overall high TOC content and its

variability within the lower mudstone facies indicate generally low sedimentation rates that varied primarily due to changes in either productivity or terrigenous (meltwater-derived) sediment. In meltwater-dominated glacial marine systems (temperate to sub-polar) the TOC content is inversely proportional to the sedimentation rate, due to the dilution effect associated with increased terrigenous sedimentation (Syvitiski et al., 1990). In polar to sub-polar systems, which are not dominated by abundant meltwater input, the TOC content is proportional to sedimentation rate as it reflects changes in surface water productivity (i.e. sea ice cover) and hence carbon flux to the seafloor. Without an independent measure of sedimentation rate it is difficult to decide which of the two conditions might apply to the basal mudstone facies. However, the TOC contents from the mudstone are relatively high (>1.0%) compared to open, glacial marine settings of temperate oceanic character and more closely approach the carbon contents of sediments from the present Antarctic glacial marine settings (1-3%; Dunbar et al., 1989; 1-1.5% Domack, 1990).

The abundance of glendonites within the section would suggest a marine origin for the organic carbon because one requirement for the formation of its precursor (ikaite) is an abundance of easily metabolized (non-refractory) organic matter (Suess et al., 1982). The other requirement for ikaite formation is a low temperature, close to freezing (Shearman and Smith, 1985). For displacive growth of ikaite within unconsolidated sediments, an undisturbed mud substrate is also needed. The presence of glendonites within the mudstone section can therefore be used as an indicator of cold bottom water conditions and undisturbed (deep water) depositional conditions.

Changes in the depositional environment are marked by the succession into the upper mudstone facies. Such changes include a decrease in glendonite abundance, increase in dropstone frequency and size, first occurrence of plant macrofossils, well preserved palynoflora, and to the southeast a well sorted very fine-grained sandstone. Locally, as in the Bonney's Plains borehole, the section fines upward in a composite fashion or, as in the Eaglehawk Neck borehole, the mudstone is succeeded by a very coarse-grained and sorted sequence of sandstones and conglomerates. These observations all point to a progressive decrease in water depths and/or an increase in the vigour of bottom currents. Sandy pebble horizons with down-mixed burrows and robust macrofauna (Figure 4) can be interpreted as indicating either periods of intensive sea-ice rafting (Gilbert, 1990) or the formation of current lags or hiati as observed in deep marine contourite deposits (Gonthier et al., 1984). Hence, the increased frequency and thickness of these layers, upward in the section, would logically indicate closer proximity to the shoreline (drop in relative sea level) or increased frequency of strong bottom currents. A drop in relative sea level is supported by the presence of littoral coarse grained sandstones and conglomerates above black mudstone in the Eaglehawk Neck borehole. The increased disturbance of the sediment/water interface prevented the formation of ikaite (glendonites) by the periodic resuspension of the upper few centimeters of sediment.

Well-sorted very fine grained sandstones found within the southeastern portions of the basin indicate the progradation of a subaqueous delta from that direction, as paleocurrents trend toward the west northwest. The sandstones are intensively bioturbated indicating a marine depositional setting rather than a fluvial deltaic environment. Channels are not observed in outcrop thus suggesting more uniform depositional conditions within the system. The distal equivalents of this sandy facies are the muddy siltstones found within the center of the basin (Figure 3). The reduction in preserved TOC and the lack of high frequency fluctuations in preserved carbon are taken as an indication of terrigenous dilution and a dominance of terrigenous sediment input over phytoplankton productivity (Figures 8 and 9). These changes in the environment were accompanied by a change toward temperate conditions. This is clearly indicated by the presence of well-preserved leaf impressions and palynoflora within the upper mudstone facies.

3.3 Sedimentation Rates

Although rates of deposition cannot be calculated directly we can assume a reasonable range for the open marine section of some 1-3 mm/yr. This is a good estimate based upon observations of sedimentation rates within open marine, continental shelf environments, of modern high latitudes and from known rates of sedimentation within ikaite bearing organic-rich muds (Suess et al., 1982). Given such rates and

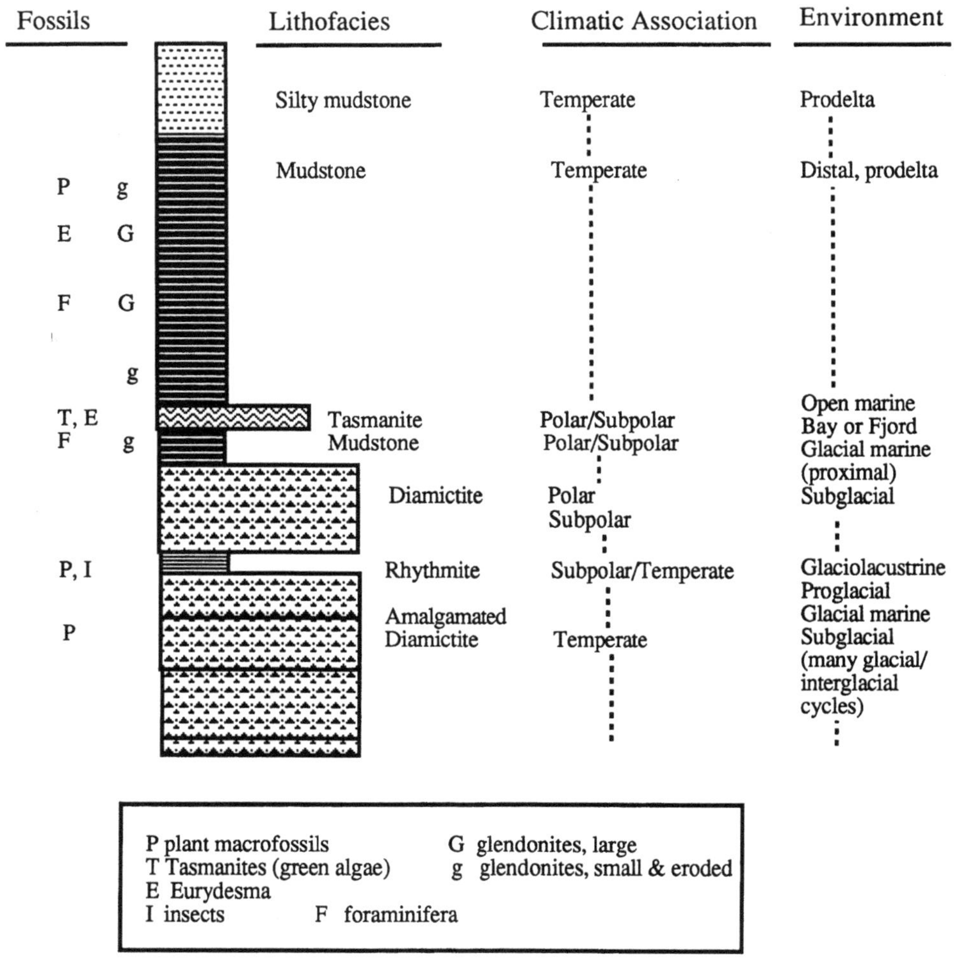

Fig. 10 Overall paleoclimatic and environmental evolution of the glacigenic portion of the Parmeener Surpergroup.

allowing for corrections due to compaction and lithfication then the maximum thickness of about 200 m (Tunbridge borehole) would represent a period of approximately 1.0 myr. The minimum thickness of 55 m (Douglas River borehole) represents a much reduced sedimentation rate as indicated by the abundance of ice-rafted material throughout the section (Figure 6). Since the amount of stratigraphic growth between the two boreholes is about x 3.5 (Figure 3) a sedimentation rate for the Douglas River borehole can be estimated to be about 0.2 to 0.8 mm/yr. This is a realistic rate for glacial marine sedimentation within a deep strait as indicated by studies in the modern Gerlache Strait, Antarctica (Harden et al., 1991).

4 CONCLUDING REMARKS

The mudstone was deposited between an underlying glacially dominated sequence and an overlying and alternating glacial marine and non-marine sequence. The long-term paleoclimatic evolution of the entire sequence is illustrated in Figure 10 where it can be seen that the transition from glacial (diamictite) to deep glacial-marine (mudstone) sedimentation represents a response of the ice sheet to relative sea level change. The relative sea-level change was brought about by either increased basin subsidence or a real rise in eustatic sea-level; the former has been suggested as a plausible mechanism for deglaciation within the Dwyka basin (Visser, 1990). Whatever the driving mechanism, the event was rapid and was not necessarily associated with a regional change in climate as both the glacigenic diamictites and deep marine mudstones are compatible with polar to sub-polar climatic settings (Figure 10). However, there must have been temperate glaciation during an earlier history of the basin because of the abundance of meltwater facies (Powell, 1990), fossil plants, and even insects (Banks and Clarke, 1987; Martini and Banks, 1989). Post-glacial climatic warming to temperate conditions took place over a period of several hundred thousand years. This was a long term climatic change unlike deglacial warming episodes (terminations) associated with the Quaternary. Information on high frequency paleoclimatic changes during the main period of mudstone deposition (estimated as some 600,000 years) could possibly be extracted from subtle changes in the character of the mudstone, such as from geochemical indices or silt/clay contents. In this regard, the cycles of TOC and glendonite abundance are intriguing and may be hinting at a paleoclimatic signal not unlike the oxygen isotope record of deep marine pelagic carbonates during the Quaternary. Further work is needed on the detailed geochemical composition of particulate organic carbon and glendonites within the boreholes.

ACKNOWLEDGEMENTS

This project was supported by the American Chemical Society (Petroleum Research Fund), The University of Tasmania, Hamilton College, and The Department of Mines and Mineral Resources, Tasmania. We would like to thank Chesley Williams and Kevin Colangelo for their assistance in the field. This is a contribution to IGCP Project #260: The Earth's Glacial Record.

REFERENCES

Anderson, J.B. and Domack, E.W., 1991. The glacial marine environment: subglacial and climatic setting. In J.B. Anderson and G.M. Ashley, (eds.) Paleoclimatic Interpretation from Glacial Marine Deposits. Geological Society of America. Special Paper, v-viii. 261:

Banks, M.R. and Clarke, M.J., 1987. Changes in the geography of the Tasmania basin in the Late Paleozoic. In G.D. McKenzie (ed.): Gondwana Six: Stratigraphy, Sedimentology, and Paleontology. American Geophysical Union, Washington D.C.: 1-14.

Calver, C.R., Clarke, M.J. and Truswell, E.M., 1983. The stratigraphy of a Late Paleozoic borehole section at Douglas River, Eastern Tasmania: A synthesis of marine macro invertebrate and palynological data. Papers and Proceedings of the Royal Society of Tasmania. 118: 137-161.

Denwer, K., 1986. Geology and Geochemistry of the Tasmanite Oil Shale. Unpublished Honours Thesis, University of Tasmania.

Domack, E.W., 1990. Laminated terrigenous sediments from the Antarctic Peninsula: the role of subglacial and marine processes. In J.A. Dowdeswell and J.D. Scourse. (eds.) Glacimarine Environments: Processses and Sediments. Geological Society Special Publication 53: 91-103.

Domack, E.W., Jull, A.J.T., Anderson, J.B. and Linick, T.W., 1991. Mid-Holocene ice sheet recession from the Wilkes Land continental shelf, East Antarctica. In M.R.A. Thomson, J.A. Crame and J.W. Thomson (eds.) Geological Evolution of Antarctica. Cambridge University Press, Cambridge, U.K.: 693-698.

Drewry, D.J. and Cooper, A.P.R., 1981. Processes and models of Antarctic glacimarine sedimentation. Annals of Glaciology, 3: 117-122.

Dunbar, R.B., Leventer, A.R. and Stockton, W.L., 1989. Biogenic sedimentation in McMurdo Sound, Antarctica. Marine Geology, 85: 331-358.

Gilbert, R., 1990. Rafting in glacimarine environments. In J.A. Dowdewell and J.D. Scourse (eds.) Glacimarine Environments: Processes and Sediments. Geological Society Special Publication 53: 105-120.

Gonthier, E.G., Faugeres, J-C. and Stow, D.A.V., 1984. Contourite facies of the Faro Drift, Gulf of Cadiz. In D.A.V. Stow and D.J.W. Piper (eds.): Fine-Grained Sediments:

Deep-Water Processes and Facies. Geological Society London, Special Publication 15: 275-292.

Harden, S.L., Demaster, D.J., and Nittrouer, C.A., 1991. Developing sediment geochronologies for high-latitude continental shelf deposits: a radiochemical approach. Marine Geology, 103: 69-97.

Haq, B.U., Hardenbol, J. and Vail, P.R., 1987. Chronology of fluctuating sea levels since the Triassic. Science, 235: 1156-1167.

Krumbein, W.C., 1941. The measurement and geological significance of shape and roundness of sedimentary particles. Journal of Sedimentary Petrology, 11: 64-72.

Martini, I.P. and Banks, M.R., 1989. Sedimentology of the cold-climate, coal-bearing, "Lower Freshwater Sequence" of Tasmania. Sedimentary Geology, 64: 25-41.

Paerl, H.W., 1988. Nuisance phytoplankton blooms in coastal, estuarine, and inland waters. Limnology and Oceanography, 33: 823-847.

Powell, R.D., 1990. Glacimarine processes at grounding-line fans and their growth to ice-contact deltas. In J.A. Dowdeswell and J.D. Scourse (eds.) Glacimarine Environments: Processes and Sediments. Geological Society Special Publication 53: 53-73.

Savdra, C., Bottjer, D. and Gorsline, D., 1984. Development of a comprehensive oxygen-deficient marine biofacies model: evidence from Santa Monica, San Pedro, and Santa Barbara Basins, California continental boarderland. American Association of Petroleum Geologists, Bulletin, 68: 1179-1192.

Shearman, D.J. and Smith, A.J., 1985. Ikaite, the parent mineral of jarrowite-type pseudomorphs. Proceedings Geological Association, 96: 305-314.

Smith, W.O., Jr. and Nelson, D.M., 1985. Phytoplankton bloom produced by a receding ice edge in the Ross Sea: spatial coherence with the density field. Science, 227: 163-166.

Smith, W.O., Jr. and Sakshaug, E., 1990. Polar Phytoplankton. In W.O. Smith, Jr. (ed.) Polar Oceanography, Part B: Chemistry, Biology, and Geology. Academic Press, San Diego: 477-526.

Suess, E., Balzer, W., Hess, E.K.F., Muller, P.J., Ungerer, P.J. and Wefer, G., 1982. Calcium carbonate hexahydrate from organic-rich sediments of the Antarctic shelf: precursor of glendonites. Science, 1216: 1128-1131.

Sullivan, C.W., McClain, C.R., Comiso, J.C. and Smith, W.O., Jr., 1988. Phytoplankton standing crops within an Antarctic ice edge assessed by satellite remote sensing. Journal of Geophysical Research, 93: 12,487-12,498.

Syvitiski, J.P.M., William, K., LeBlanc, G. and Cranston, R.E., 1990. The flux and preservation of organic carbon in Baffin Island fjords. In J.A. Dowdeswell and J.D. Scourse (eds.) Glacimarine Environments: Processes and Sediments. Geological Society of London, Special Publication 53: 177-200.

Truswell, E.M., 1978. Palynology of the Permo-Carboniferous in Tasmania. Geological Survey Bulletin 56: Tasmania Department of Mines, 31 pp.

Visser, J.N.J., 1990. Influence of basin subsidence on the disintegration of the marine ice sheet in the Dwyka Basin, South Africa. International Association of Sedimentologists, Nottingham, U. K.: abstracts of meeting.

Zeller, E.J., Dreschoff, G.S. and Turner, M. D., 1988. Impact of ice-related nutrients on glacial margin environments. Northeast Meeting of the Geological Society of America, Abstracts with Programs: p. 80.

Gondwana Eight, Findlay, Unrug, Banks & Veevers (eds) © 1993 Balkema, Rotterdam. ISBN 90 5410 304 3

Isotope composition of the ice and sub-glacial geology near the Allan Hills, Victoria Land, Antarctica

G. Faure, K. S. Wehn, J. M. Montello, E. H. Hagen, M. L. Strobel & K. S. Johnson
Department of Geological Sciences & Byrd Polar Research Center, The Ohio State University, Columbus, Ohio, USA

ABSTRACT: Rock clasts and sediment deposited by the East Antarctic ice sheet in supraglacial moraines and along the ice edge can provide information about the subglacial geology of East Antarctica. Such studies are enhanced by using the isotope composition of the ice to constrain its provenance. The $\delta^{18}O$ values of ice collected at regular intervals along surveyed lines near the Elephant and Reckling moraines on the East Antarctic ice sheet west of Victoria Land range from -42.6 to -50.4 ‰ relative to standard mean ocean water (SMOW). Ice adjacent to the Allan Hills has a significantly different isotope composition with $\delta^{18}O$ values between -35.2 and -43.0 ‰. These results indicate that the ice presently exposed at the Elephant and Reckling moraines originated from the eastern slope of the central ice divide in East Antarctica. The less negative $\delta^{18}O$ values of the ice at the Allan Hills indicate either that the ice originated from a local source close to the Transantarctic Mountains or that it formed during an interglacial stage of the Pleistocene epoch, whereas at least some of the ice at the Elephant and Reckling moraines formed during glacial stages. The rock clasts in the moraines at all three sites are composed primarily of dolerite and basalt (Ferrar Supergroup, Jurassic) and of sandstone, siltstone, and shale (Beacon Supergroup, Devonian to Triassic). Clasts of granitic rocks and of high-grade metamorphic rocks of Precambrian age were not found among about 70,000 clasts that were classified on the Elephant and Reckling moraines, but a few granite clasts (Granite Harbour Intrusives, Cambro-Ordovician) were seen in the moraines at the Allan Hills. The absence of basement rocks in the Elephant and Reckling moraines indicates that these kinds of rocks are not exposed along the flowlines of the ice arriving at the moraines, whereas granitic rocks do occur under the ice close to the Allan Hills. The presence of sedimentary rocks in the moraines and the absence of basement rocks in the Elephant and Reckling moraines implies that the Beacon Supergroup is more extensive under the ice in East Antarctica than suggested by the present outcrops in the Transantarctic Mountains and that granitic basement rocks are not exposed along the flowline from the central ice divide to the Elephant and Reckling moraines.

1. INTRODUCTION

The ice-covered areas of East and West Antarctica, together with Greenland, are the last geologically unmapped regions on Earth. Some information about the sub-ice geology of East and West Antarctica has been derived from geophysical surveys, including the airborne radar surveys of Drewry (1982), which have provided much useful information about the bottom topography, ice thickness, and surface elevation of the Antarctic ice sheets.

We propose that information about the sub-ice geology of ice-covered areas can also be derived by systematic studies of the sediment and rock clasts which the ice sheets deposit along their margins or in supraglacial moraines. We have begun such a study of Neogene and Holocene till deposited by the East Antarctic ice sheet along the western slopes of the Transantarctic

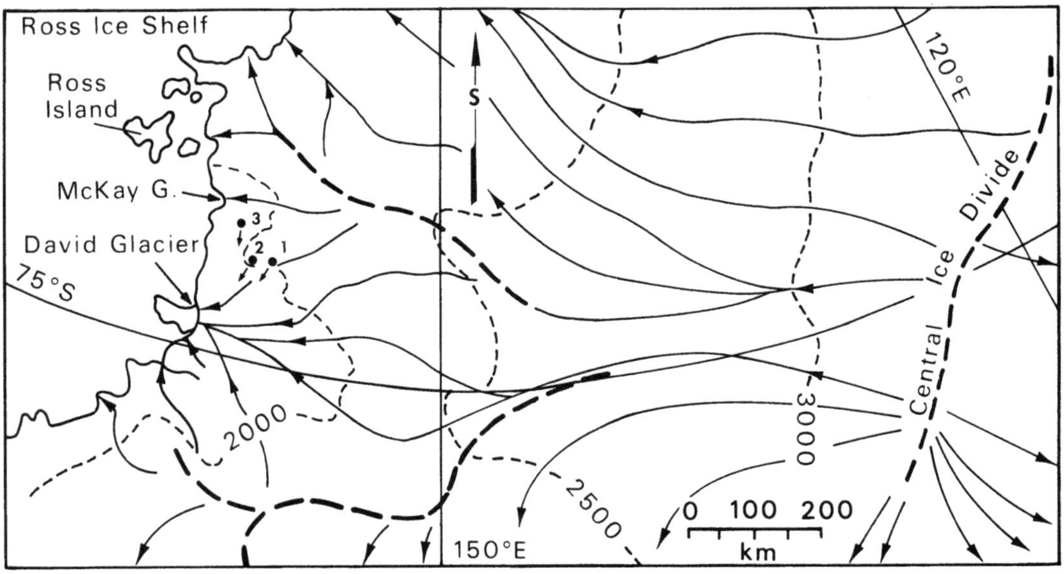

Figure 1 Ice-flow pattern in the drainage basin of the David Glacier. Locality 1) Elephant Moraine; 2) Reckling Moraine; 3) Allan Hills. The elevations of the topographic contours on the surface of the ice (thin, dashed line) are in meters. The heavy, dashed lines are ice divides (After Drewry, 1982).

Mountains. In this report we present data for till from the Elephant Moraine (76°17'S, 157°20'E, informal name), Reckling Moraine (76°15'S, 158°40'E, informal name), and from the Allan Hills (76°45'S, 159°40'E). The locations of these sites south of the ice drainage basin of the David Glacier are indicated in Figure 1.

2. GLACIOLOGICAL SETTING

The Elephant and Reckling moraines are supraglacial moraines located on a broad band of exposed ice that extends from Reckling Peak (76°16'S, 159°15'E) in a westerly direction for about 75 kilometers (km) (Faure and Taylor, 1985; Faure et al., 1987; Faure and Buchanan, 1987; Faure, 1990 a,b,c,d). Both moraines are located at an elevation of about 2,000 meters (m) above sea-level, are about 5.6 km long, up to 1.5 km wide, and are elongated in a northeasterly direction parallel to the movement of ice (Figure 1). Each moraine consists of a thin layer of rock clasts and sediment released by sublimation of basal ice that is exposed at these locations in the form of arcuate bands of black sediment-rich ice. The exposure of the basal ice is caused by the presence of sub-glacial bedrock ridges and mesas over which the ice thins to less than 100 meters (Faure, 1990b; Delisle and Sievers, 1991). The presence of multiple bands of basal ice are partly the result of repeated thrust faulting of the ice sheet in response to bedrock obstructions in its path. The response of an ice sheet to sub-glacial bedrock peaks was modelled by Robin and Millar (1982).

Both moraines consist of loose till of Holocene age described by Faure and Taylor (1985) and by Faure et al. (1987). In addition, the ice sheet at both moraines contains layers of sediment-rich ice, commonly less than 2 meters thick, which contain oval-shaped nodules of indurated but uncemented till. Some of the nodules contain striated and faceted clasts and appear to have been moulded by the movement of the ice. The till nodules probably originated from glacial sediment that was previously deposited along the western slopes of the Transantarctic Mountains. These till deposits may therefore be related to the Sirius Group of tills of Late Tertiary age defined by Mercer

(1972) based on deposits at the summit of Mt. Sirius (84°05′S, 163°15′E).

Faure and Harwood (1990) described a small suite of marine diatoms, silicoflagellates, ebridians, radiolarians, sponge spicules, chrysomonad cysts, pollen, and microclasts of diatomaceous marine sediment from nodular and loose till in the Elephant Moraine. The presence of marine fossils in the till at the Elephant Moraine confirms the existence of marine embayments west of the Transantarctic Mountains in Pliocene time postulated by Webb et al. (1984).

At the Allan Hills, the East Antarctic ice sheet flows across a bedrock ridge part of which is actually exposed. The topographic ice basin in the lee of this ridge contains several small supraglacial moraines whose origin is similar to that of the Elephant and Reckling moraines. In addition, highly compacted but uncemented basal till was deposited on bedrock in a deglaciated area at the southern end of Manhaul Bay, which is a large cirque in the Allan Hills (Cherry and Noltimier, 1985). Clasts of similarly compacted till also occur in the supraglacial moraines near the Allan Hills.

The ice-flow directions in Figure 1 indicate that the Elephant and Reckling moraines lie on ice that originated from the central ice divide of East Antarctica. Therefore, the sediment and rock clasts of these moraines must have originated from bedrock sources distributed along the flow lines extending from the central ice divide of East Antarctica to the moraines. However, the bulk of the sediment in the form of large angular boulders was probably transported for only short distances of about 10 km (Strobel and Faure, 1987). Consequently, the till of the Elephant and Reckling moraines is primarily of local origin, but also contains a component that was transported for hundreds of kilometers from the interior of East Antarctica.

3. TRANSPORT OF SEDIMENT BY ICE SHEETS

Studies in North America and northern Europe, reviewed by Strobel and Faure (1987), indicate that clasts of distinctive lithologic composition were transported by the Pleistocene ice sheets for hundreds of kilometers. The abundance of a particular type of indicator clast derived from a specific bedrock source under the ice typically decreases exponentially with increasing distance from its source measured along a flow line. However, in less than about 50 km the average abundance of the indicator clasts in till becomes constant at about 1 to 5% of all clasts for distances of up to 800 km or more, depending on the extent of the ice sheet. Strobel and Faure (1987) attributed the exponential decrease in the abundances of indicator clasts primarily to progressive dilution with clasts of other lithologies at the base of the ice sheet and defined the transport half-distance as the distance along a flow line of a glacier where the abundance of a particular variety of indicator clasts decreases by one-half.

Clasts that remain at the base of an ice sheet are ultimately destroyed by grinding and their abundance in basal till therefore should approach zero with increasing transport distance. The fact that the abundances of indicator clasts in till do not approach zero is caused by the fact that some of the clasts are transported englacially and thereby survive the trip to the terminal moraine; unless the ice sheet stagnates, in which case they are deposited in melt-down till. In this way, a small fraction of a particular variety of indicator clasts is transported over great distances until these clasts are incorporated into the sediment deposited by the ice sheet in the ablation zone. Although the model of Strobel and Faure (1987) is based on the transport of indicator clasts by the Laurentide and Scandinavian ice sheets, we assume that it also applies to sediment transported by the East Antarctic ice sheet.

4. ISOTOPE COMPOSITION OF THE ICE

The isotope composition of oxygen and hydrogen in glacial ice depends primarily on the temperature of condensation of water vapor in the clouds, on the latitude of the site of deposition, and on its elevation above sea-level (Faure, 1986). The ice-flow pattern of the East Antarctic ice sheet indicates that the ice at the Elephant and Reckling moraines and at the Allan Hills formed elsewhere and flowed to

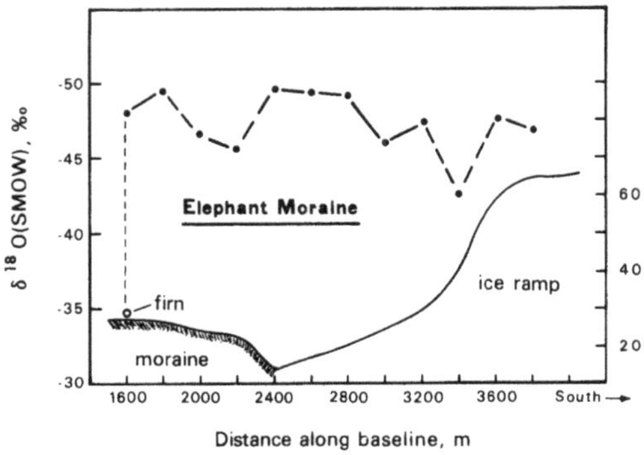

Figure 2A Variation of $\delta^{18}O$ of ice and topographic profile along a surveyed line extending from the southern end of the Elephant Moraine onto the surrounding plateau.

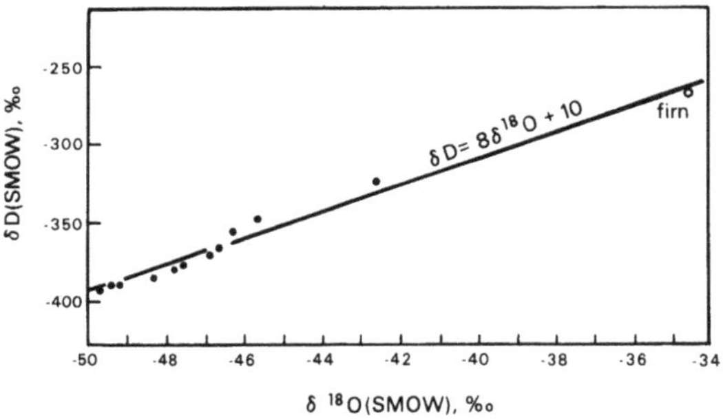

Figure 2B Correlation of δD and $\delta^{18}O$ of the ice samples from the Elephant Moraine (Figure 2A above) in relation to the meteoric water line. (Isotope analyses by Geochron Laboratory, Cambridge, Massachusetts, U.S.A.).

these sites. Therefore, the provenance of the ice at each study site was evaluated by means of its isotope composition of oxygen and hydrogen in order to provide a basis for the interpretation of the abundance of different clast lithologies.

4.1 Elephant Moraine

The ice samples from the Elephant Moraine were collected at 200 meter intervals along a surveyed line extending from the southern end of the moraine up to the level of the surrounding plateau shown on a map in Faure and Sutton (1985). The $\delta^{18}O$ values of the ice were measured by standard methods and are expressed relative to standard mean ocean water (SMOW) in accordance with the equation:

$$\delta^{18}O = \frac{(^{18}O/^{16}O)_{spl} - (^{18}O/^{16}O)_{std}}{(^{18}O/^{16}O)_{std}} \times 10^3 \quad (1)$$

The $\delta^{18}O$ values of ice along the surveyed line in the Elephant Moraine in Figure 2A range from −42.6 ‰ (station 3400 m S) to -49.7 ‰ (stations 1800 m and 2400 m S), whereas one sample of recently deposited firn at 1600 m S has $\delta^{18}O$ = -34.5 ‰ (Appendix). The ice at stations 1800, 2400, 2600 and 2800 m S is strongly depleted in ^{18}O and has an average $\delta^{18}O$ value of -49.5 ± 0.2 ‰ , whereas the other 8 samples average $\delta^{18}O$ = -46.4 ± 1.8 ‰ (the errors are one standard deviation). The variation in the isotopic composition of oxygen in the ice along this traverse may have three causes: 1) climate changes in the accumulation area; 2) structural intercalation of ice of different ages formed under different climatic conditions; 3) contamination with recent meltwater ice.

Contamination by recently deposited snow was avoided by collecting the ice samples from pits more than 10 cm deep. In addition, contamination with modern precipitation ($\delta^{18}O$ = −34.5 ‰ would have enriched the ice samples in ^{18}O and therefore would have made the $\delta^{18}O$ values less negative. There is no evidence in the data to suggest that this has occurred. Therefore, the range of variation of $\delta^{18}O$ values probably records climatic variations. However, the structural deformation of the ice sheet evident in the field near the Elephant and Reckling moraines as well as in the vicinity of the Allan Hills, suggests that the climate record in the ice has probably been disturbed by faulting and folding. Additional evidence in support of this conclusion for ice at the Reckling Moraine was presented by Faure et al. (1992).

The isotope compositions of hydrogen, expressed as δD values relative to SMOW, and the $\delta^{18}O$ values of ice from the Elephant Moraine plot along the meteoric water line in Figure 2B (Faure, 1986). This result is consistent with the relationship between the

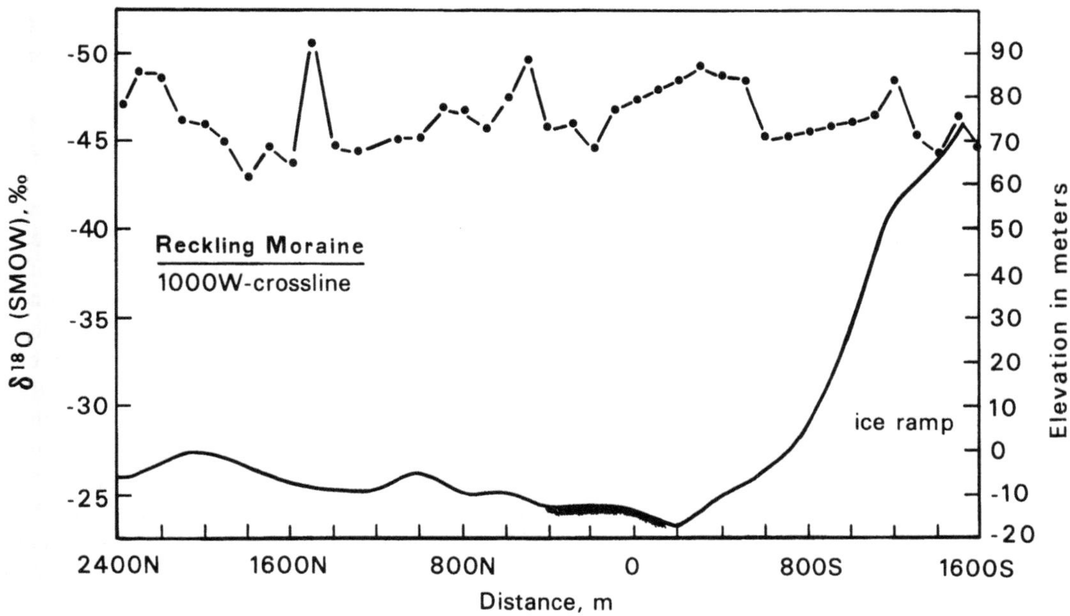

Figure 3 Variation of $\delta^{18}O$ of ice and topographic profile across the Reckling Moraine. (Isotope analyses by Geochron Laboratory, Cambridge, Massachusetts, U.S.A.).

isotope compositions of hydrogen and oxygen in meteoric precipitation.

4.2 Reckling Moraine

A suite of 41 ice samples was collected from shallow pits at 100 m intervals along a traverse across the Reckling Moraine and the adjacent ice fields (crossline 1000W, Faure et al., 1987; Faure et al., 1992). The $\delta^{18}O$ values of the ice range from -42.8 ‰ (1800 m N along 1000W) to -50.4 ‰ (1500 m N along 1000W). One firn sample has $\delta^{18}O$ = -36.1 ‰ and snow that fell on 19 December 1986 has $\delta^{18}O$ = -32.3 ‰. The $\delta^{18}O$ profile has been combined in Figure 3 with a topographic cross-section of the ice sheet along the surveyed lines. A more complete evaluation of the $\delta^{18}O$ values of the icefields adjacent to the Reckling Moraine was presented by Faure et al. (1992).

These results indicate that some of the ice at the Elephant and Reckling moraines is strongly depleted in ^{18}O compared to modern snow in the area and therefore formed at a lower average temperature than occurs in the area at the present time.

4.3 Allan Hills

The oxygen in ice at the Allan Hills is much less fractionated isotopically than oxygen in the ice at the Elephant and Reckling moraines. The $\delta^{18}O$ values of ice collected along a surveyed line leading up to the edge of the ice sheet near the Allan Hills are displayed in Figure 4. (The line is located on a map in Faure and Buchanan, 1987). The $\delta^{18}O$ values range from –35.2 ‰ to –43.0 ‰, whereas recently deposited snow has $\delta^{18}O$ = -32.0 ‰ (Appendix).

Figure 4 also contains a cross-section of the thinning edge of the ice sheet in relation to the bedrock topography based on radar-echo soundings (Faure and Buchanan, 1987). The sub-glacial bedrock ridge detected by these measurements is actually exposed a few hundred meters south of the surveyed line. The supraglacial moraine is typically located in the lee of the sub-glacial ridge much like in the Reckling and Elephant moraines. The association of supraglacial moraines with the sub-glacial bedrock topography suggests that a genetic relationship exists between them. The presence of obstructions in the path of the ice sheet causes sediment-rich basal ice to be forced to

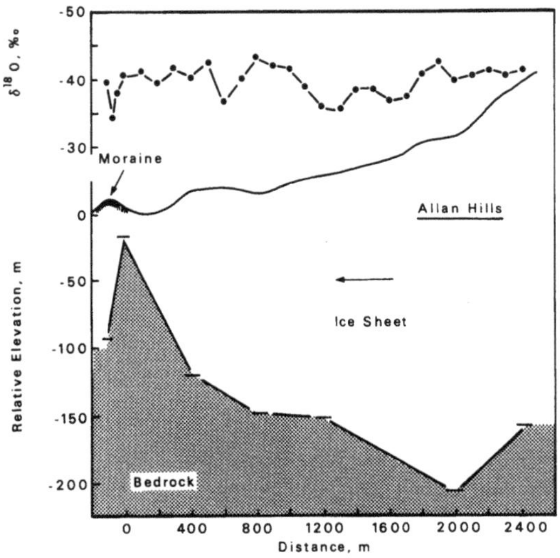

Figure 4 Variation of δ^{18}O, surface topography, and ice thickness along a surveyed line at the margin of the East Antarctic ice sheet near the Allan Hills.

the surface, provided that the thickness of the ice is low enough to permit this to happen.

5. PROVENANCE OF THE ICE

Some indication of the location and environmental conditions at the site of deposition of the ice that is now exposed in the ice fields can be derived from the work of Lorius and Merlivat (1977). These authors measured δD values of snow along a traverse from Dumont d'Urville (67°S, 147°E) on the coast of East Antarctica toward Dome C (74°30'S, 123°10'E) on the central divide. They also measured the mean temperature and the elevation above sea-level of each of 48 stations along an 833 km track. The results of their study indicate that the δD values of snow vary linearly with the mean annual temperature and that these parameters are related by the equation:

$$\delta D = 6.04\ T\,(°C) - 51 \qquad (2)$$

The δD values of ice at the Elephant Moraine can now be used to calculate average annual temperatures based on equation (2). The results indicate values ranging from -45 °C to –57 °C for the source area of the ice at the time of its formation. Under present climatic conditions such temperatures occur only along the central ice divide of East Antarctica about 1000 km from the Elephant Moraine.

The average annual temperatures of ice now located at the Reckling Moraine were calculated similarly after converting the δ^{18}O values to δD using the equation of the meteoric water line (Faure, 1986). The results indicate that the annual temperatures at the time of formation of the ice at the Reckling Moraine are virtually identical to those derived from ice at the Elephant Moraine. The present average annual temperature at the Reckling Moraine is -33 °C (based on δ^{18}O = -32.3 ‰ for modern snow). The relationship between δD of snow and topographic elevation derived by Lorius and Merlivat (1977) from their data yields an elevation of 2100 m above sea-level for the Reckling Moraine, in good agreement with the known elevation.

The average annual temperatures indicated by the δ^{18}O values of ice at the Allan Hills range from -37 °C and -47 °C compared to the present mean annual temperature of –32 °C (based on δ^{18}0 = –32.0 ‰ for modern snow). The evidence for relatively warm climatic conditions at the time of formation of the ice now located adjacent to the Allan Hills indicates that this ice either originated near the Transantarctic Mountains rather than in the interior of East Antarctica, or that it formed during the Sangoman or an even earlier interglacial stage prior to the Wisconsin glaciation. If the ice formed during the Sangoman interglacial, its age would be between 2×10^5 and 4×10^5 years.

The low temperatures of formation of the ice at the Elephant and Reckling moraines can be accounted for by assuming that it formed along the central ice divide of East Antarctica during the Wisonsin glacial stage (Faure et al., 1992). However, the ice now located at the two moraines did not originate from the crest of the ice divide because Wisconsinan ice in a core drilled at the Russian station Vostock (78°28'S, 106°48'E, 3488 m above sea-level) has δ^{18}O values of -62 ‰ (Lorius et al. 1985). Low δ^{18}O values of this magnitude have not been detected anywhere else along the margins of the

East Antarctic ice sheet. Therefore, the ice exposed in the vicinity of the Elephant and Reckling moraines probably originated along the eastern slope of the ice divide during the Wisconsin glacial stage and flowed from there to its present location.

6. SUB-GLACIAL GEOLOGY OF EAST ANTARCTICA

6.1 Elephant Moraine

The composition of rocks clasts in the Holocene till of the Elephant Moraine was determined by identifying 47,687 clasts larger than 16 mm in diameter at 230 surveyed sites throughout the moraine (Faure and Taylor, 1985; Taylor, 1986). The clasts are composed primarily of dolerite, basalt, diamicton, sandstone, and siltstone with minor amounts of till nodules, chert, black calcite, shale, and coal (Figure 5A). The dolerite clasts very probably originated from sills of the Ferrar Dolerite (Jurassic), whereas the basalt is assignable to the Kirkpatrick Basalt. Faure et al. (1984) reported an average initial $^{87}Sr/^{86}Sr$ ratio of 0.7099 ± 0.0001 for two dolerite clasts from the Elephant Moraine. This value is characteristic of the Ferrar Dolerite in the Transantarctic Mountains (Faure et al., 1991). The diamicton clasts resemble the Mawson Formation (Jurassic), which is well exposed in the Allan Hills and at Carapace Nunatak (Borns and Hall, 1969). The sandstone, siltstone, shale, and coal represent the formations of the Beacon Supergroup in the Transantarctic Mountains. Isotopic data reported by Faure et al. (1988) indicate that the black calcite and chert formed in equilibrium with glacial meltwater under the present ice sheet.

In the Elephant Moraine the clasts in the Neogene till nodules have virtually the same lithologic composition as clasts in the loose Holocene till as shown in Figure 5B (Montello, 1986). This indicates that both varieties of till have the same geologic provenance. No clasts composed of granitic gneiss, granulites, or charnockites of Archean and Early Proterozoic age, known from exposures along the coast of East Antarctica (Grew, 1978; DePaolo et al., 1982; Sheraton and Collerson, 1983), were found in the Elephant Moraine. In view of the demonstrated ability of ice sheets to transport clasts for hundreds of kilometers (Strobel and Faure, 1987), the absence of clasts composed of crystalline basement rocks means that basement rocks are not exposed along the flowline of ice arriving at the Elephant Moraine.

In spite of the absence of exposures of basement rocks along the flowline leading to the

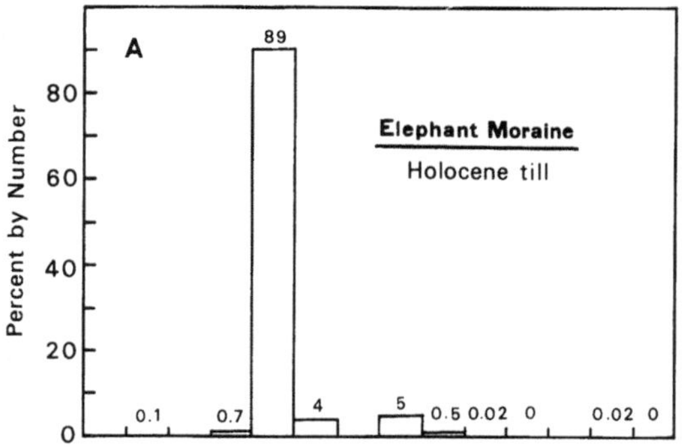

Figure 5A Lithologic composition of 47,687 rock clasts more than 16 mm in diameter in Holocene till at 230 surveyed sites on the Elephant Moraine. The abundance of different lithologies are expressed in percent by number (Taylor, 1986).

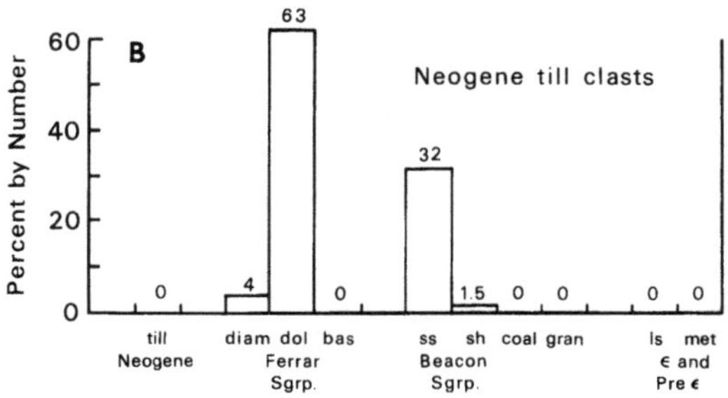

Figure 5B Lithologic composition of rock clasts (>16 mm) in Neogene till pellets at the Elephant Moraine. The abundances of different clasts lithologies abundances are expressed in percent by number (Montello, 1986). The abbreviations here and in Figures 6 and 7 are: diam = diamicton, dol = dolerite, bas = basalt, ss = sandstone, sh = shale, gran = granite, ls = limestone, met = metamorphic rocks.

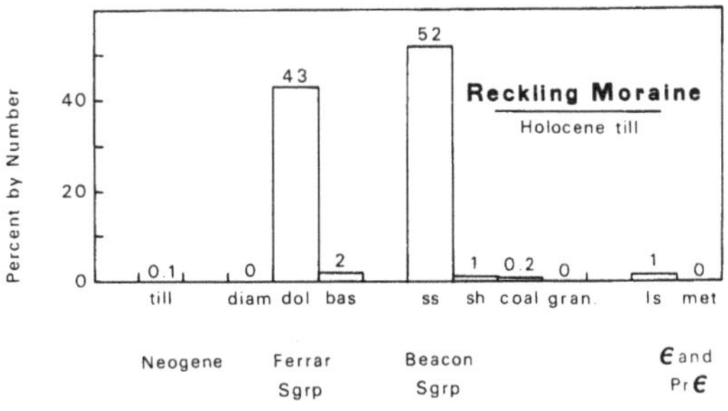

Figure 6 Lithologic composition of rock clasts (>16 mm) in Holocene till in the Reckling Moraine in percent by number.

Elephant Moraine, microcline and plagioclase grains do occur in the sand fractions of the nodular till. A Rb-Sr isochron of these feldspar grains from a nodule of Neogene till yielded a date of 660 ± 33 Ma and an initial $^{87}Sr/^{86}Sr$ ratio of 0.7075 ± 0.0011 (Montello, 1986). The date indicates that the feldspar grains originated primarily from rocks of Paleozoic age, but that a small fraction of them originated from rocks of Precambrian age. Since feldspar - bearing granitic rocks are apparently not exposed in the path of the ice, the feldspar grains in the nodular till at the Elephant Moraine probably originated from arkosic sandstones of the Beacon Supergroup.

6.2 Reckling Moraine

The pebble count at the Reckling Moraine included 21,346 rock clasts more than 16 mm in diameter collected at 78 surveyed stations. The Holocene till that forms this moraine is composed primarily of sedimentary rocks (Beacon Supergroup), almost an equal proportion of dolerite and basalt (Ferrar Supergroup), and minor amounts of nodular till, coal, and carbonate rocks (Figure 6).

No igneous and metamorphic rocks of granitic composition were found in the Reckling Moraine. The granite clast donated by L.A. Rancitelli and dated by Faure et al. (1984) by the Rb-Sr mineral-isochron method probably originated from moraines near the Allan Hills where clasts of that lithologic composition do occur and not from the Reckling Moraine as originally reported. The age of that granite clast is 496 ± 5 Ma and its initial $^{87}Sr/^{86}Sr$ ratio is 0.70715 ± 0.00037. The age of this clast is identical to that of Granite Harbour Intrusives in southern Victoria Land (Faure et al., 1974).

One sample of Neogene till from the Reckling Moraine contained 73.8% sandstone, 17.5% shale, and 8.7% dolerite in granules and clasts more than 4 mm in diameter. In addition, about 8% of these clasts were found to be striated confirming their transport in basal ice (Hagen, 1988).

6.3 Allan Hills

Clasts in loose Holocene till in supraglacial moraines near the Allan Hills consist of dolerite, sandstone, and basalt (Figure 7A). Neogene till deposited on bedrock in Manhaul Bay and till clasts on the ice are composed of the same kinds of clasts. However, sandstone clasts are more abundant than dolerite clasts (Figure 7B). None of the five samples examined in this study contained granitic clasts although a small number of clasts of coarse grained, equigranular granite containing both K-feldspar and albitic plagioclase do occur in some places in the supraglacial moraines adjacent to the Allan Hills. The presence of these rare granite clasts indicates that granite bedrock does occur under the ice in the vicinity of the Allan Hills.

7. IMPLICATIONS FOR THE GEOLOGY OF GONDWANA

The absence of clasts of granitic basement rocks in the Elephant and Reckling moraines strongly indicates that basement rocks are not exposed to the ice along the flowline leading from the central ice divide to these locations. Instead, the dominance of sedimentary rocks and of basalt/dolerite in the glacial deposits requires the conclusion that these rocks cover the crystalline basement in the area of East Antarctica drained by the ice at the Elephant and Reckling moraines. Consequently we conclude that the rocks of the Beacon Supergroup are not restricted to the Transantarctic Mountains, but extend onto the East Antarctic shield under the present ice sheet.

These results therefore suggest that the

sedimentary basins of Gondwana during the Permian and Triassic periods were more extensive than suggested by Elliot (1974) on the basis of known exposures of the Beacon Supergroup in the Transantarctic Mountains and in the vicinity of the Amery Ice Shelf in East Antarctica. However, additional studies of glacial deposits along the western slope of the Transantarctic Mountains are needed in order to test the hypothesis that sedimentary rocks of the Beacon Supergroup occur widely and throughout the area between the ice divide of East Antarctica and the Transantarctic Mountains.

Acknowledgements

We thank David Buchanan and John Schutt for their valuable assistance during several field seasons on the polar plateau and Todd Aebie and R. James Elliot for collecting the ice samples in the Reckling Moraine. This is contribution No 89 of the Laboratory for Isotope Geology and Geochemistry of The Ohio State University. The research on which this paper is based was supported by the National Science Foundation through grants DPP-8314136A01 and DPP-8714324.

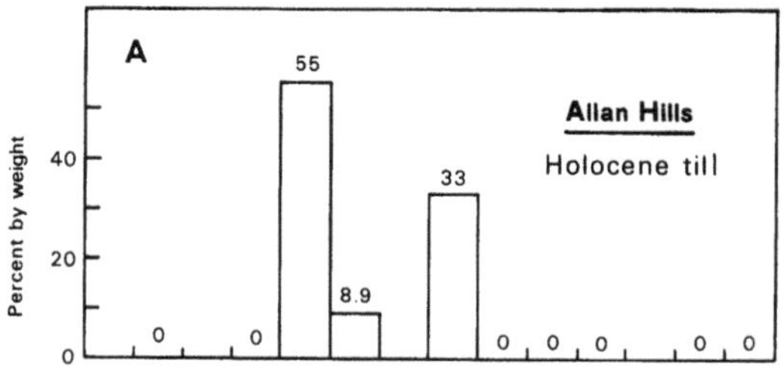

Figure 7A Lithologic composition of rock clasts in Holocene till from supraglacial moraines near the Allan Hills.

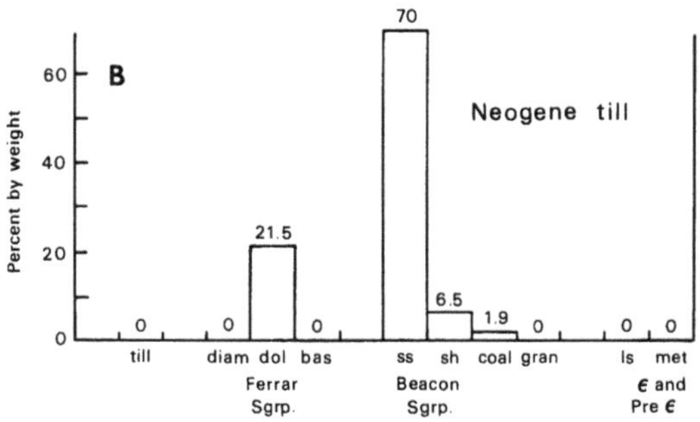

Figure 7B Lithologic composition of rock clasts in Neogene till at Manhaul Bay and in supraglacial moraines near the Allan Hills.

REFERENCES

Borns, H.W., Jr. & B.A. Hall 1969. Mawson "tillite" in Antarctica: preliminary report of a volcanic deposit of Jurassic age. Science 166: 870-872.

Cherry, E.M. & H.C. Noltimier 1985. Dating till by paleomagnetism: Allan Hills, southern Victoria Land. Antarctic Journal of the U.S. 20(5): 10.

Delisle, G. & J. Sievers 1991. Sub-ice topography and meteorite finds near the Allan Hills and the Near Western ice field, Victoria Land, Antarctica. Journal of Geophysical Research 96(E1): 15,577-15,587.

DePaolo, D.J., W.I. Manton, E.S. Grew & M. Halpern 1982. Sm-Nd, Rb-Sr and U-Th-Pb systematics of granulite facies rocks from Fyfe Hills, Enderby Land, Antarctica. Nature 298: 614-618.

Drewry, D.J. 1982. Antarctica: Glaciological and Geophysical Folio. Scott Polar Research Institute, University of Cambridge.

Elliot, D.H. 1974. Gondwana Basins of Antarctica. Third Gondwana Symposium, Canberra: 493-536.

Faure, G. 1986. Principles of isotope geology, second edition. Wiley, New York, 589 p.

Faure, G. 1990a. Origin of stranding surfaces. Lunar and Planetary Institute Technical Report No. 90-03: 11-14.

Faure, G. 1990b. Physical description of the Elephant and Reckling moraines. Lunar and Planetary Institute Technical Report No. 90-03: 24-25.

Faure, G. 1990c. Subice topography and the formation of supraglacial moraines. Lunar and Planetary Institute Technical Report No. 90-03: 32-33.

Faure, G. 1990d. Supraglacial moraines, associated meteorites and climate change. Lunar and Planetary Institute Technical Report 90-03: 34-35.

Faure, G., L.M. Jones & L.B. Owen 1974. Isotope composition of strontium and geologic

history of the basement rocks in Wright Valley, southern Victoria Land. New Zealand Journal of Geology and Geophysics 17(3): 611-627.
Faure, G., M.L. Kallstrom & T.M. Mensing 1984. Classification and age of terrestrial boulders in the Elephant and Reckling Moraines. Antarctic Journal of the U.S. 19(5): 28-29.
Faure, G. & S. Sutton 1985. Thermoluminescence of sandstone clasts of the Elephant Moraine. Antarctic Journal of the U.S. 20(5): 12-14.
Faure, G. & K.S. Taylor 1985. The geology and origin of the Elephant Moraine on the East Antarctic ice sheet. Antarctic Journal of the U.S. 20(5): 11-12.
Faure, G. & D. Buchanan 1987. Glaciology of the East Antarctic ice sheet at the Allan Hills: A preliminary interpretation. Antarctic Journal of the U.S. 22(5): 74-75.
Faure, G., M.L. Strobel & E.H. Hagen 1987. Glacial geology of the Reckling Moraine on the East Antarctic ice sheet. Antarctic Journal of the U.S. 22(5): 74-75.
Faure, G., J. Hoefs, L.M. Jones, J.F. Curtis & D.E. Pride 1988. Extreme ^{18}O depletion in calcite and chert clasts from the Elephant Moraine on the East Antarctic ice sheet. Nature 332: 352-354.
Faure, G. & D.M. Harwood 1990. Marine microfossils in till clasts of the Elephant Moraine on the East Antarctic ice sheet. Antarctic Journal of the U.S. 25(5): 23-25.
Faure, G, T.M. Mensing, L.M. Jones, J. Hoefs & E.M. Kibler 1991. Isotopic and geochemical studies of Ferrar Dolerite sills in the Transantarctic Mountains. Gondwana 7 Proceedings, 669-683. Universidade de Saõ Paulo, Brasil, 714 pp.
Faure, G., P. Grootes, D. Buchanan, & E.H. Hagen 1992. Oxygen-isotope study of the icefields surrounding the Reckling Moraine on the East Antarctic ice sheet. American Geophysical Union, Antarctic Research Series, Annual Contributions, vol. 3 (in press).
Grew, E.S. 1978. Precambrian basement at Molodezhnaya Station, East Antarctica. Geological Society America Bulletin 89: 801-813.
Hagen, E.H. 1988. Geochemical studies of Neogene till in the Transantarctic Mountains; Evidence for an extraterrestrial component. M.S. Thesis, Department of Geological Sciences, The Ohio State University, 322 p.
Lorius, C. & L. Merlivat 1977. Distribution of mean surface stable isotope values in East Antarctica: observed changes with depth in the coastal area. International Symposium on Isotopes and Impurities in Snow and Ice, 127-137. International Association of Hydrologic Sciences, Pub. 118.
Lorius, C., J. Jouzel, C. Ritz, L. Merlivat, N.I. Barkov, Y.S. Korotkevich & V.M. Kotlyakov 1985. A 150,000-year climatic record from Antarctic ice. Nature 316: 591-596.
Mercer, J.H. 1972. Some observations on the glacial geology of the Beardmore Glacier area. In R.J. Adie (ed.), Antarctic Geology and Geophysics: 427-433. Universitets-forlaget, Oslo.
Montello, J.M. 1986. The provenance of till at the Elephant Moraine on the East Antarctic ice sheet. M.S. thesis, Department of Geological Sciences, The Ohio State University, 168 p.
Robin, G. de Q & D.H.M. Millar 1982. Flow of ice sheets in the vicinity of subglacial peaks. Annals Glaciology 3: 290-294.
Sheraton, J.W. & K.D. Collerson 1983. Archaean and Proterozoic geological relationships in the Vestfold Hills-Prydz Bay area, Antarctica. BMR Journal Australian Geology and Geophysics 8: 119-128.
Strobel, M.L. & G. Faure 1987. Transport of indicator clasts by ice sheets and the transport half-distance: a contribution to prospecting for ore deposits. Journal of Geology 95: 687-697.
Taylor, K.S. 1986. Geological and geochemical study of the Elephant Moraine, Allan Hills area, Antarctica. M.S. thesis, Department of Geology, Kent State University, Ohio.
Webb, P.N., D.M. Harwood, B.C. McKelvey, J.H. Mercer & L.D. Stott 1984. Cenozoic marine sedimentation and ice-volume variation on the East Antarctic craton. Geology 12: 287-291.

Appendix*

1. Isotope composition of oxygen and hydrogen in ice along a surveyed line at the Elephant Moraine (see text)

Location	Material	δD,‰	$\delta^{18}O$,‰
1600S	white ice	-384	-48.3
1600S	firn	-266	-34.5
1800S	ice	-395	-49.7
2000S	ice	-365	-46.6
2200S	ice	-346	-45.6
2400S	ice	-393	-49.7
2600S	ice	-388	-49.4
2800S	ice	-388	-49.2
3000S	ice	-355	-46.2
3200S	ice	-376	-47.4
3400S	ice	-324	-42.6
3600S	ice	-377	-47.7
3800S	ice	-369	-46.8

2. Isotope compositions of oxygen in ice along the 1000 W crossline at the Reckling Moraine were reported by Faure et al. (1992).

3. Isotope composition of oxygen in ice along a surveyed line on the ice sheet adjacent to the Allan Hills (see text)

Location	Material	$\delta^{18}O$,‰
100W	ice	-40.9
200W	ice	-39.4
300W	ice	-41.2
400W	ice	-40.2
500W	ice	-42.3
600W	ice	-36.3
700W	ice	-39.9
800W	ice	-43.0
900W	ice	-42.1
1000W	ice	-41.3
1100W	ice	-38.6
1200W	ice	-35.7
1300W	ice	-35.2
1400W	ice	-38.4
1500W	ice	-38.4
1600W	ice	-36.7
1700W	ice	-37.0
1800W	ice	-40.7
1900W	ice	-42.2
2000W	ice	-39.8
2100W	ice	-40.4
2200W	ice	-41.3
2300W	ice	-40.6
2400W	ice	-41.0
05E	ice	-40.4
50E	ice	-37.8
50E	snow	-32.0
75E	ice	-34.4
100E	ice	-39.2

* Analyses by Geochron Laboratories, Cambridge, MA, 02139, U.S.A. relative to SMOW (standard mean ocean water)

Gondwana Eight, Findlay, Unrug, Banks & Veevers (eds) © *1993 Balkema, Rotterdam. ISBN 90 5410 304 3*

Gondwanan foreland basin along the Panthalassan margin of Antarctica

D.L.Sahagian & J.W.Collinson
Ohio State University, Columbus, Ohio, USA

ABSTRACT: Permo-Triassic rocks in the Transantarctic Mountains and Permian rocks in the Ellsworth Mountains were deposited in a foreland basin in front of the Gondwanan fold-thrust belt. Evidence supporting this hypothesis includes the thickening of comparable sections toward West Antarctica and paleocurrent and lithologic data suggesting a calc-alkaline volcanic source in West Antarctica. Gondwana basins in Victoria Land are relatively small and their size is well constrained, suggesting a width of about 200 km. A cross-sectional width of the foreland basin in the Central Transantarctic Mountains is at least 400 km. The Ellsworth Mountains-Pensacola Mountains segment of the basin, which includes folded Permian rocks, is about 100 km wide. The reconstructed 200 km Permian width of this segment, when added to the Central Transantarctic Mountains width, yields a basin width of 600 km. The Triassic width of the foreland basin, excluding the folded rocks of the Ellsworth-Pensacola Mountains, but now including the Central Transantarctic Mountains and Victoria Land, was also 600 km wide. Thus, a constant basin width was maintained as the fold-thrust belt advanced toward the craton. Since lithospheric flexural rigidity (elastic thickness) controls foreland basin width, a thermal model which describes the evolution of the Antarctic lithosphere can be used to retrodict basin width. Given an initial elastic thickness of 35 km with orogenic thickening by factors of 1.5 at 1700 Ma and ranging from 1.0 to 1.4 at 500 Ma, the model indicates a Permo-Triassic basin width of 600-770 km. The agreement between the model minimum and reconstructions based on stratigraphic observations suggests that the Beardmore/Ross orogeny at 500 Ma did not significantly thicken the lithosphere beneath the Transantarctic foreland basin.

1 INTRODUCTION

The Permo-Triassic paleo-Pacific margin of Gondwanaland was bordered by a series of foreland basins. These basins developed in front of thrust and fold belts related to convergence along the plate margin. The best known of these basins are the Sydney basin in eastern Australia (Veevers 1984) and the Karoo basin in South Africa (Tankard et al. 1982). In between these basins lay the less well exposed or preserved Transantarctic basin along the cratonic margin of East Antarctica (Collinson 1991).

The geometries of the Sydney and Karoo basins are relatively well preserved, having undergone relatively little tectonism since their formation. Strike-normal cross-sections can be directly reconstructed from geologic maps. The width of these basins can be measured directly from the fold-thrust belt to where the stratigraphic sequence abruptly thins. The depth of these basins can be established directly through stratigraphic thicknesses. The construction of a cross-section of the Transantarctic basin is much more difficult. In addition to being mostly concealed by ice, the basin has undergone at least two episodes of extension, the first associated with dolerite intrusions, and the second with uplift of the Transantarctic Mountains in Mid-Cenozoic times (Elliot 1991; Wilson, this volume).

In this paper we discuss two ways in which the general geometry of the basin can be established. The first method integrates our knowledge of the stratigraphy and sedimentology of Permo-Triassic deposits throughout the Transantarctic Mountains. The second method uses orogenic history to retrodict the evolution of lithospheric flexural rigidity and elastic thickness, which controlled the geometry of the foreland basin.

2 STRATIGRAPHIC SYNTHESIS

Foreland basins have been observed to be generally elongate and asymmetrical toward their thrust-fold belts (Allen et al. 1986). Because these basins are formed by flexural loading of the lithosphere by the thrust sheets in a thrust-fold belt, the greatest subsidence is immediately in front of the thrust-fold belt. The axis of the basin moves toward the craton as thrusting and folding impinges on the orogenic margin of the basin. This asymmetry is clear in the

Sydney and Karoo basins (Veevers 1984; Tankard et al. 1982), because the thrust-fold belts are still preserved. In both cases the sedimentary sequences are clastic wedges that thin toward the craton.

All the components of a foreland basin are found in Antarctica, but as a result of subsequent tectonic activity as well as modern ice cover, they cannot be seen together. These components include (Fig. 1): (1) the orogenic margin of folded and presumably thrusted Permian rocks in the Ellsworth and Pensacola Mountains, (2) the cratonic margin in northern Victoria Land where the Permo-Triassic sequence wedges out toward the continent, and (3) the central part of the basin in the Central Transantarctic Mountains where a 2.5 km thick and mostly complete Permo-Triassic sequence is preserved. If the axis of the Transantarctic foreland basin was approximately parallel to the margin of the East Antarctic craton, the trend of the Transantarctic Mountains is slightly oblique to this axis.

Consequently, the stratigraphy of the Transantarctic Mountains represents a time-transgressive sampling of the foreland basin sections, younging toward Victoria Land.

The geometry of the Transantarctic foreland basin can be reconstructed by projecting the ends into a strike-normal cross-section through the Central Transantarctic Mountains (Fig. 2) (Collinson 1990; 1991). The West Antarctic part of the basin must be palinspastically restored to undo the effects of considerable post-Triassic extension (Fig. 1). The Antarctic Peninsula, Thurston Island, Ellsworth-Whitmore Mountains, and Marie Byrd Land crustal blocks also were involved in the crustal extension that

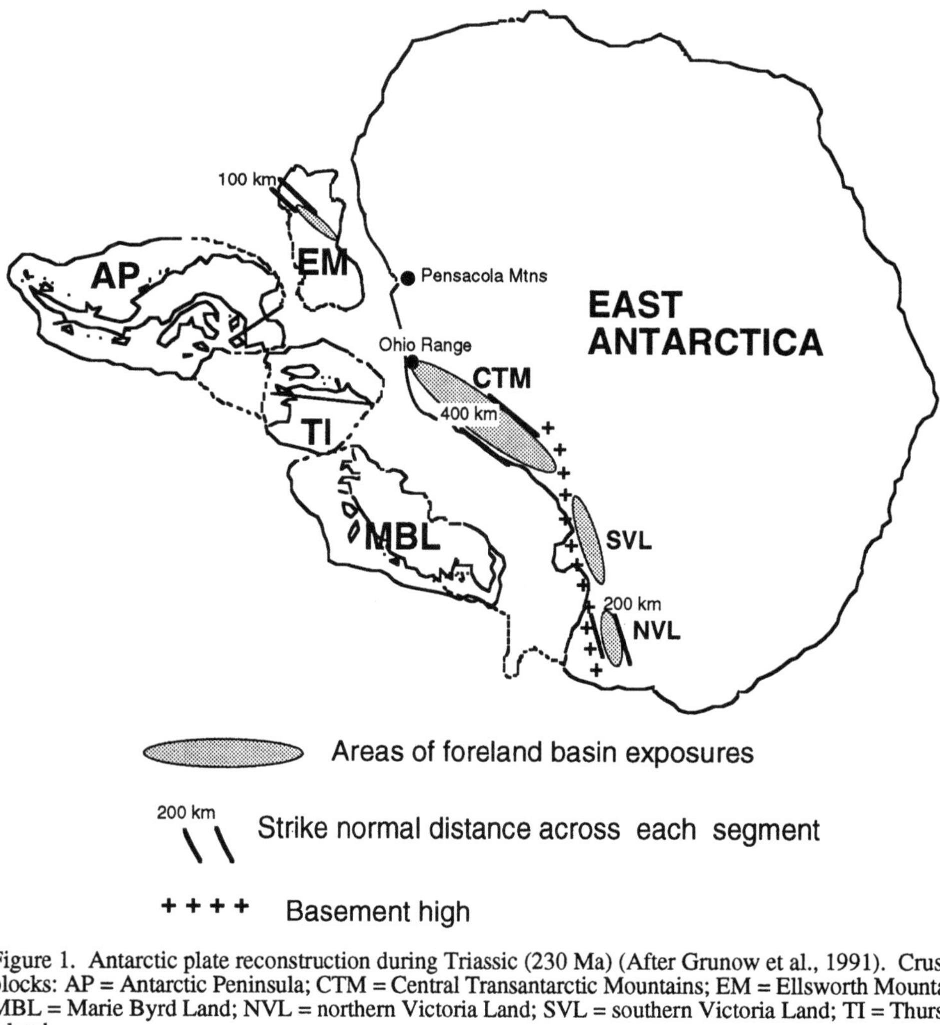

Figure 1. Antarctic plate reconstruction during Triassic (230 Ma) (After Grunow et al., 1991). Crustal blocks: AP = Antarctic Peninsula; CTM = Central Transantarctic Mountains; EM = Ellsworth Mountains; MBL = Marie Byrd Land; NVL = northern Victoria Land; SVL = southern Victoria Land; TI = Thurston Island.

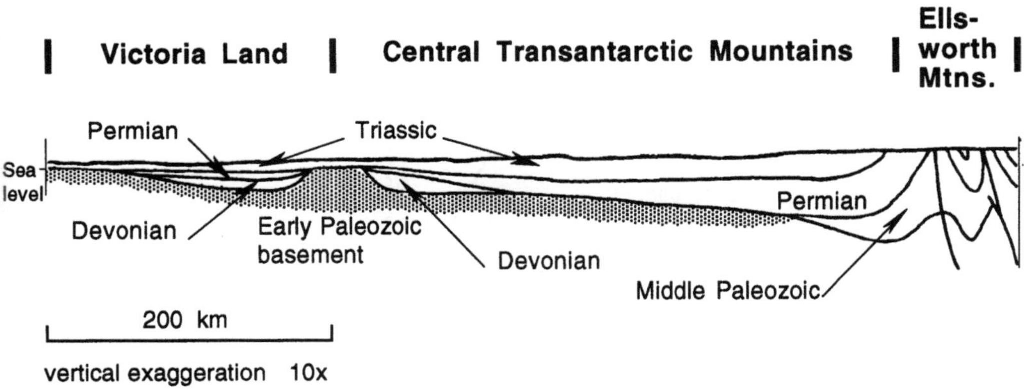

Figure 2. Diagrammatic strike normal cross-section across Transantarctic foreland basin assembled by combining segments shown in figure 1.

affected all of West Antarctica and are restored to their approximate position based on paleomagnetic data (Grunow et al. 1991).

The vertical scale of the cratonic side of the basin can be established from stratigraphic thicknesses in Victoria Land and the Central Transantarctic Mountains (Fig. 3) where stratigraphic sections are complete. Permian units in the Ellsworth Mountains are thicker than their counterparts in the Central Transantarctic Mountains (Fig. 3). The top of the section in the Ellsworth Mountains was removed by erosion, probably during folding and thrusting in the Triassic (Elliot 1975).

In order to reconstruct the width of the Transantarctic basin from the preserved and exposed pieces in the Transantarctic Mountains and environs, the widths of the individual components must be projected into the cross-section. We have assumed that the segments do not greatly overlap and that no large parts are missing. The Ellsworth Mountains crustal block cannot greatly overlap the East Antarctic craton as palinspastically restored. Unfolded Permian rocks in the northern Pensacola Mountains suggest that the folds die out toward the craton. Continuity in the stratigraphy and environmental setting of deposits from the Ellsworth Mountains to the Ohio Range and throughout the Central Transantarctic Mountains indicates that no great gap exists between these regions (Fig. 4). The post-glacial Permian sequences in Victoria Land are different than those in the Central Transantarctic Mountains (Figs. 3 and 4), probably because a cratonic high separated these areas (Collinson 1990). Volcaniclastic sandstones, which dominate the Upper Permian in the Central Transantarctic Mountains, did not override this high to reach southern Victoria Land until Middle Triassic time (Fig. 4). Differences in the Central Transantarctic Mountains and Victoria Land sequences argue against overlap and the nearly adjacent positions of these regions on the craton suggest that little if any gap exists.

We estimated the widths of the various components of the cross-section from the outcrop belts. The width of the Victoria Land part of the basin (200 km) can be seen in northern Victoria Land (Fig. 1) where coarse Permian clastics onlap the Ross high to the east and the Permo-Triassic section wedges out against the craton to the west (Collinson et al. 1986). The Ross high, which forms the eastern margin of the Victoria Land part of the basin, is near the north end of the Central Transantarctic Mountains (Fig. 1). Here the Lower Permian section is missing between basement and the Permian coal measures (Laird 1963). Permian stratigraphic units can be traced throughout the Central Transantarctic Mountains from Victoria Land to the Ohio Range. The axis of this part of the basin is well defined by Permo-Triassic paleocurrent directions, which reflect the predominantly longitudinal flow of the major stream systems. This belt is approximately 400 km wide (Fig. 1). Outcrops at both ends of the Ellsworth Mountains show that the region was covered by Permian deposits at one time (Craddock et al. 1986). This segment is about 100 km wide, but was probably twice as wide prior to Triassic folding based on fold geometry.

Reconstruction of the now disjoint basin segments on the basis of the above stratigraphic arguments results in a Triassic basin approximately 600 km wide from the inner fold-thrust belt margin to the cratonic margin. This includes the Central Transantarctic Mountains and Victoria Land segments of the basin. The greatest thickness is about 2.5 km, but an increase in the thickness of the Triassic rocks toward the outer margin of the basin suggests that thicknesses could have been 3 to 4 km, which are not particularly great for a foreland basin. During the Permian, the foreland basin was more outboard and included the Ellsworth-Pensacola Mountains and the Central Transantarctic Mountains segments. As the foreland basin migrated cratonward with folding and thrusting in the Ellsworth-Pensacola Mountains in the Triassic, Victoria Land became part of the foreland basin, maintaining a constant basin width of 600 km.

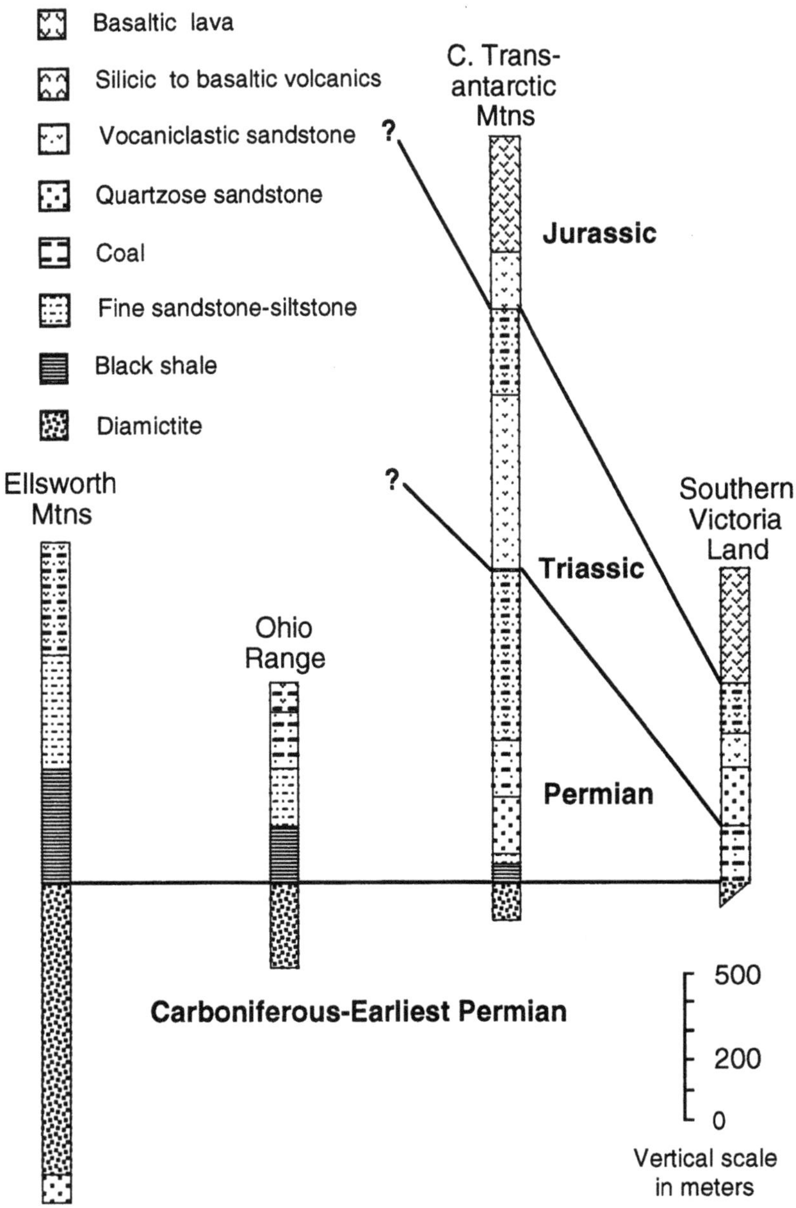

Figure 3. Representative stratigraphic sections across the Transantarctic basin.

3 THERMO-MECHANICAL EVOLUTION OF THE LITHOSPHERE

The thermal structure and flexural rigidity of the lithosphere are controlled by lithospheric thickening and thinning events, as well as by subsequent thermal reequilibration over time (Sahagian and Holland, in review; Holland et al. 1990). Thus, a region of the lithosphere having experienced a complex tectonic history is expected to have had a complex thermal history and consequent thermal structure. This thermal structure strongly controls the flexural rigidity of the lithosphere, which can become particularly large as a result of mechanical thickening involved in orogeny (Karner and Watts 1983). The width of any foreland basin is directly controlled by the flexural rigidity (commonly characterized by an elastic thickness T_e) of the lithosphere (McNutt et al.

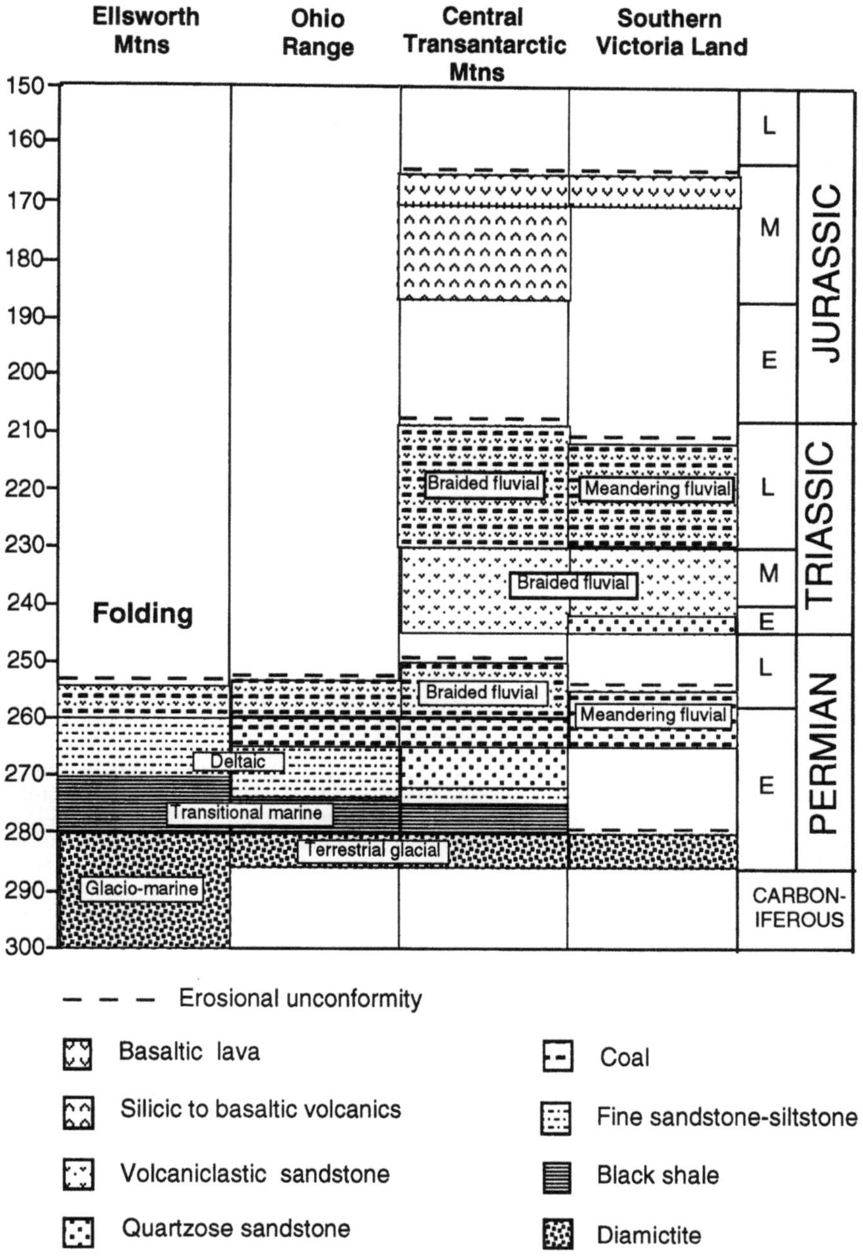

Figure 4. Chronostratigraphy of the Transantarctic basin from 300 to 150 Ma.

1988; Karner and Watts 1983). In this section, we discuss the relative importance of mechanical (tectonic thinning & thickening) effects, and thermal equilibration effects on the elastic thickness of the lithosphere. The flexural rigidity and elastic thickness of the lithosphere underlying the Transantarctic foreland basin resulting from its tectonic history prior to the Permian directly controls the width of the thrust-loaded Transantarctic foreland basin. Thus if the thermo-mechanical evolution of the lithosphere can be accounted for, then the width of the foreland basin can be modelled as a check against stratigraphic data.

The flexural characteristics of mechanically and thermally evolved lithosphere become apparent when it is loaded. In general, a history of tectonic events should result in a particular thermal structure at any point in time, with a corresponding flexural rigidity and elastic thickness. In the context of continental lithosphere loaded by thrust sheets and foreland basin sediments, the elastic thickness (as modulated by the evolving distribution of sediment load) controls foreland basin geometry and accommodation space, which results in an observable stratigraphic architecture. To understand these effects in continental lithosphere subjected to tectonic and sediment loads, we use the general model developed by Sahagian and Holland (in review) which takes into account the effects of multiple tectonic thinning and thickening events of the lithosphere, and the evolution of lithospheric thermal structure. Model predictions of flexural rigidity and lithospheric elastic thickness can be applied to basins whose geometries are not directly observable such as the Permo-Triassic Transantarctic foreland basin, which has been largely obscured by subsequent tectonic events such as uplift of the Transantarctic Mountains and rifting of West Antarctica, as well as covered by modern ice. A summary of elastic thicknesses for a variety of continental environments loaded by overthrusting (McNutt et al. 1988) shows values of T_e ranging from 5 to 105±25 km. This wide range of T_e indicates very different thermal structures, and cannot be explained solely on the basis of age. This is expected, since the previous mechanical history of the flexed lithosphere should profoundly affect T_e. As the lithosphere is physically thickened by tectonic events, the elastic thickness increases directly. An evolving thermal structure also controls elastic thickness because the lithosphere stiffens as it cools. Consequently, we consider both thermal and mechanical controls, and assess the relative importance of each at time-scales relevant to basin development (10^7-10^8 m.y.).

The rheological structure of the lithosphere is complex (Davy and Cobbold 1991; Stevenson and Cloetingh 1991), but in general the upper, cooler portions are stiffer than the relatively ductile lower portions at temperatures greater than about 450 °C (Chen 1988, Karner and Watts 1982, Kirby and Kronenberg 1987, McNutt, et al. 1988, Zoetmeijer, et al. 1990; Buck 1992). Thus, the elastic thickness (taken to reflect the depth to the 450 °C isotherm) of the lithosphere is distinct from the thermal thickness (to 1350 °C) defined by the onset of partial melting. We consider the elastic thickness as the thickness that the lithosphere would have to be (if it strictly obeyed Hooke's law) to respond to loads as observed. Between the 450 and 1350 °C isotherms, the lithosphere is taken to have a relaxation time too short to consider. However, the depth (temperature) at which the lithosphere can be considered elastic depends on the time-scale of load response. While our treatment pertains to thrust sheet and sedimentary loads formed during tens of millions of years, on time-scales of glacial variability (thousands of years) the elastic thickness approaches the thermal thickness (Bills and May 1987; Peltier 1984).

3.1 *Thermal Model*

We use the simple numerical model developed by Sahagian and Holland (in review) to describe the processes of heat flow and lithospheric thinning and thickening in response to stretching and orogenic events. Deep lithospheric heat transfer occurs by conduction only, while underlying mantle heat transfer proceeds by convection, which can be far more efficient. Since the thermal lithosphere is defined by the position of a particular isotherm (1350 °C), it is sensitive to changes in thermal structure at its base. If the mantle did not convect (e.g. thermal expansion coefficient of 0), the crust and mantle would behave as a cooling halfspace. If, on the other hand, the top of the convecting mantle were held at a specific temperature at a specific depth (plate model), the lithosphere would attain a linear, permanent equilibrium temperature profile. Mantle convection has the effect of supplying heat to the top of the convecting mantle such that the lithospheric cooling rate is reduced from what it would be for a cooling halfspace, but is not reduced to zero and is thus an intermediate case.

4 MODEL RESULTS

Flexure of the lithosphere can be treated as a problem of a flexed beam or plate, depending on the relevant geometry (Hetenyi 1946). As such, the response to loads applied to the lithosphere from thrusting of crustal sheets and from sediment deposition (Flemings and Jordan 1989) is controlled by the flexural rigidity. Foreland basins are caused by the load of an overthrusting crustal sheet on the lithosphere in front of the thrust (the foreland). This relationship is commonly found in compressional mountain belts such as the Rockies or Alps, and creates basins such as the Denver basin and the Molasse basin, respectively, into which sediments from the eroding mountains are shed (Stockmal and Beaumont 1987). A foreland basin is formed over tens of millions of years as the thrust front advances, "pushing" the basin along in front of it. In detail, as new thrust surfaces form in front of old ones, older basins (e.g. the Cretaceous basin of Utah) are incorporated into the new thrust sheets and are thrust over the new basins (e.g. Denver basin).

The formulation of the problem of foreland basin loading as plate flexure assumes that the (entire) lithosphere exhibits effectively elastic behavior on the time scale of tens of millions of years. There is some indication that the viscous relaxation time for oceanic lithosphere as well as continental lithosphere is at least several tens of millions of years (McNutt and Menard 1978; Watts 1978), so the assumption of elasticity may not be unreasonable at these time-scales. At longer time-scales, two opposing factors affect the flexural rigidity of the lithosphere. These are a cooling thermal structure in response to lithospheric heat flow (or on longer time-scales yet, to planetary cooling), balanced against viscous relaxation of the lithosphere (rate at which it becomes inelastic and deformations become permanent). The precise rate of relaxation in not well constrained, but is a function of thermal structure and applied load. However, on the time-scales of foreland basin development, the lithosphere can be considered elastic, as discussed above.

The results of applying a finite difference model to an orogenic event in the lithosphere indicate that the lithospheric thickness (to the 1350 °C isotherm) is sensitive to thermal reequilibration in response to mantle heat flow, but the elastic thickness (to the 450 °C isotherm) is relatively insensitive (Sahagian and Holland, in review). An implication of this result is that estimates of flexural rigidity during foreland basin development are sensitive only to the thickening factor, γ, and not to the time elapsed since the last orogenic event.

4.1 *History of Transantarctic Basin lithospheric thickness*

The Transantarctic basin shares a similar tectonic history to the Karoo basin and Sydney basin, to which it was joined during and prior to basin development. Following an unclear number of orogenic events after 1700 Ma and a subsequent prolonged tectonic quiescence, the East Antarctic margin underwent the Beardmore and Ross Orogenies at 700-500 Ma (Borg et al. 1990; Goodge et al. 1991). During the late Paleozoic-early Mesozoic, both the Transantarctic and Karoo basins were enveloped within a foreland basin during the Gondwanide Orogeny (Collinson 1991, Lock 1980).

The thickening factor, γ, is difficult to calculate exactly, and the available lines of evidence can be used to only approximately constrain the amount and date of thickening in response to each orogenic event. Because of the uncertainty of timing and magnitude of orogenic events between 1700 and 500 Ma (despite estimates of current crustal thickness in the McMurdo Sound area of over 40 km (Smithson 1972)), various scenarios were modelled (Figure 5). Each model run had an initial elastic thickness of 34 km and experienced various orogenic thickening events until 500 Ma, subject to thermal reequilibration until 200 Ma. Immediately following a major orogeny, the elastic thickness increases in response to continued surface heat flux, but then begins to decrease as increased lower lithospheric temperatures become evident through conduction to shallower (elastic) levels. To provide an upper limit on the elastic thickness, one model run assumed an orogeny at 500 Ma (relatively recently) with $\gamma = 1.5$; a lower limit is provided in the second run by assuming no orogeny ever. In the no-orogeny model, the elastic thickness slowly increased to 100 km. In the maximum orogenic model ($\gamma = 1.5$ at 500 Ma), the elastic thickness was approximately 160 km following the orogeny, then decreased to 158 km in the following 365 m.y. Model runs with intermediate values of compression factor or older ages of orogeny had correspondingly intermediate values of final elastic thickness (Figure 5). The relative importance of mechanical thickening events (orogeny) to thermal reequilibration can be seen from Figure 5. Orogenies of sufficient age (2 Ga) are not observable because enough time is available for subsequent thermal reequilibration. Consequently, the curves for "no orogeny" and "1.5 at 2760 Ma" are superimposed by about 1 Ga. This is in part also due to the rapid rates of thermal reequilibration exhibited by thin (<70 km) elastic lithosphere. However, at later times (500 Ma), when the elastic lithosphere is closer to its equilibrium thickness of about 100 km, thermal processes proceed much more slowly, and mechanical thickening (orogeny) is the dominant factor in controlling elastic lithospheric thickness. Thus, thermo-mechanical models (Sahagian and Holland, in review) can be used to constrain the orogenic history of the Antarctic lithosphere only for events in the last billion years or so. Events prior to that will have been thermally reequilibrated by now (and by the Permo-Triassic). A second important implication of the relative importance of mechanical thickening and thermal reequilibration is that the model results are independent of the initial value chosen for elastic thickness. We chose an initial elastic thickness of 34 km, but the elastic thickness after 1 b.y. would be indistinguishable from that corresponding to an initial elastic thickness of 50 km or 10 km.

The thermo-mechanical evolution model (Sahagian and Holland in review) can be used to retrodict the width of the foreland basin during the Gondwanide Orogeny, which is particularly helpful in Antarctica where outcrops are scarce and the basin must be reconstructed with only limited field evidence as discussed previously. The maximum Permian elastic thickness of 158 km implies a basin width of 770 km, which may be taken as an upper limit. From the non-orogenic results, a lower limit on elastic thickness of 100 km corresponds to a basin width of 500 km. Intermediate values of T_e=120 to 125 km (1.2 at 560 Ma or 1.5 at 1760 Ma) correspond to a basin width of 600 km. Measured from the fold belt to the foreland edge of the peripheral bulge, the Karoo basin is presently approximately 550 km wide (Winter and Venter 1971). Reconstructions of the Transantarctic basin suggest a basin width of approximately 600 km, corresponding most closely to our minimum estimate and also consistent with the 115±10 km of Stern and Ten Brink (1989). Thus, the geologic data, combined with our model results suggest that there was at most a very minor orogeny (lithospheric thickening event) at 700 Ma, or more possibly a larger orogeny at earlier times (1700 Ma) that affected the lithosphere of the present Transantarctic basin. The results do not preclude the possibility that there was no significant mechanical thickening of the Transantarctic lithosphere at all.

A thermo-mechanical model allows us to explore the relationship between lithospheric thickening (or thinning) events, and thermal reequilibration. These results can be compared to field observations of

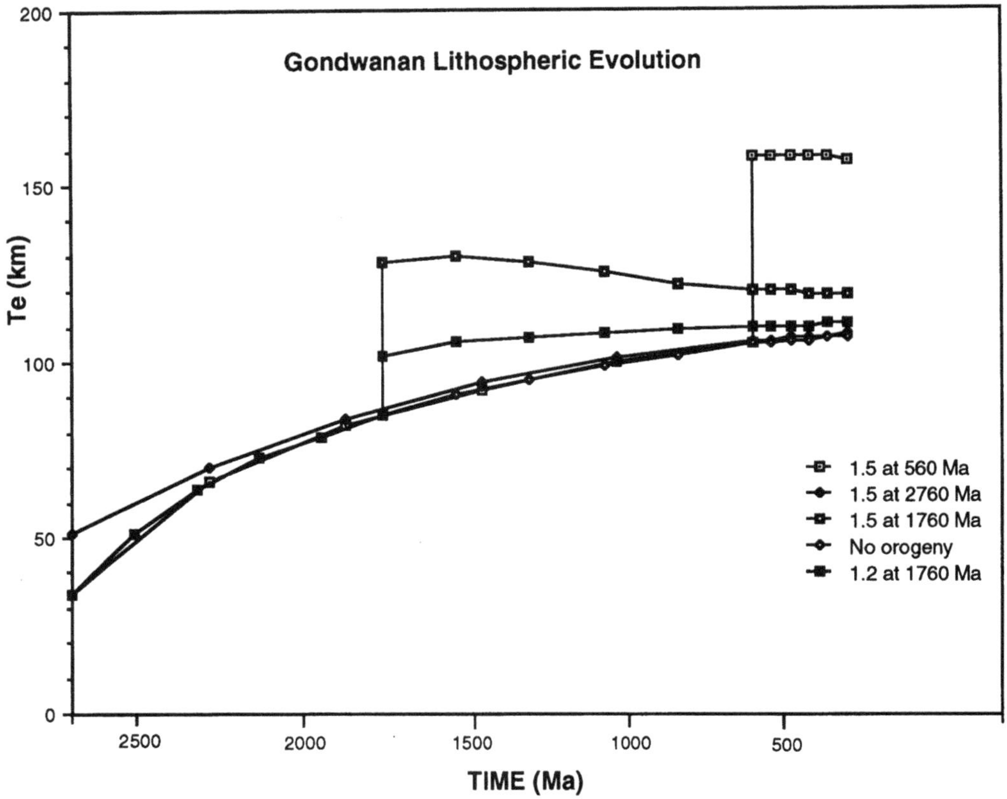

Figure 5. Modeled evolution scenarios of elastic and thermal lithospheric thicknesses in the Transantarctic basin. The initial conditions for the lithosphere are such that it has an elastic thickness of 34 km, but the final results are not sensitive to this choice of initial thickness. Because γ for the various possible orogenic events are poorly constrained as discussed in the text, various estimates were made, corresponding to a range of possible histories. For cases where the Nimrod Orogeny caused significant lithospheric thickening, elastic thickness initially increases due to surface heat flux, but then decreases in response to heating at the base of the lithosphere transferred to shallow (elastic) levels by conduction. The various thickening scenarios include: minimum estimate with simple thermal cooling only and no orogenic events; maximum estimate with orogenic event with $\gamma = 1.5$ at 560 Ma; various intermediate estimates with either smaller amounts of thickening or older thickening events. Note lag time of thermal reequilibration after major 1760 Ma event. This is due to the rising of a thermal wave in response to increased temperature at the base of the thickened lithosphere.

continental lithosphere which has been subject to flexural loads by thrust emplacement and consequent foreland basin development. If the elastic thickness of the lithosphere can be determined from foreland basin geometry, then the model can be used to constrain tectonic history to within certain limits discussed above. Conversely, if the tectonic history can be documented by some independent means, then the model can be used to determine the evolution of elastic thickness and consequent width of foreland basins which may be lost to erosion or otherwise obscured.

5 CONCLUSIONS

A concatenation of field observations indicates that during the Permian, the Transantarctic foreland basin was filled by sediments now found in the Ellsworth-Pensacola Mountains and Central Transantarctic Mountains. As the thrust sheets advanced, the Ellsworth-Pensacola Mountains were incorporated into the fold-thrust belt, and the locus of lithospheric loading migrated cratonward to include Victoria Land. Palinspastic reconstruction of basin segments indicates a constant basin width of approximately 600 km.

On the basis of model calculations, we have found that except for cases of particularly thin lithosphere, tectonic activity dominates the evolution of lithospheric flexural rigidity at time-scales relevant to foreland basin development. Given palinspastic reconstructions of the Transantarctic foreland basin, the model can be used to constrain tectonic history.

The reconstructed 600 km width suggests that there was only minor, if any, Late Proterozoic lithospheric thickening. Conversely, given tectonic history, elastic thickness at any point in time can be obtained. From this elastic thickness, the response to loads such as thrust sheets and foreland basin sedimentation can be retrodicted. Based on model runs using the best available estimates of orogenic compression factors affecting the lithosphere underlying the Transantarctic foreland basin, we retrodict a Permo-Triassic basin width of 600 km, in agreement with reconstructions based on field data.

REFERENCES

Allen, P.A., P. Homewood, & G. Williams 1986. Foreland basins: An introduction, In Allen, P. & P. Homewood (eds.), Foreland Basins. *International Association of Sedimentologists Special Publication* 8: 3-12

Bills, B. & G. M. May 1987. Lake Bonneville: constraints on lithospheric thickness and upper mantle viscosity from isostatic warping of Bonneville, Provo, and Gilbert stage shorelines. *Journal of Geophysical Research* 92: 11493-11508

Borg, S. G., D. J. DePaolo & B. M. Smith 1990. Isotopic structure and tectonics of the Central Transantarctic Mountains. *Journal of Geophysical Research* 95: 6647-6667

Buck, R. 1992. Modes of continental lithospheric extension. *Journal of Geophysical Research* 96: 20161-20178

Chen, Y. 1988. Thermal model of oceanic transform faults. *Journal of Geophysical Research* 93: 8839-8851

Collinson, J.W. 1990. Depositional setting of Late Carboniferous to Triassic biota in the Transantarctic Basin. In Taylor, T.N. & E.L. Taylor (eds.), *Antarctic paleobiology: its role in the reconstruction of Gondwana*. New York, Heidelberg, Berlin, Springer-Verlag, p. 1-114

Collinson, J.W. 1991. The palaeo-Pacific margin as seen from East Antarctica. In Thomson, M.R.A., J.A. Crame, J.W. Thomson (eds.), *Geological evolution of Antarctica*: Cambridge, Cambridge University Press, p. 199-204

Collinson, J.W., Pennington, D.C. & A. Kemp 1986. Stratigraphy and petrology of Permian and Triassic fluvial deposits in northern Victoria Land, Antarctica. In Stump, E. (ed.), Geological investigations in northern Victoria Land: *Antarctic Research Series*, 46, Washington, DC, American Geophysical Union: 211-242

Craddock, C., Webers, G.F., Rutford, R.H., Spörli, K.B. & Anderson, J.J. 1986. Geologic map of the Ellsworth Mountains, Antarctica: *Geological Society of America, Map and Chart Series* MC-57

Elliot, D.H. 1991. Triassic-Early Cretaceous evolution of Antarctica. in Thomson, M.R.A., Crame, J.A., Thomson, J.W. (eds.), *Geological evolution of Antarctica*: Cambridge, Cambridge University Press: 541-548

Davy, P. & P. Cobbold 1991. Experiments on shortening of a 4-layer model of the continental lithosphere. *Tectonophysics* 188: 1-25

Flemings, P. B. & T. E. Jordan 1989. A synthetic stratigraphic model of foreland basin development. *Journal of Geophysical Research* 94: 3851-3866

Goodge, J., S. Borg, B. Smith & V. Bennet 1991. Tectonic significance of Proterozoic ductile shortening and translation along the Antarctic margin of Gondwana. *Earth & Planetary Science Letters* 102: 58-70

Grunow, A., D. Kent, & I. Gill 1991. New Paleomagnetic data from Thurston Island: implications for the tectonics of West Antarctica and Weddell Sea opening, *Journal of Geophysical Research* 96: 17935-17954,

Hetenyi, M. 1946. *Beams on elastic foundations*. University of Michigan Press, Ann Arbor, Michigan, 255 pp

Holland, S., D. Sahagian & J. Isbell 1990. Thermal modeling of the Transantarctic foreland basin: Comparison with field data. *Eos* 71: 1599

Karner, G. D. & A. B. Watts 1982. On isostasy at Atlantic-type continental margins, *Journal of Geophysical Research* 87: 2923–2948

Karner, G. D. & A. B. Watts 1983. Gravity anomalies and flexure of the lithosphere at mountain ranges. *Journal of Geophysical Research* 88: 10449–10477

Kirby, S. H. & A. K. Kronenberg 1987. Rheology of the lithosphere: Selected topics. *Reviews of Geophysics* 25: 1219–1244

Laird, M.G. 1963. Geomorphology and stratigraphy of the Nimrod Glacier-Beaumont Bay region, southern Victoria Land, Antarctica. *New Zealand Journal of Geology and Geophysics* 6: 465-484.

Lock, B. 1980. Flat-plate subduction and the Cape Fold Belt of South Africa. *Geology*, 8: 35-39

Mayne, S., E. Nicholas, A.L. Bigg-Wither, J.S. Rasidi, & M.J. Raine 1974. Geology of the Sydney basin- A review, *Australian Bureau of Mineral Resources, Geology and Geophysics, Bulletin* 149: 129 pp

McNutt, M. K., M. Diament & M. G. Kogan 1988. Variations of elastic plate thickness at continental thrust belts. *Journal of Geophysical Research* 93: 8825-8838

McNutt, M. K. & H. W. Menard 1978. Lithospheric flexure and uplifted atolls. *Journal of Geophysical Research* 83: 1206–1212

Peltier, W. R. 1984. The thickness of the continental lithosphere. *Journal of Geophysical Research*, 89: 11303-11316

Sahagian, D.L. & S.M. Holland, in review. On the thermo-mechanical evolution of continental lithosphere. *Journal of Geophysical Research*

Smithson, S. 1972. Gravity interpretation in the Transantarctic Mountains near McMurdo. *Geological Society of America Bulletin* 83: 3437-3442

Stern, T., & U. Ten Brink 1989. Flexural uplift of the Transantarctic Mountains. *Journal of*

Geophysical Research 94: 10315-10330
Stevenson, R. & S. Cloetingh 1991. Some examples and mechanical aspects of continental lithospheric folding. *Tectonophysics* 188: 27-37
Stockmal, G. & Beaumont, C. 1987. Geodynamic models of convergent margin tectonics: the southern Canadian Cordillera and the Swiss Alps. In C. Beaumont & A.J. Tankard (eds.), Sedimentary Basins and Basin-Forming Mechanisms: *Canadian Society of Petroleum Geologists, Memoir* 12: 393-412
Tankard, A.J., M.P.A. Jackson, K.A. Eriksson, D.K. Hobday, D.R. Hunter, & W.E.L. Minter 1982. *Crustal Evolution of southern Africa: 3.8 billion years of earth history*: Springer-Verlag, New York: 523 pp
Veevers, J.J., (ed.) 1984. *Phanerozoic earth history of Australia.* Clarendon Press, Oxford: 418 pp
Watts, A. B. 1978. An analysis of isostasy in the world's oceans 1. Hawaiian-Emperor seamount chain. *Journal of Geophysical Research* 83: 5989–6004
Watts, D.R. & Bramall, A.M. 1981, Palaeomagnetic evidence for a displaced terrane in western Antarctica: *Nature* 293: 638-641
Wilson, T.J., Jurassic faulting and magamatism in the Transantarctic Mountains: implications for Gondwana breakup. *This Volume*
Winter, H. & J. Venter 1971. Lithostratigraphic correlations of recent deep boreholes in the Karoo-Cape sequence. *Proceedings & Papers of the 2nd Gondwana Symp.*, International Union of Geological Sciences Commision on Stratigraphy, Subcommission on Gondwana stratigraphy and Palaeontology : 395-408
Zoetmeijer, R., P. Desegaulx, S. Cloetingh, F. Roure & I. Moretti 1990. Lithospheric dynamics and tectonic-stratigraphic evolution of the Ebro Basin. *Journal of Geophysical Research* 95: 2701-2712

Gondwana Eight, Findlay, Unrug, Banks & Veevers (eds) © 1993 Balkema, Rotterdam. ISBN 90 5410 304 3

Gondwana basins of India occupy the middle of a 7500 km sector of radial valleys and lobes in central-eastern Gondwanaland

Ram C.Tewari
Department of Geology, D.S. College, Aligarh, U.P., India

J.J.Veevers
Australian Plate Research Group, School of Earth Sciences, Macquarie University, N.S.W., Australia

ABSTRACT: During Permian and Triassic times, the Gondwana sequence of India, lateral equivalents in southern Australia and southeastern Africa, and an upslope equivalent in coastal East Antarctica were deposited in a 7500 km sector of valleys and lobes that radiated through 180° of arc from an inferred upland in East Antarctica. Initially glacigenic, subsequently coal-measure sediments, and finally redbeds were deposited during the stage of Pangean sagging. The middle part of the sector, crossed by alluvial valleys, is underlain by brittle Precambrian basement which, during the Late Triassic and Jurassic stage of Pangean rifting, was dismembered into grabens and half-grabens. The outer parts of the sector in southeastern Australia and southern Africa underlain by flexible Paleozoic basement were crossed by sedimentary lobes in ovate basins little deformed by later rifting. In its vast scale, the Gondwana sector of radial drainage is matched in the modern world only by the drainage that radiates north from the Central Asian upland.

INTRODUCTION

Elliot (1975) and Mitra et al. (1979) suggested that the Permian sediments of East Antarctica and peninsular India were deposited in a master basin and that the Indian sediments were derived in part from Antarctica. Kent (1991) suggested that the fill of the Gondwana basins of India was derived from an uplift across coastal East Antarctica and eastern India. This region was underlain by a thermal anomaly in the asthenosphere that emerged in the Cretaceous as the Kerguelen hot spot. By considering a wider frame of reference that includes southern Australia and southeast Africa we suggest that the Gondwana basins of India occupied the middle part of a 7500 km-wide sector of radial valleys and sedimentary lobes that radiated from an upland in the interior of East Antarctica.

GONDWANA PALEO-VALLEYS

The Permian and Triassic alluvium of the Gondwana basins of India was deposited on a northwest paleo-slope in broad paleo-valleys within paleo-uplands. The example of this paleo-geography shown in Figure 1 pertains to the Early Permian deposition of the Karharbari-Barakar Formations in the Godavari, Son-Mahanadi, Damodar, and Rajmahal paleo-valleys within the Chhatisgarh, Faizabad, Chhota-Nagpur, and Monghyr-Saharsa paleo-uplands (M-S in Fig. 1). The downslope to the northwest is indicated (Casshyap and Tewari 1982) by 1) the area of marine flooding (from Bap in the northwest to the Son, Damodar, and Rajmahal paleo-valleys on the northeast) at the end of Talchir deposition, part of the glacio-eustatic Tastubian transgression probably reinforced by a lag in post-glacial isostatic rebound, and 2) the polarized northwestward fluvial paleo-currents and their northward deflection around the eastern side of the inferred Monghyr-Saharsa paleo-upland. Parts of the valleys were bound by scarps as indicated by the proximal (or piedmont) facies of the Karharbari Formation in the Damodar, Mahanadi, and Godavari paleo-valleys. Paleo-valleys locally wider than the present structural basins are suggested by the outliers at Phulbani (20°27'N, 84°15'E) and the Mailaram High, 150 km northwest of Narsapur.

The Karharbari-Barakar paleo-geography is essentially unchanged from that of the preceding Talchir Formation, as indicated (but not shown here) by ice-flow

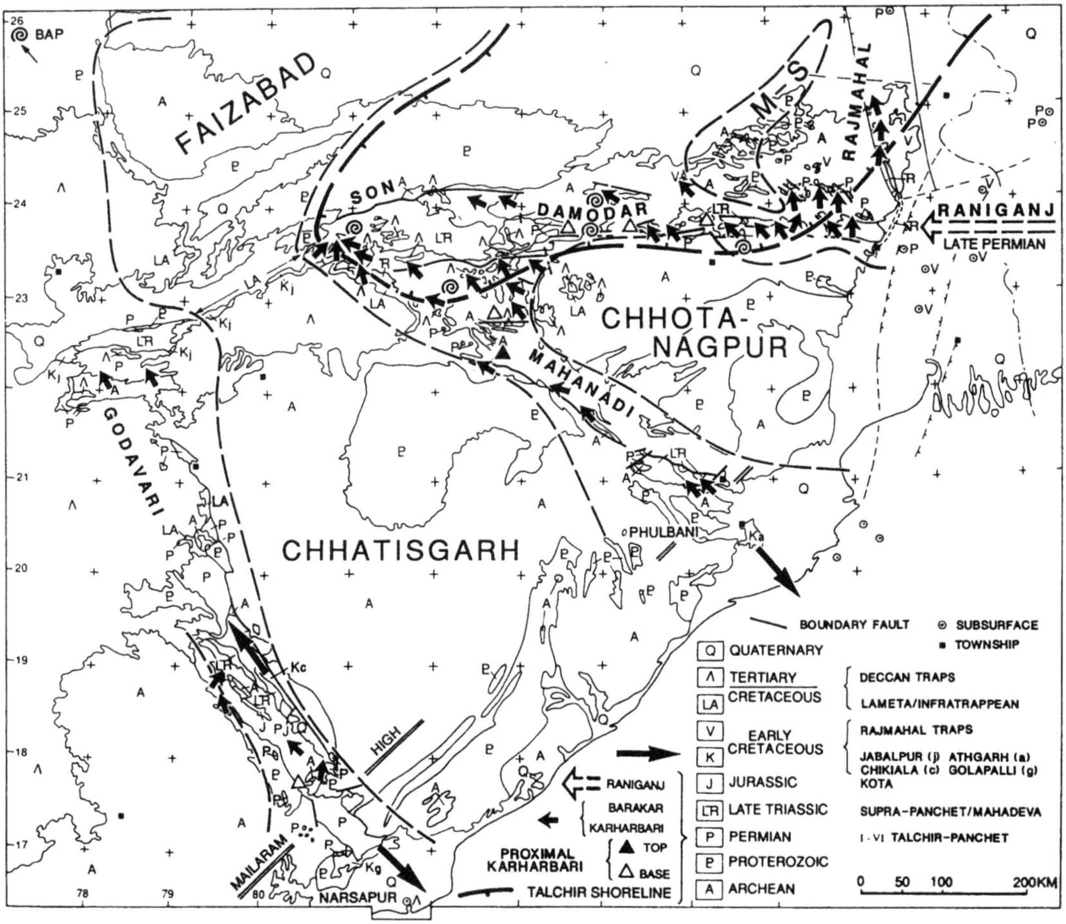

Fig. 1 Gondwana basins of India, showing paleo-currents in the Early Permian Karharbari and Barakar Formations (short full arrows) of palynozones II and [II (Tiwari and Tripathi 1988), Late Permian Raniganj Formation (projected broken arrow) of palynozone V, and Early Cretaceous Athgarh, Chikiala, and Golapalli Formations (long full arrow). Marine invertebrates (coil) indicate the Early Permian (palynozone I or Tastubian) marine ingression towards the end of the deposition of the Talchir Formation that preceded the Karharbari and Barakar Formations. Paleo-valleys (e.g. Damodar) within paleo-uplands (e.g., Chhota-Nagpur) shown on a base map from Roy (1962). Data from Veevers and Tewari (in prep.) include Karharbari-Barakar: Casshyap (1973), Casshyap and Tewari (1984), Khan (1987), Ramanamurthy (1985), and Tewari and Casshyap (1982); proximal or piedmont facies of Karharbari Formation from Mitra (1987); Raniganj: Casshyap and Kumar (1987); Athgarh: Kumar and Bhandari (1973); Chikiala: Dutta and Laha (1979); and Golapalli: Prasada Rao (1976). M-S= Monghyr-Saharsa paleo-upland.

directions from pavement striations and till fabric, and river flow from cross-dip azimuths (Casshyap 1979, Casshyap and Tewari 1982, Casshyap and Srivastava 1988, Chaudhuri and Mondal 1989, Frakes et al. 1975, Smith 1963a,b). The succeeding Late Permian Raniganj Formation had the same paleo-geography, except it was deposited by paleo-currents running from east to west along the Damodar paleo-valley (Figure 1); the paleo-geography of the Permo-Triassic Panchet Formation (Dutta and Laha 1979) and the Late Triassic Supra-Panchet/Mahadeva Formations (Tewari, personal observations 1990), though based on scanty data, reverted to that of the

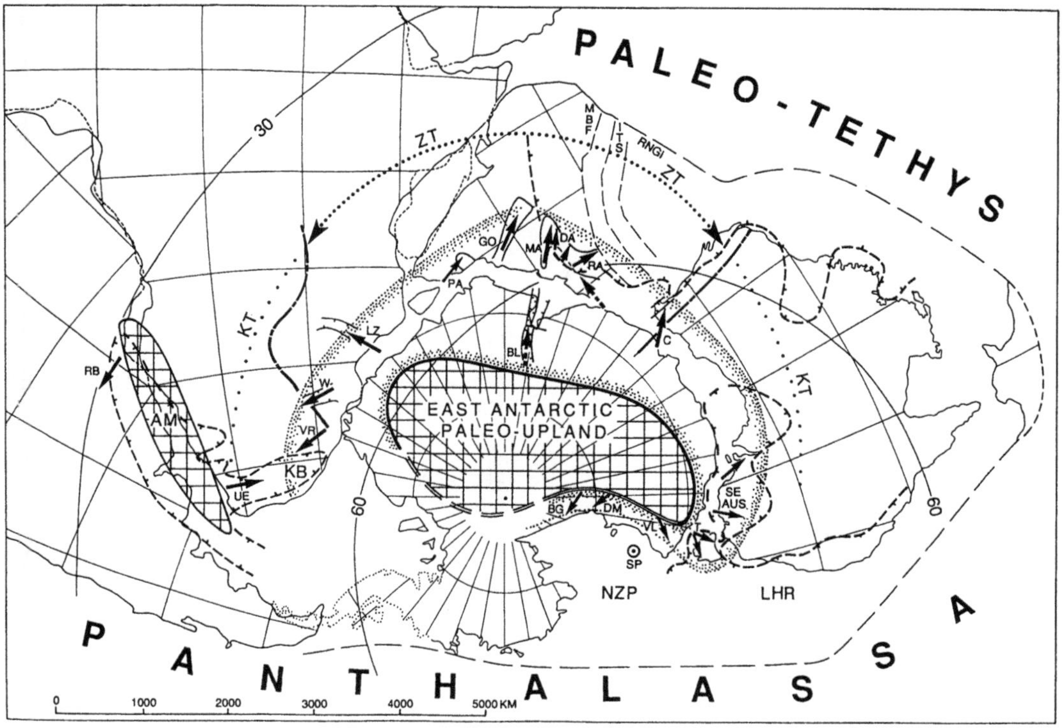

Fig. 2 Reconstruction and paleo-latitude of Early Permian southern Pangea between Paleo-Tethys and Panthalassa (C.McA. Powell and Z.X. Li in prep.). Permian and Triassic fluvial transport azimuth (full arrow) except projected Late Permian Raniganj Formation and Amery Group of Beaver Lake (broken arrow) of radial drainage (stippled) around East Antarctic paleo-upland (cross-hatched), and Early Permian (Tastubian) marine ingressions (heavy barbed line) from Paleo-Tethys and Panthalassa. The azimuth in the Beardmore Glacier area pertains to the Permian only; by the Triassic the paleo-current had reversed (Collinson 1990). Another paleo-upland (cross-hatched) is the post-glacial Atlantic Mountains (AM). BG=Beardmore Glacier (Barrett and Kohn 1975), BL=Beaver Lake (the sediment here is Late Permian and Early Triassic: Webb and Fielding 1990), C=Collie (Wilson 1989), DA=Damodar (Fig. 1), DM=Darwin Mountains (Barrett and Kohn 1975), GO=Godavari (Fig. 1), ITS=Indus-Tsangpo Suture, KB=Karoo Basin, KT=basement of the Karoo terrain (Rust 1975), LHR=Lord Howe Rise, LZ=Lower Zambezi (Rust 1975), MA=Mahanadi (Fig. 1), MBF=Main Boundary Fault, NZP=New Zealand Plateau, PA=Palar, southern India (Mitra et al. 1979), RA=Rajmahal (Fig. 1), RB=Rio Bonito (Castro 1988), RNGI=restored northern edge of Greater India, SE AUS= south-eastern Australia (Veevers 1984, p. 240), SP=South paleo-Pole, T=Tasmania (Hand 1990), UE=Upper Ecca (Ryan and Whitfield 1979), VL=Victoria Land (Walker 1983), VR=Vryheid (Ryan and Whitfield 1979), W=Waterberg (Rust 1975), and ZT=Zambezian or fault-bounded terrain (Rust 1975).

Karharbari and Barakar Formations. It is only with the deposition of the Early Cretaceous Athgarh and Golapalli Sandstones in rivers that flowed southeastward from the Mailaram High (with counter-flowing rivers that deposited the Chikiala Sandstone on the northwest flank - Figure 1) that a new paleo-geography, due to the breakup of India and Antarctica, replaced the old. And it was not until the Cretaceous/Tertiary boundary that the present southeastward slope of peninsular India was established, as shown by eruption of the Deccan Traps from a centre in the Western Ghats extending to the east coast (represented by an erosional outlier at Narsapur [16°28°N, 81°45'E] - Jaeger et al. 1989) . The modern drainage of the Godavari/Krishna Rivers, Mahanadi River, and Koel-Damodar Rivers, which give their names to the Gondwana structural basins, follows that of their Permian and Mesozoic forebears due to the endurance of the intervening uplands; but the flow is reversed because India was detached from uphill Antarctica by seafloor spreading during the Early Cretaceous and then uplifted in the western part of the peninsula during the latest Cretaceous.

7500 KM SECTOR OF RADIAL DRAINAGE

The paleo-slopes of the Gondwana sequence and its lateral equivalents define a 7500 km-wide sector of alluvial paleo-valleys that radiated through 180° of arc from an inferred upland in East Antarctica (Figure 2). Between the upland and the Panthalassan margin lay the Transantarctic foreland basin (Collinson 1990). Permian paleo-currents in Victoria Land (VL in Figure 2), the Darwin Mountains (DM), and the Beardmore Glacier (BG) point obliquely away from the inferred paleo-upland and from a saddle across the axis of the Transantarctic Basin.

A moat about the radial valleys is indicated by the areas flooded by the Tastubian transgression. Anticlockwise from Tasmania, the moat crosses 1) southeastern Australia, 2) southwestern Australia and northern India (Figure 1), 3) southern Africa, and 4) the South Atlantic margin of South America. With the regression of the sea, fluvial deposition resumed except in the Atlantic margin, which became a source upland for the Upper Ecca Group and Rio Bonito Coal Measures.

Over the whole land area shown in Figure 2, individual fingers and lobes of glacigenic (initially non-marine and finally marine), then coal measure sediment (the first deposits at Beaver Lake), and finally redbeds were deposited during the Early Permian to mid-Triassic stage of Pangean sagging (Veevers 1989, in press). The middle (Indian) part of the sector, crossed by alluvial valleys, is underlain by brittle Precambrian basement (Zambezian terrain of Rust 1975), which, during the Late Triassic and Jurassic stage of Pangean rifting, was dismembered into grabens and half-grabens. In contrast, the outer parts of the sector in southeastern Australia and southern Africa underlain by flexible Paleozoic and Precambrian basement (Karoo terrain of Rust 1975) were crossed by sediment lobes that accumulated in ovate flexured basins little deformed by later rifting.

The Gondwana sector of radial drainage and its inferred source are today matched in size (but not in gross tectonic setting) by the arc of drainage (Caspian and Aral Seas and the rivers Ob, Yenisey, Lena, Amur, Hwang-Ho, and Yangtse) about the Central Asian upland. Furthermore, the Gondwanan foreland basin of the Transantarctic Mountains is matched by the present-day foreland basin of the Persian Gulf and the Indus and Ganges Rivers on the other side of the Central Asian upland.

ACKNOWLEDGEMENTS

We thank Dr P.J. Conaghan and an anonymous referee for their comments. Supported by the Australian Research Council and the governments of India and Australia through the India-Australia Science Agreement.

REFERENCES

Barrett, P.J. and Kohn, B.P. 1975. Changing sediment transport directions from Devonian to Triassic in the Beacon Super-Group of south Victoria Land, Antarctica. In Campbell, K.S.W. (ed.): Gondwana Geology, Australian National University Press, Canberra: 15-35.

Casshyap, S.M., 1973. Paleocurrents and palaeogeographic reconstruction of the Barakar (Lower Gondwana) sandstones of Peninsular India. Sedimentary Geology 9: 283-303.

Casshyap, S.M. 1979. Patterns of sedimentation in Gondwana Basins. In Laskar, B. and Raja Rao, C.S. (eds.): Fourth International Gondwana Symposium, Hindustan Publishing Corporation, New Delhi: 525-551.

Casshyap, S.M. and Kumar, A. 1987. Fluvial architecture of the Upper Permian Raniganj coal measure in the Damodar Basin, eastern India. Sedimentary Geology 51: 181-213.

Casshyap, S.M. and Srivastava, V.K. 1987. Glacial and proglacial Talchir sedimentation in Son-Mahanadi Gondwana basin: paleogeographic reconstruction. American Geophysical Union Geophysical Monograph 41: 167-182.

Casshyap, S.M. and Tewari, R.C. 1982. Facies analysis and paleogeographic implications of a late Paleozoic glacial outwash deposit, Bihar, India. Journal of Sedimentary Petrology 52: 1243-1256.

Casshyap, S.M. and Tewari, R.C. 1984. Fluvial models of the Lower Permian coal measures of Son-Mahanadi and Koel-Damodar Valley basins, India. Special Publications of the International Association of Sedimentologists 7: 121-147.

Castro, J.C. 1988. Gondwana sequence of the Parana Basin. 7th Gondwana

Symposium, Sao Paolo, Guidebook A2, 30 pages.
Chaudhuri, S. and Mondal, K. 1989. Eotomariidae gastropod in Talchir Formation of Ramgarh Coalfield and its paleoenvironmental significance. Journal of the Geological Society of India 34: 665-668.
Collinson, J.W. 1990. Depositional setting of Late Carboniferous to Triassic biota in the Transantarctic Basin. In Taylor, T.N. and Taylor, E.L. (eds.): Antarctic Paleobiology, Springer, New York: 1-15.
Dutta, P.K. and Laha, C. 1979. Climatic and tectonic influence on Mesozoic sedimentation in Peninsular India. In Laskar, B. and Raja Rao, C.S. (eds.): Fourth International Gondwana Symposium, Hindustan Publishing Corporation, New Delhi: 563-580.
Elliot, D.H. 1975. Gondwana Basins of Antarctica. In Campbell, K.S.W. (ed.): Gondwana Geology, Australian National University Press, Canberra: 493-536.
Frakes, L.A., Kemp, E.M. and Crowell, J.C. 1975. Late Paleozoic Glaciation: Part VI, Asia. Bulletin of the Geological Society of America 86: 454-464.
Hand, S.J. 1990. Palaeogeography of Tasmania's Permo-Carboniferous glacigene deposits: a tentative reconstruction. Geological Society of Australia, Tasmanian Division (Geology in Tasmania, a generalist's influence - the M.R. Banks Symposium): 47-51.
Jaeger, J.-J., Courtillot, V. and Tapponier, P. 1989. Paleontological view of the ages of the Deccan Traps, the Cretaceous/Tertiary boundary, and the India-Asia collision. Geology 17: 316-319.
Kent, R. 1991. Lithospheric uplift in eastern Gondwana: Evidence for a long-lived mantle plume system? Geology 19: 19-23.
Khan, Z.A. 1987. Paleodrainage and paleochannel morphology of a Barakar river (Early Permian) in the Rajmahal Gondwana Basin, Bihar, India. Palaeogeography, Palaeoclimatology, Palaeoecology 58: 235-247.
Kumar, S. and Bhandari, L.L. 1973. Palaeocurrent analysis of the Athgarh Sandstone (Upper Gondwana), Cuttack District, Orissa (India). Sedimentary Geology 10: 61-75.
Mitra, N.D. 1987. Structures and tectonics of Gondwana Basins of Peninsular India. In Singh, R.M. (ed.): Proceedings of the National Seminar on Coal Resources of India, Department of Geology, Banaras Hindu University, Banaras: 30-41.
Mitra, N.D., Bandyopadhyay, S.K. and Basu, U.K. 1979. Sedimentary framework of the Gondwana sequence of eastern India and its bearing on Indo-Antarctic fit. In Laskar, B. and Raja Rao, C.S. (eds.): Fourth International Gondwana Symposium, Hindustan Publishing Corporation, New Delhi: 37-41.
Prasada Rao, R. 1976. A rational approach for exploration of oil in the Cauvery and Godavari Basins. In Avasthi, D.N. and Varadarajan, K. (eds.): General Proceedings of the workshop on coastal sedimentaries of India south of 18°N latitude, Madras: 31.
Ramanamurthy, B.V. 1985. Gondwana sedimentation in Ramagundam - Mantheni area, Godavari Valley Basin. Journal of the Geological Society of India 26: 43-55.
Roy, B.C., 1962. Geological Map of India, 6th edition, 1:2 000 000, secant conical projection. Geological Survey of India, Calcutta, 11 sheets.
Rust, I.C. 1975. Tectonic and sedimentary framework of Gondwana Basins in Southern Africa. In Campbell, K.S.W. (ed.): Gondwana Geology. Australian National University Press, Canberra: 537-564.
Ryan P.J. and Whitfield, G.G. 1979. Basinal analysis of the Ecca and lowermost Beaufort Beds and associated coal, uranium and heavy mineral beach sand occurences: Geological Society of South Africa Special Publication 6: 91-101.
Smith, A.J. 1963a. Evidence for a Talchir (lower Gondwana) glaciation: Striated pavement and boulder bed at Irai, central India. Journal of Sedimentary Petrology 33: 739-750.
Smith, A.J. 1963b. A striated pavement beneath the basal Gondwana sediments on the Ajay River, Bihar, India. Nature 198: 880.
Tewari, R.C. and Casshyap, S.M. 1982. Paleoflow analysis of Late Paleozoic Gondwana deposits of Giridih and adjoining basins and paleogeographic implications. Journal of the Geological Society of India 23: 67-79.
Tiwari, R.S. and Tripathi, A. 1988. Palynological zones and their climatic inference in the coal-bearing Gondwana of peninsular India. Palaeobotanist 36: 87-101.
Veevers, J.J. (ed.) 1984. Phanerozoic earth history of Australia: Clarendon, Oxford, 418 pages.
Veevers, J.J. 1989. Middle/Late Triassic (230±5 Ma) singularity in the stratigraphic and magmatic history of the Pangean heat anomaly: Geology 17: 784-787.

Veevers, J.J. in press. Gondwana facies of the Pangean supersequence: a review. In Findlay, R.H. (ed.): Gondwana 8 - Assembly, Evolution, and Dispersal. A.A. Balkema, Rotterdam.

Walker, B.C. 1983. The Beacon Supergroup of northern Victoria Land, Antarctica. In Oliver, R.L., James, P.R. and Jago, J.B. (eds): Antarctic Earth Science. Australian Academy of Science, Canberra: 211-214.

Webb, J. and Fielding, C. 1990. Once upon a time in the Prince Charles Mountains: ANARE (Australian National Antarctic Research Expeditions) News 61: 5-6.

Wilson, A.C. 1989. Palaeocurrent patterns in the Collie Coal Measures - the implications for sedimentation and basin models. Professional Papers of the Geological Survey of Western Australia 25: 85-91.

Gondwana Eight, Findlay, Unrug, Banks & Veevers (eds) © 1993 Balkema, Rotterdam. ISBN 90 5410 304 3

Gondwana facies of the Pangean supersequence: A review

J.J.Veevers
Australian Plate Research Group, School of Earth Sciences, Macquarie University, N.S.W., Australia

ABSTRACT: This paper reviews ideas of the Permian-Triassic Gondwana facies as the southern hemisphere component of the Pangean supersequence and its tectonic stages (Veevers, 1990a). I present a new idea on the development of an Early Permian volcanic stage on the mid-Carboniferous Kanimblan fold belt of eastern Australia and Variscan fold belt of Europe.

As a free-standing continental assemblage, Gondwanaland arose in the earliest Palaeozoic at the breakup of a Late Proterozoic supercontinent and lost identity immediately before the first deposit of the Gondwana facies.The late Paleozoic-early Mesozoic Gondwana sequence of India and its equivalents in the present southern continents were deposited as a facies of the Pangean super-sequence after Gondwanaland merged in Pangea in the late Paleozoic. The Gondwana facies owes its distinctive character not to its isolation in Gondwanaland, but to its mid- to high-latitude depositional environment in the Southern Hemisphere.

In common with the rest of the Pangean supersequence, the Gondwana facies reflects three tectonostratigraphic stages generated by the Pangean heat anomaly. Stage 1 (320-286 Ma or Late Carboniferous) is the stratigraphic lacuna that resulted from the high-standing Pangean platform that initially blocked the escape of heat from the interior. Stage 2 (286-230 Ma or Permian and Early-Middle Triassic) is the first Pangea-wide accumulation of platform sediments in sags and grabens generated by the initial loss of Pangean heat; the parts of Pangea that had undergone terminal deformation in the mid-Carboniferous - Kanimblan eastern Australia and Variscan Europe - were distinguished by the eruption of thick and widespread volcanics. Stage 3 (230-160 Ma or Late Triassic and Early-Middle Jurassic) is the Pangea-wide rifting that dissipated further heat. Compressive events in Africa, India, and Australia in the mid-Triassic or Carnian (230±5 Ma) correlate with the initial accretion to Eurasia of Cimmeria and the North and South China blocks that marks the maximum coalescence in Pangea. Incipient dispersal by rifting therefore marks a singularity in earth history at the change from a waxing to a waning Pangea.

INTRODUCTION

The Gondwana sequence of peninsular India (Figure 1, IX) comprises Permian glacigene deposits, initially marine and then nonmarine, overlain by coal measures containing the Glossopteris flora, succeeded by Triassic-Jurassic nonmarine facies containing the Dicroidium and Ptillophyllum floras (Sastry et al. 1977). Suess (1885) interpreted the restricted occurrence of the Glossopteris flora in peninsular India, southern and middle Africa, and Madagascar as due to a continuous landmass which he called Gondwana-Land. Later authors, including Wegener (1915) and Du Toit (1937), extended the landmass to include the rest of Africa, Arabia, South America, Antarctica, and Australia, and compressed its name to Gondwanaland. The distinction between sequence (Gondwana) and supercontinent (Gondwanaland), blurred by Schwarzbach (1981), was reinstated by Sengor (1983).

Wegener's (1915) Late Carboniferous reconstruction showed one continent only, Pangea, continuous across the provinces of Laurasia and Gondwanaland, a view kept in sight by paleobiogeographers (e.g., Anderson 1981; Ross and Ross 1985) and paleomagnetists (e.g., Irving 1983). Many geologists seemed to overlook Pangea by giving the unintended or unwitting impression that Gondwanaland was independent during the late Paleozoic and early Mesozoic deposition of the Gondwana sequence. This led to the paradox that the late Paleozoic and early Mesozoic heyday of Gondwanaland, denoted by the deposition of the distinctive Gondwana facies, was the very time during which Gondwanaland and the other continents ceased to exist as independent entities by their union in Pangea.

PANGEAN SUPERSEQUENCE

Figure 1 shows the platform sequences of various continents (column IX) set against the time span of Pangea. Sloss and Speed (1974) saw "nothing unique in Gondwana sediments or demanding of special circumstances other than those typical of Absaroka strata on this [North American] continent. Rather, postglacial Gondwana rocks and the tectonic milieu of their deposition in peninsular India..., South Africa..., Australia..., and Brazil... bear striking parallels with Absaroka sediments and their conditions of deposition in North America."

The parallel between the two regions of Pangea affected by terminal deformation - eastern Australia and Europe - is equally striking. In scale, sequence, magma type, and age, the extensional province of Europe (Ziegler 1988, 1989, 1990) matches that of eastern Australia, especially in its tectonic setting of deposition after the terminal Variscan (or Hercynian) deformation that accompanied the final collision with Laurentia and in the extent and variety of magma types that were emplaced across the Fennoscandian craton (Oslo Rift) and Variscan Fold Belt in the latest Carboniferous and earliest Permian (Stephanian-Autunian). According to Ziegler (1989, p. 53-55) "the Variscan fold belt became transected by a system of dextral post-orogenic wrench faults that caused its collapse [cf. the dextral shear system in Eastern Australia during the Late Paleozoic (Evans and Roberts 1980)]. Activity along the fault system was associated with crustal fracturing and widespread [1.5×10^6 km^2] volcanics of a highly variable chemistry .... [cf. 0.5×10^6 km^2 coeval mafic-felsic volcanics at the base of the Bowen-Gunnedah-Sydney Basin of Eastern Australia]. Syntectonic deposits consist of continental, often coarse, coal-bearing series" [cf. Reids Dome Beds of the Denison Trough of Eastern Australia]. Furthermore, S-type granitoids continued to be emplaced in the Moldanubian section of the Variscan orogen (Finger and Steyrer 1990)[cf. Bundarra and Hillgrove Suites of New England - Shaw and Flood 1981]. Despite their many differences in plate configuration, paleo-latitude and paleo-climate - Europe was in the equatorial zone, Eastern Australia near the South Pole - the two regions have a common sedimentary and magmatic history which I interpret as reflecting extension of hot foldbelt basement immediately after terminal deformation during the first release of Pangean heat.

GONDWANA FACIES

The Gondwana sequence of India and its equivalents in Australia (Innamincka), southern Africa (Karoo), Brazil (Delta and Delta-A)(Figure 1, IX), and Antarctica (Victoria Group; Elliot, 1975) constitute the Austral facies of the Pangea super-sequence. This facies is distinguished by initially marine and then nonmarine glacigene strata at the base, reflecting mid to high southern latitudes (Veevers and Powell 1987), and succeeded by strata, again mainly nonmarine, that contain a provincial biota separated from Boreal Pangea on the east by the equatorial oceanic gulf of Neo-Tethys and on the west by the tropical land zone across northern South America/Africa and southern North America/Europe (Figure 2). The glacigene sediment and endemic biota of the Gondwana province were isolated from the rest of Pangea not by wide continental separation but by the tropical continental zone on the west and by the oceanic gulf of Neo-Tethys on the east (Anderson 1981). That the tropical land barrier was not absolute is shown, for example, by the close similarities between the Late Triassic Maleri tetrapods of India and those of Morocco, Europe, and Texas (Chatterjee 1987).

GONDWANALAND

As a free-standing entity, Gondwanaland spanned 250 m.y. from the fragmentation of a Late Proterozoic supercontinent (Bond et al. 1984) at or about the start of the Paleozoic (570 Ma) to the coalescence of Gondwanaland and Baltica/Laurentia in the mid-Carboniferous (320 Ma). Except for the early loss of Cimmeria (Figure 2), the Gondwanaland assemblage remained intact within Pangea until incipient dispersal by rifting in the Carnian (230 Ma) and definitive dispersal by seafloor spreading in the Late Jurassic. The original

Gondwanaland elements of Pangea (South America, Africa, Antarctica, India, Australia) became independent from the Late Jurassic and Cretaceous. New combinations arose during the Cenozoic merging of South and North America, and of India and Africa/Arabia in Eurasia.

The supercontinent Gondwanaland spanned the Cambrian to the mid-Carboniferous. It commenced with breakup of a Late Proterozoic supercontinent and ended by merging in Pangea immediately before the first deposit of the Gondwana facies. Compared with the rest of the Pangean supersequence, the mid-Carboniferous to mid-Jurassic Gondwana facies owes its distinctive character not to its isolation in Gondwanaland but to its mid- to high-latitude depositional environment in the Southern Hemisphere.

TECTONOSTRATIGRAPHIC STAGES OF PANGEA

According to Veevers (1990a), the earth alternates in a supercycle 400 million years long from a single continent (Pangea) and ocean (Panthalassa) with an icehouse climate to many continents and oceans with a greenhouse climate. The supercycle is driven by the heat that accumulates beneath the insulator of Pangea, and the greenhouse is made by the excess CO_2 vented from the mantle during the faster rate of plate activity. Anderson (1982) found that the Atlantic/African high of the residual geoid coincides in location with the central part of Pangea 200 Ma ago located from Morgan's (1981) hotspot frame (Figure 2), and interpreted the residual geoid high as a relic of the heat trapped beneath the insulator of Pangea.

In common with the rest of the Pangean supersequence, the Gondwana facies reflects three tectonostratigraphic stages generated by the Pangean heat anomaly:

stage 1 (320-286 Ma or Late Carboniferous) is the stratigraphic lacuna of the platform that resulted from the high-standing Pangean platform that initially blocked the escape of heat from the interior;

stage 2 (286-230 Ma or Permian and Early-Middle Triassic) is the first Pangea-wide accumulation of sediments in sags and grabens between uplifts generated by the initial differential loss of Pangean heat; and

stage 3 (230-160 Ma or Late Triassic and Early-Middle Jurassic) is the Pangea-wide rifting that represents an accelerating loss of heat in rifts.

Stage 1 (lacuna) is general over the Gondwanaland part of the Pangean platform. Stage 2 is represented by the Talchir through Panchet succession in India, by the Dwyka through Beaufort succession in southern Africa, by the Delta succession in Brazil, by the the Victoria Group in Antarctica, and by the Early Innamincka sequence in Australia. Stage 3, Late Triassic and Jurassic sediment deposited (except in Antarctica) unconformably on Middle Triassic and older deformed rocks, is represented by the Supra-Panchet succession of India (Mitra 1987), by the Stormberg and Drakensberg succession of southern Africa (Halbich et al. 1983), and by the Middle and Late parts of the Innamincka sequence in Australia (Veevers 1984). The mode of pre-Carnian compressive deformation that generated the unconformity was as follows: a) in the Damodar River basins of India folding and transpressive faulting accompanied lateral movement of fault blocks (Mitra 1987); b) in southern Africa, the southern Karoo Basin was folded by the terminal thrusting of the Cape Fold Belt, dated by the $^{40}Ar/^{39}Ar$ method as 230±3 Ma or Carnian (Halbich et al. 1983); c) in eastern Australia, the foreland basin was folded during the terminal overthrusting of the New England Orogen (Murray et al. 1989); and d) in the northern Canning Basin of northwest Australia, the folding accompanied transpressive movement of fault blocks during a lacuna bracketed by Carnian deposits above and Ladinian deposits below (Veevers 1990b). The pinpointing of these compressive events to the mid-Triassic or Carnian (230±5 Ma) reflects a compressive stress field throughout Pangea generated by the accretion to Pangea of Cimmeria and the North and South China blocks (Sengor 1984) that marks the maximum coalescence in Pangea. Incipient dispersal during the coeval initial stage of rifting therefore marks a singularity in earth history at the change from a waxing to a waning Pangea.

SUMMARY

1. Independent Gondwanaland lasted between the breakup of a Late Proterozoic supercontinent and the late Paleozoic coalescence of the Pangean supercontinent.
2. The Gondwana facies started to be deposited after Gondwanaland lost its identity by merging in Pangea.
3. The late Paleozoic and early Mesozoic Gondwana facies was deposited in the distinctive climatic environment of the Gondwanaland province as part of the

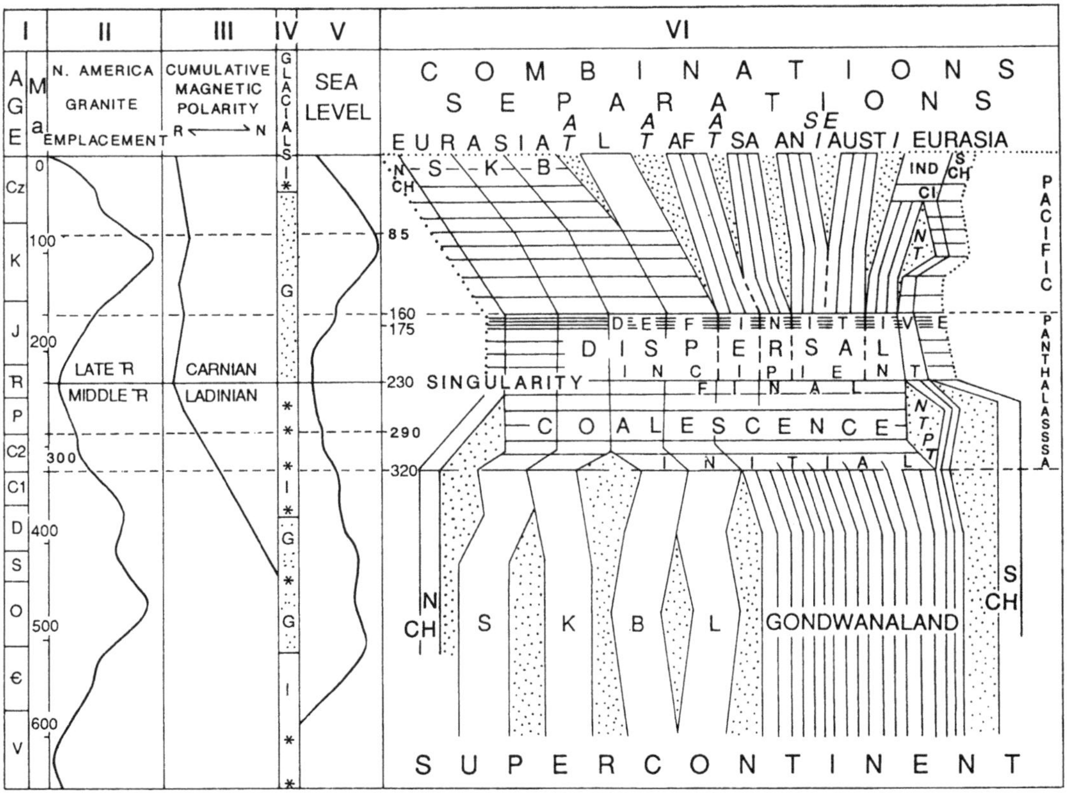

Fig. 1 Phanerozoic tectonic indices, combinations and separations of the continents and oceans, the Pangean heat store, and platform sequences. I: Timescale, after Palmer (1983). II: North American granite emplacement, from Fischer (1984); maxima to the right. III: Cumulative magnetic polarity (Nicolaysen 1985) (N normal, R reversed), normal polarity intervals represented by lines sloping to the right and reversed intervals by left sloping lines so that the dominantly reversed Permian-Carboniferous field slopes to the left, mixed polarity in latest Permian to mid-Cretaceous slopes alternately right and left, and finally right in the Cretaceous normal interval (118-84 Ma), and then slopes left during the dominantly reversed mixed polarity (since 84 Ma). IV: Glacial occurrences (asterisks) with alternating icehouse (I) and greenhouse (G) states, and V: Sea level (maxima to the right), both from Fischer (1984). VI: Continental and oceanic combinations and separations, after Briden et al. (1974), Ziegler et al. (1979), Bond et al. (1984), and Sengor (1984). Gondwanaland and subsequent fragments marked with vertical lines, Pangea and Eurasia with transverse lines, and oceans with stipple; dotted lines alongside Eurasia join up on either side. Continents: AF-Africa, AN-Antarctica, AUST-Australia, B-Baltica, CI-Cimmeria, IND-India, K-Kazakhstania, L-Laurentia, N CH-north China, S-Siberia, SA-South America, S CH-south China, oceans (italic letters) AT-Atlantic (north, central, south), I-Indian, NT-NeoTethys, PT-PaleoTethys, SEI-Southeast Indian. VII: Number of continents (full line) and incipient continents (dotted line) from 1 on left to 7 on right, and combinations. VIII: Accumulation and dissipation of Pangean heat anomaly in stages 1 through 5, rate of loss of heat (dotted line) that drives plate tectonics, emplaces volcanics, and initiates basin structure, and residual heat or balance (full line). Late Paleozoic and Mesozoic flood lavas: A-European (Wopfner 1984, Ziegler 1988) and eastern Australian (Veevers 1984, p. 240), B-Siberian traps (Renne and Basu 1991), C-Amazon (Mosmann et al. 1986),

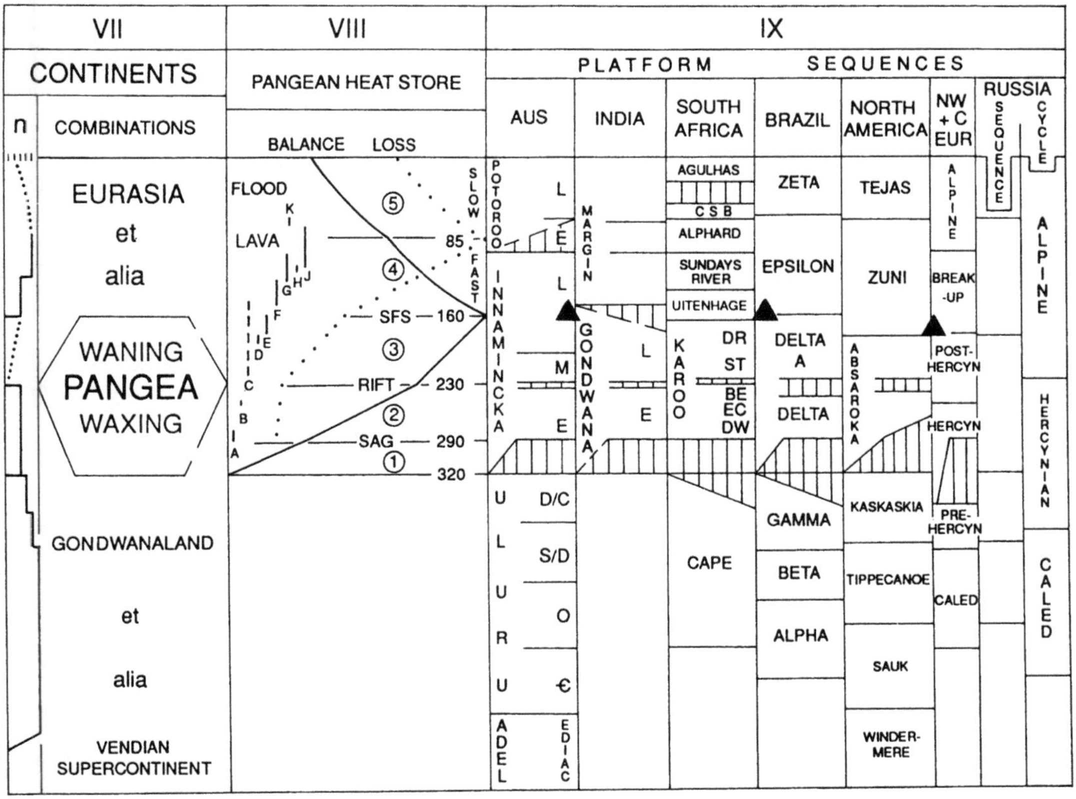

D-Karoo (Bristow and Saggerson 1983), E-Transantarctic and Tasmanian (Kyle et al. 1981, Schmidt and McDougall 1977), F-Serra Geral (Schobbenhaus 1984), G-Pacific ridge-crest (Watts et al. 1980), H-Rajmahal (Baksi 1988), J-Pacific mid-plate (Schlanger et al. 1981), K-Deccan (Jaeger et al. 1989). IX: sequences of continental platforms. Note in particular lacunas of Late Carboniferous and Middle/ Late Triassic (vertical hachured area), and Middle Jurassic onset of drift (triangles). AUS-Australia (Veevers 1984): onset of deposition on the platform of Gondwanaland is dated at 290 Ma (epi-Carboniferous or Gzelian, East Australian palynological stage 2); ADEL-Adelaidean, C-Cambrian, DC-Late Devonian/Early Carboniferous, E-Early, EDIAC-Ediacarian, L-Late, M-Middle, O-Ordovician, SD-Silurian/Middle Devonian. India: the Gondwana sequence, divided by the mid-Triassic lacuna into two parts, E-Early and L-Late, and the overlying sequence of the eastern margin. Southern Africa (Dingle et al. 1983): BE-Beaufort, CSB-Cape St Blaize, DR-Drakensberg, DW-Dwyka, EC-Ecca, ST-Stormberg. Brazil (Soares et al. 1978). North America: sequences from Sloss (1988); in the Permian Basin of Texas and New Mexico, the earliest Permian (286 Ma) Wolfcamp Series overlaps the Central Basin Platform (Sloss 1988, p. 265), the base of the Absaroka II subsequence (268 Ma) is marked by greatly accelerated rates of subsidence (Sloss 1988, p. 39). Northwest and Central Europe: main phases of plate boundary reorganization, from Ziegler (1988); in the Permian basins of Europe, subsidence was initiated in the Oslo (volcanic) Graben and renewed in the half grabens between Norway and Greenland at the same time (290 Ma, Stephanian B or Gzelian) as sagging in the south was accompanied by the eruption of flood lavas and the deposition of the Rotliegend and later successions (Ziegler 1988); HERCYN-Hercynian, CALED-Caledonian. Russian Platform: sequences (unnamed) as in North America, from Sloss (1976), cycles from Ronov et al. (1969). From Veevers (1990a), with the permission of Elsevier Science Publishers.

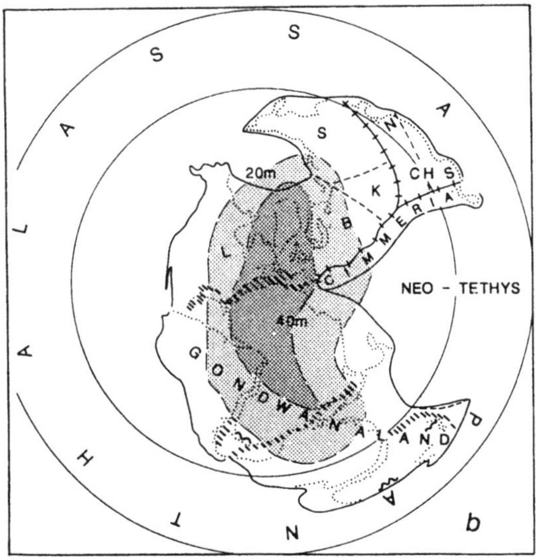

Fig. 2 Anderson's (1982) Early Jurassic (200 Ma) configuration of Pangea based on Morgan's (1981) hotspot-based reconstruction, superimposed on the present (0 Ma) Atlantic-African residual geoid high. Lambert equal-area projection of the entire sphere (Briden et al. 1974) such that the front hemisphere is bounded by the inner circle and the back hemisphere lies within the outer ring. The paleo-equator bisects Pangea and Tethys. Abbreviations of continents given in the caption of Figure 1.

Pangean supersequence that accumulated in basins generated by the Pangean tectonic regime.

ACKNOWLEDGEMENTS

This synthesis derives from studies supported by the Australian Research Council.

REFERENCES

Anderson, D.L. 1982. Hotspots, polar wander, Mesozoic convection and the geoid. Nature 297: 391-393.

Anderson, J.M. 1981. World Permo-Triassic correlations: Their biostratigraphic basis. In Cresswell, M.M. and Vella, P. (eds.): Gondwana Five, A.A. Balkema, Rotterdam: 3-10.

Baksi, A.K. 1988. reply to comment on Critical evaluation of the age of the Deccan Traps, India: Implications for flood-basalt volcanism and faunal extinctions: Geology 16: 758-759.

Bond, G.C., Nickerson, P.A., Kominz, M.A. 1984. Breakup of a supercontinent between 625 Ma and 555 Ma: New evidence and implications for continental histories. Earth and Planetary Science Letters 70: 325-345.

Briden, J.C., Drewry, G.E., Smith, A.G. 1974. Phanerozoic equal-area world maps. Journal of Geology 82: 555-574.

Bristow, W., Saggerson, E.P. 1983: A general account of Karoo vulcanicity in southern Africa. Geologische Rundschau 72: 1015-1059.

Chatterjee, S.A. 1987: A new theropod dinosaur from India with remarks on the Gondwana-Laurasia connection in the Late Triassic. In McKenzie, G.D. (ed.): Gondwana six: Stratigraphy, sedimentology, and paleontology. American Geophysical Union, Washington D.C., Monograph 41, 183-189.

Dingle, R.V., Siesser, W.G., Newton, A.R. 1983. Mesozoic and Tertiary geology of southern Africa. A.A. Balkema, Rotterdam, 375 pages.

Du Toit, A.L. 1937. Our wandering continents. Oliver and Boyd, Edinburgh, 366 pages.

Elliot, D.H. 1975. Gondwana Basins of Antarctica. In Campbell, K.S.W. (ed.): Gondwana Geology. Australian National University Press, Canberra, 493-536.

Evans, P.R. and Roberts, J. 1980. Evolution of central eastern Australia during the late Palaeozoic and early Mesozoic: Journal of the Geological Society of Australia 29: 325-340.

Finger, F. and Steyrer, H. P. 1990. I-type granitoids as indicators of a late Paleozoic convergent ocean-continent margin along the southern flank of the central European Variscan orogen. Geology 18: 1207-1210.

Fischer, A.G. 1984. The two Phanerozoic supercycles. In Berggren, W.A. and van Couvering, J.A. (eds.): Catastrophes and Earth history. Princeton University Press, Princeton: 129-150.

Halbich, I.W., Fitch, F.J. and Miller, J.A. 1983. Dating the Cape orogeny. In Sohnge, A.P.G. and Halbich, I.W. (eds), Geodynamics of the Cape Fold Belt. Geological Society of South Africa Special Publication 12: 149-164.

Irving, E. 1983. Fragmentation and assembly of the continents, mid-Carboniferous to present. Geophysical Surveys 5: 299-333.

Jaeger, J.-J., Courtillot, V. and Tapponier, P. 1989. Paleontological view of the ages of the Deccan Traps, the Cretaceous/Tertiary boundary, and the India-Asia collision. Geology 17: 316-319.

Kyle, P.R., Elliot, D.H., Sutter, J.F. 1981. Jurassic Ferrar Supergroup tholeiites from the Transantarctic Mountains, Antarctica, and their relationship to the initial fragmentation of Gondwana. In Cresswell, M.M. and Vella, P. (eds.), Gondwana Five: A.A. Balkema, Rotterdam: 283-287.
Mitra, N.D., 1987. Structures and tectonics of Gondwana Basins of Peninsular India, in Singh, R.M. (ed.), Proceedings of the National Seminar on Coal Resources of India: Department of Geology, Banaras Hindu University, 30-41.
Morgan, W.J. 1981. Hotspot tracks and the opening of the Atlantic and Indian Oceans. In Emiliani, C. (ed.): The oceanic lithosphere, v. 7 of The Sea. Wiley-Interscience, New York: 443-487.
Mosmann, R., Falkenheim, F.U.H., Goncalves, A., Nepomuceno, F. 1986. Oil and gas potential of the Amazon Paleozoic basins. In Halbouty, M.T. (ed.): Future petroleum provinces of the world: American Association of Petroleum Geologists Memoir 40: 207-241.
Murray, C.G., Scheibner, E., Walker, R.N. 1989. Regional geological interpretation of a digital coloured residual Bouguer gravity image of eastern Australia with a wavelength cut-off of 250 km. Australian Journal of Earth Sciences 36: 423-449.
Nicolaysen, L.O. 1985. On the physical basis for the extended Wilson cycle, in which most continents coalesce and then disperse again. Geological Society of South Africa Transactions 88: 562-580.
Palmer, A.R. 1983. The Decade of North American Geology 1983 geologic time scale. Geology 11: 503-504.
Renne, P.R. and Basu, A.R. 1991. Rapid eruption of the Siberian Traps flood basalts at the Permo-Triassic boundary. Science 253: 176-179.
Ronov, A.B., Migdisov, A.A. and Barskaya, N.V. 1969. Tectonic cycles and regularities in the development of sedimentary rocks and paleogeographic environments of sedimentation of the Russian Platform (an approach to a quantitative study). Sedimentology 13: 179-212.
Ross, C.A. and Ross, J.R.P. 1985. Carboniferous and Early Permian biogeography. Geology 13: 27-30.
Sastry, M.V.A., Acharyya, S.K., Shah, S.C., Satsangi, P.P., Ghosh, S.C., Raha, P.K., Singh, G. and Ghosh, R.N. 1977. Stratigraphic index of Gondwana formations of India. Geological Survey of India Miscellaneous Publication 36: 170 pages.
Schlanger, S.O., Jenkyns, H.C., Premoli-Silva, I. 1981. Volcanism and vertical tectonics in the Pacific Basin related to global Cretaceous transgressions. Earth and Planetary Science Letters 52: 435-449.
Schmidt, P.W. and McDougall, I. 1977. Palaeomagnetic and potassium-argon studies of the Tasmanian Dolerites. Journal of the Geological Society of Australia 24: 321-328.
Schobbenhaus, C. (ed.) 1984. Geologia do Brasil. Departamento Nacional da Productacao Mineral, Brasilia: 501 pages.
Schwarzbach, M. 1981. Gondwana and Gondwanaland. Geologische Rundschau 70: 497-498.
Sengor, A.M.C., 1983. Gondwana and Gondwanaland: A discussion. Geologische Rundschau 72: 397-400.
Sengor, A.M.C. 1984. The Cimmeride orogenic system and the tectonics of Eurasia. Geological Society of America Special Paper 195: 82 pages.
Shaw, S.E. and Flood, R.H. 1981. The New England Batholith, Eastern Australia: geochemical variations in time and space. Journal of Geophysical Research 86: 10530-10544.
Sloss, L.L. 1976. Areas and volumes of cratonic sediments, western North America and eastern Europe. Geology 4: 272-276.
Sloss, L.L. (ed.) 1988. Sedimentary Cover - North American Craton, U.S. Geological Society of America, Boulder, Colorado, The Geology of North America, Volume D-2, 506 pages.
Sloss, L.L. and Speed, R.C. 1974. Relationships of cratonic and continental-margin tectonic episodes. Society of Economic Paleontologists and Mineralogists Special Publication 22: 98-119.
Soares, C.S., Landim, P.M.B. and Fulfaro, V.J. 1978. Tectonic cycles and sedimentary sequences in the Brazilian intracratonic basins. Geological Society of America Bulletin 89: 181-191.
Suess, E. 1885. Das Antlitz der Erde. Temsky, Vienna: 779 pages.
Veevers, J.J. (ed.) 1984. Phanerozoic earth history of Australia. Clarendon, Oxford, 418 pages.
Veevers, J.J. 1990a. Tectonic-climatic supercycle in the billion-year plate-tectonic eon: Permian Pangean icehouse alternates with Cretaceous dispersed-continents greenhouse. Sedimentary Geology 68: 1-16.

Veevers, J.J. 1990b. Development of Australia's post-Carboniferous sedimentary basins. Petroleum Exploration Society of Australia Journal 16: 25-32.
Veevers, J.J. and Powell, C.McA. 1987. Late Paleozoic glacial episodes in Gondwanaland reflected in transgressive-regressive depositional sequences in Euramerica: Geological Society of America Bulletin 98: 475-487.
Watts, A.B., Bodine, J.H. and Ribe, N.M. 1980. Observations of flexure and the geological evolution of the Pacific Ocean basin. Nature 283: 532-537.
Wegener, A. 1915. Die Entstehung der Kontinente und Ozeane. Vieweg, Braunschweig: 367 pages.
Wopfner, H. 1984. Permian deposits of the Southern Alps as product of initial alpidic taphrogenesis. Geologische Rundschau 73: 259-277.
Ziegler, A.M., Scotese, C.R., McKerrow, W.S., Johnson, M.E. and Bambach, R.K. 1979. Paleozoic paleogeography. Annual Review of Earth and Planetary Science 7: 473-502.
Ziegler, P.A. 1988. Evolution of the Arctic-North Atlantic and the Western Tethys. American Association of Petroleum Geologists Memoir 43: 198 pages.
Ziegler, P.A. 1989. Evolution of Laurussia, a study in Late Palaeozoic plate tectonics. Kluwer, Dordrecht: 102 pages.
Ziegler P.A, 1990. Geological Atlas of Western and Central Europe. 2nd edition. Shell Internationale Petroleum Maatschappij B.V., The Hague: 239 pages, 56 enclosures.

Dispersal

Gondwana Eight, Findlay, Unrug, Banks & Veevers (eds) © 1993 Balkema, Rotterdam. ISBN 90 5410 304 3

The determination of the positions of the Gondwana continents back to the Ordovician using a method of plate path plotting

T. Frank Lee
Melbourne, Vic., Australia

ABSTRACT: If the inner core of the Earth: (i) does not lie at the centre of the Earth but lies in the Northern Hemisphere at the intersection of the 6° latitudinal plane and the line joining the North and South Magnetic Poles; and (ii) it exerts expansional forces on the remainder of the Earth's body; then it is possible to derive by trial and error a simple rule which allows plotting of the paths of the continental plates back through Phanerozoic time. The only assumption required is the assumption that the rotation/magnetic axes relationship has been approximately constant during this time and that accurately determined and dated, not-too-widely spaced palaeomagnetic poles for each plate are used. Because there are several gaps in palaeomagnetic pole data only approximate plotting can be done at present for the Gondwana continents.
Ten maps are presented showing the probable distribution of the Gondwana continents at different times from Middle Ordovician to Upper Cretaceous. A study of these maps allows an explanation of the distribution of the Palaeozoic glaciation across Gondwanaland.

INTRODUCTION

If the Earth's core were displaced from the geocentre, if the core exerted an outward radial pressure on passive surrounding material, and if the remainder of the Earth acted as a homogeneously strained body, the stress field produced by such a configuration could well be reflected by major physiographic and/or structural features on the Earth's surface. If so, major physical features of the Earth's surface should be related to one another.

Study has shown (Lee, 1989) that if the inner core is assumed to be at a particular offset position relative to the Earth's geocentre then a mathematical relationship does exist between major physiographical and tectonic features. This is because these features coincide with, or are simply mathematically related to, the axes and planes of stress derived from such a structural configuration. The following hypothesis was developed from these considerations.

HYPOTHESIS

The Earth consists of two spheres, a "core sphere" (consisting of the inner core) and a "mantle sphere" (the remainder of the Earth including the outer core). The north-south principal stress axis of the core sphere does not coincide with that of the mantle sphere but instead lies along the Earth's north-south magnetic axis (the line joining the north and south magnetic poles). The centre of the core sphere is located within the northern hemisphere at the point where the 6° latitudinal plane cuts the core sphere axis (Fig. 1a).

The core sphere exerts forces on the mantle sphere such that the mantle sphere behaves under the usual stress mechanics of an unconfined body (Hills, 1963), plus forces emanating from the centre of the core sphere to produce conical planes of stress about both halves of the two principal axes of stress which lie in the horizontal plane of the figure of stress. (Not shown in Fig. 1 as they do not enter the following discussion.)

It was accepted that a mathematical relationship between surface features,

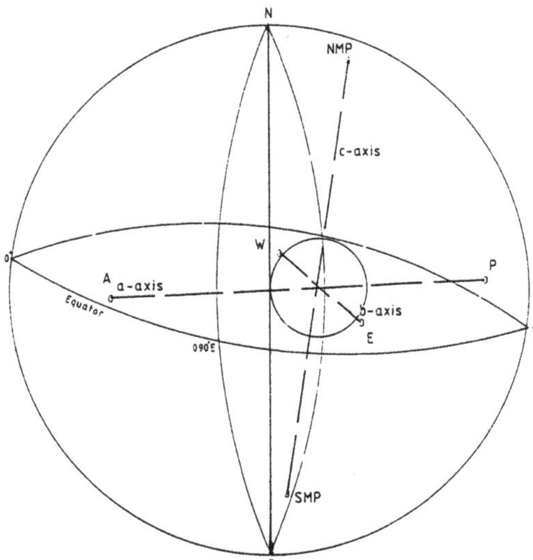

Fig.1a Sketch showing the location of the displaced inner core and its three principal stress axes (dash lines).

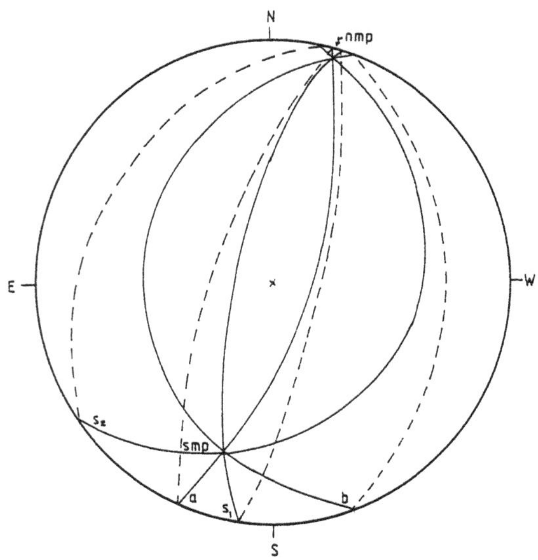

Fig.1b Stereographic projection of the four vertical planes.

even derived under the conditions postulated, did not necessarily prove that the inner core was offset. However, if the structure as defined were correct and it affected the Earth's surface then other physical properties of the Earth such as shape, seismic behaviour, magnetic properties, etc. should show coincidence with or symmetry to the determined stress axes and planes.

While studying these other properties, consideration of the polar wander paths of the North American and European continental plates during Mesozoic-Tertiary times suggested that these paths might also be related to the stress planes determined from the hypothesis. Testing showed that there is indeed a simple, common relationship when it is assumed: (a) published palaeomagnetic poles give the Earth's palaeo- rotation poles (e.g. McElhinny, 1979.); and (b) the relative positions of the magnetic poles to the rotation poles of the Earth have been approximately fixed throughout geological time. This relationship was tested for eleven other continental plates and has been found to hold (Lee, 1989; Lee, 1992). This study gave the following rule: the movements of continental plates throughout at least Phanerozoic time have been such that they have always moved parallel to one of the four vertical stress planes of the model; shown in Fig. 1b.

From the above rule it is a simple matter to develop a method (see Appendix 1) of plotting a continental plate's path back through time from the Present using palaeomagnetic poles, provided it is assumed that the Earth's outer shell has not moved relative to the Earth's rotation axis and that the plate does not rotate about its own centre (there are a few, recognisable exceptions to this latter assumption, e.g. Italy before the Cretaceous).

Where the intervals between ages of palaeomagnetic poles are greater than about 25 My, the rule cannot be applied accurately and an approximation is necessary. Empirical testing has shown (Lee, 1989) that generally the result using the approximation is not too inaccurate, but the accuracy of the palaeomagnetic pole values becomes especially important here.

Using this model for each of the Gondwana continents it has been possible to determine the plates' relationships to one another from Ordovician to Present times by first calculating the paths of the plates and then moving plate templates step by step back along the paths plotted on a globe. Figures 2 to 11 are plots of the relative positions of the plates for progressively later periods beginning at the Middle Ordovician. They are used to describe the original form and the gradual breakup of Gondwana as given in the next section.

The a and b, and the s_1 and s_2 planes shown on Fig. 1b were determined using magnetic pole positions of 65°S139°E and 73°N100°W. To assist those who may wish to draw the four vertical stress planes

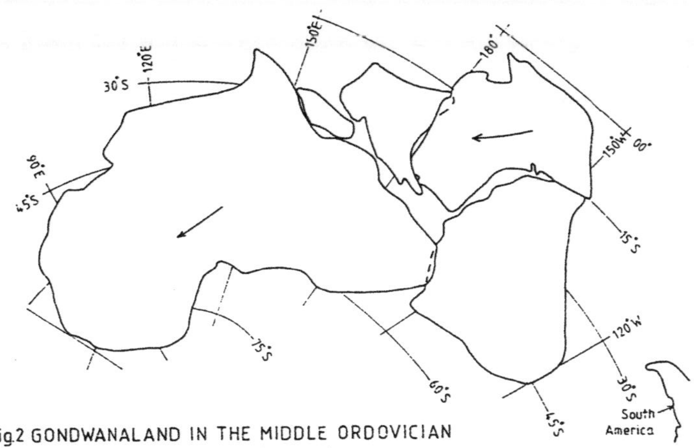

Fig.2 GONDWANALAND IN THE MIDDLE ORDOVICIAN immediately prior to break-up.

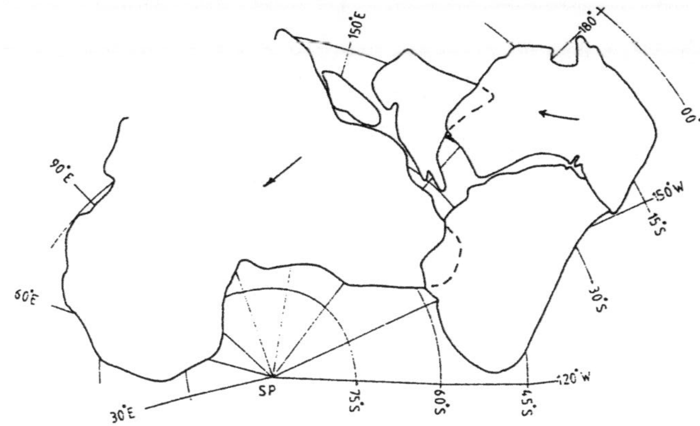

Fig.3 Estimated positions of the plates at the base of the Upper Ordovician.

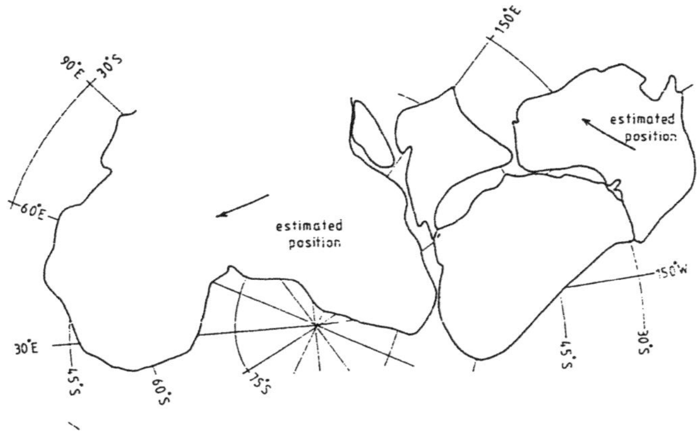

Fig.4 TOP OF ORDOVICIAN

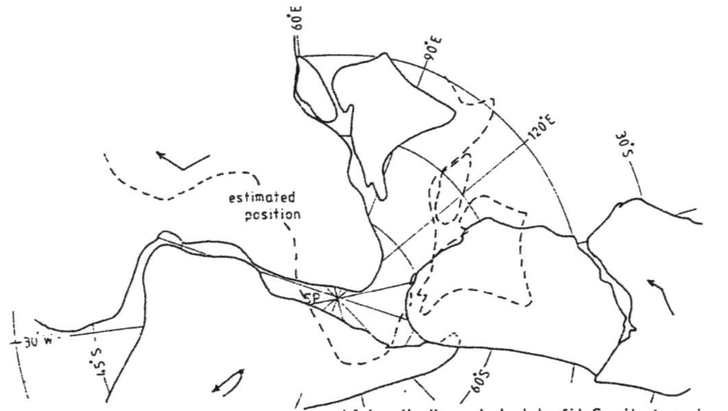

Fig.5 LOWER DEVONIAN

Africa/India rotated to fit South America.
One of the two available, very poor and greatly different Devonian poles (neither used) for Africa suggests an African position between the two plate positions shown but partly overlapping South America.

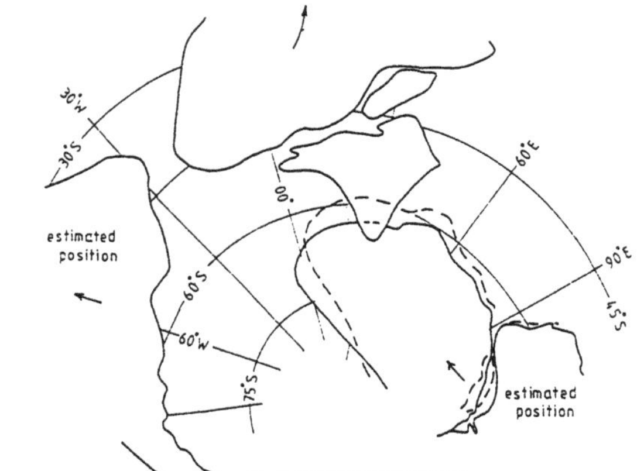

Fig.6 CARBONIFEROUS

Dashed lines show plotted outlines of slightly adjusted plates.

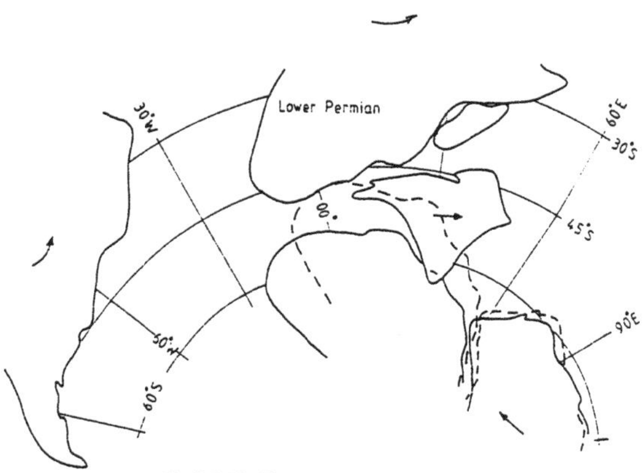

Fig.7 PERMO-CARBONIFEROUS

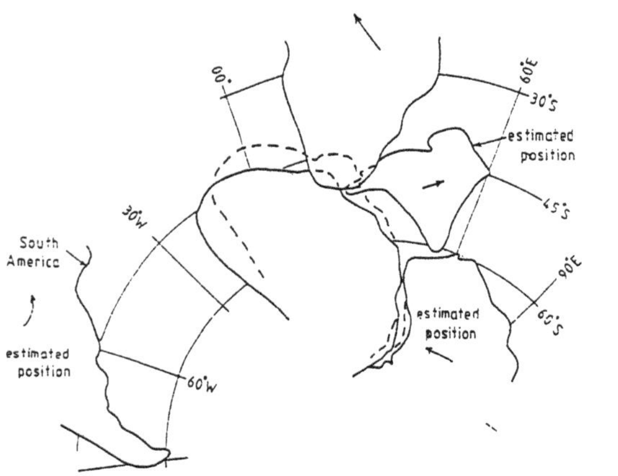

Fig.8 MIDDLE PERMIAN

on a globe it is noted here that because the core sphere is offset from the geographical centre of the Earth the planes form small circles. To draw them on the surface of a sphere one should draw circular curves through the magnetic poles and the following points:

for a-plane	6°N027°E	6°N155°W
b-plane	23½°N124°E	10½°S073°W
s_1-plane	18°N166°E	6°S022°W
s_2-plane	20°N074°E	5°S116°W

Because of doubtful palaeomagnetic poles for the various plates, and because of the scarcity of palaeomagnetic data of some plates at various times, particularly during the middle and early Palaeozoic, approximations must come into the plots of the paths of the Gondwana plates back into Ordovician times. In particular the paths of Eastern Antarctica and Africa are poorly defined. However there are both geological and palaeomagnetic data which shows that Australia and Eastern Antarctica were one plate before the beginning of the Tertiary, that Madagascar was part of the African plate before Upper Permian times, and India was part of the African plate before Upper Carboniferous times.

A DESCRIPTION OF THE BREAK-UP OF GONDWANALAND AS DEDUCED USING THE RULE

Despite the deficiency in palaeomagnetic pole data, plotting the plate paths of the Gondwana continents using the rule reveals a simple sequence of continental break-up. The break-up is described as follows.

In Middle Ordovician times Africa, Eastern Antarctica, Australia, India, and Madagascar formed a single continent. South America did not form a part of this continent unless it was connected through Tierra del Fuego by another plate.

Towards the end of the Middle Ordovician a split occurred between the Australo-Antarctica part of Gondwana and the remainder of the Gondwana plate. Both pieces began to move approximately southwestwards but parallel to different stress planes and at different speeds. Until Africa and South America collided in Lower Devonian times, this movement resulted not only in considerable distance travelled but also in relative rotation about their southern ends.

On the collision of Africa and South America there was an almost immediate rebound of South America and Africa-India-Madagascar but Australia-Antarctica continued onwards. It is possible that part of Africa amalgamated with South America. It is almost certain that the movements of these two plates were such that the relative movement between them was a sliding separation, so that final contact in possibly Middle Carboniferous times was of the Cape area of Africa with what is now the easternmost part of South America.

The continuing, now northwesterly, movement of Australia-Antarctica resulted in Antarctica striking glancingly what is now the eastern part of India, but which was then still part of the African plate, in about Middle Carboniferous times. The resultant torsional force (because the African plate was moving anticlockwise northwards) separated India and probably freed Madagascar which, due to the movement of Africa, remained in contact with Africa.

By Permo-Carboniferous times the pinching of India between Africa and Antarctica resulted in a hinging separation of India and Africa and a movement of India towards the Cape end of Africa. The hinge for rotation was the southern end of the Indian plate. As Antarctica passed India, and Africa continued to turn anticlockwise, the Indian plate received a northeasterly thrust from the African plate and began its long movement towards the Northern Hemisphere.

In Middle Permian times Africa ceased its anticlockwise movement and began

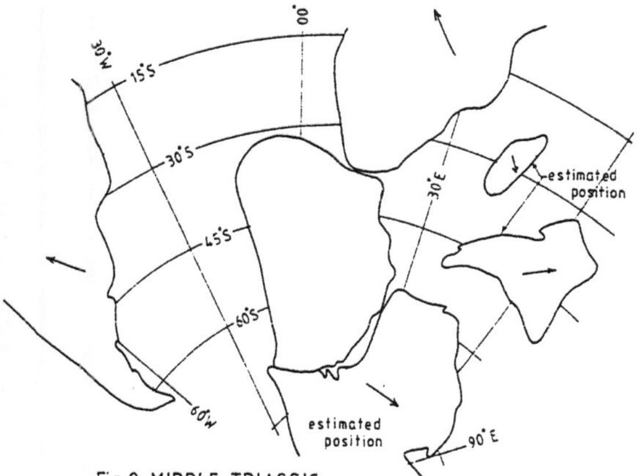

Fig.9 MIDDLE TRIASSIC

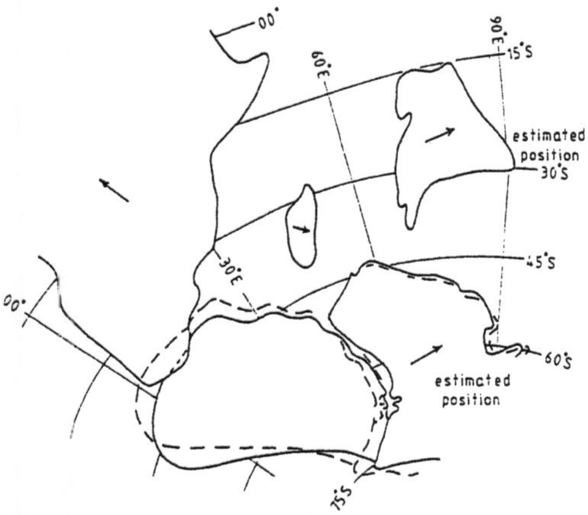

Fig.10 MIDDLE JURASSIC

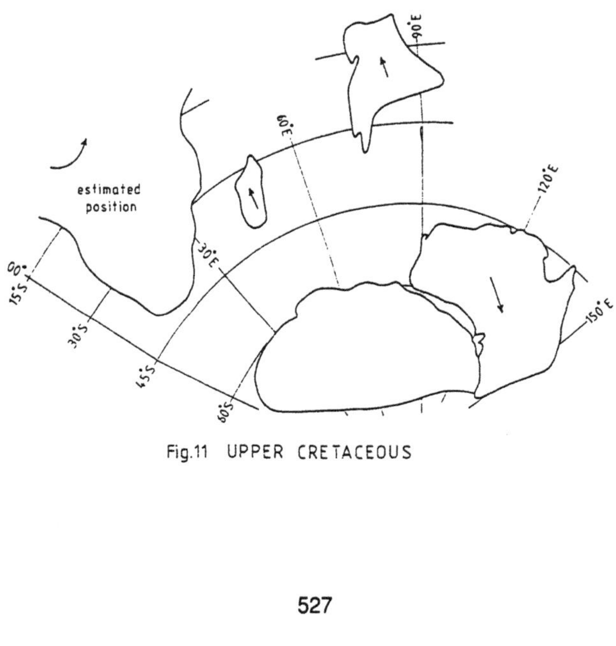

Fig.11 UPPER CRETACEOUS

to move northwest. It is at this time that Madagascar became a separate plate (see Lee,1992 for supporting geological evidence).

Australia-Antarctica continued to move approximately northwest until the then northern coast of Antarctica was near 30°S latitude in Middle Triassic times. It was still in contact with the southern tip of Africa (see p.316 of Lee, 1992 for supporting evidence).

The southern Atlantic began to open at the end of the Carboniferous and, because of the relative movements of South America and Antarctica, remained much less than half the width it is at present until Middle Triassic times. Then, with the reversal of direction of Australia-Antarctica the ocean rapidly widened and began to take its present form.

In approximately Middle Triassic times Australia-Antarctica reversed its past 250my general southwesterly movement and has kept this reversed direction (with zig-zags) to the present day. However, because of the relative movements of the various plates, Africa and Antarctica were still in contact during the Middle Jurassic, and Madagascar and India were still relatively near Australia. By Upper Jurassic times Africa and Antarctica had separated and the individual plates, because of their relative movements, rapidly parted. By Upper Cretaceous times Madagascar had attained a position only 15° of latitude south of its present position since when it has moved slowly northwards to its present position.

Australia-Antarctica was certainly still a single plate in Middle Jurassic times and had certainly broken apart some time before 40my ago, but the absence of palaeomagnetic poles for Eastern Antarctica in this time interval prevents determining the time of separation using the rule presented. Based on the rule presented and the assumption of a uniform speed of Antarctica between its positions in Middle Jurassic and Middle Tertiary times, it is estimated that separation occurred at about 70my. This agrees well with estimates given by other methods (see Paine, 1990).

References

Hills, E.S. 1963. *Elements of Structural Geology*. Norwich: Methuen & Co Ltd.

Klootwijk, C.T. 1984. A preview of Indian Phanerozoic palaeomagnetism: implications for the Indo-Asia collisions. *Tectonophysics 105:* 331-353.

Lee,T.F. 1989. *Structural analysis of plate movements as an aid to mineral exploration: an asymmetric Earth? - a hypothesis.* MApSc. thesis, Royal Melbourne Institute of Technology. (unpublished).

Lee,T.F. 1992. *An Asymmetrically Structured Earth? A Hypothesis; and its use in solving various geological problems.* Lee, Melbourne.

McElhinny,M.W. 1979. *Palaeomagnetism and Plate Tectonics.* Cambridge: Cambridge University Press.,

Paine,A.G.L(ed) 1990 *Australia, Evolution of a Continent.* Sydney: Alken Press Pty. Ltd. for Bureau of Mineral Resources, Canberra.

APPENDIX: THE METHODS OF DETERMINING A PLATE PATH AND THE MOVING OF A PLATE TO ITS POSITION IN AN EARLIER GEOLOGICAL AGE

All calculations have been done using a sphere of the Earth, a pencil, a ruler, and a pair of compasses, and this is the method which will be described. The motion of the Indian plate will be used as illustration.

(a) The palaeomagnetic poles of India (Klootwijk, 1984) back to 55-65 My are listed below. Their palaeocore poles must be determined using a stereographic net. (i.e. each pole given is assumed to be the rotation South Pole but the South Magnetic Pole (SMP) for each must be determined.)

Date My	Palaeomagnetic pole	Palaecore pole
(i)5-22.5	78½°S114½°E	54°S134°E
(ii)25-35	72½°S136°E	48°S139°E
(iii)35-45	67½°S132°E	43°S137°E
(iv)45-55	57½°S121½°E	34°S131°E
(v)55-65	40½°S120°E	17°S129°E
etc.		

(b) The a, b, s_1, and s_2 planes (Fig. 1b) are drawn on a globe of the Earth.

(c) The palaeocore pole for time (i) of the list is plotted on the sphere. It will be found to lie at a distance of 11 equatorial degrees from SMP in the direction of the East Indies, 1 equatorial degree west of the b-plane (in shorthand - 11°bE).

For the movement between the palaeomagnetic pole for time (i) and the palaeomagnetic pole for time (ii): plot the (ii) palaeocore pole; move palaeocore pole (i) to the SMP; move palaeocore pole (ii) an identical amount in direction and distance. It will be found that when plaeocore pole (i) lies at the SMP then palaeocore pole (ii) lies ½°W of the s_1-plane and 7° in the direction of the Pacific Ocean. This can be written in shorthand as 7°s_1E.

For the movement between the palaeomagnetic pole for time (ii) and the palaeomagnetic pole for time (iii): plot the (iii) palaeocore pole; move palaeocore pole (ii) to the SMP; move palaeocore pole (iii) an identical amount; measure the distance between the poles and the location of the latter pole relative to the nearest vertical plane.

The results can be tabulated as:

From My	To	Fit	Plane	Distance	Comments
SMP	5-22.5	Good	bE	11°	1°W of plane.
5-22.5	25-35	Good	s_1E	7°	½°W of plane.
25-35	55-65	Good	bE	32°	On plane.

(It will be found that the poles for times (iii), (iv), and (v) plot consecutively onto the b-curve in the direction of the East Indies - designated bE.)

The path of the Indian plate is now known back to the 55-65 My period. To move the Indian plate back to its former positions:

(d) A template of the Indian plate, with its centre marked, is placed over the present position of India on the globe of the Earth given in (b). The centres of the b- and s_1- planes circles are marked on the sphere surface.

If the plate were near the b-plane circle centre, to find the 5-22.5 position it would be sufficient to draw a line on the globe surface from the b circle centre to the plate centre, measure off 11° with a protractor in the direction of the SMP, mark a point an equal distance from the circle centre on this radius, and move the plate so that the plate centre lies over the point. This gives the position of the plate at av(5-22.5)My.

In most cases, as here, the plate centre is too far from the plane circle centre to measure the angle with a protractor. The method, using the above 11°bE as an example, is to draw with a pair of compasses an arc through the plate centre. The angle of distance (11°) is multiplied by the number sin(90°.ac/ab); where ac = distance from the b circle centre to the centre of the plate over the sphere surface, ab = distance from the b circle centre to the b-plane over the sphere surface. This gives the distance to be measured in equatorial degrees along the arc towards the SMP for location of a point x. This distance subtends an angle of 11° at the b circle centre. Move the plate along the arc to point x. The plate is now at its av(5-22.5) My position.

The construction is repeated step by step.

Gondwana Eight, Findlay, Unrug, Banks & Veevers (eds) © 1993 Balkema, Rotterdam. ISBN 90 5410 304 3

Structural development of Tanzanian Karoo Basins and the break-up of Gondwana

Helmut Wopfner
Universität zu Köln, Germany

ABSTRACT: The Karoo basins of Tanzania show abundant evidence for repeated syndepositional tectonism during Permian and Early Triassic times. These movements were essentially caused by rifting. Specific facies associations resulting from these movements include alluvial fan and braid-plain deposits, coarsening upwards sequences and scarp foot fan wedges interfingering with lacustrine sediments. Apart from initial movements responsible for the onset of Karoo deposition, major tectonic pulses occurred during the Sakmarian, in the early Artinskian, in the late Kazanian and about the Permo-Triassic boundary. Each of these events was associated with sedimentary overlap, indicating progressive widening of individual rift systems. The event near the Permo-Triassic boundary was associated with detachment mechanisms. It caused the foundering of the eastern parts of the Karoo rift system and introduced the transformation from the intracratonic rift to the pericratonic basin stage. Karoo deposition was terminated by a Middle to Late Triassic tectonic event. The orientation of Karoo grabens was northeast . Northwest trending structures formed as connecting faults between major graben systems. Similar rift systems controlled the sedimentation of Karoo rocks of western Madagascar. In the restored position of the island opposite Tanzania they form the eastern counterpart of the Tanzanian grabens. Both systems resulted from the opening of a major trans-tensional chasm on the Tethyan margin of Gondwana in late Palaeozoic times.

1 BASIN DEVELOPMENT

The formation of the Karoo succession in Tanzania commenced with the deposition of glacigene sediments in the Late Carboniferous. This depositional event was preceeded by a prolonged period of erosion and subaerial exposure extending at least from the Late Proterozoic (Bucoban) to the Late Carboniferous, a time span almost equalling that of the total Phanerozoic (450 to 500 Ma). In places where the Karoo succession rests on metamorphic rocks of the Early Proterozoic Ubendian or Usagaran foldbelts, as in the Ruhuhu and Selous basins, this interval may have been even longer (Fig. 1).

It may be reasonable to conclude that the pre-Karoo landscape was one of low relief and extensive peneplains. The creation of receptacles for the accumulation of Karoo sediments exceeding 3000 m in thickness required considerable adjustments which could have been achieved only by tectonic movements. This has already been suggested by, amongst others, Janensch (1927) and Dixey (1937). A purely glacially overdeepened morphology, causing the basin formation, as for instance envisaged by McKinlay (1954) would appear inadequate to accomodate the observed sediment accumulations of Tanzanian Karoo basins.

Within the Ruhuhu Basin, which contains the classical reference

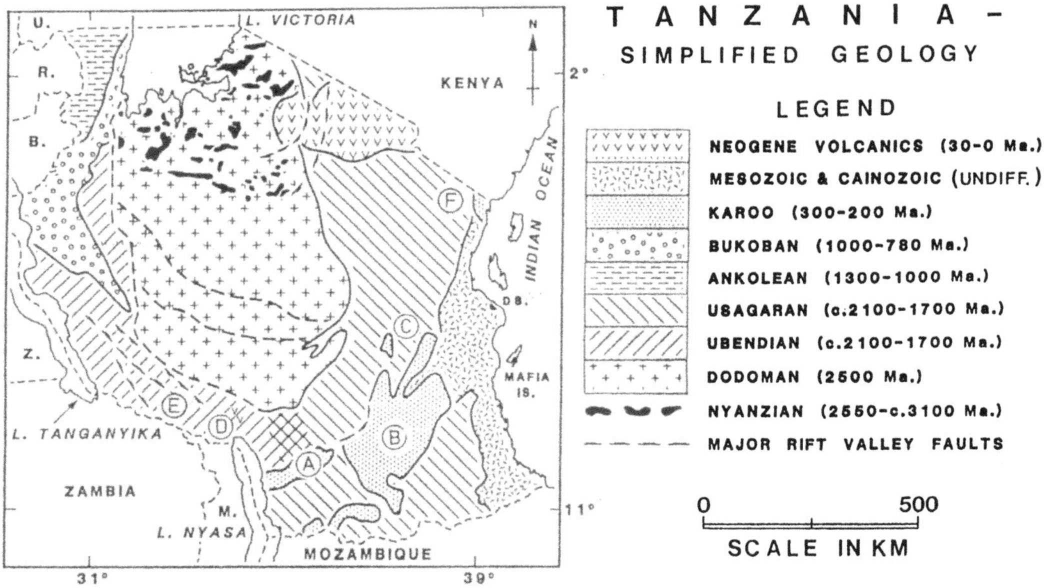

Figure 1: Distribution of major Karoo basins in relation to major Precambrian tectono-stratigraphic units of the Tanzanian craton. Basins are identified by capital letters: (A) Ruhuhu; (B) Selous; (C) Mikumi; D) Ilima; E) Galula-Rukwa; (F) Tanga.

section of the Karoo succession in Tanzania, the Permo-Carboniferous comprises four depositional sequences, each commencing with coarse, rudaceous deposits and gradually fining upwards to fine-grained siliciclastic rocks. A fifth sequence, consisting of a coarse, supermature quartz sandstone at the base and a thick succession of individually fining upwards cyclothems was laid down during the Early Triassic. The stratigraphy of the Ruhuhu Basin has been discussed recently by Kreuser et al. (1990) and that of the Selous Basin by Wopfner & Kaaya (1991). At the moment, the old symbols K1 to K8 identifying, in ascending order, the rock-units of Tanzanian Karoo basins (c.f. McKinlay, 1954) are used parallel with formation names still awaiting formal definition.

2 STRUCTURAL HISTORY

The development of Tanzanian Karoo basins was characterized by frequent changes in basin geometry, rates of subsidence and attitude. Depocentres with continuous subsidence were bordered by intermittently uplifted regions, whilst basin margins experienced sedimentary overlap. Thick, more or less continuous successions are often in proximity to condensed sections, frequently interrupted by hiati or unconformities. This reflects the complex structural history of Tanzanian Karoo basins. The relationship between tectonic movements and the development of the Karoo succession is shown in the schematic time-space diagram of Figure 2.

Indirect evidence for the existence of tectonic movements prior to and during the deposition of the glacigene Idusi Formation (K1) has been presented by Wopfner & Kreuser (1986) and this has been confirmed by geophysical data. Seismograph reflection surveys across the Selous Basin, for instance, show substantial increases in time intervals between basal Karoo reflections at the downthrown side of basement faults (Kajato, 1986).

Substantial tectonic movements occurred near the boundary between the Idusi Formation (K1) and the succeeding coal bearing Mchuchuma Formation (K2). Within the deeper

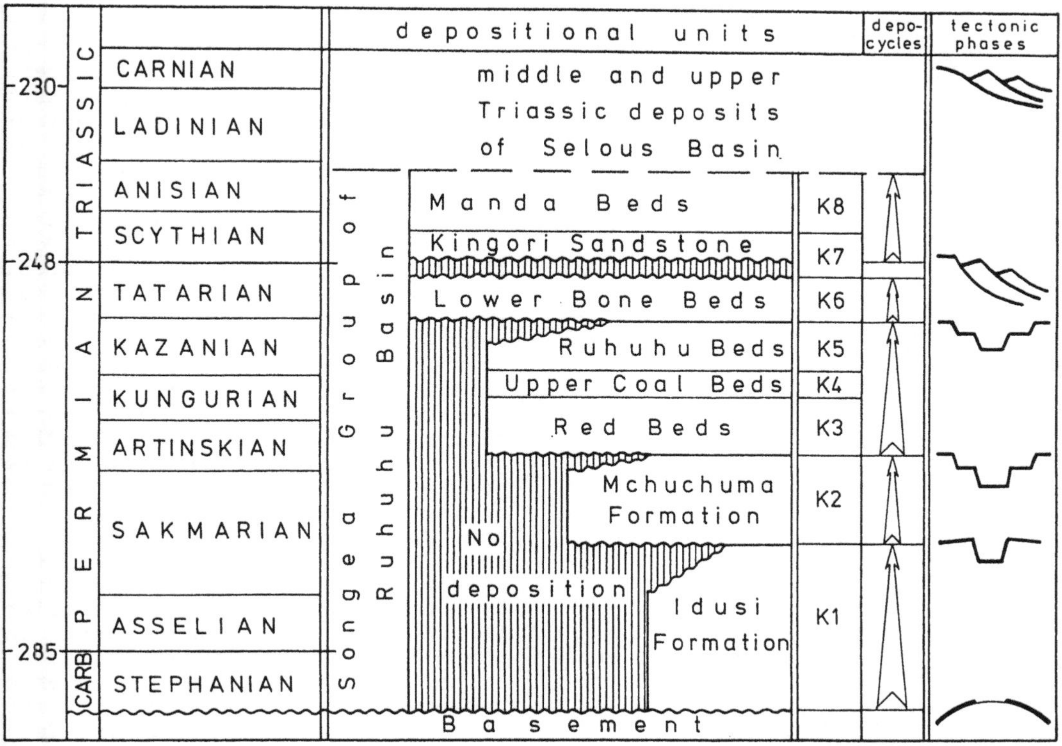

Figure 2: Generalized time-space diagram of the Karoo succession in south-western Tanzania related to syn-depositional tectonic events. Tectonic mechanisms are indicated by simplified sketches in right hand column. They show in ascending order, upwarp, rifting, rift expansion and detachment. The symbols K1 to K8 (K for Karoo) denote the old lithostratigraphic subdivision used by previous workers (e.g. McKinlay, 1954).

parts of the basins, the basal, coarse and arkosic fluviatile sandstone of the latter rests with apparent conformity on black lacustrine lutites of the upper Idusi Formation. In many instances, however, the basal K2-sandstone overlaps onto basement or rests with an angular unconformity on truncated sections of the Idusi Formation. Lag-boulders of up to 25 cm diameter may be incorporated within the base of the K2 sandstone. Good examples of that unconformity occur in the upper reaches of the Chizi River near Galula, on the southern margin of the Rukwa Basin (Wopfner & Kaaya, in press) and in the drainage area of the Mkapa River in the eastern part of the Ruhuhu Basin. From the structural geometry of the latter area, Wopfner (1990) demonstrated that rift movements were responsible for the unconformable relationship between K1 and K2. This is shown in Figure 3.

In some areas post-Idusi movements apparently continued into, or were rejuvenated during the coal depositional cycle of the Mchuchuma Formation (K2) as evidenced by coarsening upward cycles exposed along Liweta River, about 17 km south-east of Manda. Each of these cycles consists of a coal seam overlain successively by siltstone, sandstone and breccia. The breccia consists of angular clasts of up to 12 cm diameter and bears all features of a typical scarp foot fan deposit. Seven such cycles, varying in thickness from about 2 to 5 m, have been identified.

In the bore hole M1 which was drilled to basement near Mkapa, the basal sandstone of the K3 succes-

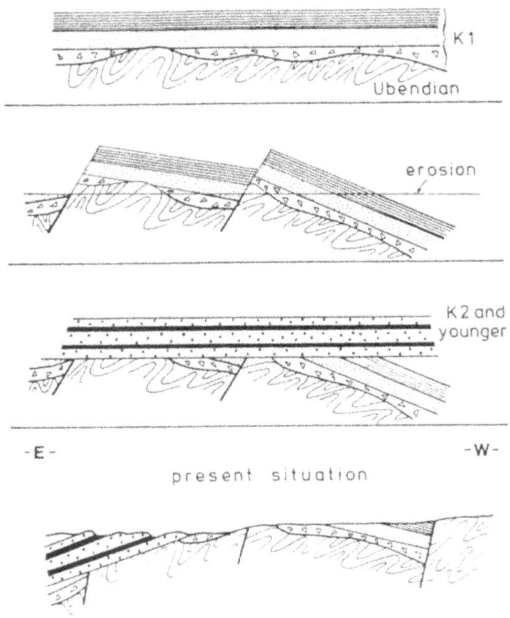

Figure 3: Reconstruction of the sequence of events which led to the present distribution of K1/K2 strata in the drainage area of Mkapa River in the eastern Ruhuhu Basin (after Wopfner, 1990).

sion (Scarp Sandstone) rests on a truncated and intensely mottled section of Idusi Formation. The coal-bearing Mchuchuma Formation is missing and this, together with the mottling, indicates prolonged aerial exposure prior to the deposition of the K3 sandstone (P.M. Semkiwa, pers. comm. 1991).

The next tectonic pulse is evidenced by marked changes in depositional environment and basin geometry near the end of the Early Permian. It coincided with the beginning of the K3 cycle. Sedimentation commenced with coarse clastics, ranging from very coarse arkoses to cobble conglomerates and was succeeded by red and green feldspar conglomerates and sandstones. These sediments also overlap the older beds and rest on basement, for example in the area north of Manda. In fact, these overstepping late Artinskian Red Beds, resting directly on basement, are the oldest exposed deposits in the Mikumi area and in the Tanga Basin (Fig. 1).

In many instances the basal clastics of the K3 depositional cycle are composed of up to 70 % of fresh microcline and plagioclase clasts with diametres up to 5 cm. The lack of rounding, the poor sorting of the sediments and cut and fill structures (Wopfner, 1989; 1990; Wopfner & Kaaya, in press) indicate that the material was dumped across the edge of a rising source area. The linear distribution of the wedges of coarse clastics and the overlapping relationship demonstrate that elevation of the source area was caused by rifting. Towards the end of this sequence the deposition of fine sediments, including coal and lacustrine mudrocks, of K4 and K5, indicate a prolonged period of tectonic tranquility in many of the Karoo basins in Tanzania. It was also during this time that the proximity of the sea is indicated by the appearance of an estuarine to brackish water bivalve fauna in the Mikumi region (Kreuser, 1983).

The fourth depositional sequence commenced near the end of the Kazanian, when the lacustrine-deltaic sedimentation of the Ruhuhu Beds (K5) was interrupted by the deposition of rudaceous scarp foot fans, alluvial fans and flood plain splays, the latter frequently containing reptilian remains. Again, substantial widening of the existing graben systems is indicated by depositional overlap. These movements were most pronounced in the Mikumi and in the northern Selous Basin. There, movements along NE-trending active faults were responsible for a pronounced asymmetry of Late Permian deposits, which grade from scarp-foot fan-conglomerates and slope-scree boulder beds in the west to alluvial fan and lacustrine deposits further east. Time equivalent finger sands demonstrate the existence of a delta which was maintained by a south-west flowing trunk drainage following the long axis of the rift (Wopfner & Kaaya, 1991; in press).

Palaeocurrent directions indicate that bilateral sediment influx and axial drainage systems were maintained throughout the Permian. This was changed by a new tectonic event near the Permian-Triassic boundary (see Figure 2). It is evidenced by a low angle unconformity between

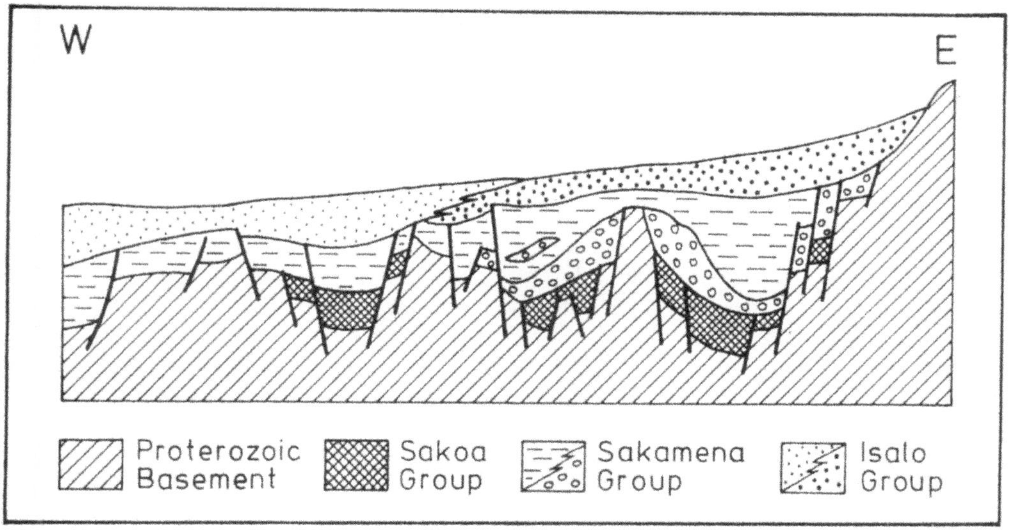

Figure 4: Schematic cross section of Morondava Basin redrawn from Cliquet (1957). This basin formed part of the eastern rift system of the Malagassy Trough. Grabens filled with deposits of the Late Carboniferous to Early Permian Sakoa Group were covered by the less intensely faulted Permo-Triassic Sakamena Group. These Karoo-eqivalents were subsequently overlapped by the Mesozoic strata of the Isalo Group.

the Late Permian K6 and the Early Triassic Kingori Sandstone (K7) and by reworked cobbles and boulders of K6 incorporated within the basal Kingori Sandstone. The supermature nature of this sandstone indicates the existence of a chemically weathered source area in the Early Triassic, but as erosion started to cut down into the unweathered crystalline rocks, supply of fresh feldspars increased, thus leading to the arkosic composition of the Manda Beds. Foreset azimuths, plotting in very narrow segments, indicate a palaeoslope with a fairly uniform inclination to the east (Markwort, 1991). As the newly elevated rift shoulder established itself as a major source of detrital material, the eastern rift margin foundered and subsided. These movements which were probably related to detachment mechanisms, initiated the transformation of the intracratonic rift to the pericratonic basin stage.

3 DISCUSSION

The structural history, comprising initial doming and rifting, rift expansion and finally detachment and foundering of the eastern side of the rift system is summarized in a schematic way on the right hand column of Figure 2. This picture, gathered from surface geological observations is supported by seismic reflection profiles across the central and northern Selous Basin (Kajato, 1986). These demonstrate the existence of syn-Karoo rifting and the progressive expansion of Karoo rifts up to and including the movements between K5 and K6 and the commencement of the foundering of the eastern rift shoulder during the Triassic.

Comparable tectonic developments to those described from Tanzanian Karoo basins also took place in Madagascar, where the Early Permian glacigene and coal-bearing graben fill of the Sakoa Group is overstepped by the Permo-Triassic Sakamena Group (Cliquet, 1957; Wright & Askin, 1987). However, as shown in Figure 4, the foundering portion of the Malagassy rift system was that facing Africa, i.e. the western part of the Morondava Basin. This makes the structure of

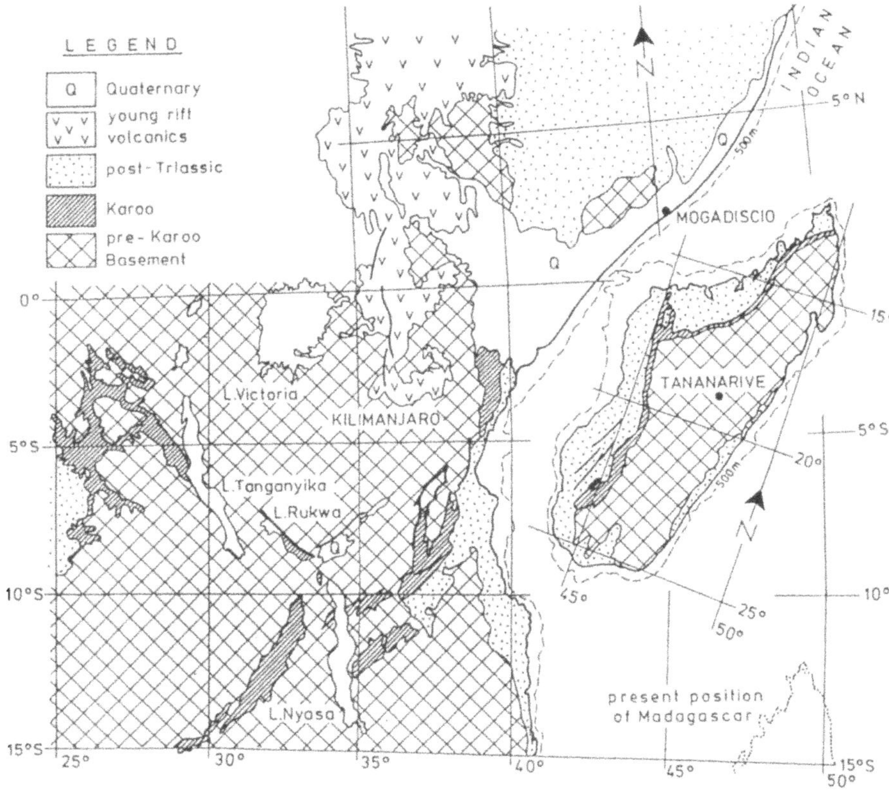

Figure 5: The Karoo basins in eastern Africa constitute an essentially northeast trending graben system. The north-west oriented Galula-Rukwa Basin apparently forms a connection with the Karoo basins of eastern Zaire. A slightly more northerly and clockwise rotated position of Madagascar than that depicted in conventional reconstructions (e.g. De Wit et al., 1988) aligns the Morondava Basin with the Karoo rift system of eastern Africa.

the Permo-Triassic basins of Madagascar almost a mirror image of that described from the Karoo basins of eastern Africa.

The dominant trend of the Tanzanian Karoo rifts was northeast (Wopfner, 1990; Kreuser et al., 1990; Wopfner & Kaaya, 1991; in press). Structures with a northwest orientation accumulated thick sedimentary successions locally (e.g. Rukwa Basin - pers. comm. E. Mbede) but on the whole they were of lesser importance. Apparently they formed as connecting or complimentary faults between major graben systems (Fig. 5).

The northeasterly trending and eastward foundering graben systems of eastern Africa may be aligned conveniently with the westward foundering rift system of Madagascar. As shown in figure 5 this results in a much better fit of geological structures than the conventional reconstruction, for instance that of De Wit et al. (1988). As Madagascar was further north than the Tanzanian Karoo basins, Permian brachiopod and cephalopod faunas could invade the Malagassy basins from the north, whereas the sea barely touched some of the estuaries of the Tanzanian rift. The marine faunas were clearly of Tethyan provenance (Teichert, 1970; Archbold, 1983) and thus demonstrate the existence of a deep incision into the Tethyan margin of Gondwana during the late Palaeozoic, as already suggested by Wopfner (1991). The model is fur-

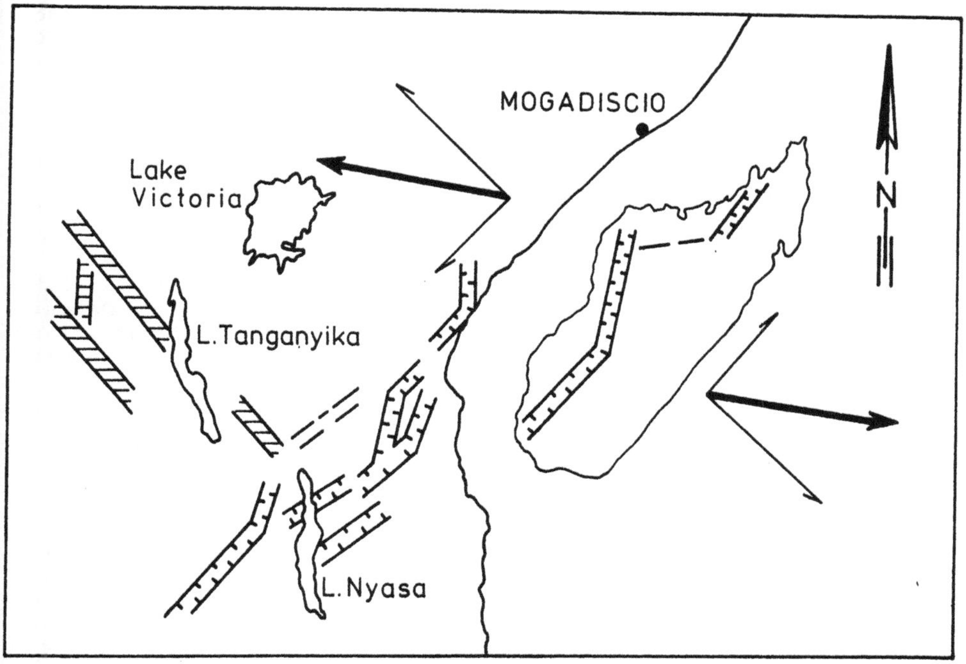

Figure 6: Observed distribution of Karoo graben complex in eastern Africa and Madagascar may be explained by a trans-tensional stress-field operating in Permo-Carboniferous and early Triassic times. Magnitudes of couplets indicated by length of arrows are only qualitative and will have changed in time. Projection and scale are the same as in Figure 5.

ther supported by the occurrence of presumed Permo-Carboniferous glacial deposits in an oil exploration well in Ogaden (J.Drong, pers.comm. 1991) and by the existence of the Early Permian brachiopod fauna of Oman (Hudson & Sudbury, 1959). The northward continuation of the western rift shoulder is further indicated by the Huqf-Haushi uplift of southern Oman (Lee, 1990). The opening of the Malagassy Trough would therefore appear to have been a concommittant event with the formation of the passive continental margin of north Oman (Blendinger et al., 1990).

The geometry of the individual graben structures shows them to be mainly of tensional origin. However, northerly deviations from the general northeast orientation of the rift system in both eastern Africa and Madagascar suggests participation of a sinistral strike-slip component (see Figure 6).

4 CONCLUSIONS

The tectonic development along the Tethyan margin of Gondwana during Permian-Triassic times differed considerably from that of its Pacific edge. Whilst the latter was governed by plate collision and the formation of foreland basins (Collinson, 1987; Powell et al., 1991) rifting and desintegration commenced along the Tethyan edge (Wopfner, 1991). It led to the partial separation of the Tibet-South China blocks (Metcalf, 1988) and to the opening of two rifts, the Westralian Trough between Australia and Greater India and the Malagassy Trough between Afro-Arabia and Madagascar-India (Wopfner, 1991). However, whilst Tethyan faunas entered the Westralian Trough rather slowly (Archbold, 1983) they quickly took possession of the Malagassy Trough, indicating a much more open and unrestricted entrance from and free connection with the Tethys (Teichert, 1970; Archbold, 1983).

5 ACKNOWLEDGEMENT

The work was supported by a grant of the Deutsche Forschungsgemeinschaft (Wo-289/2) . The attendance at the 8th Gondwana Symposium was made possible by grant 477/95/91 from the same organisation. The author records his sincere appreciation for both.

REFERENCES

Archbold, N.W. 1983. Permian marine invertebrate provinces of the Gondwana realm. - Alcheringa 7: 59-73.

Blendinger, W., Van Vliet, A. & Hughes Clarke, M.W. 1990. Updoming, rifting and continental margin development during the Late Palaeozoic in northern Oman. In: Robertson, A.H.F., Searle M.P. & Ries, A.C. (eds) The Geology and Tectonics of the Oman Region. Geological Society London Special Publication No. 49: 27-37.

Cliquet, P.L., 1957. La tectonique profonde du sud Bassin de Morondava. - Rep. 2nd Meeting East-Central & South Reg. Comm. Geol. C.C.T.A., Tananarive: 199.

Collinson, J.W. & Isbell, J.L., 1987. Evidence from the Beardmore Glacier region for a late Paleozoic/early Mesozoic foreland basin. - Antarctic Journal, Review, 22: 17-19.

De Wit, M., Jeffery, M., Bergh, H. & Nicolaysen, L. 1988. Geological map of sectors of Gondwana. American Association of Petroleum Geologists, Tulsa.

Dixey, F. 1937. The Pre-Karroo landscape of the Lake Nyasa region, and a comparison of the Karroo structural directions with those of the Rift Valley. - Q.J.Geol.Soc. London 93: 77-93.

Hudson, R.G.S. & Sudbury, M. 1959. Permian brachiopoda from south-east Arabia. Museum National d'Histoire Naturelle, VII: 19-55, Paris.

Janensch, W., 1927. Beiträge zur Kenntnis der Karru-Schichten im Östlichen Deutsch-Ostafrika. - Palaeontographica, Supplement VII: 107-142.

Kajato, H.K., 1986. The geology and hydrocarbon potential of Tanzania. - Proceedings, Oil and Gas Exploration in the SADCC region, SADCC Energy Sector Seminar Arusha, Tanzania, 26 - 29 November 1986.

Kreuser, T., 1983. Stratigraphie der Karroo-Becken in Ost-Tansania. Geologisches Institut, Universität Köln, Sonderveröffentlichung No. 45: 217.

Kreuser,T., Wopfner,H., Kaaya, C.Z., Markwort, S., Semkiwa, P.M. and Aslanidis, P. 1990. Depositional evolution of Permo-Triassic Karoo Basins in Tanzania with reference to their economic potential. Journal of African Earth Sciences 10 (1/2): 151-167.

Lee, C.W. 1990. Review of platform sedimentation in the Early and Late Permian of Oman, with particular reference to the Oman Mountains. in: Robertson, A.H.F., Searle, M.P. & Ries, A.C. (eds), The Geology and Tectonics of the Oman Region. Geological Society London, Special Publication No. 49: 39-47.

Markwort, S. 1991. Sedimentation und Diagenese der untertriadischen Karoo-Schichten des Ruhuhu Beckens, SW Tansania. Geologisches Institut, Universität Köln, Sonderveröffentlichung No. 80: 118.

McKinlay, A.C. 1954. The geology of the Ketewaka-Mchuchuma Coalfield, Njombe District. Bulletin Geolog. Survey Tanganyika 21: 46.

Metcalf, I. 1988. Origin and assembly of Southeast Asia continental terranes. Geological Society of London, Special Publication, No. 37: 101-118.

Powell, C.McA., Veevers, J.J., Collinson, J., Conaghan, P.J. & Lopez Gamundi, O.R. 1991. Late Carboniferous to Early Triassic foreland basin behind the Panthalassa margin of Gondwanaland. 8th International Symposium on Gondwana, Abstracts: 67, Hobart.

Teichert, C. 1970. Marine fossil invertebrate faunas of the Gondwana region. Proceedings and Papers, Second Gondwana Symposium. Geological Society of South Africa: 125-138.

Wopfner, H. 1989. Palaeoclimate and depositional control of the Karoo sequence in southern Tanzania. XI[e] Congres International de Stratigraphie et de Geologie

du Carbonifere Beijing 1987, Compte Rendu 4: 89-95.
Wopfner, H. 1990. Rifting in Tanzanian Karoo basins and its economic implications. 15th Colloquium on African Geology - Nancy 1990, Extended Abstracts, CIFEG - Service Documentations & Editions, Orleans.
Wopfner, H. 1991. Permo-Triassic sedimentary basins in Australia and East Africa and their relationship to Gondwanic stress pattern. In: Ulbrich, H. & Rocha-Campos, A.C. (eds), Gondwana Seven Proceedings, Sao Paulo - Brazil, 1988: 133-146.
Wopfner, H. and Kreuser, T. 1986. Evidence for late Palaeozoic glaciation in southern Tanzania. Palaeogeography, Palaeoclimatology, Palaeoecology 56: 269-275.
Wopfner, H. and Kaaya, C.Z. 1991. Stratigraphy and morphotectonics of Karoo deposits of the northern Selous Basin - Tanzania. Geological Magazine 128: 319-334.
Wopfner, H. & Kaaya, C.Z. in press. Syn-depositional Karoo rifting in the Ruhuhu and Selous Basins - Tanzania. Geology for Economic Development, Newsletter 9, UNESCO, Bruxelles.
Wright, R.P. and Askin, R.A. 1987. The Permian-Triassic boundary in the southern Morondava Basin of Madagascar as defined by plant microfossils. IUGS Gondwana Six, American Geophysical Union, Washington: 157-166.

Gondwana Eight, Findlay, Unrug, Banks & Veevers (eds) © 1993 Balkema, Rotterdam. ISBN 90 5410 304 3

Mafic and ultramafic igneous suites in the Lambert-Amery rift zone

E.V. Mikhalsky, A.V. Andronikov, B.V. Beliatsky & E.N. Kamenev
VNII Okeangeologia, St. Petersburg, Russia

ABSTRACT: Geologic setting and petrogenetic characteristics of basic and ultrabasic igneous rocks are important indications of lithosphere extension and rifting. The dyke swarms in Vestfold Hills, Mawson Escarpment, Mt.Collins, Beaver Lake area, plutons at Mt. Willing and Mt.Collins, stocks and sills in Jetty Peninsula are considered as manifestation of deep-seated thermal processes affecting the lithosphere in Proterozoic and in Phanerozoic. Proterozoic igneous rocks are mostly tholeitic known. The Phanerozoic igneous rocks are confired to alkaline series. Late Palaeozoic dykes of alkaline dolerites and alkaline lamprophyres are widespread in the Beaver Lake area. This igneous event directly predates the formation of paleorift graben filled by Permian sequence. An Cretaceous complex of K-Na alkaline-ultrabasic rocks in Jetty Peninsula intruded into Precambrian metamorphic rocks and Palaeozoic sedimentary strata. Formation of this complex is believed to record a middle stage of Gondwana breakup. Main structure features of the rift zone seem to be controlled by ancient linear structures marked by swarms trends. Ancient regional fracture structures have been revived with a new direction being developed in Phanerozoic time. We consider some of these trends to be inherited by the rift structure.

1 INTRODUCTION

The Lambert-Amery rift stretches across the East Antarctic Precambrian Shield from the sea coast for at least 700 km inland. Igneous rocks are widespread along the lenght of the rift. The vary in composition from alkaline-ultrabasic and thole iitic to acidic and form dyke swarms, small plutons, sills, stocks and rare volcanic strata (Fig. 1). Isotopic age determina- tions of mafic rocks have shown 2400, 1800, 1380, 420, 320, 120-145, 50 Ma (Tingey 1976, Sheraton & England 1980, Sheraton 1983, Collerson & Sheraton 1986, Hofmann 1986, Laiba et al. 1987). The geologic setting and petrogenetic characteristics of basic and ultrabasic igneous rocks are important indications of lithosphere extension and rifting. In this report the dyke swarms in Vestfold Hills, Mawson Escarpment, Mt.Collins, Beaver Lake area, plutons at Mt.Willing and Mt.Collins, stocks and sills in Jetty Peninsula (Fig. 2) are considered as manifestations of deep-seated thermal processes affecting the lithosphere in Proterozoic and Phanerozoic preceding to (Late Palaeozoic) and accompanying Gondwana breakup.

2 GEOLOGIC SETTING

According to Sheraton et al. (1987) and Kuehner (1987) in the high-grade Archaean metamorfic terrane of the Vestfold Hills basic dykes were intruded during three intrusive episodes: 1) sublongitudinal dykes of high-Mg tholeiitic (2400 Ma), 2) sublatitudinal tholeiitic dykes intruded at the Low/Upper Proterozoic boundary (c.1800 Ma), and 3) North-South trending tholeiitic and alkaline basic dykes of Late Proterozoic age (1380 Ma). Our field observations show that the oldest high-Mg tholeiites form not only sublongitudinal but also latitudinal dykes parallel to 1800 Ma system and often change their trend to North-Eastern (Fig. 3a). In addition to previously described dyke groups some younger systems were newly re- cognized: 1) North-Western trending tholeiitic dykes that cut the Early Proterozoic (1800 Ma) dykes but are transected by the Riphean (c.1400 Ma) tholeiites, and 2) North-North-Eastern (Fig. 3b) alkaline lamprophyres which are cut by the dated Riphean dykes but transect all older systems. These lamprophyres are of specific significance due to their alkaline character (Fig. 4) and the presence of deep-seated nodules which may be related to ancient rifting in the area.

3 PETROGRAPHY AND GEOCHRONOLOGY

The alkaline lamprophyres of Vestfold Hills contain phenocrysts of clinopyroxene ($Ca_{48}Mg_{40}Fe_{12}$), olivine and

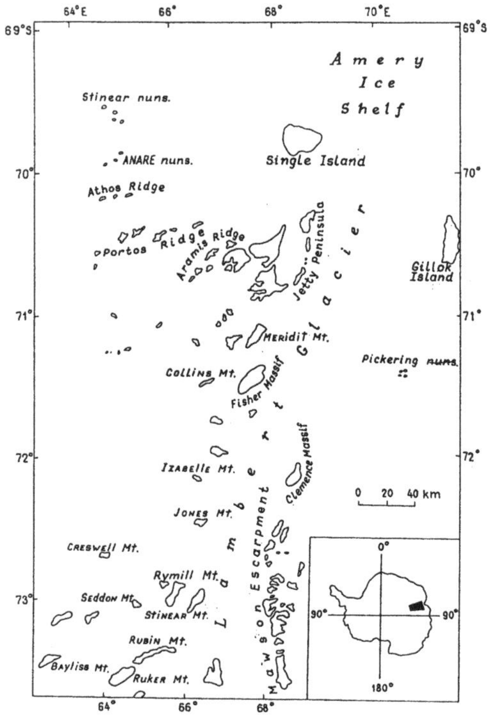

Fig.1 Schemaic map of outcrops in Prince Charles Mountains

phlogopite (f=38, MgO=18%, TiO_2=6.5%) enclosed in a relatively fresh, well crystallized groundmass of carbonate + phlogopite and/ or plagioclase. No modal alkaline minerals were recorded. Small carbonate-feldspar globules are common. Colour index varies from 55 to 95-100, and Mg# is 50-75. According to petrographical and geochemical scheme of Rock (1986), these rocks are defined as K-Na biotite - augite camptonites although less evolved varieties resemble ultramafic lamprophyres (e.g. in SiO_2, K_2O, MgO abundances). Field evidence suggests that the more salic dykes are relatively older. Some geochemical features (high Mg#, high Ni, Cr, Zr, Ba abundances, and a relatively high Rb/Sr ratio of 0.05-0.08) are consistent with primary derivation as an upper-mantle melt in source region(s) rich in incompatible elements. Zr/Nb, Zr/Y ratios are suggest a heterogeneous source region and its increase deeping. Preliminary Sm-Nd wholerock isotope data yield a Late Proterozoic reference line (c.1100-1200 Ma) (Fig. 5). The exeeding scatter of data (Table 1) probable is due to a different degree of crustal contamination or to variations in composition. Low εNd values (about -3.0) imply crustal material, assimilation, though the degree of contamination probably was not great. However the source heterogenety is very likely to be involved as well.

In places the lamprophyre dykes thicken into lensoid bodies of similar composition containing abundant small nodules of dunite, wehrlite, harzburgite, spinel and (very rarely) garnetbearing lherzolite. We suggest that these nodules have Low-Crust and Upper-Mantle origin.

In the southern part of the Mawson Escarpment numerous dykes of basic rocks cut the upper-amphibolite facies terrane. These dykes have different orientations (Fig. 3c) and compositions. At least two chemically distinct age groups have been recognized. High-Ni, Cr and Mg hyperstene- olivine-normative metatholeiites and amphi- bolites compose the earliest dykes and trend West-East or South-Western - North- Eastern. Less evolved members of this high-Mg suite have a higher Zr content that might be due to alaline character of primary melt. The younger group is represented by hyperstene-qurtz- normative metatholeiites and slightly metamorphosed dolerites making up North-South or North-Western - South-Eastern dykes. These rocks have lower Zr/Y ratios and probably were derived from another source region. Our preliminary (Table 1) isotopic Sm-Nd data are too scarce for isochron solutions, and represent two igneous events: Early Proterozoic (older 1800 Ma) (metatholeiites) and Late Proterozoic (1200-800 Ma) (Fig. 6) (dolerites). However, close geochemical affinities of both rock types and hence their possible comagmatic origin (with dolerites being unmetamorphosed equivalents) leave chances for them to be simultaneous (c.1800 Ma in this case). A model U-Pb zircon study of the high-Mg metabasites yields an age of 2400 Ma (Beliatsky, unpublished data), which probably indicates crystallisation age, but the possibility of contamination factors should not be disregarded.

A large body of metamorphosed syenites, monzonites and granosyenites is exposed at Mt.Collins in the central part of Prince Charles Mountains. These rocks are considered us to have crystallized at 1400 Ma (U-Pb zircon data) and to have undergone subsequent thermal events at 700-900 Ma (Beliatsky, unpublished data). The metasyenites are cut by numerous metamorphosed basic dykes of variable orientations (Fig. 3d) and composition (Fig. 4) and partly of alkaline character probably indicated by Zr and Ba-enrichment and low Y/Nb ratio. The presence of metasyenite intrusions and the accompaying dyke suite of Late Proterozoic age confirms a Late Precambrian extensional igneous activity in the central part of Prince Charles Mountains.

A layered gabbroid complex is exposed at Mt.Willing some 30 km to the south of Mt. Collins. The rocks are mainly gabbronorites, gabbro, norites (some are olivine-bearing) with minor hornblende gabbro and ferrogabbronorites at the uppermost part of the exposed section. The rocks are converted to amphibolites along shear zones. The cumulus assemblage changes upwards from Opx+Cpx+Pl+Ol to Pl+Cpx+Opx+ +Hrb+Mt. The Willing Massive contains some concordant or gently dissending small gabbro-anorthozite bodies of the second generation, and numerous pegmatites and granite bodies as thick as tens metres. The Willing Massive is dissending by steeply deepping mafic dykes composed of amphibolized (uralitic) dolerites and trachydolerites. Nine samples define a Sm-Nd whole-rock isochron of 1233+160 Ma (Fig. 7) corresponding probably to a crystallization age. Mineral separations (Pl, Amph) gave a Sm-Nd isochron of c.1000 Ma, indicating metamorphic reworking. Relatively high values of εNd=1.5-3.5 (Fig. 8) are thought to indicate low degree of crustal

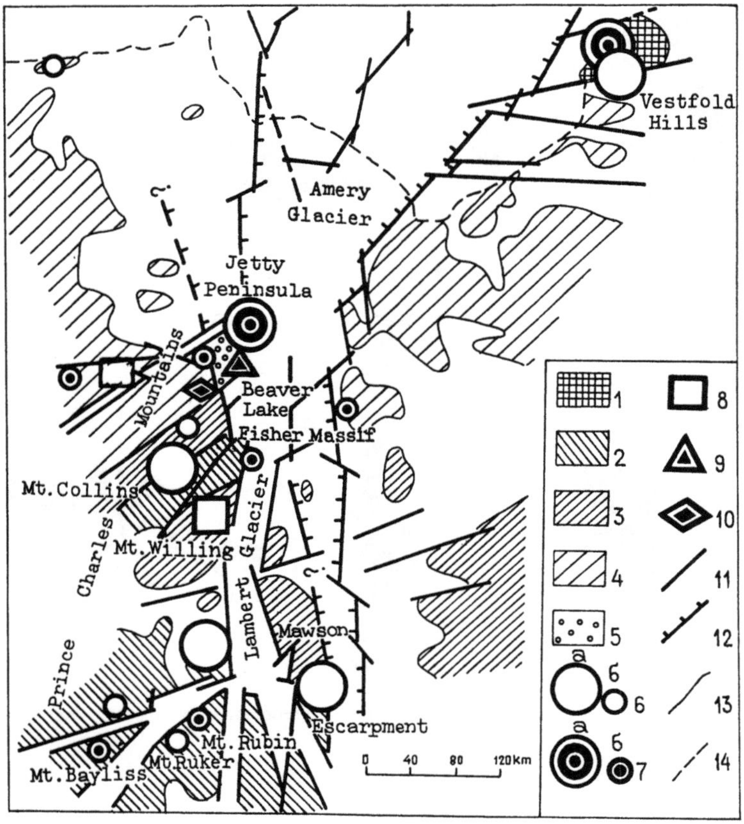

Fig.2 Locality map of basic and ultrabasic bodies in relation to basement geology

1-4 - basement geology (after Kamenev & Kameneva 1990): 1-Vestfold Hills granulites (Archaean), 2-Fisher and Rucker greenstone belts (respectively Late Proterozoic and Late Archaean), 3-transition granite-gneiss-schists belts, 4-granulite-gneiss mobile matamorphic belt (Proterozoic); 5-Permian sandstones; 6,7 - locations of dykes: 6-of tholeitic and calc-alkaline composition, 7-of alkaline composition (a-dense swarms, b-rare dykes); 8-location of gabbroid plutons; 9-location of alkaline-ultrabasic rocks; 10-location of alkaline basalt flow; 11,12 - fault zones: 11-recent, 12-of Palaeozoic(?) generation; 13-contours of bedrock above sea level; 14-coast line.

contamination and derivation from the initial melt of depleted mantle. The low εNd of one of the samples (0.17) may indicate more extensive contamination represented by the second generation of gabbro injections (Fig. 8).

Rare thin high-K lamproite-like dykes are known from the southern part of the Prince Charles Mountains (My.Bayliss, Mt.Rubin) (Sheraton & England 1980, Laiba, unpublished data). The rocks consist mainly of potassic feldspar (high-orthoclase and microcline), Ti-phlogopite and low-Ti high-K ferro richterite - magnesio-arfvedsonite groundmass enclosing phenocrysts of diopside (?) which are rarely altered. The rocks contain reworked xenoliths of country rocks, suggesting some degree of crustal contamination as to caused of the unusual amphibole composition.

The Phanerozoic igneous rocks are confined to the alkaline series. Late Palaeozoic (K-Ar age 320 Ma) (Hofmann 1986) dykes of alkaline dolerites and alkaline lamprophyres are widespread in the Beaver Lake area forming a dense North-South trending swarm in Jetty Peninsula (Fig. 3 e). This igneous event directly predates the formation of paleorift graben filled by a Permian coal-bearing sedimentary sequence. An Early Cretaceous (120- 145 Ma) (Laiba et al. 1987) complex of K-Na alkaline-ultramafic rocks in Jetty Peninsula is formed of small stocks intruding along a North-South major fault zone, sills and dykes intruded into Precambrian metamorphic rocks and Palaeozoic sedimentary strata. The stocks were formed by three successive injections of melt. The earliest olivine-melilite tuffaceous breccias were followed earlier by melilite lamprophyres and later by alkaline picrites and ultramafic foidites. Deep-seated nodules are abundant in all stocks and represented by garnet-bearing and spinel lherzolites, garnet-free dunites, harzburgites and wehrlites (Andronikov 1990). Formation of this complex is believed to record an middle stage of Gondwana breakup. Thin kimberlite dykes (Egorov & Melnik in

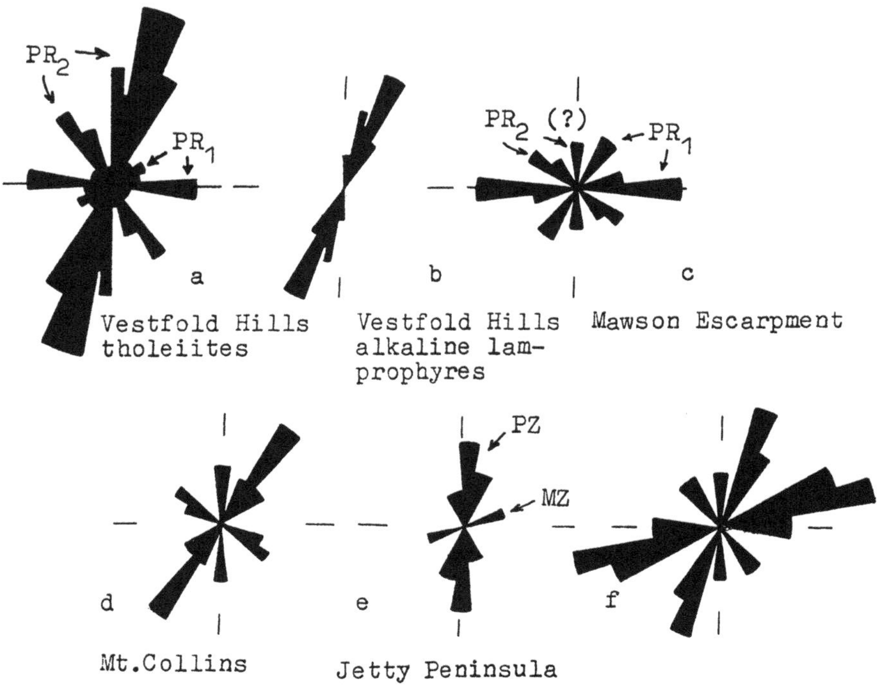

Fig.3 Rose diagrams for main dyke swarms (a-e) and post-dyke (predomonantly recent) fracture zones at Vestfold Hills (f)

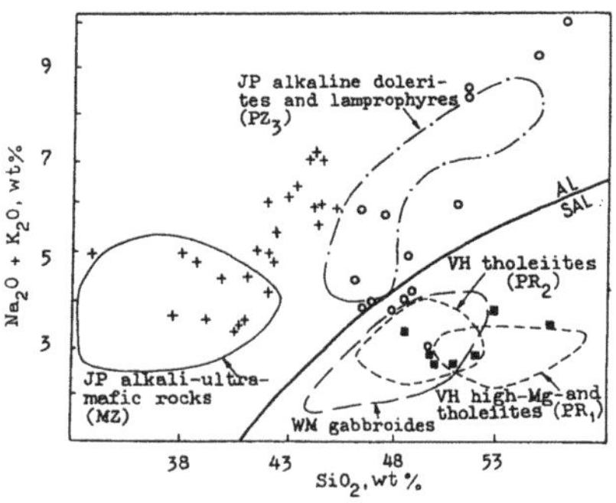

Fig.4 Diagram SiO_2 versus (Na_2O+K_2O)

Crosses - Vestfold Hills alkaline lamprophyres; boxes - Mawson Escarpment metabasites; circles - Mt.Collins metabasites. Fields of different igneous rocks at individual lacalities are shown by contours: JP - Jetty Peninsula; WM - Willing Massif; VH - Vestfold Hills. Alkaline (AL) - subalkaline (SAL) dividing line after Irvine & Baragar (1971)

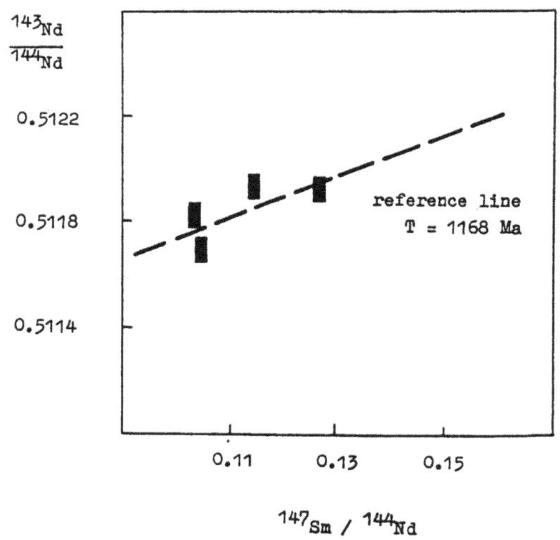

Fig.5 Sm-Nd whole-rock isotopic data for Vestfold Hills alkaline lamprophyre

press) recently discovered at Fisher Massif are probably refered to the same tectonic event. The latest igneousevent in this region is dated at 50 Ma (Tingey 1976) and is represented by a thin flow of K-rich basalts (olivine-leucite trachybasalts)at Manning Massif (northern part of Prince Charles Mountains).

4 CONCLUSION

The main structural features of the Lambert-Amery rift zone seems to be controlled by ancient linear structures marked by the trends of swarms trends. Ancient regional fractures (the earliest Proterozoic to be East-West

Table 1. Sm-Nd isotopic data for Prince Charles Mountains basic rocks

#	Sample #	Rock name	Nd ppm	Sm ppm	147Sm/144Nd	143Nd/144Nd
		VESTFOLD HILLS				
1	34539-1a	alkaline lamprophyre	67.80	14.20	0.12665	0.511917+7
2	34572-5	alkaline lamprophyre	177.98	30.54	0.10372	0.511807+9
3	34583-1	alkaline lamprophyre	91.19	15.67	0.10389	0.511706+9
4	34583-2	alkaline lamprophyre	63.64	12.06	0.11455	0.511916+11
		MAWSON ESCARPMENT				
5	33112	dolerite	18.06	5.08	0.16958	0.512478+7
			18.12	5.11	0.17049	0.512482
6	33112-2	dolerite	17.62	5.19	0.17804	0.512537+10
7	33111-12	dolerite	17.66	5.17	0.17682	0.512487+14
8	33113	dolerite	8.56	2.62	0.18476	0.512467+8
9	33110-11	metatholeiite	8.40	2.46	0.17728	0.512558+6
10	33110-7	metatholeiite	34.12	7.38	0.13079	0.511693+11
11	33106-7	metatholeiite	17.67	4.32	0.14796	0.512197+7
12	33025-3	high-Mg metabasite	13.24	3.16	0.14434	0.511099+57
		WILLING MASSIF				
13	33047-8	altered gabbro	2.27	0.68	0.18064	0.512611+10
14	33047-2	altered gabbro	2.13	0.58	0.16474	0.512523+9
15	33047-1	altered gabbro	2.87	0.95	0.20112	0.512755+15
16	34703	ferrogabbro-norite	36.99	9.01	0.14730	0.512366+8
			35.80	8.65	0.14609	0.512297+17
17	34703-1	altered ferrogabbro	36.30	8.62	0.14349	0.512297+4
			34.22	8.00	0.14122	0.512333+4
18	34336-4	Ol gabbro-norite	2.62	0.83	0.19184	0.512773+22
19	34338-1	Hb norite	30.77	9.03	0.17736	0.512496+4
		JETTY PENINSULA				
20	34MK10-10	dolerite	23.96	5.03	0.12680	0.512480+7
21	30MK40-2	dolerite	39.85	6.87	0.10419	0.512352+6
22	30MK24-3	dolerite	32.66	6.94	0.12836	0.512376+9
23	30MK57- 6	augite camptonite	43.07	7.79	0.10935	0.512297+8
24	30MK9-1	Py-Bt micromonzonite	54.26	9.38	0.10446	0.512326+7
25	30MK19	Bt micromonzonite	58.86	12.78	0.13125	0.512262+9
26	5/8-1	Bt micromonzonite	56.41	9.78	0.10485	0.512318+5
27	30MK1-4	Amph micromonzonite	60.27	10.27	0.10303	0.512211+10
		NORTHERN PRINCE CHARLES MOUNTAINS				
28	35546-2	metagabbro	10.33	2.45	0.14354	0.512213+15
		MOUNTAIN RUBIN				
29	35647-2	lamproite	130.48	16.21	0.07508	0.511879+7

Note. Constants used: 147Sm decay 6.54×10^{-12} y^{-1} ; CHUR 147Sm/144Nd=0.1967; CHUR 143Nd/144Nd=0.512636.

trending and more young Proterozoic North-South and North-North- Eastern trending) have been revived with a new East-North-Eastern direction being developed in Phanerozoic time. We consider some of these trends (mostly Proterozoic) to be inherited by rift structure. It is likely that the Gondwana breakup followed a number of preceding mantle thermal activities and crustal extension episodes which prepared the lithosphere for rifting. The earliest record of extensional igneous activity in sialic crust is represented by Proterozoic (2400-1800 Ma) tholeiites. The first rift-related magmatism is believed to have occured at c.1200 Ma when the oldest alkaline rocks were formed. In Late Palaeozoic a renewed K-Na magmatic activity began in

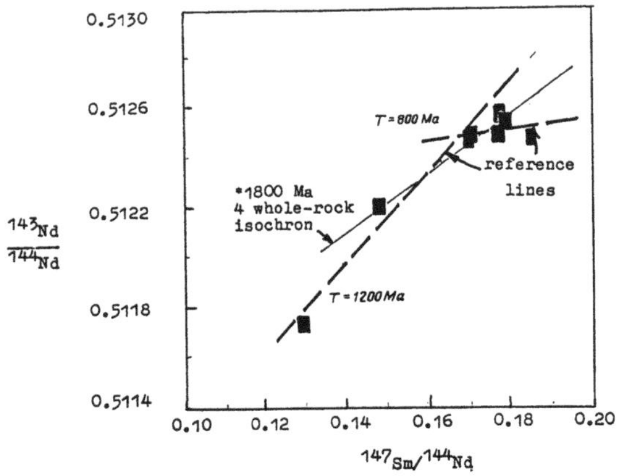

Fig.6 Sm-Nd whole-rock isotopic data for Mawson Escarpment dyke rocks

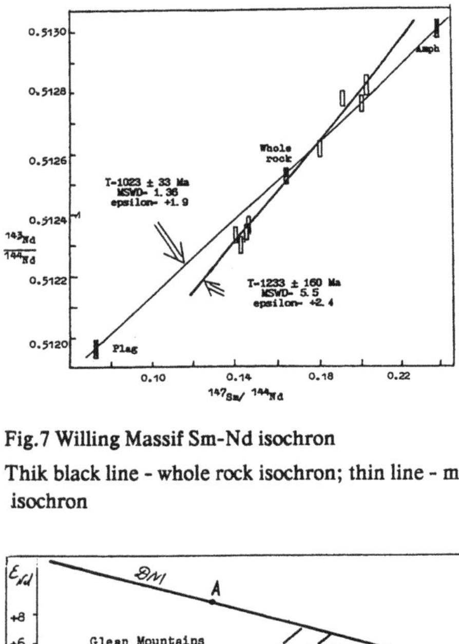

Fig.7 Willing Massif Sm-Nd isochron

Thik black line - whole rock isochron; thin line - mineral isochron

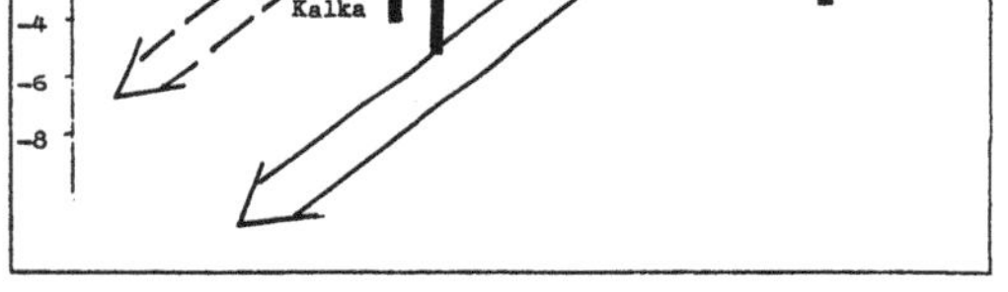

Fig.8 Nd evolutions diagram for Willing Massif gabbroides (circles) and miscellaneous layered plutons DM - depleted mantle evolution trend; A - depleted mantle source for Willing Massif rocks

relation to event 120-145 Ma. This was succeded by a period of mantle degassing which was terminated by an extrusion of Early Cenozoic K-rich basalts coeval with final opening of the Southern Ocean.

REFERENCES

Andronikov A.V. 1990. Mesozoic alkaline-ultrabasic magmatism of Jetty Peninsula (Prince Charles Mountains, East Antarctica) (in Russian). Ph.D. degree dissertation: 228, Leningrad.

Collerson K.D. & Sheraton J.W. 1986. Age and geochemical characteristics of a mafic dyke swarm in the Archaean Vestfold Block, Antarctica: inferences about Proterozoic dyke emplacement in Gondwana. Journal of Petrology, 27, 4: 853-886.

Hofmann I. 1986. Bruchtektonic und Magmatism im kristallinen Fundament der Octantarktiscen Plattform. Freiberg. Unpublished report.

Irvine T.N. & Baragar W.R.A. 1971. A guide to the chemical classificaton of the common volcanic rocks. Canad.Jour. of Earth Sci., 8: 523- 548.

Kamenev E.N. & Kameneva G.I. 1990. Structural and metallogenetic provinces in Antarctica (in Russian). In V.L.Ivanov & E.N.Kamenev (eds.), The geology and mineral resources of Antarctica: 15-40. Moscow, Nauka.

Kuehner S.M. 1987. Mafic dykes of the East Antarctic Shield: a note on the Vestfold Hills and Mawson Coast occurences. In H.C.Halls & W.F.Fahrig (eds.), Magic dyke swarms: 429-430. Geol.Assoc. of Canada spec. paper, 34.

Laiba A.A., Andronikov A.V., Egorov L.S., Fedorov L.V. 1987. Stock-like bodies and dykes of alkaline-ultramafic rocks in Jetty Peninsula (Prince Charles Mountains, East Antarctica). In V.L.Ivanov & G.E.Grikurov (eds.), Geological and geophysical research in Antarctica: 35-47. Leningrad.

Rock N.M.S. 1986. The nature and origin of ultramafic lamprophyres: Alnoites and allied rocks. J.Petrol. 27 (1): 155-196.

Sheraton J.W. 1983. Geochemestry of mafic igneous rocks of the northern Prince Charles Mountains, Antarctica. J.Geol. Soc.Austral., 30: 295-304.

Sheraton J.W. & England R.N. 1980. Highly potassic mafic dykes from Antarctica. J.Geol.Soc.Austral., 27: 129-135.

Sheraton J.W., Thomson J.W., Collerson K.D. 1987. Mafic dyke swarms of Antarctica. In H.C.Halls & W.F.Fahrig (eds.), Mafic dyke swarms: 419-432. Geol.Assoc. of Canada spec.paper, 34.

Tingey R.J. 1976. Basic and alkaline intrusive rocks in the Prince Charles Mountains. Bur.Miner. Res.Rep., 194: 73.

Gondwana Eight, Findlay, Unrug, Banks & Veevers (eds) © 1993 Balkema, Rotterdam. ISBN 90 5410 304 3

Mesozoic alkaline-ultrabasic magmatism of Jetty Peninsula

A.V. Andronikov & L.S. Egorov
VNII Okeangeologia, St.Petersburg, Russia

ABSTRACT: Detailed geological and petrological studies of small alkaline-ultrabasic intrusions have been carried out in the Jetty Peninsula region of the Prince Charles Mountains. These intrusions include six dykes, seven stocks and two sills forming a submeridional chain over 40km long. The main conclusions drawn from these studies are: 1) all the subisometric in plan intrusive bodies of Jetty Peninsula represent small polyphase stocks with analogues in Paraguay, South Africa and Australia; 2) mesozoic alkaline-ultrabasic magmatism is associated with the middle stages of Gondwana break-up (120-150 Ma); 3) all polyphase stocks have the same sequence of the emplacement: tuffaceous breccias - melilite lamprophyres - alkaline picrites and ultrabasic foidites; sills - polcenites (pyroxene free alnoites); dykes - singlephase bodies of by polcenites, alkaline picrites or ultrabasic foidites; 4) the alkaline ultrabasc rocks of Jetty Peninsula contain a great number of mantle nodules: garnet-bearing and spinel lherzolites, garnet-free harzburgites, dunites and wehrlites. Three types of mantle material are recognized: "fertile", "depleted" and enriched LREE (resulting from mantle metasomatism). The source region depth ranges from 35 to 80 km (12-27 kb) and the equilibrum temperature range of the mantle mineral assemblages stet 825-1180 C. Based on these mantle xenoliths a schematic section of the upper mantle to a depth of 80 km is presented.

1 INTRODUCTION

The complex of Mesozoic (120-150 Ma) alkaline ultrabasic rocks is recognized in northern Prince Charles Mountains, Jetty Peninsula area (Laiba et al. 1987). Alkaline ultrabasic rocks are represented by small intrusions forming a linear chain more than 40 km long, and occur at the intersection of Northern-East - Southern-West and North - South faults (Fig.1). Dykes occur the Northern-East - Southern-West faults and the stock-like intrusions are located at the intersections with the eastern msrgin of the Beaver Lake graben.

The studied intrusive bodies of alkaline-ultrabasic composition include six dykes, seven stocks and two sills. The chain of stocks, all named, extends for 14 km along the eastern flank of the Beaver Lake graben. In plan these bodies are isometric, oval or irregular, the dimensions ranging from 25x60 m to 100x180 m (Fig.2).

2 BODIES' CHARACTERISTICS

The MARTOVSKOYE body intrudes Permian sandstones and consists of two phases of injection: tuffaceous breccia and olivine-melilite melanephelinite. The presence of glassy pyroxene-free melilitite clasts, representing rapidly chilled magma, is characteristic of tuffaceous breccias from all bodies investigated. The SEVERNOYE VERKHNEYE body consists of a sequence of three igneous intrusions: tuffaceous breccia (quasy-intrusive), polcenite (pyr-oxene-free alnoite) and nepheline-melilite-pyroxene alkaline picrite. The SEVERNOYE NIZHNEYE body, like Verkhneye, occurs in crustalline basement and in partly covered by moraine and fluvio-glacial deposits. Tuffaceous breccia together with melilite-pyroxene and nepheline-melilite-pyroxene alkaline picrites make up the stock. The melilite-pyroxene picrite are the early differentiates of the final phase of injection (cumulates that separated early from magma that later gave nepheline-melilite-pyroxene rocks) and nepheline-melilite-pyroxene ones constitute the bulk of the body; similar rocks occur in the Severnoye Verkhneye body. The NOVOYE body was emplaced into the Precambrian metamorphic rocks, now forming a small surface depression, and is a tiny three-phase central intrusion with a radial outcrop pattern. The tuffaceous breccia, polcenites, pyroxene-free olivine melilitites (nepheline rich), melilite-olivine melanephelinites were intruded successively. The KONUS body was intruded into granulite facies Archean rocks. Two thirds of its area are composed of tuffaceous breccia, the rest of polcenite with alnoite schliren. The PLOSKOYE body (the only single-phase stock in Jetty Peninsula) consists of olivine melanephelinite. The extremely high concentration of xenoliths of Permian terrigenous rocks points to the small cross section of the stock. The YUZHNOYE body was also emplaced through the Permian sedimentary sequence. The greater part is composed of biotite-pyroxene alkaline picrite which is locally cut by nepheline picrite veins. Tuffaceous breccia, the first phase of injection, occurs in the northern part of the body.

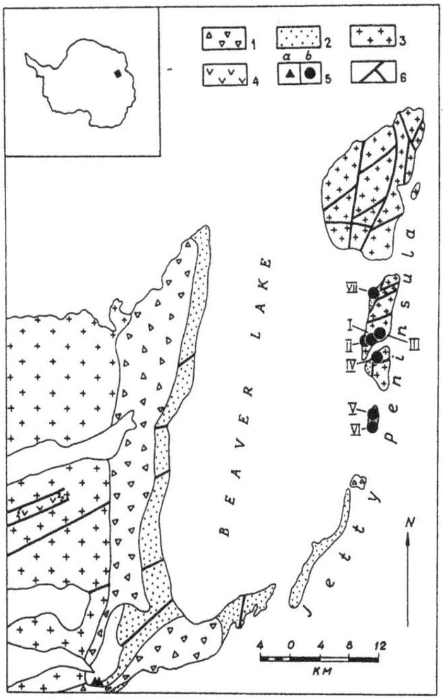

Fig.1 Schematic geological map of the Jetty Peninsula area

1-moraine; 2-Permian sandstones; 3-Precambrian crystalline basement; 4-Cenozoic alkaline basaltic rocks; 5-Cretaceous alkaline-ultrabasic rocks: a-sills, b-stocks (I-Severnoye Verchneye, II-Severnoye Nizhneye, III-Novoye, IV-Konus, V-Ploskoye, VI-Yuzhnoye, VII-Martovskoye); 6-faults

3 ROCKS' CHARACTERISTICS

3.1 General view

The dykes of alkaline-ultrabasic composition are commonly spatially associated with stocks, are singlephase and represent-

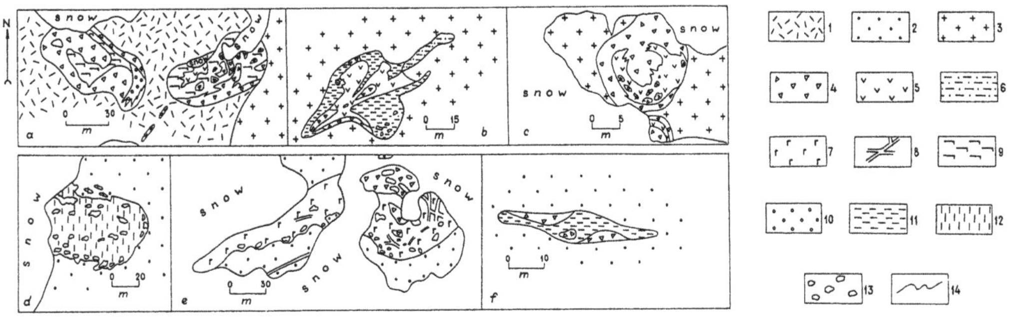

Fig.2 Schematic geological maps of the alkaline-ultrabasic rock bodies (a-Severnoye Nizhneye and Severnoye Verchneye, b-Novoye, c-Konus, d-Ploskoye, e-Yuzhnoye f-Martovskoye)

1-moraine and fluvio-glacial deposits; 2-Permian sandstones; 3-Precambrian crystalline basement; 4-tuffaceous breccia; 5-polcenites; 6-pyroxene-free olivine melilitites; 7-biotite-pyroxene alkaline picrites; 8-viens of nepheline picrites; 9-nepheline-melilite-pyroxene picrites; 10-melilite-pyroxene picrites; 11-melilite-olivine melanephelinites; 12-olivine nephelinites; 13-xenoliths of the host rocks; 14-geological boundaries ed all of the described types of magmatic rocks except for tuffaceous breccia. The sills of polcenite (Radock-1 and Radock-2) were intruded into Permian strata in the south-western part of region. The thicker (11-13 m) upper sill, Radock-1, can be traced in a cliff bordering Radock Lake for about 1 km. In both sills a symmetrical zonation is expressed by the distribution of textural variations (an axial zone; upper and lower massive zones, and two contact zones).

The alkaline-ultrabasic rocks of Jetty Peninsula classifed into 3 groups (in order of emplacement): 1) tuffaceous breccias; 2) melilite lamprophyres; 3) alkaline picrites and ultrabasic foidites. The tuffaceous breccias represent mixtures of debris derived from different host rocks with autoliths consisting of glassy pyroxene-free olivine melilitite. The matrix is formed of finely comminuted host rocks lithologies and cemented by hydrous calcite, rarely, limonite. On the basis of xenolith/autolith ratios, the breccias are divided into:

1. Autolithic (less than 15% of xenogenous material);
2. Xenolith-autolithic (15-50% of xenogenous material);
3. Autolith-xenolithic (50-85% of xenogenous material);
4. Xenolithic (more than 85% of xenogenous material).

The autolithic breccias occur in the Martovskoye body, xenolith-aitolithic in the Yuzhnoye and in the both of Severnoye bodies, autolith-xenolithic in the Novoye body, and xenolithic in the Konus body.

The polcenites and alnoites are members of the melilite lamprophyre group. They were emplaced during the second phase of magma injection in the stocks. In addition, the polcenites form the singlephase bodies: two sills and dyke. The alnoites form the schliren in the polcenites of the Konus body and are probably the products of their contamination by host rock silica (Andronikov 1990).

The alkaline picrites and ultrabasic foidites occur in the separate stocks and compositionally consisist of:

1.Biotite-pyroxene and nepheline alkaline picrite (Yuzhnoye body);

2.Nepheline-melilite-pyroxene and melilite-pyroxene alkaline picrite (Severnoye bodies);

3.Pyroxene-free olivine melilitite and melilite-olivine melanephelinite (Martovskoye, Novoye, Ploskoye bodies).

The modal compositions of the intrusive rocks are given in Table 1. A characteristic of these rocks is the hypocrystalline olivine-phyric textures caused by the presence of relatively large olivine xenocrysts and reflecting the high rate of ascent and rapid crystallization of magma; the structures of rocks are massive. The alkaline-ultrabasic rocks were subjected to secondary alteration, usually expressed as serpentinization, phlogopitization and carbonatization.

3.2 Geochemestry

All alkaline picrites are similar in their chemical composition (see Appendix, Table 1). A higher content of SiO_2 (up to 42%) and lower content of MgO (down to 18%) is present in the magmatic bodies emplaced through the Permian strata. The distinguishing petrochemical feature of the olivine melanephelinites is the lower content of MgO (12-14%), higher alkalinity (sum of alkalis up to 5.5%) and content of Al_2O_3 (up to 8-10%).

Two trends of evolution are shown in the $CaO-MgO-Al_2O_3$ ternary (Fig.3). The first

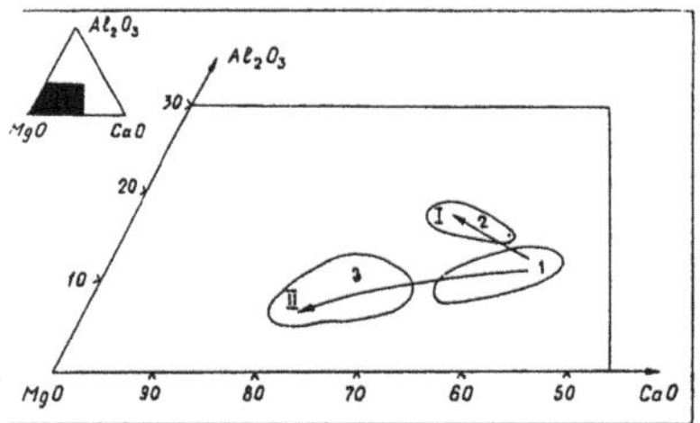

Fig.3 Ternary diagram ($MgO-Al_2O_3-CaO$) for alkaline-ultrabasic rocks of Jetty Peninsula

I,II-trends of magma evolution.
1-3-fields of compositions: 1-melilite lamprophyres; 2-ultrabasic foidites; 3-alkaline picrites

trend (I) appears to reflect the evolution from melilite lamprophyres to ultrabasic foidites. The second branch (II), defined by increasing Mg, is mainly the results of irregular contamination of the magmas by ultramafic mantle material.

The individual geochemical features of different alkaline-ultrabasic rocks are weakly expressed (Appendix, Table 2). The relatively increased content of Ti (up to 14120 ppm) in polcenites and alnoites, caused here by abundant perovskite, of Cr (up to 2166 ppm) in alkaline picrites, and relatively high content of Ni (up to 1056 ppm) and Co (up to 78.8 ppm) probably

Table 1. Modal composition of the alkaline-ultrabasic rocks of the Jetty Peninsula

	OL	CPX	MEL	NE	PHL	TMG	PER	AP	SLP	GL
Glassy pyroxene-free olivine melilitite	10-15		20-30		+	+	+			55-60
Polcenite	10-15		30-40	10-35	5-45	1-10	1-10	1-2	3-7	3-10
Alnoite	10-15	5-20	25-30	10-15	5-15	5-7	5-7	+	+	5-7
Biotite-pyroxene alkaline picrite	15-20	35-40		5-15	5-10					20-40
Nepheline picrite	10-15	50-60		7-15	10-15	5-7	2-3			5-7
Melilite-pyroxene alkaline picrite	15-20	30-40	10-20		+	7-10	+			30-40
Nepheline-melilite-pyroxene picrite	15-20	30-40	5-10	10-15	15-20	5-10				10-15
Pyroxene-free olivine melilitite	10-15		20-25	40-45	+	10-15	+			7-10
Melilite-olivine melanephelinite	10-15	8-10	7-10	50-60	7-10	7-10	2-5			3-7
Olivine melanephelinite	15-25	25-30		10-35	3-7				1-3	5-10

Note: OL-olivine, CPX-clinopyroxene, MEL-melilite, NE-nepheline, PHL-phlogopite, TMG-Ti-magnetite, PER-perovskite, AP-apatite, SLP-sulphides, GL-volcanic glass.

result from the presence of pyroxene and abundant olivine phenocrysts (Andronikov 1990). Despite the varying absolute abundances, the Ni/Co ratios are rather constant (6-14), and similar to mantle ratios. The alkaline-ultrabasic rocks of Jetty Peninsula are highly enriched is REE. The degree of fractionation (La/Yb and Ce/Yb) is variable, and lies within the range 50-94 and 59-113 for alkaline picrites and ultrabasic foidites and 22-108 and 37-108 for polcenites.

3.3 Mineralogical features

Microprobe analyses of the compositions of rock-forming minerals such as phlogopite (Mg#=80.3-86.4), nepheline (Ks=19.7-25.6%), melilite (Ak=66.3-73.3%, Sodium melilite (Sm)=21.8-30.4%), clinopyroxene (Ca=48.2-55.3, Mg=31.0-38.6, Fe=9.9-16.3, TiO_2=2.18-3.67%) and olivine (Fa=12.0-13.7%) from different alkaline-ultrabasic rocks (Andronikov 1990) demonstrate their resemblance to the minerals of similar rocks from other regoins of the world (Dawson, 1980; Jaques et al. 1986 and others).

4 DEEP-SEATED NODULES

All the intrusive bodies of the region contain irregularly distributed xenoliths of various rock types. Contacts of the alkaline-ultrabasic rocks with the inclusions indicate non-equilibrium conditions. The ultramafic mantle inclusions in their lateral parts are weakly phlogopitized and serpentinized. From petrografic study, five main varieties of deep-seated (mantle) inclusion have been defined: garnet-bearing and spinel lherzolites, garnet-free dunites, harzburgites and wehrlites. The overwhelming majority of nodules (near 60%) are lherzolites, nine tenths of which consist of garnet-free varieties. Dunites form nearly 25% of xenoliths, harzburgites 15%, and wehrlites are represented only by rare occurences.

On the basis of chemistry (Andronikov, 1990) the nodules are combined into two principal groups (Fig.4). The first ($CaO+Al_2O_3$ > 5%) is garnet-bearing and includes part of the spinel lherzolites; the second ($CaO+Al_2O_3$ < 3%) is mainly spinel lherzolites, harzburgites and some of the wehrlites. These groups are connected by the wehrlite compositions and a few of the spinel lherzolites. If changes in the composition of peridotites are regarded as the result of partial melting, the two main groups of rocks may represent respectively "fertile" and "depleted" fragments of the upper mantle. The dunites form a separate group; their compositions lie in the field of depleted rocks but have a high Fe/(Fe+Mg) ratio (0.10-0.12). We consider such

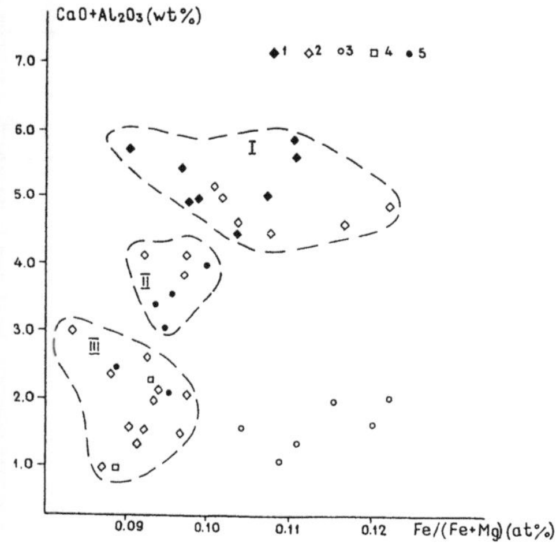

Fig.4 Diagram ($CaO+Al_2O_3$) versus Fe/(Fe+Mg) for mantle nodules

1-garnet-bearing lherzolites; 2-spinel lherzolites; 3-wehrlites; 4-harzburgites; 5-dunites.
I-"fertile" rocks; II-intermediate rocks; III-"depleted" rocks

dunites to be the refractory residue from the partial melting of peridotite.

On the basis of absolute abundances (Appendix, Table 3) (INAA method) and the patterns of REE distribution (Fig.5) several mantle sources for the range of deep-seated inclusions can be distinguished:

1. Deep-seated rocks with weak fractionation of the lanthanides and insignificant deviation (normalised abundance) from the content in standard chondrite. Such nodules can be assigned to derivation from "fertile" mantle.

2. Rocks subjected to a partial melting episode accompanied by removal of REE mainly of the cerium group. The positive slope in lanthanide distribution is characteristic of xenoliths of these rocks.

3. Inclusions that are fragments of mantle enriched in light REE and characterized by a negative slope on the chondrite normalized diagram. These rocks were affected by metasomatism with introduction of REE of the cerium group.

4. Dunites characterized by a relatively flat REE pattern with insignificant enrichement relative to chondritic values. This pattern shows some enrichement of the

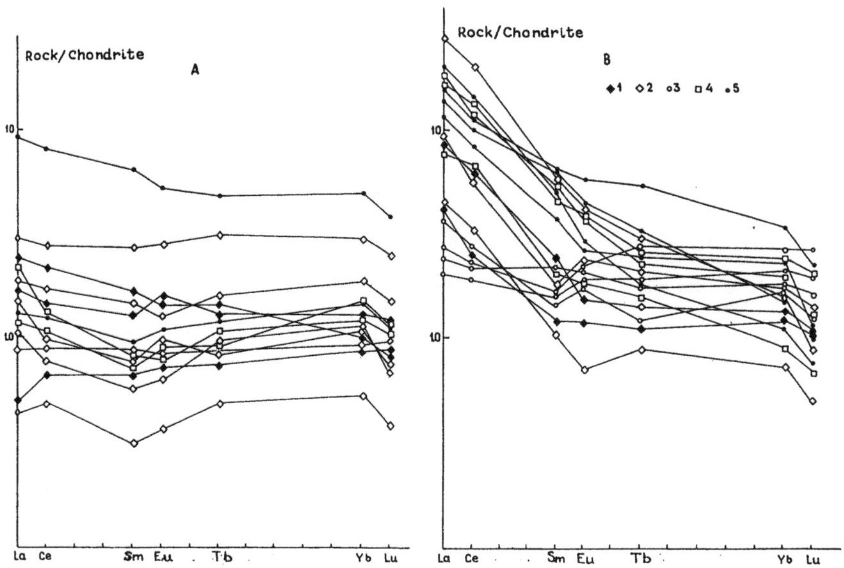

Fig.5 Chondrite-normalized REE abundance for mantle nodules (A-wiyhout cryptic metasomatism enrichement; B-with cryptic metasomatism enrychement)

1-garnet-bearing lherzolites; 2-spinel lherzolites; 3-dunites; 4-harzburgites 5-wehrlites

Table 2. Temperature and pressure in the mantle nodules

NN	Sample #	TEMPERATURE (°C) Boyd, Shairer 1964	Wells 1977	Ehrenberg 1982	Sachtleben, Seck, 1981	PRESSURE (kb) Ring-wood 1975	Mysen Boettcher 1975
		Spinel - garnet lherzolites					
1	6/1	985	1129	1155		22	
2	32D-A5	1020	1123	1151		24	
3	32D-A5-1	875	1016	1020		20	
4	32D-A6	990	1079	1101		24	
5	32D-A28	990	1065	1083		23	
6	32Y-A9	960	1050	1065		22	
7	32Y-A42				1180	27	
		Spinel lherzolites					
8	32D-A13	880	922	878	882		
9	32D-A14	910	982	970			
10	32SV-A49/2	825					
11	32K-A105/1	840	850	853			12
12	32K-A107	955	1005	1003			
13	32R-A118	980	1043	1048			18
		D u n i t e s					
14	32Y-A37	970					
		H a r z b u r g i t e s					
15	32SN-N15	905					14

Note: Samples 1-9 & 14 - from the Yuzhnoye body; 10 - from the Severnoye Verchneye body; 15 - from the Severnoye Nizhneye body; 11-12 - from the Martovskoye body; 13 - from the Radock-1 sill.

rocks by light lanthanides due to mantle metasomatism.

The chemical composition of rock-forming minerals of the nodules has been studied to establish P-T conditions of equilibration of mantle mineral assemblage. On the basis of calculation using different geothermo-barometers (Boyd & Shairer 1964; Ehrenberg 1982; Mysen & Boettcher 1975; Ringwood 1975; Sachtleben & Seck 1981; Ukhanov et al. 1988; Wells 1977) (Table 2) a schematic section of the upper mantle below Jetty Peninsula can be constructed from petro-geochemical data and evidence on the composition and the frequency of occurence of nodules. This section is limited to a depth of about 80 km (27 kb) and is characterized by vertical petrographical and geochemical heterogeneity, i.e. by changing rock composition with depth (Fig.6).

Garnet-bearing lherzolites, rocks, least of all affected by mantle metasomatism, occur from 80 km to near 60 km depth. Spinel lherzolites occur from about 60 km up to the base of the crust at a depth of 30-35 km (Ravitch et al. 1978). Throughout the entire section of this unit, numerous lenses and interlayers of dunites, and rare harzburgites and very rare wehrlites, both most likely representing the refractory residue of varying degrees of partial melting of spinel lherzolites, are developed.

Alkaline-ultrabasic magmas are commonly regarded as the result of equilibrium partial melting of garnet peridotite containing small ammounts of CO_2 and H_2O at pressures greater than 30 kb and tempera-tures of 1150-1200°C (Ringwood 1975). These conditions are commensurate with those observed at Jetty Peninsula. The tempera-ture of ascending magmas (not less than 1150° C) (Andronikov 1990) points to magma generation below the depth for formation of garnet-bearing lherzolites i.e. near the boundary of diamond and diamond-graphite facies of upper mantle (Sobolev 1974).

Near parallel REE patterns (Fig.7), with strong enrichement of light elements (INAA method) (Appendix, Table 4) and marked fractionation of lanthanides indicate common genetic features and common mantle sources for the alkaline-ultrabasic magmas. In studying the behavior of REE the

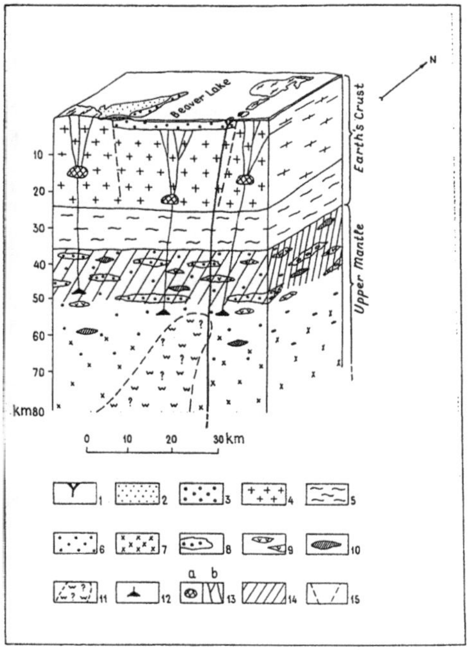

Fig.6 A composite section of lithosphere in the Beaver Lake area

1-feeders to alkaline-ultrabasic bodies; 2-5-Earth's Crust rocks: 2-moraine, 3-Permian deposits, 4-"granitic" layer, 5-"basaltic" layer; 6-11-Mantle rocks: 6-spinel lherzolites, 7-garnet-bearing lherzolites, 8-dunites, 9-harzburgites, 10-wehrlites, 11-hypothetical intra-mantle intrusion; 12-mafic magma sources; 13-a: intermediate magma chambers; b: mafic dykes; 14-zone of cryptic mantle metasomatism; 15-faults bounding the Beaver Graben

Kay-Gast model (1973), using the crystal-melt distribution coefficients was followed. During partial melting of a source rock lanthanides are partitioned melt and solid phases and the concentration of light REE in the melt is inverse related to the degree of melting. The Ce/Yb ratio is the main index of the degree of fractionation of lanthanides. Judging from this ratio in fresh polcenites (100-108), the volume of initial melting did not exceed 5%.

The active fenitization of xenoliths by Archaean and mainly Permian host rocks and the high degree of autometamorphism point to enrichement of magmas by the sodium-rich fluids. These processes are particularly intensive in polcenites from the sills (Egorov et al. 1989).

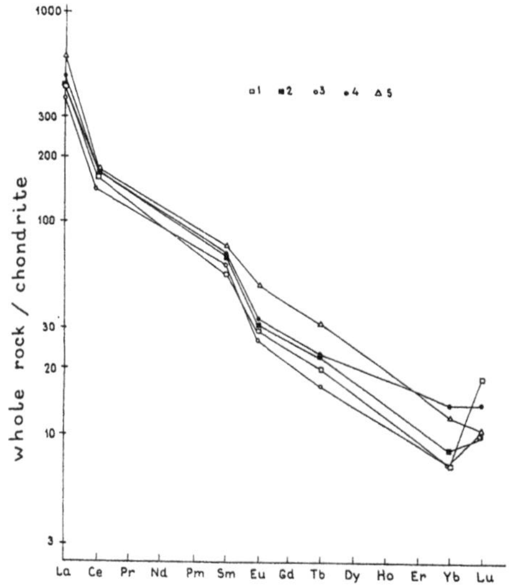

Fig.7 Chondrite-normalized REE abundances for the alkaline-ultrabasic rocks

1-biotite-pyroxene alkaline picrites; 2-nepheline-melilite-pyroxene picrites; 3-olivine melanephelinites; 4-melilite-olivine melanephelinites; 5-polcenites

5 CONCLUSION

Thus, from the studies conducted, the following conclusions can be drawn:

1. All subisometric in plan bodies of Jetty Peninsula are represented by small polyphase stocks with analogues in Paraguay, South Africa and Australia (Andronikov 1990; Egorov et al. 1989 and others).

2. Mesozoic alkaline-ultrabasic magmatism is associated with middle stages of Gondwana break-up (120-150 Ma).

3. All polyphase stocks have the same sequence of emplacement: tuffaceous breccia-melilite lamprophyre - alkaline picrite and ultrabasic foidite. The sills are composed of polcenite whereas the dykes are single-phase bodies composed of polcenite, alkaline picrite or ultrabasic foidite.

4. The alkaline ultrabasic rocks of Jetty Peninsula contain numerous mantle nodules: garnet-bearing and spinel lherzolites; garnet-free harzburgites, dunites and wehrlites. Three main types of mantle material are recognized: "fertile", "depleted" and light REE enriched (resulting from mantle metasomatism). The depth of origin of these nodules is 35-80 km (12-27 kb), the equilibration temperature of the mantle mineral assemblages was 825-1180°C. A schematic section for the upper mantle to a

depth of 80 km can be constructed for this region.

5. The alkaline-ultrabasic magmas of Jetty Peninsula were generated at a depth not less than 100 km and by small (not exceed 5%) degrees of partial melting of garnet peridotite containing small amounts of CO_2 and H_2O.

6 ACKNOWLEDGEMENTS

We would like to thank Prof. L.Kogarko for the help in determination of content and distribution of REE in nodules and for discussions on the result obtained. The authors are particulary grateful to E.Miloradovskaya who translated the text.

REFERENCES

Andronikov A.V. 1990. Mesozoic alkaline-ultrabasic magmatism of Jetty Peninsula (Prince Charles Mountains, East Antarctica) (in Russian). Author's abstract. Leningrad, 228 p.

Boyd F.R. & Shairer J.F. 1964. The system $MgSiO_3$-$CaMgSi_2O_6$. J.of Petrology 3: 275-309.

Dawson J.B. 1980. Kimberlites and their xenoliths. Berlin: Springer Verlag, 252 p.

Egorov L.S., Surina N.P., Andronikov A.V. et al. 1989. Alkaline-ultrabasic rocks of Jetty Peninsula, formation and geodinamical prerequisites of the revealing of kimberlite and carbonatite mineral deposits on Antarctic and other platform (in Russian). Open-file report. Leningrad, 256 p.

Ehrenberg S.N. 1982. Petrogenesis of garnet lherzolite and megacristalline nodules from Thumb, Navajo volcanic field. J.of Petrology 23: 507-547.

Jaques A.L., Lewis J.D., Smith C.B. 1986. The kimberlites and lamproites of Western Australia. Geol.Surv. of West.Austral. 132, 268 p.

Kay R.W. & Gast P.W. 1973. The rare-earth content and the origin of alkaline-rich basalts. J.Geol. 3: 653-683.

Laiba A.A., Andronikov A.V., Egorov L.S., Fedorov L.V. 1987. Stock-like and dyke bodies of alkaline-ultrabasic composition in Jetty Peninsula (Prince Charles Mountains, East Antarctica) (in Russian). In V.Ivanov & G.Grikurov (eds.), Geological and geophysical investigations in Antarctica: 35-47. Leningrad.

Mysen B.O. & Boettcher A.L. 1975. Melting of hydrous mantle: I.Phase relations of natural peridotites at high pressures and temperatures with controlled activities of water, carbon dioxide and hydrogen. J.of Petrology 16: 520-548.

Ravich M.G., Soloviev D.S., Fedorov L.V. 1978. Geological structure of MacRobertson Land (East Antarctica) (in Russian). Leningrad, Gidrometeoizdat, 230 p.

Ringwood A.E. 1975. Composition and Petrology of Earth's Mantle. McGrew-Hill, New-York, 618 p.

Sachtleben Th., Seck H.A. 1981. Chemical control of Al-solubility in orthopyroxene and its implications on pyroxene geothermometry. Contr.Miner.Petrol. 78: 157-165.

Sobolev N.V. 1974. Deep-seated inclusions in kimberlites and the problem of the upper mantle composition (in Russian). Novosibirsk, Nauka, 264 p.

Ukhanov A.V., Riabchikov I.D., Kharkiv A.D. 1988. Lithospheric mantle of Jakutia kimberlite province (in Russian). Moscow, Nauka, 286 p.

Wells P.R.A. 1977. Pyroxene thermometry in simple and complex system. Contr.Miner. Petrol. 62: 129-139.

APPENDIX

Table 1. Representative chemical analises of alkaline-ultrabasic rocks of Jetty Peninsula (wt.%)

	Polcenite					Alnoite			Bt-Px picrite	
	1	2	3	4	5	6	7	8	9	10
SiO_2	36.06	37.21	37.00	35.00	35.97	37.24	36.50	35.80	41.35	40.09
TiO_2	2.28	2.13	2.18	2.38	2.20	2.38	2.27	2.49	1.80	1.80
Al_2O_3	9.22	7.11	7.39	9.79	8.60	10.55	9.40	11.14	7.80	7.66
Fe_2O_3	2.57	3.97	3.56	4.29	3.77	4.03	5.76	4.23	0.39	0.22
FeO	7.32	6.68	7.06	6.26	7.51	6.18	5.11	5.69	7.89	7.95
MnO	0.23	0.18	0.17	0.22	0.19	0.21	0.18	0.21	0.19	0.19
MgO	19.76	20.45	19.26	14.23	15.71	12.61	12.73	12.41	18.56	20.57
CaO	13.02	12.82	13.60	14.63	15.27	15.05	16.48	15.37	8.85	8.56
Na_2O	2.00	2.31	2.20	2.12	1.80	1.84	1.59	1.38	2.05	2.01
K_2O	2.08	1.49	1.92	3.68	2.94	1.96	2.07	2.10	1.99	1.69
P_2O_5	0.89	0.68	0.72	1.11	0.85	1.00	1.00	1.05	0.72	0.75
Ignition losses	4.41	4.97	4.63	5.64	5.04	6.83	6.68	7.93	7.47	8.03
Total	99.84	100.00	99.69	99.35	99.85	99.88	99.77	99.80	99.06	99.52
Na_2O+ +K_2O	4.08	3.80	4.12	5.80	5.04	3.80	3.66	3.48	4.04	3.70

	Bt-Px picrite			Ne picrite		Mel-Px picrite		Ne-Mel-Px picrite		
	11	12	13	14	15	16	17	18	19	20
SiO_2	42.07	40.17	38.50	35.60	35.76	39.07	38.30	35.83	39.50	39.96
TiO_2	1.72	1.77	1.98	1.74	1.50	2.06	1.77	1.71	1.86	1.95
Al_2O_3	8.24	6.10	6.76	5.76	5.34	8.95	7.10	6.68	7.44	8.59
Fe_2O_3	0.88	4.65	3.48	1.51	2.31	2.31	4.63	2.69	3.51	3.20
FeO	7.01	5.05	5.17	6.86	6.70	6.37	5.31	6.34	6.34	6.25
MnO	0.18	0.13	0.16	0.17	0.17	0.20	0.14	0.21	0.17	0.18
MgO	18.21	22.80	22.99	23.76	23.73	18.06	19.11	23.87	21.08	21.41
CaO	7.57	9.00	9.14	9.10	8.69	10.22	11.03	10.02	10.88	10.67
Na_2O	2.50	0.58	0.86	1.48	1.42	1.63	2.00	1.67	2.10	2.34
K_2O	1.76	1.33	1.64	1.74	1.80	2.37	1.77	1.66	1.82	1.48
P_2O_5	0.78	0.69	0.75	0.66	0.63	0.66	0.63	0.64	0.65	0.63
Ignition losses	8.60	6.43	7.70	11.86	12.18	7.56	7.85	8.43	4.46	6.74
Total	99.52	98.70	99.99	100.24	100.23	99.46	99.64	99.75	100.15	100.24
Na_2O+ +K_2O	4.26	1.91	2.50	3.22	3.22	4.00	3.77	3.33	3.92	3.82

	Ne-Mel-Px picrite		Mel-Ol melanephelinite				Ol melanephelinite			
	21	22	23	24	25	26	27	28	29	30
SiO_2	37.72	37.22	35.82	36.10	35.55	38.16	36.64	38.94	38.70	40.30
TiO_2	1.74	1.57	2.26	2.41	2.11	2.20	2.30	2.38	2.28	2.21
Al_2O_3	6.67	6.53	8.18	8.14	7.70	8.66	10.09	9.81	10.20	9.60
Fe_2O_3	2.47	3.27	3.65	5.25	5.26	3.43	3.23	2.14	4.14	3.06
FeO	6.58	6.01	5.96	5.72	5.11	5.71	5.82	7.43	6.08	6.47
MnO	0.17	0.16	0.21	0.18	0.17	0.19	0.17	0.19	0.15	0.13
MgO	25.20	22.38	15.84	17.46	16.16	14.99	13.49	11.93	12.00	12.84
CaO	9.94	9.95	11.85	12.50	13.45	13.53	11.58	11.91	12.13	11.36
Na_2O	1.64	1.72	2.19	2.54	2.77	2.74	2.84	2.62	3.00	2.11
K_2O	1.20	1.34	2.66	2.13	2.38	2.28	2.32	2.10	2.10	2.24
P_2O_5	0.52	0.39	0.91	0.89	0.87	0.80	0.58	0.64	0.69	0.66
Ignition losses	6.47	9.55	9.99	6.91	7.64	7.28	10.81	9.69	8.11	8.12
Total	100.32	100.09	99.52	100.23	99.17	99.97	99.87	99.78	99.58	99.10
Na_2O+ +K_2O	2.84	3.06	4.85	4.67	5.15	5.02	5.16	4.72	5.10	4.35

Table 2. Content of trace elements in alkaline-ultramafic rocks of Jetty Peninsula (ppm)

	Polcenite (25)	Alnoite (1)	Bt-Px picrite (5)	Ne picrite (2)	Mel-Px picrite (4)	Ne-Mel-Px picrite (12)	Melanephelinite Mel-Ol (1)	Melanephelinite Ol (4)
Be	1.5-3.7	3.3	2.3-3.2	2.1-4.1	2.0-2.9	1.7-3.5	3.3	2.0-3.8
Sc	17.9-40.5	27.7	21.3-42.0	18.4-26.1	22.4-27.1	19.9-33.1	51.7	22.6-26.2
Ti	8082-14120	13200	6710-13200	8084-9366	8516-10552	7501-12678	13200	8458-12085
V	102-280	207	83-169	96-142	123-198	108-183	240	125-140
Cr	280-1611	431	941-1746	872-1656	755-1817	774-2166	917	504-700
Mn	912-1405	1353	674-1128	834-1024	790-2181	1032-1816	1266	797-1579
Co	42.1-77.8	44.8	47.4-70.3	59.8-63.7	48.0-67.3	53.3-78.8	68.2	34.6-47.9
Ni	222-815	270	671-1045	779-998	529-1061	633-1056	543	214-354
Zn	72-149	72	72-164	77-144	144-155	72-175	144	73-102
Ga	5.3-19.0	15.6	4.9-12.9	4.0-9.7	4.3-12.9	4.0-16.4	7.0	9.8-17.6
Y	11.6-40.1	15.3	10.3-26.1	18.7-24.5	12.2-19.1	11.5-23.8	22.1	10.4-13.8
Zr	111-450	199	167-285	170-249	118-176	130-217	223	144-188
Nb	56.6-109.0	86.2	53.1-102.0	56.5-100.0	55.5-69.2	49.0-74.3	91.6	58.5-76.2
In	1.1-2.1	1.6	1.0-1.7	1.2-1.4	1.0-1.8	1.2-2.0	1.9	1.1-1.6
Sn	2.4-16.5	2.4	2.4-10.5	4.8-12.3	2.4-14.7	2.4-8.2	4.8	5.8-10.9
Ba	918-9000	1500	1070-2029	1375-1593	1189-2080	926-1500	1837	1314-1588
Pb	8.6-62.1	13.9	9.3-25.9	8.5-42.9	8.0-28.1	4.0-24.6	20.3	12.0-25.8
Ni/Co	4.5-9.7	6.0	11.8-15.4	12.2-16.7	11.0-15.8	10.3-14.4	8.0	6.2-7.4

Note: quantite of analises is in the brackets

Table 3. Contents of REE in deep-seated nodules of Jetty Peninsula (ppm).

	Garnet-bearing lherzolites								Spinel lherzolites				
	1	2	3	4	5	6	7	8	9	10	11	12	13
La	1.004	0.460	1.310	1.550	2.896	0.920	0.165	0.770	0.490	1.221	0.690	0.148	1.820
Ce	1.980	0.804	2.202	3.206	5.701	2.502	0.562	1.925	1.420	2.905	1.600	0.436	3.860
Sm	0.286	0.289	0.240	0.464	0.475	0.543	0.140	0.360	0.330	0.530	0.309	0.054	0.674
Eu	0.114	0.116	0.086	0.132	0.124	0.194	0.057	0.120	0.112	0.152	0.135	0.026	0.225
Tb	0.058	0.059	0.050	0.072	0.082	0.157	0.040	0.067	0.074	0.090	0.097	0.025	0.120
Yb	0.292	0.230	0.234	0.289	0.276	0.531	0.158	0.190	0.391	0.224	0.460	0.094	0.354
Lu	0.039	0.037	0.034	0.033	0.036	0.085	0.027	0.025	0.059	0.026	0.070	0.010	0.045

	Spinel lherzolites				Harzburgites				Wehrlites		
	14	15	16	17	18	19	20	21	22	23	24
La	0.310	0.890	1.580	5.310	1.450	2.920	0.630	0.580	5.370	2.560	3.160
Ce	0.718	1.738	2.850	11.300	3.160	5.600	1.240	1.553	10.500	6.252	6.404
Sm	0.145	0.216	0.405	1.010	0.250	0.411	0.194	0.340	0.958	1.271	0.730
Eu	0.057	0.056	0.146	0.272	0.058	0.162	0.060	0.145	0.203	0.370	0.192
Tb	0.046	0.041	0.087	0.110	0.046	0.104	0.057	0.095	0.094	0.226	0.116
Yb	0.220	0.175	0.370	0.375	0.132	0.333	0.210	0.373	0.200	0.790	0.358
Lu	0.023	0.022	0.056	0.042	0.016	0.046	0.027	0.059	0.024	0.113	0.039

	Dunites				
	25	26	27	28	29
La	0.370	3.570	0.686	0.786	3.790
Ce	0.829	8.732	1.706	1.735	9.220
Sm	0.173	1.164	0.440	0.356	1.250
Eu	0.066	0.306	0.143	0.147	0.400
Tb	0.052	0.151	0.087	0.128	0.251
Yb	0.223	0.239	0.310	0.420	0.530
Lu	0.030	0.030	0.044	0.078	0.069

Table 4. Concentration of REE in alkaline-ultrabasic rocks of Jetty Peninsula (ppm)

	Bt-Px		picrite		Ne-Mel-Px picrite		Melanephelinite Ol	Mel-Ol	Polcenite			
	1	2	3	4	5	6	7	8	9	10	11	12
La	108	104	119	109	96	118	116	100	115	151	147	125
Ce	131	136	135	135	130	147	130	123	136	171	169	145
Sm	13.0	12.5	14.1	12.2	12.4	16.3	14.1	14.0	14.6	18.6	14.0	16.4
Eu	2.3	2.1	2.1	2.4	2.1	2.7	2.0	2.1	2.4	3.2	3.2	2.6
Tb	1.3	0.9	0.8	1.1	1.4	1.2	0.8	0.9	1.1	1.8	1.7	1.3
Yb	1.7	1.2	1.4	1.5	1.5	1.6	1.4	1.3	2.3	1.6	2.0	1.6
Lu	0.6	0.5	0.7	0.7	0.3	0.3	0.2	0.3	0.4	0.3	0.4	0.3
La/Yb	63.5	86.7	85.0	72.7	64.0	73.7	82.9	76.9	50.0	94.4	73.5	78.1
Ce/Yb	77.1	113.3	96.4	90.0	86.7	91.9	92.9	94.6	59.1	106.9	84.5	90.6

Gondwana Eight, Findlay, Unrug, Banks & Veevers (eds) © 1993 Balkema, Rotterdam. ISBN 90 5410 304 3

Prospects of mineral detection associated with alkaline-ultrabasic rocks (Prince Charles Mountains)

L.S.Egorov & A.V.Andronikov
VNII Okeangeologia, St.Petersburg, Russia

ABSTRACT: Manifestations of alkaline-ultrabasic rocks in the northern Prince Charles Mountains of East Antarctica were studied during the 1986-1987 summer field season. Pyrope garnets were discovered in lherzolite nodules within one stock of alkaline ultrabasite some time ago, and this discover, plus the general abundance of mantle inclusions, enabled the identification of subisometric in plan bodies with kimberlite diatremes. However, as a result of detailed geological and petrological studies it has been established that all the intrusive bodies can be classified as melilite lamprophyres, alkaline picrites and ultrabasic foidites. Elsewhere small diamond grains of no economic significance were found in similar rocks. The following conclusions are based on detailed studies of the Antarctic alkaline-ultrabasic rocks, analyses of the geological structure of the region, and correlation with similar intrusive bodies of other regions of the world: 1) the region between 71° and 73° S is the most promising for the discovery of diamonde-bearing rocks; 2) tuffaceous breccias from stock-like bodies of alkaline ultrabasic rocks in the Jetty Peninsula could contain accessory diamonds. Besides the alkaline-ultrabasic rocks there are also Early Palaeozoic dykes of lamproite-like rocks in the southern Prince Charles Mountains.

In East Antarctica the manifestations of alkaline-ultrabasic and mainly alkaline-basic magmatism are known over a vast area (Priestly Peaks in Enderby Land, Vestfold Hills in Princess Elizabeth Land, Gaussberg in Queen Mary Land and Bunger Hills in Wilkes Land) and are related predominantly to the development of rift systems (Fig.1). The manifestations of alkaline ultrabasites in the northern part of Prince Charles Mts. on Jetty Peninsula area (Fig.2) have received the most study.

In lherzolite nodules from the Yuzhnoye intrusive body essentially pyrope garnets were discovered (Andronikov 1990, Andronikov et al. 1989). This discovery and the general abundance of mantle inclusions enabled a reliable identification (for that time) of the sub-isometric in plan bodies of alkaline-ultrabasic rocks with kimberlite diatremes (Laiba et al. 1987). The geological and petrological studies carried out during 1986-87 summer field season have allowed us to identify more accurately the alkaline-ultrabasic rocks of Jetty Peninsula.

It has been found that all intrusive bodies are composed of melilite lamprophyres, alkaline picrites and ultrabasic foidites pertaining to a diatreme-dyke magmatic formation (Egorov 1989). Rocks such as these elsewhere in the world have yielded diamond grains of no industrial importance (Dovzhikov 1986, Egorov 1989, Moore & Verwoerd 1985, Neal & Nixon 1985).

There four initial prerequisities for the discovery of potential diamond-bearing areas in Prince Charles Mountains:

1. In almost all the regions of alkaline-ultrabasic magmatism elsewhere (Eastern Siberia, north of the East European Platform, Southern Africa) the bodies that are composed of alkaline-ultrabasic lamprophyres, alkaline picrites and ultrabasic foidites are found to have a spacial connection with kimberlites of the same age. All these formations are associated with systems of deep faults.

2. The similarity of the East Antarctic intrusive bodies with diamond-bearing lamproite diatrems of Western Australia both in structure and in history was noted during mapping the bodies of alkaline ultrabasites of Jetty Peninsula (Egorov & Andronikov 1989) and confirmed during further processing of the data. The studied intrusions are most similar to polyphasic bodies of analcime-bearing alkaline picri-

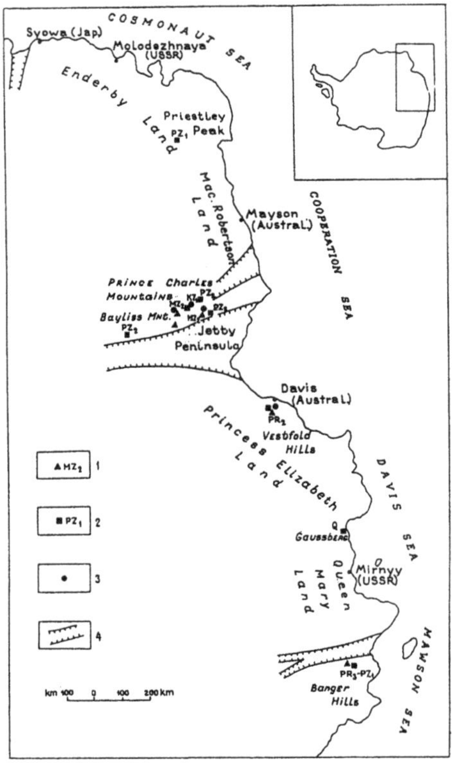

Fig.1 Scheme of the location of alkaline-ultrabasic and alkaline-basic manifestations in central part of East Antarctica (by the data: Andronikov 1990, Mikhalsky et al. 1991, Sheraton & Cundari 1980, Sheraton & England 1980, Sheraton et al. 1990, Tingey 1976)

1-alkaline-ultrabasic rocks and their age; 2-alkaline-basic rocks and their age; 3-the places of location of deep-seated inclusions; 4-rift zones

tes of the Wandagee region, which in some cases contain accessory diamonds (Jaques et al. 1986). In all both Australian and Antarctic cases the formation of the bodies began with the development of tuffaceous breccia, but continued and ended with injection of alkaline-ultrabasic magma (melilite lamprophyre and alkaline picrite (Jetty Peninsula), and olivine lamproite, alkaline picrite and fitzroite in the Wandagee region of North-Western Australia. The industrial diamonds in lamproite diatremes of North-Western Australia occur where tuffaceous breccia larmed during the first phase of injection (Jaques et al 1986).

3. In the Middle Timan region (Russia) diatremes consisting of alkaline picrites identical to East Antarctic rocks are known. These bodies are commonly polyphasic, and accessory grains of diamonds in them were detected in the tuffaceous breccias only and not in massive rocks (Dovzhikov 1986).

4. In alkaline-ultrabasic magmatites (not kimberlites) enriched with mantle inclusions, the presence of rare diamond grains has been established in some cases (Egorov 1989, Jaques et al. 1986, Moore & Verwoerd 1985, Neal & Nixon 1985). Analogous finding are quite probable in the Yuzhnoye body of Jetty Peninsula (Fig.3). The rocks of this stock contain nodules of garnet-bearing lherzolites picked up by magma at the greatest depth (around 80 km) for this region (Andronikov 1990), and numerous xenocrysts, which are the products of disintegration of these nodules. In this connection the contamination of the stock by completely disintegrated inclusions of the even deeper mantle rocks, indicates a strong probability that these rocks could

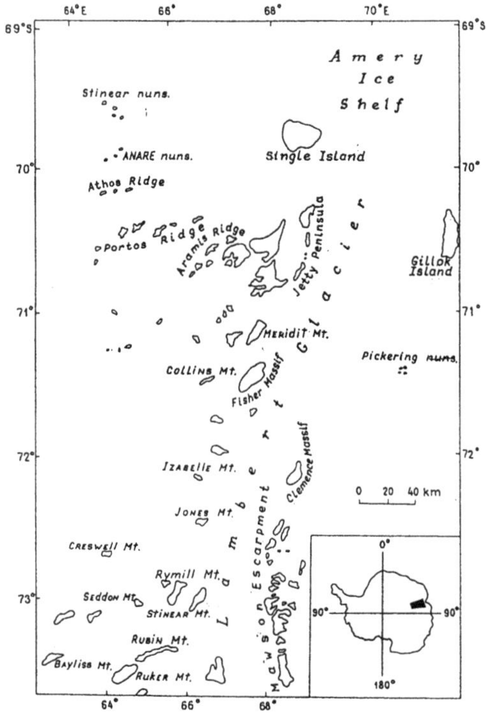

Fig.2 Schematic map of outcrops in Prince Charles Mountains

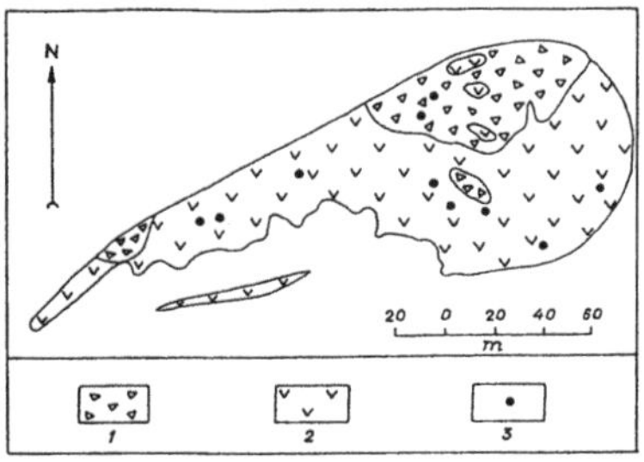

Fig.3 Schematic geological map of the Yuzhnoye body (without the snow cover)

1-tuffaceous breccia; 2-alkaline picrites; 3-garnet-bearing lherzolite nodules location

contain diamonds. Of the rocks composing the stock the tuffaceous breccias are the most prospective for diamonds, by analogy with similar formayion in other regions of the world.

On the basis of the study of alkaline-ultrabasic rocks of Jetty Peninsula with allowance for general regularites of spatial-temporal manifestations of alkaline-ultrabasic magmatism (Egorov 1989) and the geological structure of Lambert Glacier (Ravich et al 1978), the following conclusions about the prospects of detection of minerals connected with alkaline-ultrabasic rocks can be made:

1. The area southward of Jetty Peninsula between 71 and 73 latitudes is the most promising for kimberlite detection. The Fisher, Clemence and Shaw Massifs and Mawson Escarpment are of particular interest (in the summer field season of 1988-1989 the first kimberlite dyke was found by Russian geologists in the Fisher Massif).

2. Starting from the similarity in the structure and history of the formation of East Antarctic stocks with lamproite bodies of Western Australia, the tuffaceous breccias from the Antarctic bodies may contain individual diamond grains. In order to determine whether the alkaline ultrabasites in Jetty Peninsula contain diamonds it is necessary to sample those breccias representing the products of the first quesi-intrusive phase of injection.

Along with the rocks considered the manifestations of Middle-Palaeozoic (413-430 Ma) dyke lamproite-like rocks at Bayliss and Rubin Mountains (Sheraton & England 1980, Mikhalsky et al. 1991) are known. Thus it can be suggested that the native diamond potential in the region can be connected both with kimberlites and lamproites formed over a long period.

REFERENCES

Andronikov A.V. 1990. Spinel-garnet lherzolite nodules from alkaline-ultrabasic rocks of Jetty Peninsula (East Antarctica). Antarctic Science 2(4):321-330.

Andronikov A.V., Dovzhikov N.A., Shablya A.O. 1989. The pyropes of peridotite inclusions from the stock of alkaline-ultrabasic rocks of Jetty Peninsula (Prince Charles Mountains, East Antarctica) (in Russian). Inf. Bull. SAE 111: 17-21.

Dovzhikov N.A. 1986. The mineralogy and geochemical features of alkaline-ultrabasic dyke series of Middle Timan (in Russian). Author's abstract: 18 p.

Egorov L.S. 1989. Alkaline-ultrabasic magmatism in space and in time. In: Magmatic formations (petrology, geology) (in Russian). Sverdlovsk: 30-47.

Egorov L.S. & Andronikov A.V. 1989. The new data about stock-like bodies of alkaline-ultrabasic rocks of Jetty Peninsula (Mac Robertson Land, East Antarctica), obtained in 1986-1987 field season (in Russian). Inf.bull.SAE 111: 5-13.

Jaques A.L., Lewis J.D., Smith C.B. 1986. The kimberlites and lamproites of Western Australia. Geol.Surv.West.Austral. Bull.132: 268 p.

Laiba A.A., Andronikov A.V., Egorov L.S., Fedorov L.V. 1987. The stock-like and dyke bodies of alkaline-ultrabasic composition in Jetty Peninsula (Prince Charles Mountains, East Antarctica) (in Russian). In V.L.Ivanov & G.E.Grikurov (eds.), Geological and geophysical investigations in Antarctica. Leningrad: 35-47.

Mikhalsky E.V., Andronikov A.V., Beliatsky B.V. 1991. Mafic and ultramafic igneous suites in the Lambert-Amery rift zones. 6th International Symposium on Antarctic Earth Sciences, Abstracts, Tikyo: 406-411.

Moore A.E. & Verwoerd W.J. 1985. The olivine melilite-"kimberlite"-carbonate suite of Namaqualand and Bushmanland, South Africa. Transact.Geol.Soc.South Africa 88, 2: 281-293.

Neal C.R. & Nixon P.N. 1985. Spinel-garnet relationships in mantle xenoliths from Malaita alnoites, Solomon Island, South-Western Pacific. Transact.Geol.Soc.South Africa 88, 2: 347-353.

Ravitch M.G., Soloviev D.S., Fedorov L.V. 1978. Geological structure of Mac Robertson Land, East Antarctica (in Russian). Leningrad, Hydrometeoizdat: 230 p.

Sheraton J.W. & England R.N. 1980. Highly potassic mafic dykes from Antarctica.

J.Geol.Soc Austral. 27: 129-135.
Sheraton J.W. & Cundari A. 1980. Leucitites from Gaussberg, Antarctica. Contr. Miner.Petrol. 71: 417-427.
Sheraton J.W., Black L.P., McCulloch M.T., Oliver R.L. 1990. Age and origin of a compositionally varied mafic dyke swarm in the Bunger Hills, East Antarctica. Chem.Geol. 85: 215-246.
Tingey R.J. 1976. Basic and alkaline intrusive rocks in the Prince Charles Mountains. Bur.Miner.Res.Rep. 194: 73.

Gondwana Eight, Findlay, Unrug, Banks & Veevers (eds) © 1993 Balkema, Rotterdam. ISBN 90 5410 304 3

Jurassic faulting and magmatism in the Transantarctic Mountains: Implications for Gondwana breakup

Terry J.Wilson
Byrd Polar Research Center & Department of Geological Sciences, Ohio State University, Columbus, Ohio, USA

ABSTRACT: New structural mapping in the southern Victoria Land and Beardmore Glacier regions of the Transantarctic Mountains, Antarctica, has documented the regional development of Jurassic normal faulting and consistent Jurassic Ferrar dolerite dike trends. Dikes form two subperpendicular sets with northwest and east-northeast trends. Two sets of conjugate normal faults, subparallel to the dike trends, are developed in both areas. Because the subperpendicular dike and fault sets formed contemporaneously, these orientation patterns document a regionally consistent Jurassic triaxial strain field characterized by both northeast-southwest and northwest-southeast extension. The parallelism between the Jurassic structural trends within the Transantarctic Mountains and offshore basins in the Weddell and Ross embayments suggests that these basins may have initially formed in the Jurassic, within a zone of failed rifting that transected the Antarctic sector of Gondwana.

Time relations between faults and Ferrar dolerites indicate that crustal extension occurred along the Transantarctic trend before and during the Ferrar magmatism. Similar time relations and orientation patterns of Jurassic structures in adjacent sectors of Gondwana indicate that a regional tensional stress field was present within the lithosphere of the supercontinent prior to breakup. Crustal extension followed by magmatism is consistent with recent models in which rifts form where extending lithosphere encounters a region of anomalously hot mantle, possibly associated with mantle plumes.

1 INTRODUCTION

The fragmentation of the Gondwana supercontinent was preceded by extremely widespread Jurassic magmatism, generally taken as an indicator of a thermal 'event' related either to the insulating effect of the supercontinent, to back-arc phenomena associated with the active margin of Gondwana, to mantle plume activity, or to some combination of these factors (Cox 1978, 1988; Dalziel et al. 1987; White & McKenzie 1989). Crustal extension preceded and accompanied magmatism in southern Africa and South America, but in Antarctica the presence of rift structures has only been inferred from the apparent distribution of Jurassic igneous rocks (Kyle et al. 1981; Schmidt & Rowley 1986; Elliot 1991), and from structural interpretations of subglacial morphology and geophysical trends in West Antarctica (Doake et al. 1983; Kadmina et al. 1983; Jankowski et al. 1983; Maslanyj & Storey 1990). A Jurassic age has been inferred for some subglacial rift basins (Garrett et al. 1988; Kristoffersen & Hinz 1991), but elsewhere the basins may have resulted from successive episodes of Mesozoic and Cenozoic rifting (Davey 1981; Cooper et al. 1987; Maslanyj & Storey 1990; Cooper et al. 1991). Indeed, it has been argued recently that such basins may be mainly Cenozoic in age (Behrendt & Cooper 1991) and, therefore, unrelated to the earliest phase of Gondwana breakup.

The timing of magmatism relative to crustal extension is a key indicator of the evolving stress regime in the lithosphere during rifting and, thus, of the driving forces for supercontinent breakup. Where lithospheric plate stresses activate rifting, regional tectonic stresses will be present prior to and during magmatic activity. Crustal thinning by normal faulting should thus both precede and accompany magmatism, and dike planes, which represent hydraulic fractures that track the principal stress trajectories in the crust (Anderson 1951; Delaney et al. 1986), should form parallel swarms perpendicular to the least compressive stress direction. In contrast, where rifting is initiated above a rising mantle plume, magmatism may precede any crustal extension, in which case normal faults would cut intrusions and volcanic rocks, and early dike arrays would have random orientations, reflecting the absence of a regional tectonic stress field during emplacement. For example, feeder dikes for the Deccan traps of India are randomly oriented, whereas dikes that cut the basalt flows are oriented in parallel swarms, indicating that volcanism preceded crustal extension (Hooper 1991). On this basis, Hooper (1991) argued for a plume-activated rifting process during breakup of that sector of Gondwana.

I report here the initial results of a study of the time relations between Jurassic Ferrar magmatism and crustal extension in the Antarctic sector of Gondwana. Structural mapping in the southern

Victoria Land and Beardmore Glacier sectors of the Transantarctic Mountains (Figure 1) has revealed widespread Jurassic faulting and the presence of regionally consistent Jurassic dolerite dike trends. Crustal extension preceded and accompanied the Jurassic magmatism in Antarctica, and Jurassic dike patterns are similar to dike trends found in the adjacent portions of the Gondwana supercontinent. This consistent dike pattern suggests that a regionally persistent stress regime may have existed along an extensional corridor ca. 5,000 km in length during the magmatic activity that preceded breakup of the supercontinent.

2 GEOLOGIC SETTING OF THE TRANSANTARCTIC MOUNTAINS

The Transantarctic Mountains are a young mountain belt that transects the Antarctic continent (Figure 1); uplift of the present mountain chain is related to Mesozoic and Cenozoic rifting (Fitzgerald et al. 1986; Stern & ten Brink 1989; Fitzgerald & Stump 1991). The Transantarctic Mountains developed along the boundary between the Precambrian crust of cratonic East Antarctica and the mosaic of crustal blocks that make up West Antarctica (Dalziel & Elliot 1982). This boundary appears to have marked the continental margin of East Antarctica between Proterozoic and early Palaeozoic times, and a series of orogenic events affected the margin during this interval (e.g. Elliot 1975a; Borg & de Paolo 1991; Stump 1992). Early Palaeozoic uplift and erosion of the basement complex is marked by the regionally developed unconformity surface known as the Kukri Peneplain. Subsequent subsidence formed a major basin or series of basins in which the Devonian - Jurassic Beacon Supergroup clastic sequence was deposited (Elliot 1975b; Barrett 1982; Collinson 1991).

Voluminous Jurassic magmatic rocks are exposed along the length of the present site of the Transantarctic Mountains (Kyle et al. 1981; Elliot 1991). The earliest volcanic deposits are silicic tuffs and breccias of probable Early Jurassic age (Barrett et al. 1986; Elliot 1991). The extrusion of the Kirkpatrick Basalts and emplacement of the extensive sills, dikes and plugs of Ferrar Dolerite followed in the early Middle Jurassic (Elliot et al. 1985). Elliot (1991) attributed the presence of local topography during extrusion of Jurassic silicic and mafic volcanic rocks to crustal extension and uplift, possibly associated with dome formation along all or

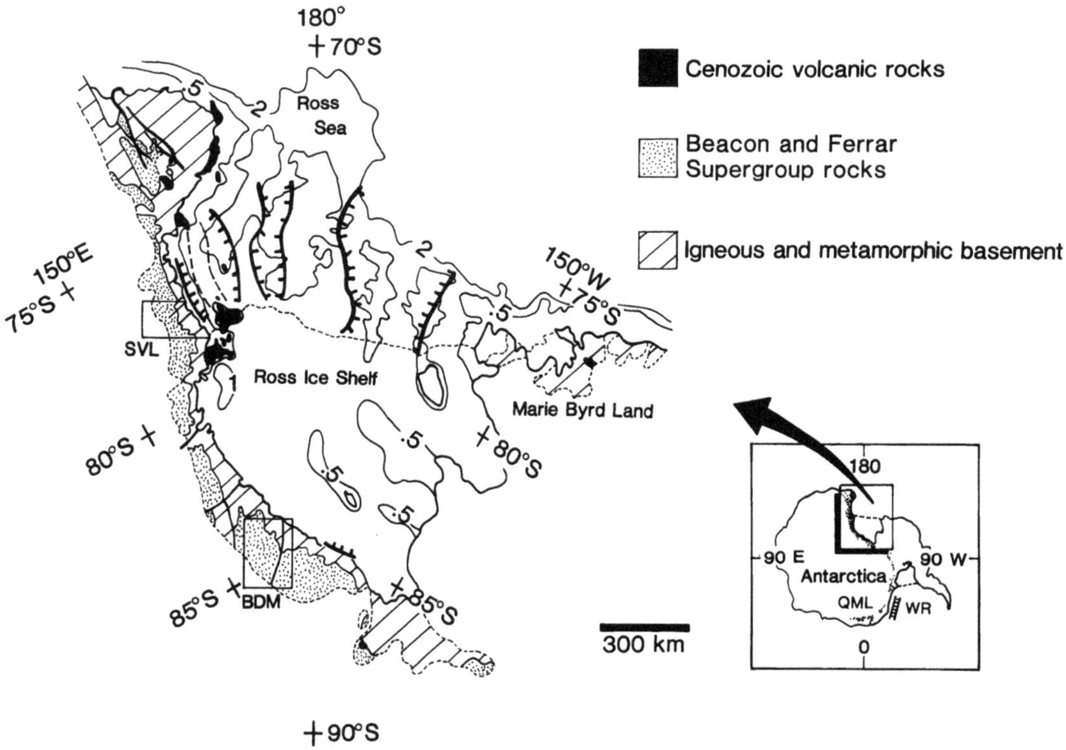

Figure 1. Simplified geologic map of the Transantarctic Mountains. Boxes mark study areas in southern Victoria Land (SVL) and the Beardmore Glacier area (BDM). Structural and morphological basins in the Ross embayment (Ross Sea and Ross Ice Shelf) indicated by hachured lines and isobaths (in kilometres), respectively. Hachures indicate downthrown side of the rift basins. QML: Queen Maud Land; WR: Weddell Rift.

part of the Transantarctic Mountains prior to Ferrar magmatism. Bradshaw (1987), in order to explain the presence of sedimentary intercalations and pillow basalts throughout the sequence, preferred the alternative that steady regional subsidence occurred along the Ferrar province throughout extrusion of the Kirkpatrick Basalts. It is thus not clear whether the Beacon basin was inverted in the Jurassic to form a proto-Transantarctic Mountain chain along the flank of a major "Transantarctic rift system" as proposed by Schmidt and Rowley (1986), or continued to subside as part of such a rift system. Uplift of the present Transantarctic Mountain chain in southern Victoria Land and in the Beardmore Glacier region began in the Eocene (Gleadow & Fitzgerald 1987; Fitzgerald & Stump 1991) and, together with extrusion of volcanic rocks, is related to renewed rifting in the West Antarctic rift system in Cenozoic times (Fitzgerald et al. 1986; Stern & ten Brink 1989; Behrendt & Cooper 1991).

3 JURASSIC STRUCTURES IN THE TRANSANTARCTIC MOUNTAINS

No regional investigation of structures associated with the Ferrar magmatic rocks has been carried out previously. The trends of Ferrar dolerite dikes have been described in some areas of southern Victoria Land (Grapes et al. 1974; Pyne 1984; Bradshaw 1987; Morrison 1989). Elsewhere in the Transantarctic Mountains, dikes have not been described or have been considered uncommon (e.g. Barrett et al. 1986). Earlier workers used the field relations between faults and Ferrar dolerite bodies to demonstrate Jurassic faulting in southern Victoria Land (Hamilton 1965; Korsch 1984; Pyne 1984; Morrison 1989) and at one locality in the Beardmore Glacier area (Collinson & Elliot 1984). However, due to the sparse post-Jurassic stratigraphic record, faults in the Transantarctic Mountains have generally only been found to be younger than Jurassic (e.g. Katz 1982; Barrett et al. 1986).

The present study concentrated on a systematic investigation of the orientation patterns of and age relations between Ferrar dolerite dikes and associated faults. The Ferrar dolerite dikes were found to be widely disseminated throughout both southern Victoria Land and the Beardmore Glacier area, but only locally do they occur in closely spaced arrays that could be described as 'swarms'. The dikes are typically 20 - 300 cm thick, and extend as far as several km in strike length. Individual dikes are made up of a series of en echelon segments, which is a geometry typical of opening-mode tensile hydraulic fractures (Pollard et al. 1982).

Throughout both of the study areas, striated, mesoscopic faults cut Beacon Supergroup strata. Most of the faults examined and measured offset Beacon strata with normal, dip-slip motion of a few meters or less. It is important to note, however, that faults with displacements of tens to hundreds of meters have been mapped on stratigraphic grounds (Hamilton 1965; Barrett & Elliot 1973; Pyne 1984; Bradshaw 1987) and are parallel to the fault trends established in this study, demonstrating that the abundant minor faults are reliable indicators of the large-scale displacement patterns. In the Beardmore Glacier area, some of the faults are associated with regionally developed, northeast-trending monoclines (Barrett & Elliot 1973). The monoclines are inferred to have formed by the draping of Beacon strata over relatively large-displacement normal faults cutting crystalline basement below.

3.1 *Time relations of faults and intrusions*

There is abundant evidence that faulting preceded dike emplacement and continued during dike emplacement. In southern Victoria Land, conjugate mesoscopic normal fault sets occur in rotated blocks of sediment floating within dolerite dikes and sills, indicating that faulting preceded dolerite emplacement (Figure 2A). In both southern Victoria Land and the Beardmore Glacier area, there is a systematic spatial association of dikes and faults. There is commonly a zone of closely spaced fractures adjacent to dike margins, resembling the marginal fracture sets termed "adjacent joints" by Delaney et al. (1986). Dip-parallel striae and normal offsets of bedding show, however, that these surfaces are faults rather than tensile fractures. Offset of strata across dikes is common. Well-developed fibrous calcite veins along dike margins record dominantly dip-slip motion (Figure 2B). The fact that these striated planes extend beyond dike terminations indicates either that pre-existing fault planes were dilated during dike injection, or that the faults were forming synchronously with dike emplacement.

During dolerite emplacement, heat initiated a flux of fluids which caused extensive alteration of surrounding host rocks (Craw & Findlay 1984) and remobilized Beacon clastic sediments to produce clastic dikes and breccia bodies (Grapes et al. 1974; Korsch 1984; Bradshaw 1987; my observations). In places such remobilized material occupies segments of mesoscopic normal fault planes, indicating that the faults formed prior to or during dike emplacement. Volcanic breccias and dikes occupy faults associated with regional monoclines in the Beardmore Glacier region, demonstrating that the monoclines are of Jurassic age. In southern Victoria Land, faults cutting dolerite are rare whereas, in the Beardmore Glacier area, normal faults cutting dolerite are common (Figure 2C), showing that some faulting continued after magmatic activity ceased.

3.2 *Geometry and kinematic interpretation of faults and dikes*

Dolerite dikes in both southern Victoria Land and around Beardmore Glacier occur in two, loosely clustered, subperpendicular systems with northwest and east-northeast trends (Figure 3). In some places, these dike sets show mutual crosscutting relations but, most commonly, the dikes merge at T-

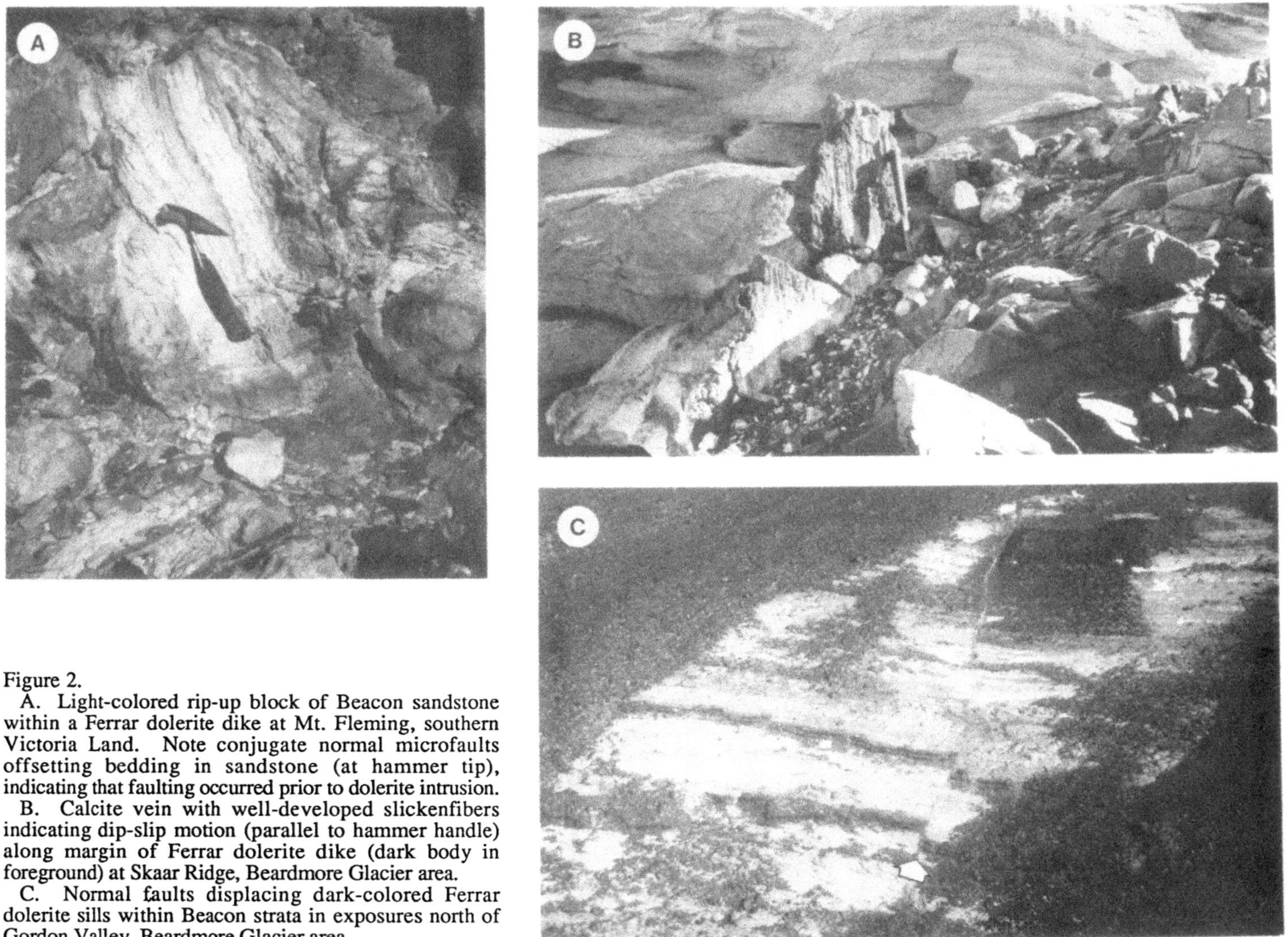

Figure 2.
A. Light-colored rip-up block of Beacon sandstone within a Ferrar dolerite dike at Mt. Fleming, southern Victoria Land. Note conjugate normal microfaults offsetting bedding in sandstone (at hammer tip), indicating that faulting occurred prior to dolerite intrusion.
B. Calcite vein with well-developed slickenfibers indicating dip-slip motion (parallel to hammer handle) along margin of Ferrar dolerite dike (dark body in foreground) at Skaar Ridge, Beardmore Glacier area.
C. Normal faults displacing dark-colored Ferrar dolerite sills within Beacon strata in exposures north of Gordon Valley, Beardmore Glacier area.

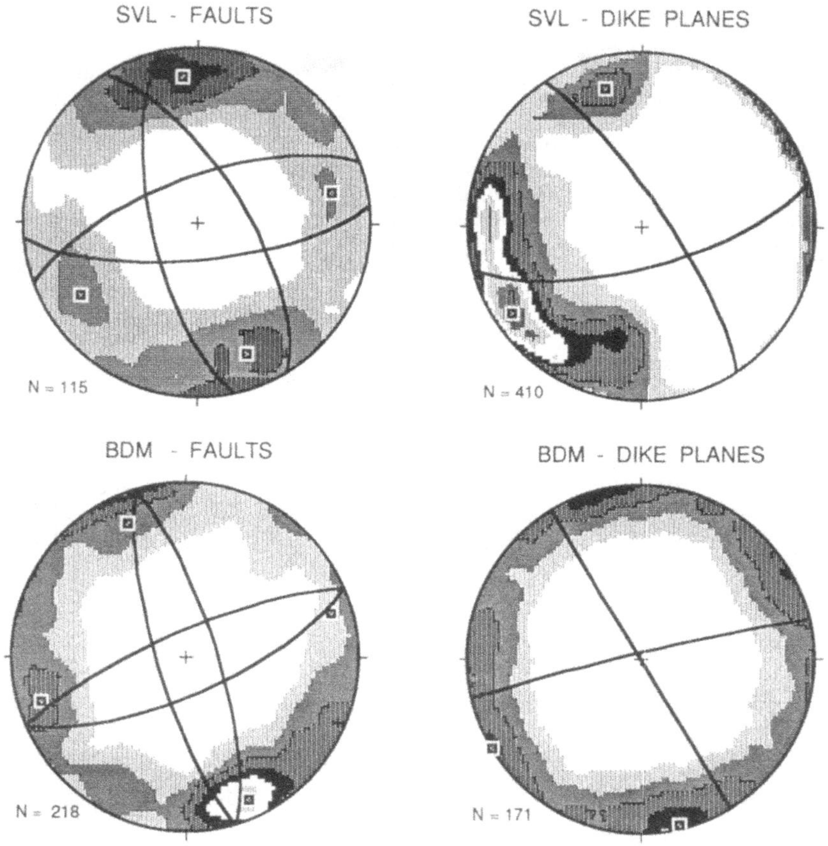

Figure 3. Summary stereoplots of Jurassic fault and dike orientations in southern Victoria Land (SVL) and the Beardmore Glacier area (BDM). Note the two conjugate pairs of normal faults and two subperpendicular dike sets in both regions. North in BDM is an average of 5° clockwise of north in SVL. Lower hemisphere, equal area projections; poles to faults and dikes contoured by the Kamb method with contour interval of 2 sigma; average poles indicated by boxes and corresponding average fault and dike planes indicated by great circles.

junctions, showing that the two subperpendicular systems are contemporaneous. Dilation perpendicular to each dike set requires extension in both northeast and north-northwest directions during dike emplacement.

North-northwest- and east-northeast-striking normal fault sets, subparallel to the dike systems, also occur in both areas (Figure 3). The fault arrays consist of two subperpendicular pairs of conjugate normal faults. Because both sets of faults have the same relative age relations with the Ferrar dikes, they must have formed contemporaneously. The development of two conjugate fault sets occurs where there is strain in each of the three principal strain directions, in contrast to the more familiar two-dimensional strain state in which a single conjugate fault set forms (Reches 1978; Aydin & Reches 1982; Krantz 1988). A Jurassic triaxial strain field is thus indicated by the simultaneous development of subperpendicular conjugate normal fault sets in the Transantarctic Mountains. Extension occurred in both northeast-southwest and northwest-southeast horizontal directions, and shortening occurred in the vertical direction. The development of the subperpendicular dike systems is consistent with this interpretation.

Due to the extensive ice cover, it is not possible to directly determine the direction of maximum extension by summation of fault displacements. The principal dike trend in southern Victoria Land is northwest (Figure 3), possibly indicating a dominant northeast extension direction, whereas in the Beardmore Glacier region the preferred orientation of dikes is to the east-northeast (Figure 3), suggesting a dominant north-northwest extension direction there. In the Beardmore Glacier area, the orientation of the regional monoclines might suggest that the northeast fault trend is the dominant structural trend and, thus, that the principal extension direction was northwest-southeast, consistent with the dike data. A regional northeast-southwest principal extension direction is suggested by the parallelism between the northwest faults and dikes and the trend of the mountains and offshore basins in the Ross embayment (Figure 4). The overall triaxial strain regime and a northeast-southwest principal extension direction are both inconsistent with the right-lateral transtensional rift kinematics proposed for the Jurassic by Schmidt and Rowley (1986), because such a strain regime would produce a uniform northeast orientation of normal faults and dike swarms in response to principal northwest-southeast extension.

4 JURASSIC DIKE TRENDS IN GONDWANA

A regionally consistent strain regime over approximately 1,000 km of strike length in the Transantarctic Mountains is documented by the correspondence of Jurassic dike and fault patterns in southern Victoria Land and around Beardmore Glacier. The total regional extent of this extensional stress field can be evaluated by comparing the results from the Transantarctic Mountains with Jurassic dike trends described by other workers in adjacent sectors of Gondwana (Figure 4). In Queen Maud Land, along the Weddell Sea margin of East Antarctica, Jurassic dolerite dikes form subperpendicular swarms that parallel two conjugate normal fault sets (Spaeth & Schull 1987; Grantham & Hunter 1991). This structural pattern is remarkably similar to that shown in the Transantarctic Mountains, indicating that a triaxial strain regime persisted through to the Weddell Sea margin, ca. 2,250 km across the continent. The dominant trend of Jurassic faults and dikes in Queen Maud Land indicates that the principal extension direction was perpendicular to the present continental margin, with subordinate extension parallel to the margin (Spaeth & Schull 1987). This is consistent with the position of the Queen Maud Land margin along the flank of the failed Jurassic Weddell Rift and related Explora volcanic wedge (Kristoffersen & Hinz 1991).

In Gondwana reconstructions, the Queen Maud Land coast of East Antarctica lies adjacent to southern Africa and, according to recent paleomagnetic data (Taylor & Shaw 1989), the Falkland Islands. In both of these regions, there are dolerite dike swarms with subperpendicular

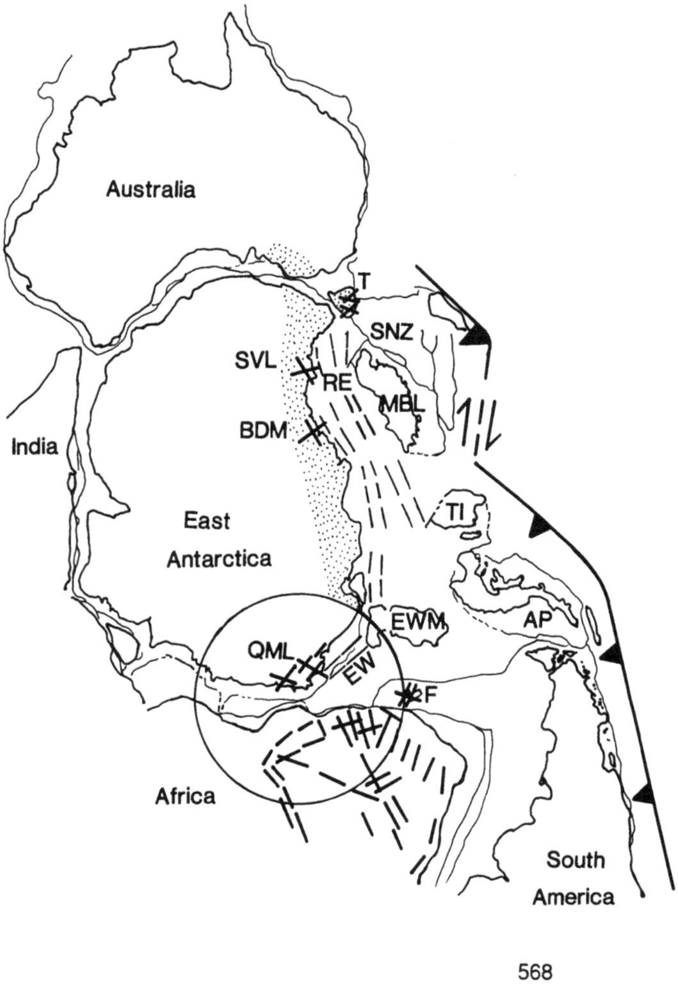

Figure 4. Average orientations of Jurassic dike planes (heavy lines) plotted on the Gondwana reconstruction of Grunow, Kent & Dalziel (1991). Circle shows position of mantle plume proposed by White & McKenzie (1989). Stipple pattern indicates Transantarctic Ferrar magmatic province and equivalent rocks in Tasmania and Australia. Dashed lines mark positions of known or inferred submarine and subglacial rift basins. T: Tasmania; SNZ: southern New Zealand block; MBL: Marie Byrd Land block; SVL: southern Victoria Land; BDM: Beardmore Glacier area; TI: Thurston Island block; AP: Antarctic Peninsula block; EWM: Ellsworth - Whitmore Mountains block; F: Falkland Islands; QML: Queen Maud Land; EW: offshore Explora Wedge forming part of the margin of the failed Weddell Rift of Kristoffersen and Hinz (1991). Dike orientations after this study, Spaeth and Schull (1987), Hunter and Reid (1987), Greenway (1972), Forsyth (1984) and Gulline (1984).

orientations similar to those in Queen Maud Land (Figure 4). Dike swarms in the Karoo province of southern Africa were emplaced between 190 - 120 Ma, but are mainly of Jurassic age (Hunter & Reid 1987). In the Falkland Islands, subperpendicular dike systems are present (Greenway 1972; Taylor & Shaw 1989), and one dike has yielded an Early Jurassic K-Ar age of 192±10 Ma (Cingolani & Varela 1976). In Tasmania, along the boundary between the Antarctic and Australian sectors of Gondwana, widespread dolerites dated at ca. 175 Ma (Schmidt & McDougall 1977) represent the continuation of the Transantarctic Ferrar magmatic province. There, the dikes show subperpendicular trends and are associated with normal faulting (e.g. Leaman 1975; Forsyth 1984; Gulline 1984). Strike-slip displacement, as indicated by striae on dike margins, has also occurred parallel to the dike trends (R. Berry, pers. comm. 1991).

Outside of the Transantarctic Mountains and Queen Maud Land, there are insufficient data on the timing and kinematics of dike emplacement and related faulting to evaluate definitively the significance of the Jurassic structural trends for the regional Gondwana stress field. Viewed in a reconstructed Gondwana framework (Figure 4), however, all of the Jurassic dike swarms occur in subperpendicular sets with similar orientations. This suggests the intriguing possibility that a homogeneous extensional stress regime persisted for over 5,000 km within Gondwana during the Jurassic, prior to Cretaceous fragmentation of the supercontinent.

5 DISCUSSION

The new structural mapping reported here shows that Jurassic faulting was widespread in the Transantarctic Mountains and that regional crustal extension was intimately associated with Ferrar magmatic activity. The Jurassic magmatism and crustal extension in Queen Maud Land was related to an early period of failed rifting in the Weddell Sea embayment (Kristoffersen & Hinz 1991; Storey 1991). The coeval bimodal magmatism in the Ellsworth-Whitmore block (Storey et al. 1988; Garrett et al. 1988) indicates that rifting in that zone is also of Jurassic age. Within the Ross embayment, the Transantarctic Mountains form the uplifted flank of an extensional province marked by major subglacial and submarine rift basins (Figure 4). Because the age of the oldest sedimentary and/or volcanic material in these basins has not yet been determined, the age of the onset of rifting in the Ross embayment is not well defined. The prevailing view, however, is that rifting occurred in the Late Cretaceous when the New Zealand - Campbell Plateau block rifted from the Marie Byrd Land margin of Antarctica and when spreading commenced between Australia and East Antarctica (e.g. Cooper et al. 1991; Tessensohn & Wörner 1991). It is important to consider, however, that the Jurassic fault and dike trends in the Transantarctic Mountains are oriented parallel (and perpendicular) to the trends of the offshore basins of the Ross embayment (Figure 4). This parallelism suggests that these structural trends may have initially formed in the Jurassic, as part of the regional crustal extension associated with Ferrar magmatism. The basins in the Ross embayment would thus be direct analogues of the failed Weddell Rift, and together would define a continuous zone of failed rifting that transects the Antarctic sector of Gondwana. The Weddell Sea end of this zone would have connected with the Karoo province; the Ross segment would have linked with the Jurassic transtensional and rift basins which formed between Australia and Antarctica prior to breakup (Willcox & Stagg 1990). This Jurassic rift system corresponds to the "Jurassic Transantarctic Rift " of Schmidt and Rowley (1986), but the triaxial strain regime documented here indicates that the rift system was dominated by crustal stretching, rather than accommodating right-lateral translation between East and West Gondwana.

Crustal extension prior to Ferrar magmatism is revealed by the field relations between faults and dikes in the Transantarctic Mountains. The development of oriented dike sets and the crosscutting relations of faults, dikes, and sills demonstrate that extension continued during magmatism and after it had ceased. The linear belt of Ferrar magmatic rocks suggests that the locus of extension and magmatism followed Precambrian and Palaeozoic structural trends along the margin of East Antarctica. On more local scales, however, the extensional structures cut preexisting trends, showing the influence of the rift-related stress field. I suggest that the correspondence of Jurassic structural patterns throughout such an extensive region within Antarctica and beyond shows that a uniform stress regime may have existed within the Gondwana lithosphere prior to breakup. The existence of a regional tectonic stress field prior to and during magmatism suggests a lithosphere-activated rifting process, in which the extensional stress regime in the Gondwana lithosphere initiated breakup of the supercontinent. The Jurassic magmatism in the Ferrar province apparently occurred in a back-arc setting cratonward of the active Gondwana plate margin (Kyle et al. 1981; Elliot 1991). According to recent Gondwana reconstructions, the position of the dolerite belt changes from ca. 1,000 km to ca. 3,000 km cratonward of the Gondwana active margin from Tasmania to the Queen Maud Land-Falkland-Karoo region (Figure 4). This divergence suggests that the position and orientation of the associated extensional province was not controlled by convergent margin geometry. It is likely, however, that the tensional stress regime within the Gondwana lithosphere originated from plate boundary forces generated along the active subducting margins of the supercontinent (Bott 1982).

White and McKenzie (1989) proposed that the impingement of a mantle plume on a zone of crustal stretching in the Gondwana lithosphere led to the widespread Jurassic magmatism of the Karoo province, the Queen Maud Land region, and the volcanic margin of the Weddell Rift (Figure 4). Based on the geochemical differences of the Ferrar magmatic rocks, White and McKenzie (1989) excluded the Transantarctic Ferrar province from a similar origin but, as pointed out by Cox (1989), the geochemical variation may reflect differences in degree of interaction between melt and the lithosphere. The Transantarctic Ferrar province extends ca. 3,500 km beyond the circumference of the proposed plume, across Antarctica and into Tasmania (Figure 4). The available geochronological data suggest that intrusion and extrusion occurred during a short period along the entire belt, at the same time as that in Queen Maud Land and as the second main pulse of magmatism in the Karoo province (Kyle et al. 1981; Cox 1988). There is thus no evidence for age progression along the belt that could suggest passage of the Gondwana lithosphere over a single plume.

In the Transantarctic province, regional crustal extension preceded Ferrar magmatism. This is comparable to the sequence proposed by Bott (1982) and White and McKenzie (1989), in which volcanism occurs after a zone of lithospheric thinning comes in contact with a region of anomalously hot mantle. However, the present exposure of Ferrar rocks suggests a narrow, elongate belt with a form and extent different from the ca. 1,000-km-radius circular or mushroom-shaped province typical of a plume-related thermal anomaly (White & McKenzie 1989). Formation of an elongate magmatic belt would imply either an aligned series of plumes with overlapping thermal anomalies, or a substantial deflection and elongation of the 'Karoo' plume, much like the "hot line" proposed by Cox (1988). Alternatively, the geometry of the mantle thermal anomaly may reflect upwelling caused by the insulating effect of the supercontinent (Anderson 1982; Gurnis 1988), rather than a mantle plume. In either case, it would appear that the relatively young and hot lithosphere along the orogenic margin of the East Antarctic craton marked a weak zone where the lithosphere was preferentially rifted in response to pre-breakup tensional stresses in the Gondwana lithosphere. It is likely that thinning of the lithosphere along this zone caused mantle decompression which served to initiate the melting that produced the Transantarctic Ferrar magmatic belt.

ACKNOWLEDGMENTS

This research was supported by National Science Foundation Grant DPP 88-16932. The logistic support of VXE-6 Squadron of the U.S. Navy and of Helicopters, NZ, Ltd., made much of the field study possible. Special thanks to Peter Braddock, Mike Gibson, Chuck Kroger and Barry Muller for assisting with the field work.

REFERENCES CITED

Anderson, D.L. 1982. Hotspots, polar wander, Mesozoic convection, and the geoid. *Nature* 297: 391-393.

Anderson, E.M. 1951. *The Dynamics of Faulting and Dyke Formation with Application to Britain.* Edinburgh: Oliver and Boyd, Ltd.

Aydin, A. & Z. Reches 1982. Number and orientation of fault sets in the field and in experiments. *Geology* 10: 107-112.

Barrett, P.J. 1982. History of the Ross Sea region during the deposition of the Beacon Supergroup 400 - 180 million years ago. *Journal of the Royal Society of New Zealand* 11: 447-458.

Barrett, P.J. & D.H. Elliot. 1973. Reconnaissance geologic map of the Buckley Island Quadrangle, Transantarctic Mountains, Antarctica. U.S. Antarctic Research Program Map, Map A-3, U.S. Geological Survey.

Barrett, P.J., D.H. Elliot & J.F. Lindsay. 1986. The Beacon Supergroup (Devonian - Triassic) and Ferrar Group (Jurassic) in the Beardmore Glacier area, Antarctica. In M.D. Turner & J.F. Splettstoesser (eds.), *Geology of the Central Transantarctic Mountains.* Washington, D.C.: American Geophysical Union: Antarctic Research Series 36: 339-428.

Behrendt, J. C. & A. Cooper 1991. Evidence of rapid Cenozoic uplift of the shoulder escarpment of the Cenozoic West Antarctic rift system and a speculation on possible climate forcing. *Geology* 19: 315-319.

Borg, S.G. & D.J. DePaolo 1991. A tectonic model of the Antarctic Gondwana margin with implications for southeastern Australia: isotopic and geochemical evidence. *Tectonophysics* 196: 339-358.

Bott, M.H.P. 1982. The mechanism of continental rifting. *Tectonophysics* 81: 301-309.

Bradshaw, M.A. 1987. Additional field interpretation of the Jurassic sequence at Carapace Nunatak and Coombs Hills, south Victoria Land, Antarctica. *New Zealand Journal of Geology and Geophysics* 30: 37-49.

Cingolani, C.A. & R. Varela 1976. Investigaciones geologicas y geochronologicas en el extremo sur de la Isla Gran Malvina, sector de Cabo Belgrano (Cape Meredith), Islas Malvinas. *Actas 6th Congreso Geologico Argentino* 1: 457-473.

Collinson, J. W. 1991. The palaeo-Pacific margin as seen from East Antarctica. In M.R.A. Thomson, J.A. Crame & J.W. Thomson (eds.), *Geological Evolution of Antarctica.* Cambridge: Cambridge University Press: 199-204.

Collinson, J.W. & D.H. Elliot. 1984. Geology of Coalsack Bluff, Antarctica. In M.D. Turner & J.F. Splettstoesser (eds.), *Geology of the Central Transantarctic Mountains.* Washington, D.C.: American Geophysical Union: Antarctic Research Series 36: 103-117.

Cooper, A.K., F.J. Davey & J.C. Behrendt 1987. Seismic stratigraphy and structure of the Victoria

Land Basin, western Ross Sea, Antarctica. In A.K. Cooper & F.J. Davey (eds.), *The Antarctic Continental Margin, Geology and Geophysics of the Western Ross Sea*. CPCEMR Earth Science Series 5B: 27-76.

Cooper, A.K., F.J. Davey & K. Hinz. 1991. Crustal extension and origin of sedimentary basins beneath the Ross Sea and Ross Ice Shelf, Antarctica. In M.R.A.Thomson, J.A. Crame & J.W. Thomson (eds.), *Geological Evolution of Antarctica*. Cambridge: Cambridge University Press: 285-291.

Cox, K.G. 1978. Flood basalts, subduction and the breakup of Gondwanaland. *Nature* 274: 47-49.

Cox, K.G. 1988. The Karoo Province. In J.D. Macdougall (ed.), *Continental Flood Basalts*. Dordrecht: Kluwer Academic Publishers: 239-272.

Cox, K.G. 1989. Hot plumes from the mantle. *Nature* 340: 341-342.

Craw, D. & R.H. Findlay 1984. Hydrothermal alteration of Lower Ordovician granitoids and Devonian Beacon Sandstone at Taylor Glacier, McMurdo Sound, Antarctica. *New Zealand Journal of Geology and Geophysics* 27: 465-475.

Dalziel, I.W.D. & D.H. Elliot 1982. West Antarctica: Problem child of Gondwanaland. *Tectonics* 1: 3-19.

Dalziel, I.W.D., B.C. Storey, S.W. Garrett, A.M. Grunow, L.D.B. Herrod & R.J. Pankhurst 1987. Extensional tectonics and the fragmentation of Gondwanaland. In M. P. Coward, J.F. Dewey & P.L. Hancock (eds.), *Continental Extensional Tectonics*. London: Special Publication of the Geological Society 28: 433-41.

Davey, F.J. 1981. Geophysical studies in the Ross Sea region. *Journal of the Royal Society of New Zealand* 11: 465-479.

Delaney, P.T., D.D. Pollard, J.I. Ziony & E.H. McKee 1986. Field relations between dikes and joints: emplacement processes and paleostress analysis. *Journal of Geophysical Research* 91: 4920-4938.

Doake, C.S.M., R.D. Crabtree & I.W.D. Dalziel 1983. Subglacial morphology between Ellsworth Mountains and Antarctic Peninsula: New data and tectonic significance. In R.L. Oliver, P.R. James & J.B. Jago (eds.), *Antarctic Earth Science*. Canberra: Australian Academy of Science: 270-273.

Elliot, D.H. 1975a. Tectonics of Antarctica: A review. *American Journal of Science* 275A: 46-106.

Elliot, D.H. 1975b. Gondwana basins of Antarctica. In K.S.W. Campbell (ed.), *Gondwana Geology*. Canberra: ANU Press: 493-536.

Elliot, D.H. 1991. Triassic - Early Cretaceous evolution of Antarctica. In M.R.A. Thomson, J.A. Crame & J.W. Thomson (eds.), *Geological Evolution of Antarctica*. Cambridge: Cambridge University Press: 541-548.

Elliot, D.H., R.J. Fleck & J.F. Sutter 1985. Potassium-argon age determinations of Ferrar Group rocks, central Transantarctic Mountains. In M.D. Turner & J.F. Splettstoesser (eds.), *Geology of the Central Transantarctic Mountains*. Washington, D.C.: American Geophysical Union: Antarctic Research Series 36: 197-224.

Fitzgerald, P.G., M. Sandiford, P.J. Barrett & A.J.W. Gleadow 1986. Asymmetric extension associated with uplift and subsidence in the Transantarctic Mountains and Ross embayment. *Earth and Planetary Science Letters* 81: 67-78.

Fitzgerald, P.G., & E. Stump 1991. Uplift history of the Transantarctic Mountains from Victoria Land (~70° S) to Scott Glacier (~86°S): Evidence from fission track analysis. In *Abstracts: Sixth International Symposium on Antarctic Earth Sciences*, Tokyo: National Institute of Polar Research: 144.

Forsyth, S.M. 1984. Geological Atlas 1:50,000 series. Sheet 68. Oatlands. Explanatory Report geological Survey of Tasmania: 182 pp.

Garrett, S.W., M. P. Maslanyj & D. Damaske 1988. Interpretations of aeromagnetic data from the Ellsworth Mountains - Thiel Mountains ridge, West Antarctica. *Journal of the geological Society, London* 145: 1009 - 1017.

Gleadow, A.J.W. & P.G. Fitzgerald 1987. Uplift history of the Transantarctic Mountains in the Dry Valleys area, southern Victoria Land, Antarctica, from apatite fission-track ages. *Earth and Planetary Science Letters* 82: 1-14.

Grantham, G.H. & D.R. Hunter 1991. The timing and nature of faulting and jointing adjacent to the Pencksokket, western Dronning Maud Land, Antarctica. In M.R.A. Thomson, J.A. Crame & J.W. Thomson (eds.), *Geological Evolution of Antarctica*. Cambridge: Cambridge University Press: 47-51.

Grapes, R.H., D.L. Reid & J.G. McPherson 1974. Shallow dolerite intrusion and phreatic eruption in the Allan Hills region, Antarctica. *New Zealand Journal of Geology and Geophysics* 17(3): 563-577.

Greenway, M.E. 1972. The Geology of the Falkland Islands. *British Antarctic Survey Scientific Report* 76: 42 pp.

Grunow, A.M., D.V. Kent & I.W.D. Dalziel 1991. New paleomagnetic data from Thurston Island: Implications for the tectonics of West Antarctica and Weddell Sea opening. *Journal of Geophysical Research* 96: 17,935-17,954.

Gulline, A.B. 1984. Geological Atlas 1:50,000 series. Sheet 83. Hobart: Explanatory Report geological Survey of Tasmania: 46 pp.

Gurnis, M. 1988. Large-scale mantle convection and the aggregation and dispersal of supercontinents. *Nature* 332: 695-699.

Hamilton, W. 1965. Diabase sheets of the Taylor Glacier Region, Victoria Land, Antarctica. *U.S. Geological Survey Professional Paper* 3456-B: 71 pp.

Hooper, P.R. 1991. The timing of crustal extension and the eruption of continental flood basalts. *Nature* 345: 246 - 249.

Hunter, D.R. & D.L. Reid 1987. Mafic dyke swarms in southern Africa. In Halls, H.C. & W.F. Fahrig (eds.), *Mafic Dyke Swarms*. Geological Association of Canada Special Paper 34: 445-457.

Jankowski, E.J., D.J. Drewry & J.C. Behrendt. 1983. Magnetic studies of upper crustal structure in West Antarctica and the boundary with East Antarctica. In R.L. Oliver, P.R. James & J.B. Jago (eds.), *Antarctic Earth Science*. Canberra: Australian Academy of Science: 197-203.

Kadmina, I.N., R.G. Kurinin, V.N. Masolov & G.E. Grikurov 1983. In R.L. Oliver, P.R. James & J.B. Jago (eds.), *Antarctic Earth Science*. Canberra: Australian Academy of Science: 498-502.

Katz, H.R. 1982. Post-Beacon tectonics in the region of Amundsen and Scott Glaciers, Queen Maud Range, Transantarctic Mountains. In C. Craddock (ed.), *Antarctic Geoscience*. Madison: University of Wisconsin Press: 827-834.

Korsch, R.J. 1984. The structure of Shapeless Mountain, Antarctica, and its relation to Jurassic igneous activity. *New Zealand Journal of Geology and Geophysics* 27: 487-504.

Krantz, R.W. 1988. Multiple fault sets and three-dimensional strain: theory and application. *Journal of Structural Geology* 10: 225-237.

Kristoffersen, Y. & K. Hinz 1991. Evolution of the Gondwana plate boundary in the Weddell Sea area. In M.R.A. Thomson, J.A. Crame & J.W. Thomson (eds.), *Geological Evolution of Antarctica*. Cambridge: Cambridge University Press: 225-230.

Kyle, P.R., D.H. Elliot & J.F. Sutter 1981. Jurassic Ferrar Supergroup tholeiites from the Transantarctic Mountains, Antarctica, and their relationship to the initial fragmentation of Gondwana. In M.M. Cresswell & P. Vella (eds.), *Gondwana Five*. Rotterdam: Balkema: 283-287.

Leaman, D.E. 1975. Form, mechanism and control of dolerite intrusion near Hobart, Tasmania. *Journal Geological Society of Australia* 9: 175-186.

Maslanyj, M. P. & B.C. Storey 1990. Regional aeromagnetic anomalies in Ellsworth Land: Crustal structure and Mesozoic microplate boundaries within West Antarctica. *Tectonics* 9: 1515-1532.

Morrison, A.D. 1989. Geological observations from Terra Cotta Mountain, upper Taylor Glacier. *New Zealand Antarctic Record* 9: 31-43.

Pollard, D.D., P. Segall & P.T. Delaney 1982. Formation and interpretation of dilatant echelon cracks. *Geological Society of America Bulletin* 93: 1291-1303.

Pyne, A.R. 1984. Geology of the Mt Fleming area, South Victoria Land, Antarctica. *New Zealand Journal of Geology and Geophysics* 27: 505-512.

Reches, Z. 1978. Analysis of faulting in three-dimensional strain field. *Tectonophysics* 47: 109-129.

Schmidt, P.W. & I. McDougall 1977. Palaeomagnetic and potassium-argon dating studies of the Tasmanian dolerites. *Journal of the Geological Society of Australia* 25: 321-328.

Schmidt, D.L. & P.D. Rowley 1986. Continental rifting and transform faulting along the Jurassic Transantarctic rift, Antarctica. *Tectonics* 5: 279-292.

Spaeth, G. & P. Schull 1987. A survey of Mesozoic dolerite dikes from western Neuschwabenland, Antarctica, and their geotectonic significance. *Polarforschung* 57: 93-113.

Stern, T.A. & U.S. ten Brink 1989. Flexural uplift of the Transantarctic Mountains. *Journal of Geophysical Research* 94: 10,315-10,330.

Storey, B.C. 1991. The crustal blocks of West Antarctica within Gondwana: reconstruction and break-up model. In M.R.A. Thomson, J.A. Crame & J.W. Thomson (eds.), *Geological Evolution of Antarctica*. Cambridge: Cambridge University Press: 587-592.

Storey, B.C., M.J. Hole, R.J. Pankhurst, I.L. Millar & W. Vennum 1988. Middle Jurassic within-plate granites in West Antarctica and their bearing on the break-up of Gondwanaland. *Journal of the Geological Society, London* 145: 999-1007.

Stump, E. 1992. The Ross orogen of the Transantarctic Mountains in light of the Laurentia-Gondwana split. *GSA Today* 2(3): 25-31.

Taylor, G.K. & J. Shaw 1989. The Falkland Islands: New palaeomagnetic data and their origin as a displaced terrane from southern Africa. In J.W. Hillhouse (ed.), *Deep Structure and Past Kinematics of Accreted Terranes*. Geophysical Monograph 50, IUGG 5: 59-72.

Tessensohn, F. & G. Wörner 1991. The Ross Sea rift system, Antarctica: structure, evolution and analogues. In M.R.A. Thomson, J.A. Crame & J.W. Thomson (eds.), *Geological Evolution of Antarctica*. Cambridge: Cambridge University Press: 273-277.

White, R.S. & D.P. McKenzie 1989. Magmatism at rift zones: the generation of volcanic continental margins and flood basalts. *Journal of Geophysical Research* 94: 7685-7729.

Willcox, J.B. & H.M.J. Stagg 1990. Australia's southern margin: a product of oblique extension. *Tectonophysics* 173: 269-281.

Gondwana Eight, Findlay, Unrug, Banks & Veevers (eds) © *1993 Balkema, Rotterdam. ISBN 90 5410 304 3*

Remnants of Gondwana in the Australian landscape

C.R.Twidale & E.M.Campbell
Department of Geology and Geophysics, University of Adelaide, S.A., Australia

ABSTRACT: Significant elements of the Australian landscape date from Mesozoic or earlier times. Australia did not separate completely from other Gondwana components until Early Tertiary times and these Mesozoic and older elements can therefore be regarded as Gondwanan. A pre-Cretaceous exhumed surface is especially well preserved, but epigene and etch forms are prominent in many areas.

During the separation and northern drift of the continent and particularly in Late Jurassic and Early Cretaceous times, the sea transgressed much of the landmass. The associated sediments covered and preserved much of the pre-existing land surface and augmented the effects of thalassostatic loading at the margins of the seas, so that subsidence of the basins caused adjacent land masses to rise. Many old land surfaces have been re-exposed at the former oceanic margins, but old epigene forms are preserved high in the relief on the uplifted blocks.

1 INTRODUCTION

Cainozoic events are responsible for most of the great plains, erosional and depositional, that characterise much Australian scenery. The widespread planation and concomitant deep weathering of the central and western sectors of the continent obliterated much that had gone before. In addition, river dissection consequent on lower sea levels of the later Cainozoic glacial periods, plus local glacial and nival modifications of the landscape, also changed the scenery. Nevertheless, substantial landscape elements survived these events. Some, like most of the lateritised and silcreted plains, plateaux and ridges are of Early-Middle Tertiary age, but some are of even greater antiquity.

2 TYPES OF SURFACE

Erosional surfaces are of three types: epigene, etch and exhumed. Epigene forms are due to epigene or subaerial processes, in most instances fluvial action, and have been exposed since inception. Weathering of such surfaces produces a regolith, the base of which, the junction between altered and fresh bedrock, is called the weathering front. Stripping of the regolith exposes the weathering front as an etch surface. Both epigene and etch forms may be covered by riverine, lacustrine, volcanic, glacial, or, most commonly marine deposits, and later as these cover materials are eroded, re-exposed, resurrected or exhumed surfaces are formed.

3 GONDWANA SURFACES AND FORMS

3.1 The age of 'Gondwana' surfaces

The component parts of Gondwana did not separate and drift apart simultaneously (see e. g. Craddock 1977; Veevers 1984; Barron 1987; Fairbridge 1989). First, in late Palaeozoic times western Gondwana (Africa, Arabia & South America) became detached from the eastern sector (Antarctica, Madagascar, India & Australia). There are signs of Early Jurassic rifting off the northwest coast of Western Australia, but the western margin of the Australian continent did not fully develop until the Early Cretaceous. The parting of Antarctica and Australia was a Late Cretaceous event, though new seafloor was not emplaced between the two until the Early Tertiary. A similar youthfulness has been suggested for the Tasman

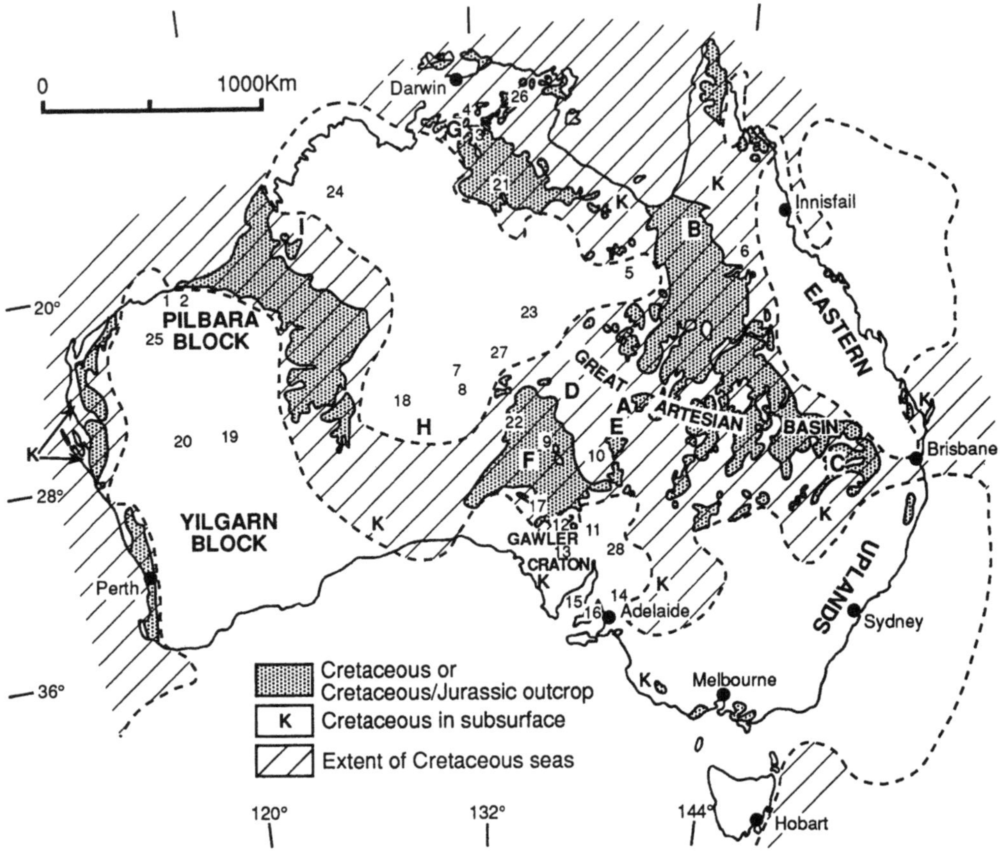

Figure 1. Localities mentioned in the text, areas covered by Cretaceous (Aptian) seas (after Frakes et al. 1987) and Jurassic-Cretaceous sediments. A Eromanga Basin; B Carpentaria Basin; C Surat Basin; D Pedirka Basin; E Warburton Basin; F Arckaringa Basin; G Daly Basin; H Officer Basin; I Fitzroy Basin. 1 Port Hedland; 2 Shay Gap; 3 Tindal Plain; 4 Pine Creek; 5 Isa Highlands; 6 Georgetown Upland and Einsleigh Upland; 7 Lake Amadeus; 8 Ayers Rock; 9 Peake and Davenport Ranges; 10 Lake Eyre; 11 Arcoona Plateau; 12 Lake Gairdner; 13 Gawler Ranges; 14 Mt Lofty Ranges; 15 Spencer Gulf; 16 St Vincent Gulf; 17 Tarcoola; 18 Blackstone Range; 19 Wiluna; 20 Meekatharra; 21 Ashburton Range; 22 Indulkana Range; 23 Forster Range; 24 Kimberley Plateau; 25 Hamersley Range; 26 Arnhem Land Plateau; 27 MacDonnell Range; 28 Flinders Ranges.

Sea. On the other hand the Carpentaria Basin (Fig. 1) opened as early as the Triassic (Smart & Senior 1980) suggesting that deformation of the continent was initiated as soon as the disintegration of Gondwana began. Thus any erosional or depositional surface or other landform of Mesozoic or earlier age can be regarded as of Gondwanan origin in the sense that not until the Early Tertiary was Australia a detached and separate entity.

3.2 Older elements

Exhumed Precambrian surfaces have been identified in several parts of the Australian Shield. Thus to the east of Lake Gairdner, the Scrubby Surface, cut in Middle Proterozoic Gawler Range Volcanics (1592 Ma) is being exhumed from beneath the 1424 Ma Pandurra Formation (Campbell & Twidale, 1991). In the Forster Range, near Barrow Creek, Northern Territory, a surface eroded in Precambrian amphibolites and gneisses is being exhumed as the overlying horizontally disposed Proterozoic quartzites are stripped away. In the Pilbara of Western Australia, eroded granitic domes in Archaean granite and gneiss are being exhumed from beneath lowermost Proterozoic and uppermost Archaean volcanics and sediments (Twidale 1986). In addition glaciated pavements of Middle Proterozoic age are reported from several parts of the northern Shield (e.g. Perry & Roberts 1968).

Pre-Middle Cambrian surfaces, exhumed and part of the modern landscape, have been described from the southern parts of the Isa Highlands (Öpik 1961) and from the eastern Kimberley region of Western Australia (Plumb & Veevers 1971; Young 1991) where many of the local ranges are in process of exhumation from beneath Cambrian basalt flows. A Cambrian age has been claimed for parts of the Ashburton Ranges in the Northern Territory by Stewart et al. (1986).

The late Palaeozoic glaciation was a significant event in Australia as in other parts of Gondwana, and pre-Permian surfaces have been reported both in unconformity and in the landscape west of the Peake and Davenport Ranges (Wopfner 1964) and west and southwest of Lake Amadeus (Wells et al. 1970). In Western Australia they have been reported from the western Officer Basin (Jackson & Van de Graaff 1981), around the Blackstone Ranges (Daniels 1974), in the Wiluna-Meekatharra region (Mabbutt 1988) and in the Fitzroy Basin (Jennings & Sweeting 1963). In addition numerous minor glaciated pavements of Permian age have been recorded from various parts of southeastern Australia, including Tasmania. But pre-Permian landscapes are not as frequent and prominent as might be expected. Perhaps ice sheets modify landforms rather than create new ones.

The lateritised summit surface of the Gulfs region in South Australia, is certainly of Mesozoic age (e.g. Campana 1958; Twidale 1976a) and may be as old as Triassic (Daily et al. 1974; Daily et al. 1979). In the southern Flinders Ranges the various Triassic basins (Boolcunda, Springfield & Leigh Creek) were filled with sediments derived from a surface of low relief but no remnants of that surface have been recognised in the present landscape.

4 THE EARLY CRETACEOUS TRANSGRESSION AND ITS IMPACTS

The separation of Australia from Antarctica was accompanied by a northerly drift of the former. Just as modern intraplate seismic events can be attributed to such movements, so the horizontal migration of the structurally heterogeneous mass caused dislocations, and in particular the buckling and bending of the crust resulting in the bowing down of some areas and the uplift of others. Later Mesozoic seas invaded these downwarped and downfaulted areas of the continent.

The Late Jurassic and Cretaceous marine trangressions were world wide and extensive. In Australia their limits varied in time (see Frakes et al. 1987) but the inundation of the Early Cretaceous (Neocomian-Aptian) in particular extended over much of the newly defined continent (Fig. 1). Only the Eastern Uplands and the western blocks, including the Yilgarn and Pilbara, the Kimberleys and Gawler Craton, and the crests of the higher ranges, stood above sea level. Some of the present uplands, such as the Isa Highlands, were inundated. The main seaway was the present Great Artesian Basin, but the ocean extended in a northwest-southeast corridor between the Yilgarn and the Pilbara on the one hand and the Kimberleys on the other. The extent of the Early Cretaceous marine transgression is even greater than suggested by the extent of surface outcrop on maps of solid geology, for Cretaceous strata underlie many basins (Fig. 1).

These Cretaceous marine transgressions had several effects. Pre-existing land surfaces were extensively buried by the associated marine sediments, and many remain preserved in unconformity. Others, particularly at the margins of the basins, were temporarily covered but have been re-exposed so that exhumed pre-Cretaceous surfaces are an important landscape element in many parts of Australia. The land masses of the time were eroded by rivers graded to Cretaceous shorelines, and these too have long been recognised as components of the present scenery. The weight of water and sediments laid down in the seas caused the underlying crust to subside, while the adjacent land masses were gradually unloaded by erosion and tended to rise isostatically. Such combinations of thalassostatic and denudational isostasy (Figure 2) have an inbuilt reinforcement mechanism (Twidale & Campbell 1991, 1992), for the greater the erosion of the upstanding blocks, the greater the isostatic response and the greater

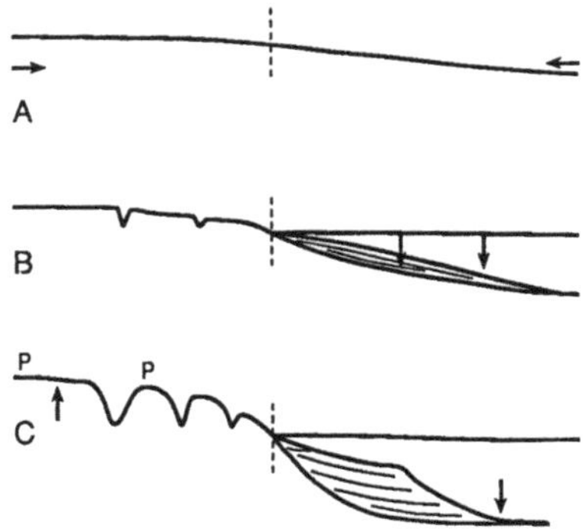

Figure 2. Schematic isostatic response of land surfaces to thalassostatic loading, to deposition and erosion and to tectonism.

the uplift; and conversely with basins. Such positive and negative blocks are regions that have, in an isostatic sense, behaved consistently over long periods. Because of uplift and stream rejuvenation, palaeosurfaces, and particularly summit plains, commonly survive as dissected remnants on the positive blocks. They are preserved in their entirety, in unconformity, in the adjacent basins.

5 THE PRE-CRETACEOUS SURFACE

The nature of exhumed pre-Cretaceous surfaces was recognised early by Jack (1915, 1931) who published clear descriptions of such surfaces marginal to the southwestern Great Artesian Basin. He pointed out that in the Tarcoola district the Precambrian rocks had been "worn down to an almost level surface" (Jack 1931, p. 6) before the Jurassic and Lower Cretaceous strata were laid down, and that to the north, adjacent to the Indulkana Range, "Prior to Upper Cretaceous time the processes of erosion appear to have planed down the old land surface to almost its present contours." (Jack 1915, p. 24).

The Early Cretaceous (Neocomian-Aptian) seas lapped onto the northernmost Flinders Ranges, where marine strata containing plant fossils indicate the position of the then shoreline (Woodard 1955; Alley & Lemon 1988). The location of Cretaceous marine beds beneath the Eyre and Frome plains immediately adjacent to the upland, as well as in Mt Babbage, on the summit surface, show that the high plain, cut across Precambrian rocks of various types and structures, is exhumed and of pre-Cretaceous age. The prominent summit surface preserved on folded Precambrian rocks in the Peake and Davenport (Denison) Ranges may be of the same type and age, and like the north Flinders surface has been dislocated by faulting (Wopfner 1968). A Late Jurassic surface is extensively exposed north and west of the Ranges in both South Australia and the Northern Territory (Jack 1931; Wopfner 1964).

In the Daly Basin of the Northern Territory the Tindal Plain is a remarkably flat karst plain cut in Cambrian strata. It carries residuals of basal Cretaceous sandstone, and as the unconformity between the Cretaceous above and the Cambrian below is coincident with the plain surface, the latter has been interpreted as exhumed and pre-Cretaceous in age (Twidale 1984). Limestone pinnacles and low hills already existed before the epicontinental Early Cretaceous seas invaded the area, for Cretaceous sediments are preserved in joint partings and as smears on the surface of the minor landforms (Twidale 1984). To the north, west of Pine Creek, the granitic plains and nubbins have been exposed as a result of stripping of the Early Cretaceous strata (Stuart-Smith et al. 1987). The exhumed surfaces of the Tindal and Pine Creek areas are the same surface, the parts differing in elevation by about 165 m as a result of faulting at the northern margin of the Daly Basin.

In the Darwin area much of the topography, including the shore platform at Stokes Bay, is exhumed from beneath the Cretaceous cover. The granite domes and nubbins of the Port Hedland area, Western Australia, as well as surfaces preserved on plateaux around Shay Gap, are also demonstrably exhumed and of pre-Cretaceous age (Twidale 1986).

In the Isa Highlands the pre-Cretaceous surface is especially prominent. Cut across Proterozoic strata and granitic rocks, the surface carries a scatter of Lower Cretaceous beds not only high in the relief but also in the piedmont of the upland (Plate 1), and it can be traced into unconformity beneath the Carpentaria Basin, whence it emerges to the east as a summit bevel, with Cretaceous remnants preserved in the Georgetown and Einsleigh uplands (Twidale 1956; Carter & Öpik 1959; Branch 1966). It may extend into the summit bevel of the Coast Ranges behind

Plate 1. Unconformity expressed as an exhumed pre-Cretaceous surface, south of Cloncurry, Isa Highlands, Queensland. A-B-C unconformity; A-B exhumed surface; D lateritised surface; E Precambrian schists; X Early Cretaceous sediments. (Twidale/CSIRO).

Plate 2. View looking southwest of part of the summit surface of the southern Gawler Ranges, South Australia, west of Yardea Homestead. Relief amplitude 100 m.

Innisfail (de Keyser 1964), but no remnants of a cover are preserved in this area.

6 CRETACEOUS LAND SURFACES

Cretaceous land surfaces have long been recognised in the Australian landscape. Hossfeld (1926) deduced that the lateritised summit surface in the northern Mount Lofty Ranges in South Australia is of Cretaceous age, a conclusion sustained by later research (e. g. Campana 1958; Brock 1971; Twidale 1976a). Craft (1932, 1933), working in the Eastern Uplands of central New South Wales, and Hills (1934) in the highlands of eastern Victoria, both concluded that the prominent summit surface throughout the upland region is of Cretaceous age. Hills' argument was based on an Early Tertiary age for the Older Basalts of Victoria, and the fact that they occupied valleys cut into the upland surface; though he later considered the surface to be even older, i.e. of Early to Middle Mesozoic age (Hills 1975). The same arguments are used today, though they are more secure as a result of the radiometric dating of the volcanic flows, which range in age from Palaeocene to

Miocene (Wellman & McDougall 1974).

Subsequently many surfaces of similar age have been recognised in various parts of the continent. In some areas, and particularly in the Eastern Uplands (see e.g. Ollier 1982) it is difficult to distinguish such epigene and etch surfaces from the contiguous exhumed elements. Similarly the summit surface prominent throughout the southern Flinders Ranges is demonstrably epigene, for although in physical continuity with the exhumed pre-Cretaceous surface of the northern Flinders, the shoreline was located around Mt Babbage. No evidence of a marine transgression has been found south of Mt Babbage, suggesting that this high plain surface may have been shaped by rivers graded to the Early Cretaceous shoreline. Such a Cretaceous age is consistent with ridge and valley topography eroded below the level of the summit surface around the northern margin of the intermontane Willochra Basin being of Early Tertiary age (Twidale & Campbell 1988).

The Hamersley Surface of the Hamersley Range, Western Australia, is demonstrably of etch type and pre-Eocene age; the local stratigraphy suggests a Cretaceous planation (Twidale et al. 1985). The higher levels of the MacDonnell Ranges (Mabbutt 1966; Twidale & Campbell 1988) and the Ashburton Surface of the central highlands of the Northern Territory (Hays 1967) appear to be of similar antiquity. The summit surface of Arnhem Land may be of Cretaceous or older age for the upland is surrounded by Cretaceous sediments and sea cliffs are preserved at the margins of the plateaux (e.g. Needham 1982). The plateau and high plain preserved in the Arcoona Plateau, South Australia, both predate the Early Cretaceous as sediments of that age occupy valleys cut into the surface. The prominent summit surface of the Gawler Ranges (Plate 2), a massif of silicic volcanic rocks in the interior of South Australia, is of etch type. The planation of the surface is of Jurassic or older age, but the stripping and exposure of the surface of low relief took place in the Early Cretaceous (Campbell & Twidale 1991).

In addition many minor Cretaceous elements have been recognised. Thus the summit bevel of Ayers Rock is part of the Cretaceous landscape (Harris & Twidale 1991).

7 ISOSTATIC ADJUSTMENTS

The pattern of Australian topography appears to have been constant for a long time. The uplands have been uplands throughout the Cainozoic. In some instances, as in the Isa Highlands and the Flinders Ranges, there is a strong suggestion that the uplands have stood in positive relief through most of Phanerozoic time. In this respect they resemble the Eastern Uplands of Australia, which remain much as they have been through the Cainozoic (e.g. Bishop 1982; Young 1983; Taylor et al. 1985; Young & McDougall 1985). The intervening basins, though changing in shape and area have consistently been structurally negative areas. The Carpentaria Basin is a simple structure, having been formed by epeirogenic movements in the early Mesozoic (Smart & Senior 1980) and having been the site of sedimentation through the remainder of Mesozoic times. The Eromanga Basin, which with the Surat and Carpentaria basins together form the Great Artesian Basin, is not only an area of later Mesozoic sedimentation but also covers such older depositional basins as the Pedirka (late Palaeozoic), Warburton (late Palaeozoic-early Mesozoic) and Arckaringa (late Palaeozoic) basins.

Lambeck (1984) attributed basin development in central Australia, and as a corollary the rise of the intervening uplands, to north-south lateral compression. Jones and Veevers (1982) associated the pronounced phases of Cainozoic rejuvenation of the Eastern Highlands with periods of active volcanism. Wellman and Greenhalgh (1988) considered the Flinders and associated ranges as being associated with compressive faulting, which is consistent with the nature of exposed fault planes bordering the uplands. Lambeck and Stephenson (1986) interpreted the rise of the Eastern Uplands as an isostatic response to erosion. On the other hand, none of these mechanisms involving block faulting, epeirogenic uplift, and isostatic adjustment seems appropriate on its own to explain the various uplands.

Tectonism must have played a major part in the evolution of the topographic pattern. The Carpentaria Basin, for example, and other components of the Great Artesian Basin were initiated as negative features by the crustal movements related to the separation of Australia and Antarctica (Smart & Senior 1980). Possibly the downfaulted gulfs of St Vincent and Spencer mark the early beginning of the splitting apart of Australia, the fracturing taking effect first in the south and extending northwards to the western side of Lake Eyre and including the Levi and Mt Margaret faults of the Peake and Davenport Ranges. On the other hand, the Tennant Creek earthquake of 1990 and other recent intraplate earthquakes are due to compression (Gordon & Lewis 1980; McCue 1990).

But this is probably not the whole story. In addition to epeirogenic movements associated with the break-up of Gondwana eustatic changes of sealevel caused the invasion of the continent by Early Cretaceous epicontinental seas. This induced isostatic responses due first to the thalassostatic loading of the basins and second to deposition in the same basins and uplift and erosion of the land. Such isostatic adjustments have long been recognised, Bourcart (1938,1950) for instance discussing continental flexure and Jessen (1943) the Randschwellen concept, which involves marine transgression causing loading and subsidence of the flooded area (Figure 2). The coastal zone of the time acts as a fulcrum, and the adjacent land rises, though, according to Jessen the interior areas may also subside in response to this see-saw effect. In terms of continental flexure the impact near the coast may have taken the form of upwarping, as for instance in the Isa Highlands and the northern Flinders Ranges, or involve block faulting as in the Gawler Ranges. The timing of these events is uncertain. There must have been delays between the loading or unloading and isostatic response. Once they had begun to rise, however, the positive blocks were dissected as a result of rejuvenation of streams grading to the lower base levels. Thus a positive feedback or reinforcement mechanism takes effect.

It seems likely that each of these three effects, tectonic displacement, thalassostatic isostasy and denudational isostasy, have operated. The isostatic responses were triggered by tectonism and marine transgression. Whether the depths of incision and the volume of sediments transferred are consistent with the suggested cause is, however, questionable (see e.g. Crittenden 1963; Morrison 1965). Overriding any such considerations, however, tectonic and isostatic adjustments may have been cushioned by the inherent strength and rigidity of the crust at a regional scale, as has been suggested in respect of the Flinders Ranges (Wellman & Greenhalgh 1988).

8 MODELS OF LANDSCAPE EVOLUTION

The landscape evolution implied is inconsistent with all the conventional models used in geomorphological interpretation (see Davis 1909; King 1942, 1953; Hack 1960). None allows for the survival of palaeoforms (see also Thornbury 1954; Brown 1980). The steady state model implies constant erosion of the entire land surface. The cyclic models allow time for landscape revival to take effect; there is a period during which rejuvenated streams incise and extend. Remnants of the "original" or initial surface persist until late into the scarp retreat cycle, but are eliminated quite early in the peneplanation sequence. In both cyclic schemes, however, the duration of the geomorphic cycle is critical. The time taken to reduce the land mass to base level obviously depends on its nature (area, elevation), on climatic conditions and on tectonic renewal, if any. Given warm humid conditions and tectonic stability, and allowing for isostatic response to erosion, it has been calculated that a land mass the size of the continental United States would be baselevelled in the order of 33 Ma (Gilluly 1955; Schumm 1963). In other words, no land surface older than Oligocene ought to survive, unless it has been exhumed.

The United States estimates are applicable to other land areas such as Australia, where, indeed, many of the palaeosurfaces recognised in the landscape are of exhumed type. Others remain in unconformity. Both either are or have been preserved by burial: neither presents problems for survival. On the other hand, many of the palaeosurfaces recorded in the landscape and of epigene or etch type are older than Oligocene. The landscapes of the central and southern Flinders epitomise such ancient epigene landscapes, and the summit surface of Ayers Rock is an example of an etch surface from which the regolith was most likely, almost certainly, eroded long before the Oligocene. Such palaeosurfaces pose problems for the conventional models of landscape evolution (e.g. Twidale 1976b; Young 1983).

The occurrence in the contemporary landscape of uncomfortably old palaeosurfaces is not peculiar to Australia. Surfaces of similar antiquity have been reported from many other Gondwanan continents, from southern Africa (e.g. King 1962; Partridge & Maud 1987; Twidale 1988, 1990), west Africa (e.g. Pugh 1954; Michel 1978), east Africa (e.g. Willis 1936), peninsular India (e.g. Demangeot 1978) and the Guyana Shield (Briceño & Schubert 1990). One possible solution to the problem posed by such ancient land surfaces is that the conventional models of landscape evolution are in error. Davis (1909) argued that water erosion and wasting together affected the entire land surface, implying that the entire surface was lowered simultaneously. Knopf

(1924) and Crickmay (1932, 1976) pointed out that some facets of the landscape, particularly on divides, appear to persist. Horton (1945) advocated a zone of nil erosion on interfluves. Crickmay's concept of Unequal Activity (see Crickmay 1976 for references), particularly where applied to streams, allows strong dissection in and near the channels, but leaves the divides untouched. Thus palaeosurface remnants are left high and dry, thus ensuring their survival (Twidale 1991).

9 CONCLUSIONS

Substantial elements of the Australian landscape are of Gondwanan origin: they had developed before the breakup of the supercontinent. Some are exhumed, but others appear to have survived as a result of a sequence of tectonism, thalassostatic and depositional isostatic subsidence, related uplift of adjacent land areas, and unequal erosion of those land masses. In this way palaeosurfaces were left high in the relief where they have persisted intrinsically unchanged.

ACKNOWLEDGEMENTS

The work reported here was carried out under the auspices of an ARC grant.

REFERENCES

Alley, N. F.; Lemon, N. M. 1988: Evidence of earliest Cretaceous (Neocomian) marine influence, northern Flinders Ranges. *Geological Survey of South Australia Quarterly Geological Notes* 106: 2-7.

Barron, E. J. 1987: Cretaceous plate tectonic reconstructions. *Palaeogeography, Palaeoclimatology, Palaeoecology* 59: 3-29.

Bishop, P. 1982: Stability or change: a review of ideas on ancient drainage in eastern New South Wales. *Australian Geographer* 15: 219-230.

Bourcart, J. 1938: Le marge continentale. *Bulletin de la Société Géologique de France* 5: 393-474.

Bourcart, J. 1950: Le théorie de la flexure continentale. *Congrès International de Géographie, Lisbon 1949*, 2: 167-190.

Branch, C. D. 1966: Volcanic cauldrons, ring complexes and associated granites of the Georgetown Inlier, Queensland. *Bureau of Mineral Resources, Geology and Geophysics Bulletin* 76.

Briceño, H.; Schubert, C., 1990: Geomorphology of the Gran Sabana, Guyana Shield. *Geomorphology*. 3: 125-141.

Brock, E. J. 1971: The denudation chronology of the Fleurieu Peninsula, South Australia. *Transactions of the Royal Society of South Australia* 95: 85-94.

Brown, E. H. 1980: Historical geomorphology - principles and practices. *Zeitschrift für Geomorphologie Supplementband* 36: 9-15.

Campana, B. 1958: The Mount Lofty- Olary Region and Kangaroo Island. In Glaessner, M. F.; Parkin, L. W. (eds): *The Geology of South Australia*. Melbourne University Press / Geological Society of Australia, Melbourne: 3-27.

Campbell, E. M.; Twidale, C. R. 1991: The evolution of bornhardts in silicic volcanic rocks in the Gawler Ranges. *Australian Journal of Earth Sciences* 38: 79-93.

Carter, E. K.; Öpik, A. A. 1959: *Geological map of northwest Queensland* (10 miles/1 inch). Bureau of Mineral Resources, Geology and Geophysics, Canberra.

Craddock, C. 1977: The evolution and fragmentation of Gondwanaland. In Laskar, B.; Raja Rao, C. S. (ed.): *Fourth International Gondwana Symposium, Calcutta, India, Papers* II: 711-719.

Craft, F. A. 1932: The physiography of the Shoalhaven Valley. *Proceedings of the Linnaean Society of New South Wales* 57: 245-260.

Craft, F. A., 1933: Surface history of the Monaro. *Proceedings of the Linnaean Society of New South Wales* 58: 229-244.

Crickmay, C. H. 1932: The significance of the physiography of the Cypress Hills. *Canadian Field Naturalist 46*: 185-186.

Crickmay, C. H. 1976: The hypothesis of unequal activity. In Melhorn, W. N.; Fleming, R. C. (eds): *Theories of landform development*. State University of New York, Binghampton: 103-109.

Crittenden, M. D., 1963: New data on the isostatic deformation of Lake Bonneville. *United States Geological Survey Professional Paper*. 454-E.

Daily, B.; Milnes, A. R.; Twidale, C. R.; Bourne, J. A. 1979: Geology and Geomorphology. In Tyler, M. J.; Twidale, C. R.; Ling, J. K. (eds): *Natural History of Kangaroo Island*. Royal Society of South Australia, Adelaide: 1-38.

Daily, B.; Twidale, C. R.; Milnes, A. R. 1974: The age of the lateritized surface on Kangaroo Island and adjacent areas of South Australia. *Journal*

of the Geological Society of Australia 21: 387-392.
Daniels, J. L. 1974: The geology of the Blackstone region, Western Australia. *Western Australian Geological Survey Bulletin* 123.
Davis, W. M. 1909: *Geographical essays.* Dover, Boston.
Demangeot, J. 1978: Les reliefs cuirassés de l'Inde du Sud. *Travaux et Documents de Géographie Tropicale* 33: 97-111.
Fairbridge, R. W. 1989: International fracture propagation in the breakup of Gondwanaland. In Le Maitre, R. W. (ed): *Pathways in Geology. Essays in Honour of Edwin Sherbon Hills.* Hills Memorial Volume Committee, Melbourne: 336-344.
Frakes, L. A.; Burger, D.; Apthorpe, M.; Wiseman, J.; Dettmann, M.; Alley, N.; Flint, R.; Gravestock, D.; Ludbrook, N.; Backhouse, J.; Skwarko, S.; Scheibnerova, V.; McMinn, A.; Moore, P. S.; Bolton, B. R.; Douglas, J. G.; Christ, R.; Wade, M.; Molinar, R. E.; McGowran, B.; Balme, B. E.; Day, R. A.: 1987: Australian Cretaceous shorelines, stage by stage. *Palaeogeography, Palaeoclimatology, Palaeoecology* 59: 31-48.
Gilluly, J. 1955: Geologic contrasts between continents and ocean basins. *Geological Society of America Special Paper* 62: 7-18.
Gordon, F. R.; Lewis, J. D. 1980: The Meckering and Calingiri earthquakes, October 1968 and March 1970. *Geological Survey of Western Australia Bulletin* 126.
Hack, J. T. 1960: Interpretation of erosional topography in humid temperate regions. *American Journal of Science* 237A: 80-97.
Harris, W. K.; Twidale, C. R. 1991: Revised age for Ayers Rock and the Olgas. *Transactions of the Royal Society of South Australia* 15: 109.
Hays, J. 1967: Land surfaces and laterites in the north of the Northern Territory. In Jennings, J. N.; Mabbutt, J. A. (eds): *Landform Studies in Australia and New Guinea.* Australian National University Press, Canberra: 182-210.
Hills, E. S. 1934: Some fundamental concepts in Victorian physiography. *Proceedings of the Royal Society of Victoria* 47: 158-174.
Hills, E. S. 1975: *Physiography of Victoria; An Introduction to Geomorphology.* Whitcombe & Tombs, Melbourne.
Horton, R. E. 1945: Erosional development of streams and their drainage basins: hydrophysical approach to quantitative morphology. *Geological Society of America Bulletin* 56: 275-370.
Hossfeld, P. S. 1926: *The geology of portions of the counties of Light, Eyre, Sturt and Adelaide.* Unpublished M. Sc. thesis, University of Adelaide.
Jack, R. L. 1915: The geology and prospects of the region to the south of the Musgrave Ranges, and the geology of the western portion of the Great Australian Artesian Basin. *Geological Survey of South Australia Bulletin* 5.
Jack, R. L. 1931: Report on the geology of the region to the north and northwest of Tarcoola. *Geological Survey of South Australia Bulletin* 15.
Jackson, M. J.; Van de Graaff, W. J. E. 1981: The geology of the Officer Basin, Western Australia. *Bureau of Mineral Resources Bulletin* 206.
Jennings, J. N.; Sweeting, M. M. 1963: The limestone ranges of the Fitzroy Basin, Western Australia. *Bonner Geographische Abhardlungen* 32.
Jessen, O. 1943: Die Randschwellen der Kontinente. *Petermanns Geographische Mitteilungen Erganz* 241.
Jones, J. G.; Veevers, J. J. 1982: A Cainozoic history of Australia's Southeast Highlands. *Journal of the Geological Society of Australia* 29: 1-12.
Keyser, F. de 1964: *Innisfail, Queensland 1:250,000 Geological Series Explanatory Notes.* Bureau of Mineral Resources Geology and Geophysics, Canberra.
King, L. C. 1942: *South African Scenery.* Oliver & Boyd, Edinburgh.
King, L. C. 1953: Canons of landscape evolution. *Geological Society of America Bulletin* 64: 721-752.
King, L. C. 1962: *Morphology of the Earth.* Oliver & Boyd: Edinburgh.
Knopf, E. B. 1924: Correlation of residual erosion surfaces in the eastern Appalachian highlands. *Geological Society of America Bulletin 35*: 633-668.
Lambeck, K. 1984: Structure and evolution of the Amadeus, Officer and Ngalia basins of central Australia. *Australian Journal of Earth Sciences* 31: 25-48.
Lambeck, K.; Stephenson, R. 1986: The post-Palaeozoic uplift history of southeastern Australia. *Australian Journal of Earth Sciences* 33: 253-270.
Mabbutt, J. A. 1966: Landforms of the western Macdonnell Ranges. In Dury, G. H. (ed.): *Essays in Geomorphology.* Heinemann, London: 83-119.

Mabbutt, J. A. 1988: Australian desert landscapes. *GeoJournal* 16: 355-369.

McCue, K. 1990: Australia's large earthquakes and Recent fault scarps. *Journal of Structural Geology* 12: 761-766.

Michel, P. 1978: Cuirassés bauxitiques et ferrugineuses d'Afrique occidentale: Aperçu chronologique. *Travaux et Documents de Géographie Tropicale* 33: 11-32.

Morrison, R. B. 1965: Quaternary geology of the Great Basin. pp. 265-285 in H. E. Wright & D. G. Frey (eds), *The Quaternary of the United States*. Princeton University Press: Princeton, New Jersey.

Needham, R. S. 1982: East Alligator, Northern Territory. *1:100,000 Geological Map Commentary*. Bureau of Mineral Resources Geology and Geophysics, Canberra.

Ollier, C. D. 1982: The Great Escarpment of eastern Australia: tectonic and geomorphic significance. *Journal of the Geological Society of Australia* 25: 13-24.

Öpik, A. A. 1961: The geology and palaeontology of the headwaters of the Burke River, Queensland. *Bureau of Mineral Resources, Geology and Geophysics Bulletin* 53.

Partridge, T. C.; Maud, R. R. 1987: Geomorphic evolution of southern Africa since the Mesozoic. *South African Journal of Geology* 90: 179-208.

Perry, W. J.; Roberts, H. G. 1968: Late Precambrian glaciated pavements in the Kimberley region, Western Australia. *Journal of the Geological Society of Australia* 15: 51-56.

Plumb, K. A.; Veevers, J. J. 1971: Cambridge Gulf *1:250,000 Geological Series Explanatory Notes Sheet SD/52-14*. Bureau of Mineral Resources Geology and Geophysics, Canberra.

Pugh, J. C. 1954: High level surfaces in the Eastern Highlands of Nigeria. *South African Geographical Journal* 36: 31-42.

Schumm, S. A., 1963: Disparity between present rates of denudation and orogeny. *United States Geological Survey Professional Paper* 454.

Smart, J.; Senior, B. R. 1980: Jurassic-Cretaceous basins of northeastern Australia. In Henderson, R. A.; Stephenson, P. J. (eds): *The Geology and Geophysics of Northeastern Australia*. Geological Society of Australia, Queensland Division, Brisbane: 315-329.

Stewart, A. J.; Blake, D. H.; Ollier, C. D. 1986: Cambrian river terraces and ridgeslopes in central Australia: oldest persisting landforms? *Science* 333: 758-761.

Stuart-Smith, P. G.; Needham, R. S.; Bagas, L.; Wallace, D. A. 1987: Pine Creek, Northern Territory *1:100,000 Geological Series Map Commentary*. Bureau of Mineral Resources Geology and Geophysics/Northern Territory Geological Survey, Canberra.

Taylor, G.; Taylor, G. R.; Binks, M.; Fondoulis, C.; Gordon, I.; Hedstrom, J.; Minello, J.; Whippy, F. 1985: Pre-basaltic topography of the northern Monaro and its implications. *Australian Journal of Earth Sciences*, 32: 65-72.

Thornbury, W. D. 1954: *Regional geomorphology of the United States*. Wiley, New York.

Twidale, C. R. 1956: Chronology of denudation in northwest Queensland. *Geological Society of America Bulletin* 67: 867-882.

Twidale, C. R. 1976a: Geomorphological evolution. In Twidale, C. R.; Tyler, M. J.; Webb, B. P. (eds): *Natural History of the Adelaide Region*. Royal Society of South Australia, Adelaide: 43-59.

Twidale, C. R. 1976b: On the survival of palaeoforms. *American Journal of Science* 276: 77-94.

Twidale, C. R. 1984: The enigma of the Tindal Plain, Northern Territory. *Transactions of the Royal Society of South Australia* 108: 95-103.

Twidale, C. R. 1986: Granite platforms and low domes: newly exposed compartments or degraded remnants? *Geografischer Annaler (Series A)* 68: 399-411.

Twidale, C. R., 1988: The missing link: planation surfaces and etch forms in southern Africa. pp. 31-46 in G. F. Dardis & B. P. Moon (eds). *Geomorphological Studies in Southern Africa*. Balkema, Rotterdam.

Twidale, C. R. 1990: The origin and implications of some erosional landforms. *Journal of Geology* 98: 343-364.

Twidale, C. R. 1991: A model of landscape evolution involving increased and increasing relief amplitude. *Zeitschrift für Geomorphologie* 35: 85-109.

Twidale, C. R.; Campbell, E. M. 1988: Ancient Australia. *GeoJournal* 16: 339-354.

Twidale, C. R.; Campbell, E. M. 1991: The Early Cretaceous marine transgression and its significance for landscape interpretation. *Australian Geographer* 22: 178-184.

Twidale, C. R.; Campbell, E. M. 1992: Some aspects of the geomorphological development of the eastern margin of the Australian Shield. *Earth Surface Processes and Landforms*. In press.

Twidale, C. R.; Horwitz, R. C.; Campbell, E. M.

1985: Hamersley landscapes of the northwest of Western Australia. *Revue de Géologie Dynamique et de Géographie Physique* 26: 173-186.

Veevers, J. J. (ed.) 1984: *Phanerozoic earth history of Australia.* Clarendon Press, Oxford.

Wayland, E. J. 1934: Peneplains and some erosional landscapes. *Geological Survey of Uganda Annual Report and Bulletin 1*: 77-79.

Wellman, P.; McDougall, I. 1974: Potassium-Argon ages on the Cainozoic volcanic rocks of New South Wales. *Journal of the Geological Society of Australia.* 21:247-272.

Wellman, P; Greenhalgh, S.A. 1988: Flinders/Mt Lofty Ranges, South Australia: their uplift, erosion and relationship to coastal structure. *Transactions of the Royal Society of South Australia* 112: 11-19.

Wells, A. T.; Forman, D. J.; Randford, L. C.; Cook, P. J. 1970: Geology of the Amadeus Basin, central Australia. *Bureau of Mineral Resources Geology and Geophysics Bulletin* 100.

Willis, B. 1936: East African plateaus and rift valleys. *Studies in Comparative Seismology Publication* 470. Carnegie Institute Washington D. C.

Woodard, G. D. 1955: The stratigraphic succession in the vicinity of Mt Babbage, South Australia. *Transactions of the Royal Society of South Australia* 78: 8-17.

Wopfner, H. 1964: Permian-Jurassic history of the western Great Artesian Basin. *Transactions of the Royal Society of South Australia* 88: 117-128.

Wopfner, H. 1968: Cretaceous sediments on the Mt Margaret Plateau and evidence for neo-tectonism. *Geological Survey of South Australia Quarterly Geological Notes* 28: 7-11.

Young, R. W. 1983: The tempo of geomorphological change. Evidence from southeastern Australia. *Journal of Geology* 91: 221-230.

Young, R. W. 1991: Structural inheritance and planation in the evolution of landforms in the East Kimberley. *Australian Journal of Earth Sciences*, In press.

Young, R. W.; McDougall, I. 1985: The age, extent and geomorphologic significance of the Sassafras basalt, southeastern New South Wales. *Australian Journal of Earth Sciences* 32: 321-331.

Gondwana Eight, Findlay, Unrug, Banks & Veevers (eds) © *1993 Balkema, Rotterdam. ISBN 90 5410 304 3*

Petrologic data on the evolution of the Great Australian Bight

Jonathan D.A.Clarke
Western Mining Corporation, Preston, Vic., Australia
Neville F.Alley
South Australian Department of Mines and Energy, Eastwood, S.A., Australia

ABSTRACT: Sedimentation in the Great Australian Bight has occurred in a variety of settings from the initial intra-cratonic rift to the modern carbonate shelf. An extensive palaeodrainage system, remnants of which are still preserved, supplied clastic sediments to the Great Australian Bight from the north-west, north, and north-east during the Mesozoic and early Tertiary. Clastic sediments were deposited in lacustrine to prograding shelf and deltaic environments. Petrological and palynological data indicate that climates were predominantly humid. High rainfall, run-off and clastic input prevented development of rift stage evaporites, and delayed establishment of a carbonate shelf until the Middle Eocene, when influence of the palaeodrainages diminished, possibly through the establishment of a less humid climatic regime. Scarcity of volcanics suggests rifting was controlled by external tectonic processes, rather than a mantle heat source. The sedimentary evolution of the conjugate Antarctic margin up to the Oligocene may partly mirror that of the Great Australian Bight.

1 INTRODUCTION

The Great Australian Bight (Figure 1) has developed from rifts in the interior of Gondwana and contrasts with the northern and eastern margins of Australia, which formed part of the Palaeo-Pacific rim of the super-continent (Coney, 1990). This paper: (1) summarises petrological and stratigraphic studies of samples collected by the Bureau of Mineral Resources's research vessel "Rig Seismic" Cruise 11 (Bureau of Mineral Resources Cruise 66), preliminary results of which were published by Willcox *et al.* (1988) and Davies *et al.* (1988); and (2) provides further data on the evolution of the Great Australian Bight.

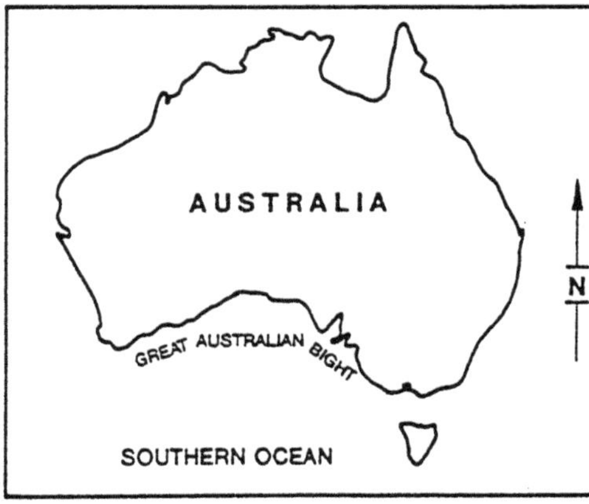

FIGURE 1. Location of study area.

2 REGIONAL ASPECTS

2.1 Setting

The Great Australian Bight (Figure 2) contains several basins and Sub-basins, with most of the sediment fill occurring in the Bight Basin (Hill, in press). The cratonic on-lap of the Bight Basin forms the Eucla Basin. Basins and Sub-basins within the Great Australian Bight are the Eyre (Bein & Taylor, 1981) and Ceduna (Falvey *et al.*, 1990) Sub-basins of the Bight Basin, Duntroon Basin (Robertson *et al.*, 1979), and Polda Basin (Nelson *et al.*, 1986).

2.2 Stratigraphy

Basement rocks in the Great Australian Bight consist, for the most part, of Precambrian crystalline rocks of the Albany-Fraser Orogen, and the Gawler and Yilgarn Cratons.

Basalts are rare in the Great Australian Bight. Prior to Bureau of Mineral Resources Cruise 66, the only known volcanic was the Middle Jurassic Wisanger Basalt on Kangaroo Island (Milnes *et al.*, 1982).

Pre-rift sediments in the Great Australian are also rare. Onshore these are confined to limited Permian platform cover (Lowry, 1970) and the unusual late

TABLE 1. Mesozoic-Cainozoic stratigraphy of the GAB (after Hill, in press; Benbow et al., in press; and Willcox et al., 1988).

AGE	BIGHT BASIN	EUCLA PLATFORM	EYRE BASIN	POLDA BASIN	DUNTROON BASIN
L Eocene-Quaternary	Eucla Group	Eucla Group	Eucla Group	Bridgewater Fm (Quarternary)	Eucla Group
Paleocene-Eocene	Immarna Group	Immarna Group	Immarna Group	Poelpena Fm	Immarna Group
Maastrichtian-Danian	Potoroo Fm	----------	----------	----------	Potoroo Fm
Turonian-Campanian	Wigunda Fm	----------	----------	----------	Wigunda Fm
Cenomanian	Platypus Fm	----------	----------	----------	Platypus Fm
Albian-Aptian	Madura & Ceduna Fm	Madura Fm	Madura Fm	----------	Ceduna, Borda, & Neptune Fm
Neocomanian	Loongana & Echidna Fm	Loongana Fm	Loongana Fm	----------	Echidna Fm
Late Jurassic	Unnamed sequence	---------- Cratons	Unnamed sequence	Polda Fm	----------

TABLE 2. Petrology of igneous rocks from the GAB.

FORMATION	SAMPLE	AGE	LITHOLOGY	MINERALOGY	ALTERATION
Eyre Terrace crystalline basement	66DR2B	Proterozoic	granite	plag-qtz-hnbe-bio	seafloor weathering & Mn encrustation
Eyre Terrace volcanics	66DR1J	Eocene-Late Cretaceous	vesicular ol alkali basalt	ol-cpx-gls	seafloor weathering & silicification
	66DR3H	Eocene-Late Cretaceous	vesicular ol alkali basalt	ol-cpx-gls	seafloor weathering & carbonation

Mineral abbreviations: alb=albite; ap=apatite; bio=biotite; cal=calcite; clay=unidentified clay minerals clinop=clinoptolite; dol=dolomite; cpx=clinopyroxene; glauc=glauconite; gls=glass; goet=goethite; hnbe=hornblende; ill=illite; kaol=kaolinite; ol=olivine; plag=plagioclase; qtz=quartz; sid=siderite.

Proterozoic sediments of the Polda Basin (Nelson *et al.*, 1986). The deep off-shore stratigraphy is not sufficiently well known to determine if any pre-rift sediments are present; all wells drilled to date have, however, passed directly from Mesozoic sediments into Precambrian basement.

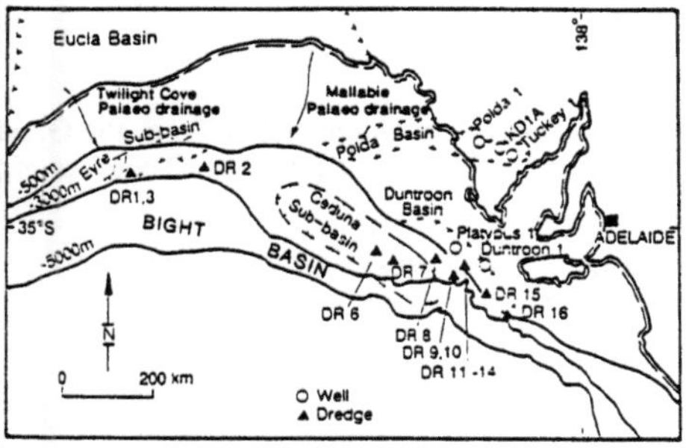

FIGURE 2. Great Australian Bight localities (after Davies *et al.*, 1988; Lowry, 1976).

The sedimentary stratigraphy of the Great Australian Bight (Table 1) was defined by Robertson *et al.* (1979), and reviewed in Willcox *et al.* (1988). The Mesozoic section of the Great Australian Bight was revised by Hill (in press), and the stratigraphy of the Polda Trough summarised by Nelson *et al.* (1986).

Earliest rift sediments are of Jurassic age, and correspond to sequence Mz3 (sequence stratigraphy of Hocking, 1990), they occur only in the deepest parts of the basin. These are overlain by two Cretaceous clastic sequences. Cretaceous sediments occur over much of the Great Australian Bight with the Early Cretaceous sequence (Mz4) being the most extensive, and representing deposition during the rift-drift transition. The Late Cretaceous sediments represent the start of the post-rift sedimentation (sequence Mz5). Three sequences are present in the Tertiary succession. The earliest (Cz1) is represented by basalts and subaerial sediments. This was followed by a relatively thin clastic Paleocene to Middle Eocene clastic sequence (lower Cz2) and later, carbonate dominated sequences (upper Cz2, Cz3 and Cz4) that extend into the Holocene.

3 PETROLOGY & PETROGRAPHY

3.1 Procedure

Samples were dredged from numerous localities on the sea floor during Cruise 66. The geology of some of these, as determined from seismic sections and dredged samples at several of these localities, is shown in Figure 3. Tables 2-5 summarise the petrology of selected samples collected during Bureau of Mineral Resources Cruise 66. Some samples from wells in the Polda and Duntroon Basins, and the Ceduna Sub-basin are also included. Petrographic studies were carried out using standard techniques for the study of rocks in thin section. Mineralogy was determined with a Phillips PW 1800 quantitative XRD machine. Whole rock compositions of igneous lithologies were determined by ICP at the School of Earth Sciences, Flinders University. Sample ages were determined using foraminifers (McGowran, 1991), nannofossils (Shafik, 1990) and palynology (Alley & Clarke, in press). Biostratigraphic and lithologic criteria were employed to match dredge samples with the established stratigraphy.

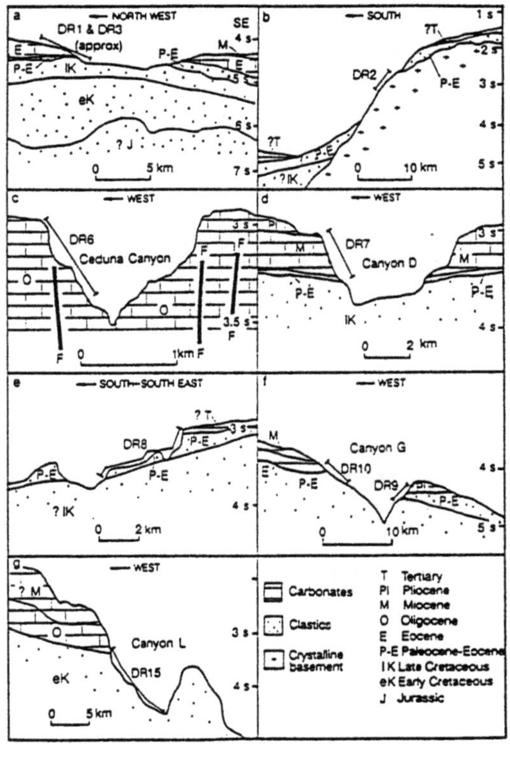

FIGURE 3. Interpreted stratigraphy of selected dredge stations from seismic and geologic data (modified from Davies *et al.*, 1988). Vertical scale in seconds.

3.2 Basement

Granodiorite samples were recovered by dredge haul 2 during Bureau of Mineral Resources Cruise 11 from a submerged scarp on the Eyre Terrace (Table 2 & 3). The granites are unfoliated and unmetamorphosed, despite their presumed Precambrian age. In thin section they contain 60 % feldspar (orthoclase 40 %, perthite 15 %, microcline 5 %), 20 % quartz (an early phase of rounded crystals within feldspars and a later relict space-filling phase), 10 % biotite, and 10 % hornblende.

Sea-floor weathering consists of partial breakdown of feldspars to sericite, and mafic minerals to chlorite and epidote. Fracturing and infill by clay and iron oxide, and development of a very thin manganese crust has also occurred.

3.3 Volcanics

Vesicular, amygdaloidal, possibly pillowed, alkaline olivine basalts (Table 2 & 3) with minor associated hyaloclastites were recovered in dredges 66DR01J and 66DR03, as part of a Coniacian to Late Eocene dredge suite. The basalt is thus most likely of Late Cretaceous or Palaeogene age. The recovered samples are probably related to the small volcanic centres interpreted by James & von der Borch (1991) on the Eyre Terrace. Similar seismic mounds were also identified as volcanic centres by Fraser & Tilbury (1979) on seismic sections from the Ceduna Terrace. A cluster of magnetic highs beneath the Ceduna Terrace (Fraser & Tilbury, 1979, Fig 6) may be related to volcanics and associated intrusions. High heat flow on the Ceduna Terrace has been attributed to buried volcanics and intrusions (Willcox *et al.*, 1988).

The basalts are porphyritic, containing phenocrysts of olivine and clinopyroxene in an originally glassy groundmass containing fine plagioclase needles. Both samples have been extensively altered, the glass to amorphous clays, and with extensive vesicle infill by either chalcedony (66DR1J) or calcite and dolomite (66DR3H). Partial breakdown of olivines and pyroxenes to iron oxides and carbonates has occurred. Comparison of whole rock and glass assays indicates higher whole rock silica in both samples and calcium in 66DR3H. Amygdaloidal silica and carbonate from alteration on the sea-floor is the most likely source of these differences.

3.4 Carbonates

Carbonate sediments are common only in the Tertiary succession (Table 4). Cretaceous calcareous sediments (Table 5) were recovered from the Eyre Sub-basin and Duntroon Basin. Eyre Sub-basin samples comprise sideritic dolomite possibly formed as diagenetic nodules (66DR03D). Duntroon Basin samples are either glauconitic calcareous siltstone or fenestral, calcareous mudstone (66DR12F). The calcareous siltstone is interpreted as being deposited in relatively deep environments, while the fenestral mudstone is interpreted as having been deposited on coastlines distal to fluvial input. Based on a limited number of samples (Table 4), it appears that although most samples younger that Eocene were deposited in deep water, episodes of shallower deposition occurred on what is now the continental slope of the outer Great Australian Bight until the Miocene. This is demonstrated by evidence of meteoric daigenesis in a sample (66DR07G) of Miocene limestone (see below).

TABLE 3. Igneous rock composition from the Great Australian Bight (66DR1J glass after Davies *et al.*, 1988).

Major (weight %)	66DR1J	66DR1J (glass)	66DR3H	66DR2B (@3)
SiO_2	69.05	50.60	59.74	66.05
TiO_2	2.25	2.25	1.69	1.13
Al_2O_3	12.36	16.20	9.47	14.92
Fe_2O_3	0.04	8.55	7.69	4.81
MnO	0.09	0.13	0.08	0.36
MgO	4.54	5.01	4.57	0.62
CaO	6.32	8.07	11.02	2.56
Na_2O	2.12	5.07	2.73	4.13
K_2O	2.02	3.30	2.16	3.64
P_2O_5	1.20	0.78	0.85	0.43
TOTAL	99.99	99.96	100.00	98.65
Minor (ppm)				
Ba	359.0	—	214.0	1659.7
Ce	67.1	—	20.9	173.0
Co	53.5	—	82.6	171.8
Cr	202.0	—	275.0	12.4
Cu	36.5	—	33.6	10.1
La	66.4	—	60.8	83.7
Nd	45.9	—	30.3	75.4
Ni	81.8	—	248.0	41.0
Sr	568.0	—	587.0	44.1
V	211.0	—	162.0	96.2
Zn	144.0	—	123.0	89.2
Zr	281.0	—	209.0	69.8

Shelf carbonates are fine-grained bioclastic lime mudstone and wackestone. One sample of coarse-grained coralline limestone (66DR7G) was also recovered. Reef facies were not sampled, nor are they known from the on-shore Eucla Basin (Lowry, 1970; Benbow *et al.*, in press). Deeper-water facies are represented by pelagic chalks. Both facies are dominated by fossils with calcite mineralogy, calcareous nannofossils, foraminifers, and sponge spicules being the most common. Mollusc, echinoderm and coral fragments are rare. Siliceous elements, particularly sponge spicules, but also radiolaria and diatoms, are common in many samples.

Diagenesis has been mainly limited to simple compaction and weak cementation by micro-crystalline calcite during burial. Late partial dolomitisation is present in some chalks. Evidence of meteoric diagenesis is confined to two samples, 66DR07G

(Miocene) and 66DR12F (Late Cretaceous). The coralline limestone, 66DR07G has undergone dissolution of molluscs and corals with subsequent partial infill of mouldic porosity by drusy calcite, characteristic of meteoric phreatic environments (James & Choquette, 1984). Sample 66DR12F has partial infill of fenestrae by drusy calcite.

Many samples contain evidence of weathering on the sea-floor. Dissolution of spicules and mollusc fragments is common, some moulds have been partly in-filled by glauconite, silica, and oxides of iron and manganese. Case-hardening of exposed surfaces is evident in some samples, as well as encrustation by iron and manganese oxides. Syn-depositional bioturbation and later boring are common.

Partially silicified carbonates and nodular cherts are common in the Eocene and Oligocene rocks. The silica is predominantly microcrystalline quartz, chalcedony and minor clinoptilite. No evidence of the detrital volcanic material, the silica source suggested by Jones & Fitzgerald (1986) for silicified spicular sediments of similar age near Adelaide, was identified in these rocks. The common remains of siliceous organisms were the most likely source for secondary silica. Global precipitation of silica during Eocene and Oligocene times was attributed by McGowran (1989) to combined palaeoclimatic and oceanographic conditions. Cherts from the Great Australian Bight correspond with those in the Wilson Bluff Limestone and Port Willunga Formation in the Eucla Basin and St Vincent Basin respectively. None are time-equivalents of the unusual silica facies of the Blanche Point Formation (Jones & Fitzgerald, 1986) in the St Vincent Basin near Adelaide. The presence of common siliceous spicules in Late Cretaceous clastic sediments from the Great Australian Bight indicates that conditions favourable for the proliferation of siliceous organisms were common-place, prior to the Late Eocene.

3.5 Clastics

Sandstones from the Great Australian Bight contain grains of plutonic and metamorphic provenances, indicating derivation from Precambrian cratons, or possibly the metamorphic and igenous rocks of the Kanmantoo trough (Tables 4 & 5). A minor volcanoclastic component occurs in the eastern part of the Breat Australian Bight. Recycled palynomorphs (Alley & Clarke, in press), and second cycle quartz grains show significant input from older sediments. Possible candidates include Proterozoic and Palaeozoic sediments of the Officer Basin. Reworking of some older rift related sediments has also occurred. Recovered clastics are predominantly of Cretaceous age (Table 5), and one sample of Pliocene mudstone was also recovered (66DR09A, Table 4, age data from McGowran 1991). Contamination of Cretaceous samples by Eocene palynomorphs (Alley & Clarke, in press), suggest that thin outcrops of Eocene clastics which shed palynomorphs down-slope are present at most dredge sites. These were not, however, sampled directly, apart from one phosphatised sample (DR11F) described below in section 3.6.

Clastic sediments would have been transported down the numerous palaeodrainage systems radial to the Great Australian Bight from the north and west (van de Graaff *et al.*, 1977, Benbow *et al.*, in press). Feldspars are predominantly albite, together with minor K-feldspar. Clays are predominantly kaolinite and illite. Clinoptilite occurs in a few samples. Fossils present include wood fragments, radiolaria, calcareous and agglutinating forams, molluscs, and calcareous and siliceous sponge spicules.

Most clastics are only weakly indurated, although exceptions include iron or pyrite-cemented sandstone, and samples of the Platypus, Madura, and Ceduna Formations. The iron-cemented sample 66DR08G, consists of a gravelly rizoconcretion (? Cretaceous or Paleocene) with a meniscus cement of haematite, indicating subaerial, vadose cementation. Glauconite is a common constituent, occurring as detrital grains, fossil infills, and as replacement of existing grains (most evident in 66DR01E, where clasts of microbored limestone have been glauconitised).

Burial processes, mainly compaction, predominate. Where present, feldspars are largely unaltered, except in the Polda Formation samples, where *in situ* partial alteration to sericite has occurred.

Evidence of sea-floor weathering has been confined to partial oxidation of pyrite, and partial dissolution of siliceous spicules.

3.6 Phosphorites

Two samples of phosphatised sediment were recovered (66DR01I and 66DR11F). The first is of Maastrichtian age, and the second Tertiary (both from Alley & Clarke, in press). The two samples consist of siliceous sand or silt grains in a calcareous matrix which has undergone subsequent phosphatisation. In 66DR01I, the phosphate has formed a band displacing the original sediment. The phosphate in 66DR11F however, forms a nodule replacing the original carbonate matrix. Both specimens are interpreted as having formed as hard-grounds on the Late Cretaceous to Paleocene disconformity.

4 EVOLUTION

4.1 Introduction

The above petrologic and petrographic data allows an expansion of the environmental history of the evolving

TABLE 4. Petrology of Tertiary sediments from the GAB. Age data after Alley & Clarke (in press), Shafik (1990), & McGowran (1991).

FORMATION	SAMPLE	AGE	LITHOLOGY	MINERALOGY	PROVENANCE	FOSSILS	ENVIRONMENT	DIAGENESIS
Pliocene	66DR9A	Pliocene	mudstone	----------	----------	----------	?shelf	compaction
sequence	66DR9B	Pliocene	chalk	cal	----------	cal f cal spic nan	pelagic	compaction
Nullarbor Lmst equivalent	66DR7G	Middle Miocene	coralline lmst	cal	---------- / ----------	cal f coral moll ech	shelf	seafloor cement, meteoric dissolution & partial infill
(Miocene)	66DR11B	Early Miocene	chalk	cal	----------	cal f cal & sil spic nan	pelagic	compact & cement
Abrakurrie	66DR6B	Late Oligocene	chalk	cal	----------	cal f cal & sil spic nan	pelagic	compact & cement
Lmst equiv.	66DR14B	Early Oligocene	chalk	cal	----------	cal f cal & sil spic nan	pelagic	compact & cement
(Oligocene)	66DR14E	Early Oligocene	cherty lmst	cal+qtz	----------	cal f cal & sil spic nan	pelagic	sil mob, compact & cement
	66DR15C	Oligocene	glauc chalk		----------	cal f cal & sil spic nan	pelagic shelf	glauc formation, sil mob, compact, sil & calc cement
	66DR16A	Oligocene	glauc dol lmst	cal+dol+glauc	----------	cal f ech	shelf	glauc formation, compact, sil & calc cement, dolomitisation
	66DR12C	Eocene/Early Oligocene	chalk	cal	----------	cal f cal & sil spic nan	pelagic	compact & cement, arag dissolution, glauc formation
	66DR14C	Eocene-Oligocene	chalk	cal	----------	cal f moll	pelagic	compact & cement, arag dissolution, Mn encrustation
Wilson Bluff equivalent	66DR1A	Late Eocene	chalk	cal	----------	cal f cal & sil spic nan	pelagic	compact & cement, arag Lmst dissolution, Mn encrustation
(Eocene)	66DR14D	Late Eocene	glauc chalk	cal+glauc	----------	cal f cal & sil spic nan	pelagic	glauc formation, compact, silica diss, calc & sil cement
	66DR1D	Middle Eocene	cherty chalk	qtz+cal+ tr clinop	----------	cal f dia sil spic nan	pelagic	compact, silica diss, calc cal & sil cement
	66DR1B	Eocene	cherty chalk	qtz+tr clinop	----------	dia cal f rad sil spic nan	pelagic shelf	glauc formation, compact, sil diss, calc & sil cement
	66DR8A	Early Eocene	slty glauc lmst	cal qtz glauc	----------	cal f moll nan	shelf	compact & cement, glauc formation
	66DR11G	Early Tertiary	lmst	cal	----------	cal f	shelf	compact & cement, arag dissolution
Immarna Group	66DR8G	? Paleocene	ferruginous gravelly sst	qtz+goet	plut-meta-volc, 1st & 2nd cycle	----------	fluvial	vadose Fe cementation
	66DR11F	Tertiary	phosphatic calcareous sst	qtz+cal+ap	plut-meta, 1st & 2nd cycle	----------	marine	compact & calc cement, phosphatisation

Fossil abbreviations: agg f=aggultinating forams; bur=burrows; cal spic=calcareous sponge spicules; cal f=calcareous forams; dia=diatoms; ech=echinoderms; moll=molluscs; nan=nannofossils; rad=radiolaria; sil spic=siliceous sponge spicules; wood=coalified wood.

TABLE 5. Mesozoic sediments from the GAB. Age data after Clarke & Alley (in press), Shafik (1990), and Hill (in press).

FORMATION	SAMPLE	AGE	LITHOLOGY	MINERALOGY	PROVENANCE	FOSSILS	ENVIRONMENT	DIAGENESIS
Potoroo Fm	66DR1E	L Maas-E Danian	glauconitic sst	qtz+alb+cal+ glauc	plut-meta, 1st cycle	agg f	shal marine	clast boring & glauconit-isation, calc cement
	66DR1H	L Maas-E Danian	organic siltstone	qtz+clinop	plut-meta, 1st & 2nd cycle	moll sil spic nan	shal marine	compaction & cementation
	66DR7F	L Maas-E Danian	siltstone	qtz+cal	plut-meta, 1st & 2nd cycle	cal f cal & sil spic	shal marine	compact & cementation, silica dissolution
	66DR12D	L Maas-E Danian	organic mudstone	qtz+kaol	----------------	wood	marg marine	compaction & cementation
	66DR7E	L Cret-E Tert	organic mudstone	qtz+kaol	meta, 1st & 2nd cycle	------------	marg marine	authigenic glauc formation, compaction & cementation
	66DR1I	Maas-E Danian	phosphatic calc siltstone	qtz+cal+ap+alb	plut-meta, 1st cycle	------------	?marg marine	shal burial phosphatisation, compact & calc cement
	66DR8F	Maas-E Danian	coaly siltstone	qtz+kaol	plut-meta, 1st cycle	wood bur	marg marine	compaction & cementation
	66DR9C	Maas-E Danian	organic mudstone	qtz+alb+?ill	meta, 1st & 2nd cycle	------------	marg marine	compact & cement,
	66DR12E	Maas-E Danian	coaly siltstone	qtz+?clay	meta, 1st & 2nd cycle	agg f wood cal & sil spic rad	marg marine	compact & cement, pyrite precip, seafloor pyrite ox authigenic pyrite growth
	66DR12F	Maas-E Danian	fenestral pyritic, calc mudstone	qtz+cal+?clay	sed-meta, 1st & 2nd cycle	cal f rad sil spic	peritidal	compaction & cementation, partial meteoric infill
	66DR3A	Maas E Danian	organic mudstone	qtz+clinop	plut-meta, 1st & 2nd cycle	moll nan	shal marine	compaction & cementation
	66DR3D	Maastrichtian	sideritic, dolomitic lime mdst	qtz+sid+tr dol	----------------	cal spic bur	peritidal	cementation of early dol, seafloor Fe encrustation
	66DR8D	Cretaceous	gravelly slst	qtz+kaol	plut-meta, 1st & 2nd cycle	------------	marg marine	compaction & cementation
Wigunda Fm	66DR3C	Campanian	glauc intraclastic conglomerate	qtz+cal+alb	plut-meta, 1st & 2nd cycle	moll shark tooth	shal marine	burial, calc cement, partial pyritisation
Platypus Fm	Platypus 1, 3889 m	Early Cretaceous	silty sandstone	qtz+clay	plut-meta, 1st cycle	------------	delta front	compaction & cementation
Ceduna/ Madura Fm	Duntroon 1, 2759 m	Early Cretaceous	intraclastic, feldspathic sst	qtz+alb+clay	plut-met, 1st cycle	------------	delta top	compaction & cementation
	66DR16B	Early Cretaceous	glauc calc slst	qtz+cal+glauc	-----------	cal f cal & sil spic rad moll	deep shelf	glauc formation, silica mob, compact, sil & calc cement
Polda Fm	KD 1, 550 m	Middle Jurassic	gravelly feld sst	qtz+alb+clay+cal	plut-meta-volc, 1st & 2nd cycle	------------	fluvial	compaction
	Tuckey 1, 208 m	Middle Jurassic	slst & gravel sst	qtz+alb+clay	plut-meta, 1st & 2nd cycle	------------	fluvial	compaction
	Polda 1, 113 m	Middle Jurassic	feld sandy slst	qtz+alb+clay	plut-meta-volc, 1st cycle	------------	fluvial	compaction

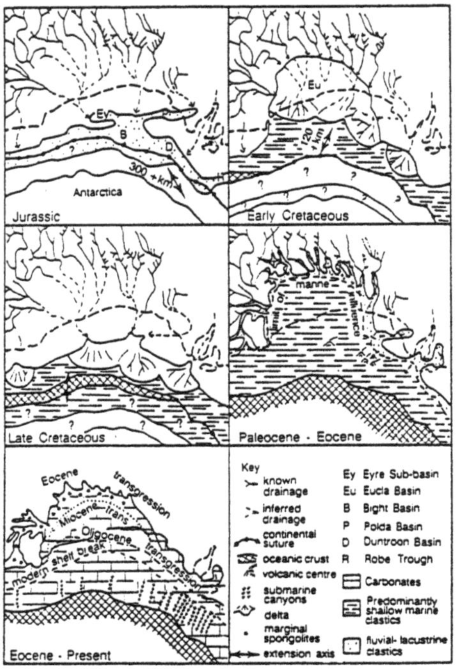

FIGURE 4. Palaeogeographic evolution of the Great Australian Bight (from Veevers, 1984; Scotese *et al.*, 1988, Bureau of Mineral Resources Palaeogeographic Group, 1990).

Great Australian Bight, previously outlined by Willcox & Stagg (1990), Stagg & Willcox (1988) and Deighton *et al.* (1976). The palaeogeographic evolution of the Great Australian Bight is shown schematically in Figure 4. Significant depositional events occurred in the Jurassic, Early and Late Cretaceous, and the Eocene. Veevers *et al.* (1991) document regional reorganisation of sea-floor spreading patterns in the Late Jurassic, Early Cretaceous, Early Paleocene, and Middle Eocene. This suggests that changes in tectonic regime in the Great Australian Bight and southern margin were reflections of the larger global pattern.

4.2 Jurassic (Mz3)

Earliest rift-related sedimentation in the Polda Basin, Eyre Sub-basin and (possibly) Ceduna Sub-basin occurred during the Late Jurassic (Willcox & Stagg, 1990; Stagg & Willcox, 1988). Oblique extension gave rise to wrench-controlled grabens in which fluvial and lacustrine sediments were deposited. Minor basaltic volcanism occurred in the eastern part of the Great Australian Bight region. Palaeodrainage incision began during this time as a result of uplift of the cratonic margin. The numerous palaeodrainage systems along the margins of the Eucla Basin coalesced into two main systems, the Twilight Cove and the Mallabie systems. The Twilight Cove system entered the Great Australian Bight from the north-west through the Eyre Sub-basin, while the Mallabie flowed into the Great Australian Bight from the north (both after Lowry, 1976). Veevers (1984) suggested that the Palaeo-Cooper drainage system also entered the Great Australian Bight, and the Palaeo-Murray supplied clastic sediments to the Duntroon Basin, this was, however, disputed by Stephenson & Brown (1989) and is not supported by palaeocurrent data from the Eromanga Basin. Veevers (1984) argued the palaeodrainages marginal to the Eucla Basin appeared insufficent to supply the volume of Mesozoic clastics in the Great Australian Bight. If the Palaeo-Cooper and Palaeo-Murray did not enter the Great Australian Bight, alternative sources of sediment must be found. The presence of reworked Late Permian palynomorphs (Alley & Clarke, in press) suggests that very large volumes of Permian sediments may have been reworked into the Great Australian Bight together with earlier rift-related sediments, and that these older sediments may have been much more extensive, and thicker, than indicated by their present distribution.

4.3 Early Cretaceous (Mz4)

Oblique extension changed to normal extension during the Neocomian, and continued through to the Aptian. This interval marks the rift-early drift transition, and contains several possible break-up unconformities (terminology of Falvey, 1974). Non-marine to shallow marine sediments of the Loongana Sandstone were deposited disconformably across wide areas of the developing rift. This event may represent the earliest appearance of oceanic crust in the basin (Willcox & Stagg, 1990). Veevers (1984) suggested seaway connections from the Great Australian Bight to the epiric seas of the Canning Basin to the north-east and the Eromanga Basin to north-west during the highest Early Cretaceous sea-level, which he dated as Aptian. This was disputed by the Bureau of Mineral Resources Palaeogeographic Group (1990), which interpreted connections as existed to the Canning Basin only, which formed earlier, during the Cenomanian.

Deepening of the rift resulted in the development of a series of prograding deltaic and clastic shelf sediments. Initial basin drowning and fore-delta sedimentation is represented by the progradational systems tract of the Albian to Aptian Madura and Ceduna Formations. Calcareous units occurred in areas of low clastic influx. The overlying Cenomanian Platypus Formation forms the delta topset succession. The down-lap surface at the top of the Platypus Formation represents a 96 Ma hiatus, coincident with the main separation of Australia and Antarctica (Veevers, 1986).

4.4 Late Cretaceous (Mz5)

A second cycle of clastic sediments was deposited across the unconformity following break-up. Sedimentation was less extensive, and confined to the Ceduna Sub-basin and western Duntroon Basins. The Wigunda Formation resulted from shelf and delta progradation during the Turonian and Campanian (Robertson *et al.*, 1979). Marly sediments were deposited between individual delta systems in near-shore and shore-line environments. The progradation of non-marine and near-shore facies in the Maastrichtian led to the deposition of the overlying Potoroo Formation in near-shore environments. Diapirs, most likely deltaic mudlumps, locally penetrate the Late Cretaceous succession (Falvey *et al.*, 1990). The high organic content, including both wood and comminuted organic material, suggest a humid palaeoclimate. This is borne out by the palynological evidence, which consists of podocarp and araucarian dominated rain forest taxa (Alley & Clarke, in press).

4.5 Paleocene to Middle Eocene (Cz1 & lower Cz2)

Deposition of the Immarna Group (Cz1 and lower Cz2) was diachronous from the Great Australian Bight Basin (?Paleocene) and Eucla Basin (Middle to Late Eocene), forming lowstand to transgressional systems tracts.

Sequence Cz1 is represented by largely by basalts and phosphatic hard grounds on Maastrichtian and Tertiary sediments. A subaerial gravelly sandstone (66DR8G) may represent subaerial deposition in latest Cretaceous to Paleocene times in the Ceduna Sub-basin prior to the earliest Paleocene transgression. Basaltic eruptions and associated minor intrusions occurred from small volcanic centres on the Ceduna and Eyre Sub-basins in during the Paleocene. The eruptions occurred in a shallow marine setting.

The lower part of sequence Cz2 was not sampled directly during cruise 66. The presence of thin organic-rich Early Eocene clastic sediments is inferred from the presence of palynomorph contamination in many Cretaceous sediments (Alley & Clarke, in press).

A major contraction in the extent of clastic facies occurred during the Eocene, with reduced clastic input. Off-shore clastic facies are extremely thin, compared with the thick sequences of the Mesozoic. The reduction in clastic sediments indicates diminished palaeodrainage flow, perhaps the first indication of the increasing aridity that led to the present climatic regime of much of Australia.

4.6 Middle Eocene to Holocene (upper Cz2-4)

A series of major Middle to Late Eocene transgessions led to deposition of marginal marine to marine sediments high up the marginal palaeodrainage channels of the Eucla Basin (Jones, 1990). This Late Eocene transgression represents the most extensive marine on-lap in the Great Australian Bight (upper Cz2). Near-shore sediments were organic-rich, and subaerial facies are present in marginal palaeodrainages. Clastic sediments pass offshore into the carbonates of the Wilson Bluff and Toolina Limestones.

The change from clastic to carbonate sedimentation coincided with the onset of rapid sea-floor spreading (Shafik, 1990), but is more likely related to a major reduction in clastic input, allowing the accumulation of skeletal carbonates off-shore. The temperate carbonates of the Great Australian Bight form a distally steepened shelf or ramp (James & von der Borch, 1991; James & Bone, 1991) of sequences upper Cz2-4. Petrographic evidence for the progressive deepening of the outer margin of the shelf from the Oligocene onwards accords with the nannofossil record, documented by Shafik (1990). The Ceduna and Eyre Terraces thus comprise drowned portions of the platform margin complex (Kendall & Schlager, 1981). The accelerated subsidence of the Great Australian Bight during the Oligocene coincides with the final separation of Australian an Antarctica along the south Tasman Rise at 32 Ma (Willcox & Stagg, 1990). Subsequent shelf derived terrigenous and carbonate sediments were channelled along submarine canyons (Davies *et al.*, 1988; von der Borch, 1968). Sedimentation away from the continental shelf and canyons was predominantly pelagic, with localised turbidite deposition. Carbonate deposition, apart from a brief interval of clastic sedimentation during the Pliocene, has continued to the present.

5 ANTARCTIC COMPARISONS

The eastern Antarctic continental margin (Veevers, 1987) is poorly documented geologically compared with the conjugate southern margin of Australia. Despite this it is clear that the sequences share a similar sedimentary history until the Oligocene (Wannesson, 1990). The histories of the two margins diverge after this time, with a southern Australian carbonate shelf contrasting with the glaciogene sediments of the eastern Antarctic continental margin. The pre-Oligocene sediments of the southern Australian margin, and their sequence stratigraphy, may be used as analogues of those occursing along eastern Antarctica, pending more detailed geological data from the later.

6 CONCLUSIONS

The evolution of the Great Australian Bight is generally similar to other Australian marginal basins (*c.f.* Falvey, 1974, Veevers, 1984). However, unlike other southern Australian marginal basins, such as the Otway (Douglas *et al.*, 1988) or West Tasmania (Hinz *et al.*, 1987) Basins, the Great Australian Bight has had only minimal volcanic activity. Unlike the eastern or western Australian margins, pre-rift sediments are limited to localised platform cover. Both features may be a reflection of low heat flow prior to break-up. The significance of wrench tectonics in the early history of the Great Australian Bight suggests that rifting was initiated by processes external to the region, rather than by local thermal uplift and subsequent rifting.

In contrast to the well-documented margins of the Atlantic (Bond & Kominz, 1988), the Great Australian Bight lacks evaporite and carbonates during the early and middle stages of its history, and reefs were absent once the carbonate-dominated margin was established from the Eocene onwards. The absence of evaporites during the rift phase of margin evolution may be linked to a humid palaeoclimate resulting in high rainfall and run-off and clastic influx from rapidly eroding Permian and older Mesozoic sediments. The high terrigenous flux down the palaeodrainages prevented deposition of carbonates until the Eocene, when the rainfall may have begun to diminish, and palaeodrainage flow decreased. The absence of reefs may result from the high palaeolatitude and consequent non-tropical sea temperatures, as bioherm-building communities were present from the Eocene onwards (James, 1983). These differences illustrate the need to use a wide range of examples to illustrate the evolution of passive margins (Edwards & Santogrossi, 1989).

ACKNOWLEDGMENTS

The authors wish to thank the Bureau of Mineral Resources and all crew members of the R/V "Rig Seismic" for the opportunity to participate in Cruise 66. From Flinders University M. Abbot performed ICP analyses and J Lorenzin prepared many of the thin sections. At Kambalda, J. Clout allowed use of research facilities, T. Laffan and F. Raeside processed the samples. Figures were drafted by R. Clout and T. Richards. Additional thanks are due to A. van Bentum, A. Bancroft, and D. Inglesby for proofreading. The paper was reviewed by S. Shafik and D. Haig, formatting was by A Lisci. Many thanks are also due to the editor for his persistance and patience.

REFERENCES

Alley, N. F. & J. D. A. Clarke in press. Cretaceous-Tertiary palynology and stratigraphy of sediments from the Great Australian Bight. *Bureau of Mineral Resources Journal of Australian Geology and Geophysics*

Bein, J., & M. L. Taylor 1981. The Eyre Sub-basin: recent exploration results. *Australian Petroleum Exploration Association Journal* 21:91-98.

Benbow, M. C., J. M. Lindsay & N. F. Alley in press. Bight and Duntroon Basins, Eucla Basin and Paleodrainage. *In* Johns, R. J. (ed). The Geology of South Australia. *South Australian Department of Mines and Energy Special Publication.*

Bond, G. C. & M. A. Kominz 1988. Evolution of thought on passive continental margins from the origin of geosynclinal theory (~1860) to the present. *Geological Society of America Bulletin* 100: 1909-1933.

Bureau of Mineral Resources Palaeogeographic Group 1990. Australia: Evolution of a Continent. *Bureau of Mineral Resources, Canberra.*

Coney, P. J. 1990. Terranes, tectonics and the Pacific rim. *Abstracts of the 1990 Pacific Rim Conference* p 19-30.

Davies, H. L., J. D. A. Clarke, H. M. J. Stagg, S. Shafik, B. M. McGowran, N. F. Alley & J. B. Willcox 1988. Maastrichtian and younger sediments from the Great Australian Bight. *Bureau of Mineral Resources Report* 288.

Deighton, I., D. A. Falvey, & D. J. Taylor 1976. Depositional Environments and Geotectonic Framework: South Australian Continental Margin. *Australian Petroleum Exploration Association Journal* 16: 25-36.

Douglas, J. G., C. Abele, S. Benedek, M. E. Dettmann, P. R. Kenley, C. R. Lawrence, T. H. V. Rich, & P. V. Rich 1988. Mesozoic. *In* Douglas, J. G. & J. A. Ferguson (eds). Geology of Victoria. *Victorian Division Geological Society of Australia* 213-250.

Edwards, J. D., & P. A. Santogrossi 1989. Divergent/Passive Margin Basins. *American Association of Petroleum Geologists Memoir* 48.

Falvey, D. A. 1974. The development of continental margins in plate tectonic theory. *Australian Petroleum Exploration Association Journal* 14: 95-106.

Falvey, D. A., P. A. Symonds, J. B. Colwell, J. B. Willcox, J. F. Marshall, P. E. Williamson, & H. M. J. Stagg 1990. Australia's deepwater frontier petroleum basins and play types. *Australian Petroluem Exploration Association Journal* 30: 239-262.

Fraser, A. J., & L. A. Tilbury 1979. Structure and stratigraphy of the Ceduna Terrace region, Great Australian Bight Basin. *Australian Petroleum Exploration Association Journal* 19: 53-65.

Hill, A. J. in press. Bight and Duntroon Basins. *In* Johns, R. J. (ed). The Geology of South Australia. *South Australian Department of Mines and Energy Special Publication.*

Hinz, K., J. B. Willcox, M. Whitcar, H.-H. Kudrass, N. F. Exon, & D. A. Feary 1987. The West Tasmania margin: An underrated Petroleum Province. *In* Glenie, R. C. (ed). *Second southeastern Australia Oil Exploration Symposium PESA Symposium, 1985*, Melbourne, p 395-410.

Hocking, R. M. 1990. Eucla Basin. *In* Geology and Mineral Resources of Western Australia. *Western Australia Geological Survey Memoir* 3: 548-561.

James, N. P. 1983. Reef Environment. *In* Scholle, P. A., D. G. Bebout, & C. H. Moore (eds). Carbonate Depositional Environments. *American Association of Petroleum Geologists Memoir* 33: 345-362.

James, N. P., & Y. Bone 1991. Origin of a cool-water, Oligo-Miocene deep shelf limestone, Eucla Platform, southern Australia. *Sedimentology* 38: 323-341.

James, N. P., & P. W. Choquette 1984. Diagenesis 9. Limestones - the meteoric diagenetic environment. *Geoscience Canada* 11(4): 161-194.

James, N. P., & C. C. von der Borch 1991. The carbonate shelf edge off southern Australia; A prograding open platform margin. *Geology* 19: 1005-1009.

Jones, B. G. 1990. Cretaceous and Early Tertiary sedimentation on the western margin of the Eucla Basin. *Australian Journal of Earth Sciences* 37 (3): 317-329.

Jones, J. B., & M. J. Fitzgerald 1986. Silica rich layering at Blanche Point, South Australia. *Australian Journal of Earth Sciences* 33 (4): 529-551.

Kendall, C. G. St. C., & W. Schlager 1981. Carbonates and relative change in sealevel. *Marine Geology* 44: 181-212.

Lowry, D. C. 1970. Geology of the Western Australian part of the Eucla Basin. *Geological Survey of Western Australia Bulletin* 122.

Lowry, D. C. 1976. Eucla Basin. *In* Leslie, R.B., H. J. Evans, & C. L. Knight (eds). *Australasian Institute of Mining and Metallurgy Monograph* 7 Economic Geology of Australia and Papua New Guinea - 3 Petroleum p 95-98.

McGowran, B. 1989. Silica burp in the Eocene ocean. *Geology* 17: 857-860.

McGowran, B 1991. Maastrichtian and early Cainozoic, southern Australia: planktonic foraminiferal biostratigraphy. *In* Williams, M. A., De Deckker, P., & A. P. Kershaw. (eds). The Cainozoic in Australia: A Re-appraisal of the Evidence. *Geological Society of Australia Special Publication* 18: 79-98.

Milnes, A. R., B. J. Cooper, & J. A. Cooper. 1982. The Jurassic Wisanger Basalt of Kangaroo Island, South Australia. *Transactions of the Royal Society of South Australia* 106 (1): 1-13.

Nelson, R. G., T. N. Crabb, & R. A. Gerdes. 1986. A review of geophysical exploration in the Polda basin, South Australia. *APEA Journal* 26: 319-333.

Robertson, C. S., D. K. Cronk, E. Nicholas, S. J. Mayne, & D. G. Townsend. 1979. A review of petroleum exploration and prospects in the Great Australian Bight region. *Bureau of Mineral Resources Record* 1979/20.

Scotese, C. R., L. M. Gahagan, & R. I. Larson 1988. Plate tectonic reconstructions of the Cretaceous and Cenozoic ocean basins. *Tectonophysics* 155: 27-48.

Shafik, S 1990. The Maastrichtian and Early Tertiary record of the Great Australian Bight Basin and its onshore equivalents on the Australian southern margin: a nannofossil study. *Bureau of Mineral Resources Journal of Australian Geology and Geophysics* 11: 473-497.

Stagg, H. M. J., & J. B. Willcox 1988. Seismic facies distribution in the Central Great Australian Bight. *Palaeogeography, Sealevel and Climate, 17th Bureau of Mineral Resources Research Symposium extended abstracts* 29-38.

Stephenson, A. E., & C. M. Brown 1989. The ancient Murray river system. *Bureau of Mineral Resources Journal of Australian Geology and Geophysics* 11(2/3): 387-395.

Van de Graaff, W. F. E; R. W. A. Crowe, J. A. Bunting, & M. J. Jackson 1977. Relict Early Cenozoic drainage in arid Western Australia. *Zeitschrift fur Geomorphologie N. F.* 21:(4): 379-400.

Veevers, J. J. 1984 (ed). Phanerozoic earth history of

Australia. *Clarendon Press*, Oxford, 418p.

Veevers, J. J. 1986. Breakup of Australia and Antarctica estimated as mid-Cretaceous (95 +/- 5 Ma) from magnetic and seismic data from the continental margin. *Earth and Planetary Science Letters* 77: 91-99.

Veevers, J. J. 1987. The Conjugate Continental Margins of Antarctica and Australia. *In* Eitreim, S. L., & M. A. Hampton (eds). The Antarctic Continental Margin: Geology and Geophysics of Offshore Wilkes Land. *Circum-Pacific Council for Energy and Mineral Resources Earth Science Series* 5A: 5-73.

Veevers, J. J; C McA Powell, & S. R. Roots 1991. Review of seafloor spreading around Australia. 1. Synthesis of the patterns of spreading. *Australian Journal of Earth Science* 38 (4): 373-390

Von der Borch, C. C. 1968. Southern Australian submarine canyons, their distribution and ages. *Marine Geology* 6: 267-279.

Wannesson, J. 1990. Geology and Petroleum Potential of the Adelie Coast Margin, East Antarctica. *In* St. John, B., (ed). Antarctica as an Exploration Frontier-Hydrocarbon Potential, Geology, and Hazards. *American Association of Petroleum Geologists Studies in Geology #31* 77-87.

Willcox, J. B., H. M. J. Stagg, & H. L. Davies, 1988. Rig Seismic Research Cruises 10 & 11: geology of the central Great Australian Bight Region. *Bureau of Mineral Resources Report* 286.

Willcox, J. B., H. M. J. Stagg 1990. Australia's southern margin: a product of oblique extension. *Tectonophysics* 173: 269-281.

Gondwana Eight, Findlay, Unrug, Banks & Veevers (eds) © 1993 Balkema, Rotterdam. ISBN 90 5410 304 3

The architecture of Gondwana rifting in southeastern Australia: Evidence from apatite fission track thermochronology

David A. Foster & Andrew J.W. Gleadow
Victorian Institute of Earth and Planetary Sciences, Department of Geology, La Trobe University, Bundoora, Vic., Australia

ABSTRACT: Apatite fission track data indicate that the boundary between the Kanmantoo and Lachlan Fold Belts in western Victoria, Australia was reactivated as a transfer fault during Mesozoic rifting. Reactivation of this zone of crustal weakness probably determined the location of the Tasman Fracture Zone of the southeast Indian Ridge, controlled the shape of the Otway Basin, and influenced the Cretaceous denudation history of western Victoria. When Australia and Antarctica are reconstructed along the trace of the Tasman Fracture Zone the reactivated tectonic boundary in Victoria lines up with the Leap Year Fault in northern Victoria Land. The existence of this tectonic boundary and its effect in deflecting the southeast Indian Ridge to the south of Tasmania probably prevented Cretaceous rifting in Bass Strait from becoming sea-floor spreading. Without this transfer structure Tasmania would have remained attached to Antartica rather than Australia.

1. INTRODUCTION

Zones of crustal weakness along tectonic boundaries are commonly reactivated during continental break-up. Reactivation of such zones influences the initial pattern of rifting and leads to the development of oceanic fracture zones (Wilson, 1965). In this paper we present apatite fission track evidence for Mesozoic reactivation of an early Palaeozoic tectonic boundary between the Kanmantoo and Lachlan Fold Belts in southeastern Australia. This tectonic boundary was exploited during rifting between Australia and Antarctica and probably determined the location of the Tasman Fracture Zone, which substantially offsets the southeast Indian Ridge.

During Palaeozoic times southeastern Australia was the locus of large scale accretion as the terranes of the Tasman Orogen were joined to the Australian craton (e.g., Coney et al., 1990). Following amalgamation along the Pacific margin of Gondwana, Australia began to rift away from Antarctica at ~160 Ma, leading to initial opening of the southeast Indian ocean at ~96 Ma (Veevers et al., 1991). Rifting in western Victoria formed the Otway Basin, which contains up to 10 km of Late Jurassic to Tertiary sediments (Megallaa, 1986). Rifting along the eastern margin of Australia also began at ~95 Ma (Veevers et al., 1991). Lister et al. (1991) suggested that the eastern and southern rifted margins of Australia represent upper and lower plates, respectively, of asymmetric lithospheric detachments. They also noted an abrupt change in extensional style along the western Otway Basin, the area of this study.

In southeastern Australia the Tasman Orogen consists of the Kanmantoo and Lachlan Fold Belts (VandenBerg, 1978). The Kanmantoo Fold Belt extends from the Adelaide Fold Belt in South Australia into western Victoria (Fig. 1) and was a late Precambrian to Cambrian passive margin sequence deformed during the Cambro-Ordovician Delamerian Orogeny. The Lachlan Fold Belt extends east from the Kanmantoo Fold Belt to the eastern continental margin in southeastern Australia and was amalgamated during a series of Late Ordovician to Devonian orogenies. The nature of the basement rocks of these two fold belts is unclear. The exposure of transitional MORB/BAB tholeiites in the deepest parts of the sections and isotopic constraints from granitic rocks that do not require the

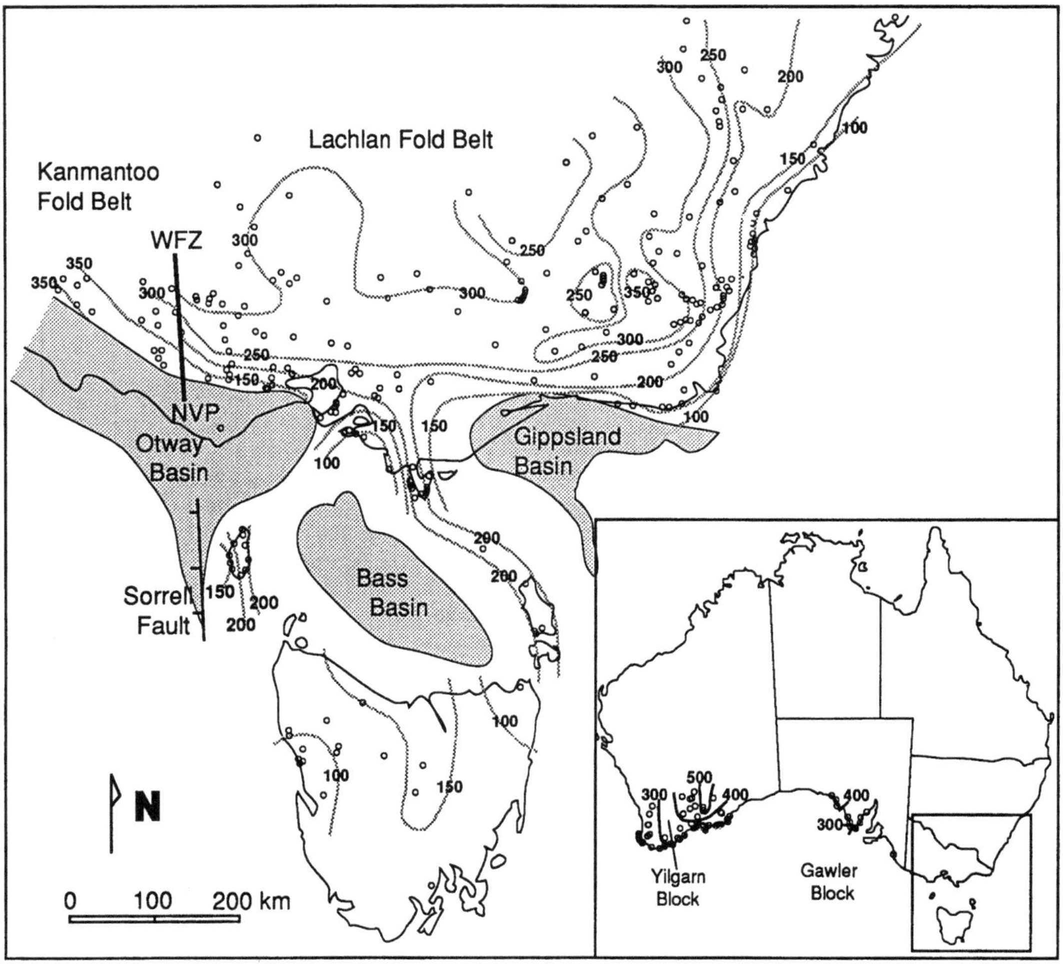

Fig. 1. Map of southeastern Australia with apatite fission track ages and smoothed fission track age contours [modified from Dumitru et al. (1991) with unpublished data from Foster and Gleadow]. The solid line north of the Otway Basin is located along an abrupt change in fission track age trends correlated with the Woorndoo Fault Zone (WFZ). The inset map shows the location of the larger scale map and apatite fission track ages from the southern margins of Western Australia and South Australia from Ferguson (1981). NVP, Newer Volcanic Province.

presence of Proterozoic crust (Gray, 1990; C. Wilson et al., submitted) suggest an ensimatic origin, though some continental basement may exist such as that exposed in Tasmania (Coney et al., 1990).

The lithospheric boundary between the Kanmantoo and Lachlan Fold Belts, in Victoria, is located east of the Grampians Mountains (Fig. 1) (VandenBerg, 1978; Coney et al., 1990). This boundary, referred to as the Woorndoo Fault Zone, is defined by: a belt of thrust-bounded late Proterozoic to Cambrian greenstones (Mt. Stavely Belt) and their north-northwest-trending subsurface continuation defined by their geophysical signatures (Coney et al., 1990); a major isotopic discontinuity in the basalts of the Newer Volcanic Province which separates low from high initial $^{87}Sr/^{86}Sr$ ratios in the west and east, respectively (Gray et al., 1990); and the boundary between highly sodic ca. 400 Ma plutons to the west and more typical I-type plutons of similar age to the east

(Chappell et al., 1988). The surface expression of the boundary between the fold belts may be the Avoca Fault to the east of the Woorndoo Fault Zone (Wilson et al., submitted). However, thrust-bounded packages of Kanmantoo Group rocks between the Avoca and Woorndoo Faults probably represent only the allochthonous, upper crustal overlap between the two fold belts, similar to the large allochthons in the Appalachian Mountains that were transported long distances beyond deep crustal ramps (Cook et al., 1979). The geochemical and isotopic data from western Victoria indicate that rocks east and west of the Stavely Belt overlie fundamentally different lithospheric blocks. Even though these data do not identify a single fault they probably mark the subsurface boundary between the two fold belts.

Late Precambrian to Palaeozoic rocks in southeastern Australia may have been continuous with the Wilson, Bowers, and Robertson Bay terranes (Fig. 2) of northern Victoria Land, Antarctica, prior to rifting (e.g., Oliver, 1972; Harrington et al., 1973; Burret and Findlay, 1984; Stump et al., 1986; Findlay, 1987), based on the following evidence: 1. similarities between the Ross Orogeny of the Wilson terrane and the Delamerian Orogeny in Australia; 2. the resemblance of Cambrian mafic volcanic rocks of the Sledgers Group and the Mt. Stavely Group in western Victoria; 3. Similarities between graywacke turbidites of the Robertson Bay Group and the Admiralty Intrusives in the Robertson Bay terrane, and the St. Arnaud Beds and Mt Cole granitic suite in the western Lachlan Fold Belt.

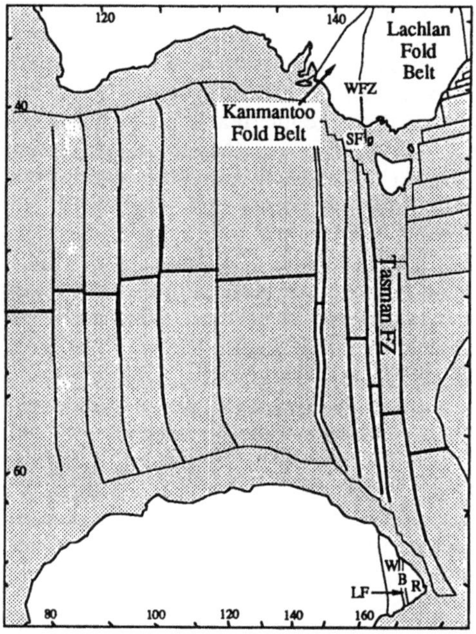

Fig. 2. Map showing Palaeozoic terranes within the formally adjacent margins of Australia and Antarctica, and major features of the southeast Indian Ridge and fracture zones (FZ) of the southeast Indian Ocean (modified from Foster and Gleadow, 1992). Terranes and faults: B, Bowers terrane; LF, Leap Year Fault; R, Robertson Bay terrane; SF, Sorrel Fault; W, Wilson terrane; WFZ, Woorndoo Fault Zone.

2. APATITE FISSION TRACK DATA

Apatite fission track analysis, which is based on the accumulation of radiation damage from the spontaneous fission of ^{238}U, is a powerful tool for elucidating the low temperature histories of rocks. This technique reveals the thermal history of rock masses below ~120°C from apparent age and length measurements of fission tracks in apatite (e.g., Gleadow et al., 1983; Dumitru et al., 1991).

Figure 1 shows contours of apatite fission track ages of pre-Mesozoic rocks in southern Australia (Gleadow and Lovering, 1978a,b; Ferguson, 1981; Moore et al., 1986; Dumitru et al., 1991; Foster and Gleadow, 1992). For much of southeastern Australia there is a consistent pattern of young apatite ages (<100 Ma) along the rifted continental margins and older ages (>300 Ma) 100-300 km inland. The youngest fission track ages (80-96 Ma) occur south of the Sydney Basin, New South Wales; at Cape Woolamai and the Mornington Peninsula in Victoria; and throughout Tasmania. Mean lengths of confined fission tracks for the suite of samples from southeastern Australia suggest maximum Cretaceous palaeotemperatures of <60°C for exposed rocks in the inland areas and ≥120°C in the coastal areas (Moore et al., 1986; Dumitru et al., 1991).

The exposure of rocks that were at temperatures of ≥120°C in Cretaceous time requires significant denudation of a continental flexure that was produced by rifting (Moore et al., 1986). To quantify the amount of denudation requires knowledge of the geothermal gradient during the rifting episode. Such temperatures are consistent with ≥2 km of denudation in the coastal areas if the geotherms were high (30°-50°C/km), or ~4-5 km of denudation if the geothermal gradient remained normal (~25°-30°C/km) (Moore et al., 1986; Dumitru et al. 1991). Elevated geothermal gradients during rifting are suggested by: Cretaceous aged argon loss from K-feldspar samples in Tasmania (Sawka et al., 1990), virtrinite reflectance data from the Sydney Basin (Middleton and Schmidt, 1982), and the variation in apparent apatite age with elevation in the southeastern highlands, which gives a gradient of 35°-40°C/km (B. Kohn, pers. com., 1992). In addition, Late Cretaceous igneous rocks are present in southeastern Australia (McDougall and Wellman, 1976) which are probably the surface manifestation of deeper crustal magmatism and underplating which occurred during rifting (Drummond and Collins, 1986; Lister et al., 1991).

Low average elevations inland from the coast suggest that significantly less continental flexure and denudation occurred in western Victoria and southern Australia than along the eastern margin. However, young apatite ages are present along the margin of the sedimentary basins as far west as ~143° longitude in western Victoria (Fig. 1). Apatites from Devonian granitoids adjacent to the eastern part of the northern margin of the Otway Basin yield fission track ages as young as ~100-125 Ma; within 60-100 km further inland apatite ages increase to >300 Ma. Mean confined track lengths for the younger and older samples are 13-14 μm with standard deviations of 1-2 μm (Fig. 3). The moderately long track lengths for the >300 Ma samples reflect steady cooling after crystallization and residence at <60°C since the Palaeozoic, whereas mean tracks lengths of 13-14 μm in the younger samples indicate relatively rapid cooling following Mesozoic rifting. The intermediate aged samples (~140-280 Ma) yield broad and locally bimodal track length distributions with mean track lengths of ~11.4-12.1 μm and standard deviations >2 μm due to partial annealing during Cretaceous time (Fig. 3). These data suggest Cretaceous temperatures of ~100° to ≥120°C adjacent to sedimentary basins and <60°C inland.

The trend towards young ages adjacent to the marginal sedimentary basins changes abruptly in western Victoria along the boundary between

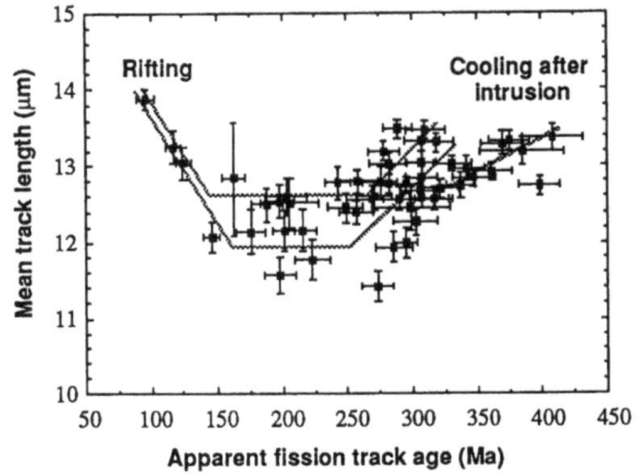

Fig. 3. Plot showing the relationship between apatite fission track age and the mean length of confined fission tracks for samples of Palaeozoic basement rocks in western Victoria.

the Kanmantoo and Lachlan Fold Belts. West of this abrupt change no decrease in apatite age toward the northern margin of the Otway Basin is observed (Fig. 1), in fact there is a slight increase. In that area apatite fission track ages >350 Ma occur within 5 km of the Otway Basin margin. Apatites from Lower Devonian and Cambro-Ordovician plutons in these areas yield fission track ages of ~300-400 Ma with mean track lengths of 12.5-13.5 μm and standard deviations of ~1.3 μm (Fig. 3). These ages are more similar to apparent apatite ages along the margins of southern and western Australia (Ferguson, 1981) (Fig. 1) and indicate that plutons exposed there have been at temperatures below ~60°C since Palaeozoic time.

Different apatite age trends in eastern and western parts of western Victoria can only be due to contrasting thermal histories of exposed basement rocks. If it were the result of similar thermal histories superimposed upon apatites with different initial ages there would be a larger component of short fission tracks in the >350 Ma samples. The variations cannot be attributed to different sample elevations because no more than ~200 m separates samples with apparent ages of <150 Ma from those with ages of >300 Ma (Foster and Gleadow, 1992). The estimated difference in palaeotemperature between the two areas is ≥50°C, which indicates that 1-2 km more material was denuded during rifting east of the age discontinuity. This difference in amount of denudation is consistent with the preservation of superficial Permian glacial sediments west of the Woorndoo Fault Zone (Bowen and Thomas, 1988) and the removal of Silurian Grampians Group sediments east of the Woorndoo Fault Zone.

3. DISCUSSION AND CONCLUSIONS

The discontinuity in low temperature thermal histories of basement rocks exposed in western Victoria suggests that there were fundamental differences in the style of Mesozoic rifting across this region. The different denudational histories east and west of the Woorndoo Fault Zone suggest it was reactivated as a transfer fault (e.g., Gibbs, 1984) separating crust that was strongly flexed and subsequently eroded from that which was not.

If the boundary between the Kanmantoo and Lachlan Fold Belts is the Avoca Fault and not the Woorndoo Fault Zone, as suggested by Wilson et al. (submitted), then it could be argued that the apatite discontinuity does not lie along any tectonic boundary. However, the geochemical, isotopic, and geophysical data indicate deep crustal differences across and parallel to the Woorndoo Fault Zone. It is possible that this fault zone lies over a major accretionary ramp in the deep crust and its reactivation during rifting was similar to the inversion of major thrust ramps as normal faults in fold and thrust belts (e.g., Price, 1981) (Fig. 4).

Although the exact nature of reactivation of the Woorndoo Fault Zone is still unresolved, rifting seems to have exploited anisotropies between Kanmantoo and Lachlan crustal blocks. The apatite age discontinuity intersects the Otway Basin where structural trends change from northwest to northeast and along the projection of the Sorrel Fault, the major basin bounding fault (Megallaa, 1986) (Figs. 1). This suggests that transfer movement on the Woorndoo Fault Zone controlled the shape of the Otway Basin. The fault zone and apatite age discontinuity also line up with the Tasman Fracture Zone which offsets the southeast Indian Ridge (Fig. 2). We suggest that the location of the Tasman Fracture Zone was determined by the boundary between the Kanmantoo and Lachlan Fold Belts.

The Tasman Fracture Zone extends as a continuous feature from southern Australia across the Southern Ocean to Antarctica where

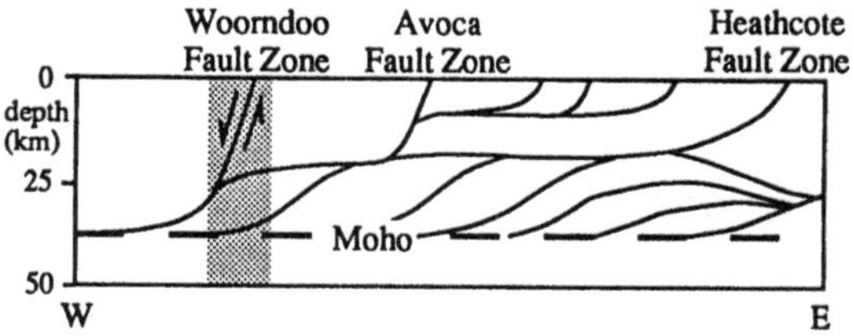

Fig. 4. Possible crustal section of western Victoria showing major Palaeozoic thrusts (modified from Gray et al., 1991) and the boundary between lithospheric blocks of the Kanmantoo and Lachlan Fold Belts (shaded area). Reactivation of the Woorndoo Fault Zone is shown by the arrows.

it intersects northern Victoria Land at the Leap Year Fault of the Bowers terrane (Fig. 2) (Royer and Sandwell, 1989). Reconstruction of Australia and Antarctica along the trace of the Tasman Fracture Zone brings the Grampians-Stavely belt and Bowers terrane into alignment, which is consistent with the geologic correlations of Stump et al. (1986) and Findlay (1987). This method of reconstruction produces an improved fit between Australia and Antarctica which does not require large lateral movements of Tasmania during rifting as have previously been suggested (e.g., Harrington et al., 1973).

The extension of this transfer fault along the west coast of Tasmania was probably the major reason why rifting failed to propagate completely through Bass Strait between Tasmania and Victoria. Final rifting was deflected to the south leaving Tasmania part of Australia and not part of north Victoria Land. The widespread exposure of rocks that were at temperatures of >100° and locally >200°C during rifting, as indicate by 150-80 Ma apatite apparent ages (Fig. 1) and disturbed Ar systematics of K-feldspar (Swaka et al., 1990), are consistent with Tasmania being close to the axis of extension. The thermochronological data indicate that more material was denuded from Tasmania during rifting than most of mainland southeastern Australia. This deeper level of erosion could explain the extensive exposure of Precambrian rocks in western Tasmania and may have been the result of denudation and uplift following large scale magmatic underplating of the crust coincident with the intrusion of Jurassic dolerite (see Burret and Martin, 1989).

REFERENCES

Bowen, R.L. & G.A.Thomas 1988. Permian. In Douglas, J.G. & J.A. Ferguson, (eds.), *Geology of Victoria*: 195-212. Melbourne: Geological Society of Australia.

Burrett, C.F. & E.L. Martin (eds.), 1989. *Geology and Mineral Resources of Tasmania*, Hobart: Geological Society of Australia.

Burret, C.F. & R.H. Findlay 1984. Cambrian and Ordovician conodonts from the Robertson Bay Group, Antarctica and their tectonic significance. *Nature* 307: 723-726.

Chappell, B.W., A.J.R. White, & R. Hine 1988. Granite provinces and basement terranes in the Lachlan fold belt, southeastern Australia. *Australian Journal of Earth Science* 35: 505-521.

Coney, P.J., A.E. Edwards, R. Hine, F. Morrison & D. Windrim 1990. The regional tectonics of the Tasman orogenic system, eastern Australia. *Journal of Structural Geology* 12: 519-543.

Cook, F.A., D.S. Albaugh, L.D. Brown, S. Kaufman, J.E. Oliver & R.D. Hatcher 1979. Thin skinned tectonics in the crystalline southern Appalachians: COCORP seismic-reflection profiling of the Blue Ridge and Piedmont. *Geology* 7: 563-567.

Drummond, B.J. & C.D.N. Collins 1986. Seismic evidence for underplating of the lower continental crust of Australia. *Earth and Planetary Science Letters* 79: 361-372.

Dumitru, T.A., K.C. Hill, D.A. Coyle, I.R. Duddy, D.A. Foster, A.J.W. Gleadow, P.F. Green, B.P. Kohn, G.M. Laslett & A.J. O'Sullivan 1991. Fission track thermochronology: application to continental rifting of south-eastern Australia. *Australian Petroleum Exploration Association Journal* 31: 131-142.

Ferguson, K.U. 1981. Fission track dating of shield areas, Australia; relationships between tectonic and thermal histories and fission track age distributions. Msc thesis University of Melbourne.

Findlay, R.H. 1987. A review of the problems important for interpretation of the Cambro-Ordovician palaeogeography of northern Victorian Land (Antarctica), Tasmania, and New Zealand. In McKenzie, G.D. (ed.). *Gondwana Six: Structure, Tectonics, and Geophysics, Geophysical Monograph 40*: 49-66. Washington: American Geophysical Union.

Foster, D.A. & A.J.W. Gleadow 1992. Reactivated tectonic boundaries and implications for the reconstruction of southeastern Australia and northern Victoria Land, Antarctica, *Geology* 20: 267-270.

Gibbs, A.D. 1984. Structural evolution of extensional basin margins. *Journal of Geological Society of London* 141: 609-620.

Gleadow, A.J.W. & J.F. Lovering 1978a. Fission track geochronology of King Island, Bass Strait, Australia: relationship to continental rifting. *Earth and Planetary Science Letters* 37: 429-437.

Gleadow, A.J.W. & J.F. Lovering 1978b. Thermal history of granitic rocks from western Victoria: a fission-track dating study. *Geological Society of Australia Journal* 25: 323-340.

Gleadow A.J.W., I.R. Duddy & J.F. Lovering 1983. Fission track analysis: a new tool for the evaluation of thermal histories and hydrocarbon potential, *Australian Petroleum Exploration Association Journal* 23: 93-102.

Gray, C.M. 1990. A strontium isotopic traverse across the granitic rocks of southeastern Australia: Petrogenetic and tectonic implications. *Australian Journal of Earth Science* 37: 331-349.

Gray, C.M., M. Tatsumoto, K.R. Ludwig & R.C. Price 1990. Correlated Sr, Nd and Pb isotopic systems and chemical composition in basalts of the Newer Volcanics, Victoria, Australia. *Geological Society of Australia Abstracts* 27: 41.

Gray, D.R., C.J.L. Wilson & T.J. Barton 1991. Intracrustal detachments and implications for crustal evolution within the Lachlan fold belt, southeastern Australia. *Geology* 19: 574-577.

Harrington, H.J., K.L. Burns & B.R. Thompson 1973. Gambier-Baconsfield and Gambier-Sorell fracture zones and the movements of plates in the Australia-Antarctica-New Zealand region. *Nature Physical Science* 245: 109-112.

Lister, G.S., M.A. Etheridge & P.A. Symonds 1991. Detachment models for the formation of passive continental margins. *Tectonics* 10: 1038-1064.

McDougall, I. & P. Wellman 1976. Potassium-argon ages for some Australian Mesozoic igneous rocks. *Geological Society of Australia Journal* 23: 1-9.

Megallaa, M. 1986. Tectonic development of Victoria's Otway Basin - a seismic interpretation. In Glenie, R.C. (ed.), *Second South-Eastern Australia Oil Exploration Symposium, Technical Papers*: 201-218. Melbourne: Petroleum Exploration Society of Australia.

Middleton, M.D. & P.W. Schmidt 1982. Palaeothermometry of the Sydney Basin. *Journal of Geophysical Research* 86: 3998-4008.

Moore M.E., A.J.W. Gleadow & J.F. Lovering 1986. Thermal evolution of rifted continental margins: new evidence from fission tracks in basement apatites from southeastern Australia. *Earth and Planetary Science Letters* 78: 255-270.

Oliver, R.L. 1972. Some aspects of Antarctic-Australian geological relationships. In Adie, R.J. (ed.), *Antarctic Geology and Geophysics*: 858-864. Oslo: Univeristetsforlaget.

Price, R.A. 1981. The Cordilleran foreland thrust and fold belt in the southern Canadian Rocky Mountains. *Geological Society of London Special Publication* 9: 427-448

Royer, J & D.T. Sandwell 1989. Evolution of the Eastern Indian Ocean since the Late Cretaceous: constraints from Geosat altimetry. *Journal of Geophysical Research* 94: 13755-13782.

Stump, E., A.J.R. White & S.G. Borg 1986. Reconstruction of Australia and Antarctica: evidence from granites and recent mapping. *Earth and Planetary Science Letters* 79: 348-360.
Sawka, W.N., M.T. Heizler, R.W. Kistler & B.W. Chappell 1990. Geochemistry of highly fractionated I- and S-type granites from the tin-tungsten province of western Tasmania. *Geological Society of America Special Paper* 246: 161-179.
Vandenberg, A.H.M. 1978. The Tasman fold belt system in Victoria. *Tectonophysics* 48: 267-297.
Veevers, J.J., C.M Powell & S.R. Roots 1991. Review of seafloor spreading around Australia. I. synthesis of the patterns of spreading. *Australian Journal of Earth Science* 38: 373-389.
Wilson, C.J.L., T.M. Will, R.A. Cayley & S. Chen (submitted). Geologic framework and tectonic evolution in western Victoria, Australia. *Tectonophysics*.
Wilson, T.J. 1965. A new class of faults and their bearing on continental drift. *Nature* 207: 343-347.

ACKNOWLEDGEMENTS

We thank Kevin Hill, Rod Brown, Dave Coyle, Ian Duddy, Paul Green and Barry Kohn for comments and contributions. The Australian Research Council, La Trobe University, and Australian Institute of Nuclear Science and Engineering provided financial support.

Gondwana Eight, Findlay, Unrug, Banks & Veevers (eds) © 1993 Balkema, Rotterdam. ISBN 90 5410 304 3

Reconnaissance apatite fission-track data from the East Antarctic Shield

D.C.Arne & P.R.Kelly
Department of Geology, University of Melbourne, Parkville, Vic., Australia
R.W. Brown & A.J.W.Gleadow
Department of Geology, La Trobe University, Bundoora, Vic., Australia

ABSTRACT: The low-temperature (i.e.<~120°C) thermal history of the East Antarctic Shield has been assessed by apatite fission-track analysis of 42 Precambrian outcrop samples. Fission-track ages for these samples range from 466 ± 20 Ma to 153 ± 3 Ma and mean track lengths vary from 14.40 ± 0.11 μm to 12.41 ± 0.17 μm. The variability in the relationship between apatite fission track age and mean track length in data from various geographical regions suggests differences in styles of cooling during the Phanerozoic. Samples from the continental margin east of Prydz Bay are inferred to have cooled during the late Palaeozoic and early Mesozoic, between ~300 and 200 Ma, while samples from the Prince Charles Mountains and from the continental margin west of Prydz Bay are inferred to have cooled more recently in the Mesozoic, at ~150 Ma. Evidence for cooling during the Mesozoic at ~150 Ma is attributed to exhumation associated with the early stages of Gondwana break-up. Cooling during the late Palaeozoic and early Mesozoic may have been associated with an earlier phase of aborted rifting. These preliminary apatite fission track data suggest that the Lambert Graben has had an important influence on the Phanerozoic exhumation history of the East Antarctic Shield, with the region to the west of the Lambert Graben showing the greatest amount of uplift and erosion related to the early stages of Gondwana break-up.

1 INTRODUCTION

The East Antarctic Shield is formed predominantly of Precambrian crystalline basement rocks, many of which were metamorphosed at high grade during the Proterozoic (Craddock, 1972). During the early Palaeozoic much of the East Antarctic Shield was affected by widespread intrusive activity and heating at ~500 Ma, indicated by disturbed K-Ar and Rb-Sr isotopic systems (e.g. Tingey, 1982). The cooling history of the East Antarctic Shield following this event is poorly understood.

A suite of 42 Precambrian basement samples was assembled at the University of Melbourne for reconnaissance apatite fission-track analysis in the late 1970s in an effort to constrain the sub-120°C thermal history of the East Antarctic Shield. The more recent measurement of confined track length distributions in these samples allows a more detailed thermal history interpretation and indicates that the East Antarctic Shield has been affected by discrete cooling episodes at various time during the Phanerozoic; these new data are reported in this paper.

2 APATITE FISSION TRACK ANALYSIS

Apatite fission-track analysis is a well established method for determining the sub-120°C thermal history of various rock units and regions (Green et al., 1989). In particular, apatite fission track analysis has provided useful information concerning the cooling history of passive continental margins. Studies of basement apatite grains have demonstrated the significance of cooling associated with Cretaceous rifting along the southeast margin of Australia (Moore et al., 1986), the northeastern and southern margins of the United States (Miller and Duddy, 1989; Arne, 1992), and the southern margin of Africa (Brown et al., 1990). Evidence for cooling associated with uplift

Table 1. Apatite fission track data

Sample number	Location	Approximate position	Number of grains	Fluence (10^{15}ncm^{-2})	ρs (X10^6/cm^2)	ρi (X10^6/cm^2)	r	Fission-track age (Ma)	Mean track length (µm)	Standard deviation (µm)
East of Prydz Bay										
7910-108	Grove Mtns.	72° S 75° E	5	7.62	2.92 (732)	4.61 (579)	0.87	287 ± 17	13.34 ± 0.16	1.58 (100)
7910-101	Long Sound	68° S 78° E	7	3.45	0.81 (185)	0.78 (89)	0.95	215 ± 10	14.40 ± 0.11	1.12 (100)
7910-103	Club Lake	68° S 78° E	6	3.42	1.27 (387)	0.98 (150)	0.99	263 ± 6	14.18 ± 0.16	1.60 (100)
7910-82	Bunger Hills	67° S 101° E	6	3.51	2.00 (319)	1.73 (138)	0.83	243 ± 13	13.01 ± 0.21	2.05 (100)
7810-06	Nelly Island	67° S 111° E	8	6.86	1.09 (314)	1.45 (210)	0.99	305 ± 8	13.81 ± 0.16	1.64 (100)
7810-25	Lloyd Island	67° S 111° E	8	7.25	0.56 (224)	0.92 (183)	0.94	265 ± 10	13.78 ± 0.18	1.83 (100)
7810-28	Chapel Island	67° S 111° E	5	7.01	0.69 (325)	1.27 (299)	0.99	228 ± 7	13.88 ± 0.20	2.00 (100)
7810-41	Clarke Peninsula	67° S 111° E	5	6.94	3.60 (648)	5.57 (501)	0.99	268 ± 6	12.93 ± 0.17	1.71 (100)
7810-102	Mitchell Peninsula	67° S 111° E	5	7.09	1.08 (463)	1.92 (411)	0.97	239 ± 7	14.03 ± 0.16	1.56 (100)
									13.82 ± 0.16	1.61 (100)
7910-100	Chick Island	67° S 121° E	6	8.45	1.94 (399)	2.74 (282)	0.94	354 ± 28	13.57 ± 0.12	1.17 (100)
7810-324	Petrel Island	67° S 140° E	5	6.78	0.97 (211)	1.38 (151)	0.99	283 ± 9	13.87 ± 0.13	1.10 (71)
7810-315	Lion Island	67° S 140° E	11	35.8	0.62 (249)	2.82 (563)	0.91	466 ± 20	14.20 ± 0.12	1.18 (101)
7810-01	Commonwealth Bay	67° S 143° E	10	8.87	2.17 (902)	4.41 (915)	0.82	261 ± 7	13.42 ± 0.16	1.58 (100)
7810-04	Commonwealth Bay	67° S 143° E	6	7.17	1.75 (569)	2.32 (377)	0.98	322 ± 10	12.97 ± 0.16	1.58 (100)
Prince Charles Mountains/ Prydz Bay										
7722-2	Larry Rocks	70° S 72° E	6	9.28	1.75 (489)	2.91 (414)	0.99	332 ± 10	12.88 ± 0.16	1.60 (100)
7722-3	Landing Bluff	70° S 72° E	5	7.89	2.21 (307)	3.75 (260)	0.96	278 ± 9	12.63 ± 0.20	2.01 (100)
7722-6	Loewe Massif	71° S 68° E	4	9.18	3.71 (382)	11.25 (643)	0.85	164 ± 7	14.04 ± 0.14	1.41 (100)
7722-10	Moore Pyramid	71° S 65° E	6	7.81	1.84 (418)	3.89 (442)	0.98	221 ± 7	13.73 ± 0.17	1.70 (100)
7722-11	Mt. Willing	72° S 67° E	7	8.62	1.50 (438)	2.29 (335)	0.93	335 ± 9	13.48 ± 0.16	1.58 (100)
7722-12	Mt. Collins	71° S 67° E	7	8.37	1.68 (478)	2.62 (373)	0.99	319 ± 7	13.05 ± 0.18	1.77 (100)

The number of tracks measured is given in parentheses.

Table 1. Continued

Sample number	Location	Approximate position	Number of grains	Fluence ($10^{15}ncm^{-2}$)	ρs ($X10^6/cm^2$)	ρi ($X10^6/cm^2$)	r	Fission-track age (Ma)	Mean track length (μm)	Standard deviation (μm)
7722-18	Clemence Massif	72° S 69° E	7	7.57	1.64 (423)	3.05 (394)	0.57	243 ± 13	13.67 ± 0.23	2.35 (100)
7722-19	Shaw Massif	72° S 67° E	7	8.76	3.80 (666)	6.10 (535)	0.97	324 ± 10	12.41 ± 0.17	1.70 (100)
7722-20	Mawson Escarpment	74° S 68° E	5	8.66	2.12 (524)	5.30 (654)	0.97	208 ± 6	13.09 ± 0.19	1.86 (100)
7910-95	Depot Peak	69° S 64° E	6	8.91	2.05 (801)	4.17 (814)	0.98	262 ± 14	13.69 ± 0.13	1.34 (100)
7910-98	Mt. Woinarski	71° S 67° E	8	8.68	0.97 (444)	2.29 (524)	0.98	220 ± 15	13.59 ± 0.20	2.01 (100)
									13.60 ± 0.17	1.70 (100)
7910-99	Lichen Island	69° S 70° E	8	8.56	1.22 (620)	2.05 (521)	0.99	303 ± 19	13.98 ± 0.18	1.81 (100)
7910-106	Wilsons Bluff	74° S 67° E	6	7.91	1.29 (652)	1.69 (427)	0.97	358 ± 23	14.08 ± 0.13	1.30 (100)
West of Prydz Bay										
7910-114	Molodezhnaya	68° S 46° E	6	3.39	1.70 (757)	2.26 (504)	0.99	153 ± 3	13.51 ± 0.17	1.75 (100)
7910-117	Molodezhnaya	68° S 46° E	6	3.32	0.88 (398)	1.14 (260)	0.98	153 ± 4	14.21 ± 0.15	1.49 (100)
R61203	Enderby Land	?	8	7.34	0.53 (236)	1.28 (285)	0.99	183 ± 4	13.49 ± 0.18	1.84 (99)
7722-01	Cook Mtn.	68° S 57° E	9	9.38	1.04 (669)	2.89 (929)	0.93	203 ± 11	13.86 ± 0.19	1.90 (100)
7722-24	Mt. Mueller	67° S 56° E	7	8.35	3.15 (830)	5.58 (734)	0.94	282 ± 9	12.84 ± 0.18	1.78 (100)
7910-91	Kitchenside Rocks	68° S 47° E	10	9.73	1.43 (888)	2.56 (793)	0.95	308 ± 16	13.45 ± 0.19	1.90 (100)
7910-92	Wheeler Rocks	66° S 56° E	10	3.48	3.10 (243)	3.57 (140)	0.87	182 ± 9	14.17 ± 0.13	1.33 (99)
7910-116	Proclamation Island	66° S 54° E	7	3.36	1.45 (429)	0.98 (145)	0.95	296 ± 13	13.83 ± 0.17	1.74 (100)
7722-15	Chapman Ridge	67° S 61° E	13	9.93	2.57 (1475)	4.79 (1375)	0.97	317 ± 14	12.96 ± 0.20	2.02 (100)
7722-16	Scullin Monolith	68° S 67° E	7	8.97	1.62 (476)	4.61 (675)	0.94	190 ± 6	13.44 ± 0.19	1.87 (101)
7722-17	Painter Peak	68° S 63° E	7	8.87	1.57 (153)	4.03 (197)	0.77	206 ± 12	13.13 ± 0.20	2.02 (101)
7722-21	Scullin Monolith	68° S 67° E	5	8.56	1.89 (589)	5.39 (838)	0.98	181 ± 6	13.38 ± 0.17	1.66 (100)
7722-22	Mawson	67° S 63° E	6	7.49	1.12 (357)	3.19 (510)	0.91	158 ± 6	14.22 ± 0.14	1.43 (100)
7722-23	Evans Island	68° S 63° E	11	36.4	0.75 (422)	5.92 (16669)	0.98	275 ± 9	13.44 ± 0.21	1.86 (80)
7910-93	Baillieu Peak	68° S 61° E	8	9.13	0.75 (385)	1.30 (332)	0.99	315 ± 24	13.28 ± 0.22	2.28 (104)

The number of tracks measured is given in parentheses.

marginal to continental rifts has also been reported adjacent to the Ross Embayment (Gleadow and Fitzgerald, 1987) and the Red Sea (Bohannon et al., 1989; Omar et al., 1989). Although apatite fission-track data has not previously been reported from the East Antarctic Shield, data from northern Victoria Land, to the east of the study area, have been presented by Stump et al. (1990).

Details of the 42 samples analysed are presented in Table 1. Approximate locations of the sample are shown in Figure 1. The samples were first pulverized in a jaw crusher and then ground to fine sand size in a rotary crushing mill. The crushed material was washed to remove fines and processed by conventional heavy liquid and magnetic separation techniques to concentrate the heavy minerals. Apatite separates were prepared for fission-track analysis by the external detector method (Gleadow, 1981) following the procedure outlined by Green (1986). Neutron irradiations were carried out in a well-thermalised neutron flux at the X-7 facility of the Australian Nuclear Science and Technology Organization's Hifar reactor, Lucas Heights. This facility has a Cd ratio for Au of 125. The irradiations employed a nominal fluence of 10^{16} ncm^{-2} monitored using NBS standard glass SRM 612 included at either end of the irradiation canister. As the induced track densities for the standard glasses from either end of the irradiation canisters showed a significant difference, a unique standard glass density was calculated for each sample in the irradiation by assuming a linear fluence gradient.

Track densities were generally determined for between 5 and 10 apatite grains in each sample and the data pooled to calculate a fission track age with a 1σ uncertainty using the "conventional analysis" described by Green (1981). Fission track ages were calculated using a personal calibration factor for analyst PRK, similar to the zeta factor recommended by Hurford (1990), determined from replicate analyses of Fish Canyon Tuff, Durango and Mt. Dromedary age standards. Horizontal confined lengths were measured for 100 tracks where possible following the later technique described by Green et al. (1986).

Analytical results for all samples are presented in Table 1. Apatite fission track ages range from 153 ± 3 Ma for a sample from Molodezhnaya station to 466 ± 20 Ma for a sample from Dumont d'Urville station (Figure 1). Mean lengths for all samples range from 12.41 ± 0.17 μm to 14.40 ± 0.11 μm. The track length distributions are typically dominated by tracks 12 to 15 μm in length and are negatively skewed, with a subordinate population of tracks less than 12 μm in length. Uncertainties in both the fission-track ages and mean track lengths are quoted at 1σ in Table 1 and in Figure 1.

3 INTERPRETATION OF THERMAL HISTORY

Regional temperatures during the early Palaeozoic were sufficiently high enough to affect Rb-Sr isotopic systems (e.g. Tingey, 1982) and so were too hot for the retention of fission tracks in Precambrian basement apatite grains (>120°C). Thus the apatite fission track age data presented in Table 1 indicate cooling through temperatures below ~120°C since the early Palaeozoic. The range in observed apatite fission track ages also suggests that cooling during the Phanerozoic was prolonged and extended into the Mesozoic. Whether cooling occurred gradually following the early Palaeozoic or resulted from discrete periods of cooling cannot be determined from the fission track age data alone.

The relationship between fission track age and mean track length for samples from three different geographical regions is shown in Figure 2. Samples from the continental margin east of Prydz Bay generally have fission track ages in the range ~350 to 200 Ma. Mean track lengths over this range increase from ~13 to 14 μm, suggesting that the samples cooled from a range of temperatures involving varying degrees of track retention during the late Palaeozoic and early Mesozoic, between ~300 and 200 Ma. A single sample from Dumont d'Urville (7810-315) has an apatite fission track age of 466 ± 20 Ma and a mean track length of 14.20 ± 0.12 μm, indicating that it must have cooled rapidly below ~120°C during the early Palaeozoic.

Apatite fission-track ages of the samples from the Prince Charles Mountains range from ~350 Ma to ~150 Ma, whereas mean track lengths vary between ~12.5 to 14 μm (Figure 2). The data show generally that those samples (7910-106, 7722-6) with the longest mean track lengths also have the oldest and youngest fission-track ages, whereas those (e.g. 7722-3, 772-19) with fission-track ages in the range

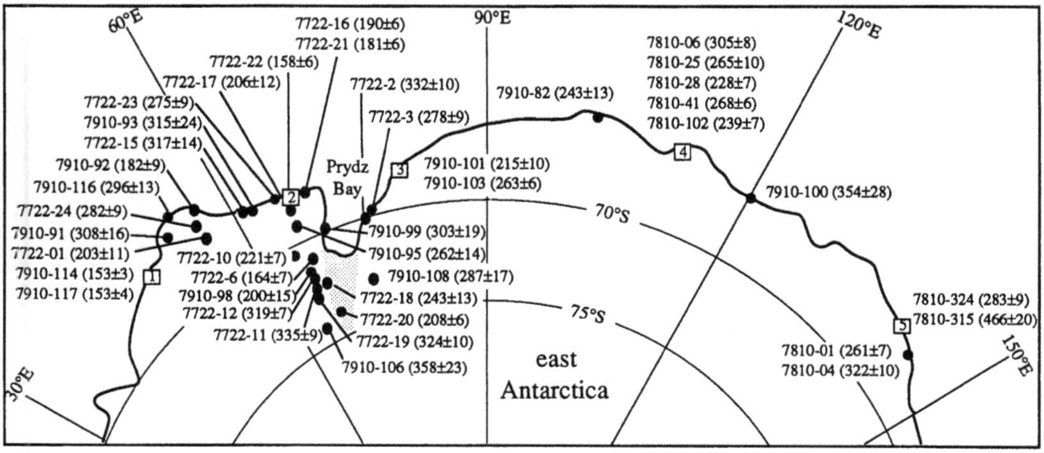

Fig. 1 Location of samples from the East Antarctic Shield. Apatite fission-track ages are given in parentheses. Antarctic base localities are numbered as follows: 1 - Molodezhnaya, 2 - Mawson, 3 - Davis, 4 - Casey, 5 - Dumont d'Urville. The position of the Lambert Graben is shaded.

~300 to 200 Ma have the shortest mean track lengths. Such a U-shaped trend is consistent with varying degrees of track retention prior to cooling in the Mesozoic, at ~150 Ma (e.g. Green, 1986).

Samples 7722-11, 7722-12, 7910-99, 7910-106 with apatite fission-track ages in the range 300 to 350 Ma and mean track lengths >13 μm are inferred to have cooled well below ~120°C at some time during the late Palaeozoic and show minimal evidence of subsequent track annealing. In contrast, sample 7722-6 with a fission-track age of 164 ± 7 Ma and a mean track length of 14.04 ± 0.14 μm is inferred to have been severely annealed prior to cooling at ~150 Ma. Samples 7722-3, 7722-10, 7722-18, 7722-20, 7910-95, 7910-98 with intermediate fission track ages between ~300 and 200 Ma are interpreted to contain a discrete population of track lengths shortened prior to cooling in the Mesozoic and a population of longer, unannealed tracks formed since cooling.

The remaining samples from the continental margin west of Prydz Bay have apatite fission-track ages in the range ~300 Ma to 150 Ma (Figure 2). Mean track lengths in these samples vary between ~13 and 14 μm, with a general trend of increasing mean track length with decreasing fission-track age. The interpretation of these data is similar to that inferred for samples from the Prince Charles Mountains, except that all samples from west of Prydz Bay appear to have been severely to moderately annealed prior to cooling at ~150 Ma and no direct evidence of earlier cooling episodes is preserved.

4 DISCUSSION

Although this study is of a reconnaissance nature, there appear to be clear differences in apatite fission-track data from different geographic regions. More detailed sampling in specific areas is required to determine patterns of fission track age and length variation with elevation (e.g. Arne, 1991) and until such studies are complete the findings of the present study should be considered as preliminary.

However, these data suggest differences in style of cooling for the continental margin east and west of Prydz Bay, which forms the northern extension of the Lambert Graben (Stagg, 1985; Wellman and Tingey, 1976). We infer that the samples from east of Prydz Bay underwent major cooling during late Palaeozoic and early Mesozoic times, perhaps in response to early phase(s) of aborted rifting related to mantle plume activity (Kent, 1991). There is little evidence for any cooling associated with the Cretaceous breakup between east Antarctica and Australia during the mid-Cretaceous (Veevers et al., 1991). We infer a similar conclusion from apatite fission track data from the western part of the Wilson terrane in northern Victoria Land and from western Victoria (Stump et al, 1990). Samples from the continental margin west of Prydz Bay and some samples from the Prince Charles Mountains show a major phase of cooling initiated during the Mesozoic, at ~150 Ma. Cooling at that time could well have been associated with the separation of east Antarctica and India during the Mesozoic (Veevers et al., 1991).

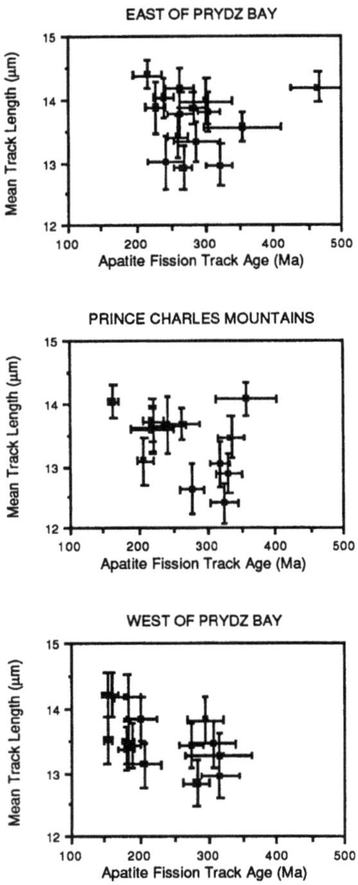

Fig. 2 Summary of apatite fission track data from three regions of the East Antarctic Shield.

Evidence for a late Palaeozoic phase of cooling is preserved in samples from the Prince Charles Mountains. Recent detailed sampling in the northern Prince Charles Mountains indicates that the range in apatite fission-track age between ~300 and 100 Ma for this area is generally related to differences in the elevation from which the samples were collected (Arne, 1991). Here the oldest apatite fission-track ages are preserved in samples from the highest elevations. Thus differences between the apatite fission-track data from the East Antarctic Shield west of Prydz Bay and the Prince Charles Mountains probably are a result of sampling the various erosional levels in the two regions.

The differences between the cooling histories of samples from west of Prydz Bay, including those from the Prince Charles Mountains, and samples from east of Prydz Bay probably indicate significant differences in the timing of exhumation during the Phanerozoic. Such a conclusion further suggests that the Lambert Graben forms a major tectonic feature that has had a significant influence on the uplift and cooling history of the East Antarctic Shield during the Phanerozoic.

5 CONCLUSIONS

Apatite fission-track analysis of a reconnaissance suite of samples indicates differences in the Phanerozoic exhumation history of the East Antarctic Shield to the east and west of the Lambert Graben. Samples from west of the Lambert Graben cooled during late Palaeozoic times and during the Cretaceous break-up of Gondwana, whereas samples from east of the Lambert Graben cooled during late Palaeozoic to early Mesozoic times and show little evidence for Cretaceous cooling.

ACKNOWLEDGEMENTS

The authors are indebted to Professor John Lovering of Flinders University, Adelaide and Bob Tingey of the Bureau of Mineral Resources, Canberra for providing sample material collected during various Australian National Antarctic Research Expeditions. The Australian Institute of Nuclear Science and Engineering is also thanked for a grant providing neutron irradiations. DCA would like to acknowledge the Australian Antarctic Science Advisory Committee (ASAC) for financial support as well as Geotrack International for technical support. PRK was supported by a grant from the Australian Research Grants Council at the time this study was initiated.

REFERENCES

Arne, D.C. 1991: Evidence from apatite fission track analysis for Early Cretaceous uplift and erosion of the northern Prince Charles Mountains, Antarctica (Abst.). 8th International Symposium on Gondwana: 4-5.

Arne, D.C.1992: Evidence for regional mid-Cretaceous cooling in the Ouachita Mountain Fold Belt and Arkoma Basin of Arkansas from apatite fission track analysis. American Association of Petroleum Geologists Bulletin : 392-402.

Bohannon, R.G.; Naeser, C.W.; Schmidt, D.L.; Zimmermann, R.A. 1989: The timing of uplift, volcanism, and rifting peripheral to the Red Sea: a case for passive rifting. Journal of Geophysical Research 94 :1683-1701.

Brown, R.D.; Rust, D.J.; Summerfield, M.A.; Gleadow, A.J.W.; De Wit, M.C.J. 1990: An Early Cretaceous phase of accelerated erosion on the south-western margin of Africa: evidence from apatite fission track analysis and the offshore sedimentary record. Nuclear Tracks and Radiation Measurements 17 : 339-350.

Craddock, C. 1972: Antarctic Tectonics. In Aide, R.J. (ed.): 'Antarctic Geology and Geophysics'. Universitetsforlaget, Oslo: 449-455.

Gleadow, A.J.W. 1981: Fission track dating methods: what are the real alternatives ? Nuclear Tracks 5 : 3-14.

Gleadow, A.J.W.; Fitzgerald, P.G. 1987: Uplift history and structure of the Transantarctic Mountains: new evidence from fission track dating of basement apatites in the Dry Valleys area, southern Victoria Land. Earth and Planetary Science Letters 82 : 1-14.

Green, P.F. 1981: A new look at statistics in fission track dating. Nuclear Tracks 5 : 77-86.

Green, P.F. 1986: On the thermo-tectonic evolution of Northern England: evidence from fission track analysis. Geological Magazine 123 : 493-506.

Green, P.F.; Duddy, I.R.; Gleadow, A.J.W.; Tingate, P.R.; Laslett, G.M. 1986: Thermal annealing of fission tracks in apatite: 1 - a qualitative description. Chemical Geology (Isotope Geoscience Section) 79 : 237-253.

Green, P.F.; Duddy, I.R.; Gleadow, A.J.W.; Hegarty, K.A.; Laslett, G.M.; Llovering, J.F. 1989: Thermal annealing of fission tracks in apatite: 4 - quantitative modelling techniques and extension to geologic timescales. Chemical Geology (Isotope Geoscience Section) 79 : 155-182.

Hurford, A.J. 1990: Standardization of fission track dating calibration: recommendation by the Fission Track Working Group of the I.U.G.S. Subcommission on Geochronology. Chemical Geology (Isotope Geoscience Section) 80 : 171-178.

Kent, R. 1991: Lithospheric uplift in eastern Gondwana: evidence for a long-lived mantle plume system ? Geology 19: 19-23.

Omar, G.I.; Steckler, M.S.; Buck, W.R.; Kohn, B.P. 1989: Fission-track analysis of basement apatites at the western margin of the Gulf of Suez rift: evidence for synchroneity of uplift and subsidence. Earth and Planetary Science Letters 94 : 316-328.

Miller, D.S.; Duddy, I.R. 1989: Early Cretaceous uplift and erosion of the northern Appalachian Basin, New York, based on apatite fission track analysis. Earth and Planetary Science Letters 93 : 35-49.

Moore, M.E.; Gleadow, A.J.W.; Lovering, J.F. 1986: Thermal evolution of rifted continental margins: new evidence from fission tracks in basement apatites from southeastern Australia. Earth and Planetary Science Letters 78 : 255-270.

Stagg, H.M.J. 1985: The structure and origin of Prydz Bay and MacRoberston Shelf, East Antarctica. Tectonophysics 114 : 315-340.

Stump, E.; Fitzgerald, P.F.; Gleadow, A.J.W. 1990: Comparison through fission-track analysis of portions of Australia and Antarctica adjacent prior to continental drift. Nuclear Tracks and Radiation Measurements 17 : 359-365.

Tingey, R.J., 1982: The geological evolution of the Prince Charles Mountains - an Antarctic Archean cratonic block. In Craddock, C. (ed.) 'Antarctic Geoscience' University of Wisconsin: 455-464.

Veevers, J.J.; Powell, C. McA.; Roots, S.R. 1991: Review of seafloor spreading around Australia. I. Synthesis of the patterns of spreading. Australian Journal of Earth Sciences 38: 373-389 .

Wellman, P.; Tingey, R.J. 1976: Gravity evidence for a major crustal fracture in eastern Antarctica. Journal of the Bureau of Mineral Resources, Geology and Geophysics, Australia 1 : 105-108.

Gondwana Eight, Findlay, Unrug, Banks & Veevers (eds) © 1993 Balkema, Rotterdam. ISBN 90 5410 304 3

Thermo-tectonic influences on the Sydney Basin during the breakup of Gondwana

M.F. Middleton
Department of Exploration Geophysics, Curtin University of Technology, Perth, W.A., Australia

ABSTRACT: The Sydney Basin formed during the late Palaeozoic and early Mesozoic as a foreland basin. The basin has relatively high rank coals (about 1 percent vitrinite reflectance near the surface along its eastern boundary). Quantitative modelling suggests that the thermal history that is most consistent with the tectonic history of the basin, and observed palaeotemperature data, is that of a short burst of high heat flow for about 5 Ma. Thermal conductivity measurements were made for the first time on shales of the Wianamatta Group, the youngest known pre-breakup sequence in the basin, and yielded very low values in the range 1.3-2.1 W/m°K. A thermal model is propopsed that entails 1) a thickness of about 800-900 m of missing section, and 2) a short period of maximum heat flow of about 182 mW/m², caused by a regional deep crustal intrusion which is deeper than 10 km and is about 10 km thick.

1 INTRODUCTION

The Sydney Basin (Fig. 1) formed during the late Palaeozoic and early Mesozoic as a foreland basin (Herbert, 1980). During this time, fluvial to marine deposition occurred in the basin, with significant coal swamps forming in the Permian. Deposition probably ceased during the Late Triassic, although the youngest known sequence from this period of deposition, the Wianamatta Group, is dated as Middle Triassic. Little deposition probably occurred during the Jurassic and Early Cretaceous. At about 75 Ma, continental breakup and seafloor spreading occurred along the basin's present day eastern margin (Shaw, 1978). The basin has been strongly intruded by igneous rocks during the Triassic and Jurassic, and again during the Tertiary when significant flows also occurred (Wellman and McDougall, 1974; Crawford et al., 1980). It is well documented in the literature that foreland basins have well defined and predictable temperature and organic maturity patterns (Beaumont, 1981; Beaumont et al., 1985). The Sydney Basin, therefore, has been shown to have similarly regular and predictable patterns of coal rank (Shibaoka et al., 1973).

The aim of the present paper is to review the causes of the high coal rank near the basin's eastern margin. Contours of surface vitrinite reflectance (a quantitative

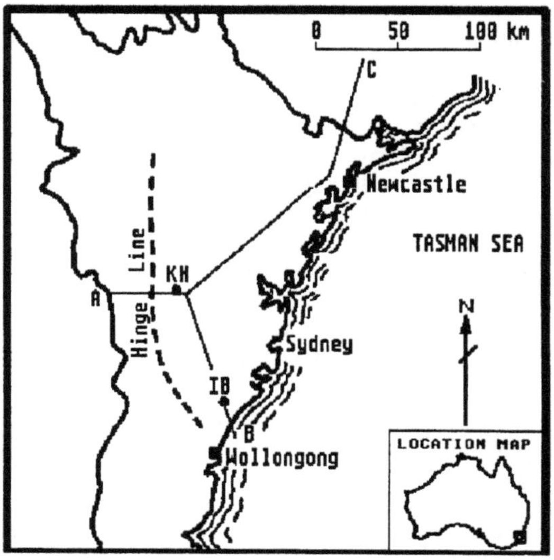

Figure 1: Location diagram showing the Sydney Basin; sections AB and AC, which are discussed in the text, are also shown.

index of coal rank) show a region greater than 1 percent between Sydney and Wollongong (Fig.2). These contours show that this region was the hottest part of the basin, and must have been buried in the past. However, there is considerable controversy about the depth of burial, i.e. the thickness of the missing section. Various workers believe that the missing sequence was relatively thin (less than a kilometre) and high heat flow caused the high coal rank, whereas other workers have provided evidence of a thick missing sequence (2000 m or more) based on quantitative modelling of thermal maturity indicators (vitrinite reflectance, fluid inclusion homogenization temperatures, palaeomagnetic overprint signatures).

This study investigates quantitatively these aspects further. More thermal conductivity measurements have been made on the top two preserved sequences: the Hawkesbury Sandstone and the Wianamatta Shale. Further, the near surface thermal gradient caused by a regional igneous intrusion into the crust beneath the basin is also quantitatively modelled.

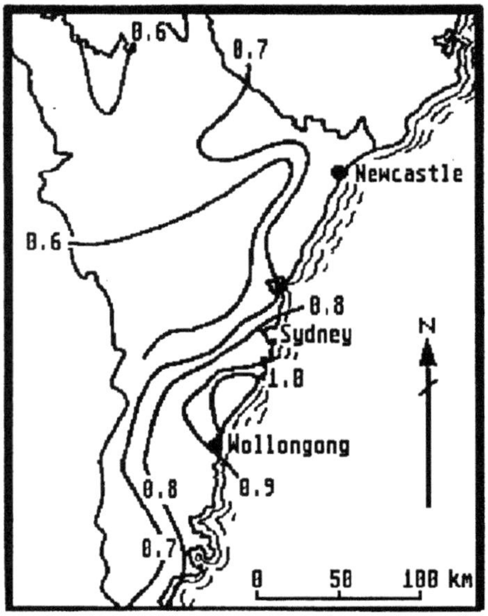

Figure 2: Contours of surface vitrinite reflectance from Middleton and Schmidt (1982).

2 HIGH HEAT FLOW ORIGIN OF THE COAL RANK

Diessel (1975) suggested that the high coal rank was due more to high heat flow from igneous activity rather than a significantly thick missing sequence. These views were supported by Facer et al. (1979, 1980), who suggested that the high heat flow might also be enhanced by radiogenic heat generation in underlying basement rocks, as well as basin sediments derived from them.

Finally, Branagan (1983) rejected a thick missing sequence essentially on the grounds:

1. That there is little evidence of a thick section being deposited off the eastern margin of the Sydney Basin in the Tasman Sea, and

2. Erosion of diatremes suggest loss of no more than 1100-1400 m of sediment (comprising 800-1000 m of unknown sequence) near the coast at Sydney; he suggested high heat flow as the cause of the high coal rank.

No quantitative models of the thermal structure of the Sydney Basin have been reported in the literature by the supporters of the high heat flow origin of the coal rank. Rather they quote relatively high radiogenic heat generation values for some of the surrounding igneous rocks of the Lachlan Fold Belt, and have accumulated a large number of thermal conductivity values. The difficulty is that extrapolation of these values to a missing sequence gives thicknesses of 2000 m or more.

3 A THICK MISSING SEQUENCE AS THE CAUSE OF HIGH COAL RANK

Middleton (1983a), and Middleton and Schmidt (1983) proposed that the high coal rank was due to a significant thickness of missing section. They suggested that in excess of 2000 m previously existed near the coast in the regions of surface reflectance in excess of 1 percent. The calculations of the thickness of the missing sequence were essentially based on a quantitative model of the behaviour of vitrinite reflectance with temperature and time. They also calculated that the maximum geothermal gradient in the high vitrinite reflectance regions of the basin was in the range 50-60 °C/km.

Recent work by Eadington and Hamilton (1991) and Bai et al. (1991) have supported these findings and they have quoted missing section in the range 1800-2100 m and maximum heat flow of about 86 mW/m² (2.1 HFU). Their calculations are based on a chemical kinetic model of vitrinite reflectance (Sweeney and Burnham, 1990) and is significantly different from the model used by Middleton (1983a) and Middleton and Schmidt (1983).

4 PALAEOTEMPERATURE FROM VITRINITE REFLECTANCE

Considerable insight into the problem can be obtained by superimposing the observed vitrinite reflectance trends onto a cross-section across the Sydney Basin (Fig. 3). Vitrinite reflectance trends can be deduced from over 50 well and sample localities throughout the basin (Middleton, 1983b). The cross-section (AB in Fig 1) passes through the locations KH (Kurrajong Heights Bore), IB (Illawarra Bores) and coal mines (not shown). The isoreflectance trends seen on the cross-section (Fig. 3) are representative of the large volume of data published by Middleton (1983b). The occurrence of the 1 percent vitrinite reflectance contour at the surface is clearly seen to be confined to near the eastern margin of the basin. The 1.4 percent contour almost parallels the higher 1 percent contour, suggesting a similar geothermal gradient across most of the central and eastern part of the basin.

The vitrinite reflectance contours can be converted directly into temperature using the

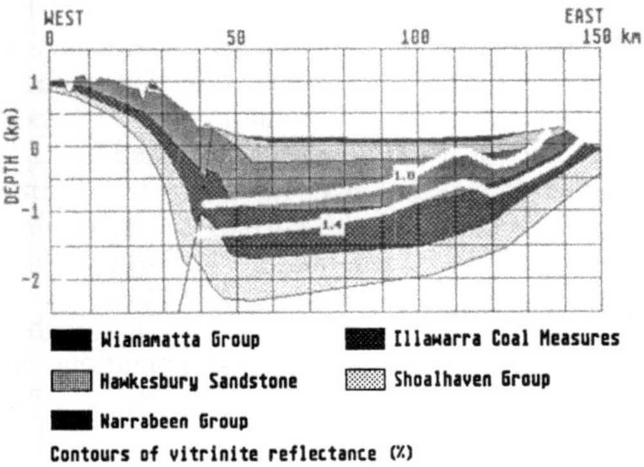

Figure 3: Cross-section AB with contours of 1 percent and 1.4 percent vitrinite relectance shown.

vitrinite reflectance-temperature-time nomogram of Sweeney and Burnham (1990) shown in Figure 4. The conversion from vitrinite reflectance to temperature is made under the assumptions that the temperature:(i) is applied rapidly, (ii) maintained for a given period of time, and (iii) removed rapidly. These assumptions are approximately realised in the case of an intrusion if the temperature is maintained for about 5 Ma (see below), and in the case of rapid burial and uplift with a long period of maximum burial in between. Middleton (1983a) showed that in the latter case the coalification is dominated by the long period at maximum burial. Figure 5a shows the same cross-section as in Figure 3 with the vitrinite reflectance converted to maximum temperature, which has been maintained for 5 Ma (to approximate a deep regional igneous intrusion). Figure 5b shows the cross-section with maximum temperature maintained for 150 Ma (to approximate the burial history model of Middleton (1983a)). The possible range of maximum surface palaeotemperatures in the east of the basin is seen from Figure 5 to vary from 121 °C to 143 °C. Further, the geothermal gradient determined from the two temperature contours shown in Figure 5 is 52 °C/km in both the 5 Ma maximum temperature case and the 150 Ma maximum temperature case.

The thickness of the missing sequence can be calculated by continuing this geothermal gradient to the palaeosurface of that sequence, which is assumed to have a temperature of 20 °C. In the case of the 5 Ma maximum temperature (Fig. 5a), we find 2.37 km of missing sequence, and in the case of the 150 Ma maximum temperature (Fig. 5b), we find 1.94 km of missing sequence. Thus, on the basis of this argument, we find support for the near 2 km of missing sequence, as predicted by Middleton and Schmidt (1983), Eadington and Hamilton (1991) and Bai et al. (1991).

5 THERMAL EFFECT OF A DEEP REGIONAL INTRUSION

All authors agree on one point, and that is that there has been a period of high heat flow probably associated with continental breakup during the Cretaceous, roughly in the period 75 Ma to 100 Ma. It is important therefore

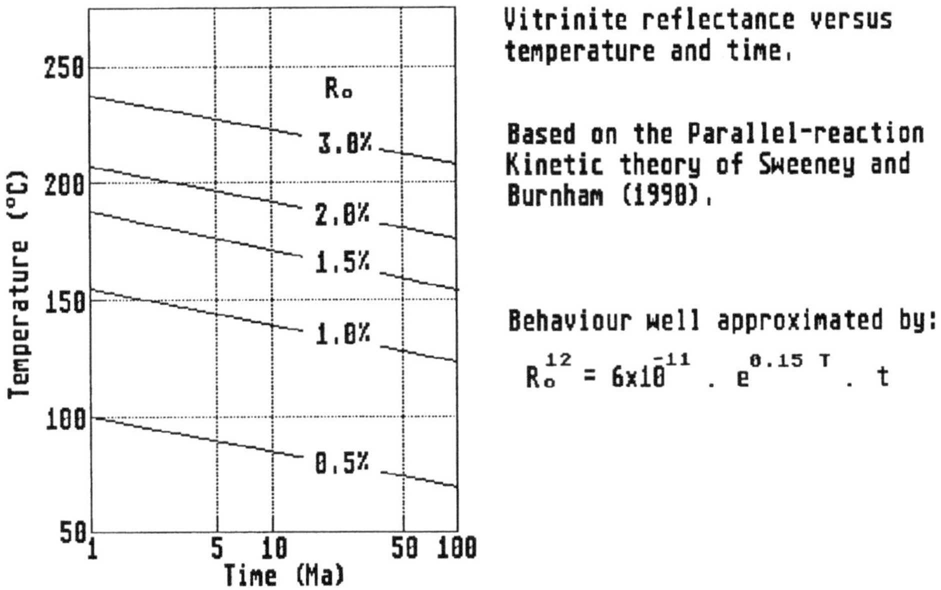

Figure 4: A nomogram showing the relationship of vitrinite reflectance, temperature and time; the nomogram is based on the chemical kinetic model of Sweeney and Burnham (1990).

to have some quantitative estimate of the effect of heat emanating from a deep intrusive source that may have been emplaced during continental breakup. A model is proposed of a slab of magma 10 km thick, 1000 km long, 70 km wide and of temperature 1100 °C, which is intruded at a range of depths between 10 and 20 km beneath the earth's surface. The 10 km thickness is based on the isostatic calculation for the intrusion to cause the change of elevation of about 1000 m across the hinge zone (see in Figures 1 and 3). The 70 km x 1000 km dimensions are to approximate such a slab being injected along the eastern coast of Australia during continental breakup. The hot slab is assumed to cool immediately upon intrusion by conduction. The theory to determine temperatures in the upper 1000 m of the earth directly above the centre of the slab is well known in the literature (Carslaw and Jaeger, 1959), and is not described herein.

The model is shown schematically in Figure 6. Also shown in Figure 6 are curves of the geothermal gradient (measured between 1.0 and 1.5 km from the surface) versus time after intrusion for depths to the top of the intrusive slab of 10, 15 and 20 km. The thermal effect of the intrusion is clearly seen to be that of a rapid pulse of heat within 10 Ma of the magma being intruded.

The curves in Figure 6 can be used in a quantitative sense to provide some information about the postulated intrusive body. To match the geothermal gradient of 52 °C/km, observed in Figure 5, the top of the intrusion would have to be between 10 and 15 km. It is worth noting that a similar geothermal gradient could be produced by a deeper intrusion that was maintained at about 1100 °C for several million years by convection, rather than cooling immediately upon intrusion as assumed in this model. Nevertheless, it is possible for a deep regional intrusion to cause sufficiently high geothermal gradients (about 52 °C/km) within 1000 m of the surface to satisfy the observations in Figure 5. Further, the intrusion model shows that the 5 Ma scenario (Figure 5b) is probably a reasonable approximation of the thermal conditions of such a heat pulse.

Although high heat flow as a cause of the high coal rank in the eastern part of the basin now has some quantitative support, we are still left with the problem of the thickness

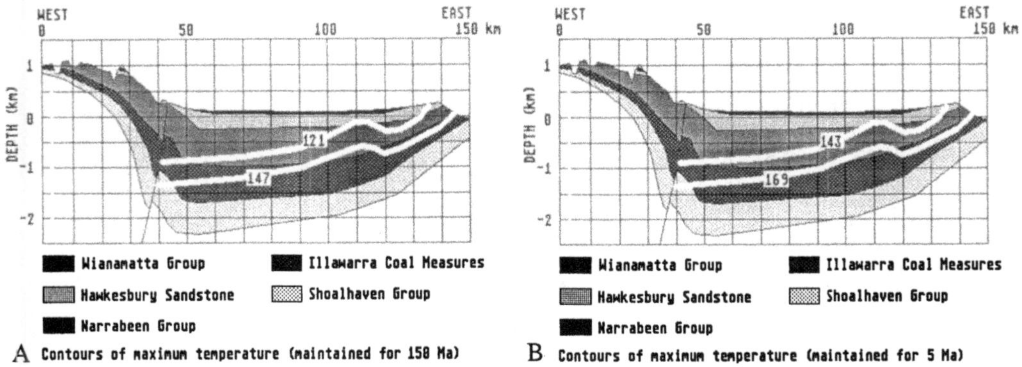

Figure 5: Cross-section AB with contours of maximum temperature corresponding to the contours of 1 percent and 1.4 percent vitrinite reflectance: a) assuming that the maximum temperature was maintained for 150 Ma, b) assuming that the maximum temperature was maintained for 5 Ma.

of the vanished sequence. The scenario in Figure 5b still requires some 2.37 km of missing sequence.

6 THERMAL CONDUCTIVITY

The other critical component in thermal analysis, and heretobefore generally neglected, is thermal conductivity. As heat flow is the product of thermal conductivity and the thermal gradient, thermal conductivity has an important part to play in constraining the various proposed thermal models.

A significant suite of thermal conductivity measurements of Sydney Basin sediments has been made and published in the literature, and are summarised by Facer et al. (1980). These values essentially show the Illawarra Coal Measures and Narrabeen Group sediments to be in the range 3.0 to 3.5 W/m°K, the Hawkesbury Sandstone to be high with a mean about 4.5 W/m°K; no data have been acquired for the Wianamatta Group. This study supplemented these data, and the current thermal conductivity information available for the basin are summarised in Figure 7.

These data can be used to convert the 52°C/km geothermal gradient observed in Figure 5 to an approximate heat flow. As the geothermal gradient in Figure 5 was measured across the Narrabeen Group and the Hawkesbury Sandstone, the pertinent rock thermal conductivities are in the range 2.7 and 4.5 W/m°K. A near median value of 3.5 W/m°K is assumed to approximate the maximum heat flow in the basin, which is calculated to be 182 mW/m^2 (~4.4 HFU). This is substantially greater than that proposed in the modelling of Bai et al. (1991).

The other important finding of the present study of thermal conductivity is that the previously untested Wianamatta Group shales have very low values in the range 1.3 to 2.1 W/m°K. These values are consistent with sediments of clay or feldspar mineralogy that have relatively high porosity (Palciauskas, 1986), and thus have not been buried deeper than about 1000 m.

Similar values have been obtained by Detrich et al. (1986) for poorly consolidated deep ocean sediments at depths shallower than about 1000 m. Isolated clay-rich shale lenses in the dominantly sand and carbonate rich sediments of the Uinta Basin are also found to have thermal conductivity values in this range (Brigaud et al., 1990).

The Wianamatta Group thermal conductivity data provide a vital clue to the thermal history of the Sydney Basin. Essentially, these data are telling us:

1. That the Wianamatta Group was not buried deeper than 1000 m
2. The missing sequence, if argillaceous, probably has thermal

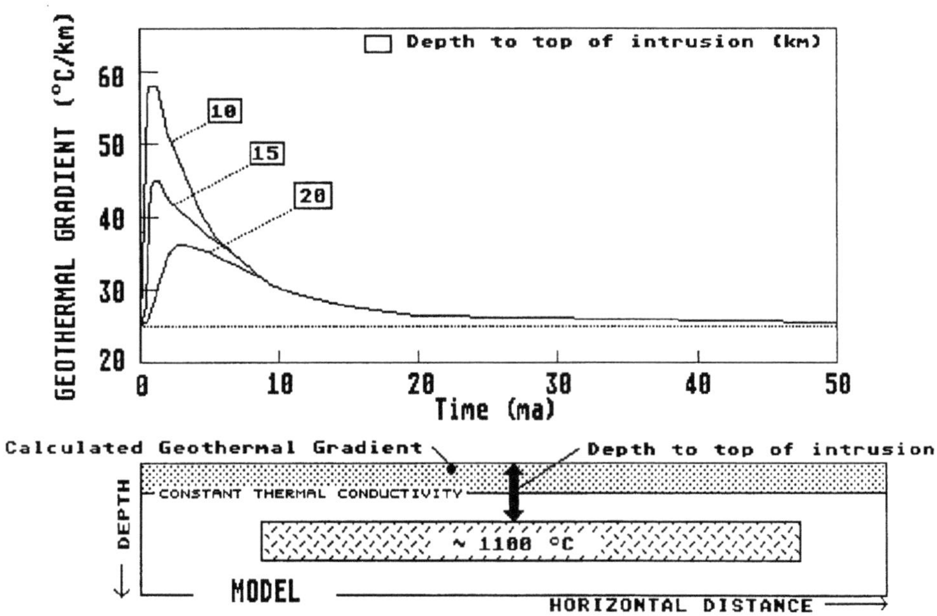

Figure 6: (Upper) Curves of geothermal gradient versus time after intrusion for the model (shown below) with different depths to the top of the intrusion; the curves support the maximum temperature scenario of Figure 5b, which has a geothermal gradient of 52 °C/km maintained for about 5 Ma.

conductivities in the same range or less.

The second point is most important in the calculation of the thickness of the missing sequence, as previous estimates are based on assuming similar thermal properties to the sediments beneath the Hawkesbury Sandstone. In fact, the data of Palciauskas (1986) shows that clay-rich sediments exhibit a fairly rapid increase of thermal conductivity from about 1.05 W/m°K to about 1.5 W/m°K in the initial 1000 m of burial, and then a gradual increase to about 2 W/m°K at greater depths.

Further, if the missing sequence as well as the Wianamatta Group consisted of poorly consolidated argillaceous sediments, weathering of this sequence would be relatively easy and possibly explains why most of it has since vanished.

STRATIGRAPHY		LITHOLOGY	THERMAL CONDUCTIVITY (W/m°K)
Basalt			2.4
Missing Sequence			
Wianamatta Group			1.3-2.1*
Hawkesbury Sandstone			3.8-4.5*
Narrabeen Group	BALD HILL CL		2.7
	BULGO SST		3.7
	STANWELL P CL		3.0
	SCARBO'H SST		3.4
	WOMBARRA SH		3.2
	COALCLIFF SST		3.3
Illawarra Coal Measures	BULLI COAL		
			2.9
	BALGOWNI COAL		
			3.1
	WONGAWILLI C		

* Measurements made for this study

Figure 7: Table of thermal conductivities of sediments from the Sydney Basin; the data have been obtained from the literature and have been supplemented by this study. Mean values are shown where no range is given.

7 THERMAL MODEL

A thermal model is now proposed based on the assumption that the missing sequence consists dominantly of shale with clay and feldspar mineralogy. Thus the model assumes that the missing sequence has a thermal conductivity 1.25 W/m°K in its initial 500 m and 1.50 W/m°K for its remaining thickness. The maximum heat flow through the basin is assumed to be 182 mW/m², as determined from the data in Figures 5 and 7.

The model is shown in Figure 8, with the assumed thermal parameters. The temperature at the top of the Hawkesbury Sandstone is 143 °C and at the top of the Illawarra Coal Measures is 167 °C, which are consistent with data in Figure 5b.

Furthermore, the thickness of missing sequence is approximately 850 m, which is consistent with the thickness suggested by Branagan (1983) based on the erosion of diatremes within the basin.

This model is extented to the cross-section AB in Figure 9, which shows the calculated thickness of missing sequence across the basin. This is also consistent with the diatreme based thicknesses of Branagan (1983), who suggested about 800 to 1000 m of missing section near the coast and decreasing to several hundred metres approaching the basin hinge line (see Fig. 1).

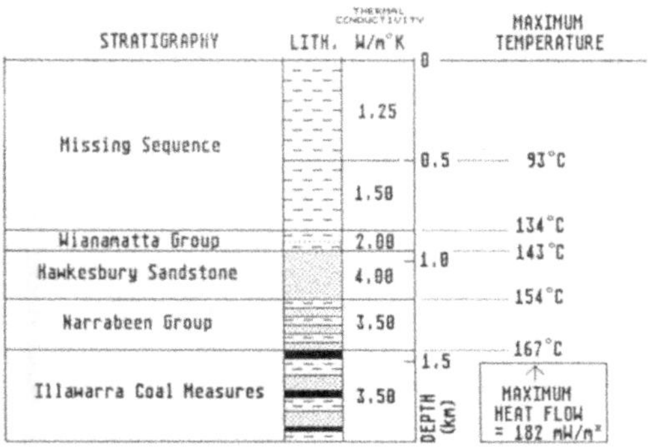

Figure 8: The thermal model proposed as a result of this study; the principal advance of this model is the use of lower thermal conductivities of the Wianamatta Group and the missing sequence, which significantly reduces the thickness of missing sequence.

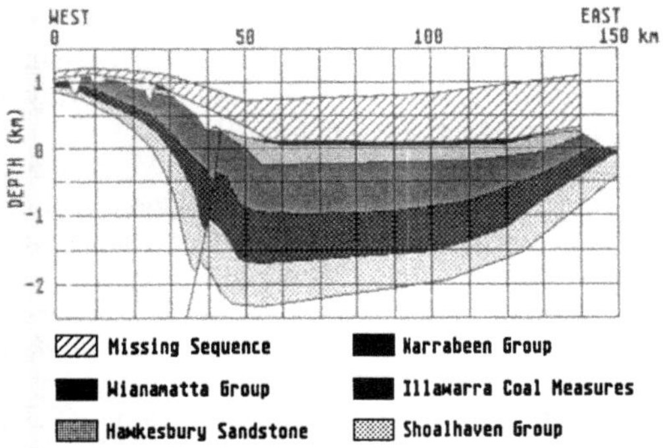

Figure 9: The cross-section AB with the newly modelled thickness of missing sequence added to the top of the existing preserved section.

8 CONCLUSION

This study of the thermal and tectonic history of the Sydney Basin suggests that substantially less missing sequence existed in the basin than some previous authors had calculated. The principal difference between the present thermal model for the basin, and that proposed previously by Middleton (1983a), and Middleton and Schmidt (1983) is that the thermal conductivity of the missing sequence for the present model is significantly less than the underlying rocks.

The study also showed quantitatively that a deep regional intrusion, associated with continental breakup and seafloor spreading in the Tasman Sea, is capable of producing the observed geothermal gradients in the preserved sediments of the Sydney Basin. The existence of a deep intrusion beneath the basin has been debated in the literature for some years on the basis of the bouguer gravity signature of the basin (Qureshi, 1984; Leaman, 1990). The result of this debate is that no major intrusion appears to lie above about 10 km depth, which is not inconsistent with the modelling in Figure 6. Further gravity studies are required to determime if a deeper intrusion, of more regional extent than that proposed by Qureshi (1984), is a viable model; currently the intrusive slab forms a component of the "isostatic function" used by Qureshi (1984) and Leaman (1990). A summary of the tectonic history of the basin is now proposed as follows (Fig. 10):

1. The basin formed as a foreland basin during the Permian to Late Triassic (including a maximum of about 800-900 m of totally missing sequence), with overthrusting from the north and attendant downward mantle flow.

2. A period of uplift occurred some time between the Early Jurassic and Early Cretaceous in response to the mantle rebound associated with cessation of downward convective flow (mechanism described by Middleton, 1990); at this time the Eastern

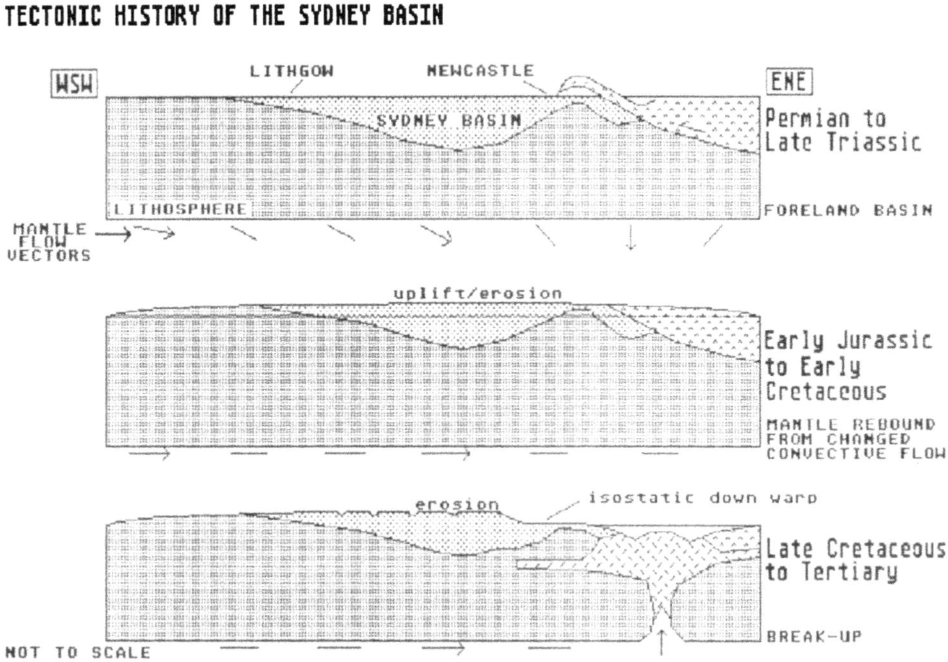

Figure 10: Tectonic history of the Sydney Basin; location is section AC in Figure 1. The basin formed as a foreland basin during the Permian to Late Triassic; uplift occurred between the Early Jurassic and Early Cretaceous, due to mantle rebound from changed convective flow; continental breakup occurred in the Late Cretaceous, and erosion occurred during the Tertiary with the opening of the Tasman Sea.

Highlands began to form, but little erosion occurred in this elevated plateau prior to continental breakup (because of the distance to the existing coast, some several hundred kilometres to the east). The proposed uplift complements the isostatic rebound origin of the Eastern Highlands proposed by Lambeck and Stevenson (1986). However, as suggested above, the uplift, or part of the uplift, may have occurred when the major collisional tectonism waned, some 50 Ma later than proposed by Lambeck and Stevenson (1986).

3. Continental breakup occurred in the Late Cretaceous with attendant intrusive igneous activity; a large wedge of mantle-derived material is assumed to have been intruded under the Sydney Basin to the east of the hinge line, which caused an isostatic down warp of about 1000 m in response to the altered deep crustal density. The deep intrusive body under the eastern part of the basin elevated the heat flow to about 182 mW/m^2 (4.4 HFU), and caused the elevated coal rank in that part of the basin.

4. Subsequent to continental breakup, the missing sequence was gradually eroded, and redeposited in the Tasman Sea.

REFERENCES

Beaumont, C., 1981. Foreland basins. *Geophysical Journal of the Royal Astronomical Society* 65: 291-329.

Beaumont, C; Boutilier, R.; Mackenzie, A.S.; Rullkötter, J, 1985. Isomerization and aromatization of hydrocarbons and paleothermometry and burial history of Alberta foreland basin. *American Association of Petroleum Geologists Bulletin 69*: 546- 566.

Bai, G.; Keene, J.; Eadington, P.J.; Hamilton, P.J.; Macdougall, I. 1991. Fluid flow history of E. Triassic Narrabeen Group sandstones of the Southern Sydney Basin. *The Australian Society of Exploration Geophysicists 8th Conference and Exhibition*, and *The Geological Society of Australia Exploration Symposium, 1991*, "Exploration in a Changing Environment", Abstracts No. 30: 122.

Branagan, D.F. 1983. The Sydney Basin and its vanished sequence. *Journal of the Geological Society of Australia* 30. 75-84.

Brigaud, F.; Chapman, D.S.; Le Douaran, S. 1990. Estimating thermal conductivity in sedimentary basins using lithologic data and geophysical well logs. *American Association of Petroleum Geologists Bulletin* 74: 1459-1477.

Carslaw, H.S. and Jaeger, J.C. 1959. Conduction of heat in solids. *Oxford University Press*, Oxford.

Crawford, E.A.; Herbert, C.; Taylor, G.; Helby, R.; Morgan, R.; Ferguson, J. 1980. Diatremes of the Sydney Basin. In Herbert, C. and Helby, R. (eds): 'A Guide to the Sydney Basin'. *Geological Survey of New South Wales Bulletin* No. 26: 294-323.

Detrick, R.S.; Von Herzen, R.P.; Parsons, B.; Sandwell, D.; Dougherty, M. 1986. Heat flow observations on the Bermuda Rise and thermal models of midplate swells. *Journal of Geophysical Research* 91: 3701-3723.

Diessel, C.F.K. 1975. Coalification trends in the Sydney Basin, New South Wales. In Campbell, K.S.W. (ed.): 'Gondwana Geology'. *Australian National University Press*, Canberra: 295-309.

Eadington, P.J. and Hamilton, P.J. 1991. Integrating time specific and time dependent temperature indicators for thermal histories of sedimentary basins. *The Australian Society of Exploration Geophysicists 8th Conference and Exhibition*, and *The Geological Society of Australia Exploration Symposium, 1991*, "Exploration in a Changing Environment", Abstracts No. 30: 119- 120.

Facer, R.A.; Cook, A.C.; Beck, A.E. 1980. Thermal properties and coal rank in rocks and coal seams of the southern Sydney Basin, New South Wales: A palaeogeothermal explanation of coalification. *International Journal of Coal Geology* 1: 1-17.

Facer, R.A.; Hutton, A.C.; Frost, D.J. 1979. Heat generation by siliceous igneous rocks of the basement and its possible influence on coal rank in the Sydney Basin, New South Wales. *Proceedings of the Linnean Society of New South Wales* 104: 95- 109.

Herbert, C. 1980. 2. Depositional development of the Sydney Basin. In Herbert, C. and Helby, R. (eds): 'A Guide to the Sydney Basin'. *Geological Survey of New South Wales Bulletin* No. 26: 11- 52.

Lambeck, K. and Stevenson, R. 1986: The post-Palaeozoic uplift history of south-eastern Australia. *Australian Journal of Earth Sciences* 33: 253-270.

Leaman, D.E. 1990. The Sydney Basin: Composition of basement. *Australian Journal of Earth Sciences* 37: 107-108.

Middleton, M.F. 1983a. Tectonic history from vitrinite reflectance. *Geophysical Journal of the Royal Astronomical Society* 68: 121-132.

Middleton, M.F. 1983b: Coal rank trends in eastern Australian Permian coal basins. *Investigation Report 141*, Institute of Energy and Earth Resources, CSIRO.

Middleton, M.F. 1989. A model for the formation of intracratonic seg basins. *Geophysical Journal International* 99: 665-676.

Middleton, M.F. and Schmidt, P.W. 1982. Paleothermometry of the Sydney Basin. *Journal of Geophysical Research* 87: 5351-5359.

Palciauskas, V.V. 1986. Models for thermal conductivity and permeability in normally compacting basins. In Burrus, J. (ed.): 'Thermal Modeling in Sedimentary Basins'. *Gulf Publishing Company, Houston*: 323-336.

Qureshi, I.R. 1984. Wollondilly-Blue Mountains Gravity Gradient and its bearing on the origin of the Sydney Basin. *Australian Journal of Earth Sciences* 31: 293-302.

Shaw, R.D. 1978. Seafloor spreading in the Tasman Sea: A Lord Howe Rise-Eastern Australia reconstruction. *Australian Society of Exploration Geophysicists Bulletin* 9: 75-81.

Shibaoka, M.; Bennett, A.J.R.; Gould, K.W., 1973. Diagenesis of organic matter and occurrence of hydrocarbons in some Australian sedimentary basins. *Australian Petroleum Exploration Association Journal* 13: 73-80.

Sweeney, J.J. and Burnham, A.K. 1990. Evaluation of a simple model of vitrinite reflectance based on chemical kinetics. *American Association of Petroleum*

Geolobists Bulletin 74: 1557-1570.
Wellman, P. and McDougall, I. 1974. Potassium-argon ages on the Cainozoic volcanic rocks of New South Wales. *Journal of the Geological Society of Australia* 21: 247-272.

Gondwana Eight, Findlay, Unrug, Banks & Veevers (eds) © 1993 Balkema, Rotterdam. ISBN 90 5410 304 3

Author index

www.ingramcontent.com/pod-product-compliance
Ingram Content Group UK Ltd.
Pitfield, Milton Keynes, MK11 3LW, UK
UKHW052050230926
475702UK00025B/1019
* 9 7 8 9 0 5 4 1 0 3 0 4 2 *